CATALYTIC METHODS IN ASYMMETRIC SYNTHESIS

CATALYTIC METHODS IN ASYMMETRIC SYNTHESIS

Advanced Materials, Techniques, and Applications

EDITED BY

MICHELANGELO GRUTTADAURIA
FRANCESCO GIACALONE

Department of Molecular and Biomolecular Sciences (STEMBIO)
Section of Organic Chemistry
University of Palermo
Palermo, Italy

WILEY

A JOHN WILEY & SONS, INC., PUBLICATION

Published by John Wiley & Sons, Inc., Hoboken, New Jersey
Published simultaneously in Canada

Library of Congress Cataloging-in-Publication Data

Gruttadauria, Michelangelo.
Catalytic methods in asymmetric synthesis : advanced materials, techniques, and applications / Michelangelo Gruttadauria, Francesco Giacalone.
p. cm.
ISBN 978-0-470-64136-1 (hardback)
1. Asymmetric synthesis. 2. Catalysis. I. Giacalone, Francesco. II. Title.
QD262.G78 2011
541'.395–dc22

2011006415

Printed in the United States of America

oBook ISBN: 978-1-118-08799-2
ePDF ISBN: 978-1-118-08797-8
ePub ISBN: 978-1-118-08798-5

10 9 8 7 6 5 4 3 2 1

This book is dedicated to Prof. Renato Noto

CONTENTS

PREFACE xi

FOREWORD xiii

CONTRIBUTORS xv

I NEW MATERIALS AND TECHNOLOGIES: SUPPORTED CATALYSTS, SUPPORTS, SELF-SUPPORTED CATALYSTS, CHIRAL IONIC LIQUID, SUPERCRITICAL FLUIDS, FLOW REACTORS, AND MICROWAVES 1

1 **RECYCLABLE STEREOSELECTIVE CATALYSTS** **3**

Carlos M. Monteiro, Alexandre F. Trindade, Pedro M. P. Gois, and Carlos A. M. Afonso

2 **RECYCLABLE ORGANOCATALYSTS IN ASYMMETRIC REACTIONS** **83**

Michelangelo Gruttadauria, Francesco Giacalone, and Renato Noto

3 **SYNTHESIS AND CHARACTERIZATION OF SUPPORTED CHIRAL CATALYSTS** **177**

Carmela Aprile, Hermenegildo Garcia, and Paolo P. Pescarmona

4 SYNTHESIS OF CHIRAL CATALYSTS SUPPORTED ON ORGANIC POLYMERS 209
Tor Erik Kristensen and Tore Hansen

5 SELF-SUPPORTED CHIRAL CATALYSTS 257
Hongchao Guo and Kuiling Ding

6 CATALYSIS WITH CHIRALLY MODIFIED METAL SURFACES: SCOPE AND MECHANISMS 291
Angelo Vargas, Cecilia Mondelli, and Alfons Baiker

7 CHIRAL IONIC LIQUIDS FOR ASYMMETRIC REACTIONS 323
Annie-Claude Gaumont, Yves Génisson, Frédéric Guillen, Viacheslav Zgonnik, and Jean-Christophe Plaquevent

8 ASYMMETRIC REACTIONS IN FLOW REACTORS 345
Munawwer Rasheed, Simon C. Elmore, and Thomas Wirth

9 ASYMMETRIC CATALYTIC SYNTHESIS IN SUPERCRITICAL FLUIDS 373
Tomoko Matsuda

10 MICROWAVE-ASSISTED TRANSITION METAL-CATALYZED ASYMMETRIC SYNTHESIS 391
Luke R. Odell and Mats Larhed

II RECENT ADVANCES IN ORGANOCATALYTIC, ENZYMATIC, AND METAL-BASED MEDIATED ASYMMETRIC SYNTHESIS 413

11 RECENT ADVANCES ON STEREOSELECTIVE ORGANOCATALYTIC REACTIONS. ORGANOCATALYTIC SYNTHESIS OF NATURAL PRODUCTS AND DRUGS 415
Monika Raj and Vinod K. Singh

12 RECENT ADVANCES IN BIOCATALYSIS APPLIED TO ORGANIC SYNTHESIS 491
Gonzalo De Gonzalo, Iván Lavandera, and Vicente Gotor

13 PEPTIDES FOR ASYMMETRIC CATALYSIS 529
Matthias Freund and Svetlana B. Tsogoeva

14 SILICATE-MEDIATED STEREOSELECTIVE REACTIONS CATALYZED BY CHIRAL LEWIS BASES 579
Maurizio Benaglia, Stefania Guizzetti, and Sergio Rossi

15 RECENT ADVANCES IN THE METAL-CATALYZED STEREOSELECTIVE SYNTHESIS OF BIOLOGICALLY ACTIVE MOLECULES 625
Catalina Ferrer, Xavier Verdaguer, and Antoni Riera

16 STEREOSELECTIVE NITROGEN HETEROCYCLE SYNTHESIS MEDIATED BY CHIRAL METAL CATALYSTS 671
Sherry R. Chemler

INDEX 689

PREFACE

Catalytic asymmetric synthesis is the "art" of promoting the exclusive achievement of an enantiomer over another with the help of substoichiometric amounts of a proper catalyst. This frontier field of research experiences, day after day, an exponential growth, and in the meantime it evolves in a parallel manner in thousands of laboratories all over the world. This continued evolution has led in the last decade to the astounding discovery of a new foundational pillar in the field. In fact, similar to the well-established biocatalysis and organometallic catalysis fields, in the last decade a third column called organocatalysis is strongly helping chemists with new additional tools in the preparation of three-dimensional chemical structures. However, due to fast developments in the field, with new concepts and methods almost daily being discovered, it is difficult to carefully describe it in a given moment.

In this book we tried to take an instant picture of the most recent methods, applications, and techniques exploited in asymmetric synthesis. We gathered an authoritative list of worldwide experts in their corresponding areas of interest and asked them to contribute to this project. In order to cover the more important aspects of asymmetric synthesis that in the last recent years have received considerable interest from the scientific community, the book has been divided in two parts.

In the first part new materials and technologies are collected. Chapters 1 and 2 focus on the recycling of stereoselective catalysts and organocatalysts, respectively, which paves the way to more sustainable processes. Chapters 3 and 4 are devoted to the synthesis and characterization of covalently supported chiral catalysts on both inorganic and polymeric organic supports. The most important aspect is that these materials can be easily separated from the reaction mixture and reused several times without affecting their efficiency. On the other hand, very recently metal-organic supramolecular polymers, constituted by self-complementary units or by orthogonal supramolecular building blocks, have emerged as powerful microporous heterogeneous catalysts; these are highlighted in Chapter 5. Next, chirally modified metals employed in the asymmetric hydrogenation of prochiral double bonds with special emphasis on the mechanistic aspects of the processes involved are discussed in Chapter 6. Then, in Chapters 7–10, asymmetric catalysis in alternative green reaction media such as ionic liquids or supercritical fluids will be thoroughly addressed as well as the use of enabling technologies such as continuous flow or microwave-assisted reactors.

In the second part the most relevant and recent advances regarding the three pillars of asymmetric catalysis will be described. Particularly, much attention will be devoted to the synthesis of natural products mediated by organocatalysts, metal-free organic compounds of relatively low molecular weight and simple structure (Chapter 11), as well as to the use of the more structurally complex enzymes (Chapter 12) or natural and synthetic peptides (Chapter 13), meant as simplified versions of biocatalysts, in asymmetric catalysis.

Chapter 14 covers chiral silicon-based Lewis bases, which play an important role as promoters of a large variety of stereoselective reactions. Finally, the last two chapters cover the most recent and innovative advances in the field of chiral metal catalysis with special emphasis on the applications for the synthesis of biologically active molecules (Chapter 15) and the stereoselective nitrogen heterocycle synthesis, since nitrogen heterocycles play a central role in many biologically active molecules.

This book will serve to introduce the reader to the wide field of asymmetric catalysis, giving him or her an insight into the current status of the area. Moreover, the presence in one book of two interconnected and complementary aspects may allow teachers to give a wider overview of the topic and, at the same time, give an advantage to the students.

We would like to acknowledge once again all the contributors, the efforts of whom have made the publication of this book come true. We want to thank also Jonathan Rose and all the people at Wiley-Blackwell that supported us during the whole adventure.

Michelangelo Gruttadauria
Francesco Giacalone

Palermo
March 2011

FOREWORD

There will always be a need for organic synthesis. New compounds will always be required for evaluation as pharmaceuticals, agrochemicals, dyestuffs, materials, and for a host of other purposes. In the context of organic synthesis, the need for asymmetric synthesis to provide efficient access to enantiomerically enriched materials is a supreme challenge and the development of catalytic processes for asymmetric synthesis is at the forefront of advances in this area.

Modern advances in catalytic asymmetric synthesis include not only the recognition and application of organic catalysts together with improved ligands and transition-metal based catalysts, but also the introduction of solid supported catalysts, flow systems, and homochiral organic liquids. Advances in biotechnology are also providing improved and more generally applicable enzymic processes for asymmetric synthesis. The need for innovative partition and extraction procedures has led to the use of supercritical fluids and fluorous reagents and solvents. Microwave heating can also provide much faster reactions than conventional heating and the demand for environmentally acceptable processes requires the minimisation of waste whether from reagents or solvents and so the use of recyclable catalysts and atom efficiency are of paramount importance not least for industrial processes.

So this is a large and rapidly evolving field!

In this timely and very broadly based volume of recent advances in asymmetric synthesis, technological as well as chemical advances are presented. The use of solid supported and recyclable catalysts is discussed with a range of polymer and other solid supports considered in detail. The question of how to convert a homogeneous catalyst into a heterogeneous one is addressed together with procedures for the characterisation of solid supported catalysts. The benefits of different catalyst supports are also presented alongside self-supported catalysts, chiral ionic liquids, and chirally modified metal surfaces. Asymmetric synthesis in flow systems and in supercritical fluids in considered together with microwave heating for asymmetric catalysis. Finally recent advances in asymmetric catalysis are presented in the context of many different types of reaction.

This book will be useful not only to experts in the field but also to all synthetic organic chemists involved in asymmetric synthesis of chiral compounds. Post-graduate students and post-doctoral researchers will also find it an invaluable introduction to an important and burgeoning field.

School of Chemistry
The University of Manchester
April 2011

ERIC J. THOMAS

CONTRIBUTORS

Carlos A. M. Afonso, CQFM, Centro de Química-Física Molecular and IN—Institute of Nanosciences and Nanotechnology, Instituto Superior Técnico, 1049-001 Lisboa, Portugal; iMed.UL, Faculdade de Farmácia da Universidade de Lisboa, Av. Prof. Gama Pinto, 1649-003 Lisboa, Portugal.

Carmela Aprile, Facultés Universitaires Notre-Dame de la Paix (FUNDP), 61 rue de Bruxelles, B-5000 Namur, Belgium.

Alfons Baiker, Department of Chemical and Applied Biosciences, ETH Zurich, Switzerland, Wolfgang-Pauli-Str. 10, ETH Hönggerberg, HCI E 133, CH-8093 Zürich, Switzerland.

Maurizio Benaglia, Dipartimento di Chimica Organica e Industriale, Università degli Studi di Milano, via Golgi 19, 20133 Milano, Italy.

Sherry R. Chemler, Department of Chemistry, The State University of New York at Buffalo, Buffalo, NY 14260.

Kuiling Ding, State Key Laboratory of Organometallic Chemistry, Shanghai Institute of Organic Chemistry, Chinese Academy of Sciences, 345 Lingling Road, Shanghai 200032, China.

Simon C. Elmore, Cardiff University, School of Chemistry, Main Building, Park Place, Cardiff CF10 3AT, United Kingdom.

Catalina Ferrer, Unitat de Recerca en Síntesi Asimètrica (URSA-PCB), Institute for Research in Biomedicine (IRB Barcelona), and Departament de Química Orgànica, Universitat de Barcelona, Baldiri Reixac 10, 08028 Barcelona, Spain.

Matthias Freund, Department of Chemistry and Pharmacy, Chair of Organic Chemistry I, University of Erlangen-Nuremberg, Henkestrasse 42, 91054 Erlangen, Germany.

Hermenegildo Garcia, (b) Instituto de Tecnología Quimica CSIC-UPV, Av. de los Naranjos s/n, Universidad Politécnica de Valencia, 46022 Valencia, Spain.

Annie-Claude Gaumont, Laboratoire de Chimie Moléculaire et Thioorganique, UMR CNRS 6507, INC3M FR 3038, ENSICAEN & Université de Caen, 14050 Caen, France.

Yves Génisson, CNRS-UMR 5068, Synthèse et Physicochimie de Molécules d'Intérêt Biologique, Université Paul Sabatier, 118 route de Narbonne, 31062 Toulouse cedex 9, France.

Francesco Giacalone, Department of Molecular and Biomolecular Sciences (STEMBIO), Section of Organic Chemistry, University of Palermo, Viale delle Scienze, Ed. 17, 90128 Palermo, Italy.

Pedro M. P. Gois, iMed.UL, Faculdade de Farmácia da Universidade de Lisboa, Av. Prof. Gama Pinto, 1649-003 Lisboa, Portugal.

Gonzalo de Gonzalo, Departamento de Química Orgánica e Inorgánica, Instituto Universitario de Biotecnología de Asturias, University of Oviedo, 33006 Oviedo, Spain.

Vicente Gotor, Departamento de Química Orgánica e Inorgánica, Instituto Universitario de Biotecnología de Asturias, University of Oviedo, 33006 Oviedo, Spain.

Michelangelo Gruttadauria, Department of Molecular and Biomolecular Sciences (STEMBIO), Section of Organic Chemistry, University of Palermo, Viale delle Scienze, Ed. 17, 90128 Palermo, Italy.

Frédéric Guillen, CNRS-UMR 6014, COBRA, IRCOF, Université de Rouen, rue Tesnière, 76821 Mont-Saint-Aignan, France.

Stefania Guizzetti, Dipartimento di Chimica Organica e Industriale, Università degli Studi di Milano, via Golgi 19, 20133 Milano, Italy.

Hongchao Guo, Department of Applied Chemistry, China Agricultural University, 2 Yuanmingyuan West Road, Beijing 100193, China.

Tore Hansen, Department of Chemistry, University of Oslo, P.O. Box 1033 Blindern, NO-0315 Oslo, Norway.

Tor Erik Kristensen, Department of Chemistry, University of Oslo, NO-0315 Oslo, Norway.

Mats Larhed, Department of Medicinal Chemistry, Organic Pharmaceutical Chemistry, BMC, Uppsala University, Box 574, SE-75123 Uppsala, Sweden.

Iván Lavandera, Departamento de Química Orgánica e Inorgánica, Instituto Universitario de Biotecnología de Asturias, University of Oviedo, 33006 Oviedo, Spain.

Tomoko Matsuda, Department of Bioengineering, Graduate School of Bioscience and Biotechnology, Tokyo Institute of Technology, 4259 Nagatsuta, Midori-ku, Yokohama, Japan, 226-8501.

Cecilia Mondelli, Department of Chemical and Applied Biosciences, ETH Zurich, Switzerland, Wolfgang-Pauli-Str. 10, ETH Hönggerberg, HCI E 133, CH-8093 Zürich, Switzerland.

Carlos M. Monteiro, CQFM, Centro de Química-Física Molecular and IN—Institute of Nanosciences and Nanotechnology, Instituto Superior Técnico, 1049-001 Lisboa, Portugal.

Renato Noto, Department of Molecular and Biomolecular Sciences (STEMBIO), Section of Organic Chemistry, University of Palermo, Viale delle Scienze, Ed. 17, 90128 Palermo, Italy.

Luke R. Odell, Department of Medicinal Chemistry, Organic Pharmaceutical Chemistry, BMC, Uppsala University, Box 574, SE-75123 Uppsala, Sweden.

Paolo P. Pescarmona, Centre for Surface Chemistry and Catalysis, K.U. Leuven, Kasteelpark Arenberg 23, 3001 Heverlee, Belgium.

Jean-Christophe Plaquevent, CNRS-UMR 5068, Synthèse et Physicochimie de Molécules d'Intérêt Biologique, Université Paul Sabatier, 118 route de Narbonne, 31062 Toulouse cedex 9, France.

Monika Raj, Indian Institute of Science Education and Research, Bhopal Transit Campus: ITI Campus (Gas Rahat) Building, Govindpura, Bhopal—460 023 India.

Munawwer Rasheed, Cardiff University, School of Chemistry, Main Building, Park Place, Cardiff CF10 3AT, United Kingdom.

Antoni Riera, Unitat de Recerca en Síntesi Asimètrica (URSA-PCB), Institute for Research in Biomedicine (IRB Barcelona), and Departament de Química Orgànica, Universitat de Barcelona, Baldiri Reixac 10, 08028 Barcelona, Spain.

Sergio Rossi, Dipartimento di Chimica Organica e Industriale, Università degli Studi di Milano, via Golgi 19, 20133 Milano, Italy.

Vinod K. Singh, Indian Institute of Science Education and Research, Bhopal Transit Campus: ITI Campus (Gas Rahat) Building, Govindpura, Bhopal—460 023 India.

Alexandre F. Trindade, CQFM, Centro de Química-Física Molecular and IN—Institute of Nanosciences and Nanotechnology, Instituto Superior Técnico, 1049-001 Lisboa, Portugal.

Svetlana B. Tsogoeva, Department of Chemistry and Pharmacy, Chair of Organic Chemistry I, University of Erlangen-Nuremberg, Henkestrasse 42, 91054 Erlangen, Germany.

Angelo Vargas, Department of Chemical and Applied Biosciences, ETH Zurich, Switzerland, Wolfgang-Pauli-Str. 10, ETH Hönggerberg, HCI E 133, CH-8093 Zürich, Switzerland.

Xavier Verdaguer, Unitat de Recerca en Síntesi Asimètrica (URSA-PCB), Institute for Research in Biomedicine (IRB Barcelona), and Departament de Química Orgànica, Universitat de Barcelona, Baldiri Reixac 10, 08028 Barcelona, Spain.

Thomas Wirth, Cardiff University, School of Chemistry, Main Building, Park Place, Cardiff CF10 3AT, United Kingdom.

Viacheslav Zgonnik, CNRS-UMR 5068, Synthèse et Physicochimie de Molécules d'Intérêt Biologique, Université Paul Sabatier, 118 route de Narbonne, 31062 Toulouse cedex 9, France.

PART I

NEW MATERIALS AND TECHNOLOGIES: SUPPORTED CATALYSTS, SUPPORTS, SELF-SUPPORTED CATALYSTS, CHIRAL IONIC LIQUID, SUPERCRITICAL FLUIDS, FLOW REACTORS, AND MICROWAVES

CHAPTER 1

RECYCLABLE STEREOSELECTIVE CATALYSTS

CARLOS M. MONTEIRO, ALEXANDRE F. TRINDADE, PEDRO M. P. GOIS, AND CARLOS A. M. AFONSO

1.1. Introduction 3
1.2. Chiral phosphines 5
  1.2.1. Hydrogenation 5
  1.2.2. Hydroformylation 13
  1.2.3. Cycloaddition 14
  1.2.4. Allylic substitution 15
1.3. Chiral alkaloids 17
1.4. Bisoxazolines 24
  1.4.1. Py-BOX ligands 24
  1.4.2. BOX ligands 26
  1.4.3. Aza-BOX and phe-BOX ligands 33
1.5. Salen-type ligands 35
1.6. Enzymes 47
1.7. Chiral diamines, diols, and aminoalcohols 51
  1.7.1. TsDPEN-type ligands 51
  1.7.2. Chiral aminoalcohols 55
  1.7.3. BINOL-type ligands 59
1.8. Conclusions 64
References 64

1.1. INTRODUCTION

Asymmetric catalysis constitutes an important subject, generating thousands of published works every year. Still, the application of such methodologies in

Catalytic Methods in Asymmetric Synthesis: Advanced Materials, Techniques, and Applications, First Edition. Edited by Michelangelo Gruttadauria and Francesco Giacalone.

the chemical industry is rather limited due to the high cost of the chiral ligands and/or noble metals used in such transformations. Additionally, sometimes the final products contain high levels of metal contamination derived from catalysts descomplexation or degradation phenomena, which can became a serious drawback if the metal is toxic, particularly for the pharmaceutical and food industries. For these reasons, there are still advantages to using the chiral building blocks readily available in nature or by applying resolution of optical isomers [1].

Stereochemical and chemical efficiency of a certain transformation are, in principle, better reproduced and predicted in homogeneous catalysis than in heterogeneous catalysis. The presence of the heterogeneous support in a reaction vessel can create, in some cases, unpredictable results (negative vs. novel positive effects) [2]. The choice of the heterogeneous support for the catalysis is a crucial decision. Some properties like high thermal, chemical and physical stabilities, chemical inertia, and homogeneous-like behavior are highly desirable. Furthermore, the catalyst is easily recovered using just filtration or extraction techniques that are impossible to be applied in homogeneous catalysis [3].

Amorphous and ordered silicas, clays, and highly cross-linked polymers are the standard supports to heterogenize a homogeneous catalyst. The principal immobilization mechanisms consist of ligand grafting, metal coordination, microencapsulation, electrostatic interactions, and ion exchange [3] (see also Chapters 3 and 4).

It is possible to combine the advantages of homogeneous and heterogeneous systems by running reactions with catalysts that have been chemically linked to soluble macromolecules like oligomeric and/or low cross-linked soluble polymers, poly(ethylene glycol) (PEG), and dendritic structures. The supported homogeneous catalyst can be precipitated at the end of the reaction by addition of a cosolvent and recovered like a heterogeneous system [3].

The reutilization of asymmetric catalysts and the reactions media was possible using greener solvents like water, ionic liquids (ILs), PEG, perfluorinated solvents, and supercritical CO_2 ($scCO_2$), which constitute alternatives to volatile organic solvents. Water appears as the cheapest solvent, bearing unique characteristics that differ from the others solvents: it is cheaper, most abundant in nature, and proven to have some unexpected beneficial effects in organic transformations [3, 4].

The need to transfer the asymmetric catalysis methodology to large-scale synthesis technology is a crucial goal for synthetic organic scientists worldwide. There are many contributions toward achieving this goal in the literature [2, 5]. In 2002, Chan et al. combined all these type of transformations or chiral ligands [6], or self-supported heterogeneous catalysts [7], or solvent-free transformations [8]. In 2009, Trindade et al. [3] updated the earlier Chan et al. work [2], covering a broader number of transformations and all type of catalyst recycling methodologies. Other reviews were published in the literature, where the majority focused on catalyst immobilization for both chiral and achiral organic transformations, but they do not cover all types of reported catalysts immobilization processes [9]. From the reviews that focus only on enantioselective catalysis, some cover only one transformation performed with heterogeneous catalysts [10].

This chapter provides an overview of our selection transformations and all types of catalyst (except organocatalysts) recycling methodologies described up to March 2010.

At the beginning of each paragraph, there is a small description of the ligand in question and its applications in asymmetric catalysis. In the text, maximum efforts were made to identify the best-reported catalyst recycle system using heterogeneous catalysts, homogeneous catalysts, and alternative reaction media. Also, the method for recovery, in terms of reactivity and enantioselectivity, will be highlighted, giving special attention to the recycling process. This recycling process will be analyzed both in terms of executability and efficiency. In terms of efficiency, at times the reader may encounter $Y_e(x) = \ldots$ % and $ee_e(x) = \ldots$ %. These percentages should be read as the yield (Y_e) or enantioselectivity (ee_e) erosion at run x. The percentages are calculated according to the following expressions:

$$Y_e(x) = \frac{Y(1) - Y(x)}{Y(1)} \qquad ee_e(x) = \frac{ee(1) - ee(x)}{ee(1)}$$

1.2. CHIRAL PHOSPHINES

The wide synthetic applications of phosphine catalysts have motivated many investigations on enantioselective reactions, particularly hydrogenation, hydroformylation, cycloaddition, and allylic substitution. The search for new, particularly designed chiral phosphorus catalysts has become a goal of several research groups. Despite the high activity and enantioselectivity achieved by complexing these ligands with Ru, Rh, and Ir, their air sensitivity and high cost represent a drawback. To circumvent these limitations, more stable and recyclable catalysts have been developed.

1.2.1. Hydrogenation

Rhodium(I) and chiral phosphine catalysts are widely used in asymmetric alkene hydrogenation due to their efficiency in the preparation of enantiopure compounds with excellent atom economy. Generally, methyl α-acetamidoacrylate (MAA), methyl α-acetamido cinnamate (MAC), and dimethyl itaconate are the model substrates chosen by the authors to perform asymmetric hydrogenation reactions (Fig. 1.1).

Ding and Wang prepared a self-supported heterogeneous catalyst for enamines hydrogenation from an Rh complex $[Rh(Cod)_2BF_4]$ and a bis-phosphoamidate derived from 1,1′-bi-2-naphthol (BINOL) (Monophos) **1**.

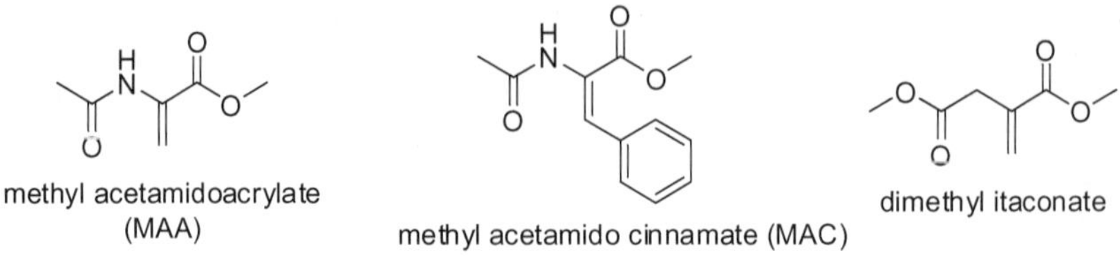

FIGURE 1.1. Main substrates used for asymmetric hydrogenation assays.

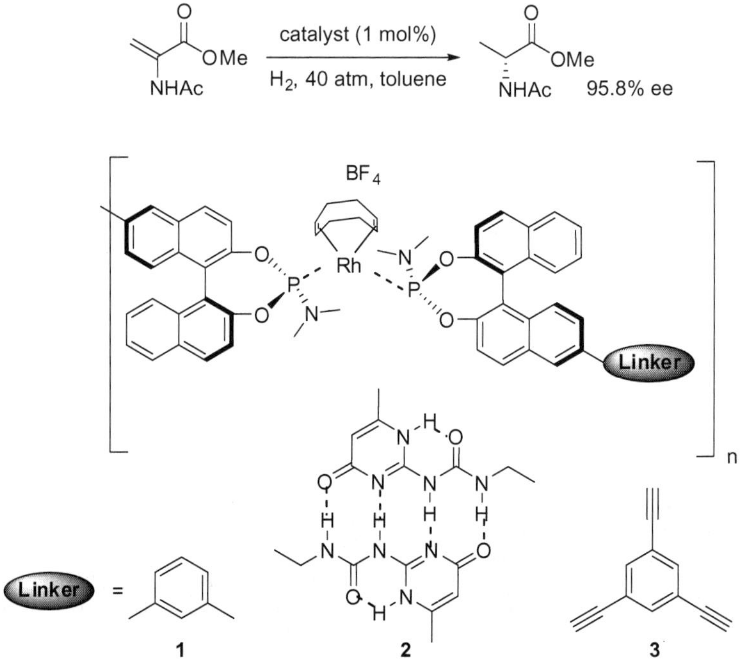

SCHEME 1.1. Hydrogenation reaction with self-supported heterogeneous catalyst.

Excellent enantioselectivities were obtained with aromatic and alkylic enamines with this heterogeneous catalyst (equivalent to a homogeneous ligand) (Scheme 1.1). The catalyst was recycled five times with a minimal 5% enantioselectivity erosion [11]. Two years later, Ding and Wang prepared a new self-supported polymer based on the same homogeneous catalyst (catalyst **2**). This polymeric complex proved to be an efficient heterogeneous catalyst for enamines hydrogenation reactions [substrate/catalyst (S/C) = 100, 40 atm H_2, 91%–96% enantiomeric excess (ee)]. It could be recovered by filtration and reused with minimal ee erosion [12].

More recently, the same authors developed a continuous-flow system using another self-supported catalyst **3** based on the same catalyst. Under experimental conditions (2 atm H_2), the α-dehydroamino acid methyl esters could be continuously hydrogenated in >99% conversion (93–94 μmol/h) in 97% ee for a total of 144 hours and an overall yield of 2.52 g (15.75 mmol). The immobilized catalyst also showed high enantioselectivity (96%–97% ee) but was observed with lower reactivity in a batch recycling system during 10 cycles [13].

In 2000, de Rege et al. performed the heterogenization of a [((*R*,*R*)-Me-DuPHOS)Rh(COD)]$^+$ catalyst in mesoporous crystalline material (MCM)-41. This heterogeneous catalyst proved to be as selective as the free catalyst (up to 99% ee) for the asymmetric hydrogenation of amidoacrylic acids and proved to be recyclable without loss of activity or enantioselectivity [14].

The chiral ligands (*R*)-2,2′-bis(diphenylphosphino)-1,1′-binaphthyl (BINAP) [15], (*R*,*R*)-Me-DuPhos [16] and (2*S*,4*S*)-BDPP [17], and carbon-activated [18] supports were studied under Augustine et al.'s immobilization methodology [19]. Generally, the heterogeneous catalysts were found to be less active than their homogeneous partners under batchwise conditions but provided comparable levels of enantioselectivities (generally above 90% ee). The high degree of recyclability of these catalysts was also demonstrated (up to 11 cycles).

Hutchings et al. showed that achiral [Rh(cod)$_2$]BF$_4$ could be immobilized in mesoporous MCM-41 by ion exchange, followed by chiral ligand coordination [(*R*,*R*)-Me-DuPhos]. Alternatively, they also immobilized directly the chiral rhodium complex. Both types of heterogeneous catalyst proved to be as efficient as their homogeneous analogs (up to 99% ee) and could be reused nine times without enantioselectivity erosion in methyl itaconate hydrogenation (S/C = 250, 80 psi H_2) [20].

Luna et al. grafted a Ru-(BINAP) (dpea) complex on the surface of amorphous $AlPO_4$ support to achieve its recycling. It was tested in the successive liquid-phase enantioselective hydrogenation of dimethyl itaconate, MAA, and MAC (in dichloromethane, S/C = 45.4, 6.8 atm H_2 at 50–70°C, Table 1.1). This heterogeneous catalyst **4** (Table 1.1) provided the respective products with excellent enantioselectivities (99% ee) over several days and worked continuously (up to 10 cycles) [21].

Other chiral phosphines immobilization in heterogeneous supports, such as acidic aluminosilacte (AlTUD-1) [22], phosphotungstic acid (PTA) on alumina [23], and carbon nanotubes [24], were reported. Excellent enantioselectivity (>96% ee) was obtained in this transformation, with good recyclability.

Fan et al. prepared and tested a chiral dendritic ligand **5** bearing a Pyrphos moiety linked to its core for the asymmetric hydrogenation of α-acetamido cinnnamic acid. After in situ catalyst formation [reaction between Rh(Cod)$_2$BF$_4$ and the dendritic ligand **5**, Fig. 1.2], it furnished phenylalanines with excellent ee's (>97% vs. 99% ee for nonsupported ligands) with all dendritic generations. The catalyst was recovered by filtration and reused with constant ee. However, the conversion decreased considerably upon recycling (run 1: 94%, run 2: 55%) [25]. Similar chiral dendritic ligand was tested with analogous results, but improving in terms of reutilization. A drastic decrease of conversion was only observed on the fourth cycle [26].

TABLE 1.1. Successive Hydrogenation Rates of Different Substrates

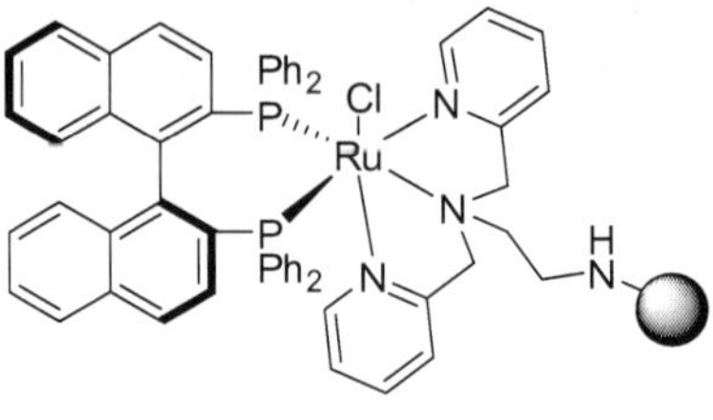

4

Run	Substrate	Temperature (°C)	Reaction Rate (μmol/s)	TOF (h^{-1})	Half Reaction Time (h)	Reaction Time (h)
1	Dimethyl itaconate	70.0	0.248	2.53	10.0	46.0
2	Dimethyl itaconate	55.0	0.062	0.63	76.0	144.0
3	Dimethyl Itaconate	70.0	0.023	0.24	96.0	216.0[a]
4	MAA	70.0	1.023	10.45	2.2	9.0
5	MAC	70.0	0.039	0.40	96.0	288.0

[a] After 1473 hours (2 months approximately) continuously working, TON = 408.

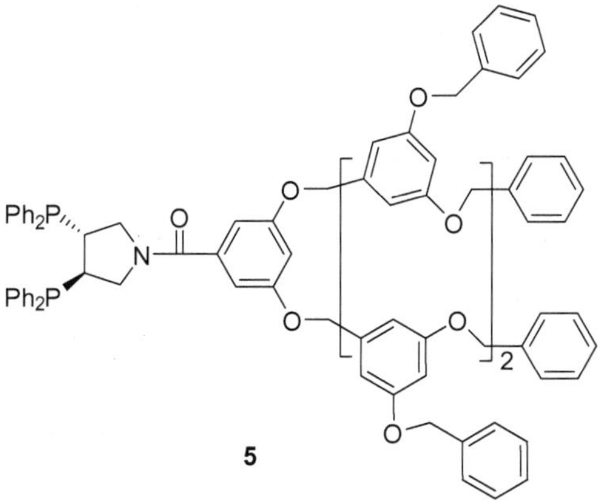

5

FIGURE 1.2. Structure of the dendritic ligand **5**.

Feng et al. reported the use of wet ILs as biphasic reaction media for MAA and MAC asymmetric hydrogenation [27]. The mixture of [omim]BF_4/water (2 mL of IL to 2–3 mL of cosolvent) proved to be the best reaction media for MAA, affording higher levels of enantioselectivity and conversion (up to >99% ee) than the traditional methanol/*i*-propanol system (catalyst loading of 1 mol%), for all ferrocene-containing catalysts (Fig. 1.3). This biphasic system proved to be very stable, and only at the sixth cycle was an extension on the reaction time required to always achieve complete conversion with a consistent 99% ee. For more apolar substracts such as MAC, the organic

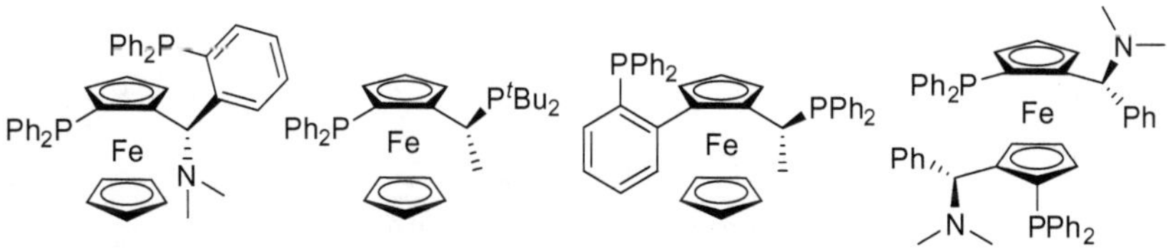

Taniaphos Josiphos Walphos Mandyphos

FIGURE 1.3. Chiral biphosphine ligands for asymmetric hydrogenation reaction in [omim]BF_4/water.

TABLE 1.2. Asymmetric Hydrogenation of Enamides

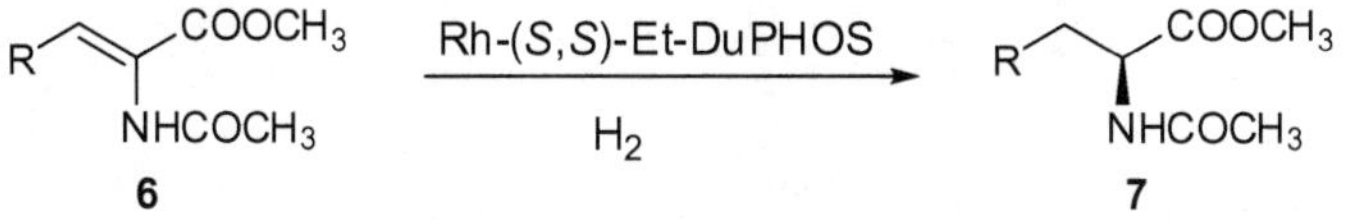

Entry	Substrate (R)	Solvent	System (v/v)	ee (%)[a]
1	**6a** (C_6H_5)	PEG600/MeOH	(1:3)	97.6
2	**6a** (C_6H_5)	PEG600/MeOH	(1:4)	98.1 (98.6)
3	**6b** (3-Cl-C_6H_4)	PEG600/MeOH	(1:3)	97.0 (98.6)
4	**6c** (4-Cl-C_6H_4)	PEG600/MeOH	(1:3)	97.1 (98.8)
5	**6d** (4-CH_3O-C_6H_4)	PEG600/MeOH	(1:3)	98.6 (99.0)
6	**6e** (H)	PEG600/MeOH	(1:3)	98.2 (99.6)

[a] Data in parentheses were obtained by using MeOH as solvent under otherwise identical conditions.

cosolvent toluene was added to the wet IL in order to achieve high enantioselectivity.

Chan et al. showed that PEG could be added to methanol to conduct ruthenium- and rhodium-mediated hydrogenation of 2-arylacrylic acids and enamides (Table 1.2). This solvent mixture proved to be as efficient as the original solvent used in this transformation, providing excellent levels of enantioselectivity (quantitative yields, up to 98.6% ee). Furthermore, it added the possibility of recycling the homogeneous catalysts during nine efficient cycles [28].

Chiral phosphines also have a great potential for asymmetric hydrogenations and reductions of ketones (Scheme 1.2). Several chiral phosphines were described mixed with 1,2-diphenylethylenediamine (DPEN)-type ligands type with very good results. Some of the supported ligands provided excellent levels of chiral induction (up to 98% ee).

One of the most successful immobilizations techniques was carried out by Noyori et al. when they anchored a BINAP derivative to a polystyrene polymer. The ruthenium complex **8** (Fig. 1.4) with this ligand and DPEN were used to hydrogenate acetoacetates with 97% ee during 10 consecutive cycles.

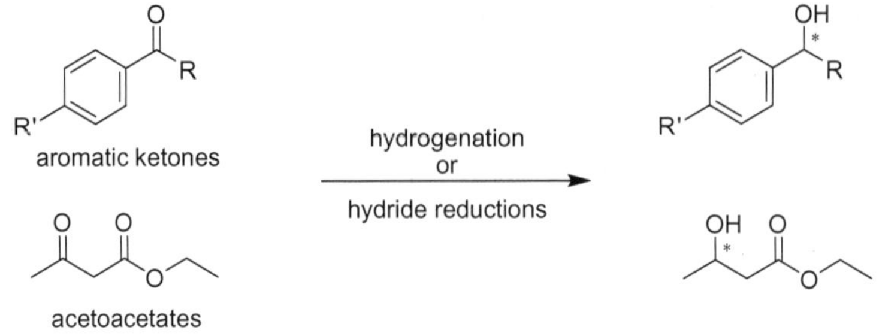

SCHEME 1.2. General substrates and methods for hydrogenation/reduction of ketones.

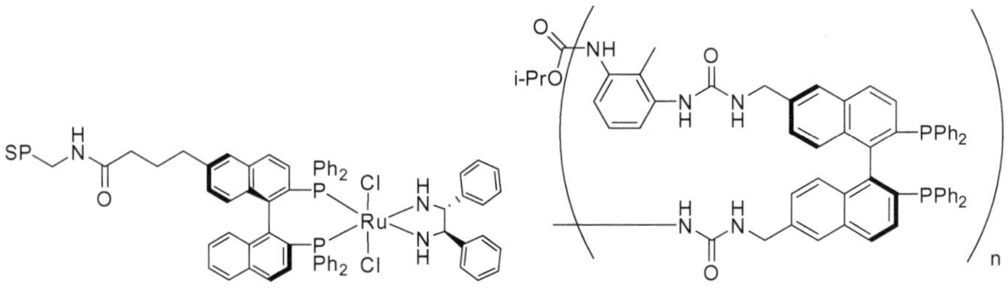

8 9

FIGURE 1.4. Structures of ruthenium complex **8** and ligand **9**.

Lemaire et al. immobilized diAm-BINAP in a polymer copolymer using di-isocyanates as the linker. The most rigid copolymer ligand **9** (Fig. 1.4) was used in ruthenium-mediated hydrogenation of acetoacetates with excellent performance and near-perfect selectivity (S/C = 1000, 40 atm hydrogen pressure, 50°C, 16 hours, 99% ee). This heterogeneous catalyst (in methanol) was reused three times without any change on its performance [29].

Lin et al. were interested in preparing hybrid materials containing organic linkers and metal nodes. The catalyst Ru(4,4′-$(PO_3H_2)_2$-BINAP) (DPEN)Cl_2 coordinated to a zirconium salt (ZrOt-Bu_4) was synthesized for such purposes. With this heterogeneous catalyst in hand, the hydrogenation of aromatic ketones was conducted with higher enantioselective than its homogeneous counterpart and achieved excellent performance (0.1 mol%, 700 psi of H_2, 20 hours to give up to 99.2% ee). Furthermore, the catalyst was reused seven times [30].

Bergens et al. reported the first polymeric asymmetric catalyst prepared by ring-opening metathesis polymerization (ROMP) of *trans*-$RuCl_2(Py)_2$((*R,R*)-Norphos)) and cycloctene. The living ends of the polymer were cross-linked with dicyclopentadiene and the pyridine ligands with chiral DPEN ligand and deposited on $BaSO_4$ (catalyst **10**). It was tested in 1′-acetonaphthone hydrogenation (S/C = 500, 4 atm H_2, Scheme 1.3), being reused up to 10 cycles with

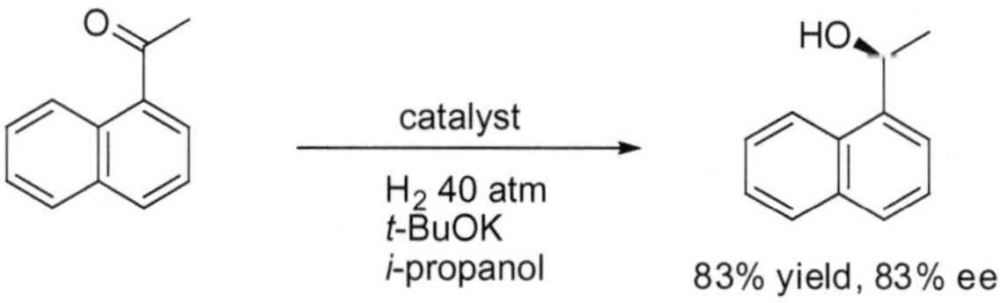

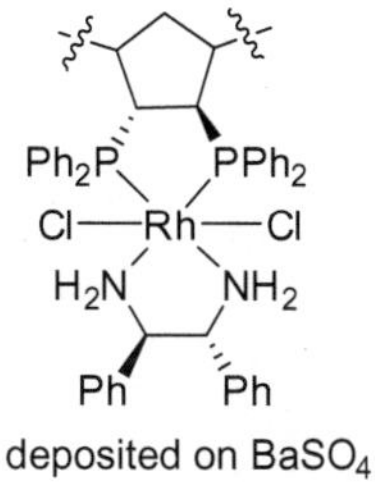

10

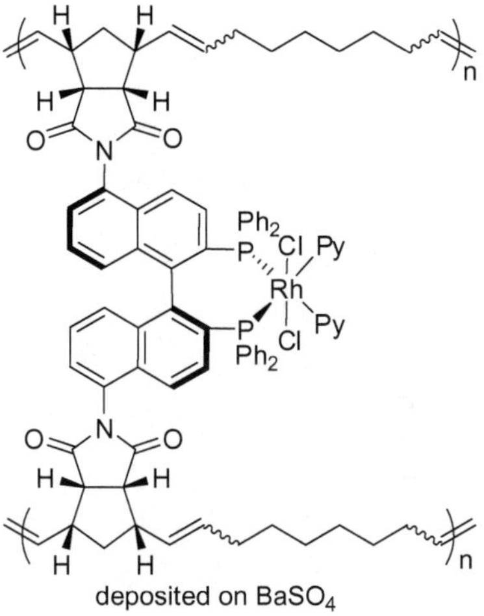

11

SCHEME 1.3. Hydrogenation of aromatic ketones using chiral Rh catalyst deposited in $BaSO_4$.

the activity reaching 100% and the optical yield maintaining above 80% ee. Despite this catalyst being less reactive than the homogeneous one, it proved to be by far more selective (83% vs. 48% ee) [31].

The enantioselectivity reached 95% ee when the bisphosphine ligand used was exchanged by a BINAP derivative **11**. Furthermore, this catalyst could be reused 25 times without any detrimental effect on the conversion and enantioselectivity using 0.1 mol% of catalyst [32].

Kantam et al. reported the application of nanocrystalline copper oxide/BINAP as a catalyst for asymmetric hydrosilylation of ketones in the absence of an inorganic base. The secondary alcohols can be obtained in very good yields and enantioselectivities after hydrolysis with tetrabutylammoniumfluoride (TBAF) (up to 99% ee). This heterogeneous catalyst was reused three times without any yield or enantioselectivity erosion [33].

For homogeneous ruthenium-mediated hydrogenation of acetoacetates, Guerreiro et al. presented a PEG-supported BINAP ligand. Excellent enantioselectivity (99% ee) was obtained with ligand **12** (Fig. 1.5) and was recycled four times with consistent efficiency [34].

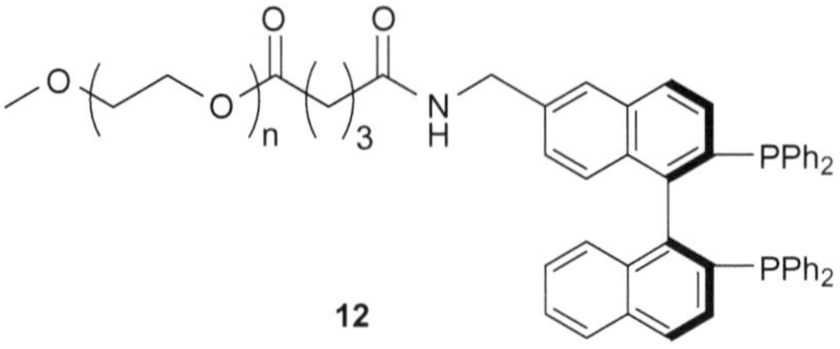

12

FIGURE 1.5. Structure of ligand **12**.

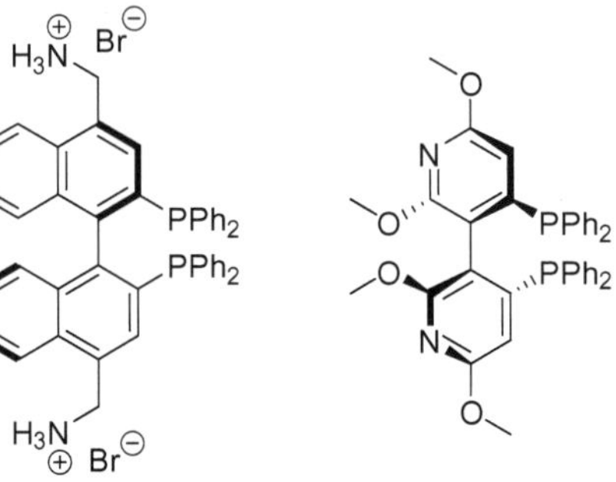

13 14

FIGURE 1.6. Structures of ligands **13** and **14**.

Zhu et al. prepared a new IL containing the *N*-n-butylpyridinium cation and the 1-carbadodecaborate anion ($BPCB_{10}H_{12}$). This new IL was applied in acetophenone and ethyl formate hydrogenation using a rhodacarborane as catalyst precursor and BINAP. Quantitative conversion was observed after 8 hours with high ee (>99%), turnover frequency (TOF) (>239/h) during six cycles [35].

Lemaire et al. applied BINAP ligand **13** (Fig. 1.6) and its 4,4′-analog in the ruthenium-mediated hydrogenation of acetoacetates (S/C = 1000 and 40 atm H_2). In a two-phase protocol, acetoacetates were added to an aqueous solution of the catalyst for hydrogenation to the respective alcohol in 99% ee (both ligands gave identical ee comparable to the original homogeneous protocol). At the end of each cycle, the product was extracted with pentane and the aqueous phase could be reused for at least eight cycles without any decrease of selectivity [36].

While looking for air-stable iridium catalyst for asymmetric imine reduction or hydrogenation, Chan et al. found that a complex with the ligand P-Phos **14** (Fig. 1.6) possessed those characteristics [37]. This catalyst was used to quantitatively hydrogenate quinolines in high enantioselectivities (90%–92% ee, 700 psi H_2, S/C = 200). To make the catalytic system more sustainable, the reaction was carried out in poly(ethylene glycol) dimethyl ether (DMPEG), providing the same levels of efficiency, especially for a biphasic mixture with

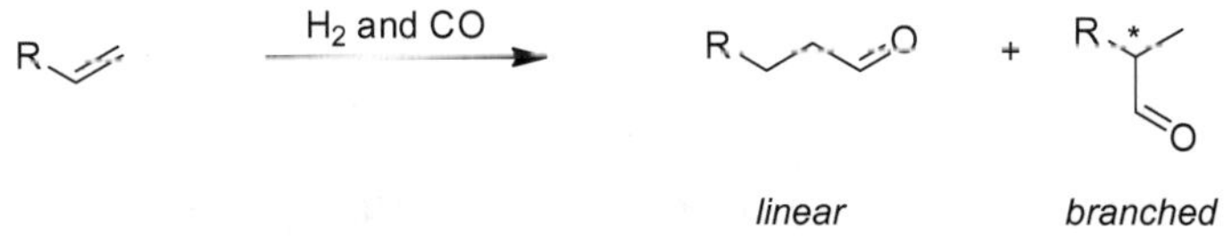

SCHEME 1.4. Hydroformylation reaction.

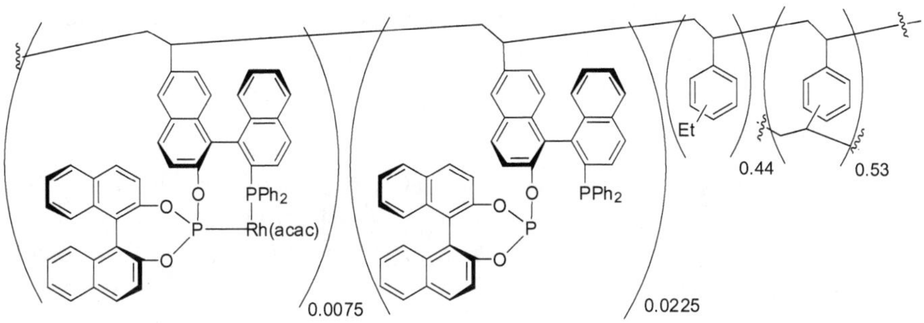

15

FIGURE 1.7. Structure of catalyst **15**.

hexane. The immobilized catalyst was reused seven times without any negative impact on the reaction enantioselectivity. This protocol was also extended to other chiral diphosphine ligands [38].

1.2.2. Hydroformylation

The hydroformylation reaction constitutes one of the most successful organometallic reactions in organic synthesis history with high atom economy (Scheme 1.4). In 1993, Nozaki and co-workers [39] combined an unsymmetrical phosphine–phosphite ligand, (*R,S*)-BINAPHOS, and a rhodium (I) catalyst to afford chiral aldehydes with excellent enantioselectivities from prochiral olefins. Since then, these catalysts, (*R,S*)-BINAPHOS-Rh(I) (acac), became (and still are) the most potent catalysts for asymmetric hydroformylation. Several efforts have been made to make this reaction more sustainable via recycling approaches. However, the immobilized catalysts showed lower catalytic activity than the free catalyst.

In order to be used in continuous vapor-flow column reactors, Shibahara and co-workers packed the catalyst **15** (Fig. 1.7) together with sea sand in a stainless column. This column reactor was used to hydroformylate 3,3,3-trifluoropropene. The authors observed that high selectivities and enantioselectivities (iso/normal: 95:5, up to 90% ee), which characterize a homogeneous protocol, were maintained despite the TOF remaining at lower levels due to the aldehydes' strong affinity for the catalyst (9 vs. 64) [40] (see also Chapter 9).

The need to obtain a general protocol for the hydroformylation of gaseous, liquid, or solid olefins motivated the authors to apply *sc*CO_2 as a solvent flow. Styrene was injected portionwise into the column reactor under the flow of syngas and *sc*CO_2. When the styrene hydroformylation was carried out at a total pressure of 120 atm, selectivity, conversion, and ee's remained constant for seven reaction cycles [80% iso aldehyde, 90% conversion, 85% ee, $Y_e(7) = 11\%$,and $ee_e(7) = 0\%$)]. This last apparatus was successfully applied to synthesize a library of optically active aldehydes, using a sequential injection protocol, with similar results compared to a homogeneous procedure.

1.2.3. Cycloaddition

Metal-based asymmetric catalysts act as Lewis acids, activating the dienophile in cycloadditions reactions. The main strategies earlier developed relied on the direct ligand linkage with recoverable support, where the several Lewis acids could be coordinated.

Hoveyda et al. immobilized a chiral phosphinimine ligand in a Wang resin to achieve its recovery and recycling [41]. This ligand **17** was applied in Danishefsky's diene, and arylimines cycloaddition reaction was catalyzed by Ag(I)/ligand in air with undistilled tetrahydrofuran (THF) as the solvent. Using 5 mol% of catalyst, 96% yield and 86% ee were obtained, and ligand reutilization was achieved in five cycles. Under the same conditions, the non-supported protocol (ligand **16**) furnishes the respective arylenamine with 98% yield and 95% ee (Scheme 1.5).

Furuno et al. reported the use of new rare earth catalysts with chiral BINAP-type phosphorates [42]. The new catalysts showed enhanced solubility in dichloromethane (without the aid of any additive) and were tested in hetero

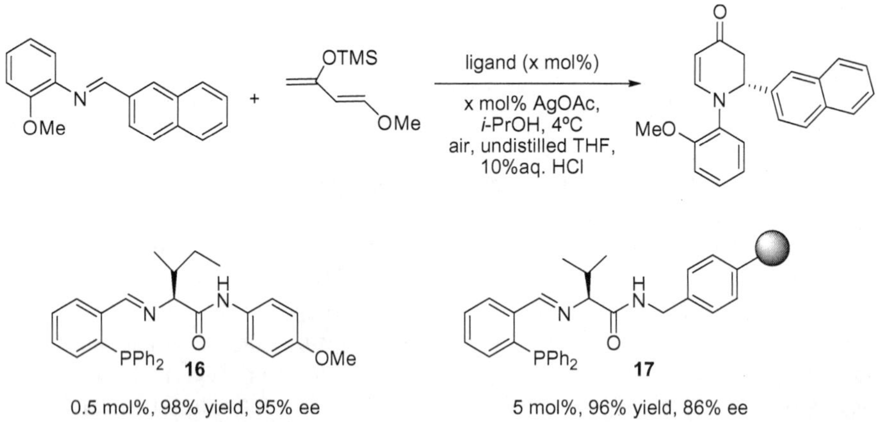

SCHEME 1.5. Asymmetric cycloaddition reactions promoted by chiral Ag–phosphinimine complex immobilized in Wang resin.

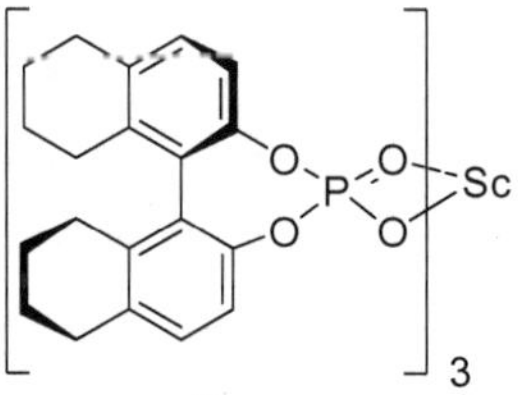

18

FIGURE 1.8. Structure of catalyst **18**.

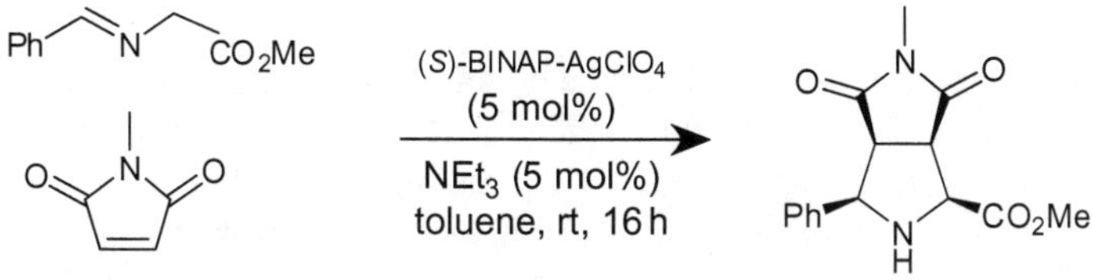

SCHEME 1.6. Asymmetric cycloaddition reactions promoted by heterogeneous Ag-BINAP catalyst.

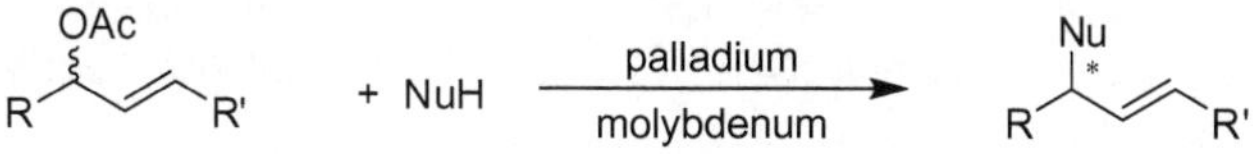

SCHEME 1.7. Asymmetric allylic substitution.

Diels–Alder reaction of Danishefsky's diene with aromatic aldehydes. Heavy lanthanides such as Yb and Er have a tendency to afford higher ee's (up to 98%) than lighter ones (up to 91%). The catalyst **18** (Fig. 1.8) proved to be the most robust combination and showed excellent performance (99% yield and 93% ee). It was also recovered by precipitation with diethyl ether and was reused up to three cycles with low yield (4%) erosion and ee (6%) erosion.

Nájera et al. discovered that the (*S*)-BINAP-$AgClO_4$ complex was insoluble in toluene, and it promoted the 1,3-dipolar cycloaddition reaction between azomethine ylides and maleimides (Scheme 1.6). The respective products were obtained in good-to-excellent yields (82%–90%), excellent *endo* selectivity (98:2), and excellent ee's (up to >99%). Due to its insolubility it was almost quantitatively recovered and could be reused efficiently up to five cycles [43].

1.2.4. Allylic Substitution

The asymmetric allylic substitution reaction (Scheme 1.7) constitutes an important tool to build chiral building blocks for several compounds with economical interest. This reaction is traditionally performed using palladium catalysts in the presence of expensive chiral mono- and bis-dentate phosphines.

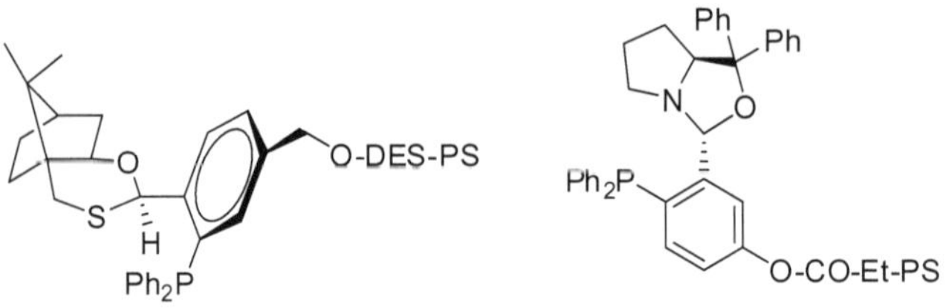

19 **20**

FIGURE 1.9. Structures of ligands **19** and **20**.

Owing to a heterogeneous approach, Nakano et al. grafted a chiral phosphinooxathiane ligand to a diethylsylilchloride polystyrene support. The ligand **19** (Fig. 1.9) promoted quantitative benzylamine condensation with allylic acetate in 99% ee. The heterogenous Pd/ligand complex was reused twice with 30 and 20% yield and enantioselectivity erosion, respectively [44].

Better results for the alkylation of allylic acetate were obtained when a chiral phosphonioxazolidine ligand **20** (Fig. 1.9) was heterogenized using the same strategy. In this case, the best support was PS–Et–COOH, providing the respective product in a 99% yield and 99% ee. Unfortunately, if no palladium was added to the next run, a 57% yield and 33% enantioselectivity erosion were observed [45].

Ding et al. prepared a new type of C_2-symmetric bisphosphine ligand bearing a cyclobutane backbone and tested it on palladium-promoted allylic alkylation and amination reaction [46]. Due to the excellent results obtained with those new ligands, considerable efforts were made to link them to soluble PEG polymers in order to accomplish their recovery at the end of each reaction. In fact, one of those polymer-supported ligands (ligand **21**) was found to be more enantioselective than its nonsupported partner (ligand **22**) (Scheme 1.8).

The supported catalyst was quantitatively recovered by filtration and reused three times in allylic alkylation and eight times in allylic amination reaction, though with some yield (11%–18%) and ee (6%–8%) erosion.

Mino et al. attached two perfluoroalkyl chains on a diaminophosphine ligand to achieve its reuse together with the palladium catalyst (Scheme 1.9) [47]. The Pd/perfluoro ligand **23** complex was tested for the allylic alkylation reaction of racemic (*E*)-1,3-diphenyl-3-acetoxyprop-1-ene with dimethylmalonate, still proving some of the qualities of the parent nonfluorinated ligand **24**. The reaction became slower, requiring a higher temperature (30°C) to achieve quantitative conversion with 90% ee. Using 5 mol% of catalyst loading, it was possible to recycle both ligand and palladium efficiently, up to six cycles with a nearly unchanged outcome.

Herein, the main strategies for chiral phosphines catalyst recycling reported are ILs, perfluoruous techniques, and polymer-supported ligands (mainly heterogeneous). There is a tendency to use heterogeneous methods to the detriment of homogeneous methods, especially when the heterogenization

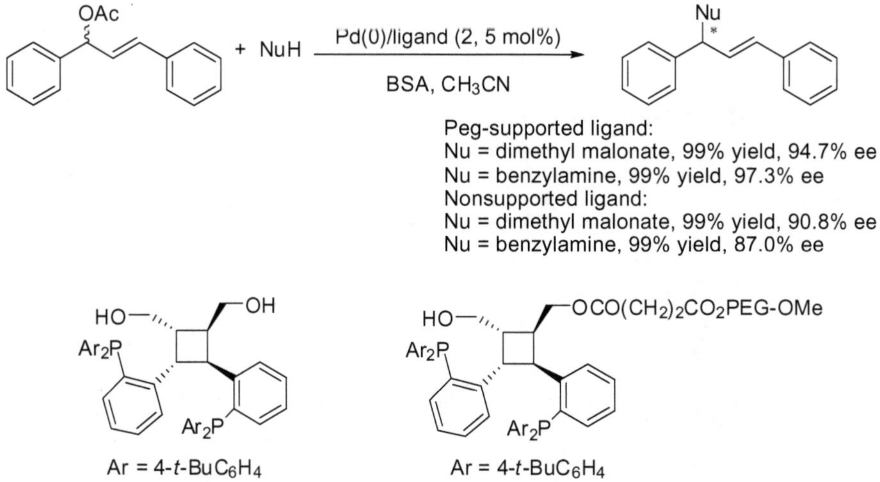

21 22

SCHEME 1.8. Palladium-catalyzed allylic alkylation promoted by C_2-symmetric bisphosphine ligand immobilized in PEG.

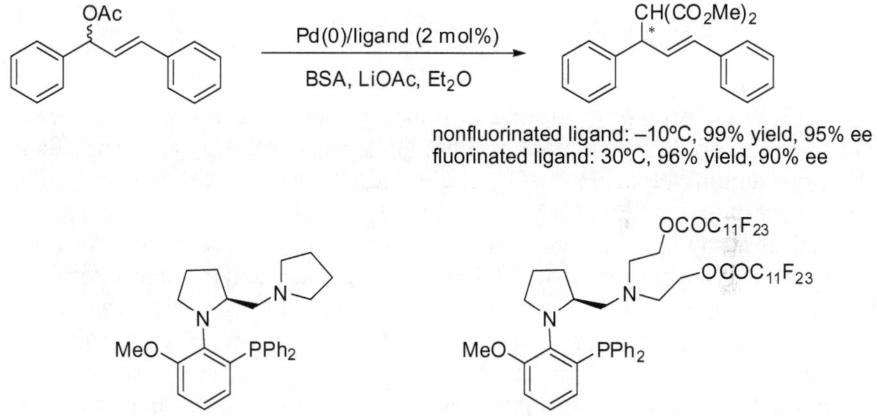

23 24

SCHEME 1.9. Asymmetric allylic alkylation promoted by Pd/perfluoro diaminophosphine complex.

does not involve the chiral ligand modification. Nevertheless, in some reactions the reutilization approaches still needs to be improved.

1.3. CHIRAL ALKALOIDS

Osmium-catalyzed asymmetric dihydroxylation (AD) of olefins is a robust methodology to conceive a wide range of enantiomerically pure vicinal diols

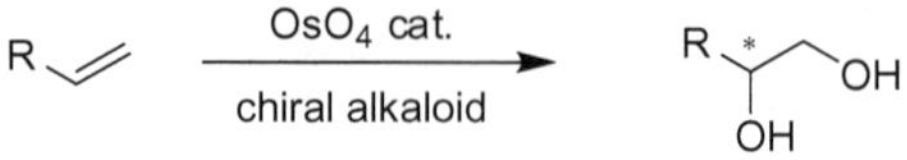

SCHEME 1.10. Asymmetric dihydroxylation reaction of alkenes.

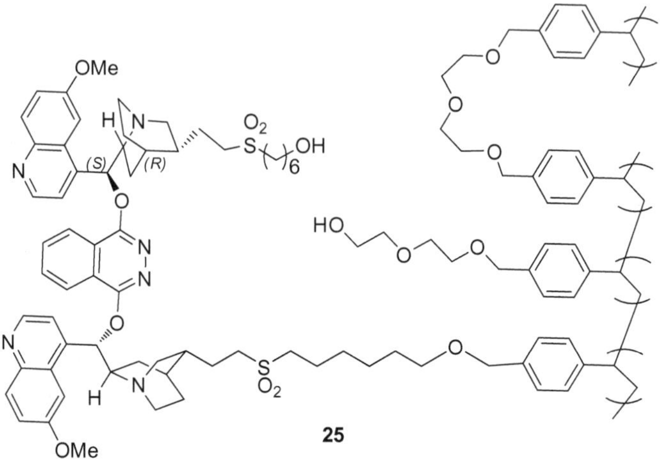

25

FIGURE 1.10. Structures of the supported ligand **25**.

[48]. This transformation became an enantioselective catalysis event when Markó et al. applied *N*-methylmorpholine-*N*-oxide (NMO) as a cooxidant, allowing osmium tetroxide (OsO_4) and a chiral cinchona alkaloid ligand to be used in catalytic amounts (Scheme 1.10) [49].

Although AD is a very important transformation, it uses expensive reactants (OsO_4 and a cinchona alkaloid) and a very toxic (and in some cases volatile) osmium component. To overcome these drawbacks, some alternatives have been developed, such as the use of $K_2OsO_2(OH)_4$ in place of OsO_4, and recovery and reuse of the Os ligand with high efficiency and without osmium contamination in the final product. Several strategies were developed for this purpose, namely by anchoring the ligand to a soluble or insoluble polymer matrix to an inorganic solid support (silica), or by immobilization of an osmium component through microencapsulation or ion exchange techniques [2–6c].

Salvadori et al. synthesized the supported ligand **25** (Fig. 1.10) with a copolymer architecture. When submitted to the same AD conditions it proved to be very efficient for typical substrates, with conversions up to 100% affording the respective diols in high ee's (87%–99%). Under recycle assays, using styrene as a starting material, a small amount of OsO_4 (0.35% mol) was added to account for losses during the recovery of the ligand at the beginning of each new cycle. The conversion and optical purity were analyzed at 1.5 hours of reaction time. The enantioselectivity fell in the 87%–91% range while the conversion fell in the 72%–94% range during 12 recycles [50].

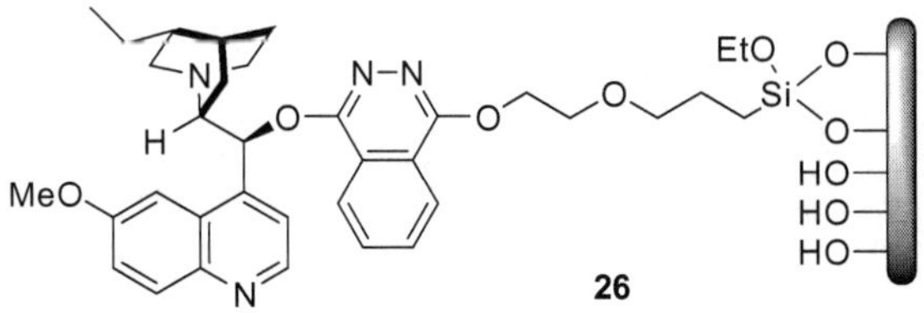

26

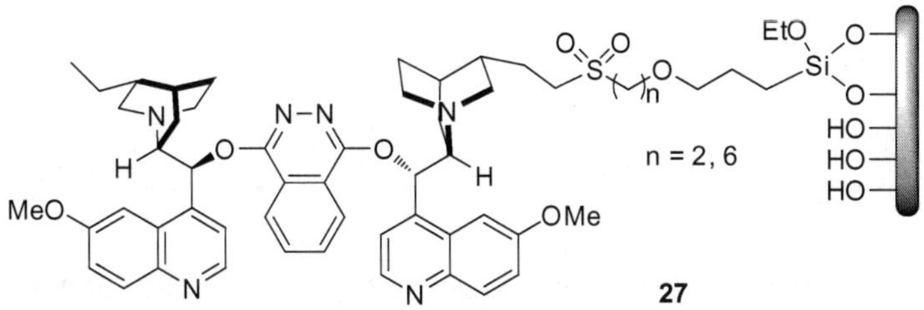

27

FIGURE 1.11. Structures of silica-supported *Cinchona* ligands **26** and **27**.

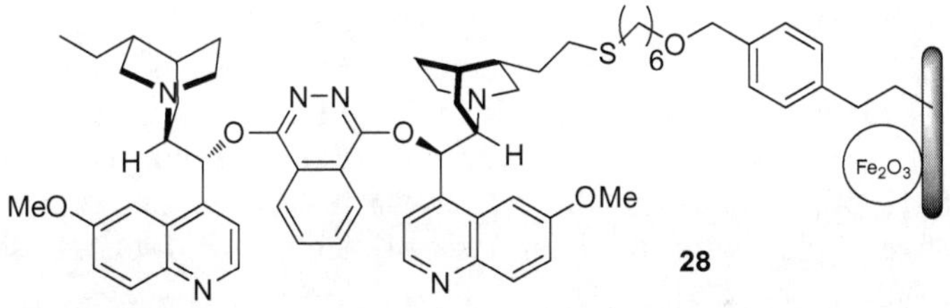

28

FIGURE 1.12. Structures of Fe_3O_4-supported $(DHQ)_2PHAL$ ligand **28**.

Silica was the preferred inorganic material used for this kind of *Cinchona* ligands immobilization. Lee and co-workers achieved high yields (up to 97%) and ee's (>99%) with catalysts **26** and **27** (Fig. 1.11) using *trans*-stilbene as a starting material; moreover, recycling experiments were successful for six runs [51].

Lee et al. prepared mesocellular silica foam with ordered mesoporous walls containing magnetic nanoparticles (γ-Fe_2O_3) and $(DHQ)_2PHAL$ ligand grafted in the pore surface (ligand **28**, Fig. 1.12) [52]. This ligand was tested for AD reaction using $K_3Fe(CN)_6$ as cooxidant (1 mol% catalyst). Several aromatic olefins were successfully dihydroxylated in high yields (up to 95%) and excellent ee's (up to 99.5%). At the end of each cycle, the supported ligand was separated by decantation with the aid of a magnet (magnetic separation) and reused seven times without any loss of efficiency. Unfortunately, it was necessary to add more osmium tetroxide at the end of each reaction to maintain a high level of conversion.

TABLE 1.3. AD Reaction with Heterogenized OsO_4

	$K_2OsO_4.2H_2O$-$Na_2WO_4.2H_2O$		LDH-OsW	
Olefin	Yield (%)	ee (%)	Yield (%)	ee (%)
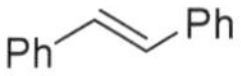	85	99	93	99
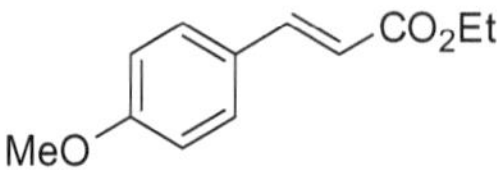	82	99	89	99
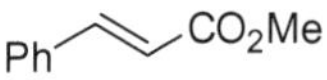	91	99	92	99

Following their work in ion exchange techniques with double-layered hydroxides, Choudary et al. designed a bifunctional catalyst composed of OsO_4^-–WO_4^- for one-pot synthesis of chiral diols via *N*-oxidation–AD reaction. This heterogeneous catalyst was found to be more active under AD conditions than its homogeneous analog, providing diols with comparable or even higher yields and ee's (*N*-methylmorpholine (NMM), Table 1.3). Furthermore, it provided the advantage of easy separation from organic products and uses lower amounts of the chiral alkaloid ligand $(DHD)_2$PHAL as well [53]. This methodology offers diols with higher yields and ee's than the Kobayashi PSresin-MC Os system [2, 54]. In addition, this system was easily recycled and reused for at least five cycles with minimal loss of activity and enantioselectivity [$Y_e(5) = 2\%$, $ee_e(5) = 0\%$].

Motivated by these results, the authors idealized and developed a trifunctional catalyst based on $PdCl_4^-$–OsO_4^-–WO_4^- for tandem Heck coupling–*N*-oxidation–AD reactions, using the same ion exchange technology. The desired diol was obtained in 85% yield with 99% ee. The LDH-PdOsW catalyst was recovered quantitatively by filtration, while the chiral ligand could be recovered (up to 95% recovery) by a simple acid/base extraction. The recovered catalyst was submitted to recycling assays without a significant drop in activity [$Y_e(5) = 3.5\%$] for at least five runs. To prove the synthetic utility of this trifunctional catalyst, the authors applied it in the synthesis of two well-known drugs, Diltiazem and Taxol side chain, minimizing in this way the number of unit operations needed in the process. The success of this kind of catalyst prompted the filing of several patents [55].

One of the major advantages of soluble polymer-bounded catalyst resides in the absence of diffusion problems of the reactants, which traditionally characterizes heterogeneous systems.

Bolm and Gerlach developed a hydroquinidinediether (DHQD) ligand attached to methoxy (MeO)-PEG via an aryl spacer (linked to 4,6-bis-(9-*O*-dihydrochininyl)pyrimidine (PYR) segment, **29**, Fig. 1.13) [56]. They successfully dihydroxylated several vinylic and aliphatic olefins in high yields and high

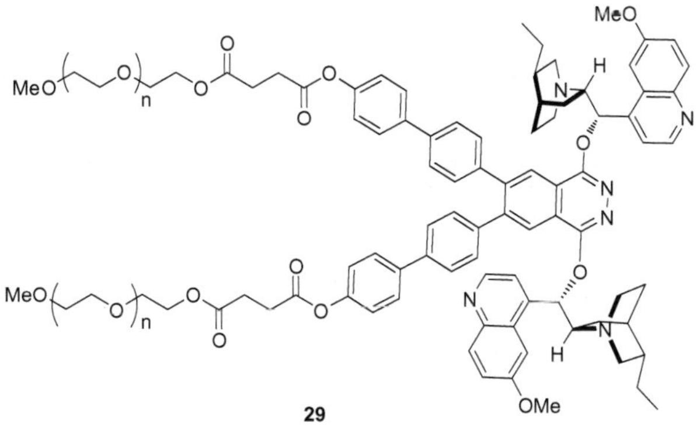

29

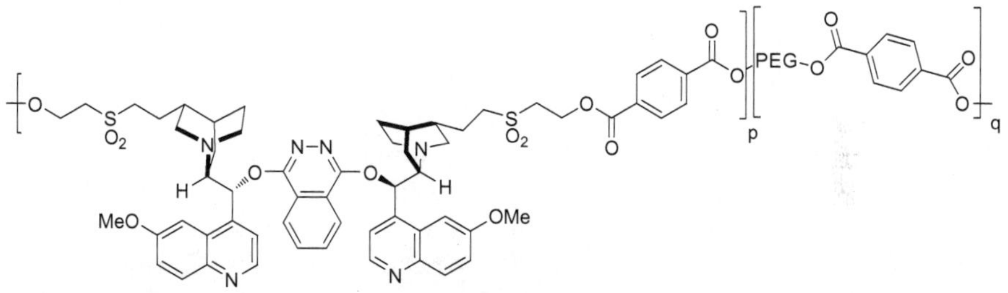

30

FIGURE 1.13. Structures of ligands **29** and **30**.

ee's—95%–99% and 87%–90% ee, respectively. This system was reused six times without a detrimental change on its enantioselectivity [$ee_e(6) = 2\%$].

Zhang et al. reported the immobilization of cinchona alkaloids in PEG-4000 [57] (Mw = 4000) instead of PEG-OMe [58] as reported in previous work [58]. The ligand **30** (Fig. 1.13) was always recovered in good yields (95%) and reused four times without changes on the activity (high yields up to 87%, up to 99% ee).

Taking advantage of greater affinity of the alkaloid-OsO_4 complex to the room temperature ionic liquids (RTILs) phase, Branco and Afonso started using RTILs to scope the substitution of the standard organic solvent (^{t}BuOH) used in AD reactions. In this way, they studied two different solvent systems: a biphasic RTIL/H_2O system and a monophasic RTIL/H_2O/^{t}BuOH system, using two different ligands [$(DHQD)_2$PHAL and $(DHQD)_2$PYR]. Using [bmim]PF_6, both systems gave similar or even higher results when compared with the traditional ^{t}BuOH/H_2O system [$K_2OsO_2(OH)_4$ (0.5 mol%), $(DHQD)_2$PYR/PHAL (1 mol%), $K_3Fe(CN)_6$ (3 eq), solvent, room temperature during 24 hours, Table 1.4] [59].

Once again, the use of $K_2OsO_2(OH)_4$ as oxidant proved to be less efficient, affording diols with lower chemical yield and ee's than OsO_4. The authors believe that this decrease was well compensated for by using a less toxic oxidant.

TABLE 1.4. Comparative Dihydroxylations Reactions Run in Different Reaction Media

		$^tBuOH/H_2O$		$RTIL/H_2O$		$RTIL/^tBuOH/H_2O$	
Olefin	Ligand	Yield (%)	ee (%)	Yield (%)	ee (%)	Yield (%)	ee (%)
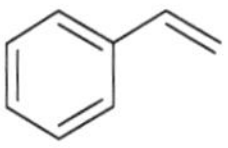	PHAL	88	97	87	62	86	94
	PYR	91	93	86	75	90	89
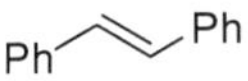	PHAL	89	96	87	98	92	99
	PYR	94	87	81	96	79	77
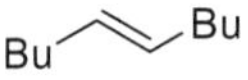	PHAL	94	94	69	87	96	92
	PYR	79	75	52	63	92	96

Both systems could be easily recycled as the products were extracted with diethyl ether. The biphasic system was efficiently reused for nine runs without loss of activity and enantioselectivity [$Y_e(9) = 10\%$, $ee_e(9) = 4\%$ and total turnover number (TON) = 1334], while the monophasic system proved to be a little more robust since it was successfully recycled 10 times [$Y_e(10) = 12.5\%$, $ee_e(10) = 9\%$ and total TON = 1720], affording higher yields. In all assays, the osmium content on the extracted phase was analyzed as being under 4% on average from its initial amount (0.5 mol%). Later, the authors achieved comparable results without the slow addition of olefin and a chiral ligand by using chiral ILs based on the combination of guanidinium cation and readily available chiral anion such as that derived from quinic acid and lactic acid [60].

Song et al. developed a simple method to achieve both ligand $(QN)_2$PHAL and osmium tetroxide recycling without the aid of any solid or polymeric support, nor with the use of an alternative reaction media such as ILs or perfluorophases [61]. In a simple and traditional homogeneous protocol, sucrose was added to the reaction media prior to the AD reaction. In the end, only the AD product could be extracted to an ethereal phase while the ligand and osmium products remained in the water phase. This retention was due to hydrogen bonding between the ligand and the sucrose (sugar, Fig. 1.14). It should be remembered that the applied ligand also undergoes AD and becomes more soluble in water. This system could be reused three times with an average activity erosion of 15%.

More recently, Kim et al. developed a similar strategy to recover the catalyst under traditional homogeneous conditions (Table 1.5) [62]. At the end of the reaction, osmium tetroxide was mainly in the organic phase (as neutral OsO_4) due to the large excess of cooxidant employed. Addition of a low-weight olefin like ethylvinylether returns the reduced specie (OsO_4^{2-}) to the aqueous phase, which will be retained when all the cooxidant is consumed. Solvent extraction allowed the immobilization of 89% of the osmium catalyst in water, which can be used in eight consecutive cycles with minimal yield erosion.

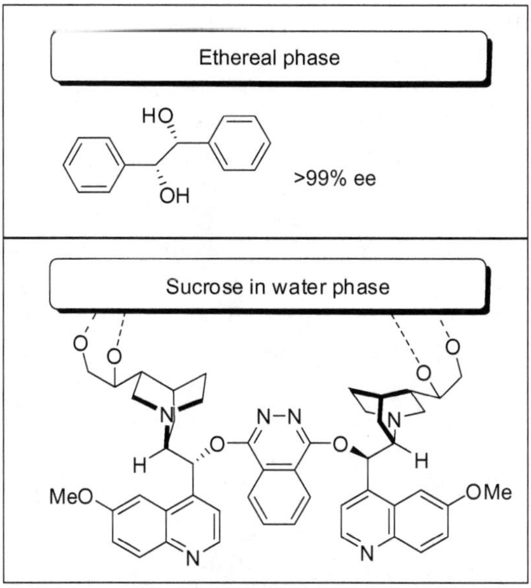

FIGURE 1.14. Catalytic system recycling mechanism.

TABLE 1.5. AD Using Chemoentrapment Strategy

Entry	Olefin	Extracting Solvent	Yield (%)	ee (%)
1	Styrene	Ethyl acetate	97	96.2
2	Styrene	TBME	97	97
3	α-Methylstyrene	TBME	96	89.6
4	*trans*-Stilbene	TBME	96	>99.5

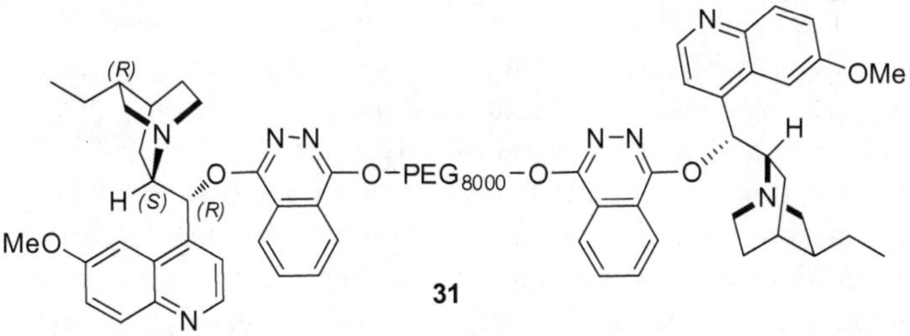

31

FIGURE 1.15. Structure of the PEG_{8000}-supported ligand **31**.

Xu et al. developed a methodology of preparing a family of new PEG-bound bis-cinchona alkaloid ligands. This new family was evaluated on cinnamic esters with amino-hydroxylation in order to achieve the ideal polymer length. The ligand that had the biggest polymeric chain (PEG_{8000}, ligand **31**, Fig. 1.15) behaved better in terms of efficiency (90% yield, 99% ee) and allowed the capture of osmium during the recycling procedure (precipitation upon the addition of ethyl ether). Despite recyclability, it achieved five cycles, but decreasing reactivity in each run was observed [63].

In previous years, some of the methods for AD reactions had already afforded diols with excellent yields and enantiopurity. In this way, AD studies became more focused on obtaining the optimal method, where both the osmium and ligand can be recovered and reused with high efficiency. These efforts have been directed not only in terms of a large number of successful recycles, but also in lowering the catalyst systems loadings. In contrast, less extended approaches have been reported for the parent asymmetric aminohydroxylation (AA) reaction.

1.4. BISOXAZOLINES

The bisoxazoline ligands (BOX ligands) rose as one of the most important family of ligands in asymmetric catalysis, which are easily synthesized to take advantage of the chirality nature of amino acids. Their principal role in catalysis is to act as a ligand for copper catalysts, being applied in a wide range of transformations in which it could behave as Lewis acids or as carbene generators [64]. Depending on the link between bisoxazoline rings, it is possible to have Aza-BOX (nitrogen), BOX (dimethyl methylene), Phe-BOX (phenyl), and Py-BOX (pyridine) ligand families. It is true that bisoxazolines are not as expensive as chiral phosphines; however, the reutilization of Cu-bisoxazolines catalysts can make their industrial application more viable and prevent high degrees of copper contamination on the final products.

1.4.1. Py-BOX Ligands

Regarding Py-BOX ligands, several methods were applied to achieve the recycling of the catalytic system. The main benchmark transformations used to test the efficiency of each system are alkyne addition to imines, ruthenium-catalyzed cyclopropanations, indium-catalyzed aldehydes allylation, and lanthanides-catalyzed silylcyanation of aldehydes.

Afonso et al. demonstrated that RTILs can be applied as solvent media for a Cu(I)-promoted phenylacetylene in addition to *N*-benzylidene-aniline. The transformation occurred with the same efficiency compared with toluene, but some trouble appeared when the catalytic system was recycled (Scheme 1.11) [65]. Hexane was not polar enough to fully extract the reaction product from the IL, while diethyl ether was able to accomplish such task. Unfortunately, 25% of catalyst was also extracted to the ethereal phase.

Li et al. achieved an identical recyclability when the same transformation was conducted in water in the presence of a surfactant (stearic acid, Scheme 1.11) [66]. Under certain conditions the hydrolysis of Py-BOX ligands in the presence of Lewis acids [$Ce(OTf)_4$] or in acidic conditions has been observed [67]. No signs of decomposition were reported on this example.

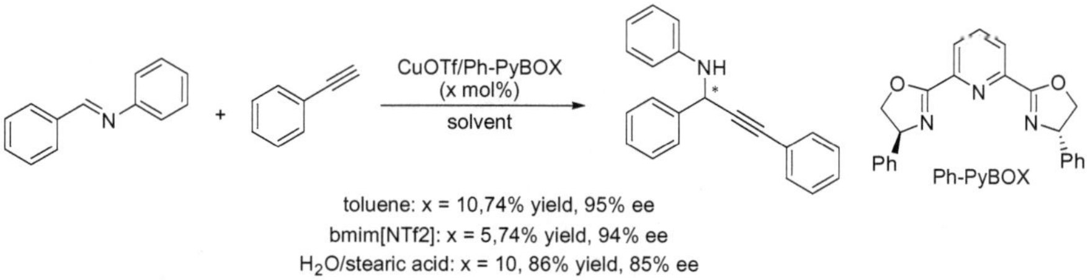

SCHEME 1.11. Asymmetric addition of alkynes to imines promoted by Cu-Ph-PyBOX complex in ILs and water/stearic acid medium.

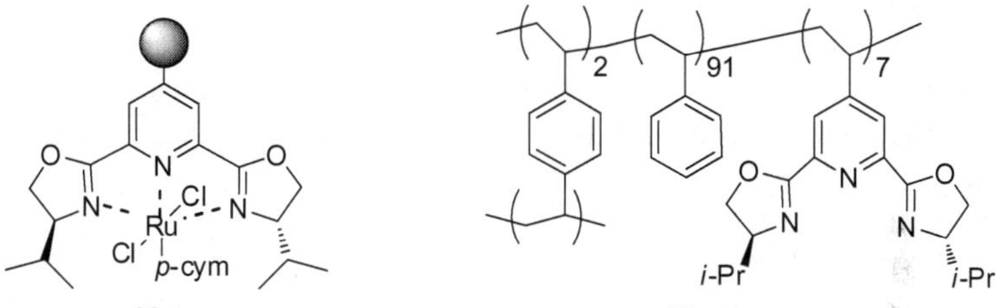

FIGURE 1.16. Structures of immobilized Ru-PyBOX catalyst **32** and polymeric ligand **33**.

Alternatively, O'Leary and co-workers immobilized a Cu(I)- and Cu(II)-Ph-PyBOX triflate complex on silica due to electrostatic interactions, but the catalyst's reusability was studied only during three cycles [68]. The immobilization of the Py-BOX ligand in a Wang resin did not provide a very recyclable system for the same transformation [69]. The reactions performed in IL provided the higher levels of enantiomeric induction, but the water/stearic acid system was more efficiently recyclable.

A series of works were conducted for the reutilization of a Py-BOX ligand with soluble and insoluble copolymers with high cross-linking. Mayoral et al. immobilized the Ru-PyBOX catalyst **32** (Fig. 1.16) to a polystyrene polymer by copolymerization. With this catalyst, the respective cyclopropanes (derived from styrene and ethyl diazoacetate) were isolated with high *trans* selectivity (85:15 *trans/cis*), in which the major diasteriomer was obtained in 85% ee's with 30% yields. The lack of reactivity could be explained by the presence of several inaccessible catalyst sites inside the polymer chains. In the third consecutive cycle, both activity and enantioselectivity dropped about 50% [70]. Recently, the authors prepared a monolithic polystyrene resin bearing the Ru-PyBOX to be used under continuous flow conditions [71].

The polymeric ligand **33** (Fig. 1.16) with a lower cross-linking degree provided an efficient catalyst (52%–70% yields, *trans/cis* selectivity of 90:10 and enantioselectivities of around 85%–91% for the *trans*-cyclopropanes) [72]. This catalytic system was reused consistently during four cycles;

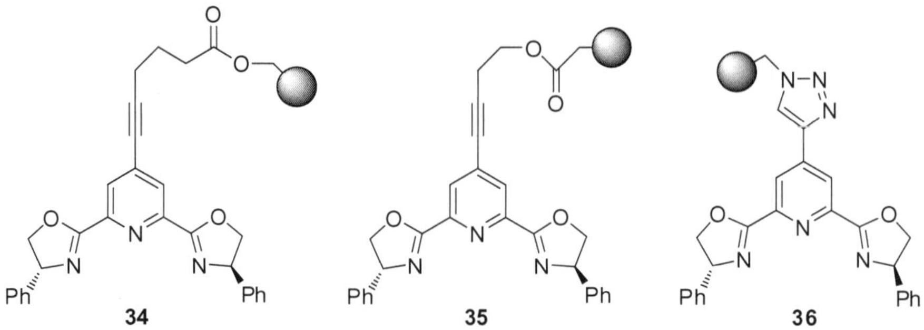

FIGURE 1.17. Structures of heterogeneous Ph-PyBOX ligands **34**–**36**.

however, the recovery protocol required inert conditions. Grafting the ligand in Merrifield resin proved slightly less efficient than catalysts prepared from copolymerization.

Although low cross-linking polymerization is the best method for the immobilization of the Ru-PyBOX system (when looking at the overall efficiency of the reaction: yield and enantioselectivity), microencapsulation appears as a competitive alternative since no ligand modification is required [73]. Both polystyrene and Ru-PyBOX are soluble in dichloromethane (the solvent used in the cyclopropanation reaction), but when hexane was added, the polymer has a tendency to microencapsulate the catalyst, precipitating together and allowing its recovery.

In(III)-PyBOX was used as a catalyst for allylation of ketones and aldehydes in a mixture of [hmim]PF_6/dichloromethane, and its reutilization was achieved up to four cycles with consistent yield and slightly decreasing ee [$ee_e(4) = 12\%$] [74]. In 2003, the Ph-PyBOX ligand was immobilized to TentaGel resins. Both heterogeneous Ph-PyBOX ligands **34** and **35** (20 mol%) (Fig. 1.17) were tested in the catalytic asymmetric silylcyanation of benzaldehyde in the presence of 10 mol% of $YbCl_3$, being slightly less enantioselective than the homogeneous analog. Curiously, both polymeric ligands showed an outstanding lifetime since they can be used in 30 consecutive cycles without any change in the reaction outcome (enantioselectivity and conversion). However, the catalytic system (ytterbium/ligand complex) was only efficiently reused during four cycles, due to metal-leaching phenomena during the filtration procedures [75]. More recently, the same ligand was grafted to polystyrene resin, bearing an azide group via click chemistry [76]. Ligand **36** (Fig. 1.17) was tested in the cyanosilylation of aromatic aldehydes promoted by ytterbium and lutetium. High conversions (68%–87%) and high enantioselectivities (67%–78% ee) were obtained for this reaction in four consecutive cycles (10 mol% of catalyst, acetonitrile/dichloromethane 3:2) [77].

1.4.2. BOX Ligands

Cu-BOX catalysts have been addressed for recycling for cyclopropanation, carbonyl-ene, Mukaiyama aldol, and cyclo addition reactions. The enantioselectivity of a cyclopropanation reaction catalyzed by Cu-BOX catalysts can be extremely sensitive to two parameters: the presence of halide and/or water impurities and the alterations of the bite angle in the BOX ligand. Davies et al. showed that the presence of 5% of halide impurities on an RTIL's phase was more than enough to deactivate completely the copper catalyst for cyclopropanation reactions [78]. Due to their synthesis strategy, RTILs in general can have halide impurities that could prevent a smooth cyclopropanation reaction in this media. Furthermore, since RTILs are in general hygroscopic, that limits their application in this transformation. It is known that a BOX bearing two methyl groups attached in the C-bridge atom provides cyclopropanes with higher enantioselectivities compared with BOX ligands bearing benzyl groups [64]. This is due to the alteration of the bite angle of the ligand.

The first studies for BOX ligand immobilization in polymeric structures were conducted using BOX ligands bearing two styril moyeties linked in the bridge carbon atom that could be polymerized and copolymerized with styrene (and cross-linker, ligands **37** and **38**, Fig. 1.18) [79]. The polymeric ligands were tested in the benchmark cyclopropanation reaction (styrene cyclopropanation with diazoacetate), affording the respective cyclopropanes with an efficiency similar to the one observed with homogeneous nonsupported ligands [37:63 *trans/cis* selectivity, 78%:72% ee (*trans/cis*)].

The rapid immobilization of such ligands in polystyrene was accomplished using only one flexible linker that separated the ligand from the polymer backbone (Scheme 1.12) [80]. Not only did the immobilization have a minor effect on the enantioselectivity of the reaction, but the polymeric catalyst (Cu/BOX **39**, Fig. 1.19) could be recycled four times without any efficiency erosion.

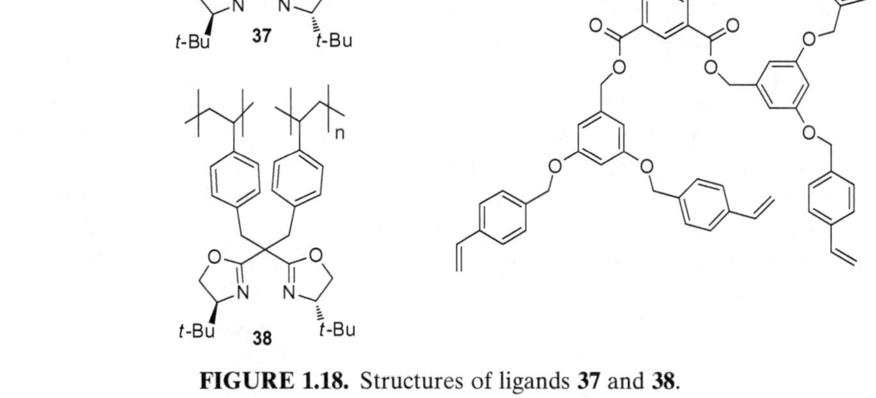

FIGURE 1.18. Structures of ligands **37** and **38**.

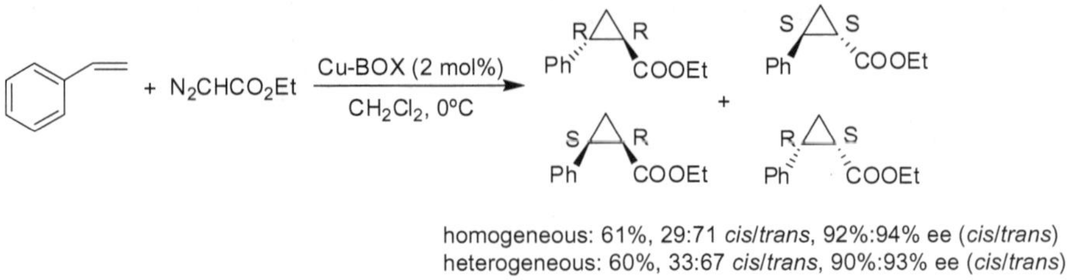

SCHEME 1.12. Benchmark cyclopropanation reaction promoted by copper/polymer-supported BOX complex.

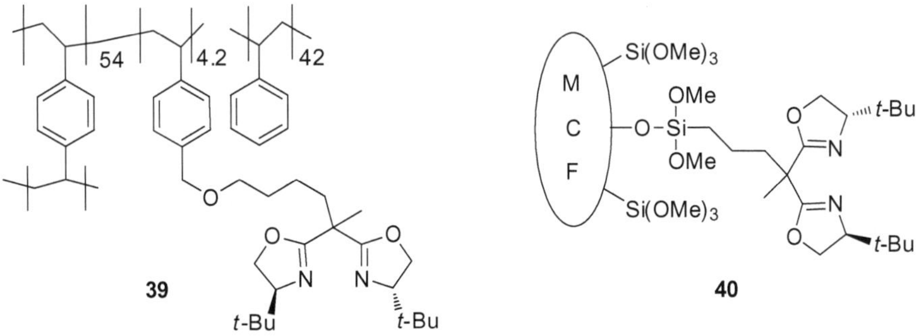

39

40

FIGURE 1.19. Structures of ligands **39** and **40**.

As an alternative to organic polymeric supports, Ying et al. grafted a chiral BOX ligand to the surface of siliceous mesocellular foams (MCFs) using only one silanized linker [81]. It was necessary to protect free silanols groups with trimethylsilane in order to achieve excellent enantioselectivities. The heterogeneous ligand **40** (Fig. 1.19) was applied in styrene cyclopropanation, providing cyclopropanes in 80% yield with excellent enatiomeric excesses (65:35 *trans*/*cis*, 95%:92% ee *trans*/*cis*, comparable with the homogeneous counterpart). The catalyst loading could be decreased to 0.2 mol% without compromising the reaction outcome. The Cu/ligand **40** was reused for 12 consecutive reactions without losing activity or enantioselectivity. Once again, monotethered Box ligands proved to be more selective than bis-tethered ones [82].

Sometimes, the immobilization of a catalyst can alter considerably the reaction selectivity compared with the one observed in homogeneous solution. This phenomenon revealed itself when a Cu-BOX complex was immobilized in laponite containing a supported IL film. This heterogeneous catalyst was tested in a benchmark styrene cyclopropanation reaction with diethyl diazoacetate. Despite the moderate success, a totally different *trans*/*cis* ratio was disclosed when compared with the same reaction run in IL media. It was also observed that the thickness of the film had a crucial influence on the selectivity [83].

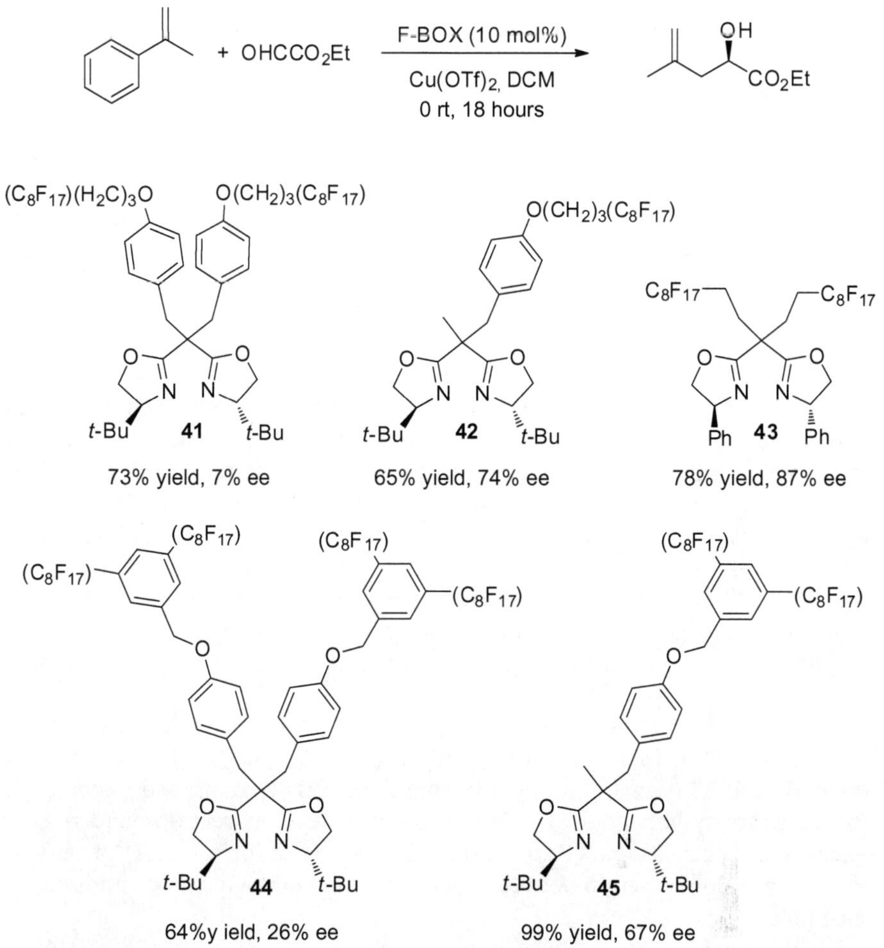

SCHEME 1.13. Asymmetric carbonyl-ene reaction promoted by Cu–perfluoroBox complexes.

The carbonyl-ene reaction catalyzed by copper-BOX complexes also proved to be considerably dependent on the BOX bite angle. In Scheme 1.13, it is possible to compare the evolution of enantioselectivity in the function of the fluorous tail linked to the BOX ligand [84].

The BOX **43** provided higher levels of enantioselectivity compared with other fluorinated BOX (**41–45**, Scheme 1.13). The catalyst was reused efficiently at least five times with the aid of fluorous reverse phase silica gel [85].

Kanemasa et al. attached the BOX ligand to gold nanoparticles. Despite the excellent enantioselectivities obtained during five cycles, the recovery protocol is rather complex, which could represent an important drawback [86].

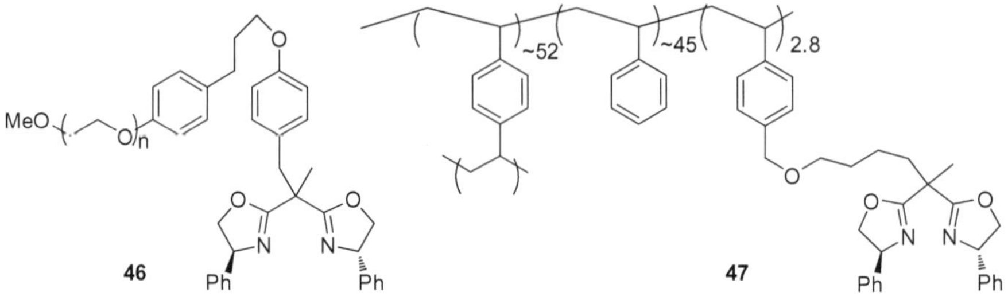

46 47

FIGURE 1.20. Structures of ligands **46** and **47**.

For this transformation, the best results were obtained when the BOX was grafted to PEG and copolymerized with styrene/divinylbenzene. The benchmark carbonyl-ene reaction between α-methylstyrene and glyoxylate afforded the respective product in 96% yield with 95% ee (10 mol% Cu/ligand **46**, Fig. 1.20). The supported catalyst was precipitated with diethyl ether and reused twice with marginal loss of efficiency [2]. The copolymer **47** (Fig. 1.20) in a 10 mol% loading provided 96% yield with 90% ee for the same reaction. In the recycling assays, consistent 90% ee was obtained during five cycles. However, the reaction time required to achieve high conversions increased gradually in each cycle [87].

Ligand **46** was studied in copper-catalyzed Mukaiyama aldol reactions in water (20 mol%). Low chemical yield was accounted for due to the low aldehyde solubility in water, but the enantioselectivity remained unchanged. The catalytic system could be 85% recovered with extractions and reused twice upon pretreatment [88]. Ligand **42** was tested for a Cu(II)-promoted Mukaiyama aldol reaction between silylketene and ethylglyoxalate and reused once [84b].

When the BOX ligand was immobilized on mesoporous silica SBA-15 (ligand **48**, Fig. 1.21) and in the core of dendritic structures (ligand **49**, Fig. 1.21), it proved to be at least as efficient as the nonsupported counterpart. The copper/**48** catalyst was easily recovered and reused five times for nitro-Mannich with gradual loss of enantioselectivity and diastereoselectivity. The catalyst heterogenization had a positive impact on the reaction enantioselectivity surpassing the homogeneous catalyst [89]. The copper/**49** catalyst proved to be more efficient than the nonsupported catalyst in the Mukaiyama aldol reaction, affording 1,3-dydroxyketone with 78% yield in 64% ee and 2.2:1 *syn*/*anti* selectivity (Scheme 1.14) [90].

In the following works, the application of IL enhanced ecyclability. It appears that ILs act as a protecting agent toward metal leaching that would lead to loss of catalytic activity.

Doherty et al. showed that ligand **50** (Fig. 1.21) provided the same levels of enantioselectivity (90% ee) in a Cu(II)-catalyzed Mukaiyama aldol reaction

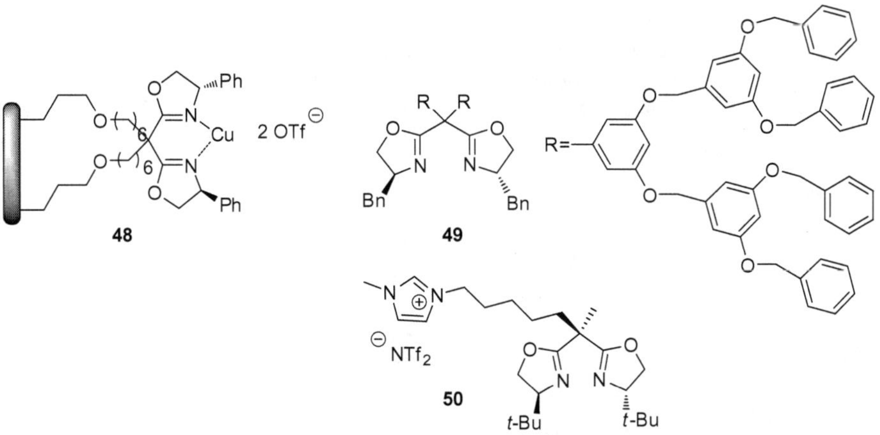

FIGURE 1.21. Structures of catalyst **48** and ligands **49** and **50**.

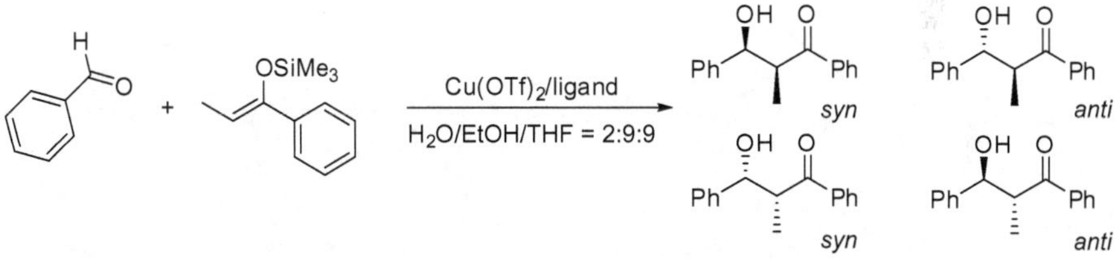

SCHEME 1.14. Mukaiyama aldol reaction catalyzed by copper in the presence of dendritic Box ligand.

between methyl pyruvate and 1-methoxy-1-trimethylsilyloxypropene in [emim]NTf_2 that are observed in dichloromethane at −78°C. The catalytic activity of this system was retained over three cycles. The electrostatic immobilization of the catalytic system was achieved when the silica treated with IL was used as an immobilizing agent. The reusability in this case reached eight cycles with excellent enantioselectivity [91].

The same ligand **50** was also explored in the copper-catalyzed Diels–Alder reaction between *N*-acryloyloxazolidinone and cyclopentadiene in [emim] NTf_2, providing identical levels of recyclability (nine cycles). An accelerating effect was observed that allowed more demanding substrates to be scoped [92].

Kim et al. observed that the same transformation catalyzed by Cu-Inda-BOX ligand can have its enantioselectivity enhanced up to 94% ee when run in [bmim]SbF_6 instead of in dichloromethane (8 mol% catalyst, 3°C, 10 minutes, Table 1.6). Furthermore, the authors show that the catalytic system and IL could be recycled 16 times with residual enantioselectivity erosion (88%–92% yields, 10 minutes) [93].

TABLE 1.6. Asymmetric Diels–Alder Reaction Between Cyclopentadiene and *N*-acryloyloxazolidinone with Recycled [Bmim]SbF_6 Containing IndaBOX–Cu Complex

Run	*Endo*/*Exo*	*Endo* ee (%)
1	97:3	94
6	96:4	88
10	95:5	86
15	93:7	84
17	93:7	84

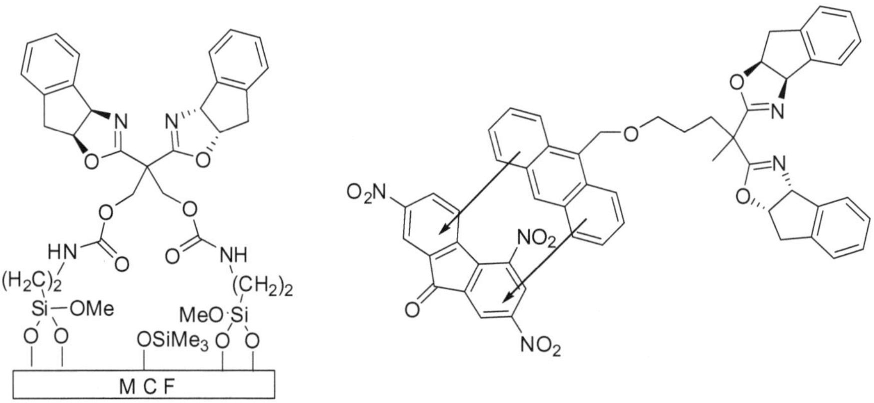

51 **52**

FIGURE 1.22. Structures of ligands **51** and **52**.

Other immobilization strategies were applied to the Cu-Inda-BOX catalyst, such as grafting in siliceous materials, copolymerization with methylene diphenyl isocyanate, and immobilization in MCFs [94]. The best results were obtained in the heterogenization on MCFs after capping the free silanols groups (ligand **51**, Fig. 1.22). The enantioselectivity of the Cu(II)-catalyzed cycloaddition of 3-cryloyl-2-oxazolidinone with cyclopentadiene reached 88% ee with quantitative conversion in 8 hours and 96% *endo* selectivity. The catalyst was reused at least four times without any observable efficiency loss [95].

Chollet et al. reported a very peculiar way to recycle a Cu-Inda-BOX complex by preparing a charge transfer complex that could be recovered by precipitation. Ligand **52** (Fig. 1.22) was applied in Cu(II)-promoted cycloaddition of 3-acryloyl-2-oxazolidinone with cyclopentadiene, providing the respective product with quantitative conversion, 93% de, and 84% ee (Scheme 1.15). The system was reused with an extraordinary longevity (11 recycles). Surprisingly, the catalyst became more efficient as it was being reused since the reaction time reduced from 44 hours (run 1) to 0.5 hours (run 11) [96].

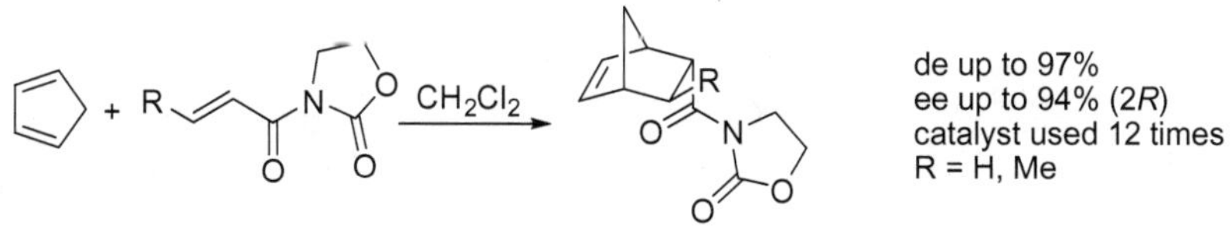

SCHEME 1.15. Asymmetric cycloaddition reactions promoted by Cu–Box complex immobilized in a charge transfer complex.

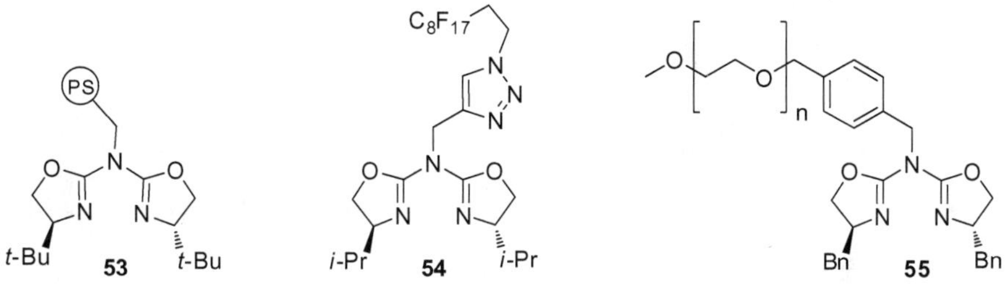

FIGURE 1.23. Structures of ligands **53–55**.

1.4.3. Aza-BOX and Phe-BOX Ligands

Several methods were reported in the literature to achieve the reutilization of Aza-BOX ligands. In a first method, polystyrene-bound Aza-BOX ligand **53** (Fig. 1.23) was prepared by grafting and coplymerization to be tested for Cu(I)-catalyzed cyclopropanations of styrene with ethyl diazoacetate [97]. The supported ligands derived from grafting proved to have a large number of catalyst active centers accessible to the substrate molecules, providing very active catalysts. The respective cyclopropanes were isolated in enantioselectivities comparable or even higher than those provided by homogeneous and nonsupported catalysts for several olefins (as an example, for styrene: 94%, 1 mol% catalyst, up to 99% ee). The catalyst could be recycled up to five cycles for reactive alkenes [98].

Alternatively, Cu-AzaBOX catalyst immobilization can be achieved by the following methods: electrostatic interactions in laponite (clay), nafion-silica [98, 99], retention in the [emim][OTf] phase, immobilization in PEG, and self-supported polymers. From these contributions, the highlights go to the immobilization IL, PEG, and in a self-supported polymer. The recyclability with IL strategy was considerably higher because the enantioselectivity of the isolated *trans* product remained at 90% ee during eight cycles [100, 101].

The MeO-PEG Aza-BOX ligand **55** (Mw = 5000) as a homogeneous-supported catalyst proved to be at least as efficient as its nonsupported analog in copper-catalyzed cyclopropanation of styrene and 1,1-diphenyl ethylene with ethyl diazoacetate. The enantioselectivity observed for the cyclopropanes

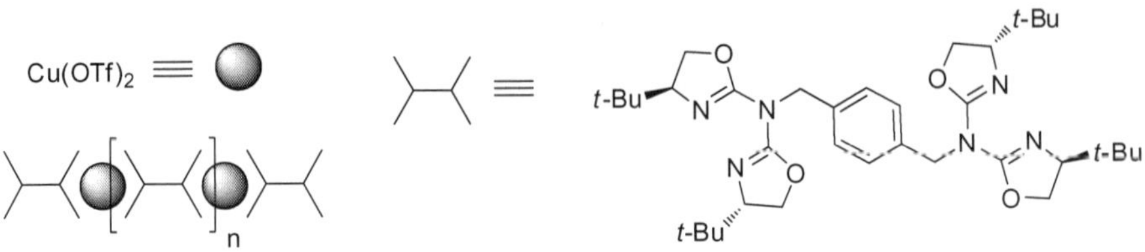

56

FIGURE 1.24. Structure of self-supported polymeric Aza-BOX catalyst **56**.

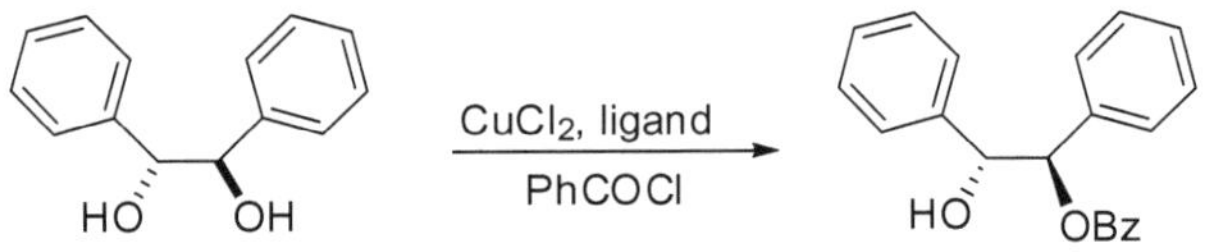

SCHEME 1.16. Desymmetrization of 1,2-diols with copper- and polymer-supported aza-BOX.

with this catalyst rose to 90% ee and its reutilization reached nine consecutive cycles [102].

In 2008, the application of the self-supported polymeric Aza-BOX catalyst **56** (Fig. 1.24) in the cyclopropanation of styrene (which curiously behaved as a heterogeneous and homogeneous catalyst) was reported. During the catalysis, the heterogeneous polymer becomes soluble, perhaps due to substrate-induced depolymerization and, at the end of the reaction, it precipitates. The recyclability of this catalyst was demonstrated during 14 cycles (>95% ee) [103].

The ligand **55** was also applied for the kinetic resolution (KR) of racemic 1,2-diols via copper-catalyzed benzoylation (Scheme 1.16) [104]. The product was obtained in 41% yield and >99% ee. The catalyst was recovered in high yields (>95%) and reused up to five cycles without efficiency variations. The nonsupported version of ligand **55** was also anchored to other supports like Merrifield resin, PEG chains, and superparamagnetic magnetite/silica nanoparticles using click chemistry developed by the Sharpless group [76]. From this list of supports, it was possible to achieve the same recyclability with only superparamagnetic magnetite/silica nanoparticles [105]. Recently, this ligand was also immobilized in a dendritic structure [106].

Interestingly, it was possible to recycle fluorous-tagged Aza-BOX **54** without the need to use perfluorous solvents. The homogeneous catalyst prepared with ligand **54** could be precipitated with diethyl ether (when dichloromethane is used as solvent) and recycled with similar efficiency as for **55** [107].

The Phe-BOX immobilization was conducted by Weissberg and Portnoy, via preparation of the Rh-Phe-BOX catalysts **57** (Fig. 1.25) by solid-phase synthetic chemistry. The preliminary results of the application of **57** in benzaldehyde enantioselective allylation indicates that there is still room for improvement in the recovery and reutilization efficiency [108].

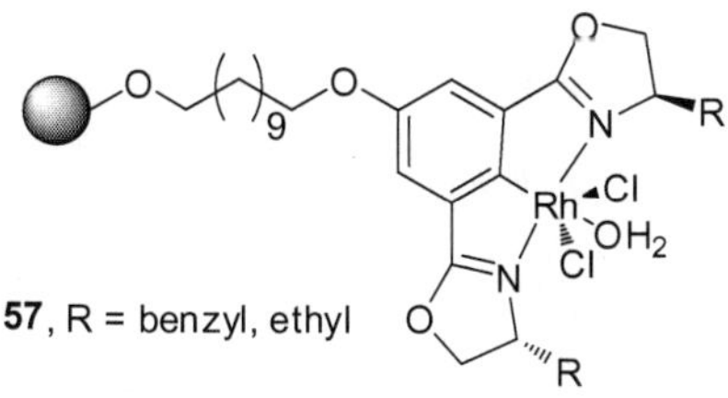

FIGURE 1.25. Structure of Rh-Phe-BOX catalysts **57**.

In conclusion, Phe-BOX ligands were considerably less studied for recycling strategies than BOX, Py-BOX, and Aza-BOX. There are several strategies to achieve the immobilization of the catalystic systems and their consequent reutilization. From such strategies in general the best results are obtained when the immobilization is done in a robust support that does not alter the nature of the catalyst. It is also noteworthy that for immobilization by electrostatic interaction, the utilization of ILs and an ionic-tagged catalyst provided excellent results when the BOX ligands are applied in Lewis acid catalysis.

1.5. SALEN-TYPE LIGANDS

The family of salen-based complexes has applications in several synthetic methodologies; however, the epoxidation of alkenes, KR of epoxides, and cyanosylilation of aldehydes are among the most well developed. The number of works based on catalyst reutilization with salen complexes is high. The aim of this section is to highlight some successful examples. For a more complete view on the topic, consult a recent review [2].

This section will be structured by reaction. In the asymmetric epoxidation of olefins, the main supports chosen to immobilize the catalyst are inorganic. The recycling of the catalytic system with this strategy was conducted by coordination of the support to the metal, grafting of ligand moiety, or electrostatic interaction.

Nguyen et al. prepared a coordination polymer by reaction of a [bis(catechol) salen]Mn(III) complex with several di- and trivalent metal ions. The polymer prepared with copper was found to be insoluble in several organic solvents and a good heterogeneous catalyst to perform the asymmetric epoxidation of 2,2,-dimethyl-2*H*-chromene with 2-(*tert*-butylsulfonyl)iodosylbenzene (79% yield, 76% ee). The catalyst **58** (Fig. 1.26) could be reused during 10 cycles with minimal yield erosion (10%) and constant enantioselectivity [109].

A series of works were published that used functionalized siliceous materials to which the Mn-salen complex could be coordinated. Li et al. grafted phenolic groups on the mesoporous surface of siliceous MCM-41 heterogeneous support [110]. The chiral Mn-salen complex was then anchored into this

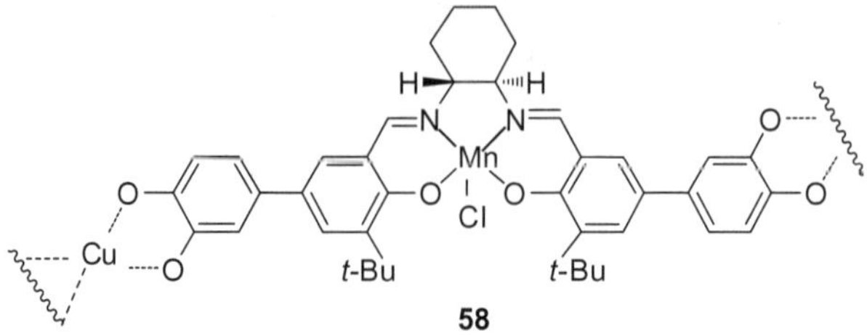

58

FIGURE 1.26. Structure of catalyst **58**.

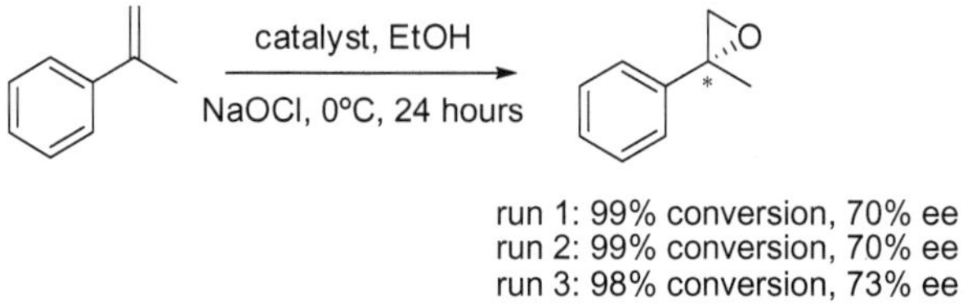

SCHEME 1.17. Recycling of Mn-salen catalyst anchored in modified MCM-41 (phenolic groups).

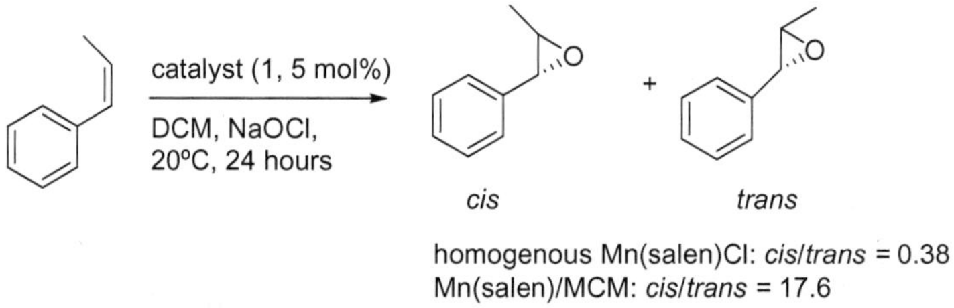

SCHEME 1.18. Pore effect in the selectivity of *cis*-β-methylstyrene epoxidation.

modified support through axial complexation of manganese on the phenoxy groups. This catalyst was tested for α-methylstyrene epoxidation and gave better results in terms of enantioselectivity (67%–73% ee) than the homogeneous Mn-salen complex (56% ee, Scheme 1.17). Interestingly, alkenes that do not fit inside the cavities (such as 1-phenylcyclohexene) were not epoxidized: a proof that all Mn-salen complexes were strongly anchored and behaved like a heterogeneous catalyst. The catalyst was reused three times without any loss of activity and enantioselectivity.

The same methodology was extended to create a monolayer of phenyl sulfonic groups on the surface of several inorganic supports (MCM-41, SBA-15, and activated silica) [111]. The same behavior in terms of enantioselectivity was detected. When the Mn-salen–silica catalyst was reused, consistent 76% ee was achieved during five cycles [$Y_e(5) = 30\%$]. Interestingly, when the olefin was changed to *cis*-β-methylstyrene, a curious selectivity behavior was detected with anchored Mn-salen catalysts. In fact, on going from homogeneous environment to low-sized pores, the *cis*/*trans* ratio changed from 0.38 to 17.6 (Scheme 1.18).

TABLE 1.7. The Recycles of Mn-Salen in the Asymmetric Epoxidation of α-Methylstyrene

Run	Time	Conversion (%)	ee (%)	TOF × 10^{-4}(s^{-1})
1	24	>99	>99	2.31
5	24	96	>99	2.22
6	24	95	99	2.20
10	24	90	91	2.08
11	24	80	85	1.85

The supported catalyst was found to be more selective for *cis* epoxide than the homogeneous catalysts that showed a higher tendency to generate the *trans* isomer. The sulfonic and phenoxy groups were also prepared in the surface of highly cross-linked polystyrene resins [112] and phenoxy-modified zyrconiumpoly-(styrene-phenylvinylphosphanate)phosphate [113]. The first new catalysts proved to be easier to handle than the silica-based ones, providing the same results. Regarding the second family, the coordination of a supported amine to the Mn-salen allowed a recyclability of 11 cycles (99% ee) for epoxidation α-methylstyrene (Table 1.7) [114].

The authors found that the increase in enantioselectivity was due to a positive pore effect since catalysts anchored in the external surface did not provide enantioselectivities as high as those catalysts immobilized inside nanopores [115]. It was also reasoned that the inversion on the *cis*-β-methylstyrene epoxidation selectivity was due to the effect on the rotation motions necessary to achieve the *trans* transition state caused by the lack of space inside the nanopore. *N*-oxide groups were also scoped as ligands to anchor Mn-salen inside MCM-41 [116].

Chemical functionalization of silica with ligand moieties and functionalization of salen ligand with silanes were also tested to achieve recyclable systems. Kureshy et al. functionalized the surface of MCM-41 and SBA-15 with a reactive 3-aminopropyltriethoxysilane, allowing the salen complex to be grafted and therefore immobilized on the support [117]. Both heterogeneous catalysts **59** (Fig. 1.27) proved to be nearly as efficient as the nonsupported analog, and quantitative yields were achieved with good-to-excellent levels of enantioselectivities (chromenes, 5 mol%, up to 94% ee). Furthermore, they were successfully recycled with minimal yield erosion and consistent ee up to four cycles. In the case of styrene derivatives, the heterogenized catalyst provided the respective epoxides with higher enantioselectivity [118]. More recently, Liu et al. studied the pore effect for this type of immobilization. They observed that siliceous supports, with bigger pores, facilitated diffusion phenomena enhancing the reactivity, but with a penalty on the enantioselectivity of the epoxides [119].

The authors also studied the immobilization of Mn(III)-salen complex in MCM-48. The enantioselectivity and reactivity of the immobilized catalyst **60** (Fig. 1.27) sometimes surpassed those obtained with the homogeneous

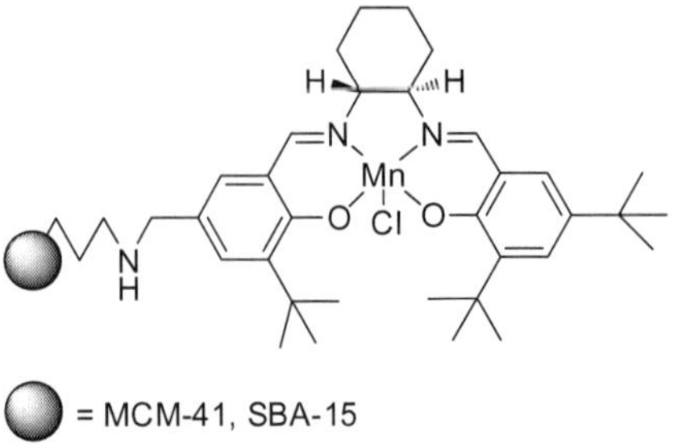

59

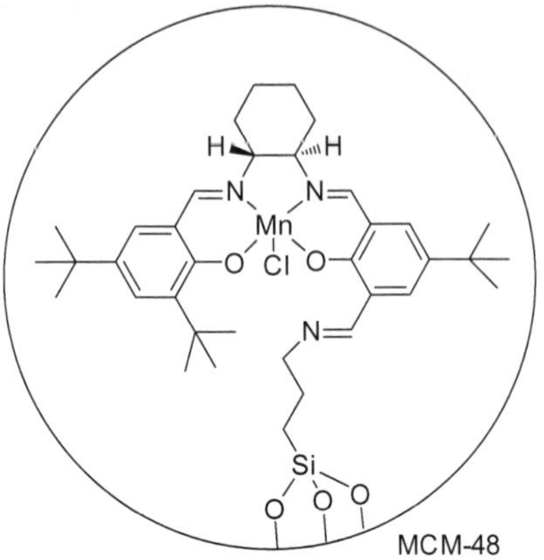

60

FIGURE 1.27. Structures of catalysts **59** and **60**.

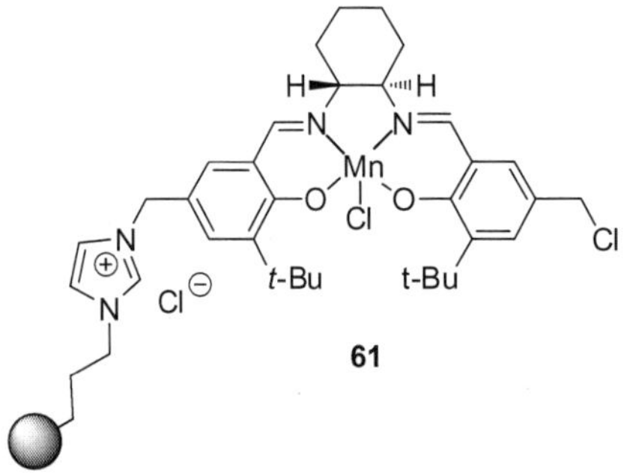

61

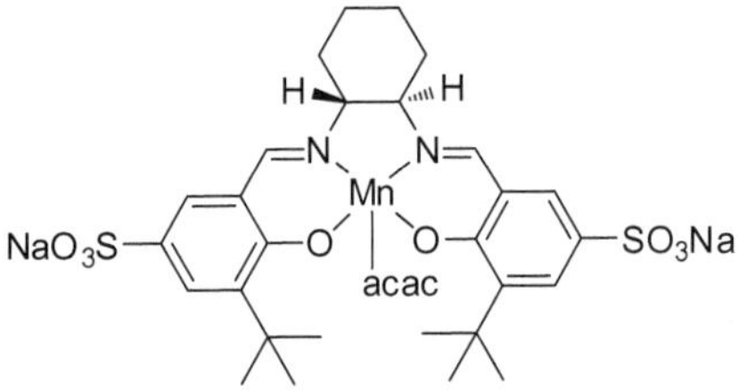

62

FIGURE 1.28. Structures of catalysts **61** and **62**.

complex, and it could be recycled up to three cycles [120]. A few years later, the authors anchored this ligand on several siliceous supports (MCM-41, SBA-15, amorphous silica, and MCM-48) using an ionic linker (imidazolium chloride salt). In this case, the catalyst **61** (Fig. 1.28) was recycled five times with success [121]. The immobilization of the Mn(III)-salen complexes in MCM-48 modified with imidazolium-based ILs units treated with [bmim]PF_6 was also analyzed. This heterogeneous catalyst was tested in asymmetric epoxidation of unfunctionalized olefins, where it was found to be stable, recyclable (three cycles), and exhibited comparable activity and considerable higher enantioselectivities than those obtained by the homogeneous counterparts (e.g., >99%, 92% ee for α-methylstyrene) [122].

Hydrotalcite-like materials [123, 124] and clays [125] were also used to immobilize the ionic Mn-salen complex by electrostatic interaction. In the work of Choudary et al., the catalyst **62** (Fig. 1.28) was immobilized on layered double hydroxide (LDH) and cationic resin with beneficial effects on catalyst recycling, namely in terms of enantioselectivity (which remains unchanged up to five cycles). Epoxidation of cyclic olefins affords excellent levels of enantioselectivity (>94% ee), while acyclic ones like styrene were converted to the respective epoxide in moderate ee (48% ee) [126].

Liese et al. reported the application of a chiral Mn(III)-salen complex covalently anchored to a hyperbranched polyglycerol polymer as an efficient recyclable catalyst for asymmetric epoxidation of chromenes (96% conversion, 96% ee). Epoxidation of 6-cyano-2,2-dimethylchromene, using catalyst **63** (Fig. 1.29), was carried out in a continuously operated chemical membrane reactor for 20 residence times, and steady conversions up to 70% as well as steady ee's of up to 92% were achieved (total TON = 240) [127].

Seebach and co-workers prepared insoluble polymers using ligand **64** (Fig. 1.29) as a cross-linker. The application of this ligand in asymmetric epoxidation of styrene was successful, namely in terms of recyclability (it reached 10 cycles with consistent 77%–80% ee) [128].

Until now, all the methods described herein have a heterogeneous nature. The following contributions are in homogeneous catalysis. Kureshy et al. showed that dimeric Mn(III)-salen complexes **65** and **66** (Fig. 1.30) are able

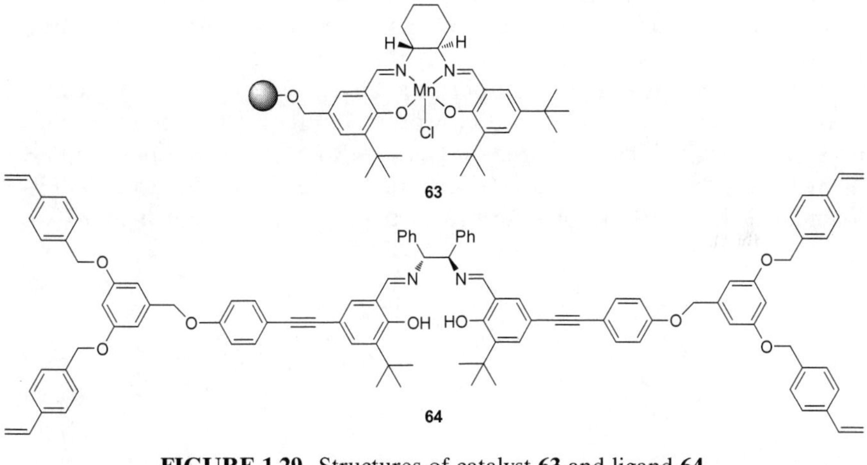

FIGURE 1.29. Structures of catalyst **63** and ligand **64**.

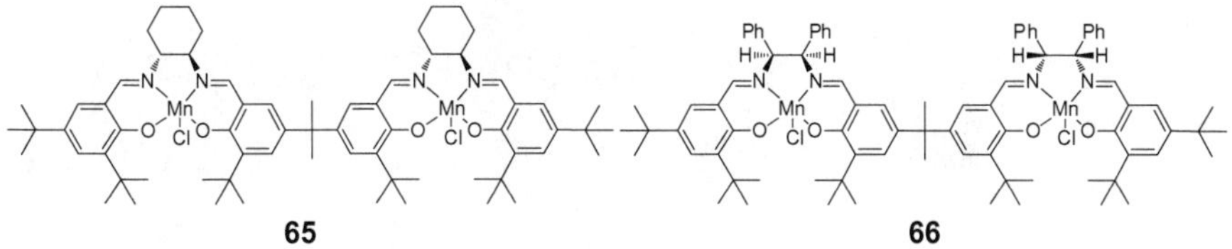

FIGURE 1.30. Structures of dimeric Mn(III)-salen complexes **65** and **66**.

to catalyze the epoxidation of several alkenes (in the presence of a 4-PhPy-oxide and NaOCl) with the same efficiency described for Jacobsen's catalysts; however, the first was recoverable by precipitation and reused for five cycles without any efficiency erosion [129]. Later, the authors described a greener protocol where urea hydrogen peroxide was substituted by NaOCl [130].

In 2004, the asymmetric epoxidation of dihydronaphthalene in the presence of a Mn–Katsuki-type salen catalyst in IL media ([bmim]PF_6, 2.5 mol%, 93% ee) was performed. The IL phase bearing the catalyst was reused eight times [$Y_e(9) = 25\%$, $ee_e(9) = 13\%$] [131].

The KR of epoxides was explored almost at the same time as the asymmetric epoxidation of alkenes. In this transformation, the synergetic effects of two near Co-salen units provide high levels of enantioselectivity and in principle should be taken into account when the catalyst is immobilized.

Yang et al. demonstrated for the first time that cooperative activation effect (between two metallic units) can be enhanced in the nanocage of mesoporous materials like SBA-16. The Co(III)-salen catalyst confined to nanocages exhibited a significantly higher activity and enantioselectivity for the resolution of racemic aromatic epoxides (S/C up to 12,000) compared with the homogeneous catalyst. Furthermore, the system retained its activity for more than eight cycles (with regeneration) [132]. Recently, a polymeric chiral Co-salen was anchored in the walls of SBA-15 pretreated with dimethylcarbonate. This heterogeneous catalyst proved to have an efficiency similar to the homogeneous version and was reused efficiently three times [133].

To take advantage of mechanism cooperativity of nearby Co-salen centers in dendritic structures, Weck and co-workers grafted dendrons to insoluble polystyrene resins. The heterogeneous catalyst **67** (Fig. 1.31) could be applied in hydrolytic kinetic resolution (HKR) of terminal epoxides with catalysts loadings as low as 0.04 mol% furnishing the respective products in 99% ee.

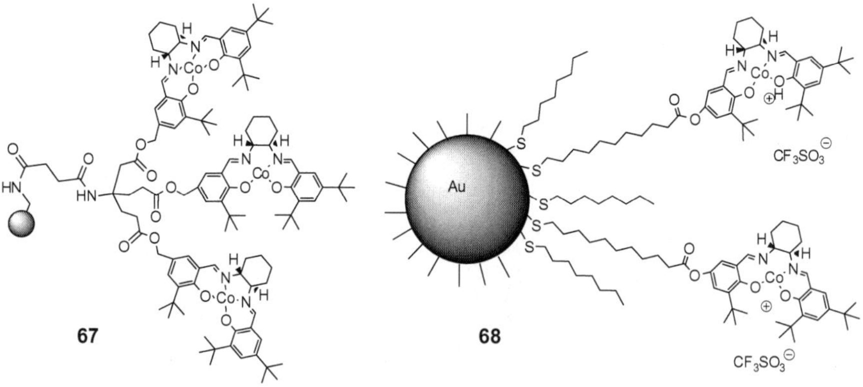

67 **68**

FIGURE 1.31. Structures of catalysts **67** and **68**.

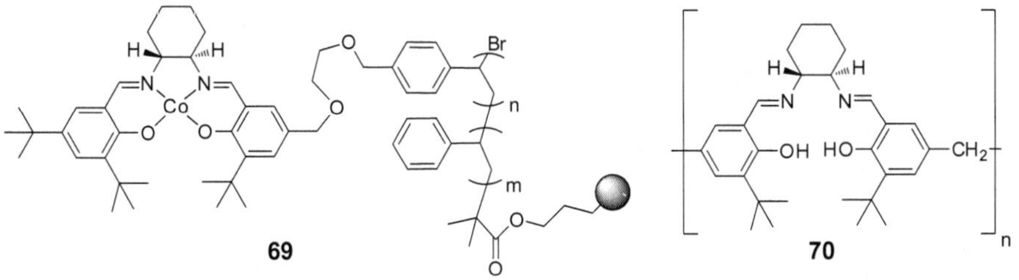

FIGURE 1.32. Structures of catalyst **69** and ligand **70**.

The efficiency of reutilization of the catalytic system was demonstrated during five cycles [134].

Jacobsen and Belser prepared gold colloids coated with a monolayer of *n*-octanethiolates and thiolates with a chiral Co(III)-salen catalyst and tested it as the catalyst for the HKR of hexene-1-oxide. Due to cooperative effects, the catalyst **68** (Fig. 1.31) exhibited a significant acceleration relative to the homogeneous system (>99.9% ee epoxide, 0.01% mol% catalyst). Recovery of the immobilized catalyst was possible by simple filtration, and after catalyst reoxidation, it was reused seven times without any loss of activity and enantioselectivity [135].

Alternatively, it is possible to have identical efficiency if the catalyst is immobilized in polymeric units with considerable flexibility. Co-salen with a flexible pendant group was immobilized in a polystyrene residue anchored in silica. Catalyst **69** (Fig. 1.32) was applied efficiently during five cycles in the resolution of epichlorohydrin. The respective epoxide was isolated in 99% ee after a consistent conversion of 55% [136]. Recently, silica has been substituted by a magnetic nanoparticle with the same success [137].

Kwon and Kim tested a family of polymeric (linear and cross-linked) salen ligands for Co(II)-mediated KR of terminal epoxides [138]. Using 0.5 mol% of Co/salen **70** (Fig. 1.32), racemic epoxides were resolved with excellent enantioselectivity (>99% ee for both diol and epoxide). The counteranion played an important role in the efficiency of this immobilized catalyst, allowing its recyclability (tetrafluoroborate and hexafluorophosphate were the best). These catalysts were reused up to seven cycles with extraordinary consistency (44% yield and 99% ee for all cycles) without any need for reactivation at the end of each reaction. Recently, Kim et al. showed that group 13 salts ($AlCl_3$, $GaCl_3$, and $InCl_3$) can enhance the catalytic and enantioselectivity of polymeric Co(III)-salen catalysts [139].

Song et al. showed that KR of terminal alkylic epoxides (with water) could proceed smoothly in the presence of poly-salen-Co(III) complexes to furnish excellent conversions (>49%) and high enantioselectivities (up to 98%, Scheme 1.19) [140]. HKR of more reactive epoxides such as epychlorohydrin featured a different behavior with the formation of a unique phase. The free terminal amine from the catalyst **71** probably reacted with the epoxide, improving

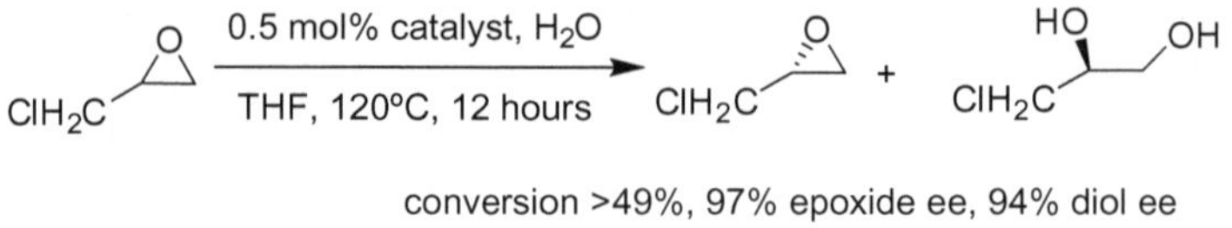

SCHEME 1.19. Racemic epichlorohydrin resolution with polymeric Co-salen catalyst.

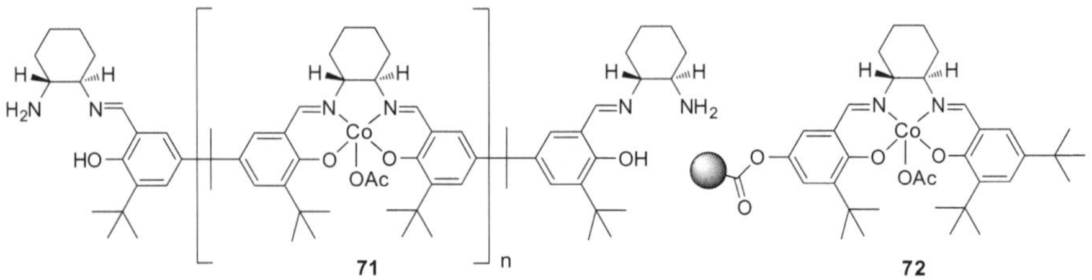

71

72

FIGURE 1.33. Structures of catalysts **71** and **72**.

its solubility in water, allowing an easier way to achieve the catalyst recovery. The recovered polymer catalyst **71** (Fig. 1.33) showed good activity and selectivity.

Weberskirch et al. built a self-assembled nanoreactor to perform the HKR of aromatic epoxides in pure water. The Co(III)-salen complex was anchored to an amphyphilic copolymer that formed micellar structures in water. The local high concentration of catalyst and the reasonable low concentration of water inside the micellar reactor allowed for the resolution of several types of aromatic epoxides with excellent enantioselectivitites (up to 51% conversion, 99% ee of epoxide, 96% ee of diol). The polymeric catalyst **72** (Fig. 1.33) could be separated and reused in four consecutive cycles without loss of enantioselectivity. The increase of the reaction time is due to incomplete regeneration of the catalyst [141].

Song et al. performed the HKR of racemic epoxides with water in the presence of a Jacobsen catalyst (Co-salen) in a mixture of IL and tetrahydrofuran ([bmim]PF_6/THF: 1:4) [142]. The KR of epichlorohydrins in water was achieved with excellent enantioselectivities (>99% ee for the epoxide and 92% ee for the diol with 0.1 mol% of catalyst). Furthermore, the authors disclosed that the catalyst did not follow deactivation to cobalt (II) species, which is a characteristic of these reactions when run in organic media. In addition, the catalyst was found to be retained in the ionic phase during product removal procedures (distillation and water/IL extraction), allowing its recyclability and reutilization (up to 10 cycles with consistent activity and enantioselectivity with consecutively faster reactions). When the reaction was run in pure IL, the use of an additive (p-nitrobenzoic acid) was necessary in order to have an active catalyst for asymmetric aminolytic KR of terminal epoxides [143]. In the end of the reaction, no further additive was required to maintain an active catalytic system during seven cycles.

Homogeneous protocols have a tendency to afford equivalent results compared with traditional protocols, while heterogeneous- or macromolecules-supported catalysts should be aware of mechanistic evidence (synergistic effects by two near metal units) in its design so as to achieve higher levels of efficiency (compared with traditional method) and recyclability. Interestingly, in general, the procedures described uses at least identical (if not lower) amounts of catalyst loading compared with traditional homogeneous reactions. This is outstanding because traditionally the immobilization, especially in heterogeneous supports, required the application of higher levels of catalyst loading to achieve the same reactivity. Perhaps the supports tend to enhance the number of synergetic interactions improving the reactivity. In case of epoxidation reactions, the pore effect can be considered an important discovery in this field.

In case of the silylcyanation of aldehydes, Corma and collaborators described a very interesting study by comparing different protocols in order to achieve VOsalen catalysts reutilization. The first contribution focuses on the immobilization of the catalyst in the walls of a structured inorganic mesoporous materials by a postsynthesis treatment [144]. However, the immobilization proved to be detrimental for the ee of benzaldehyde silylcyanation. One year later, VOsalen was immobilized by five distinct methods: by covalent bonding to an ionic moiety (catalyst **73**), by grafting in silica (catalyst **74**), in single-wall nanotubes (catalyst **75**), in activated carbon (catalyst **76**) (Fig. 1.34), and finally, by retention in an IL phase ([beim]PF_6 and [emim]PF_6) [145]. The catalyst **74** and VOsalen in the IL ([emim]PF_6 or [beim]PF_6) afforded higher enantioselectivities (85% and 89% ee, respectively) [146]. In the case of catalyst **75** and catalyst **76**, the low enantioselectivity observed could be due to achiral interference of the solid support. The homogeneous VOsalen catalyst immobilized in IL presented a higher recyclability performance [Ye(5) = 2%] [146, 147].

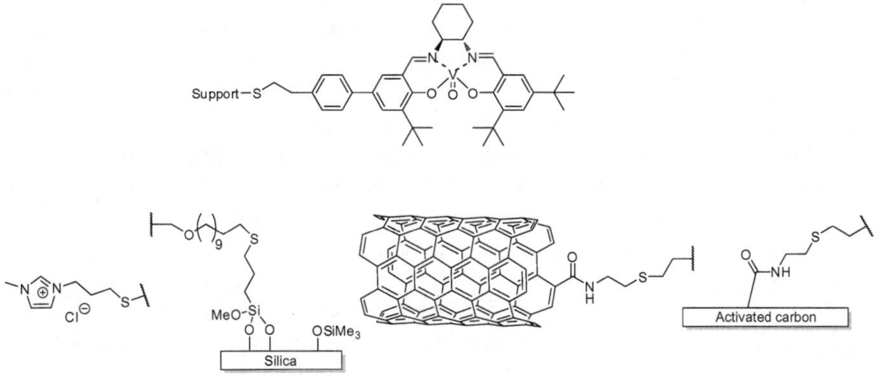

73 74 75 76

FIGURE 1.34. Structures of catalysts **73–76**.

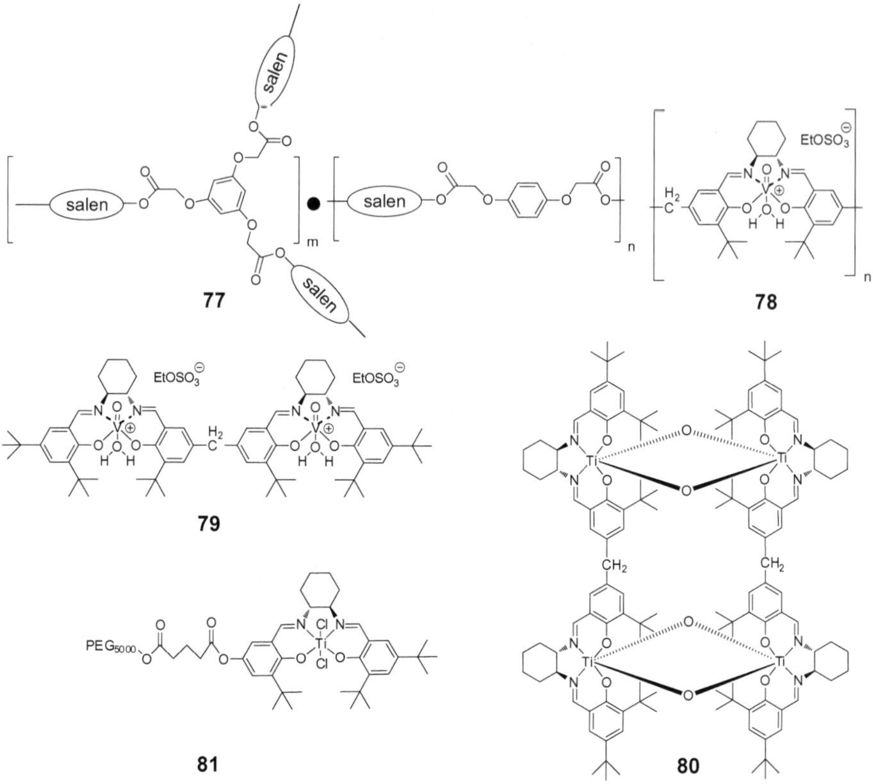

FIGURE 1.35. Structures of catalysts **77–81**.

Polymeric supports and dimeric catalysts were also studied in this transformation, and those that are homogeneous relied on the precipitation to achieve the reutilization. Zheng et al. prepared a polymeric Ti(IV)salen catalyst **77** (Fig. 1.35) for the synthesis of the *O*-acetylcyanohydrins from KCN, Ac_2O, and aldehydes. The polymeric catalysts were prepared from coplymerization of a di- or tribenzaldehydes with a chiral diamine to obtain supported catalysts with several degrees of cross-linking [148]. The polymeric ligand with a cross-linking ratio of 25:100 was reused six times with minimal conversion [run 1: 99%, $Ce(7) = 4\%$] and ee erosion [run 1: 80% ee, $ee_e(7) = 7\%$]. The lower observation arises from the difficulty of the catalyst to form titanium oxo-bridges between two near units due to a high cross-linking degree.

Khan et al. prepared a linear polymeric VOsalen possessing a methylene group as spacer (tri/di ratio of 0:100). The catalyst **78** (Fig. 1.35) proved to be as reactive as the previous catalysts and provided higher levels of enantioselectivity (97% yield and 94% ee). Furthermore, it tends to precipitate in apolar solvents, allowing its reutilization (up to four efficient cycles) [149].

The dimeric VOsalen and Ti(IV)salen catalysts were also studied under identical conditions. Both catalysts showed improved efficiency, where conversions up to 99% with enantioselectivities up to 96% were obtained for several benzaldehydes. Furthermore, catalysts **79** and **80** (Fig. 1.35) were recovered by precipitation (with hexane) and reused up to four cycles with consistent enantioselectivity [150].

It is known that the strategies based on catalyst immobilization in soluble macromolecules require precipitation techniques to achieve their recovery. In some cases, those techniques do not furnish the immobilized catalyst in quantitative yields. An example of such complex catalysts are those immobilized in PEG. Anyanwu and Venkataraman showed that a PEG-supported Ti-salen complex **81** (Fig. 1.35) could be recovered and reused for silylcyanation of benzaldehyde (0.1 mol% of catalyst) at least five times without any loss of activity (95% conversion, 86% ee) [151]. These results are comparable with those obtained with the nonsupported catalyst analog. The recovery process involved a Soxhlet-dialysis protocol, where the Soxhlet chamber was refilled with fresh dichloromethane every 20 minutes. After 38 hours, about 98% of cyanohydrins were separated from the PEG-supported Ti-salen complex.

Besides the reactions already addressed in this section, where salen ligands are involved, it is possible to apply this ligand in a Diels–Alder reaction, with the resolution of secondary alcohols and Michael additions. The following examples constitute highlights of the reutilization of salen complexes in these transformations.

Schulz and co-workers applied a distinct immobilization technique to recycle a Cr-salen catalyst based on induced electropolymerization. The resulting polymer **82** (Fig. 1.36) was then applied to the hetero Diels–Alder reaction [152] between Danishefsky's diene and heptanal and the Henry reaction [153]. In both cases, the catalyst immobilization proved to be a bit detrimental for the reaction enantioselectivity. The authors showed that despite some yield fluctuation, this system was very robust and achieved a recyclability of 20 cycles in the cyclization reaction [154].

Heckel and Seebach immobilized the Cr-salen complex in a pore-controlled glass silica gel support and tested it in a hetero Diels–Alder reaction of Danishefsky's diene with caproaldehyde ($HOCC_5H_{11}$) [155]. The supported

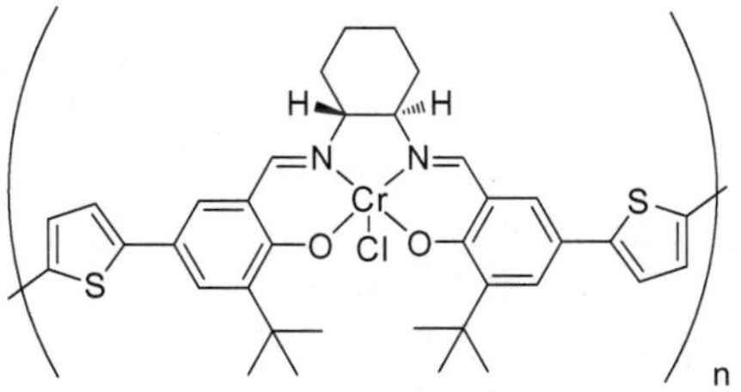

82

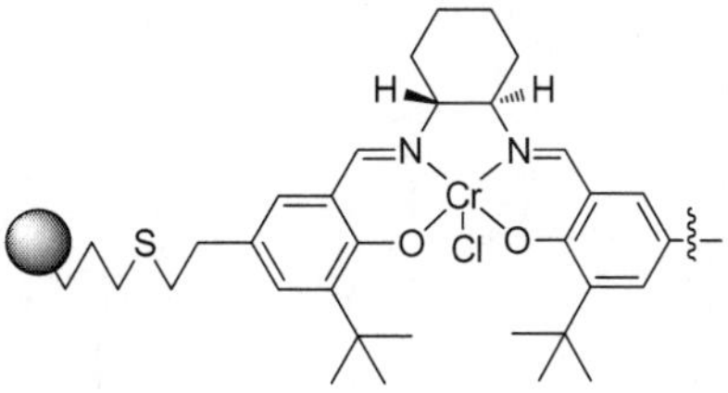

83

FIGURE 1.36. Structures of catalysts **82** and **83**.

catalyst **83** (Fig. 1.36) proved to be almost as efficient as Jacobsen's catalyst, yielding the respective cycloadduct in 91% conversion and 80% ee (vs. 97% conversion with 83% ee). The catalyst could be recovered by filtration and reused up to 10 cycles, and while the conversion showed some fluctuations (91%–55%), the enantioselectivity increased with the number of runs up to a certain level (84%).

Madhavan and Weck tested AlCl(salen) immobilized in a poly(norbonene) polymeric matrix with a relatively flexible linker in the asymmetric addition of cyanide to α,β-insaturated imides (Scheme 1.20). The homogeneous polymeric catalyst **84** (Fig. 1.37) provided the respective products in excellent enantioselectivities during four cycles. Kinetic studies indicated that the activity of the polymeric catalyst is significantly higher than that of the unsupported analog, suggesting an enhancement effect of the bimetallic pathway due to the catalytic sites' proximity [156].

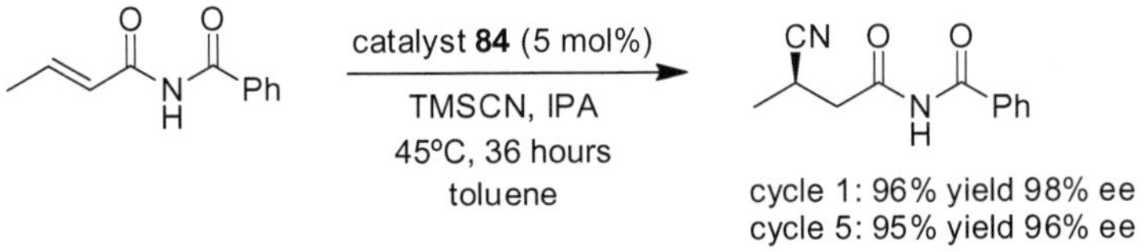

SCHEME 1.20. Asymmetric Michael addition catalyzed by AlCl(salen) anchored to polynorbonene resin.

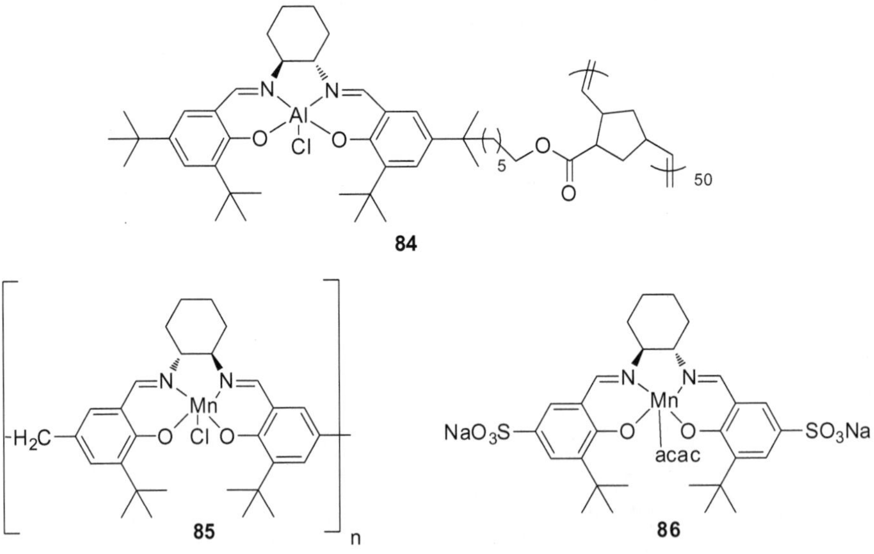

FIGURE 1.37. Structures of catalysts **84–86**.

Kureshy et al. used a chiral Mn(II)-salen polymer to separate a mixture of racemic secondary alcohols by oxidation in the presence of $PhI(OAc)_2$ and KBr (dichloromethane/water 1:2). The complex selectively oxidizes one enantiomer to the respective ketone, furnishing enantiopure unreacted alcohol (>99%). The catalyst **85** (Fig. 1.37) could be recovered by precipitation and reused. Its activity slowly decreased over five runs [Ye(5) = 18%], while the alcohol was always isolated with high enantioselectivity (95%–96% ee) [157]. Kantam et al. immobilized a sulfonate derivative of Mn-salen catalyst **86** (Fig. 1.37) in resin-bearing tetraalkyl ammonium groups, which could be used with identical efficiency during four cycles [158].

There are already a few works on the immobilization of chiral vanadyl and titanium salen complexes for cyanosilylation reported so far. The catalysts supported on polymers (by grafting or polymerization) have some tendency to perform better than those supported on inorganic materials.

1.6. ENZYMES

Enzymes are an important and versatile class of asymmetric catalysts (see also Chapter 12). They have an important role in organic synthesis when it comes to asymmetric induction or resolution of racemic compounds, namely sulfoxides, alcohols, and amines in mild and nonaqueous conditions. For industrial scale applications, the immobilization of enzymes has normal advantages since there is a control of the active site, and it allows reutilization of the catalyst or its application in a continuous process [159].

The KR of racemic alcohols and amines has one limitation: 50% is the maximum yield. In this way, dynamic kinetic resolution (DKR) allows total conversion of a racemic mixture to only one optical isomer using the resolving catalyst and a new catalyst that will racemize the remaining alcohol. Both DKR catalysts are extremely expensive, and therefore highly desirable recycling protocols should be developed (Scheme 1.21) [160].

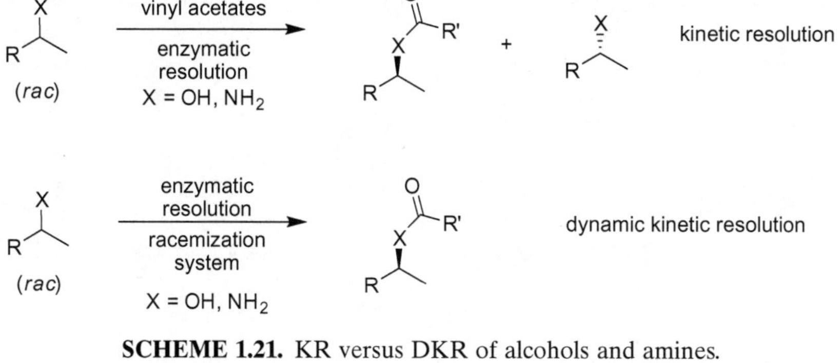

SCHEME 1.21. KR versus DKR of alcohols and amines.

Since the pioneering work by Allen and Williams [161], Pamies and Bäckvall envisaged the possibility of performing the reutilization of the catalytic system used in alcohol DKR [162]. *Candida antarctica* lipase B was recovered by filtration and recycled more than three times in the KR of β-hydroxynitriles (providing enantioselectivities up to >99% and 50% conversion). The racemization ruthenium catalyst (4 mol%) was also reused three times. Unfortunately, the recovered enzyme from the DKR did not retain any activity when tested under KR conditions. Therefore, the catalytic system could only be recycled in a two-step manner (KR followed by racemization) instead of one-pot DKR, which is limited by several purifications tasks.

Kim et al. reported the first ionic liquid-coated enzyme (ILCE) for secondary racemic alcohol resolution in the presence of *Pseudomonas cepacia* lipase (CPL). The enzyme was coated into [PPMIM]PF_6 [PPMIM = 1-(3′-phenylpropyl)-3-methylimidazolium], which becomes liquid above 53°C. The catalyst preparation protocol was very simple. The powder enzyme was added to the liquid [PPMIM]PF_6 and stirred until a uniform heterogeneous solution was formed, then it was cooled down to room temperature. The resulting solid was broken down to small-sized particles and used to resolve several secondary alcohols. The ILCE provided higher enantioselectivities than the native enzyme (twofold enhancement, Scheme 1.22) [163].

The authors observed that fresh ILCE showed reduced activities in the first cycle due to some diffusion difficulties, which were relieved in the second cycle when the particle size was decreased (100% relative activity of native enzyme is achieved). After five cycles, 93% of the native enzyme activity was retained.

Iborra et al. tested the possibility of having a flow system for 1-phenylethanol racemic resolution. The main idea consisted of immobilizing an enzyme (*C. antarctica* lipase B) in an IL that acted as a stationary phase, and a racemic mixture could be flown through the biocatalyst using *sc*CO_2 as the mobile phase [164]. More polar IL such as [emim]NTf_2 tended to give higher selectivity (86.3% vs. 84.8%) and half-life time (24 vs. 22 cycles) when compared with the less polar [bmim]NTf_2. The (*R*)-1-phenyl ethyl propionate was always obtained in conversions higher than 35% and 99.9% ee's (Table 1.8). The use of IL as the stationary phase provided a protective effect to the biocatalyst; therefore, higher lifetimes are observed (about twofold) when compared with enzyme adsorbed on celite.

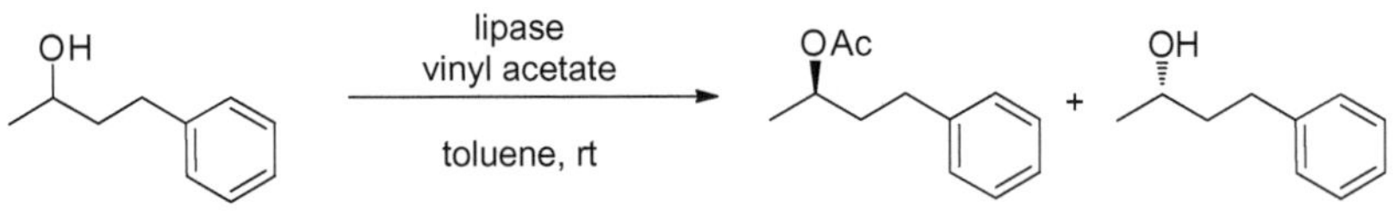

	ee_s	ee_p	E
native	54.5%	98.7%	265
ILCE	29.5%	99.5%	532

SCHEME 1.22. KR of secondary alcohols with IL-coated *P. cepacia* lipase.

TABLE 1.8. Activity and Operational Stability Parameters of Free *C. antarctica* Lipase B Dissolved in ILs for Continuous (R)-1-Phenylethyl Propionate Synthesis in scCO2 at 15 MPa

IL	Temperature (°C)	Specific Activity (*U*/ mg Enz.)	Selectivity (%)	ee (%)	Half-Life Time (cycles)
[EMIM]	50	1.6 ± 0.3	86.3 ± 1.3	>99.9	24
[Tf_2N]	100	1.1 ± 0.1	95.2 ± 1.5	>99.9	16
[BMIM]	50	1.7 ± 0.2	84.8 ± 3.2	>99.9	22
[Tf_2N]	100	0.6 ± 0.1	88.1 ± 4.6	>99.9	8
None	50	0.2 ± 0.02	81.5 ± 2.9	>99.9	10

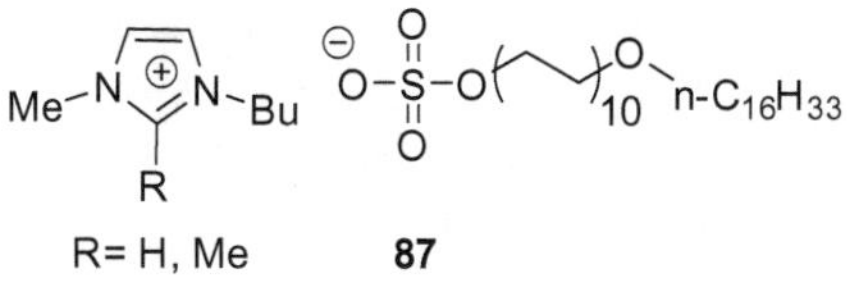

87

FIGURE 1.38. Structure of IL **87**.

Matsuda et al. applied the same flow methodology reported by Iborra et al [164], but did not use IL to immobilize *C. antarctica* Novozym 435 [165]. The KR of racemic 1-phenylethanol was conducted with this technology for more than 3 days of operation under supercritical condition (12.9–13 MPa of CO_2 at 42°C) and resulting in quantitative transformation of 221 g to (*S*)-1-phenylethanol with 99% ee and (*R*)-(1-phenylethyl) acetate with 99% ee using 1.73 g of the immobilized enzyme. This new flow system proved to be much more efficient than the batch protocol.

Itoh et al. showed that [bdmim]BF_4 was a suitable reaction media to perform the enzymatic resolution of 5-phenyl-1-penten-3-ol and methyl mandelate alcohols with Novozym 345. The products and acetaldehyde could be efficiently separated from the IL phase and enzyme, allowing its reuse up to 10 cycles [166]. In 2006, the authors described the application of several types of imidazolium-type ILs derived from poly(oxyethyleneglycol)alkyl sulfate **87** (Fig. 1.38) as an additive or coating material for lipase (*Burkholderia cepacia* lipase and *C. rugosa* lipase). It was disclosed that this material accelerated the enzymatic resolution of about 500- to 1000-fold, maintaining excellent levels of enantioselectivity (up to 99% ee). Both stabilized enzymes were reused four times without loss of activity [167].

Lourenco and Afonso developed a new method to achieve the separation of both enantiomers of racemic mixtures of secondary alcohols without application of laborious chromatographic separations. The separation was accomplished by using an ionic acylating agent that remains in the IL medium ([bmim]PF_6) while the unreacted *S*-enantiomer was removed by extraction.

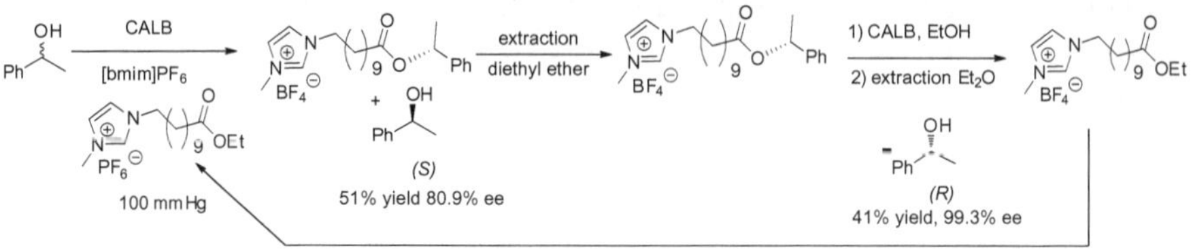

SCHEME 1.23. Protocol for separation of secondary alcohols enantiomers without using any chromatography separation.

Next, the reacted *R*-enantiomer as ionic ester was transesterified with ethanol by the same enzyme to be also removed by extraction. This protocol could be used for four consecutive cycles for secondary alcohol resolution (Scheme 1.23) [168]. Later, the same authors performed the same approach using ethyl myristate as an anchoring agent and reaction medium and isolation of each enantiomer by distillation without the need of an organic solvent for the overall process [169].

De Vos et al. immobilized a palladium catalyst in alkaline-earth supports and found that this catalyst could also promote racemization of chiral amines. Combined with Novozym 435, achiral benzylic and alkylic amines were resolved in excellent yields (64%–98%) with near-perfect enantioselectivity (99% ee, Pd/$BaSO_4$ 5 mol%, 100 mg/mmol enzyme). The one-pot heterogeneous catalytic system was reused twice without efficiency erosion. The support for the palladium catalyst proved to have an important role in controlling the formation of by-products [170]. Recently, the authors discovered that microwave irradiation could be used to accelerate the racemization rate without affecting the catalytic activity of Novozym 345 [171].

Meanwhile, the same authors discovered that the Raney Ni catalyst can be used to replace palladium as the racemization agent. In the case of aliphatic amines, racemization and enzymatic resolution could be combined in one pot, resulting in an efficient DKR. However, to achieve efficient DKR for benzylic amines, it was required that the racemization and resolution reactions be physically separated [172].

Leaching problems due to desorption events could be an important issue when working with adsorbed catalysts in oxidation of sulfides. Therefore, Strukul et al. immobilized chloroperoxidase (CPO) in a microporous silica gel cage to prevent leaching events and catalyst deactivation with hydrogen peroxide (Scheme 1.24). Two distinct samples were prepared [173]. Sample A has 3.7 CPO units per 0.3 g of catalyst and gave lower enantioselectivity (73%–92% ee) than sample B, which has 123 CPO units per 0.3 g (>99% ee) (Scheme 1.24).

In terms of reactivity, sample B reacted faster despite not always giving higher conversions. Sample A failed to be reused in a second cycle (dramatic drops on reactivity and enantioselectivity were detected), while sample B was

SCHEME 1.24. Asymmetric sulfide oxidation in the presence of CPO immobilized in silica cages.

efficiently applied in four consecutive cycles affording chiral sulfoxides in >99% ee with only 6% yield erosion.

As mentioned before, the CPO desnaturation promoted by hydrogen peroxide together with the known low solubility of organic compounds in water constitutes the most important drawbacks of CPO utilization in industrial applications. To overcome this limitation, Spreti et al. reported the utilization of PEG polymers on CPO-catalyzed asymmetric sulfide oxidation [174]. The addition of PEG to the reaction mixture solves both of these problems: They increase the solubility of organic substrates in water and allow the enzyme to retain more of its initial activity. Under optimized addition of oxidant, it was possible to achieve 100% yield with >99% ee (in the case of thioanisole oxidation). Unfortunately, the chemical yield dropped about 15% on each recycle.

Wang et al. reported for the first time the performance of immobilization of CPO in magnetic nanoparticles with an iron oxide and polymer shell in the sulfoxidation of thioanisole. The covalently bound CPO with a long spacer showed the sulfoxidation activity and enantioselectivity identified as those observed for free CPO. The thick polymer shell significantly increased the stability of the nanobiocatalyst. A great improvement was achieved when compared with CPO on other solid supports, giving excellent enantioselectivity (99% ee) and no loss of activity after recycling 11 times [175].

Enzymes showed to be an excellent catalyst in KR and asymmetric oxidation. The immobilization and use of alternative media represent the main approach for reutilization in batch or continuous systems.

1.7. CHIRAL DIAMINES, DIOLS, AND AMINOALCOHOLS

In this part, the goal is to provide insights in how chiral diamines, 1,2-amino alcohols, and diols can be recovered and reused together with the metal catalyst.

1.7.1. TsDPEN-Type Ligands

In the family of chiral diamine ligands, (1S,2S)-(+)-*N*-(4-toluenesulfonyl)-1,2-diphenylethylenediamine (TsDPEN) appears as one of the most well-known ligands that are applied for the reduction of ketones and imines [176]. This ligand was recycled with the aid of several types of homogeneous and heterogeneous supports and alternative reaction media.

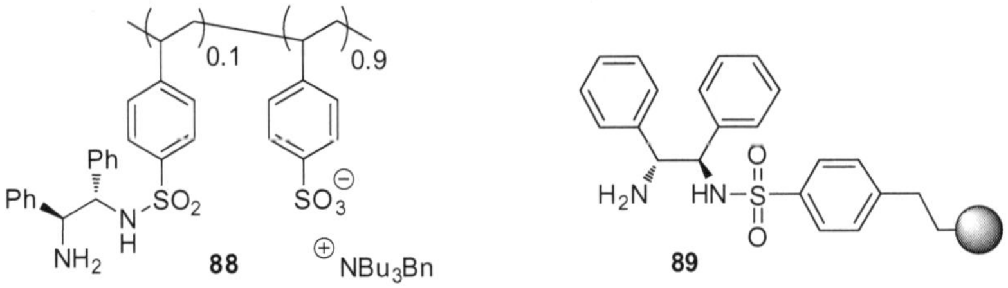

FIGURE 1.39. Structures of ligands **88** and **89**.

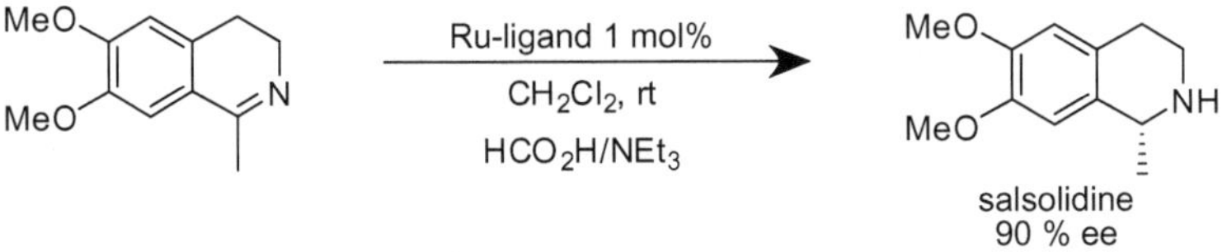

SCHEME 1.25. Reduction of aromatic imines in the presence of RuTsDPEN catalyst immobilized in silica.

In case of heterogeneous systems, the immobilization of TsDPEN ligand in polystyrene was conducted by grafting [177] and coplymerization [178]. Both methods did not have a negative impact on the reaction efficiency. However, the coplymerized ligand **88** (Fig. 1.39) provided a more recyclable system for the ruthenium-mediated reduction of acetophenone in water. The presence of a styrene derivative monomer bearing a sulfonic group in the polymer structure proved to be crucial for the polymer swelling in water and to obtain 98% ee. This heterogeneous catalyst was recycled up to five cycles with the same outcome of enantioselectivity.

Alternatively, it was demonstrated that it is possible to have active ruthenium catalysts anchored in amorphous silica gel, MCM-41, SBA-15, modified silica [179], functionalized SBA-16 [180], and MCFs [181] (ligand **89**, Fig. 1.39). All supports performed well in the recycling tests, with the best being amorphous silica gel (10 cycles), followed by silica MCFs (six cycles), and functionalized SBA-16 (six cycles). Recently, magnetic MCFs were used as support, allowing a recyclability of nine cycles [182].

Shortly after, the authors showed that ligand **89** anchored in silica gel can be applied in the reduction of aromatic ketones in water in the presence of TBAB as an additive and sodium formate. Curiously, the water induced an acceleration effect on this transformation, and the high degree of reusability was conserved [183].

Another application of TsDPEN ligand **89**, anchored in siliceous MCFs, is the reduction of salsolidine to prepare 1-substituted 1,2,3,4-tetrahydroisoquinolines (applied in the treatment of Parkinson's disease). The product was obtained in high yields (95%–100%) and consistent 90% ee over six consecutive cycles (Scheme 1.25).

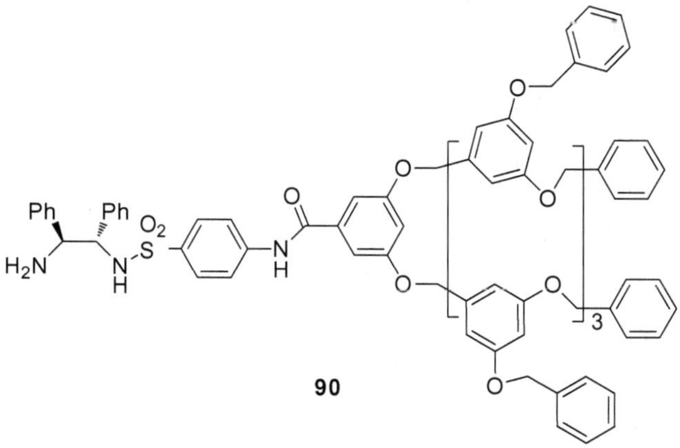

90

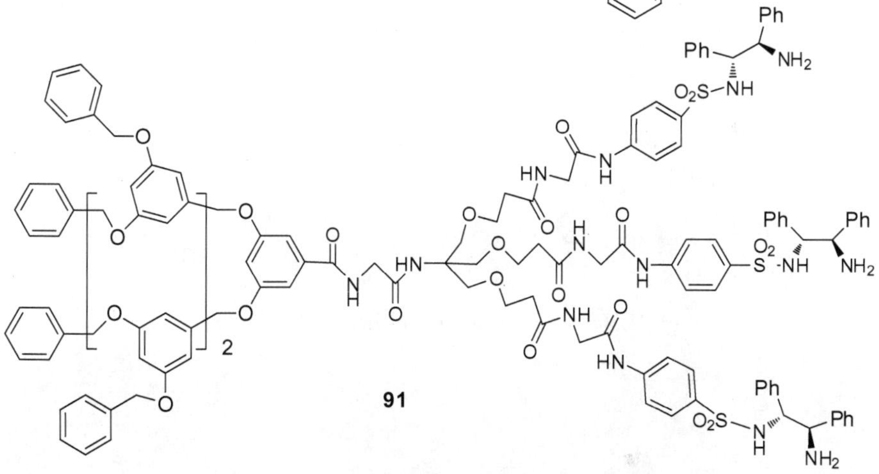

91

FIGURE 1.40. Structures of ligands **90** and **91**.

Regarding homogeneous catalysis, the ligand was linked to dendritic structures and to PEG polymers. Chen et al. showed that rhodium/ligand **90** (Fig. 1.40) can induce excellent levels of enantioselectivity in the reduction of aromatic ketones in water. The recyclability achieved five cycles with 25% yield erosion [184].

The authors also studied the effect of having the ligand anchored in the periphery of the dendrimer in ruthenium-mediated reduction of acetophenones [185]. This catalyst proved to be as selective as the nonsupported one (97.7% ee), but it was difficult to achieve an effective recycling. To overcome this limitation, the authors prepared a hybrid dendrimer (ligand **91**, Fig. 1.40) that was recycled twice [186]. Higher recyclability (14 and 7 cycles) were observed when the two TsDPEN-PEG ligands **92**[187, 188] (Table 1.9) and **93** [189] (Fig. 1.41), respectively, were studied using sodium formate as the hydride source in water.

TABLE 1.9. Asymmetric Transfer Hydrogenation of Ketones With Ru-PTsDPEN by HCOONa in Water

Ketone	Temperature (°C)	Time (h)	Conversion (%)	ee (%)
Acetophenone (acp)	22	8	>99	93
2′-chloro-acp	22	13	>99	91
3′-methoxy-acp	22	18	98	90
4′-methyl-acp	22	18	>99	86
1′-acetonaphthone	22	36	85	88
2′-acetonaphthone	22	36	>99	94
2-acetylfuran	22	18	98	91
1-tetralone	22	18	>99	94

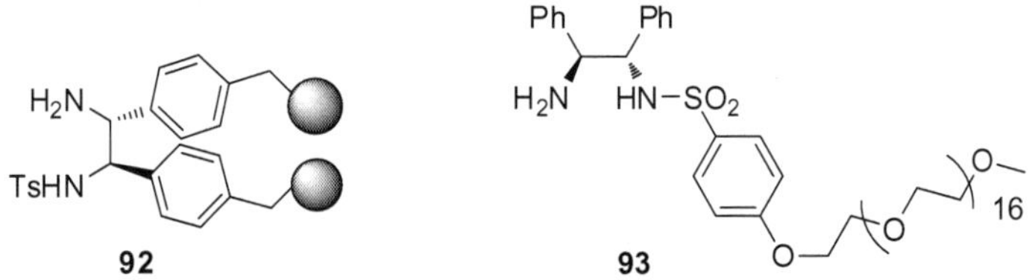

FIGURE 1.41. Structures of ligands **92** and **93**.

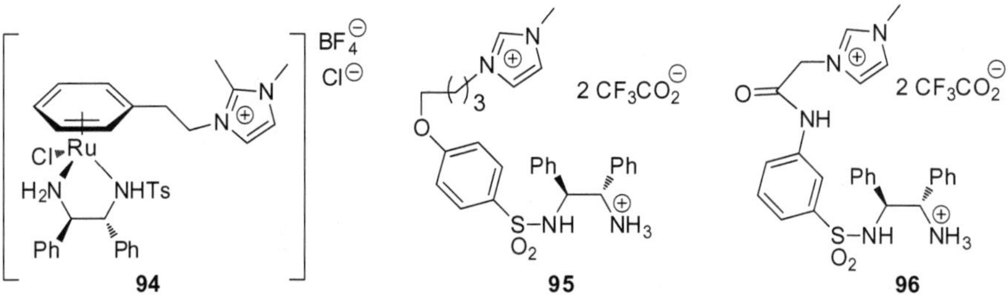

FIGURE 1.42. Structures of catalyst **94** and ligands **95** and **96**.

Another possibility is to attach ionic tags to the ligands such as imidazolium rings (catalyst **94** [190], ligand **95** [191], and ligand **96** [192]), or hydrophilic sulfonic groups (ligand **97**) [193] (Fig. 1.42). However, better results were described when surfactants or PEG were applied as additives for aqueous phase reactions. Zhu et al. disclosed a biphasic system (water/dichloromethane) that enables the Ru-TsDPEN catalyst recovery and recycle, with aid of a mixture of surfactants (sodium dodecyl sulfate (SDS)/cetyl trimethylammonium bromide (CTAB): 2:1). This system allowed an increase of activity, chemoselectivity, and enantioselectivity (S/C = 100, up to 99%) in aromatic

TABLE 1.10. Asymmetric Hydrogenation of Ketones Catalyzed by Ru-TsDPEN with HCOONa in PEG-H_2O

Entry	Ketone	Time (h)	Conversion (%)	ee (%)
1	Acetophenone	3	98	95
2	4′-methoxy-acp	8	91	94
3	1′-acetonaphthone	8	99	93
4	1-Indanone	6	99	95

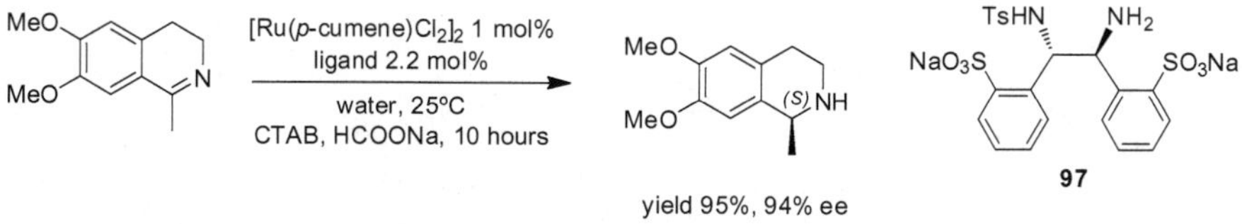

SCHEME 1.26. Reduction of aromatic imines in water/surfactant media.

ketones reduction with sodium formate. The hydrophobic catalyst remained inside the micelles while the products were found mainly in the dichloromethane phase. The aqueous phase was efficiently reused six times in this transformation [194]. Chan and co-workers showed that a solvent mixture of PEG_{400}/water (3:1) was a suitable media to efficiently run reduction of aromatic ketones with sodium formate in the presence of RuTsDPEN (S/C = 100) (Table 1.10). The reaction products were extracted with hexane and the catalyst could be reused up to 15 cycles with consistent enantioselectivity (95% ee) [195].

The use of surfactants in water was also explored in salsolidine synthesis. Working above the micellar critical concentration of CTAB, it was possible to enhance the enantioselectivity of the reaction [196]. A large number of arylimines were quantitatively reduced with excellent enantioselectivity (S/C = 100, generally above 95% ee, Scheme 1.26). At the end of the reaction, the products were extracted with a diethyl ether/hexane (1:1) mixture, leaving the catalyst Ru/ligand **97** in the aqueous phase. The catalytic system could be reused over eight consecutive times with consistent 94% ee, though with some yield erosion [$Y_e(8) = 9\%$].

The TsDPEN ligand has high reliability and efficiency in enantioselective reductions of ketones and imines which lead to several reutilization studies.

1.7.2. Chiral Aminoalcohols

An important class of chiral ligands are 1,2-aminoalcohols that can be prepared from the respective amino acid. Two of the most important applications of these ligands are in boranes-mediated reduction of ketones and the alkylation of aldehydes.

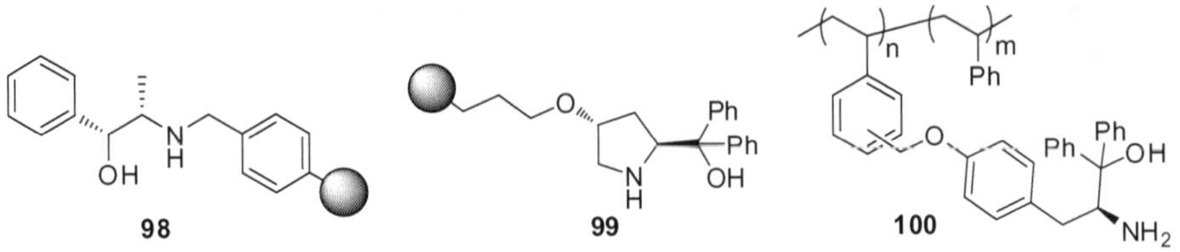

FIGURE 1.43. Structures of 1,2-aminoalcohol ligands **98–100**.

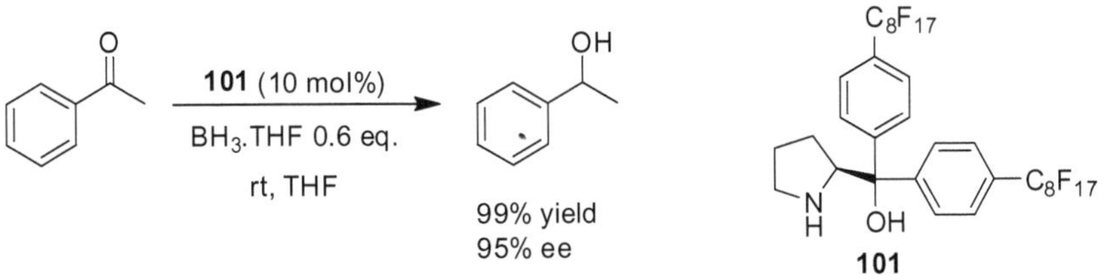

SCHEME 1.27. Reduction of aromatic ketones with borane in the presence of a perfluorinated-diphenylprolinol catalyst **101**.

Starting with borane reductions, Sandee et al. applied a heterogeneous ruthenium catalyst as the solid phase of a flow reactor for the reduction of acetophenone. The immobilization of an ephedrine derivative **98** (Fig. 1.43) in a silica support allowed the efficiency of the system to be retained after 1 week of work (95% conversion, 90% ee) [197].

Meanwhile, ligands **99** [198] and **100** [199] (Fig. 1.43) were immobilized in silica and polystyrene, respectively. The respective oxazoborolidine were reused, taking advantage of nanofiltration technology for the asymmetric reduction of acetophenone. In the case of ligand **100**, the reactions were carried out in a continuously operated membrane reactor equipped with a nanofiltration membrane. The chiral alcohols were obtained in good-to-excellent ee and space–time yield [up to 99% ee and 1.4 kg/($L_{reactor\ volume}$ × day)].

Soós et al. studied the possibility of applying extraction techniques of perfluorous compounds. The authors synthesized a prolinol catalyst bearing two perfluoroalkyl chains (ligand **101**, Scheme 1.27). It was coordinated in situ to BH_3. THF promoted the reduction of aromatic ketones. Using only 10 mol% of catalyst with 0.6 eq. borane, quantitative yields were achieved with an excellent 95% ee for the acetophenone reduction. The perfluorinated ligand (not the oxazaborolidine) was quantitatively recovered with a solid–liquid extraction methodology and reused up to three cycles without any yield and ee erosion [200].

Dialkyl zinc addition to aldehydes was, in the past, one of the most explored reactions under the umbrella of recyclable stereoselective catalysts. A considerable number of strategies were tested and a representative fraction of those works are described here.

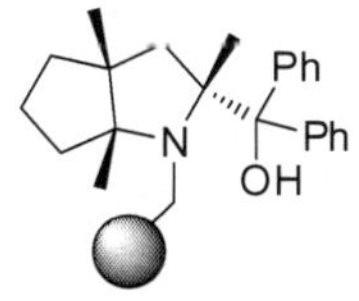

102

FIGURE 1.44. Structure of heterogeneous polymeric ligand **102**.

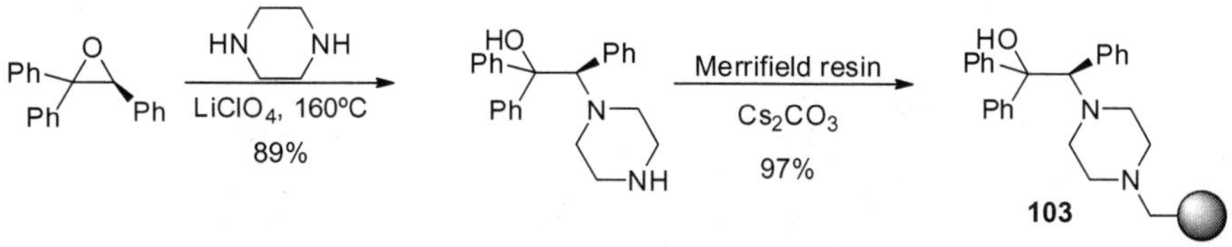

103

SCHEME 1.28. Preparation of (*R*)-2-piperidino-1,1,2-triphenylethanol anchored in a Merrifield resin.

Martens and co-workers immobilized a chiral amino alcohol by two different strategies in polystyrene polymers (grafting or coplymerization) [201]. The second immobilization strategy provided the heterogeneous polymeric ligand **102** (Fig. 1.44). The appropriate polymerization conditions provided a monolith with the desired morphology and properties that could be applied in a stationary monolithic column. This feature allows the design of a continuous-flow system where the reaction mixture is continuously passed through the monolith for 24 hours. This new protocol provided (*R*)-1-phenylpropanol in 85% with 99% ee in the first cycle. Further reutilization up to four cycles showed no loss of activity and enantioselectivity. A continuous-flow system is a very appealing solution for industrial application because the need for catalyst separation was suppressed, and the products can be easily isolated just by removing the solvent, thereby increasing the overall efficiency.

Pericàs et al. reported in 2003 the synthesis of (*R*)-2-piperidino-1,1,2-triphenylethanol covalently bonded to polystyrene polymeric support (five steps) [202]. More recently, Castellnou et al. [203] reported a simplified procedure for such supported ligand synthesis (Scheme 1.28). The supported ligand **103** was tested in the direct diethylzinc addition to benzaldehydes, where 94%–95% ee were obtained in high yield using only 2 mol% of catalytic resin at 0°C after 4 hours. The catalyst could be recovered and reused efficiently in five consecutive batches (resin 4 mol%, toluene, 0°C). Recently, such polymeric ligand was efficiently applied under a continuous-flow process. The high catalytic activity of this system allowed complete conversion of substrates with the use of stoichiometric reagent ratios and unprecedented short residence times (less than 3 minutes) [204].

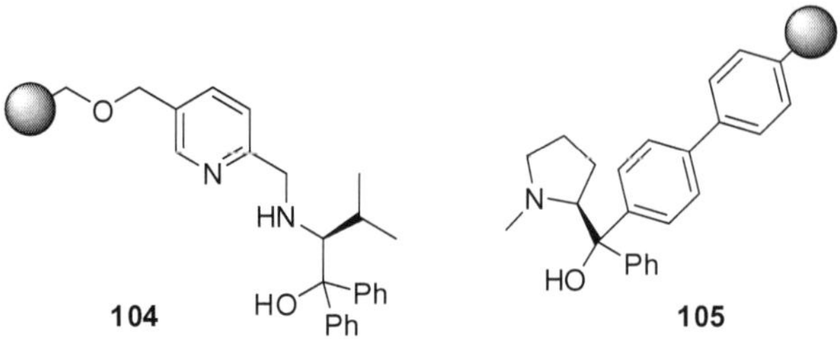

FIGURE 1.45. Structures of heterogeneous polymeric ligands **104** and **105**.

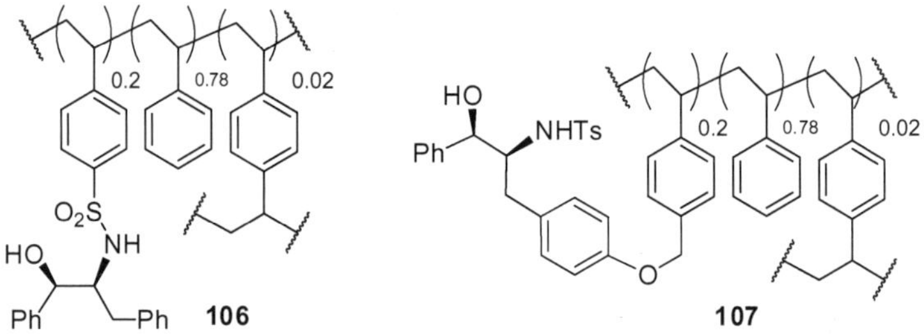

FIGURE 1.46. Structures of ligands **106** and **107**.

The first examples described proved to be very efficient, and their applications in flow reactors were accomplished. The next five examples were studied under batch conditions. Gros et al. described the application of a pyridine-based tridentate chiral ligand in enantioselective direct addition of diethylzinc to benzaldehydes supported on a Merrifield resin. This heterogeneous ligand **104** (Fig. 1.45) provided the respective phenylpropanols with excellent levels of enantioselectivities (up to 93% ee) and proved to be fully recyclable during five cycles [205]. Hodge et al. applied an *N*-methyl-diphenylprolinol grafted in polystyrene beads as catalyst for alkylzinc addition to benzaldehydes. Despite the high levels of catalyst loading required to obtain levels of enantioselectivities comparable with the homogeneous system (78%–94% ee), catalyst **105** (Fig. 1.45) was recycled efficiently nine times [206].

Gau et al. copolymerized a chiral styrene-substituted *N*-sulfonylated amino alcohol with styrene and divinylbenzene to achieve its recycling. The Ti/ligand **106** (Fig. 1.46) complex was tested in the diethyl zinc addition to aldehydes, which resulted in excellent levels of enantioselectivity. The ligand, recovered by filtration, was reused several times with a small erosion of enantioselectivity in each cycle (about 3%) [207]. Recently, the authors described the ligand **107** (Fig. 1.46) that is more reactive and more reusable than **106** (nine efficient recycles) [208].

Xu et al. immobilized *N*-sulfonylated amino alcohols in the amorphous silica to be used in titanium-mediated ethyl zinc addition to aldehydes. In the first tests, it was discovered that it was necessary to pretreat the free silinols

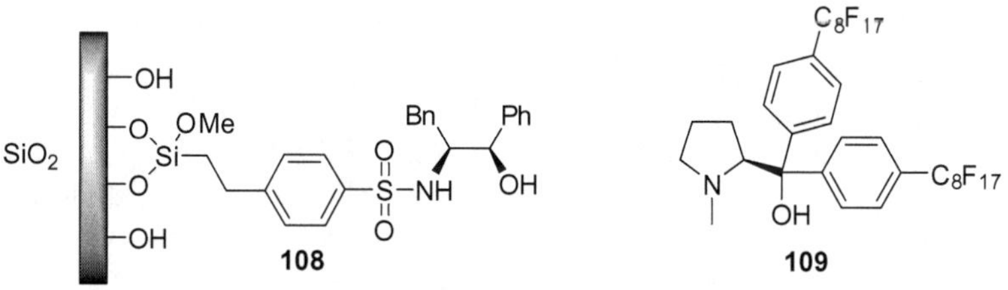

FIGURE 1.47. Structures of ligands **108** and **109**.

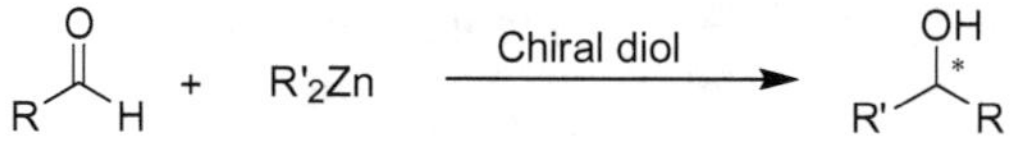

SCHEME 1.29. Asymmetric dialkylzinc addition to aldehydes.

with titanium isopropoxide to achieve any reaction (98% yield, 80% ee, substrate: benzaldehyde). The catalyst **108** (Fig. 1.47) showed an impressive recycling ability, since it was reused nine times without any significant decrease of reactivity and enantioselectivity [209].

Heterogeneous supports furnished a more robust catalyst than those prepared for homogeneous catalysis, in which soluble polystyrene polymers [210] and dentritic [211, 212] structures were examined as supports. However, when dealing with homogeneous catalysis it is crucial to highlight the work of Kim et al. In this work was described a method for the modification of pyrrolidinylmethanol derivatives to be applied in asymmetric addition of diethyl zinc to aldehydes [213]. Under biphasic conditions (FC-72/hexane, 40°C), the desired product was obtained in 92% ee. The ligand was recovered by lowering the temperature phase to 0°C, where a phase separation occurred. The authors claimed that stable **109** Li Et_2Zn (Fig. 1.47) forced the ligand to stay in the FC-72 phase. This protocol allows the catalytic system to have a long lifetime (about nine runs), affording (*S*)-1-phenylpropanol with high ee's.

As was observed in the previous section, a considerable amount of work has been described for dialkyl zinc additions to aldehydes from which very good results were highlighted in this section.

1.7.3. BINOL-Type Ligands

Over the past 20 years, BINOL ligands have become some of the most broadly used ligands in catalytic asymmetric reactions, such as the formation of new carbon–carbon bonds or on oxidation reactions. Mainly, they act as a ligand in titanium-promoted (Lewis acid) catalysis.

The asymmetric dialkylzinc addition aldeydes can be conducted in Ti/BINOL complexes (Scheme 1.29). BINOL can be immobilized in soluble and insoluble polymers with considerable success in this transformation. Several polymeric ligands provided high levels of enantioselectivity (up to 98% ee) and proved to be recyclable [2].

An organosilane substituted (*S*)-BINOL was anchored in the surface of two siliceous materials, MCM-41, MCF, and SBA-15 by Abdi et al. [214]. These heterogeneous catalysts were applied to a Ti(IV)-mediated diethyl zinc addition to aldehydes where MCF readily proved to be the most efficient support. In the preliminary results, low enantioselectivities were observed caused by free silinol groups present in the supports, which interact with titanium. When those groups were protected with tetramethylsilane, enantioselectivities increased to 94% ee (comparable with free BINOL). The ligand **110** (Fig. 1.48) was recovered by simple filtration and reused up to four efficient cycles with some yield (27%) and enantioselectivity (9%) erosion before regeneration.

Ma et al. prepared Frechet-type dendritic BINOL ligands to be applied in titanium-promoted diethylzinc addition to benzaldehydes. Excellent yields and enantioselectivities were achieved with this ligand **111** (Fig. 1.48), sometimes surpassing the nonimmobilized BINOL system (up to 89% ee). Ligand **111** could be recovered by precipitation with methanol addition and reused twice without yield and enantioselectivity erosion [215].

Moreau et al. immobilized BINOL derivates for titanium-promoted ethyl zinc addition to benzaldehyde in an IL [216]. The immobilized ligand **112** (Fig. 1.48) proved to be as effective as the BINOL itself, affording (*S*)-1-

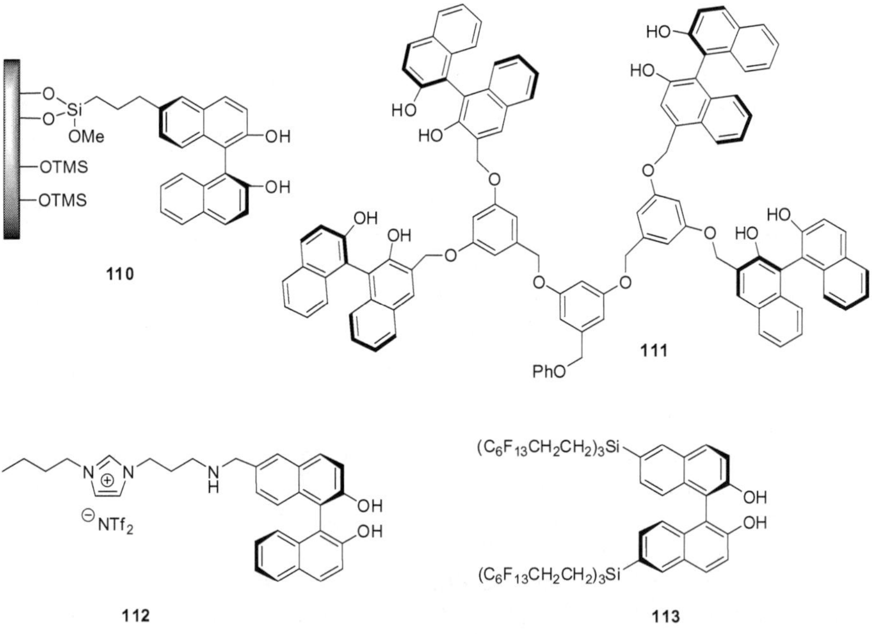

FIGURE 1.48. Structures of BINOL-type ligands **110–113**.

phenylpropanol with 82% ee's. Its recovery was achieved by extractions and filtration. It was reused three times with constant enantioselectivity (81%–82% ee's).

Curran et al. reported in a previous work that the ligand **113** (Fig. 1.48) with a fluorine content of 61.2% provided results comparable to the original system (81%–89% and 80%–83% ee vs. 97% and 85% ee) in titanium-promoted diethylzinc addition to aldehydes using a biphasic solvent system of toluene/hexane/FC-72. Despite its fluorine content, it was not possible to prevent some leaching to the organic phase during the liquid–liquid extraction. The reutilization problems were solved by using a solid-phase extraction technique. The reaction mixture was passed through a fluorous reverse-phase silica-gel column and eluted with acetonitrile. The retained ligand **113** was recovered using FC-72 and was reused up to four cycles without any efficiency loss [217].

The carbonyl-ene coupling mediated by a titanium or copper catalyst allows the formation of a new C–C bond with a high level of enantioselectivity. Titanium-based catalysts applied in this chemistry use BINOL-type ligands, which are quite expensive. Therefore, it is necessary to achieve reutilization of such catalysts to boost their industrial application.

Sasai et al. reported the first example of a chiral self-assembled structure, formed through the self-assembly of chiral multidentate ligands with a metal. This polymer acted as an asymmetric catalyst in carbonyl-ene reactions providing high enantioselectivities (20 mol%). The catalyst 114 (Fig. 1.49) was tested for enantioselective carbonyl-ene reaction of ethyl glycoxalate and prop-1-en-2-ylbenzene, affording the respective product after 98 hours in 88% yield with 88% ee. While the enantioselectivity remained unchanged for more than four cycles, the reactivity suffered a considerable decrease [218].

Wang et al. showed that the enantioselectivity could be increased to 95% ee using only 1 mol% of catalyst 115 (Fig. 1.49), if it does not have an oxo-bridge between the two titanium centers. Further ligand derivatization allowed the discovery of a new insoluble polymeric catalyst that provided quantitative yields with 97% ee. This new catalyst was also reused four times, but in this case both yield (19%) and enantioselectivity (27%) decreased [219].

Using the same polymeric-catalyst approach, Sasai et al. reported the first example of a chiral self-assembled structure, formed through self-assembly of chiral multidentate ligands with a aluminum instathat that acted as an asymmetric catalyst for Michael additions in high enantioselectivity [218].

This insoluble polymeric catalyst **116** (Fig. 1.49) showed a similar performance (Scheme 1.30) to the homogeneous monomer ALB [Al-Li-bis(binaphthoxide), 12 hours, quantitative yield, 97% ee]. At the end of the fourth cycle, significant erosion on the activity (~30%) and enantioselectivity (~20%) was observed.

Loh et al. performed the cycloaddition of acroleins with butadienes and cyclopentadiene in [bmim]PF_6 in the presence of (*S*)-BINOL-In(III) and allyltributyl stannane. This system performed better in terms of yields (78%–92%) in the IL media than in dichloromethane and furnished excellent

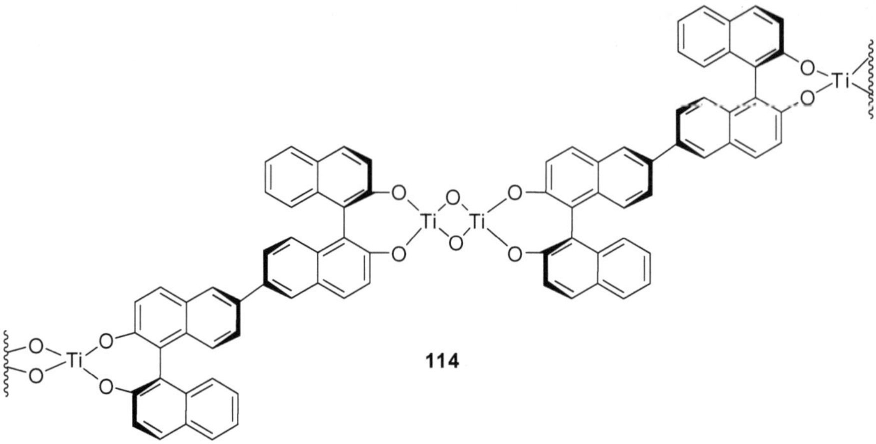

114

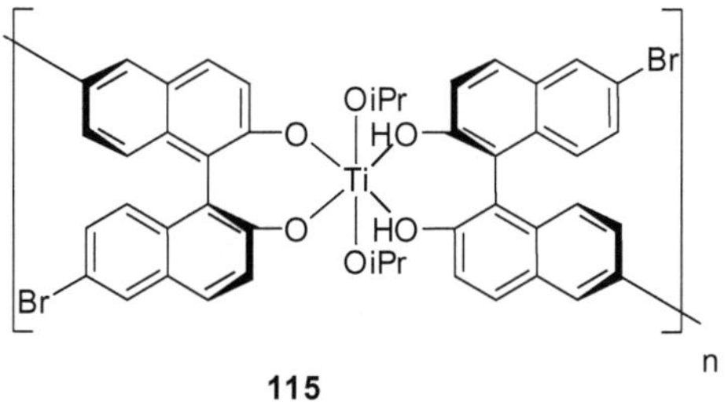

115

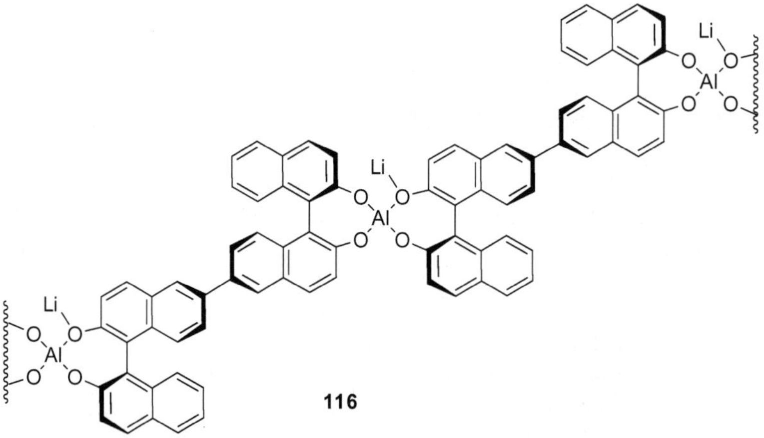

116

FIGURE 1.49. Structures of catalysts **114–116**.

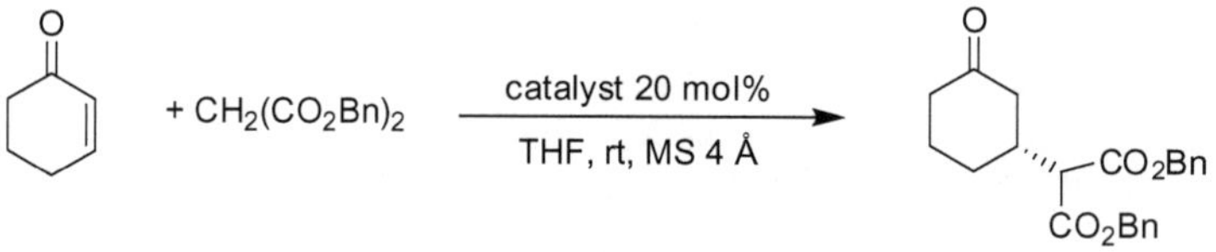

48 hours, 86% yield, 96% ee

SCHEME 1.30. Asymmetric Michael addition catalyzed by insoluble self-assembled catalyst.

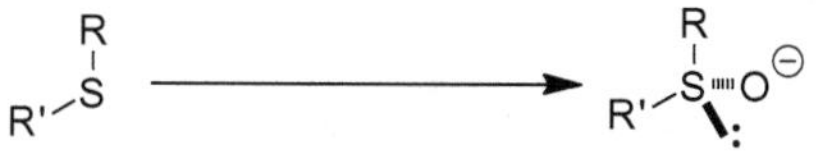

SCHEME 1.31. Asymmetric sulfide oxidation to sulfoxides.

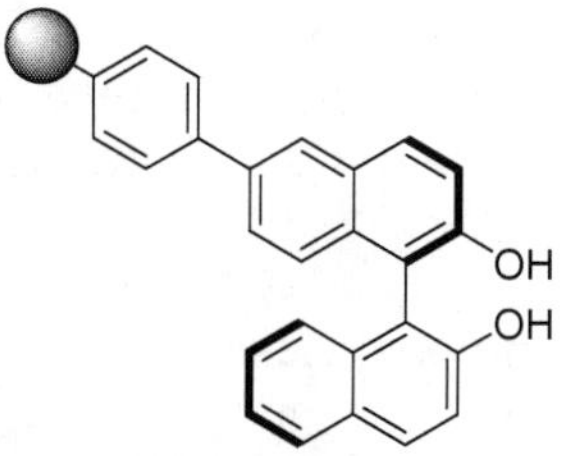

117

FIGURE 1.50. Structure of polymeric BINOL ligand **117**.

enantioselectivities (up to 98% ee). Furthermore, the catalytic system remained in the IL while the products were extracted enclosing the reutilization. In the seventh cycle, the yield eroded 5% while the enantioselectivity fell 12% [220].

The reutilization of BINOL catalysts was also described in asymmetric allylation [221], aldol-type reactions [222], or in Claisen–Schmidt condensation [223], but the results were not competitive with other approaches.

Several methodologies have been applied for the oxidation of the sulfide moiety to chiral sulfoxides (Scheme 1.31), from which the highlight here goes to the Ti/BINOL system.

Hodge et al. disclosed a straightforward method to immobilize a BINOL unit in polystyrene beads via a Suzuki coupling. Such heterogeneous ligands were coordinated to titanium tetraisopropoxide and tested in oxidation of prochiral thioethers (10 mol% titanium loading, ee up to 91%). The authors observed that retreatment of polymeric BINOL **117** (Fig. 1.50) with a titanium complex achieved four runs with consistent enantioselectivities [224].

Sahoo et al. immobilized chiral Ti/BINOL onto IL-modified SBA-15. The immobilized catalyst presented a high enantioselectivity in oxidation of thioanisole (99.2% and 62% yield). The supported catalyst was recycled seven times without any loss of enantioselectivity [225].

A wide range of protocols and reactions were explored using BINOL systems. Generally, BINOL complexes give good-to-excellent enantioselectivities and conversions. Furthermore, several efficient immobilization and polymerization approaches were shown. However, more effort should be made in the area of reutilization since some protocols showed low recyclability and erosion of efficiency through cycles.

1.8. CONCLUSIONS

In the past several years, remarkable progress has been observed in the development of efficient asymmetric synthetic methodologies that allow catalyst reuse for a diverse range of organic reactions under homogeneous or heterogeneous conditions, or by using alternative reaction media such as water, ILs, perfluorinated solvents and catalysts, supercritical CO_2, and membrane separation technology. This area has been established as one of hottest areas in organic chemistry. However, at this stage, the following directions may be considered:

(1) the development of more efficient protocols that allow less leaching of precious (as well as toxic in many cases) metals, minimal use of volatile organic solvents, and the use of water as reaction media;
(2) more developments on the less explored methodologies based on the use of microreactors and membrane-based technologies;
(3) exploration of efficient methodologies for catalyst reuse for some (bio) catalysts.

One of the most important parameters that help to identify the limitations and potentialities of a recycling protocol using organometallic catalysis is the degree of metal leaching. This phenomenon is the main cause for the decrease of reactivity upon recycling in most of the protocols described herein, and this parameter is not always determined.

REFERENCES

[1] (a) Breuer, M., Ditrich, K., Habicher, T., Hauer, B., Kesseler, M., Sturmer, R., Zelinski, T. (2004). Industrial methods for the production of optically active intermediates. *Angew. Chem. Int. Ed.*, *43*, 788–824; (b) Ikunaka, M. (2003). A process in need is a process indeed: scalable enantioselective synthesis of chiral compounds for the pharmaceutical industry. *Chem. Eur. J.*, *9*, 379–388; (c) Hawkins, J. M., Watson, T. J. N. (2004). Asymmetric catalysis in the pharmaceutical industry. *Angew. Chem. Int. Ed.*, *43*, 3224–3228; (d) Cole-Hamilton, D. J. (2003).

Homogeneous catalysis—new approaches to catalyst separation, recovery, and recycling. *Science*, *299*, 1702–1706.

[2] Fan, Q. H., Li, Y. M., Chan, A. S. C. (2002). Recoverable catalysts for asymmetric organic synthesis. *Chem. Rev.*, *102*, 3385–3465.

[3] Trindade, A. F., Gois, P. M. P., Afonso, C. A. M. (2009). Recyclable stereoselective catalysts. *Chem. Rev.*, *109*, 418–514.

[4] Lindstrom, U. M. (2002). Stereoselective organic reactions in water. *Chem. Rev.*, *102*, 2751–2771.

[5] (a) Benaglia, M., Puglisi, A., Cozzi, F. (2003). Polymer-supported organic catalysts. *Chem. Rev.*, *103*, 3401–3429; (b) Corma, A., Garcia, H. (2003). Lewis acids: from conventional homogeneous to green homogeneous and heterogeneous catalysis. *Chem. Rev.*, *103*, 4307–4365; (c) Dickerson, T. J., Reed, N. N., Janda, K. D. (2002). Soluble polymers as scaffolds for recoverable catalysts and reagents. *Chem. Rev.*, *102*, 3325–3343; (d) Bergbreiter, D. E. (2002). Using soluble polymers to recover catalysts and ligands. *Chem. Rev.*, *102*, 3345–3383; (e) McMorn, P., Hutchings, G. J. (2004). Heterogeneous enantioselective catalysts: strategies for the immobilisation of homogeneous catalysts. *Chem. Soc. Rev.*, *33*, 108–122; (f) Bell, A. T. (2003). The impact of nanoscience on heterogeneous catalysis. *Science*, *299*, 1688–1691; (g) Rolison, D. R. (2003). Catalytic nanoarchitectures—the importance of nothing and the unimportance of periodicity. *Science*, *299*, 1698–1701.

[6] (a) Xia, Q. H., Ge, H. Q., Ye, C. P., Liu, Z. M., Su, K. X. (2005). Advances in homogeneous and heterogeneous catalytic asymmetric epoxidation. *Chem. Rev.*, *105*, 1603–1662; (b) Baleizao, C., Garcia, H. (2006). Chiral salen complexes: an overview to recoverable and reusable homogeneous and heterogeneous catalysts. *Chem. Rev.*, *106*, 3987–4043; (c) Kobayashi, S., Sugiura, M. (2006). Immobilization of osmium catalysts for asymmetric dihydroxylation of olefins. *Adv. Synth. Catal.*, *348*, 1496–1504.

[7] Ding, K. L., Wang, Z., Wang, X. W., Liang, Y. X., Wang, X. S. (2006). Self-supported chiral catalysts for heterogeneous enantioselective reactions. *Chem. Eur. J.*, *12*, 5188–5197.

[8] Walsh, P. J., Li, H. M., de Parrodi, C. A. (2007). A green chemistry approach to asymmetric catalysis: solvent-free and highly concentrated reactions. *Chem. Rev.*, *107*, 2503–2545.

[9] (a) Pozzi, G., Shepperson, I. (2003). Fluorous chiral ligands for novel catalytic systems. *Coord. Chem. Rev.*, *242*, 115–124; (b) Muzart, J. (2006). Ionic liquids as solvents for catalyzed oxidations of organic compounds. *Adv. Synth. Catal.*, *348*, 275–295; (c) Afonso, C. A. M., Branco, L. C., Candeias, N. R., Gois, P. M. P., Lourenco, N. M. T., Mateus, N. M. M., Rosa, J. O. N. (2007). Efficient catalyst reuse by simple dissolution in non-conventional media. *Chem. Commun.*, 2669–2679; (d) Parvulescu, V. I., Hardacre, C. (2007). Catalysis in ionic liquids. *Chem. Rev.*, *107*, 2615–2665; (e) Chen, J., Spear, S. K., Huddleston, J. G., Rogers, R. D. (2005). Polyethylene glycol and solutions of polyethylene glycol as green reaction media. *Green Chem.*, *7*, 64–82; (f) Andrade, C. K. Z., Alves, L. M. (2005). Environmentally benign solvents in organic synthesis: current topics. *Curr. Org. Chem.*, *9*, 195–218; (g) Song, C. E. (2004).). Enantioselective chemo- and bio-catalysis in ionic liquids. *Chem. Commun.*, 1033–1043; (h) Song, C. E., Yoon, M. Y., Choi, D. S. (2005). Significant improvement of catalytic efficiencies in ionic liquids. *Bull. Korean Chem. Soc.*, *26*, 1321–1330; (i) Jain, N., Kumar, A., Chauhan, S., Chauhan,

S. M. S. (2005). Chemical and biochemical transformations in ionic liquids. *Tetrahedron*, *61*, 1015–1060; (j) Lee, S. G. (2006). Functionalized imidazolium salts for task-specific ionic liquids and their applications. *Chem. Commun.*, 1049–1063; (k) Corma, A., Garcia, H. (2006). Silica-bound homogenous catalysts as recoverable and reusable catalysts in organic synthesis. *Adv. Synth. Catal.*, *348*, 1391–1412; (l) Mery, D., Astruc, D. (2006). Dendritic catalysis: major concepts and recent progress. *Coord. Chem. Rev.*, *250*, 1965–1979; (m) Helms, B., Frechet, J. M. J. (2006). The dendrimer effect in homogeneous catalysis. *Adv. Synth. Catal.*, *348*, 1125–1148; (n) Bergbreiter, D. E., Sung, S. D. (2006). Liquid/liquid biphasic recovery/reuse of soluble polymier-supported catalysts. *Adv. Synth. Catal.*, *348*, 1352–1366; (o) Cozzi, F. (2006). Immobilization of organic catalysts: when, why, and how. *Adv. Synth. Catal.*, *348*, 1367–1390; (p) Benaglia, M. (2006). Recoverable and recyclable chiral organic catalysts. *New J. Chem.*, *30*, 1525–1533.

[10] Heitbaum, M., Glorius, F., Escher, I. (2006). Asymmetric heterogeneous catalysis. *Angew. Chem. Int. Ed.*, *45*, 4732–4762.

[11] Wang, X. W., Ding, K. L. (2004). Self-supported heterogeneous catalysts for enantioselective hydrogenation. *J. Am. Chem. Soc.*, *126*, 10524–10525.

[12] Shi, L., Wang, X. W., Sandoval, C. A., Li, M. X., Qi, Q. Y., Li, Z. T., Ding, K. L. (2006). Engineering a polymeric chiral catalyst by using hydrogen bonding and coordination interactions. *Angew. Chem. Int. Ed.*, *45*, 4108–4112.

[13] Shi, L., Wang, X. W., Sandoval, C. A., Wang, Z., Li, H. J., Wu, J., Yu, L. T., Ding, K. L. (2009). Development of a continuous-flow system for asymmetric hydrogenation using self-supported chiral catalysts. *Chem. Eur. J.*, *15*, 9855–9867.

[14] de Rege, F. M., Morita, D. K., Ott, K. C., Tumas, W., Broene, R. D. (2000). Non-covalent immobilization of homogeneous cationic chiral rhodium-phosphine catalysts on silica surfaces. *Chem. Commun.*, 1797–1798.

[15] Augustine, R. L., Tanielyan, S. K., Mahata, N., Gao, Y., Zsigmond, A., Yang, H. (2003). Anchored homogeneous catalysts: the role of the heteropoly acid anchoring agent. *Appl. Catal. A*, *256*, 69–76.

[16] Brandts, J. A. M., Berben, P. H. (2003). Application of immobilized rhodium catalyst precursors in enantio- and chemoselective hydrogenation reactions. *Org. Process Res. Dev.*, *7*, 393–398.

[17] Zsigmond, A., Undrala, S., Notheisz, F., Szollosy, A., Bakos, J. (2006). Substituent effects in enantio selective hydrogenations catalyzed by immobilized Rh complexes. *Appl. Catal. A*, *303*, 29–34.

[18] (a) Augustine, R. L., Goel, P., Mahata, N., Reyes, C., Tanielyan, S. K. (2004). Anchored homogeneous catalysts: high turnover number applications. *J. Mol. Catal. A Chem.*, *216*, 189–197; (b) Barnard, C. F. J., Rouzaud, J., Stevenson, S. H. (2005). Straightforward preparation of highly enantioselective anchored chiral homogeneous catalysts. *Org. Process Res. Dev.*, *9*, 164–167.

[19] Augustine, R., Tanielyan, S., Anderson, S., Yang, H. (1999). A new technique for anchoring homogeneous catalysts. *Chem. Commun.*, 1257–1258.

[20] Hems, W. P., McMorn, P., Riddel, S., Watson, S., Hancock, F. E., Hutchings, G. J. (2005). Asymmetric hydrogenation using chiral Rh complexes immobilised with a new ion-exchange strategy. *Org. Biomol. Chem.*, *3*, 1547–1550.

[21] Bautista, F. M., Caballero, V., Campelo, J. M., Luna, D., Marinas, J. M., Romero, A. A., Romero, I., Serrano, I., Llobet, A. (2006). Heterogeneization of a new

Ru(II) homogeneous asymmetric hydrogenation catalyst containing BINAP and the N-tridentate bpea ligand, through covalent attachment on amorphous AlPO4 support. *Top. Catal.*, *40*, 193–205.

[22] Simons, C., Hanefeld, U., Arends, I., Sheldon, R. A., Maschmeyer, T. (2004). Noncovalent anchoring of asymmetric hydrogenation catalysts on a new mesoporous aluminosilicate: application and solvent effects. *Chem. Eur. J.*, *10*, 5829–5835.

[23] Simons, C., Hanefeld, U., Arends, I., Maschmeyer, T., Sheldon, R. A. (2006). Comparison of supports for the electrostatic immobilisation of asymmetric homogeneous catalysts. *J. Catal.*, *239*, 212–219.

[24] Xing, L., Xie, J. H., Chen, Y. S., Wang, L. X., Zhou, Q. L. (2008). Simply modified chiral diphosphine: catalyst recycling via non-covalent absorption on carbon nanotubes. *Adv. Synth. Catal.*, *350*, 1013–1016.

[25] Yi, B., Fan, Q. H., Deng, G. J., Li, Y. M., Qiu, L. Q., Chan, A. S. C. (2004). Novel chiral dendritic diphosphine ligands for Rh(I)-catalyzed asymmetric hydrogenation: remarkable structural effects on catalytic properties. *Org. Lett.*, *6*, 1361–1364.

[26] Yi, B., He, H. P., Fan, Q. H. (2010). Synthesis and application of peripherally alkyl-functionalized dendritic pyrphos ligands: homogeneous-supported catalysts for enantioselective hydrogenation. *J. Mol. Catal. A Chem.*, *315*, 82–85.

[27] Pugin, B., Studer, M., Kuesters, E., Sedelmeier, G., Feng, X. (2004). Mixtures of ionic liquids and water as a medium for efficient enantioselective hydrogenation catalyst recycling. *Adv. Synth. Catal.*, *346*, 1481–1486.

[28] Zhou, H. F., Fan, Q. H., Tang, W. J., Xu, L. J., He, Y. M., Deng, G. J., Zhao, L. W., Gu, L. Q., Chan, A. S. C. (2006). Polyethylene glycol as an environmentally friendly and recyclable reaction medium for enantioselective hydrogenation. *Adv. Synth. Catal.*, *348*, 2172–2182.

[29] Saluzzo, C., Lamouille, T., Herault, D., Lemaire, M. (2002). Polymer-supported catalysts: enantioselective hydrogenation and hydrogen transfer reduction. *Bioorg. Med. Chem. Lett.*, *12*, 1841–1844.

[30] Hu, A. G., Ngo, H. L., Lin, W. B. (2003). Chiral porous hybrid solids for practical heterogeneous asymmetric hydrogenation of aromatic ketones. *J. Am. Chem. Soc.*, *125*, 11490–11491.

[31] Ralph, C. K., Akotsi, O. M., Bergens, S. H. (2004). A reusable polymeric asymmetric hydrogenation catalyst made by ring-opening olefin metathesis polymerization. *Organometallics*, *23*, 1484–1486.

[32] Ralph, C. K., Bergens, S. H. (2007). A highly reusable catalyst for enantioselective ketone hydrogenation. Catalyst-organic frameworks by alternating ROMP assembly. *Organometallics*, *26*, 1571–1574.

[33] Kantam, M. L., Laha, S., Yadav, J., Likhar, P. R., Sreedhar, B., Choudary, B. M. (2007). Asymmetric hydrosilylation of prochiral ketones catalyzed by nanocrystalline copper(II) oxide. *Adv. Synth. Catal.*, *349*, 1797–1802.

[34] Guerreiro, P., Ratovelomanana-Vidal, V., Genet, J. P., Dellis, P. (2001). Recyclable diguanidinium-BINAP and PEG-BINAP supported catalysts: syntheses and use in Rh(I) and Ru(II) asymmetric hydrogenation reactions. *Tetrahedron Lett.*, *42*, 3423–3426.

[35] Zhu, Y. H., Carpenter, K., Bun, C. C., Bahnmueller, S., Ke, C. P., Srid, V. S., Kee, L. W., Hawthorne, M. F. (2003). *R*)-Binap-mediated asymmetric hydrogenation with a rhodacarborane catalyst in ionic-liquid media. *Angew. Chem. Int. Ed.*, *42*, 3792–3795.

[36] Berthod, M., Saluzzo, C., Mignani, G., Lemaire, M. (2004). 4,4′ and 5,5′-DiamBINAP as a hydrosoluble chiral ligand: syntheses and use in Ru(II) asymmetric biphasic catalytic hydrogenation. *Tetrahedron Asymmetry*, *15*, 639–645.

[37] Xu, L. K., Lam, K. H., Ji, J. X., Wu, J., Fan, Q. H., Lo, W. H., Chan, A. S. C. (2005). Air-stable Ir-(P-Phos) complex for highly enantioselective hydrogenation of quinolines and their immobilization in poly(ethylene glycol) dimethyl ether (DMPEG). *Chem. Commun.*, 1390–1392.

[38] Chan, S. H., Lam, K. H., Li, Y. M., Xu, L., Tang, W., Lam, F. L., Lo, W. H., Yu, W. Y., Fan, Q., Chan, A. S. C. (2007). Asymmetric hydrogenation of quinolines with recyclable and air-stable iridium catalyst systems. *Tetrahedron Asymmetry*, *18*, 2625–2631.

[39] Sakai, N., Mano, S., Nozaki, K., Takaya, H. (1993). Highly enantioselective hydroformylation of olefins catalyzed by new phosphinephosphite-Rh(I) complexes. *J. Am. Chem. Soc.*, *115*, 7033–7034.

[40] Shibahara, F., Nozaki, K., Hiyama, T. (2003). Solvent-free asymmetric olefin hydroformylation catalyzed by highly cross-linked polystyrene-supported (*R,S*)-BINAPHOS-Rh(I) complex. *J. Am. Chem. Soc.*, *125*, 8555–8560.

[41] Josephsohn, N. S., Snapper, M. L., Hoveyda, A. H. (2003). Efficient and practical Ag-catalyzed cycloadditions between arylimines and the Danishefsky diene. *J. Am. Chem. Soc.*, *125*, 4018–4019.

[42] Furuno, H., Kambara, T., Tanaka, Y., Hanamoto, T., Kagawa, T., Inanaga, J. (2003). Highly enantioselective homogeneous catalysis of chiral rare earth phosphates in the hetero-Diels-Alder reaction. *Tetrahedron Lett.*, *44*, 6129–6132.

[43] Nájera, C., de Gracia Retamosa, M., Sansano, J. M. (2007). Recoverable (*R*)- and (*S*)-Binap-Ag(I) complexes for the enantioselective 1,3-dipolar cycloaddition reaction of azomethine ylides. *Org. Lett.*, *9*, 4025–4028.

[44] Nakano, H., Takahashi, K., Suzuki, Y., Fujita, R. (2005). Polymer-supported chiral phosphinooxathiane ligands for palladium-catalyzed asymmetric allylations. *Tetrahedron Asymmetry*, *16*, 609–614.

[45] Nakano, H., Takahashi, K., Fujita, R. (2005). Polymer-supported chiral phosphinooxazolidine ligands for palladium-catalyzed asymmetric allylic alkylations and Diels-Alder reactions. *Tetrahedron Asymmetry*, *16*, 2133–2140.

[46] Zhao, D. B., Sun, J., Ding, K. L. (2004). New types of soluble polymer-supported bisphosphine ligands with a cyclobutane backbone for Pd-catalyzed enantioselective allylic substitution reactions. *Chem. Eur. J.*, *10*, 5952–5963.

[47] Mino, T., Sato, Y., Saito, A., Tanaka, Y., Saotome, H., Sakamoto, M., Fujita, T. (2005). Synthesis, reaction, and recycle of fluorous palladium catalysts for an asymmetric allylic alkylation without using fluorous solvents. *J. Org. Chem.*, *70*, 7979–7984.

[48] Kolb, H. C., Vannieuwenhze, M. S., Sharpless, K. B. (1994). Catalytic asymmetric dihydroxylation. *Chem. Rev.*, *94*, 2483–2547.

[49] Jacobsen, E. N., Markó, I., Mungall, W. S., Schroder, G., Sharpless, K. B. (1988). Asymmetric dihydroxylation via ligand-accelerated catalysis. *J. Am. Chem. Soc.*, *110*, 1968–1970.

[50] Mandoli, A., Pini, D., Fiori, M., Salvadori, P. (2005). Asymmetric dihydroxylation with recoverable cinchona alkaloid derivatives: a warning note and an improved, insoluble polymer-bound ligand 5 architecture. *Eur. J. Org. Chem.*, 1271–1282.

[51] Lee, H. M., Kim, S. W., Hyeon, T., Kim, B. M. (2001). Asymmetric dihydroxylation using heterogenized cinchona alkaloid ligands on mesoporous silica. *Tetrahedron Asymmetry*, *12*, 1537–1541.

[52] Lee, D., Lee, J., Lee, H., Jin, S., Hyeon, T., Kim, B. M. (2006). Filtration-free recyclable catalytic asymmetric dihydroxylation using a ligand immobilized on magnetic mesocellular mesoporous silica. *Adv. Synth. Catal.*, *348*, 41–46.

[53] Choudary, B. M., Chowdari, N. S., Madhi, S., Kantam, M. L. (2003). A trifunctional catalyst for one-pot synthesis of chiral diols via Heck coupling-N-oxidation-asymmetric dihydroxylation: application for the synthesis of diltiazem and Taxol side chain. *J. Org. Chem.*, *68*, 1736–1746.

[54] Kobayashi, S., Endo, M., Nagayama, S. (1999). Catalytic asymmetric dihydroxylation of olefins using a recoverable and reusable polymer-supported osmium catalyst. *J. Am. Chem. Soc.*, *121*, 11229–11230.

[55] (a) Choudary, B. M., Chowdary, N. S., Madhi, S., Kantam, M. L. (2003). US6780810; (b) Choudary, B. M., Chowdary, N. S., Kantam, M. L., Raghavan, K. V., Reddy, C. V. (2002). US6387033 B1; (c) Choudary, B. M., Chowdary, N. S., Madhi, S., Kantam, M. L., Jyothi, K. (2004). US6706901 B1.

[56] Bolm, C., Gerlach, A. (1998). Polymer-supported catalytic asymmetric sharpless dihydroxylations of olefins. *Eur. J. Org. Chem.*, 21–27.

[57] Kuang, Y. Q., Zhang, S. Y., Wei, L. L. (2003). A soluble block copolymer-supported bis-cinchona alkaloid ligand for the asymmetric dihydroxylation of olefins. *Synth. Commun.*, *33*, 3545–3550.

[58] Cheng, S. K., Zhang, S. Y., Wang, P. A., Kuang, Y. Q., Sun, X. L. (2005). Homogeneous catalytic asymmetric dihydroxylation of olefins induced by an efficient and recoverable polymer-bound ligand QN-AQN-OPEG-OMe. *Appl. Organometallic Chem.*, *19*, 975–979.

[59] Branco, L. C., Afonso, C. A. M. (2002). Catalytic asymmetric dihydroxylation of olefins using a recoverable and reusable OsO_4 in ionic liquid [bmim] [PF_6]. *Chem. Commun.*, 3036–3037.

[60] Branco, L. C., Gois, P. M. P., Lourenco, N. M. T., Kurteva, V. B., Afonso, C. A. M. (2006). Simple transformation of crystalline chiral natural anions to liquid medium and their use to induce chirality. *Chem. Commun.*, 2371–2372.

[61] Kwueon, E. K., Choi, D. S., Choi, H. Y., Lee, Y. J., Jo, C. H., Hwang, S. H., Park, Y. S., Song, C. E. (2005). Osmium tetroxide-(QN) (2)PHAL-sugar: a new recyclable catalyst system for asymmetric dihydroxylation of olefins. *Bull. Korean Chem. Soc.*, *26*, 1839–1842.

[62] Lee, D., Lee, H., Kim, S., Yeom, C. E., Kim, B. M. (2006). A novel chemoentrapment approach for supportless recycling of a catalyst: catalytic asymmetric dihydroxylation. *Adv. Synth. Catal.*, *348*, 1021–1024.

[63] Yang, X. W., Liu, H. Q., Xu, M. H., Lin, G. Q. (2004). A highly efficient and practical new PEG-bound bi-cinchona alkaloid ligand for the catalytic asymmetric aminohydroxylation of alkenes. *Tetrahedron Asymmetry*, *15*, 1915–1918.

[64] Desimoni, G., Faita, G., Jørgensen, K. A. (2006). C2-Symmetric chiral bis(oxazoline) ligands in asymmetric catalysis. *Chem. Rev.*, *106*, 3561–3651.

[65] Rosa, J. N., Santos, A. G., Afonso, C. A. M. (2004). Enantioselective addition of alkynes to imines in ionic liquids. *J. Mol. Catal. A Chem.*, *214*, 161–165.

[66] Liu, J. T., Liu, B., Jia, M., Li, X. S., Chan, A. S. C. (2007). Asymmetric addition of alkynes to imines in water catalyzed with a recyclable Cu(I)-bis(oxazoline) and stearic acid system. *Tetrahedron Asymmetry*, *18*, 396–399.

[67] Loh, T.-P., Zhou, J.-R. (1999). An enantioselective allylation reaction of aldehydes in an aqueous medium. *Tetrahedron Lett.*, *40*, 9115–9118.

[68] McDonagh, C., O'Conghaile, P., Gebbink, R., O'Leary, P. (2007). Electrostatic immobilisation of copper(I) and copper(II) bis(oxazolinyl)pyridine catalysts on silica: application to the synthesis of propargylamines via direct addition of terminal alkynes to imines. *Tetrahedron Lett.*, *48*, 4387–4390.

[69] Weissberg, A., Halak, B., Portnoy, M. (2005). The first solid-phase synthesis of bis(oxazolinyl)pyridine ligands. *J. Org. Chem.*, *70*, 4556–4559.

[70] Cornejo, A., Fraile, J. M., Garcia, J. I., Garcia-Verdugo, E., Gil, M. J., Legarreta, G., Luis, S. V., Martinez-Merino, V., Mayoral, J. A. (2002). The first immobilization of pyridine-bis(oxazoline) chiral ligands. *Org. Lett.*, *4*, 3927–3930.

[71] Burguete, M. I., Cornejo, A., Garcia-Verdugo, E., Gil, M. J., Luis, S. V., Mayoral, J. A., Martinez-Merino, V., Sokolova, M. (2007). Pybox monolithic miniflow reactors for continuous asymmetric cyclopropanation reaction under conventional and supercritical conditions. *J. Org. Chem.*, *72*, 4344–4350.

[72] Cornejo, A., Fraile, J. M., Garcia, J. I., Gil, M. J., Luis, S. V., Martinez-Merino, V., Mayoral, J. A. (2005). A flexible and versatile strategy for the covalent immobilization of chiral catalysts based on pyridinebis(oxazoline) ligands. *J. Org. Chem.*, *70*, 5536–5544.

[73] Cornejo, A., Fraile, J. M., Garcia, J. I., Gil, M. J., Martinez-Merino, V., Mayoral, J. A. (2005). Reversible microencapsulation of pybox-Ru chiral catalysts: scope and limitations. *Tetrahedron*, *61*, 12107–12110.

[74] (a) Lu, J., Ji, S. J., Loh, T. P. (2005). Enantioselective allylation of aldehydes catalyzed by chiral indium(III) complexes immobilized in ionic liquids. *Chem. Commun.*, 2345–2347; (b) Lu, J., Ji, S. J., Teo, Y. C., Loh, T. P. (2005). Highly enantioselective allylation of aldehydes catalyzed by indium(III)-PYBOX complex. *Org. Lett.*, *7*, 159–161; (c) Lu, J., Ji, S. J., Teo, Y. C., Loh, T. P. (2005). Asymmetric allyltributylstannane addition to ketones catalyzed by chiral PYBOX-In(III) complex immobilized in ionic liquid. *Tetrahedron Lett.*, *46*, 7435–7437.

[75] Lundgren, S., Lutsenko, S., Jonsson, C., Moberg, C. (2003). Polymer-supported pyridine-bis(oxazoline). Application to ytterbium-catalyzed silylcyanation of benzaldehyde. *Org. Lett.*, *5*, 3663–3665.

[76] Kolb, H. C., Finn, M. G., Sharpless, K. B. (2001). Click chemistry: diverse chemical function from a few good reactions. *Angew. Chem. Int. Ed.*, *40*, 2004–2021.

[77] Tilliet, M., Lundgren, S., Moberg, C., Levacher, V. (2007). Polymer-bound pyridine-bis(oxazoline). Preparation through click chemistry and evaluation in asymmetric catalysis. *Adv. Synth. Catal.*, *349*, 2079–2084.

[78] Davies, D. L., Kandola, S. K., Patel, R. K. (2004). Asymmetric cyclopropanation in ionic liquids: effect of anion and impurities. *Tetrahedron Asymmetry*, *15*, 77–80.

[79] (a) Burguete, M. I., Fraile, J. M., Garcia, J. I., Garcia-Verdugo, E., Luis, S. V., Mayoral, J. A. (2000). Polymer-supported bis(oxazoline)-copper complexes as

catalysts in cyclopropanation reactions. *Org. Lett.*, *2*, 3905–3908; (b) Diez-Barra, E., Fraile, J. M., Garcia, J. I., Garcia-Verdugo, E., Herrerias, C. I., Luis, S. V., Mayoral, J. A., Sanchez-Verdu, P., Tolosa, J. (2003). Polymer immobilization of bis(oxazoline) ligands using dendrimers as cross-linkers. *Tetrahedron Asymmetry*, *14*, 773–778; (c) Burguete, M. I., Diez-Barra, E., Fraile, J. M., Garcia, J. I., Garcia-Verdugo, E., Gonzalez, R., Herrerias, C. I., Luis, S. V., Mayoral, J. A. (2002). Improvement of ligand economy controlled by polymer morphology: the case of polymer-supported bis(oxazoline) catalysts. *Bioorg. Med. Chem. Lett.*, *12*, 1821–1824.

[80] Mandoli, A., Orlandi, S., Pini, D., Salvadori, P. (2003). A reusable, insoluble polymer-bound bis(oxazoline) (IPB-box) for highly enantioselective heterogeneous cyclopropanation reactions. *Chem. Commun.*, 2466–2467.

[81] Lee, S. S., Hadinoto, S., Ying, J. Y. (2006). Improved enantioselectivity of immobilized chiral bisoxazolines by partial precapping of the siliceous mesocellular foam support with trimethylsilyl groups. *Adv. Synth. Catal.*, *348*, 1248–1254.

[82] Lee, S. S., Ying, J. Y. (2006). Siliceous mesocellular foam-supported chiral bisoxazoline: application to asymmetric cyclopropanation. *J. Mol. Catal. A Chem.*, *256*, 219–224.

[83] Castillo, M. R., Fousse, L., Fraile, J. M., Garcia, J. I., Mayoral, J. A. (2007). Supported ionic-liquid films (SILF) as two-dimensional nanoreactors for enantioselective reactions: surface-mediated selectivity modulation (SMSM). *Chem. Eur. J.*, *13*, 287–291.

[84] (a) Annunziata, R., Benaglia, M., Cinquini, M., Cozzi, F., Pozzi, G. (2003). Synthesis of perfluoroalkyl-substituted bis(oxazolines) as ligands for catalytic enantioselective reactions. *Eur. J. Org. Chem.*, 1191–1197; (b) Simonelli, B., Orlandi, S., Benaglia, M., Pozzi, G. (2004). New perfluoroalkyl-substituted bisoxazolines as chiral ligands in asymmetric Cu-II-catalyzed reactions. *Eur. J. Org. Chem.*, 2669–2673.

[85] Kolodziuk, R., Goux-Henry, C., Sinou, D. (2007). Recyclable copper catalysts in the enantioselective glyoxylate-ene reaction catalyzed by chiral fluorous-tagged bis(oxazolines). *Tetrahedron Asymmetry*, *18*, 2782–2786.

[86] Ono, F., Kanemasa, S., Tanaka, J. (2005). Reusable nano-sized chiral bisoxazoline catalysts. *Tetrahedron Lett.*, *46*, 7623–7626.

[87] Mandoli, A., Orlandi, S., Pini, D., Salvadori, P. (2004). Insoluble polystyrene-bound bis(oxazoline): batch and continuous-flow heterogeneous enantioselective glyoxylate-ene reaction. *Tetrahedron Asymmetry*, *15*, 3233–3244.

[88] Benaglia, M., Cinquini, M., Cozzi, F., Celentano, G. (2004). Enantioselective catalysis in water: Mukaiyama-aldol condensation promoted by copper complexes of bisoxazolines supported on poly(ethylene glycol). *Org. Biomol. Chem.*, *2*, 3401–3407.

[89] Lee, A., Kim, W., Lee, J., Hyeon, T., Kim, B. M. (2004). Heterogeneous asymmetric nitro-Mannich reaction using a bis(oxazoline) ligand grafted on mesoporous silica. *Tetrahedron Asymmetry*, *15*, 2595–2598.

[90] Yang, B. Y., Chen, X. M., Deng, G. J., Zhang, Y. L., Fan, Q. H. (2003). Chiral dendritic bis(oxazoline) copper(II) complexes as Lewis acid catalysts for enantioselective aldol reactions in aqueous media. *Tetrahedron Lett.*, *44*, 3535–3538.

[91] Doherty, S., Goodrich, P., Hardacre, C., Parvulescu, V., Paun, C. (2008). Efficient heterogeneous asymmetric catalysis of the Mukaiyama aldol reaction by silica- and ionic liquid-supported Lewis acid copper(II) complexes of bis(oxazolines). *Adv. Synth. Catal.*, *350*, 295–302.

[92] Doherty, S., Goodrich, P., Hardacre, C., Knight, J. G., Nguyen, M. T., Parvulescu, V. I., Paun, C. (2007). Recyclable copper catalysts based on imidazolium-tagged bis(oxazolines): a marked enhancement in rate and enantioselectivity for Diels-Alder reactions in ionic liquid. *Adv. Synth. Catal.*, *349*, 951–963.

[93] Yeom, C. E., Kim, H. W., Shin, Y. J., Kim, B. M. (2007). Chiral bis(oxazoline)-copper complex catalyzed Diels-Alder reaction in ionic liquids: remarkable reactivity and selectivity enhancement, and efficient recycling of the catalyst. *Tetrahedron Lett.*, *48*, 9035–9039.

[94] (a) Rechavi, D., Lemaire, M. (2002). Enantioselective catalysis of Diels-Alder reactions by heterogeneous chiral bis(oxazoline) catalysts. *J. Mol. Catal. A Chem.*, *182*, 239–247; (b) Rechavi, D., Albela, B., Bonneviot, L., Lemaire, M. (2005). Understanding the enantioselectivity of a heterogeneous catalyst: the influence of ligand loading and of silica passivation. *Tetrahedron*, *61*, 6976–6981; (c) Tanaka, S., Tada, M., Iwasawa, Y. (2007). Enantioselectivity promotion by achiral surface functionalization on SiO_2-supported Cu-bis(oxazoline) catalysts for asymmetric Diels-Alder reactions. *J. Catal.*, *245*, 173–183.

[95] Lancaster, T. M., Lee, S. S., Ying, J. Y. (2005). Effect of surface modification on the reactivity of MCF-supported IndaBOX. *Chem. Commun.*, 3577–3579.

[96] (a) Chollet, G., Rodriguez, F., Schulz, E. (2006). A new method for recycling asymmetric catalysts via formation of charge transfer complexes. *Org. Lett.*, *8*, 539–542; (b) Chollet, G., Guillerez, M. G., Schulz, E. (2007). Reusable catalysts for the asymmetric Diels-Alder reaction. *Chem. Eur. J.*, *13*, 992–1000; (c) Chollet, G., Didier, D., Schulz, E. (2010). Reusable chiral bis(oxazoline)-copper complexes immobilized by donor-acceptor interactions on insoluble organic supports. *Catal. Commun.*, *11*, 351–355.

[97] Werner, H., Herrerias, C. I., Glos, M., Gissibl, A., Fraile, J. M., Perez, I., Mayoral, J. A., Reiser, O. (2006). Synthesis of polymer bound azabis(oxazoline) ligands and their application in asymmetric cyclopropanations. *Adv. Synth. Catal.*, *348*, 125–132.

[98] Fraile, J. M., Perez, L., Mayoral, J. A., Reiser, O. (2006). Multipurpose box- and azabox-based immobilized chiral catalysts. *Adv. Synth. Catal.*, *348*, 1680–1688.

[99] (a) Fernandez, M. J., Fraile, J. M., Garcia, J. I., Mayoral, J. A., Burguete, M. I., Garcia-Verdugo, E., Luis, S. V., Harmer, M. A. (2000). Enantioselective cyclopropanation reactions promoted by immobilized bis(oxazoline)-copper complexes. *Top. Catal.*, *13*, 303–309; (b) Fraile, J. M., Garcia, J. I., Herrerias, C. I., Mayoral, J. A., Reiser, O., Socuellamos, A., Werner, H. (2004). The role of binding constants in the efficiency of chiral catalysts immobilized by electrostatic interactions: the case of azabis(oxazoline)Copper complexes. *Chem. Eur. J.*, *10*, 2997–3005; (c) Fraile, J. M., Garcia, J. I., Herrerias, C. I., Mayoral, J. A., Harmer, M. A. (2004). Bis(oxazoline)-copper complexes supported by electrostatic interactions: scope and limitations. *J. Catal.*, *221*, 532–540.

[100] Fraile, J. M., Garcia, J. I., Herrerias, C. I., Mayoral, J. A., Gmough, S., Vaultier, M. (2004). Comparison of the immobilization of chiral bis(oxazoline)-copper complexes onto anionic solids and in ionic liquids. *Green Chem.*, *6*, 93–98.

[101] Fraile, J. M., Garcia, J. I., Herrerias, C. I., Mayoral, J. A., Reiser, O., Vaultier, M. (2004). The importance of complex stability for asymmetric copper-catalyzed cyclopropanations in [emim][OTf] ionic liquid: the bis(oxazoline)-azabis(oxazoline) case. *Tetrahedron Lett.*, *45*, 6765–6768.

[102] Glos, M., Reiser, O. (2000). Aza-bis(oxazolines): new chiral ligands for asymmetric catalysis. *Org. Lett.*, *2*, 2045–2048.

[103] García, J. I., López-Sánchez, B., Mayoral, J. A. (2008). Linking homogeneous and heterogeneous enantioselective catalysis through a self-assembled coordination polymer. *Org. Lett.*, *10*, 4995–4998.

[104] Gissibl, A., Finn, M. G., Reiser, O. (2005). Cu(II)-aza(bisoxazoline)-catalyzed asymmetric benzoylations. *Org. Lett.*, *7*, 2325–2328.

[105] Schätz, A., Hager, M., Reiser, O. (2009). Cu(II)-Azabis(oxazoline)-complexes immobilized on superparamagnetic magnetite@silica-nanoparticles: a highly selective and recyclable catalyst for the kinetic resolution of 1,2-diols. *Adv. Funct. Mater.*, *19*, 2109–2115.

[106] Gissibl, A., Padie, C., Hager, M., Jaroschik, F., Rasappan, R., Cuevas-Yanez, E., Turrin, C. O., Caminade, A. M., Majoral, J. P., Reiser, O. (2007). Synthesis and application of phosphorus dendrimer immobilized azabis(oxazolines). *Org. Lett.*, *9*, 2895–2898.

[107] Rasappan, R., Olbrich, T., Reiser, O. (2009). Combining fluorous and triazole moieties for the tagging of chiral azabis(oxazoline) ligands. *Adv. Synth. Catal.*, *351*, 1961–1967.

[108] Weissberg, A., Portnoy, M. (2003). Unprecedented preparation of pincer bis(oxazolinyl)phenyl ligands on solid support and their use in the first heterogeneously-catalyzed enantioselective allylation of aldehydes. *Chem. Commun.*, 1538–1539.

[109] Cho, S. H., Gadzikwa, T., Afshari, M., Nguyen, S. T., Hupp, J. T. (2007). [Bis(catechol) salen]Mn-III coordination polymers as support-free heterogeneous asymmetric catalysts for epoxidation. *Eur. J. Inorg. Chem.*, 4863–4867.

[110] (a) Xiang, S., Zhang, Y. L., Xin, Q., Li, C. (2002). Enantioselective epoxidation of olefins catalyzed by Mn (salen)/MCM-41 synthesized with a new anchoring method. *Chem. Commun.*, 2696–2697; (b) Zhang, H. D., Xiang, S., Xiao, J. L., Li, C. (2005). Heterogeneous enantioselective epoxidation catalyzed by Mn(salen) complexes grafted onto mesoporous materials by phenoxy group. *J. Mol. Catal. A Chem.*, *238*, 175–184.

[111] (a) Zhang, H. D., Xiang, S., Li, C. (2005). Enantioselective epoxidation of unfunctionalised olefins catalyzed by Mn(salen) complexes immobilized in porous materials via phenyl sulfonic group. *Chem. Commun.*, 1209–1211; (b) Zhang, H., Li, C. (2006). Asymmetric epoxidation of 6-cyano-2,2-dimethylchromene on Mn(salen) catalyst immobilized in mesoporous materials. *Tetrahedron*, *62*, 6640–6649.

[112] Zhang, H. D., Zhang, Y. M., Li, C. (2005). Asymmetric epoxidation of unfunctionalized olefins catalyzed by Mn(salen) axially immobilized onto insoluble polymers. *Tetrahedron Asymmetry*, *16*, 2417–2423.

[113] (a) Ren, W. S., Fu, X. K. (2009). Chiral Mn (III) salen complexes covalently bonded on zirconium oligostyrenylphosphonate-phosphates as catalysts for enantioselective epoxidation of nonfunctionalized alkenes. *J. Mol. Catal. A*

Chem., *312*, 40–47; (b) Ren, W. S., Fu, X. K., Bao, H. B., Bai, R. F., Ding, P. P., Sui, B. L. (2009). Enantioselective epoxidation of unfunctionalized olefins catalyzed by chiral salen Mn(III) catalyst immobilized on zirconium oligostyrenylphosphonate-phosphate. *Catal. Commun.*, *10*, 788–793; (c) Zou, X. C., Fu, X. K., Li, Y. D., Tu, X. B., Fu, S. D., Luo, Y. F., Wu, X. J. (2010). Highly enantioselective epoxidation of unfunctionalized olefins catalyzed by chiral Jacobsen's catalyst immobilized on phenoxy-modified zirconium poly(syrene-phenylvinylphosphonate)phosphate. *Adv. Synth. Catal.*, *352*, 163–170.

[114] Gong, B., Fu, X., Chen, J., Li, Y., Zou, X., Tu, X., Ding, P., Ma, L. (2009). Synthesis of a new type of immobilized chiral salen Mn(III) complex as effective catalysts for asymmetric epoxidation of unfunctionalized olefins. *J. Catal.*, *262*, 9–17.

[115] Zhang, H. D., Zhang, Y. M., Li, C. (2006). Enantioselective epoxidation of unfunctionalized olefins catalyzed by the Mn(salen) catalysts immobilized in the nanopores of mesoporous materials. *J. Catal.*, *238*, 369–381.

[116] Kureshy, R. I., Ahmad, I., Khan, N. U. H., Abdi, S. H. R., Singh, S., Pandia, P. H., Jasra, R. V. (2005). New immobilized chiral Mn(III) salen complexes on pyridine *N*-oxide-modified MCM-41 as effective catalysts for epoxidation of nonfunctionalized alkenes. *J. Catal.*, *235*, 28–34.

[117] Kureshy, R. I., Ahmad, I., Khan, N. H., Abdi, S. H. R., Pathak, K., Jasra, R. V. (2005). Encapsulation of a chiral Mn-III(salen) complex in ordered mesoporous silicas: an approach towards heterogenizing asymmetric epoxidation catalysts for non-functionalized alkenes. *Tetrahedron Asymmetry*, *16*, 3562–3569.

[118] Kureshy, R. I., Ahmad, I., Khan, N. H., Abdi, S. H. R., Pathak, K., Jasra, R. V. (2006). Chiral Mn(III) salen complexes covalently bonded on modified MCM-41 and SBA-15 as efficient catalysts for enantioselective epoxidation of nonfunctionalized alkenes. *J. Catal.*, *238*, 134–141.

[119] Yu, K., Gu, Z., Ji, R., Lou, L. L., Ding, F., Zhang, C., Liu, S. (2007). Effect of pore size on the performance of mesoporous material supported chiral Mn(III) salen complex for the epoxidation of unfunctionalized olefins. *J. Catal.*, *252*, 312–320.

[120] Yu, K., Lou, L. L., Lai, C., Wang, S. J., Ding, F., Liu, S. X. (2006). Asymmetric epoxidation of unfunctionalized olefins catalyzed by Mn(III) salen complex immobilized on MCM-48. *Catal. Commun.*, *7*, 1057–1060.

[121] Lou, L. L., Yu, K., Ding, F., Peng, X. J., Dong, M. M., Zhang, C., Liu, S. X. (2007). Covalently anchored chiral Mn(III) salen-containing ionic species on mesoporous materials as effective catalysts for asymmetric epoxidation of unfunctionalized olefins. *J. Catal.*, *249*, 102–110.

[122] Lou, L. L., Yu, K., Ding, F., Zhou, W., Peng, X. J., Liu, S. X. (2006). An effective approach for the immobilization of chiral Mn(III) salen complexes through a supported ionic liquid phase. *Tetrahedron Lett.*, *47*, 6513–6516.

[123] Bhattacharjee, S., Dines, T. J., Anderson, J. A. (2004). Synthesis and application of layered double hydroxide-hosted catalysts for stereoselective epoxidation using molecular oxygen or air. *J. Catal.*, *225*, 398–407.

[124] Bhattacharjee, S., Anderson, J. A. (2006). Novel chiral sulphonato-salen-manganese (III)-pillared hydrotalcite catalysts for the asymmetric epoxidation of styrenes and cyclic alkenes. *Adv. Synth. Catal.*, *348*, 151–158.

[125] (a) Kureshy, R. I., Khan, N. U. H., Abdi, S. H. R., Ahmad, I., Singh, S., Jasra, R. V. (2004). Dicationic chiral Mn(III) salen complex exchanged in the interlayers of

montmorillonite clay: a heterogeneous enantioselective catalyst for epoxidation of nonfunctionalized alkenes. *J. Catal.*, *221*, 234–240; (b) Kureshy, R. I., Khan, N. H., Abdi, S. H. R., Ahmad, I., Singh, S., Jasra, R. V. (2003). Immobilization of dicationic Mn(III) salen in the interlayers of montmorillonite clay for enantioselective epoxidation of nonfunctionalized alkenes. *Catal. Lett.*, *91*, 207–210.

[126] Choudary, B. M., Ramani, T., Maheswaran, H., Prashant, L., Ranganath, K. V. S., Kumar, K. V. (2006). Catalytic asymmetric epoxidation of unfunctionalised olefins using silica, LDH and resin-supported sulfonato-Mn(salen) complex. *Adv. Synth. Catal.*, *348*, 493–498.

[127] Beigi, M., Haag, R., Liese, A. (2008). Continuous application of polyglycerol-supported salen in a membrane reactor: asymmetric epoxidation of 6-cyano-2,2-dimethylchromene. *Adv. Synth. Catal.*, *350*, 919–925.

[128] Sellner, H., Karjalainen, J. K., Seebach, D. (2001). Preparation of dendritic and non-dendritic styryl-substituted salens for cross-linking suspension copolymerization with styrene and multiple use of the corresponding Mn and Cr complexes in enantioselective epoxidations and hetero-Diels-Alder reactions. *Chem. Eur. J.*, *7*, 2873–2887.

[129] Kureshy, R. I., Khan, N. U. H., Abdi, S. H. R., Singh, S., Ahmad, I., Jasra, R. V., Vyas, A. P. (2004). Enantioselective epoxidation of nonfunctionalized alkenes catalyzed by recyclable new homochiral dimeric Mn(III) salen complexes. *J. Catal.*, *224*, 229–235.

[130] Kureshy, R. I., Singh, S., Khan, N. U. H., Abdi, S. H. R., Ahmed, I., Bhatt, A., Jasra, R. V. (2006). Environment friendly protocol for enantioselective epoxidation of non-functionalized alkenes catalyzed by recyclable homochiral dimeric Mn(III) salen complexes with hydrogen peroxide and UHP adduct as oxidants. *Catal. Lett.*, *107*, 127–130.

[131] Smith, K., Liu, S. F., El-Hiti, G. A. (2004). Use of ionic liquids as solvents for epoxidation reactions catalysed by a chiral Katsuki-type salen complex: enhanced reactivity and recovery of catalyst. *Catal. Lett.*, *98*, 95–101.

[132] (a) Yang, H. Q., Zhang, L., Su, W. G., Yang, Q. H., Li, C. (2007). Asymmetric ring-opening of epoxides on chiral Co(Salen) catalyst synthesized in SBA-16 through the ship in a bottle strategy. *J. Catal.*, *248*, 204–212; (b) Yang, H. Q., Zhang, L., Zhong, L., Yang, Q. H., Li, C. (2007). Enhanced cooperative activation effect in the hydrolytic kinetic resolution of epoxides on (Co[salen]) catalysts confined in nanocages. *Angew. Chem. Int. Ed.*, *46*, 6861–6865.

[133] Kawthekar, R. B., Lee, Y. H., Kim, G. J. (2009). Asymmetric catalysis by chiral salen complexes immobilized on mesoporous materials having modified pore channel system by dimethylcarbonate. *J. Porous Mater.*, *16*, 367–378.

[134] Goyal, P., Zheng, X., Weck, M. (2008). Enhanced cooperativity in hydrolytic kinetic resolution of epoxides using poly(styrene) resin-supported dendronized Co-(salen) catalysts. *Adv. Synth. Catal.*, *350*, 1816–1822.

[135] Belser, T., Jacobsen, E. N. (2008). Cooperative catalysis in the hydrolytic kinetic resolution of epoxides by chiral [(salen)Co(III)] complexes immobilized on gold colloids. *Adv. Synth. Catal.*, *350*, 967–971.

[136] Gill, C. S., Venkatasubbaiah, K., Phan, N. T. S., Weck, M., Jones, C. W. (2008). Enhanced cooperativity through design: pendant Co(III)-salen polymer brush

catalysts for the hydrolytic kinetic resolution of epichlorohydrin (salen-bis(salicylidene)ethylenediamine dianion. *Chem. Eur. J.*, *14*, 7306–7313.

[137] Gill, C. S., Long, W., Jones, C. W. (2009). Magnetic nanoparticle polymer brush catalysts: alternative hybrid organic/inorganic structures to obtain high, local catalyst loadings for use in organic transformations. *Catal. Lett.*, *131*, 425–431.

[138] Kwon, M. A., Kim, G. J. (2003). Synthesis of polymeric salen complexes and application in the enantioselective hydrolytic kinetic resolution of epoxides as catalysts. *Catal. Today*, *87*, 145–151.

[139] Lee, K. Y., Kawthekar, R. B., Kim, G. J. (2007). Synthesis of chiral intermediates catalyzed by new chiral polymeric (salen) cobalt complexes bearing Lewis acidic metal halides. *Bull. Korean Chem. Soc.*, *28*, 1553–1561.

[140] Song, Y. M., Yao, X. Q., Chen, H. L., Bai, C. M., Hu, X. Q., Zheng, Z. (2002). Highly enantioselective resolution of terminal epoxides using polymeric catalysts. *Tetrahedron Lett.*, *43*, 6625–6627.

[141] Rossbach, B. M., Leopold, K., Weberskirch, R. (2006). Self-assembled nanoreactors as highly active catalysts in the hydrolytic kinetic resolution (HKR) of epoxides in water. *Angew. Chem. Int. Ed.*, *45*, 1309–1312.

[142] (a) Song, C. E., Oh, C. R., Roh, E. J., Choo, D. J. (2000). Cr(salen) catalysed asymmetric ring opening reactions of epoxides in room temperature ionic liquids. *Chem. Commun.*, 1743–1744; (b) Oh, C. R., Choo, D. J., Shim, W. H., Lee, D. H., Roh, E. J., Lee, S., Song, C. E. (2003). Chiral Co(III) (salen)-catalysed hydrolytic kinetic resolution of racemic epoxides in ionic liquids. *Chem. Commun.*, 1100–1101.

[143] Kureshy, R. I., Prathap, K. J., Agrawal, S., Kumar, M., Khan, N. U. H., Abdi, S. H. R., Bajaj, H. C. (2009). Highly efficient recyclable Co-III-salen complexes in the catalyzed asymmetric aminolytic kinetic resolution of aryloxy/terminal epoxides for the simultaneous production of *N*-protected 1,2-amino alcohols and the corresponding epoxides in high optical purity. *Eur. J. Org. Chem.*, 2863–2871.

[144] Baleizao, C., Gigante, B., Das, D., Alvaro, M., Garcia, H., Corma, A. (2003). Synthesis and catalytic activity of a chiral periodic mesoporous organosilica (ChiMO). *Chem. Commun.*, 1860–1861.

[145] Baleizao, C., Gigante, B., Garcia, H., Corma, A. (2004). Chiral vanadyl salen complex anchored on supports as recoverable catalysts for the enantioselective cyanosilylation of aldehydes. Comparison among silica, single wall carbon nanotube, activated carbon and imidazolium ion as support. *Tetrahedron*, *60*, 10461–10468.

[146] Baleizao, C., Gigante, B., Garcia, H., Corma, A. (2002). Ionic liquids as green solvents for the asymmetric synthesis of cyanohydrins catalysed by VO(salen) complexes. *Green Chem.*, *4*, 272–274.

[147] Baleizao, C., Gigante, B., Garcia, H., Corma, A. (2003). Chiral vanadyl Schiff base complex anchored on silicas as solid enantioselective catalysts for formation of cyanohydrins: optimization of the asymmetric induction by support modification. *J. Catal.*, *215*, 199–207.

[148] Huang, W., Song, Y. M., Bai, C. M., Cao, G. Y., Zheng, Z. (2004). Asymmetric addition of KCN and Ac_2O to aldehydes catalyzed by recyclable polymeric salen-Ti(IV) complexes. *Tetrahedron Lett.*, *45*, 4763–4767.

[149] (a) Khan, N. U. H., Agrawal, S., Kureshy, R. I., Abdi, S. H. R., Mayani, V. J., Jasra, R. V. (2006). Asymmetric addition of trimethylsilyl cyanide to aldehydes promoted by chiral polymeric vanadium(V) salen complex as an efficient and recyclable catalyst. *Tetrahedron Asymmetry*, *17*, 2659–2666; (b) Khan, N. U. H., Agrawal, S., Kureshy, R. I., Abdi, S. H. R., Mayani, V. J., Jasra, R. V. (2007). Easily recyclable polymeric V(V) salen complex for the enantioselective *O*-acetyl cyanation of aldehydes. *J. Mol. Catal. A Chem.*, *264*, 140–145.

[150] Khan, N. U. H., Agrawal, S., Kureshy, R. I., Abdi, S. H. R., Mayani, V. J., Jasra, R. V. (2006). Asymmetric synthesis of *O*-acetylcyanohydrins by reaction of aldehydes with NaCN/KCN catalyzed by recyclable chiral dimeric titanium(IV)/vanadium(V) salen complexes. *Eur. J. Org. Chem.*, 3175–3180.

[151] Anyanwu, U. K., Venkataraman, D. (2005). Soxhlet-dialysis: a method to recover soluble polymer supported catalysts. *Green Chem.*, *7*, 424–425.

[152] Mellah, M., Ansel, B., Patureau, F., Voituriez, A., Schulz, E. (2007). Electropolymerized Cr-salen complexes for the heterogeneous asymmetric hetero Diels-Alder reaction. *J. Mol. Catal. A Chem.*, *272*, 20–25.

[153] Zulauf, A., Mellah, M., Schulz, E. (2009). New chiral thiophene–salen chromium complexes for the asymmetric Henry reaction. *J. Org. Chem.*, *74*, 2242–2245.

[154] Zulauf, A., Mellah, M., Guillot, R., Schulz, E. (2008). Chromium-thiophene-salen-based polymers for heterogeneous asymmetric hetero-Diels-Alder reactions. *Eur. J. Org. Chem.*, 2118–2129.

[155] Heckel, A., Seebach, D. (2002). Enantioselective heterogeneous epoxidation and hetero-Diels-Alder reaction with Mn- and Cr-salen complexes immobilized on silica gel by radical grafting. *Helv. Chim. Acta.*, *85*, 913–926.

[156] Madhavan, N., Weck, M. (2008). Highly active polymer-supported (salen)Al catalysts for the enantioselective addition of cyanide to alpha,beta-unsaturated imides. *Adv. Synth. Catal.*, *350*, 419–425.

[157] (a) Kureshy, R. I., Ahmad, I., Pathak, K., Khan, N. U. H., Abdi, S. H. R., Prathap, J. K., Jasra, R. V. (2007). Easily recyclable chiral polymeric Mn(III) salen complexes for oxidative kinetic resolution of racemic secondary alcohols. *Chirality*, *19*, 352–357; (b) Pathak, K., Ahmad, I., Abdi, S. H. R., Kureshy, R. I., Khan, N. U. H., Jasra, R. V. (2007). Oxidative kinetic resolution of racemic secondary alcohols catalyzed by recyclable chiral dimeric Mn(III) salen catalysts. *J. Mol. Catal. A Chem.*, *274*, 120–126.

[158] Kantam, M. L., Ramani, T., Chakrapani, L., Choudary, B. M. (2007). Oxidative kinetic resolution of racemic secondary alcohols catalyzed by resin supported sulfonato-Mn(salen) complex in water. *J. Mol. Catal. A Chem.*, *274*, 11–15.

[159] Hanefeld, U., Gardossi, L., Magner, E. (2009). Understanding enzyme immobilisation. *Chem. Soc. Rev.*, *38*, 453–468.

[160] Ghanem, A., Aboul-Enein, H. Y. (2005). Application of lipases in kinetic resolution of racemates. *Chirality*, *17*, 1–15.

[161] Allen, J. V., Williams, J. M. J. (1996). Dynamic kinetic resolution with enzyme and palladium combinations. *Tetrahedron Lett.*, *37*, 1859–1862.

[162] Pamies, O., Bäckvall, J. E. (2002). Efficient lipase-catalyzed kinetic resolution of and dynamic kinetic resolution of beta-hydroxy nitriles. Correction of absolute configuration and transformation to chiral beta-hydroxy acids and gamma-amino alcohols. *Adv. Synth. Catal.*, *344*, 947–952.

[163] Lee, J. K., Kim, M. J. (2002). Ionic liquid-coated enzyme for biocatalysis in organic solvent. *J. Org. Chem.*, *67*, 6845–6847.

[164] Lozano, P., de Diego, T., Carrie, D., Vaultier, M., Iborra, J. L. (2002). Continuous green biocatalytic processes using ionic liquids and supercritical carbon dioxide. *Chem. Commun.*, 692–693.

[165] Matsuda, T., Watanabe, K., Harada, T., Nakamura, K., Arita, Y., Misumi, Y., Ichikawa, S., Ikariya, T. (2004). High-efficiency and minimum-waste continuous kinetic resolution of racemic alcohols by using lipase in supercritical carbon dioxide. *Chem. Commun.*, 2286–2287.

[166] Itoh, T., Nishimura, Y., Ouchi, N., Hayase, S. (2003). 1-Butyl-2,3-dimethylimidazolium tetrafluoroborate: the most desirable ionic liquid solvent for recycling use of enzyme in lipase-catalyzed transesterification using vinyl acetate as acyl donor. *J. Mol. Catal. B Enzymol.*, *26*, 41–45.

[167] Itoh, T., Matsushita, Y., Abe, Y., Han, S. H., Wada, S., Hayase, S., Kawatsura, M., Takai, S., Morimoto, M., Hirose, Y. (2006). Increased enantioselectivity and remarkable acceleration of lipase-catalyzed transesterification by using an imidazolium PEG-alkyl sulfate ionic liquid. *Chem. Eur. J.*, *12*, 9228–9237.

[168] Lourenco, N. M. T., Afonso, C. A. M. (2007). One-pot enzymatic resolution and separation of sec-alcohols based on ionic acylating agents. *Angew. Chem. Int. Ed.*, *46*, 8178–8181.

[169] Monteiro, C. M., Lourenco, N. M. T., Afonso, C. A. M. (2010). Separation of secondary alcohols via enzymatic kinetic resolution using fatty esters as reusable acylating agents. *Tetrahedron Asymmetry*, *21*, 952–956.

[170] Parvulescu, A. N., Jacobs, P. A., De Vos, D. E. (2007). Palladium catalysts on alkaline-earth supports for racemization and dynamic kinetic resolution of benzylic amines. *Chem. Eur. J.*, *13*, 2034–2043.

[171] Parvulescu, A. N., Van der Eycken, E., Jacobs, P. A., De Vos, D. E. (2008). Microwave-promoted racemization and dynamic kinetic resolution of chiral amines over Pd on alkaline earth supports and lipases. *J. Catal.*, *255*, 206–212.

[172] Parvulescu, A. N., Jacobs, P. A., De Vos, D. E. (2008). Heterogeneous Raney nickel and cobalt catalysts for racemization and dynamic kinetic resolution of amines. *Adv. Synth. Catal.*, *350*, 113–121.

[173] Trevisan, V., Signoretto, M., Colonna, S., Pironti, V., Strukul, G. (2004). Microencapsulated chloroperoxidase as a recyclable catalyst for the enantioselective oxidation of sulfides with hydrogen peroxide. *Angew. Chem. Int. Ed.*, *43*, 4097–4099.

[174] Spreti, N., Germani, R., Incani, A., Savelli, G. (2004). Stabilization of chloroperoxidase by polyethylene glycols in aqueous media: kinetic studies and synthetic applications. *Biotechnol. Prog.*, *20*, 96–101.

[175] Wang, W., Xu, Y., Wang, D. I. C., Li, Z. (2009). Recyclable nanobiocatalyst for enantioselective sulfoxidation: facile fabrication and high performance of chloroperoxidase-coated magnetic nanoparticles with ion oxide core and polymer shell. *J. Am. Chem. Soc.*, *131*, 12892–12893.

[176] Gladiali, S., Alberico, E. (2006). Asymmetric transfer hydrogenation: chiral ligands and applications. *Chem. Soc. Rev.*, *35*, 226–236.

[177] (a) Li, Y. Z., Li, Z. M., Li, F., Wang, Q. R., Tao, F. G. (2005). Preparation of polymer-supported Ru-TsDPEN catalysts and use for enantioselective synthesis of (*S*)-

fluoxetine. *Org. Biomol. Chem.*, *3*, 2513–2518; (b) Arakawa, Y., Chiba, A., Haraguchi, N., Itsuno, S. (2008). Asymmetric transfer hydrogenation of aromatic ketones in water using a polymer-supported chiral catalyst containing a hydrophilic pendant group. *Adv. Synth. Catal.*, *350*, 2295–2304.

[178] Arakawa, Y., Haraguchi, N., Itsuno, S. (2006). Design of novel polymer-supported chiral catalyst for asymmetric transfer hydrogenation in water. *Tetrahedron Lett.*, *47*, 3239–3243.

[179] Liu, P. N., Gu, P. M., Wang, F., Tu, Y. Q. (2004). Efficient heterogeneous asymmetric transfer hydrogenation of ketones using highly recyclable and accessible silica-immobilized Ru-TsDPEN catalysts. *Org. Lett.*, *6*, 169–172.

[180] Yang, H. Q., Li, J., Yang, J., Liu, Z. M., Yang, Q. H., Li, C. (2007). Asymmetric reactions on chiral catalysts entrapped within a mesoporous cage. *Chem. Commun.*, 1086–1088.

[181] Huang, X. H., Ying, J. Y. (2007). Asymmetric transfer hydrogenation over Ru-TsDPEN catalysts supported on siliceous mesocellular foam. *Chem. Commun.*, 1825–1827.

[182] Li, J., Zhang, Y., Han, D., Gao, Q., Li, C. (2009). Asymmetric transfer hydrogenation using recoverable ruthenium catalyst immobilized into magnetic mesoporous silica. *J. Mol. Catal. A Chem.*, *298*, 31–35.

[183] (a) Liu, P. N., Deng, J. G., Tu, Y. Q., Wang, S. H. (2004). Highly efficient and recyclable heterogeneous asymmetric transfer hydrogenation of ketones in water. *Chem. Commun.*, 2070–2071; (b) Liu, P. N., Gu, P. M., Deng, J. G., Tu, Y. Q., Ma, Y. P. (2005). Efficient heterogeneous asymmetric transfer hydrogenation catalyzed by recyclable silica-supported ruthenium complexes. *Eur. J. Org. Chem.*, 3221–3227.

[184] Jiang, L., Wu, T. F., Chen, Y. C., Zhu, J., Deng, J. G. (2006). Asymmetric transfer hydrogenation catalysed by hydrophobic dendritic DACH-rhodium complex in water. *Org. Biomol. Chem.*, *4*, 3319–3324.

[185] Chen, Y. C., Wu, T. F., Deng, J. G., Liu, H., Cui, X., Zhu, J., Jiang, Y. Z., Choi, M. C. K., Chan, A. S. C. (2002). Multiple dendritic catalysts for asymmetric transfer hydrogenation. *J. Org. Chem.*, *67*, 5301–5306.

[186] Chen, Y. C., Wu, T. F., Jiang, L., Deng, J. G., Liu, H., Zhu, J., Jiang, Y. Z. (2005). Synthesis of dendritic catalysts and application in asymmetric transfer hydrogenation. *J. Org. Chem.*, *70*, 1006–1010.

[187] Li, X. G., Chen, W. P., Hems, W., King, F., Xiao, J. L. (2004). Asymmetric transfer hydrogenation of ketones with a polymer-supported chiral diamine. *Tetrahedron Lett.*, *45*, 951–953.

[188] Li, X. G., Wu, X. F., Chen, W. P., Hancock, F. E., King, F., Xiao, J. L. (2004). Asymmetric transfer hydrogenation in water with a supported Noyori-Ikariya. *Catal. Org. Lett.*, *6*, 3321–3324.

[189] Liu, J. T., Zhou, Y. G., Wu, Y. N., Li, X. S., Chan, A. S. C. (2008). Asymmetric transfer hydrogenation of ketones with a polyethylene glycol bound Ru catalyst in water. *Tetrahedron Asymmetry*, *19*, 832–837.

[190] Geldbach, T. J., Dyson, P. J. (2004). A versatile ruthenium precursor for biphasic catalysis and its application in ionic liquid biphasic transfer hydrogenation: conventional versus task-specific catalysts. *J. Am. Chem. Soc.*, *126*, 8114–8115.

[191] Kawasaki, I., Tsunoda, K., Tsuji, T., Yamaguchi, T., Shibuta, H., Uchida, N., Yamashita, M., Ohta, S. (2005). A recyclable catalyst for asymmetric transfer hydrogenation with a formic acid-triethylamine mixture in ionic liquid. *Chem. Commun.*, 2134–2136.

[192] Zhou, Z., Sun, Y. (2009). Water-soluble chiral aminosulfonamides as ligands for ruthenium(II)-catalyzed asymmetric transfer hydrogenation. *Catal. Commun.*, *10*, 1685–1688.

[193] Ma, Y. P., Liu, H., Chen, L., Cui, X., Zhu, J., Deng, J. E. (2003). Asymmetric transfer hydrogenation of prochiral ketones in aqueous media with new water-soluble chiral vicinal diamine as ligand. *Org. Lett.*, *5*, 2103–2106.

[194] Wang, F., Liu, H., Cun, L. F., Zhu, J., Deng, J. G., Jiang, Y. Z. (2005). Asymmetric transfer hydrogenation of ketones catalyzed by hydrophobic metal-amido complexes in aqueous micelles and vesicles. *J. Org. Chem.*, *70*, 9424–9429.

[195] Zhou, H. F., Fan, Q. H., Huang, Y. Y., Wu, L., He, Y. M., Tang, W. J., Gu, L. Q., Chan, A. S. C. (2007). Mixture of poly(ethylene glycol) and water as environmentally friendly media for efficient enantioselective transfer hydrogenation and catalyst recycling. *J. Mol. Catal. A Chem.*, *275*, 47–53.

[196] Wu, J. S., Wang, F., Ma, Y. P., Cui, X. C., Cun, L. F., Zhu, J., Deng, J. G., Yu, B. L. (2006). Asymmetric transfer hydrogenation of imines and iminiums catalyzed by a water-soluble catalyst in water. *Chem. Commun.*, 1766–1768.

[197] Sandee, A. J., Petra, M. G. I., Reek, J. N. H., Kamer, P. C. J., van Leeuwen, P. W. N. M. (2001). Solid-phase synthesis of homogeneous ruthenium catalysts on silica for the continuous asymmetric transfer hydrogenation reaction. *Chem. Eur. J.*, 1202–1208.

[198] Felder, M., Giffels, G., Wandrey, C. (1997). A polymer-enlarged homogeneously soluble oxazaborolidine catalyst for the asymmetric reduction of ketones by borane. *Tetrahedron Asymmetry*, *8*, 1975–1977.

[199] Giffels, G., Beliczey, J., Felder, M., Kragl, U. (1998). Polymer enlarged oxazaborolidines in a membrane reactor: enhancing effectivity by retention of the homogeneous catalyst. *Tetrahedron Asymmetry*, *9*, 691–696.

[200] Dalicsek, Z., Pollreisz, F., Gomory, A., Soós, T. (2005). Recoverable fluorous CBS methodology for asymmetric reduction of ketones. *Org. Lett.*, *7*, 3243–3246.

[201] (a) Burguete, M. I., Garcia-Verdugo, E., Vicent, M. J., Luis, S. V., Pennemann, H., von Keyserling, N. G., Martens, J. (2002). New supported beta-amino alcohols as efficient catalysts for the enantioselective addition of diethylzinc to benzaldehyde under flow conditions. *Org. Lett.*, *4*, 3947–3950; (b) Luis, S. V., Altava, B., Burguete, M. I., Collado, M., Escorihuela, J., Garcia-Verdugo, E., Vicent, M. J., Martens, J. (2003). Preparation and optimization of polymer-supported and amino alcohol based enantioselective reagents and catalysts. *Ind. Eng. Chem. Res.*, *42*, 5977–5982.

[202] Castellnou, D., Sola, L., Jimeno, C., Fraile, J. M., Mayoral, J. A., Riera, A., Pericàs, M. A. (2005). Polystyrene-supported (*R*)-2-piperazino-1,1,2-triphenylethanol: a readily available supported ligand with unparalleled catalytic activity and enantioselectivity. *J. Org. Chem.*, *70*, 433–438.

[203] Pericàs, M. A., Castellnou, D., Rodriguez, I., Riera, A., Sola, L. (2003). Tail-tied ligands: an immobilized analogue of (*R*)-2-piperidino-1,1,2-triphenylethanol with intact high catalytic activity and enantioselectivity. *Adv. Synth. Catal.*, *345*, 1305–1313.

[204] Pericas, M. A., Herrerias, C. I., Sola, L. (2008). Fast and enantioselective production of 1-aryl-1-propanols through a single pass, continuous flow process. *Adv. Synth. Catal.*, *350*, 927–932.

[205] Kelsen, V., Pierrat, P., Gros, P. C. (2007). Enantioselective ethylation of aldehydes using a regenerable polymer-supported *N*-picolylvalinol tridentate ligand. *Tetrahedron.*, *63*, 10693–10697.

[206] Kell, R. J., Hodge, P., Nisar, M., Watson, D. (2002). Towards more chemically robust polymer-supported chiral catalysts for the reactions of aldehydes with dialkylzincs. *Bioorg. Med. Chem. Lett.*, *12*, 1803–1807.

[207] Hui, X. P., Chen, C. A., Wu, K. H., Gau, H. M. (2007). Polystyrene-supported *N*-sulfonylated amino alcohols and their applications to titanium(IV) complexes catalyzed enantioselective diethylzinc additions to aldehydes. *Chirality.*, *19*, 10–15.

[208] Chen, C. A., Wu, K. H., Gau, H. M. (2008). Synthesis and application of polymer-supported *N*-sulfonylated aminoalcohols as enantioselective catalysts. *Polymer.*, *49*, 1512–1519.

[209] Huang, L. N., Hui, X. P., Xu, P. F. (2006). Asymmetric addition of diethylzinc to benzaldehyde catalyzed by silica-immobilized titanium(IV) complexes of *N*-sulfonylated amino alcohols. *J. Mol. Catal. A Chem.*, *258*, 216–220.

[210] Forrat, V. J., Ramon, D. J., Yus, M. (2006). Polymer supported trans-1-phenylsulfonylamino-2-isoborneolsulfonylaminocyclohexane ligand for the titanium catalyzed organozinc addition to ketones. *Tetrahedron Asymmetry*, *17*, 2054–2058.

[211] Sato, I., Kodaka, R., Hosoi, K., Soai, K. (2002). Highly enantioselective addition of dialkylzincs to aldehydes using dendritic chiral catalysts with flexible carbosilane backbones. *Tetrahedron Asymmetry*, *13*, 805–808.

[212] El-Shehawy, A. A., Sugiyama, K., Hirao, A. (2008). Application of well-defined chain-end-functionalized polystyrenes with dendritic chiral ephedrine moieties as reagents for highly catalytic enantioselective addition of dialkylzines to aldehydes. *Tetrahedron Asymmetry*, *19*, 425–434.

[213] Park, J. K., Lee, H. G., Bolm, C., Kim, B. M. (2005). Asymmetric diethyl- and diphenylzinc additions to aldehydes by using a fluorine-containing chiral amino alcohol: a striking temperature effect on the enantioselectivity, a minimal amino alcohol loading, and an efficient recycling of the amino alcohol. *Chem. Eur. J.*, *11*, 945–950.

[214] (a) Pathak, K., Bhatt, A. P., Abdi, S. H. R., Kureshy, R. I., Khan, N. H., Ahmad, I., Jasra, R. V. (2006). Enantioselective addition of diethylzinc to aldehydes using immobilized chiral BINOL-Ti complex on ordered mesoporous silicas. *Tetrahedron Asymmetry*, *17*, 1506–1513; (b) Pathak, K., Ahmad, I., Abdi, S. H. R., Kureshy, R. I., Khan, N. U. H., Jasra, R. V. (2008). The synthesis of silica-supported chiral BINOL: application in Ti-catalyzed asymmetric addition of diethylzinc to aldehydes. *J. Mol. Catal. A Chem.*, *280*, 106–114.

[215] Yin, L., Li, R., Wang, F. S., Wang, H. L., Zheng, Y. F., Wang, C. F., Ma, J. T. (2007). The synthesis of dendritic BINOL ligands and their applications in the asymmetric addition of diethylzinc to benzaldehyde. *Tetrahedron Asymmetry*, *18*, 1383–1389.

[216] Gadenne, B., Hesemann, P., Moreau, J. J. E. (2005). Easily recoverable BINOL ligand with ionic tag for asymmetric catalysis. *Tetrahedron Asymmetry*, *16*, 2001–2006.

[217] (a) Nakamura, Y., Takeuchi, S., Okumura, K., Ohgo, Y., Curran, D. P. (2002). Recyclable fluorous chiral ligands and catalysts: asymmetric addition of diethylzinc to aromatic aldehydes catalyzed by fluorous BINOL-Ti complexes. *Tetrahedron.*, *58*, 3963–3969; (b) Nakamura, Y., Takeuchi, S., Ohgo, Y. (2003). Enantioselective carbon-carbon bond forming reactions using fluorous chiral catalysts. *J. Fluorine Chem.*, *120*, 121–129.

[218] Takizawa, S., Somei, H., Jayaprakash, D., Sasai, H. (2003). Metal-bridged polymers as insoluble multicomponent asymmetric catalysts with high enantiocontrol: an approach for the immobilization of catalysts without using any support. *Angew. Chem. Int. Ed.*, *42*, 5711–5714.

[219] Wang, X. S., Wang, X. W., Guo, H. C., Wang, Z., Ding, K. L. (2005). Self-supported heterogeneous titanium catalysts for enantioselective carbonyl-ene and sulfoxidation reactions. *Chem. Eur. J.*, *11*, 4078–4088.

[220] Fu, F., Teo, Y. C., Loh, T. P. (2006). Catalytic enantioselective Diels-Alder reaction in ionic liquid via a recyclable chiral In(III) complex. *Org. Lett.*, *8*, 5999–6001.

[221] Fawcett, J., Hope, E. G., Stuart, A. M., West, A. J. (2005). Recycling of a perfluoroalkylated BINOL ligand using fluorous solid-phase extraction. *Green Chem.*, *7*, 316–320.

[222] Choudary, B. M., Ranganath, K. V. S., Pal, U., Kantam, M. L., Sreedhar, B. (2005). Nanocrystalline MgO for asymmetric Henry and Michael reactions. *J. Am. Chem. Soc.*, *127*, 13167–13171.

[223] Choudary, B. M., Kantam, M. L., Ranganath, K. V. S., Mahendar, K., Sreedhar, B. (2004). Bifunctional nanocrystalline MgO for chiral epoxy ketones via Claisen-Schmidt condensation-asymmetric epoxidation reactions. *J. Am. Chem. Soc.*, *126*, 3396–3397.

[224] Yuan, X. Y., Li, H. Y., Hodge, P., Kilner, M., Tastard, C. Y., Zhang, Z. P. (2006). An easy synthesis of robust polymer-supported chiral 1,1′-bi-(2-naphthol)s (BINOLs): application to the catalysis of the oxidation of prochiral thioethers to chiral sulfoxides. *Tetrahedron Asymmetry*, *17*, 2401–2407.

[225] Sahoo, S., Kumar, P., Lefebvre, F., Halligudi, S. B. (2009). Synthesis of chiral sulfoxides by enantioselective sulfide oxidation and subsequent oxidative kinetic resolution using immobilized Ti-binol complex. *J. Catal.*, *262*, 111–118.

CHAPTER 2

RECYCLABLE ORGANOCATALYSTS IN ASYMMETRIC REACTIONS

MICHELANGELO GRUTTADAURIA, FRANCESCO GIACALONE, AND RENATO NOTO

2.1.	General aspects	84
2.2.	Asymmetric epoxidation	88
	2.2.1. Covalently linked on insoluble supports: Polystyrene and silica	88
	2.2.2. Covalently linked to soluble support: PEG	90
	2.2.3. Noncovalently linked: Silica	90
	2.2.4. Ionic liquid-anchored catalysts	91
2.3.	Asymmetric synthesis of α-amino acids via PTC	92
	2.3.1. Covalently linked on insoluble supports: Polystyrene	92
	2.3.2. Covalently linked to soluble support: PEG	95
	2.3.3. Chiral ammonium polymers	96
	2.3.4. Dendrimer and fluorous-modified catalysts	97
	2.3.5. Noncovalently linked PTC catalysts	97
2.4.	Desymmetrization	100
	2.4.1. Covalently linked on insoluble supports: Polystyrene, magnetite, and silica	100
2.5.	Aldol reaction	102
	2.5.1. Covalently linked on insoluble supports: Polystyrene, methacrylates, and silica	102
	2.5.2. Covalently linked to soluble support: PEG	114
	2.5.3. Dendrimers	115
	2.5.4. Ionic liquids as solvents	115

Catalytic Methods in Asymmetric Synthesis: Advanced Materials, Techniques, and Applications, First Edition. Edited by Michelangelo Gruttadauria and Francesco Giacalone.

2.5.5. Noncovalently linked organocatalysts: Ionic liquid-modified silica gels; alumina support; polyoxometalate support 117
2.5.6. Ionic liquid-tagged organocatalysts 120
2.5.7. Fluorous organocatalysts 126
2.5.8. Noncovalently supported organocatalysts 126
2.5.9. Nonsupported organocatalysts 130
2.6. Michael reaction 131
2.6.1. Covalently linked on insoluble supports: Polystyrene and silica 131
2.6.2. Covalently linked on soluble supports: PEG 138
2.6.3. Dendrimers 140
2.6.4. Ionic liquid-anchored organocatalysts 141
2.6.5. Fluorous organocatalysts 147
2.6.6. Nonsupported organocatalysts 148
2.7. Ketone reduction 149
2.8. Diels–Alder reaction 150
2.9. Friedel–Craft-type reaction 153
2.10. Asymmetric reduction 155
2.11. Miscellaneous 158
2.11.1. Allyltrichlorosilane addition to C=O 158
2.11.2. Aza-Morita–Baylis–Hillman reaction 159
2.11.3. Robinson annulation 160
2.12. Multisupported catalyst-mediated reactions: The ultimate goal in asymmetric synthesis 160
2.13. Conclusions 162
References 162

2.1. GENERAL ASPECTS

After the term "organocatalytic" was coined in 2000 [1], there has been a tremendous interest in this field especially for asymmetric reactions [2]. Actually, several important (now called organocatalytic) reactions were reported in the nineteenth century. Almost parallel to the development of organocatalytic reactions, homogeneous chiral organocatalysts were immobilized in order to obtain heterogeneous materials that could be easily manipulated. One of the first approaches was the use of polymer-supported asymmetric synthesis. Indeed, polymers in insoluble bead form offer a great advantage in recovery of the supported chiral catalyst by filtration. As an example, which will be discussed in detail later, the epoxidation of α,β-unsaturated carbonyl compounds, reported by Juliá and Colonna, had the major drawback of furnishing a system that was difficult to separate and recycle because of the semisolid paste-like poly(amino acid). The usefulness of supporting an organocatalyst, in this case, is due to an easier workup of the reaction and the recovery of the catalyst. Generally, the main reason that prompted chemists to immobilize chiral organocatalysts lies in the synthetic advantages for larger scale production of cheap, easily available, and recoverable materials that can be reused for several cycles.

While recovery can be easily achieved by simple filtration when insoluble catalysts are used, the possibility of catalyst reuse could be limited by deactivation process. Two aspects are, then, of primary importance for a recyclable catalyst: recoverability and reusability. When an organocatalyst fulfills the above aspects, then it could be defined as recyclable.

Usually, a recyclable organocatalyst is a supported heterogeneous material, but this is not always true. Recyclable organocatalysts can be linked to a homogeneous support and then recovered by other different strategies such as extraction or precipitation.

Generally speaking, a recyclable organocatalyst is useful when: (a) sophisticated and synthetically time-consuming catalysts are used; (b) organocatalysts in up to 20–30 mol%, or even higher loading, are used; and (c) the catalyst recovery can be carried out by means of simple methods like solvent extraction or, more likely, by simple filtration, resulting in a simpler workup of the reactions with beneficial effects under the process sustainability point of view. Employing a nonrecyclable organocatalyst under a homogeneous condition, a tedious workup procedure is often required to purify the product. The possible contamination of the product by the catalyst may also restrict its use in larger scale applications.

It should be noted that immobilization of an organocatalyst is more expensive than the use of a nonimmobilized organocatalyst, because several synthetic steps may be necessary for the immobilization procedure. To counterbalance this point, the supported organocatalyst should be easily recovered (>95%) and reused many times with unchanged activity and selectivity. Moreover, immobilization allows the use of supported organocatalyst derivatives in different solvents, and in general enables the exploration of new solubility profiles for the immobilized catalytic species. Finally, immobilization gives the possibility of exploring modifications of the properties of the supported catalysts by employing specific characteristics of the support.

In recent years, a great deal of work has been focused on supported chiral organic catalysts that have been the subject of many reviews [3] and books [4].

Chiral organocatalysts can develop two main tasks: on the one hand, they are responsible for the activation of the electrophile or nucleophile of the reaction (or both of them in the case of bifunctional catalysts) and, on the other, they are liable for the induction of the enantioselectivity of the reaction (see Fig. 2.1 for some examples). The organocatalyst behaves as a shield by preferentially blocking one of the two prochiral faces of the substrate (which usually has a prochiral Csp^2 center, at least in the transition state), making possible the reaction with the corresponding electrophile or nucleophile takes place from the unshielded side.

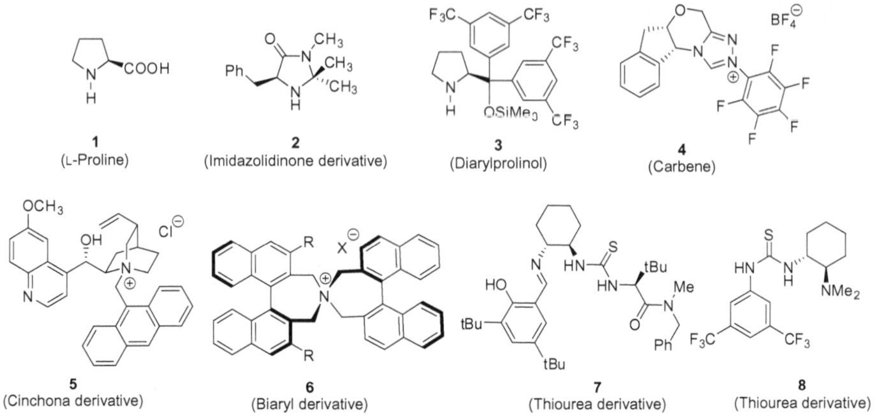

1 (L-Proline) 2 (Imidazolidinone derivative) 3 (Diarylprolinol) 4 (Carbene)

5 (Cinchona derivative) 6 (Biaryl derivative) 7 (Thiourea derivative) 8 (Thiourea derivative)

FIGURE 2.1. Structures of commonly used organocatalysts.

The substrate may be activated by the organocatalyst either by covalent linking or by noncovalent interactions. In the former case we may find two large classes of catalysts: secondary amines (**1**–**3**) and carbenes (**4**). On the other hand, activation of substrates by means of noncovalent bonds may take place through ionic interactions, as in the case of *Cinchona*-based phase transfer catalysts (**5**) or binaphthyl-derivatives (**6**), or by means of hydrogen bonds, in this case employing thiourea derivatives (**7**–**8**).

The exact role of the organocatalyst is of primary importance because its knowledge drives the choice of the "anchoring" point in the organocatalytic molecules. Usually, it is better to avoid a modification of the reacting group(s). Moreover, because economic aspects must be taken into consideration, it should be better to use less expensive molecules as starting materials for the preparation of the supported organocatalyst. Cozzi has reported a good example regarding the immobilization of proline [3c]. Supported proline derivatives can be obtained starting from 4-hydroxy-proline using the hydroxyl group as the anchoring point, or starting from proline using the carboxylic group, via amide formation, as the anchoring group (Fig. 2.2a). From an economic point of view, the latter approach is more convenient since proline is cheaper than 4-hydroxy-proline, but, on the other hand, the former immobilization approach allows the synthesis of more active and selective organocatalysts when compared with simple prolinamide derivatives.

The choice of the anchoring point depends on the structure of the organocatalyst and from the availability of the starting materials. MacMillan imidazolidinone (Fig. 2.2b) can be anchored through the 4-position of the phenyl ring (starting material: tyrosine) or through the R group on the N-3 position (starting material: substituted amine). Phase transfer catalysts based on *Cinchona* alkaloids can be anchored through the hydroxyl group or the C=C double bond or by the proper substituent on an N atom or through the oxygen atom

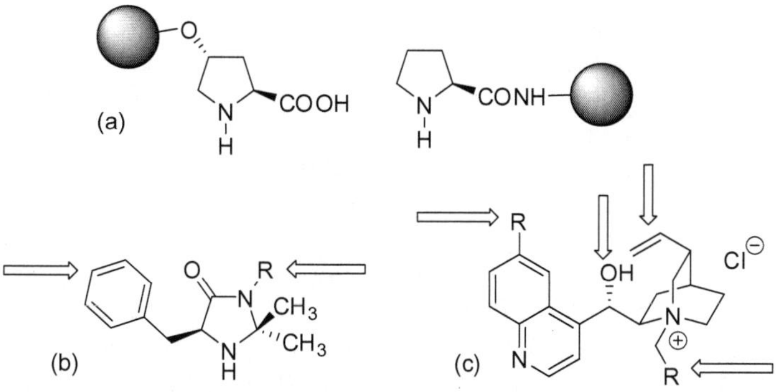

FIGURE 2.2. Anchoring points for (a) proline, (b) MacMillan's imidazolidinone, and (c) *Cinchona*-type catalysts.

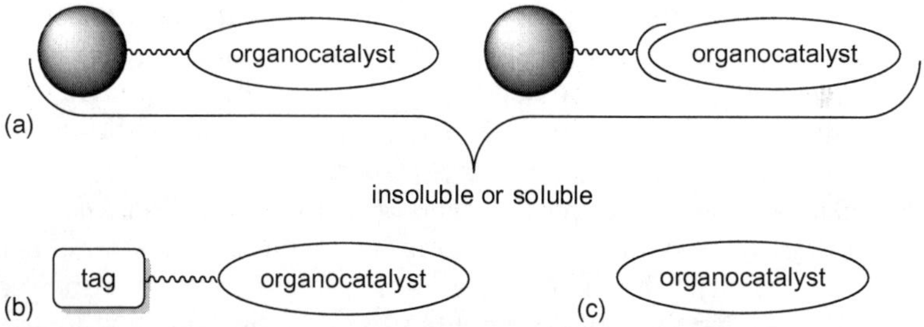

FIGURE 2.3. Classes of recyclable organocatalysts: (a) linked to a polymer support, (b) tagged with an appropriate organic moiety, and (c) nonsupported.

of the quinine or quinidine alkaloids (Fig. 2.2c). Very different results can be observed just by changing the anchoring point.

Another aspect that must be taken into consideration is the synthetic approach when a polymer-supported catalyst is prepared. Two main strategies can be developed: the postmodification of a suitable support, or a bottom-up synthesis by copolymerization of the proper monomers (see Chapter 4). Synthetic applications of recyclable chiral organocatalysts cover reactions of paramount interest such as C=C bond formation, oxidation, and phase transfer catalysis (PTC). In addition, recyclable chiral organocatalysts can be divided in the following sections (Fig. 2.3): (a) organocatalysts linked to a polymer support (covalently linked or not covalently linked, insoluble or soluble); (b) tagged organocatalysts (tag: ionic liquid moieties, fluorous moieties, etc.); and (c) nonsupported recyclable organocatalysts.

In several cases, with the goal of obtaining a more active, stereoselective, and recyclable material, all these strategies have been investigated, beginning from the same organocatalyst. For example, a poly-(L)-leucine (pLL) catalyst or proline-based catalysts have been: (a) covalently linked to insoluble supports such as polystyrene resins or silica, supported on soluble polymers, or supported onto inorganic polymers through simple physical absorption; (b) tagged to imidazolium moieties; and (c) employed as recyclable nonsupported catalysts.

The main advantage of covalently linked insoluble organocatalysts is their easy recovery, whereas soluble supported organocatalysts have the main disadvantage of being recovered after precipitation, which could be not quantitative. On the other hand, soluble supported organocatalysts and tagged organocatalysts could provide higher levels of enantioselectivity, owing to their ability to mimic the activity and stereoselectivity of the corresponding unsupported catalysts. However, this expectation is not always true, and in many cases, lower enantioselectivities were observed.

On the other hand, soluble supported organocatalysts and tagged organocatalysts are easily characterized using standard techniques, whereas polymer-supported insoluble organocatalysts need more complex technologies for their characterization. The ultimate goal could be reached with nonsupported recyclable organocatalysts, which present the great advantage of no additional costs for the immobilization procedure. In this case, the disadvantage could be represented by the recovery procedure that could be tedious and not quantitative.

In this chapter, recyclable chiral organocatalysts will be discussed from the point of view of their synthetic applications, highlighting the kind of support and their reusability. The immobilization strategies for metal-based and organic-based catalysts will be discussed in the next two chapters.

2.2. ASYMMETRIC EPOXIDATION

2.2.1. Covalently Linked on Insoluble Supports: Polystyrene and Silica

Asymmetric epoxidation of α,β-unsaturated ketones has been extensively studied, and several important procedures have been developed in the past decade because optically active epoxides are highly useful intermediates and building blocks for the synthesis of biologically active compounds [5]. Previously, to set up the immobilization procedure for pLL recovery, the best strategy was the use of a two-phase system, comprising a substrate and a peroxide donor—such as a urea-H_2O_2 complex in an organic solvent containing a base (diazabicycloundecene)—in the presence of insoluble pLL. This methodology allowed for the recycling of catalyst by simple decantation; however, losses of catalyst amounted to approximately 10% per run. In addition, after some time the catalyst had to be reactivated before reuse [6]. In the early 1990s, Itsuno et al. reported the use of poly(styrene-co-divinylbenzene)-supported poly(amino acid) **9** as an efficient chiral catalyst in the epoxidation of α,β-unsaturated carbonyl compounds with hydrogen peroxide to yield

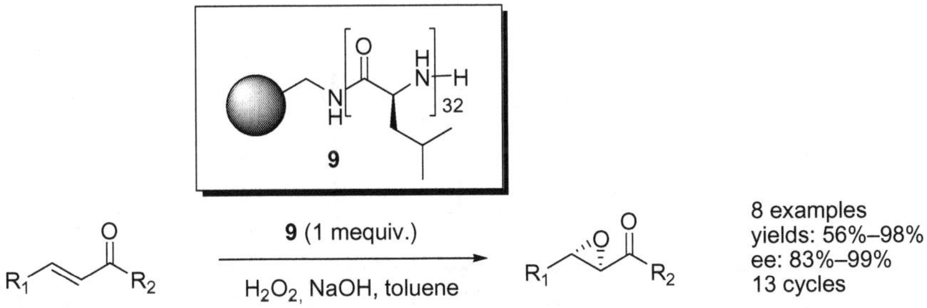

SCHEME 2.1. Asymmetric epoxidation mediated by polystyrene-supported poly-(L)-leucine (pLL).

optically active epoxy ketones in high enantioselectivities up to 99% (Scheme 2.1). The catalytic material was synthesized starting from a 2% cross-linked polystyrene resin having aminomethyl functionality, which was used as initiator for *N*-carboxyanhydride of L-leucine (L-Leu-NCA) polymerization. The resin, having a degree of functionalization of 0.32 and a degree of polymerization of 32, gave good-to-high enantiomeric excess (ee) values. Moreover, the resin was recycled 12 times with unaffected results [7].

Poly-L-leucine, immobilized on cross-linked aminomethyl polystyrene, was used in a fixed-bed reactor for the asymmetric epoxidation of chalcone in *t*-butyl methyl ether and DABCO-H_2O_2 as oxidant in excellent yield and ee [8].

Silica gel was used for covalent anchorage of poly-L-leucine catalyst for asymmetric epoxidation of chalcone. Immobilization was achieved using (3-aminopropyl)triethoxysilane as linker but, interestingly, different outcomes in recycling resulted by changing the anchorage strategy. In a first attempt, silica was functionalized through the treatment with (3-aminopropyl) triethoxysilane and then the polymerization of L-Leu-NCA was carried out. This material gave decreased enantioselectivity upon recycling [9].

In a second attempt, polymerization of L-Leu-NCA was initiated by (3-aminopropyl)triethoxysilane, and then this polymer was grafted onto a silica gel surface. In this case, after six cycles no decrease in yield (89%–93%) and enantioselectivity (95%–97%) was observed in the asymmetric epoxidation of benzalacetophenone, whereas the average loss of catalyst amounted to approximately 2% per run. The lower performances of the former catalyst can be ascribed to the negative influence of the Si-OH groups on the polymerization. Besides –NH_2, Si-OH on a silica gel surface could also initiate the polymerization of L-leucine NCA, causing a decreased average chain length, which influenced the properties of the catalyst. Indeed, a comparison among three catalysts prepared following the latter approach, each having a degree of polymerization of 10, 30, and 50, showed that the catalyst possessing the shorter chain afforded the epoxide in lower yield and enantioselectivity [10].

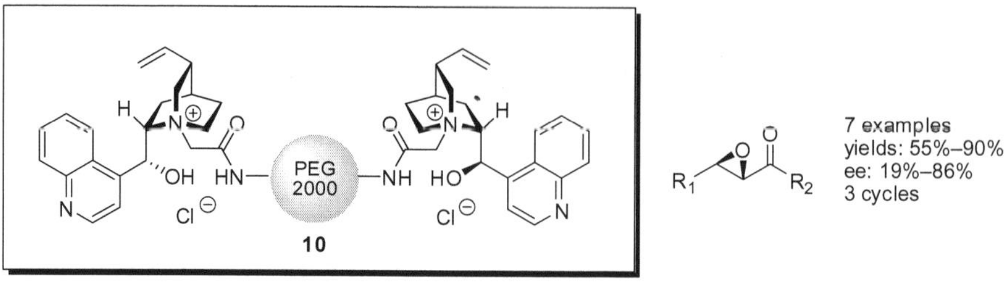

FIGURE 2.4. PEG-supported cinchonidine **10** for the asymmetric epoxidation of chalcones.

2.2.2. Covalently Linked to Soluble Support: PEG

Although a pLL catalyst is usually the catalyst of choice for asymmetric epoxidation of chalcones, good results were also obtained with other catalysts. Dimeric cinchonidine catalyst anchored on poly(ethylene glycol) (PEG) 2000 (**2a**) was chosen among the corresponding dimeric cinchonine and quinine PEG-supported catalysts for the asymmetric epoxidation of chalcones (Fig. 2.4). When catalyst **10** was employed in the presence of *t*-BuOOH in CH_2Cl_2/KOH, 1 M at 0°C, epoxides were obtained in good yields, whereas ee values ranged from low (19%) to good (86%). Catalyst **10** was recovered by precipitation with diethyl ether and reused up to three times with only a 3% decrease in yield and ee [11].

2.2.3. Noncovalently Linked: Silica

Adsorption on silica gel was one of the first approaches for pLL recovery. Such an immobilization strategy is certainly one of the simplest approaches for obtaining a recyclable material since the procedure is based on mixing unmodified amorphous silica gel and pLL followed by filtration. This very simple methodology furnished a robust catalyst, which was employed in several epoxidation reactions with a high level of enantioselectivity and without loss of activity and enantioselectivity after several cycles [12]. Epoxidation with silica-adsorbed pLL was successfully employed in the synthesis of the nonsteroidal anti-inflammatory agent (*S*)-Fenoprofen [13]. An improved protocol for the preparation and activation of the pLL catalyst for subsequent immobilization on silica gel was later reported [14]. Polymerization of L-Leu-NCA in toluene, in the presence of 1,3-diaminopropane as initiator, was carried out at 0°C to room temperature (rt) for 1 day instead of the 10 days required when L-Leu-NCA polymerization was carried out in tetrahydrofurane (THF). Further studies revealed that in addition to pLL, poly*neo*-pentylglycine (pLN) immobilized on silica gel also furnished excellent results (Scheme 2.2) [15].

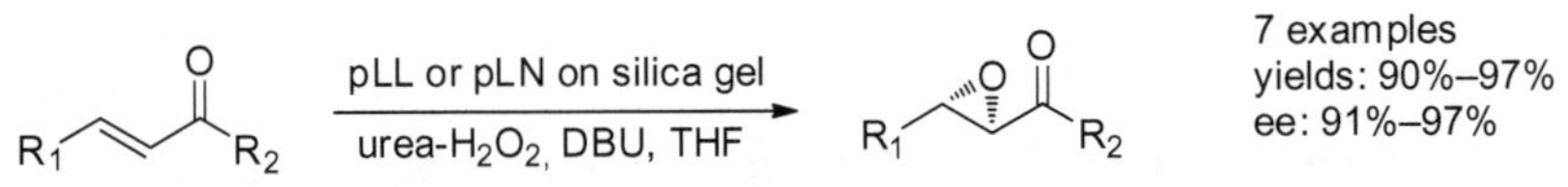

SCHEME 2.2. Asymmetric epoxidation mediated by pLL or pLN on silica gel.

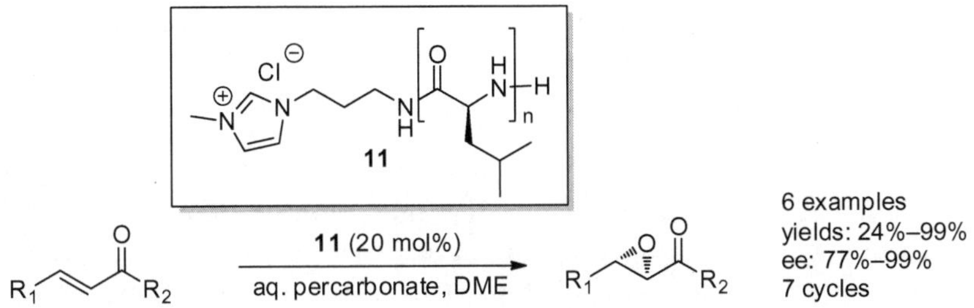

SCHEME 2.3. Asymmetric epoxidation mediated by ionic liquid-supported pLL.

2.2.4. Ionic Liquid-Anchored Catalysts

Ionic liquid-anchored pLL catalyst **11** was prepared by polymerization of L-Leu-NCA initiated by 1-(3-aminopropyl)-3-methyl-imidazolium chloride with an initiator/L-Leu-NCA ratio of 1:30. The asymmetric epoxidation of chalcones was carried out in dimethoxyethane (DME) using sodium percarbonate as oxidant and base in the presence of 20 mol% of catalyst **11** (Scheme 2.3). Use of the lower catalytic loading of **11** furnished the epoxide of chalcone with lower enantioselectivity. Interestingly, when the traditional catalyst *n*-$BuNH_2$-pLL was used, the enantioselectivity and yield were much lower than when catalyst **11** was employed. The use of Br and BF_4 as anion instead of Cl did not improve the outcome of the reaction.

Catalyst **11** was used for the asymmetric epoxidation of six chalcones producing the corresponding products in a very short time (15 minutes), and in high yield and enantioselectivity, except when an *O*-substituted substrate was employed. The high activity of the imidazolium pLL catalyst was ascribed to the presence of the quaternary ammonium moiety, which could provide a good phase transfer ability so that it can transfer more hydroperoxide from the aqueous phase to the reactive sites. Moreover, the ionic liquid tag allowed recycling of the catalyst, which was successfully used in seven consecutive runs without loss in activity and enantioselectivity. Another interesting aspect that deserves attention was the fact that the molecular weight of catalyst **11** is much lower than those of other inorganic or organic supported catalysts. This means that the weight percent of the catalyst is much less [16].

2.3. ASYMMETRIC SYNTHESIS OF α-AMINO ACIDS VIA PTC

2.3.1. Covalently Linked on Insoluble Supports: Polystyrene

PTC uses catalytic amounts of phase transfer agents that facilitate interphase transfer of species, making reactions possible between reagents in two immiscible phases. PTC is used widely in the synthesis of various chiral organic chemicals in both liquid–liquid and solid–liquid systems due to its simple experimental operation, mild reaction conditions, inexpensive and environmentally benign reagents and solvents, and the possibility it offers to conduct large-scale preparations [17].

The PTC methodology applied to the asymmetric alkylation of glycine and alanine Schiff bases is one of the most simple and easily scalable procedures for the synthesis of α-amino acids [17a]. Quaternized *Cinchona* alkaloids allowed high levels of enantioselection to be obtained using a very simple procedure. Many efforts have been made for the development of an easily reusable PTC methodology. One of the most investigated approaches was based on the attachment of the alkaloid-derived PTC to a solid support. First attempts using *N*-methylephedrine and (*R*)-*N,N*-dimethyl-α-methylbenzylamine, and even *Cinchona* alkaloids immobilized on cross-linked chloromethylated polystyrene for different asymmetric transformations under PTC conditions, gave low ee's. Further, deeper studies were carried out, mainly on polymer-supported *Cinchona* catalysts linked through the nitrogen atom (Fig. 2.5, type I). Fewer examples were reported on *Cinchona* catalysts linked through the secondary OH group (Fig. 2.6, type II), whereas scattered examples are reported on *Cinchona* catalyst linked through the vinyl moiety (Fig. 2.6, type III).

The most investigated catalysts were those of type I. The nature of the support, the length, and the bulkiness of the linker as well as its electronic nature were examined for these kinds of catalysts.

Nájera et al. screened several supported cinchonidine catalysts of type I (**12–14**, **19–24**). They employed several supports such as polystyrene [1% divinylbenzene (DVB)] (**12–14**), polymer-bound triphenylchloromethane (1% DVB) (**19**), JandaJel (2% cross-linked) (**20**), Wang-Br resin (1% DVB) (**21**), polyethyleneglycol-polystyrene copolymer (1% DVB) (**22**, **23**), and a mercaptomethyl polystyrene (**24**) (Scheme 2.4). A comparison between these materials indicated that the best catalyst was the polystyrene-supported cinchonidine **12**. Higher ee's were obtained when a hydroxyl group was present in the alkaloid moiety, their *O*-allylated counterparts giving lower enantioselectivities. Catalyst **12** was used with a set of bromides to give the corresponding alkylated glycine derivatives (Scheme 2.5) [18, 19].

The four alkaloids—cinchonine, cinchonidine, quinine, and quinidine—were supported on polystyrene resin through a spacer having one, four, six, and eight carbon atoms. Screening of these catalytic materials indicated that the enantioselectivity was moderately altered by the length of the spacer and

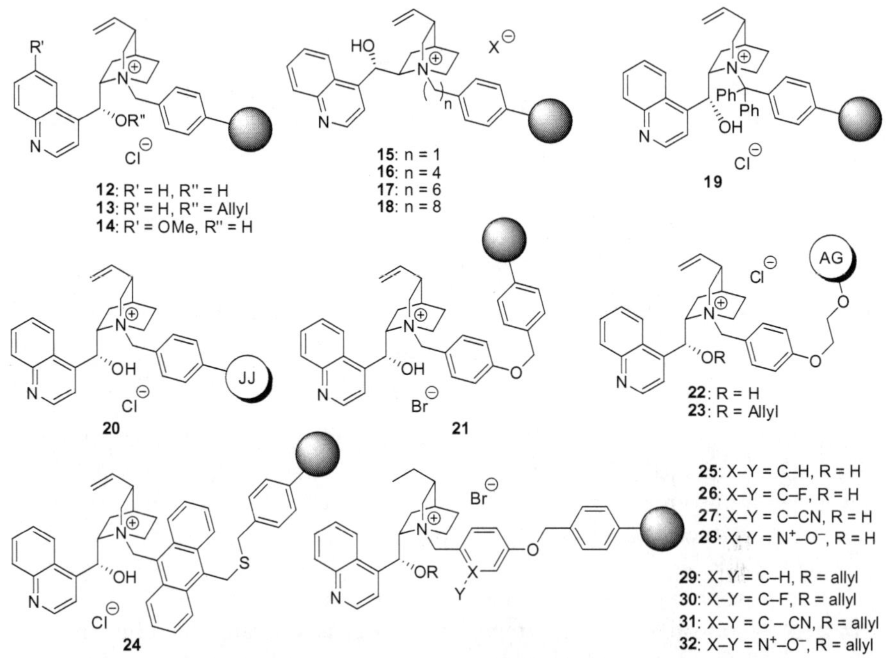

FIGURE 2.5. Type I supported *Cinchona* catalysts.

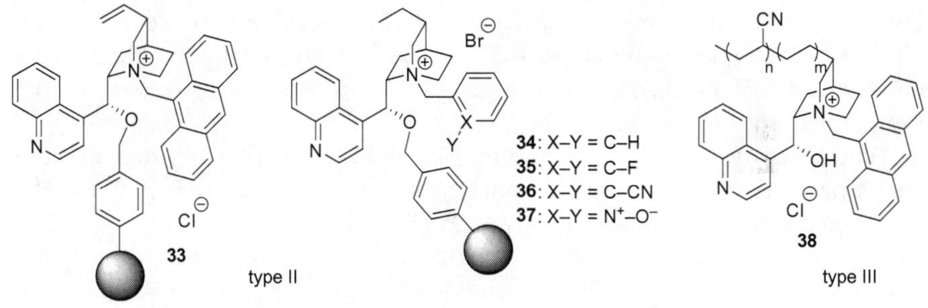

FIGURE 2.6. Type II and type III supported *Cinchona* catalysts.

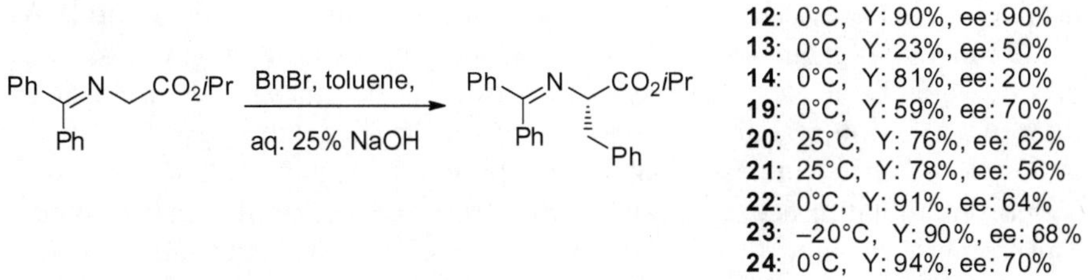

SCHEME 2.4. Asymmetric benzylation of *i*-propylglycinate-benzophenone Schiff base.

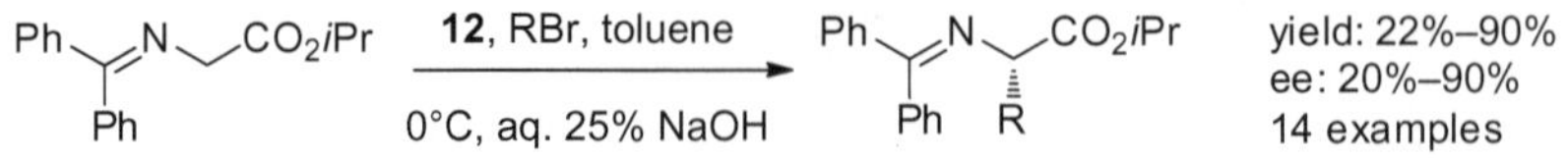

SCHEME 2.5. Asymmetric alkylation of *i*-propylglycinate-benzophenone Schiff base.

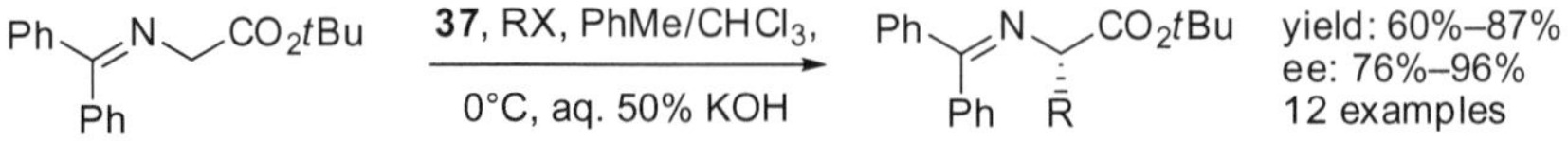

SCHEME 2.6. Asymmetric alkylation of *tert*-butylglycinate-benzophenone Schiff base.

strongly dependent on the alkaloid. Among the four alkaloids, cinchonine gave higher enantioselectivities. Further experiments with this alkaloid and three different spacers (**16–18**) showed that the best performances were observed with catalyst **16** (yields: 60%–80%, ee: 37%–81% *R*, four examples) [20].

While it has been proposed that the steric bulkiness of the N^+-arylmethyl group is responsible for the high enantioselective catalytic efficiency of *Cinchona* PTCs, other authors demonstrated that electronic effects of the N^+-arylmethyl group in the catalyst are also responsible for high enantioselectivity. This result prompted authors to prepare the supported version of these catalysts. Catalysts **25–32** were tested in the asymmetric benzylation of glycine imine ester. While no great differences were observed between non-*O*-allylated **25–28** and *O*-allylated **29–32** catalysts, electronically modified molecules **26–28** and **30–32** gave improved ee's up to 91% with respect to the unmodified ones (**25** and **29**) [21].

Type II catalysts were supported on Merrifield resin. Two approaches were developed: the use of the bulky 9-anthracenylmethyl group (**33**) [22] and the use of electronically modified catalysts **35–37** [23]. All the four *Cinchona* alkaloids were grafted to polystyrene support and quaternized with the N^+-anthracenylmethyl group. Only cinchonidine catalyst **33** gave benzylation of the *tert*-butylglycinate-benzophenone Schiff base in high ee (93%) but working at low temperature (−40°C), which is less practical for large-scale production. On the other hand, catalysts **35–37** worked nicely at 0°C to give the benzylated product in >90% ee. The electronic effect of the 2′-fluoro, 2′-cyano and 2′-*N*-oxide functional groups in **35–37** was evident since catalyst **34** gave the same product in 69% ee. Resin **37** was chosen as the best catalyst and employed in a set of alkylation reactions with excellent results (Scheme 2.6).

A type III catalyst, **38**, was prepared by *N*-alkylation with 9-chloromethylanthracene of an acrylonitrile and cinchonidine copolymer. This catalyst reached a 71% ee and 74% yield in the benzylation of *iso*-propylglycinate-benzophenone Schiff base with 50% KOH at −20°C in toluene/chloroform [24].

2.3.2. Covalently Linked to Soluble Support: PEG

PEG was explored as support for *Cinchona* alkaloids immobilization to create applications in the enantioselective synthesis of α-amino acids. The ee values were strongly dependent on the kind of alkaloid, solvent, and type of linker and its position.

The four alkaloids—cinchonine, cinchonidine, quinine, and quinidine—were supported on a PEG ($M_W = 5000$) trough of a 4-carboxybenzyl moiety. These catalysts were screened in the benzylation of a *tert*-butylglycinate-benzophenone Schiff base. Only the cinchonine-supported catalyst **39** (Fig. 2.7) afforded the benzylated product in good optical purity (81%) working at 0°C. Moreover, the reaction was strongly influenced by the solvent. While toluene gave the highest enantioselectivity, dichloromethane, chloroform, or THF gave an almost racemic product. In the latter cases, the PEG chain could act as a host molecule for the potassium cation, as crown ether, promoting the achiral benzylation [25].

Cinchonidine was again the alkaloid of choice in the synthesis of catalyst **40** (Fig. 2.7). The reactions were carried out at rt. The benzylated product was obtained in slightly lower ee value. Alkylation with allyl and ethyl

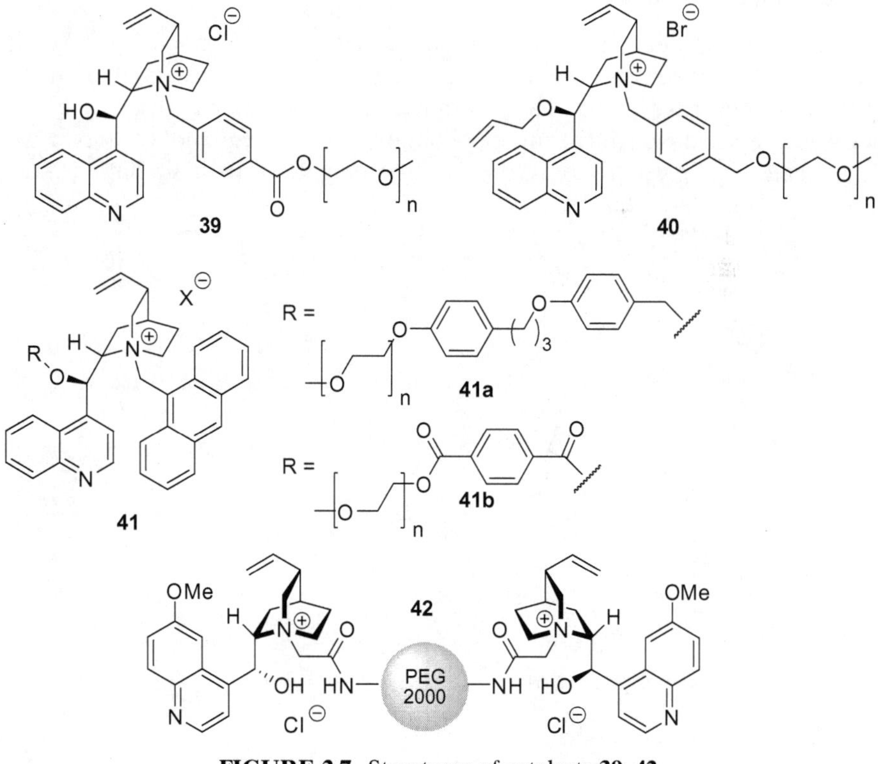

FIGURE 2.7. Structures of catalysts **39–42**.

bromide gave products in moderate ee values. The catalyst was recovered and was reused in another cycle with a sensible decrease in activity and stereoselectivity [26].

To improve the stereoselectivity, the hydroxy group was chosen as anchorage sites on the PEG, whereas the more bulky 9-anthracenylmethyl group was used (**41a–b**) (Fig. 2.7). Catalyst **41a** gave the benzylated product in lower ee than the unsupported catalyst. Again, the PEG chain promoted the achiral alkylation, lowering the stereoselectivity of the reaction. More interestingly, recycling studies revealed that the catalyst, both in the supported and unsupported form, was intrinsically unstable, causing a decreased activity and enantioselectivity upon recycling [27]. A moderate ee value (62%) was also observed when other linkers were employed, such as in catalyst **41b** [25]. When the bulky 9-anthracenylmethyl group was replaced by the diacetamido-PEG moiety (catalyst **42**), high enantioselectivities were obtained. Moreover, the dimeric-PEG-supported cinchonidine catalyst **42** was more stable when being used in four consecutive runs with a small decrease in activity and almost unchanged enantioselectivity [28]. In addition, catalyst **42** gave better results when employed in water instead of in usual organic solvents.

2.3.3. Chiral Ammonium Polymers

The asymmetric benzylation of the *tert*-butylglycinate-benzophenone Schiff base was also carried out by employing optically active quaternary ammonium polymers that have main chain chirality. The cinchonidine-based polymer **43** was prepared, starting from cinchonidine and an equimolar amount of the dihalide followed by polycondensation (Fig. 2.8). In a similar fashion,

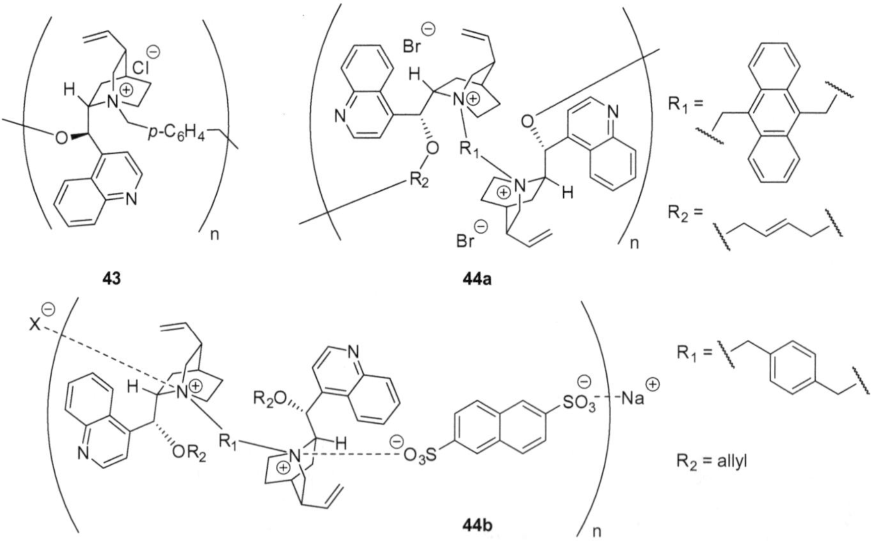

FIGURE 2.8. Cinchonidine-based polymers **43–44**.

bis(quaternary ammonium salts) **44** were also synthesized (Fig. 2.8) [29]. Catalyst **43** reached a good ee value (80%) when used in 10 mol% at 0°C. Several catalysts of type **44** were also screened in the same reaction, displaying ee values in the range 69%–86%. In particular, catalyst **44a** gave the benzylated product in high ee value (86%) working at 0°C, whereas catalyst **44b** afforded the benzylated product in higher ee value (94%) at –20°C (Table 2.1). The chiral polymer **44a** was easily recovered after precipitation in hexane, whereas **44b** was not soluble in the organic solvent used. Their reuse in two further cycles indicated that these materials maintained both the activity and stereoselectivity.

2.3.4. Dendrimer and Fluorous-Modified Catalysts

Dendrimers-supported cinchonidine (first, second, and third generation) were tested as phase transfer catalysts for the alkylation of an *iso*-propylglycinate-benzophenone Schiff base. Enantioselectivities were not high (28%–72%). Thanks to their high molecular weight, dendritic catalysts were retained by dialysis membranes, while reactants flowed through the membrane pores. While the second-generation catalysts gave decreased activity and enantioselectivity after three cycles, the third-generation catalysts maintained activity and enantioselectivity, although they were lower than the former catalyst. The different behavior in recycling could be ascribed to the higher molecular weight, which prevented leaching of the catalyst out of the membrane [30].

The highly fluorinated catalyst **45** (Fig. 2.9) was designed as a recyclable chiral phase transfer catalyst, as fluorous phase separation techniques for the recovery of fluorinated catalysts have been found to be most useful in recently advanced catalyst recovery techniques. Under the reaction condition adopted, the aqueous KOH/toluene biphasic system, catalyst **45** became heterogeneous. Nevertheless, **45** was found to promote the phase-transfer alkylation of the *tert*-butylglycinate-benzophenone Schiff base with various alkyl halides with good enantioselectivity (87%–93% ee) in 82%–93% yield. After the reaction, catalyst **45** was easily recovered by simple extraction with FC-72 as a fluorous solvent and could be utilized for the next run (three cycles) without any loss of reactivity and selectivity [31]. In Table 2.1 are the collected performances displayed by catalysts **39–45** in the synthesis of α-amino acids via PTC.

2.3.5. Noncovalently Linked PTC Catalysts

In 2008, Itsuno et al. reported the first example of the immobilization of quaternary ammonium salt through an ionic interaction [32].

Two approaches were developed for the ionic immobilization: (a) polymerization of a chiral quaternary ammonium sulfonate monomer and

TABLE 2.1. Performances for Catalysts 39–45 Employed in the Asymmetric Synthesis of α-Amino Acids Via PTC

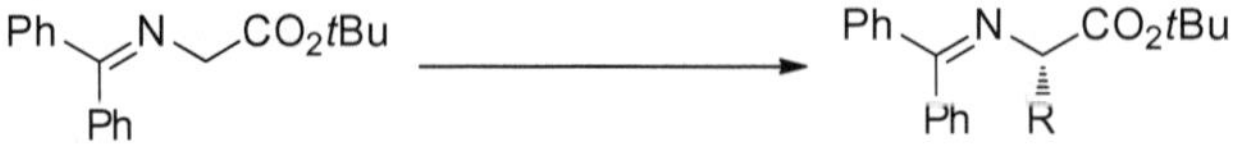

Entry	Catalyst (mol%) Solvent Base	R	Yields (%)	ee (%)	# Cycles	Ref.
1	**39** (10) Toluene, 0°C 50% KOH	Bn	84	81	—	[25]
2	**40** (10) Toluene, rt 50% KOH	Bn	83	73	2	[26]
		Cinnamyl	87	71		
		Allyl	77	47		
		Ethyl	62	36		
3	**41a** (10) DCM, −78°C Solid CsOH	Bn	75	64	Decomp.	[27]
4	**41b** (10) Toluene, 0°C Solid KOH	Bn	83	62	—	[25]
5	**42** (10) Water, rt 1M KOH	Bn	98	83	4	[28]
		2-CH_3-Bn	84	82		
		3-CH_3-Bn	86	90		
		4-CH_3-Bn	91	85		
		2-Cl-Bn	94	97		
		3-Cl-Bn	83	92		
		4-Cl-Bn	82	91		
		Allyl	77	86		
		Ethyl	82	90		
		Methyl	79	83		
6	**43** (10) Toluene/$CHCl_3$ 0°C, 50% KOH	Bn	78	80	—	[29]
7	**44a** (10) Toluene/$CHCl_3$ 0°C, 50% KOH	Bn	83	86	3	[29]
8	**44b** (10) Toluene/$CHCl_3$ −20°C, 50% KOH	Bn	91	94	3	[29]
9	**45** (3) Toluene 0°C, 50% KOH	Bn	82	90	3	[31]
		Allyl	81	90		
		4-CH_3-Bn	82	92		
		4-F-Bn	93	93		
		Ethyl	83	87		

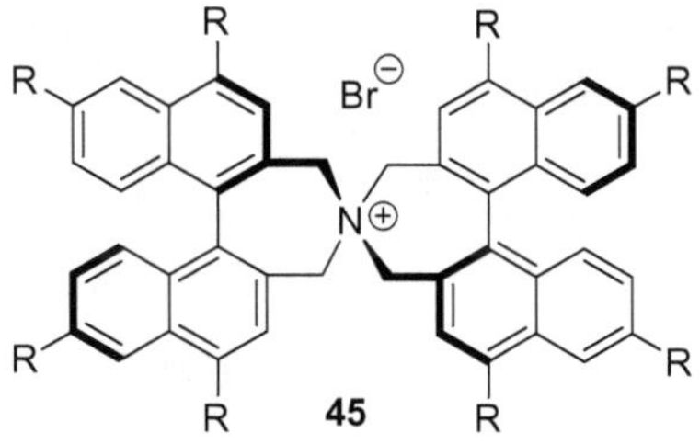

$R = SiMe_2(CH_2CH_2C_8F_{17})$

FIGURE 2.9. Fluorinated Maruoka's catalyst **45**.

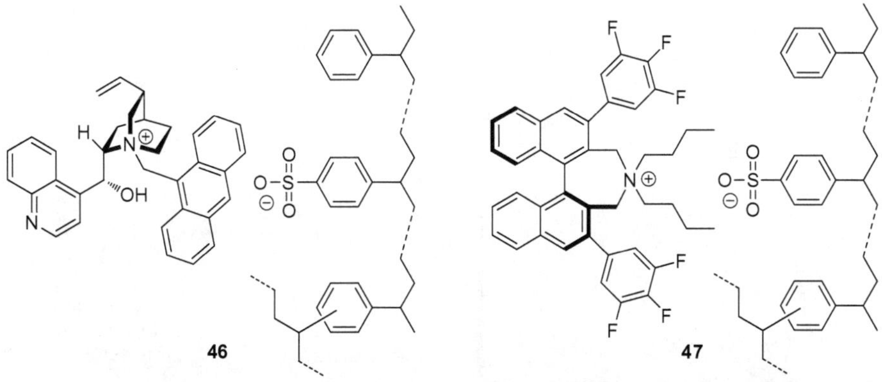

FIGURE 2.10. Electrostatically immobilized PTC catalysts **46–47**.

(b) immobilization of a chiral quaternary ammonium salt onto a sulfonated polymer through an ion-exchange reaction.

Chiral polymeric organocatalyst **46** (Fig. 2.10) was prepared following the above approaches and successfully employed in the alkylation of *N*-diphenylmethylene glycine *tert*-butyl ester, which afforded the expected products in very high enantioselectivities. Organocatalyst **47** (Fig. 2.10) was prepared by an ion-exchange strategy and represents the only way to obtain the polymer-supported Maruoka catalyst without modification of its original structure. Benzylation of the glycine ester derivative employing only 1 mol% of catalyst **47** afforded the benzylated product high yield and excellent enantioselectivity (98%). Catalysts **46** and **47** were fully recyclable without loss in activity and enantioselectivity. In both cases, no elimination of the chiral quaternary ammonium moieties was observed in solution. The high recoverability was due to the strong affinity between the sulfonate anion and the ammonium cation. In Table 2.2, data obtained with catalysts **46** and **47** are summarized.

TABLE 2.2. Performances for Electrostatically Immobilized Catalysts 46–47 Employed in the Asymmetric Synthesis of α-amino Acids Via PTC

Entry	Catalyst (mol%) Solvent Base	R	Yields (%)	ee (%)	# Cycles	Ref.
1	**46** (10)	Bn	84	94	3	[32]
	Toluene, 0°C	4-CH_3-Bn	90	97		
	50% KOH	4-Br-Bn	95	91		
		Allyl	85	98		
		Me	45	70		
2	**47** (1)	Bn	84	98	3	[32]
	Toluene, 0°C					
	50% KOH					

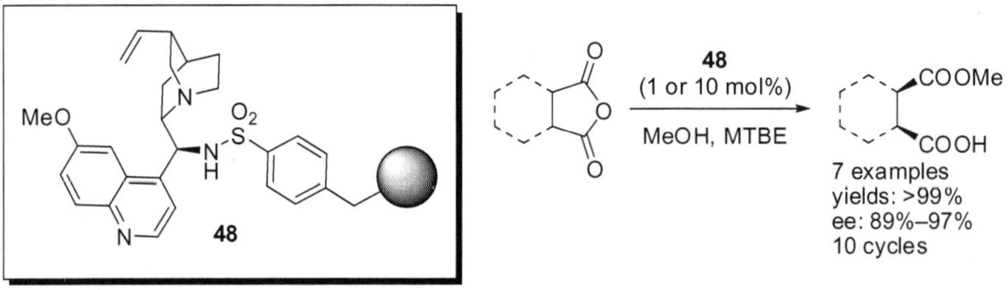

SCHEME 2.7. Desymmetrization of cyclic anhydrides.

2.4. DESYMMETRIZATION

2.4.1. Covalently Linked on Insoluble Supports: Polystyrene, Magnetite, and Silica

Several kinds of organocatalysts were employed in desymmetrization reactions. The polymer-supported *Cinchona* bifunctional sulfonamide catalyst **48** was synthesized by suspension copolymerization and successfully employed in the methanolic desymmetrization of cyclic anhydrides (Scheme 2.7). The hemiesters were obtained in excellent yields and high ee's. In addition, recycling experiments carried out with 10 mol% or 1 mol% of catalyst **48** gave reproducible results of up to 10 cycles [33].

Pyrrolidine derivative **49** supported on JandaJel was employed as catalyst in the kinetic resolution of several cyclic secondary alcohols (Scheme 2.8). *Trans*-2-phenylcyclohexanol was benzoylated, at −78°C, in five consecutive runs to give the ester with unchanged yields (34%–36%) and ee (94%–96%). Kinetic resolution of acyclic alcohols was less successful [34].

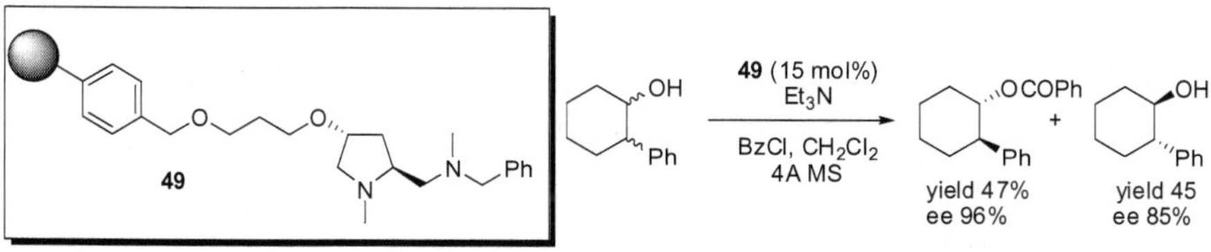

SCHEME 2.8. Kinetic resolution of cyclic secondary alcohols.

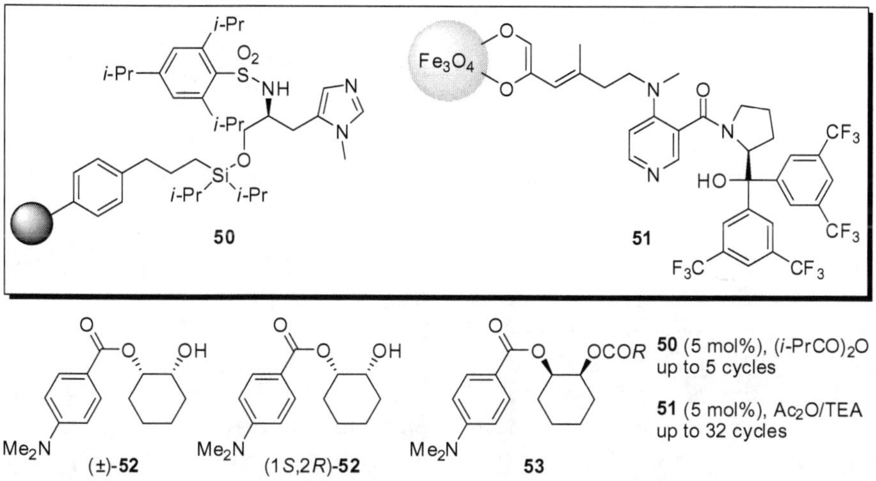

FIGURE 2.11. Supported catalysts **50** and **51** employed in the chiral resolution of alcohol **52**.

A polystyrene-supported artificial acylase **50**, which originated from natural L-histidine, was developed as a recyclable catalyst for the kinetic resolution of racemic alcohols (Fig. 2.11). Resolution of racemic alcohol **52** at rt gave (1*S*, 2*R*)-**52** in 82%–86% ee and ester (1*R*, 2*S*)-**53** (R = *i*-Pr) in 62%–64% ee after a 42%–44% conversion in five consecutive runs [35].

Another class of chiral organocatalysts, which has been successfully designed for kinetic resolution of *sec*-alcohols, is represented by chiral 4-dimethylaminopyridine (DMAP) analogs. Desymmetrization of alcohol **52** was also carried out using the magnetite-supported chiral DMAP catalyst **51** (Fig. 2.11). Supported magnetite catalysts can be easily removed from the reaction mixture by exposure to an external magnetic field. Catalyst **51** was synthesized from magnetite nanoparticles treated with *N*-methyl dopamine hydrochloride followed by S_NAr reaction with the chiral-substituted chloropyridine [36]. Asymmetric acylation reaction on racemic alcohol **52** was carried out in the presence of 5 mol% of catalyst **51** and Ac_2O/triethylamine (TEA). The ee of the recovered alcohol was in the range 82%–99% after 59%–72% conversion in 20 consecutive runs. The same batch of catalyst was used for the

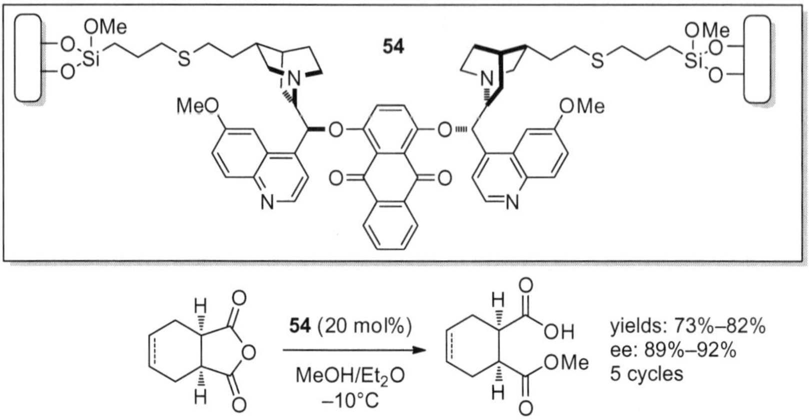

SCHEME 2.9. Desymmetrization of *meso*-cyclic anhydrides.

kinetic resolution of a set of *sec*-alcohols of variable steric and electronic characteristics. The recovered alcohols displayed ee values in the range 70%–93% after a conversion of 62%–87%. More importantly, catalyst **51** was reusable for at least 32 cycles.

1,4-Bis(dihydroquinidinyl)anthraquinone supported on silica gel (**54**) was a recyclable catalyst for the desymmetrization of *meso*-cyclic anhydrides (Scheme 2.9) [37]. Catalyst **54** was superior than the corresponding supported 1,4-bisdihydroquinidine derivative. The rigidity of the active sites was crucial for the enantioselectivity; in fact, the supported bis-derivatives were superior to the corresponding supported monoderivatives. Unfortunately, hindered cyclic anhydrides gave low yields.

2.5. ALDOL REACTION

2.5.1. Covalently Linked on Insoluble Supports: Polystyrene, Methacrylates, and Silica

Asymmetric intermolecular aldol reaction affording *β*-hydroxy ketones—important building blocks for the synthesis of polyfunctional compounds and natural products—is one of the most important C–C bond-formation methods in organic synthesis [38]. In this light, a lot of work has been done on supported proline and proline derivatives for both the direct asymmetric aldol reaction and the asymmetric Michael reaction. It is likely that Takemoto et al. reported the first polystyrene-supported proline organocatalyst in 1985 [39]. *Trans*-4-hydroxy-L-proline was linked to a polystyrene resin (**55**) and employed in the intramolecular aldol reaction, affording the cyclic product in low yield and low enantioselectivity. Until the rediscovery of the proline-catalyzed aldol reaction, no further studies were carried out on this topic. In 2006, Barbas et al.

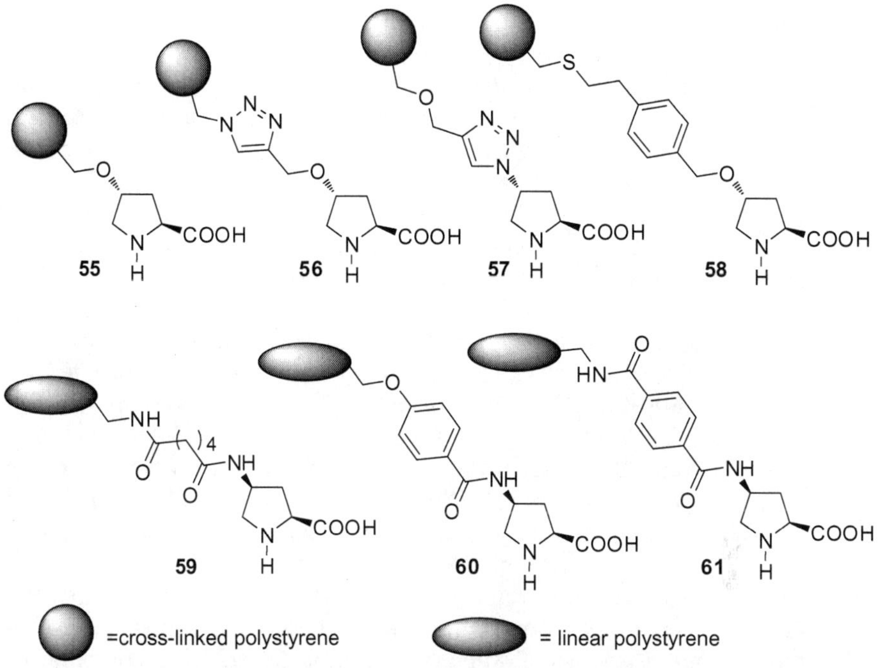

FIGURE 2.12. Resin-supported prolines **55–61**.

and Hayashi et al. reported the use of amphiphilic proline derivatives for asymmetric aldol reactions in water [40, 41].

The hydrophobic moieties present in the amphiphilic proline derivatives could be replaced with a hydrophobic resin. The more outstanding results were obtained using *trans*-4-hydroxy-L-proline as a starting material by anchoring the proline moiety through the hydroxy group in the 4-position. Several kinds of linkers have been used, and results were found to be dependent on their nature. In 2006, Pericás et al. reported the synthesis and the use of catalyst **56** (Fig. 2.12). Hydroxy-L-proline was supported to the resin through a 1,2,3-triazole linker by using the click chemistry strategy. Catalyst **56** was prepared by a 1,3-dipolar cycloaddition of an azide-substituted Merrifield resin with an *O*-propargyl hydroxyl-proline derivative [42]. Aldol reaction between four ketones and aromatic aldehydes was carried out in the presence of catalyst **56**. The screening of solvents [H_2O, dimethylformamide (DMF), dimethylsulfoxide (DMSO), and their mixtures] showed that better performances were obtained in water. Resin **56** was employed in 10 mol% using water-soluble DiMePEG (10 mol%) as additive to facilitate the diffusion to the resin. Increased yields were obtained using this additive. The insoluble support enabled the recovery by filtration. Resin **56** was used for several cycles with unchanged results. Moreover, a comparison between **56** and the corresponding unsupported monomer showed the beneficial contribution of the polymer backbone. The monomer afforded the aldol products in lower diastereomeric ratios and ee values.

The same resin was employed in the α-aminoxylation of aldehydes and ketones [43]. Enantioselectivities were excellent (>96%), whereas in several cases yields were determined by nuclear magnetic resonance (NMR) since the chromatographic purification process induced partial N-O cleavage. The usefulness of this catalyst was evident since reaction rates were improved compared to native proline. Recycling investigations showed that resin could be reused three times without loss in enantioselectivity and slowly diminishing yields.

Interestingly, a similar linker in resin **57** (Fig. 2.12) gave a different behavior. Resin **57** was again obtained by click chemistry between 4-azido-L-proline and an *O*-propargyl-substituted Merrifield resin [44]. This resin was compared with catalysts **55–56** and other two resins having the same linker as **56** but anchored on PS-PEG NovaBioSyn and on Argopore resins. In sharp contrast to the latter materials, which produced multiphase systems when the aldol reaction between cyclohexanone and benzaldehydes was carried in water, catalyst **57** perfectly swelled in water. With the latter resin, a gel-like single phase, containing up to 24% in weight of water, was formed. This behavior arose from the formation of a hydrogen bond network connecting the proline and a 1,2,3-triazole unit. The role of water was demonstrated by a comparative experiment. An anhydrous sample of **57** was used for an aldol reaction. In a parallel experiment, the same reaction was carried out with the same resin swollen in water and used in dichoromethane, after the excess water was removed. In the latter case, the reaction was faster and more stereoselective than the corresponding reaction carried out with anhydrous **57**. Catalyst **57** was used in the aldol reaction between cyclohexanone and cyclopentanone and several arylaldehydes. In addition, a cross-aldol reaction of propanal—which afforded the aldol product in a 5:1 *anti*/*syn* ratio and 97% ee—was reported. Resin **57** was recycled and reused five times without appreciable loss in yield and in stereoselectivity.

Supported organocatalysts **58** (Fig 2.12) were prepared by a different synthetic approach [45]. The key step was the thiol-ene coupling reaction between a mercaptomethyl polystyrene resin and a styrene-proline derivative. This approach easily afforded the Merrifield-substituted proline catalyst **58** having a longer hydrophobic linker compared to resin **55**. Interestingly, solvent screening showed that the reaction took place only in the presence of water. Other solvents, such as DMSO, DMF, $CHCl_3$, and dioxane did not promote the reaction. Addition of water to these solvents promoted the reaction, although in lower yield. Alcoholic solvents did not promote the reaction, with the exception of methanol, which did not give a good yield. Resin **58**, which works only in the presence of water, and resin **57**, which gave improved performances when expanded in water, can be considered mimics of the natural class I aldolase enzymes, which use an enamine-mediated mechanism in water.

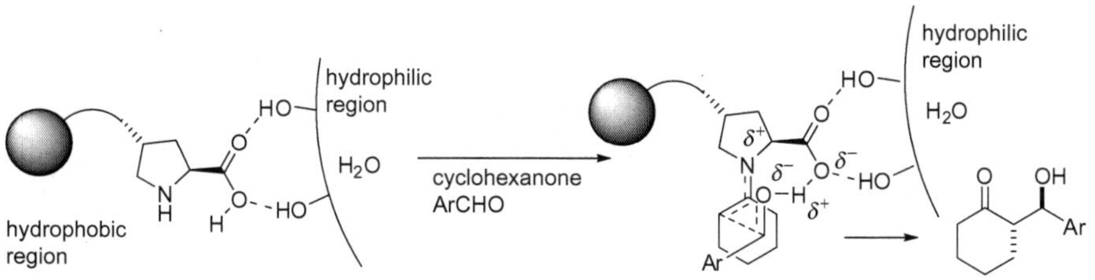

FIGURE 2.13. Proposed favored transition state for the aldol reaction catalyzed by PS-supported proline.

In the case of resin **58**, the higher activity, diastereoselectivity, and enantioselectivity observed were explained in the following model in Figure 2.13. Water molecules lying in the hydrophilic outer region forced hydrophobic arylaldehydes into the restricted hydrophobic inner pocket. In this way, water promoted the reaction and increased the stereoselectivity. The insoluble nature of the catalyst allowed its recovery by filtration and reuse for five consecutive runs without loss in activity and stereoselectivity. Interestingly, the same catalyst was found to be active and recyclable in both the α-selenenylation of aldehydes [45a] and Baylis–Hillman reactions, although no enantioselectivity was displayed [46].

In addition to cross-linked polystyrene resins, linear polystyrene support (M_W = ca. 5000, f = 0.60 mmol/g) was also employed. Organocatalysts **59–61** were prepared using this support and were used in DMF/water (15:1) or in ketone/water mixtures, employing a very large excess of ketone in each case [47]. Because of the solubility of the catalyst, the recovery was performed by addition of diethyl ether followed by filtration of the precipitate. Reuse of these catalysts gave comparable ee values and slowly diminishing yields. A comparison of these data indicates that cross-linked polystyrene proline resins behaved better than linear-supported proline resins in the catalytic performances and recycling.

As stated before, two strategies can be followed in order to obtain supported organocatalysts: (a) immobilization into a preformed support and (b) immobilization by in situ polymerization. In the case of proline immobilization, several examples of the first strategy have been described above. The second approach was deeply investigated by Hansen et al. [48]. Polymer-supported organocatalysts were prepared in a bottom-up fashion where methacrylic functional monomers were prepared in an entirely nonchromatographic manner and subsequently copolymerized with suitable monomers to give cross-linked polymers beads. The synthetic methodology employed was easily scalable; resins **62–64** were prepared by suspension copolymerization by using 2 mol% of the proper cross-linker while resin **66** was prepared by dispersion polymerization (Fig. 2.14). These catalysts were screened by using the aldol reaction between cyclohexanone and 4-nitrobenzaldehyde. No significant

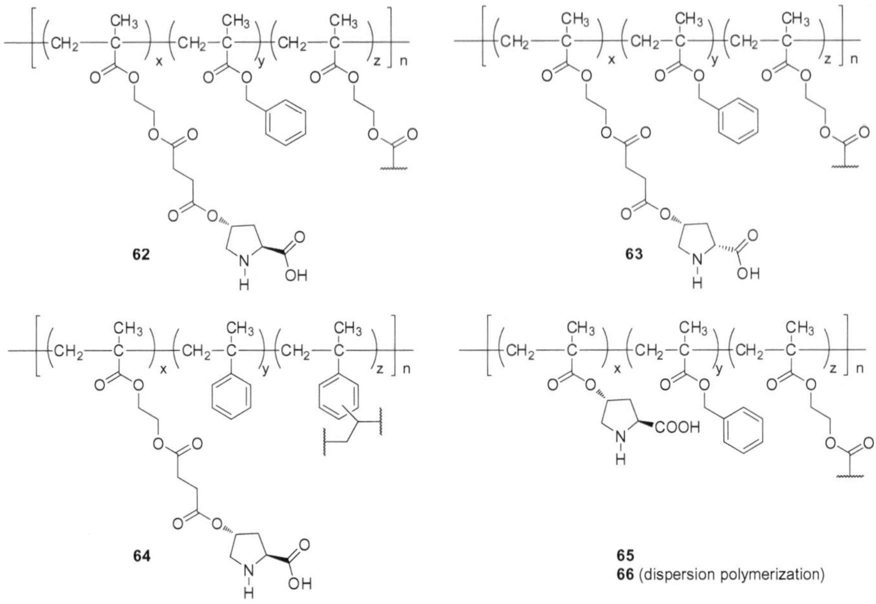

FIGURE 2.14. Proline-based resins prepared from their corresponding monomers.

differences were observed in yields, diastereoselectivity, and enantioselectivity when resins **62** and **64–66** were compared. Even supports that did not contain any form of linker (**65** and **66**) showed good results. Moreover, catalyst **64** retained its activity at only 1 mol% of catalyst loading. Further investigations were carried out using catalysts **62** and **63**. Aldol reaction between cyclohexanone and arylaldehydes, carried out in water, afforded the corresponding products good yields and stereoselectivities. Moreover, C-2 epimeric *cis*-catalyst **63** exhibited a selectivity that was only modestly lower than that of *trans*-catalyst **62**, and nearly identical for several substrates. These acrylic supports were easily recycled by filtration, and catalyst **64** was reused five times with unaffected results. The results obtained with resin-supported prolines are shown in Table 2.3.

In order to achieve higher activity and stereoselectivity as well as broader applicability of proline-based organocatalyst, several studies have been carried out using proline-based peptide catalysts. In the examples reported so far, proline was immobilized through the hydroxy group on C-4. The other strategy for proline immobilization relies on the anchorage through the COOH group via amide formation (Fig. 2.15, Type I). On the other hand, connection to the support via the hydroxy group on C-4 (Fig. 2.15, Type II) can be also followed. Looking at the first approach (Type I), several di- and tripeptides linked on several supports (PS-PEG, TentaGel, PS) were prepared.

TABLE 2.3. Performances for Proline-Supported Resins 56–63 Employed in the Asymmetric Aldol Reaction or in the α-aminoxylation of Aldehydes

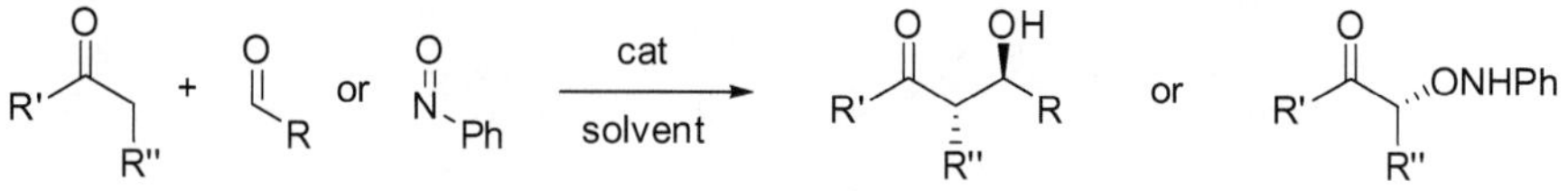

Entry	Catalyst (mol%) Solvent	Product	Yields (%)	d.r. (*anti*/*syn*) ee (%)	# Cycles	Ref.
1	**56** (10) H_2O	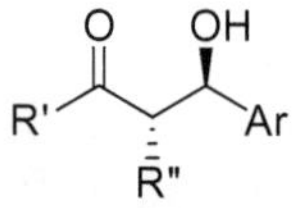 10 examples	45–97	58:42–98:2 45–97	4	[42]
2	**56** (10–20) DMF	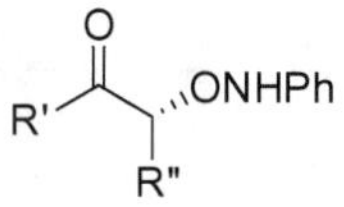 13 examples	35–86	— 96–99	3	[43]
3	**57** (10) H_2O	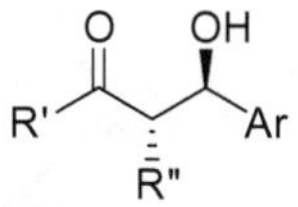 8 examples	16–99	84:16–97:3 94–>99	5	[44]
4	**58** (10) H_2O	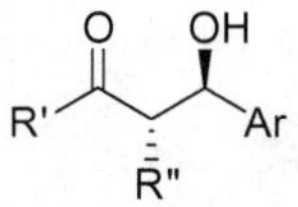 19 examples	33–99	68:32–>99:1 66–98	5	[45]
5	**59** (5) DMF/H_2O 15:1	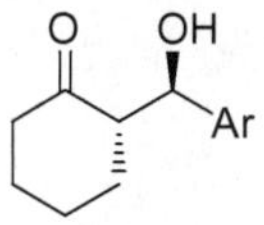 7 examples	69–94	83:17–93:7 88–94	6	[47]
6	**60** (5) DMF/H_2O 15:1	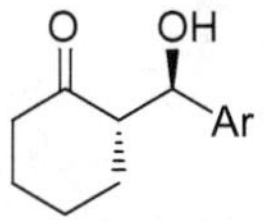 7 examples	62–72	87:13–91:9 86–98	6	[47]
7	**61** (5) DMF/H_2O 15:1	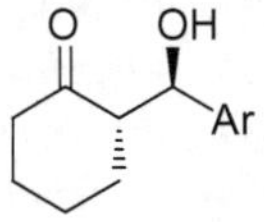 7 examples	60–81	83:17–93:7 82–95	—	[47]

(*Continued*)

TABLE 2.3. (*Continued*)

Entry	Catalyst (mol%) Solvent	Product	Yields (%)	d.r. (*anti*/*syn*) ee (%)	# Cycles	Ref.
8	**62** (10) H_2O	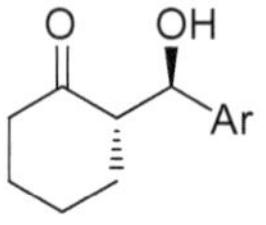 8 examples	7–88	91:9–95:5 89–99	5 With cat. **64**	[48]
9	**63** (10) H_2O	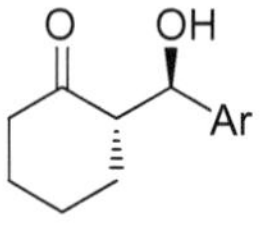 8 examples	13–91	93:7–96:4 86–98	—	[48]

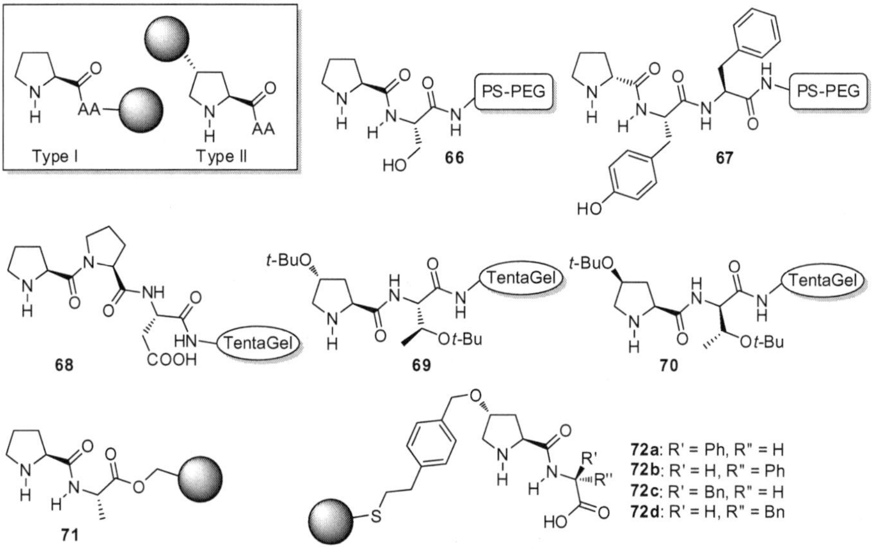

FIGURE 2.15. Resins-supported peptides.

Organocatalysts **66–71** (Fig. 2.15) were synthesized following this idea. Davis et al. synthesized several di- and tripeptides of type H-Pro-[AA_1]-[AA_2]-NH-support using a Novasyn TG amino resin, an amine-terminated PEG-polystyrene graft copolymer that was chosen for its compatibility with a wide range of solvents. Catalysts were tested in the aldol reaction between acetone and 4-nitrobenzaldehyde. The best outcome was observed with organocatalyst **66** in 13 mol% at −25°C (conv. >98%; ee 82%) [49].

Tripeptide **67** (20 mol%) was tested in the same reaction in water at rt to give the aldol product in only 33% ee. However, when the reaction was carried out in the presence of 20 mol% of $ZnCl_2$ at 0°C in water/THF, the ee value reached 73%. Reuse of such catalyst over five runs gave reproducible values. The stereochemical outcome of the reaction was driven by C-2 configuration of the proline moiety. Indeed, aldol products with *S* configuration were obtained [50].

Tripeptide H-L-Pro-L-Pro-L-Asp-NH_2 was a powerful catalyst for the aldol reaction between acetone and aldehydes [51]. Such tripeptide was covalently immobilized onto four different supports (polystyrene with an ε-aminocaproic acid as spacer, polyacrilamide, TentaGel, and poly(ethyleneglycol-acrylamide) [PEGA]). These catalytic materials were used in 1 mol% in the presence of *N*-methylmorpholine as a base and evaluated in the reaction between acetone and 4-nitrobenzaldehyde. The optimal catalyst was a TentaGel-supported tripeptide **68** with a loading of 0.1–0.2 mmol/g. This catalyst was used in the aldol reaction between acetone and five aldehydes to give comparable results to those obtained with a nonsupported peptide. The catalyst was isolated by filtration and reused up to eight cycles. Unfortunately, activity decreased during recycles [52].

TentaGel-supported catalysts **69** and **70** found interesting application in the synthesis of optically active chromanones. The use of a linear *tert*-butyl ether [Ser(*t*Bu)] as a second amino acid, instead of the branched *tert*-butyl ether [Thr(*t*Bu)], as well as the absence of the O*t*Bu group on C-4, caused a lower ee. The configuration of chiral centers in **69** and **70** and the position of R^3 (Scheme 2.10) dictated the configuration of the obtained chromanones. The reactions gave complete diastereoselectivity and generally high enantioselectivity (>90%). Moreover, these catalysts were employed in preliminary assays in a series of aldol, Michael, Robinson annulation, and Mannich reactions to provide final products in moderate-to-good yields and high ee's. More interestingly, over 40 recycles were carried out without loss in efficiency [53].

Yan and Wang reported the synthesis of several dipeptide-supported organocatalysts on Merrifield resin of the type H-Pro-[AA_1]-[AA_2]-COO-support. Screening of these catalysts indicated resin **71** as the most promising. It was used in several aldol reactions between ketones and

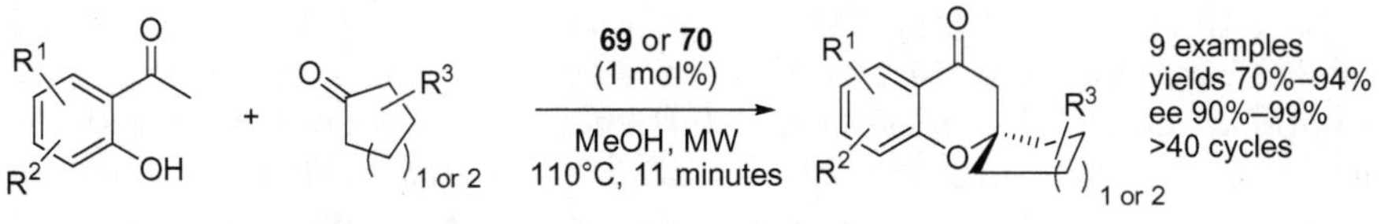

SCHEME 2.10. Asymmetric synthesis of chromanones.

arylaldehydes under neat conditions. This material was easily recovered and used in seven consecutive runs with unchanged results [54]. Dipeptide-supported organocatalysts **72a–d** are representative of Type II and were linked through the C-4 of hydroxyproline. Several catalysts of the type support-Pro-(AA) were prepared and used in the aldol reaction between cyclic ketones and arylaldehydes in water. A comparison between **72a**, **72b**, **72c**, and **72d** showed that the presence of a D-amino acid (resins **72b**,**d**) gave better stereoselectivity than L-amino acid (resins **72a**,**c**). In particular, D-phenylglycine derivative **72d** gave the highest stereoselectivity. This catalyst was employed in several aldol reactions in up to nine cycles with unchanged results. These supported dipeptide catalysts were more active than the unsupported catalysts. In fact, pristine dipeptides H-Pro-(AA) were active only in the presence of a base and a surfactant. The base furnished the corresponding more soluble carboxylate, whereas the surfactant allowed a better interaction with the hydrophobic reagents. Thanks to the hydrophobic backbone, catalysts **72** worked as in the model reported in Figure 2.13, avoiding the need for such additives [55]. In Table 2.4 are the collected performances displayed by supported peptides catalysts in the asymmetric aldol reaction.

While simple prolinamide was not a good catalyst for the aldol reaction, prolinamides of type **73** showed good activity and selectivity (Fig. 2.16). The poor activity of prolinamide was ascribed to the insufficient acidity of the CONH hydrogen. The hydroxyl group in **73** may form a second hydrogen bond to the carbonyl oxygen in the aldol transition state, reinforcing the effect of the NH and restoring activity. After the discovery of several highly active and stereoselective prolinamides, Gruttadauria et al. prepared two polystyrene-supported prolinamides (**74** and **75**) (Fig. 2.16) [56, 57]. These organocatalysts were prepared by postmodification strategy using the same methodology used for resins **58** and **72**. Again, the basilar role of water was evident since no reaction or very poor yields were observed using organic solvents. Good yields were observed when water was employed as a reaction medium, but the use of a mixture of chloroform and water gave the highest yields. This fact could be ascribed to the good swelling properties of chloroform, while water is able to create a concentrated organic phase on the hydrophobic resin, increasing both the activity and stereoselectivity of the catalyst. Catalysts **74** and **75** were employed in the aldol reaction using acetone and several cyclic ketones with arylaldehydes. Enantio- and diastereoselectivities were excellent. The recycling of these catalysts indicated a drop in activity; however, a simple treatment with formic acid fully restored the activity. Applying such regeneration methodology, catalysts **74** and **75** were used in up to 22 consecutive runs without a decrease in performance. Most likely, the excess of ketone with the formation of the corresponding imidazolidinone deactivated the catalysts. Therefore, catalysts were regenerated, treated with formic acid in order to hydrolyze the imidazolidinone. Diastereomeric catalysts **76** and **77** (Fig. 2.16) were prepared via acrylic copolymerization and used in aldol reactions

TABLE 2.4. Summary of Aldol Reactions Carried Out with Supported Peptides

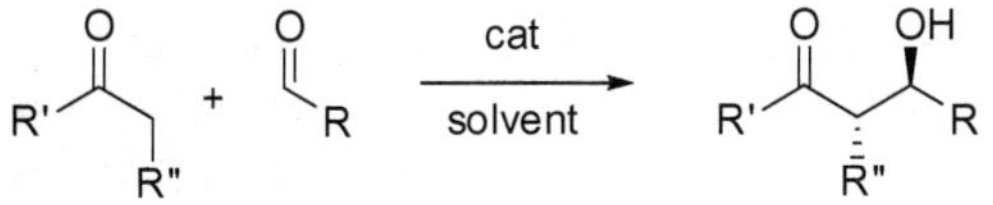

Entry	Catalyst (mol%) Solvent	Product	Yields (%)	d.r. (*anti*/*syn*) ee (%)	# Cycles	Ref.
1	**67** (20) + $ZnCl_2$ (20) THF/H_2O	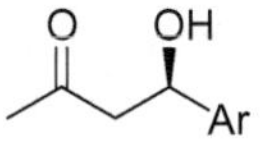 5 examples	50–100	— 71–84	5	[50]
2	**68** (5) + NMM (5) Neat	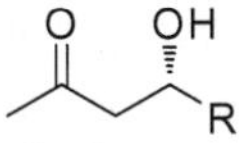 5 examples	30–93	— 72–80	8	[52]
3	**71** (10) Neat	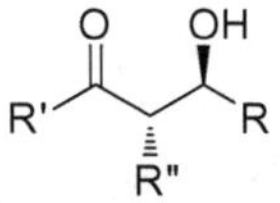 14 examples	52–98	20:80–90:10 75–95	7	[54]
4	**72d** (20) H_2O	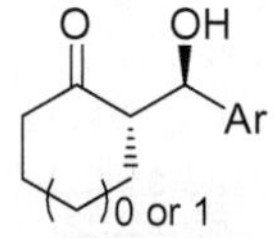 9 examples	34–98	50:50–97:3 64–99	9	[55]

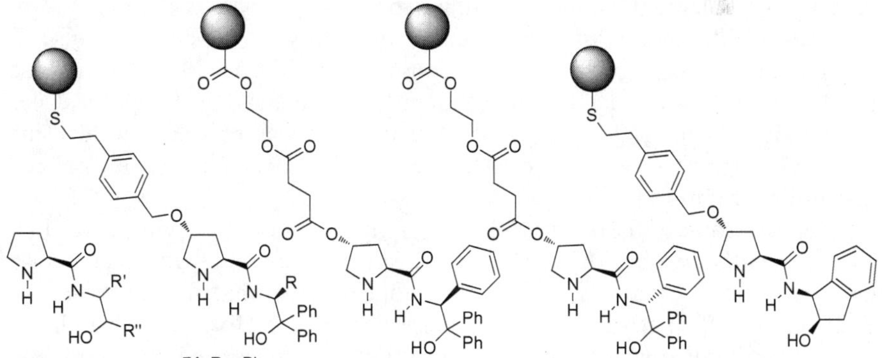

FIGURE 2.16. Structure of supported prolinamides **74–78**.

between acetone and arylaldehydes. Products were obtained in excellent enantioselectivity. Generally, catalyst **77**, having a *cis*-relationship, gave a somewhat lower enantioselectivity, but for some derivatives, the selectivity was almost identical. Differently from catalysts **74** and **75**, which worked in chloroform and water, catalysts **76** and **77** worked well in just water. This difference could be due to the more hydrophilic nature of the support. Catalyst **76** was used in five consecutive cycles while still maintaining activity and selectivity [48].

Catalyst **78** afforded aldol products in lower diastereo- and enantioselectivity, compared to the other supported prolinamides. However, supported prolinamide **78** was much more stereoselective than the corresponding unsupported prolinamide, which displayed low enatioselectivities (12%–42% ee) when employed in water. The usefulness of the immobilization approach was evident because it provided a more stereoselective and recyclable catalyst [55] (Table 2.5).

Amorphous and mesoporous silica gels were employed for proline immobilization. Moreover, for proline anchorage, two strategies were applied: grafting on preformed silica gels or synthesis via sol-gel processes. Proline linked through a urea moiety in *trans* configuration, both on MCM-41 and SiO_2 (**79**), gave poor results in ee values and recycling (Fig. 2.17) [58]. Using the same urea moiety proline, having a *cis* configuration, was linked to the silica gel. Among several different supports, the catalyst MCM-41 **80** gave the best results (Fig. 2.17). Such a catalyst was used in DMSO or toluene for the aldol reaction between hydroxyacetone and five aldehydes. In general, yields were higher using homogeneous conditions. Only *iso*-butyraldehyde gave the same ee value both under homogeneous and heterogeneous conditions. With cyclohexyl carboxaldehyde and benzaldehyde, MCM-41 **80** gave a lower ee value. Interestingly, the last two aldehydes provided diastereoselectivity complementary to the homogeneous catalyst. This result was ascribed to interactions of the reactants with the solid support. Because of the harsh conditions required for condensation (90°C), MCM-41 **80** was also used in DMSO with the assistance of microwave heating. Reaction times were reduced and yields increased. The catalyst was recycled twice using *iso*-butyraldehyde showing decreased yield (45%–40%) and an unchanged diastereomeric ratio [59].

Later, the same authors reported further investigations on asymmetric aldol reactions catalyzed by MCM-41 **80** [60]. Solvent screening for the reaction between 4-nitrobenzaldehyde and dioxanone showed that the reaction proceeded more efficiently in the hydrophilic polar solvent. However, the addition of small amount of water (up to 5 eq.) to the reaction carried out in toluene had a positive effect on the rate and the stereoselectivity. An excess of water (10 or 20 eq.) produced a drastic drop of reactivity. Optimized procedures were employed to furnish useful intermediates for the synthesis of azasugars. In addition to MCM-41 **80**, catalyst MCM-41 **81**—in which the urea group was replaced by a carbamate group with *trans* configuration in the proline ring—

TABLE 2.5. Summary of Aldol Reactions Carried Out with Supported Prolinamides 74–78

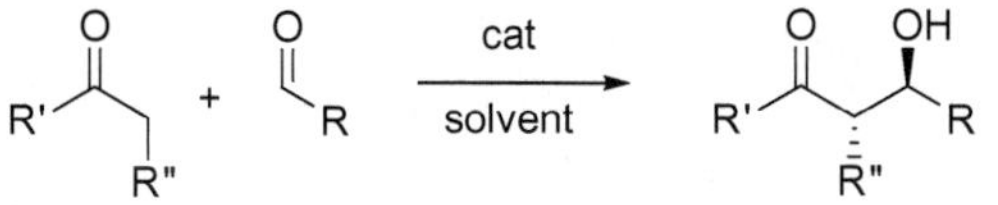

Entry	Catalyst (mol%) Solvent	Product	Yields (%)	d.r. (*anti*/*syn*) ee (%)	# Cycles	Ref.
1	**74** (10) $CHCl_3$/ H_2O	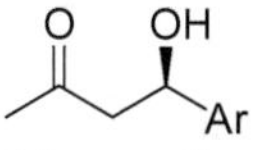 14 examples	30–98	— 89–97	11	[56]
2	**74** (10) $CHCl_3$/ H_2O	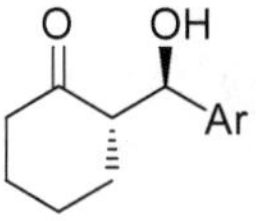 14 examples	4–97	96:4–98:2 95–99	11	[56]
3	**75** (10) $CHCl_3$/ H_2O	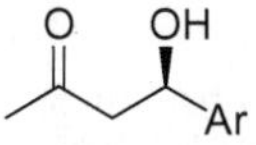 14 examples	14–98	— 86–96	16	[57]
4	**75** (10) $CHCl_3$/ H_2O	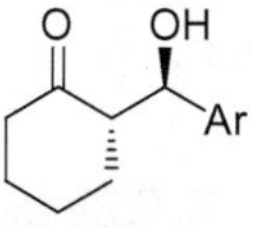 6 examples	31–95	91:9–98:2 80–98	22	[57]
5	**76** (10) H_2O	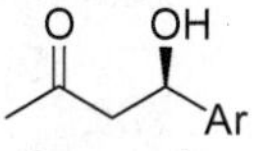 6 examples	29–90	— 90–99	5	[48]
6	**77** (10) H_2O	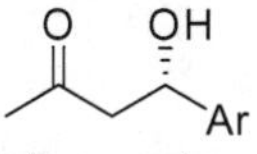 6 examples	47–88	— 82–99	—	[48]
7	**78** (10) H_2O	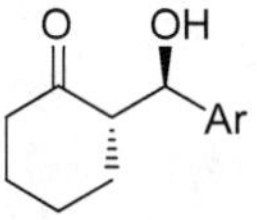 5 examples	31–99	81:19–89:11 75–86	6	[55]

was used. The latter catalyst gave lower diastereoselectivities and ee values. Only one recycle was performed using 4-nitrobenzaldehyde in formamide as a solvent. A small decrease both in yield and ee was observed.

The same catalyst was prepared by other authors using a bottom-up approach via the sol-gel process in the presence of a variable amount of

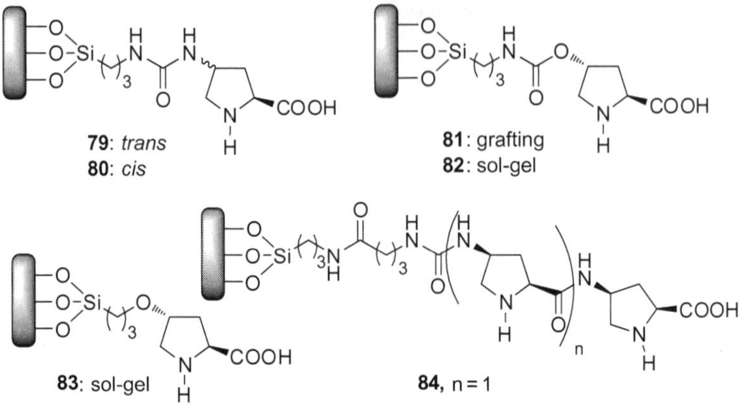

FIGURE 2.17. Silica-supported prolines.

tetraethoxysilane with or without porogen. This methodology was also applied for the synthesis of catalyst **83**. Silica gels with high surface areas were obtained (352–989 m^2/g) (Fig. 2.17). However, catalysts **82** and **83** displayed low enantioselectivity and recycling properties in the aldol reaction between acetone and 4-nitrobenzaldehyde. The low enantioselectivity can be ascribed to the competition of the acidic silanols with the COOH-proline moieties for the positioning of the aldehyde. The decrease activity in recycling was explained in terms of inhibition of the catalytic sites [61].

It appears evident that, whatever the immobilization procedure, simple proline immobilized on silica gel gave catalysts that displayed the worst performances compared with the polystyrene-supported proline catalysts. Better performances were obtained by employing a silica gel-supported proline-proline dipeptide catalyst **84** [62]. Catalyst **84** was chosen among four catalysts having the same linker and from one to four proline units ($n = 1–3$). Aldol reactions between acetone and arylaldehydes in the presence of catalyst **84** furnished products in high yields (55%–97%) and enantioselectivities (64%–96%). Moreover, after five cycles, no decrease in activity and enantioselectivity was observed.

2.5.2. Covalently Linked to Soluble Support: PEG

Immobilization of (2*S*, 4*R*)-4-hydroxyproline on PEG-500 monomethyl ether by means of a succinate spacer gave a recyclable soluble catalyst that promoted the enantioselective aldol condensation between acetone and hydroxyacetone with several aldehydes. The same catalyst was also used in the synthesis of the Wieland–Mischler ketone and in the Mannich reaction.

The PEG-proline catalyst gave a similar yield and enantioselectivity compared to nonsupported proline. Recovery of catalyst was achieved by precipitation with diethyl ether and filtration (70%–80% yield). It was reused three

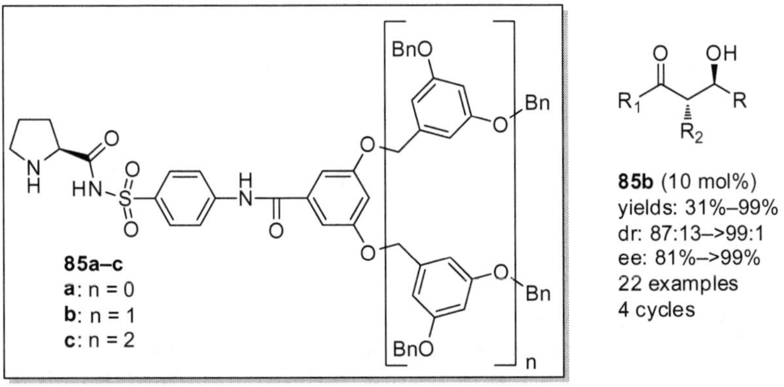

FIGURE 2.18. Dendrimer-supported prolinamides **85a–c**.

times in the aldol reaction between acetone and 4-nitrobenzaldehyde, giving the same ee value but decreasing yield (68%–51%) [63, 64].

2.5.3. Dendrimers

In 2006, Wang et al. reported the synthesis of chiral dendritic catalysts **85a–c** derived from proline-*N*-sulfonamide (Fig. 2.18) [65]. These catalytic materials were tested in the aldol reaction between cyclohexanone and 4-nitrobenzaldehyde in the presence of water as reaction medium. The best result was obtained with catalyst **85b**, which worked well in other aldol reactions.

Good yields and excellent stereoselectivities were obtained. Dendritic materials were recovered by precipitation and filtration. Different solvents were tested in order to find the optimal conditions for recovery. Four consecutive cycles were carried out, giving reproducible excellent yields and stereoselectivities, recovering about 95% of catalyst in each cycle.

2.5.4. Ionic Liquids as Solvents

$[Bmim]PF_6$ was used to confine proline (30 mol%) as a catalyst in the aldol reaction between acetone and several substituted benzaldehydes [66]. Yields and ee ranged from 55% to 94% and from 63% to 82%, respectively. Recycling experiments (up to three cycles) showed decreased yields and selectivities. Employing this ionic liquid, proline was used also in 1–5 mol% with good results in the reaction between acetone and 4-trifluoromethylbenzaldehyde. High yield (91%), diastereomeric ratio (d.r.) (20:1, *anti*/*syn*), and ee (93%) were obtained in the reaction between the above aldehyde and cyclohexanone. Several ionic liquids were investigated as a medium for the aldol reaction between acetone and benzaldehydes [67]. $[Bmim]PF_6$ was selected as reaction

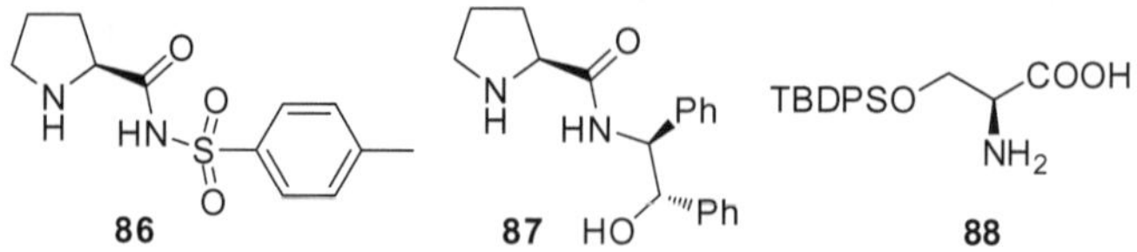

FIGURE 2.19. Catalysts employed in ionic liquids.

medium, giving aldol products in good yields (58%–83%) and ee's (67%–89%). Recycling experiments (four cycles) showed a slight decrease in yield (58–52%) and selectivity (71%–67%). Better results were obtained in the cross-aldol reaction [68]. Using proline (5 mol%) in [bmim]PF_6/DMF 1.5:1 at 4°C for 15–17 hours, good yields (69%–78%), and excellent stereoselectivities (d.r.: 3:1–>19:1; ee: 99%–>99%) were obtained.

This methodology was also applied to the direct assembly of pyranoses with good results. Several recycling experiments were carried out. In all the cases, no decrease both in yields and stereoselectivities was observed.

An enantioselective aldol reaction between acetone and aromatic aldehydes using *N*-toluenesulfonyl-proline **86** in [bmim]PF_6 was also reported (Fig. 2.19). Results were comparable to those obtained by simple proline in ionic liquid. Recycling studies were also performed [69].

The direct aldol reaction between acetone or butanone and several aldehydes was carried out using prolinamide **87** (20 mol%) in [bmim]BF_4 at 0°C [70]. Good yields and excellent enantioselectivities were obtained. Noticeably, ee values were higher than those obtained by the same catalyst in acetone at −20°C. Recycling experiments carried out using 4-trifluoromethylbenzaldehyde and acetone showed no decrease in enantioselectivity after four cycles but did show a drop in activity in the last cycle (79%–41%).

The silyloxy-L-serine **88** (10 mol%) in [bmim]BF_4 catalyzed the aldol reaction between cyclohexanone, cyclopentanone, and acetone with aromatic aldehydes in good yields (40%–86%), d.r. ratio up to 88/12, and enantioselectivity (41%–90%). The catalyst was recovered by extraction and reused, but a drop in activity was observed in the fourth cycle [71]. When catalyst **88** mediated the aldol reaction of TBSO-protected hydroxyacetone (TBS, *tert*-butyldimethylsilyl) with aromatic aldehydes in [bmim]BF_4, the drop in activity was observed in the third cycle. Such decreased activity could be due to the leaching of the catalyst to either layer during the extraction procedure [72].

Ionic liquids may be used both as solvents and catalysts. The chiral ionic liquid 1-ethyl-3-methylimidazolium-(*S*)-2-pyrrolidinecarboxylic acid salt acted as a catalyst (30 mol%) for the aldol reaction between several ketones and aromatic aldehydes in [bmim]BF_4. Catalyst and solvent were reused after recovery of the product by extraction. Recycling indicated a decreased stereoselectivity and activity. The poor results in recycling experiments together with the high catalyst loading makes this approach less interesting [73].

2.5.5. Noncovalently Linked Organocatalysts: Ionic Liquid-Modified Silica Gels; Alumina Support; Polyoxometalate Support

After the first applications of the use of ionic liquids as reaction media for asymmetric aldol reactions appeared [66, 67], Gruttadauria et al. reported the first example of immobilized ionic liquid as support for an organocatalyst (L-proline) [74]. The supported ionic liquid phase (SILP) serves as the reaction phase in which the homogeneous catalyst is dissolved. Since ionic liquids are still expensive, it is desirable to minimize the amount of utilized ionic liquid in a process while at the same time allowing for an easy recovery of the catalyst. Moreover, it is of great interest to have a material behaving as a bulk ionic liquid even when it is covalently attached as a monolayer or adsorbed multilayer on a surface. This new class of advanced materials could share the properties of true ionic liquids and the advantages of a solid support. In order to develop this approach, three different materials were prepared and tested in aldol reactions between acetone and several aldehydes: (a) proline (30 mol%) adsorbed on ionic liquid-modified silica gels; (b) proline (30 mol%) adsorbed on ionic liquid-modified silica gels with additional adsorbed ionic liquid; and (c) proline (30 mol%) adsorbed on unmodified silica gel containing adsorbed ionic liquid (Fig. 2.20).

Preliminary investigations carried out with these materials—using the reaction between acetone and benzaldehyde as model reaction—gave significant results. Proline immobilized on unmodified silica gel cointaining adsorbed ionic liquid [Bmim]BF_4 (type c) gave the aldol product in low optical purity. On the other hand, proline immobilized on covalently modified silica gel with or without adsorbed ionic liquid (type b or a) gave a higher ee value. In the case of material **89**/[Bmim]BF_4/pro [silica **89** containing adsorbed ionic liquid plus proline (see Fig. 2.21)], a better yield was obtained. These results indicated that the surface of silica gel must be modified by a covalently attached ionic liquid for a better outcoming of the reaction. Free acidic hydroxy groups on the surface of silica gel (see type c) probably participate in the coordinaton of the aldehydes, and then play the same role of the intramolecular hydrogen bond furnished by the proline carboxylic group. Further tests with a small set of aldehydes indicated that material **89**/[Bmim]

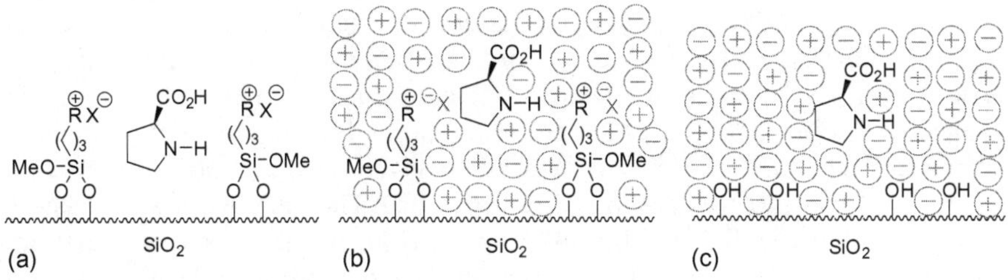

FIGURE 2.20. Different modes for L-proline immobilization on ionic liquid-modified silica gel.

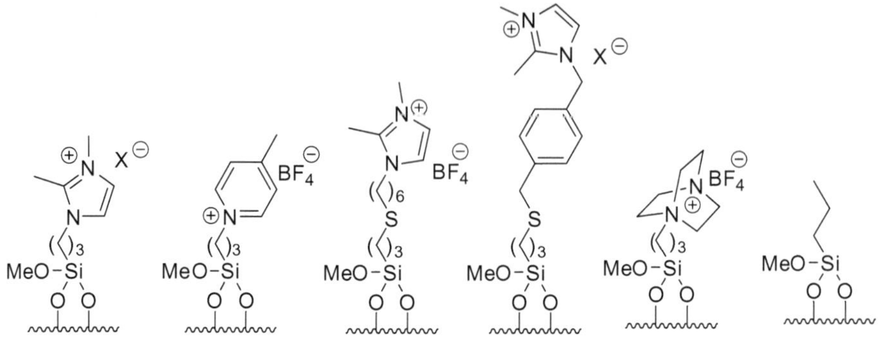

FIGURE 2.21. Structure of ionic liquid-modified silica gels.

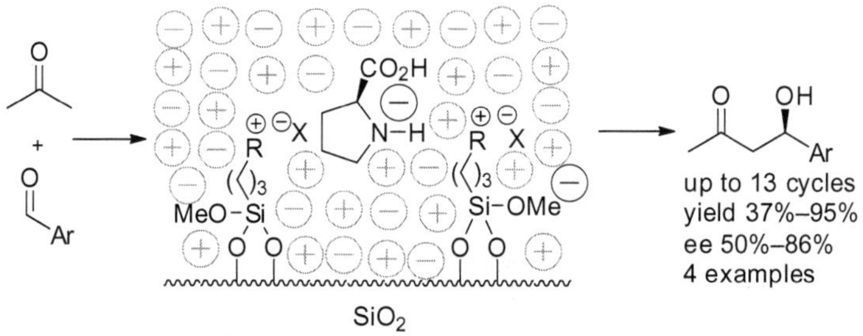

SCHEME 2.11. Aldol reactions performed with **1**/[Bmim]BF_4/pro system.

BF_4/pro gave good yields and ee values and could be used for up to four consecutive runs [75].

The surface of silica gel was modified with a set of anchored ionic liquids moieties (**89–97**) (Fig. 2.21). The use of the last two modified silica gels (**97–98**) as support for proline gave catalytic materials, which afforded aldol products in low optical purity, while the remaining modified silica gels gave high ee value. On a whole, the above results indicated that, in order to have high ee values, the surface of the silica gel must be modified with a covalently attached aromatic ionic liquid phase. Recycling studies were also carried out using two catalytic systems, **92**/pro and **89**/[Bmim]BF_4/pro. The catalytic system **92**/pro was not highly recyclable, while **89**/[Bmim]BF_4/pro system was successfully used for six cycles. Moreover, the support **89** was easily recovered and recharged with fresh [Bmim]BF_4/proline. Thus, the regenerated catalytic system was used for up to 13 cycles with unchanged results (Scheme 2.11).

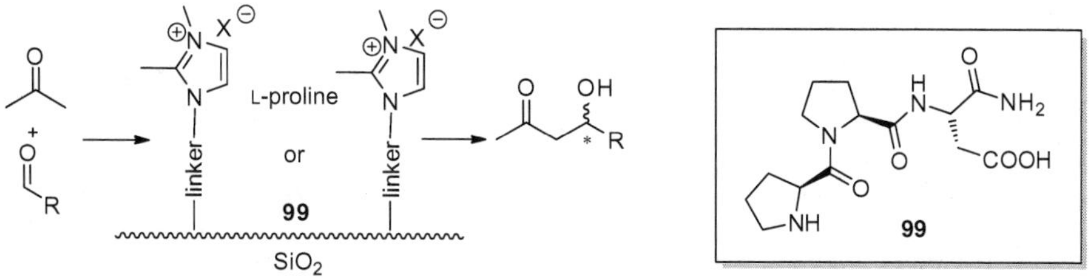

L-Proline: * = *R*; yields: 50%–98%, ee 66%–95%, 9 cycles, 7 examples
99 : * = *S*; yields 99%–57% I–IV cycle; ee 72%–75%, T = 25°C, support **94**
yields 91%–42% I–IV cycle; ee 82%–83%, T = –20°C, support **94**

SCHEME 2.12. Aldol reactions performed with **93–96**/pro and **94/99** systems.

Further studies were carried out using new ionic liquid-modified silica gels **93–96** (Scheme 2.12) [76].

Proline (30 mol%) was adsorbed on supports **93–96** from a methanol solution. Unlike the previous investigations, no additional adsorbed ionic liquid was used. Support **94** was the best one, as it was recyclable up to nine times without the need for regeneration. Aldol products were obtained in 50%–98% yield and 66%–95% ee. In particular, the ee values observed with some aldehydes were comparable to those obtained in pure ionic liquid and were higher than those obtained in pure acetone. These results suggested that the covalently attached monolayer of ionic liquid behaved as a bulk ionic liquid medium, but with the advantage of avoiding the need of expensive ionic liquids as solvents. Tripeptide H-Pro-Pro-Asp-NH_2 **99** (5 mol%) was supported on silica gels **94–96**, and the catalytic materials were tested in the reaction between acetone and 4-nitrobenzaldehyde, both at rt and at –20°C. The catalytic material **94/99** gave the best results being used in four consecutive runs with slightly decreased enantioselectivity with respect to the unsupported **99** when used at rt. Reproduction of good enantioselectivity was displayed in all reaction cycles, but usually decreased conversions were observed in the third and fourth cycle.

Aldol reactions were carried out in the presence of alkaline Al_2O_3 in the presence of 20 mol% of the dipeptide Pro-Trp as catalyst and 20 mol% of *N*-methylmorpholine or 1,4-diazabicyclo[2.2.2]octane as additive at –20°C. Interestingly, this approach allowed the use of equivalent molar amounts of aldehydes and ketones (300 mg of alkaline Al_2O_3 per 0.5 mmol of aldehyde plus 0.5 mmol of ketone). Among simple amino acids such as proline, thiazolidine-4-carboxylic acid, and tryptophan, as well as the dipeptides Val-Val, Pro-Phe, and Pro-Trp, the most effective catalytic activities were displayed by the latter dipeptide. The use of other solid media (neutral Al_2O_3, silica gel, bentonite) showed lower performances. Aldol products with cyclic ketones were obtained in high enantioselectivity (74%–98%, 13 examples), whereas acetone afforded aldols in 41%–94% ee (three examples). The use of a solid medium allowed the recycling of the organocatalyst by simply washing the

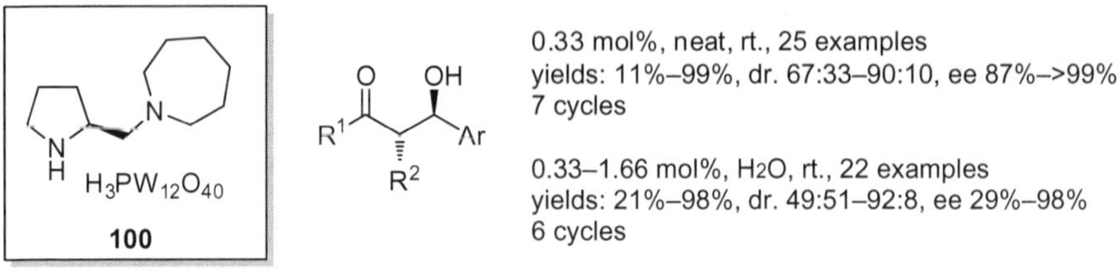

FIGURE 2.22. POM-supported catalyst **100** and its catalytic activity.

reaction mixture with ethyl acetate and leaving most of the catalyst on the alkaline Al_2O_3. The catalytic system was reused (three cycles) without decrease of activity by the addition of 10 mol% catalyst and base in the catalytic system before reuse each time. However, this approach constituted an obvious limit since the recovery is not quantitative, and fresh catalyst/base must be added in every cycle [77].

Another simple approach of noncovalently linked organocatalysts is based on the use of chiral amine–polyoxometalate (POM) hybrids [78]. These materials can be prepared by the slow addition of POM acid into a solution of the chiral amine in dry tetrahydrofuran followed by removal of the solvent. The resulting powders were washed with ethyl ether and dried. These materials were well dispersed in water, showing uniform micelle-like aggregates with 0.4 µm mean diameter. This surfactant-like property encouraged their use in water. After the screening of five chiral pyrrolidine diamines and seven POM acids, (*S*)-1-(pyrrolidin-2-ylmethyl)azepane supported on $H_3PW_{12}O_{40}$ (**100**) gave the best results and was employed in several aldol reactions both under neat conditions and in water (Fig. 2.22).

Under both conditions, aldol products were obtained with high levels of stereoselectivity and, more important, catalyst **100** worked efficiently with 0.33 mol% loading, which represented a very low amount for a supported organocatalyst. When catalyst **100** was used under neat conditions, it was recovered by precipitation with diethyl ether. The recovered catalyst was used in seven consecutive runs while still maintaining high enantioselectivity, but with a reduced activity (first cycle: 87% yield, 24 hours; seventh cycle: 60% yield, 30 hours). In the case of reactions performed under aqueous conditions, the product was extracted with diethyl ether and the catalyst directly used for the next run. In this case, the catalyst showed a more marked decrease in activity likely due to the slow aggregations of POM hybrid for extending reuse.

2.5.6. Ionic Liquid-Tagged Organocatalysts

Organocatalysts can be modified with ionic liquid moieties, which may help in recycling procedures. In addition, ionic liquid-modified organocatalysts can present enhanced stereoselectivity since the ionic liquid tags could act as chiral-induction groups. Ionic liquid-anchored organocatalysts may be more

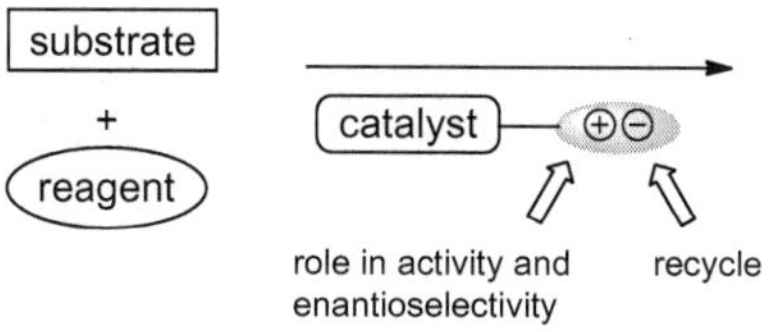

FIGURE 2.23. Roles of ionic liquid tags.

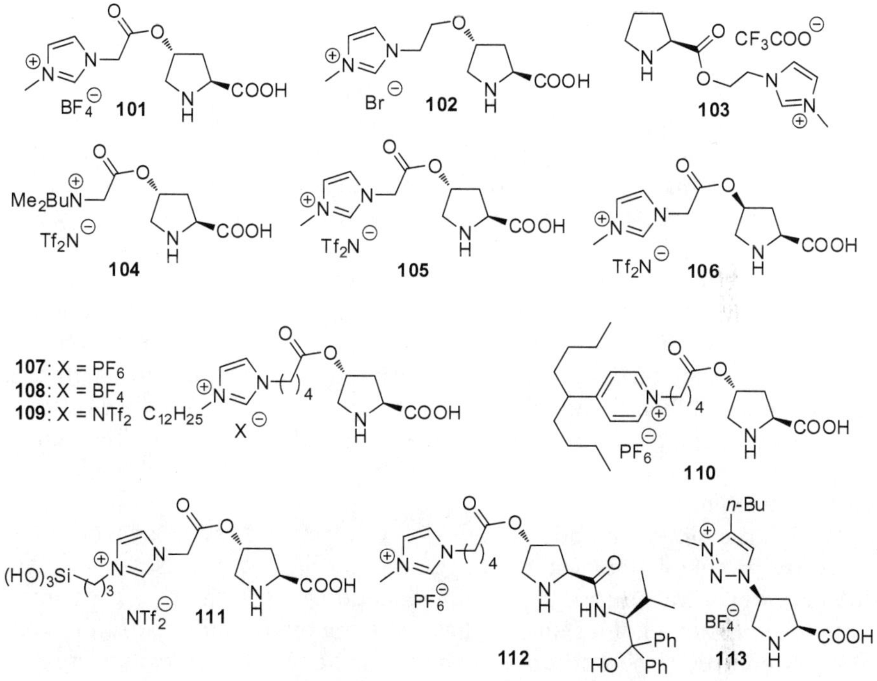

FIGURE 2.24. Ionic liquid-tagged proline-based organocatalysts **101–113**.

active with respect to the solid-supported ionic liquid organocatalysts because of their homogeneous nature (Fig. 2.23).

Solubilities of ionic liquid-anchored organocatalysts can be tuned by an accurate choice of cations and anions groups present in their structure. This approach allows for phase separation from organic as well as aqueous media. Following the proper extraction procedure, it could be possible to separate the catalytic molecule from the product and reuse it.

Catalysts **101–112** (Fig. 2.24) were prepared with the same strategy—nucleophilic displacement of the halogen atom followed by anion exchange and deprotection—as depicted in Scheme 2.13. Catalyst **113** was prepared following a longer procedure that involves the synthesis of the corresponding (2*S*, 4*S*)-4-azido-proline.

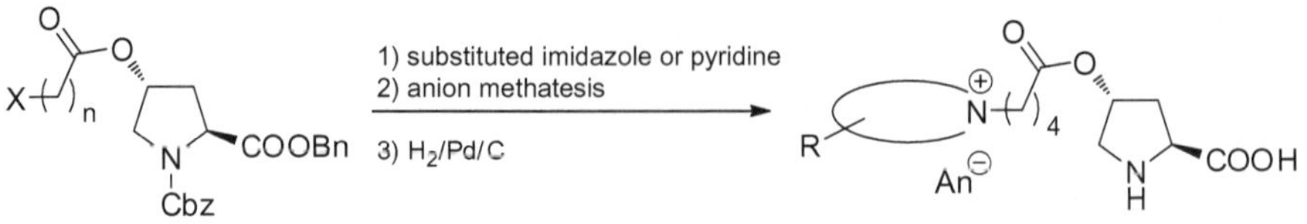

SCHEME 2.13. General scheme for the synthesis of 4-acyloxy ionic liquid-anchored prolines.

In 2006, after the development of the supported ionic liquid asymmetric catalysis [74], the first example of an organocatalyst (**101**) anchored to an ionic liquid was reported [79]. Aldol reactions performed in pure ketone employing organocatalyst **101** (30 mol%) gave quite comparable or superior results to those obtained in DMSO. Moreover, catalyst **101** showed higher activity and gave better enantioselectivities in neat acetone when compared to native L-proline. These results suggested that the ionic liquid-supported proline **101** was a more efficient and stereoselective organocatalyst than proline. In addition, the recyclability of catalyst **101** was also examined. After the reaction was completed, the mixture was concentrated, rinsed twice with dichloromethane, centrifuged, and then decanted. The dichloromethane solutions were concentrated to give the aldol product, whereas the residue contained the catalyst **101**. Four cycles were carried out with almost unchanged yield and enantioselectivity.

While the ionic liquid moiety in catalyst **101** was anchored to 4-hydroxyproline with an ester linkage, an ether linkage was used in catalyst **102** [80], which was in turn used in ionic liquid as a solvent, thus combining the advantages of a homogeneous phase with the opportunity to exploit solubility differences in the workup step. The amount of catalyst **102** was decreased with respect to **101** being used in 10 mol% and the reactions were carried out in [Bmim]BF_4. Reactions between acetone and several substituted benzaldehydes afforded the aldol products in good to high yields and high enantioselectivity. Moreover, both catalyst **102** and [Bmim]BF_4 were recovered and reused for six cycles with minor decreases in yields, but always with reproducible ee values.

The ionic liquid tag was also introduced as an ester derivative of simple L-proline (**103**) [81]. Catalyst **103** was used in the aldol reaction between acetone and several aldehydes. The same catalyst was previously reported by Miao and Chan and was used in the presence of *N*-methyl morpholine, producing an aldol product in low ee value [79]. Aldol products were obtained in good yields and good-to-high ee values. Moreover, the catalyst was recovered after evaporation of the acetone and extraction with diethyl ether, which afforded the aldol product, leaving the catalyst as residue. The catalyst was used four times with unchanged yields and enantioselectivity. Further studies were carried out by using ionic liquid-supported prolines **104** and **105** [82]. While catalyst **101** was employed in large amount (30 mol%), catalysts **104** and **105** were screened in lower amount (1–5 mol%). Four ionic liquids were used as solvents: two imidazolium-based ionic liquids, [Bmim]Tf_2N and [Bmim]Tf_2O, and two butylmethylpyrrolidinium-based ionic liquids, [Bmpy]Tf_2N and [Bmpy]Tf_2O. Preliminary tests were carried out on the reaction between acetone and 4-nitrobenzaldehyde. The best result was obtained when catalyst **104** was used in 5 mol% with 10 eq. of acetone in [Bmim]Tf_2N. However, the drawback of this procedure was represented by the difficult recovery and reuse of the catalyst at the concentration level used. After three cycles a dramatic drop in yield was observed. Several reactions were carried out, affording aldol products in better ee value with respect to native proline in DMSO.

Although recycling experiments were unsatisfactory, this work demonstrated that the ionic liquid-anchored organocatalysts can be used in lower catalytic amount than "classical" 30 mol% usually used for the proline-catalyzed aldol reactions.

In order to improve the stereoselectivities observed using catalyst **104** in ionic liquid, further studies were presented that employed catalyst **105** both in ionic liquid and water as reaction media. Passing from the use of **105** under homogeneous conditions in [Bmim]Tf_2N to heterogeneous conditions in water, both enantio- and diastereoselectivities increased significantly (ee increase up to 19%) [83].

Again, the important role that water plays in stereoselective organocatalytic reactions was evidenced [84]. Using catalyst **105**, several aldol reactions were carried out, affording products in high stereoselectivities in the case of cyclohexanone and cycloheptanone, whereas cyclopentanone gave an aldol product in lower stereoselectivity. In the presence of a small amount of water (ca. 4% v/v), aliphatic aldehydes (*iso*-butyraldehyde and cyclohexanecarboxaldehyde) also afforded the corresponding aldol products in good yields and excellent stereoselectivities. Catalyst **105** was recovered after the removal of the water/ketone mixture and extraction. After five cycles, a 10% decrease in yield was observed while stereoselectivities remained unchanged.

The possibility of having an organocatalyst that could work in lower catalytic amounts prompted the same authors to search for new structures and new reaction conditions than those previously reported. They reported that ionic liquid-anchored proline **106** having a 1,4-*cis*-configuration—unlike that of compound **105**, which has a 1,4-*trans*-configuration—was a powerful catalyst for the aldol reaction [85]. Moreover, instead of ionic liquid, water was again found to be a much better reaction medium. *Cis*-catalyst **106** was synthesized, considering that the ionic tag and its Tf_2N^- counterion are in spatial proximity to the reactive centers during the rate-determining enamine addition step. If an internal electrostatic stabilization of the transition state was in action, a reaction rate enhancement should be observed. Authors referred to this hypothesis as a "*cis* effect."

Using the reaction between cyclohexanone and 4-nitrobenzaldehyde as a model reaction, catalyst **106** was tested by employing different catalytic loadings (from 5 to 0.1 mol%) and different amounts of water (0.25, 1.2, 3.2, and 32 eq.). A comparison between catalysts **105** and **106** in 0.5 mol% catalytic loading and in the presence of 1.2 eq. of water was made. After 5 hours, both catalysts afforded the aldol product in excellent optical purity (ee: >99%, *anti/syn* 92:8–94:6), but with marked different activity (yield: **105**, 37%; **106**, 87%). These results indicated that ionic liquid-anchored catalyst **106** was highly efficient. It was then used in the aldol reaction with several aromatic aldehydes in 0.1–2 mol% and with aliphatic aldehydes in 10 mol% with excellent results.

Since the use of water as reaction medium allows, in many cases, products to be obtained with higher optical purity, it might be expected that more hydrophobic structural analogs of catalysts **101** would possess a much higher activity and stereoselectivity when used in water. To test this hypothesis, two new ionic liquid-anchored proline derivatives (**107** and **108**) bearing long-chain hydrocarbon groups at the imidazolium nitrogen atoms were prepared [86]. Compounds **107** and **108** had quite different solubilities in water. Tetrafluoroborate **108** gave a clear 5% aqueous solution, whereas hexafluorophosphate **107** gave a suspension under the same conditions. In the presence of the water-soluble catalyst **108**, the reaction did not take place. This result resembles those of aldol reactions catalyzed by water-soluble amino acids in water: low activity and stereoselectivity. On the other hand, good results were obtained with water-unsoluble catalyst **107**. These reactions afforded aldol products in low to high yield and high stereoselecivity but, disappointingly, the activity of catalyst **107** was not high. In fact, it was used in 30 mol%.

Recycling experiments were carried out. The aldol products were recovered by extraction with diethyl ether and the residue reused. After five cycles the catalyst **107** retained its activity and stereoselectivity.

Another highly hydrophobic ionic liquid-anchored compound was reported by the same authors and used as catalysts for the aldol reaction in water. Such catalysts were based on a pyridinium cation bearing a long hydrophobic group in the 4-position (**110**). In addition, compound **109**, carrying the (bistrifluoromethylsulfonyl)imide, anion was prepared and tested [87]. This catalyst, used in 15 mol%, gave excellent results in terms of yield and stereoselectivity in the reaction between cyclohexanone (3 eq.) and 4-nitrobenzaldehyde, even if a dramatic drop in activity was observed in the third cycle. Compound **110** showed excellent results, and it was used in eight consecutive runs without loss in activity and stereoselectivity. In addition to catalyst **110**, 4-alkylpyridinium-anchored-serine and threonine were prepared and were used in 20 mol%. These catalysts gave the corresponding aldol products in lower yield and stereoselectivity. Then, catalyst **110** was employed for a set of reactions with good results. Compared with catalyst **106**, which was used in a small amount (2–0.1 mol%) with excellent results, catalyst **110** was much less active

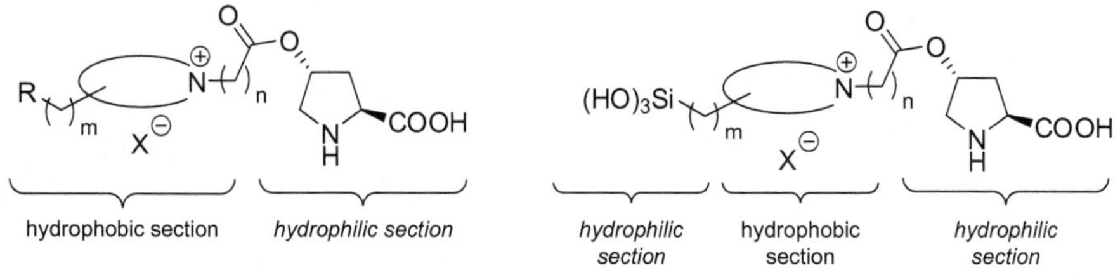

114 **115**

FIGURE 2.25. General structure of molecules of type **114** and **115**.

(15 mol%). It could be interesting to investigate if inversion of configuration at C-4 could enhance its activity.

Looking at the data reported in the literature, the most useful ionic liquid-anchored proline derivatives for aldol reactions in water could be represented in the general formula **114** (Fig. 2.25). The key factor for the successful application of these molecules is the hydrophobicity of the ionic liquid moiety. However, it has been questioned that maximizing the hydrophobicity of this part of the molecule could provide an inefficient recycling procedure at the extraction stage since the catalyst could be extracted in the same solvent. In order to overcome this possible drawback, a new type of ionic liquid-anchored organocatalyst with the correct amphiphilicity profile should be designed.

It has been proposed that molecules of type **115** (Fig. 2.25) could help in the recycling procedure [88]. The presence of the NTf_2 anion created the hydrophobic core of catalyst **111**, which made it active and stereoselective for its use in water, whereas the hydrophilic sections made possible its extraction in water. Catalyst **111** was used in 10 mol% in water to give aldol products in good yields and high-to-excellent stereoselectivity. Indeed, at the end of the reaction, excess of ketone and water were removed under reduced pressure and the solid residue partitioned between water and diethyl ether. Catalyst **111** was polar enough to be miscible with water while aldol products were extracted in the organic phase. The aqueous phase containing catalyst **111** was directly used for the next run. However, this approach did not give excellent results; in fact, in the fourth and fifth cycles, yield decreased while a decrease of the enantioselectivity was observed in the fifth cycle. Replacement of the NTf_2 anion in **111** with the chloride ion furnished a molecule completely soluble in water, which did not catalyze the aldol reaction between cyclohexanone and 4-nitrobenzaldehyde.

Since supported prolinamides showed high performances as catalysts in aldol reactions in water, ionic liquid-anchored prolinamide **112** was synthesized and tested in the above reactions [89].

Preliminary tests indicated that organocatalyst **112** was more active and stereoselective than the corresponding bromide salt. Indeed, bromide salt produced a clear 7% aqueous solution at rt, whereas the hexafluorophosphate **112**/H_2O mixture was a suspension under the same conditions. Using catalyst **112** in water at 3°C, several aldol reactions with cyclic ketones were carried out, affording products in high yields and stereoselectivities. Several methyl alkyl ketones were also employed in the reaction with 4-nitrobenzaldehyde at rt to give aldol products in high enantioselectivity. Recycling investigations showed a drop in activity in the fourth cycle, still maintaining the same level of stereoselectivity.

Catalyst **113** was employed in 20 mol% under neat conditions to afford aldol products in high yields and stereoselectivities. Products were recovered by extraction with cyclohexane leaving the catalyst as an oil, which was concentrated and reused. After five cycles, although stereoselectivities remained unchanged, yield decreased due to a possible leaching of the catalyst in the extraction procedure [90]. All the performances of ionic liquid-tagged organocatalysts **101–107** and **110–113** are collected in Table 2.6.

2.5.7. Fluorous Organocatalysts

Fluorous sulfonamides **116** [91] and **117** [92] were investigated as recyclable catalysts for the asymmetric aldol reaction (Fig. 2.26). Because of the hydrophobicity of such molecules, aldol reactions were carried out in water or brine. Catalyst **116** was employed in 10 mol% in water without additives, whereas catalyst **117** needed the additional use of trifluoroacetic acid (TFA). Both catalysts were recovered by fluorous solid-phase extraction (F-SPE) in >90% yield. Stereoselectivity was maintained upon recycling, whereas a loss in activity was observed, especially for **116**. Indeed, to maintain the yield, reaction times were progressively increased.

2.5.8. Noncovalently Supported Organocatalysts

The main advantage of noncovalently supported organocatalysts is represented by the easier immobilization approach, which does not require additional synthetic steps. Two examples are presented: host–guest complexes between *β*-cyclodextrin and 4-substituted proline derivatives (Fig. 2.27) and entrapment into montmorillonite (Scheme 2.14). Catalyst **118** was easily obtained by refluxing a solution of (4*S*)-phenoxyproline and *β*-cyclodextrin and subsequent removal of the solvent [93]. Catalyst **119** was formed employing an adamantane proline derivative [94].

Both catalysts were used in 10 mol% loading. Catalyst **118** was used in the aldol reaction of acetone and substituted benzaldehydes, whereas catalyst **119** was used in the aldol reaction of cyclohexanone and substituted benzaldehydes. In the latter case, the use of the more hydrophobic ketone allowed the use of water as a reaction medium. Recycling experiments indicated no loss in activity and stereoselectivity after four cycles. In both cases, aldol products were obtained in good to high yields and stereoselectivity.

TABLE 2.6. Summary of Aldol Reactions Carried Out with Ionic Liquid Tagged Proline-Based Organocatalysts 101–107, 110–113

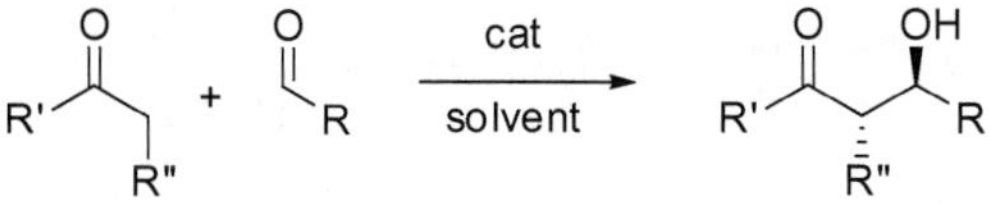

Entry	Catalyst (mol%) Solvent	Product	Yields (%)	d.r. (*anti*/*syn*) ee (%)	# Cycles	Ref.
1	**101** (30) acetone	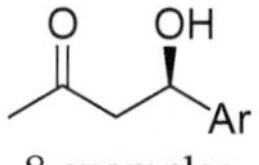 8 examples	40–92	— 64–85	4	[79]
2	**101** (30) DMSO	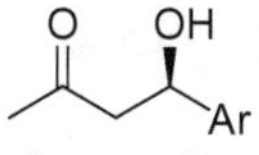 8 examples	46–83	— 60–87	—	[79]
3	**102** (10) [Bmim] BF_4	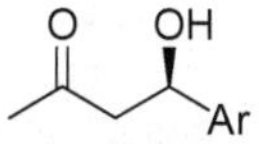 14 examples	53–94	— 64–93	6	[80]
4	**103** (10) Acetone	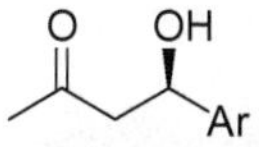 6 examples	54–94	— 67–92	4	[81]
5	**104** (5) [Bmim] Tf_2N	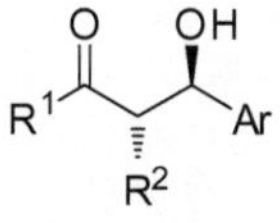 7 examples	35–78	67:33–80:20 75–94	3	[82]
6	**105** (5) H_2O	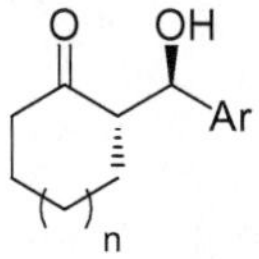 11 examples	25–98	58:42–98:2 80–99	5	[82]
7	**106** (0.1–10) H_2O	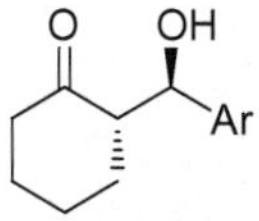 13 examples	44–>99	80:20–>99:1 70–>99	—	[85]

(*Continued*)

TABLE 2.6. (*Continued*)

Entry	Catalyst (mol%) Solvent	Product	Yields (%)	d.r. (*anti*/*syn*) ee (%)	# Cycles	Ref.
8	**107** (30) H_2O	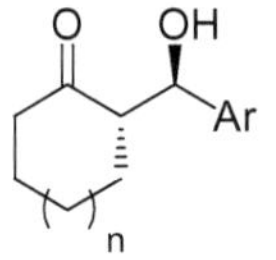 13 examples	20–>95	84:16–97:3 80–>99	5	[86]
9	**110** (15) H_2O	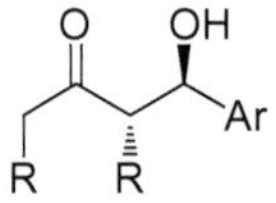 10 examples	38–97	80/20–98/2 92–>99	8	[87]
10	**111** (10) H_2O	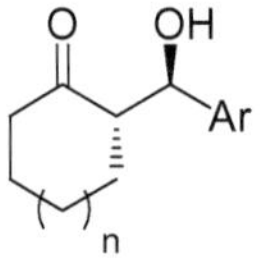 10 examples	25–98	70:30–>99:1 82–>99	5	[88]
11	**112** (1–5) H_2O	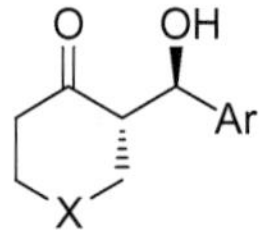 8 examples	64–99	62/38–99/1 92–99	4	[89]
12	**112** (5) H_2O	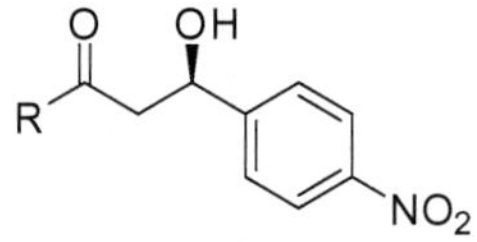 5 examples	48–95	— 82–97	—	[89]
13	**113** (20) Neat	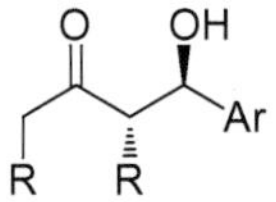 8 examples	84–99	67:33–99:1 76–99	5	[90]

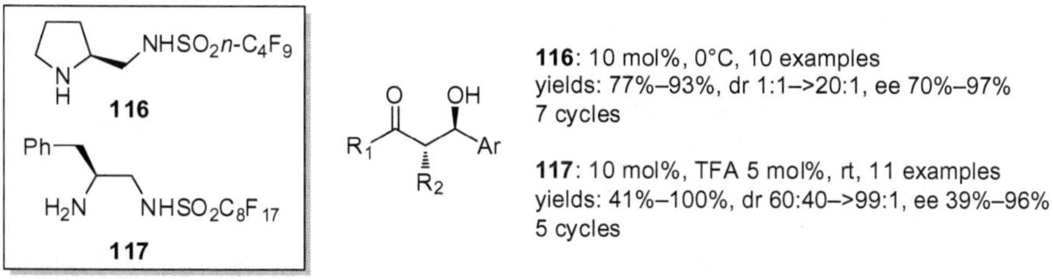

FIGURE 2.26. Fluorinated organocatalysts for aldol reaction and their performances.

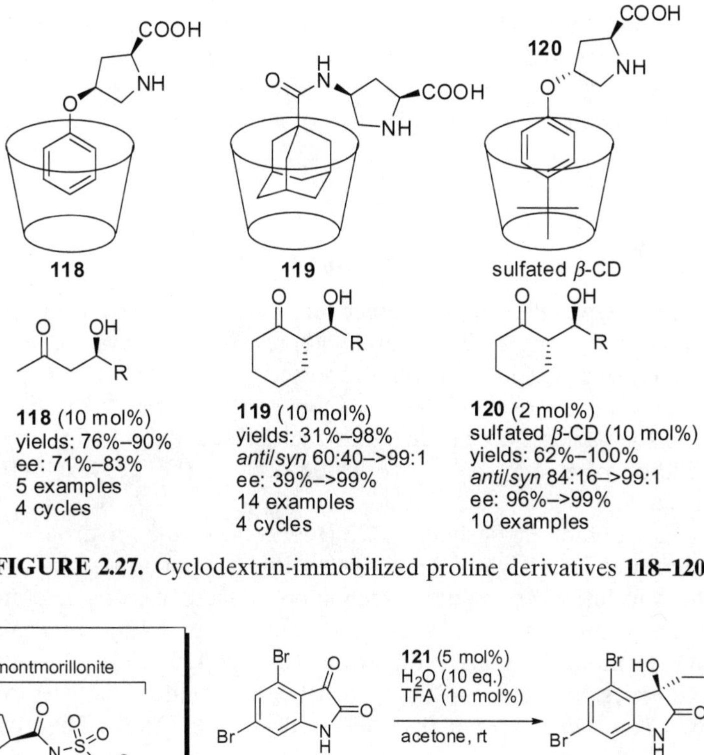

FIGURE 2.27. Cyclodextrin-immobilized proline derivatives **118–120**.

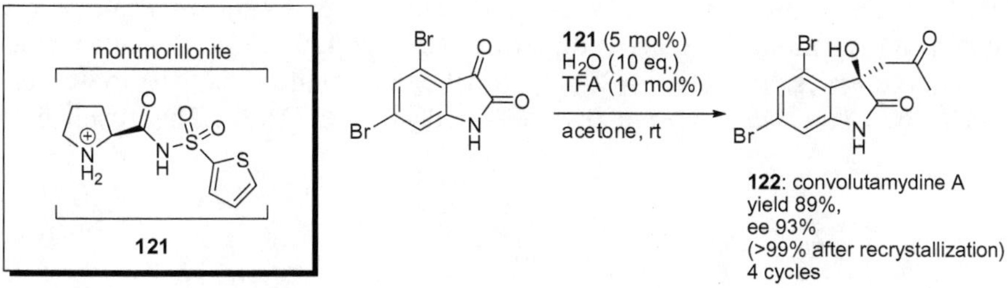

SCHEME 2.14. Asymmetric synthesis of convolutamydine A.

Trans-4-(4-*tert*-butylphenoxy)-proline **120** was used as a catalyst (2 mol%) in the aldol reaction between cyclohexanone and several substituted benzaldehydes in water in the presence of sulfated *β*-cyclodextrin (10 mol%). The authors concluded that the aldol reaction occurred in the water phase where organocatalyst **120** resided with sulfated *β*-CD. Noticeably, this approach allowed the use of stoichiometric amount of cyclohexanone but, disappointingly, no recycling studies were carried out [95]. Entrapment of *N*-(2-thiophenesulfonyl)prolinamide into montomorillonite was achieved by simple ion-exchange reaction [96].

The catalytic material **121** was used in 5 mol% in the presence of water (10 eq.) and TFA (10 mol%) (Scheme 2.14). Several aldol reactions between various isatins with acetone or acetaldehyde were carried out, affording products in high yields and enantioselectivity (up to 96% ee). This approach was successfully employed in the asymmetric synthesis of convolutamydine A (**122**) (Scheme 2.14), which was obtained without column chromatography purification and in excellent ee value after recrystallization. The catalytic material **121**, which was recovered by filtration, retained its activity and enantioselectivity for four cycles. In a further recycling investigation on a different isatin, catalyst **121**, some degree of activity loss was observed in the third and fourth cycles.

2.5.9. Nonsupported Organocatalysts

Here we report several examples of recoverable nonsupported catalysts, which can be recovered using three different approaches: decantation, extraction, and chromatography. Although nonsupported catalysts are less expensive—since no additional costs related to their synthesis are required for the immobilization—procedures such as extraction and chromatography represent an additional cost in the recovery step because the recovery may be not quantitative.

L-Proline amphiphilic derivative **123** (Fig. 2.28) formed hydrogels above a concentration of 2 mM after dissolution in hot water and sudden cooling at 25°C with sonication (1 minute). Such supramolecular catalytic system was used for the aldol reaction between cyclohexanone and 4-nitrobenzaldehyde. When the reaction was carried out with hydrogel **123** (20 mol%) in toluene with large excess of ketone (20 eq.) at 5°C, the enantioselectivity observed (88%) was much higher than that obtained at 25°C (18%). This result was

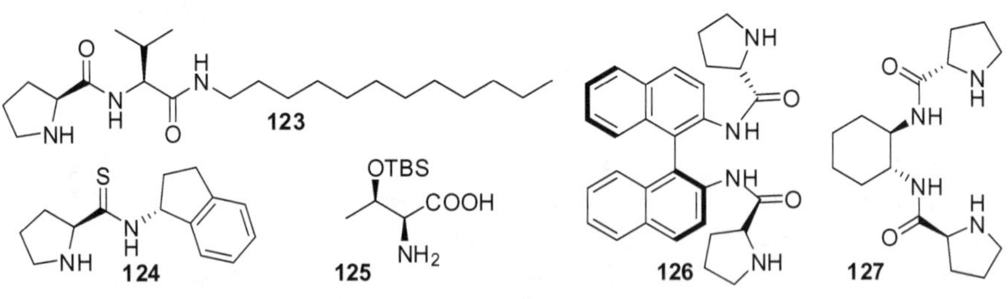

FIGURE 2.28. Recyclable molecular organocatalysts.

ascribed to the restricted mobility of the reactants as well as to the structural rigidification of the supramolecular aggregates. After decantation of the organic layer, the hydrogel **123** was reused for an additional two cycles with unchanged results (yield >99%, *anti/syn* 92:8, ee 90%) [97].

Good results were observed when prolinethioamide **124** (5 mol%) was used under neat conditions in the aldol reaction of 4-nitrobenzaldehyde with a set of ketones. This catalyst was easily recovered, but not quantitatively, after extractive acid/base workup [98].

Silyloxy-protected L-threonine **125** catalyzed the aldol reaction between cyclohexanone and aromatic aldehydes in water, in 2 mol% loading, affording products in high enantioselectivity (91%–99%) and good to high d.r. ratio (4:1–19:1). This catalyst was recovered by extraction and reused in up to three runs with unchanged results [99].

(S_a)-Binam-L-prolinamide **126** (5 mol%) was used in combination with benzoic acid (10 mol%) under three different methods in the aldol reaction between several ketones and aldehydes to afford product in useful levels of stereoselectivity. In this case, recovery of the catalyst was possible after extraction and column chromatography. Catalyst **126** was recovered in 86% yield [100]. Catalyst **127** was used in 10 mol% in [bmim]BF_4/H_2O (1:1 v/v) and did not reach useful levels of enantioselectivity. After extraction of the product, the reaction was repeated in the same system, and after five cycles, stereoselectivity remained unchanged, although reaction times were increased to maintain the same levels of yield [101].

A different approach consisted of the use of PEG-400 as reaction medium for the reaction between acetone and aldehydes using native proline (10 mol%). The reactions were faster than in DMSO, giving good isolated yields (58%–94%). In several cases, ee values were lower than those obtained using proline in DMSO.

This methodology allowed the recovery of proline. At the end of each cycle the product was extracted with diethyl ether while PEG + proline was reused for the subsequent cycle. After 10 cycles, a slowly decreasing yield was observed (94%–84%), while enantioselectivity was maintained [102].

2.6. MICHAEL REACTION

2.6.1. Covalently Linked on Insoluble Supports: Polystyrene and Silica

Supported organocatalysts on polystyrene, such as pyrrolidine derivatives and diarylprolinols, were used in the asymmetric Michael reaction. Catalysts in Figure 2.29 were prepared following the postmodification approach. Pericás et al. prepared and tested catalysts **128** and **129** under several conditions (neat; solvents: water, toluene, DMF; additives: TFA, DiMePEG). The best conditions were found when catalyst **128** was employed in water/DiMePEG with

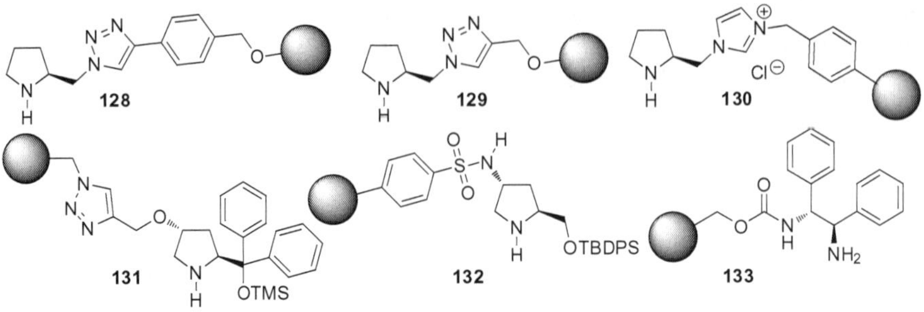

FIGURE 2.29. Polystyrene-supported organocatalysts for Michael reaction obtained by resin postmodification.

good results and used in three consecutive runs [103]. On the other hand, Miao and Wang soon after reported that catalyst **129** gave excellent performances when used under neat conditions in the presence of TFA [104]. This catalyst was used in 10 consecutive runs without decrease in activity and stereoselectivity. In both catalysts, the triazole ring had the roles of linker and of shielding of the two faces of the enamine from the nitrostyrene acceptor approach. In catalyst **130**, the latter roles were played by the imidazolium moiety. The Michael addition of cyclohexanone to nitrostyrene proceeded smoothly, giving high yield and excellent enantioselectivity, in polar and apolar solvents, except those from diethyl ether, dichloromethane, and tetrahydrofuran. However, neat conditions slightly improved the yield. Other anions such as BF_4^- and PF_6^- had no effects on activity and stereoselectivity. Catalyst **130** displayed broad substrate applicability with high yields and stereoselectivities. The recyclability was tested in up to eight consecutive trials without significant loss in activity and stereoselectivity [105].

The triazole linker was also used in the α,α-diarylprolynol silyl ether catalyst **131** (Fig. 2.29), which displayed catalytic activity and enantioselectivity comparable to the best homogeneous catalysts in the Michael addition of aldehydes to nitroolefins. Under the adopted conditions, only linear aldehydes underwent Michael addition, whereas bulky aldehydes or ketones were unreactive. Such substrate selectivity must result from the topology of the reaction cavity and, in this sense, material **131** behaved as an enzyme mimic. Upon use of catalyst **131**, deactivation occurred due to the hydrolysis of the labile silyl ether. To overcome such deactivation, it was found that a treatment with trimethylsilyl *N,N*-dimethylcarbamate in hexane/acetonitrile, followed by washing to remove the excess of silylating agent, led to the selective protection of the hydroxyl group with full recovery of catalytic activity. Regeneration of the catalyst after each cycle allowed the use of **131**, in recycling experiments, up to six cycles with unchanged results [106].

The Michael addition of several aldehydes and ketones to nitrostyrenes was also carried out by using supported organocatalysts **132**. It was prepared by a reaction between MP-sulfonyl resin and *O*-TBDPS-protected *trans*-4-aminoprolinol. In this case, the bulky $CH_2OTBDPS$ group shielded the *si*-face of the enamine double bond, whereas the NH sulfonamide proton could provide the hydrogen bond with the acceptor nitrostyrene, which activated the latter effectively. Catalyst **132** was successfully employed and recycled up to six times [107].

In addition to pyrrolidine derivatives, primary amines were also used as catalysts for Michael addition reactions. Wang polystyrene resin having polymer-bound benzylic hydroxyl groups was used to immobilize chiral primary amines. Catalyst **133** was prepared by anchoring the amine through a carbamate linker. As for catalyst **132**, catalyst **133** could simultaneously activate the nucleophile and the electrophile. The primary amine moiety could act as a Lewis base, whereas the carbamate hydrogen atom could act as an H-bond donor. Catalyst **133** gave better results compared with other catalysts having two H-bond donors in the proximity of the amine (urea linker) and catalysts having different tether lengths to the support. Moreover, the use of benzoic acid as an additive strongly enhanced the activity. Catalyst **133** was tested only in three Michael addition reactions with good results and was used in three consecutive runs [108]. All the data obtained with the above catalysts are summarized in Table 2.7.

In addition to pyrrolidine-based catalysts, thiourea-based catalysts have been largely employed for Michael reactions. A chiral thiourea-based catalyst was bound to a carboxypolystyrene HL resin and TentaGel carboxy resin, but these catalysts showed lower catalytic activities as compared to the corresponding soluble catalyst [109].

Diarylprolinols **134–137** were prepared following the bottom-up approach (Fig. 2.30). Studer et al. reported the immobilization of homogeneous catalysts into polymer nanofiber via nitroxide-mediated polymerization followed by the electrospinning process. Such methodology was applied to the synthesis of catalyst **134** (Mn = 7500 g/mol; polydispersity index = 1.30). This fibrous catalyst was tested in the asymmetric Michael addition of dimethyl malonate to cinnamaldehyde in EtOH. The chemical yield was 42% and the ee 92%. Immobilization caused a decreased activity since the corresponding homogeneous catalyst afforded the same product in 84% yield and 94% ee. Recycling experiments (10 cycles) showed that the initial activity was maintained only in the second run and then decreased in the following runs. The decreased activity was not ascribed to the desilylation process, but to small changes in the macrostructure of the fibrous system [110].

Standard suspension copolymerization of a C-4 methacrylate-substituted α,α-diarylprolinol with methyl methacrylate and 2 mol% of ethyleneglycol dimethylacrylate (EGDMA) gave microporous polymer beads **135a**, useful for reaction in nonpolar solvents. By substituting the methyl methacrylate/EGDMA with a 50:50 mixture of PEG400 methyl ester methacrylate and PEG600 dimethacrylate, a swellable macroporous resin **135b** for reactions in more polar solvents (lower alcohols, MeCN, aqueous solvents) was prepared [48].

TABLE 2.7. Performances for Supported Organocatalysts Employed in the Asymmetric Michael Reaction

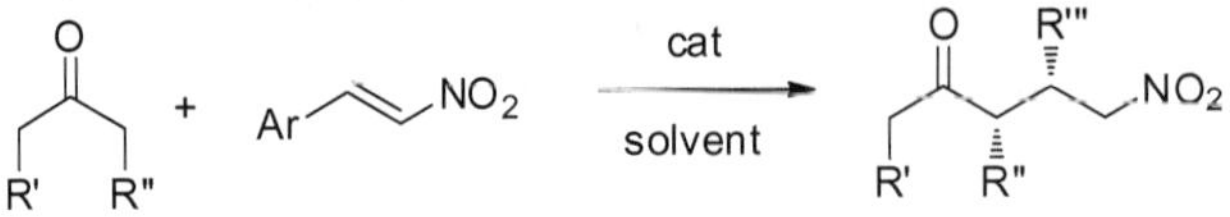

Entry	Catalyst (mol%) Solvent	Product	Yields (%)	d.r. (syn/anti) ee (%)	# Cycles	Ref.
1	**128** (10) DiMePEG (10) H_2O	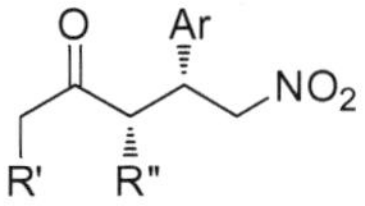 14 examples	40–85	89:11–>99:1 26–>99	3	[103]
2	**129** (10) TFA (2.5) Neat	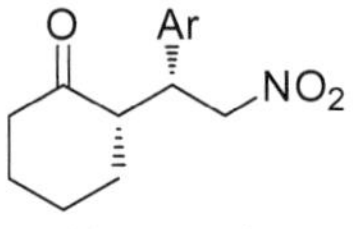 10 examples	85–97	97:3–>99:1 88–>99	10	[104]
3	**130** (10) Neat	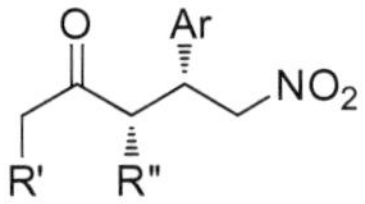 22 examples	69–98	75:25–>99:1 51–>99	8	[105]
4	**131** (10) CH_2Cl_2	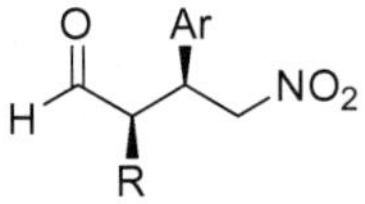 8 examples	44–98	75:25–>99:1 90–>99	6	[106]
5	**132** (10) *p*NO_2BA (10) H_2O	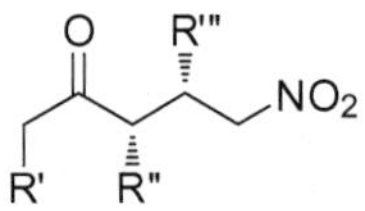 27 examples	39–100	5:95–97:3	6	[107]
6	**133** (30) BA (10) Toluene	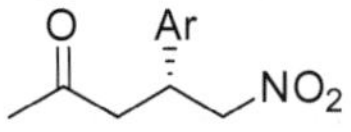 3 examples	28–60	— 83–92	3	[108]

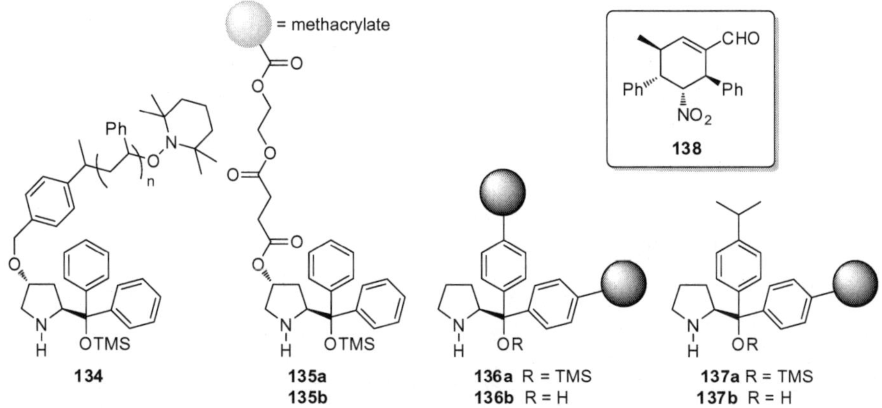

FIGURE 2.30. Supported organocatalysts for Michael reaction obtained with the bottom-up approach.

The cross-linked catalyst **136a** was prepared by suspension copolymerization of the Boc-protected monomer (having two styril moieties) with styrene followed by deprotection and treatment with a silylating agent. The pendant-bound catalyst **137a** was prepared in a similar way, by suspension copolymerization of the Boc-protected monomer (having one styril moiety) with styrene/DVB (2 mol%) [111].

Catalysts **135a**, **136a**, and **137a** were benchmarked in the asymmetric Enders cascade reaction that preceded a tetrasubstituted cyclohexene carbaldehyde with four stereogenic centers (**138**) via an enamine and iminium triple cascade reaction. Reactions were carried out in toluene, and only catalysts **135a** and **137a** were active. It is likely that the more constrained steric environment in **136a** is less compatible with the construction of the bulky three-component product, whereas the use of **135b** in MeCN, as a solvent, afforded only an intermediate product. Cross-linked catalyst **136a** was also employed in several Michael addition reactions of aliphatic aldehydes to nitrostyrene in benzene, giving high yields (>80%) and high stereoselectivities (*syn*/*anti*, >95/5; ee, >87%). This catalyst was slightly superior to **137a** and to the unsupported catalyst. Unfortunately, activity of catalyst **136a** dropped upon reuse even after complete deprotection and careful reprotection. These results are in agreement with the fact that diarylprolinol systems are not so robust for reuse. Only the activity of catalyst **131** was fully restored after regeneration. For the Michael addition reaction, simple pyrrolidine-based catalysts (**128–130**) appear more robust.

Cinchona alkaloids were also employed as nonionic catalysts. In these cases, the two most commonly used connection sites to the polystyrene backbone were the double bond and the oxygen atom at C-9.

Two different spacers were used to connect the alkaloid to the polystyrene resin (1% DVB cross-linked) in **139** and **140** (Fig. 2.31). These materials were

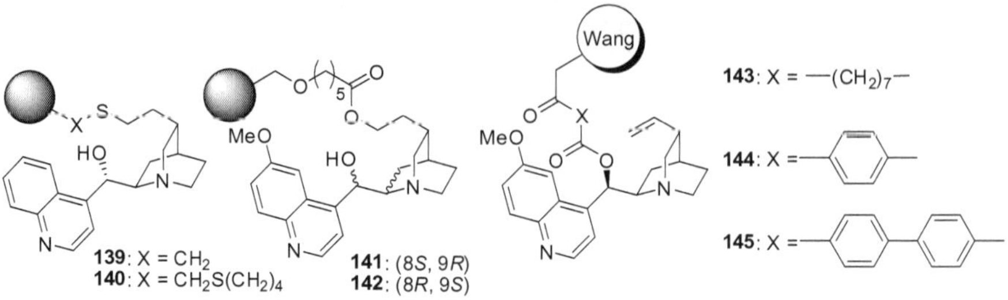

FIGURE 2.31. Resin-supported *Cinchona* alkaloids **139–145** for the Michael reaction.

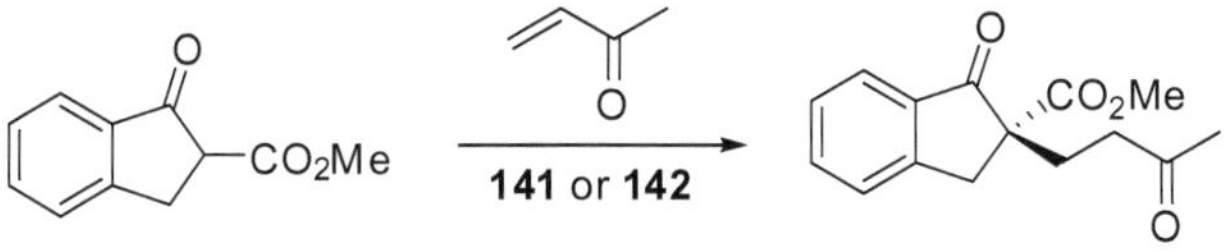

SCHEME 2.15. Michael addition reaction.

tested in the addition of thiols and thiolbenzoic acid to cyclohexenone and nitrostyrene. Catalyst **139** was the most interesting, achieving a 45% ee value in the reaction between 4-thiocresol and cyclohexenone. This catalyst was recovered and reused in three consecutive cycles with unchanged results [112]. The quinine-based catalyst **141** and the quinidine-based catalyst **142** were used, among others catalysts, in the Michael reaction reported in Scheme 2.15.

Catalyst **141** afforded the product in 87% ee and 85% yield, whereas catalyst **142** gave lower performances (39% ee). Other catalysts having spacers of different length and nature also gave lower ee's. Interestingly, no matter the alkaloid, the configuration of the product was *R* [113].

Lectka et al. reported that benzoylquinine (10 mol%) catalyzed the reaction between phenylketene and an imino ester to give the corresponding *β*-lactam in high ee and d.r. Under homogeneous conditions the catalyst was not easily recovered. The column asymmetric catalysis methodology was then applied, which allowed the catalyst to perform more than 60 cycles. In order to get the supported catalyst, the oxygen atom at C-9 was used to link quinine to the Wang resin by means of three different spacers to give catalysts **143–145** (Fig. 2.31). The three linkers were different in length and shape. The idea was that short or "floppy" linkers would allow close proximity of the catalytic moiety with the resin and thus lower the rate and affect the selectivity in an unpredictable way.

Catalyst **145**, having the more rigid and longer linker, gave the highest d.r. (13:1) and high ee (90%), very similar to catalyst **144**, which afforded the *β*-lactam in 93% ee and 10:1 d.r. Catalyst **143**, having the longer and "floppy" linker, gave a much lower diastereoselectivity (2:1) with 87% ee. Interestingly,

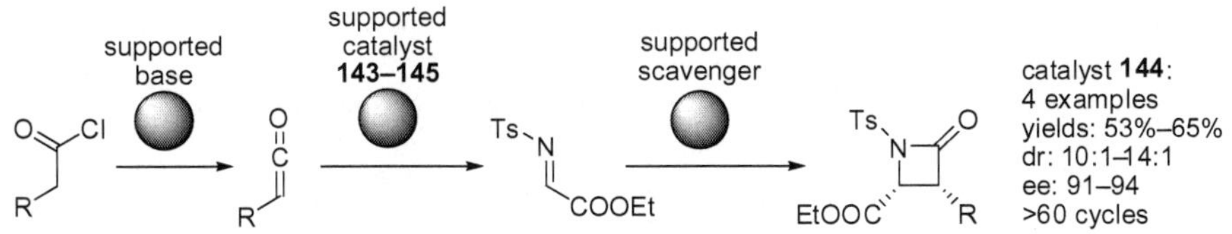

SCHEME 2.16. Asymmetric synthesis of β-lactams.

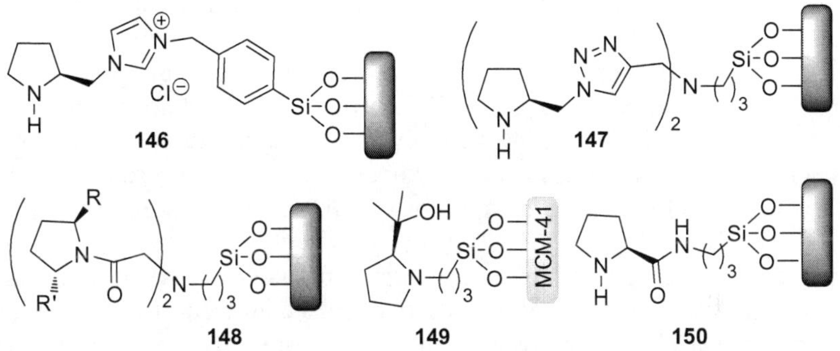

FIGURE 2.32. Catalysts for Michael addition reactions: catalysts **146–149** were prepared via grafting; catalyst **150** was prepared via the sol-gel process.

it was necessary to run a small number of cycles, starting from fresh resins, before consistent results were obtained. Once the resins were aged through 5–10 cycles, they were employed with no significant loss in yield and selectivity for over 60 cycles. The column asymmetric catalysis methodology with catalyst **144** was used to give the β-lactams good yields and high stereoselectivity (Scheme 2.16) [114].

Chiral organocatalysts supported on silica gels were deeply tested as recyclable materials for Michael addition reactions. Again, the two possible approaches—grafting and sol-gel processes—were investigated. Following the grafting approach, several organocatalysts were synthesized (**146–149**) (Fig. 2.32). Usually, a chiral pyrrolidine group represents the chiral moiety in the catalyst. In compounds **146** and **147**, the imidazolium and the 1,2,3-triazole moieties played both the roles of linker and of shielding of the one of the two faces of the intermediate enamine.

Organocatalyst **146** was benchmarked in the addition of cyclohexanone to nitrostyrene in a set of solvents. In each case, stereoselectivity was very high, whereas the best activity was observed when the reaction was carried out under neat conditions. Then, employing 10 mol% of catalyst **146**, several Michael reactions were carried out, affording the final products in high yields and stereoselectivities (Scheme 2.17). Only cyclopentanone and acetone gave lower stereoselectivity. Moreover, after six consecutive runs the activity and stereoselectivity of **146** did not decrease [115].

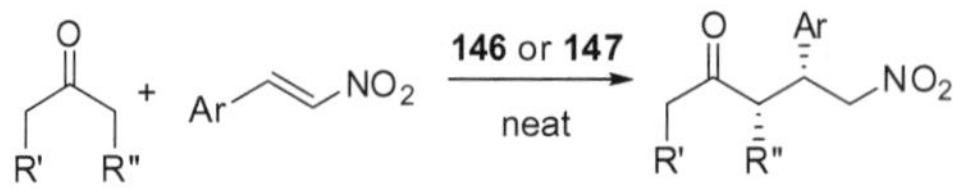

146 (10 mol%): 16 examples, yields: 84%–94%; dr: 75:25–>99:1; ee: 41%–>99%; 6 cycles

147 (10 mol%): 14 examples, yields: 55%–98%; dr: 95:5–99:1; ee: 40%–93%; 4 cycles

SCHEME 2.17. Michael addition reaction.

Slightly lower performances were obtained when catalyst **147** was employed under similar conditions. However, recycling (four cycles) confirmed the reusability of such systems [116].

Chiral pyrrolidine-based catalysts of type **148** were supported on silica, MCM-41, and delaminated zeolite, ITQ-2, whereas a prolinol-based catalyst was supported on MCM-41 (**149**). These catalytic materials were tested in the Michael addition of ethyl 2-oxocycloalkenecarboxylates to acrolein and their activity and selectivity were compared with the corresponding unsupported catalysts [117].

In all cases, much higher conversions were obtained with supported samples in comparison with corresponding homogeneous counterparts. Depending on the support, variable enantioselectivities were obtained, from very low to moderate (5%–60%). The best results were observed with catalyst **148** supported on MCM-41 and **149**. The enantioselectivities displayed by these catalysts were higher compared to those obtained by the same authors when chinconine and chinconidine supported on MCM-41 were used [118]. L-Proline was immobilized onto mesoporous silica starting from (*S*)-*N*-(3-triethoxysilylpropyl)pyrrolidine-2-carboxamide via a sol-gel approach under microwave irradiation. Three different proline loadings, with respect to SiO_2, were used (2.5%, 5.0%, and 7.5%). These catalysts were employed in the Michael addition of diethyl malonate to cyclohexenone. Catalyst **150**, having the highest proline loading, displayed better conversion and higher enantioselectivity (78% ee). In addition, each supported proline catalyst (2.5%, 5.0%, and 7.5%) gave Michael adduct in much higher enantioselectivity compared to the homogeneous proline. Such enhanced enantioselectivity was explained by the confinement effect due to the pore structure of the material. Catalyst **150** was used twice, showing the same enantioselectivity and slightly decreased activity [119].

2.6.2. Covalently Linked on Soluble Supports: PEG

While a thiourea-based catalyst supported on cross-linked polystyrene resins led to low yields in several Michael addition reactions, much better results were obtained when the same catalyst was supported on soluble PEG support [109].

The PEG-supported catalyst **151**, with a purity of about 80%, was used in the Michael reaction of nitrostyrene with diethyl malonate to give the adduct in good yield and slightly lower enantioselectivity compared with the

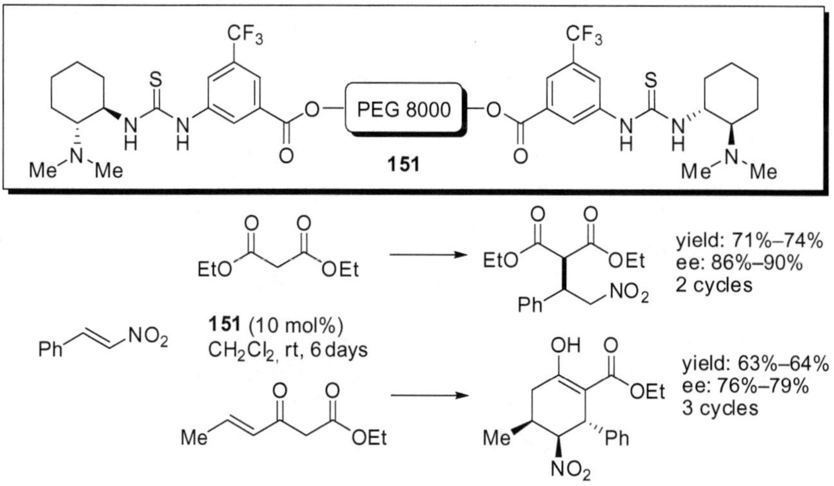

SCHEME 2.18. Michael addition reaction.

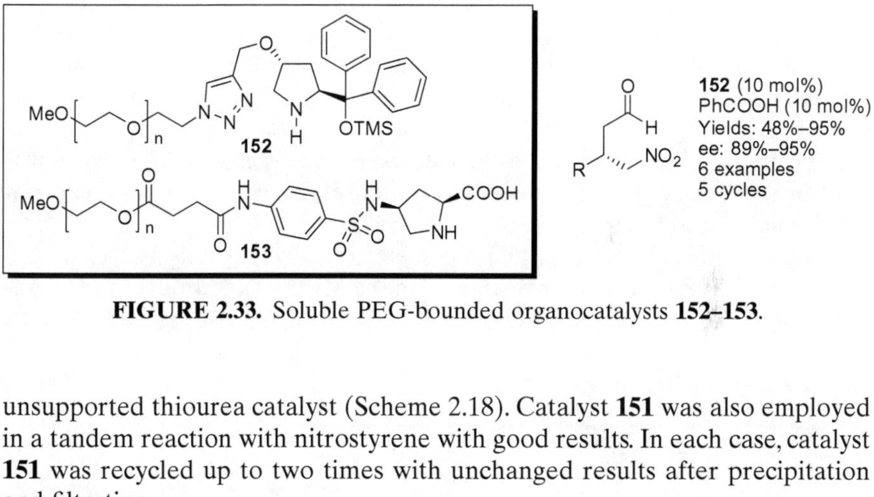

FIGURE 2.33. Soluble PEG-bounded organocatalysts **152–153**.

unsupported thiourea catalyst (Scheme 2.18). Catalyst **151** was also employed in a tandem reaction with nitrostyrene with good results. In each case, catalyst **151** was recycled up to two times with unchanged results after precipitation and filtration.

TMS-protected diarylprolinol supported on MeOPEG 5000 through a 1,2,3-triazole linker **152** was used in the Michael addition of nitromethane to α,β-unsaturated aldehydes (Fig. 2.33) [120].

The catalytic moiety of such molecules was the same of the polystyrene-supported catalyst **131** (Fig. 2.29). Diarylprolinol **152** was tested in six Michael reactions to afford the corresponding adducts in good yields and high enantioselectivities. The reactions were carried out in methanol in the presence of benzoic acid as additive. In water, the fully soluble catalyst **152** proved to be inefficient. After five cycles, although the enantioselectivity was maintained, the activity of catalyst **152** decreased. Such recycling experiments shed some light on the loss of catalytic activity of diarylprolinol silyl ethers. Usually, deactivation is attributed to a desilylation process. However, recycling experiments carried out with a methyl-protected diarylprolinol showed a similar loss in activity. Taking advantage of the solubility of the MeOPEG-supported catalyst, NMR experiments were carried out showing that deactivation ensues from unreleased product. This observation allowed the development of a catalyst regeneration procedure. A simple stirring of the α,β-unsaturated aldehyde restored the catalyst's initial activity.

Proline sulfonamide **153** supported on PEG-5000 was used in several Michael additions between ketones or aldehydes and nitroolefins. The recovery was performed by precipitation and filtration (average recovery yields: 80%–90%). The recycling study, performed using cyclohexanone as a ketone, showed a dramatic decrease both in yield (94%–24%) and in enantioselectivity (60%–<10%) after four cycles [121].

2.6.3. Dendrimers

Pyrrolidine-triazole based dendritic catalysts **154a–c** (first, second, and third generation) were investigated in the Michael addition of ketones to nitroolefins under neat conditions in the presence of TFA as cocatalyst (Fig. 2.34). The third-generation catalyst **154c** provided the best results in terms of catalytic activity and stereoselectivity. Recycling of the catalyst was performed by removing the unreacted cyclohexanone and extraction of the product with diethyl ether. This procedure left the catalyst in the reaction vessel, which was used in the next run. After six cycles, stereoselectivity was maintained but activity decreased [122].

Recycling properties of amphiphilic dendritic organocatalysts based on a chiral pyrrolidine core were investigated in the Michael addition of cyclohexanone and *trans*-β-nitrostyrene. Usually, soluble dendrimer-supported catalysts

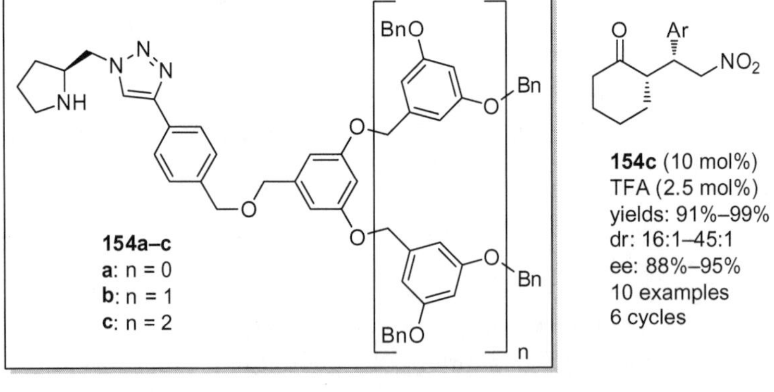

FIGURE 2.34. Dendrimer-supported catalysts **154a–c**.

can be recovered by precipitation or by membrane dialysis. However, such approaches cannot be applied in the case of these amphiphilic dendritic materials, which were recovered by partition between MeOH and heptanes. After five cycles, the second- and third-generation dendritic catalysts gave unchanged activity and stereoselectivity [123].

2.6.4. Ionic Liquid-Anchored Organocatalysts

Organocatalysts **155** and **156** were used in 15 mol% in the presence of trifluoroacetic acid as an additive (5 mol%) and were employed in several Michael addition reactions, affording the final products in excellent stereoselectivity in the case of cyclohexanone (Fig. 2.35 and Table 2.8). Cyclopentanone, acetone, *iso*-butyraldehyde, and *iso*-valeraldehyde gave adducts with lower stereoselectivity (ee values 43%–89%). Recycling experiments, carried out using catalyst

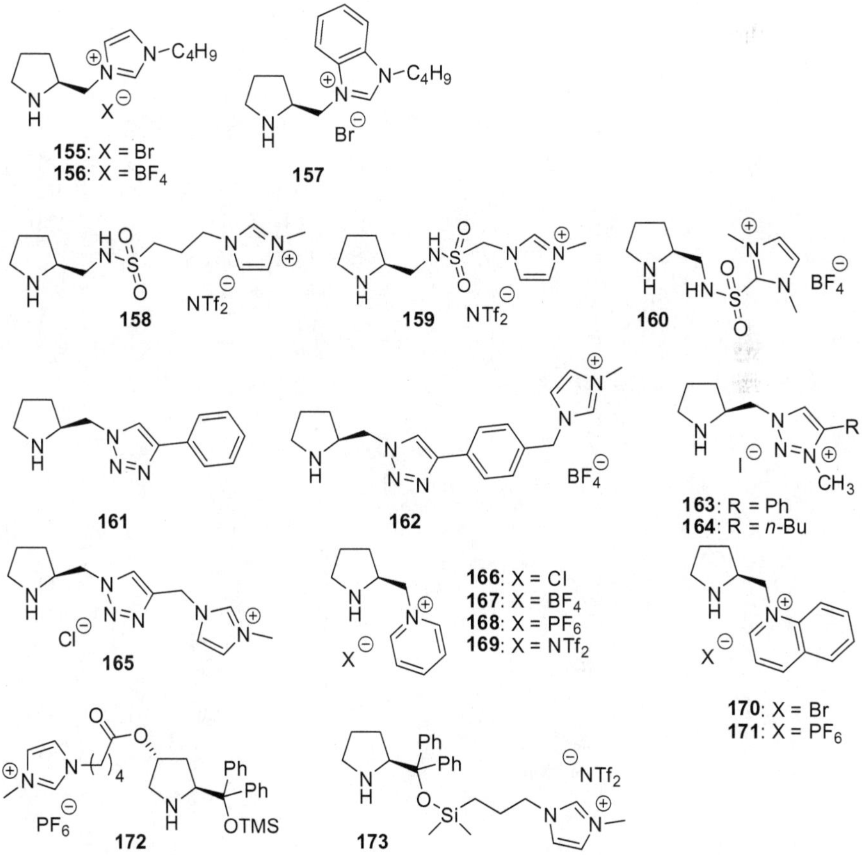

FIGURE 2.35. Ionic liquid-tagged organocatalysts **155–173** for the Michael reaction.

TABLE 2.8. Summary for Ionic Liquid-Tagged Organocatalyzed Michael Reaction

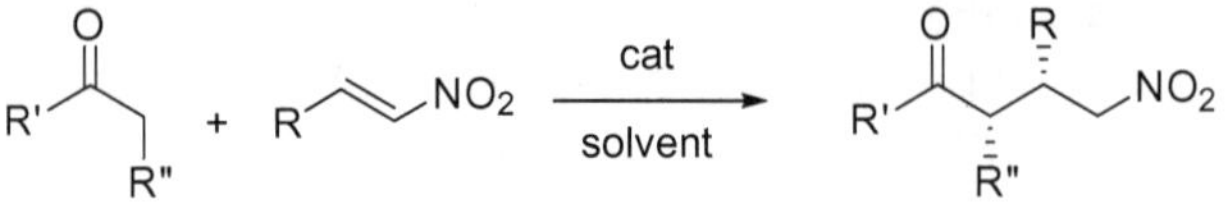

Entry	Catalyst (mol%) Solvent	Product	Yields (%)	d.r. (*syn*/*anti*) ee (%)	# Cycles	Ref.
1	**155** or **156** (15) TFA (5)	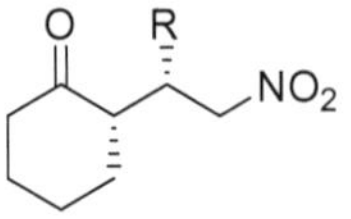 7 examples	76–100	96:4–99:1 92–99	4	[124]
2	**157** (15) Salicylic acid (5)	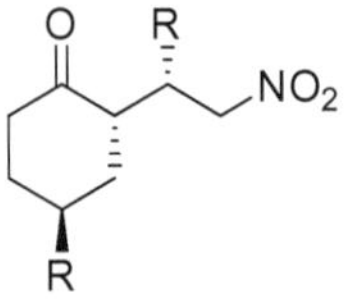 16 examples	61–99	4:1–10:1 93–99	4	[125]
3	**158** (20) Et_2O	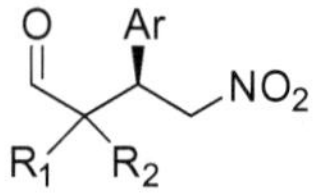 8 examples	29–64	89:11–97:3 64–82	3	[126]
4	**159** (20) MeOH	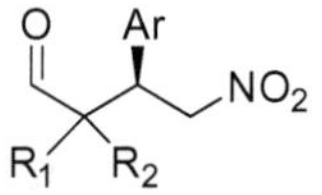 8 examples	91–99	78:22–94:6 33–46	5	[127]
5	**160** (10) *i*-PrOH	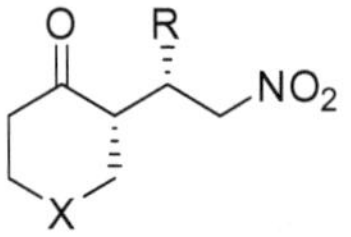 10 examples	83–97	92:8–99:1 80–99	5	[128]
6	**162** (15) TFA (5)	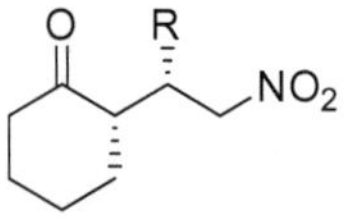 12 examples	94–99	92/8–49/1 85–97	4	[131]

Entry	Catalyst (mol%) Solvent	Product	Yields (%)	d.r. (*syn*/*anti*) ee (%)	# Cycles	Ref.
7	**163** or **164** (10) TFA (2 or 2.5)	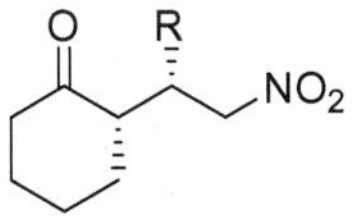 4 examples	93–98	95:5–99:1 83–99	4	[132]
8	**165** (10) TFA (5) EtOH	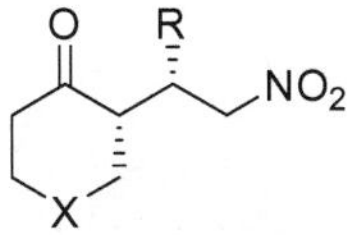 15 examples	85–95	99:1–>99:1 91–>99	8	[133]
9	**167** (15) TFA (5)	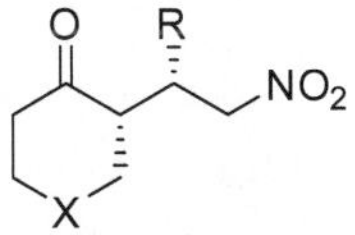 8 examples	90–100	94:6–>99:1 92–98	3	[134]
10	**170** or **171** (10) [bmim]BF_4	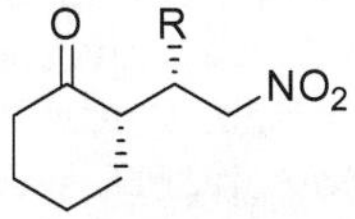 13 examples	93–95	91:9–96:4 75–>99	5	[135]
11	**172** (10) EtOH	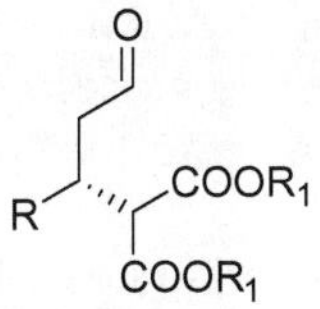 10 examples	93–98	– 76–96	6	[136]
12	**173** (1–2) DCM or H_2O	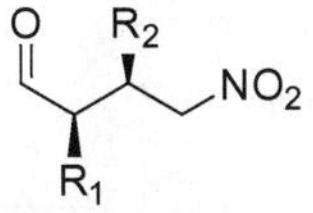 13 examples	65–99	89:11–98:21 97–>99	—	[137]

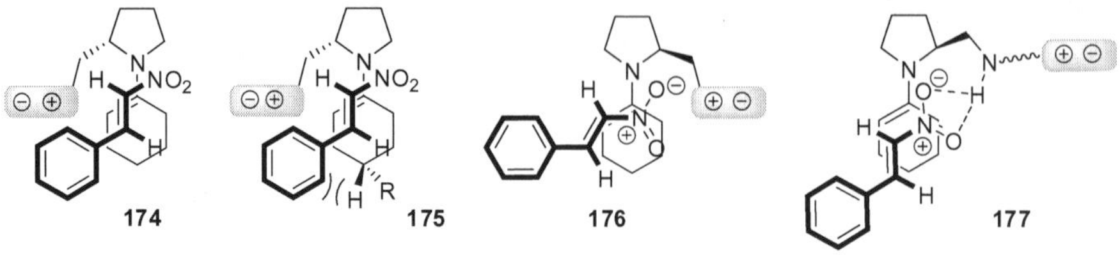

FIGURE 2.36. Nitrostyrene approach models for cyclohexanone enamine donors.

156, showed no loss in stereoselectivity after four cycles but showed a loss in activity [124].

The high stereoselectivity observed when cyclohexanone was used was explained with model **174** (Fig. 2.36), in which the ionic liquid moiety would effectively shield the *Si* face of the enamine double bond in the ketone donor and the reaction would occur through a *Re-Re* approach. For aldehyde donors, *anti*-enamine would be formed and the reactions would occur through a *Si-Si* approach.

Ionic liquid-anchored pyrrolidine organocatalysts were also employed for the enantioselective desymmetrizations of prochiral ketones via asymmetric Michael addition reactions to nitrostyrenes [125]. Screening experiments on several ionic liquid-anchored organocatalysts revealed better performances than were displayed by organocatalyst **157**. Moreover, screening of additives revealed that salicylic acid was the best one. Under these conditions (**157** 10mol%, salicylic acid 5mol%, no added solvent), several Michael addition reactions between prochiral ketones and nitrostyrenes were carried out. Apart from ketones carrying OH, Br, and CN groups, which were unreactive, in all the other examples, high yields and stereoselectivities were observed. These latter were explained by invoking the plausible model **175**. Organocatalyst **157** was recovered by precipitation and reused up to four cycles with unchanged stereoselectivity but diminished activity.

In the reported examples, pyrrolidine-based ionic liquid-anchored catalysts, employed for Michael addition reactions, did not contain acidic hydrogens able to form a hydrogen bond with substrates (nitrostyrenes). In order to investigate this type of organocatalyst, a chiral pyrrolidine moiety including a protonic functionality, linked to an imidazolum ionic liquid, was designed [126].

The introduction of the ionic liquid moiety would help in the recovery of the organocatalyst. Compound **158** was designed with this aim and synthesized starting from (*S*)-2-amino-1-*N*-Boc-pyrrolidine and 3-chloropropanesulfonyl chloride. Preliminary experiments were carried out on the Michael reaction of *iso*-butyraldehyde and *trans*-*β*-nitrostyrene. Various solvents were examined at rt and the best results were obtained when catalyst **158** (20mol%) was used in less polar solvents such as Et_2O or $CHCl_3$. Moreover, working at lower temperature (4°C), a small increase in enantioselectivity was observed (82% at 4°C; 78% at rt), although reaction times were quite long (6 days). The catalyst was recovered by precipitation and reused twice with unchanged yield and enantioselectivity. Then, using the optimized reaction conditions, several Michael additions of aldehydes to *trans*-β-nitrostyrenes were carried out. Adducts were obtained in moderate yields (29%–64%), good enantioselectivity (64%–82%), and high diastereoselectivity (89:11–97:3 *syn*/*anti*) [126]. Adduct with cyclohexanone was obtained in low yield (38%) and good enantioselectivity (88%), although lower than that obtained with catalyst **156**.

A comparison between the catalytic systems **156** (15 mol%)/TFA (5 mol%) and **158** (20 mol%) indicated that the former was more active and selective. The shielding role of the imidazolium moiety probably plays a major role compared to the coordinating role of the acidic N–H hydrogen bond. A more active organocatalyst compared to **158** was later reported by the same authors [127]. Following a similar synthetic strategy for organocatalyst **158**, ionic liquid-anchored organocatalyst **159** was prepared and used for the Michael addition of aldehydes to *trans*-β-nitrostyrenes and the results compared to those obtained using catalyst **158**. Michael reactions catalyzed by **159** were carried out in methanol at rt, instead of in diethyl ether at 4°C for catalyst **158**. Reaction times were significantly shortened (2 days for **159**, 6 days for **158**) and yields markedly increased. However, enantioselectivities observed with **159** were lower than those observed with **158**. Catalyst **159** gave better results compared to **158** when used for the addition of α,α-disubstituted aldehydes. Recycling experiments showed a decreased activity in the fifth cycle. Again, reaction with cyclohexanone was not satisfactory (yield: 40%; ee: 90%).

The stereochemical outcome of the reactions catalyzed by **158** and **159** is believed to be based on the acidity of the N–H bond. In order to increase the N–H acidity, a new type of ionic liquid-anchored pyrrolidine sulfonamide organocatalyst (**160**) was synthesized [128].

The increased acidity is ascribed to the sulfonyl group, which, in turn, is linked to the C-2 position of the electron withdrawing the imidazolium cation. In addition, the C-2 position is more electronegative than the C-4 or C-5 positions. The increased acidity will result in stronger hydrogen bonds that are formed in the transition states of these reactions. As a result, enhancement of the catalytic activity and selectivity, without the need of additives, could be expected. Also, the closer presence of the imidazolium moiety with respect to organocatalysts **158** and **159** could introduce more steric bulk in proximity to the catalytic site, leading to an improved stereoselectivity.

Addition reactions of six membered cyclic ketones to nitrostyrenes gave products in high yields and stereoselectivities. *Iso*-butyraldehyde also gave good results while acetone gave good yield but low enantioselectivity. Cyclopentanone gave low yield. Catalyst **160** was recovered (>90%) by phase separation and reused for five cycles. Stereoselectivity was maintained, but activity decreased in the fourth and fifth cycles.

Pyrrolidine-based organocatalyst **161** gave excellent results when used in 10 mol% in chloroform or water without additives [129] or under neat conditions in 10 mol% and in the presence of TFA (2.5 mol%) [130]. However, no recycling experiments were carried out. In order to get a recyclable catalyst able to maintain the high activity of **161**, several ionic liquid-anchored organocatalysts based on this structure were synthesized.

When catalyst **162** [131] was used in chloroform or water without additives, poor results were obtained, showing the different activity, due to the presence of the imidazolium tetrafluoroborate moiety, compared with catalyst **161**. Other reaction conditions were screened. Finally, it was found that catalyst **162** could be successfully employed under neat conditions (15 mol%) in the presence of TFA (5 mol%). Using these reaction conditions, excellent results were obtained in the addition of cyclohexanone to nitroolefins. Lower stereoselectivities were observed when cyclopentanone (ee *syn* 81%; *syn*/*anti* 5:3) and acetone (ee 40%) were employed. Catalyst **162** was recovered by precipitation with diethyl ether and reused for four consecutive runs without loss in activity and stereoselectivity. In catalyst **162** the 1,2,3-triazole unit serves as linker while the imidazolium unit acts as phase tag for recycling. Other authors synthesized the ionic liquid-anchored organocatalysts **163–164** in which the 1,2,3-triazolium unit does not play the role of linker but could act as in the analogs imidazolium compounds **155–156** [132].

Several Michael addition reactions of ketones to nitrostyrenes were carried out with good results. Cyclopentanone, acetone, *iso*-butyraldehyde, and *iso*-valeraldehyde gave lower stereoselectivities (ee 36%–82%) Recycling investigations (four cycles) with organocatalyst **163** showed that enantioselectivities were unsatisfactory in the last two cycles. Disappointingly, organocatalysts **163–164** are not directly comparable with **155–156** since the anions are different; however, on the whole, the latter showed better performances.

Astonishing results were observed in the Michael addition of six membered ketones to nitrostyrenes catalyzed by organocatalyst **165** [133], which was prepared through the "click chemistry" procedure. Preliminary tests indicated organocatalyst **165** was the most promising when used in 10 mol% in addition to TFA (5 mol%) in EtOH at 10°C for 36 hours compared to the corresponding BF_4 and PF_6 salts. Products were obtained in excellent yields and stereoselectivities. Moreover, this compound was used for eight consecutive cycles without loss in activity and stereoselectivity.

Simple pyridinium ionic liquid-anchored pyrrolidine organocatalysts were easily synthesized starting from (*S*)-(2-aminomethyl)-1-*N*-Boc-pyrrolidine and Zincke's salt, followed by deprotection and anion exchange [134]. Following this procedure, four organocatalysts (**166–169**) were prepared (Fig. 2.35). All organocatalysts **166–169** (15 mol%) catalyzed the asymmetric Michael addition of cyclohexanone to *trans*-*β*-nitrostyrene, but, overall, compounds **166** and **167** containing Cl and BF_4 anions gave the best performances with high yields, high diastereo- and enantioselectivity. The addition of TFA (5 mol%) increased the reaction rate without decreasing enantioselectivity. Then, organocatalysts **166** (at rt) and **167** (at 4°C) were used for the reaction between cyclohexanone, or dihydro-2H-pyran-4(3H)-one, and nitrostyrenes to afford the final adducts in high yields and stereoselectivities. Other donors (*iso*-butyraldehyde, acetone, and cyclopentanone) were also used for the reaction with *trans*-β-nitrostyrene, affording the product with modest stereoselectivities. Catalysts were recovered by precipitation, and recycling experiments (three cycles) carried out with **167** showed a decreased yield and steroselectivity in the third cycle.

Catalysts **170–171** were used in [bmim]BF_4 and recovered after extraction of the product in diethyl ether. After five cycles, activity and stereoselectivity were maintained [135].

As reported above, the stereoselectivities observed with cyclic ketones were explained by invoking the approach model **174** (Fig. 2.36). Other authors also invoked the occurrence of the approach model **176** (Fig. 2.36), in which the possible attraction between ionic liquid moiety and nitro group of the substrate should also contribute to the enantioselectivity observed. Alternatively, for organocatalysts bearing N-H acidic hydrogens, the approach model **177** (Fig. 2.36) can also explain the observed stereoselectivity.

Michael reactions were also carried out using ionic liquid-anchored α,α diphenyl-(*S*)-prolinol derivatives. Two main examples were reported, the first one carrying the ionic liquid tag in the 4-position of the proline moiety (**172**) and the second one carrying the ionic liquid tag as substituent in the *O*-TMS group (**173**). The ionic liquid-anchored α,α-diphenyl-(*S*)-prolinol **172** was used as a catalyst in the Michael reaction between α,β-enals and dialkyl malonates [136].

This catalyst worked well in 96% EtOH at 4°C in 10 mol% loading. Use of water (50 eq.) or neat conditions gave slightly decreased performances. Several Michael reactions were carried out, affording the final product in high yields and enantioselectivities. The catalyst was recovered after solvent evaporation and phase separation and used for six consecutive cycles. Although enantioselectivity was unaffected, activity decreased in the fifth and sixth cycles.

α,α-Diphenyl-(*S*)-prolinol **173** was tested in the reaction between propanal and nitrostyrene in 16 different solvents always affording the adduct with excellent enantioselectivity (99%) and good diastereoselectivity (85:15) [137]. Dichloromethane and water were selected as solvents of choice because of the higher activity in these media. Catalyst loading was tested in dichloromethane, down to 0.25 mol% with improved stereoselectivity (>99.5% ee, 96:4 d.r.) and high yield (88%). Then, several Michael addition reactions were carried out using catalyst **173** in 1–2 mol% in four different conditions (dichloromethane or water with or without benzoic acid as an additive). Adducts were obtained in high yields and diastereoselectivities and excellent enantioselectivities. The data regarding ionic liquid-tagged organocatalysts are reported in Table 2.8.

2.6.5. Fluorous Organocatalysts

The fluorous (*S*)-pyrrolidine catalyst **116** (Fig. 2.26), also employed for aldol reaction, successfully promoted the Michael addition of cyclic ketones with

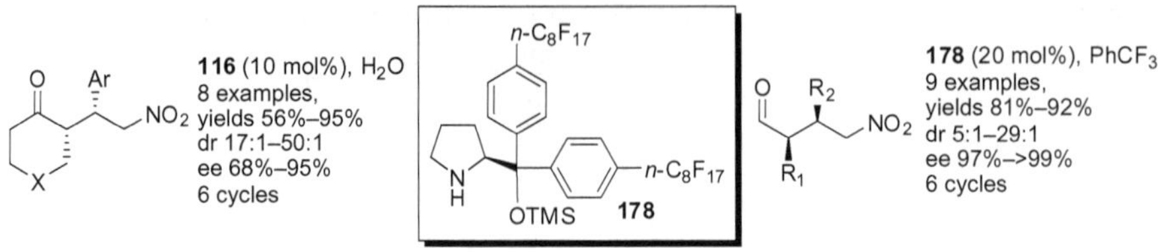

FIGURE 2.37. Structure of fluorinated catalyst **178** and its activity and that of **116** in the Michael reaction.

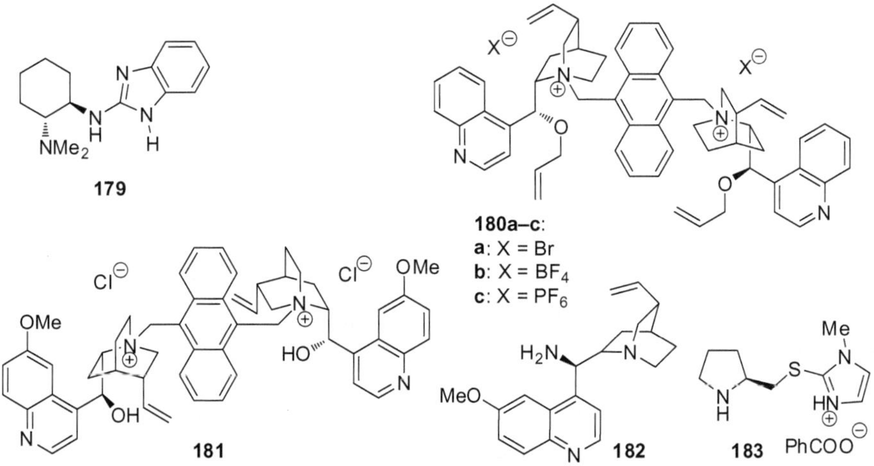

FIGURE 2.38. Recyclable molecular organocatalysts for Michael reaction.

nitroolefins in water. Nonanal and *iso*-butyraldehyde gave adducts in 81% and 86% ee, respectively [138]. Aldehydes were largely employed in the Michael addition catalyzed by (*S*)-diphenylpyrrolinol silyl ether **178** (20 mol%) (Fig. 2.37). Reactions were carried out in $PhCF_3$, affording adducts with high levels of stereoselectivity. As usual, these catalysts were recovered by F-SPE. While stereoselectivities were maintained, reaction times were progressively increased to obtain a reproducible yield [139].

2.6.6. Nonsupported Organocatalysts

A few examples of nonsupported organocatalysts for Michael addition reactions were also reported. Three recovery approaches were applied: extraction, precipitation of the catalyst and precipitation of the product, and recovery of the soluble catalyst. An acid–base extractive workup was used for the recovery of chiral 2-aminobenzimidazole **179** in 94% yield (Fig. 2.38). Such a catalyst was used for the addition of 1,3-dicarbonyl compounds to nitroalkenes with a high level of enantioselectivity (ee 87%–96%) [140].

Dimeric anthracenyldimethyl-derived *Cinchona* catalysts **180a–c** and **181** were recovered by precipitation with diethyl ether. Ammonium salts **180a–c** were used in 5 mol% at low temperatures (–35°C or –55°C) as phase transfer catalysts in the Michael addition reaction of *N*-diphenylmethyleneglycine *tert*-butyl ester to electron-poor olefins. The ee values were dependent on the nature of the anion [141].

The quinidine-derived dimeric ammonium salt **181** was used in the Michael addition of cyclic *β*-keto esters to electron-poor olefins in 30 mol% at –40°C. Products were obtained in 51%–94% ee values [142]. Catalyst **182** was used in 15 mol% in the Michael addition of 1,3-diaryl-1,3-propanedione to 13 different nitroolefins to give the corresponding adducts in high yields (86%–97%) and enantioselectivities (94%–>99%). These reactions were carried out in diethyl ether. The Michael adducts were not very soluble in such solvents. Then, pure products were isolated by filtration, and the catalyst, which was kept in the filtrate, could be reused directly for the next cycle. After seven cycles, catalyst **182** kept its stereoselectivity, while increased reaction times were needed [143].

Another strategy for achieving a recyclable unsupported chiral organocatalyst is represented by the use of PEG as a reaction medium in which the catalyst is dissolved. The pyrrolidinyl-thioimidazolium salt **183** formed a host–guest complex with PEG-800 and was used in 20 mol% in the reaction between cyclic ketones and nitroolefins to afford the adducts in high yields (70%–97%) and high stereoselectivities (*syn*/*anti* 90:10–97:3; ee 88%–99%, 13 examples). The use of PEG-800 increased activity and stereoselectivity when compared with DMSO or other solvents, and it allowed recycling. After extraction of the product with hexane/ether (6:1), the complex **183**/PEG-800 was reused up to seven times with unchanged stereoselectivity but with prolonged reaction times [144].

2.7. KETONE REDUCTION

Unprotected supported diarylprolinols such as **136b** and **137b** (Fig. 2.30) were employed as Corey–Bakshi–Shibata (CBS) catalysts for the enantioselective ketone reduction. In this case, the cross-linked catalyst **136b** afforded almost identical enantioselectivities to those of the homogeneous catalyst, whereas the pendant-bound catalyst **137b** showed reduced activity [145, 146].

The postmodification approach was the strategy of choice for the synthesis of catalysts **184** and **185** (Fig. 2.39). Starting from polystyrene beads containing phenylboronic acid residues and employing the Suzuki coupling reaction, diarylprolinols **184** and **185** were prepared. Catalyst **184** was a mixture of double-bonded and single-bonded diarylprolinol. Both catalysts were tested in the CBS reduction of prochiral ketones in 30 mol%, giving comparable results to those observed with unsupported diarylprolinol in 5 mol%. Moreover, catalyst **184** was recycled 14 times, whereas catalyst **185** was recycled eight times [147].

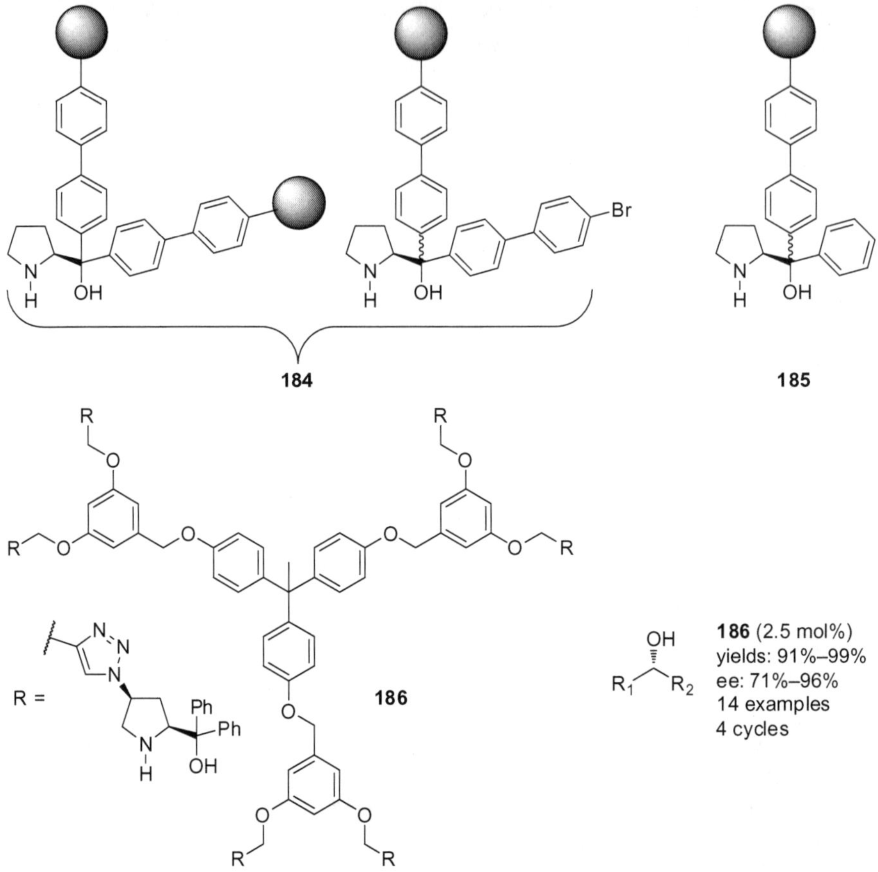

FIGURE 2.39. Supported diarylprolinols employed in ketone reductions.

Asymmetric borane reduction of prochiral ketones was accomplished using 1,2,3-triazole-linked dendrimers. In particular, catalyst **186** was used in 2.5 mol% in the presence of $BH_3{\cdot}Me_2S$ to afford chiral alcohols in high yields and enantioselectivities. Recovery of the catalyst was performed by precipitation, and its reuse indicated no significant loss in performances after four cycles [148].

2.8. DIELS–ALDER REACTION

In 2000, MacMillan et al. discovered that imidazolidinone **2** (Fig. 2.1) was able to asymmetrically catalyze the Diels–Alder reaction between cinnamaldehyde and cyclopentadiene [1]. Since then, the immobilized version of MacMillan imidazolidinone was frequently employed to obtain recyclable organocatalysts for the above reaction. The three main kinds of recyclable catalysts (covalently supported, noncovalently supported, and nonsupported) were investigated. Despite the many efforts that were devoted to the realization of recyclable organocatalysts based on MacMillam imidazolidinone, in several cases these materials lost their activity after few cycles. In other cases, only one reuse was carried out, which did not allow a good evaluation of the recyclability of the catalytic materials. Benaglia et al. showed that ^{1}H NMR of the recovered samples of **187** indicated extensive degradation after three cycles [149]. Prolonged exposure of supported imidazolidinone to the acidic reaction medium can lead to catalyst degradation, resulting in the observed decreased yields upon recycling. Moreover, further investigations pointed out that catalyst degradation could be induced by the presence of the reagents [150].

The chiral imidazolidinone was supported by covalent linkage to JandaJel (**188**), silica (**189**) [151], siliceous mesocellular foam (MCF) (**190**), and polymer-coated MCF [152] and PEG (**187**) [149] (Fig. 2.40).

Materials **187–190** were prepared following the postmodification approach, whereas catalyst **191** was prepared following the bottom-up approach by acrylic copolymerization by using a 50:50 mixture of feedstock PEG 400 methyl ether methacrylate and PEG 600 dimethacrylate.

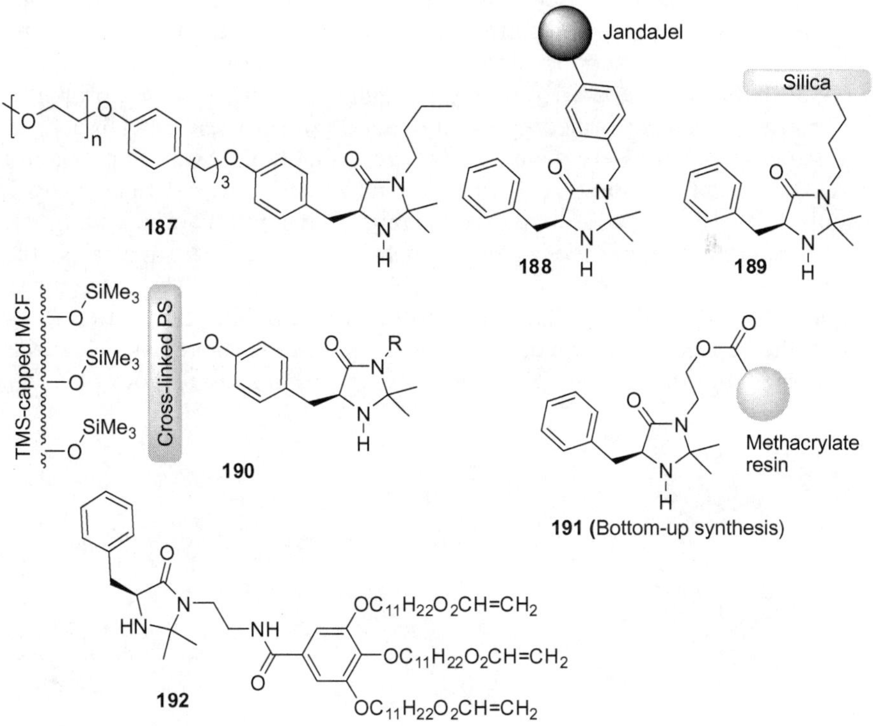

FIGURE 2.40. Catalysts used in the asymmetric Diels–Alder reaction.

Recovery of the soluble PEG-supported catalyst was not quantitative (70%–80%), whereas the insoluble catalysts were recovered by filtration. More in-depth studies on recycling were carried out on catalysts **187** and **191**. Activity or stereoselectivity dropped after few cycles. Imidazolidinone monomer **192** was used for a different immobilization approach via the acid-induced liquid crystal self-assembly and subsequent photopolymerization. Protonation not only activated the catalyst but also simultaneously induced lyotropic (i.e., amphiphilic) liquid crystal (LLC) behavior. The resulting LLC assemblies (**192**/*n*HX) were radically photo-cross-linked to afford nanostructured solid organic catalysts with very good reactive site accessibility (ca. 70%) and retention of the high enantioselectivities exhibited by the parent chiral imidazolidinone (as HCl salt) catalyst in solution. Such material was employed as phosphate, sulfate, and tartrate salts in the Diels–Alder reaction of crotonaldeyde and cyclopentadiene in MeCN/H_2O, which gave similar stereoselectivities than the unmodified MacMillan catalyst, but in lower yield. These materials were recycled at least once with unchanged stereoselectivities and slightly reduced yields [153].

Immobilization of the chiral imidazolidinone was also achieved by using the noncovalent linkage approach (Fig. 2.41). Three main examples were described: (a) the catalyst was entrapped into montmorillonite [154]; (b) the polymer-supported imidazolidinone was prepared by an ion-exchange reaction with polymer-supported sulfonic acid [155]; and (c) the imidazolidinone was immobilized as a supported ionic liquid catalyst [156].

Catalyst **193** was prepared by simple ion exchange with Na^+-montmorillonite. The recovery was performed by filtration, and after five cycles both activity and stereoselectivity were maintained. A very simple preparation procedure was also needed for catalyst **194**. This material was used twice with unchanged results. In the latter two examples, no catalyst leaching was observed. In the case of material **195**, in order to avoid leaching of the MacMillan catalyst, the reactions were carried out in the hydrophobic *t*-amyl alcohol. This catalyst was prepared by adsorption on silica gel and recovered by filtration. After seven cycles, the ee values decreased, especially for the *endo* product. The yield was maintained, but the reaction time was increased from 22 to 56 hours.

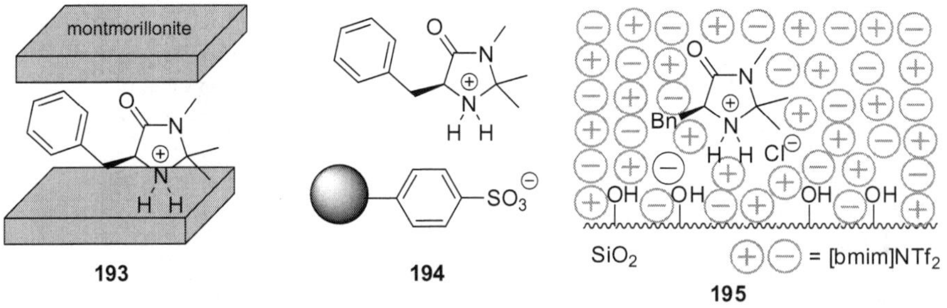

FIGURE 2.41. Noncovalently supported catalysts **193–195**.

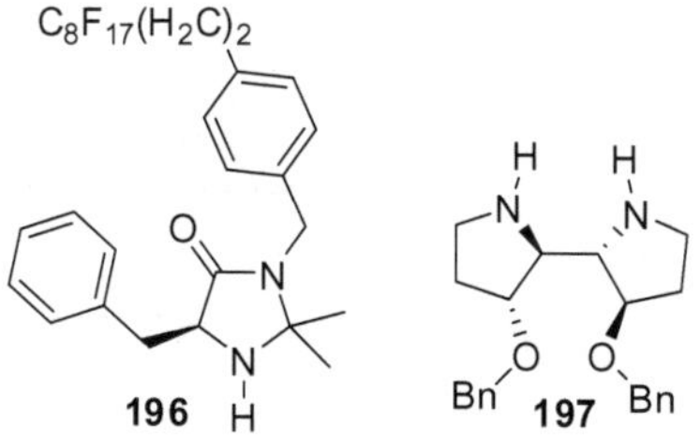

FIGURE 2.42. Organocatalysts used for Diels–Alder reactions.

Unmodified MacMillan imidazolidinone **2** was used as a catalyst for Diels–Alder reactions between cyclopentadiene and cinnamaldehyde in [Bmim]PF_6/water 95:5. Use of ionic liquid allowed recovery and reuse of the chiral catalyst, but both yield and enantioselectivities of *endo* and *exo* products dropped in the fourth cycle [157].

The fluorous imidazolidinone derivative **196** (Fig. 2.42) was successfully used in several Diels–Alder reactions and recovered by F-SPE but not quantitatively (84%). Its reuse in a subsequent cycle did not show a decrease in activity and stereoselectivity [158]. However, the fluorous functionalization allowed a higher yield, higher stereoselectivity catalyst recovery yield, and purity of the recovered catalyst compared to the unmodified MacMillan imidazolidinone.

Asymmetric Diels–Alder reactions between cyclopentadiene and various α,β-unsaturated aldehydes were carried out in the presence of C_2-symmetric bipyrrolidine **197** as $HClO_4$ salt. Reactions were performed in water without an additional cosolvent. The use of water allowed the recovery of the catalyst by simple extraction of the products with diethyl ether. Catalyst **197** remained in the water phase and was used in the next cycle. This procedure allowed the recovery of the catalyst without any modification. However, in the fifth cycle a great drop in activity was observed [159].

2.9. FRIEDEL–CRAFT-TYPE REACTION

Peptide catalyst **198**, supported on PS-PEG-$CH_2CH_2NH_2$ resin having a polyleucine linker, gave interesting results both in the Friedel–Crafts-type alkylation of *N*-methyl indole and *N*-methyl pyrrole to nitrostyrenes (Scheme 2.19) [160], and in the asymmetric transfer hydrogenation in the presence of the Hantzsch ester (see Section 2.10). The former reactions were carried out in aqueous media (THF/H_2O 1:2) in the presence of TFA. Preliminary results, obtained with proline linked to PS-PEG-$CH_2CH_2NH_2$ with the polyleucine linker, indicated that the use of THF as the sole solvent did not promote the reaction, whereas increasing the amount of water (THF/H_2O 1:2) resulted in enhanced reaction rate (59% conversion). The acceleration could be ascribed

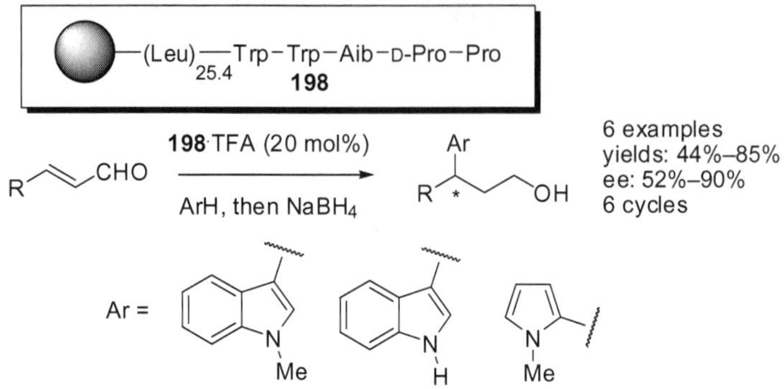

SCHEME 2.19. Asymmetric Friedel–Crafts-type alkylation of *N*-methyl indole and *N*-methyl pyrrole to nitrostyrenes.

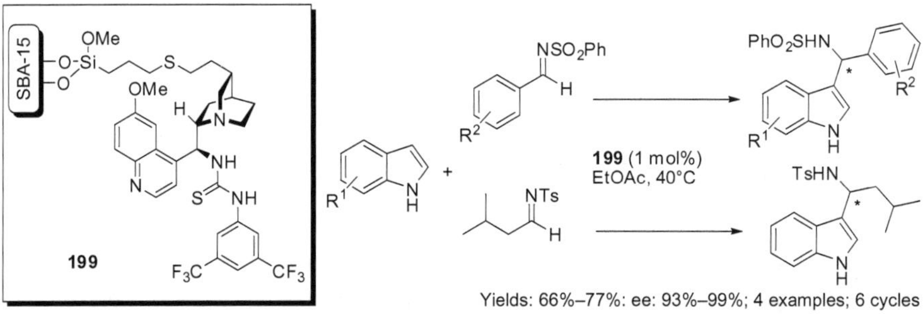

SCHEME 2.20. Asymmetric Friedel–Craft reaction of imines with indoles.

to the increased interactions between hydrophobic substrates and the hydrophobic polyleucine chain. However, ee values were low. When catalyst **198**/TFA was used in THF/H_2O 1:2, high yield and higher ee values were observed. The role of the polyleucine linker was evident since catalyst **198** without the linker promoted the reaction in low yield and enantioselectivity. Then, the Pro-D-Pro-Aib-Trp-Trp unit constituted the catalytically active site and effectively shielded one face of the iminium ion intermediate by the rigid peptide framework. The polyleucine chain played a role in increasing activity and stereoselectivity. Six reactions were carried out and the products were isolated with useful levels of enantioselectivity. The catalyst was recovered by filtration, and up to six cycles were performed with reproducible results.

The asymmetric Friedel–Craft reaction of imines with indoles was carried out employing the 9-thiourea *epi*-quinine catalyst **199** supported on mesoporous SBA-15 silica gel (Scheme 2.20) [161].

The immobilization was achieved starting from the SBA-15 support that was treated with trimethoxy-3-mercaptopropylsilane. The catalyst was anchored via a radical addition reaction in the presence of α,α'-azobisisobutyronitrile

(AIBN). The Friedel–Craft products were obtained in high enantioselectivities and good yields albeit with a long reaction time (5 days). Interestingly, the role of the immobilized 9-thiourea *epi*-quinine catalyst was evident because of the enhancement in the yield and the selectivity obtained with the supported catalyst compared with the homogeneous catalyst. Catalyst **199** was considered recyclable because no decrease in enantioselectivity was evident after four cycles. A decrease in the yield and enantioselectivity appeared in the fifth cycle. After washing **199** with toluene, acetone, ether, and hexane, the enantioselectivity was restored even if the yield was lower with respect to the first cycle.

The Friedel–Craft reaction between *N*-methyl pyrrole and cinnamaldehyde was carried out with the **190** (Fig. 2.40) catalyst, affording the product in 72% yield and 71% ee after two cycles.

2.10. ASYMMETRIC REDUCTION

Metal-free organocatalysis is now emerging as a novel synthetic strategy with the ambition to replace, whenever possible, the traditional transition metal catalysis for the asymmetric reduction of prochiral ketimines and C=C double bonds. As discussed in depth in Chapter 14, silicate-mediated stereoselective reactions catalyzed by chiral Lewis acids are currently receiving greater attention. In this context, recyclable catalysts for silicate-mediated reactions are now the subject of a new research field. Asymmetric reduction of ketimines was accomplished using trichlorosilane in the presence of supported amino acid-derived formamide catalysts or supported chiral Brønsted acid. Reduction with trichlorosilane was performed using polystyrene catalysts **200a,b** (Scheme 2.21). In addition to catalysts **200a,b**, which displayed better performances, other supports (Wang, TentaGel, Marshall, and extended Merrifield resins) were employed. Catalyst **200a** was used in a higher amount (25 mol%) than **200b** (15 mol%).

Each resin was used in chloroform in six consecutive runs with reproducible ee's (81%–82%). The enantioselectivity was lower (76%–77%), with only the first runs suggesting that "conditioning" of both the catalysts was required to attain their optimal performances. The supported catalysts **200a,b** exhibited lower enantioselectivity (ca. 10% ee) than their soluble counterparts. Control experiments carried out with resins lacking the catalytic formamide moiety indicated that the support did catalyze the reaction, though at a considerable lower rate. The polymeric backbone affected the selectivity in an adverse way, and it was more evident when the reactions were performed in toluene. Catalyst **200b** was used for further reactions in 30 mol% loading. It was tested up to six consecutive cycles with different ketimines in each cycle. Products were isolated in 62%–82% yield and 77%–81% ee [162]. Previously, the same authors reported the use of fluorous-tagged catalysts of type **200**. Three different catalysts (**200c–e**) were tested, having X = $(CH_2)_2C_6F_{13}$ and R = H, Me, *i*-Pr (Scheme 2.21). A comparison between these catalysts indicated that

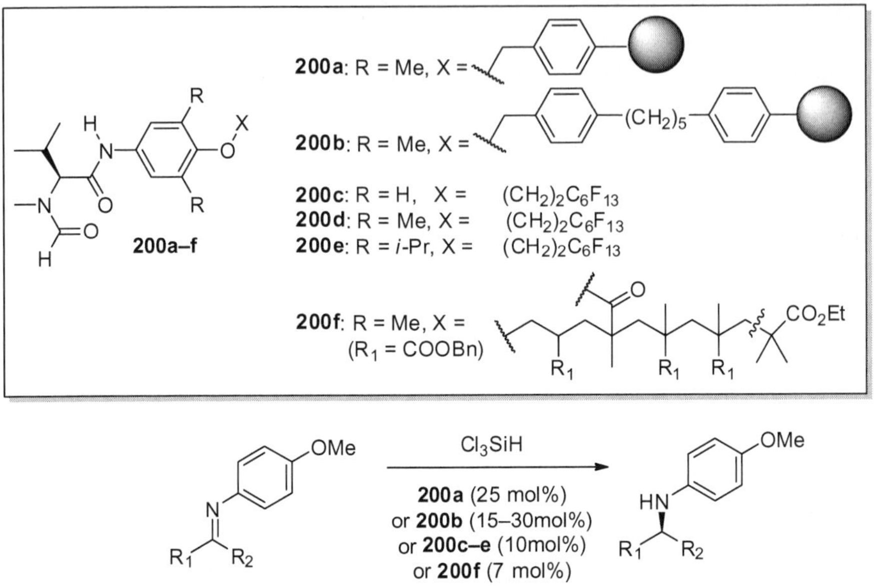

200b in $CHCl_3$: yields 62%–87%, ee 78%–82%, 6 cycles, 6 examples
200d in toluene: yields 68%–90%, 86%–92% ee, 4 cycles, 4 examples
200f in toluene: yields 70%–90%, ee 55%–91%, 5 cycles,10 examples

SCHEME 2.21. Asymmetric reduction of ketimines.

catalysts **200d,e** gave higher enantioselectivity. Catalyst **200c** was used in 5–10mol% in toluene with four ketimines furnishing products in 68%–90% yield and 86%–92% ee. Interestingly, performances of fluorous-tagged organocatalysts matched those observed with the corresponding untagged catalysts. Recycling studies were carried out on the three catalysts **200c–e**. While catalysts **200c,d** displayed reduced enantioselectivity and catalyst recovery upon reuse, catalyst **200e** displayed unchanged enantioselectivity after five cycles and recovered yield in the range 99%–88%. Each catalyst was recovered by filtration through a pad of fluorous silica [163].

The major difference between the latter two approaches was that heterogeneous polystyrene-supported catalysts **200a,b** were separated easily by filtration but had a negative impact on the enantioselectivity, whereas homogeneous fluorous catalysts did not have a negative impact on the enantioselectivity but recovery was not quantitative. Soluble polymer-supported catalyst **200f** had as its main advantage, with respect to the latter catalytic systems, the increased enantioselectivity compared to the heterogeneous catalysts and the almost quantitative recovery after precipitation with methanol. Catalyst **200f** was prepared by copolymerization of a 5:95 methacrylate-derived catalyst and benzyl methacrylate mixture, in the presence of CuI, 2,2′-bipyridine and ethyl 2-bromoisobutyrate. The average molecular weight was approximately 6000g/mol with a fairly narrow distribution of molecular mass for the

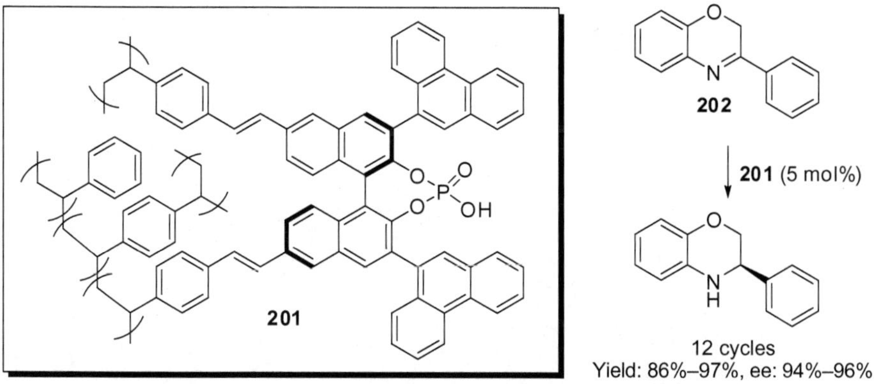

SCHEME 2.22. Asymmetric transfer hydrogenation of C=N bond.

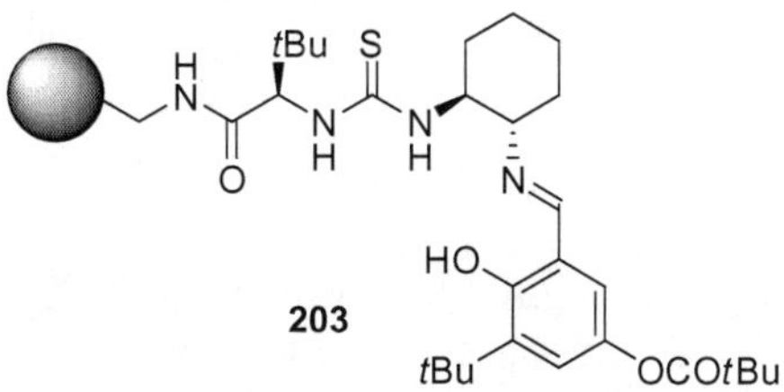

FIGURE 2.43. Structure of catalyst **203**.

polymer. Polymer **200f** was employed in 7 mol% for the reduction of 10 imines in high yields (up to 90%) and ee values (up to 91%). Recycling studies (five cycles) were carried out using **200f** in 20 mol% with unchanged activity and enantioselectivity [164].

The polymer-supported chiral Brønsted acid **201** was employed for the asymmetric transfer hydrogenation of the benzoxazine **202** in the presence of a Hantzsch ester (Scheme 2.22). Catalyst **201** was prepared following the bottom-up approach by copolymerization of styrene, DVB, and a styrenic monomer of the chiral Brønsted acid. The catalytic activity was evaluated in multiple transfer hydrogenation reactions. Polymer **201** was used for 12 consecutive cycles without loss in activity and stereoselectivity, and results were comparable to those of the corresponding homogeneous catalyst. The catalytic material was prepared in the form of a polymer stick to be incorporated in an open reaction container and easily recovered in a tea bag approach. A less bulky catalyst carrying the linkage for immobilization in 3,3′-positions of the 1,1′-binaphthol (BINOL) skeleton and carrying no substituents in the 6,6′-positions was less enantioselective [165].

Imines are also very useful starting materials for the Strecker reaction, which allows the asymmetric synthesis of α-amino acid derivatives. The polystyrene-bound catalyst **203** (Fig. 2.43) was used in 10 consecutive runs in the Strecker reaction of pivaladimine under preparative conditions. Only a

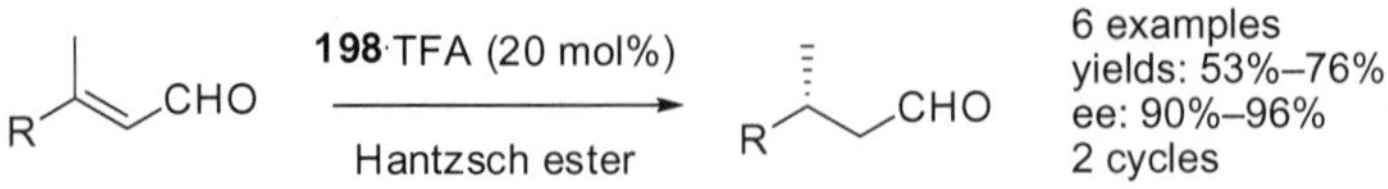

SCHEME 2.23. Asymmetric transfer hydrogenation of C=C bond.

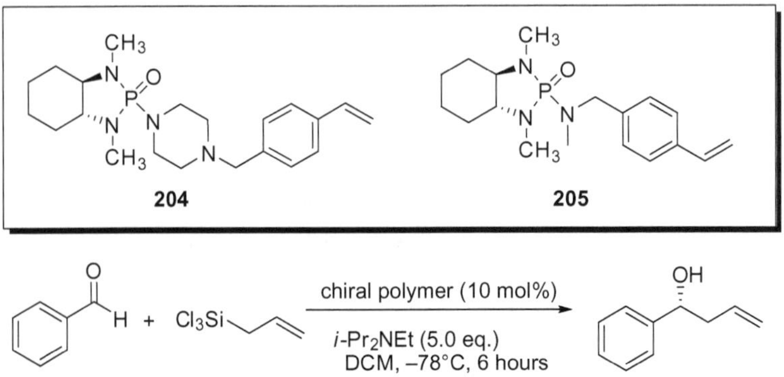

SCHEME 2.24. Asymmetric allylation of C=O group.

slight reduction in enantioselectivity (from 96% to 93%) was observed using the supported catalyst **3b** compared to the unsupported one [166].

Asymmetric transfer hydrogenation of C=C double bonds was performed using the resin-supported peptide **198** (see Scheme 2.19) in aqueous media (THF/H_2O 1:2) in the presence of TFA (Scheme 2.23). Catalyst **198** was the best among several supported di-, tri-, and tetra-peptides. The length of the chain was changed, and the catalyst having 25.4 leucine residues enhanced the reaction. Again, as for the Friedel–Craft reaction (see Section 2.9), the polyleucine chain was introduced with the aim of generating a hydrophobic environment around the prolyl residue that increased both yield and enantioselectivity when **198** was used in aqueous medium. Indeed, catalyst **198** without the polyleucine linker displayed lower performances. The PS-PEG support was essential for avoiding aggregation and sedimentation of the hydrophobic unsupported catalyst [167, 168].

2.11. MISCELLANEOUS

2.11.1. Allyltrichlorosilane Addition to C=O

Asymmetric allylation of C=O group promoted by chiral Lewis acids is another topic of great interest (see Chapter 14). Despite their usefulness, recyclable organocatalysts for such reactions have been very scarcely investigated, which leaves enough room for further research. The only significant report was published in 2005 by Oyama et al. who reported the polymerization of two styrenic chiral phosphoramides (**204** and **205**) (Scheme 2.24). The polymerization reactions were carried out in the absence of styrene or with styrene in various ratios with respect to **204** and **205** (25:75, 50:50, 75:25) in the presence of AIBN as radical initiator (80°C, 24 hours) [169].

Homopolymer obtained from **205** displayed the highest activity (84% yield) and enantioselectivity (63% ee). Moreover, these values were 2.4 times better and 1.4 higher, respectively, than those obtained with the corresponding unsupported catalysts. The principal reason for the high activity and selectivity of the polymeric catalysts was ascribed to the favorable multiple coordination of the phosphoramide units induced by the polymer effect. In addition, homopolymer and polymers obtained from **205** with variable amounts of styrene afforded similar results, indicating that at least 37% of chiral phosphoramide units are sufficient for multiple coordination. On the other hand, homopolymer and polymers obtained from **204** with variable amounts of styrene gave products in decreasing yield when the amount of **204** decreased in the polymer. Disappointingly, no data were presented about the recoverability and reusability of the catalytic systems. Finally, no further studies have been reported to date, making this field very attractive for new developments.

2.11.2. Aza-Morita–Baylis–Hillman Reaction

Another scattered example on supported chiral Lewis bases regards the use of dendritic chiral phosphine organocatalysts **206** for the asymmetric aza-Morita–Baylis–Hillman reaction of *N*-sulfonated imines with methyl vinyl ketone, ethyl vinyl ketone, and acrolein (Scheme 2.25) [170].

Screening of solvents indicated THF as the best reaction medium, while the second-generation catalyst **206** bearing benzyl groups gave the highest yield and enantioselectivity. Products were obtained in high yields and enantioselectivities, and results were comparable with those obtained with the

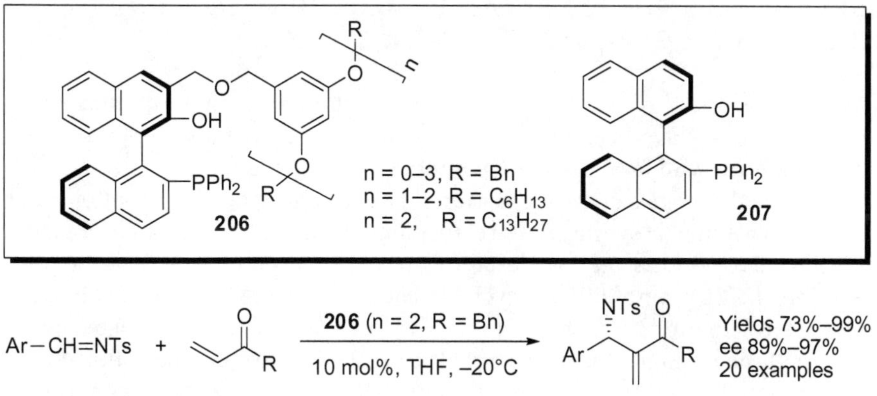

SCHEME 2.25. Asymmetric aza-Morita–Baylis–Hillman reaction of *N*-sulfonated imines.

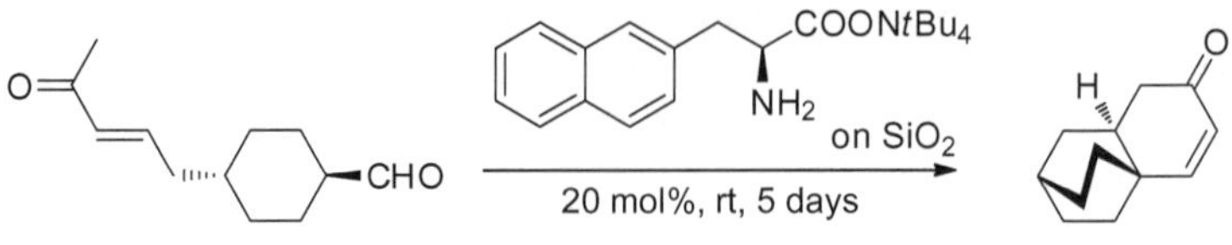

SCHEME 2.26. Asymmetric Robinson annulation.

corresponding unsupported catalyst **207**. Using the latter catalyst, both yields and enantioselectivities were lower. In addition, recovered catalyst **207** showed that most of it was oxidized to the corresponding phosphine oxide, and the catalytic activity disappeared completely when the recovered compound was reused. On the other hand, the recovered dendritic catalyst retained its activity and stereoselectivity in the second cycle. Most likely, the large branched polyether carbon chain in the dendrimer-supported organocatalyst retarded the oxidation of the phosphorus atom during the reaction, and therefore, the recovered supported organocatalyst could be successfully reused. The major drawback of such a catalytic system was due to the not-quantitative recovery of the catalyst (75%), which was obtained after the removal of the solvent and the washing of the residue with hexane/ether (8:1) to extract the product and unreacted ketone.

2.11.3. Robinson Annulation

The tetrabutylammonium salt of 2-naphthyl-substituted alanine amino acid gave the annulated product in 97% ee in 31% yield. When the same catalyst was adsorbed on silica gel, the yield increased up to 84% while maintaining the ee value. The immobilization allowed recycling for three cycles with no decrease in activity and enantioselectivity (Scheme 2.26) [171].

2.12. MULTISUPPORTED CATALYST-MEDIATED REACTIONS: THE ULTIMATE GOAL IN ASYMMETRIC SYNTHESIS

Over the years, a plethora of new chiral metal-based catalysts and chiral organocatalysts have been developed and successfully used in a great number of asymmetric syntheses (Part II, Chapters 11–16). At the same time, new materials and technologies (Part I, Chapter 1–10) are under development in order to make asymmetric synthesis more appealing from a practical point of view.

These two aspects may coalesce in the asymmetric domino (or cascade) reactions [172] promoted by recyclable catalysts. The use of recyclable catalysts for asymmetric domino reactions is worth being developed, since it could represent the ultimate goal in asymmetric synthesis and could be coupled with new materials and technologies, thereby allowing access to an enormous number of molecules (usually complex) with high levels of stereocontrol and in an efficient, atom-economical manner.

In this chapter, three examples of supported organocatalysts for domino reactions have been described (catalysts **135a**, **136a**, and **137a**, Section 2.6.1) but, actually, this intriguing field remains to be explored. Moreover, this field is not limited only to organocatalytic reactions, but is also open to metal-catalyzed reactions or metal/organocatalytic-based or enzyme-based reactions. In the reported examples, a single supported catalyst has been used to mediate one-pot reactions involving both iminium and enamine catalysis. The use of a single catalyst for multiple catalytic cycles is certainly advantageous, but the same catalyst could be optimal just for one catalytic cycle and less active for other cycles.

In 2008, Fréchet et al. reported the use of soluble star polymers with highly branched noninterpenetrating catalytic cores for one-pot multicomponent asymmetric cascade reactions [173]. They investigated the MacMillan imidazolidinone-mediated nucleophilic addition of *N*-methyl indole to 2-hexenal (via iminium catalysis), followed by a chiral pyrrolidine-catalyzed Michael addition of the product (via enamine catalysis) to methyl vinyl ketone (activated by H-bond catalysis) (Fig. 2.44). The catalysts were encapsulated within star polymers, which cannot penetrate each other's core, thereby maintaining their catalytic integrity. MacMillan's imidazolidinone was retained within the core during catalysis by electrostatic interactions, whereas the chiral pyrrolidine catalyst was covalently linked within the core of the star polymer.

This approach could be further developed by using a combination of otherwise incompatible or stereo-incompatible catalysts. Moreover, if an efficient

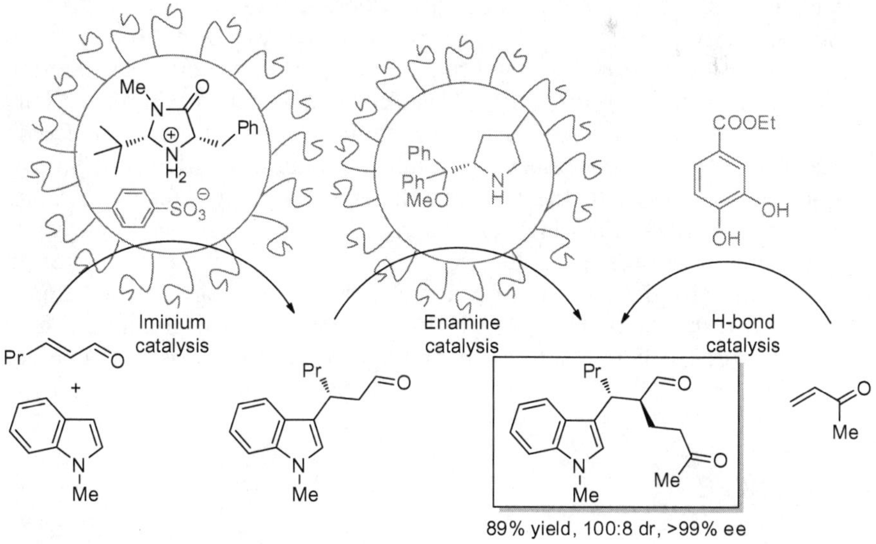

FIGURE 2.44. Noninterpenetrating star polymers for one-pot cascade asymmetric catalysis.

recovery and reuse of such catalytic materials can be done, it certainly will represent the most advanced asymmetric catalytic technology.

Finally, the new concept of "waste as catalyst/cocatalyst" [174] is briefly worth mentioning. This strategy is based on the employment of the waste generated in the upstream step of a cascade reaction as the catalyst or cocatalyst for the downstream step of the reaction. As a first example of this approach, it has been reported that Ph_3PO generated as waste in the Wittig reaction can be used as a useful additive in chiral Lewis acid-catalyzed reactions since it is capable of improving reactivity and selectivity. Indeed, Ph_3PO has been recycled in a tandem Wittig–cyanosilylation reaction in the presence of a chiral salen complex to afford products in 65%–93% ee. This approach is in its very early stages, and further investigations might lead to the use of supported chiral catalysts capable of being reused many times.

2.13. CONCLUSIONS

In the present chapter, the most innovative and advanced methodologies for recovery and reuse of chiral organic catalysts have been reported. Homogeneous chiral organic catalysts are currently finding a plethora of synthetic applications, and interest in this field is believed to increase in the near future. Although industrial application of chiral organic catalysts is still in its infancy, the development of this field is certainly of great interest for the entire community. At the same time, the development of highly recyclable chiral organic catalysts represents a step forward. The great advantages of this approach are due to the easier handling and workup of the reaction, to the possibility of using the catalysts many times, to the possibility of exploring new reaction conditions, and exploring the influence of the support (in the case of supported chiral organic catalysts). These aspects are useful in order to achieve not only recyclable materials but also to achieve even higher catalytic performances in terms of activity and stereoselectivity. Finally, the approaches developed in this chapter can be joined to the flow reaction methodology, which allows the application of a recyclable chiral organic catalyst to a continuous-flow system. As an example, Pericás et al. developed the use of a polystyrene-supported proline catalyst for the asymmetric Mannich reaction of aldehydes and ketones with the *N*-(p-methoxyphenyl)ethyl glyoxylate imine in a continuous-flow apparatus with excellent results [175]. This approach will be discussed in more depth in Chapter 8.

REFERENCES

[1] Ahrendt, K. A., Borths, C. J., MacMillan, D. W. C. (2000). New strategies for organic catalysis: the first highly enantioselective organocatalytic Diels-Alder reaction. *J. Am. Chem. Soc.*, *122*, 4243–4244.

[2] (a) Dalko, P. I. (2007). *Enantioselective Organocatalysis*. Wiley-VCH Verlag, Weinheim, Germany; (b) Berkessel, A., Gröger, H. (2005). *Asymmetric Organocatalysis: From Biomimetic Concepts to Applications in Asymmetric Synthesis*. Wiley-VCH, Weinheim, Germany; (c) Pellissier, H. (2007). Asymmetric organocatalysis. *Tetrahedron*, *63*, 9267–9331; (d) Dondoni, A., Massi, A. (2008). Asymmetric organocatalysis: from infancy to adolescence. *Angew. Chem. Int. Ed.*, *47*, 4638–4660; (e) Melchiorre, P., Marigo, M., Carlone, A., Bartoli, G. (2008). Asymmetric aminocatalysis-Gold rush in organic chemistry. *Angew. Chem. Int. Ed.*, *47*, 6138–6171; (f) List, B., Ed. (2010). Asymmetric organocatalysis. *Top. Curr. Chem.*, *291*. Springer, Dordrecht, South Africa.

[3] (a) Gruttadauria, M., Giacalone, F., Noto, R. (2008). Supported proline and proline-derivatives as recyclable organocatalysts. *Chem. Soc. Rev.*, *37*, 1666–1688; (b) Benaglia, M., Puglisi, A., Cozzi, F. (2003). Polymer-supported organic catalysts. *Chem. Rev.*, *103*, 3401–3430; (c) Cozzi, F. (2006). Immobilization of organic catalysts: when, why, and how. *Adv. Synth. Catal.*, *348*, 1367–1390; (d) Benaglia, M. (2006). Recoverable and recyclable chiral organic catalysts. *New J. Chem.*, *30*, 1525–1533; (e) Trindade, A. F., Gois, P. M. P., Afonso, C. A. M. (2009). Recyclable stereoselective catalysts. *Chem. Rev.*, *109*, 418–514; (f) Bergbreiter, D. E., Tian, J., Hongfa, C. (2009). Using soluble polymer supports to facilitate homogeneous catalysis. *Chem. Rev.*, *109*, 530–582; (g) Lu, J., Toy, P. H. (2009). Organic polymer supports for synthesis and for reagent and catalyst immobilization. *Chem. Rev.*, *109*, 815–838.

[4] (a) Cole-Hamilton, D. J., Tooze, R. (Eds.). (2006). *Catalyst Separation, Recovery and Recycling. Chemistry and Process Design*. Springer, Dordrecht, South Africa; (b) Albericio, F., Tulla-Puche, J. (Eds.) (2008). *The Power of Functional Resins in Organic Synthesis*. Wiley-VCH, Weinheim, Germany; (c) Benaglia, M. (Ed.) (2009). *Recoverable and Recyclable Catalysts*. Wiley-VCH, Weinheim, Germany.

[5] (a) Bonini, C., Righi, G. (2002). A critical outlook and comparison of enantioselective oxidation methodologies of olefins. *Tetrahedron*, *58*, 4981–5021; (b) Xia, Q.-H., Ge, H.-Q., Ye, C.-P., Liu, Z.-M., Su, K.-X. (2005). Advances in homogeneous and heterogeneous catalytic asymmetric epoxidation. *Chem. Rev.*, *105*, 1603–1662; (c) Wong, O. A., Shi, Y. (2009). Chiral ketone and iminium catalysts for olefin epoxidation. *Top. Curr. Chem.*, *291*, 201–232.

[6] (a) Flisak, J. R., Gombatz, K. J., Holmes, M. M., Jarmas, A. A., Lantos, I., Mendelson, W. L., Novack, V. J., Remich, J. J., Snyder, L. (1993). A practical, enantioselective synthesis of SK&F 104353. *J. Org. Chem.*, *58*, 6247–6254; (b) Allen, J. V., Bergeron, S., Griffiths, M. J., Mukherjee, S., Roberts, S. M., Williamson, N. M., Wu, L. E. (1998). Juliá-Colonna asymmetric epoxidation reactions under non-aqueous conditions: rapid, highly regio- and stereo-selective transformations using a cheap, recyclable catalyst. *J. Chem. Soc. Perkin Trans. 1*, 3171–3179.

[7] Itsuno, S., Sakakura, M., Ito, K. (1990). Polymer-supported poly(amino acids) as new asymmetric epoxidation catalyst of α,β-unsaturated ketones. *J. Org. Chem.*, *55*, 6047–6049.

[8] Cappi, M. W., Flood, R. W., Roberts, S. M., Skidmore, J., Williamson, N. M., Chen, W.-P., Liao, Y.-W., Smith, J. A. (1998). New procedures for the Juliá-Colonna asymmetric epoxidation: synthesis of (+)-clausenamide. *Chem. Commun.*, 1159–1160.

[9] Yi, H., Zou, G., Li, Q., Chen, Q., Tang, J., He, M. (2005). Asymmetric epoxidation of α,β-unsaturated ketones catalyzed by silica-grafted poly-(l)-leucine catalysts. *Tetrahedron Lett.*, *46*, 5665–5668.

[10] Yang, F., He, L.-M., Yi, H., Zou, G., Tang, J., He, M.-Y. (2007). An efficient silica-grafted poly-(l)-leucine catalyst for asymmetric epoxidation of α,β-unsaturated ketone. *J. Mol. Catal. A Chem.*, *273*, 1–4.

[11] Lv, J., Wang, X., Liu, J., Zhang, L., Wang, Y. (2006). Catalytic asymmetric epoxidation of chalcones under poly(ethylene glycol)-supported *Cinchona* ammonium salt catalyzed conditions. *Tetrahedron Asymmetry*, *17*, 330–335.

[12] Geller, T., Roberts, S. M. (1999). A new procedure for the Juliá-Colonna stereoselective epoxidation reaction under non-aqueous conditions: the development of a catalyst comprising polyamino acid on silica (PaaSiCat). *J. Chem. Soc. Perkin Trans. 1*, 1397–1398.

[13] Carde, L., Davies, H., Geller, T. P., Roberts, S. M. (1999). PaaSicats: powerful catalysts for asymmetric epoxidation of enones. Novel syntheses of α-arylpropanoic acids including (*S*)-fenoprofen. *Tetrahedron Lett.*, *40*, 5421–5424.

[14] Baars, S., Drauz, K.-H., Krimmer, H.-P., Roberts, S. M., Sander, J., Skidmore, J., Zanardi, G. (2003). Development of the Juliá-Colonna asymmetric epoxidation reaction: Part 1. Preparation and activation of the polyleucine catalyst. *Org. Proc. Res. Dev.*, *7*, 509–513.

[15] Dhanda, A., Drauz, K.-H., Geller, T., Roberts, S. M. (2000). PaaSiCats: novel polyamino acid catalysts. *Chirality*, *12*, 313–317.

[16] Qiu, W., He, L., Chen, Q., Luo, W., Yu, Z., Yang, F., Tang, J. (2009). Imidazolium-modified poly(l-leucine) catalyst: an efficient and recoverable catalyst for Juliá-Colonna reactions. *Tetrahedron Lett.*, *50*, 5225–5227.

[17] (a) O'Donnell, M. J. (2004). The enantioselective synthesis of α-amino acids by phase-transfer catalysis with achiral Schiff base esters. *Acc. Chem. Res.*, *37*, 506–517; (b) Hashimoto, T., Maruoka, K. (2007). Recent development and application of chiral phase-transfer catalysts. *Chem. Rev.*, *107*, 5656–5682.

[18] Chinchilla, R., Mazón, P., Nájera, C. (2004). Polystyrene-anchored *Cinchona* ammonium salts: easily recoverable phase-transfer catalysts for the asymmetric synthesis of α-amino acids. *Adv. Synth. Catal.*, *346*, 1186–1194.

[19] Chinchilla, R., Mazón, P., Nájera, C. (2000). Asymmetric synthesis of α-amino acids using polymer-supported *Cinchona* alkaloid-derived ammonium salts as chiral phase-transfer catalysts. *Tetrahedron Asymmetry*, *11*, 3277–3281.

[20] Thierry, B., Plaquevent, J.-C., Cahard, D. (2001). New polymer-supported chiral phase-transfer catalysts in the asymmetric synthesis of α-amino acids: the role of a spacer. *Tetrahedron Asymmetry*, *12*, 983–986.

[21] Shi, Q., Lee, Y.-J., Song, H., Cheng, M., Jew, S., Park, H., Jeong, B.-S. (2008). Electronically modified polymer-supported *Cinchona* phase-transfer catalysts for asymmetric synthesis of α-alkyl-α-amino acid derivatives. *Chem. Lett.*, *37*, 436–437.

[22] Thierry, B., Perrard, T., Audouard, C., Plaquevent, J.-C., Cahard, D. (2001). Solution- and solid-phase approaches in asymmetric phase-transfer catalysis by *Cinchona* alkaloid derivatives. *Synthesis*, 1742–1746.

[23] Shi, Q., Lee, Y.-J., Kim, M.-Y., Park, M.-K., Lee, K., Song, H., Cheng, M., Jeong, B.-S., Park, H., Jew, S. (2008). Highly efficient polymer supported phase-transfer

catalysts containing hydrogen bond inducing functional groups. *Tetrahedron Lett.*, *49*, 1380–1383.

[24] Chinchilla, R., Mazón, P., Nájera, C. (2004). Polymer-supported *Cinchona* alkaloid-derived ammonium salts as recoverable phase-transfer catalysts for the asymmetric synthesis of α-amino *acids*. *Molecules*, *9*, 349–364.

[25] Thierry, B., Plaquevent, J.-C., Cahard, D. (2003). Poly(ethylene glycol) supported *Cinchona* alkaloids as phase transfer catalysts: application to the enantioselective synthesis of α-amino acids. *Tetrahedron Asymmetry*, *14*, 1671–1677.

[26] Kim, D. Y., Huh, S. C. (2004). Poly(ethylene glycol) supported chiral quaternary ammonium salts as phase-transfer catalysts for catalytic enantioselective synthesis of α-amino acids. *Bull. Korean Chem. Soc.*, *25*, 347–348.

[27] Danelli, T., Annunziata, R., Benaglia, M., Cinquini, M., Cozzi, F., Tocco, G. (2003). Immobilization of catalysts derived from *Cinchona* alkaloids on modified poly(ethylene glycol). *Tetrahedron Asymmetry*, *14*, 461–467.

[28] Wang, X., Yin, L., Yang, T., Wang, Y. (2007). Synthesis of new dimeric-PEG-supported *Cinchona* ammonium salts as chiral phase transfer catalysts for the alkylation of Schiff bases with water as the solvent. *Tetrahedron Asymmetry*, *18*, 108–114.

[29] (a) Itsuno, S., Paul, D. K., Ishimoto, M., Haraguchi, N. (2010). Designing chiral quaternary ammonium polymers: novel type of polymeric catalysts for asymmetric alkylation reaction. *Chem. Lett.*, *39*, 86–87; (b) Itsuno, S., Paul, D. K., Salam, M. A., Haraguchi, N. (2010). Main-chain ionic chiral polymers: synthesis of optically active quaternary ammonium sulfonate polymers and their application in asymmetric catalysis. *J. Am. Chem. Soc.*, *132*, 2864–2865.

[30] Guillena, G., Kreiter, R., van de Coevering, R., Gebbink, R. J. M. K., van Koten, G., Mazón, P., Chinchilla, R., Nájera, C. (2003). Chiroptical properties and applications in PTC of new dendritic cinchonidine-derived ammonium salts. *Tetrahedron Asymmetry*, *14*, 3705–3712.

[31] Shirakawa, S., Ueda, M., Tanaka, Y., Hashimoto, T., Maruoka, K. (2007). Design of binaphthyl-modified symmetrical chiral phase-transfer catalysts: substituent effect of 4,4′,6,6′-positions of binaphthyl rings in the asymmetric alkylation of a glycine derivative. *Chem. Asian J.*, *2*, 1276–1281.

[32] Arakawa, Y., Haraguchi, N., Itsuno, S. (2008). An immobilization method of chiral quaternary ammonium salts onto polymer supports. *Angew. Chem. Int. Ed.*, *47*, 8232–8235.

[33] Youk, S. H., Oh, S. H., Rho, H. S., Lee, J. E., Lee, J. W., Song, C. E. (2009). A polymer-supported *Cinchona*-based bifunctional sulfonamide catalyst: a highly enantioselective, recyclable heterogeneous organocatalyst. *Chem. Commun.*, 2220–2222.

[34] Clapham, B., Cho, C.-W., Janda, K. D. (2001). A polymer-supported proline-based diamine catalyst for the kinetic resolution of racemic secondary alcohols. *J. Org. Chem.*, *66*, 868–873.

[35] Kosugi, Y., Akakura, M., Ishihara, K. (2007). Kinetic resolution of racemic alcohols catalyzed by minimal artificial acylases derived from L-histidine. *Tetrahedron*, *63*, 6191–6203.

[36] Gleeson, O., Tekoriute, R., Gun'ko, Y. K., Connon, S. J. (2009). The first magnetic nanoparticle-supported chiral DMAP analogue: highly enantioselective acylation and excellent recyclability. *Chem. Eur. J.*, *15*, 5669–5673.

[37] Kim, H. S., Song, Y.-M., Choi, J. S., Yang, J. W., Han, H. (2004). Heterogeneous organocatalysis for the asymmetric desymmetrization of *meso*-cyclic anhydrides using silica gel-supported bis-*Cinchona* alkaloids. *Tetrahedron*, *60*, 12051–12057.

[38] (a) Machajewski, T. D., Wong, C.-H. (2000). The catalytic asymmetric aldol reaction. *Angew. Chem. Int. Ed.*, *39*, 1352–1375; (b) Alcaide, B., Almendros, P. (2003). The direct catalytic asymmetric cross-aldol reaction of aldehydes. *Angew. Chem. Int. Ed.*, *42*, 858–860; (c) Palomo, C., Oiarbide, M., Garcia, J. M. (2002). The aldol addition reaction: an old transformation at constant rebirth. *Chem. Eur. J.*, *8*, 36–44; (d) Mahrwald, R. (Ed.). (2004). *Modern Aldol Reactions*, Vols. *1, 2*. Wiley-VCH, Weinheim, Germany.

[39] Kondo, K., Yamano, T., Takemoto, K. (1985). Asymmetric Robinson cyclization reaction catalyzed by polymer-bound L-proline. *Makromol. Chem.*, *186*, 1781–1785.

[40] Mase, N., Nakai, Y., Ohara, N., Yoda, H., Takabe, K., Tanaka, F., Barbas, III, C. F. (2006). Organocatalytic direct asymmetric aldol reactions in water. *J. Am. Chem. Soc.*, *128*, 734–735.

[41] Hayashi, Y., Sumiya, T., Takahashi, J., Gotoh, H., Urushima, T., Shoji, M. (2006). Highly diastereo- and enantioselective direct aldol reactions in water. *Angew. Chem. Int. Ed.*, *45*, 958–961.

[42] Font, D., Jimeno, C., Pericàs, M. A. (2006). Polystyrene-supported hydroxyproline: an insoluble, recyclable organocatalyst for the asymmetric aldol reaction in water. *Org. Lett.*, *8*, 4653–4656.

[43] Font, D., Bastero, A., Sayalero, S., Jimeno, C., Pericàs, M. A. (2007). Highly enantioselective α-aminoxylation of aldehydes and ketones with a polymer-supported organocatalyst. *Org. Lett.*, *9*, 1943–1946.

[44] Font, D., Sayalero, S., Bastero, A., Jimeno, C., Pericàs, M. A. (2008). Toward an artificial aldolase. *Org. Lett.*, *10*, 337–340.

[45] (a) Giacalone, F., Gruttadauria, M., Mossuto Marculescu, A., Noto, R. (2007). Polystyrene-supported proline and prolinamide. Versatile heterogeneous organocatalysts both for asymmetric aldol reaction in water and α-selenenylation of aldehydes. *Tetrahedron Lett.*, *48*, 255–259; (b) Gruttadauria, M., Giacalone, F., Mossuto Marculescu, A., Lo Meo, P., Riela, S., Noto, R. (2007). Hydrophobically directed aldol reactions: polystyrene-supported L-proline as recyclable catalyst for direct asymmetric aldol reaction in the presence of water. *Eur. J. Org. Chem.*, 4688–4698.

[46] Giacalone, F., Gruttadauria, M., Mossuto Marculescu, A., D'Anna, F., Noto, R. (2008). Polystyrene-supported proline as recyclable catalyst in the Baylis-Hillman reaction of arylaldehydes and methyl or ethyl vinyl ketone. *Catal. Commun.*, *9*, 1477–1481.

[47] (a) Liu, Y.-X., Sun, Y.-N., Tan, H.-H., Liu, W., Tao, J.-C. (2007). Linear polystyrene anchored L-proline, new recyclable organocatalysts for the aldol reaction in the presence of water. *Tetrahedron Asymmetry*, *18*, 2649–2656; (b) Liu, Y.-X., Sun, Y.-N., Tan, H.-H., Tao, J.-C. (2008). Asymmetric aldol reaction catalyzed by new recyclable polystyrene-supported L-proline in the presence of water. *Catal. Lett.*, *120*, 281–287.

[48] Kristensen, T. E., Vestli, K., Jakobsen, M. G., Hansen, F. K., Hansen, T. (2010). A General approach for preparation of polymer-supported chiral organocatalysts via acrylic copolymerization. *J. Org. Chem.*, *75*, 1620–1629.

[49] Andreae, M. R. M., Davis, A. P. (2005). Heterogeneous catalysis of the asymmetric aldol reaction by solid-supported proline-terminated peptides. *Tetrahedron Asymmetry*, *16*, 2487–2492.

[50] Akagawa, K., Sakamoto, S., Kudo, K. (2005). Direct asymmetric aldol reaction in aqueous media using polymer-supported peptide. *Tetrahedron Lett.*, *46*, 8185–8187.

[51] Krattiger, P., Kovasy, R., Revell, J. D., Ivan, S., Wennemers, H. (2005). Increased structural complexity leads to higher activity: peptides as efficient and versatile catalysts for asymmetric aldol reactions. *Org. Lett.*, *7*, 1101–1103.

[52] Revell, J. D., Gantenbein, D., Krattiger, P., Wennemers, H. (2005). Solid-supported and pegylated H-Pro-Pro-Asp-NHR as catalysts for asymmetric aldol reactions. *Biopolymers*, *84*, 105–113.

[53] Carpenter, R. D., Fettinger, J. C., Lam, K. S., Kurth, M. J. (2008). Asymmetric catalysis: resin-bound hydroxyprolylthreonine derivatives in enamine-mediated reactions. *Angew. Chem. Int. Ed.*, *47*, 6407–6410.

[54] Yan, J., Wang, L. (2008). Merrifield resin supported dipeptides: efficient and recyclable organocatalysts for asymmetric aldol reactions under neat reaction conditions. *Synthesis*, 2065–2072.

[55] Gruttadauria, M., Salvo, A. M. P., Giacalone, F., Agrigento, P., Noto, R. (2009). Enhanced activity and stereoselectivity of polystyrene-supported proline-based organic catalysts for direct asymmetric aldol reaction in water. *Eur. J. Org. Chem.*, 5437–5444.

[56] Gruttadauria, M., Giacalone, F., Mossuto Marculescu, A., Noto, R. (2008). Novel prolinamide-supported polystyrene as highly stereoselective and recyclable organocatalyst for the aldol reaction. *Adv. Synth. Catal.*, *350*, 1397–1405.

[57] Gruttadauria, M., Giacalone, F., Mossuto Marculescu, A., Salvo, A. M. P., Noto, R. (2009). Stereoselective aldol reaction catalyzed by a highly recyclable polystyrene supported substituted prolinamide catalyst. *ARKIVOC*, *viii*, 5–15.

[58] Dhar, D., Beadham, I., Chandrasekaran, S. (2003). Proline and benzylpenicillin derivatives grafted into mesoporous MCM-41: novel organic-inorganic hybrid catalysts for direct aldol reaction. *Proc. Indian Acad. Sci. (Chem. Sci.)*, *115*, 365–372.

[59] Calderón, F., Fernández, R., Sánchez, F., Fernández-Mayoralas, A. (2005). Asymmetric aldol reaction using immobilized proline on mesoporous support. *Adv. Synth. Catal.*, *347*, 1395–1403.

[60] Doyagüez, E. G., Calderón, F., Fernández, R., Sánchez, F., Fernández-Mayoralas, A. (2007). Asymmetric aldol reaction catalyzed by a heterogenized proline on a mesoporous support. The role of the nature of solvents. *J. Org. Chem.*, *72*, 9353–9356.

[61] Zamboulis, A., Rahier, N. J., Gehringer, M., Cattoën, X., Niel, G., Bied, C., Moreau, J. J. E., Wong Chi Man, M. (2009). Silica-supported L-proline organocatalysts for asymmetric aldolisation. *Tetrahedron Asymmetry*, *20*, 2880–2885.

[62] Yan, J., Wang, L. (2009). Asymmetric aldol reactions catalyzed by efficient and recyclable silica-supported proline-based peptides. *Chirality*, *21*, 413–420.

[63] Benaglia, M., Celentano, G., Cozzi, F. (2001). Enantioselective aldol condensations catalyzed by poly(ethylene glycol)-supported proline. *Adv. Synth. Catal.*, *343*, 171–173.

[64] Benaglia, M., Cinquini, M., Cozzi, F., Puglisi, A., Celentano, G. (2002). Poly(ethylene glycol)-supported proline: a versatile catalyst for the enantioselective aldol and iminoaldol reactions. *Adv. Synth. Catal.*, *344*, 533–542.

[65] Wu, Y., Zhang, Y., Yu, M., Zhao, G., Wang, S. (2006). Highly efficient and reusable dendritic catalysts derived from *N*-prolylsulfonamide for the asymmetric direct aldol reaction in water. *Org. Lett.*, *8*, 4417–4420.

[66] Kotrusz, P., Kmentová, I., Gotov, B., Toma, Š., Solčaniová, E. (2002). Proline-catalysed asymmetric aldol reaction in the room temperature ionic liquid [bmim] PF6. *Chem. Commun.*, 2510–2511.

[67] Loh, T.-P., Feng, L.-C., Yang, H. Y., Yang, J.-Y. (2002). L-Proline in an ionic liquid as an efficient and reusable catalyst for direct asymmetric aldol reactions. *Tetrahedron Lett.*, *43*, 8741–8743.

[68] Córdova, A. (2004). Direct catalytic asymmetric cross-aldol reactions in ionic liquid media. *Tetrahedron Lett.*, *45*, 3949–3952.

[69] Mečiarová, M., Toma, Š., Berkessel, A., Koch, B. (2006). *Lett. Org. Chem.*, *3*, 437–441.

[70] Guo, H.-M., Cun, L.-F., Gong, L.-Z., Mi, A.-Q., Jiang, Y.-Z. (2005). Asymmetric direct aldol reaction catalyzed by an L-prolinamide derivative: considerable improvement of the catalytic efficiency in the ionic liquid. *Chem. Commun.*, 1450–1452.

[71] Teo, Y.-C., Chua, G.-L. (2008). A recyclable non-immobilized siloxy serine organocatalyst for the asymmetric direct aldol reaction. *Tetrahedron Lett.*, *49*, 4235–4238.

[72] Yong, F.-F., Poh, C.-Y., Chua, G.-L., Teo, Y.-C. (2010). Organocatalytic asymmetric *syn*-selective direct aldol reaction in ionic liquid. *Chem. Lett.*, *39*, 490–492.

[73] Qian, Y., Zheng, X., Wang, Y. (2010). A green and efficient asymmetric aldol reaction catalyzed by a chiral anion modified ionic liquid. *Eur. J. Org. Chem.*, 3672–3677.

[74] Gruttadauria, M., Riela, S., Lo Meo, P., D'Anna, F., Noto, R. (2004). Supported ionic liquid asymmetric catalysis. A new method for chiral catalysts recycling. The case of proline-catalyzed aldol reaction. *Tetrahedron Lett.*, *45*, 6113–6116.

[75] Gruttadauria, M., Riela, S., Aprile, C., Lo Meo, P., D'Anna, F., Noto, R. (2006). Supported ionic liquids. New recyclable materials for the L-proline-catalyzed aldol reaction. *Adv. Synth. Catal.*, *348*, 82–92.

[76] Aprile, C., Giacalone, F., Gruttadauria, M., Mossuto Marculescu, A., Noto, R., Revell, J. D., Wennemers, H. (2007). New ionic liquid-modified silica gels as recyclable materials for L-proline or H-Pro-Pro-Asp-NH_2-catalyzed aldol reaction. *Green Chem.*, *9*, 1328–1334.

[77] Lei, M., Xia, S., Wang, J., Ge, Z., Cheng, T., Li, R. (2010). An efficient dipeptide-catalyzed direct asymmetric aldol reaction of equimolar reactants in solid media. *Chirality*, *22*, 580–586.

[78] Li, J., Hu, S., Luo, S., Cheng, J.-P. (2009). Chiral amine-polyoxomethalate hybrids as recoverable asymmetric esamine catalysts under neat and aqueous conditions. *Eur. J. Org. Chem.*, 132–140.

[79] Miao, W., Chan, T. H. (2006). Ionic-liquid-supported organocatalyst: efficient and recyclable ionic-liquid-anchored proline for asymmetric aldol reaction. *Adv. Synth. Catal.*, *348*, 1711–1718.

[80] Zhou, L., Wang, L. (2007). Chiral ionic liquid containing L-proline unit as a highly efficient and recyclable asymmetric organocatalyst for aldol reaction. *Chem. Lett.*, *36*, 628–629.

[81] Zhuo, C., Yong, L., Hui, X., Chang-gang, H., Xian, D. (2008). Ionic liquid-supported proline as catalyst in direct asymmetric aldol reaction. *Russ. J. Org. Chem.*, *44*, 1807–1810.

[82] Lombardo, M., Pasi, F., Easwar, S., Trombini, C. (2007). An improved protocol for the direct asymmetric aldol reaction in ionic liquids, catalysed by onium ion-tagged prolines. *Adv. Synth. Catal.*, *349*, 2061–2065.

[83] Lombardo, M., Pasi, F., Easwar, S., Trombini, C. (2008). Direct asymmetric aldol reaction catalyzed by an imidazolium-tagged *trans*-4-hydroxy-L-proline under aqueous biphasic conditions. *Synlett*, 2471–2474.

[84] Gruttadauria, M., Giacalone, F., Noto, R. (2009). Water in stereoselective organocatalytic reactions. *Adv. Synth. Catal.*, *351*, 33–57.

[85] Lombardo, M., Easwar, S., Pasi, F., Trombini, C. (2009). The ion tag strategy as a route to highly efficient organocatalysts for the direct asymmetric aldol reaction. *Adv. Synth. Catal.*, *351*, 276–282.

[86] Siyutkin, D. E., Kucherenko, A. S., Struchkova, M. I., Zlotin, S. G. (2008). A novel (*S*)-proline-modified task-specific chiral ionic liquid—an amphiphilic recoverable catalyst for direct asymmetric aldol reactions in water. *Tetrahedron Lett.*, *49*, 1212–1216.

[87] Siyutkin, D. E., Kucherenko, A. S., Zlotin, S. G. (2009). Hydroxy-α-amino acids modified by ionic liquid moieties: recoverable organocatalysts for asymmetric aldol reactions in the presence of water. *Tetrahedron*, *65*, 1366–1372.

[88] Lombardo, M., Easwar, S., De Marco, A., Pasi, F., Trombini, C. (2008). A modular approach to catalyst hydrophobicity for an asymmetric aldol reaction in a biphasic aqueous environment. *Org. Biomol. Chem.*, *6*, 4224–4229.

[89] Siyutkin, D. E., Kucherenko, A. S., Zlotin, S. G. (2010). A new (*S*)-prolinamide modified by an ionic liquid moiety—a high performance recoverable catalyst for asymmetric aldol reactions in aqueous media. *Tetrahedron*, *66*, 513–518.

[90] Shah, J., Khan, S. S., Blumenthal, H., Liebscher, J. (2009). 1,2,3-Triazolium-tagged prolines and their application in asymmetric aldol and Michael reactions. *Synthesis*, 3975–3982.

[91] Zu, L., Xie, H., Li, H., Wang, J., Wang, W. (2008). Highly enantioselective aldol reactions catalyzed by a recyclable fluorous (*S*)-pyrrolidine sulfonamide on water. *Org. Lett.*, *10*, 1211–1214.

[92] Miura, T., Imai, K., Ina, M., Tada, N., Imai, N., Itoh, A. (2010). Direct asymmetric aldol reaction with recyclable fluorous organocatalyst. *Org. Lett.*, *12*, 1620–1623.

[93] Shen, Z., Ma, J., Liu, Y., Jiao, C., Li, M., Zhang, Y. (2005). β-Cyclodextrin-immobilized (4*S*)-phenoxy-(*S*)-proline as a catalyst for direct asymmetric aldol reactions. *Chirality*, *17*, 556–558.

[94] Liu, K., Häussinger, D., Woggon, W.-D. (2007). Aldol reactions in water using a β-cyclodextrin-binding proline derivative. *Synlett*, 2298–2300.

[95] Huang, J., Zhang, X., Armstrong, D. W. (2007). Highly efficient asymmetric direct stoichiometric aldol reactions on/in water. *Angew. Chem. Int. Ed.*, *46*, 9073–9077.

[96] Hara, N., Nakamura, S., Shibata, N., Toru, T. (2010). Enantioselective aldol reaction using recyclable montmorillonite-entrapped *N*-(2-thiophenesulfonyl)prolinamide. *Adv. Synth. Catal.*, *352*, 1621–1624.

[97] Rodríguez-Llansola, F., Miravet, J. F., Escuder, B. (2009). A supramolecular hydrogel as a reusable heterogeneous catalyst for the direct aldol reaction. *Chem. Commun.*, 7303–7305.

[98] Almaşi, D., Alonso, D. A., Nájera, C. (2008). Prolinamides versus prolinethioamides as recyclable catalysts in the enantioselective solvent-free inter- and intramolecular aldol reactions. *Adv. Synth. Catal.*, *350*, 2467–2472.

[99] Wu, X., Jiang, Z., Shen, H.-M., Lu, Y. (2007). Highly efficient threonine-derived organocatalysts for direct asymmetric aldol reactions in water. *Adv. Synth. Catal.*, *349*, 812–816.

[100] Guillena, G., del Carmen Hita, M., Nájera, C., Viózquez, S. F. (2008). A highly efficient solvent-free asymmetric direct aldol reaction organocatalyzed by recoverable (*S*)-binam-L-prolinamides. ESI-MS Evidence of the enamine-iminium formation. *J. Org. Chem.*, *73*, 5933–5943.

[101] Kucherenko, A. S., Syutkin, D. E., Zlotin, S. G. (2008). Asymmetric aldol condensation in an ionic liquid-water system catalyzed by (*S*)-prolinamide derivatives. *Russ. Chem. Bull. Int. Ed.*, *57*, 591–594.

[102] Chandrasekhar, S., Ramakrishna Reddy, N., Shameen Sultana, S., Narsihmulu, C., Venkatram Reddy, K. (2006). L-Proline catalysed asymmetric aldol reactions in PEG-400 as recyclable medium and transfer aldol reactions. *Tetrahedron*, *62*, 338–345.

[103] Alza, E., Cambeiro, X. C., Jimeno, C., Pericás, M. A. (2007). Highly enantioselective Michael additions in water catalyzed by a PS-supported pyrrolidine. *Org. Lett.*, *9*, 3717–3720.

[104] Miao, T., Wang, L. (2008). Polystyrene-immobilized pyrrolidine as a highly stereoselective and recyclable organocatalyst for asymmetric Michael addition of cyclohexanone to nitroolefins. *Tetrahedron Lett.*, *49*, 2173–2176.

[105] Li, P., Wang, L., Wang, M., Zhang, Y. (2008). Polymer-immobilized pyrrolidine-based chiral ionic liquids as recyclable organocatalysts for asymmetric Michael additions to nitrostyrenes under solvent-free reaction conditions. *Eur. J. Org. Chem.*, 1157–1160.

[106] Alza, E., Pericás, M. A. (2009). A highly selective, polymer-supported organocatalyst for the Michael additions with enzyme-like behavior. *Adv. Synth. Catal.*, *351*, 3051–3056.

[107] Chuan, Y., Chen, G., Peng, Y. (2009). Highly enantioselective Michael addition reactions in water catalyzed by an insoluble MPS-supported 4-sulfonamidyl prolinol *tert*-butyldiphenylsilyl ether. *Tetrahedron Lett.*, *50*, 3054–3058.

[108] Tuchman-Shukron, L., Portnoy, M. (2009). Polymer-supported highly enantioselective catalyst for nitro-Michael addition: tuning through variation of the number of H-bond donors and spacer length. *Adv. Synth. Catal.*, *351*, 541–546.

[109] Miyabe, H., Tuchida, S., Yamauchi, M., Takemoto, Y. (2006). Reaction of nitroorganic compounds using thiourea catalysts anchored to polymer support. *Synthesis*, 3295–3300.

[110] Röben, C., Stasiak, M., Janza, B., Greiner, A., Wendorff, J. H., Studer, A. (2008). Immobilization of oligostyrene-prolinol conjugates into polystyrene via electrospinning and applications of these fibers in catalysis. *Synthesis*, 2163–2168.

[111] Varela, M. C., Dixon, S. M., Lam, K. S., Schore, N. E. (2008). Asymmetric epoxidation, Michael addition, and triple cascade reaction using polymer-supported prolinol-based auxiliaries. *Tetrahedron*, *64*, 10087–10090.

[112] Hodge, P., Khoshdel, E., Waterhouse, J. (1985). Michael additions catalysed by *Cinchona* alkaloids bound via their vinyl groups to preformed crosslinked polymers. *J. Chem. Soc. Perkin Trans. 1*, 2327–2331.

[113] Alvarez, R., Hourdin, M.-A., Cavé, C., Angelo, J., Chaminade, P. (1999). New polymer-supported catalysts derived from *Cinchona* alkaloids: their use in the asymmetric Michael reaction. *Tetrahedron Lett.*, *40*, 7091–7094.

[114] Hafez, A. M., Taggi, A. E., Dudding, T., Lectka, T. (2001). Asymmetric catalysis on sequentially-linked columns. *J. Am. Chem. Soc.*, *123*, 10853–10859.

[115] Li, P., Wang, L., Zhang, Y., Wang, G. (2008). Silica gel supported pyrrolidine-based chiral ionic liquid as recyclable organocatalyst for asymmetric Michael addition to nitrostyrenes. *Tetrahedron*, *64*, 7633–7638.

[116] Zhao, Y.-B., Zhang, L. W., Wu, L.-Y., Zhong, X., Li, R., Ma, J.-T. (2008). Silica-supported pyrrolidine-triazole, an insoluble, recyclable organocatalyst for the enantioselective Michael addition of ketones to nitroalkenes. *Tetrahedron Asymmetry*, *19*, 1352–1355.

[117] Fuerte, A., Corma, A., Sánchez, F. (2005). Heterogenised chiral amines as environmentally friendly base catalysts for enantioselective Michael addition. *Catal. Today*, *107–108*, 404–409.

[118] Corma, A., Iborra, S., Rodríguez, I., Sánchez, F. (2002). MCM-41 heterogenized chiral amines as base catalysts for enantioselective Michael reaction. *Catal. Lett.*, *82*, 237–242.

[119] Prasetyanto, E. A., Lee, S.-C., Jeong, S.-M., Park, S.-E. (2008). Chiral enhancement in diethyl malonate addition by morphosynthesized L-proline mesoporous silica. *Chem. Commun.*, 1995–1997.

[120] Mager, I., Zeitler, K. (2010). Efficient, enantioselective iminium catalysis with an immobilized, recyclable diarylprolinol silyl ether catalyst. *Org. Lett.*, *12*, 1480–1483.

[121] Gu, L., Wu, Y., Zhang, Y., Zhao, G. (2007). A new class of efficient poly(ethylene-glycol)-supported catalyst based on proline for the asymmetric Michael addition of ketones to nitrostyrenes. *J. Mol. Catal. A Chem.*, *263*, 186–194.

[122] Lv, G., Jin, R., Mai, W., Gao, L. (2008). Highly efficient and recoverable dendritic organocatalyst from click chemistry for the asymmetric michael addition of ketones to nitroolefins without the use of organic solvent. *Tetrahedron Asymmetry*, *19*, 2568–2572.

[123] Lo, C.-M., Chow, H.-F. (2009). Structural effects on the catalytic, emulsifying, and recycling properties of chiral amphiphilic dendritic organocatalysts. *J. Org. Chem.*, *74*, 5181–5191.

[124] Luo, S., Mi, X., Zhang, L., Liu, S., Xu, H., Cheng, J.-P. (2006). Functionalized chiral ionic liquids as highly efficient asymmetric organocatalysts for Michael addition to nitroolefins. *Angew. Chem. Int. Ed.*, *45*, 3093–3097.

[125] Luo, S., Zhang, L., Mi, X., Qiao, Y., Cheng, J.-P. (2007). Functionalized chiral ionic liquid catalyzed enantioselective desymmetrizations of prochiral ketones via asymmetric Michael addition reaction. *J. Org. Chem.*, *72*, 9350–9352.

[126] Ni, B., Zhang, Q., Headley, A. D. (2007). Functionalized chiral ionic liquid as recyclable organocatalyst for asymmetric Michael addition to nitrostyrenes. *Green Chem.*, *9*, 737–739.

[127] Zhang, Q., Ni, B., Headley, A. D. (2008). Asymmetric Michael addition reactions of aldehydes with nitrostyrenes catalyzed by functionalized chiral ionic liquids. *Tetrahedron*, *64*, 5091–5097.

[128] Ni, B., Zhang, Q., Dhungana, K., Headley, A. D. (2009). Ionic liquid-supported (ILS) (*S*)-pyrrolidine sulfonamide, a recyclable organocatalyst for the highly enantioselective Michael addition to nitroolefins. *Org. Lett.*, *11*, 1037–1040.

[129] Yan, Z.-Y., Niu, Y.-N., Wei, H.-L., Wu, L.-Y., Zhao, Y.-B., Liang, Y.-M. (2006). Combining proline and "click chemistry": a class of versatile organocatalysts for the highly diastereo- and enantioselective Michael addition in water. *Tetrahedron Asymmetry*, *17*, 3288–3293.

[130] Luo, S., Xu, H., Mi, X., Li, J., Zheng, X., Cheng, J.-P. (2006). Evolution of pyrrolidine-type asymmetric organocatalysts by "click" *chemistry. J. Org. Chem.*, *71*, 9244–9247.

[131] Wu, L.-Y., Yan, Z.-Y., Xie, Y.-X., Niu, Y.-N., Liang, Y.-M. (2007). Ionic-liquid-supported organocatalyst for the enantioselective Michael addition of ketones to nitroolefins. *Tetrahedron Asymmetry*, *18*, 2086–2090.

[132] Yacob, Z., Shah, J., Leistner, J., Liebscher, J. (2008). *S*)-Pyrrolidin-2-ylmethyl-1,2,3-triazolium salts—ionic liquid supported organocatalysts for enantioselective Michael additions to β-nitrostyrenes. *Synlett*, 2342–2344.

[133] Miao, T., Wang, L., Li, P., Yan, J. (2008). A highly efficient and recyclable ionic liquid anchored pyrrolidine catalyst for enantioselective Michael additions. *Synthesis*, 3828–3834.

[134] Ni, B., Zhang, Q., Headley, A. D. (2008). Pyrrolidine-based chiral pyridinium ionic liquids (ILs) as recyclable and highly efficient organocatalysts for the asymmetric Michael addition reactions. *Tetrahedron Lett.*, *49*, 1249–1252.

[135] Xu, D.-Q., Wang, B.-T., Luo, S.-P., Yue, H.-D., Wang, L.-P., Xu, Z.-Y. (2007). Pyrrolidine-pyridinium based organocatalysts for highly enantioselective Michael addition of cyclohexanone to nitroalkenes. *Tetrahedron Asymmetry*, *18*, 1788–1794.

[136] Maltsev, O. V., Kucherenko, A. S., Zlotin, S. G. (2009). *O*-TMS-α,α-diphenyl-(*S*)-prolinol modified with an ionic liquid moiety: a recoverable organocatalyst for the asymmetric Michael reaction between α,β-enals and dialkyl malonates. *Eur. J. Org. Chem.*, 5134–5137.

[137] Lombardo, M., Chiarucci, M., Quintavalle, A., Trombini, C. (2009). Highly efficient ion-tagged catalyst for the enantioselective Michael addition of aldehydes to nitroalkenes. *Adv. Synth. Catal.*, *351*, 2801–2806.

[138] Zu, L., Wang, J., Li, H., Wang, W. (2006). A recyclable fluorous (*S*)-pyrrolidine sulfonamide promoted direct, highly enantioselective Michael addition of ketones and aldehydes to nitroolefins in water. *Org. Lett.*, *8*, 3077–3079.

[139] Zu, L., Li, H., Wang, J., Yu, X., Wang, W. (2006). Highly enantioselective aldehyde-nitroolefin Michael addition reactions catalyzed by recyclable fluorous (*S*) diphenylpyrrolinol silyl ether. *Tetrahedron Lett.*, *47*, 5131–5134.

[140] Almaşi, D., Alonso, D. A., Gómez-Bengoa, E., Nájera, C. (2009). Chiral 2-aminobenzimidazoles as recoverable organocatalysts for the addition of 1,3-dicarbonyl compounds to nitroalkenes. *J. Org. Chem.*, *74*, 6163–6168.

[141] Chinchilla, R., Mazón, P., Nájera, C., Ortega, F. J., Yus, M. (2005). Enantioselective Michael reactions promoted by recoverable dimeric anthryl-derived *Cinchona* phase-transfer *catalysts. ARKIVOC*, *vi*, 222–232.

[142] Tarí, S., Chinchilla, R., Nájera, C. (2009). Enantioselective Michael reaction of β-keto esters organocatalyzed by recoverable *Cinchona*-derived dimeric ammonium salts. *Tetrahedron Asymmetry*, *20*, 2651–2654.

[143] Tan, B., Zhang, X., Chua, P. J., Zhong, G. (2009). Recyclable organocatalysis: highly enantioselective Michael addition of 1,3-diaryl-1,3-propanedione to nitroolefins. *Chem. Commun.*, 779–781.

[144] Xu, D. Q., Luo, S. P., Wang, Y. F., Xia, A. B., Yue, H. D., Wang, L. P., Xu, Z. Y. (2007). Organocatalysts wrapped around by poly(ethylene glycol)s (PEGs): a unique host–guest system for asymmetric Michael addition reactions. *Chem. Commun.*, 4393–4395.

[145] Price, M. D., Sui, J. K., Kurth, M. J., Schore, N. E. (2002). Oxazaborolidines as functional monomers: ketone reduction using polymer-supported Corey, Bakshi, and Shibata catalysts. *J. Org. Chem.*, *67*, 8086–8089.

[146] Degni, S., Wilén, C.-E., Rosling, A. (2004). Highly catalytic enantioselective reduction of aromatic ketones using chiral polymer-supported Corey, Bakshi, and Shibata catalysts. *Tetrahedron Asymmetry*, *15*, 1495–1499.

[147] Kell, R. J., Hodge, P., Snedden, P., Watson, D. (2003). Towards more chemically robust polymer-supported chiral catalysts: α,α-diphenyl-L-prolinol based catalysts for the reduction of prochiral ketones with borane. *Org. Biomol. Chem.*, *1*, 3238–3243.

[148] Niu, Y.-N., Yan, Z.-Y., Li, G.-Q., Wei, H.-L., Gao, G.-L., Wu, L.-Y., Liang, Y.-M. (2008). 1,2,3-Triazole-linked dendrimers as a support for functionalized and recoverable catalysts for asymmetric borane reduction of prochiral ketones. *Tetrahedron Asymmetry*, *19*, 912–920.

[149] Benaglia, M., Celentano, G., Cinquini, M., Puglisi, A., Cozzi, F. (2002). Poly(ethylene glycol)-supported chiral imidazolidin-4-one: an efficient organic catalyst for the enantioselective Diels-Alder cycloaddition. *Adv. Synth. Catal.*, *344*, 149–152.

[150] Puglisi, A., Benaglia, M., Cinquini, M., Cozzi, F., Celentano, G. (2004). Enantioselective 1,3-dipolar cycloadditions of unsaturated aldehydes promoted by a poly(ethylene glycol)-supported organic catalyst. *Eur. J. Org. Chem.*, 567–573.

[151] Selkälä, S. A., Tois, J., Pihko, P. M., Koskinen, A. M. P. (2002). Asymmetric organocatalytic Diels-Alder reactions on solid support. *Adv. Synth. Catal.*, *344*, 941–945.

[152] Zhang, Y., Zhao, L., Lee, S. S., Ying, J. Y. (2006). Enantioselective catalysis over chiral imidazolidin-4-one immobilized on siliceous and polymer-coated mesocellular foams. *Adv. Synth. Catal.*, *348*, 2027–2032.

[153] Pecinovsky, C. S., Nicodemus, G. D., Gin, D. L. (2005). Nanostructured, solid-state organic, chiral Diels-Alder catalysts via acid-induced liquid crystal assembly. *Chem. Mater.*, *17*, 4889–4891.

[154] Mitsudome, T., Nose, K., Mizugaki, T., Jitsukawa, K., Kaneda, K. (2008). Reusable montmorillonite-entrapped organocatalyst for asymmetric Diels-Alder reaction. *Tetrahedron Lett.*, *49*, 5464–5466.

[155] Haraguchi, N., Takemura, Y., Itsuno, S. (2010). Novel polymer-supported organocatalyst via ion exchange reaction: facile immobilization of chiral imidazolidin-4-one and its application to Diels-Alder reaction. *Tetrahedron Lett.*, *51*, 1205–1208.

[156] Hagiwara, H., Kuroda, T., Hoshi, T., Suzuki, T. (2010). Immobilization of MacMillan imidazolidinone as Mac-SILC and its catalytic performance on sustainable enantioselective Diels-Alder cycloaddition. *Adv. Synth. Catal.*, *352*, 909–916.

[157] Park, J. K., Sreekanth, P., Kim, B. M. (2004). Recycling chiral imidazolidin-4-one catalyst for asymmetric Diels-Alder reactions: screening of various ionic liquid. *Adv. Synth. Catal.*, *346*, 49–52.

[158] Chu, Q., Zhang, W., Curran, D. P. (2006). A recyclable fluorous organocatalyst for Diels-Alder reactions. *Tetrahedron Lett.*, *47*, 9287–9290.

[159] Ma, Y., Jin, S., Kan, Y., Zhang, Y. J., Zhang, W. (2010). Highly active asymmetric Diels-Alder reactions catalyzed by *C*2-symmetric bipyrrolidines: catalyst recycling in water medium and insight into the catalytic mode. *Tetrahedron*, *66*, 3849–3854.

[160] Akagawa, K., Yamashita, T., Sakamoto, S., Kudo, K. (2009). Friedel-Crafts-type alkylation in aqueous media using resin-supported peptide catalyst having polyleucine. *Tetrahedron Lett.*, *50*, 5602–5604.

[161] Yu, P., He, J., Guo, C. (2008). 9-Thiourea *Cinchona* alkaloid supported on mesoporous silica as a highly enantioselective, recyclable heterogeneous asymmetric catalyst. *Chem. Commun.*, 2355–2357.

[162] Malkov, A. V., Figlus, M., Kočovsky, P. (2008). Polymer-supported organocatalysts: asymmetric reduction of imines with trichlorosilane catalyzed by an amino acid-derived formamide anchored to a polymer. *J. Org. Chem.*, *73*, 3985–3995.

[163] Malkov, A. V., Figlus, M., Stončius, S., Kočovsky, P. (2007). Organocatalysis with a fluorous tag: asymmetric reduction of imines with trichlorosilane catalyzed by amino acid-derived formamides. *J. Org. Chem.*, *72*, 1315–1325.

[164] Malkov, A. V., Figlus, M., Prestly, M. R., Rabani, G., Cooke, G., Kočovsky, P. (2009). Soluble polymer-supported organocatalysts: asymmetric reduction of imines with trichlorosilane catalyzed by an amino acid derived formamide anchored to a soluble polymer. *Chem. Eur. J.*, *15*, 9651–9654.

[165] Rueping, M., Sugiono, E., Steck, A., Theissmann, T. (2010). Synthesis and application of polymer-supported chiral Brønsted acid organocatalysts. *Adv. Synth. Catal.*, *352*, 281–287.

[166] Sigman, M. S., Vachal, P., Jacobsen, E. N. (2000). A general catalyst for the asymmetric Strecker reaction. *Angew. Chem. Int. Ed.*, *39*, 1279–1281.

[167] Akagawa, K., Akabane, H., Sakamoto, S., Kudo, K. (2008). Organocatalytic asymmetric transfer hydrogenation in aqueous media using resin-supported peptide having a polyleucine tether. *Org. Lett.*, *10*, 2035–2037.

[168] Akagawa, K., Akabane, H., Sakamoto, S., Kudo, K. (2009). Asymmetric transfer hydrogenation in aqueous media catalyzed by resin-supported peptide having a polyleucine tether. *Tetrahedron Asymmetry*, *20*, 461–466.

[169] Oyama, T., Yoshioka, H., Tomoi, M. (2005). Polystyrenes with chiral phosphoramide substituents as Lewis base catalysts for asymmetric addition of allyltrichlorosilane: enhancement of catalytic performance by polymer effect. *Chem. Commun.*, 1857–1859.

[170] Liu, Y., Shi, M. (2008). Dendritic chiral phosphine Lewis bases-catalyzed asymmetric aza-Morita-Baylis-Hillman reaction of *N*-sulfonated imines with activated olefins. *Adv. Synth. Catal.*, *350*, 122–128.

[171] Li, P., Yamamoto, H. (2009). Amino acid salt catalyzed intramolecular Robinson annulation. *Chem. Commun.*, 5412–5414.

[172] (a) Pellissier, H. (2006). Asymmetric domino reactions. Part B: reactions based on the use of chiral catalysts and biocatalysts. *Tetrahedron*, *62*, 2143–2173; (b) Enders, D., Grondal, C., Hüttl, M. R. M. (2007). Asymmetric organocatalytic domino reactions. *Angew. Chem. Int. Ed.*, *46*, 1570–1581; (c) Yu, X., Wang, W. (2008). Organocatalysis: asymmetric cascade reactions catalysed by chiral secondary amines. *Org. Biomol. Chem.*, *6*, 2037–2046; (d) Simons, C., Hanefeld, U., Arends, I. W. C. E., Maschmeyer, T., Sheldon, R. A. (2007). Towards catalytic cascade reactions: asymmetric synthesis using combined chemo-enzymatic catalysts. *Top. Catal.*, *40*, 35–44; (e) Chapman, C. J., Frost, C. G. (2007). Tandem and domino catalytic strategies for enantioselective synthesis. *Synthesis*, 1–21.

[173] Chi, Y., Scroggins, S. T., Fréchet, J. M. J. (2008). One-pot multi-component asymmetric cascade reactions catalyzed by soluble star polymers with highly branched non-interpenetrating catalytic cores. *J. Am. Chem. Soc.*, *130*, 6322–6323.

[174] Cao, J.-J., Zhou, F., Zhou, J. (2010). Improving the atom efficiency of the Wittig reaction by a "waste as catalyst/co-catalyst" strategy. *Angew. Chem. Int. Ed.*, *49*, 4976–4980.

[175] Alza, E., Rodríguez-Escrich, C., Sayalero, S., Bastero, A., Pericás, M. A. (2009). A Solid-supported organocatalyst for highly stereoselective, batch, and continuous-flow Mannich reactions. *Chem. Eur. J.*, *15*, 10167–10172.

CHAPTER 3

SYNTHESIS AND CHARACTERIZATION OF SUPPORTED CHIRAL CATALYSTS

CARMELA APRILE, HERMENEGILDO GARCIA,
AND PAOLO P. PESCARMONA

3.1. Introduction 178
3.2. Types and features of solid supports 179
3.2.1. Structured porous materials 179
3.2.2. Layered materials 182
3.2.3. Unstructured materials 182
3.3. How to convert a homogeneous catalyst into a heterogeneous one 183
3.3.1. Synthesis 183
3.3.2. Covalent immobilization 183
3.3.2.1. Postsynthetic approach 183
3.3.2.2. Cosynthetic strategies 188
3.3.3. Noncovalent immobilization 190
3.3.3.1. Immobilization through electrostatic interactions 191
3.3.3.2. Entrapment 192
3.3.3.3. Adsorption 194
3.4. Characterization of supported chiral catalysts 196
3.4.1. Nuclear magnetic resonance 196
3.4.2. Electron paramagnetic resonance 198
3.4.3. Infrared spectroscopy 198
3.4.4. UV-Vis-NIR spectroscopy 200
3.4.5. X-ray diffraction 202
3.4.6. N_2 adsorption–desorption isotherms 203
3.4.7. Elemental analysis 203
3.4.8. Other techniques 204
References 206

Catalytic Methods in Asymmetric Synthesis: Advanced Materials, Techniques, and Applications, First Edition. Edited by Michelangelo Gruttadauria and Francesco Giacalone.

3.1. INTRODUCTION

The synthesis of enantiomerically pure compounds is generally achieved using chiral catalysts, which exploit their own chirality to induce that of the products. A vast class of chiral catalysts is represented by chiral complexes or chiral compounds that act as homogeneous catalysts. Homogenous catalysts find limited industrial applications, mainly due to the difficulty in separating the often expensive catalysts from the reaction medium. To solve this issue, many attempts have been made to immobilize the chiral catalysts on suitable supports in order to transform the homogeneous system into a heterogeneous catalyst. Heterogeneous catalysts are advantageous because they can be easily separated from the reaction mixture and can be recycled, which also implies a more straightforward purification of the products. Many examples of recyclable chiral catalysts are reported in Chapters 1 and 2. In this chapter, the general synthetic strategies and characterization procedures of supported chiral catalysts will be discussed. In Chapter 4, the synthesis of chiral catalysts supported on organic polymers is reported.

The world of heterogeneous catalysis is decidedly fascinating and perennially relevant. It is worth noting that more than 90% of the chemical manufacturing industries use catalysis in at least one step of their processes. The science and technology of heterogeneous catalysis are clearly of central and practical importance, also because the intrinsic characteristics of the selected solid such as porosity and surface area play a crucial role in the catalytic performances.

Traditionally, the two fields of homogeneous and heterogeneous catalysis are considered as separate domains, but there are different ways in which a homogeneous catalyst can be converted into a heterogeneous one, which helps to bridge the gap between these two fields and create an optimum balance between the respective advantages and disadvantages. The drawbacks of the process of "heterogenization" of a chiral homogeneous catalyst are predominantly related to the catalytic activity, since both enantioselectivity and yield may suffer a decrease when a solid support is used. The lower enantioselectivity can be due to the distortion of the structure of the immobilized organic moiety or to the steric hindrance that precludes the catalyst from reaching the transition state geometry achieved in homogeneous phase. The activity can diminish due to the decreased accessibility of the active sites located within the support. Particular attention should also be given to the possible leaching of the organic moiety from the solid supports. In spite of these problems, all the recent advances in heterogeneous asymmetric catalysis demonstrate that is possible to obtain materials with good activity and selectivity, comparable with those of homogeneous systems, while maintaining the easy recovery and recycling properties of the solid support.

The main focus of this chapter is on reviewing the various supports that have been used for the immobilization of chiral catalysts. Due to the large and constantly increasing number of solids that have been already used or that constitute suitable candidates for this purpose, an extensive description with a comprehensive list of all the possible solid supports would be difficult to compile and to read. Instead, we chose to use a systematic approach to review the various types of supported chiral catalysts. In the next section of this chapter, we present a general description of the most common families of materials used as supports. In the third section, we review the different synthetic approaches used for the preparation of the heterogeneous chiral catalysts. Finally, in the fourth section we discuss the most relevant techniques used for the characterization of the supported chiral catalysts.

3.2. TYPES AND FEATURES OF SOLID SUPPORTS

The characteristics of the solid support and the way in which the active compound can be immobilized on it have a large influence on the performance of the heterogeneous catalyst.

To give a general perspective, an ideal solid support should present the following features [1]:

- Chemical, thermal, and mechanical stability
- Large surface area
- Good amount of functionalization sites
- Fast mass transport of reactants and products to and from the active sites

There are many types of solids that fulfill the above-mentioned requirements and, therefore, are good candidates to act as solid supports. A general classification based on the structural features of the material leads to the definition of three main categories of supports:

- Structured porous materials
- Layered materials
- Unstructured materials

In the next section, we will describe the synthesis of supported chiral catalysts that will be described in terms of the type of interaction between the support and the enantioselective catalyst that has to be supported. All the materials that can be used for this purpose belong to one of the categories presented here. For the sake of conciseness, only selected and representative examples of each category will be given.

3.2.1. Structured Porous Materials

Structured porous materials present a porous structure in which the cavities have defined sizes and are organized either in a crystalline framework (as in zeolites) or in an amorphous structure with a long-range order (as in MCM-41)

[2]. These materials can be further classified as micro-, meso-, and macroporous solids on the basis of the dimension of their pores. According to the International Union of Pure and Applied Chemistry (IUPAC) notation, microporous solids are those with a pore size distribution (p) between 0 and 2 nm; mesoporous materials have pores in the range $2\,nm < p < 50\,nm$; and macroporous materials display larger pores with $p > 50\,nm$. The synthesis of structured materials with a regular pore size distribution represents an extremely interesting field of research. By using the appropriate structure-directing agent, the dimension of the pores can be tuned from angstroms to microns in an almost continuous spectrum of materials. The flexibility of this approach is very attractive from the scientific and industrial point of view because the structure of the material is directly connected with its properties and consequently, with its possible applications.

The most representative classes of structured porous materials are the following:

(a) Zeolites are the best known and most widely used class of microporous materials for many applications, including as support for immobilized chiral catalysts. Zeolites are aluminosilicates with a crystalline microporous structure that contains channels and cages. Zeolites can be described as an assembly of polyhedral cages and/or channels built from tetrahedral units of silica and alumina. The physical and chemical properties of these solids are largely dependent on both structure and composition. In Fig. 3.1 three examples of zeolite frameworks are presented.

(b) Metal-organic frameworks (MOFs) (Fig. 3.2) are a class of microporous materials that has been attracting growing interest. While the framework is completely inorganic in zeolites, MOFs are characterized by the combination of organic and inorganic moieties, which generates a two- or three-dimensional structure often characterized by a very higher surface area (up to $4500\,m^2/g$).

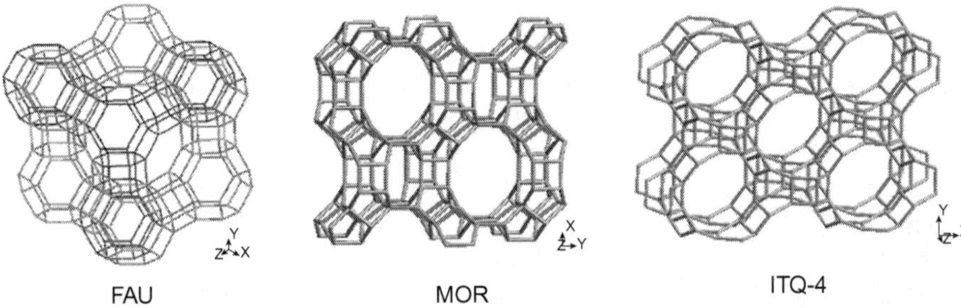

FIGURE 3.1. Structure of some typical zeolites: Faujasite (FAU), Mordenite (MOR), and ITQ-4.

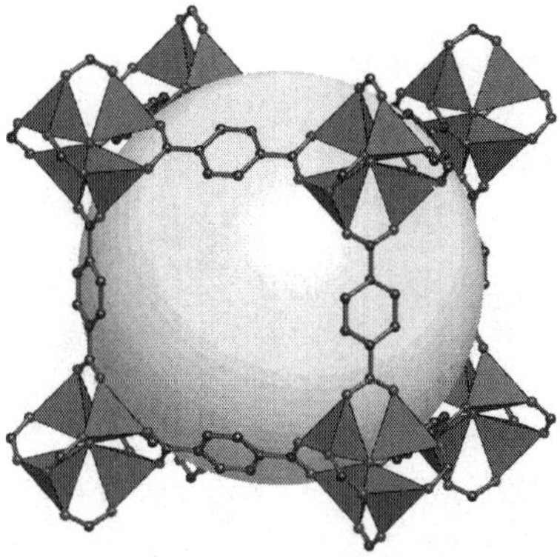

FIGURE 3.2. Schematic representation of MOF-5.

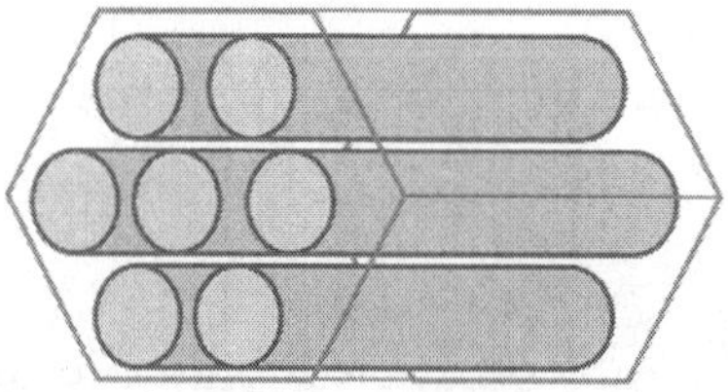

FIGURE 3.3. Schematic representation of an MCM-41.

(c) The most common structured mesoporous materials are MCM-41, MCM-48, SBA-15, and SBA-16 (Fig. 3.3). To obtain the ordered array of the mesopores characteristic of these materials, a templating agent such as a surfactant or a triblock copolymer must be used during the synthesis. The template can be easily removed by calcination or extraction. The extraction procedure is used to preserve the organic part of the material when a mixed organic–inorganic solid is formed either through postfunctionalization or by a cosynthetic approach (see Section 3.3). MCM-41-like materials have been extensively used as solid supports for chiral homogeneous catalyst as described in Section 3.3.

(d) Materials with ordered macroporores can be obtained following the templating approach by using emulsion or the sphere templating method [3]. Macroporous carbon can be synthesized starting from a phenol resin used to fill the internal void of opals, followed by a mild temperature treatment and by the removal of the silica matrix using HF and pyrolysis at 1000°C. Macroporous silicates bearing organic functionalities can be prepared in a similar way as structured mesoporous solids, but using colloidal crystals instead of the compounds used as templates in the synthesis of ordered mesoporous materials (see Section 3.2.1.c). The colloidal crystal templating procedure is also employed to synthesize porous polymers [3].

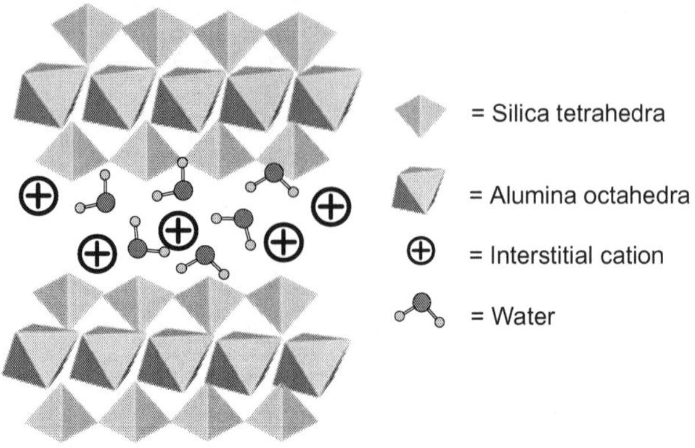

FIGURE 3.4. Structure of an aluminosilicate clay.

3.2.2. Layered Materials

Clays and layered double hydroxides (LDHs) are the most used layered materials (LMs) for the heterogenization of chiral catalysts, and the only ones that show some confinement effects (see Section 3.3). Clays (both natural and synthetic) are aluminosilicates or magnesiosilicates constituted by a combination of tetrahedral and octahedral layers of oxides. Isomorphic substitutions such as Al by Si, Mg by Al, or Li by Mg generate in the solid a charge deficiency that is compensated by the inclusion of monovalent cations (lithium, sodium, ammonium) in the lamellar spaces (Fig. 3.4) [4]. These cations can subsequently be exchanged with the chiral compounds to form the heterogeneous enantioselective catalyst. These solids can be considered as nanostructured since the distance between layers is on the scale 0.3 and 1.5 nm. Examples of clays used as catalyst supports include bentonite, K10- and KSF-montmorillonites, and hectorite. The LDHs are another class of LMs that are finding application as support for chiral catalysts. These materials are a class of synthetic ionic inorganic compounds with a similar structure to clays and with the general formula $[M^{II}_{1-x}M^{II}_{x}\ (OH)_2]^{x+}[A^{n-}]_{x/n}{\cdot}yH_2O$ [5]. Their structure is based on a series of layers, where a divalent metal cation is located in the center of oxygen octahedron, and two-dimensional infinite layers are formed by edge-sharing of the octahedra. The partial substitution of divalent cations by trivalent ones generates a positive charge on the layers that is balanced by anions or molecules of solvent. These anions can be exchanged with chiral compounds to generate the supported enantioselective catalysts.

3.2.3. Unstructured Materials

Amorphous solids that do not present any regular pore size distribution or structural organization can also be used as supports for chiral catalysts with good results. The high surface area is, in many cases [e.g., amorphous carbon,

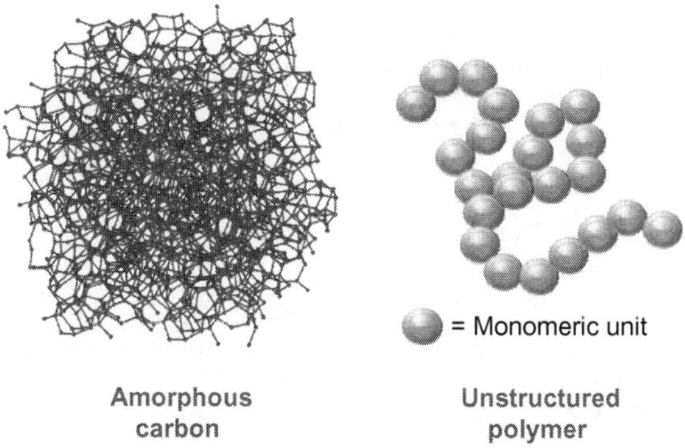

FIGURE 3.5. Schematic representations of two, largely used, amorphous solids.

silica, unstructured polymers (Fig. 3.5)], a sufficient feature to guarantee a high catalytic activity of the heterogeneous catalyst. In general, the synthesis of unstructured materials can be achieved using similar methods to those employed for structured porous materials (see Section 3.2.1), simply avoiding the use of the expensive structure directing agent.

3.3. HOW TO CONVERT A HOMOGENEOUS CATALYST INTO A HETEROGENEOUS ONE

3.3.1. Synthesis

As mentioned in the introduction, one of the major problems hindering the industrial application of homogeneous enantioselective catalysis is represented by the difficult separation of the catalyst from the reaction media in an easy and inexpensive way. One of the most convenient strategies to circumvent this problem is the immobilization of the compound acting as a catalyst on a solid support. There are different approaches to immobilize a chiral compound and various classifications have been proposed by different authors. Here, a classification based on the interaction between the chiral catalyst and the support is proposed. This interaction can be divided into two main categories, based on covalent and noncovalent immobilization.

3.3.2. Covalent Immobilization

Covalent immobilization is based on the covalent bonding of the chiral catalyst on the solid support and can be further divided into postsynthetic and cosynthetic strategies.

3.3.2.1. Postsynthetic Approach. This approach implies the grafting of the chiral moiety on a preformed solid support. In most of the cases, a linker is

FIGURE 3.6. Scheme of the anchoring of a homogeneous chiral compound to a support through a linker.

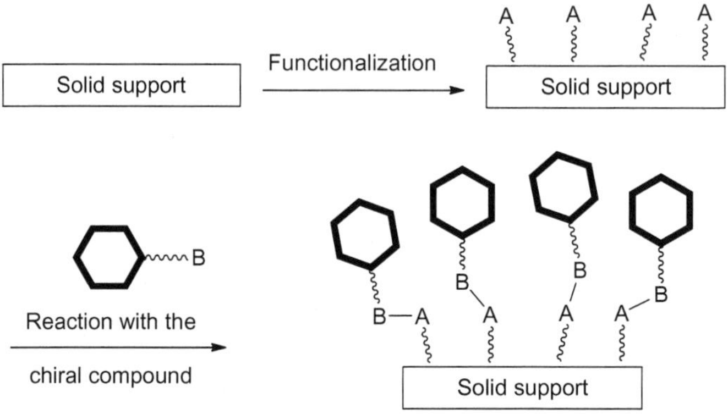

FIGURE 3.7. General procedure for postsynthetic covalent immobilization.

used to allow the covalent bonding of the chiral catalyst to the support. Typical examples of materials that have been used for this purpose are constituted by inorganic mesoporous materials such as MCM-41 or SBA-15, carbon nanotubes, and polymers, while the linker is usually represented by an organic molecule with a double functionality that can react from one end with the solid and from the other end with the chiral catalyst (Fig. 3.6). The features of the linker are of crucial importance because the nature, as well as the length of the linker, can have an influence on the activity and selectivity of the catalyst [6]. Therefore, the selection or design of the linker should be performed carefully.

One of the most common strategies within the postsynthetic approach is represented by the prefunctionalization of the support followed by the anchoring of the chiral compound (Fig. 3.7).

As an example of the general applicability of the grafting procedure, a work in which a chiral vanadyl salen complex (Fig. 3.8) was anchored using a postsynthetic approach onto supports of different nature is presented. In this work, Garcia et al. compared the activity of the same homogeneous catalyst anchored on amorphous high-surface-area SiO_2, delaminated ITQ2 material, and mesoporous MCM-41 [6]. In another work of the same group, single-wall nanotubes (SWNTs) were used as support using a modified version of the synthetic approach employed with the silica-based supports [7]. The same 1-mercaptopropyltrimethoxy silane linker was used with all the silica-based

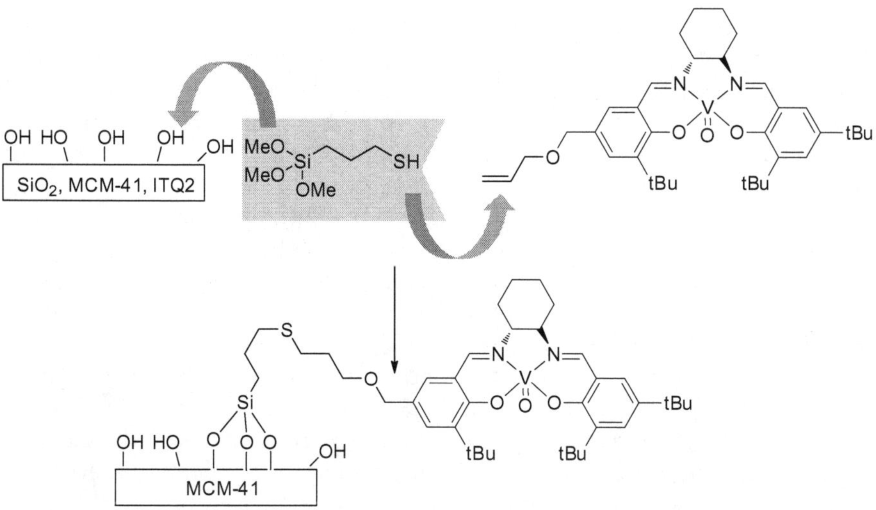

FIGURE 3.8. Synthetic approach followed for the preparation of the chiral vanadyl salen complex supported on MCM-41.

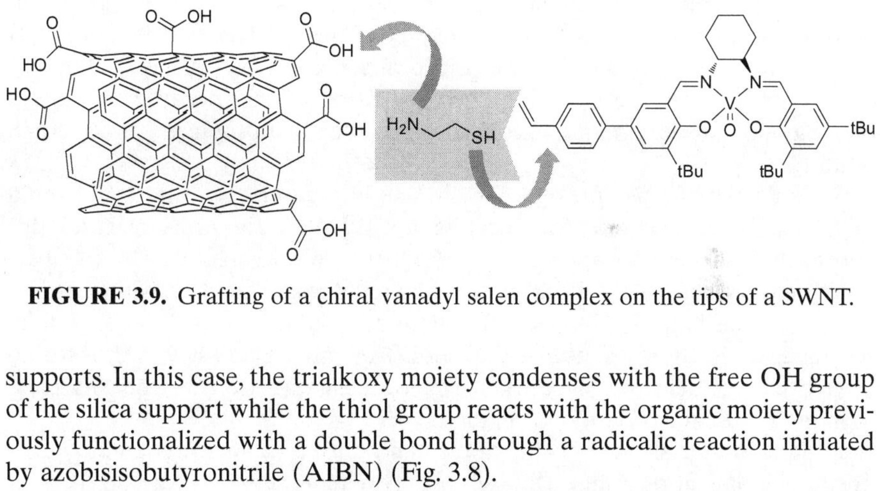

FIGURE 3.9. Grafting of a chiral vanadyl salen complex on the tips of a SWNT.

supports. In this case, the trialkoxy moiety condenses with the free OH group of the silica support while the thiol group reacts with the organic moiety previously functionalized with a double bond through a radicalic reaction initiated by azobisisobutyronitrile (AIBN) (Fig. 3.8).

These solids were used to catalyze the cyanosilylation of benzaldehyde [4]. The best catalytic activity is given by the solid constituted by amorphous silica. Due to the need of a silylation of the free OH group, which can have a negative influence on the catalysis, the authors explain the results in terms of a more difficult complete silylation of the MCM-41 and ITQ2 supports due to sterical hindrance. Amorphous silica was found to be the most suitable support in terms of both activity and selectivity, when compared with carbon composites like SWNTs (Fig. 3.9) or amorphous carbon, for the enantioselective addition of trimethylsilyl cyanide to benzaldehyde catalyzed by chiral vanadyl salen complexes.

In the case of SWNTs, the postfunctionalization can be obtained using the free carboxylic groups on the tips and/or on the walls that become accessible after a purification procedure [7]. Amorphous carbon can be functionalized following the same synthetic strategy by using the surface carboxylic groups.

Comparison with the homogeneous vanadyl salen in imidazolium ionic liquid under the same reaction conditions shows that the pure chiral complex gives a better yield and selectivity than the silica-supported catalyst. However, traces of vanadyl salen were found in the products after extraction, indicating that the separation of the homogeneous catalyst was not complete. Therefore, the authors concluded the vanadyl salen complex anchored on silanized silica was the catalyst of choice.

A different type of inorganic support is constituted by nanoparticles. Nanoparticles are considered a separate class of supports because it is difficult to classify them in terms of their interaction with the organic chiral moiety. In the case of metal oxide nanoparticles, the presence of covalent bond with the linker can be claimed. This is the case of proline supported on magnetite [8], in which azide-functionalized magnetite nanoparticles were treated with a proline derivative in the presence of Cu(I) to catalyze the cycloaddition reaction (Fig. 3.10).

Excellent yields were found for the *N*-arylations of *N*-heterocycles, and recycling studies showed a good reusability of the catalyst with less than 5% deactivation after the fourth cycle [8]. In other cases—when metal nanoparticles are used—the classification becomes more difficult because the nature of the bond can be a matter of controversy, such as in the case of the Au–S bond.

As an example, we present the first report describing the use of gold nanoparticles as a support for chiral ligands [9]. As in the previous cases, this approach is of general applicability to all the chiral substrates that can be functionalized with a thiol group. Gold colloids were modified with a monolayer of octanethiol by the organic-phase reduction of $AuCl_4^-$ with sodium borohydride in the presence of the thiol. The chiral catalyst was introduced after exchange reaction (Fig. 3.11). This catalyst was used for the Sharpless asymmetric dihydroxylation of alkenes.

Organic materials such as polymers can be used as supports employing a postfunctionalization route. The use of polymers as supports for chiral molecules will be extensively discussed in the following chapter. Here, we report

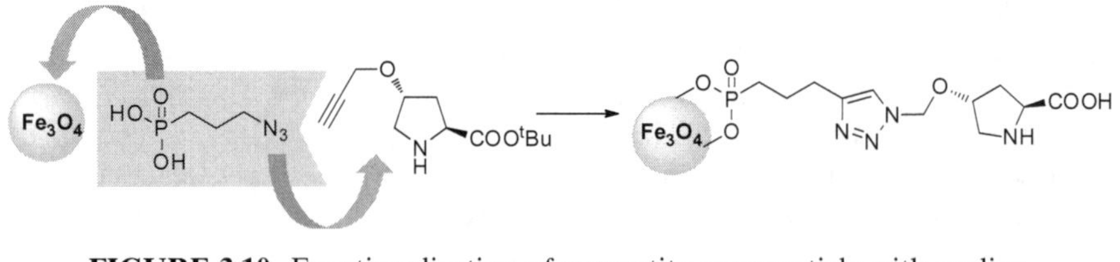

FIGURE 3.10. Functionalization of magnetite nanoparticle with proline.

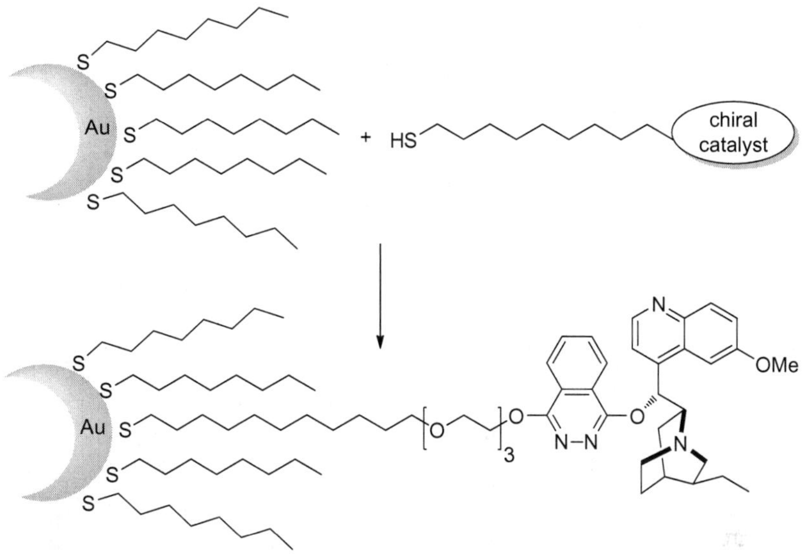

FIGURE 3.11. Representation of the structure of gold nanoparticle-supported dihydroquinidine ligands.

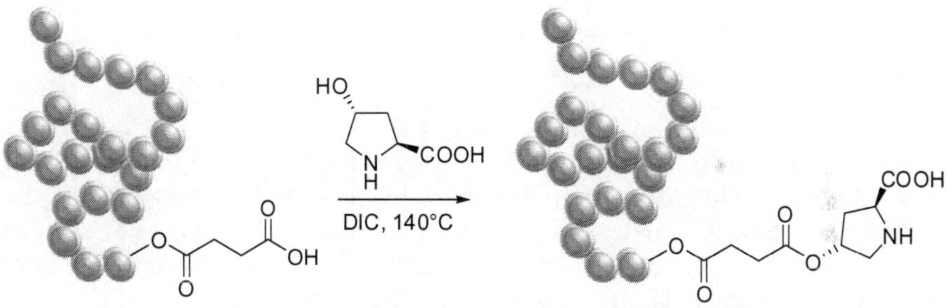

SCHEME 3.1. PEG functionalized with proline.

just one example in which poly(ethylene glycol) (PEG)-500 monomethyl ether by means of a succinate spacer was used to immobilize proline (Scheme 3.1). The as-synthesized homogeneous recyclable catalyst promotes the enantioselective aldol condensation of acetone and hydroxyacetone with various aldehydes. The PEG-supported catalyst gives a high level of stereocontrol [up to 98% enantiomeric excess (ee)], very similar to those obtained with the non-supported catalyst. The solid can also be easily recovered and recycled, supporting the validity of the use of polymers as solid supports [10].

Chiral complexes can be generated also in situ on the inorganic supports following a multistep grafting procedure. Compared to the above-mentioned

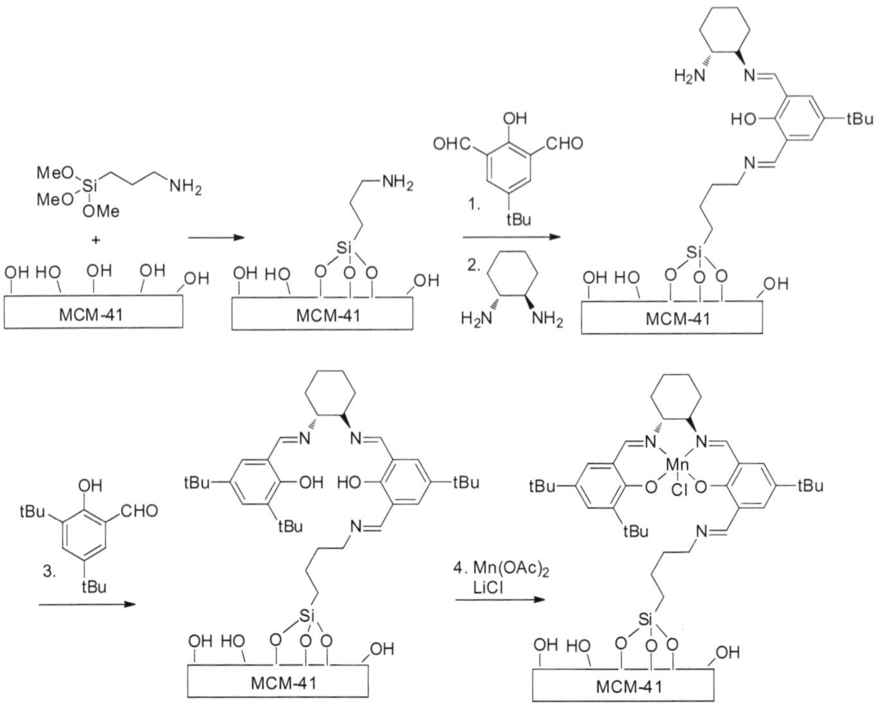

SCHEME 3.2. Synthetic route followed for the grafting of a manganese-salen complex on MCM-41.

procedures, in which the ligand is previously synthesized and then anchored in a one-step synthesis, the main drawback of the multistep procedure is the possibility of having a material in which by-products of the synthesis, such as incompletely formed metal complexes, are present. Particularly negative from the point of view of the enantioselectivity would be to have a solid with a population of uncomplexed transition metal that could catalyze the reaction, but not promote enantioselectivity. In Scheme 3.2, an example is reported in which a chiral Mn-salen complex is anchored on a MCM-41 through a multi-step synthetic strategy [11].

3.3.2.2. Cosynthetic Strategies. In contrast to the grafting procedures, in which the chiral catalyst is anchored on a previously synthesized support, in the cosynthetic strategy both the chiral catalyst (previously modified by adding a specific functional group) and the molecular precursors of the support react in a one-pot procedure to build together the chiral heterogeneous catalyst. This approach is normally applied to silica-made materials, such as periodic mesoporous organosilicas (PMOs) or silsesquioxanes derivatives, to polymers, and, in some recent applications, to MOFs. Following this approach, a more uniform distribution of active sites can be obtained, and a completely new

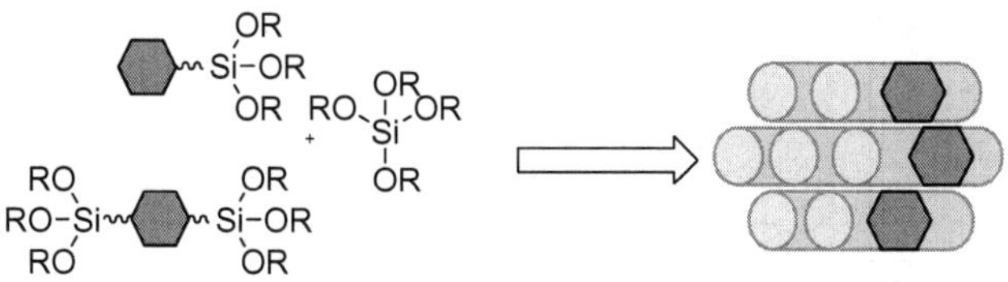

FIGURE 3.12. General procedure for the synthesis of a PMO.

material is synthesized in which the different hydrophobic/hydrophilic interactions, as well as the textural and porosity characteristics, can be tuned by varying the relationship between organic and inorganic precursors. On the other hand, some relevant parameters of the support such as the pore size distribution and the surface area can be negatively affected by the presence of an "external" component in the synthesis gel.

In the case of materials constituted by silica, as in the case of mesoporous MCM-41 or related materials, the organic moiety bearing trialkoxy functionalities is directly introduced in the synthesis gel via self- or co-condensation with tetraethyl orthosilicate (TEOS) or another silicon source in a basic or acid medium. Due to the mild reaction conditions involved in this approach, the sol-gel chemistry becomes one of the most promising routes to obtain the organic–inorganic hybrids materials. The general procedure followed for this approach is illustrated in Figure 3.12.

In the case of structured mesoporous materials, a templating agent is also added to the synthesis gel in order to obtain a defined pore size distribution and high surface area.

Moreau et al. [12] compared the catalytic activity of similar heterogeneous catalysts obtained by reacting a chiral silylated diaminocyclohexane with TEOS in the presence or absence of a templating agent. They found that the surface area of the material prepared by using the templating agent was more then 10 times higher that the corresponding solid obtained without the aid of the template. The higher surface area had a positive influence on the catalytic behavior of the solid; the material complexed with rhodium (Rh) exhibits a good catalytic activity and selectivity comparable to the homogeneous molecular complex.

PMO with an MCM-41-like structure was synthesized following a cosynthetic approach by Garcia and co-workers [13].

A vanadyl Schiff base complex was modified by introducing two trialkoxy silyl groups (Scheme 3.3) and then used in combination with TEOS as a silicon source and cetyltrimetylammonium bromide as the structure-directing agent. This approach allows the direct synthesis of solid materials in which the catalyst is connected to the silicate walls. The catalytic activity of VO(salen) αPMO after extraction of the template was tested for the room-temperature cyanosilylation of aldehydes in chloroform using trimethylsilyl cyanide as the reagent. High conversion (80%) and essentially complete selectivity to the

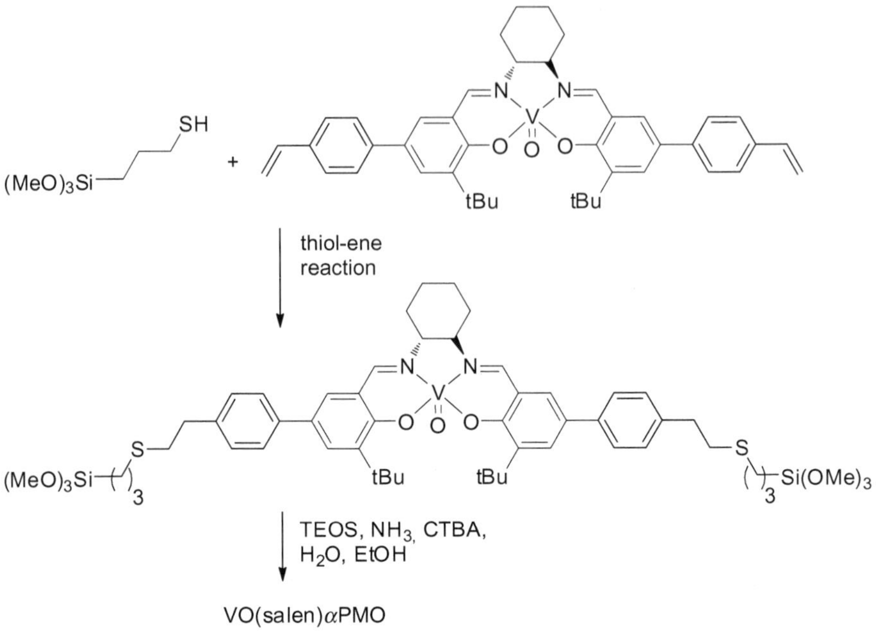

SCHEME 3.3. Vanadyl complex functionalized with two trimethoxysilicon groups.

silylated cyanohydrin was achieved, although the ee was lower (30%) than that observed in solution. In addition, studies carried out on the corresponding achiral catalyst showed no leaching, and the catalyst maintained the initial activity after consecutive runs, proving the validity and demonstrating the advantages of the cosynthetic approach.

A different synthetic approach should be followed for the preparation of chiral metal-organic frameworks (CMOFs). In this case, the organic moiety should be synthesized with some typical functionalities such as a carboxylic or an amino group that can be subsequently used to interact with the metal center. Hupp and co-workers nicely described a cosynthetic approach for the synthesis of a microporous MOF featuring a chiral Mn-salen (Fig. 3.13). This CMOF was successfully used in the asymmetric epoxidation of olefins [14]. More detailed examples are reported in Chapter 5.

3.3.3. Noncovalent Immobilization

The noncovalent immobilization is an extensively applied approach for the preparation of heterogeneous catalysts, starting from the homogeneous chiral molecules. In this case, the immobilization of the chiral catalyst leading to the formation of the chiral heterogeneous catalyst can be achieved by electrostatic interaction, inclusion, or adsorption procedures.

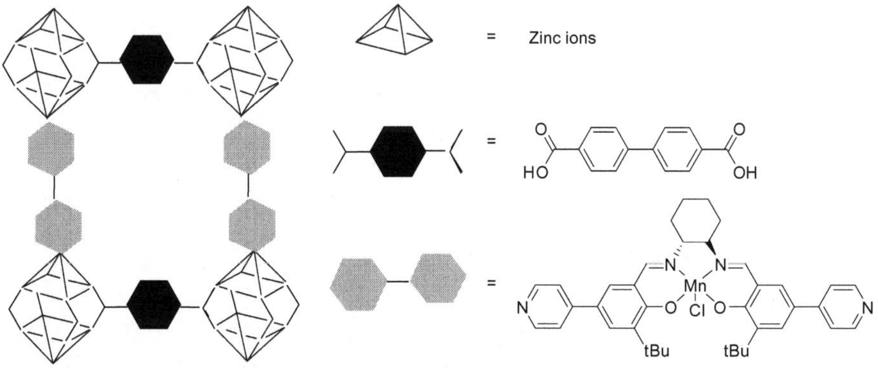

FIGURE 3.13. Schematic representation of the CMOF synthesized by Hupp and co-workers in Reference 13.

3.3.3.1. Immobilization Through Electrostatic Interactions. Noncovalent immobilization by means of electrostatic interaction can be employed if the chiral active molecules are positively (or negatively) charged and the solid support presents cationic (or anionic) species that can be exchanged with the ionic chiral compounds. LMs exhibit most of the characteristics of a suitable solid support for the immobilization of chiral catalysts by electrostatic interactions: high surface area, intercalation, and high ion-exchange capacity. LMs exist both in the form of cationic clays (such as hectorite or montmorillonite) and anionic compounds (such as hydrotalcite-like LDH; see Section 3.2.2). Several methods have been proposed for the synthesis of heterogeneous LDH catalysts intercalating a chiral organic molecule or complex. One of them involves the direct inclusion in the inorganic host by stirring a solution of a freshly prepared anionic organic catalyst with the desired LDH. A reorganization of the ionic counterparts (both organic and inorganic) occurs during the process, which allows for the incorporation of the organocatalyst into the inorganic matrix. An example of this approach is reported by He and co-workers. They successfully prepared an Mg/Al LDH in which proline was intercalated [15]. Different materials were prepared following this approach, and one of them was tested for the asymmetric aldol reaction between acetone and benzaldehyde with similar results to L-proline. With the LDHs intercalated with proline (intercalation yields of 59%), high yield (89%) and enantioselectivity (94%) were obtained. It is worth noting that this is the highest enantioselectivity obtained for this reaction.

Another synthetic strategy is represented by the coprecipitation approach. As an example, Choudary and co-workers compared the previously described ion-exchange strategy with the direct synthesis by coprecipitation [16]. The two different approaches were employed for the immobilization of two anionic organocatalysts such as proline and binol. The heterogeneous catalysts were tested for C–C bond formation reactions. Aldol condensation, Henry and

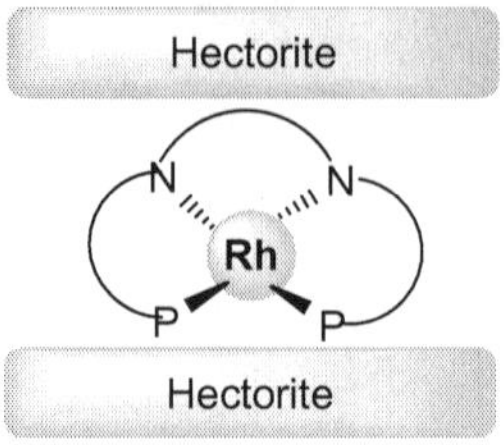

FIGURE 3.14. Hectorite intercalating a rhodium chiral complex.

Michael reactions, and cyanosilylation were also performed. The materials showed good yields but very low enantioselectivity.

Similar to the intercalation of anionic compounds in LDHs, cationic clays can be used for the intercalation of cationic chiral complexes by ion exchange. The first noncovalent immobilization of this type reported in literature involved the use of different cationic clays (hectorite, bentonite, nontronite, haloysite) in which the $[(PNNP)\text{-}Rh(cod)]^+$ complex (where PNNP is a diiminophosphine ligand, and cod is the cyclooctadiene molecule) was intercalated by direct ion exchange. The catalysts were used for the selective hydrogenation of amino acids precursors [17]. The type of clay was found to be extremely important for the catalytic performance of the chiral catalyst. For example, hectorite was the best host in the case of cinnamic substrates. Hectorite-based support was also very active for the hydrogenation of 2-acetamidoacrylic acid with a high enantioselectivity (72%) and good recycling properties.

Figure 3.14 shows a schematic representation of the heterogeneous catalyst in the case of hectorite as solid support [17].

Ion-exchange resins are another class of extensively used supports for the noncovalent immobilization of chiral organocatalysts. Some examples are reported in the previous chapters. From a synthetic point of view, their preparation is quite simple because they are based on ion exchange. A discussion on the general approaches for the synthesis of the polymers will be described in more detail in the next chapter. Just one example is provided here, which describes the first reported immobilization on polymers. In this case, the selected support was a sulfonated poly(styrene-divinylbenzene) copolymer that was used to immobilize the chelate $[Rh(COD)(Ph\text{-}\beta\text{-glup})]^+BF_4^-$, to give the catalyst reported in Figure 3.15 [18].

The catalyst was tested for the selective hydrogenation of amino acids, and excellent enantioselectivities were found. Even if the catalytic activity was lower than the corresponding homogeneous Rh-complex, the high stability in the recycling tests made this catalyst very promising and opened the door for following studies.

3.3.3.2. Entrapment. In this method, the homogeneous catalyst remains trapped inside the structure of the porous support without the need for any

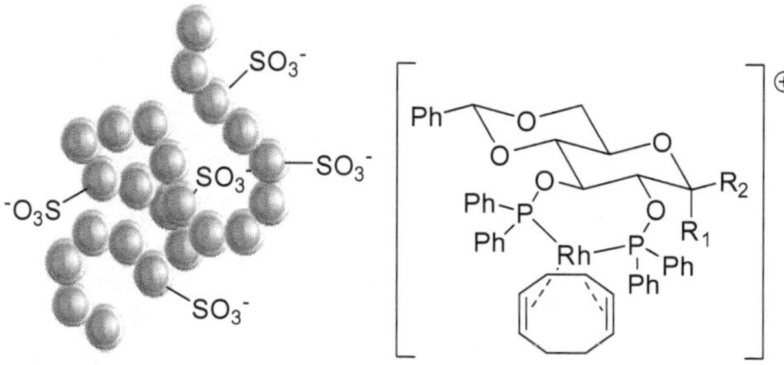

FIGURE 3.15. Cationic rhodium(I) chelate immobilized on a polystyrene polymer functionalized with benzenesulphonate moieties.

strong physical or chemical interaction. This is generally achieved by using a support with a uniform pore size distribution. For a successful immobilization, the size of the active catalyst must be bigger than the pores of the host to prevent leaching during the catalytic process. The entrapment methods can be classified in the function of the support used: entrapment into a flexible polymeric matrix and entrapment into a rigid inorganic framework. In this chapter, only the part concerning the inorganic hosts will be described.

The entrapment into a rigid inorganic framework can be performed by synthesizing the active catalyst directly in the pores of the inorganic support. Alternatively, the inorganic matrix can be built around the preformed complex. In the first case, the synthetic approach is based on the principle that the precursors of the final active catalyst are either flexible or small enough to pass through the pores of the supports. The assembly of the precursors inside the structured material will generate a bigger compound that remains trapped into the pores. For this procedure, called the "ship-in-a-bottle" approach, the material of choice is represented by zeolites due to their small apertures and internal cavities.

Zeolite NaY, for example, was successfully used for the ship-in-a-bottle inclusion of rhodium complexes $[Rh(COD)L]^+$ via the flexible ligand method [19]. The heterogeneous catalyst was used for the enantioselective hydrogenation of (Z)-methyl-α-acetamidocinnamate, exhibiting higher activity and enantioselectivity than the homogeneous Rh-complex. A nice example in which the metal complex is formed inside the cages in a limited number of synthetic steps, starting from the small organic precursors, is depicted in Figure 3.16. In general, in this approach the metal is introduced inside the pores of the support by ion exchange—implying that the selected support should be charged—or by preadsorption of a labile metal complex followed by the reaction with the small precursor to form the final catalyst inside the pores [20].

In the example reported in Figure 3.16, the catalytic activity of the chiral catalyst was tested for the epoxidation of different prochiral alkenes, and the results were compared with those obtained under homogeneous conditions.

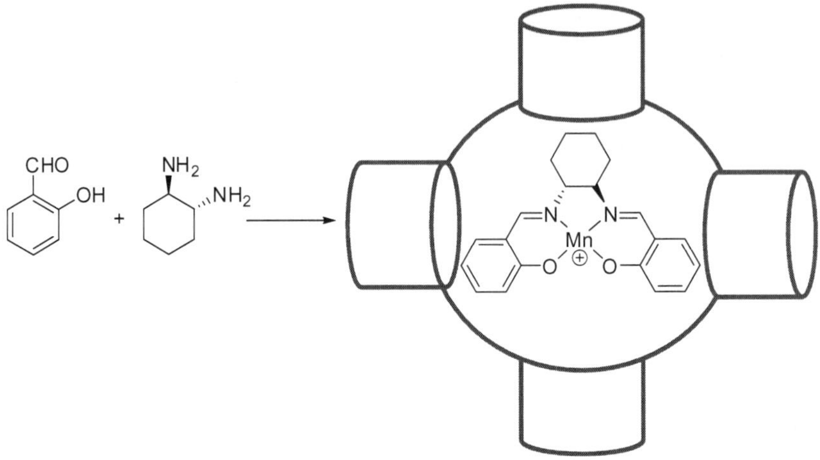

FIGURE 3.16. Schematic representation of the "ship-in-a-bottle" approach.

The Mn-complex trapped in the cage of Zeolite Y showed a catalytic activity similar to that of the chloride complex in homogeneous phase.

An alternative approach for entrapment involves building the support around the preformed complex. In this case, the first and fundamental requisite is that the catalyst must be stable under the conditions of support preparation. Some disadvantages related to this method are the difficulty in obtaining a material with a well-defined pore distribution, the presence of different types of encapsulation sites, and the possibility that part of the catalyst will remain on the external surface after washing or Soxhlet extraction, with consequent leaching during the catalyst tests. This was, for example, the case of the [(1,1′-Binaphthalene-2,2′-bis(diphenylphosphine)) (BINAP)Ru(p-cumene) Cl]Cl, [(2,3-isopropylidene-2,3-dihydroxy-1,4-bis(diphenylphosphino)butane) (DIOP)Rh(cod)Cl], and [(*N*-(t-butoxycarbonyl)-4-(diphenylphosphino)-2-[(diphenyl-phosphino)methyl]pyrrolidine)) (BPPM)Rh(cod)Cl] complexes trapped in the pores of a silica matrix by using a sol-gel approach [21]. Although the catalyst was washed several times, a substantial leaching was found during the enantioselective hydrogenation of itaconic acid in THF.

3.3.3.3. Adsorption. Noncovalent immobilization by adsorption can occur either by physisorption, in which only Van der Waals interactions are involved, or by chemisorption, in which stronger interactions such as those of hydrogen bonds take place. In the case of physisorption, when different supports were compared for the adsorption of chiral metal complexes, silica was often the host that gave the best results. In some cases, the hydrophobic character of the support can play a fundamental role for the adsorption properties of the selected solid. For example, a higher loading of the neutral [(BPPM)Rh(cod) Cl] is achieved by immobilization on methylated silica than by immobilization on the corresponding nonmethylated silica, and thus, less hydrophobic host is

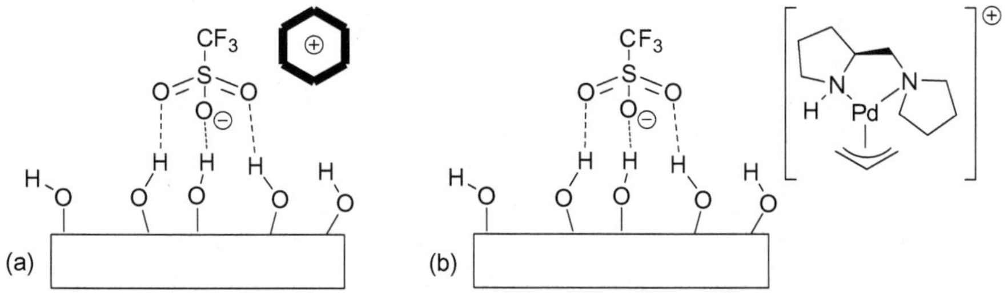

FIGURE 3.17. (a) Immobilization of a generic cation compound on the surface of a solid bearing free hydroxyl groups; (b) synthetic approach followed by Raja and co-workers [24].

needed. The catalyst was used for the selective hydrogenation of a DOPA precursor with good results but lower recycling capacity [22]. Mesoporous silicates substituted with Al, Ga, or Fe were also successfully used for the impregnation of a chiral Mn(III)-salen complex. The catalysts were used for selective epoxidation reactions with results comparable with those obtained in homogeneous conditions [23].

Supports with free OH groups can also be used for the immobilization through hydrogen bond formation. In this case, the chiral catalyst can be directly immobilized on the support, or a "linker" can be used (Fig. 3.17a).

In this case, the linker does not react with the support nor with the substrate, but it can interact with the support through a hydrogen bond. At the same time, it can provide a suitable counterion for a cationic chiral catalyst. Following this approach, different types of cationic substrates can be trapped on the surface of the material. For example, complexes of chiral diamine with different metallic centers such as Rh and Pd (Fig. 3.17b) were immobilized on a series of silicas with different surface areas (from 300 to 700 m^2/g) and a range of mesoporosity from 4 to 25 nm [24]. The enantioselectivity for hydrogenation reactions increased after the immobilization, probably due to structural restriction induced by the support.

Defining whether a supporting method is truly based on chemisorption is often a matter of controversy because the formation of hydrogen bonds, as well as the presence of the sulfur–gold interaction (see the examples concerning the use of gold nanoparticles as support in Section 3.3.2.1), is not considered as a chemisorption technique by some authors. However, it is unlikely to find a system in which a molecule is adsorbed on a solid support, while unmistakably excluding the presence of some other kind of interaction.

To conclude this section we would like to reiterate that the identification of a catalyst as a heterogeneous one is not always an easy task. There are cases in which the catalyst can be located at the interface between homogeneous and heterogeneous catalysis. Solid supports like short SWNTs or systems in which the chiral catalyst is dissolved in a supported layer of ionic liquids can be cited as representative examples of this category [25, 26].

3.4. CHARACTERIZATION OF SUPPORTED CHIRAL CATALYSTS

In the previous sections of this chapter, we reviewed the different approaches that can be employed to synthesize enantioselective heterogeneous catalysts and the different types of supports that can be used as hosts for chiral compounds. Once an enantioselective heterogeneous catalyst has been prepared by the immobilization of chiral compounds on a chosen support, it is essential to characterize the catalytic material with a suitable set of analytical and spectroscopic techniques. This section describes the most widely used approaches for the characterization of these catalysts. For the most part, the characterization of supported chiral catalysts has mainly three different purposes:

(1) Establishing whether the chiral catalyst was successfully supported with the selected immobilization technique and whether the structure of the supported compound is identical to that of the free chiral catalyst;
(2) Establishing whether the support is affected by the presence of the chiral catalyst;
(3) Establishing the catalytic activity and stability of the supported chiral catalyst under the reaction conditions: this implies ascertaining the integrity of the material after the catalytic test and determining if no leaching occurred.

The techniques used for each of these purposes are summarized in Table 3.1 and are presented and discussed in the rest of this section.

3.4.1. Nuclear Magnetic Resonance

Nuclear magnetic resonance (NMR) spectroscopy is a versatile technique for the characterization of immobilized chiral catalysts. NMR is based on the absorption of an electromagnetic radiation by a nucleus with a nonzero spin in an external magnetic field. The frequency of the absorbed radiation provides information about the chemical environment of the nucleus and, thus, about the compound. Solid-state NMR spectroscopy is often used to demonstrate

TABLE 3.1. Characterization of Supported Chiral Catalysts

Question	Characterization Techniques
Was the chiral compound successfully immobilized? What is the structure of the supported catalyst?	NMR; EPR; IR; UV-Vis-NIR; ICP-AES; combustion analysis; XPS; EXAFS; XANES; CV; TGA
Was the support affected by the presence of the supported chiral catalyst?	XRD; N_2 isotherms; TEM
Is the supported chiral catalyst stable under the reaction conditions?	ICP-AES; ICP-MS; AAS

the presence of the ligand of the chiral complex and its interaction with the support in the heterogeneous catalyst. Many ligands are based on a carbon backbone, and can be studied by means of ^{13}C NMR. For example, the integrity of the salen ligand in a vanadium-salen complex immobilized on a silica support could be confirmed by the ^{13}C NMR spectrum of the material [27]. Similarly, solid-state ^{13}C and ^{29}Si cross-polarization magic-angle spinning (CP MAS) NMR were used in combination with liquid-phase ^{13}C and ^{29}Si NMR to confirm the incorporation of the chiral diamine moiety in hybrid organic–inorganic chiral catalysts prepared by the sol-gel hydrolytic condensation of silyl-substituted chiral diamine-rhodium complexes with TEOS [28]. As another example, solid-state ^{29}Si MAS NMR has been used to confirm the successful incorporation of a chiral vanadyl Schiff base complex having two terminal trimethoxysilyl groups peripheral to the ligand into a PMO material, leading to an efficient enantioselective heterogeneous catalyst for the cyanosilylation of carbonyl groups. The ^{29}Si NMR spectrum of the material shows the presence of T_3 silicon atoms [–CH_2–Si(OSi≡)$_3$] originating from the complete condensation reaction of the T_0 groups [–CH_2–Si(OR)$_3$] of the ligand: this represents clear evidence that the Si atoms of the ligand are covalently attached to the silicate framework [13]. Chiral ligands often contain heteroatoms that can be monitored by means of NMR spectroscopy. ^{31}P MAS NMR has been used to study the synthesis of rhodium-diphosphine complexes encapsulated in Al-MCM-41: The NMR data confirmed the absence of a free phosphine ligand, suggesting the successful immobilization of the chiral complex [29]. NMR spectroscopy can also provide more specific information concerning the interaction between the ligand and the support. This was exemplified by a study on cationic chiral rhodium-phosphine catalysts with triflate counter ions supported by noncovalent immobilization on mesoporous MCM-41 and on silica gel [30]. The ^{31}P and ^{19}F NMR spectra of the free and bound complexes suggested that it is the triflate counterion that interacts strongly with the support. The observed broadening of the signals in the ^{31}P and ^{19}F NMR spectra of the complexes supported on MCM-41 is consistent with the restricted mobility expected for immobilized complexes (Fig. 3.18).

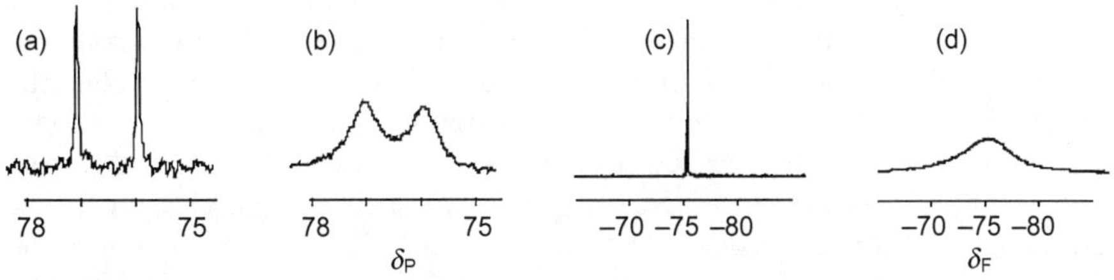

FIGURE 3.18. (a) ^{31}P NMR of the free Rh-complex in CH_2Cl_2 and (b) after the addition of MCM-41; (c) ^{19}F NMR of the same complex in CH_2Cl_2; and (d) after the addition of MCM-41 (reproduced from Reference [30]).

3.4.2. Electron Paramagnetic Resonance

Electron paramagnetic resonance (EPR), also known as electron spin resonance (ESR), is a spectroscopic technique that allows the characterization of species containing one or more unpaired electrons. Many transition metal ions present in chiral complexes have unpaired electrons and, therefore, lend themselves to be studied with this technique. An example of the application of EPR spectroscopy to the characterization of enantioselective heterogeneous catalysts is represented by the study of a chiral catalyst based on a zeolite Y containing Cu and Na ions, chirally modified with a bis(oxazoline) ligand [31]. EPR spectroscopy provided the first direct experimental evidence for the formation of square planar and square pyramidal complexes of Cu(II) with bis(oxazoline) inside the pores of the Y zeolite. This EPR study allowed the investigation of the coordination and oxidation state of the copper-containing catalyst, estimation of the percentage of copper that was successfully complexed, and acquisition of useful information about the mechanism of the catalytic reaction. Another example of the use of EPR in the characterization of supported chiral catalysts is given by the evaluation of the V-V interatomic distance in vanadium complexes with chiral tridentate Schiff base ligands supported on SiO_2. The supported complex was used as a catalyst for the oxidative coupling of 2-napthol: EPR was employed to study the interaction of the reagents (oxygen and 2-napthol) with the vanadium catalyst, providing useful information about the reaction mechanism [32]. EPR spectroscopy can also be employed to determine the oxidation state of a metal in a chiral material, provided that the metal ion is paramagnetic (and thus EPR active) in at least one of the oxidation states. This approach has been used to study the ratio between Mn^{II} and Mn^{III} in manganese Schiff base complexes encapsulated in zeolite Y [33]. Through the comparison with the results of cyclic voltammetry (CV) (see Section 3.4.8), these EPR data suggested the inhomogeneous distribution of the two ions between the external layer of the zeolite and in its bulk.

3.4.3. Infrared Spectroscopy

Infrared (IR) spectroscopy provides information about the vibrational modes of the analyzed material. Since different groups have different vibrational modes that correspond to absorption of radiation at characteristic wavelengths, specific regions of the IR spectrum can be often used as a fingerprint to ascertain the identity of the analyzed material. IR spectroscopy is generally measured using Fourier transform (FT) to process the spectral signal: the technique is then referred to as Fourier transform infrared (FT-IR) spectroscopy. In the case of supported chiral catalysts, IR spectroscopy proves very useful in studying whether the chiral catalyst has been successfully supported and whether the heterogenization process caused any structural change in the chiral compound. These questions can be easily answered by comparing the spectrum of

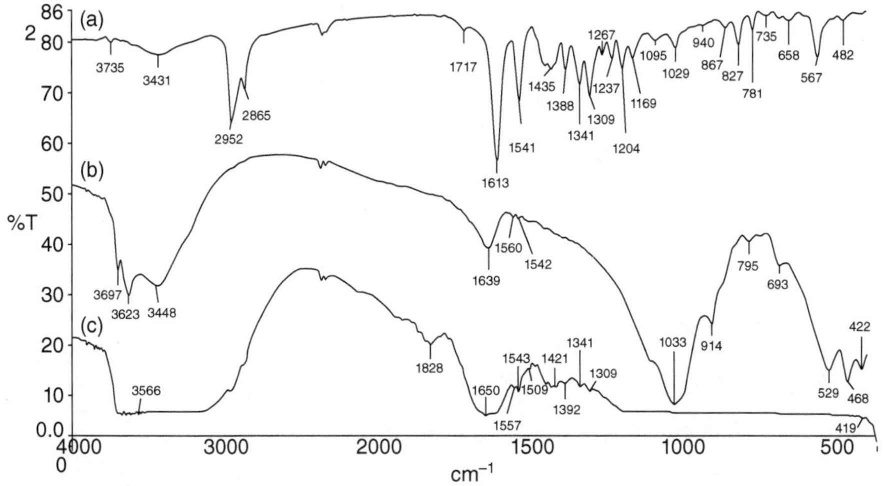

FIGURE 3.19. IR spectra of the Mn-salen complex **(A)**; of the montmorillonite clay (**B**); of the complex intercalated in the clay (**C**). The bands at 2962 and 2865 cm^{-1} ($C_{aliphatic}$–H) and the broad band in the range 1750–1550 cm^{-1} in spectrum **C** suggest the presence of the Mn-salen complex in the clay matrix (reproduced from Reference [34]).

the free chiral compound with that of the supported catalyst. For these reasons, FT-IR spectroscopy is one of the most widely used techniques in the characterization of supported chiral catalysts, though the depth of information achieved with this technique is inferior to that obtained with NMR (see Section 3.4.1). A typical application of IR spectroscopy is in the study of the integrity of chiral complexes after their heterogenization by intercalation in the interlayer space of LMs, such as clays and LDHs. Examples include dicationic chiral Mn^{III}-salen complexes exchanged in the interlayers of montmorillonite clay as heterogeneous enantioselective catalysts for the epoxidation of alkenes (Fig. 3.19) [34] and chiral sulfonato-salen-Mn^{III} complexes intercalated into a Zn-Al LDH [35]. IR spectroscopy can also provide in-depth information about the interaction of the chiral compound with the support. For example, by determining which of the vibrational modes of vanadium Schiff base complexes immobilized on silica (see Section 3.4.2) were modified and which were unaffected in the supported material compared to the free complex, it was possible to gain information about the interaction of the chiral compounds with the silica support [32]. The application of IR spectroscopy to the characterization of supported chiral catalysts is not limited to the study of the integrity of the chiral compounds. An example in this sense is provided by the use of IR spectroscopy in the characterization of vanadyl salen complexes (both achiral and chiral) covalently anchored to SWNTs (see Section 3.3.2.1). The successful anchoring of the chiral complex to the nanotube was monitored by FT-IR: the reaction of the –SH function of the modified nanotube with the

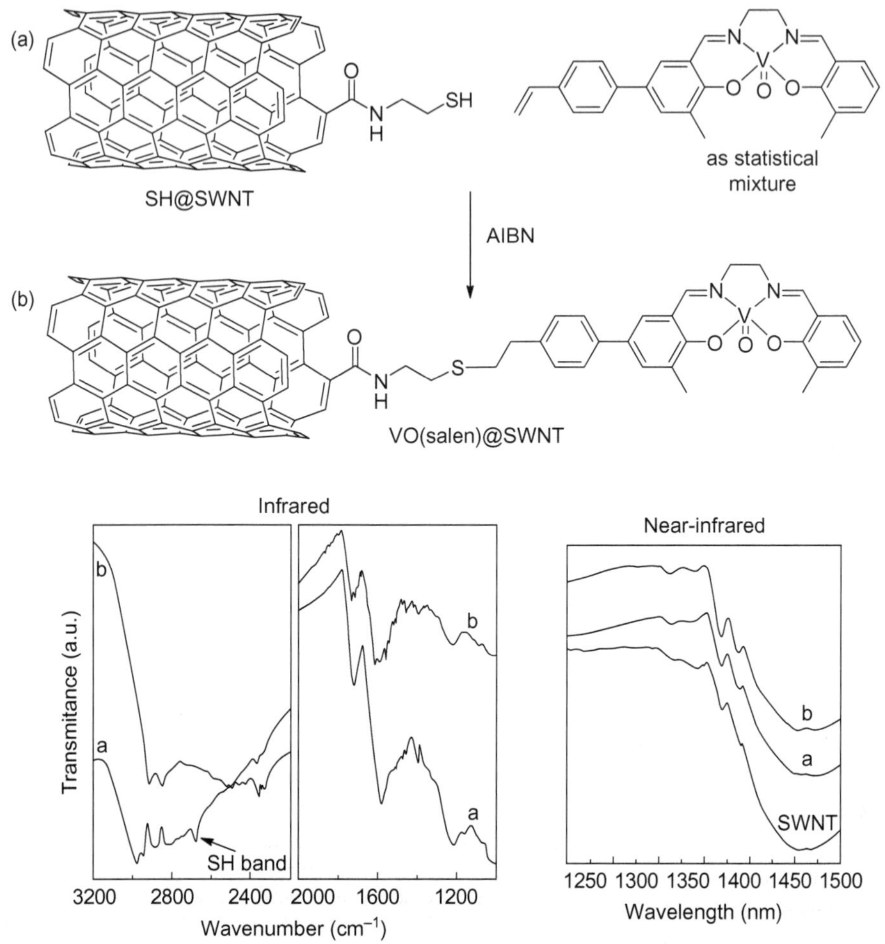

FIGURE 3.20. Infrared and near-infrared spectra of (a) SH-functionalized SWNT support and (b) of the supported catalyst (reproduced from Reference [7]).

terminal double bond of the complex was demonstrated by the disappearance of the SH-stretching band, while the signal of the amide group at 1726 cm^{-1} was preserved (Fig. 3.20) [7].

3.4.4. UV-Vis-NIR Spectroscopy

Electronic transitions in chemical compounds correspond to radiation in the ultraviolet, visible, and near-infrared (UV-Vis-NIR) region of the electromagnetic spectrum. These transitions take place at characteristic wavelengths. Therefore, UV-Vis-NIR spectroscopy can be used in a similar way to that described above for IR spectroscopy. These two types of spectroscopy are

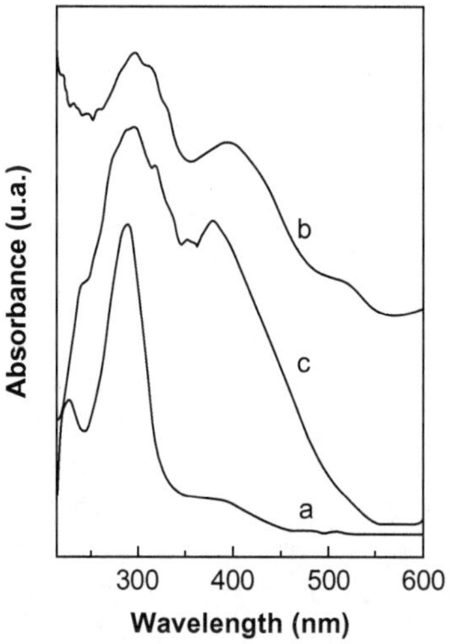

FIGURE 3.21. (a) UV-Vis absorption spectra of the free vanadyl salen complex and (b) diffuse reflectance UV-Vis spectra of the mesoporous organosilica material containing the vanadyl salen complex before and (c) after extraction of the template (reproduced from Reference [13]).

often used in combination for the characterization of the integrity of chiral catalysts within the supporting host. Examples of the complementary use of UV-Vis and IR spectroscopies are given by the characterization of the supported chiral catalysts obtained by intercalation of chiral complexes within LMs mentioned in Section 3.4.3. For all these materials, UV-Vis spectroscopy provided supporting evidence that the complex was not significantly affected by the intercalation [34, 35]. Similarly, UV-Vis spectroscopy was used to confirm the presence of the vanadyl salen complexes in the PMO material discussed in Section 3.4.1. This analysis could also prove that only a fraction of the complex was lost after extraction of the template, which has to be removed to free the pores of the material for catalytic application (Fig. 3.21). UV-Vis can also indicate that a supported complex is different from the free complex, as in the case of the vanadium Schiff base complexes immobilized on silica (mentioned in Sections 3.4.2 and 3.4.3): The appearance of a band at ~550 nm in the UV-Vis spectrum, assigned to a d-d transition of the vanadium species, suggested that the symmetry of the supported vanadyl complexes became distorted [32]. Although the study of electronic transitions is mainly performed in the UV-Vis region of the electromagnetic spectrum, NIR spectroscopy can also provide valuable information. This technique was used to complement the FT-IR results in the characterization of the vanadyl salen complexes anchored to SWNTs described in Section 3.4.3. Analysis in the NIR region showed that the functionalization of the nanotube and the consecutive grafting of the vanadyl salen complex did not alter the structure of the nanotube (Fig. 3.20) [7].

3.4.5. X-Ray Diffraction

Powder X-ray diffraction (XRD) is widely used to investigate the crystalline structure or the long-range order in solids. In the case of the solids used as supports for the immobilization of chiral catalysts, XRD can be employed in the characterization of the material before and after the heterogenization of the catalytic species to determine whether the process causes any structural change in the material used as support. XRD analysis proved particularly informative in the characterization of heterogeneous chiral catalysts prepared by immobilization of chiral complexes by intercalation in the interlayer space of LMs, such as cationic clays (e.g., hectorite and montmorillonite) and anionic, hydrotalcite-like LDHs. The successful intercalation generally causes an expansion of the spacing between the layers, which can be easily monitored by XRD. For example, the intercalation of the cationic Rh-DIOP chiral complex into the cationic clay hectorite was performed by cation exchange and resulted in an increase of the basal spacing to 1.33 nm, which is in agreement with the molecular size of the Rh-DIOP$^+$ complex [36]. In a similar way, a stereoselective and reusable heterogeneous epoxidation catalyst was prepared by intercalation of a chiral sulfonato-salen-Mn^{III} complex into a Zn-Al LDH host: The increase in the basal spacing observed by XRD was considered to be strong evidence of the successful intercalation [35]. XRD can also be used to characterize the structural properties of porous materials containing chiral moieties. For example, the XRD pattern of the PMO material discussed in Section 3.4.1 displays the characteristic (100), (110), and (200) reflections corresponding to the hexagonal array of channels of MCM-41-like materials. This ordered structure was preserved after extraction of the template using acidified ethanol (Fig. 3.22) [13].

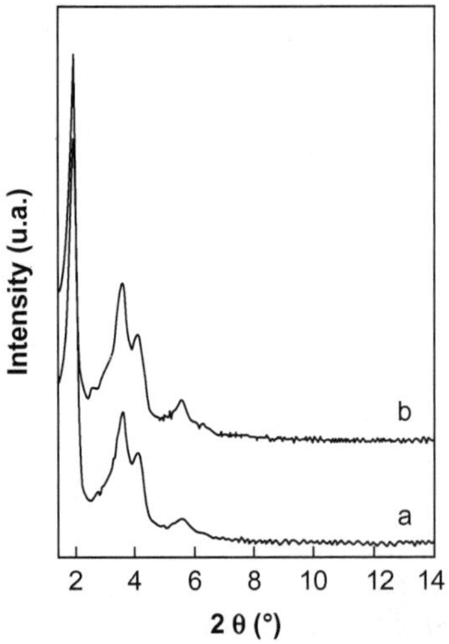

FIGURE 3.22. (a) XRD patterns of the PMO material with vanadium-salen moieties before and (b) after the extraction of the template (reproduced from Reference [13]).

3.4.6. N_2 Adsorption–Desorption Isotherms

The measurement of N_2 adsorption–desorption isotherms is commonly used to elucidate the textural properties of solids, and particularly of porous materials. These measurements provide information about the specific surface area of the material and about the size distribution and volume of its pores. Porous materials are often used as supports for chiral catalysts (see Section 3.2.1) because their cavities offer a high surface area and a suitable location to immobilize the active species. The presence of immobilized chiral catalysts generally causes a decrease in the specific surface area and in the volume and size of the pores of the support. These changes can be monitored by the measurement and analysis of the N_2 adsorption–desorption isotherms. Therefore, this type of analysis can provide information on the effect of the immobilization on the support and, indirectly, on the degree of loading of the chiral catalyst. An example of this application of N_2 adsorption–desorption isotherms is given by the study of chiral Cr-salen and Mn-salen complexes immobilized on an aminopropyl-functionalized ordered mesoporous MCM-41 material. The N_2 isotherms of the material obtained after grafting of the complexes showed a decrease in specific surface area and in the size and volume of the mesopores, suggesting the successful anchoring of the complexes inside the pores [37]. Similarly, a decrease in specific surface area and mesopore volume of the MCM-41 support was observed upon the noncovalent immobilization of the chiral rhodium-phosphine complex described in Section 3.4.1 [30].

N_2 and Ar adsorption–desorption isotherms can also be used to characterize the textural properties of porous materials containing chiral moieties. For example, these techniques were used to determine the specific surface area ($900\,m^2/g$) and the pore diameter (4.2 nm) of the PMO material discussed in Section 3.4.1 [23].

3.4.7. Elemental Analysis

Elemental analysis by means of atomic spectroscopy is routinely used to evaluate whether leaching of the catalytically active species from the support occurred. The approach commonly employed consists of analyzing the presence of metal ions of the chiral complex in solution after the liquid-phase catalytic reaction. The absence of metal ions in the solution is considered a sufficient proof of the absence of leaching of the chiral catalyst [38]. However, it is important to keep in mind that each metal has a different detection limit and also that the leached metal species can be inefficient to promote the reaction. Generally, the analysis is performed using inductively coupled plasma atomic emission spectroscopy (ICP-AES), also known as inductively coupled optical emission spectroscopy (ICP-OES). In this technique, plasma is used to decompose the analytes into atoms, which then emit radiation due to the relaxation of electrons from excited states to lower energy states. Each element emits radiation with a characteristic wavelength. As an alternative to the measurement of the emitted radiation, the atomization by ICP can be combined with analysis by mass spectrometry (ICP-MS). The equipment for this type of analysis is much more expensive, but its detection limits are much lower as compared to ICP-AES. Therefore, ICP-MS can prove particularly advantageous for analyzing elements that have a relatively high detection limit with ICP-AES [39] (e.g., the detection limit of Rh with ICP-AES is 20 ng/mL, compared with 0.2 ng/mL for Os and Mn, while the detection limit of Rh with ICP-MS is 0.0003 ng/mL) [40]. Other elemental analysis techniques that can be used for studying the leaching of metal complexes are based on atomization with a flame or in a graphite furnace combined with analysis by atomic absorption spectroscopy (AAS) [41].

Elemental analysis by atomic spectroscopy can also be used to measure the amount of a selected metal present into a solid and, thus, to estimate the loading of immobilized chiral catalyst on the support and their decrease upon reuse [34, 37]. However, atomic spectroscopy is not suitable for the analysis of C, H, N, and S. Elemental analysis can be performed by combustion analysis if the supported chiral compound does not contain a metal or, more generally, if we need to obtain information about the organic part of the chiral compound [26]. The sample is treated at high temperatures (around 1000°C) in the presence of O_2 and with suitable catalysts. Under these conditions, the organic part of the material oxidizes to CO_2, H_2O, N_2, and SO_2. The gas mixture is separated by gas chromatography, and each component is quantified by IR absorbance or with a thermal conductivity detector. Combustion analysis allows for the estimation of the loading of the immobilized organic compounds, though it should be kept in mind that the accuracy of this quantification can be affected by the relative error of the results of this type of analysis and by the possible presence of other organic species (e.g., solvents).

3.4.8. Other Techniques

The techniques reviewed in the previous paragraphs are those most widely employed and generally the most informative in the characterization of supported chiral catalysts. Other techniques can be used for this purpose, although their application is less widespread, mainly because they provide relevant information only for specific systems or because the equipment is less easily available [e.g., extended X-ray absorption fine structure (EXAFS) measurements are generally performed at synchrotron radiation laboratories]:

- X-ray photoelectron spectroscopy (XPS) is a spectroscopic technique based on the ejection of electrons from inner shells of an element upon irradiation with a beam of X rays. XPS can be used for a quantitative analysis of the elemental composition of the surface layer of a material, and to determine the chemical and electronic state of surface atoms. EXAFS and X-ray absorption near edge structure (XANES) are closely related spectroscopic techniques based on the measurement of the X-ray absorption coefficient of a material. EXAFS provides information about the structure of the analyzed material, particularly for systems for which diffraction techniques such as XRD are not suitable, while XANES can be used to determine the average oxidation state and the coordination environment of the absorbing atoms in the sample. These spectroscopic techniques were used in the detailed study of the vanadium Schiff base complexes immobilized on silica mentioned in Sections 3.4.2–3.4.4. XANES showed that the V=O bond in the chiral complex remained unchanged when the compound was supported on SiO_2. XANES in combination with XPS indicated that the oxidation state of vanadium (4+) was not affected by the coordination with 2-napthol, which was used as reagent in the catalytic test of the material. Finally, EXAFS provided information about the local structure of the complex around the vanadium center [32].

- CV is an electroanalytical technique based on the measurement of the electrical current generated in an electrochemical cell as a function of a difference of potential applied with a triangular profile in repeated cycles. This type of measurement provides information on the redox behavior of chemical compounds and on the kinetics of electrode reactions in the study of supported chiral catalysts. CV has been employed in combination with spectroscopic techniques (EPR, IR, UV-Vis) to study Mn Schiff base complexes encapsulated in zeolite Y in which manganese can be present in different oxidation states (see also Section 3.4.2) [33, 42].
- Transmission electron microscopy (TEM) is a type of microscopy based on the use of a beam of electrons. Since TEM pictures allow visualization of materials at the nanoscale, this technique is a very powerful tool in the characterization of many solids, including materials used as support for chiral catalysts (e.g., MCM-41, nanotubes). However, TEM is less useful for characterizing the materials obtained after supporting the chiral catalysts because the immobilized compounds are generally not visualized by TEM. Among the few reported applications of TEM in the characterization of supported chiral catalysts there are the visualization of the increased spacing between the layers of clays intercalated with an Rh-DIOP complex [36] and the monitoring of the aggregation of magnetic Fe_3O_4-nanoparticles used as support for chiral catalysts [43].
- Thermogravimetric analysis (TGA) is an analytical technique based on the measurement of the weight loss of a material as a function of the temperature at which the material is heated. TGA has been used in the study of supported chiral catalysts to confirm the incorporation of the complex in the host, for example, Mn-salen complexes intercalated in LDHs [35]. However, TGA alone does not provide sufficient evidence of the successful heterogenization and should always be complemented by other characterization techniques.

REFERENCES

[1] Benaglia, M. (2009). *Recoverable and Recyclable Catalysts*. John Wiley and Sons, Chicester, UK.

[2] Fraile, J. M., Garcia, J. I., Mayoral, J. A. (2008). Recent advances in the immobilization of chiral catalysts containing bis(oxazolines) and related ligands. *Coord. Chem. Rev.*, *252*, 624–646.

[3] Stein, A. (2000). Spheres templating methods for periodic mesoporous solids. *Micropor. Mesopor. Mat.*, *44–45*, 227–239.

[4] Fraile, J. M., Garcia, J. I., Mayoral, J. A. (2009). Noncovalent immobilization of enantioselective catalysts. *Chem. Rev.*, *109*, 360–417.

[5] Drits, V. A., Sokolova, T. N., Sokolova, G. V., Cherkashin, V. I. (1987). New members of the hydrotalcite-manasseite group. *Clays Clay Miner.*, *35*, 401–417.

[6] Baleizão, C., Gigante, B., Garcia, H., Corma, A. (2003). Chiral vanadyl Schiff base complex anchored on silicas as solid enantioselective catalysts for formation of cyanohydrins: optimization of the asymmetric induction by support modification. *J. Catal.*, *215*, 199–207.

[7] Baleizão, C., Gigante, B., Garcia, H., Corma, A. (2004). Vanadyl salen complexes covalently anchored to single-wall carbon nanotubes as heterogeneous catalysts for the cyanosilylation of aldehydes. *J. Catal.*, *221*, 77–84.

[8] Chouhan, G., Wang, D., Alper, H. (2007). Magnetic nanoparticle-supported proline as a recyclable and recoverable ligand for the CuI catalyzed arylation of nitrogen nucleophiles. *Chem. Commun.*, *45*, 4809–4811.

[9] Li, H., Luk, Y.-Y., Mrksich, M. (1999). Catalytic asymmetric dihydroxylation by gold colloids functionalized with self-assembled monolayers. *Langmuir*, *15*, 4957–4959.

[10] (a) Benaglia, M., Cinquini, M., Cozzi, F., Pugliesi, A., Celentano, G. (2002). Poly(ethylene glycol)-supported proline: a versatile catalyst for the enantioselective aldol and iminoaldol reactions. *Adv. Synth. Catal.*, *344*, 533–542; For more examples regarding supported proline-based catalysts: (b) Gruttadauria, M., Giacalone, F., Noto, R. (2008). Supported proline and proline-derivatives as recyclable organocatalysts. *Chem. Soc. Rev.*, *37*, 1666–1688.

[11] Li, C. (2004). Chiral synthesis on catalysts immobilized in microporous and mesoporous materials. *Catal. Rev.*, *46*, 419–492.

[12] Moreau, J. J. E., Vellutini, L., Wong Chi Man, M., Bied, C. (2001). New hybrid organic–inorganic solids with helical morphology via H-bond mediated sol–gel hydrolysis of silyl derivatives of chiral (R,R)- or (S,S)-diureidocyclohexane. *J. Am. Chem. Soc.*, *123*, 1509–1510.

[13] Baleizão, C., Gigante, B., Das, D., Alvaro, M., Garcia, H., Corma, A. (2004). Periodic mesoporous organosilica incorporating a catalytically active vanadyl Schiff base complex in the framework. *J. Catal.*, *223*, 106–113.

[14] Cho, S.-H., Ma, B., Nguyen, S. T., Hupp, J. T., Albrecht-Schmitt, T. E. (2006). A metal-organic framework material that functions as an enantioselective catalyst for olefin epoxidation. *Chem. Commun.*, (24), 2563–2565.

[15] An, Z., Zhang, W., Shi, H., He, J. (2006). An effective heterogeneous L-proline catalyst for the asymmetric aldol reaction using anionic clays as intercalated support. *J. Catal.*, *241*, 319–327.

[16] Choudary, B. M., Kavita, B., Chowdari, N. S., Sreedhar, B., Kantam, M. L. (2002). Layered double hydroxides containing chiral organic guests: synthesis, characterization and application for asymmetric C-C bond-forming reactions. *Catal. Lett.*, *78*, 373–377.

[17] Mazzei, M., Marconi, W., Riocci, M. (1980). Asymmetric hydrogenation of substituted acrylic acids by Rh′-aminophosphine chiral complex supported on mineral clays. *J. Mol. Catal.*, *9*, 381–387.

[18] Selke, R. (1986). Phosphinites of carbohydrates as chiral ligands for asymmetric synthesis catalyzed by complexes. Part III. Immobilization of cationic rhodium(I) chelates of phenyl 4,6-*O*-(*R*)-benzylidene-2,3-*O*-bis(diphenylphosphino)-β-D-glucopyranoside on cation exchangers for hydrogenation of dehydroamino acids. *J. Mol. Catal.*, *37*, 227–234.

[19] Zsigmond, A., Bogar, K., Notheisz, F. (2003). Comparative study of "ship-in-a-bottle" and anchored heterogenized Rh complexes. *J. Catal.*, *213*, 103–108.

[20] Sabater, M. J., Corma, A., Domenech, A., Fornes, V., Garcia, H. (1997). Chiral salen manganese complex encapsulated within zeolite Y: a heterogeneous enantioselective catalyst for the epoxidation of alkenes. *Chem. Commun.*, (14), 1285–1286.

[21] Gelman, F., Avnir, D., Schumann, H., Blum, J. (1999). Sol-gel entrapped chiral rhodium and ruthenium complexes as recyclable catalysts for the hydrogenation of itaconic acid. *J. Mol. Catal. A Chem.*, *146*, 123–128.

[22] Ishizuka, N., Togashi, M., Inoue, M., Enomoto, S. (1987). Asymmetric hydrogenation of 3,4-methylenedioxy-α-acetamidocinnamic acid using newly developed silica gel-supported chiral rhodium(I)-phosphine complexes. *Chem. Pharm. Bull.*, *35*, 1686–1690.

[23] Frunza, L., Kosslick, H., Landmesser, H., Hoeft, E., Fricke, R. (1997). Host/guest interactions in nanoporous materials. I. The embedding of chiral salen manganese(III) complex into mesoporous silicates. *J. Mol. Catal. A Chem.*, *123*, 179–187.

[24] Raja, R., Thomas, J. M., Jones, M. D., Johnson, B. F. G., Vaughan, D. E. W. (2003). Constraining asymmetric organometallic catalysts within mesoporous supports boosts their enantioselectivity. *J. Am. Chem. Soc.*, *125*, 14982–14983.

[25] Gruttadauria, M., Riela, S., Aprile, C., Lo Meo, P., D'Anna, F., Noto, R. (2006). Supported ionic liquids. New recyclable materials for the L-proline-catalyzed aldol reaction. *Adv. Synth. Catal.*, *348*, 82–92.

[26] Aprile, C., Giacalone, F., Gruttadauria, M., Marculescu, A., Noto, R., Revell, J. D., Wennemers, H. (2007). New ionic liquid -modified silica gels as recyclable materials for L-proline- or H-Pro-Pro-Asp-NH_2-catalyzed aldol reaction. *Green Chem.*, *9*, 1328–1334.

[27] Heckel, A., Seebach, D. (2002). Enantioselective heterogeneous epoxidation and hetero-Diels-Alder reaction with Mn- and Cr-salen complexes immobilized on silica gel by radical grafting. *Helv. Chim. Acta.*, *85*, 913–926.

[28] Adima, A., Moreau, J. J. E., Wong Chi Man, M. (1997). Chiral organic-inorganic solids as enantioselective catalytic materials. *J. Mater. Chem.*, *7*, 2331–2333.

[29] Wagner, H. H., Hausmaan, H., Hölderich, W. H. (2001). Immobilization of rhodium diphosphine complexes on mesoporous Al-MCM-41 materials: catalysts for enantioselective hydrogenation. *J. Catal.*, *203*, 150–156.

[30] de Rege, F. M., Morita, D. K., Ott, K. C., Tumas, W., Broene, R. D. (2000). Non-covalent immobilization of homogeneous cationic chiral rhodium-phosphine catalysts on silica surfaces. *Chem. Commun.*, 1797–1798.

[31] Traa, Y., Murphy, D. M., Farley, R. D., Hutchings, G. H. (2001). An EPR study on the enantioselective aziridination properties of a CuNaY zeolite. *Phys. Chem. Chem. Phys.*, *3*, 1073–1080.

[32] Tada, M., Kojima, N., Izumi, Y., Taniike, T., Iwasawa, Y. (2005). Chiral self-dimerization of vanadium complexes on a SiO_2 surface for asymmetric catalytic coupling of 2-naphthol: structure, performance, and mechanism. *J. Phys. Chem. B*, *109*, 9905–9916.

[33] Domenech, A., Formentin, P., Garcia, H., Sabater, M. J. (2000). Combined electrochemical and EPR studies of manganese Schiff base complexes encapsulated within the cavities of zeolite Y. *Eur. J. Inorg. Chem.*, *6*, 1339–1344.

[34] Kureshy, R. I., Khan, N. H., Abdi, S. H. R., Ahmad, I., Singh, S., Jasra, R. V. (2004). Dicationic chiral Mn(III) salen complex exchanged in the interlayers of montmorillonite clay: a heterogeneous enantioselective catalyst for epoxidation of nonfunctionalized alkenes. *J. Catal.*, *221*, 234–240.

[35] Bhattacharjee, S., Anderson, J. A. (2004). Synthesis and characterization of novel chiral sulfonato-salen-manganese(III) complex in a zinc-aluminium LDH host. *Chem. Commun.*, *5*, 554–555.

[36] Sento, T., Shimazu, S., Ichikuni, N., Uematsu, T. (1999). Asymmetric hydrogenation of itaconates by hectorite-intercalated Rh-DIOP complex. *J. Mol. Catal. A*, *137*, 263–267.

[37] Kureshy, R. I., Ahmad, I., Khan, N. H., Abdi, S. H. R., Singh, S., Pandia, P. H., Jasra, R. V. (2005). New immobilized chiral Mn(III) salen complex on pyridine N-oxide-modified MCM-41 as effective catalysts for epoxidation of nonfunctionalized alkenes. *J. Catal.*, *235*, 28–34.

[38] Choudary, B. M., Pal, U., Kantam, M. L., Ranganath, K. V. S., Sreedhar, B. (2006). Asymmetric epoxidation of olefins by manganese(III) complexes stabilized on nanocrystalline magnesium oxide. *Adv. Synth. Catal.*, *348*, 1038–1042.

[39] Stephenson, P., Licence, P., Ross, S. K., Poliakoff, M. (2004). Continuous catalytic asymmetric hydrogenation in supercritical CO_2. *Green Chem.*, *6*, 521–523.

[40] Harris, D. C. (2007). *Quantitative Chemical Analysis*, 7th ed. W.H Freeman & C, New York. Chapter 21.

[41] Simons, C., Hanefeld, U., Arends, I. W. C., Maschmeyer, T., Sheldon, R. A. (2006). Comparison of supports for the electrostatic immobilization of asymmetric homogeneous catalysts. *J. Catal.*, *239*, 212–219.

[42] Domenech, A., Formentin, P., Garcia, H., Sabater, M. J. (2002). On the existence of different zeolite-associated topological redox isomers. Electrochemistry of the Y zeolite-associated Mn(Salen)N_3 complex. *J. Phys. Chem. B*, *106*, 574–582.

[43] Hu, A., Yee, G. T., Lin, W. (2005). Magnetically recoverable chiral catalysts immobilized on magnetite nanoparticles for asymmetric hydrogenation of aromatic ketones. *J. Am. Chem. Soc.*, *127*, 12486–12487.

CHAPTER 4

SYNTHESIS OF CHIRAL CATALYSTS SUPPORTED ON ORGANIC POLYMERS

TOR ERIK KRISTENSEN AND TORE HANSEN

4.1. Introduction 209
4.2. The preparation of chiral catalysts supported on organic polymers through postmodification 211
4.2.1. Polymer resins for immobilization of chiral catalysts 211
4.2.2. The preparation of chiral organometallic catalysts supported on organic polymers through postmodification 213
4.2.3. The preparation of chiral organocatalysts supported on organic polymers through postmodification 220
4.3. The preparation of chiral catalysts supported on organic polymers according to bottom-up strategies 226
4.3.1. The character of radical polymerization from the synthetic point of view 226
4.3.2. The nature of different heterophase polymerizations 228
4.3.3. The preparation of chiral organometallic catalysts supported on organic polymers through bottom-up strategies 230
4.3.4. The preparation of chiral organocatalysts supported on organic polymers through bottom-up strategies 237
4.4. Conclusions and outlook 244
References 245

4.1. INTRODUCTION

Among supported chiral catalysts, those prepared using organic polymers as their base are currently both the most developed and the ones most actively

Catalytic Methods in Asymmetric Synthesis: Advanced Materials, Techniques, and Applications, First Edition. Edited by Michelangelo Gruttadauria and Francesco Giacalone.

pursued [1]. Organic polymers have, by nature of being organic materials (and if appropriately prepared), a natural conformity with the organic solvents that tend to be the medium for most asymmetric transformations using chiral catalysts. In addition, their hydrophilic–hydrophobic balance is most often easily adjusted, providing leverage for the preparation of catalytic systems that can operate under a wide range of solvent polarity. This includes the aqueous conditions that are of major importance in modern biomimetic asymmetric synthesis. After all, the enzymes that mediate the catalytic transformations that take place in biological systems are, in essence, organic polymers (peptides) operating under aqueous conditions. By striking a balance between the hydrophilic peptide backbone (reinforced if necessary with segments containing polar side chains) with hydrophobic side chains, this biopolymer can create favorable hydrophobic microenvironments (if necessary) in aqueous solution. It is only natural that scientists are attempting to copy the properties of these remarkable biopolymers by using synthetic organic polymers. In the future, we will probably regard the polymeric backbones themselves as opportunities and not only as "inert scaffolds" for anchoring substrates, as, historically, they have tended to be treated. Organic polymers do consequently appear to be the most adaptable of the solid supports for immobilization of chiral catalysts simply because of the near-endless structural variations they offer and the ease at which they often can be synthesized.

The expedient syntheses of polymer-supported chiral catalysts are certainly of foremost importance for the reasons discussed above. However, the synthetic avenues available for their production are probably undervalued as the focus within this field of research tends to be heavily biased toward the utilization, and not preparation, of polymer-supported catalysts. This is unfortunate, as the greatest obstacle for the widespread utilization of polymer-supported reagents and catalysts is undoubtedly their prohibitive costs. This chapter will address the synthetic strategies available for preparation of polymer-supported chiral catalysts instead of the catalytic chemistry of the finished systems (which is covered in great detail in Chapters 1–3).

However, for certain applications, some constraints enforced on the properties of the finished systems as a result of the use for which they were intended will severely affect, and possibly limit, the possibilities available during their synthesis.

Although several classifications of the strategies available for synthesis of polymer-supported chiral catalysts are potentially useful, a helpful distinction can be made as to where in the overall sequence we introduce the chiral catalyst as related to the macromolecular synthesis forming the core "skeleton" of the final immobilized catalyst. If the catalyst is introduced after the macromolecular synthesis (polymerization), it will be referred to as a postmodification strategy. If the catalyst (or at least its main components) is introduced before (or rather during or part of the macromolecular synthesis), it will be referred to as a bottom-up strategy. As presented later in this chapter, the second of these two strategies usually consists of some sort of copolymerization and will often be referred to as a copolymerization strategy. This approach is therefore modular at the monomeric, rather than at the polymeric, level. Although there certainly will be differences of opinion, it can, with at least some sort of confidence, be stated that the former of these two strategies (postmodification) has been slightly dominant in the preparation of polymer-supported catalysts. This is connected with the fact that it is simply the strategy that outsources the macromolecular synthesis (generally, few synthetic organic chemists have received adequate training in preparative polymer science), and also because postmodification is an integral part of solid phase peptide synthesis. However, for research conducted by scientists within materials science, a bottom-up-type philosophy is often preferred as the research then tends to focus on functional materials per se rather than their eventual utilization. Although it is rather entrenched by now, this artificial partition is probably not truly rational science but rather a likely result of the traditional and "invisible" boundaries existing between different fields of chemistry. Still, both strategies can actually address the same problems, and a truly interdisciplinary approach is surely the most useful, as these two strategies each have their strengths and weaknesses and are highly complementary. The traditional notion of the superiority of classical postmodification has more historical foundations than factual ones, as we will see. This is perhaps especially pronounced with the advent of chiral organocatalysts in the past decade, something that has boosted copolymerization strategies because of the robustness and simplicity of many organocatalysts, as well as the need for efficient preparations since organocatalysts, to a certain extent, tend to be utilized at substantial catalyst loadings.

The remainder of this chapter is devoted to an analysis of the preparation of covalently bound polymer-supported chiral catalysts by using the postmodification/bottom-up partition of synthetic strategies, focusing especially on examples that provide useful insight on the preparation rather than only on the performance of the immobilized catalysts [2].

4.2. THE PREPARATION OF CHIRAL CATALYSTS SUPPORTED ON ORGANIC POLYMERS THROUGH POSTMODIFICATION

4.2.1. Polymer Resins for Immobilization of Chiral Catalysts

Extensive reviews dealing with available polymer resins for the immobilization of different types of substrates are now available [3,4]. As part of a current microreview, we have also contributed with a short, although relatively comprehensive, assessment of the most important polymer resins that have been reported and that are commercially available [2]. Though numerous polymer resins are presently available, prices regrettably continue to be prohibitively high for most of them. Despite the fact that such prices can be defended when the immobilized substrates are high-valued peptides or oligonucleotides, they can prove problematic for many chiral catalysts, as these can sometimes be relatively cheap, and they sometimes may be needed at considerable catalyst loadings (this refers, of course, more strongly to organocatalysts). As a result of this, most researchers tend to use the archetypical Merrifield resin [5] because of convenience and/or cost, even though other polymeric supports (ones that often have more favorable characteristics for certain uses) are available.

Since postmodification strategies for preparation of chiral catalysts supported on organic polymers are usually synonymous with the purchase of prefabricated polymer products as a starting material, only a brief outline will be presented here. Their detailed preparation will not be explored here but can be located elsewhere [2]. In the section on bottom-up strategies later in this chapter, the characteristics of these preparations will be dealt with more thoroughly, as they form the cornerstone of the bottom-up philosophy. Radical polymerization in particular has certain characteristics connected with it that are immensely useful from the synthetic point of view; a fact that, unfortunately, is somewhat underappreciated.

Organic polymers for use in the immobilization of catalysts can be of a homogenously soluble linear type or of a cross-linked type (the latter very often in the form of spherical polymer beads). The cross-linked types are especially convenient and will be of particular importance for this chapter, but the synthetic strategies presented can normally be readily adapted for both types. The cross-linked polymer beads can be of two major morphological types: microporous or macroporous [1, 6]. Microporous polymer beads contain low levels of cross-linker (a few percent) and do not have any appreciable porosity in their dry state; their interior is accessible only after swelling in a reaction environment containing a suitable solvent. The swelling is accompanied by a significant expansion in volume to a gel-like state, and these resins are therefore often referred to as gel type. Macroporous (or macroreticular) polymer beads, on the other hand, contain a large proportion of cross-linker (all the way up to 50% or more) and retain most of their porosity even in a dry state. Unlike microporous beads, they can take up solvent with little or no change in volume, but generally have a lower loading capacity as a result of their larger and permanent pore structure. The high degree of cross-linker can make these polymer beads more rigid and useful for packaging in columns, but can also make them, for example, brittle and vulnerable to magnetic stirring, a clear disadvantage in organic synthesis. Microporous beads, on the other hand, require less care in handling and are usually less fragile. However, the performance of microporous polymeric structures varies to a large extent on the solvent used because of the accompanying swelling. As a result, finding the right type of polymer backbone to match a required set of reaction conditions is a major challenge for their utilization. Macroporous structures can be used with almost any solvent, irrespective of whether or not it is a good solvent for the non-cross-linked polymer, and they do not have diffusional limitations on reaction rates. Unfortunately, the greater heterogeneity of the resultant reactions mixtures when using macroporous structures is frequently detrimental for reactions in organic synthesis. Microporous supports firmly remain the most important.

The chemistry of cross-linked polymer beads, which form the largest type of polymer supports, can very roughly be divided into four major types (this certainly encompasses only the most important ones, though): styrenic, acrylic, polyetheral, or hybrids of more than one of these (either through grafting or cross-linking) [1, 2, 4]. Purely styrenic networks like the Merrifield resin have the advantage of being chemically quite inert, but they suffer from poor swelling in polar solvents and suffer badly in that they have few handles for functionalization, meaning that the chloromethylated polystyrene is one of very few starting points for derivatization. Acrylic networks (PEGA, CLEAR, etc.), on the other hand, are both readily functionalized and are compatible with a wide range of solvent polarity. However, they are chemically less robust (the acrylic esters more so than the acrylic amides, naturally). Polyetheral networks (POEPS, POEPOP, SPOCC, ChemMatrix, etc.) usually contain mainly poly(ethyleneglycol) (PEG) networks. They are both chemically the most inert and have the most favorable swelling characteristics, but they are also by far the most expensive. By making styrenic-polyetheral hybrids (TentaGel, Champion, ArgoGel, etc.) by either grafting or cross-linking polystyrene with PEG, both favorable swelling characteristics and chemical inertness can be obtained simultaneously by combining the properties of the two materials. Again, these more specialized supports are significantly more expensive because of their more intricate synthesis [7]. Clever and economic preparations of chemically inert and readily expandable microporous resins (and maintaining this property in a selection of solvents spanning a broad range of polarity) are still in much need, something that, to a certain extent, is undermining the efficiency of postmodification strategies for immobilization of chiral catalysts. Hopefully, these issues can be addressed in the future, and the situation would improve significantly if polymer-supported methodologies were adapted by a larger proportion of the scientific community.

4.2.2. The Preparation of Chiral Organometallic Catalysts Supported on Organic Polymers Through Postmodification

In general, organometallic species for asymmetric catalysis consist of transition metal atom(s) surrounded by chiral ligands to induce asymmetry. As these metals more often than not belong to the more precious ones like palladium, platinum, rhodium, and so on, efficient immobilization and subsequent recycling have been of interest for decades. In 1973, Kagan and co-workers reported the first preparation of a polymer-supported chiral complex, a supported rhodium catalyst for asymmetric hydrogenation [8].

In this approach (Scheme 4.1), a phosphinated polymer resin is prepared by postmodification, and the rhodium is introduced in the final step. Chloromethylated Merrifield resin was modified into an aldehyde resin **2**,

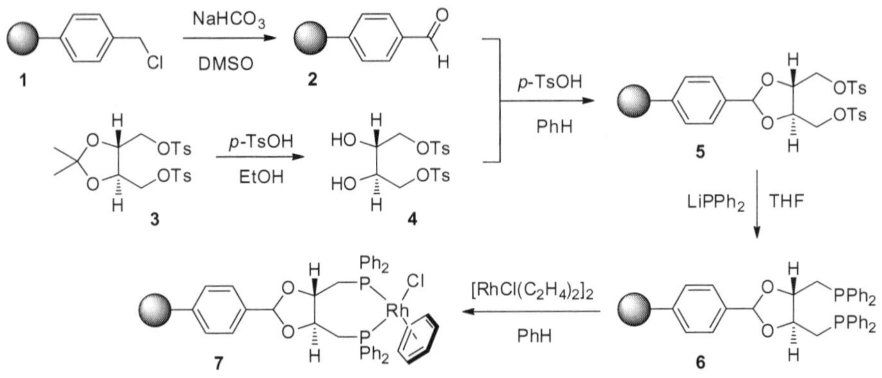

SCHEME 4.1. The first immobilization of a chiral organometallic complex.

and an immobilized DIOP (*O*-isopropylidene-2,3-dihydroxy-1,4-bis (diphenylphosphino)butane) ligand analog **6** was prepared from this by binding diol **4** (obtained by ethanolysis of acetonide **3**) to resin **2** through acetalization and subsequent phosphination with lithium diphenylphosphide in tetrahydrofuran (THF). The final immobilized rhodium complex **7** was obtained by treatment of **6** with $[RhCl(C_2H_4)_2]_2$ in benzene. This catalyst exhibited varying results in asymmetric reductions (and could be reused), but its synthesis via a postmodification approach by immobilization of chiral phosphines onto the Merrifield resin, followed by final attachment of the metal center by way of a simple metal salt, formed a synthetic strategy that has been a cornerstone for postmodification approaches up to the present. However, the efficiency and selectivity of these early catalysts must be considered rather poor (at least by today's standards), and the 1970s and 1980s came to be somewhat dominated by the impressive work of Stille and co-workers who actually obtained better results through a bottom-up strategy, where the polymeric matrix could be better tuned according to the reaction conditions, providing improved catalytic performance (this work is detailed in Section 4.3.3).

The development of polymer-supported chiral organometallic complexes has unquestionably mirrored the development of the chiral phosphine ligands that have pushed the field of asymmetric catalysis forward. The chelating bisphosphine ligands like the DIOP just described (developed by Kagan and co-workers in the early 1970s) [9], the DIPAMP (1,2-bis(*o*-anisylphenylphosphino)ethane) developed by Knowles et al. in the mid- to late 1970s (and used so successfully in the industrial synthesis of L-DOPA by Monsanto) [10], and finally the BINAP (2,2′-bis(diarylphosphino)-1,1′-binaphtyl) family of ligands pioneered by Noyori et al. in the 1980s [11] totally reshaped the landscape of asymmetric catalysis [12]. Both Knowles and Noyori received the 2001 Nobel Prize in chemistry together with Sharpless for their work in this area. The chemical structures of these phosphine ligands,

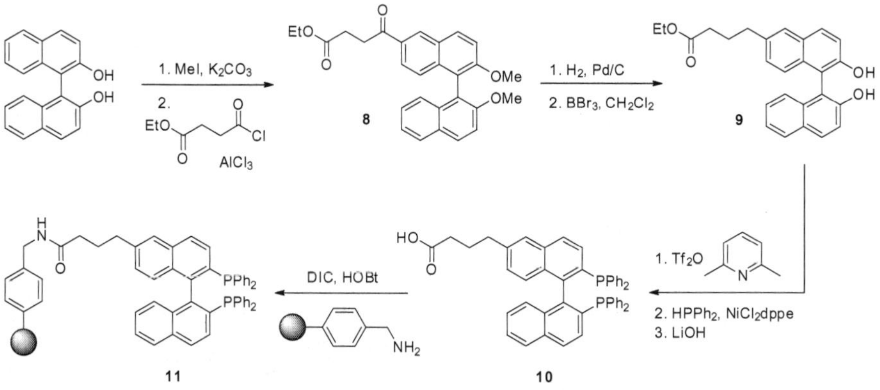

SCHEME 4.2. Polystyrene-supported BINAP.

especially the naphthalene structural elements of the BINAP family, obviously put their likely constraints on the strategies used for their immobilization [13]. Underscoring this point is the successful polymer-supported BINAP reported in 1998 by Bayston and co-workers (Scheme 4.2) [14].

Protection and regioselective Friedel–Craft acylation of BINOL gave BINOL ether **8**, which, after ketone reduction and cleavage of the methyl ethers, gave functionalized BINOL **9**. Conversion to the ditriflate and nickel-catalyzed double phosphinylation followed by acid saponification yielded the appropriately functionalized BINAP **10**. This BINAP was linked to aminomethylated polystyrene through a straightforward peptide coupling to give supported BINAP **11**. An active ruthenium hydrogenation catalyst was formed from **11**, (COD)Ru(bis-methallyl), and HBr in acetone for 1 hour followed by evaporation of solvent in vacuo. This pseudo-C_2-symmetric BINAP catalyst displayed excellent results in asymmetric reductions in THF-MeOH and could be conveniently filtered away after reaction and reused [14].

Although there are 25 years separating the polymeric immobilizations presented in Schemes 4.1 and 4.2, the overall synthetic strategies bear a striking resemblance. However, the crucial step in the synthesis of supported BINAP **11** is the clever regioselective acylation to give BINOL ether **8**, providing a point of polymer attachment at the BINOL that is located at some distance away from the final catalytic center. Exactly the same route was followed by Noyori and co-workers in 2001 when they prepared an analogous polymer-bound BINAP-diamine ruthenium catalyst for asymmetric reduction of ketones [15]. It was even commercialized by then and used by other researchers [16], a testament to how the overall attractiveness of an immobilization strategy can hinge critically on one crucial transformation. This will be a recurrent theme in this chapter.

As previously seen, the BINAP ligands have fairly limited possibilities for a quick functionalization on which to efficiently link them to polymeric

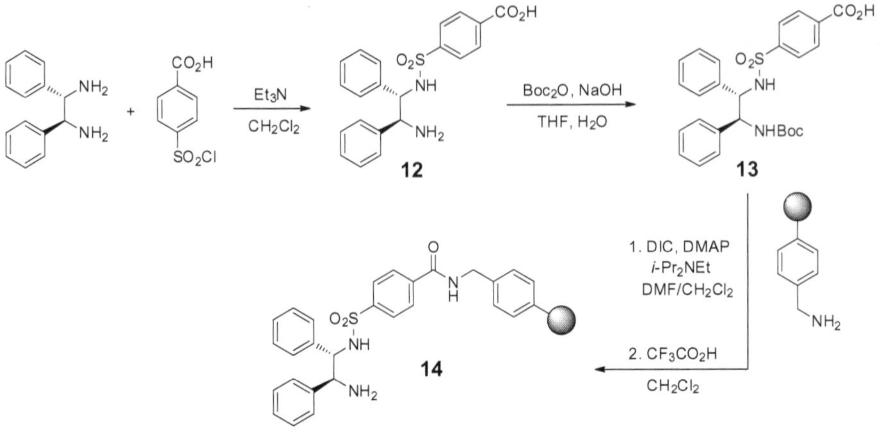

SCHEME 4.3. Polystyrene-supported TsDPEN ligand.

supports. However, other ligands of such BINAP metal assemblies may be easier to use as the point of attachment. In 2003, Itsuno and co-workers reported a polystyrene-supported BINAP-diamine ruthenium complex for asymmetric hydrogenation of acetophenone [17]. In this approach, an analog of the diphenylethylenediamine (DPEN) ligand bearing *p*-hydroxyfunctionalities was used as starting material. Upon simple deprotonation of the phenolic functionalities with NaH and using it in a nucleophilic displacement of the standard chloromethylated Merrifield resin, polymer-supported DPEN was obtained in an expedient manner. This supported ligand was then used together with BINAP to give a supported ruthenium complex. In 1989, prior to this research, a polymer-supported version of the closely related Noyori ligand *N*-(*p*-tolylsulfonyl)-1,2-diphenylethylenediamine (TsDPEN) was reported, prepared according to the outline in Scheme 4.3 [18].

A sulfonylation of DPEN, followed by Boc-protection, gave sulfonamide **13**. In an analogous manner to that described previously for the supported BINAP catalyst **11** (Scheme 4.2), a peptide coupling to aminomethylated Merrifield resin gave polymer-supported ligand **14** after deprotection. The same ligand was also immobilized in the same manner to amino-functionalized TentaGel [18]. In 2005, Chinese researchers reported a very similar immobilization of the same ligand [19]. However, in their procedures, short linkers were introduced in the preparation, and the resulting ruthenium catalysts were used successfully as part of an enantioselective synthesis of (*S*)-fluoxetine.

For asymmetric hydrogenations in water, Swedish researchers have studied phosphine ligands bound to hydrophilic poly(acrylic acid) [20]. The secondary amine of 4-diphenylphosphino-2-(diphenylphosphinomethyl)pyrrolidine (PPM) ligands were immobilized through straightforward peptide coupling with aqueous poly(acrylic acid). This conveniently prepared macroligand was then treated with an aqueous rhodium salt to give catalytically

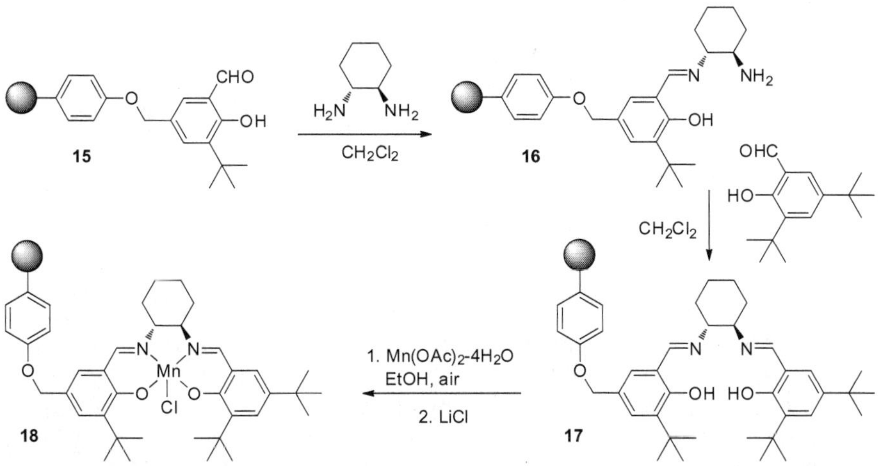

SCHEME 4.4. Polymer-supported chiral Mn^{III}(salen) epoxidation catalyst.

active catalysts used for the asymmetric hydrogenation of enamide precursors in water or under biphasic conditions (H_2O–EtOAc).

The chiral salen complexes form a very important family of organometallic catalysts for asymmetric transformations, and especially then for asymmetric epoxidation [21]. They have consequently been natural candidates for polymeric immobilization. While many of the reported procedures for preparation of polymer-supported salen catalysts belong to the bottom-up type of immobilizations, the easily adaptable nature of the salen ligands has allowed for a good deal of leverage for the synthetic strategies that are employed. Sherrington and co-workers reported several supported chiral Mn^{III}(salen) alkene epoxidation catalysts in 1998 [22]. A representative procedure is given in Scheme 4.4. In what can probably best be described as a combination of a bottom-up and postmodification approach, a series of specialized cross-linked resins, containing a structural moiety like that shown for **15**, were prepared by suspension copolymerization. The salen ligand was prepared from **15** through a series of postmodications involving a diimine preparation of **17** from cyclohexanediamine via **16**. The final introduction of manganese gave supported catalysts like **18** (both styrenic and methacrylic resins were prepared, both porous and gel like). The catalysts were used successfully in epoxidations with *m*-chloroperoxybenzoic acid (MCPBA) in CH_2Cl_2, with *N*-methylmorpholine-*N*-oxide (NMO) as an activator. A very similar strategy to one just described involved an on-bead synthesis of the salen ligand, and was reported by American researchers in the same year [23].

A more conventional postmodification was reported by the Jacobsen et al. for preparation of Co(salen) complexes for hydrolytic kinetic resolution of terminal epoxides (Scheme 4.5) [24]. A prefabricated phenolic salen **19** was immobilized on cross-linked hydroxymethylpolystyrene prederivatized as the 4-nitrophenyl carbonate. After metal insertion and oxidation, supported

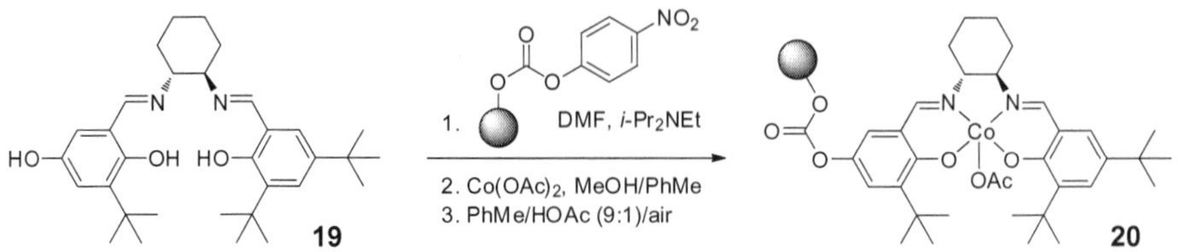

SCHEME 4.5. Polymer-supported chiral Co^{III}(salen) catalyst.

Co^{III}(salen) **20** was obtained. The preparation of salen **19** and its immobilization could be efficiently combined by using the crude material from the preparation of salen **19**. Reger and Janda also used the practical nature of phenolic salen **19** for their synthesis of a polymer-supported Mn(salen) catalyst [25]. In their approach, the phenol of salen **19** was reacted with glutaric anhydride, and the resulting carboxylic acid salen was coupled to either PEG, hydroxy-functionalized linear polystyrene, the cross-linked Merrifield resin, or the more expandable JandaJel resin through esterification with *N,N′*-dicyclohexylcarbodiimide (DCC)/4-dimethylaminopyridine (DMAP). The final catalysts were obtained by standard metal insertion and oxidation. In analogy to BINAP, it can be seen that some particularly useful intermediates (like salen **19**) are common for several independently reported procedures, highlighting the importance of identifying such central building blocks.

The introduction of chiral salen ligands via their diamine-derived moiety has also been reported and follows much of the same postmodification methodology described above: using a peptide coupling between a precursor with a pendant carboxylic acid and an amino-functionalized cross-linked resin for the immobilization [26]. A British research group reported in 2002 a polymer-supported Katsuki-type Mn(salen) complex by anchoring the salen to the support via a linker introduced on the naphtol part of the catalyst, thereby combining typical BINOL and salen type chemistry [27].

Chiral dirhodium(II) complexes are important catalysts for asymmetric cyclopropanations and C–H insertions [28]. Except for a very successful catalyst reported very recently (to be detailed in Section 4.3.3), reported polymer-supported dirhodium complexes are almost invariably prepared through postmodifications schemes. Two procedures have been dominant. Doyle and co-workers reported the synthesis of polymer-supported versions of dirhodium carboxamidates such as dirhodium(II) tetrakis(methyl 2-oxapyrrolidine-(5*S*)-carboxylate) = Rh_2(5*S*-MEPY)$_4$ by simple ligand exchange of Rh_2(5*S*-MEPY)$_4$ with carboxamidate-bearing resins (Scheme 4.6) [29].

Pyroglutamate was anchored to either hydroxy-functionalized TentaGel through standard esterification with carbodiimide to give **21** or chloromethylated Merrifield resin through nucleophilic displacement to give **23**. Ligand exchange in refluxing chlorobenzene then gave dirhodium catalysts **22** and **24**. The same research group later made catalysts with mixed chiral ligands [29b]. In the other general approach to polymer-supported dirhodium(II) catalysts,

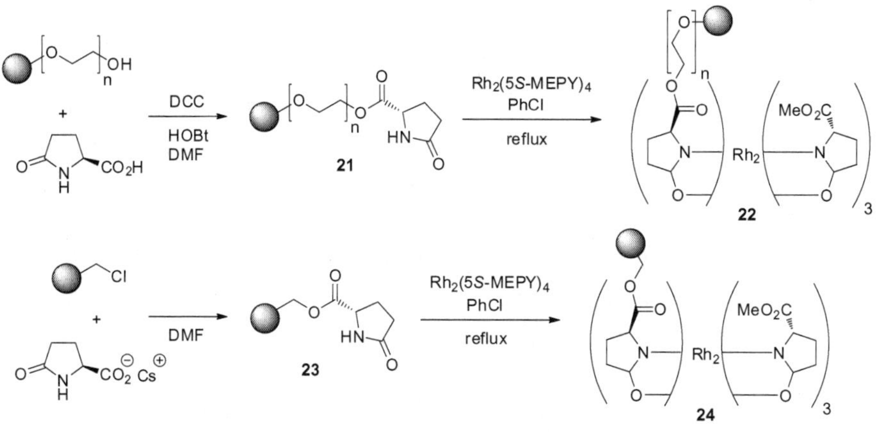

SCHEME 4.6. Polymer-supported dirhodium(II) carboxamidates.

Davies et al. have reported several heterogenized dirhodium(II) tetraprolinates such as dirhodium(II) tetrakis(((*S*)-*N*-dodecylbenzenesulfonyl)prolinate) = $Rh_2(S\text{-DOSP})_4$ by simply treating pyridine-functionalized and highly cross-linked polystyrenes with the standard catalysts in unmodified form, immobilizing them through a combined axial ligand coordination and encapsulation [30]. A more recent approach to supported dirhodium(II) catalysts has much in common with the procedure of Doyle and co-workers, but this time using a carboxylic acid bearing polystyrene resin to facilitate a carboxylate interchange reaction [31]. These researchers also used a bottom-up approach to immobilized dirhodium(II) catalysts.

The chiral bisoxaolines (Box) type ligands (especially PyBox, *t*BuBox, and PhBox) form an important family of ligands for a diverse range of synthetically useful transformations [32]. Like salen ligands, there exists quite a broad selection of opportunities for polymeric immobilization, both according to postmodification and bottom-up approaches (the latter is described in Section 4.3.3) [33]. Moberg and co-workers have published two related postmodification approaches to polymer-supported PyBox ligands (Scheme 4.7) [34].

Chelidamic acid dimethyl ester **25** was used as starting material and converted to bromo-PyBox derivative **28** via amide **27**. In their first report, a Sonogashira coupling with an alkyne-terminated carboxylic acid gave PyBox **29**, a useful derivative that was easily anchored to hydroxy-functionalized TentaGel resin to give supported PyBox **30** through esterification [34a]. In a second report, the then-popular copper-catalyzed azide-alkyne cycloaddition was utilized (Scheme 4.7) [34b]. Alkyne PyBox **32** was obtained from the same bromo-PyBox derivative **28** through Sonogashira coupling and silyl deprotection. This useful "alkyne anchor" was then used for the following immobilization through an azide-alkyne cycloaddition to give "click"-PyBox resin **33**. The use of this key azide-alkyne cycloaddition will be a recurrent theme in the

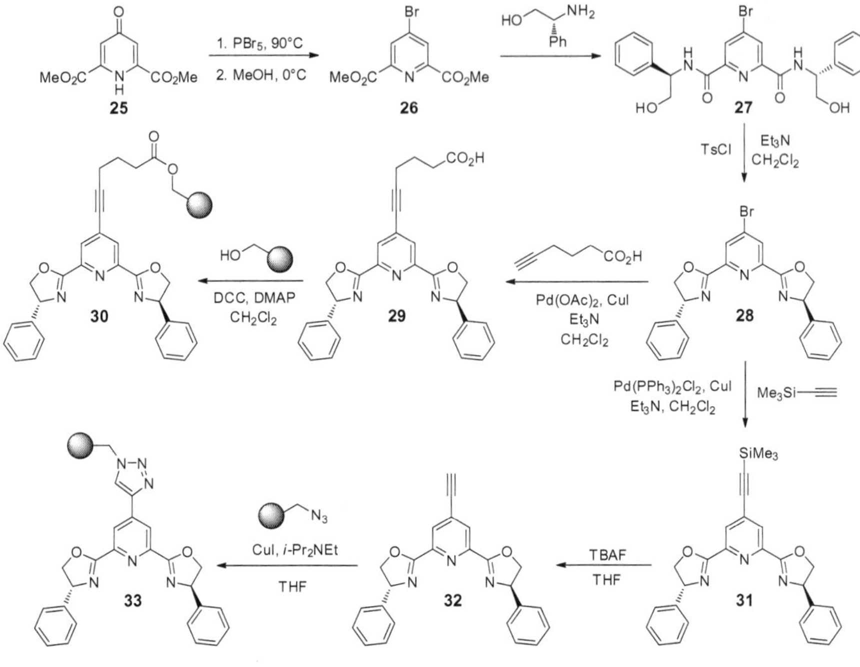

SCHEME 4.7. Polymer-supported PyBox.

more recent immobilization strategies presented, as it offers a useful alternative to the condensation reactions (ester and amide formation) that historically have dominated the postmodification strategies. Moberg et al. have also reported other polymer-supported oxazoline ligands before the two disclosures already described [35].

For more polymer-supported ligands and metal complexes, see some of the general references in this chapter [1, 36].

4.2.3. The Preparation of Chiral Organocatalysts Supported on Organic Polymers Through Postmodification

With the renaissance of organocatalysis in 2000, a great deal of effort has been directed toward the preparation of recyclable polymer-supported organocatalysts [37]. Recently, we presented a thorough treatment of the synthetic strategies for the preparation of polymer-supported chiral enamine and iminium organocatalysts, using the bottom-up/postmodification partition [2]. Many of the important chiral organocatalysts for asymmetric synthesis can be roughly classified as belonging to the enamine–iminium type (typically derived from amino acids/peptides), catalysts working by noncovalent interactions (hydrogen bonding), chiral phase-transfer catalysts (typically derived from *Cinchona* alkaloids), and chiral Brønsted acids. This crude classification will form the basis for the analysis of chiral organocatalysts in this chapter.

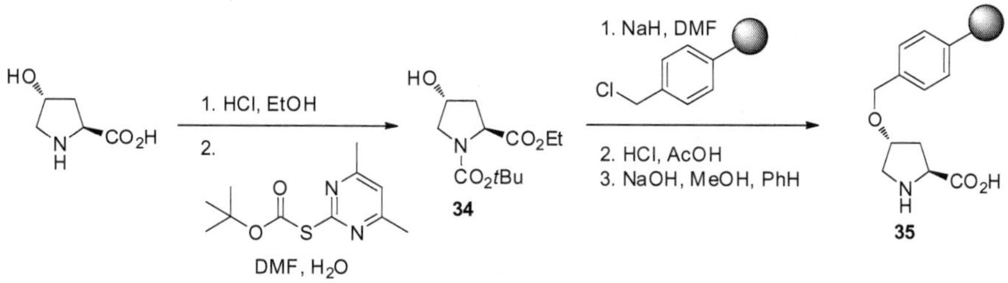

SCHEME 4.8. The first polymer-supported proline.

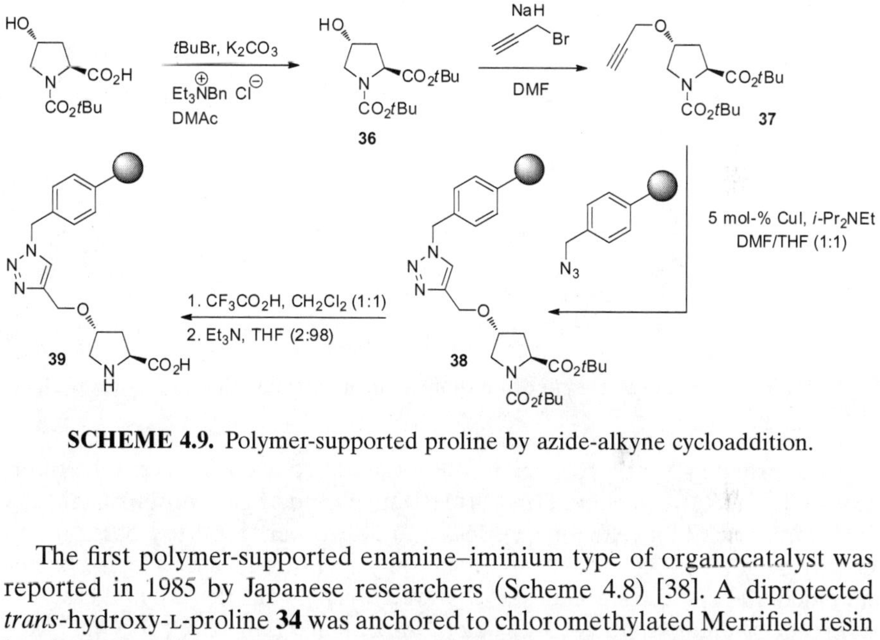

SCHEME 4.9. Polymer-supported proline by azide-alkyne cycloaddition.

The first polymer-supported enamine–iminium type of organocatalyst was reported in 1985 by Japanese researchers (Scheme 4.8) [38]. A diprotected *trans*-hydroxy-L-proline **34** was anchored to chloromethylated Merrifield resin by alkylation without using any sort of linker, providing proline resin **35** after deprotection. This catalyst was only partially successful in asymmetric intramolecular aldol reactions, and this report did not seem to have kindled more research into this area. However, after 2000, research on chiral organocatalysts supported on organic polymers was naturally reignited. The initial work focused on anchoring proline to PEG [2, 39] and provided momentum to the later work on prolines on crosslinked resins, especially the work of Pericàs and co-workers [40] and Gruttadauria and co-workers [41] from 2006 onward. These researchers used classical postmodification approaches using modified Merrifield resins (Schemes 4.9 and 4.10). In the first of these reports, Pericàs and co-workers modified diprotected hydroxyproline **36** by alkylation and anchored the resultant alkyne proline **37** to azido-modified Merrifield resin with the help of a copper-catalyzed azide-alkyne cycloaddition, giving proline resin **39** after some deprotections [40a]. Gruttadauria and co-workers prepared

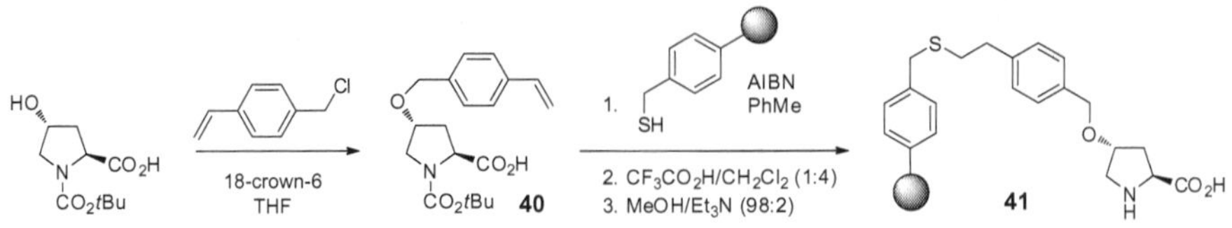

SCHEME 4.10. Polymer-supported proline by thiol-ene addition.

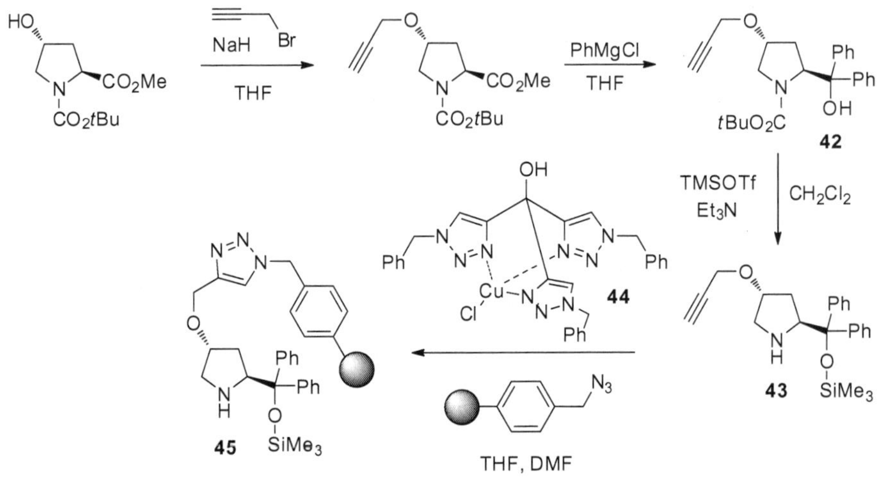

SCHEME 4.11. Polymer-supported diphenylprolinol silyl ether organocatalyst.

styrenic proline **40** and anchored it to thiol-modified Merrifield resin by thiol-ene addition to give proline resin **41** [41]. Both **39** and **41** gave promising results in asymmetric aldol reactions. Pericàs and co-workers reported other resins closely related to **39** in 2008 by anchoring an azido-proline to an alkyne-modified Merrifield resin, catalysts that were even more effective and that had improved swelling characteristics [40c]. Unfortunately, all these procedures are only useful for making moderate quantities of material, a problem when dealing with cheap catalysts such as proline.

Some typical proline derivatives, such as the prolineamides, have also traditionally been prepared by postmodification of cross-linked polystyrene. Gruttadauria and co-workers have utilized a strategy based on the thiol-ene coupling analogous to that of Scheme 4.10 to immobilize several prolineamides [42]. In the preparation of supported versions of the more valuable diarylprolinol silyl ether organocatalysts, the reported procedures are more balanced between postmodification and bottom-up approaches, perhaps due to the longer history of polymer-supported diarylprolinol ligands because of their use in the Corey–Bakshi–Shibata (CBS) asymmetric reduction. A successful postmodification approach was reported by Pericàs and co-workers in 2009 (Scheme 4.11) [43]. In comparison to their immobilization of proline, a

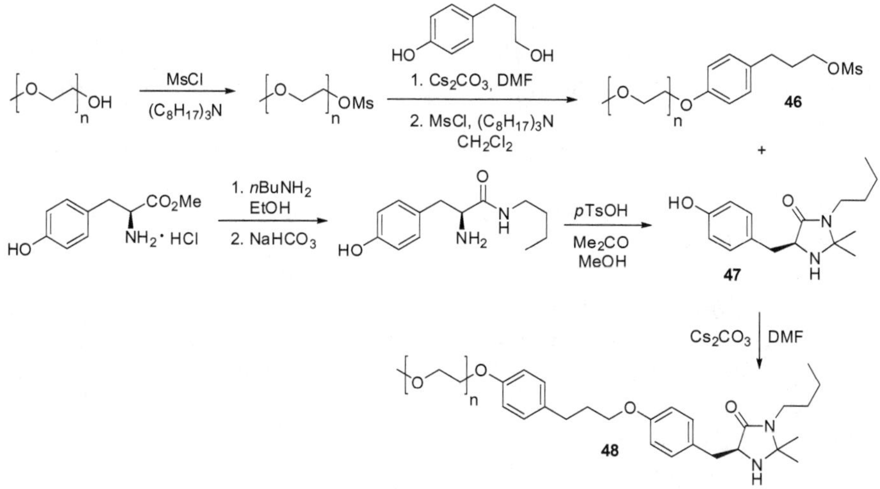

SCHEME 4.12. PEG-supported MacMillan imidazolidinone.

diphenylprolinol alkyne precursor **42** was prepared from doubly protected hydroxyproline, and then silylated and anchored to azido-modified Merrifield resin by an azide-alkyne cycloaddition (using a custom-made copper catalyst **44**) to give catalyst **45**. Other researchers have used the same key intermediate **43** and a similar azide-alkyne cycloaddition in a recently published work on a PEG-linked diphenylprolinol silyl ether [44].

The MacMillan-type imidazolidinones, together with proline, were among the first organocatalysts to be reported in polymer-supported versions. Like proline, the preparation of PEG-anchored imidazolidinones was among the first to surface in the literature in 2002 [45]. This synthesis is detailed in Scheme 4.12 and is a typical PEG approach to supported catalysts. First, a suitable PEG derivative is prepared (here by using a monomethyl ether of PEG as a starting material and transforming it into PEG-derivative **46** through mesylation, alkylation with a hydroxyalkylphenol, and a final mesylation). An imidazolidinone catalyst precursor **47** was prepared from tyrosine by formation of its *n*-butyl amide and subsequent ring closure to the imidazolidinone with acetone. Alkylation of PEG-derivative **46** with imidazolidinone **47** gave PEG-imidazolidinone catalyst **48**.

Also in 2002, Finnish researchers reported the first MacMillan imidazolidinone supported on cross-linked supports [46]. In this more specialized approach, the imidazolidinone was constructed in an on-bead fashion, using JandaJel as the cross-linked support. In 2005 and 2006, other more specialized polymer-related immobilizations of MacMillan imidazolidinones, such as procedures involving lyotropic liquid crystal assemblies followed by polymerization [47a] or polymer-coated mesocellular foams [47b], were reported. Very recently, a simple ion-exchange reaction between a standard MacMillan imidazolidinone and a sulfonated cross-linked styrene support has also been published [48], but similarly to the more specialized approaches, it probably adds only negligible generic knowledge useful for polymeric immobilization in general.

As for more advanced enamine–iminium organocatalysts, Wennemers et al. in Switzerland have had a long-standing interest in the tripeptides H-D-Pro-Pro-Glu-NH_2 and H-D-Pro-Pro-Asp-NH_2 as efficient catalysts for conjugate additions [49]. As part of these studies, a range of polymer-supported versions of such tripeptides were reported, prepared by using conventional peptide coupling protocols on suitably functionalized versions of several resins [50]. Other polymer-supported peptides have also been reported [2]. Postmodification strategies built on traditional solid-phase peptide chemistry has also been pursued by Jacobsen et al. in the preparation of polymer-supported chiral (thio)ureas [51]. Such strategies probably have limited value on larger scales, and it would be interesting to see additional research into this area because of the especially promising results obtained with these types of organocatalysts, something that hopefully can be successfully extended to polymer-supported versions at a more affordable cost than currently possible. The same is also true for typical polymer-supported versions of Juliá–Colonna epoxidations catalysts [52].

Cinchona-derived catalysts are definitely some of the chiral catalysts with the longest history with regard to polymeric immobilization, and only a very brief introduction can be provided here [53]. Among the *Cinchona* alkaloids, the four most prominent are quinine, quinidine, cinchonine, and cinchonidine, all having essentially identical core structures. Because of the terminal olefin that is present in all four of these *Cinchona* alkaloids, a structural element that does not take part in the catalytic activities (and that, in addition, is actually located at some distance from the crucial functionalities), syntheses of supported *Cinchona* alkaloids have often tended to use this as the natural place of attachment to the polymer backbone. Since olefins play such a prominent role in polymer chemistry through vinylic radical polymerization, many strategies for the immobilization of *Cinchona* alkaloids have naturally followed bottom-up schemes (some will be described later). Among them is one of the first reported polymer-supported *Cinchona* alkaloids from 1971 [54].

Polymer-supported quinine derivatives prepared by postmodification have been known since at least 1977 [55]. However, unlike many or most other polymer-supported catalysts, the pioneering era here was probably more weighted toward the bottom-up type of copolymerization schemes; postmodifications gained a stronger foothold later on. In 1983, Hodge and co-workers reported polymer-supported quaternary ammonium salts (catalysts for asymmetric Michael reactions) derived from *Cinchona* and *Ephedra* alkaloids by simple alkylation of chloromethylated Merrifield resin with the appropriate alkaloids or alkaloid derivatives [56]. Such simple derivatization of resins has been reported all the way up to the present [57]. In addition, simple immobi-

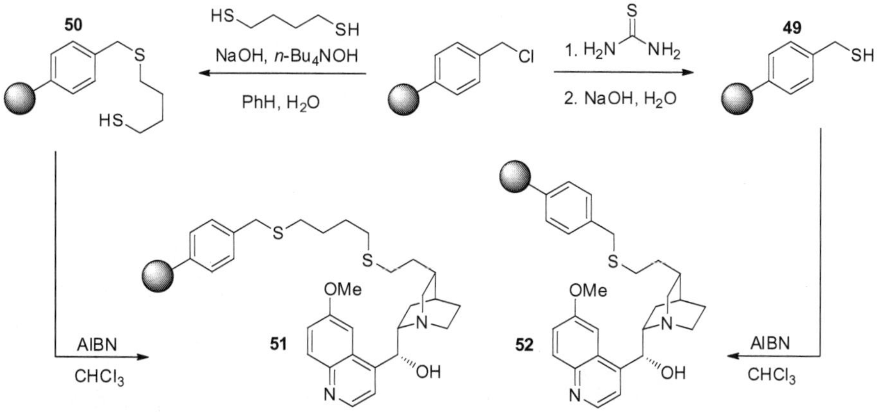

SCHEME 4.13. Polymer-supported quinine by thiol-ene addition.

lization of these *Cinchona* derivatives through their secondary alcohol for preparation of polymer-supported chiral phase-transfer catalysts is also nearly equally simple and expedient [57d]. It is, of course, less general than a linkage through the olefinic functionality as the secondary alcohol does take part in some catalysis promoted by these alkaloids. As such, the 1985 disclosure by Hodge and co-workers, where they immobilize *Cinchona* alkaloids directly through their vinylic bond onto thiol-modified and preformed cross-linked polystyrenes, is of more interest [58]. Their synthesis is outlined in Scheme 4.13. Two thiol-modified Merrifield resins were prepared from standard chloromethylated resin by treatment with either thiourea/aqueous NaOH to give **49**, or with butanedithiol with base under phase-transfer conditions to give **50**. A *Cinchona* alkaloid such as quinine could then be directly immobilized through a thiol-ene coupling. The thiol-ene coupling is together with the copper catalyzed azide-alkyne cycloaddition probably the reaction that comes closest to fulfilling the constraints laid down in the now well-known "click" concept introduced by Sharpless. Reactions using them are now often rebranded as "thiol-ene click chemistry" [59]. However, it has also been used successfully for decades in the preparation of polymer-supported chiral catalysts. Procedures such as that presented in Scheme 4.13 can definitely compete in efficiency and elegance with the best of modern procedures for preparation of polymer-supported quinines.

Another interesting strategy for the preparation of polymer-supported quinine was reported in 1999 [60]. The synthesis of the simplest supported catalyst is depicted in Scheme 4.14. Silyl-protected quinine **53** was converted to **54** by a hydroboration protocol. Immobilization through a rather harsh alkylation of chloromethylated Merrifield resin then gave quinine resin **55**, which is useful in asymmetric Michael reactions.

The three options for linkage of the *Cinchona* alkaloids are through their alcohol, tertiary amine, or vinyl functionalities, but for quinine and quinidine

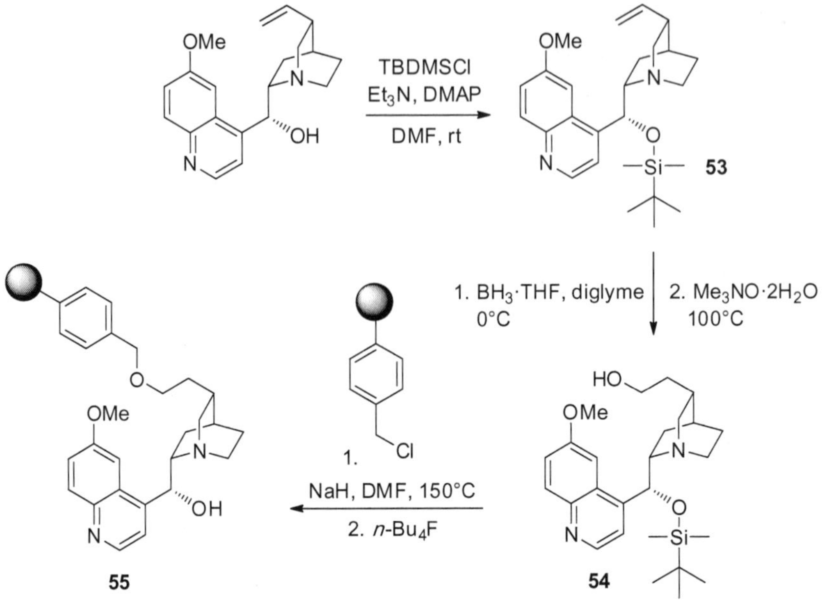

SCHEME 4.14. Polymer-supported quinine via hydroboration.

there is also the fourth possibility of linkage through the methoxy group on the quinoline system. This is the least common way of doing it, and Sherrington and co-workers reported such a procedure in 2008 [61]. *Tert*-butyldimethylsilyl (TBDMS)-protected quinine was demethylated with L-Selectride (lithium tri-*sec*-butylborohydride) in THF, and a linker was introduced by alkylation of the resulting phenol with chloroethoxyethanol-K_2CO_3-NaI in MeCN. This derivative, which contains a linker with a pendant alcohol, can easily be anchored to resins and deprotected with *n*-Bu_4NF in a conventional manner.

In summary, the long and important history of *Cinchona* alkaloids has also made the chemistry connected to their polymeric immobilization especially rich. Others will follow later in this chapter under bottom-up strategies.

4.3. THE PREPARATION OF CHIRAL CATALYSTS SUPPORTED ON ORGANIC POLYMERS ACCORDING TO BOTTOM-UP STRATEGIES

4.3.1. The Character of Radical Polymerization From the Synthetic Point of View

As detailed in the introduction to this chapter, when we refer to bottom-up strategies for preparation of chiral catalysts supported on organic polymers, we mean essentially the integration of the macromolecular synthesis and catalyst immobilization. This integration offers some significant advantages, but demanding in that it requires a comprehension of macromolecular principles. First of all, since macromolecular synthesis and catalyst immobilization are combined, the overall sequence is shortened when analyzed in its entirety, something that is seldom done when analyzing postmodification strategies because the prefabricated support is then usually purchased and, therefore, perceived as and masked as mere "starting material" (something that is revealed if actual costs are evaluated). Second, since the reactions that underpin macromolecular synthesis—such as radical polymerizations—are usually exceedingly efficient (or else they could not form any basis for the preparation of polymers), catalyst incorporation (via the functional monomer they are part of), and thereby catalyst loadings, is easily controlled and adjusted. Third, the freedom of choice with regard to what comonomers are included during polymerization provides useful leverage in the regulation of the swelling characteristics for cross-linked supports or solubility properties for linear polymers.

The above-mentioned benefits of what are essentially copolymerization strategies are easily recognized and well known. What is less known is the fact that the chemical reaction that underpins the vast majority of preparations of polymeric materials (for instance, the free radical polymerization of vinylic compounds) is a very useful reaction to integrate in the overall synthetic sequence because of its experimental simplicity and chemical tolerance [62]. The extremely favorable kinetics of the propagation step in a free radical polymerization are very useful from the synthetic point of view as it tends to outcompete nearly all other side reactions. In effect, this means that free radical polymerization is actually one of the most selective of the large assortment of reactions in organic synthesis. In addition, as polar entities such as alcohols, carboxylic acids, amines and the like (entities that very often necessitate the use of protective groups) generally have polar heteroatom–hydrogen bonds that are vulnerable to heterolytic scission; they are unusually nonperceptive toward homolytic cleavage. As such, the free radical polymerization is unusually robust toward the presence of polar and reactive functional groups (entities that typically are problematic to handle through a synthetic sequence) and can therefore also take place in "green" solvents such as water and alcohols. It also has perfect atom economy and requires virtually no auxiliary reagents—only a small amount (approximately 1 weight% or less) of a cheap free radical initiator (typically azo-compounds or peroxides).

Beginning earlier, but particularly since the mid-1990s, a number of more sophisticated forms of radical polymerization, variously described as either controlled, living, or mediated radical polymerization, have been developed [63]. The exact terminology has been the subject of some debate, but the general character of these forms of radical polymerizations is such that reversible formation of radicals from dormant species is possible, and this permits the preparation of block copolymers by the sequential addition of different monomers. In addition, their controlled character, made possible by the equilibrium between the dormant forms and radicals able to undergo chain growth, means that these techniques enable the preparation of polymer samples with both molecular weight control and narrow molecular weight distributions. The most dominant methods of controlled/living radical polymerization are nitroxide-mediated polymerization (NMP), atom transfer radical polymerization (ATRP), and radical polymerization by the reversible addition-fragmentation chain transfer (RAFT) process. Undoubtedly, the arrival of these methodologies is definitely poised to completely reshape the landscape of radical polymerization in the future. Their widespread application to polymer-supported organocatalysts is eagerly awaited. We will detail several examples later in this chapter. However, when viewed from the perspective of organic synthesis instead of materials science, they have one major drawback (in addition to the purely experimental inconvenience they offer when compared to free radical polymerization): Introduction of supplementary reagents for controlled radical polymerization lowers the functional group tolerance and robustness of the reaction toward both impurities (such as residual inhibitors) and experimental conditions (such as moisture and inert atmosphere) significantly when compared to free radical polymerization, in addition to a more problematic workup and purification. For now, free radical polymerization remains the workhorse of macromolecular synthesis, but as limitations will certainly be overcome in the future, techniques for controlled radical polymerization will unquestionably be an integral part of bottom-up schemes for polymeric immobilization.

4.3.2. The Nature of Different Heterophase Polymerizations

A free radical polymerization can be undertaken in a number of homogeneous or heterogeneous physical systems for polymerization. The conceptually simplest type is a bulk or mass polymerization where the polymerization is carried out in neat monomer. This mode of polymerization has certain advantages (especially on an industrial scale), such as the absence of solvents and a minimum of contamination of product, but it is inherently difficult to control. The exothermic nature of radical polymerization combined with the rather high activation barrier associated with initiator decomposition make temperature control critical, but this is complicated by the lack of heat dissipation in bulk and the increase of viscosity. In the worst-case scenario, a disastrous runaway reaction may occur. By performing a solution polymerization, where a diluted polymerization is carried out in an inert solvent, many of the disadvantages can be overcome. The diluted nature of the solution polymerization makes both heat control and stirring easier. The most common problems associated with solution polymerization are chain transfer to the solvent and removal of solvents during purification. Solution polymerization is the most common method for the laboratory preparation of linear polymers, and the purification is usually carried out by simply pouring the reaction mixture into a nonsolvent for the polymer. The precipitated polymer is then purified by successive dissolutions and precipitations. From the purely practical laboratory point of view, synthesis by solution polymerization can be inconvenient because of the (often) rather impractical sticky nature of the precipitated polymers. In order to avoid problems with sticky precipitates, cross-linked supports are popular when prepared in appropriate beaded form, as they are easily filtered out after reaction and obtained as a convenient powder when dried.

There are several heterogeneous polymerization systems that can be utilized for preparation of beaded and cross-linked resins by bottom-up strategies using radical copolymerization. The most important are suspension polymerization (also referred to as droplet, bead, or pearl polymerization), emulsion polymerization (latex polymerization), and several types of precipitation polymerizations (nonstabilized powder/granular polymerizations or stabilized-type dispersion polymerization) [64]. Of these, suspension polymerization is by far the dominant one, as it (if correctly carried out) provides the polymer product in the form of finely granulated and spherical beads (in the size range from approximately 20 μm up to as high as 2 mm) [63] that are remarkably convenient to handle in synthetic procedures, as they are easily decanted, filtered, and dried.

In a normal suspension polymerization, a discontinuous phase of monomer droplets containing an oil-soluble initiator is simply suspended in water by agitation and polymerized by heating. The droplets (or polymer particles after polymerization) are stabilized by the addition of non-micelle-forming emulsifiers to the aqueous phase. This has important implications for synthetic bottom-up strategies using this technique because monomers destined for suspension polymerization cannot have a high degree of water solubility. If all monomers are highly water soluble and poorly oil soluble, an inverse suspension polymerization can be carried out. In this mode of operation, an aqueous phase of monomers and a water-soluble initiator is suspended in an organic solvent as the continuous phase (often halocarbon-containing ones to increase its density relative to the suspended aqueous phase) in the presence of typical water-in-oil stabilizers. Of course, copolymerization of monomers with very different solubility profiles can obviously be difficult. However, a very useful aspect of suspension polymerization is the fact that if carried out successfully and taken to completion, the polymer composition is a near-direct reflection of the initial monomeric composition, meaning that catalysts loadings are effortlessly controlled by appropriately mixing functional monomers and comonomers in the desired ratio.

In a dispersion polymerization [65], an initially homogeneous medium of monomer(s), an organic solvent (that is a poor solvent for the polymer), free radical initiator, and particle stabilizer are prepared. Although homogeneous at the onset, the reaction mixture becomes heterogeneous after a short period of time as nucleation and formation of primary particles occur when the polymers start to precipitate. The precipitated particles coagulate on one another and are stabilized by adsorption of stabilizer. The polymer is swollen by the polymerization medium and/or monomer. The polymerization therefore proceeds largely within the particles as they absorb the monomer from the continuous phase. Unlike most precipitation polymerizations, dispersion polymerization leads to spherical and homogeneous particles (these can also be of a very uniform size). The particles are generally in the size range of 1–10 μm, thus bridging the gap in size range between emulsion polymerization (giving particles smaller than 1 μm) and suspension polymerization (giving particles larger than 10 μm). From the purely synthetic point of view, a useful property of dispersion polymerization is that unlike suspension copolymerization, it makes possible the combination of monomers with very different solubility properties. In a medium such as methanol or ethanol, both hydrophobic monomers like styrene can be combined with very polar monomers. Unfortunately, this comes at a cost as its precipitation mechanism means that the composition of the final copolymer is now no longer necessarily a reflection of the monomeric composition at all. In effect, catalyst loadings are then unpredictable and unknown prior to analysis. Consequently, this method typically requires more adjustment and is consequently less general than suspension polymerization.

4.3.3. The Preparation of Chiral Organometallic Catalysts Supported on Organic Polymers Through Bottom-Up Strategies

The preparation of chiral organometallic complexes bound to organic polymers by copolymerization has a long and intriguing history. A distinguished pioneer in bottom-up strategies for preparation of polymer-supported chiral organometallic catalysts, as well as one of the definitive pioneers of the field altogether, was the late John Kenneth Stille (famous as the father of the celebrated Stille reaction and tragically killed in a plane crash in 1989). Stille, together with co-workers, developed a series of polymer-supported chiral transition metal complexes during the 1970s and 1980s [66]. Starting with polymer-supported DIOP catalysts in the mid-1970s, the work in the 1980s by these researchers also included polymer-supported PPM catalysts [66]. By preparing functional monomers such as **56** and **59** (often on impressively robust scale), polymer-supported phosphine ligands were readily available by copolymerization (Scheme 4.15).

The first reports utilized a precipitation-type polymerization in benzene with hydroxyethyl methacrylate (HEMA) and ethyleneglycol dimethacrylate (EGDMA) to give cross-linked supports that swelled well in polar solvents [66a,b]. Later on, copolymerization with methyl vinyl ketone or a more conventional suspension copolymerization in toluene–water to give spherical beaded supports was also reported [66]. As examples, ditosylstyrene **56** was copolymerized with HEMA and EGDMA to give polymer-supported DIOP ligand **58** after phosphinylation (Scheme 4.15). As an alternative, a functional monomer already phosphinylated could be assembled completely prior to the polymerization. Acrylic PPM monomer **59** was prepared by robust procedures and could be copolymerized with HEMA–EGDMA in benzene to give PPM resin **60** or suspension copolymerized with styrene and divinylbenzene (DVB)

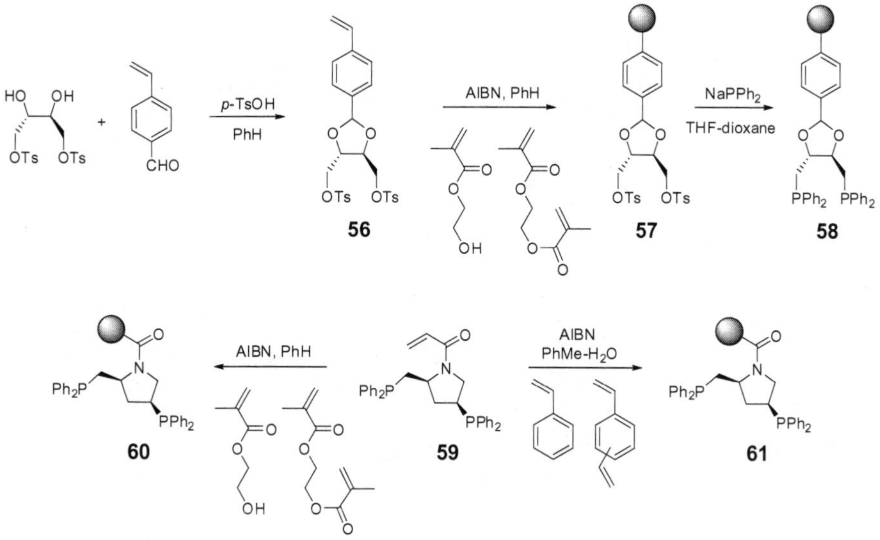

SCHEME 4.15. Polymer-supported DIOP- and PPM-type ligands.

to give PPM resin **61**. In all cases, final treatment with transition metal salts (Rh, Pt) provided the active catalysts; these catalysts could be isolated after reaction by filtration. As so correctly pointed out by Stille with regard to his methodology (and really capturing the essence of copolymerization strategies), the incorporation of ligand-bearing functionalities into the polymer chains can be controlled, and the nature of the polymer backbone, whether that be polar or nonpolar, can be varied depending on the selection of comonomers [66h]. This gave the system a natural flexibility with regard to matching a set of required reaction conditions.

As mentioned earlier, the BINAP ligands are difficult to functionalize efficiently, but due to the importance of these ligands, some novel bottom-up strategies have been devised over the years. As an example of this, Lemaire and co-workers reported a polyurea strategy for immobilization of BINAP in 2000 [67].

In their synthesis, outlined in Scheme 4.16, BINOL is converted to **62** by bromination and triflation. Dicyano-derivative **63** could be obtained directly from **62** with CuCN in NMP at high temperature. A special phosphinylation gave **64** without destroying the cyano functionalities, and a standard reduction then gave (*S*)-diam-BINAP **65** as monomeric building block. Polymerization with toluene 2,6-diisocyanate gave (*S*)-poly-NAP **66**, a soluble polymer from which active catalysts in asymmetric hydrogenations were obtained after the final addition of ruthenium salts. Although very interesting, it can be seen from this procedure, as well as from the postmodification strategy presented earlier in this chapter, that expedient functionalization of the BINAP skeleton in a manner that is suited for polymeric immobilization remains challenging. Other

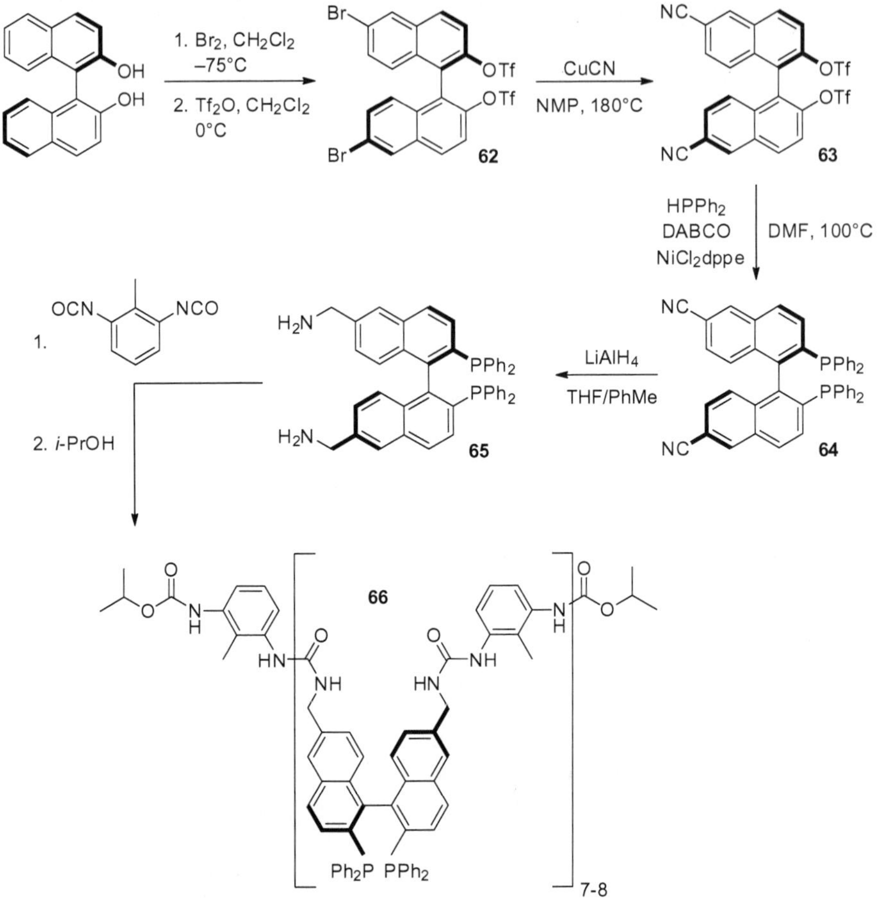

SCHEME 4.16. Polymer-supported BINAP.

researchers have presented schemes that are connected to the one presented in Scheme 4.16 by making amine-functionalized BINAP and polymerizing it with terephthaloyl chloride and PEG [68]. Poly(BINAP) made by a Suzuki-coupling-type polymerization has also been reported [69].

Like the strategies for preparation of polymer-supported BINAP that have just been discussed, several polymeric immobilizations of DPEN ligands according to bottom-up-type strategies have utilized polymerizations with diisocyanates to give polyureas or polythioureas [70]. After all, this is natural as DPEN is a diamine and as such can be polymerized even in unmodified form. Lemaire and co-workers have also developed a novel copolymer with amino functionalities for asymmetric transfer reduction of acetophenone (as its ruthenium complex) by copolymerizing optically pure glycidyl methacrylate with EGDMA and then introducing amines by epoxide ring opening [71].

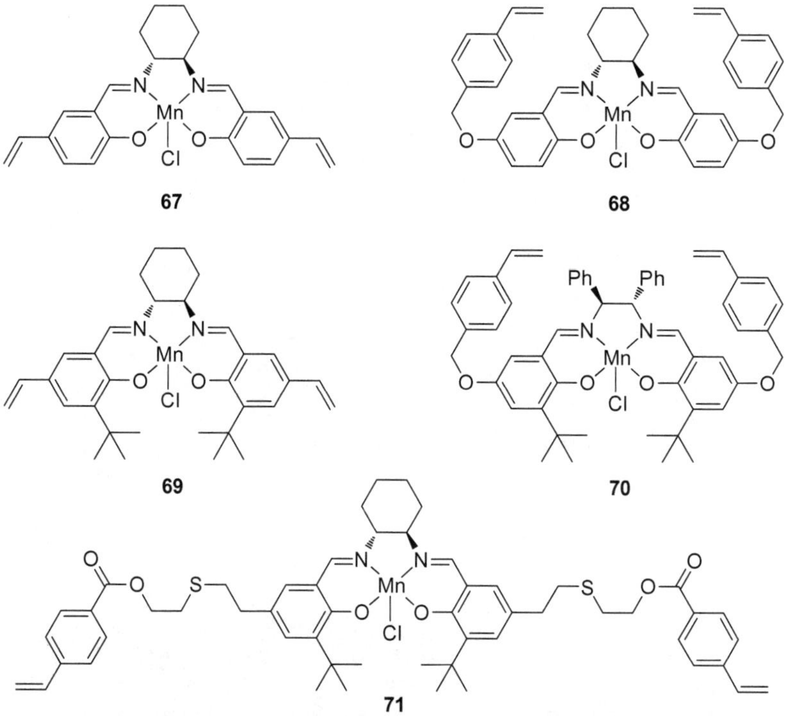

FIGURE 4.1. Functional symmetric salen monomers.

Just as for postmodification, salen ligands offer substantial opportunities for clever bottom-up immobilization. Since chiral salens are essentially the imines of salicylaldehydes and chiral diamines, usually distyrenic functional monomers such as **67–71** (Fig. 4.1) are prepared from the diamine and appropriately functionalized salicylaldehyde and then copolymerized with EGDMA or styrene/DVB to give polymer-supported salen complexes for catalytic asymmetric epoxidation of alkenes [72].

A typical preparation of such a functional monomer is shown for styrenic salen **70** in Scheme 4.17 [72f]. Hydroquinone **72** is protected and converted to **74** by Lewis acid-catalyzed coupling with paraformaldehyde. After deprotection, alkylation with 4-vinylbenzyl chloride gave salicylaldehyde **76**, which again gave salen **77** with DPEN. The functional monomer **70** was then obtained by the standard procedure for preparation of Mn^{III}(salens). The problem with distyrenic monomers such as **67–71** (Fig. 4.1) is that they give rise to very cross-linked polymers, as they are in essence cross-linkers (toluene is generally used as the porogen in these polymerizations). This may affect the catalytic properties adversely, but is a result of the symmetric nature of salen. In addition, it can be difficult to determine the proportion of the catalytic sites that are actually available in such a highly cross-linked and macroporous

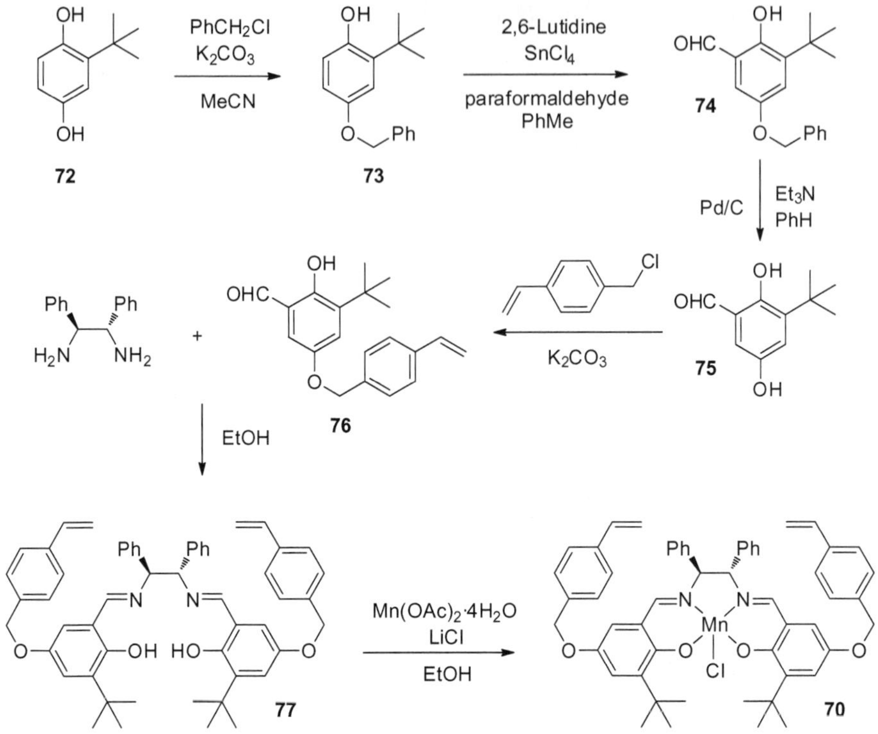

SCHEME 4.17. Typical preparation of functional salen monomer.

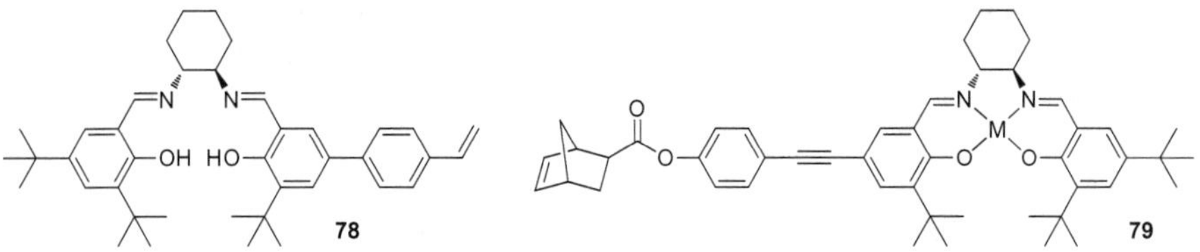

FIGURE 4.2. Functional unsymmetric salen monomers.

network. This difficulty, in effect, removes one of the strongest incentives for undertaking bottom-up approaches to polymeric immobilization: the easy estimation of catalyst loading.

The synthesis of a monofunctional and unsymmetric salen is very difficult, since both condensation reactions of the diamine in salen proceed at comparable rates. A condensation using two different salicylaldehydes inevitably produces a statistical mixture of products, and only one of these will be the desired unsymmetrical variant. Weck and co-workers have developed methodology for doing just this and have used it for the preparation of such monofuntional and unsymmetric salen ligands as **78** and **79** (Fig. 4.2) [73].

This is made possible by a two-step condensation of the monoammonium salt of the starting diamine, first with a conventional salicylaldehyde and then with a styrenic salicylaldehyde. Such clever solutions and the development of a key methodology like this for the different families of catalysts are at the absolute core for the advancement of the field of polymer-supported chiral catalysts in its entirety. More efforts should be devoted in the future to the pursuit of generic and useful methodology for preparation of key catalyst precursors for polymer-supported catalysts. The functional monomers **78** and **79** are readily polymerized, either by means of free radical polymerization for styrenic salen **78** or ring-opening metathesis polymerization (ROMP) for norbornene-functionalized salen complex **79** [73].

As described in a previous section, chiral dirhodium(II) complexes are valuable catalysts for asymmetric cyclopropanations and C–H insertions in particular and have historically been prepared almost exclusively through postmodification schemes [29–31]. Very recently, Hashimoto and co-workers reported a very successful strategy for immobilization a chiral dirhodium(II) carboxylate catalyst [74]. Their synthesis is outlined in Scheme 4.18.

A chiral dirhodium(II) complex with a pendant functionality was prepared by a ligand exchange between *tert*-leucine derivative **80** and dirhodium(II) tetrakis(*N*-phthaloyl-(*S*)-tert-leucinate) = $Rh_2(S\text{-}PTTL)_4$ (**81**) in refluxing chlorobenzene to give complex **82**. Alkylation of the phenolic functionality of

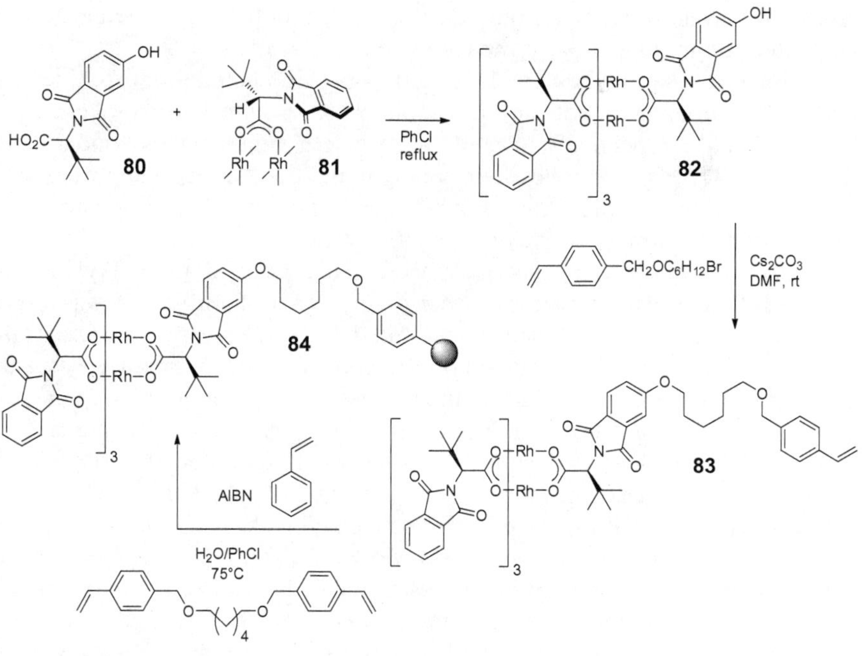

SCHEME 4.18. Polymer-supported dirhodium(II) carboxylate catalyst.

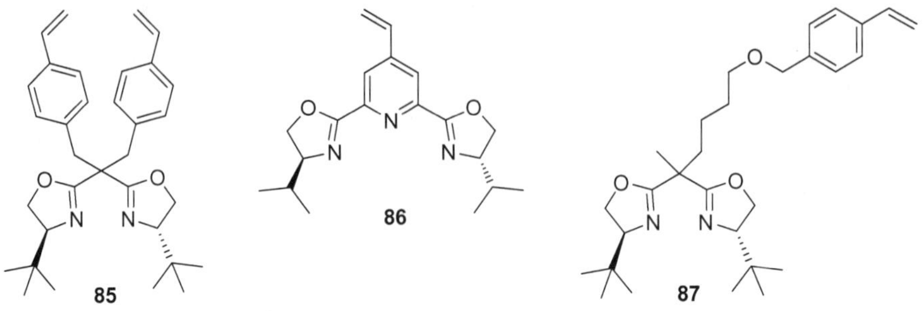

FIGURE 4.3. Functional Box monomers.

82 with 6-(4-vinylbenzyloxy)bromohexane gave styrenic dirhodium(II) complex **83**. A suspension copolymerization of **83**, styrene, and a flexible 1,6-bis(4-vinylbenzyloxy)hexane cross-linker in chlorobenzene-water gave resin **84** as homogeneous polymer beads. This resin exhibited excellent swelling characteristics in several organic solvents and was a powerful catalyst in enantioselective C–H insertion reactions of diazoesters. The catalyst had low leaching levels of rhodium and full retention of catalytic activity and selectivity during an impressive 100 sequential applications. This is an interesting example of how bottom-up strategies can sometimes be applied with spectacular results to systems almost entirely dominated by postmodification schemes. The fine-tuning of catalyst loadings and adjustment of comonomers as well as cross-linkers, which is so easily encompassed in such a bottom-up scheme, greatly facilitated the modification of the catalytic performance.

As for the salen ligands, the Box ligands provide ample opportunities for undertaking interesting bottom-up immobilization schemes. Mayoral et al. and Salvadori et al., respectively, have reported several functional Box monomers, such as **85–87** (Fig. 4.3), and their copolymerization to give supported Box ligands useful for the preparation of polymer-supported copper or ruthenium complexes for asymmetric cyclopropanations [75–77].

Bifunctional monomers such as **85** were prepared by double alkylation of the methylene bridge of precursor with 4-vinylbenzyl chloride and was either homopolymerized into a highly cross-linked support or copolymerized with styrene (in any case, either toluene or a toluene–dodecanol mixture was used as a porogenic agent) [75]. As highlighted previously, monofunctional monomers are more useful building blocks as the degree of cross-linking is then more easily controlled. PyBox monomer **86** is such a monomer and could be prepared by a rather lengthy procedure starting from chelidamic acid [76]. It was copolymerized with styrene/DVB and treated with $[RuCl_2(p\text{-cymene})]_2$ to give active cyclopropanations catalysts. Monomer **87** is particularly useful, as it is both monofunctional and contains a linker. It has been copolymerized with styrene/DVB and treated with $Cu(OTf)_2$ in THF to give catalysts for asymmetric Mukaiyama aldol reactions or cyclopropanations [77]. Unfortunately, as for PyBox **86**, the overall synthesis of **87** must be considered rather cumbersome. There still exist plenty of opportunities for the development of expedient syntheses of Box monomers useful as building blocks in copolymerization schemes.

4.3.4. The Preparation of Chiral Organocatalysts Supported on Organic Polymers Through Bottom-Up Strategies

Concurrent with the pioneering efforts by Kagan and Stille on the preparation of polymer-supported chiral organometallic catalysts in the 1970s and 1980s, several Japanese research groups very actively pursued polymer-supported *Cinchona* derivatives from 1971 onward [54, 78]. These researchers developed novel bottom-up schemes for polymeric immobilization of *Cinchona* alkaloids that have been of interest even in present times. The first report from 1971 prepared a quinine polymer by simple acylation of the alcohol in quinine with methacryloyl chloride and free radical homopolymerization in benzene [54]. This polymer was meant for studies in medicinal chemistry, but a publication from 1977 by other researchers at the same university describe the same *O*-acylation of *Cinchona* alkaloids, their homo- and copolymerization, and the use of such chiral polymers in the asymmetric addition of methanol to phenylmethylketene to give α-phenylpropionate [78a]. However, the most influential work on polymer-supported *Cinchona* alkaloids conducted in this time period was the extensive investigations of Kobayashi and Iwai from 1978 and later [78b–i]. Their method is probably the simplest immobilization of *Cinchona* alkaloids reported and is depicted in Scheme 4.19 for quinine. It was reported for the first time in 1978 and is simply the free radical copolymerization of quinine and acrylonitrile. It was the first successful vinyl polymerization of a *Cinchona* alkaloid [78b]. The method is remarkably expedient, and it highlights the functional tolerance of free radical polymerization as the other reactive functionalities of quinine remained untouched. The copolymer **88**, a light yellow powder soluble in dimethylformamide (DMF) and dimethyl sulfoxide (DMSO) and insoluble in common organic solvents, was utilized as catalyst in several asymmetric Michael reactions during the 1980s.

In 1987, another group of Japanese researchers reported a useful extension of the simple copolymers reported by Kobayashi and Iwai [78j]. By performing

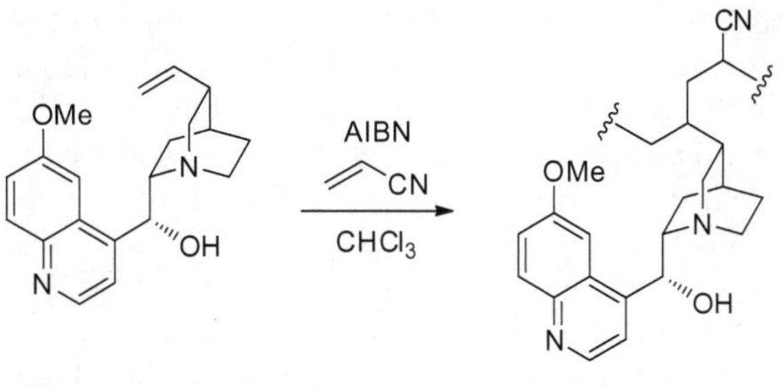

88

SCHEME 4.19. The copolymerization of quinine and acrylonitrile.

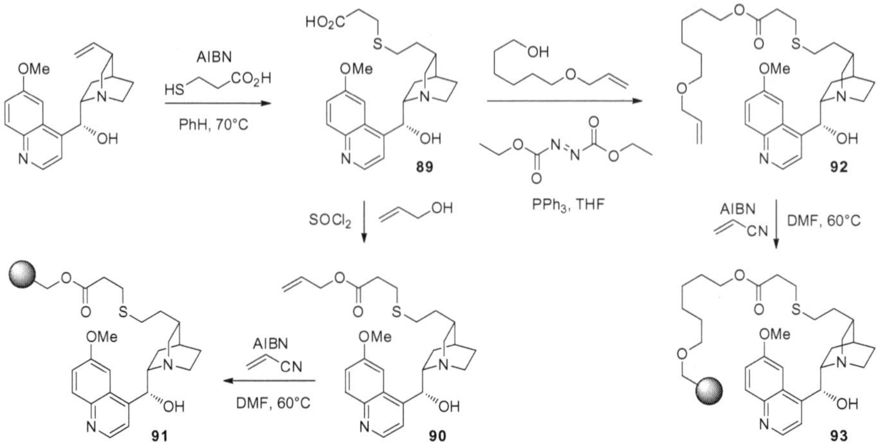

SCHEME 4.20. Polymer-supported quinine by copolymerization.

clever thiol-ene chemistry, *Cinchona* monomers such as **90** and **92** were prepared (Scheme 4.20). By reacting quinine with 3-mercaptopropionic acid in a thiol-ene coupling, carboxylic acid **89** was obtained in excellent yield. This could either be converted to ester **90** in allyl alcohol or reacted in a Mitsunobu reaction with an alcohol obtained from 6-chloro-1-hexanol and allyl alcohol to give ester **92**. Both **90** and **92** are useful vinyl monomers equipped with spacer groups that were subsequently copolymerized with acrylonitrile in DMF to give copolymers **91** and **93**. Alternatively, a very closely related strategy starting with mercaptoethanol was used to prepare an analogous acrylic quinine monomer. All these polymers were tested in asymmetric Michael reactions. Although more than 20 years have passed since the procedure in Scheme 4.20 was reported, it stands even today as remarkably elegant and modern. The thiol-ene coupling is, of course, included in the illustrious and contemporary "click" concept, but it is, in essence, a fully chain-transferred free radical polymerization that has been used reliably for decades. It can therefore be argued by analogy that a free radical polymerization, perhaps, is more or less as close as you can come to the strict definitions of a "click" reaction.

As trained polymer chemists would be quick to point out, simple allylic C=C bonds in derivatives like the *Cinchona* alkaloids or monomers such as **90** or **92** (Scheme 4.20) normally do not participate effectively in free radical polymerization because of a competitive abstraction of allylic hydrogens that would result in allylic radicals, something that will halt further chain propagation and can actually retard or inhibit additional growth. As already pointed out by Kobayashi and Iwai in their seminal work in 1978 [78b], acrylonitrile has the highest copolymerizability with the vinylic *Cinchona* alkaloids while methacrylonitrile, *N*-vinylcarbazole, and acrylamide also effectively copolymerized, but to a lesser extent. The reason for this is that only comonomers that form highly electron-deficient free radicals during polymerization (such as acrylonitrile) can effectively copolymerize with simple allylic species. This is a lesson that is easily forgotten and has been the source of some controversy. In the late 1980s, Sharpless and co-workers first reported their important methodology for asymmetric dihydroxylation of olefins by using *Cinchona*-derived materials [79]. In the years to follow, the field of polymer-supported *Cinchona* derivatives surged and numerous publications followed, nearly all of which used bottom-up approaches like that in Schemes 4.19 and 4.20, both by Sharpless and co-workers [80], as well as by other research groups [81]. Indeed, convenient copolymerizations were an attractive methodology. The problem was that several of these reports during the 1990s utilized the counterintuitive (in the sense that it should not work) copolymerization of the olefin in *Cinchona* derivatives directly with monomers like methyl methacrylate and EGDMA [81i,j]. Sherrington and co-workers (in a publication legitimately labeled "a cautionary tale") were subsequently able to repeat some of this work and prove that the *Cinchona* derivatives were, in fact, not covalently bound and could be removed from other polymeric material when samples were adequately purified [82]. The important lesson is that the simple immobilization of *Cinchona* alkaloids by direct copolymerization can only be successfully carried out when using suitable comonomers such as acrylonitrile.

With the renewed interest surrounding organocatalysis from 2000 onward, several enamine–iminium organocatalysts immediately became attractive targets for polymeric immobilization. However, unlike the *Cinchona* derivatives, the early phase in the research on polymer-supported organocatalysts was dominated by postmodification procedures. Several of these have been detailed earlier in this chapter. However, there has recently been reported quite a few successful polymer-supported prolines, diarylprolinol silyl ethers, MacMillan imidazolidinones, and chiral Brønsted acids prepared by bottom-up procedures [83–91].

Several research groups have developed the set of functional proline monomers **94–103** depicted in Figure 4.4 [83–86]. Monomers **94–97** can be prepared on a robust multigram scale in only one step from *trans*-4-hydroxy-L-proline or *cis*-4-hydroxy-D-proline by a selective *O*-acylation in CF_3CO_2H with the appropriate acid chlorides [83]. Hydrophilic monomers **94/95** can be directly polymerized in water to linear high-load polymers or dispersion copolymerized with benzyl methacrylate and EGDMA to give catalytically highly active polymer granulates. The more hydrophobic derivatives **98** and **99** are suited for suspension copolymerization, and such polymer beads have proven very successful in asymmetric aldol reactions. Prior to deprotection, such polymer beads (from **98** to **99**) can also be easily converted to prolineamides by peptide coupling with amino alcohols, making multigram preparations of polymer-supported prolineamides possible. Unlike the postmodification procedures detailed earlier in this chapter, monomers such as **94–99** all gave polymer products on a useful scale for the first time and have been utilized in gram-scale organocatalytic transformations. Other researchers have developed

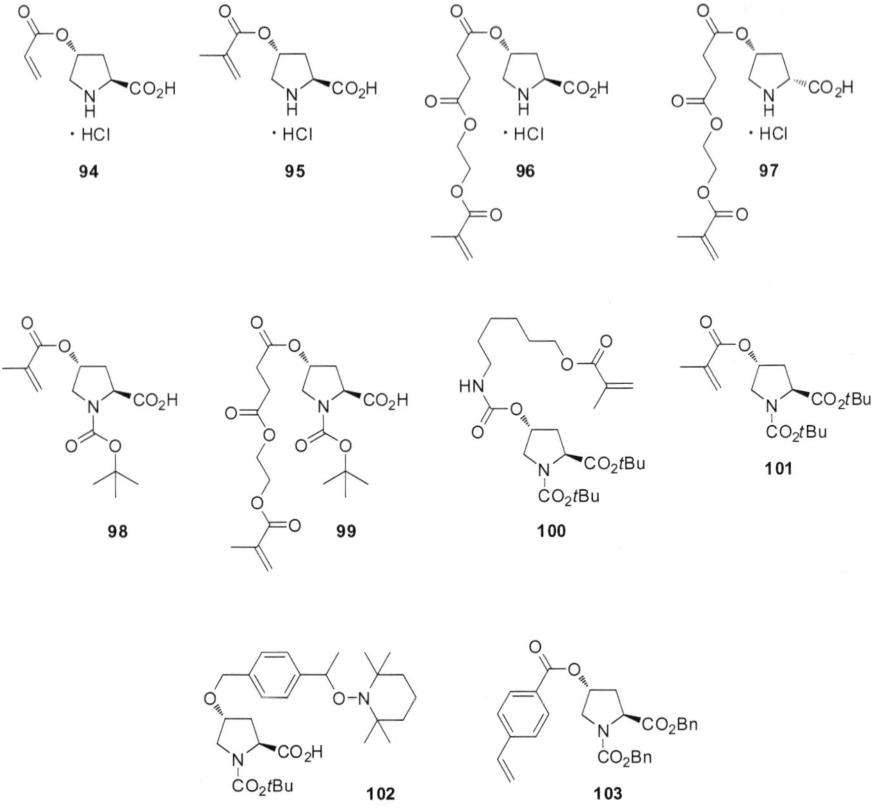

FIGURE 4.4. Reported proline monomers for bottom-up procedures.

syntheses of protected monomers **100** and **101** and used these functional monomers to make catalytically active homopolymers [84]. Nitroxide monomer **102** has been used to prepare polymer nanofibers through NMP polymerization and subsequent electrospinning, but catalytic activity of these fibers has not been reported [85]. Very recently, styrenic proline monomer **103** has been utilized in NMP- or RAFT-mediated copolymerizations to give polymers with a controlled/adjustable structure and good catalytic activity in asymmetric aldol reactions [86]. Methodologies for controlled radical polymerization such as NMP, RAFT, and ATRP open up completely new possibilities for the preparation of tailor-made macromolecular structures that can have huge potential and impact in asymmetric catalysis. It will be interesting to see whether these methodologies will be generally adopted in the future.

Unlike proline, the first reported syntheses of polymer-supported diarylprolinol silyl ether organocatalysts used bottom-up-type copolymerization schemes. Schore and co-workers reported polymer-supported diarylprolinol silyl ethers in 2008 [87], founded upon on procedures used by the same research

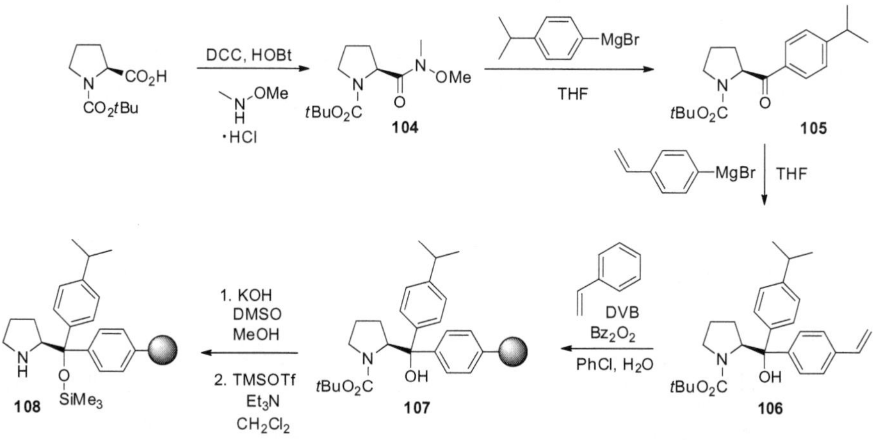

SCHEME 4.21. Styrenic polymer-supported diarylprolinol silyl ether.

group to prepare polymer-supported CBS catalysts [88]. Two types of styrenic diarylprolinol silyl ether catalysts were reported: one cross-link-bound catalyst prepared from a distyrenic precursor and the pendant-bound catalyst **108** whose synthesis is detailed in Scheme 4.21. Weinreb amide **104** was prepared via diimide coupling and converted to ketone **105** in a standard Weinreb ketone synthesis. Reaction with a styrenic Grignard reagent gave diarylprolinol **106**, and subsequent suspension copolymerization, deprotection, and a final silylation furnished resin **108**. Preparation of the cross-link-bound catalyst is analogous but simpler, as a Boc-protected proline methyl ester is directly converted to the corresponding difunctional monomer using the same styrenic Grignard reagent. These catalysts proved effective in several organocatalytic reactions, but unfortunately their styrenic backbone made them partly incompatible with the polar solvents that are so important in diarylprolinol silyl ether-mediated organocatalysis.

Scalability and compatibility with polar solvents were key issues that were addressed in 2010 by the introduction of acrylic monomer **112** (Scheme 4.22) [83c]. This monomer is easily prepared on >10 gram scale by treating the ethyl ester hydrochloride of hydroxyproline directly with excess PhMgBr to give diphenylhydroxyprolinol **109**, followed by a selective *O*-acylation in CF_3CO_2H with acid chloride **110** to give compound **111** (notably leaving both the unprotected amine and tertiary alcohol untouched). Silylation then gave acrylic monomer **112**. While normal hydrophobic polymer beads were easily obtained by standard suspension copolymerization with methyl methacrylate and EGDMA, polymer beads compatible with polar solvents were obtained by a novel suspension copolymerization with PEG methyl ether methacrylate and PEG dimethacrylate as a cross-linker (conducted in the normal mode because of the more hydrophobic nature of PEG derivatives at elevated temperatures).

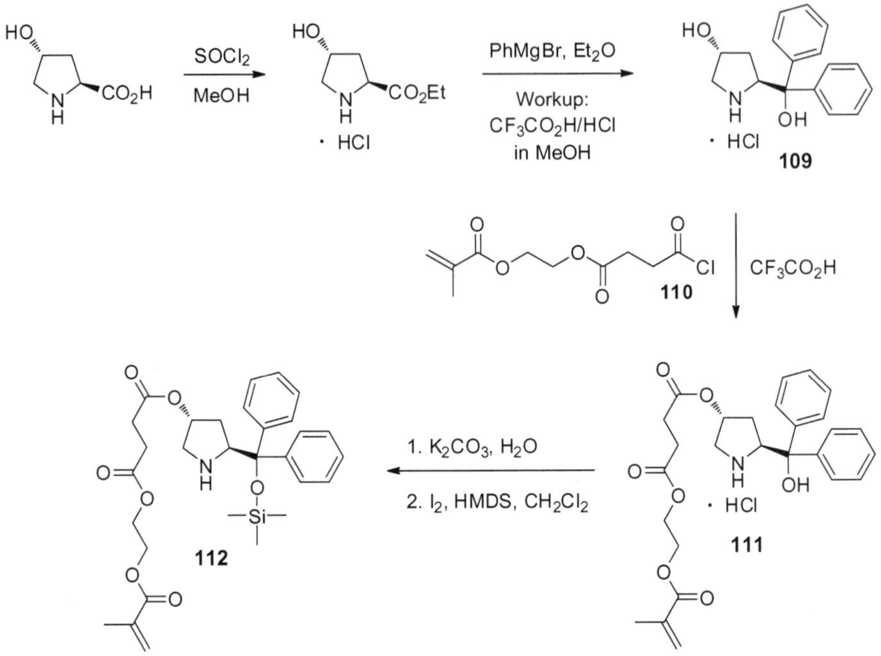

SCHEME 4.22. Acrylic diarylprolinol silyl ether monomer.

NMP-mediated polymerization has also been used for preparation of polymer-supported diphenylprolinol silyl ether organocatalysts. German researchers used a methodology very similar to that used for their proline monomer **102** (Fig. 4.4) to immobilize diphenylprolinol silyl ether by NMP polymerization and electrospinning [89]. The catalyst had moderate success in asymmetric Michael reactions. As mentioned, the diarylprolinol silyl ether organocatalysts are closely related to the diphenylprolinol catalysts used in CBS reduction. More interesting procedures are likely to be reported in the future, perhaps using more of the chemistry on polymer-supported CBS-type catalysts that has not been detailed here [90].

While MacMillan imidazolidinones have been immobilized on several occasions using conventional postmodification, bottom-up-type strategies are rare for this family of catalysts. This is probably connected to the fact that such catalysts have shown only moderate success in recycling as a result of their labile nature. One large-scale approach to polymer-supported MacMillan imidazolidinones is shown in Scheme 4.23 [83c]. By treating phenylalanine methyl ester hydrochloride with ethanolamine under neat conditions, amide **113** was obtained after a simple alkaline workup. This amide was cyclized with acetone under the azeotropic removal of water to give imidazolidinones **114**, and a selective *O*-acylation in $MeSO_3H$ with methacryloyl chloride gave methacrylic imidazolidinone **115**. This functional monomer was suspension copoly-

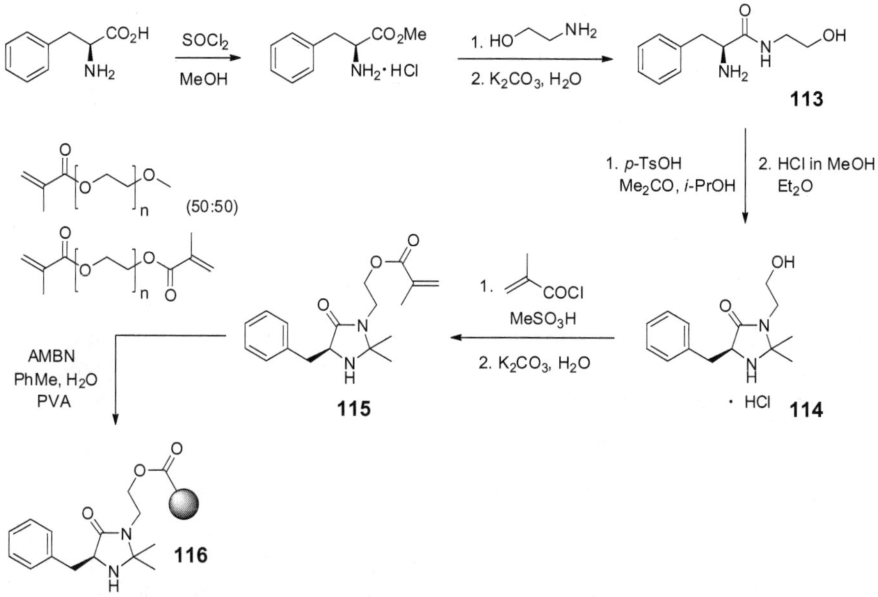

SCHEME 4.23. Acrylic polymer-supported MacMillan imidazolidinone.

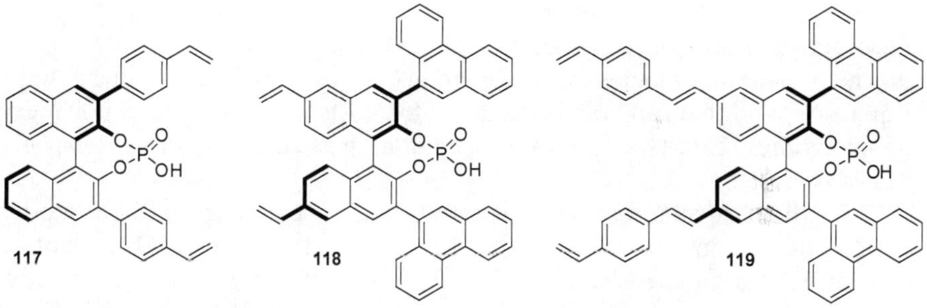

FIGURE 4.5. Functional monomers for polymer-supported chiral Brønsted acids.

merized with PEG methyl ether methacrylate and PEG dimethacrylate to give beaded resin **116**. This PEG-acrylic resin was compatible with polar solvents and exhibited catalytic selectivities comparable to the standard monomeric catalyst in asymmetric Diels–Alder reactions in MeCN-H_2O, but like other supported imidazolidinones, it also suffered from poor recycling capabilities.

The first reported immobilization of a catalyst belonging to the chiral Brønsted acids was recently reported by Rueping and co-workers in Germany [91]. The researchers prepared the advanced functional monomers **117–119** (Fig. 4.5) and copolymerized them with styrene and DVB to give small polymer sticks that could be isolated after completion of a test reaction by pulling the container out of the reaction vessel in a tea-bag-like fashion. These polymer-supported Brønsted acids were tested successfully in a transfer hydrogenation of a quinoline using Hantzsch dihydropyridine as a hydride source. The catalyst was easily isolated after the reaction and could be effectively recycled at least 10 times. It could also undertake catalysis in a variety of solvents. Unfortunately, syntheses of functional monomers **117–119** are very arduous processes that require many steps and multiple chromatographic purifications. Therefore, the final crucial steps are conducted on a minute scale and do not currently offer the same advantageous scalability as other bottom-up approaches to chiral organocatalysts. Nevertheless, the favorable recycling properties of this family of catalysts make them very attractive targets for efficient polymeric immobilization.

4.4. CONCLUSIONS AND OUTLOOK

In this chapter, an analysis of the syntheses leading up to polymer-supported chiral catalysts has been presented, with the clear goal to bring much needed attention to the task of identifying useful and hopefully generic methodology for solving these synthetic issues. Polymer-supported catalysis has been around for decades, and its basic principles have been proven beyond doubt. However, cost issues linger over the entire field, and researchers are probably overly concerned with bringing new catalysts to the literature rather than developing truly helpful methodologies that are generic (at least for the family of catalysts in question) and that can help bring down the costs to a point where polymer-supported chiral catalysts can be tools in common use rather than a specialized field of research.

As for the particular families of catalysts and the comparison of postmodification and bottom-up strategies that have been at the core of this chapter, some useful lessons can be extracted. Both overall strategies of polymeric immobilizations have been in continuous use since the 1970s. Notable highlights in the pioneering era of the 1970s and 1980s are definitely the comprehensive work of Stille on polymer-supported chiral organometallic complexes and Kobayashi's work on polymer-supported *Cinchona* alkaloids. It is notable that both utilized copolymerization strategies, and the *Cinchona*–acrylonitrile copolymerization remains today the method of choice for the immobilization of these compounds. As its pioneer Stille pointed out, a copolymerization strategy has a natural flexibility in the adjustment of catalyst loadings, the degree and type of cross-linking, and in the fine-tuning of the solubility or swelling characteristics of the polymeric materials. That said, it does require a substantial footing in materials science, and postmodification strategies have prevailed when innovation from researchers using bottom-up schemes has been low or when central participants have left the field. With the reappraisal of organocatalysis at the turn of the millennium, new possibilities opened up.

Because these catalysts often operate under mild enzyme-mimetic reaction conditions, more adaptable and more comprehensive acrylic chemistry could be introduced, and that has helped to bring out their true potential. Recent successful bottom-up approaches to chiral Brønsted acids within organocatalysis and dirhodium(II) carboxylates within organometallic catalysts have further strengthened copolymerization strategies in the past couple of years. Bottom-up strategies will probably require even more interdisciplinary cooperation in the years to come if techniques for controlled radical polymerization are to live up to their true potential.

For future work within organometallic chiral catalysts, there is still a need to tackle the core of the BINAP catalysts in a more expedient manner as they (unlike salen, Box, or DPEN-type catalysts) are more limited by virtue of their aromatic nature. Within organocatalysis, catalysts working by noncovalent interactions such as chiral thiourea catalysts or chiral Brønsted acids are especially attractive as they probably have more robust recycling capabilities and are more highly valued than several of the most ordinary types of enamine–iminium catalysts. Regardless of individual opportunities, we believe all researchers in life sciences and material sciences alike can hope that some useful generic methodology and lessons for polymeric immobilization are acquired along the way.

REFERENCES

[1] For general introductions, see: (a) De Vos, D. E., Vankelecom, I. F. J., Jacobs, P. A. (Eds.) (2000). *Chiral Catalyst Immobilization and Recycling*. Wiley-VCH, Weinheim, Germany; (b) Fan, Q.-H., Li, Y.-M., Chan, A. S. C. (2002). Recoverable catalysts for asymmetric organic synthesis. *Chem. Rev.*, *102*, 3385–3466; (c) Buchmeiser, M. R., Ed. (2003). *Polymeric Materials in Organic Synthesis and Catalysis*. Wiley-VCH, Weinheim, Germany; (d) Tulla-Puche, J., Albericio, F. (Eds.) (2008). *The Power of Functional Resins in Organic Synthesis*. Wiley-VCH, Weinheim, Germany; (e) Benaglia, M., Ed. (2009). *Recoverable and Recyclable Catalysts*. John Wiley & Sons, Chichester; (f) Trindade, A. F., Gois, P. M. P., Afonso, C. A. M. (2009). Recyclable stereoselective catalysts. *Chem. Rev.*, *109*, 418–514.

[2] For a recent treatment of this analysis for polymer-supported chiral organocatalysts, see: Kristensen, T. E., Hansen, T. (2010). Polymer-cupported chiral organocatalysts: synthetic strategies for the road toward affordable polymeric immobilization. *Eur. J. Org. Chem.*, 3179–3204.

[3] For an historical account, see: (a) Hudson, D. (1999). Matrix assisted synthetic transformations: a mosaic of diverse contributions. I. The pattern emerges. *J. Comb. Chem.*, *1*, 333–360; (b) Hudson, D. (1999). Matrix assisted synthetic transformations: a mosaic of diverse contributions. II. The pattern is completed. *J. Comb. Chem.*, *1*, 403–457.

[4] Lu, J., Toy, P. H. (2009). Organic polymer supports for synthesis and for reagent and catalyst immobilization. *Chem. Rev.*, *109*, 815–838. For more references, see Reference [2] and references cited therein.

[5] Divinylbenzene cross-linked and chloromethylated polystyrene prepared in beaded form through suspension polymerization in water, pioneered by Merrifield; see: Merrifield, R. B. (1963). Solid phase peptide synthesis I: the synthesis of a tetrapeptide. *J. Am. Chem. Soc.*, *85*, 2149–2154.

[6] Odian, G. (2004). *Principles of Polymerization*, 4th ed.. John Wiley & Sons, Hoboken, NJ.

[7] Preparation, properties and references for these different polymer materials for use in supported catalysis is readily available in the general references of this chapter, especially in References [1–4].

[8] Dumont, W., Poulin, J.-C., Kagan, H. B. (1973). Asymmetric catalytic reduction with transition metal complexes. II. Asymmetric catalysis by a supported chiral rhodium complex. *J. Am. Chem. Soc.*, *95*, 8295–8299.

[9] (a) Dang, T.-P., Kagan, H. (1971). The asymmetric synthesis of hydratropic acid and amino-acids by homogeneous catalytic hydrogenation. *J. Chem. Soc. Chem. Commun.*, 481; (b) Kagan, H., Dang, T.-P. (1972). Asymmetric catalytic reduction with transition metal complexes. I. Catalytic system of rhodium(I) with (-)-2,3-0-isopropylidene-2,3-dihydroxy-1,4-bis(diphenylphosphino)butane, a new chiral diphosphine. *J. Am. Chem. Soc.*, *94*, 6429–6433.

[10] (a) Knowles, W. S., Sabacky, M. J., Vineyard, B. D., Weinkauff, D. J. (1975). Asymmetric hydrogenation with a complex of rhodium and a chiral bisphosphine. *J. Am. Chem. Soc.*, *97*, 2567–2568; (b) Vineyard, B. D., Knowles, W. S., Sabacky, M. J., Bachman, G. L., Weinkauff, D. J. (1977). Asymmetric hydrogenation. Rhodium chiral bisphosphine catalyst. *J. Am. Chem. Soc.*, *99*, 5946–5952; (c) Knowles, W. S. (1983). Asymmetric hydrogenation. *Acc. Chem. Res.*, *16*, 106–112.

[11] (a) Miyashita, A., Yasuda, A., Takaya, H., Toriumi, K., Ito, T., Souchi, T., Noyori, R. (1980). Synthesis of 2,2'-bis(diphenylphosphino)-1,1'-binaphthyl (BINAP), an atropisomeric chiral bis(triaryl)phosphine, and its use in the rhodium(I)-catalyzed asymmetric hydrogenation of alpha.-(acylamino)acrylic acids. *J. Am. Chem. Soc.*, *102*, 7932–7934; (b) Miyashita, A., Takaya, H., Souchi, T., Noyori, R. (1984). 2, 2′-bis(diphenylphosphino)-1,1′-binaphthyl(binap):a new atropisomeric bis(triaryl) phosphine. synthesis and its use in the Rh(l)-catalyzed asymmetric hydrogenation of α-(acylamino)acrylic acids. *Tetrahedron*, *40*, 1245–1253.

[12] For more on the development of chiral phosphine ligands, see: Nagel, U., Albrecht, J. (1998). The enantioselective hydrogenation of N-acyl dehydroamino acids. *Top. Catal.*, *5*, 3–23.

[13] Several ionic organometallic complexes can be immobilized by simple ion exchange on ion exchangers, but they will not be detailed (as will not immobilization by adsorption or entrapment). See References [1] and [12].

[14] Bayston, D. J., Fraser, J. L., Ashton, M. R., Baxter, A. D., Polywka, M. E. C., Moses, E. (1998). Preparation and use of a polymer supported BINAP hydrogenation catalyst. *J. Org. Chem.*, *63*, 3137–3140.

[15] Ohkuma, T., Takeno, H., Honda, Y., Noyori, R. (2001). Asymmetric hydrogenation of ketones with polymer-bound BINAP/diamine ruthenium catalysts. *Adv. Synth. Catal.*, *343*, 369–375.

[16] (a) Fujii, A., Sodeoka, M. (1999). Palladium diaqua and hydroxo complexes with polymer-supported BINAP ligands and their use for catalytic enantioselective

reactions. *Tetrahedron Lett.*, *40*, 8011–8014; (b) Chapuis, C., Barthe, M., Laumer, J.-Y. (2001). Synthesis of citronellal by Rh(I)-catalysed asymmetric isomerization of N,N-diethyl-substituted geranyl- and nerylamines or geraniol and nerol in the presence of chiral diphosphino ligands, under homogeneous and supported conditions. *Helv. Chim. Acta.*, *84*, 230–242.

[17] Itsuno, S., Tsuji, A., Takahashi, M. (2003). Synthesis of enantiopure 1,2-diamine attached to cross-linked polystyrene and its application to an insoluble catalyst for asymmetric hydrogenation. *Tetrahedron Lett.*, *44*, 3825–3828.

[18] Bayston, D. J., Travers, C. B., Polywka, M. E. C. (1998). Synthesis and evaluation of a chiral heterogeneous transfer hydrogenation catalyst. *Tetrahedron Asymmetry*, *9*, 2015–2018.

[19] Li, Y., Li, Z., Li, F., Wang, Q., Tao, F. (2005). Preparation of polymer-supported Ru-TsDPEN catalysts and use for enantioselective synthesis of (*S*)-fluoxetine. *Org. Biomol. Chem.*, *3*, 2513–2518.

[20] (a) Malmström, T., Andersson, C. (1996). A novel chiral water-soluble phosphine ligand based on a water-soluble acrylic acid salt. *Chem. Commun.*, 1135–1136; (b) Malmström, T., Andersson, C. (1999). Enantioselective hydrogenation in water catalysed by rhodium phosphine complexes bound to polyacrylic acid. *J. Mol. Catal. A Chem.*, *139*, 259–270.

[21] For a review of asymmetric processes catalyzed by chiral salen metal complexes, see: Larrow, J. F., Jacobsen, E. N. (2004). Asymmetric processes catalyzed by chiral (salen)metal complexes. *Top. Organomet. Chem.*, *6*, 123–152.

[22] (a) Canali, L., Cowan, E., Deleuze, H., Gibson, C. L., Sherrington, D. C. (1998). Remarkable matrix effect in polymer-supported Jacobsen's alkene epoxidation catalysts. *Chem. Commun.*, 2561–2562; (b) Canali, L., Cowan, E., Deleuze, H., Gibson, C. L., Sherrington, D. C. (2000). Polystyrene and polymethacrylate resin-supported Jacobsen's alkene epoxidation catalyst. *J. Chem. Soc. Perkin Trans. 1*, 2055–2066.

[23] Angelino, M. D., Laibinis, P. E. (1998). Synthesis and characterization of a polymer-supported salen ligand for enantioselective epoxidation. *Macromolecules*, *31*, 7581–7587.

[24] Annis, D. A., Jacobsen, E. N. (1999). Polymer-supported chiral Co(salen) complexes: synthetic applications and mechanistic investigations in the hydrolytic kinetic resolution of terminal epoxides. *J. Am. Chem. Soc.*, *121*, 4147–4154.

[25] Reger, T. S., Janda, K. D. (2000). Polymer-supported (salen)Mn catalysts for asymmetric epoxidation: a comparison between soluble and insoluble matrices. *J. Am. Chem. Soc.*, *122*, 6929–6934.

[26] Song, C. E., Roh, E. J., Yu, B. M., Chi, D. Y., Kim, S. C., Lee, K.-J. (2000). Heterogeneous asymmetric epoxidation of alkenes catalysed by a polymer-bound (pyrrolidine salen)manganese(III) complex. *Chem. Commun.*, 615–616.

[27] Smith, K., Liu, C.-H. (2002). Asymmetric epoxidation using a singly-bound supported Katsuki-type (salen)Mn complex. *Chem. Commun.*, 886–887.

[28] Evans, P. A., Ed. (2005). *Modern Rhodium-Catalyzed Organic Reactions*. Wiley-VCH, Weinheim, Germany.

[29] (a) Doyle, M. P., Timmons, D. J., Tumonis, J. S., Gau, H.-M., Blossey, E. C. (2002). Preparation and catalytic properties of immobilized chiral dirhodium(II) carboxamidates. *Organometallics*, *21*, 1747–1749; (b) Doyle, M. P., Yan, M., Gau, H.-M.,

Blossey, E. C. (2003). Catalysts with mixed ligands on immobilized supports. Electronic and steric advantages. *Org. Lett.*, *5*, 561–563.

[30] (a) Nagashima, T., Davies, H. M. L. (2002). Catalytic asymmetric cyclopropanation using bridged dirhodium tetraprolinates on solid support. *Org. Lett.*, *4*, 1989–1992; (b) Davies, H. M. L., Walji, A. M. (2003). Asymmetric intermolecular C-H activation, using immobilized dirhodium tetrakis((S)-N-(dodecylbenzenesulfonyl)- prolinate) as a recoverable catalyst. *Org. Lett.*, *5*, 479–482; (c) Davies, H. M. L., Walji, A. M., Nagashima, T. (2004). Simple strategy for the immobilization of dirhodium tetraprolinate catalysts using a pyridine-linked solid support. *J. Am. Chem. Soc.*, *126*, 4271–4280; (d) Davies, H. M. L., Walji, A. M. (2005). Universal strategy for the immobilization of chiral dirhodium catalysts. *Org. Lett.*, *7*, 2941–2944.

[31] (a) Lloret, J., Estevan, F., Bieger, K., Villanueva, C., Ubeda, M. A. (2007). Immobilized chiral ortho-metalated dirhodium(II) compounds as catalysts in the asymmetric cyclopropanation of styrene with ethyl diazoacetate. *Organometallics*, *26*, 4145–4151; (b) Lloret, J., Stern, M., Estevan, F., Sanau, M., Ubeda, M. A. (2008). *Ortho*-Metalated dirhodium(II) catalysts immobilized on a polymeric cross-linked support by copolymerization. Study of their catalytic activity in the asymmetric cyclopropanation of styrene with ethyl diazoacetate. *Organometallics*, *27*, 850–856.

[32] See: (a) Basak, R. (2003). Bisoxazoline (BOX) ligand-metal complexes: an emerging chiral catalyst. *Synlett*, 1223–1224 and references cited therein. For a review of PyBox ligands, see: (b) Desimoni, G., Faita, G., Quadrelli, P. (2003). Pyridine 2,6-bis(oxazolines), helpful ligands for asymmetric catalysts. *Chem. Rev.*, *103*, 3119–3154.

[33] For a review of heterogeneous bis(oxazoline) ligands, see: Rechavi, D., Lemaire, M. (2002). Enantioselective catalysis using heterogeneous bis(oxazoline) ligands: which factors influence the enantioselectivity? *Chem. Rev.*, *102*, 3467–3494.

[34] (a) Lundgren, S., Lutsenko, S., Jönsson, C., Moberg, C. (2003). Polymer-supported pyridine-bis(oxazoline). Application to ytterbium-catalyzed silylcyanation of benzaldehyde. *Org. Lett.*, *5*, 3663–3665; (b) Tilliet, M., Lundgren, S., Moberg, C., Levacher, V. (2007). Polymer-bound pyridine-bis(oxazoline). Preparation through click chemistry and evaluation in asymmetric catalysis. *Adv. Synth. Catal.*, *349*, 2079–2084.

[35] (a) Hallman, K., Macedo, E., Nordström, K., Moberg, C. (1999). Enantioselective allylic alkylation using polymer-supported palladium catalysts. *Tetrahedron Asymmetry*, *10*, 4037–4046; (b) Hallman, K., Moberg, C. (2001). Polymer-bound bis(oxazoline) as a chiral catalyst. *Tetrahedron Asymmetry*, *12*, 1475–1478.

[36] Leadbeater, N. E., Marco, M. (2002). Preparation of polymer-supported ligands and metal complexes for use in catalysis. *Chem. Rev.*, *102*, 3217–3274.

[37] For relevant reviews, see: (a) Benaglia, M., Puglisi, A., Cozzi, F. (2003). Polymer-supported organic catalysts. *Chem. Rev.*, *103*, 3401–3429; (b) Cozzi, F. (2006). Immobilization of organic catalysts: when, why, and how. *Adv. Synth. Catal.*, *348*, 1367–1390; (c) Benaglia, M. (2006). Recoverable and recyclable chiral organic catalysts. *New J. Chem.*, *30*, 1525–1533; (d) Altava, B., Burguete, I., Luis, S. V. (2008). Polymer-supported organocatalysts. In: J. Tulla-Puche, F. Albericio (Eds.), *The Power of Functional Resins in Organic Synthesis*. Wiley-VCH, Weinheim, Germany,

pp. 247–308; (e) Gruttadauria, M., Giacalone, F., Noto, R. (2008). Supported proline and proline-derivatives as recyclable organocatalysts. *Chem. Soc. Rev.*, *37*, 1666–1688; (f) Gruttadauria, M., Giacalone, F., Noto, R. (2009). Water in stereoselective organocatalytic reactions. *Adv. Synth. Catal.*, *351*, 33–57; (g) Fraile, J. M., Garcia, J. I., Mayoral, J. A. (2009). Noncovalent immobilization of enantioselective catalysts. *Chem. Rev.*, *109*, 360–417; (h) Bergbreiter, D. E., Tian, J., Hongfa, C. (2009). Using soluble polymer supports to facilitate homogeneous catalysis. *Chem. Rev.*, *109*, 530–582; (i) Benaglia, M. (2009). Recoverable organic catalysts. In: M. Benaglia (Ed.), *Recoverable and Recyclable Catalysts*. John Wiley & Sons, Chichester, pp. 301–340.

[38] Kondo, K., Yamano, T., Takemoto, K. (1985). Asymmetric Robinson cyclization reaction catalyzed by polymer-bound L-proline. *Makromol. Chem.*, *186*, 1781–1785.

[39] (a) Benaglia, M., Celentano, G., Cozzi, F. (2001). Enantioselective aldol condensations catalyzed by poly(ethylene glycol)-supported proline. *Adv. Synth. Catal.*, *343*, 171–173; (b) Benaglia, M., Cinquini, M., Cozzi, F., Puglisi, A., Celentano, G. (2002). Poly(ethylene glycol)-supported proline: a versatile catalyst for the enantioselective aldol and iminoaldol reactions. *Adv. Synth. Catal.*, *344*, 533–542; (c) Benaglia, M., Cinquini, M., Cozzi, F., Puglisi, A., Celentano, G. (2003). *J. Mol. Catal. A*, *204–205*, 157–163.

[40] (a) Font, D., Jimeno, C., Pericàs, M. A. (2006). Polystyrene-supported hydroxyproline: an insoluble, recyclable organocatalyst for the asymmetric aldol reaction in water. *Org. Lett.*, *8*, 4653–4655; (b) Font, D., Bastero, A., Sayalero, S., Jimeno, C., Pericàs, M. A. (2007). Highly enantioselective α-aminoxylation of aldehydes and ketones with a polymer-supported organocatalyst. *Org. Lett.*, *9*, 1943–1946; (c) Font, D., Sayalero, S., Bastero, A., Jimeno, C., Pericàs, M. A. (2008). Toward an artificial aldolase. *Org. Lett.*, *10*, 337–340; (d) Alza, E., Rodríguez-Escrich, C., Sayalero, S., Bastero, A., Pericàs, M. A. (2009). A Solid-supported organocatalyst for highly stereoselective, batch, and continuous-flow Mannich reactions. *Chem. Eur. J.*, *15*, 10167–10172.

[41] (a) Giacalone, F., Gruttadauria, M., Marculescu, A. M., Noto, R. (2007). Polystyrene-supported proline and prolinamide. Versatile heterogeneous organocatalysts both for asymmetric aldol reaction in water and α-selenenylation of aldehydes. *Tetrahedron Lett.*, *48*, 255–259; (b) Gruttadauria, M., Giacalone, F., Marculescu, A. M., Meo, P. L., Riela, R., Noto, R. (2007). Hydrophobically directed aldol reactions: polystyrene-supported L-proline as recyclable catalyst for direct asymmetric aldol reaction in the presence of water. *Eur. J. Org. Chem.*, 4688–4698; (c) Giacalone, F., Gruttadauria, M., Marculescu, A. M., D'Anna, F., Noto, R. (2008). Polystyrene-supported proline as recyclable catalyst in the Baylis-Hillman reaction of arylaldehydes and methyl or ethyl vinyl ketone. *Catal. Commun.*, *9*, 1477–1481.

[42] (a) Gruttadauria, M., Giacalone, F., Marculescu, A. M., Noto, R. (2008). Novel prolinamide-supported polystyrene as highly stereoselective and recyclable organocatalyst for the aldol reaction. *Adv. Synth. Catal.*, *350*, 1397–1405; (b) Gruttadauria, M., Giacalone, F., Marculescu, A. M., Salvo, A. M. P., Noto, R. (2009). Stereoselective aldol reaction catalyzed by a highly recyclable polystyrene supported substituted prolinamide catalyst. *ARKIVOC*, *VIII*, 5–15; (c) Gruttadauria, M., Salvo, A. M. P., Giacalone, F., Agrigento, P., Noto, R. (2009). Enhanced activity and stereoselectivity of polystyrene-supported proline-based

organic catalysts for direct asymmetric aldol reaction in water. *Eur. J. Org. Chem.*, 5437–5444.

[43] Alza, E., Pericàs, M. A. (2009). A highly selective, polymer-supported organocatalyst for the Michael additions with enzyme-like behavior. *Adv. Synth. Catal.*, *351*, 3051–3056.

[44] Mager, I., Zeitler, K. (2010). Efficient, enantioselective iminium catalysis with an immobilized, recyclable diarylprolinol silyl ether catalyst. *Org. Lett.*, *12*, 1480–1483.

[45] (a) Benaglia, M., Celentano, G., Cinquini, M., Puglisi, A., Cozzi, F. (2002). Poly(ethylene glycol)-supported chiral imidazolidin-4-one: an efficient organic catalyst for the enantioselective Diels-Alder cycloaddition. *Adv. Synth. Catal.*, *344*, 149–152; (b) Puglisi, A., Benaglia, M., Cinquini, M., Cozzi, F., Celentano, G. (2004). Enantioselective 1,3-dipolar cycloadditions of unsaturated aldehydes promoted by a poly(ethylene glycol)-supported organic catalyst. *Eur. J. Org. Chem.*, 567–573.

[46] Selkälä, S. A., Tois, J., Pihko, P. M., Koskinen, A. M. P. (2002). Asymmetric organocatalytic Diels-Alder reactions on solid support. *Adv. Synth. Catal.*, *344*, 941–945.

[47] (a) Pecinovsky, C. S., Nicodemus, G. D., Gin, D. L. (2005). Nanostructured, solid-state organic, chiral Diels-Alder catalysts via acid-induced liquid crystal assembly. *Chem. Mater.*, *17*, 4889–4891; (b) Zhang, Y., Zhao, L., Lee, S. S., Ying, J. Y. (2006). Enantioselective catalysis over chiral imidazolidin-4-one immobilized on siliceous and polymer-coated mesocellular foams. *Adv. Synth. Catal.*, *348*, 2027–2032.

[48] Haraguchi, N., Takemura, Y., Itsuno, S. (2010). Novel polymer-supported organocatalyst via ion exchange reaction: facile immobilization of chiral imidazolidin-4-one and its application to Diels-Alder reaction. *Tetrahedron Lett.*, *51*, 1205–1208.

[49] For a recent summary, see: Wiesner, M., Wennemers, H. (2010). Peptide catalyzed conjugate addition reactions of aldehydes to nitroolefins. *Synthesis.*, 1568–1571 and references cited therein.

[50] Revell, J. D., Gantenbein, D., Krattiger, P., Wennemers, H. (2006). Solid-supported and pegylated H-Pro-Pro-Asp-NHR as catalysts for asymmetric aldol reactions. *Biopolymers*, *84*, 105–113.

[51] (a) Sigman, M. S., Jacobsen, E. N. (1998). Schiff base catalysts for the asymmetric Strecker reaction identified and optimized from parallel synthetic libraries. *J. Am. Chem. Soc.*, *120*, 4901–4902; (b) Sigman, M. S., Vachal, P., Jacobsen, E. N. (2000). A general catalyst for the asymmetric Strecker reaction. *Angew. Chem. Int. Ed.*, *39*, 1279–1281.

[52] Berkessel, A., Gasch, N., Glaubitz, K., Koch, C. (2001). Highly enantioselective enone epoxidation catalyzed by short solid phase-bound peptides: dominant role of peptide helicity. *Org. Lett.*, *3*, 3839–3842.

[53] For more details, see the general references of this chapter, such as References [1, 2, 37].

[54] Yamauchi, K., Kinoshita, M., Imoto, M. (1971). Vinyl polymerization. 272. Synthesis and polymerization of 9-*O*-methacryloylquinine. *Bull. Chem. Soc. Jpn.*, *44*, 3186–3187.

[55] Hermann, K., Wynberg, H. (1977). Polymer-bound Cinchona alkaloids as catalysts in the Michael reaction. *Helv. Chim. Acta.*, *60*, 2208–2212.

[56] Hodge, P., Khoshdel, E., Waterhouse, J. (1983). Michael reactions catalyzed by polymer-supported quaternary ammonium-salts derived from cinchona and ephedra alkaloids. *J. Chem. Soc. Perkin Trans. 1*, 2205–2209.

[57] (a) Zhengpu, Z., Yongmer, W., Zhen, W., Hodge, P. (1999). Asymmetric synthesis of α-amino acids using polymer-supported chiral phase transfer catalysts. *React. Funct. Polym.*, *41*, 37–43; (b) Chincilla, R., Mazón, P., Nájera, C. (2000). Asymmetric synthesis of α-amino acids using polymer-supported *Cinchona* alkaloid-derived ammonium salts as chiral phase-transfer catalysts. *Tetrahedron Asymmetry*, *11*, 3277–3281; (c) Chincilla, R., Mazón, P., Nájera, C. (2004). Polystyrene-anchored *Cinchona* ammonium salts: easily recoverable phase-transfer catalysts for the asymmetric synthesis of α-amino acids. *Adv. Synth. Catal.*, *346*, 1186–1194; (d) Shi, Q., Lee, Y.-J., Kim, M.-J., Park, M.-K., Lee, K., Song, H., Cheng, M., Jeong, B.-S., Park, H. G., Jew, S.-S. (2008). Highly efficient polymer supported phase-transfer catalysts containing hydrogen bond inducing functional groups. *Tetrahedron Lett.*, *49*, 1380–1383.

[58] Hodge, P., Khoshdel, E., Waterhouse, J., Fréchet, J. M. J. (1985). Michael additions catalysed by *Cinchona* alkaloids bound via their vinyl groups to preformed cross-linked polymers. *J. Chem. Soc. Perkin Trans. 1*, 2327–2331.

[59] For a recent treatment of "click" chemistry in biotechnology and materials science, see: Lahann, J., Ed. (2009). *Click Chemistry for Biotechnology and Materials Science*. Wiley, Chichester, U. K..

[60] Alvarez, R., Hourdin, M.-A., Cavé, C., Angelo, J., Chaminade, P. (1999). New polymer-supported catalysts derived from *Cinchona* alkaloids: their use in the asymmetric Michael reaction. *Tetrahedron Lett.*, *40*, 7091–7094.

[61] Besenius, P., Cormack, P. A. G., Liu, J., Otto, S., Sanders, J. K. M., Sherrington, D. C. (2008). Tailored polymer-supported templates in dynamic combinatorial libraries: simultaneous selection, amplification and isolation of synthetic receptors. *Chem. Eur. J.*, *14*, 9006–9019.

[62] For an introduction to free radical (co)polymerization, see ref. 6. See also: Moad, G., Solomon, D. H. (2006). *The Chemistry of Radical Polymerization*, 2nd ed. Elsevier, Oxford.

[63] For an introduction to controlled radical polymerization, see Reference [62]. See also: Matyjaszewski, K., Davis, T. P. (Eds.) (2002). *The Handbook of Radical Polymerization*. Wiley, Hoboken, NJ.

[64] For an introduction to suspension, emulsion and dispersion polymerization, see: Arshady, R. (1992). Suspension, emulsion, and dispersion polymerization: a methodological survey. *Colloid Polym. Sci.*, *270*, 717–732.

[65] For dispersion polymerization and particles made by this technique, see: (a) Barrett, K. E. J., Ed. (1975). *Dispersion Polymerization in Organic Media*. Wiley, London; (b) Kawaguchi, S., Ito, K. (2005). Dispersion polymerization. *Adv. Polym. Sci.*, *175*, 299–328; (c) Jayachandran, K. N. N., Chatterji, P. R. (2001). Preparation of linear and cross-linked polymer microspheres by dispersion polymerization. *J. Macromol. Sci. Polymer Rev.*, *C41*, 79–94.

[66] (a) Takaishi, N., Imai, H., Bertelo, C. A., Stille, J. K. (1976). Transition metal catalyzed asymmetric organic synthesis via polymer bound chiral ligands. Synthesis of R amino acids and hydratropic acid by hydrogenation. *J. Am. Chem. Soc.*, *98*, 5400–5402; (b) Takaishi, N., Imai, H., Bertelo, C. A., Stille, J. K. (1978). Transition metal catalyzed asymmetric organic syntheses via polymer-attached optically active phosphine ligands. Synthesis of R amino acids and hydratropic acid by

hydrogenation. *J. Am. Chem. Soc.*, *100*, 264–268; (c) Masuda, T., Stille, J. K. (1978). Transition metal catalyzed asymmetric organic syntheses via polymer-attached optically active phosphine ligands. Synthesis of R amino acids by hydrogenation with a polymer catalyst containing optically active alcohol sites. *J. Am. Chem. Soc.*, *100*, 268–272; (d) Baker, G. L., Fritschel, S. J., Stille, J. R., Stille, J. K. (1981). Transition-metal-catalyzed asymmetric organic synthesis via polymer-attached optically active phosphine ligands. 5. Preparation of amino acids in high optical yield via catalytic hydrogenation. *J. Org. Chem.*, *46*, 2954–2960; (e) Baker, G. L., Fritschel, S. J., Stille, J. K. (1981). Transition-metal-catalyzed asymmetric organic synthesis via polymerattached optically active phosphine ligands. 6. Asymmetric hydrogenation with polymer catalysts containing optically active pendent alcohols. *J. Org. Chem.*, *46*, 2960–2965; (f) Sybert, P. D., Bertelo, C., Bigelow, W. B., Varaprath, S., Stille, J. K. (1981). Synthesis of (R)-cyclohexylmethyl-p-styrylphosphine and the preparation of polymer-bound optically active phosphines. *Macromolecules*, *14*, 502–506; (g) Stille, J. K., Parrinello, G. (1983). Asymmetric hydroformylation of styrene by polymer-supported catalysts—platinum tin chloride on polymer-bound chiral phosphines. *J. Mol. Catal.*, *21*, 203–210; (h) Stille, J. K. (1984). Polymer-supported chiral catalysts in asymmetric synthesis. *J. Macromol. Sci., Chem.*, *A21*, 1689–1693; (i) Deschenaux, R., Stille, J. K. (1985). Transition-metal-catalyzed asymmetric organic-synthesis via polymer-attached optically-active phosphine-ligands .13. Asymmetric hydrogenation with polymer catalysts containing primary and chiral secondary pendant alcohols. *J. Org. Chem.*, *50*, 2299–2302; (j) Parrinello, G., Deschenaux, R., Stille, J. K. (1986). Platinum-catalyzed asymmetric hydroformylation with a polymer-attached optically-active phosphine ligand. *J. Org. Chem.*, *51*, 4189–4195; (k) Stille, J. K. (1989). Organic-synthesis via polymer-supported transition-metal catalysts. *React. Polym.*, *10*, 165–174.

[67] (a) Halle, R. t., Colasson, B., Schulz, E., Spagnol, M., Lemaire, M. (2000). "Diam-BINAP"; a highly efficient monomer for the synthesis of heterogeneous enantioselective catalysts. *Tetrahedron Lett.*, *41*, 643–646; (b) Halle, R., Schulz, E., Spagnol, M., Lemaire, M. (2000). Heterogeneous catalytic hydrogenation of olefinic substrates by *poly*-NAP. *Tetrahedron Lett.*, *41*, 3323–3326.

[68] Fan, Q.-H., Deng, G.-J., Chen, X.-M., Xie, W.-C., Jiang, D.-Z., Liu, D.-S., Chan, A. S. C. (2000). A highly effective water-soluble polymer-supported catalyst for the two-phase asymmetric hydrogenation: preparation and use of a PEG-bound BINAP ligand. *J. Mol. Catal. A Chem.*, *159*, 37–43.

[69] Yu, H.-B., Hu, Q.-S., Pu, L. (2000). Synthesis of a rigid and optically active poly(BINAP) and its application in asymmetric catalysis. *Tetrahedron Lett.*, *41*, 1681–1686.

[70] (a) Gamez, P., Dunjic, B., Pinel, C., Lemaire, M. (1995). "Molecular imprinting effect" in the synthesis of immobilized rhodium complex catalyst (IRC cat). *Tetrahedron Lett.*, *36*, 8779–8782; (b) Locatelli, F., Gamez, P., Lemaire, M. (1998). Molecular imprinting of polymerised catalytic complexes in asymmetric catalysis. *J. Mol. Catal. A Chem.*, *135*, 89–98; (c) Touchard, F., Fache, F., Lemaire, M. (2000). Polythioureas: main chain chiral polymers in hydride transfer hydrogenation. *Eur. J. Org. Chem.*, 3787–3792.

[71] Rolland, A., Hérault, D., Touchard, F., Saluzzo, C., Duval, R., Lemair, M. (2001). Enantiopure poly(glycidyl methacrylate-*co*-ethylene glycol dimethacrylate):

a new material for supported catalytic asymmetric hydrogen transfer reduction. *Tetrahedron Asymmetry*, *12*, 811–815.

[72] (a) De, B. B., Lohray, B. B., Dhal, P. K. (1993). Polymer supported transition metal complexes for catalytic epoxidation of olefins. *Tetrahedron Lett.*, *34*, 2371–2374; (b) De, B. B., Lohray, B. B., Sivaram, S., Dhal, P. K. (1994). Synthesis of catalytically active polymer-bound transition metal complexes for selective epoxidation of olefins. *Macromolecules*, *27*, 1291–1296; (c) De, B. B., Lohray, B. B., Sivaram, S., Dhal, P. K. (1995). Enantioselective epoxidation of olefins catalyzed by polymer-bound optically active Mn(III)-salen complex. *Tetrahedron Asymmetry*, *6*, 2105–2108; (d) Minutolo, F., Pini, D., Petri, A., Salvadori, P. (1996). Heterogeneous asymmetric epoxidation of unfunctionalized olefins catalyzed by polymer-bound (salen)manganese complexes. *Tetrahedron Asymmetry*, *7*, 2293–2302; (e) Minutolo, F., Pini, D., Salvadori, P. (1996). Polymer-bound chiral (salen)Mn(III) complex as heterogeneous catalyst in rapid and clean enantioselective epoxidation of unfunctionalised olefins. *Tetrahedron Lett.*, *37*, 3375–3378; (f) De, B. B., Lohray, B. B., Sivaram, S., Dhal, P. K. (1997). Polymeric catalysts for chemo- and enantioselective epoxidation of olefins: new cross-linked chiral transition metal complexing polymers. *J. Polym. Sci. A Polym. Chem.*, *35*, 1809–1818.

[73] (a) Zheng, X., Jones, C. W., Weck, M. (2006). Poly(styrene)-Supported Co–Salen Complexes as Efficient Recyclable Catalysts for the Hydrolytic Kinetic Resolution of Epichlorohydrin. *Chem. Eur. J.*, *12*, 576–583; (b) Holbach, M., Weck, M. (2006). Modular approach for the development of supported, monofunctionalized, salen catalysts. *J. Org. Chem.*, *71*, 1825–1836.

[74] Takeda, K., Oohara, T., Anada, M., Nambu, H., Hashimoto, S. (2010). A polymer-supported chiral dirhodium(II) complex: highly durable and recyclable catalyst for asymmetric intramolecular CH insertion reactions. *Angew. Chem. Int. Ed.*, *49*, 6979–6983.

[75] (a) Burguete, M. I., Fraile, J. M., Garcia, J. I., Garcia-Verdugo, E., Luis, S. V. (2000). Polymer-supported Bis(oxazoline)-copper complexes as catalysts in cyclopropanation reactions. *Org. Lett.*, *2*, 3905–3908; (b) Burguete, M. I., Fraile, J. M., Garcia-Verdugo, E., Luis, S. V., Martinez-Merino, V., Mayoral, J. A. (2005). Polymer-supported bis(oxazolines) and related systems: toward new heterogeneous enantioselective catalysts. *Ind. Eng. Chem. Res.*, *44*, 8580–8587.

[76] Cornejo, A., Fraile, J. M., Garcia, J. I., Garcia-Verdugo, E., Gil, M. J., Legarreta, G., Luis, S. V., Martinez-Merino, V., Mayoral, J. A. (2002). The first immobilization of pyridine-bis(oxazoline) chiral ligands. *Org. Lett.*, *4*, 3927–3930.

[77] (a) Orlandi, S., Mandoli, A., Pini, D., Salvadori, P. (2001). The aldol addition reaction: an old transformation at constant rebirth. *Angew. Chem. Int. Ed.*, *40*, 2519–2521; (b) Mandoli, A., Orlandi, S., Pini, D., Salvadori, P. (2003). A reusable, insoluble polymer-bound bis(oxazoline) (IPB-box) for highly enantioselective heterogeneous cyclopropanation reactions. *Chem. Commun.*, 2466–2467.

[78] (a) Yamashita, T., Yasueda, H., Miyauchi, Y., Nakamura, N. (1977). Asymmetric reactions .2. Asymmetric synthesis of methyl α-phenylpropionate by means of chiral polymers. *Bull. Chem. Soc. Jpn.*, *50*, 1532–1534; (b) Kobayashi, N., Iwai, K. (1978). Functional polymers. 1. Poly(cinchona alkaloid-co-acrylonitrile)s. New polymer catalysts for asymmetric synthesis. *J. Am. Chem. Soc.*, *100*, 7071–7072; (c) Kobayashi, N., Iwai, K. (1980). Functional polymers. 6. Unusual catalysis of polymeric cinchona alkaloids in asymmetric reaction. *Tetrahedron Lett.*, *21*, 2167–2170;

(d) Kobayashi, N., Iwai, K. (1980). Functional polymers. III. Asymmetric Michael reactions catalyzed by cinchona alkaloid–acrylonitrile copolymers. *J. Polym. Sci., Polym. Chem. Ed.*, *18*, 923–932; (e) Kobayashi, N., Iwai, K. (1980). Functional polymers. 4. Asymmetric addition of dodecanethiol to isopropenyl methyl ketone catalyzed by cinchona alkaloid-acrylonitrile copolymers. *Macromolecules*, *13*, 31–34; (f) Kobayashi, N., Iwai, K. (1981). Functional polymers. 7. On the C(3)-control of stereochemistry in asymmetric reactions catalyzed by polymeric cinchona alkaloids. *Polym. J.*, *13*, 263–271; (g) Kobayashi, N., Iwai, K. (1981). Functional polymers. 8. Asymmetric catalysis by quininium salts anchored to a polystyrene matrix. *Makromol. Chem. Rapid Commun.*, *2*, 105–108; (h) Kobayashi, N., Iwai, K. (1982). Functional polymers. 9. Asymmetric catalysis by new cinchona alkaloid derivatives—effect of C(3) substituent on asymmetric induction. *J. Polym. Sci. Polym. Chem. Ed.*, *20*, 85–90; (i) Kobayashi, N. (1984). Asymmetric reactions catalyzed by polymeric cinchona alkaloids. *Brit. Polym. J.*, *16*, 205–211; (j) Inagaki, M., Hiratake, J., Yamamoto, Y., Oda, J. (1987). Asymmetric induction in the base-catalyzed reactions using polymer-supported quinines with spacer groups. *Bull. Chem. Soc. Jpn.*, *60*, 4121–4126; (k) Sera, A., Takagi, K., Katayama, H., Yamada, H., Matsumoto, K. (1988). High-pressure asymmetric Michael additions of thiols, nitromethane, and methyl oxoindancarboxylate to enones. *J. Org. Chem.*, *53*, 1157–1161.

[79] Wai, J. S. M., Markó, I., Svendsen, J. S., Finn, M. G., Jacobsen, E. N., Sharpless, K. B. (1989). A mechanistic insight leads to a greatly improved osmium-catalyzed asymmetric dihydroxylation process. *J. Am. Chem. Soc.*, *111*, 1123–1125. Some other seminal publications are given in Reference [80].

[80] Kim, B. M., Sharpless, K. B. (1990). Heterogeneous catalytic asymmetric dihydroxylation: use of a polymer-bound alkaloid. *Tetrahedron Lett.*, *31*, 3003–3006.

[81] (a) Pini, D., Petri, A., Nardi, A., Rosini, C., Salvadori, P. (1991). Heterogeneous catalytic asymmetric dihydroxylation of olefins with the OsO_4/Poly(9-O-Acylquinine-co-acrylonitrile) system. *Tetrahedron Lett.*, *32*, 5175–5178; (b) Lohray, B. B., Thomas, A., Chittari, P., Ahuja, J. R., Dhal, P. K. (1992). Asymmetric catalytic dihydroxylation of alkenes on polymer support: scope and limitation. *Tetrahedron Lett.*, *33*, 5453–5456; (c) Pini, D., Petri, A., Salvadori, P. (1993). A new cross-linked polymer for the heterogeneous catalytic asymmetric dihydroxylation of alkenes. *Tetrahedron Asymmetry*, *4*, 2351–2354; (d) Lohray, B. B., Nandanan, E., Bhushan, V. (1994). Unprecedented reactivity and selectivity in heterogeneous asymmetric catalytic dihydroxylation of alkenes. *Tetrahedron Lett.*, *35*, 6559–6562; (e) Pini, D., Petri, A., Salvadori, P. (1994). Heterogeneous catalytic asymmetric dihydroxylation of olefins: a new polymeric support and a process improvement. *Tetrahedron*, *50*, 11321–11328; (f) Petri, A., Pini, D., Salvadori, P. (1995). Heterogeneous enantioselective dihydroxylation of aliphatic olefins: a comparison between different polymeric cinchona alkaloid derivatives. *Tetrahedron Lett.*, *36*, 1549–1552; (g) Song, C. E., Roh, E. J., Lee, S., Kim, I. O. (1995). Polymeric cinchona alkaloids for the heterogeneous catalytic asymmetric dihydroxylation of olefins: the influence of the polymer backbone polarity on the compatibility between polymer support and reaction medium. *Tetrahedron Asymmetry*, *6*, 2687–2694; (h) Petri, A., Pini, D., Rapaccini, S., Salvadori, P. (1995). Synthesis of optically active diols using an efficient polymer bound cinchona alkaloid derivative. *Chirality*, *7*, 580–585; (i) Song, C. E., Yang, J. W., Ha, H. J., Lee, S. (1996). Efficient and practical polymeric

catalysts for heterogeneous asymmetric dihydroxylation of olefins. *Tetrahedron Asymmetry*, *7*, 645–648; (j) Nandanan, E., Sudalai, A., Ravindranathan, T. (1997). New polymer supported Cinchona alkaloids for heterogeneous catalytic asymmetric dihydroxylation of olefins. *Tetrahedron Lett.*, *38*, 2577–2580; (k) Salvadori, P., Pini, D., Petri, A. (1997). In defense of the catalytic asymmetric cis dihydroxylation of olefins utilizing insoluble polymeric ligands. *J. Am. Chem. Soc.*, *119*, 6929–6930; (l) Park, Y. S., Jo, C. H., Choi, H. Y., Kwon, E. K., Song, C. E. (2004). Osmylated macroporous cinchona alkaloid resins: highly efficient and recyclable catalysts for asymmetric dihydroxylation of olefins. *Bull. Korean Chem. Soc.*, *25*, 1671–1675.

[82] Canali, L., Song, C. E., Sherrington, D. C. (1998). Polymer-supported bis-cinchona alkaloid ligands for asymmetric dihydroxylation of alkenes—a cautionary tale. *Tetrahedron Asymmetry*, *9*, 1029–1034.

[83] (a) Kristensen, T. E., Hansen, F. K., Hansen, T. (2009). The selective O-acylation of hydroxyproline as a convenient method for the large-scale preparation of novel proline polymers and amphiphiles. *Eur. J. Org. Chem.*, 387–395; (b) Kristensen, T. E., Vestli, K., Fredriksen, K. A., Hansen, F. K., Hansen, T. (2009). Synthesis of Acrylic Polymer beads for solid-supported proline-derived organocatalysts. *Org. Lett.*, *11*, 2968–2971; (c) Kristensen, T. E., Vestli, K., Jakobsen, M. G., Hansen, F. K., Hansen, T. (2010). A General approach for preparation of polymer-supported chiral organocatalysts via acrylic copolymerization. *J. Org. Chem.*, *75*, 1620–1629.

[84] Doyagüez, E. G., Parra, F., Corrales, G., Fernández-Mayoralas, A., Gallardo, A. (2009). New hydroxyproline based methacrylic polybetaines: synthesis, pH sensitivity and catalytic activity. *Polymer*, *50*, 4438–4446.

[85] Stasiak, M., Röben, C., Rosenberger, N., Schleth, F., Studer, A., Greiner, A., Wendorff, J. H. (2007). Design of polymer nanofiber systems for the immobilization of homogeneous catalysts—preparation and leaching studies. *Polymer*, *48*, 5208–5218.

[86] (a) Evans, A. C., Skey, J., Wright, M., Qu, W., Ondeck, C., Longbottom, D. A., O'Reilly, R. K. (2009). Functional and tuneable amino acid polymers prepared by RAFT polymerization. *J. Polym. Sci. A Polym. Chem.*, *47*, 6814–6826; (b) Evans, A. C., Lu, A., Ondeck, C., Longbottom, D. A., O'Reilly, R. K. (2010). Organocatalytic tuneable amino acid polymers prepared by controlled radical polymerization. *Macromolecules*, *43*, 6374–6380.

[87] Varela, M. C., Dixon, S. M., Lam, K. S., Schore, N. E. (2008). Asymmetric epoxidation, Michael addition, and triple cascade reaction using polymer-supported prolinol-based auxiliaries. *Tetrahedron*, *64*, 10087–10090.

[88] (a) Price, M. D., Sui, J. K., Kurth, M. J., Schore, N. E. (2002). Oxazaborolidines as functional monomers: ketone reduction using polymer-supported Corey, Bakshi, and Shibata catalysts. *J. Org. Chem.*, *67*, 8086–8089; (b) Varela, M. C., Dixon, S. M., Price, M. D., Merit, J. E., Berget, P. E., Shiraki, S., Kurth, M. J., Schore, N. E. (2007). Comparison of enantioselective reductions using bead and monolith "disk" polymer formulations of CBS catalysts. *Tetrahedron*, *63*, 3334–3339.

[89] Röben, C., Stasiak, M., Janza, B., Greiner, A., Wendorff, J. H., Studer, A. (2008). Immobilization of oligostyrene-prolinol conjugates into polystyrene via electrospinning and applications of these fibers in catalysis. *Synthesis*, 2163–2168.

[90] For more polymer-supported diarylprolinols, see: (a) Itsuno, S., Kamahori, K., Watanabe, K., Koizumi, T., Ito, K. (1994). Polymer-supported chiral Lewis acids as

asymmetric catalysts for diels-alder reactions of methacrolein with cyclopentadiene. *Tetrahedron Asymmetry*, *5*, 523–526; (b) Itsuno, S., Watanabe, K., Koizumi, T., Ito, K. (1995). Asymmetric diels-alder reaction of cyclopentadiene with methacrolein using polymer-supported chiral catalysts. *React. Polym.*, *24*, 219–227; (c) Felder, M., Giffels, G., Wandrey, C. (1997). A polymer-enlarged homogeneously soluble oxazaborolidine catalyst for the asymmetric reduction of ketones by borane. *Tetrahedron Asymmetry*, *8*, 1975–1977; (d) Degni, S., Wilén, C.-E., Leino, R. (2001). Immobilization of chiral ligands on polymer fibers by electron beam induced grafting and applications in enantioselective catalysis. *Org. Lett.*, *3*, 2551–2554; (e) Wöltinger, J., Bommarius, A. S., Drauz, K., Wandrey, C. (2001). The chemzyme membrane reactor in the fine chemicals industry. *Org. Proc. Res. Dev.*, *5*, 241–248; (f) Kell, R. J., Hodge, P., Snedden, P., Watson, D. (2003). Towards more chemically robust polymer-supported chiral catalysts: α,α-diphenyl-L-prolinol based catalysts for the reduction of prochiral ketones with borane. *Org. Biomol. Chem.*, *1*, 3238–3243.

[91] Rueping, M., Sugiono, E., Steck, A., Theissmann, T. (2010). Synthesis and application of polymer-supported chiral Brønsted acid organocatalysts. *Adv. Synth. Catal.*, *352*, 281–287.

CHAPTER 5

SELF-SUPPORTED CHIRAL CATALYSTS

HONGCHAO GUO AND KUILING DING

5.1.	Introduction	257
5.2.	Enantioselective transesterification	259
5.3.	Enantioselective hydrogenations	260
5.4.	Enantioselective epoxidation	270
5.5.	Enantioselective sulfoxidation	274
5.6.	Michael addition	277
5.7.	Carbonyl-ene reaction	278
5.8.	Addition of diethylzinc to aldehydes	278
5.9.	Ring-Opening reaction of epoxides	282
5.10.	Miscellaneous	285
5.11.	Summary and outlook	285
Acknowledgments		286
References		286

5.1. INTRODUCTION

Chiral catalysts have played an extremely important role in the catalytically asymmetric synthesis of enantiomerically enriched chiral compounds, which have attracted much attention from both the academic and industrial areas. Numerous efficient chiral catalysts have been developed in the past several decades and have demonstrated high enantioselectivity in a wide range of homogeneous asymmetric catalytic reactions [1–5]. However, their practical applications in industry are usually less successful, owing to the high costs of both the chiral ligands and the noble metals [6, 7]. In addition, trace amounts of the toxic metals that are leached from the catalysts can contaminate the

Catalytic Methods in Asymmetric Synthesis: Advanced Materials, Techniques, and Applications, First Edition. Edited by Michelangelo Gruttadauria and Francesco Giacalone.

product and are difficult to remove, which also seriously hinders the widespread applications of the chiral homogeneous organometallic catalysts in the practical synthesis of value-added fine chemicals, particularly in pharmaceutical production. Therefore, the development of procedures for easy recovery and efficient recycling of chiral catalysts would be highly desirable, but it is usually difficult to achieve these in a homogeneous catalytic process. In order to circumvent the difficulties presented by homogeneous chiral catalysts, various approaches to immobilize homogeneous asymmetric catalysts for heterogeneous enantioselective reactions have been developed, most frequently by anchoring the chiral catalyst to a support that can be either an inorganic solid or an organic polymer [8–13]. However, classical immobilization with various prefabricated supports often suffered from negative effects such as reduced catalytic activity and/or selectivity as a result of the poor accessibility, random anchoring, or disturbed geometry of the active sites in the solid matrix.

Following the rapid development of numerous metal-organic frameworks (MOFs) with microporous structures over the past 20 years and the exploration of their applications, including catalysis [14–21], homochiral metal-organic polymers (or oligomers) have emerged as an effective tool for the immobilization of chiral catalysts within the past decade. The basic strategy for the immobilization is to use multitopic chiral ligands and metal centers to form functional metal-organic assemblies via coordination bonds, which generally have an infinitely extended structure and can be insoluble in certain solvents, allowing them to be used as heterogeneous catalysts (Scheme 5.1). Two types of metal-organic polymers have been developed for this purpose. In the most straightforward way, an enantiopure ditopic (or polytopic) chiral ligand is copolymerized with a catalytically active metal ion to give a homochiral coordination polymer that may show activities in enantioselective catalysis. Alternatively, the chiral bridging ligands are designed to bear two types of orthogonal functional groups, among which one serves as the primary functional group to interact with the network-forming metal for the formation of the extended structure, and the other is responsible for attaching a metal for catalysis. In this case, there are two distinct types of metals: one responsible for polymer preparation and the other for catalysis. With regard to the linker-bridged chiral ligand, the linker moiety may be composed by either covalent bonds or noncovalent interactions such as hydrogen bonding or ligand-to-metal coordination. Using the above strategy, a homogeneous molecular catalyst could be incorporated

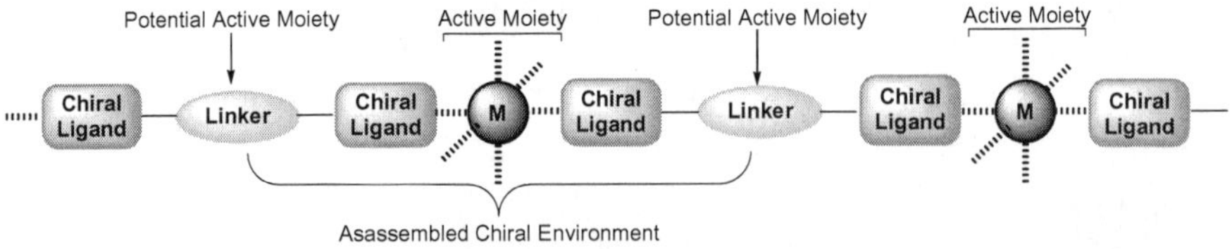

SCHEME 5.1. Schematic representation of the self-supported chiral catalysts.

into a polymeric solid in which the metal centers are integrated at the metal nodes or located on the backbones. These, in turn act as the active sites to catalyze the asymmetric reactions in the chiral environment of the assembly. Depending on the coordination geometry of the metal, as well as the bonding preference of the bridging ligand, polymeric structures with a one-dimensional (1D) chain, a two-dimensional (2D) layer, or a three-dimensional (3D) network can be generated with uniform active sites. By judicious tailoring, the linkers and the chiral units of the polytopic ligand, the chiral environment, and other useful properties such as porosity can be fine-tuned for the specific application of these homochiral metal-organic assemblies in asymmetric catalysis. Since these homochiral metal-organic assemblies were utilized as immobilized homogeneous catalysts without using any external supports, they are called self-supported chiral catalysts (Scheme 5.1).

Since the pioneering contribution in 2000 of Kim and co-workers [22], considerable progress has been achieved in the development of self-supported chiral catalysts for heterogeneous asymmetric catalysis [23]. Herein, the self-supported chiral catalysts and their application in asymmetric catalysis in the past decade are discussed as classified by the reaction types.

5.2. ENANTIOSELECTIVE TRANSESTERIFICATION

Although the earliest attempts to use coordination polymers for heterogeneous catalysis can be dated back to the 1980s [24], it was not until 2000 when Kim and co-workers reported the first examples of homochiral metal-organic polymers catalyzed enantioselective reactions [22]. An enantiopure tartaric acid-derived ligand with both carboxylic and pyridyl functional groups as the coordination sites was designed and synthesized. This multitopic chiral ligand was treated with $Zn(NO_3)_2$ in H_2O/MeOH to afford a crystalline homochiral coordination polymer with the formula $\{[Zn_3O(L)_6]\cdot 2H_3O\cdot 12H_2O\}_n$ (**1**) (L = 4-aminopyridine amide derivative of tartaric acid) (Scheme 5.2).

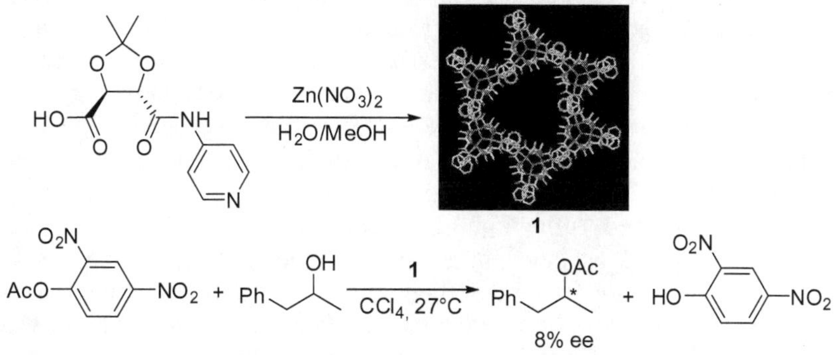

SCHEME 5.2. Synthesis of homochiral MOF **1** and its catalytic application in enantioselective transesterification of racemic 1-phenyl-2-propanol.

Based on the crystallographic analysis of the structure, the coordination polymer was found to have regular chiral 1D channels in which some dangling pyridyl groups point toward the interiors of the walls. With this homochiral polymer as a heterogeneous catalyst, the kinetic resolution of racemic 1-phenyl-2-propanol via transesterification was performed in CCl_4 at 27°C to provide the corresponding product in 8% enantiomeric excess (ee) (Scheme 5.2). Although the enantioselectivity is very modest, the feasibility of using a homochiral metal-organic polymer for enantioselective heterogeneous catalysis had been demonstrated for the first time. Kim's work represents an important breakthrough in heterogeneous asymmetric catalysis, which triggered further attention to the development of homochiral metal-organic hybrids for heterogeneous asymmetric catalysis.

5.3. ENANTIOSELECTIVE HYDROGENATIONS

Asymmetric hydrogenation of prochiral olefins, ketones, and imines has been established as one of the most reliable and powerful methods for the production of optically active compounds. Among all kinds of chiral catalysts, Ru and Rh catalysts with bidentate phosphines or monodentate phosphines as the chiral ligand are perhaps the most thoroughly studied systems and have been widely used in the homogeneous hydrogenation of a variety of substrates with excellent activity and enantioselectivity. It is thus of great significance to explore their heterogenization for practical applications.

Lin and co-workers studied the application of chiral porous zirconium phosphonates bearing Ru-BINAP [25] [BINAP = 2,2′-bis(diphenylphosphino)-1,1′-binaphthyl] moieties as heterogeneous catalysts in the highly enantioselective hydrogenation of β-keto esters [26]. The metal phosphonate hydrid material was prepared by using a molecular building-block approach, and it contains pendant chiral chelating bisphosphanes on a robust framework of metal phosphonates to generate anchored metal phosphane complexes, which were used in the heterogeneous asymmetric hydrogenation. The chiral bisphosphane ligands **2** containing phosphoric acid groups at the 4- or 6-positions of the binaphthyl backbone were synthesized with nickel-catalyzed phosphonation as the key step. Treatment of these ligands with $\{[Ru(benzene)Cl_2]_2\}$ in dimethylformamide (DMF) at 100°C led to the formation of [Ru(**2**)$(DMF)_2Cl_2$] intermediates, which were further treated with $Zr(O^tBu)_4$ in refluxing methanol to afford the chiral zirconium phosphonates **4a-b**, bearing pendant chiral bisphosphane-chelated ruthenium sites, and with a formula of {Zr[Ru(**2**)$(DMF)_2Cl_2$]}·2MeOH as amorphous solids (Scheme 5.3). In the resulting porous solids, zirconium(II) and ruthenium(II) have different functions. The former is responsible for phosphate backbone formation, and the latter works as the active metal site for the catalysis. Unfortunately, the detailed structures of the polymers were not clear because of their amorphous nature. As a result of the high porosity in the structures of **4a,b**, both polymers have been

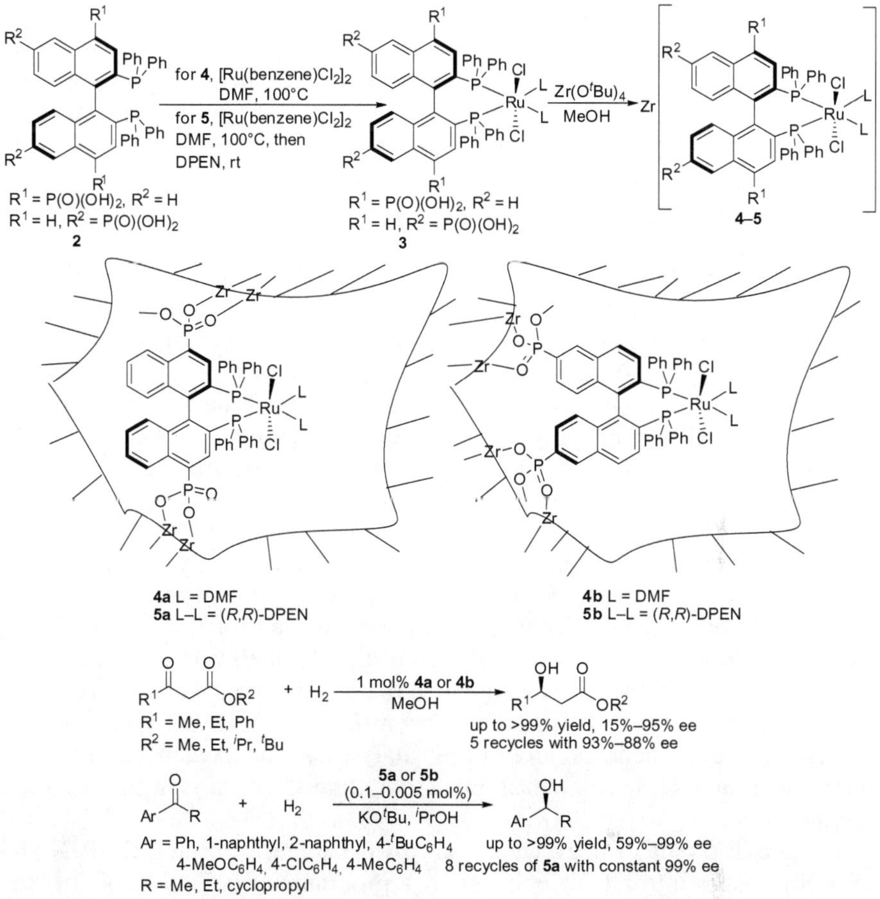

SCHEME 5.3. The preparation of Ru-bearing chiral porous zirconium phosphonates **4a,b** and **5a,b**, and their applications in enantioselective hydrogenation of β-keto esters and aromatic ketones.

demonstrated to be highly active catalysts in the asymmetric hydrogenation of β-keto esters (Scheme 5.3). The corresponding reduction products were obtained in 91.7%–95% ee by using **4b** as a heterogeneous catalyst, which was reused for five cycles in the hydrogenation of methyl acetoacetate without significant loss of enantioselectivity (93%–88% ee). In contrast, **4a** provided the same products in only modest ee values (15%–78%). The heterogeneous nature of the catalysis was evidenced with less than 0.01% ruthenium leaching in each run by direct current plasma (DCP) analysis and the inactivity of the supernatants for the catalysis.

In order to find excellent heterogeneous catalysts for the hydrogenation of simple aromatic ketones, Lin and co-workers further extended the molecular building-block approach to the preparation of two similar chiral porous Zr phosphonates (**5a**,**b**), in which Ru-BINAP-DPEN [27, 28] (DPEN = 1,2-diphenylethylenediamine) moieties were incorporated into the resulting hybrid solids by treatment of [Ru(**2**)$(DMF)_2Cl_2$] intermediates for **4** with DPEN at room temperature [29]. By using these hybrid catalysts with the built-in Ru-BINAP-DPEN entities as active sites, excellent activity and high enantioselectivity were obtained in the asymmetric hydrogenation of various aromatic ketones (Scheme 5.3). Under a 0.1 mol% of **5a**, the hydrogenation reactions were performed in isopropanol at room temperature in the presence of potassium *tert*-butoxide to afford the corresponding hydrogenation product in 90.6%–99.2% ee with full conversion. The results are comparable or even superior to those obtained with the homogeneous counterpart of the parent ruthenium complex. Particularly, even with 0.005 mol% of catalyst **5a**, the hydrogenation of 1-acetonaphthone can still be accomplished in 40 hours with 98.6% ee and full conversion [turnover frequency (TOF) ≈ $500\,h^{-1}$]. As a comparison, catalyst **5b** is also highly active for the hydrogenation of aromatic ketones; however, in most cases, **5b** can only give moderate enantioselectivity (59%–95% ee), similar to that of its parent Ru-BINAP-DPEN homogeneous catalyst. As a heterogeneous catalyst in the asymmetric hydrogenation of 1-acetonaphthone, catalyst **5a** could be easily recovered from the reaction mixture without rigorous exclusion of air, and reused for up to eight times without any loss of enantioselectivity (all 99% ee), albeit with some decrease (from 100% to 85%) in product yields in the last two runs. For each run, less than 0.2% leaching of Ru metal has been detected by DCP spectroscopic analysis. The intrinsic instability of the catalyst and the oxygen sensitivity of the ruthenium hydride complexes may account for the decline in activity after multiple runs.

Using a different strategy, Ding and co-workers incorporated Ru-BINAP-DPEN moieties into a liner polymer by a programmed self-assembly of two different multitopic ligands with Ru metallic ions [30]. In this polymer, the ruthenium(II) ions not only serve as the metallic nodes in polymer backbone formation, but also function as the active sites for the catalysis. The chiral bis-BINAPs were treated sequentially with $\{[(C_6H_6)RuCl_2]_2\}$ and the chiral bridging bis-DPEN, to afford the heterogenized Noyori-type catalysts **6a** and **6b**, respectively, by a spontaneous heterocoordination. **6b** was found to be an efficient heterogeneous catalyst in the asymmetric hydrogenation of a variety of aromatic ketones. The reactions were carried out in isopropanol at room temperature with potassium *tert*-butoxide as the base, providing the corresponding products in 100% conversion with 94%–98% ee (Scheme 5.4), which are comparable to those obtained by using the homogeneous counterpart of **6**. In contrast, **6a** only provided an ee value of 78% in the hydrogenation of acetophenone under the otherwise identical conditions. Upon completion of the hydrogenation, the catalyst could be separated from the reaction mixture by simple filtration under an Ar atmosphere (Ru leaching in product <0.1 ppm). The heterogeneous nature of the present catalytic system was evidenced by complete inactivity of the supernatant of catalyst **6b** for the hydrogenation. Even using 0.01 mol% of **6b** in the hydrogenation of acetophenone, the hydro-

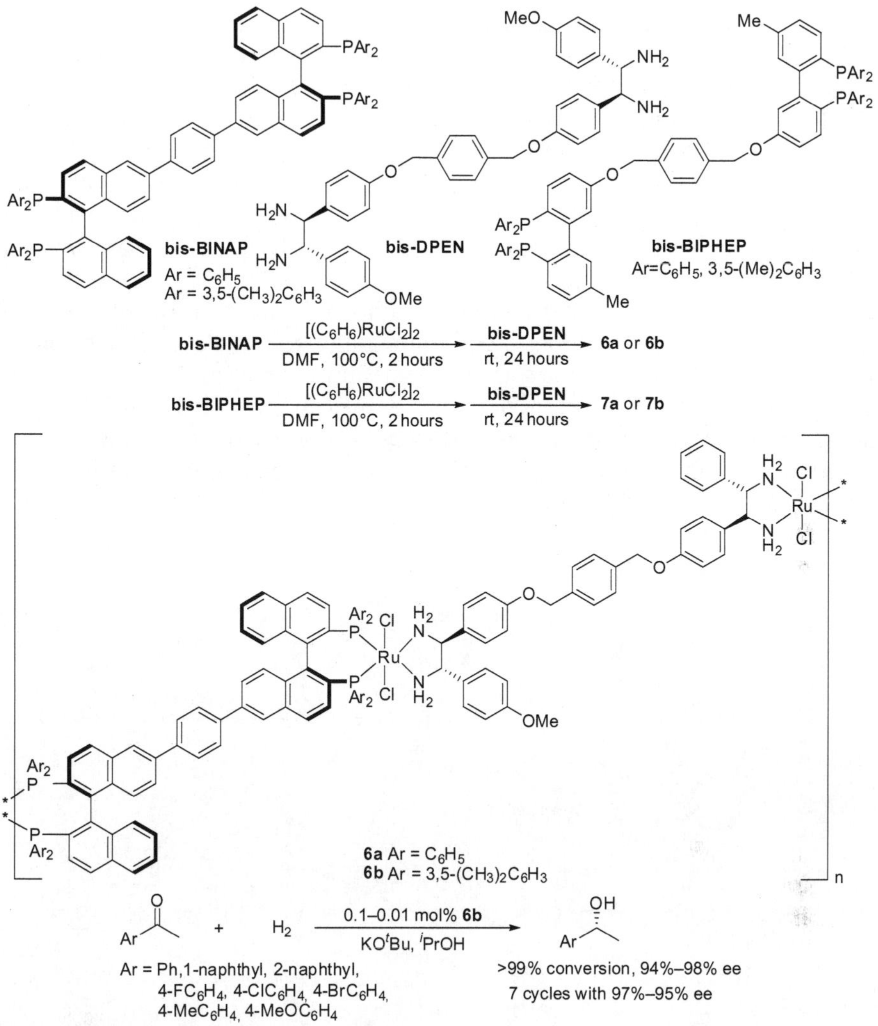

SCHEME 5.4. Self-supported heterogeneous catalysts by programmed assembly of Noyori's Ru-catalysts for enantioselective hydrogenations of ketones.

genation product could still be obtained without significant loss in either yield or enantioselectivity (95% ee). This catalyst was recycled for seven runs with nearly unchanged enantioselectivity and yield. In a subsequent work, Ding and co-workers prepared heterogenized Noyori-type catalysts **7a** and **7b** by using achiral bis-2,2′-bis(diphenylphosphino)-1,1′-biphenyl (BIPHEP) in combination with the chiral bis-DPEN and Ru salt according to a similar procedure (Scheme 5.4) [31]. Using the resulting chiral solid as a heterogeneous catalyst, up to 87% ee was obtained in the asymmetric hydrogenation of aromatic ketones. The catalyst was reused for four times in the hydrogenation of acetophenone with some decline of selectivity and activity.

Since Feringa's MonoPhos/Rh system [32, 33] is well known as an excellent chiral catalyst for asymmetric hydrogenation, its heterogenization has been studied by Wang and Ding using the self-supporting strategy [34]. For the Rh(I)-catalyzed alkene hydrogenation with monophosphoramidite ligands, mechanistic studies have revealed that the active species contains two monodentate phosphorous ligand moieties binding to one Rh center. Thus, the incorporation of active P-Rh-P sites into the coordination polymer could be achieved by assembly of an appropriate polytopic phosphorus ligand with an Rh(I) precursor. Indeed, addition of a solution of $[Rh(cod)]BF_4$ in dichloromethane into the solution of the ditopic bis-phosphoramidites (Bis-MonoPhos) in toluene led to the isolation of Rh-containing polymer **8a–c** as amorphous orange solids. The solids **8a–c** were insoluble in toluene and thus used as heterogeneous catalysts in the asymmetric hydrogenation of *β*-aryl- or alkyl-substituted dehydro-*α*-amino acid and enamide derivatives, affording the corresponding amino acid derivatives and secondary amine derivative in >99% conversion and 94%–97% ee (Scheme 5.5). Particularly, all the three catalysts

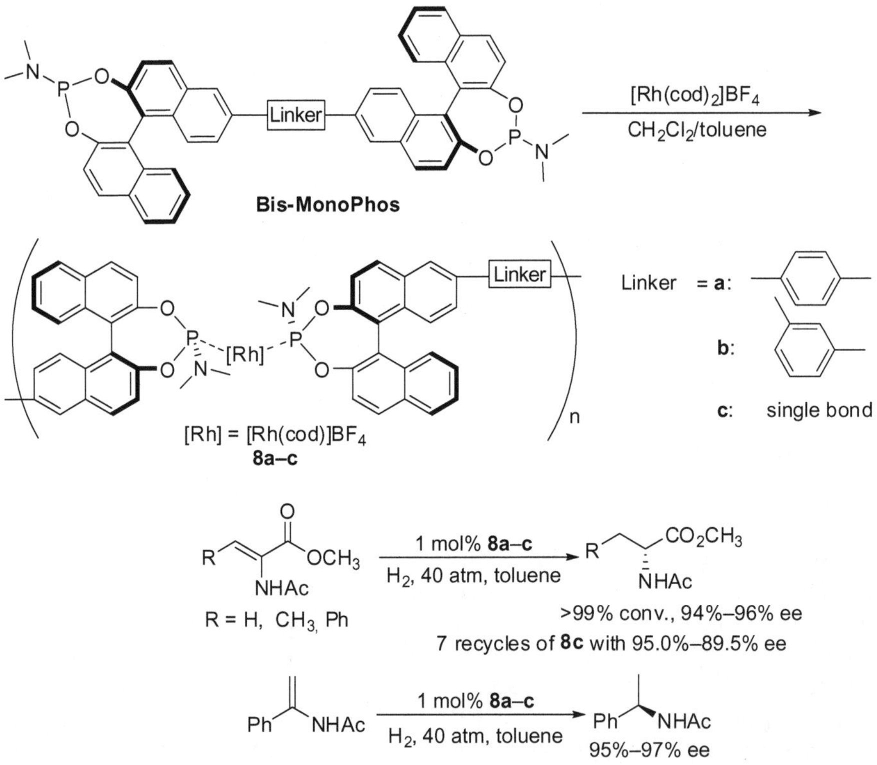

SCHEME 5.5. Heterogeneous catalysis of enantioselective hydrogenations with self-supported MonoPhos/Rh catalysts.

gave better enantioselectivity than that of their homogeneous counterpart (95%–97% vs. 89% ee) in the asymmetric hydrogenation of the enamide. Upon the completion of the hydrogenation, the catalyst was isolated from the reaction mixture by filtration, and the supernatant of the catalyst was shown as inactive in the hydrogenation. In addition, Rh leaching was negligible as detected by the inductively coupled plasma (ICP) spectroscopy, further confirming the heterogeneous nature of the catalysis. In the hydrogenation of methyl 2-acetamidoacrylate, catalyst **8c** was reused for seven runs with full conversion and a slight decrease in enantioselectivity (95%–89.5% ee).

Although covalent bonding was commonly employed for the synthesis of the multitopic chiral ligand, the process is arduous and time consuming. Following the above work, Ding and co-workers further demonstrated that noncovalent interactions such as hydrogen bonding could be used to make ditopic chiral ligands for the construction of chiral supramolecular metal-organic polymeric catalysts [35]. Here, the key is the design and synthesis of a chiral ligand bearing complementary or self-complementary hydrogen bonding recognition motifs. Ureido-4[1H]-ureidopyrimidone, which is among the most commonly used self-complementary hydrogen bonding units, was utilized as a hydrogen bonding unit and tethered on a MonoPhos ligand motif. The ligand was treated with $[Rh(cod)_2]BF_4$ to generate the supramolecular metal-organic polymer **9** as a yellow solid, which is insoluble in nonpolar organic solvents such as toluene (Scheme 5.6). ^1H- and solid-state ^{31}P

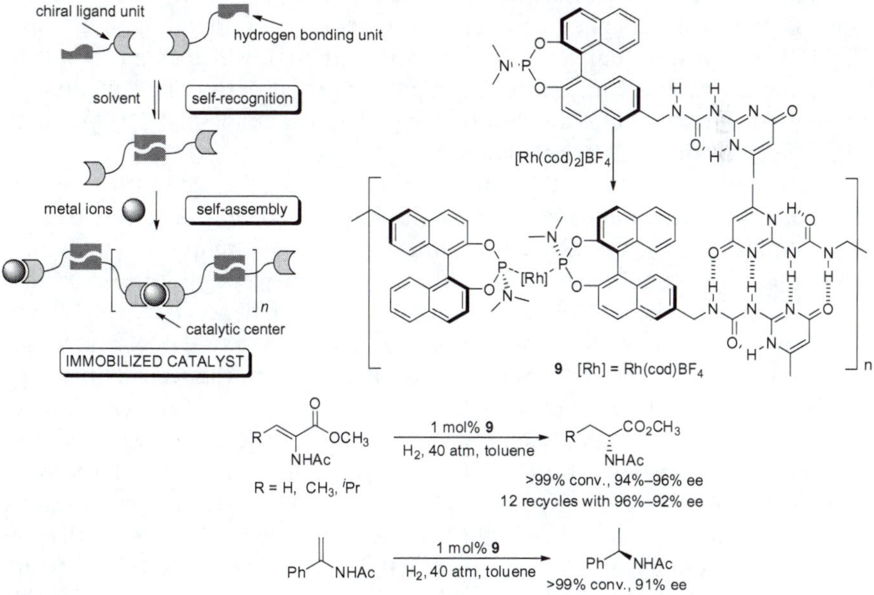

SCHEME 5.6. Homochiral supramolecular metal-organic assembly for the heterogeneous catalysis of asymmetric hydrogenations.

cross-polarization magic-angle spinning nuclear magnetic resonance (CP-MAS NMR) analyses suggested that hydrogen bonding and metal-to-ligand coordination noncovalent interactions are orthogonal to each other, which guaranteed the formation of supramolecular polymers with proper function. With **9** as the chiral catalyst, the heterogeneous hydrogenation of dehydro-α-amino acid derivatives and *N*-(1-phenylvinyl)acetamide was performed in toluene at 25°C under 40 atm of H_2 to provide the products in 91%–96% ee with >99% conversion, which are comparable to the catalytic performance of its homogeneous counterpart. After the catalyst was removed by simple filtration, the supernatant of **9** showed inactivity in the hydrogenation of the substrate under otherwise identical conditions. Moreover, no metal leaching was detected in either the filtrated organic solution or the isolated product by using ICP spectroscopy analyses within the detection limit of the instrument (1 ppm), further confirming the heterogeneous nature of the catalytic system. The filtration-recovered catalyst **9** was reused in the hydrogenation of (*Z*)-methyl 2-acetamidobut-2-enoate for 10 times to give the corresponding product with uniformly full conversion and in nearly constant 96%–92% ee. However, the catalyst reactivity deteriorated with consecutive hydrogenations, as indicated by the lower TOF for each run. It is likely that the absence of a hydrogen atmosphere during product/catalyst separation leads to partial catalyst decomposition.

Besides H-bonding interaction, the metal–ligand coordination is also noncovalent in nature and has also been elegantly used by Ding and co-workers in the construction of an immobilized Feringa's MonoPhos/Rh catalyst containing two types of metal centers [36]. In this case, a bifunctional heteroditopic ligand was designed to bear two orthogonal metal-ligating units (2,2′:6′,2″-terpyridine unit and Feringa's MonoPhos), which might result in bimetallic assemblies with interesting catalytic properties upon sequential or one-pot reaction(s) with two different metals ions, either or both of which might be catalytically active (Scheme 5.7).

The bifunctional heteroditopic ligand was readily synthesized with a copper-free Sonogashira coupling reaction as the key step. This ligand was treated sequentially with Fe^{II} and Rh^{I} salts to furnish the programmed assembly of a class of chiral bimetallic self-supported catalysts **10a–g**, which were examined in the asymmetric hydrogenation of methyl α-acetamidoacrylate. The hydrogenation reactions were run in toluene at a hydrogen pressure of 40 atm under a catalyst loading of 1 mol% (Scheme 5.8). It was found that the counterions (Cl^-, SO_4^{2-}, PF_6^-, ClO_4^-, BF_4^-, or $SO_3CF_3^-$) of the catalysts have significant impact on the activity and enantioselectivity. For example, **10c–g** having weakly coordinating counterions such as PF_6^-, ClO_4^-, BF_4^-, or $SO_3CF_3^-$ afforded full conversion and 94%–97% ee, comparable to that of their homogeneous counterpart MonoPhos/Rh (97% ee) under otherwise identical conditions. Nevertheless, **10a** with the counterion Cl^- led to no activity at all, whereas **10b** with the counterion SO_4^{2-} gave full conversion but with a slightly lowered ee (88%). It was believed that during the assembling of the catalyst **10**, the counter anions are located at the outer coordination sphere of the Fe(II)/tpy-bridged ligand,

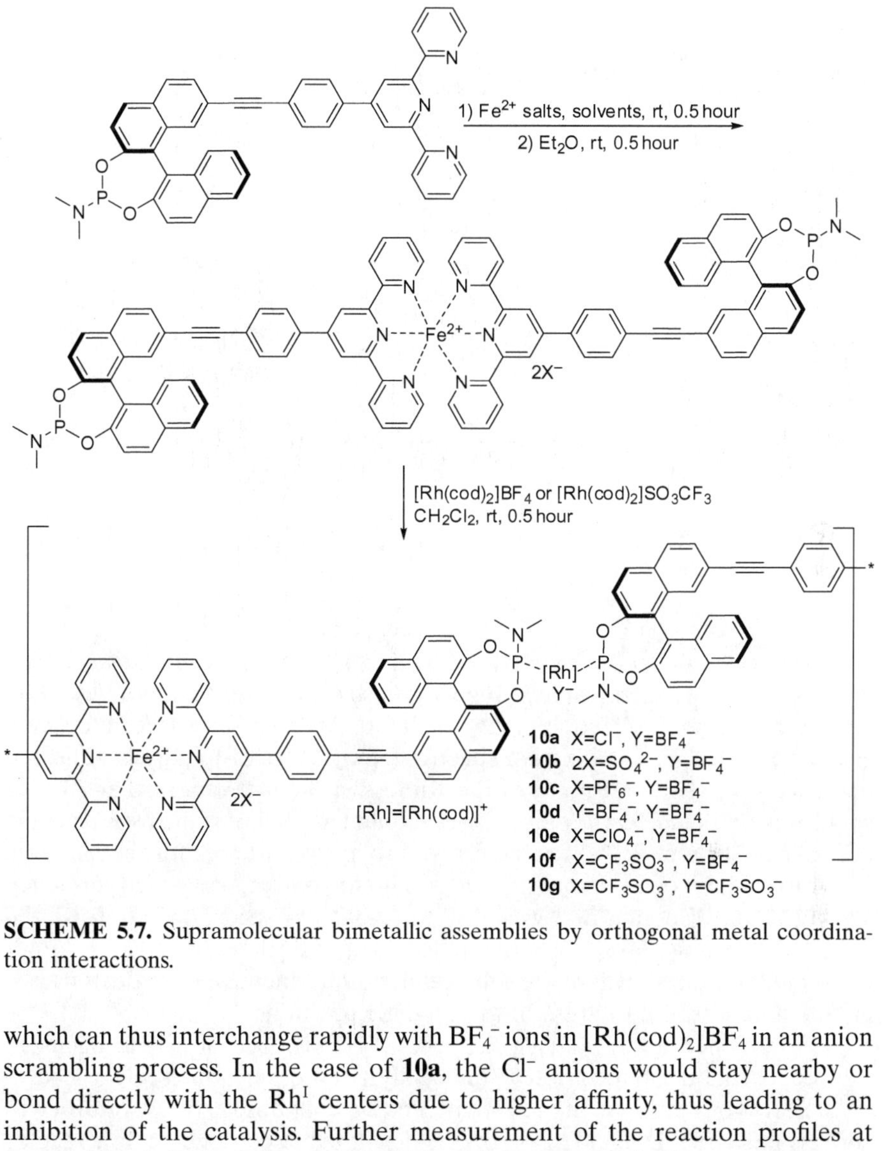

SCHEME 5.7. Supramolecular bimetallic assemblies by orthogonal metal coordination interactions.

which can thus interchange rapidly with BF_4^- ions in $[Rh(cod)_2]BF_4$ in an anion scrambling process. In the case of **10a**, the Cl^- anions would stay nearby or bond directly with the Rh^I centers due to higher affinity, thus leading to an inhibition of the catalysis. Further measurement of the reaction profiles at 1 atm of H_2 indicated that the identities of the anions also have an influence on the catalytic activity. With **10g** as the heterogeneous catalyst, asymmetric hydrogenation of α-dehydroamino acid, enamide, and itaconic acid derivatives were performed in toluene at room temperature under 40 atm of H_2 to afford the corresponding products with full conversion in 90%–97% ee, which were comparable or even superior to those obtained with their homogeneous counterpart $(MonoPhos)_2/RhI$. Upon completion of the reaction, the catalyst could be easily separated from the reaction mixture by filtration. In the asymmetric hydrogenation of methyl α-acetamidoacrylate, the catalyst **10e** and **10g** could

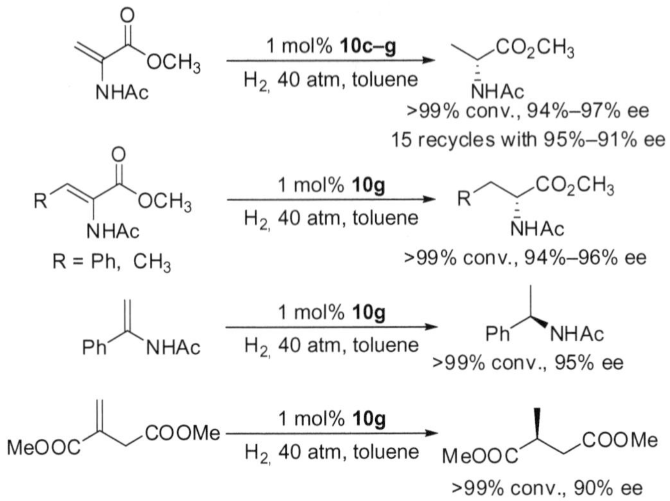

SCHEME 5.8. Heterogeneous asymmetric hydrogenations catalyzed by homochiral bimetallic assemblies **10**.

be reused more than 10 times without significant loss in the enantioselectivity or activity. In particular, catalyst **10g** was reused for 15 runs with only slightly deteriorated enantioselectivity (95%–91% ee). In the case of **10e**, inductively coupled plasma atomic emission spectroscopy (ICP-AES) analysis indicated that except for 1.7% (18 ppm) of the Rh loss in the first run, further Rh loss could not be detected (<1 ppm) in subsequent recycling runs. Iron leaching ranged from 9 ppm (0.8%) for the first run to less than 1 ppm for the 12th run. Low-molecular-weight Rh species trapped in the solid matrix, which contained weakly bound rhodium and was easy to leach, was speculated to cause the initial Rh leaching. Filtration tests indicated that further conversions do not occur at all after removal of the solid catalyst, thus unequivocally demonstrating the heterogeneous nature of the catalytic reactions.

As mentioned above, during the recovery and reuse of a heterogenized catalyst, partial decomposition of the catalyst can occur, which might lead to a deterioration in the catalytic activity and/or enantioselectivity. In order to address this issue, Ding and co-workers developed a continuous-flow reactor packed with a self-supported chiral catalyst for the heterogeneous asymmetric hydrogenation of α-dehydroamino acid and enamide derivatives [37]. Several multi-1,1′-bi-2-naphthol (BINOL) ligands **11a–i** with various linker geometries including linear, bent, trigonal-planar, and tetrahedral linkers were synthesized, and upon treatment with $[Rh(cod)_2]BF$ in dichloromethane, afforded the self-supported catalysts **12** as amorphous solids that were completely insoluble in toluene (Scheme 5.9). Catalysts **12a–i** were tested as the heterogeneous catalysts in the asymmetric hydrogenation of a wide range of α-dehydroamino acids and 2-aryl enamides to deliver the corresponding products

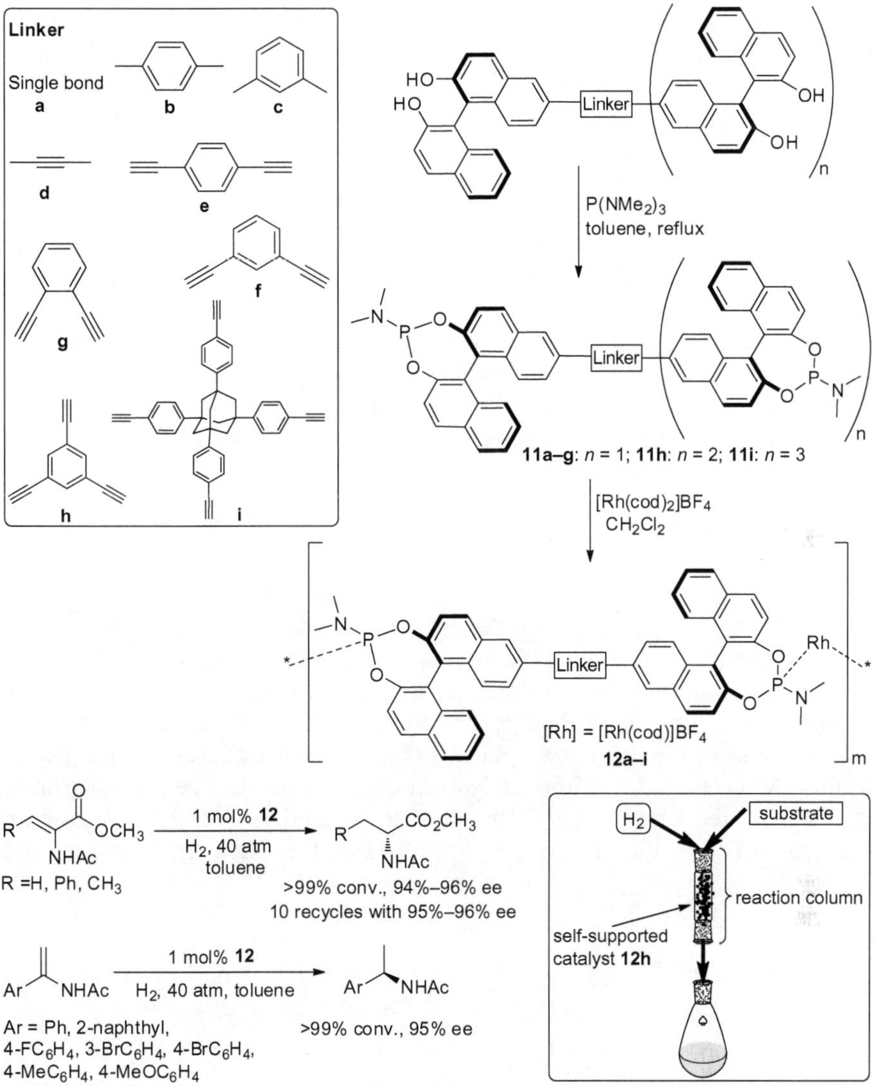

SCHEME 5.9. Continuous-flow system containing metal-organic polymer **12** for the heterogeneous catalysis of asymmetric hydrogenations.

in full conversion with excellent enantioselectivities (93%–98% and 90%–98% ee, respectively). The linker moiety of the multitopic ligand in **12** influenced the catalytic activity significantly, even though all of them demonstrated excellent enantioselectivity in the catalysis. Based on the reaction profiles measured under steady-state conditions, **12h** and **12i** demonstrated the highest catalytic activity with TOF reaching up to 95 and 97 h^{-1} under 2 atm H_2, respectively. In the asymmetric hydrogenation of methyl α-acetamidobut-2-enoate,

12h was recovered from the reaction mixture by simple cannula filtration and recycled 10 times without any loss of enantioselectivity (95%–96% ee), albeit with a decline in TOF of approximately 20% per cycle (Scheme 5.9). Furthermore, a continuous-flow system was set up to solve the problem of catalyst partial deactivation during the batch operations in catalyst recycling. The system was equipped with a reaction column as the key unit, which was packed with an activated carbon/**12h** mixture as the stationary-phase catalyst. The substrate solution and hydrogen gas were fed at a controlled rate into one end of the column, while the hydrogenation product was collected at the other end. The continuous-flow system was demonstrated to be successful in the asymmetric hydrogenation of methyl *α*-acetamidobut-2-enoate, and it ran continuously for a total of 144 hours to afford the hydrogenation product in >99% conversion and constant 96%–97% ee values. The total Rh leaching in the product solution was 1.7% of that in the original catalyst **12h**.

5.4. ENANTIOSELECTIVE EPOXIDATION

In 2005, Ding and co-workers reported the heterogenization of Shibasaki's BINOL/La [38, 39] catalyst for the enantioselective catalysis of epoxidation of *α*,*β*-unsaturated ketones [40]. Various multi-BINOL ligands with diversified linker geometries were treated with a THF solution of $La(O^iPr)_3$ in the presence of triphenylphosphine oxide, to give the heterogenized poly-BINOL/La assemblies (**13a–i**) as amorphous precipitates, which were subsequently used as heterogeneous chiral catalysts in the enantioselective epoxidation of *α*,*β*-unsaturated ketones with cumene hydroperoxide (CMHP) as an oxidant (Scheme 5.10). In the presence of 5 mol% **13b**, the oxidation reactions of a

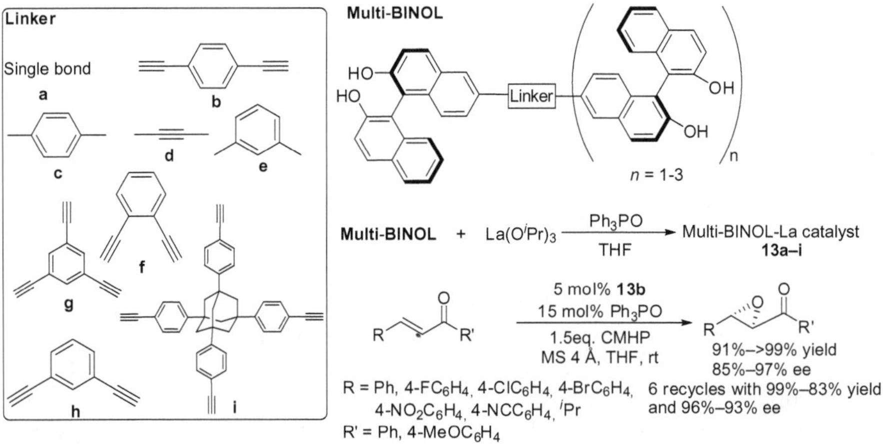

SCHEME 5.10. Self-supported Shibasaki's BINOL/La catalysts in heterogeneous asymmetric epoxidation of *α*,*β*-unsaturated ketones.

variety of α,β-unsaturated ketones proceeded smoothly in THF at room temperature, to provide the corresponding product in 91%–>99% yield with 85%–97% ee, which are comparable to those obtained using the homogeneous counterpart. The linker moieties of the bridging ligands exhibited a significant impact on the enantioselectivity and activity of the reactions. In the reaction of chalcone (R = R′ = Ph), the enantioselectivity varied between 82% and 95% ee, depending on the length or spatial orientation of the linker in the poly-BINOLs, indicating that the supramolecular structures of the assemblies have a significant influence on the catalytic processes. While the catalyst **13b** containing 1,4-diethynylbenzene linker afforded 97.6% ee, the catalyst **13d** bearing the shorter ethynyl linker only resulted in 82.9% ee. The heterogeneous nature of the catalysis was testified by the observations that the supernatant of **13b** in THF displayed no catalytic activity in the epoxidation of the chalcone (R = R′ = Ph) under the same experimental conditions, and only a negligible amount (0.093 ppm) of La(III) was detected in the above-mentioned supernatant by ICP spectroscopic analysis. By simple filtration, catalyst **13b** can be readily recovered and reused for six cycles without significant loss of enantioselectivity (from 96.5% to 93.2% ee). Moreover, the leaching of lanthanum in each recycling run was determined to be less than 0.4 ppm.

Ding and co-workers have also reported the use of chiral self-supported BINOL–Zn catalysts for the heterogeneous asymmetric epoxidation of enones, with CMHP as the oxidant [41]. The multi-BINOL ligands **14a–g** with linear or bent linkers were treated with two molar equivalents of diethylzinc in hexane/Et_2O, leading to the formation of self-supported BINOL–Zn catalysts **15** as white amorphous solids that are virtually insoluble in all organic solvents tested. With **15a** as a heterogeneous chiral catalyst, the enantioselective epoxidations of α,β-unsaturated ketones with CMHP as oxidant were performed in Et_2O to provide the corresponding products in 61%–99% yield and 73%–91% ee (Scheme 5.11). Catalyst screening revealed a significant influence of the bridging ligand on the enantioselectivity and activity of the reactions. In general, increasing the length of the spacer to some extent seems to be beneficial for the catalytic reactivity, while introducing substituents bulkier than H at the 3,3′-positions of the BINOL motifs obviously deteriorated the activity and enantioselectivity. The catalyst recycling was examined in the epoxidation of chalcone. Once the reaction had finished, the solid catalyst **15a** was easily separated from the reaction mixture by filtration under argon. The reactants and the solvent Et_2O were recharged into the reaction flask, and the catalyst was reused for five runs. However, a significant loss of reactivity (80%–38% yield) and enantioselectivity (92%–54% ee) after the second run was observed for unknown reasons.

Nguyen and co-workers reported the design and synthesis of a microporous MOF compound bearing chiral salen-Mn moieties, which was used in heterogeneous catalytic asymmetric epoxidation of 2,2-dimethyl-2H-chromene (Scheme 5.12) [42]. Metalloligand **16** bearing the chiral salen-Mn

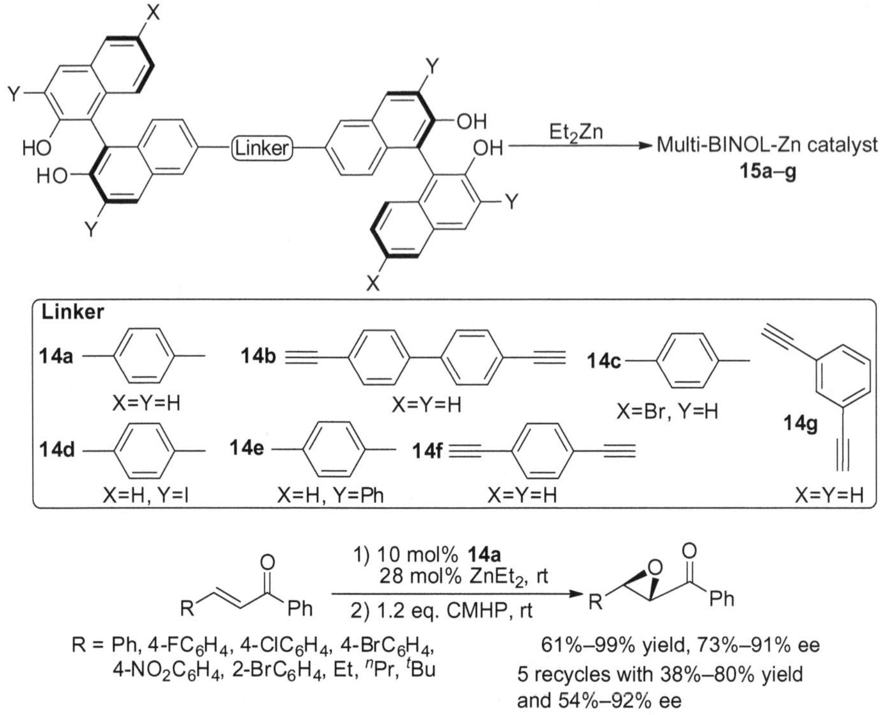

SCHEME 5.11. Self-supported BINOL/Zn catalysts for asymmetric epoxidation of α,β-unsaturated ketones.

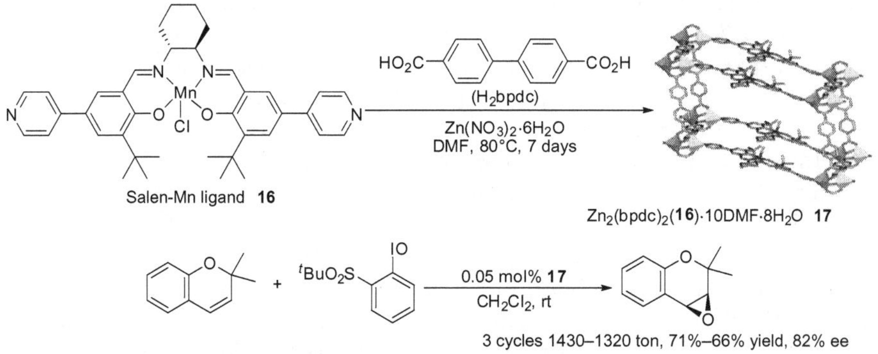

SCHEME 5.12. Homochiral salen-Mn MOF for asymmetric epoxidation of 2, 2-dimethyl-2H-chromene.

moiety was used as the ditopic bridging ligand. Such molecules, upon solvothermal treatment with $Zn(NO_3)_2{\cdot}6H_2O$ and H_2bpdc in DMF, afforded MOF **17** [Zn_2(bpdc)$_2$(**16**)$\cdot$10DMF$\cdot$8H_2O] as a crystalline solid, which was found insoluble in water and some common organic solvents (ethanol, acetonitrile, acetone, chloroform, and DMF). X-ray single crystallographic analysis indicated that **17** adopts an open framework structure, wherein the Mn(III) sites are accessible via the channels of this microporous solid. The catalytic properties of **17** was examined in the enantioselective epoxidation of 2,2-dimethyl-2*H*-chromene employing 2-(tertbutylsulfonyl) iodosylbenzene as the oxidant. The corresponding epoxide was obtained in 82% ee, a value almost rivaling the 88% ee of its homogeneous counterpart (Scheme 5.12). It is worth noting that the immobilized catalyst **17** is chemically more robust than the free salen-Mn(III) catalyst. Control experiments indicated that the free salen-Mn(III) catalyst is initially highly effective, but loses much of its activity after the first few minutes and is essentially inactive after a few hours of reaction. In contrast, the framework-immobilized catalyst **17** exhibited nearly constant reactivity during 3.4 hours of the catalytic process. It was proposed that the immobilization might prevent deactivating encounters of the active species and the degradation of the salen moiety by oxidation, and hence extended the catalyst lifetime. Catalyst **17** was recovered by centrifugation and reused for three cycles in constant ee values, albeit with some decrease in the yield of the third run. After each cycle, 4%–7% of the manganese initially present in the framework material was detected leaching into the solution, presumably caused by the inherent weakness of the Zn(II)–N(pyridine) bond, even though the small quantity of dissolved manganese did not show any catalytic activity.

In an effort to improve the stability of the immobilized salen-Mn (III) catalyst against leaching, Nguyen et al. further made use of a new chiral metalloligand, [bis(catechol)salen]Mn(III), which in principle can form stronger chelates with appropriate metal ions such as Cu(II) through the bidentate catechol moieties, as a building unit to prepare a series of coordination polymers as exemplified by **18** (Scheme 5.13) [43]. Coordination polymerization of [bis(catechol)-salen]Mn(III) with several di- and trivalent metal ions led to the formation of a series of metal-linked (salen)Mn polymers [M = Cu(II), Cr(III), Mn(II), Fe(III), Co(II), Ni(II), Zn(II), Cd(II), or Mg(II)] as amorphous powders. These were insoluble in water and in common organic solvents and thus were tested as heterogeneous catalysts in the asymmetric epoxidation of 2,2-dimethyl-2H-chromene. Generally, the oxidation products can be obtained in good yields with varying enantioselectivities (20%–76% ee). The Cu-linked salen-Mn polymer **18** can be readily recovered from the reaction mixture by centrifugation and decantation, and reused up to 10 times with a slight decrease in activity (79%–70% yield) and no loss in enantioselectivity (75%–76% ee). ICP analysis of the supernatant showed that 3.1% leaching of the manganese and 4.7% leaching of the copper occurred during the first two cycles, but the metal loss was much ameliorated during the ensuing runs, and essentially no leaching was detected in the last stages of recycling. Although the rate of heterogeneous catalytic epoxidation using insoluble catalyst **18** was found to be slower than that using its homogeneous counterpart, the lifetime of the former is significantly longer than that of the latter, and a total turnover number (TON) greater than 2000 was obtained within 3 hours using the polymer catalyst **18**.

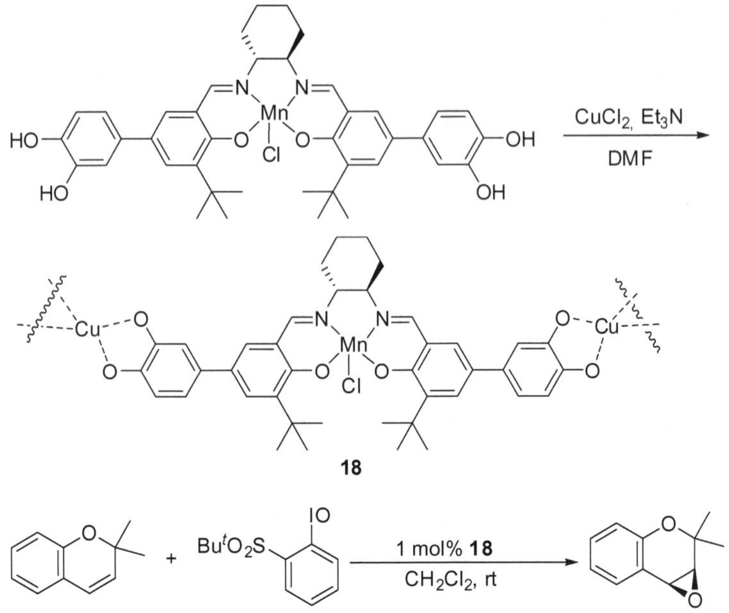

reused for 10 times with 79%–70% yield (total TON > 2000) and 76%–75% ee

SCHEME 5.13. Self-supported salen-Mn catalyst **18** for asymmetric epoxidation of 2, 2-dimethyl-2H-chromene.

5.5. ENANTIOSELECTIVE SULFOXIDATION

Ding and co-workers also reported the application of Ti–Poly-BINOL homochiral metal-organic polymers in the heterogeneous catalysis of asymmetric sulfoxidation reaction [44]. The homochiral Ti–Poly-BINOL polymers **19a–c** were prepared as amorphous orange or red powders by treatment of bridged BINOL ligands with $Ti(O^iPr)_4$ in 1:1 molar ratio in CCl_4, followed by addition of 40 eq. of water (relative to ligand). In the presence of the self-supported catalysts **19a–c**, and with CMHP as the oxidant, the oxidation of a variety of aryl alkyl sulfides proceeds with excellent enantioselectivity in CCl_4, providing the corresponding chiral sulfoxides with 96.4%–>99.9% ee in around 40% yield (Scheme 5.14). The heterogeneous nature of the oxidation reaction was confirmed by the fact that no detectable titanium (<0.1 ppm) was leached into the organic solution based on the ICP spectroscopic analyses of the mother liquid. The supernatant of **19a** in CCl_4 afforded a racemic oxidation product in a way similar to the control experiment in the absence of catalyst under otherwise identical conditions. The filtration was performed in air to recover the solid catalyst **19a**, which was reused in the sulfoxidation reaction of thioanisole for eight cycles that covered a period of more than 1 month. After eight runs, the enantioselectivity remained at the same level as

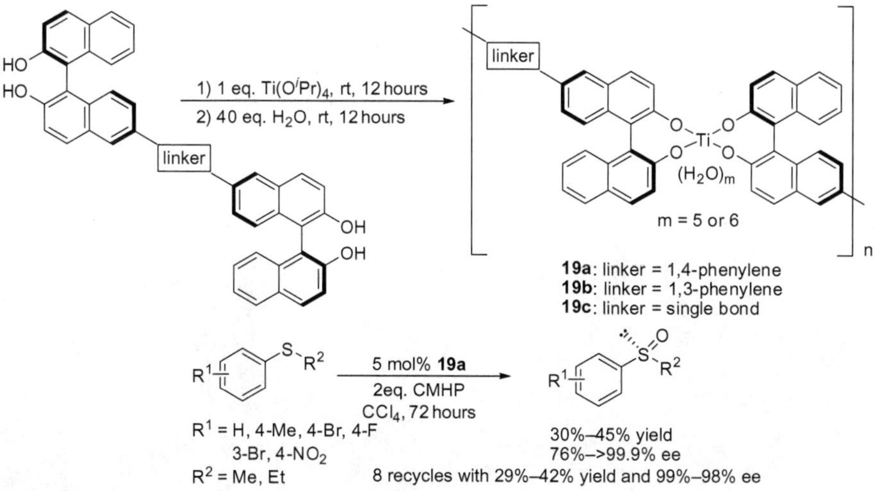

SCHEME 5.14. Self-supported BINOL-Ti catalysts for heterogeneous catalysis of asymmetric sulfoxidations.

that in the first run (99% ee), and no significant deterioration in activity was observed.

Fedin and co-workers reported the heterogeneous catalytic activity of a homochiral metal-organic polymer of zinc for the asymmetric oxidation of thioethers to sulfoxides [45]. The homochiral polymeric material, formulated as $[Zn_2(bdc)(\text{L-lac})(DMF)]\cdot(DMF)$ (**20**), was prepared as large rod-shaped colorless crystals by heating a DMF solution of $Zn(NO_3)_2$, L-lactic acid (L-H_2lac), and 1,4-benzenedicarboxylic acid (H_2bdc) in a 2:1:1 ratio in a Teflon-lined stainless steel vessel at 110°C for 2 days. The crystallographic analysis indicated that the compound **20** has a 3D MOF structure, and chiral pores of roughly 5 Å in diameter are interconnected in three directions (Scheme 5.15). By heating the as-synthesized **20** at 90°C for 3 hours, some of the guest DMF molecules were removed from the pores to afford a partially evacuated material **20** with permanent porosity, which can mediate highly size- and chemoselective catalytic oxidation of thioethers to sulfoxides by urea hydroperoxide (UHP) or H_2O_2. Smaller thioethers afforded higher conversions and better sulfoxidation selectivity than larger substrates. In all cases the chemoselectivity toward the sulfoxide formation is good to excellent (83%–100%). The remarkable size selectivity suggests that the reaction should primarily occur inside the uniform micropores of the framework solid, attesting to the heterogeneous nature of the catalysis. As a heterogeneous catalyst, each formula unit of **20** can be used for at least 30 catalytic cycles without loss of oxidation selectivity. Unfortunately, no asymmetric induction was observed in the catalytic sulfoxidations even though the catalyst has a homochiral structure. However, enantioselective sorption of the resulting racemic sulfoxide products

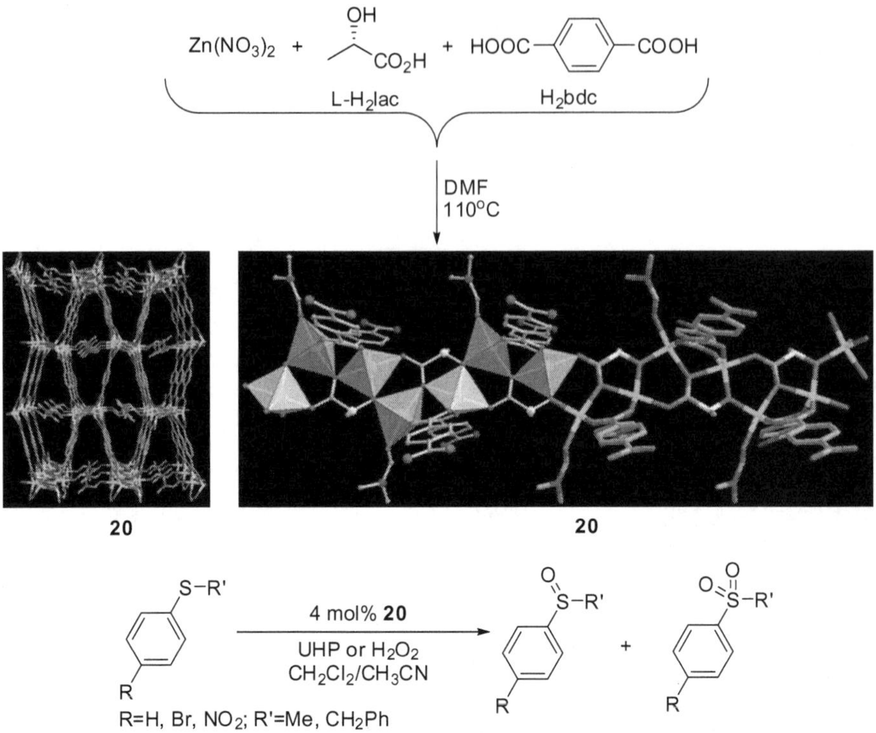

SCHEME 5.15. The homochiral Zn-MOF for the catalytic sulfoxidation and chromatographic resolution of sulfoxides.

through the chiral pore, which occurs simultaneously with the catalytic process, allowed the preferential sorption of one enantiomer from the racemic mixture with about 20% ee in the solid phase and leaving the corresponding enantiomer in excess in the solution phase. When the sorption experiments had been finished, the adsorbed guest molecules were extracted to regenerate the porous solid, which was reused in the next cycles of sorption experiments, without remarkable loss of its sorption capability. Based on the above observation, the authors developed a chiral chromatographic column by using the enantiopure porous MOF **20** as the stationary phase for the separation of racemic mixtures of chiral alkyl aryl sulfoxides [46]. A chiral chromatographic column was furnished by charging a suspension of the polymer in 10% solution of DMF in CH_2Cl_2 into a glass tube (8 mm inner diameter) with enough length. Using this column, the sulfoxide PhSOMe demonstrated a clear peak resolution and its enantiomers can be separated completely. When the catalytic oxidation of sulfides was combined simultaneously with the chromatographic enantioselective resolution of the resulting sulfoxide, using a microporous homochiral

MOF as the heterogeneous catalyst as well as the chiral stationary phase, the synthesis of enantiomerically pure sulfoxides was achieved in a unique one-pot process. A mixture of the sulfide (PhSMe) and H_2O_2 in the corresponding organic solvent was loaded onto the top of the above column. Using CH_2Cl_2/CH_3CN mixture as the eluent, enantiopure (*R*)- or (*S*)-enantiomer of the sulfoxide (PhSOMe) could be collected separately with relatively high yields at the other end of the column.

5.6. MICHAEL ADDITION

Al-Li-bis(binaphthoxide) (ALB), as a chiral heterobimetallic complex from the reaction of $LiAlH_4$ with two molar equivalents of BINOL, can promote certain asymmetric reactions in high efficiency and enantioselectivity [47]. The activity of this bifunctional catalyst depends on the synergistic cooperation between the two types of active sites. Therefore, maintaining the structural integrity of the active motif has to be considered during immobilization of this catalyst. The conventional approach involving the random introduction of ligand and functional units onto a sterically irregular polymer backbone often depresses the activity of homogeneous counterpart. A self-supported strategy uses multidentate ligands with ligating groups at the opposite ends of the molecular skeleton to form insoluble metal-bridged polymers, which were employed as the heterogeneous catalyst. This may provide a promising solution for the immobilization of these kinds of multicomponent asymmetric catalysts. With this in mind, Sasai and co-workers designed and synthesized some bis-BINOL ligands, which afforded insoluble aluminum-bridged homochiral polymers **21a–d** upon treatment with $LiAlH_4$ and BuLi in THF. Under a loading of 20 mol% heterogeneous catalysts **21a–d**, the Michael reaction between 2-cyclohexenone and dibenzyl malonate afforded the Michael adduct in good yields and up to 96% ee (for **21d**) (Scheme 5.16) [48], which is comparable to 97% ee for a homogeneous ALB catalyst. Compared with **21d**, heterogeneous catalysts **21a** and **21b** gave the Michael adduct much less satisfactory ee values (6% and 17% ee, respectively). The authors proposed that the bent shape of ligands **21a** and **21b** might facilitate the formation of unsuitable aggregates, whereas ligands **21c** and **21d** with the phenolic hydroxyl groups being situated at the opposite ends of the skeletons could form supramolecular assemblies favorable for high enantioselectivity. As a comparison, polystyrene-supported ALB gave the same Michael adduct in essentially racemic form. Obviously, the self-supporting strategy may provide a viable way for immobilization of multicomponent asymmetric catalysts. The heterogeneous nature of the catalysis was evidenced by the inactivity of supernatant solution. The catalyst **21d** was recovered by removing the supernatant with a syringe under argon and reused for five cycles with gradual decline in yield (88%–59%) and in ee values (96%–77%).

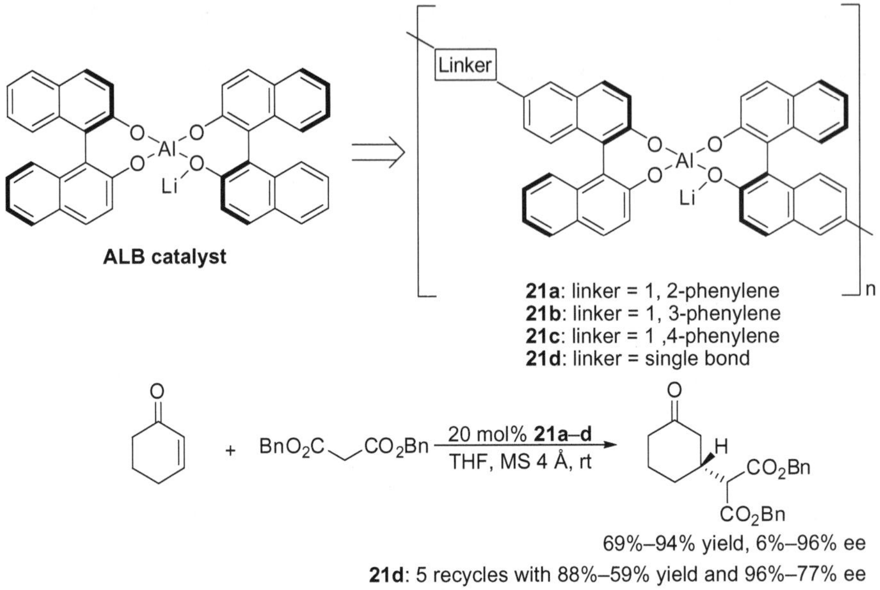

SCHEME 5.16. Al-bridged hybrid polymers for enantioselective Michael addition.

5.7. CARBONYL-ENE REACTION

Sasai et al. [48] and Ding et al. [44, 49] independently reported several homochiral bis-BINOL Ti polymers as self-supported catalysts for the asymmetric carbonyl-ene reaction. The bridged bis-BINOLs were treated with $Ti(O^iPr)_4$ in dichloromethane to afford chiral titanium-bis-BINOL polymers **22a–e** (**22a–d** from Ding et al.'s lab, and **22e** from Sasai et al.'s lab) as amorphous solids, which were then examined as immobilized catalysts for the carbonyl-ene reaction of α-methylstyrene with ethyl glyoxylate (Scheme 5.17). Under the optimized conditions, the α-hydroxyester can be obtained in high yield and excellent enantioselectivity (up to 98% ee). The nature of linkers between the two BINOL units in the bis-BINOL ligands remarkably influenced the enantioselectivity of the catalysis, presumably as a result of the differences in the supramolecular structures of the assemblies. Sasai et al.'s catalyst **22e** could be recovered by simple filtration and reused up to five consecutive runs with yields ranging from 66% to 88% and with ee values ranging from 88% to 92%. Ding et al.'s catalyst **22d** can also be reused five times, with a gradual decline in yields (87%–70%) and in ee values (97%–70%).

5.8. ADDITION OF DIETHYLZINC TO ALDEHYDES

Recently, Lin and co-workers reported the design and synthesis of several crystalline homochiral MOFs compounds and studied their network-structure-

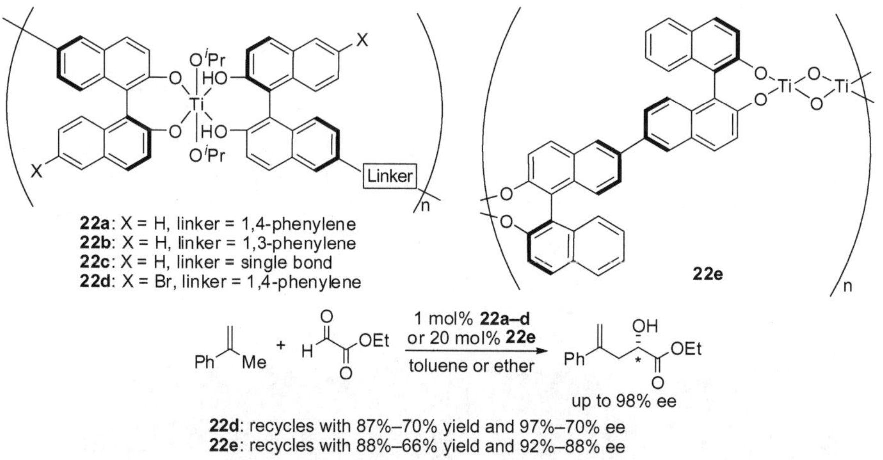

SCHEME 5.17. Self-supported BINOL-Ti catalysts for enantioselective carbonyl-ene reaction **22a–e**.

dependent catalytic activities in the enantioselective diethylzinc addition to aldehydes (Schemes 5.18 and 5.19) [50, 51]. The chiral porous MOFs were created by using a series of chiral bridging ligands containing orthogonal functional groups, that is, the primary bipyridyl groups used as metal-connecting units in forming the extended networks, and the orthogonal secondary chiral BINOL groups used to attach the catalytic metals. The crystalline MOFs (**23a–c**) were synthesized in good-to-high yields by slow diffusion of diethyl ether into a mixture of ligand and the corresponding Cd(II) salt in MeOH/DMF (Scheme 5.18), and they possess 1D chiral channels in their framework structures based on single-crystal X-ray diffraction analyses. Even after the guest solvent molecules have been removed, they still maintain permanent porosity as determined by CO_2 adsorption isotherm measurements. These solids (**23a–c**) were treated with $Ti(O^iPr)_4$ to afford the Ti(IV)-loaded active catalysts (**23a–c-Ti**), which were used in the diethylzinc addition to aldehydes (Scheme 5.19). Both **23a-Ti** and **23b-Ti** could catalyze the addition of diethylzinc to a range of aromatic aldehydes to give the corresponding product in high conversions and moderate-to-excellent enantioselectivity (45%–93% ee), but **23c-Ti** was found to be ineffective in promoting the reaction even though it also possesses chiral channels bearing binaphthyl hydroxyl groups on its walls as in **23a-Ti** and **23b-Ti**. The structural examination suggests that the environment around those dihydroxy groups in **23** is too crowded to allow the smooth coordination reaction of $Ti(O^iPr)_4$ and the binolate functionality, indicating that the detailed structure of the framework solid is critically important for the catalysis. When the evacuated catalyst **23a-Ti** was used, a remarkable size selectivity toward the aldehyde substrates was observed. The largest dendritic aldehyde did not undergo the addition reaction, and the G1-dendritic

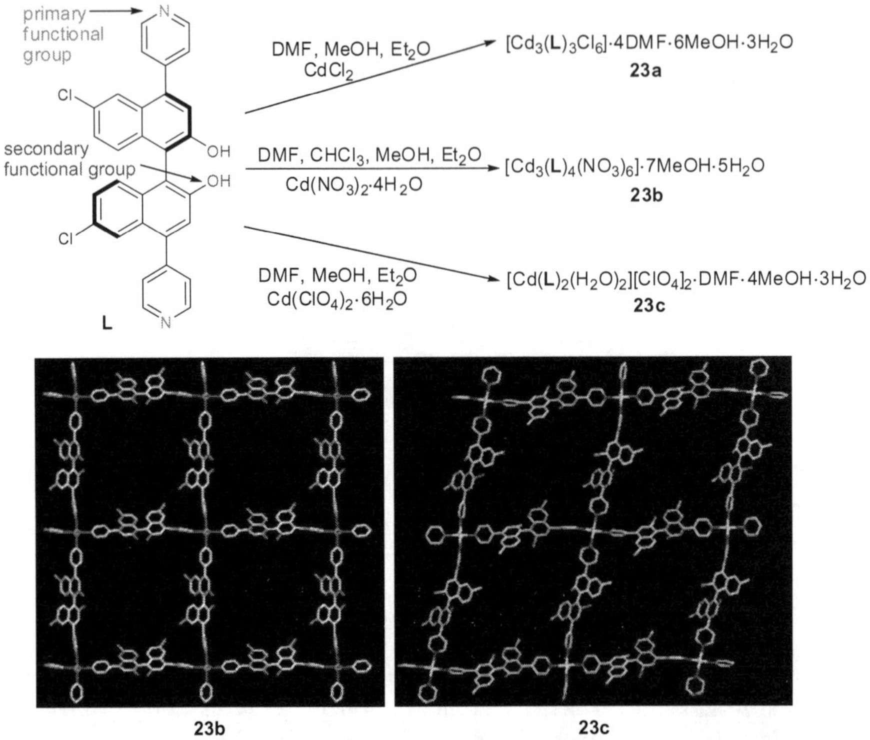

SCHEME 5.18. Preparation of crystalline MOF solids **23**.

aldehydes gave a lower conversion and enantioselectivity than G0-dendritic and other aromatic aldehydes. This size selectivity also demonstrated the true heterogeneous nature of the catalysis, but no catalyst or the framework recovery was mentioned by authors.

By using a molecular building-block approach, Lin and co-workers prepared chiral porous Zr phosphonates bearing BINOL moieties, which were subsequently used in combination with $Ti(O^iPr)_4$ for the heterogeneous catalysis of asymmetric additions of diethylzinc to aromatic aldehydes (Scheme 5.20) [52]. Treatment of BINOL-derived bisphosphonic acids with $Zr(O^nBu)_4$ in refluxing butanol afforded the chiral zirconium phosphonates **24** as amorphous solids. By drying at 80°C in vacuo, some included solvent was removed to give the evacuated hybrid polymers, which were treated with $Ti(O^iPr)_4$ to generate active catalysts for the additions of diethylzinc to aromatic aldehydes. Using a catalyst of 20–50 mol%, the heterogeneous addition reactions were carried out in toluene at room temperature to provide the corresponding chiral secondary alcohols in good conversions, albeit with enantioselectivities (up to 72% ee) somewhat lower than that of their homogeneous counterpart.

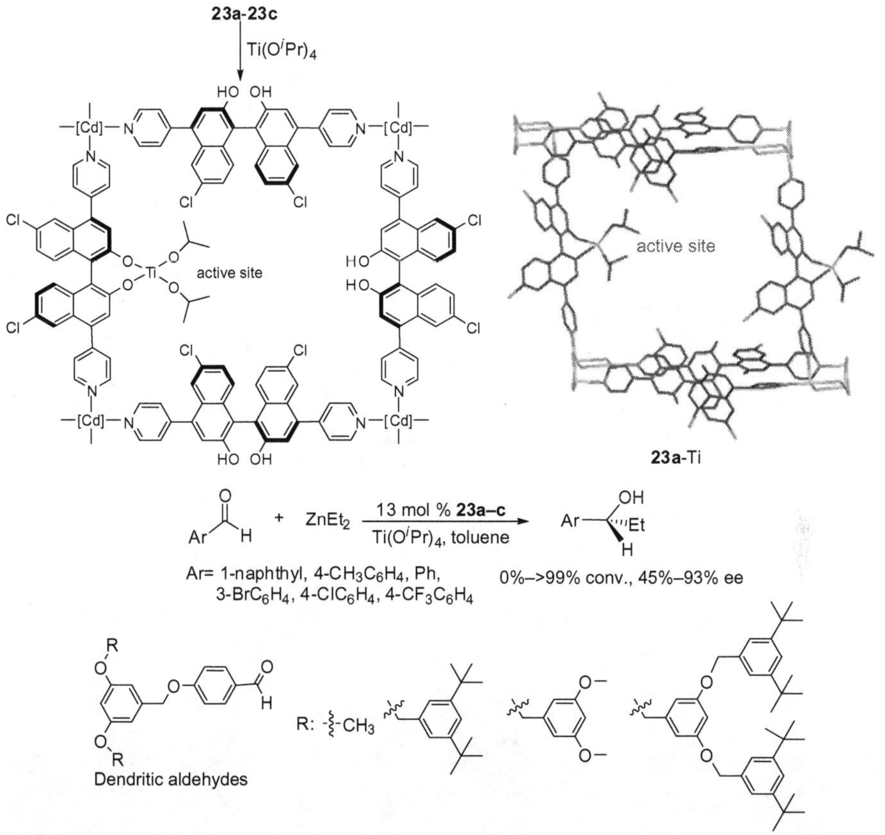

SCHEME 5.19. Application of MOF solids **23** in Ti (IV)-catalyzed $ZnEt_2$ addition to aromatic aldehydes.

In 2006, Harada and Nakatsugawa reported the use of a BINOLate/Ti(O^{i}Pr)$_4$ catalyst **25** for asymmetric addition of diethylzinc to aldehydes [53]. The rigid chiral tris-BINOL ligand was treated with Ti(O^{i}Pr)$_4$ to give the aggregate **25** as an amorphous solid, which was examined as a heterogeneous catalyst in the diethylzinc addition to aldehydes (Scheme 5.21). In the presence of a catalytic amount of **25**, the addition of diethylzinc to aldehydes proceeded smoothly in toluene/hexane to afford the corresponding products in >95% conversion and 74%–94% ee. After the reaction, the solid catalyst can be completely recovered by centrifugation. Control experiments revealed the supernatant of the reaction mixture can display a lower reactivity and less selectivity, thus indicating that the catalysis is only partially heterogeneous. For the reaction of benzaldehyde, the solid **25** can be recovered from the reaction mixture and reused for six cycles without loss of the activity (>98% conversion) or enantioselectivity (72%–75% ee).

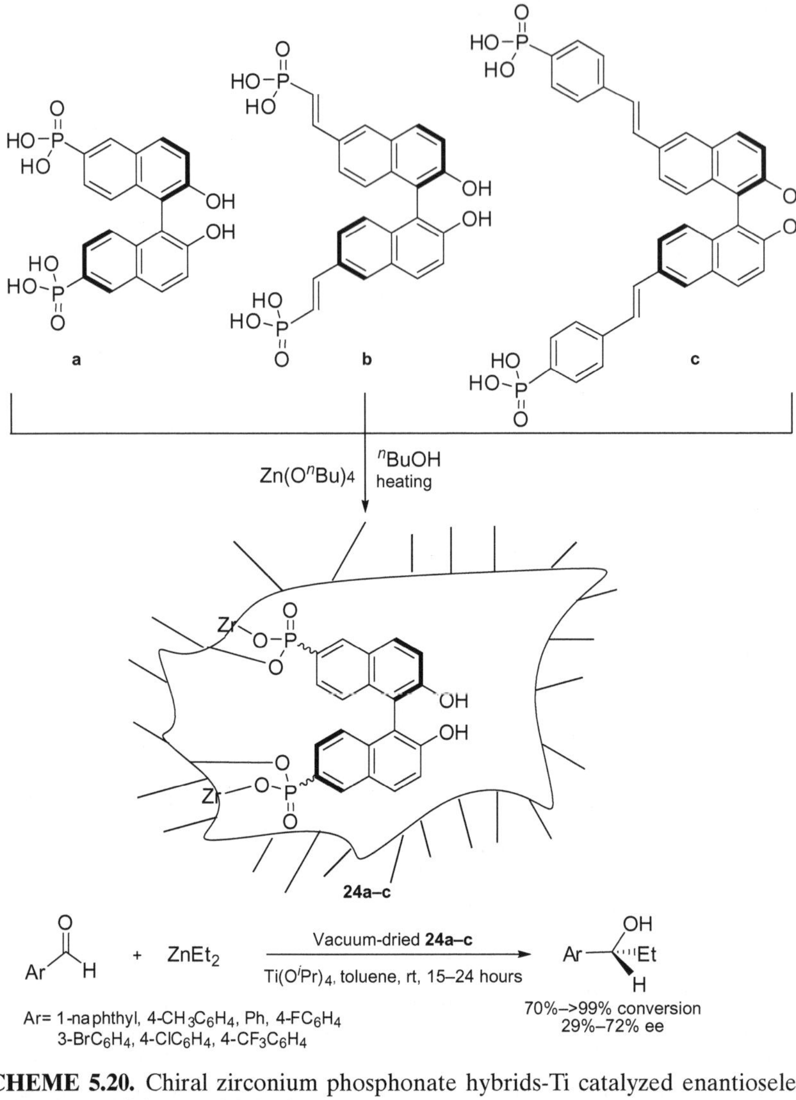

SCHEME 5.20. Chiral zirconium phosphonate hybrids-Ti catalyzed enantioselective diethylzinc addition to aldehydes.

5.9. RING-OPENING REACTION OF EPOXIDES

In 2008, Tanaka et al. reported a novel chiral MOF [Cu_2(5,5′-BDA)$_2$] (**26**) (5,5′-H_2BDA = 2,2′-dihydroxy-1,1′-binaphthalene-5,5′-dicarboxylic acid) as a heterogeneous catalyst for the asymmetric ring-opening reaction of an epoxide with amine (Scheme 5.22) [54]. Enantiopure 5,5′-H_2BDA was treated with copper nitrate in aqueous/methanol solution by slow diffusion of *N*,*N*-

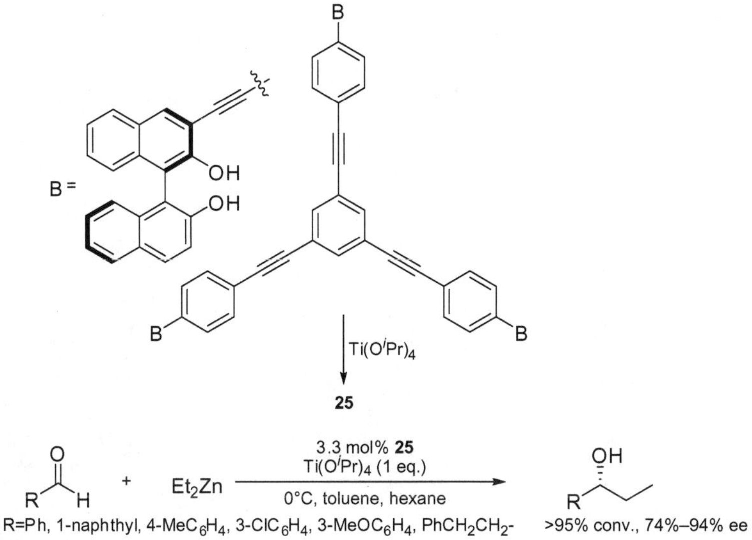

SCHEME 5.21. Tris-BINOL-Ti **25** catalyzed enantioselective diethylzinc addition to aldehydes.

dimethylaniline at room temperature to give crystalline $[Cu_2(5,5'\text{-}BDA)_2(H_2O)_2]\cdot MeOH\cdot 2H_2O$ in 33% yield. Single crystal X-ray diffraction analysis showed that **26** has a chiral MOF structure featuring a 2D dinuclear copper square grid coordination network, with each pair of Cu(II) ions bridging to four carboxylate groups. The void space of **26** is occupied by MeOH and H_2O guest molecules through hydrogen bonding, which upon evacuation in vacuo gave an amorphous solid that was tested as a heterogeneous catalyst in asymmetric ring-opening reactions of epoxides with aromatic amines. With 5 mol% of the catalyst, optically enriched β-amino alcohols were obtained in 3%–54% yields and 0%–50% ee's after 24–48-hour reactions at room temperature in toluene or under solvent-free conditions. After the reaction, the catalyst was recovered by simple filtration and was reused in the next cycles of the reaction without appreciable loss of reactivity and enantioselectivity.

In 2008, Rosseinsky and co-workers described a Brønsted acid-functionalized porous homochiral MOF solid, which was used as the heterogeneous catalyst in asymmetric ring-opening reaction of *cis*-2,3-epoxybutane with methanol (Scheme 5.23) [55]. The homochiral MOF solids Cu(L-asp) $bpe_{0.5}(HCl)(H_2O)$ (**27**) and $[Ni(\text{L-asp})bipy_{0.5}(HCl)_{0.9}(MeOH)_{0.5}]$ (**28**) were prepared by protonation of the amino acid-based open frameworks Cu(asp) $(bpe)_{0.5}$-(guests) and $[Ni(\text{L-asp})bpy_{0.5}]_n$ (L-asp = L-aspartate), respectively, with HCl in Et_2O under anhydrous conditions. Solid **27** was found catalytically more active than **28** in the methanolysis of rac-propylene oxide (PO), perhaps as a result of the nonporous nature of the latter. The ring-opening

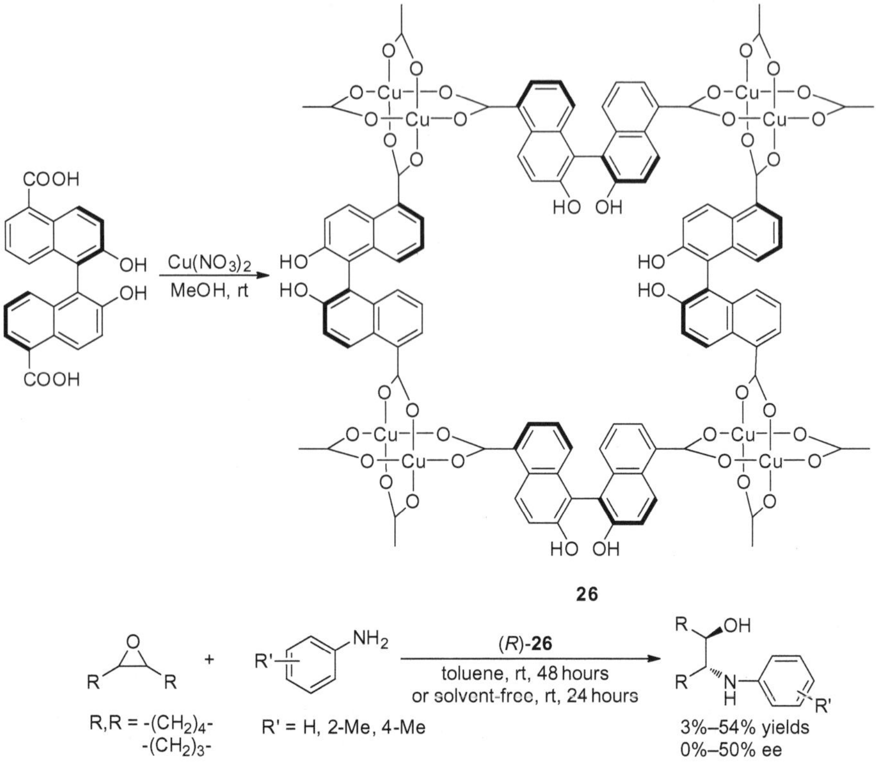

SCHEME 5.22. Chiral MOF **26** catalyzed asymmetric ring-opening reaction of epoxides with amines.

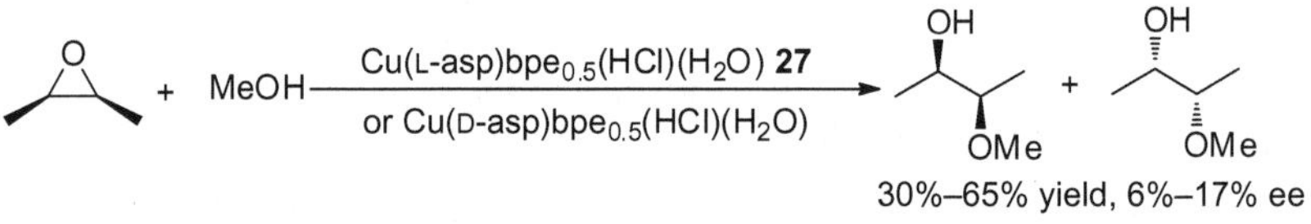

SCHEME 5.23. Homochiral MOF Cu(L-asp)bpe$_{0.5}$(HCl)(H_2O) **27** catalyzed asymmetric ring-opening reaction of *cis*-2,3-epoxybutane with methanol.

reaction performed by using **27** was demonstrated to be unambiguously heterogeneous, as the filtered supernatant of **27** was inactive for this reaction, precluding the possible catalysis by leached HCl. The impurity and parent framework were found inactive in control experiments, testifying the unique activity of the protonated material **27.** When a significantly bulkier epoxide (2,3-epoxypropyl)-benzene was used as the reactant, the turnover of methanolysis was effectively zero. This observation shows that the catalysis occurring in the pores is crucial for the conversion, and the catalysis with **27** has

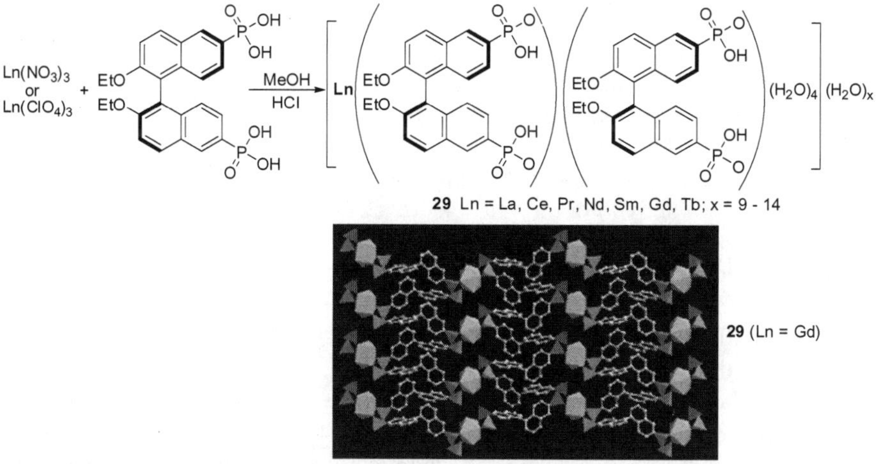

SCHEME 5.24. Homochiral lanthanide bisphosphonates.

displayed a shape selectivity toward the epoxides. After the reaction, catalyst **27** can be recovered with an essentially intact framework. Catalyst **27** was also tested in the enantioselective methanolysis of the *meso* epoxide, *cis*-2,3-epoxybutane, to give the corresponding amino alcohol in moderate yield with modest ee values (≤17%).

5.10. MISCELLANEOUS

In 2001, Lin and co-worker reported the synthesis and structural study of a class of homochiral porous lamellar lanthanide bisphosphonates **29**, as well as the application of **29e** [Ln(III) = Sm(III)] in heterogeneous catalysis of the cyanosilylation of aldehydes and ring opening of *meso*-carboxylic anhydrides (Scheme 5.24) [56]. Good reactivities were observed in the tested reactions. However, the products were essentially racemic (<5% ee) in all cases. In the chiral separation of racemic *trans*-1,2-diaminocyclohexane, the ammonia-treated **29e** can afford (*S*,*S*)-1,2-diaminocyclohexane in 13.6% ee in the beginning fractions and (*R*,*R*)-1,2-diaminocyclohexane in 10% ee in the end fractions, respectively.

5.11. SUMMARY AND OUTLOOK

During the past decade, a range of homochiral metal-organic polymers with diverse structures have been designed and prepared through coordination assembly of modular polytopic/polyfunctional ligands and metal ions. The homochiral metal-organic polymers typically contain rich functionalities within or pendant from their backbones, and were used successfully as self-supported chiral catalysts in several kinds of heterogeneous asymmetric catalytic reactions. In some cases, the self-supported chiral catalysts have demonstrated highly catalytic activity and excellent enantioselectivity that can effectively rival or even be superior to their corresponding homogeneous counterparts and can be easily recovered and reused for several times without significant loss of activity or enantioselectivity. The impressive results achieved within a short period suggest that the research on self-supported chiral catalysts has become an emerging and promising area in heterogeneous asymmetric catalysis and will stimulate further interests and activities from both academia and industry.

Despite all these achievements, however, much more research remains to be done in the future. At present, the scope of reactions that are amenable to self-supported chiral catalysts is still quite limited. The application of new types of self-supported catalysts in chemically interesting and industrially useful reactions needs to be studied. In addition, the stability of the catalysts has to be improved to ensure the catalysts are recycled and the leaching of metals is prevented. Although some polymers can be obtained as single crystals, most homochiral metal-organic polymers are amorphous, which hinders the structure elucidation of the catalysts and the further understanding of the molecular details of the heterogeneous asymmetric catalysis. Therefore, development of well-defined systems, which can be amenable to spectroscopic and crystallographic analysis, is highly desired for structural elucidation of the catalyst and mechanistic study of the catalytic cycle.

The self-supported catalysts are readily accessible from bridging chiral ligands and metal ions through a molecular building-block approach, which should allow for the fine-tuning of the catalysts for specific applications. Thus, the remarkable asymmetric induction coupled with the catalytic efficiency and the enormous chemical and structural diversity of the self-supported chiral catalysts imply that much more catalytic applications can be expected in the future.

ACKNOWLEDGMENTS

The authors are greatly indebted to their co-workers whose names are cited in the references for their fruitful dedication.

REFERENCES

[1] Noyori, R. (1994). *Asymmetric Catalysis in Organic Synthesis*. Wiley-Interscience, New York.

[2] Ojima, I. (Ed.) (2000). *Catalytic Asymmetric Synthesis*, 2nd ed. Wiley-VCH, New York.

[3] Jacobsen, E. N., Pfaltz, A., Yamamoto, H. (Eds.) (1999). *Comprehensive Asymmetric Catalysis*, Vols. I-III. Springer, Berlin, Germany.

[4] Yamamoto, H., Ishihara, K. (Eds.) (2008). *Acid Catalysis in Modern Organic Synthesis*, Vols. 1–2. Wiley-VCH, Weinheim, Germany.

[5] Börner, A. (Ed.) (2008). *Phosphorus Ligands in Asymmetric Catalysis*, Vols. 1–3. Wiley-VCH, Weinheim, Germany.

[6] Blaser, H. U., Schmidt, E. (2004). *Asymmetric Catalysis on Industrial Scale: Challenges, Approaches and Solutions*. Wiley-VCH, Weinheim, Germany.

[7] Sheldon, R. A. (1993). *Chirotechnology: Industrial Synthesis of Optically Active Compounds*. Dekker, New York.

[8] Ding, K., Uozumi, Y. (Eds.) (2008). *Handbook of Asymmetric Heterogeneous Catalysis*. Wiley-VCH, Weinheim, Germany.

[9] De Vos, D. E., Vankelecom, I. F. J., Jacobs, P. A. (Eds.) (2000). *Chiral Catalyst Immobilization and Recycling*. Wiley-VCH, Weinheim, Germany.

[10] Bergbreiter, D. E. (2002). Using soluble polymers to recover catalysts and ligands. *Chem. Rev.*, *102*, 3345–3383.

[11] Fan, Q. H., Li, Y. M., Chan, A. S. C. (2002). Recoverable catalysts for asymmetric organic synthesis. *Chem. Rev.*, *102*, 3385–3465.

[12] Heitbaum, M., Glorius, F., Escher, I. (2006). Asymmetric heterogeneous catalysis. *Angew. Chem. Int. Ed.*, *45*, 4732–4762.

[13] Dioos, B. M. L., Vankelecom, I. F. J., Jacobs, P. A. (2006). Aspects of immobilisation of catalysts on polymeric supports. *Adv. Synth. Catal.*, *348*, 1413–1446.

[14] Forster, P. M., Cheetham, A. K. (2003). Hybrid inorganic–organic solids: an emerging class of nanoporous catalysts. *Top. Catal.*, *24*, 79–86.

[15] Janiak, C. (2003). Engineering coordination polymers towards applications. *Dalton Trans.*, 2781–2804.

[16] Kitagawa, S., Kitaura, R., Noro, S. (2004). Functional porous coordination polymers. *Angew. Chem. Int. Ed.*, *43*, 2334–2375.

[17] Schuth, F. (2005). Engineered porous catalytic materials. *Annu. Rev. Mater. Res.*, *35*, 209–238.

[18] Rowsell, J. L. C., Yaghi, O. M. (2004). Metal-organic frameworks: a new class of porous materials. *Micropor. Mesopor. Mater.*, *73*, 3–14.

[19] Batten, S. R. (2001). Coordination polymers. *Curr. Opin. Solid State Mater. Sci.*, *5*, 107–114.

[20] James, S. L. (2003). Metal-organic frameworks. *Chem. Soc. Rev.*, *32*, 276–288.

[21] Rosseinsky, M. J. (2004). Recent developments in metal–organic framework chemistry: design, discovery, permanent porosity and flexibility. *Micropor. Mesopor. Mater.*, *73*, 15–30.

[22] Seo, J. S., Whang, D., Lee, H., Jun, S. I., Oh, J., Jeon, Y. J., Kim, K. (2000). A homochiral metal-organic porous material for enantioselective separation and catalysis. *Nature (London)*, *404*, 982–986.

[23] For reviews on the application of homochiral metal-organic polymers in heterogeneous asymmetric catalysis, see: (a) Kesanli, B., Lin, W. (2003). Chiral porous coordination networks: rational design and applications in enantioselective processes. *Coord. Chem. Rev.*, *246*, 305–326; (b) Lin, W. (2005). Homochiral porous

metal-organic frameworks: why and how. *J. Solid State Chem.*, *178*, 2486–2490; (c) Ngo, H. L., Lin, W. (2005). Hybrid organic-inorganic solids for heterogeneous asymmetric catalysis. *Top. Catal.*, *34*, 85–92; (d) Ding, K., Wang, Z., Wang, X., Liang, Y., Wang, X. (2006). Self-supported chiral catalysts for heterogeneous enantioselective reactions. *Chem. Eur. J.*, *12*, 5188–5197; (e) Ding, K. (2006). Development of homogeneous and heterogeneous asymmetric catalysts for practical enantioselective reactions. *Pure Appl. Chem.*, *78*, 293–301; (f) Jayaprakash, D., Takizawa, S., Arai, T., Sasai, H. (2006). Development of efficient methods for the immobilisation of multicomponent asymmetric catalysts. *J. Exp. Nanosci.*, *1*, 477–510. For a highlight on self-supported chiral catalyst, see: (g) Dai, L. X. (2004). Chiral metal-organic assemblies—a new approach to immobilizing homogeneous asymmetric catalysts. *Angew. Chem. Int. Ed.*, *43*, 5726–5729. For the comprehensive reviews on self-supported catalysts, see: (h) Wang, Z., Chen, G., Ding, K. (2009). Self-supported catalysts. *Chem. Rev.*, *109*, 322–359; (i) Fraile, J. M., García, J. I., Mayoral, J. A. (2009). Noncovalent immobilization of enantioselective catalysts. *Chem. Rev.*, *109*, 360–417.

[24] Efraty, A., Feinstein, I. (1982). Catalytic hydrogenation and isomerization of 1-hexene with [RhCl(CO)(1,4-$(CN)_2C_6H_4$)]n in the dark and under irradiation. *Inorg. Chem.*, *21*, 3115–3118.

[25] Noyori, R., Ohkuma, T., Kitamura, M., Takaya, H., Sayo, N., Kumobayashi, H., Akutagawa, S. (1987). Asymmetric hydrogenation of .beta.-keto carboxylic esters. A practical, purely chemical access to .beta.-hydroxy esters in high enantiomeric purity. *J. Am. Chem. Soc.*, *109*, 5856–5858.

[26] Hu, A., Ngo, H. L., Lin, W. (2003). Chiral, porous, hybrid solids for highly enantioselective heterogeneous asymmetric hydrogenation of beta-keto esters. *Angew. Chem. Int. Ed.*, *42*, 6000–6003.

[27] Ohkuma, T., Ooka, H., Hashiguchi, S., Ikariya, T., Noyori, R. (1995). Practical enantioselective hydrogenation of aromatic ketones. *J. Am. Chem. Soc.*, *117*, 2675–2676.

[28] Ohkuma, T., Ooka, H., Ikariya, T., Noyori, R. (1995). Preferential hydrogenation of aldehydes and ketones. *J. Am. Chem. Soc.*, *117*, 10417–10418.

[29] Hu, A., Ngo, H. L., Lin, W. (2003). Chiral porous hybrid solids for practical heterogeneous asymmetric hydrogenation of aromatic ketones. *J. Am. Chem. Soc.*, *125*, 11490–11491.

[30] Liang, Y., Jing, Q., Li, X., Shi, L., Ding, K. (2005). Programmed assembly of two different ligands with metallic ions: generation of self-supported Noyori-type catalysts for heterogeneous asymmetric hydrogenation of ketones. *J. Am. Chem. Soc.*, *127*, 7694–7695.

[31] Liang, Y., Wang, Z., Ding, K. (2006). Generation of self-supported Noyori-type catalysts using achiral bridged-BIPHEP for heterogeneous asymmetric hydrogenation of ketones. *Adv. Synth. Catal.*, *348*, 1533–1538.

[32] van den Berg, M., Minnaard, A. J., Schudde, E. P., Van Esch, J., de Vries, A. H. M., de Vries, J. G., Feringa, B. L. (2000). Highly enantioselective rhodium-catalyzed hydrogenation with monodentate ligands. *J. Am. Chem. Soc.*, *122*, 11539–11540.

[33] Pena, D., Minnaard, A. J., de Vries, J. G., Feringa, B. L. (2002). Highly enantioselective rhodium-catalyzed hydrogenation of β-dehydroamino acid derivatives using monodentate phosphoramidites. *J. Am. Chem. Soc.*, *124*, 14552–14553.

[34] Wang, X., Ding, K. (2004). Self-supported heterogeneous catalysts for enantioselective hydrogenation. *J. Am. Chem. Soc.*, *126*, 10524–10525.

[35] Shi, L., Wang, X., Sandoval, C. A., Li, M., Qi, Q., Li, Z., Ding, K. (2006). Engineering a polymeric chiral catalyst by using hydrogen bonding and coordination interactions. *Angew. Chem. Int. Ed.*, *45*, 4108–4112.

[36] Yu, L., Wang, Z., Wu, J., Tu, S., Ding, K. (2010). Directed orthogonal self-assembly of homochiral coordination polymers for heterogeneous enantioselective hydrogenation. *Angew. Chem. Int. Ed.*, *49*, 3627–3630.

[37] Shi, L., Wang, X., Sandoval, C. A., Wang, Z., Li, H., Wu, J., Yu, L., Ding, K. (2009). Development of a continuous-flow system for asymmetric hydrogenation using self-supported chiral catalysts. *Chem. Eur. J.*, *15*, 9855–9867.

[38] Bougauchi, M., Watanabe, S., Arai, T., Sasai, H., Shibasaki, M. (1997). Catalytic asymmetric epoxidation of α,β-unsaturated ketones promoted by lanthanoid complexes. *J. Am. Chem. Soc.*, *119*, 2329–2330.

[39] Watanabe, S., Arai, T., Sasai, H., Bougauchi, M., Shibasaki, M. (1998). The first catalytic enantioselective synthesis of cis-epoxyketones from cis-enones. *J. Org. Chem.*, *63*, 8090–8091.

[40] Wang, X., Shi, L., Li, M., Ding, K. (2005). Heterogenization of Shibasaki's binol/La catalyst for enantioselective epoxidation of alpha,beta-unsaturated ketones with multitopic binol ligands: the impact of bridging spacers. *Angew. Chem. Int. Ed.*, *44*, 6362–6366.

[41] Wang, H., Wang, Z., Ding, K. (2009). Self-supported BINOL–Zn catalysts for heterogeneous enantioselective epoxidation of (E)-α,β-unsaturated ketones. *Tetrahedron Lett.*, *50*, 2200–2203.

[42] Cho, S. H., Ma, B., Nguyen, S. T., Hupp, J. T., Albrecht-Schmitt, T. E. (2006). A metal–organic framework material that functions as an enantioselective catalyst for olefin epoxidation. *Chem. Commun.*, 2563–2565.

[43] Cho, S.-H., Gadzikwa, T., Afshari, M., Nguyen, S. T., Hupp, J. T. (2007). [Bis(catechol) salen]Mn^{III} coordination polymers as support-free heterogeneous asymmetric catalysts for epoxidation. *Eur. J. Inorg. Chem.*, 4863–4867.

[44] Wang, X., Wang, X., Guo, H., Wang, Z., Ding, K. (2005). Self-supported heterogeneous titanium catalysts for enantioselective carbonyl-ene and sulfoxidation reactions. *Chem. Eur. J.*, *11*, 4078–4088.

[45] Dybtsev, D. N., Nuzhdin, A. L., Chun, H., Bryliakov, K. P., Talsi, E. P., Fedin, V. P., Kim, K. A. (2006). Homochiral metal-organic material with permanent porosity, enantioselective sorption properties, and catalytic activity. *Angew. Chem. Int. Ed.*, *45*, 916–920.

[46] Nuzhdin, A. L., Dybtsev, D. N., Bryliakov, K. P., Talsi, E. P., Fedin, V. P. (2007). Enantioselective chromatographic resolution and one-pot synthesis of enantiomerically pure sulfoxides over a homochiral Zn–organic framework. *J. Am. Chem. Soc.*, *129*, 12958–12959.

[47] Shibasaki, M., Sasai, H., Arai, T. (1997). Asymmetric catalysis with heterobimetallic compounds. *Angew. Chem. Int. Ed.*, *36*, 1236–1256.

[48] Takizawa, S., Somei, H., Jayaprakash, D., Sasai, H. (2003). Metal-bridged polymers as insoluble multicomponent asymmetric catalysts with high enantiocontrol: an approach for the immobilization of catalysts without using any support. *Angew. Chem. Int. Ed.*, *42*, 5711–5714.

[49] Guo, H., Wang, X., Ding, K. (2004). Assembled enantioselective catalysts for carbonyl-ene reactions. *Tetrahedron Lett.*, *45*, 2009–2012.

[50] Wu, C. D., Hu, A., Zhang, L., Lin, W. (2005). A homochiral porous metal–organic framework for highly enantioselective heterogeneous asymmetric catalysis. *J. Am. Chem. Soc.*, *127*, 8940–8941.

[51] Wu, C. D., Lin, W. (2007). Heterogeneous asymmetric catalysis with homochiral metal-organic frameworks: network-structure-dependent catalytic activity. *Angew. Chem. Int. Ed.*, *46*, 1075–1078.

[52] Ngo, H. L., Hu, A., Lin, W. (2004). Molecular building block approaches to chiral porous zirconium phosphonates for asymmetric catalysis. *J. Mol. Catal. A Chem.*, *215*, 177–186.

[53] Harada, T., Nakatsugawa, M. (2006). Immobilization of a BINOLate-titanium catalyst by use of aggregation phenomenon. *Synlett*, 321–323.

[54] Tanaka, K., Odaa, S., Shiro, M. (2008). A novel chiral porous metal–organic framework: asymmetric ring opening reaction of epoxide with amine in the chiral open space. *Chem. Commun.*, 820–822.

[55] Ingleson, M. J., Barrio, J. P., Bacsa, J., Dickinson, C., Park, H., Rosseinsky, M. J. (2008). Generation of a solid Brønsted acid site in a chiral framework. *Chem. Commun.*, 1287–1289.

[56] Evans, O. R., Ngo, H. L., Lin, W. (2001). Chiral porous solids based on lamellar lanthanide phosphonates. *J. Am. Chem. Soc.*, *123*, 10395–10396.

CHAPTER 6

CATALYSIS WITH CHIRALLY MODIFIED METAL SURFACES: SCOPE AND MECHANISMS

ANGELO VARGAS, CECILIA MONDELLI, AND ALFONS BAIKER

6.1. Introduction to chirally modified metal surfaces 291
6.2. Asymmetric reactions at chirally modified metal surfaces 293
 6.2.1. Hydrogenation of C=O bonds 293
  6.2.1.1. Activated ketones 293
  6.2.1.2. Nonactivated ketones 297
  6.2.1.3. Deactivated ketones 297
 6.2.2. Hydrogenation of C=C bonds 298
  6.2.2.1. Unsaturated carboxylic acids and esters 298
  6.2.2.2. Unsaturated ketones 300
  6.2.2.3. Pyrones 300
  6.2.2.4. Aromatic compounds 300
 6.2.3. Hydrogenation of C=N bonds 301
6.3. The development of a model for the *Cinchona* alkaloid modified platinum asymmetric hydrogenation 301
6.4. Conclusions 307
References 307

6.1. INTRODUCTION TO CHIRALLY MODIFIED METAL SURFACES

Enantioselective catalysis features the amplification of asymmetric information, and is acknowledged as a premier methodology for the synthesis of pure enantiomers [1,2]. The growing demand for enantiomerically pure compounds

Catalytic Methods in Asymmetric Synthesis: Advanced Materials, Techniques, and Applications, First Edition. Edited by Michelangelo Gruttadauria and Francesco Giacalone.

has driven the investigation of chemical catalytic systems that can provide high enantioselectivites and high recyclability of the catalyst [1, 2]. Heterogeneous asymmetric catalysis can deliver such advantages by shifting the paradigms of the lowering of the transition state, chiral recognition, and chiral bias to a solid–liquid interface. After a reaction, the solid catalyst can be removed by simple filtration, thus allowing a renewal of the cycle [3, 4]. Only few such catalytic systems have been discovered. The two best known are the *Cinchona* alkaloid-modified metal system and related catalysts, and tartaric-acid modified nickel [3, 5]. Amino acids modified palladium and platinum, and menthyl modified Pt/Sn, are other minor asymmetric heterogeneous catalysts that are still in the early phases of investigation. In addition, the synthetic modifier Naphthylethylamine (NEA) and some of its derivatives have also been investigated within the framework of the study of *Cinchona* alkaloid surface modification. The *Cinchona* alkaloid-modified metal system is by far the most extensively investigated among the above-mentioned catalysts [5–7]. In the original version proposed by Orito et al. [5, 8], pretreatment of a platinum catalyst in cinchonidine (CD)–ethanol solutions under reflux conditions led to high enantioselectivities in the hydrogenation of ethyl pyruvate. Later, it was shown that simple addition of a *Cinchona* alkaloid to the reaction medium of the above hydrogenation reaction led to the asymmetric reduction of the keto-carbonyl moiety, with enantioselectivities exceeding 90% in the formation of ethyl lactate [9]. Since this original research, the methodology has been extended to a broad class of prochiral substrates, as will be shown in the following sections. In particular, the asymmetric hydrogenation of α-keto esters, α-fluoro ketones, and α-diketones on chirally modified platinum and rhodium has been the subject of numerous investigations [5]. It was, until recently, believed that the CD-modified Pt system was limited to activated prochiral ketones, that is, ketones bearing an electron-withdrawing group in α-position to the keto-carbonyl moiety. Recently, very encouraging results have shown the power of this methodology for the asymmetric hydrogenation of acetophenone, albeit in a slightly modified manner. From a mechanistic point of view, the chiral modification of platinum by means of alkaloids of the *Cinchona* family has been thoroughly investigated using attenuated total reflection-infrared (ATR-IR) spectroscopy and computational modeling, and it is now widely accepted that the phenomenon of chiral amplification occurs on asymmetric sites generated by the adsorption of the alkaloid on the surface of the metal. In brief, the reduction of the prochiral keto-carbonyl group occurs in the proximity of the adsorbed alkaloid, and its orientation toward surface hydrogen is efficiently biased by a 1:1 binding to the organic modifier (1:1 model).

This chapter will be organized as follows: The first part will show the development of the field in terms of the scope reached by chirally modified metals in the asymmetric hydrogenation of prochiral double bonds, and the second part will illustrate the mechanistic aspects of chiral modification, referring in particular to the *Cinchona* alkaloid modified platinum system. This latter part elucidates the evolution of our understanding of metal chiral modification and shows how mechanistic insight can pave the way to the tailoring of increasingly sophisticated metal-organic catalytic interfaces.

6.2. ASYMMETRIC REACTIONS AT CHIRALLY MODIFIED METAL SURFACES

In this section, an overview of the applicability of chirally modified metals as solid enantioselective catalysts is given. The focus is the main field of investigation, namely the enantioselective hydrogenation of C=O and C=C bonds. In addition, a short paragraph mentions the attempts to achieve asymmetric hydrogenation of C=N bonds. Enantioselectivities reported in the following tables and in the text are, in most cases, not thoroughly optimized.

6.2.1. Hydrogenation of C=O Bonds

The enantioselective hydrogenation of prochiral ketones to chiral alcohols is the most thoroughly investigated chemical transformation in the field of heterogeneous asymmetric catalysis. We can distinguish three cases of the enantioselective hydrogenation of (1) activated ketones, (2) nonactivated ketones, and (3) deactivated ketones. The first case comprises the widest variety of substrates and is characterized by ketones bearing an electron-withdrawing group in an α-position to the carbonyl moiety. The second case refers to a broad range of β-keto-compounds and to simple linear ketones. The third case is typically represented by acetophenone and introduces the additional problem of selectivity between the saturation of the phenyl group and the keto-carbonyl group.

6.2.1.1. Activated Ketones. Over the years, the asymmetric hydrogenation of α-ketoesters, α-ketoacids, α-ketoamides, α-diketones, α-ketoacetals, and α-fluorinated ketones by means of modified supported metal catalysts has been investigated [10–12]. Supported Pt modified by a *Cinchona* alkaloid, preferentially CD or *O*-methyl-cinchonidine (MeOCD), was the most successful catalyst system, while other chirally modified metals [13–15] such as Rh [16–22], Ru [23, 24], Pd [25–30], and Ir [31–33] generally achieved inferior results. Reactions are usually performed under mild conditions, at or slightly below room temperature, at 1–10 bar [34–36] (rarely up to 100 bar), and the preferred media are toluene, acetic acid, and chlorinated solvents [10–12, 37]. Under optimal conditions the typical reactant/modifier molar ratio is in the range of 150–1800, demonstrating the efficient chiral multiplication inherent to the catalyst system.

After the pioneering work on alkyl pyruvates, a number of studies have been devoted to broadening the pool of substrates suitable for this kind of reaction. Transformation of α-ketoesters to the corresponding alcohols by the Pt-*Cinchona* system resulted in enantioselectivities up to 96%–98%

enantiomeric excess (ee) [34, 36, 38–43]. CD and quinidine (QD) were shown to be the best performing modifiers for the hydrogenation of ethyl pyruvate, with both attaining 98% ee in optimized conditions [34]. Steric and electronic effects of the groups introduced either in the structure of the modifier or in the pyruvate skeleton have been explored (Table 6.1). On the modifier side, while the performance of CD was rather independent from the substrate [44], methoxy- and phenoxy-substituted quinine (QN) and PhOCD showed bulkiness-related variations of ee [36, 43, 45]. On the substrate side, substituent effects on the phenyl group in α to the keto-carbonyl were investigated. Electron-releasing methoxy groups led to the highest ee, while electron-withdrawing CF_3 groups yielded the lowest ee with all three modifiers.

Enantioselective hydrogenation of ketopantolactone (KPL), the cyclic corresponding form of ethyl pyruvate, to (*R*)-pantolactone on CD-modified Pt was first reported in 1987 [46]. The only moderate enantioselectivity achieved at that time could be markedly increased by further optimization of the reaction conditions up to 92% ee (Table 6.1) [47]. Later, 1-Naphthyl-1,2-ethanediol (NED) [48] and quaternary cinchonidinium salts [49] that gave racemic products in pyruvate hydrogenation were shown to be rather selective in KPL hydrogenation. Peptide-based modifiers built on the Trp unit have also been reported to yield moderate ee values in this latter reaction [50]. The best results achieved within the other classes of activated ketones are illustrated in Table 6.2 and show the successful broadening of substrates for enantioselective hydrogenation. In particular, the hydrogenation of α-ketoacids achieved very remarkable ee values on alumina-supported Pt modified with *O*-methylated CD [51, 52]. Catalytic data obtained with the use of solvents of different polarity, of K or Na salts of the α-ketoacids, and of quaternary cinchonidinium salts, have suggested the establishment of acid–base-type substrate-modifier interactions [51, 52]. Due to the relevance of such phenomena, the only suitable reaction media for the hydrogenation of this class of ketones are polar reaction media [51, 52].

Cinchona-modified Pt also induces high enantioselectivity in the hydrogenation of α-diketones. The best example is butane-2,3-dione (Table 6.2) [53], but hexane-3,4-dione [54], 1-phenyl-1,2-propanedione [55], and cyclohexane-1,2-dione were also converted with ee values in the range of 80%–90% [56]. For α-diketones, the enantioselectivity was explained on the basis of a combination of enantioselective reaction and kinetic resolution [53, 57], and a complex reaction route was proposed. However, the chemical yield was generally low (<30% usually) and limited the synthetic potential of the method. Other Pt group metals [14, 33] and other *Cinchona* derivatives or analogous chiral amino alcohols [58, 59] have been alternatively tested, but the results were even inferior to the Orito system.

Hydrogenation of α-ketoamides to the corresponding chiral alcohols constitutes an exception to the applicability of the CD-modified Pt/Al_2O_3 system, with only moderate enantioselectivities [60–62] being attained. On the other

TABLE 6.1. Enantioselective Hydrogenation of C=O Double Bonds: Activated Ketones

Substrate class	Substrate		Modifier ee (%)	
α-ketoesters	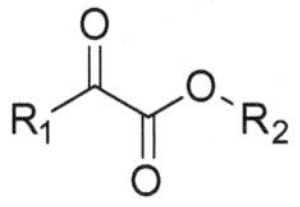	**CD**	**QN**	**PhOCD**
	R_1, R_2			
	Me, Et	80(*R*)	21(*R*)	21(*S*)
	t-Bu, Et	56(*R*)	12(*R*)	41(*S*)
	Ph, Et	86(*R*)	79(*R*)	57(*S*)
	Ph, *t*-Bu	95(*R*)	89(*R*)	78(*S*)
	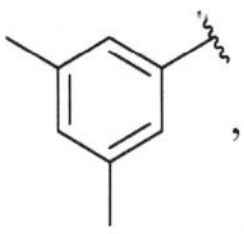 , Et	92(*R*)	75(*R*)	52(*S*)
	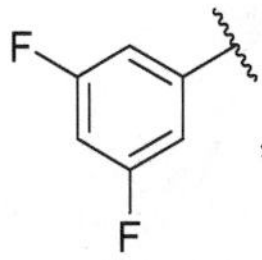 , Et	87(*R*)	76(R)	68(S)
	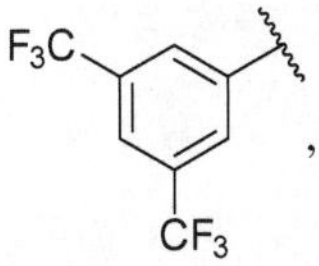 , Et	66(*R*)	47(*R*)	48(*S*)
	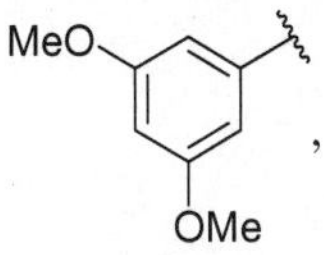 , Et	94(*R*)	84(*R*)	73(*S*)
	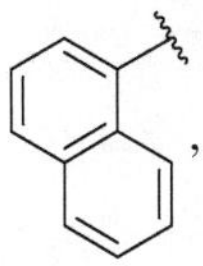 , Et	86(*R*)	60(*R*)	31(*S*)
α-ketoesters	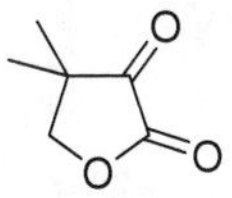	**CD**	***N,O*-DmeCD·Cl**	**NED**
		92(*R*)	44(*S*)	25(*R*)

CD, cinchonidine; QN, quinine; PhOCD, *O*-phenyl-cinchonidine; *N,O*-DmeCD·Cl, *N,O*-dimethylcinchonidine chloride; NED, 1-naphthyl-1,2-ethanediol.

TABLE 6.2. Enantioselective Hydrogenation of C=O Double Bonds: Activated Ketones

Substrate class	Substrate	Modifier eeA (%)
α-ketoacids	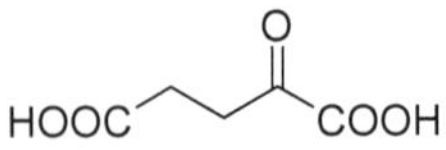	**MeODHCD** 92%
α-diketones	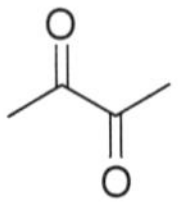	**CD** 89%
Imidoketones	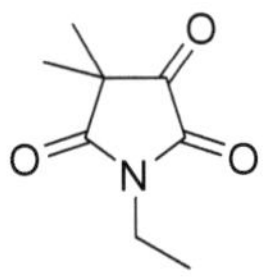	**CD** 91%
α-ketoacetals	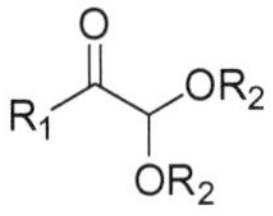	**MeODHCD**
	R_1, R_2	
	Me, Me	96%
	Me $(CH_2)_3$	97%
	Me, Et	91%
	Me, Bu	85%
	Ph, Me	89%
	$PhO(CH_2)_3$, Me	93%
	$Me_2NOC(CH_2)_3$, Me	80%
α-fluorinated ketones	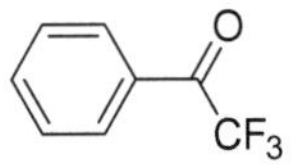	**CD** 92%
	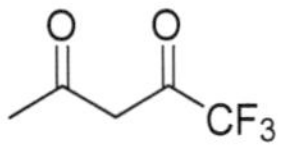	**PNEA** 93%
	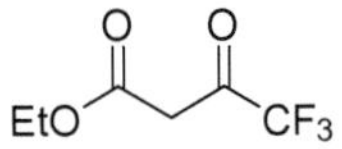	**MeOCD** 96%
	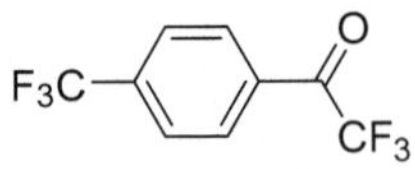	**CD** 81%

MeODHCD, 9-*O*-methyl-10,11-dihydrocinchonidine; PNEA, pantoyl-naphthylethylamine.

hand, the CD–platinum catalyst offers an attractive route for the reduction of cyclic imidoketones such as pyrrolidine-2,3,5-triones (Table 6.2) [63, 64].

Hydrogenation of α-keto acetals (Table 6.2) belongs to the most selective reactions on chirally modified Pt [65–68], and the enantioselection maintained at good levels even in the presence of additional functional groups [66].

Concerning α-fluorinated ketones, the performance of the Pt–*Cinchona* system in the hydrogenation of these substrates varies strongly depending on the structure of the chiral modifier employed and the nature of substrate hydrogenated. CD, MeOCD, and pantoyl-naphthylethylamine (PNEA) achieved the highest ee for some fluorinated β-diketones [69, 70]. The use of MeOCD afforded diversified and also limited enantioselectivities in the hydrogenation of fluorinated aromatic ketones [71–74], β-diketones [69, 70, 75], and β-ketoesters [76, 77]. CD remains the best modifier for the hydrogenation of some fluorinated aromatic ketones. The reduction of aliphatic fluorinated ketones is moderately selective [73] and complicated by solvent effects [78, 79]. Furthermore, enantioselectivities and conversions [70, 71, 76] were shown to be strongly substrate dependent.

6.2.1.2. Nonactivated Ketones. For the hydrogenation of β-functionalized and unfunctionalized ketones to alcohols, the Raney Ni-tartaric acid (TA)-NaBr is the best heterogeneous enantioselective catalyst. This system was the result of extensive research initiated by Japanese scientists several decades ago and further developed over the years.

TA conferred the best enantioselective properties to the heterogeneous catalyst with respect to other chiral modifiers such as hydroxy and amino acids [80–90]. Regarding the metal component, many efforts have been devoted to enhancing the activity of Ni, which is intrinsically lower compared to that of the Pt group metals, and remarkable improvements in the catalyst preparation technique have achieved better results [91–93].

Still, the application of elevated pressures and temperatures is required to attain reasonable reaction rates [94]. TA-modified Ni is used primarily in the hydrogenation of β-ketoesters [95, 96], but other suitable substrates are β-ketoalcohols and β-ketoethers (68%–70% ee), β-ketosulfones (67%–71% ee), and β-diketones [83]. For the latter class, kinetic resolution enhanced the enantioselectivity of the first reaction step up to 90%–91% ee to the (*R*,*R*)-diol. Ni–TA is also the only efficient catalyst in the asymmetric hydrogenation of aliphatic unfunctionalized ketones, but higher enantioselectivities require more than a stoichiometric amount of an achiral acid additive [86, 97–100].

6.2.1.3. Deactivated Ketones. The hydrogenation of unfunctionalized aromatic ketones is a demanding reaction. Not only is the hydrogenation of the keto-functionality not favored, but hydrogenation of the aromatic moiety becomes competitive. This class is best represented by acetophenone. Early attempts to hydrogenate this substrate by a CD-modified Pt cluster supported on MCM-41 attained only moderate results. Additionally, the catalyst suffered from consistent deactivation [101]. Platinum-modified with an organotin compound [102] and Pd/C in the presence of (*S*)-proline [103] were even less selective, with ee values slightly above 20%. The enantioselection improved by the ring substitution of acetophenone. The reduction of 3,4-dimethoxyacetophenone on Pt/SiO_2 modified by a chiral organotin compound gave moderate ee [104], and high ee was obtained in the hydrogenation of 3,5-bis(trifluoromethyl)acetophenone [105] on CD-modified Pt/Al_2O_3. The latter reaction, including some other acetophenone derivatives [106], resembles the hydrogenation of activated ketones, but in this case, the carbonyl group is activated by electron-withdrawing groups at the aromatic ring. More recently, the development of innovative chirally modified metals allowed for very significant progress in the hydrogenation of both substituted and unsubstituted acetophenones. A few groups reported the use of silica or hydroxyapatite-supported Ir modified with *Cinchona* alkaloids and/or chiral amines [such as (1*S*,2*S*)-diphenylethylenediamine (DPEN)], or alumina-supported Ru in the presence of (1*S*,2*S*)-DPEN and a phosphine, or even simple acetophenone, as efficient catalysts that allow for high conversion and enantioselectivities of aromatic ketones and ee values up to 80% [107–111]. Furthermore, Chinese scientists reported that the addition of a phosphine to the CD-Ir/Al_2O_3 system is able to boost the performance of the catalyst to full conversion at enantioselectivities of about 90% ee for the hydrogenation of acetophenone [112].

6.2.2. Hydrogenation of C=C Bonds

Chirally modified metals are effective only in the hydrogenation of functionalized olefins and aromatic compounds, and the best choice is usually supported Pd modified by a *Cinchona* or vinca alkaloid [113–115]. We will therefore focus on this restricted substrate pool. Four main classes can be identified: (1) unsaturated carboxylic acids and esters, (2) unsaturated ketones, (3) pyrones, and (4) aromatic compounds. The more prominent examples in the asymmetric hydrogenation within these systems are illustrated in Table 6.3, and details on the structures and efficient catalytic systems developed and employed are given in the following sections.

6.2.2.1. Unsaturated Carboxylic Acids and Esters. Among this group we have to distinguish between aromatic and aliphatic unsaturated carboxylic acids. Concerning the former case, we actually consider unsaturated acids with an aryl substituent in the β-position. The first report on the hydrogenation of α,β-unsaturated carboxylic acids dates back to the early 1960s [116] and continuous improvement (mainly by Nitta [117–122]) resulted in a highly selective transformation of some aryl-substituted acids (Table 6.3) [123, 124]. The only useful catalyst is *Cinchona*-modified Pd; other metals [125–127] and modifiers [128, 129] are poorly selective. Compared to the hydrogenation of ketones by the CD–platinum system, hydrogenation of α,β-unsaturated carboxylic acids

TABLE 6.3. Enantioselective Hydrogenation of C=C Double Bonds

Substrate	Modifier ee (%)
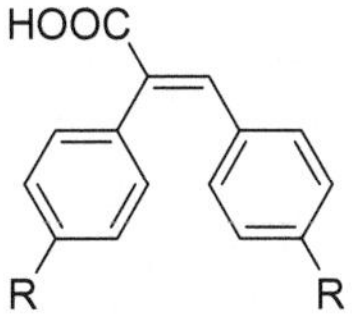	**CD** R=H 81% R=OMe 92%
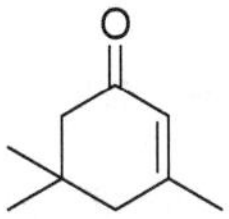	**DHVIN** 55%
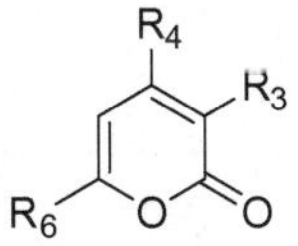	**CD/CN**
R_3, R_4, R_6	
H, OH, CH_3	85% (*S*)[a]
H, OCH_3, CH_3	94% (*R*)
H, OCH_2CH_3, CH_3	85% (*R*)
CH_3, OCH_3, CH_3	75% (*R*)
H, OCH_3, styryl	89% (*R*)
H, OCH_3, 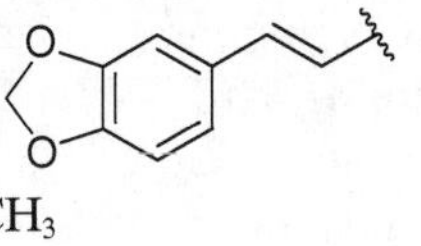	90% (*R*)
H, CH_3, CH_3	—
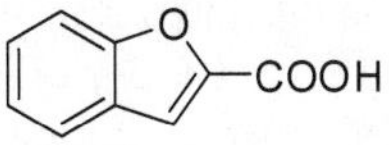	**CD** 50%

[a] Modifier CD.
DHVIN, dihydroapovincaminic acid ethyl ester.

requires a smaller substrate/modifier ratio of 25 [130] and addition of a base (e.g., benzylamine) in an almost equivalent amount. The effect of the latter is to deprotonate the substrate, weakening the adsorption of the product on Pd [131]. This method is applied in the synthesis of L-DOPA via reduction of the corresponding cinnamic acid derivative [132, 133]. Reported enantioselectivities in the hydrogenation of aliphatic α,β-unsaturated carboxylic acids are less impressive [134], with the ee varying in the range between 20% and 66% depending on the functionalization of the C=C bond [25, 135–138]. A major reason for this is the rapid isomerization of the alkenoic acid on the Pd surface, a side reaction that is accelerated by the amine-type modifier [139, 140]. Good enantioselectivities could be achieved only when the double bond migration is slow (internal C=C bond), or it is not possible at all (α-phenyl-cinnamic acid derivatives). An additional functional group in the molecule in the α-position to the C=C group, interacting with the alkaloid, makes the Pd–*Cinchona* system still effective, as it is for enol esters [141], enamines [142], and *N*-acetyl dehydroamino acid esters [143–145].

6.2.2.2. Unsaturated Ketones. The majority of the data reported in the literature refers to reactions carried out in the presence of a stoichiometric amount of proline [146–152]. The best ee's achieved in the hydrogenation of α,β-unsaturated ketones involve chirally modified Pd and are around 50%.

Concerning the hydrogenation of isophorone (a rather extensively studied reaction), Tungler's group [146–149] reported that the role of proline is to act as a chiral auxiliary and to form an adduct with isophorone that is hydrogenated diastereoselectively. Interestingly, it has recently been proposed that the origin of enantioselection is the kinetic resolution of the racemic hydrogenation product by interaction with proline in the homogeneous phase, and the Pd surface is not involved in the enantiodifferentiating step [153]. Besides proline, only the vinca alkaloid derivative (−)-dihydroapovincaminic (DHVIN) acid ethyl ester, among other investigated modifiers [154], is similarly efficient in the hydrogenation of isophorone (Table 6.3) [30, 155–161]. CD-modified Pd black is the choice for the hydrogenation of an exocyclic alkenone, 2-benzylidene-1-benzosuberone, to the corresponding saturated ketone (54% ee) [162].

6.2.2.3. Pyrones. Hydrogenation of the pseudoaromatic 4-hydroxy, 4-alkoxy, and 4-methyl derivatives of 2-pyrones to the corresponding dihydro- and tetrahydropyrones is highly selective on *Cinchona*-modified Pd. Structural effects within the substrate have been explored, and Table 6.3 illustrates the best substituting groups [163]. Partial hydrogenation of the 4-hydroxy derivative was slow and is complicated by the saturation of the quinoline ring of the *Cinchona* alkaloid [164]. Replacement of the acidic OH function by a methyl or methoxy group eliminated this difficulty, affording good yields and ee's to the corresponding dihydropyrones in short times. Only in the case that stabilization by the methoxy group at the 4-position was missing, the separation between the uptakes of 2 eq. of hydrogen was poor and the tetrahydropyrone formed with 99% diastereomeric excess (de).

6.2.2.4. Aromatic Compounds. Among the metals tested for the enantioselective hydrogenation of aromatic compounds, Rh [165, 166] and Ni [167–169] were barely successful, and the ee remained in the single-digit region. *Cinchona*-modified Pd exhibited more promising performances for the hydrogenation of furan and benzofuran carboxylic acids, although the enantioselectivities remained at a medium level (Table 6.3) [170]. The reactions are relatively slow, and the competing hydrogenation of the quinoline ring of CD necessitates high modifier/substrate (M/S) ratios (2–15 mol%). *Cinchona* alkaloids and other modifiers that possess an aromatic ring seemed not suitable for demanding hydrogenation reactions, particularly not on Pd and Rh, which are highly active for this side reaction [171, 172].

6.2.3. Hydrogenation of C=N Bonds

Enantioselective saturation of C=N bonds is a challenging reaction, which finds limited interest on a synthetic viewpoint. Therefore, only a few attempts have been made to hydrogenate such bonds, and the success was rather poor and limited to the use of chirally modified Pd. To cite a few examples, low ee was achieved in the hydrogenation of a Schiff base on Pd/SiO_2 modified by L-alaninol or L-phenylalaninol and in the reduction of an *N*-alkyl-α-iminoester on *Cinchona*-modified Pd/Al_2O_3 [173]. The enantioselectivity was slightly higher in some other reactions, but only when the modifier was applied in a stoichiometric amount [144, 174].

6.3. THE DEVELOPMENT OF A MODEL FOR THE *CINCHONA* ALKALOID MODIFIED PLATINUM ASYMMETRIC HYDROGENATION

The preceding sections show that the scope of the catalysis by chirally modified metals has been extended to Pt, Pd, Ni, Rh, and Ir metals, utilizes *Cinchona* alkaloids, modified *Cinchona* alkaloids, peptides, or synthetic modifiers as chiral amplifiers, and can be used to asymmetrically hydrogenate ketones (activated or not) as well as C=C double bonds. Though the scope is broad, the mechanistic aspects have been thoroughly investigated by means of atomistic simulations and spectroscopy only for the platinum–*Cinchona* system, while the other mentioned systems suffer from a lack of mechanistic insight. Nonetheless, the investigation patterns used for the study of the platinum–*Cinchona* system can, in principle, be used to gain insight into other catalytic chirally modified surfaces. We will therefore focus on the development of the models used for the rationalization of this catalytic system.

Early work aimed at understanding the functioning of chiral platinum surface modification that proceeded via the synthesis of chiral molecules bearing features similar to those of the *Cinchona* alkaloids, and identified three critical features of the surface modifier appropriate to rationalize the occurrence of enantioselectivity. These are shown in Figure 6.1: (1) an anchoring group, that is, an aromatic moiety that allows the binding of the chiral molecule to the metal; (2) a stereogenic region, providing the chiral environment for the asymmetric reaction; and (3) a tertiary amine bound to the anchoring moiety [175]. Previous investigations had already shown that *N*-methylation of the quinuclidine nitrogen led to a complete loss of enantioselectivity, which indicated the crucial role of the amine group for enantioselection [176]. Spectroscopic studies showed that hydrogen bonding between adsorbed

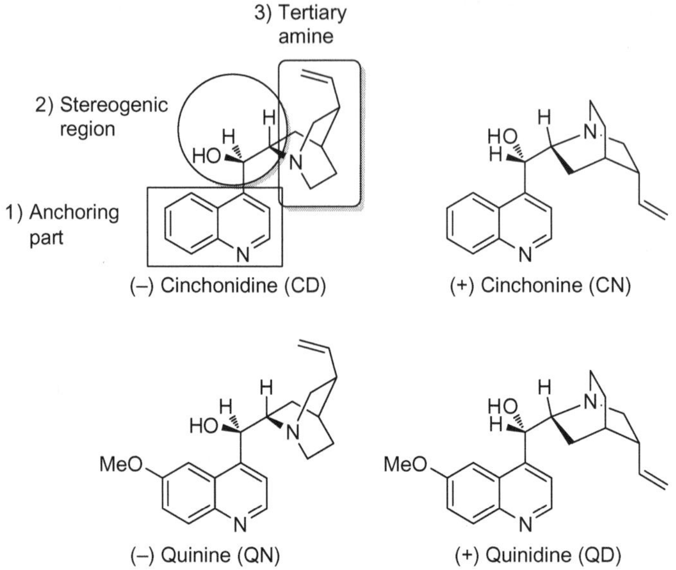

FIGURE 6.1. Structures of the cinchona alkaloids and subdivision of the structure of CD in the anchoring part, the stereogenic region, and tertiary amine.

alkaloid and KPL occurs at the metal surface [177]. The insight provided by the subdivision of the modifier into functional structural units, each having a defined role in the process of enantiodifferentiation, led to the design of a novel synthetic surface modifier, and, importantly, to the recognition that surface chiral sites can be rationally generated and used to perform asymmetric catalysis [178]. Computational modeling was applied to provide further submolecular understanding. The first molecular modeling investigations showed that the CD can bind a prochiral substrate by means of hydrogen bonding to the protonated tertiary amine, thus selectively adjusting its position toward the activated hydrogen present on the surface [179–183]. The Pro(*R*) binding was shown to be more favorable than the Pro(*S*) binding, in agreement with the observed enantioselectivity. Conceptually this early computational model portrayed an extreme simplification of the real system, since it neglected the surface and solvent, made use of a classical force field, and basically only considered the unconstrained binding between substrate and modifier. Nevertheless, this first simple model contained an essential element: chiral recognition between substrate and modifier. In addition, it expressed the concept that such recognition could be modeled to obtain qualitative and quantitative insight. Successive work would have a starting point for further improvements. Early refinements of the computational methods included the passage from classical potentials to a quantum chemical description of the substrate/modifier interaction (and therefore a better description of their interaction potential), and included the setting of planar constraints to the anchoring moiety to simulate the effect of chemisorption to platinum [184, 185]. The model was applied both to CD and to synthetic modifiers, and predicted the correct sense of enantiodifferentiation with excellent quantitative agreement [184, 185]. One of the aspects of this approach was that it considered the enantiodifferentiation process as the result of the sole enthalpic difference between Pro(*R*) and Pro(*S*) binding modes of the modifier. More strongly bound configurations would result in higher surface equilibrium concentration and therefore in enantiodiscrimination. The 1:1 interaction model, computationally implemented and refined, was still limited to an enthalpic difference. On the other hand, the well-known phenomenon of rate acceleration typically observed pointed toward the occurrence of kinetic resolution.

Since the number of available metal sites was reduced by the presence of strong adsorbates (modifiers), the presence of the surface modifiers on platinum should have, in principle, generated slower kinetics of the hydrogenation reaction, while as a matter of fact, the opposite was observed. *Cinchona* alkaloids accelerated the rate of hydrogenation, showing the important role of the modifier not only as a shape-differentiating surface-binding agent, but also as a kinetic accelerator, which raised the question as to whether the observed enantiodiscrimination also had kinetic origin. In other words, the question was whether the modifier was selectively reducing the transition state energy of one of the possible reaction pathways. It was then observed that a correlation existed between the stabilization energy of the Hartree–Fock keto-carbonyl orbitals and the reaction rate of hydrogenation of the same group [186, 187]. The more the α-substituent to the keto-carbonyl lowered the energy of its frontier orbitals, the faster was its hydrogenation kinetics. It was also calculated that a similar stabilization effect could arise due to the hydrogen bonding between a keto-carbonyl group and the quinuclidine tertiary amine of a *Cinchona* alkaloid. Given the correlation between orbital energy and hydrogenation rate, it resulted that the binding of a ketone to a *Cinchona* alkaloid would lead to an increase in the platinum surface hydrogenation rate of the carbonyl moiety. Furthermore, the orbital stabilization followed the potential energy, so that the most stable interaction complex [the Pro(*R*)] resulted in larger orbital stabilization than that of the less stable [the Pro(S)], thus predicting that the *R* configuration of the product would be produced at a faster rate than the *S* configuration, as by experimental observation (Fig. 6.2c) [186, 187]. This frontier orbital energy model was successively applied to investigate the kinetic effects of the enantioselective hydrogenation of α-hydroxyketones [188–191]. Recently, a more classical interpretation of the kinetic effects of this catalytic system has also been proposed: rate acceleration is, in fact, a well-known phenomenon in enzyme catalysis, and Page and Jencks had already observed that the specific binding process reduces the free energy of activation of reactions occurring in chiral binding sites [192]. In the same manner, the *Cinchona*-alkaloid modified platinum can be interpreted as a biomimetic structure where specific binding occurs, with corresponding increase of local concentration and therefore increase in rate [193].

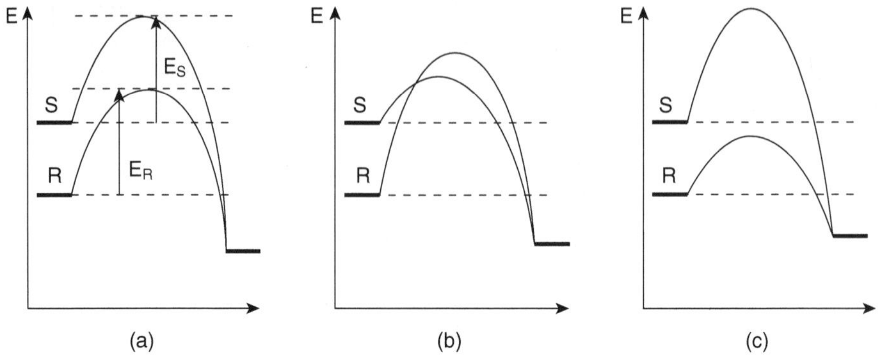

FIGURE 6.2. Thermodynamic and kinetic factors influencing enantioselectivity. If the diastereomeric complex leading to the *R* enantiomer is more stable, three cases can be distinguished: (a) *R* and *S* can be formed with the same rate, or (b) *R* can be formed with a slower or (c) with a faster rate with respect to *S*.

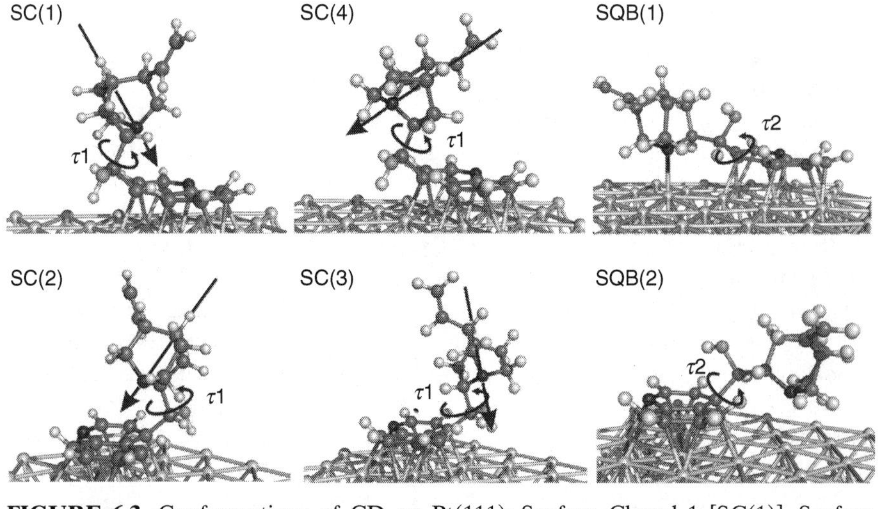

FIGURE 6.3. Conformations of CD on Pt(111): Surface Closed 1 [SC(1)], Surface Open 4 [SO(4)], Surface Quinuclidine Bound 1 [SQB(1)], Surface Closed 1 [SC(2)], Surface Open 3 [SO(3)], and Surface Quinuclidine Bound 2 [SQB(2)]. Positioning and hydrogenation of the substrate imply the pointing of the quinuclidine nitrogen toward the surface, from where the activated hydrogen will be delivered.

Formation of surface sites critically depends on the adsorption of the modifier, and since *Cinchona* alkaloids have a complex conformational behavior [194–197], the topology of the sites formed upon interaction with the metal also shows conformational complexity. These were first investigated using ATR-IR spectroscopy [198, 199], and successively those adsorption modes that appeared more relevant in catalysis were also investigated through modeling studies, with the explicit inclusion of the metal surface [193, 200, 201]. Chemisorption modes of CD on platinum are illustrated in Figure 6.3.

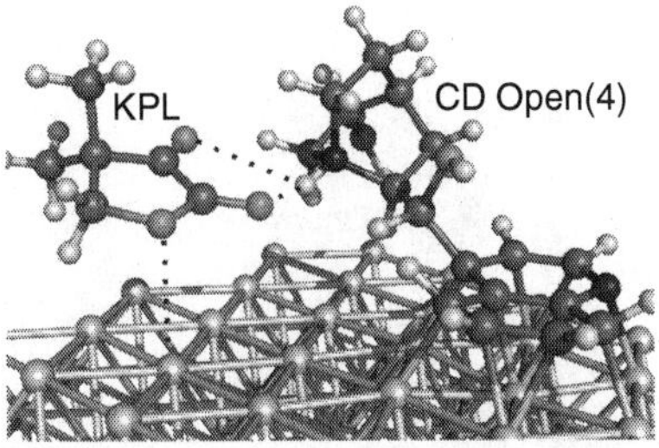

FIGURE 6.4. Docking structure of KPL in the chiral pocket of adsorbed Open(4) CD.

By rotation of the dihedral angles $\tau 1$ and $\tau 2$, CD can assume several conformations, of which only those where the quinuclidine nitrogen can point toward the surface are active. Blocking of an *anti*-Open(3) conformation in α-isocinchonidine showed the relevance of the Open conformations with respect to the Closed ones [202].

Further modeling studies identified the docking structures of CD and KPL (Fig. 6.4). The chiral pocket formed by an adsorbed *Cinchona* alkaloid and by the metal surface can adjust an incoming substrate and selectively position it in order to expose a Pro(*R*) face toward the surface hydrogen. The most relevant elements are the adsorption mode of the alkaloid, the conformational flexibility, and the ability of the tertiary amine to collect hydrogen from the surface [203] and form hydrogen bonding interactions with the substrate.

A key aspect in the chemistry of chirally modified surfaces lies in the binding of the modifier to the metal surface. Efficient metal surface modification occurs only in the presence of stable chiral sites, and their disruption, typically due to the hydrogenation of the anchoring moiety, leads to a reduction of the enantiodifferentiation [204]. Once the aromatic group is totally or partially saturated, the adsorption strength of the alkaloid decreases with its consequent desorption from the surface. The importance of an anchoring moiety in generating tunable catalytic systems is underlined by the observation that different alkaloids or alkaloid ethers may have different adsorption energies, which is at the origin of the so-called nonlinear effects. In practice, two modifiers with different adsorption strengths, if present in the same reaction medium, can compete for surface sites, inducing a bias in enantioselection proportional to their difference in adsorption energy and concentration. If the two modifiers induce opposite enantioselectivities, the catalytic system has been shown to act as a chemical device able to switch its enantioselective properties by simple addition of a competing modifier [177, 205–207]. In particular, *O*-phenyl-cinchonidine and *O*-[3,5-bis(trifluoromethyl)phenyl]cinchonidine afford opposite enantiomers in the hydrogenation of KPL just by opening and closing the chiral space with a simple conformational rearrangement of the phenyl moiety (Fig. 6.5) [207].

On the side of the submolecular tailoring of chiral sites, the tunability of the chiral environment generated by modified *Cinchona* alkaloids was shown

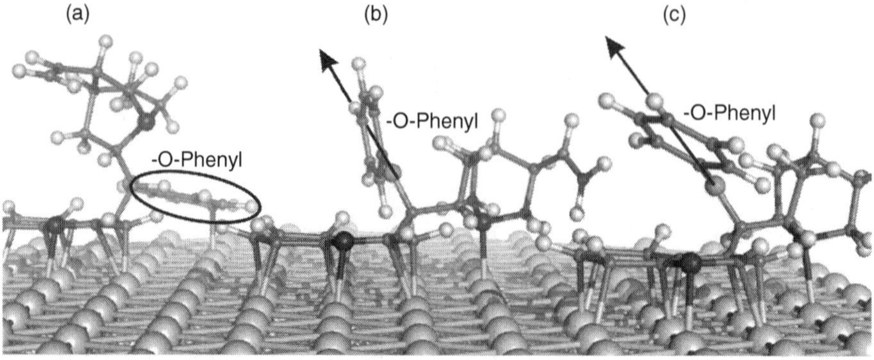

FIGURE 6.5. Adsorption of *O*-phenyl-cinchonidine on Pt(111): the phenyl group can be (a) chemisorbed on the metal, (b) inside the chiral pocket, or (c) outside the chiral pocket. Surface conformations (a) and (b) dominate, interfering with the docking of the substrate, and resulting in an inversion of enantioselectivity compared to CD.

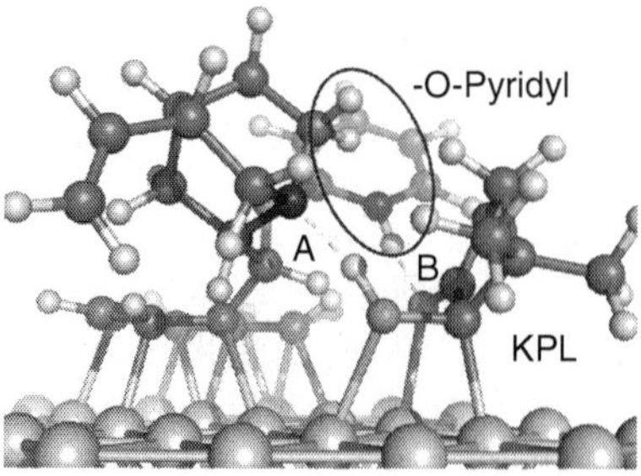

FIGURE 6.6. *O*-pyridoxy CD binds KPL in the chiral pocket with a double hydrogen bonding interaction: (A) through the quinuclidine nitrogen and (B) through the *O*-pyridyl group. This second binding does not occur for the *O*-phenyl-cinchonidine. Therefore, inversion of enantioselectivity is observed.

by combined spectroscopic and theoretical investigation of several CD ethers [205–209], among which are the already introduced *O*-phenyl-cinchonidine, and *O*-pyridyl-cinchonidine [210]. The latter is of particular relevance since within such systems, the *O*-pyridyl moiety contributes to the binding in the chiral pocket, thus inverting the enantioselectivity as compared to the *O*-phenyl-cinchonidine (Fig. 6.6). In fact, the phenyl group is not able to generate a binding interaction as the *O*-pyridyl moiety; on the contrary, it generates steric repulsion [210]. The example of *O*-pyridyl-cinchonidine shows that guiding functional groups can be added to a chiral surface scaffold in order to affect the selective binding of a prochiral substrate. Such investigations illustrate the principle that the selectivity of a surface modifier can be adjusted or fine-tuned by chemically controlling binding interactions; for example, hydrogen bonding in the case of the *O*-pyridyl moiety, and repulsive interactions, as in the case of the sterical hindrance induced by the phenyl group of *O*-phenyl-cinchonidine.

6.4. CONCLUSIONS

Chirally modified metal surfaces are excellent catalysts for the asymmetric hydrogenation of a wide range of ketones and C=C double bonds. The general principle under which such systems operate is the chiral modification of a catalytically active metal surface, by means of an organic molecule. The metal activates molecular hydrogen, while the chiral modifier, typically a complex organic molecule, furnishes the scaffold for enantiodiscrimination. A prochiral substrate enters a chiral pocket generated by a surface-bound organic molecule, and is selectively hydrogenated on the metal surface. Such catalysts are, in all respects, bifunctional, and allow design and tailoring by tuning the relative activity of metal and modifier. Metals can be chosen according to their reduction potential that must allow for fast and selective substrate hydrogenation, and at the same time must be sufficiently mild so as not to destroy the organic chiral sites. Modifiers, on the other hand, must generate suitable surface chiral sites, be able to discriminate the faces of a prochiral substrate, and at the same time must be stable in the reaction condition used. The scope of such reaction systems has constantly grown in the past years, and for many applications it can already challenge the more established homogeneous catalysis asymmetric hydrogenation strategies.

REFERENCES

[1] Halpern, J., Trost, B. M. (2004). Asymmetric catalysis. *Proc. Natl. Acad. Sci. U.S.A.*, *101*, 5347–5347.

[2] Trost, B. M. (2004). Asymmetric catalysis: an enabling science. *Proc. Natl. Acad. Sci. U.S.A.*, *101*, 5348–5355.

[3] Mallat, T., Diezi, S., Baiker, A. (2008). Chiral surfaces and enantioselective catalysis. In: G. Ertl, H. Knözinger, F. Schüth, J. Weitkamp (Eds.), *Handbook of Heterogeneous Catalysis*. Wiley-VCH, Weinheim, Germany, pp. 3603–3626.

[4] Heitbaum, M., Glorius, F., Escher, I. (2006). Asymmetric heterogeneous catalysis. *Angew. Chem. Int. Ed.*, *45*, 4732–4762.

[5] Mallat, T., Orglmeister, E., Baiker, A. (2007). Asymmetric catalysis at chiral metal surfaces. *Chem. Rev.*, *107*, 4863–4890.

[6] Blaser, H. U. (2009). Cinchona alkaloids as chirality transmitters in metal-catalyzed asymmetric reductions. In: C. E. Song (Ed.), *Cinchona Alkaloids in Synthesis & Catalysis*. Wiley-VCH, Weinheim, Germany, pp. 13–28.

[7] Ma, Z. F. Z. (2009). Chiral modification of catalytic surfaces. In: U. S. Ozkan (Ed.), *Design of Heterogeneous Catalysts*. Wiley-VCH, Weinheim, Germany, pp. 113–140.

[8] Orito, Y., Imai, S., Niwa, S., Nguyen, G. H. (1979). Asymmetric hydrogenation of methyl benzoylformate using platinum-carbon catalysts modified with cinchonidine. *J. Syn. Org. Chem. Jpn.*, *37*, 173–174.

[9] Wehrli, J. T. (1989) Diss *ETH Zürich* Nr. 8833—Enantioselektive Hydrierung von Alpha-ketoestern mit Modifizierten Platin Katalysatoren.

[10] Bartók, M. (2006). Heterogeneous catalytic enantioselective hydrogenation of activated ketones. *Curr. Org. Chem.*, *10*, 1533–1567.

[11] Murzin, D. Y., Maki-Arvela, P., Toukoniitty, E., Salmi, T. (2005). Asymmetric heterogeneous catalysis: science and engineering. *Catal. Rev.*, *47*, 175–256.

[12] Studer, M., Blaser, H. U., Exner, C. (2003). Enantioselective hydrogenation using heterogeneous modified catalysts: an update. *Adv. Synth. Catal.*, *345*, 45–65.

[13] Blaser, H. U., Jalett, H. P., Monti, D. M., Reber, J. F., Wehrli, J. T. (1988). Modified heterogeneous platinum catalysts for the enantioselective hydrogenation of alpha-ketoesters. *Stud. Surf. Sci. Catal.*, *153*(41), 153–163.

[14] Toukoniitty, E., Maki-Arvela, P., Neyestanaki, A. K., Salmi, T., Murzin, D. Y. (2002). Continuous hydrogenation of 1-phenyl-1,2-propanedione under transient and steady-state conditions: regioselectivity, enantio selectivity and catalyst deactivation. *Appl. Catal. A*, *235*, 125–138.

[15] Minder, B., Schürch, M., Mallat, T., Baiker, A. (1995). Chiral nitrogen-compounds as new modifiers for the enantioselective hydrogenation of ethyl pyruvate. *Catal. Lett.*, *31*, 143–151.

[16] Xiong, W., Ma, H. X., Hong, Y. Y., Chen, H., Li, X. J. (2005). Enantioselective hydrogenation of ethyl pyruvate catalyzed by alumina support rhodium modified with quinine. *Tetrahedron Asymmetry*, *16*, 1449–1452.

[17] Balàzsik, K., Török, B., Szakonyi, G., Bartók, M. (1999). Homogeneous and heterogeneous asymmetric reactions. Part X: enantioselective hydrogenations over K-10 montmorillonite supported noble metal catalysts with immobilized modifier. *Appl. Catal. A*, *182*, 53–63.

[18] Huang, Y. L., Chen, J. R., Chen, H., Li, R. X., Li, Y. Z., Min, L. E., Li, X. J. (2001). Enantioselective hydrogenation of ethyl pyruvate catalyzed by PVP-stabilized rhodium nanoclusters. *J. Mol. Catal. A Chem.*, *170*, 143–146.

[19] Huang, Y. L., Li, Y. Z., Hu, J. Y., Cheng, P. M., Chen, H., Li, R. X., Li, X. J., Yip, C. W., Chan, A. S. C. (2002). Enantioselective hydrogenation of ethyl pyruvate catalyzed by polyvinylpyrrolidone-stabilized and supported rhodium nanocluster. *J. Mol. Catal. A Chem.*, *189*, 219–224.

[20] Ma, H. X., Chen, H., Zhang, Q., Li, X. J. (2003). Enantioselective hydrogenation of ethyl pyruvate catalyzed by TS-1 supported rhodium nanoclusters. *J. Mol. Catal. A Chem.*, *196*, 131–135.

[21] Faraoni, M. B., Ayala, A. D., Vetere, V., Casella, M. L., Ferretti, O. A., Podesta, J. C. (2005). Synthesis of chiral organotins suitable for the preparation of asymmetric heterogeneous catalysts. *Appl. Organomet. Chem.*, *19*, 465–472.

[22] Maris, M., Mallat, T., Baiker, A. (2005). Rhodium-catalyzed enantioselective hydrogenation of ketopantolactone. *J. Mol. Catal. A Chem.*, *242*, 151–155.

[23] Maris, M., Ferri, D., Konigsmann, L., Mallat, T., Baiker, A. (2006). Why are alpha-hydroxycarboxylic acids poor chiral modifiers for Pt in the hydrogenation of ketones? *J. Catal.*, *237*, 230–236.

[24] Bhaduri, S., Darshane, V. S., Sharma, K., Mukesh, D. (1992). Carbonyl cluster derived polystyrene supported platinum for asymmetric hydrogenation of alpha-ketoesters. *J. Chem. Soc. Chem. Comm.*, 1738–1740.

[25] Hall, T. J., Johnston, P., Vermeer, W. A. H., Watson, S. R., Wells, P. B. (1996). Enantioselective hydrogenation catalysed by palladium. *Stud. Surf. Sci. Catal.*, *101*, 221–230.

[26] Tungler, A., Fodor, K., Mathe, T., Sheldon, R. A. (1997). Enantioselective hydrogenation of ethyl pyruvate and isophorone over modified Pt and Pd catalysts. In: H. U. Blaser, A. Baiker, R. Prins (Eds.), *Heterogeneous Catalysis and Fine Chemicals IV*. Elsevier, Amsterdam, pp. 157–165.

[27] Mallat, T., Szabo, S., Schurch, M., Gobel, U. W., Baiker, A. (1997). Enantioselective hydrogenation of carbonyl compounds over Pd-Pt/alumina catalysts. *Catal. Lett.*, *47*, 221–227.

[28] Török, B., Balàzsik, K., Kun, I., Szöllösi, G., Szakonyi, G., Bartók, M. (1999). Homogeneous and heterogeneous asymmetric reactions. Part 13. Clay supported noble metal catalysts in enantioselective hydrogenations. In: I. Kiricsi, G. PalBorbely, J. B. Nagy, H. G. Karge (Eds.), *Porous Materials in Environmentally Friendly Processes*. Elsevier, Amsterdam, pp. 515–522.

[29] Collier, P. J., Iggo, J. A., Whyman, R. (1999). Preparation and characterisation of solvent-stabilised nanoparticulate platinum and palladium and their catalytic behaviour towards the enantioselective hydrogenation of ethyl pyruvate. *J. Mol. Catal. A Chem.*, *146*, 149–157.

[30] Sipos, E., Tungler, A., Bitter, I. (2003). Solvent and support effects in enantioselective hydrogenation of isophorone with (*S*)-alpha,alpha-diphenyl-2-pyrrolidinemethanol modified palladium. *React. Kinet. Catal. Lett.*, *79*, 101–109.

[31] Simons, K. E., Ibbotson, A., Johnston, P., Plum, H., Wells, P. B. (1994). Enantioselective hydrogenation. 3. Methyl pyruvate hydrogenation catalyzed by alkaloid-modified iridium. *J. Catal.*, *150*, 321–328.

[32] Zuo, X. B., Liu, H. F., Yue, C. (1999). Enantioselective hydrogenation of methyl pyruvate over polymer-stabilized and supported iridium clusters. *J. Mol. Catal. A Chem.*, *147*, 63–72.

[33] Marzialetti, T., Fierro, J. L. G., Reyes, P. (2005). Iridium-supported catalyst for enantioselective hydrogenation of 1-phenyl-1,2-propanedione: the effects of the addition of promoter and the modifier concentration. *Catal. Today.*, *107–108*, 235–243.

[34] Huck, W. M., Mallat, T., Baiker, A. (2003). Non-linear effect of modifier composition on enantioselectivity in asymmetric hydrogenation over platinum metals. *Adv. Synth. Catal.*, *345*, 255–260.

[35] LeBlond, C., Wang, J., Liu, J., Andrews, A. T., Sun, Y. K. (1999). Highly enantioselective heterogeneously catalyzed hydrogenation of alpha-ketoesters under mild conditions. *J. Am. Chem. Soc.*, *121*, 4920–4921.

[36] Szöri, K., Balàzsik, K., Felföldi, K., Bartók, M. (2006). Study of enantioselective hydrogenation of bulky esters of phenylglyoxylic acid on Pt-CD and Pt-beta-ICN chiral catalysts: steric effect of ester groups and inversion of enantio selectivity. *J. Catal.*, *241*, 149–154.

[37] Baiker, A., Blaser, H. U. (1997). Enantioselective Catalysts and Reactions. In: G. Ertl, H. Knözinger, J. Weitkamp (Eds.) *Handbook of Heterogeneous Catalysis*. VCH, Weinheim, Germany, p. 2422.

[38] Zuo, X. B., Liu, H. F., Liu, M. H. (1998). Asymmetric hydrogenation of alpha-ketoesters over finely dispersed polymer-stabilized platinum clusters. *Tetrahedron Lett.*, *39*, 1941–1944.

[39] Balàzsik, K., Szöri, K., Felföldi, K., Törok, B., Bartók, M. (2000). Asymmetric synthesis of alkyl 5-oxotetrahydrofuran-2-carboxylates by enantioselective hydrogenation of dialkyl 2-oxoglutarates over cinchona modified Pt/Al_2O_3 catalysts. *Chem. Commun.*, 555–556.

[40] Szöri, K., Sutyinszki, M., Felföldi, K., Bartók, M. (2002). Heterogeneous asymmetric reactions Part 28. Efficient and practical method for the preparation of (*R*)- and (*S*)-alpha-hydroxy esters by the enantioselective heterogeneous catalytic hydrogenation of alpha-ketoesters. *Appl. Catal. A*, *237*, 275–280.

[41] Sutyinszki, M., Szöri, K., Felföldi, K., Bartók, M. (2002). Heterogeneous asymmetric reactions. 29. Enantioselective hydrogenation of ethyl benzoylformate over dihydrocinchonidine-modified platinum-alumina catalyst in acetic acid. *Catal. Lett.*, *81*, 281–284.

[42] Török, B., Balàzsik, K., Török, M., Felföldi, K., Bartók, M. (2002). Heterogeneous asymmetric reactions 20. Effect of ultrasonic variables on the enantiodifferentiation in cinchona-modified platinum-catalyzed sonochemical hydrogenations. *Catal. Lett.*, *81*, 55–62.

[43] Diezi, S., Reimann, S., Bonalumi, N., Mallat, T., Baiker, A. (2006). Steric and electronic effects in the enantioselective hydrogenation of activated ketones on platinum: directing effect of ester group. *J. Catal.*, *239*, 255–262.

[44] Szöri, K., Török, B., Felföldi, K., Bartók, M. (2000). Enantioselective hydrogenation of alpha-ketoesters over a Pt/Al_2O_3 catalyst: effects of steric constraints on the enantioselection. In: M. E. Ford (Ed.), *Catalysis of Organic Reactions*. M. Dekker, New York, p. 489.

[45] Bonalumi, N., Vargas, A., Ferri, D., Burgi, T., Mallat, T., Baiker, A. (2005). Competition at chiral metal surfaces: fundamental aspects of the inversion of enantioselectivity in hydrogenations on platinum. *J. Am. Chem. Soc.*, *127*, 8467–8477.

[46] Niwa, S., Imamura, F., Otsuka, K. (1987). JPN Patent 62-158268, Nissan Chemical Industries, Tokyo.

[47] Schürch, M., Künzle, N., Mallat, T., Baiker, A. (1998). Enantioselective hydrogenation of ketopantolactone: effect of stereospecific product crystallization during reaction. *J. Catal.*, *176*, 569–571.

[48] Marinas, A., Mallat, T., Baiker, A. (2004). 1-Naphthyl-1,2-ethanediol as a new chiral modifier of platinum in the enantio selective hydrogenation of activated ketones. *J. Catal.*, *221*, 666–669.

[49] Orglmeister, E., Mallat, T., Baiker, A. (2005). Quaternary ammonium derivatives of cinchonidine as new chiral modifiers for platinum. *J. Catal.*, *233*, 333–341.

[50] Mondelli, C., Vargas, A., Santarossa, G., Baiker, A. (2009). Fundamental aspects of the chiral modification of platinum with peptides: asymmetric induction in hydrogenation of activated ketones. *J. Phys. Chem. C*, *113*, 15246–15259.

[51] Blaser, H. U., Jalett, H. P. (1993). Enantioselective hydrogenation of alpha-ketoacids using platinum catalysts modified with cinchona alkaloids. In: M. Guisnet, J. Barbier, J. Barrault, C. Bouchoule, D. Duprez, G. Perot, C. Montassier (Eds.), *Heterogeneous Catalysis and Fine Chemicals III*. Elsevier, Amsterdam, pp. 139–146.

[52] Felföldi, K., Szöri, K., Bartók, M. (2003). Heterogeneous asymmetric reaction—Part 35. Enantioselective hydrogenation of 2-oxoglutaric acid over cinchona modified Pt/Al_2O_3 catalysts. *Appl. Catal. A*, *251*, 457–460.

[53] Slipszenko, J. A., Griffiths, S. P., Johnston, P., Simons, K. E., Vermeer, W. A. H., Wells, P. B. (1998). Enantioselective hydrogenation—V. Hydrogenation of butane-2,3-dione and of 3-hydroxybutan-2-one catalysed by cinchona-modified platinum. *J. Catal.*, *179*, 267–276.

[54] Vermeer, W. A. H., Fulford, A., Johnston, P., Wells, P. B. (1993). New enantioselective reactions catalyzed by cinchonidine-modified platinum. *J. Chem. Soc. Chem. Comm.*, 1053–1054.

[55] Toukoniitty, E., Maki-Arvela, P., Nieminen, V., Salmi, T., Murzin, D. Y. (2003). Towards a molecular understanding of asymmetric heterogeneous catalysis: hydrogenation of 1-phenyl-1,2-propanedione. *Kinet. Catal.*, *44*, 562–571.

[56] Sonderegger, O. J., Bürgi, T., Baiker, A. (2003). Asymmetric hydrogenation of cyclohexane-1,2-dione over cinchonidine-modified platinum. *J. Catal.*, *215*, 116–121.

[57] Studer, M., Okafor, V., Blaser, H. U. (1998). Hydrogenation of butane-2,3-dione with heterogeneous cinchona modified platinum catalysts: a combination of an enantioselective reaction and kinetic resolution. *Chem. Commun.*, 1053–1054.

[58] Lindholm, A., Maki-Arvela, P., Toukoniitty, E., Pakkanen, T. A., Hirvi, J. T., Salmi, T., Murzin, D. Y., Sjoholm, R., Leino, R. (2002). Hydrosilylation of cinchonidine and 9-*O*-TMS-cinchonidine with triethoxysilane: application of 11-(triethoxysilyl)-10,11-dihydrocinchonidine as a chiral modifier in the enantioselective hydrogenation of 1-phenylpropane-1,2-dione. *J. Chem. Soc. Perk. Trans. 1*, 2605–2612.

[59] Busygin, I., Toukoniitty, E., Leino, R., Murzin, D. Y. (2005). Effect of modifier structure in asymmetric 1-phenylpropane-1,2-dione hydrogenation. *J. Mol. Catal. A Chem.*, *236*, 227–238.

[60] Wang, G. Z., Mallat, T., Baiker, A. (1997). Enantioselective hydrogenation of alpha-ketoamides over Pt/Al_2O_3 modified by cinchona alkaloids. *Tetrahedron Asymmetry*, *8*, 2133–2140.

[61] Cederbaum, F., Lamberth, C., Malan, C., Naud, F., Spindler, F., Studer, M., Blaser, H. U. (2004). Synthesis of substituted mandelic acid derivatives via enantioselective hydrogenation: homogeneous versus heterogeneous catalysis. *Adv. Synth. Catal.*, *346*, 842–848.

[62] Sonderegger, O. J., Bürgi, T., Limbach, L. K., Baiker, A. (2004). Enantio selective reduction of isatin derivatives over cinchonidine modified Pt/alumina. *J. Mol. Catal. A Chem.*, *217*, 93–101.

[63] Künzle, N., Szabo, A., Schürch, M., Wang, G. Z., Mallat, T., Baiker, A. (1998). Enantioselective hydrogenation of a cyclic imidoketone over chirally modified Pt/Al_2O_3. *Chem. Commun.*, 1377–1378.

[64] Szabo, A., Kunzle, N., Mallat, T., Baiker, A. (1999). Enantioselective hydrogenation of pyrrolidine-2,3,5-triones over the Pt-cinchonidine system. *Tetrahedron Asymmetry*, *10*, 61–76.

[65] Török, B., Felföldi, K., Balàzsik, K., Bartók, M. (1999). New synthesis of a useful C3 chiral building block by a heterogeneous method: enantioselective hydrogenation of pyruvaldehyde dimethyl acetal over cinchona modified Pt/Al_2O_3 catalysts. *Chem. Commun.*, 1725–1726.

[66] Studer, M., Burkhardt, S., Blaser, H. U. (1999). Enantioselective hydrogenation of alpha-keto acetals with cinchona modified Pt catalyst. *Chem. Commun.*, 1727–1728.

[67] Ostgard, D. J., Hartung, R., Krauter, J. G. E., Seebald, S., Kukula, P., Nettekoven, U., Studer, M., Blaser, H. U. (2005). The effects of Pt/Al_2O_3 structure and stability on the O-methyl-10,11-dihydrocinchonidine modified enantioselective hydrogenation of activated ketones. In: J. R. Sowa (Ed.), *Catalysis of Organic Reactions*. Taylor & Francis, Boca Raton, FL, pp. 553–559.

[68] Felföldi, K., Balàzsik, K., Bartók, M. (2003). Heterogeneous asymmetric reactions Part 32. High enantioselectivities in the hydrogenation of activated ketones on cinchona alkaloid modified platinum-alumina catalysts. *J. Mol. Catal. A Chem.*, *202*, 163–170.

[69] Diezi, S., Hess, M., Orglmeister, E., Mallat, T., Baiker, A. (2005). An efficient synthetic chiral modifier for platinum. *Catal. Lett.*, *102*, 121–125.

[70] Diezi, S., Hess, M., Orglmeister, E., Mallat, T., Baiker, A. (2005). Chemo and enantioselective hydrogenation of fluorinated ketones on platinum modified with (*R*)-1-(1-naphthyl)ethylamine derivatives. *J. Mol. Catal. A Chem.*, *239*, 49–56.

[71] Varga, T., Felföldi, K., Forgó, P., Bartók, M. (2004). Heterogeneous asymmetric reactions—Part 38. Enantioselective hydrogenation of fluoroketones on Pt-alumina catalyst. *J. Mol. Catal. A Chem.*, *216*, 181–187.

[72] von Arx, M., Mallat, T., Baiker, A. (2001). Platinum-catalyzed enantioselective hydrogenation of aryl-substituted trifluoroacetophenones. *Tetrahedron Asymmetry*, *12*, 3089–3094.

[73] von Arx, M., Mallat, T., Baiker, A. (2001). Heterogeneous enantioselective hydrogenation of trifluoromethyl ketones. In: D. C. Sherrington, A. P. Kybett (Eds.), *Supported Catalysts and Their Applications*. Royal Society of Chemistry, Cambridge, pp. 247–254.

[74] Mallat, T., Bodmer, M., Balker, A. (1997). Enantioselective hydrogenation of trifluoroacetophenone over cinchonidine-modified platinum. *Catal. Lett.*, *44*, 95–99.

[75] Hess, R., Diezi, S., Mallat, T., Baiker, A. (2004). Chemo- and enantioselective hydrogenation of the activated keto group of fluorinated beta-diketones. *Tetrahedron Asymmetry*, *15*, 251–257.

[76] von Arx, M., Mallat, T., Baiker, A. (2002). Highly efficient platinum-catalyzed enantio selective hydrogenation of trifluoroacetoacetates in acidic solvents. *Catal. Lett.*, *78*, 267–271.

[77] Szöri, K., Szöllösi, G., Bartók, M. (2006). Dynamic kinetic resolution over Cinchona-modified platinum catalyst: hydrogenation of racemic ethyl 2-fluoroacetoacetate. *Adv. Synth. Catal.*, *348*, 515–522.

[78] Vargas, A., Hoxha, F., Bonalumi, N., Mallat, T., Baiker, A. (2006). Steric and electronic effects in enantioselective hydrogenation of ketones on platinum modified by cinchonidine: directing effect of the trifluoromethyl group. *J. Catal.*, *240*, 203–212.

[79] Felföldi, K., Varga, T., Forgó, P., Bartók, M. (2004). Enantioselective hydrogenation of trifluoromethylcyclohexyl ketone on cinchona alkaloid modified Pt-alumina catalyst. *Catal. Lett.*, *97*, 65–70.

[80] Izumi, Y. (1983). Modified Raney-Nickel (MRNi) catalyst—heterogeneous enantio-differentiating (asymmetric) catalyst. *Adv. Catal.*, *32*, 215–271.

[81] Sachtler, W. M. (1985). Asymmetric sites on heterogeneous catalysts. In: R. L. Augustine (Ed.), *Catalysis of Organic Reactions*. M. Dekker, New York, p. 189.

[82] Tai, A., Harada, T. (1986). Asymmetrically modified nickel catalysts. In: Y. Iwasawa (Ed.), *Tailored Metal Catalysts*. D. Reidel, Dordrecht, South Africa, p. 265.

[83] Tai, A., Sugimura, T. (2000). Modified Ni catalysts for enantiodifferentiating hydrogenation. In: D. E. De Vos, I. F. J. Vankelecom, P. A. Jacobs (Eds.), *Chiral Catalyst Immobilization and Recycling*. Wiley-VCH, Weinheim, Germany, p. 173.

[84] Sugimura, T. (1999). Recent progress in tartaric acid-modified Raney nickel system for enantio-differentiating hydrogenation. *Catal. Surv. Jpn.*, *3*, 37–42.

[85] Osawa, T., Harada, T., Takayasu, O. (2006). Asymmetrically modified nickel catalyst for the enantio-differentiating hydrogenation of prochiral ketones. *Curr. Org. Chem.*, *10*, 1513–1531.

[86] Osawa, T., Harada, T., Tai, A. (1997). Enantio-differentiating hydrogenation of prochiral ketones over modified nickel. *Catal. Today.*, *37*, 465–480.

[87] Klabunovskii, E. I. (1991). On some general regularities in the action of enantioselective hydrogenation catalysts. *J. Organomet. Chem.*, *417*, 181–191.

[88] Klabunovskii, E. I. (1996). Catalytic asymmetric synthesis of beta-hydroxyacids and their esters. *Russ. Chem. Rev.*, *65*, 329–344.

[89] Keane, M. A. (1997). Interaction of optically active tartaric acid with a nickel-silica catalyst: role of both the modification and reaction media in determining enantioselectivity. *Langmuir*, *13*, 41–50.

[90] Tai, A. (2003). Asymmetrically modified nickel catalysts. In: D. G. Morrell (Ed.), *Catalysis of Organic Reactions*. Dekker, New York, p. 191.

[91] Bartek, L., Kluson, P., Cerveny, L. (2005). *In situ* chiral modification of nickel catalysts for enantioselective hydrogenation of methyl acetoacetate. *Collect. Czech. Chem. Commun.*, *70*, 1642–1652.

[92] Osawa, T., Hagino, Y., Harada, T., Takayasu, O. (2006). Studies of the effects of the modification conditions on the hydrogenation rate for the enantio-differentiating hydrogenation of methyl acetoacetate over a tartaric acid-NaBr-modified nickel catalyst. *Catal. Lett.*, *112*, 57–61.

[93] Osawa, T., Yoshino, K., Takimoto, K., Takayasu, O., Harada, T. (2006). Enantio-differentiating hydrogenation of methyl acetoacetate over modified Raney nickel catalysts prepared by two-step modifications. *Catal. Lett.*, *112*, 167–171.

[94] Sugimura, T., Watanabe, J., Nakagawa, S., Okuyama, T. (2006). Enantiomer-differentiating hydrogenation of methyl 3-cyclopropyl-2-methyl-3-oxopropanoate over tartaric acid-modified nickel—performance of heterogeneous catalyst in dynamic kinetic resolution. *J. Mol. Catal. A Chem.*, *248*, 233–236.

[95] Sugimura, T., Nakagawa, S., Tai, A. (2002). Over 98% optical yield achieved by a heterogeneous catalysis. Substrate design and analysis of enantio-differentiating factors of tartaric acid-modified Raney nickel hydrogenation. *Bull. Chem. Soc. Jpn.*, *75*, 355–363.

[96] Osawa, T., Harada, T., Takayasu, O. (2000). Progress of enantio-differentiating hydrogenation of prochiral ketones over asymmetrically modified nickel catalysts and a newly proposed enantio-differentiation model. *Top. Catal.*, *13*, 155–168.

[97] Osawa, T., Harada, T., Tai, A. (1990). Enantio-differentiating hydrogenation of 2-alkanones over asymmetrically modified nickel-catalyst and its application to the preparation of optically pure 2-alkanols. *J. Catal.*, *121*, 7–17.

[98] Harada, T., Kawamura, T., Haikawa, S., Osawa, T. (1994). Enantioface-differentiating hydrogenation of the C=O double-bond using a modified Raney-Nickel—role of the carboxylic-acid added to the reaction system. *J. Mol. Catal.*, *93*, 211–219.

[99] Osawa, T., Harada, T., Tai, A. (1994). Mode of enantioface-differentiation in the hydrogenation of 2-alkanone over an asymmetrically modified nickel-catalyst. *J. Mol. Catal.*, *87*, 333–342.

[100] Osawa, T., Nakagawa, Y., Harada, T., Takayasu, O. (2006). Studies of the effect of pivalic acid on the hydrogenation rate and the enantio-differentiating ability for the hydrogenation of 2-octanone over a tartaric acid-NaBr-modified nickel catalyst. *Catal. Lett.*, *112*, 163–166.

[101] Basu, S., Mapa, M., Gopinath, C. S., Doble, M., Bhaduri, S., Lahiri, G. K. (2006). MCM-41-supported platinum carbonyl cluster-derived catalysts for asymmetric and nonasymmetric hydrogenation reactions. *J. Catal.*, *239*, 154–161.

[102] Vetere, V., Faraoni, M. B., Santori, G. F., Podesta, J., Casella, M. L., Ferretti, O. A. (2004). New approach toward the synthesis of asymmetric heterogeneous catalysts for hydrogenation reactions. *J. Catal.*, *226*, 457–461.

[103] Tungler, A., Tarnai, T., Mathe, T., Petro, J. (1991). Enantioselective hydrogenation of acetophenone in the presence of s-proline. *J. Mol. Catal.*, *67*, 277–282.

[104] Vetere, V., Faraoni, M. B. E., Santori, G. F., Podesta, J. U., Casella, M. L., Ferretti, O. A. (2005). Study of the racemic and enantioselective hydrogenation of acetophenone and 3,4-dimethoxyacetophenone using platinum-based organotin catalysts. *Catal. Today*, *107–108*, 266–272.

[105] Hess, R., Vargas, A., Mallat, T., Burgi, T., Baiker, A. (2004). Inversion of enantioselectivity in the platinum-catalyzed hydrogenation of substituted acetophenones. *J. Catal.*, *222*, 117–128.

[106] Hess, R., Mallat, T., Baiker, A. (2003). Enantioselective hydrogenation of aromatic ketones: structural effects. *J. Catal.*, *218*, 453–456.

[107] Zhang, D. L., Yang, C. F., Feng, J., Fu, H. Y., Chen, H., Li, R. X., Li, X. J. (2009). Asymmetric hydrogenations of acetophenone and its derivatives over Ir/HAP modified by (1S,2S)-DPEN. *Acta Phys-Chim. Sin.*, *25*, 2039–2044.

[108] Marzialetti, T., Oportus, M., Ruiz, D., Fierro, J. L. G., Reyes, P. (2008). Enantio selective hydrogenation of 1-phenyl-1,2-propanedione, ethyl pyruvate and acetophenone on Ir/SiO_2 catalysts effect of iridium loading. *Catal. Today*, *133*, 711–719.

[109] Yang, C. F., Jiang, H. Y., Feng, J., Fu, H. Y., Li, R. X., Chen, H., Li, X. J. (2009). Asymmetric hydrogenation of acetophenone catalyzed by cinchonidine stabilized Ir/SiO_2. *J. Mol. Catal. A Chem.*, *300*, 98–102.

[110] Cheng, H. Y., Hao, J. M., Wang, H. J., Xi, C. Y., Meng, X. C., Cai, S. X., Zhao, F. Y. (2007). R,R)-DPEN-modified Ru/gamma-Al_2O_3—an efficient heterogeneous catalyst for enantioselective hydrogenation of acetophenone. *J. Mol. Catal. A Chem.*, *278*, 6–11.

[111] Tang, B., Xiong, W., Liu, E. R., Jia, Y., Wang, J. B., Chen, H., Li, X. J. (2008). Asymmetric hydrogenation of aromatic ketones catalyzed by (1S,2S)-DPEN-modified Ru-PPh_3/gamma-Al_2O_3 catalyst. *Tetrahedron Asymmetry*, *19*, 1397–1401.

[112] Jiang, H. Y., Yang, C. F., Li, C., Fu, H. Y., Chen, H., Li, R. X., Li, X. J. (2008). Heterogeneous enantioselective hydrogenation of aromatic ketones catalyzed by Cinchona- and phosphine-modified iridium catalysts. *Angew. Chem. Int. Ed.*, *47*, 9240–9244.

[113] Mallat, T., Baiker, A., (2001). Enantioselective hydrogenation with solid catalysts. In: R. A. Sheldon, H. van Bekkum (Eds.), *Fine Chemicals through Heterogeneous Catalysis*. Wiley-VCH, Weinheim, Germany, p. 449.

[114] Tungler, A., Fogassy, G. (2001). Catalysis with supported palladium metal, selectivity in the hydrogenation of C=C, C=O and C=N bonds, from chemo- to enantioselectivity. *J. Mol. Catal. A Chem.*, *173*, 231–247.

[115] Sipos, E., Tungler, A., Fogassy, G. (2004). New substrates and modifiers in the enantio selective heterogeneous catalytic hydrogenation of the C=C double bond. *J. Mol. Catal. A Chem.*, *216*, 171–180.

[116] Padgett, R. E., Beamer, R. L. (1964). Stereospecific hydrogenations using palladium-on-silica gel catalysts. *J. Pharm. Sci.*, *53*, 689–690.

[117] Perez, J. R. G., Malthete, J., Jacques, J. (1985). Asymmetric hydrogenation of prochiral cinnamic-acids in the presence of palladium on activated carbon and of chiral bases. *C.R. Acad. Sci. II*, *300*, 169–172.

[118] Nitta, Y., Ueda, Y., Imanaka, T. (1994). Enantioselective hydrogenation of (*E*)-alpha-phenylcinnamic acid over cinchonidine-modified palladium catalysts. *Chem. Lett.*, *23*, 1095–1098.

[119] Nitta, Y., Kobiro, K., Okamoto, Y. (1997). Enantioselective hydrogenation of (E)-alpha-phenylcinnamic acid on cinchonidine-modified palladium catalysts: influence of support. In: H. U. Blaser, A. Baiker, R. Prins (Eds.), *Heterogeneous Catalysis and Fine Chemicals IV*. Elsevier, Amsterdam, pp. 191–198.

[120] Nitta, Y. (2000). Enantioselective hydrogenation of (*E*)-alpha-phenylcinnamic acid with cinchonidine-modified Pd/TiO_2: influence of solvents and additives. *Top. Catal.*, *13*, 179–185.

[121] Nitta, Y., Kubota, T., Okamoto, Y. (2001). Effect of palladium dispersion on the enantioselective hydrogenation of alpha,beta-unsaturated acids with modified Pd/TiO_2 catalysts. *Bull. Chem. Soc. Jpn.*, *74*, 2161–2165.

[122] Nitta, Y., Kubota, T., Okamoto, Y. (2004). Solvent effect on the structure sensitivity in enantioselective hydrogenation of alpha, beta-unsaturated acids with modified palladium catalysts. *J. Mol. Catal. A Chem.*, *212*, 155–159.

[123] Nitta, Y., Watanabe, J., Okuyama, T., Sugimura, T. (2005). Activation-temperature dependence in enantioselective hydrogenation of unsaturated carboxylic acids over cinchonidine-modified Pd/C catalysts. *J. Catal.*, *236*, 164–167.

[124] Sugimura, T., Watanabe, J., Okuyama, T., Nitta, Y. (2005). The effect of the substrate structure on the stereoselectivity in an asymmetric hydrogenation of unsaturated carboxylic acids over cinchonidine-modified palladium catalyst. *Tetrahedron Asymmetry*, *16*, 1573–1575.

[125] Smith, G. V., Cheng, J. J., Song, R. Z. (1996). Enantioselective hydrogenation of prochiral C=C bonds over noble metal catalysts supported by beta-cyclodextrin polymer. In: M. R.E. (Ed.), *Catalysis of Organic Reactions*. Marcel Dekker, New York, pp. 479–483.

[126] Bartók, M., Wittmann, G., Göndös, G., Smith, G. V. (1987). Homogeneous and heterogeneous catalytic asymmetric reactions .1. Asymmetric hydrogenation of

the prochiral C=C bond on a modified Raney Ni catalyst. *J. Org. Chem.*, *52*, 1139–1141.

[127] Bartók, M., Wittmann, G., Bartók, G. B., Göndös, G. (1990). Homogeneous and heterogeneous catalytic asymmetric reactions .4. Hydrogenation of the Ni complexes of cinnamic acid salts over MRNI catalyst. *J. Organomet. Chem.*, *384*, 385–395.

[128] Testa, A. C., Augustine, R. L. (2000). Chiral modifiers for supported metals. In: M. E. Ford (Ed.), *Catalysis of Organic Reactions*. M. Dekker, New York, p. 465.

[129] Jackson, S. D., Watson, S. R., Webb, G., Wells, P. B., Young, N. C. (2000). Enantioselective hydrogenation of unsaturated carboxylic acids over palladium modified by a substituted binaphthylene. In: M. E. Ford (Ed.), *Catalysis of Organic Reactions*. M. Dekker, New York, p. 477.

[130] Ferri, D., Bürgi, T., Baiker, A. (2002). *In situ* ATR-IR study of the adsorption of cinchonidine on Pd/Al2O3: differences and similarities with adsorption on Pt/Al2O3. *J. Catal.*, *210*, 160–170.

[131] Nitta, Y. (1999). Importance of product desorption in enantioselective hydrogenation of (*E*)-alpha-phenylcinnamic acid with a cinchonidine-modified Pd/TiO_2 catalyst: effect of additives. *Chem. Lett.*, *28*, 635–636.

[132] Valdes, R. H., Puzer, L., Gomes, M., Marques, C., Aranda, D. A. G., Bastos, M. L., Gemal, A. L., Antunes, O. A. C. (2004). Production of L-DOPA under heterogeneous asymmetric catalysis. *Catal. Commun.*, *5*, 631–634.

[133] Nitta, Y., Shibata, A. (1998). Enantioselective hydrogenation of (*E*)-alpha-phenylcinnamic acid on Pd/TiO_2 catalyst modified by cinchona alkaloids: effect of modifier structure. *Chem. Lett.*, *27*, 161–162.

[134] Sugimura, T., Watanabe, J., Uchida, T., Nitta, Y., Okuyama, T. (2006). Highly enantioselective hydrogenation of alpha-Alkyl-beta-arylpropenoic acids over cinchonidine-modified palladium catalyst. *Catal. Lett.*, *112*, 27–30.

[135] Borszeky, K., Mallat, T., Baiker, A. (1999). Asymmetric hydrogenation of indene carboxylic acids: stereochemistry of hydrogen addition. *Tetrahedron Asymmetry*, *10*, 4781–4789.

[136] Kun, I., Török, B., Felföldi, K., Bartók, M. (2000). Heterogeneous asymmetric reactions Part 17. Asymmetric hydrogenation of 2-methyl-2-pentenoic acid over cinchona modified Pd/Al_2O_3 catalysts. *Appl. Catal., A.*, *203*, 71–79.

[137] Bisignani, R., Franceschini, S., Piccolo, O., Vaccari, A. (2005). The solvent effect in the enantioselective hydrogenation of (*E*)-2-methyl-2-butenoic acid with cinchonidine doped Pd/Al_2O_3. *J. Mol. Catal. A Chem.*, *232*, 161–164.

[138] Casagrande, A., Franceschini, S., Lenarda, M., Piccolo, O., Vaccari, A. (2006). Cinchonidine doped Pd catalysts in the enantioselective hydrogenation of (*E*)-2-methyl-2-butenoic acid. *J. Mol. Catal. A Chem.*, *246*, 263–267.

[139] Borszeky, K., Mallat, T., Baiker, A. (1999). Palladium-catalysed enantioselective hydrogenation of alkenoic acids. Role of isomerization. *Catal. Lett.*, *59*, 95–97.

[140] Solladie-Cavallo, A., Hoernel, F., Schmitt, M., Garin, F. (2003). Heterogeneous asymmetric hydrogenation and deuteration of 2-methyl-2-pentenoic acid over Pd supported catalysts: proton/deuterium exchange. *J. Mol. Catal. A Chem.*, *195*, 181–188.

[141] Szöri, K., Szöllösi, G., Felföldi, K., Bartók, M. (2005). A novel asymmetric heterogeneous catalytic reaction: hydrogenation of ethyl 2-acetoxyacrylate on cinchonidine modified Pd and Pt catalyst. *React. Kinet. Catal. Lett.*, *84*, 151–156.

[142] Blaser, H. U., Hönig, H., Studer, M., Wedemeyer-Exl, C. (1999). Enantioselective synthesis of ethyl nipecotinate using cinchona modified heterogeneous catalysts. *J. Mol. Catal. A Chem.*, *139*, 253–257.

[143] Caulfield, N. J., McMorn, P., Wells, P. B., Compton, D., Soars, K., Hutchings, G. J. (2003). Enantioselective hydrogenation of N-acetyl dehydrophenylalanine methyl ester (NADPME) and some related compounds over alkaloid-modified palladium. In: S. D. Jackson, J. S. J. Hargreaves, D. Lennon (Eds.), *Catalysis in Application*. RSC Publishing, Cambridge, pp. 284–289.

[144] Colston, N. J., Wells, R. P. K., Wells, P. B., Hutchings, G. J. (2005). Unexpected inversion in enantioselectivity in the hydrogenation *N*-acetyl dehydrophenylalanine methyl ester using cinchona-modified Pd/Al_2O_3 catalyst. *Catal. Lett.*, *103*, 117–120.

[145] Coulston, N. J., Wells, R. P. K., Wells, P. B., Hutchings, G. J. (2006). Enantioselective hydrogenation of *N*-acetyl dehydrophenylalanine methyl ester using cinchonine-modified Pd/TiO_2 catalysts. *Catal. Today*, *114*, 353–356.

[146] Tungler, A., Kajtar, M., Mathe, T., Toth, G., Fogassy, E., Petro, J. (1989). Enantioselective hydrogenation of alpha-beta-unsaturated ketones. *Catal. Today*, *5*, 159–171.

[147] Tungler, A., Tarnai, T., Deak, A., Kemeny, S., Gyori, A., Mathe, T., Petro, J. (1993). Kinetic comparison of enantioselective hydrogenations. In: M. Guisnet, J. Barbier, J. Barrault, C. Bouchoule, D. Duprez, G. Perot, C. Montassier (Eds.), *Heterogeneous Catalysis and Fine Chemicals III*. Elsevier, Amsterdam, pp. 99–106.

[148] Tóth, G., Kovàcs, A., Tarnai, T., Tungler, A. (1993). Stereochemistry of the enantioselective reductive alkylation of proline with ketones. *Tetrahedron Asymmetry*, *4*, 331–338.

[149] Fogassy, G., Tungler, A., Lévai, A., Tóth, G. (2002). Enantioselective hydrogenation of exocyclic alpha,beta-unsaturated ketones Part II. Hydrogenation in the presence of (*S*)-proline. *J. Mol. Catal. A Chem.*, *179*, 101–106.

[150] Mhadgut, S. C., Bucsi, I., Török, M., Török, B. (2004). Sonochemical asymmetric hydrogenation of isophorone on proline modified Pd/Al_2O_3 catalysts. *Chem. Commun.*, 984–985.

[151] Pintar, A., Malacea, R., Pinel, C., Fogassy, G., Besson, M. (2004). *In situ* monitoring of catalytic three-phase enantio selective hydrogenation using FTIR/ATR spectroscopy. *Appl. Catal. A*, *264*, 1–12.

[152] Mhadgut, S. C., Török, M., Esquibel, J., Török, B. (2006). Highly asymmetric heterogeneous catalytic hydrogenation of isophorone on proline modified base-supported palladium catalysts. *J. Catal.*, *238*, 441–448.

[153] McIntosh, A. I., Watson, D. J., Burton, J. W., Lambert, R. M. (2006). Heterogeneously catalyzed asymmetric C=C hydrogenation: origin of enantioselectivity in the proline-directed Pd/isophorone system. *J. Am. Chem. Soc.*, *128*, 7329–7334.

[154] Tungler, A., Màthé, T., Tarnai, T., Fodor, K., Tóth, G., Kajtàr, J., Kolossvàry, I., Herényi, B., Sheldon, R. A. (1995). (−)-Dihydro-apovincaminic acid ethyl-ester, preparation and use as a chiral modifier in enantioselective heterogeneous catalytic hydrogenations. *Tetrahedron Asymmetry*, *6*, 2395–2402.

[155] Farkas, G., Fodor, K., Tungler, A., Màthé, T., Tóth, G., Sheldon, R. A. (1999). New chiral auxiliaries in enantioselective heterogeneous catalytic hydrogenations: (−) and (+)-dihydro-apovincaminic acid. Comparison with (−)-dihydro-apovincaminic acid ethyl ester. III. *J. Mol. Catal. A Chem.*, *138*, 123–127.

[156] Farkas, G., Sipos, E., Tungler, A., Sàrkàny, A., Figueiredo, J. L. (2001). Enantioselective hydrogenation of isophorone over Pd catalysts in the presence of (−)-dihydroapovincaminic acid ethyl ester—the effect of reduction method of Pd blacks on the enantiomeric excess. *J. Mol. Catal. A Chem.*, *170*, 101–107.

[157] Sipos, E., Farkas, G., Tungler, A., Figueiredo, J. L. (2002). Enantioselective hydrogenation of isophorone with titania supported Pd catalysts modified by (−)-dihydroapovincaminic acid ethyl ester effect of the support and the reduction method. *J. Mol. Catal. A Chem.*, *179*, 107–112.

[158] Tungler, A., Nitta, Y., Fodor, K., Farkas, G., Màthé, T. (1999). Comparison of chiral modifiers in the Pd catalysed hydrogenation of phenylcinnamic acid and isophorone. *J. Mol. Catal. A Chem.*, *149*, 135–140.

[159] Sipos, E., Tungler, A., Bitter, I., Kubinyi, M. (2002). (*S*)-Alpha,alpha-diphenyl- and (*S*)-alpha,alpha-dinaphthyl-2-pyrrolidinemethanol as chiral modifiers in asymmetric heterogeneous catalytic hydrogenation of isophorone. *J. Mol. Catal. A Chem.*, *186*, 187–192.

[160] Sipos, E., Tungler, A., Bitter, I. (2003). (*S*)-Proline based chiral modifiers. *J. Mol. Catal. A Chem.*, *198*, 167–173.

[161] Sipos, E., Fogassy, G., Tungler, A., Samant, P. V., Figueiredo, J. L. (2004). Enantioselective hydrogenations with highly mesoporous carbon supported Pd catalysts. *J. Mol. Catal. A Chem.*, *212*, 245–250.

[162] Fogassy, G., Tungler, A., Levai, A. (2003). Enantio selective hydrogenation of exocyclic alpha,beta-unsaturated ketones—Part III. Hydrogenation with Pd in the presence of cinchonidine. *J. Mol. Catal. A Chem.*, *192*, 189–194.

[163] Huck, W. R., Mallat, T., Baiker, A. (2002). Heterogeneous enantioselective hydrogenation of 2-pyrones over cinchona-modified palladium. *New J. Chem.*, *26*, 6–8.

[164] Huck, W. R., Mallat, T., Baiker, A. (2000). Transient behavior of the enantioselective hydrogenation of a hydroxymethylpyrone. *Catal. Lett.*, *69*, 129–132.

[165] Nasar, K., Fache, F., Lemaire, M., Beziat, J. C., Besson, M., Gallezot, P. (1994). Stereoselective Reduction of Disubstituted Aromatics on Colloidal Rhodium. *J. Mol. Catal.*, *87*, 107–115.

[166] Nasar, K., Besson, M., Gallezot, P., Fache, F., Lemaire, M. (1995). Enantio- and diastereoselective reduction of disubstituted aromatics. In: G. Jannes, V. Dubois (Eds.), *Chiral Reactions in Heterogeneous Catalysis*. Springer, pp. 141–145.

[167] Laine, R. M., Hum, G., Wood, B. J., Dawson, M. (1980). Asymmetric hydrogenation of 2-methylpyridine to 2-methylpiperidine. Raney-nickel modified with a molecular template. *Stud. Surf. Sci. Catal.*, *7B* (Pt. B, New Horiz. Catal.), 1478–1479.

[168] Nakagawa, S., Haruna, N., Acosta, D. E., Endo, T., Sugimura, T., Tai, A. (1999). Enantio-differentiating hydrogenation of methyl 2-furoylacetate and its analogs over tartaric acid-modified Raney nickel. *Chem. Lett.*, *28*, 1055–1056.

[169] Haruna, N., Acosta, D. E., Nakagawa, S., Yamaguchi, K., Tai, A., Okuyama, T., Sugimura, T. (2004). Asymmetric hydrogenation of furan-containing ketones over tartaric acid-modified Raney nickel catalyst. *Heterocycles*, *62*, 375–386.

[170] Maris, M., Huck, W. R., Mallat, T., Baiker, A. (2003). Palladium-catalyzed asymmetric hydrogenation of furan carboxylic acids. *J. Catal.*, *219*, 52–58.

[171] Huck, W. R., Mallet, T., Baiker, A. (2000). Potential and limitations of palladium-cinchona catalyst for the enantioselective hydrogenation of a hydroxymethyl-pyrone. *J. Catal.*, *193*, 1–4.

[172] Sonderegger, O. J., Ho, G. M. W., Bürgi, T., Baiker, A. (2005). Enantioselective hydrogenation of aromatic ketones over cinchona-modified rhodium: a new opportunity? *J. Catal.*, *230*, 499–506.

[173] Szöllösi, G., Kun, I., Bartók, M. (2001). Heterogeneous asymmetric reactions. Part 24. Heterogeneous catalytic enantioselective hydrogenation of the C=N group over cinchona alkaloid modified palladium catalyst. *Chirality*, *13*, 619–624.

[174] Borszeky, K., Mallat, T., Aeschiman, R., Schweizer, W. B., Baiker, A. (1996). Enantioselective hydrogenation of pyruvic acid oxime to alanine on Pd/alumina. *J. Catal.*, *161*, 451–458.

[175] Baiker, A. (1997). Progress in asymmetric heterogeneous catalysis: design of novel chirally modified platinum metal catalysts. *J. Mol. Catal. A Chem.*, *115*, 473–493.

[176] Blaser, H. U., Jalett, H. P., Monti, D. M., Baiker, A., Wehrli, J. T. (1991). Enantioselective hydrogenation of ethyl pyruvate—effect of catalyst and modifier structure. In: R. K. Grasselli, A. W. Sleight (Eds.), *Structure-Activity and Selectivity Relationships in Heterogeneous Catalysis*. Elsevier, Amsterdam, pp. 147–155.

[177] Bonalumi, N., Burgi, T., Baiker, A. (2003). Interaction between ketopantolactone and chirally modified Pt investigated by attenuated total reflection IR concentration modulation spectroscopy. *J. Am. Chem. Soc.*, *125*, 13342–13343.

[178] Schurch, M., Heinz, T., Aeschimann, R., Mallat, T., Pfaltz, A., Baiker, A. (1998). Design of new modifiers for the enantioselective hydrogenation of ethyl pyruvate. *J. Catal.*, *173*, 187–195.

[179] Baiker, A., Mallat, T., Minder, B., Schwalm, O., Simons, K. E., Weber, J. (1995). Enantioselective hydrogenation of alpha-ketoesters: a molecular view on the enantio-differentiation. In: G. Jannes, V. Dubois (Eds.), *Chiral Reactions in Heterogeneous Catalysis*. Plenum Press, New York, pp. 95–103.

[180] Schwalm, O., Minder, B., Weber, J., Baiker, A. (1994). Enantioselective hydrogenation of alpha-ketoesters over Pt/alumina modified with cinchonidine—theoretical investigation of the substrate-modifier interaction. *Catal. Lett.*, *23*, 271–279.

[181] Schwalm, O., Weber, J., Margitfalvi, J., Baiker, A. (1993). Ab-initio and semiempirical investigations of the complexation of methyl pyruvate by ammonia and the ammonium cation. *J. Mol. Struct.*, *297*, 285–293.

[182] Schwalm, O., Weber, J., Minder, B., Baiker, A. (1994). Theoretical investigation of the enantioselective hydrogenation of alpha-ketoesters over Pt alumina modified with cinchonidine. *Int. J. Quantum Chem.*, *52*, 191–197.

[183] Schurch, M., Schwalm, O., Mallat, T., Weber, J., Baiker, A. (1997). Enantioselective hydrogenation of ketopantolactone. *J. Catal.*, *169*, 275–286.

[184] Burgi, T., Baiker, A. (2000). Model for enantioselective hydrogenation of alpha-ketoesters over chirally modified platinum revisited: influence of alpha-ketoester conformation. *J. Catal.*, *194*, 445–451.

[185] Vargas, A., Burgi, T., Baiker, A. (2001). Model of reactant-modifier interaction in enantioselective hydrogenation of ethyl pyruvate on platinum-cinchona catalysts: extension to synthetic chiral modifiers. *J. Catal.*, *197*, 378–384.

[186] Vargas, A., Burgi, T., Baiker, A. (2002). Origin of the rate acceleration in enantioselective hydrogenation of alpha-functionalised ketones over cinchona alkaloid modified platinum. *New J. Chem.*, *26*, 807–810.

[187] Vargas, A., Bürgi, T., von Arx, M., Hess, R., Baiker, A. (2002). Relation between electronic structure of alpha-substituted ketones and their reactivity in racemic and enantioselective platinum-catalyzed hydrogenation. *J. Catal.*, *209*, 489–500.

[188] Busygin, I., Taskinen, A., Nieminen, V., Toukoniitty, E., Stillger, T., Leino, R., Murzin, D. Y. (2009). Experimental and theoretical analysis of asymmetric induction in heterogeneous catalysis: diastereoselective hydrogenation of chiral alpha-hydroxyketones over Pt catalyst. *J. Am. Chem. Soc.*, *131*, 4449–4462.

[189] Nieminen, V., Taskinen, A., Toukonitty, E., Hotokka, M., Murzin, D. Y. (2006). One-to-one reactant-modifier interactions in enantio- and diastereoselective hydrogenation of chiral alpha-hydroxyketones on Pt(111). *J. Catal.*, *237*, 131–142.

[190] Taskinen, A., Toukoniitty, E. (2005). Ab initio study of solvent effects on reactant-modifier complexes in enantioselective hydrogenation. *Catal. Today.*, *100*, 373–377.

[191] Toukoniitty, E., Nieminen, V., Taskinen, A., Paivarinta, J., Hotokka, M., Murzin, D. Y. (2004). A combined experimental and theoretical study of 1-phenylpropane-1,2-dione hydrogenation over heterogeneous cinchonidine-modified Pt catalyst. *J. Catal.*, *224*, 326–339.

[192] Page, M. I., Jencks, W. P. (1971). Entropic contributions to rate accelerations in enzymic and intramolecular reactions and chelate effect. *Proc. Natl. Acad. Sci. U.S.A.*, *68*, 1678.

[193] Vargas, A., Santarossa, G., Iannuzzi, M., Baiker, A. (2008). Chiral recognition on catalytic surfaces: theoretical insight in a biomimetic heterogeneous catalytic system. *J. Phys. Chem. C*, *112*, 10200–10208.

[194] Dijkstra, G. D. H., Kellogg, R. M., Wynberg, H. (1990). Conformational study of cinchona alkaloids—a combined NMR and molecular-orbital approach. *J. Org. Chem.*, *55*, 6121–6131.

[195] Dijkstra, G. D. H., Kellogg, R. M., Wynberg, H., Svendsen, J. S., Marko, I., Sharpless, K. B. (1989). Conformational study of cinchona alkaloids—a combined NMR, molecular mechanics, and X-ray approach. *J. Am. Chem. Soc.*, *111*, 8069–8076.

[196] Burgi, T., Baiker, A. (1998). Conformational behavior of cinchonidine in different solvents: a combined NMR and ab initio investigation. *J. Am. Chem. Soc.*, *120*, 12920–12926.

[197] Urakawa, A., Meier, D. M., Rugger, H., Baiker, A. (2008). Conformational behavior of cinchonidine revisited: a combined theoretical and experimental study. *J. Phys. Chem. A*, *112*, 7250–7255.

[198] Ferri, D., Burgi, T. (2001). An *in situ* attenuated total reflection infrared study of a chiral catalytic solid-liquid interface: cinchonidine adsorption on Pt. *J. Am. Chem. Soc.*, *123*, 12074–12084.

[199] Ferri, D., Burgi, T., Baiker, A. (2001). Chiral modification of platinum catalysts by cinchonidine adsorption studied by *in situ* ATR-IR spectroscopy. *Chem. Commun.*, 1172–1173.

[200] Vargas, A., Baiker, A. (2006). First principles study of the conformations of cinchonidine on a Pt(111) surface. *J. Catal.*, *239*, 220–226.

[201] Vargas, A., Burgi, T., Baiker, A. (2004). Adsorption of cinchonidine on platinum: a DFT insight in the mechanism of enantioselective hydrogenation of activated ketones. *J. Catal.*, *226*, 69–82.

[202] Bartok, M., Felfoldi, K., Torok, B., Bartok, T. (1998). A new cinchona-modified platinum catalyst for the enantioselective hydrogenation of pyruvate: the structure of the 1:1 alkaloid-reactant complex. *Chem. Commun.*, 2605–2606.

[203] Vargas, A., Ferri, D., Baiker, A. (2005). DFT and ATR-IR insight into the conformational flexibility of cinchonidine adsorbed on platinum: proton exchange with metal. *J. Catal.*, *236*, 1–8.

[204] Schmidt, E., Ferri, D., Vargas, A., Baiker, A. (2008). Chiral modification of Rh and Pt surfaces: effect of rotational flexibility of cinchona-type modifiers on their adsorption behavior. *J. Phys. Chem. C*, *112*, 3866–3874.

[205] Bonalumi, N., Vargas, A., Ferri, D., Baiker, A. (2007). Chirally modified platinum generated by adsorption of cinchonidine ether derivatives: towards uncovering the chiral sites. *Chem. Eur. J.*, *13*, 9236–9244.

[206] Bonalumi, N., Vargas, A., Ferri, D., Baiker, A. (2007). Catalytic chiral metal surfaces generated by adsorption of *O*-Phenyl derivatives of cinchonidine. *J. Phys. Chem. C.*, *111*, 9349–9358.

[207] Vargas, A., Ferri, D., Bonalumi, N., Mallat, T., Baiker, A. (2007). Controlling the sense of enantioselection on surfaces by conformational changes of adsorbed modifiers. *Angew. Chem. Int. Ed.*, *46*, 3905–3908.

[208] Diezi, S., Mallat, T., Szabo, A., Baiker, A. (2004). Fine tuning the "chiral sites" on solid enantio selective catalysts. *J. Catal.*, *228*, 162–173.

[209] Diezi, S., Szabo, A., Mallat, T., Baiker, A. (2003). Inversion of enantioselectivity in the hydrogenation of ketopantolactone on platinum modified by ether derivatives of cinchonidine. *Tetrahedron Asymmetry*, *14*, 2573–2577.

[210] Hoxha, F., Konigsmann, L., Vargas, A., Ferri, D., Mallat, T., Baiker, A. (2007). Role of guiding groups in cinchona-modified platinum for controlling the sense of enantiodifferentiation in the hydrogenation of ketones. *J. Am. Chem. Soc.*, *129*, 10582–10590.

CHAPTER 7

CHIRAL IONIC LIQUIDS FOR ASYMMETRIC REACTIONS

ANNIE-CLAUDE GAUMONT, YVES GÉNISSON, FRÉDÉRIC GUILLEN, VIACHESLAV ZGONNIK, AND JEAN-CHRISTOPHE PLAQUEVENT

7.1. Introduction 323
7.2. Synthesis of CILs 325
7.2.1. General strategies 325
7.2.2. Chiral cations 326
7.2.2.1. Ammonium cations 326
7.2.2.2. Imidazolium cations 327
7.2.2.3. Pyridinium cations 328
7.2.2.4. Other heterocycle-based cations 328
7.2.3. Chiral anions 329
7.2.4. Doubly CILs 331
7.3. CILs as reaction media and chiral reagents 331
7.4. Miscellaneous applications 336
7.5. Conclusion and prospects 338
Acknowledgments 339
References 339

7.1. INTRODUCTION

In the modern context of the synthesis of polyfunctional complex molecules, including pharmaceuticals, the search for new and versatile methods in stereoselective synthesis remains a major challenge. Indeed, and maybe contrarily to generally accepted ideas, the majority of asymmetric reactions

Catalytic Methods in Asymmetric Synthesis: Advanced Materials, Techniques, and Applications, First Edition. Edited by Michelangelo Gruttadauria and Francesco Giacalone.

are not general enough to ensure excellent stereoselectivities in most real cases. Chiral reagents and catalysts that are quite efficient (and sometimes excellent) for model compounds in model reactions often fail to give good results (even from the reactivity point of view) when applied to the construction of a complex, polyfunctional compound bearing labile moieties [1]. As a consequence, the quest for new methods, new reagents, and even new concepts in asymmetric synthesis is still a topic of growing importance. Obviously, the third-millennium research focuses on the discovery of perfectly catalytic, reliable, predictable, cheap, and versatile methods for the asymmetric synthesis of such complex targets, as generally are drugs and bioactive molecules.

Among the different approaches in asymmetric synthesis, chemists generally distinguish stoichiometric and catalytic methods. Usually, these methods are classified in four generations, depending on the origin of the selectivity [2]: the first-generation method uses the preexistent chirality, starting from materials obtained from the chiral pool; second-generation methods generate new stereogenic units by diastereoselective synthesis, thanks to a chiral auxiliary temporarily bonded to the prostereogenic substrate; and third- and fourth-generation methods are enantioselective and use, respectively, stoichiometric and catalytic chiral reagents.

Nevertheless, other possibilities exist beyond this paradigm, such as submitting prochiral substrates to chiral forces like circularly polarized light [3] or carrying asymmetric reactions in chiral media. Indeed, many attempts were described in this last field even at the very beginning of organic stereochemistry [4]. In any event, this approach was almost discarded as soon as chiral reagents and catalysts became impressively efficient, that is, in the beginning of the 1980s. To be certain, no real success was reported for asymmetric induction under the influence of chiral media, except for Seebach et al.'s cosolvents [5]. However, this special topic has benefited from a strong renewal of interest in the last few years, due to the discovery and development of the new class of neoteric solvents known as ionic liquids (ILs).

These solvents consist entirely of ionic species and are often liquid at room temperature. They display fascinating properties, which make them of fundamental interest to all chemists. Indeed, both the thermodynamics and kinetics of reactions carried out in ILs are different from those performed in conventional solvents, leading to a shining series of convincing examples of the increase in chemical yields, chemo-, regio-, and stereoselectivities [6].

ILs are sometimes classified in three generations [7], the more recent (third generation) being composed of task-specific ILs (i.e., designed for a specific application) and of chiral ionic liquids (CILs), which will be the main topic of this chapter. Indeed, in this growing field of organic chemistry in ILs, a novel, fascinating, and promising aspect consists in the study of these neoteric solvents in asymmetric synthesis. Indeed, it has been previously suggested that the intrinsic organization of a chiral solvent could exert a positive influence for chirality transfer to the asymmetric reaction [8]. In the example depicted

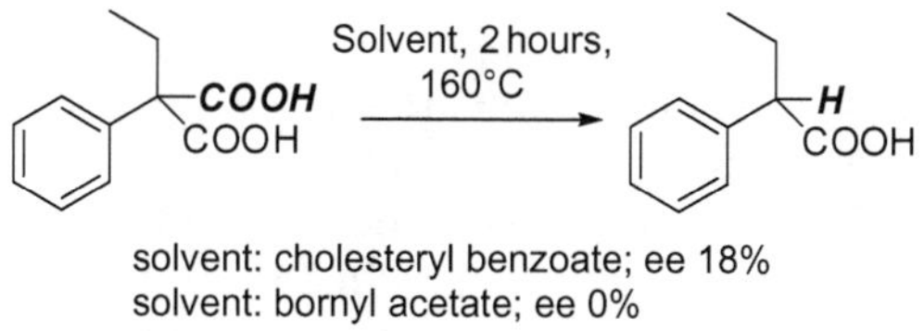

SCHEME 7.1. Decarboxylation–reprotonation of a prochiral malonate.

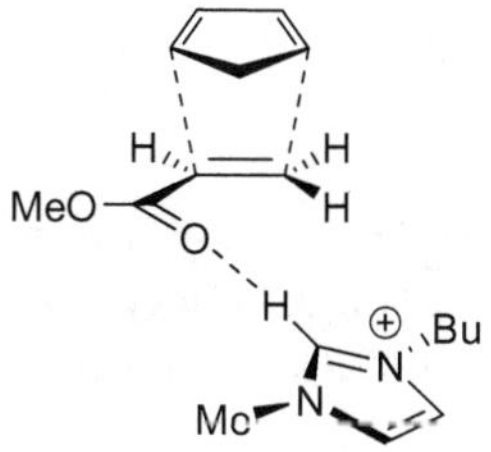

FIGURE 7.1. Proposed transition state of Diels–Alder reaction in IL.

in Scheme 7.1, the decarboxylation–reprotonation of a prochiral malonate is much more enantioselective when carried out in a liquid crystal solvent than in a molecular isotropic one.

As ILs are known to present a highly ordered and organized internal structuration [9], one can expect that chirality transfer could be more efficient in ILs than in any molecular solvent. In addition, ILs, which consist entirely of salts, can be designed and synthesized from many chiral starting materials, thus leading to an almost limitless number of solvents. In some cases, as depicted in Figure 7.1, a strong interaction could occur between the IL and the reactants in the transition state [10], hopefully with an efficient enantioselectivity if the solvent is chiral.

Literature about CILs has been reviewed recently by us [11–13] and others [14–19]. Main concepts and examples in the synthesis of CILs are given in Section 7.2, and their use as chiral solvents and chiral reagents are given in Section 7.3.

Ionic-tagged organocatalysts are beyond the scope of this chapter and are discussed in Chapter 2.

7.2. SYNTHESIS OF CILS

7.2.1. General Strategies

CILs reported thus far possess either a chiral anion, or more frequently, a stereogenic unit in the cationic part. In a few examples, the salts exhibit chirality on both the anion and the cation counterparts. The chiral element can be central, axial, or even planar. The main strategy used to prepare CILs involves the direct use of a chiral starting material, coming from the chiral pool, to prepare either the chiral anion or chiral cation.

From a synthetic point of view, the cation is often designed first, and the desired anion is then introduced by a metathetical exchange. The main cations encountered are ammonium, imidazolium, pyridinium, and to a lesser extent, a few other heterocycles such as oxazolinium, thiazolinium, and imidazolinium. Amino acid derivatives are the preferred starting material followed by amines and alkaloids, terpenes, and hydroxy acid derivatives [13]. In a few examples, CILs have been prepared by asymmetric synthesis. As an example of the wide diversity that can be attended by using the chiral pool, amino acids are a textbook case. Indeed, the CILs can be constructed without modification of the amino acid residue, through the modification of the side chain and preservation of the amino acid moiety, with alteration of one function (acidic or basic), or with polyfunctional modification of both amine and acid functions.

The first example dealing with chiral anions was reported by Seddon et al. using a lactate anion [20]. A real breakthrough was disclosed by Ohno et al. 6 years later, using an imidazolium hydroxide as a starting material in a straightforward approach, which involves neutralization with an amino acid [21]. Recent developments take into account both the ecological and economic requirements [22, 23], using renewable resources as starting materials.

7.2.2. Chiral Cations

Chiral ammonium or nitrogen-containing heterocyclic cations used for the preparation of CILs, except for very few examples, are usually made from the chiral pool, which contains the natural or easily available synthetic chiral synthons.

7.2.2.1. Ammonium Cations. Chiral ammonium ILs are mostly prepared either by alkylation (or protonation) of a chiral amine, or alkylation of an achiral amine with a chiral substituent (Fig. 7.2).

The simplest method of obtaining a chiral ammonium salt from a chiral amine is the protonation of the amino group. Treatment of free [24] or *C*-protected [25] amino acids with the appropriate Brønsted acid allows for a very straightforward and fast access to ionic species. *C*-Protection (as an ester)

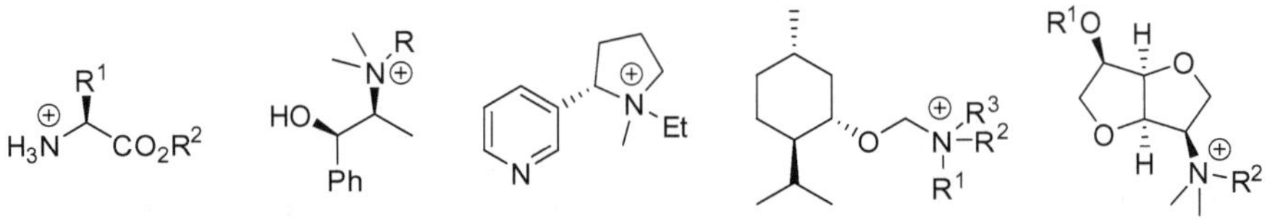

FIGURE 7.2. Examples of chiral ammonium ILs.

is required to lower the melting point of the resulting salts below 100°C. However, these ILs suffer from an obvious sensitivity to (even slightly) basic conditions.

More stable quaternary ammonium cations are prepared by alkylation of ephedrine [26]. Further studies have shown that the length of the alkyl chain has a strong influence on the melting point of the salt [27]. Other chiral ammonium cations were prepared from (*S*)-nicotine [28] using a similar strategy.

On the other hand, chiral synthons bearing an appropriate leaving group can be substituted by a tertiary amine (or by a primary or secondary amine with subsequent quaternization). The reaction of various tertiary amines with commercial (−)-chloromethylmenthylether affords, after anion metathesis, the desired ILs [29]. Reaction of primary amines with isosorbide (after monoprotection and activation as a benzenesulfonyl derivative of the remaining hydroxyl group), followed by quaternization, deprotection, and anion metathesis, leads to isosorbide-derived ILs [30].

7.2.2.2. Imidazolium Cations. Chiral imidazolium cations are obtained by substitution of a chiral synthon having an adequate leaving group by an alkylimidazole (or by imidazole with a subsequent alkylation step), by construction of the imidazole ring on a chiral substrate, or by alkylation of a chiral molecule bearing an imidazole group (Fig. 7.3).

Several chiral alcohols such as (*S*)-ethyl lactate [27] or (3*R*)-citronellol [31] were substituted by imidazole (or an alkylimidazole) after conversion of the hydroxyl group to an appropriate leaving group (tosylate, triflate, or halogen). By using Mitsunobu conditions, a chiral alcohol can be directly substituted by an imidazole [32]. Reduction of the carboxyl group of proline afforded prolinol. Substitution of the alcool function of prolinol through tosylate activation followed by alkylation of the nitrogen imidazole afforded, after anion metathesis, the expected chiral pyrrolidine-substituted imidazolium ILs [33]. Reaction of (−)-chloromethylmenthylether [34] and menthylchloroacetate [35] with an alkylimidazole, followed by anion metathesis, afforded the desired menthol-derived ILs. A carbohydrate-based IL was prepared by anomeric substitution of a phosphate arabinoside with *N*-methylimidazole hydrochloride [36].

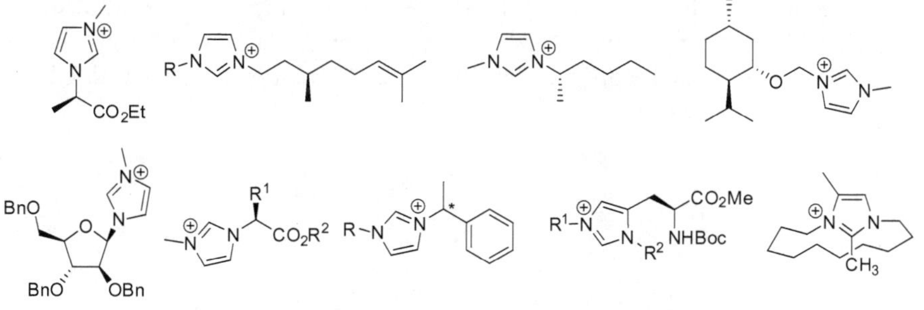

FIGURE 7.3. Examples of chiral imidazolium ILs.

Two-step double substitution of a dibromoalkane with 2,4-dimethylimidazole afforded a racemic cyclophane-type imidazolium salt with planar chirality [37].

A primary amine can be incorporated into an imidazole ring by condensation with ammonia, glyoxal, and an aldehyde (usually formaldehyde). This strategy has been used starting from amino acids or from α-methylbenzylamine [38].

A third possibility is the use of a chiral starting material bearing an imidazole group. For instance (*S*)-histidine, a natural α-amino acid having an imidazole side chain, can be dialkylated to give "symmetrical" imidazolium salts [39]. In order to obtain unsymmetrically substituted histidiniums, histidine methyl ester was first protected as a cyclic urea, alkylated on *N*-1, which was opened in the presence of *t*-butanol and the liberated *N*-3 position alkylated [40].

7.2.2.3. Pyridinium Cations. As for ammonium or imidazolium cations, pyridiniums can be obtained by substitution of a chiral substrate bearing an appropriate leaving group (Fig. 7.4) [31]. However, the use of pyridine-specific reactions such as the Zincke reaction [41] or Kröhnke condensation can also be used to prepare pyridinium ILs. Chiral derivatives having a stereogenic center in the α-position were obtained by reaction of α-methylbenzylamine with Zincke's salt obtained from pyridine and 1-chloro-2,4-dinitrobenzene followed by anion metathesis [42]. Pinene-fused pyridinium salts were prepared from (+)-pinocarvone by Kröhnke condensation followed by alkylation and anion metathesis [43].

Axially CILs were prepared by enantioselective synthesis from pyridine-4-carbaldehyde [44]. Enantioselective dehydrobromination of a dibrominated cyclic acetal, followed by cross-coupling reaction and quaternarization of the pyridine, gave the desired ILs.

7.2.2.4. Other Heterocycle-Based Cations. These chiral heterocyclic cations where the asymmetric center is part of the heterocycle are usually obtained from chiral difunctional compounds (mostly amino acid derivatives). (*S*)-Valinol, easily obtained from (*S*)-valine, can be converted into a chiral oxazoline and, after alkylation and anion metathesis, into the corresponding oxazolinium IL (Fig. 7.5) [26, 45]. However, oxazoliniums are reported to be unstable in aqueous media, leading to ring opening. The much more robust thiazolinium salts can be prepared from valinol by treatment with a dithioester

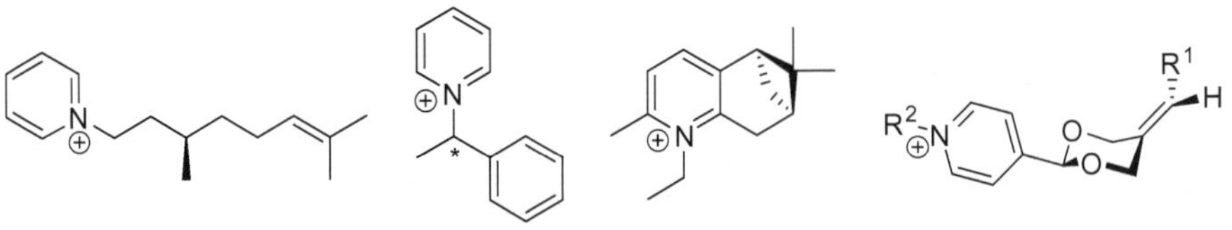

FIGURE 7.4. Examples of chiral pyridinium ILs.

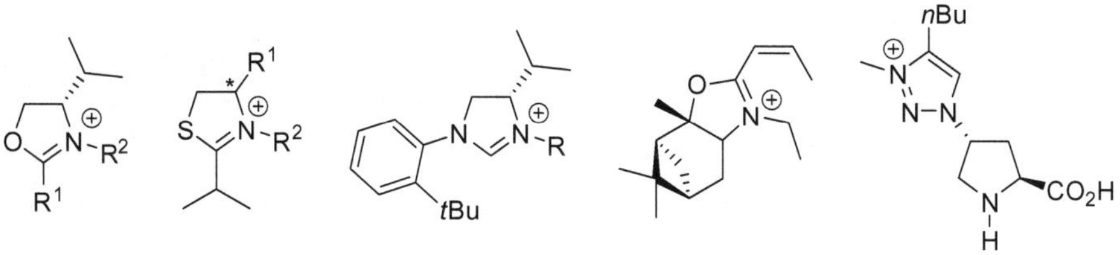

FIGURE 7.5. Examples of other heterocycle-based CILs.

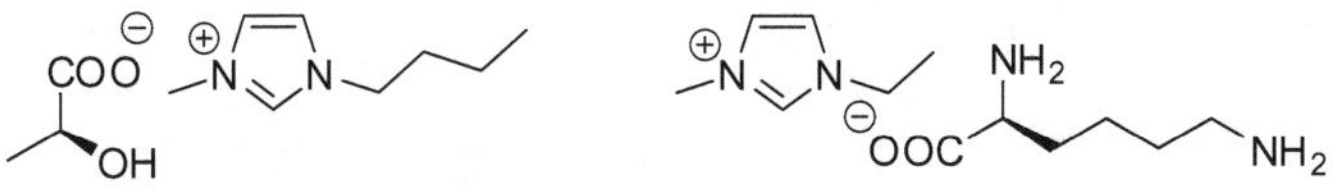

FIGURE 7.6. Examples of chiral anion-based ILs.

followed by mesylation of the intermediate hydroxythioamide, alkylation, and anion metathesis [46, 47]. Imidazoliniums can be prepared by condensation of trimethylorthoformiate with the diamine obtained after reaction of *N*-Boc-valine with 2-*t*-Bu-aniline and reduction of the amide [48].

Reaction of α-pinene-derived 3-aminohydroxypinane with the appropriate acid, followed by alkylation and anion metathesis, afforded pinane-fused ILs [49].

Triazolium-based ILs were prepared starting from (2*S*,4*R*)- and (2*S*,4*S*)-4-hydroxyproline, using Huisgen cyclization followed by alkylation [50]. These ILs found application as recyclable proline-based organocatalysts (see Chapter 2).

7.2.3. Chiral Anions

Due to their relative ease of access, ILs embedding a chiral anion have been known for more than 10 years. As a matter of fact, the first CIL ever described was a lactate salt reported by Seddon et al. in 1999 (Fig. 7.6) [20]. The latter was obtained by a metathesis between the corresponding sodium carboxylate and [Bmim]Cl in acetone followed by a filtration to remove NaCl. As part of a study on Diels–Alder cycloaddition in ILs, this entity was proposed as an alternative reaction medium.

Despite this early example, it was only in 2005 that Ohno et al. described the next examples of ILs having a chiral anion (Fig. 7.6) [21]. This group introduced a breakthrough by using an imidazolium hydroxide as starting material in a straightforward approach. This quite unstable basic precursor was obtained as an aqueous solution from the corresponding imidazolium bromide by means of an anion exchange resin. Its subsequent neutralization with a series of amino acids led to a library of 20 CILs containing an [Emim] cation. These salts displayed glass transition temperature (T_g) values ranging from −65°C to

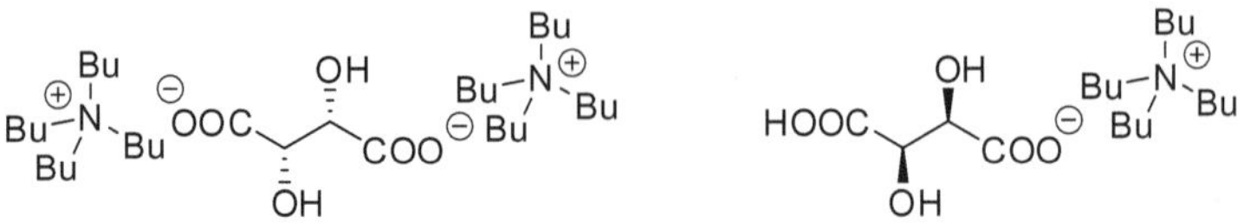

FIGURE 7.7. Examples of chiral tartrate and hemitartrate-based ILs.

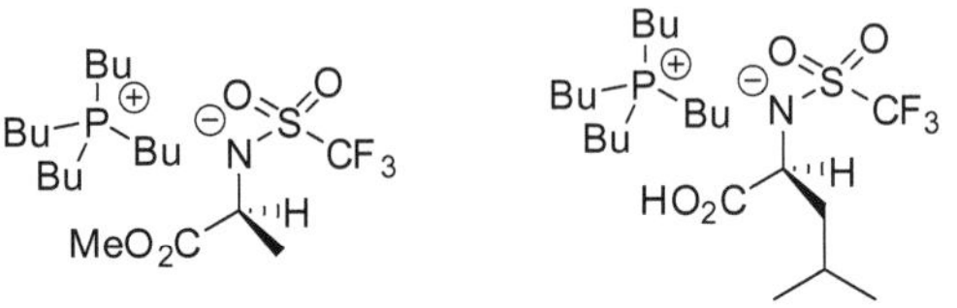

FIGURE 7.8. Examples of chiral anion ILs.

6°C, as well as good thermal stability (>173°C). Ionic conductivities were also measured and correlated to the structure of the ILs.

In 2006, Maschmeyer et al. described the use of a similar approach to prepare a collection of 23 ILs deriving from chiral amino acids and carboxylic acids (Fig. 7.7) [51]. The one-step approach relied on the neutralization of commercially available tetrabutylammonium hydroxide with the selected acids. These derivatives, described as liquids or oils at room temperature, showed good stereochemical integrity but low thermal stability (<110°C). Several tartrate and hemitartrate salts were featured in this study.

Also in 2006, Fukumoto and Ohno described a series of ILs based on modified amino acids as chiral anions wherein the negative charge was this time centered on the nitrogen atom (Fig. 7.8) [52]. After formation of the methyl esters, the treatment with trifluoromethanesulfonic anhydride led to the corresponding sulfonamides that were isolated after standard purification. Finally, deprotonation with either [Bu_4P]OH or [Bmim]OH afforded the expected ILs. Starting from three aliphatic amino acids, six derivatives were thus prepared and fully characterized, including with regard to their stereochemical integrity. Most of these salts behaved as room temperature ILs with high decomposition temperature (ca. 250°C). The T_g and viscosity values were observed to gradually increase with respect to the starting amino acid alkyl chain length. These salts proved to be more hydrophobic ILs than the corresponding standard NTf_2 congeners.

Fukumoto and Ohno further explored the unique phase separation with water of such entities that displayed a lower critical solution temperature [53]. Thus, a series of related ILs bearing a carboxylic acid moiety instead of the methyl ester was prepared (Fig. 7.8). The salt derived from leucine displays a room temperature IL behavior (T_g –50°C). Upon cooling below 22°C, this IL proved totally miscible with water, whereas warming back to 25°C induced a complete phase separation over few minutes. The phase separation

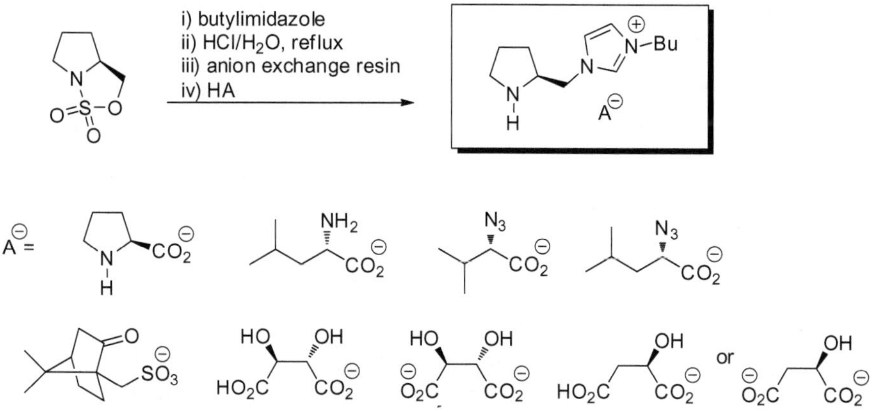

SCHEME 7.2. Doubly CILs.

temperature could be controlled by reproducibly varying the water content and the lipophilicity of the phosphonium cation.

7.2.4. Doubly CILs

Luo et al. developed a combinatorial approach to generate a library of functionalized ILs, including doubly chiral derivatives (Scheme 7.2) [54]. In this particular case, a proline-derived chiral cation was first prepared by the ring opening of a cyclic sulfamidate with 1-butylimidazole to yield a sulfamic zwitterion. Removal of the sulfonyl group by acidic treatment followed by ion-exchange resin led to a hydroxide salt that was neutralized with various chiral amino or hydroxy carboxylic and sulfonic acids. These derivatives, described as viscous liquids with low T_g values (from −37°C to −67°C) and good thermal stability (>163°C), proved to be useful in the organocatalysis of aldol condensations.

Another example of a doubly CIL bearing both a chiral imidazolium cation along with a camphorsulfonato anion has previously been described by Machado and Dorta [55].

7.3. CILS AS REACTION MEDIA AND CHIRAL REAGENTS

One of the most obvious applications of CILs in asymmetric synthesis consists in their use as a solvent or cosolvent, hopefully able to induce a significant asymmetric induction during an enantioselective process. The main advantages of such a strategy would be to benefit from the unique properties of ILs as neoteric solvents, combined with a general and nonsubstrate-dependent chiral induction. If the first attempts in the field were almost unsuccessful [20], a shining series of successes has been published since 2004. In this section are discussed the most successful and promising approaches in the field.

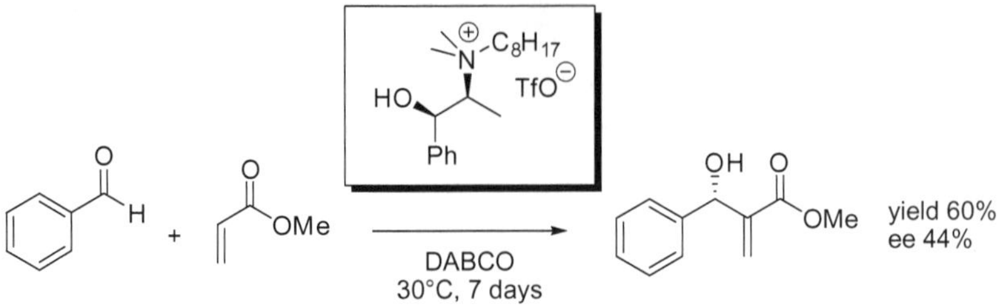

SCHEME 7.3. Asymmetric Baylis–Hillman reaction in CIL.

The first convincing example was described by Vo-Thanh and co-workers [56]. The authors showed that the use of CILs derived from *Ephedra* alkaloids induced an enantioselective Baylis–Hillman reaction with more than 40% enantiomeric excess (ee) (Scheme 7.3). Although moderate, this asymmetric induction was significant enough to demonstrate the real potential of CILs as new tools for asymmetric synthesis. It is worth stressing that most of the previous approaches using molecular chiral solvents led to no or very poor enantioselectivity [4]. It was verified that the observed ee did not come from traces of native alkaloid present in the reaction mixture.

Since this seminal finding, numerous other examples have been disclosed, some of them yielding very impressive enantioselectivities, almost equal to those observed in more classical asymmetric synthetic methods such as the use of chiral reagents or catalysts.

For example, Leitner and co-workers described in 2006 a related enantioselective aza-Baylis–Hillman reaction with up to 84% ee in CIL, using nucleophilic Lewis bases as chiral promotors (Scheme 7.4a) [57]. Based on mechanistic insights, they designed new chiral anions able to interact with the prostereogenic transition state (Scheme 7.4b). Indeed, the Brønsted acidic moieties in alcohol and acid functions act as anchoring points between the CIL and carbonyl groups in the reactants. Very interestingly, the authors demonstrated that asymmetric induction occurred only when the CIL was used as a solvent, and not if diluted into any organic molecular solvent. Again, one can stress that the high observed ee's are very similar to classical asymmetric conditions.

Studies involving transition metal catalysis, a type of chemistry perfectly fitting with the properties of ILs, were published at the same time. Afonso and co-workers prepared CILs bearing chiral anions along with the [*dmg*] cation (tetra-*n*-hexyl-dimethylguanidinium) that is less prone to give crystalline salts [58]. These new CILs were tested as the sole chiral source for rhodium-catalyzed intramolecular carbenoid C–H insertion (ee up to 27%) (Scheme 7.5) and enantioselective Sharpless dihydroxylation using osmium catalysts. In the latter case, a very impressive stereoselectivity was obtained (85% ee, Scheme 7.6). Even using catalytic amount of CIL, the authors observed a significant asymmetric induction (ca. 40% ee).

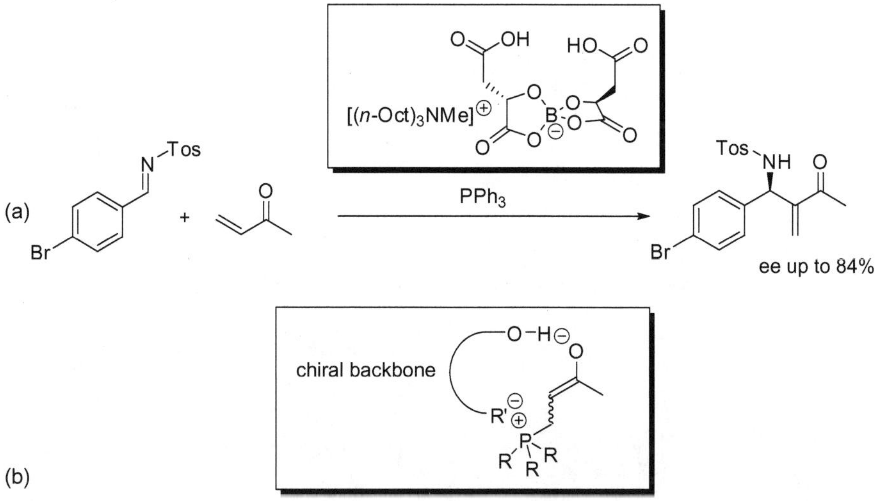

SCHEME 7.4. (a) Asymmetric aza-Baylis–Hillman reaction in CIL; (b) proposed bifunctional interaction of the zwitterionic intermediate of the aza-Baylis–Hillman reaction with the chiral anion of an IL containing a hydrogen bond donor.

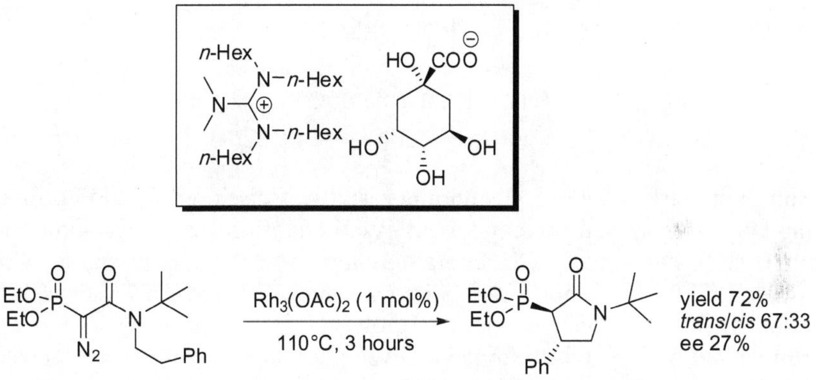

SCHEME 7.5. Asymmetric catalytic Rh(II) carbenoid C–H insertion in CIL.

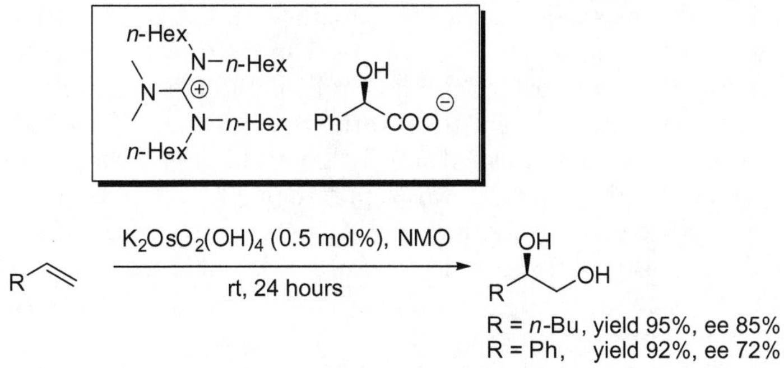

SCHEME 7.6. Asymmetric Sharpless dihydroxylation in CIL.

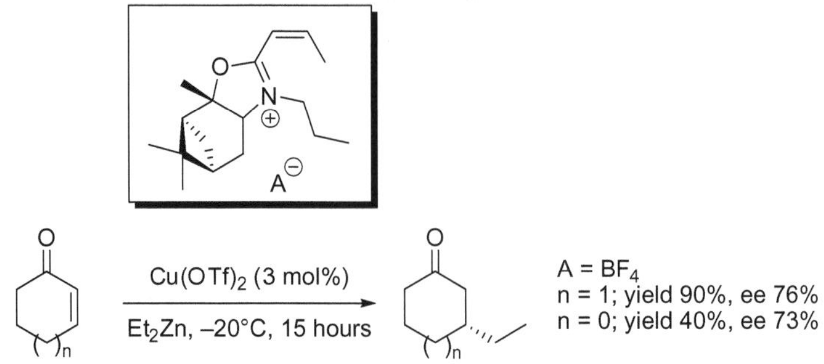

SCHEME 7.7. Asymmetric 1,4 addition of diethylzinc to enones in CIL.

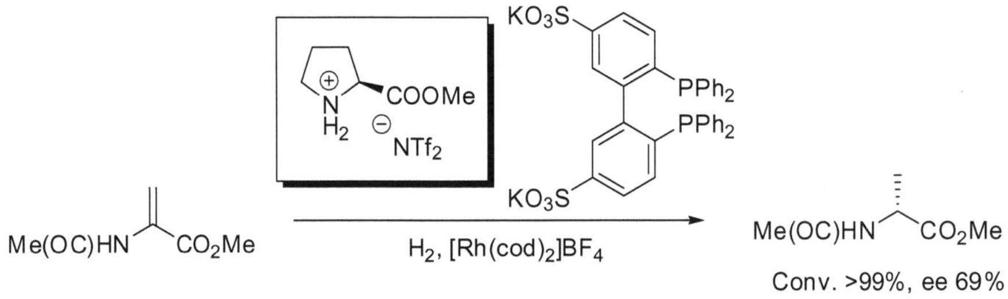

SCHEME 7.8. Asymmetric catalytic C=C hydrogenation in CIL.

Simultaneously, another outstanding result was disclosed by Malhotra and Wang [49], who studied the copper-catalyzed enantioselective 1,4 addition of diethylzinc to enones, in a CIL containing a chiral oxazolinium cation, which was prepared from a pinene backbone (Scheme 7.7). When using the CIL as a cosolvent in high concentration (35 mol%), ee's as high as 76% and 73% were reached with cyclohexenone and cyclopentanone as substrates. The role of the achiral counteranion (A = BF_4, PF_6) was examined, showing significant differences depending on its nature.

These first examples dealing with the enantioselective construction of C–C bonds as well as enantioselective oxidations demonstrated the real potency of using CILs for some very useful and generally exploited reactions in organic synthesis. Other studies dealing with powerful reactions belonging to the classic toolbox of chemists were published shortly thereafter. For example, one of the most elegant applications of CILs was reported in 2007 by Leitner et al. in the field of asymmetric catalytic hydrogenation [59]. Herein, the CILs based on amino acids such as proline served as a chiral relay able to "twist" an achiral tropos ligand in a chiral manner. Thus, the rhodium catalyst induced significant stereoselectivity during the hydrogenation of acrylates (Scheme 7.8).

Anionic sulfonated ligands were used because of their enhanced solubility, and the authors highlighted the positive role of added basic additives such as triethylamine. Also, a careful study allowed recycling of the catalytic system using extraction of the reaction product with supercritical carbon dioxide.

More recently, the same group extended the concept to the related use of racemic instead of achiral ligands [60]. The authors investigated the enantioselective hydrogenation of dimethylitaconate. Racemic as well as enantiopure BINAP were used as ligands for rhodium catalysis with proline-derived CIL as chiral cosolvent. In all cases, similar stereoselectivities were observed.

CILs can also be used as chiral reagents, in which case the IL is modified during the course of the reaction. For example, Bao and Wang [61] used several chiral tribromide ILs as chiral media and reactant for the synthesis of bromoester from benzaldehyde and *meso*-2,3-butanediol (Scheme 7.9).

A rather different approach in the field of asymmetric hydrogenation was disclosed by Wasserscheid and co-workers [62, 63]. In this case, the heterogeneous catalytic hydrogenation by means of ruthenium/C was performed on a prochiral imidazolium salt under the influence of a CIL bearing a chiral anion (Scheme 7.10). Stereoselectivity was quite high (80% ee), and was assumed to be due to the strong electrostatic interaction between the prochiral substrate and the chiral anionic part of the CIL, used as a cosolvent in methanol in order to decrease viscosity.

The authors were able to demonstrate that no chiral modification of the solid catalyst occurred. Extension of the method using ethanol as a

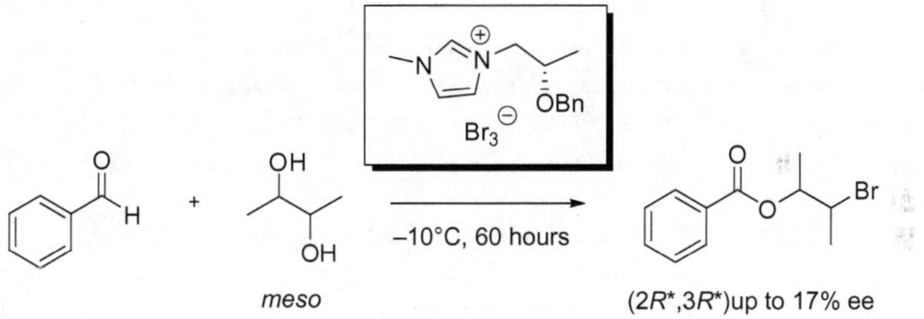

SCHEME 7.9. Asymmetric synthesis of bromoester in CIL.

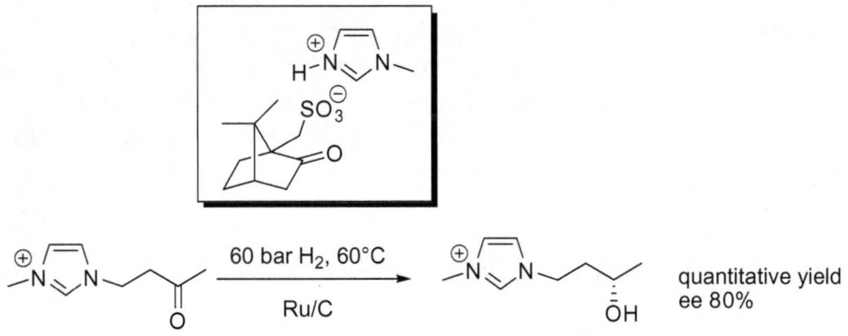

SCHEME 7.10. Asymmetric catalytic C=O hydrogenation in CIL.

solvent and improving the amount of CIL allowed them to reach 94% enantioselectivity [63].

Other asymmetric reactions were published in recent years. They will not be detailed here, either because of lower selectivities than in the results reported above, or because the actual role of the CIL remains obscure (as for example in the case of reactions using another chiral reagent or starting from a chiral substrate). Nevertheless, these approaches contribute to novel findings and insights in the field of asymmetric synthesis and must be cited as very promising tools for the development of the area, such as enantioselective photoisomerizations [34], kinetic resolution with a lipase [28], palladium-catalyzed Heck reactions [64], Michael additions [65], phase transfer catalysis [66], or diastereoselective aza-Diels–Alder reactions [30, 67].

Besides their recent development as new chiral media for organic synthesis, a few reports rely on the use of CILs as chiral solvents for stereoselective polymerization. As an example, in the use of CILs such as [mbmim*][PF_6] [68] or 3-(2-ethoxy-1-methyl-2-oxoethyl)-1-methylimidazolium [PF_6] or [BF_4] [69] as solvents in the atom transfer radical polymerization (ATRP) of acrylic monomers, a strong effect on polymer tacticity was noticed. In the free radical polymerization of vinyl monomers [35], Ma et al. have used chiral CILs such as (1-(−)-menthoxycarbonylmethylene-3-methylimidazolinium [PF_6] and 1-(−)-menthoxycarbonylmethylene-3-hexadecylimidazolinium [PF_6]) in reverse ATRP. The reaction was performed at 80°C using AIBN as an initiator and $CuCl_2$ complexes with bipyridine as catalysts. Although the monomer conversion was similar to that observed under bulk conditions, a lower polydispersity was obtained in CILs (Mw/Mn < 1.20), to compare with 2.07 in bulk conditions. This excellent result was explained by the high solubility of the catalyst complex in the medium, which accounted for a highly controlled polymerization.

At the end of this section, some conclusions can be drawn. Impressive reports have been published in which very high asymmetric inductions are described. Hence, there is no more room for doubt that the use of CILs as chiral media for various applications (asymmetric synthesis, polymerization reactions) is a realistic and powerful approach. Interestingly, this chemistry seems versatile enough to ensure high level of stereoselectivity in different useful organic reactions, like C–C bond formation, oxidations, and reductions, that is, the classic toolbox of organic chemistry. Nevertheless, one can observe that the CILs used in these studies were more or less designed for the special reaction described and thus fail to present generality. Certainly, most studies in the future will focus on this point.

7.4. MISCELLANEOUS APPLICATIONS

In this last section, we wish to emphasize that CILs (as are ILs in general) are much more than just a new class of solvents. Indeed, their unique properties allow many new applications, besides their original contribution to asymmetric reactions.

For example, since 2002, the use of CILs as chiral shift reagents in nuclear magnetic resonance (NMR) has been reported. First initiated by Wasserscheid et al. [26] and almost simultaneously by Gaumont et al. [46] and Saigo et al. [37], the research in this field demonstrated that many CILs were able to induce the shifts of enantiomers of Mosher acid used as a model compound, thus allowing the determination of ee's. Structurally different classes of CILs were used for this purpose, all of them giving promising results. Ephedrinium [26], thiazolinium [46], imidazolinium [48], and even chirally planar salts [37, 70] allowed NMR resolution for compounds not easy to resolve by means of more usual shift reagents such as chiral lanthanide salts. More recent reports, neither giving really new insights nor results, have recently been reviewed [12, 17, 18].

Another type of analytical resolution may also be carried out using CILs, that is, chiral chromatography [71]. This topic has also been recently reviewed [12, 18] and appears to us to be beyond the scope of this chapter.

An even more interesting topic in the field, which is currently under investigation, is a new approach for preparative resolution using CILs. New approaches in the field of liquid phase resolution methods may circumvent the problem of handling large amounts of solid materials. Enantioselective liquid–liquid extraction (ELLE) combines in a single technique the advantages of enantiomeric recognition and solvent extraction. Complexes of the chiral extractant with the enantiomers result from various intermolecular interactions, such as ion pairing, hydrogen bonding, π–π interactions, dipole attractions, and Van der Waals forces. ELLE processes also imply the use of two immiscible phases. A comprehensive review of this topic has recently been published [72]. ELLE, when combined with multistage countercurrent cascade technology, is one of the most promising ways in the field of new chiral separation technologies because of its scalability and recyclability. Improvements in the field will depend on the discovery of new highly selective chiral selectors, which should be easily prepared and recycled.

CILs are very promising candidates for this last role: indeed, their use both as solvent and as chiral selector in ELLE process was recently reported by Tang et al., with ee values up to 50% for single-step extraction [73]. Enantioselective enrichment of racemic amino acids was achieved through ligand exchange mechanism by means of the copper complex of CILs, which were prepared by combination of alkylimidazolium cation and prolinate anion (Scheme 7.11). In the ELLE process, the CIL played the role of the acceptor phase, and the organic solvent played that of the donor phase. The efficiency and the enantioselectivity of the extraction system were dependent on different factors, including CIL structure, copper ion concentration, organic phase, and amino acid concentration, and varied from near 0% up to 50% ee.

We are currently examining a complementary ELLE process in liquid phase, with the CILs playing the simultaneous role of enantiomer recognition center and solvent [74].

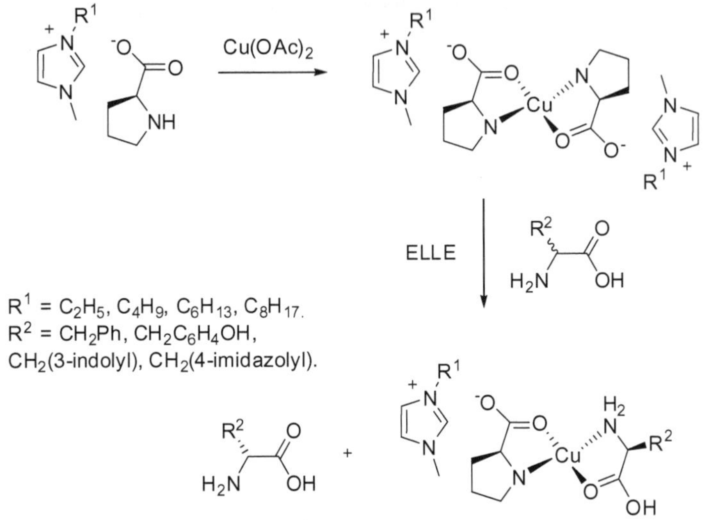

SCHEME 7.11. ELLE process in CIL.

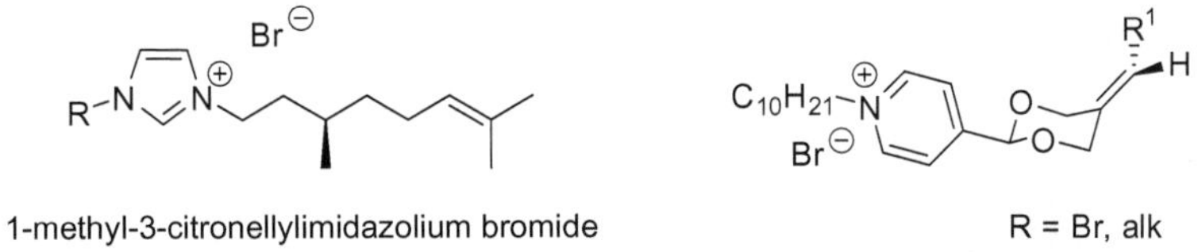

FIGURE 7.9. Examples of CILs with liquid crystal properties.

Finally, we wish to draw attention to the fact that CILs are not only new chiral media as exemplified in earlier sections of this chapter, but may also constitute a novel class of chiral materials. For example, some literature reports the construction of CILs with liquid crystal properties. These new structures were either obtained from the chiral pool [31] or from asymmetric synthesis [44] and are depicted in Figure 7.9.

7.5. CONCLUSION AND PROSPECTS

Making chiral compounds is an ever-growing target in organic synthesis. Various strategies have been explored which use a chiral environment. Since the first experiment conducted by Seebach in 1975 [5, 75], it has been recognized that chiral solvents can exert an influence on the enantioselectivity of a reaction. Then, various groups have focused their work on the solvent-mediated chirality transfer [76–78]. However, due to the moderate enantioselectivities that were obtained, this topic has been neglected for more than 20 years. It is only since the beginning of the twenty-first century that a renewal of interest in this methodology has appeared in the literature thanks to ILs, an exciting class of new solvents. Indeed, asymmetric induction by CILs has become an emerging field of research, and impressive examples of highly enantioselective reactions have been described [11, 14, 18], including the asymmetric Baylis–Hillman and aza-Baylis–Hillman reactions, developed by Vo-Thanh et al. [56] and Leitner et al. [57], and the Michael addition of ketones to nitrostyrenes reported by Luo and co-workers [33].

Although the field of CILs is still in its infancy, it has a very promising future. The main advantages are related to the fact that CILs are readily available from chiral sources such as naturally occurring amino acids and other compounds. However, it is important to recall that many starting materials for the synthesis of ILs such as pyridine, halogenoalkanes, and other starting materials come from petroleum feedstock that are neither green nor sustainable. Renewable resources should represent a valid alternative to synthetic new ILs in the future, an alternative able to take into account both the ecological and economic requirements. Additionally, ILs can be designed on purpose for a specific application, which is often the key for an efficient transfer of chirality. However, the main issue now is to go from a specific case to generality. Assuming that the dissymmetric solvation of intermediates is weakly sensitive to the structure of the substrate, the chirality transfer could thus result from the formation of macromolecular structures through various noncovalent interactions such as hydrogen bondings, electrostatic interactions, and van der Waals attractions. This would open the route to a completely new—and as yet unwritten—story in the field of asymmetry.

ACKNOWLEDGMENTS

We want to express our gratitude to INTENANT for supporting part of this study and to the European Union for the grant FP7-NMP2-SL2008-214129.

REFERENCES

[1] For a discussion, see for example: Ternois J., Guillen F., Plaquevent J.-C., Coquerel G. (2007). Asymmetric synthesis of modafinil and its derivatives by enantioselective oxidation of thioethers: comparison of various methods including synthesis in ionic liquids. *Tetrahedron Asymmetry*, *18*, 2959–2964.

[2] Stephenson, G. R., Genêt, J. P. (1996). *Advanced Asymmetric Synthesis*. Chapman & Hall, London, pp. 1–8.

[3] Moradpour, A., Nicoud, J. F., Balavoine, G., Kagan, H. B., Tsoucaris, G. (1971). Photochemistry with circularly polarized light. The synthesis of optically active hexahelicene. *J. Am. Chem. Soc.*, *93*, 2353–2354.

[4] For a short history of asymmetric induction in chiral solvents, see: Gaumont, A.-C., Génisson, Y., Guillen, F., Plaquevent, J.-C. (2011). De Pasteur aux liquides ioniques chiraux, petite histoire de l'induction asymétrique promue par le solvant. *L'actualité chimique*, *348–349*, 84–89.

[5] Seebach, D., Kalinowski, H.-O., Langer, W., Crass, G., Wilka, E.-M. (1990). Chiral media for asymmetric solvent inductions. *Org. Synth. Coll.*, *7*, 41.

[6] Wasserscheid, P., Welton, T. (2008). *IonicLiquids in Synthesis*, 2nd ed. Wiley-VCH, Weinheim, Germany.

[7] Freemantle, M. (2010). *An Introduction to Ionic Liquids*. RCS Publishing, Cambridge.

[8] Verbit, L., Halbert, T. R., Patterson, R. B. (1975). Asymmetric decarboxylation of ethylphenylmalonic acid in a cholesteric liquid crystal solvent. *J. Org. Chem.*, *40*, 1649–1650.

[9] See for example: Hardacre, C., Holbrey, J. D., Nieuwenhuyzen, M., Youngs, T. G. A. (2007). Structure and solvation in ionic liquids. *Acc. Chem. Res.*, *40*, 1146–1155.

[10] Aggarwal, A., Lancaster, N. L., Sethi, A. R., Welton, T. (2002). The role of hydrogen bonding in controlling the selectivity of Diels–Alder reactions in room-temperature ionic liquids. *Green Chem.*, *4*, 517–520.

[11] Baudequin, C., Baudoux, J., Levillain, J., Cahard, D., Gaumont, A.-C., Plaquevent, J.-C. (2003). Ionic liquids and chirality: opportunities and challenges. *Tetrahedron Asymmetry*, *14*, 3081–3093.

[12] Baudequin, C., Brégeon, D., Levillain, J., Guillen, F., Plaquevent, J.-C., Gaumont, A.-C. (2005). Chiral ionic liquids, a renewal for the chemistry of chiral solvents? Design, synthesis and applications for chiral recognition and asymmetric synthesis. *Tetrahedron Asymmetry*, *16*, 3921–3945.

[13] Plaquevent, J.-C., Levillain, J., Guillen, F., Malhiac, C., Gaumont, A.-C. (2008). Ionic liquids: new targets and media for α-amino acid and peptide chemistry. *Chem. Rev.*, *108*, 5035–5060.

[14] Ding, J., Armstrong, D. W. (2005). Chiral ionic liquids: synthesis and applications. *Chirality*, *17*, 281–292.

[15] Chen, X., Li, X., Hu, A., Wang, F. (2008). Advances in chiral ionic liquids derived from natural amino acids. *Tetrahedron Asymmetry*, *19*, 1–14.

[16] Winkel, A., Reddy, P. V. G., Wilhelm, R. (2008). Recent advances in the synthesis and application of chiral ionic liquids. *Synthesis*, 999–1016.

[17] Patil, M. L., Sasai, H. (2008). Recent developments on chiral ionic liquids: design, synthesis, and applications. *Chem. Rec.*, *8*, 98–108.

[18] Bica, K., Gaertner, P. (2008). Applications of chiral ionic liquids. *Eur. J. Org. Chem.*, 3235–3250.

[19] Prechtl, M. H. G., Scholten, J. D., Neto, B. A. D., Dupont, J. (2009). Application of chiral ionic liquids for asymmetric induction in catalysis. *Curr. Org. Chem.*, *13*, 1259–1277.

[20] Earle, M. J., McCormac, P. B., Seddon, K. R. (1999). Diels-Alder reactions in ionic liquids—a safe recyclable alternative to lithium perchlorate-diethyl ether mixtures. *Green Chem.*, *1*, 23–25.

[21] Fukumoto, K., Yoshizawa, M., Ohno, H. (2005). Room temperature ionic liquids from 20 natural amino acids. *J. Am. Chem. Soc.*, *127*, 2398–2399.

[22] Handy, S. T. (2003). Greener solvents: room temperature ionic liquids from biorenewable sources. *Chem. Eur. J.*, *9*, 2938–2944.

[23] Imperato, G., König, B., Chiappe, C. (2007). Ionic green solvents from renewable resources. *Eur. J. Org. Chem.*, 1049–1058.

[24] Tao, G.-H., He, L., Sun, N., Kou, Y. (2005). New generation ionic liquids: cations derived from amino acids. *Chem. Commun.*, 3562–3564.

[25] Tao, G.-H., He, L., Liu, W.-S., Xu, L., Xiong, W., Wang, T., Kou, Y. (2006). Preparation, characterization and application of amino acid-based green ionic liquids. *Green Chem.*, *8*, 639–646.

[26] Wasserscheid, P., Bösmann, A., Bolm, C. (2002). Synthesis and properties of ionic liquids derived from the chiral pool. *Chem. Commun.*, 200–201.

[27] Jodry, J. J., Mikami, K. (2004). New chiral imidazolium ionic liquids: 3D-network of hydrogen bonding. *Tetrahedron Lett.*, *45*, 4429–4431.

[28] Kitazume, T. (2001). Optically active ionic liquid, Patent US2001031875.

[29] Pernak, J., Feder-Kubis, J. (2005). Synthesis and properties of chiral ammonium-based ionic liquids. *Chem. Eur. J.*, *11*, 4441–4449.

[30] Van Buu, O. N., Vo-Thanh, G. (2007). Synthesis of novel chiral ammonium-based ionic liquids derived from isosorbide and their applications in an asymmetric aza Diels-Alder reaction. *Lett. Org. Chem.*, *4*, 158–167.

[31] Tosoni, M., Laschat, S., Baro, A. (2004). Synthesis of novel chiral ionic liquids and their phase behavior in mixtures with smectic and nematic liquid crystals. *Helv. Chim. Acta.*, *87*, 2742–2749.

[32] Kim, E. J., Ko, S. Y., Dziadulewicz, E. K. (2005). Mitsunobu alkylation of imidazole: a convenient route to chiral ionic liquids. *Tetrahedron Lett.*, *46*, 631–633.

[33] Luo, S., Mi, X., Zhang, L., Liu, S., Xu, H., Cheng, J.-P. (2006). Functionalized chiral ionic liquids as highly efficient asymmetric organocatalysts for Michael addition to nitroolefins. *Angew. Chem. Int. Ed.*, *45*, 3093–3097.

[34] Ding, J., Desikan, V., Han, X., Xiao, T. L., Ding, R., Jenks, W. S., Armstrong, D. W. (2005). Use of chiral ionic liquids as solvents for the enantioselective photoisomerization of dibenzobicyclo[2.2.2]octatrienes. *Org. Lett.*, *7*, 335–337.

[35] Ma, H.-Y., Wan, X.-H., Chen, X.-F., Zhou, Q.-F. (2003). Design and synthesis of novel chiral ionic liquids and their application in free radical polymerization of methyl methacrylate. *Chin. J. Polym. Sci.*, *21*, 265–270.

[36] Plaza, P. G. J., Bhongade, B. A., Singh, G. (2008). Synthesis of chiral carbohydrate ionic liquids. *Synthesis*, 2973–2976.

[37] Ishida, Y., Miyauchi, H., Saigo, K. (2002). Design and synthesis of a novel imidazolium-based ionic liquid with planar chirality. *Chem. Commun.*, 2240–2241.

[38] Bao, W., Wang, Z., Li, Y. (2003). Synthesis of chiral ionic liquids from natural amino acids. *J. Org. Chem.*, *68*, 591–593.

[39] Hannig, F., Kehr, G., Fröhlich, R., Erker, G. (2005). Formation of chiral ionic liquids and imidazol-2-ylidene metal complexes from the proteinogenic amino acid *L*-histidine. *J. Organomet. Chem.*, *690*, 5959–5972.

[40] Guillen, F., Brégeon, D., Plaquevent, J.-C. (2006). (*S*)-Histidine: the ideal precursor for a novel family of chiral aminoacid and peptidic ionic liquids. *Tetrahedron Lett.*, *47*, 1245–1248.

[41] Despite previous claims, attempts to use the Zincke methodology for the construction of structurally related imidazolium cations have been unsuccessful so far. Pastre, J. C., Correia, C. R. D., Génisson, Y. (2008). Non-occurrence of a Zincke-like process upon treatment of 1-(2,4-dinitrophenyl)-3-methylimidazolium chloride with a chiral primary amine. *Green Chem.*, *10*, 885–889.

[42] Patrascu, C., Sugisaki, C., Mintogaud, C., Marty, J.-D., Génisson, Y., Lauth de Viguerie, N. (2004). New pyridinium chiral ionic liquids. *Heterocycles*, *63*, 2033–2041.

[43] Drahonovsky, D., Labat, G. C., Sevcik, J., von Zelewsky, A. (2005). Pinene-fused chiral *N*-ethylpyridinium room temperature molten salts. *Heterocycles*, *65*, 2169–2179.

[44] Baudoux, J., Judeinstein, P., Cahard, D., Plaquevent, J.-C. (2005). Design and synthesis of novel ionic liquid/liquid crystals (IL2Cs) with axial chirality. *Tetrahedron Lett.*, *46*, 1137–1140.

[45] Winkel, A., Wilhelm, R. (2009). New chiral ionic liquids based on imidazolinium salts. *Catal. Commun.*, *11*, 276–279.

[46] Levillain, J., Dubant, G., Abrunhosa, I., Gulea, M., Gaumont, A.-C. (2003). Synthesis and properties of thiazoline based ionic liquids derived from the chiral pool. *Chem. Commun.*, 2914–2915.

[47] Brégeon, D., Levillain, J., Guillen, F., Plaquevent, J.-C., Gaumont, A.-C. (2008). Thiazolinium and imidazolium chiral ionic liquids derived from natural amino acid derivatives. *Amino Acids*, *35*, 175–184.

[48] Clavier, H., Boulanger, L., Audic, N., Toupet, L., Mauduit, M., Guillemin, J.-C. (2004). Design and synthesis of imidazolinium salts derived from (*L*)-valine. Investigation of their potential in chiral molecular recognition. *Chem. Commun.*, 1224–1225.

[49] Malhotra, S. V., Wang, Y. (2006). Application of chiral ionic liquids in the copper catalyzed enantioselective 1,4-addition of diethylzinc to enones. *Tetrahedron Asymmetry*, *17*, 1032–1035.

[50] Shah, J., Khan, S. S., Blumenthal, H., Liebscher, J. (2009). 1,2,3-Triazolium-tagged prolines and their application in asymmetric aldol and Michael reactions. *Synthesis*, 3975–3982.

[51] Allen, C. R., Richard, P. L., Ward, A. J., van de Water, L. G. A., Masters, A. F., Maschmeyer, T. (2006). Facile synthesis of ionic liquids possessing chiral carboxylates. *Tetrahedron Lett.*, *47*, 7367–7370.

[52] Fukumoto, K., Ohno, H. (2006). Design and synthesis of hydrophobic and chiral anions from amino acids as precursor for functional ionic liquids. *Chem. Commun.*, 3081–3083.

[53] Fukumoto, K., Ohno, H. (2007). LCST-type phase changes of a mixture of water and ionic liquids derived from amino acids. *Angew. Chem. Int. Ed.*, *46*, 1852–1855.

[54] Zhang, L., Luo, S. Z., Mi, X. L., Liu, S., Qiao, Y. P., Xu, H., Cheng, J. P. (2008). Combinatorial synthesis of functionalized chiral and doubly chiral ionic liquids and their applications as asymmetric covalent/non-covalent bifunctional organocatalysts. *Org. Biomol. Chem.*, *6*, 567–576.

[55] Machado, M. Y., Dorta, R. (2005). Synthesis and characterization of chiral imidazolium salts. *Synthesis*, *15*, 2473–2475.

[56] Pégot, B., Vo-Thanh, G., Loupy, A. (2004). First application of chiral ionic liquids in asymmetric Baylis–Hillman reaction. *Tetrahedron Lett.*, *45*, 6425–6428.

[57] Gausepohl, R., Buskens, P., Kleinen, J., Bruckmann, A., Lehmann, C. W., Klankermayer, J., Leitner, W. (2006). Highly enantioselective aza-Baylis-Hillman reaction in a chiral reaction medium. *Angew. Chem. Int. Ed.*, *45*, 3689–3692.

[58] Branco, L. C., Gois, P. M. P., Lourenço, N. M. T., Kurteva, V. B., Afonso, C. A. M. (2006). Simple transformation of crystalline chiral natural anions to liquid medium and their use to induce chirality. *Chem. Commun.*, *22*, 2371–2372.

[59] Schmitkamp, M., Chen, D., Leitner, W., Klankermayer, J., Franciò, G. (2007). Enantioselective catalysis with tropos ligands in chiral ionic liquids. *Chem. Commun.*, 4012–4014.

[60] Chen, D., Schmitkamp, M., Franciò, G., Klankermayer, J., Leitner, W. (2008). Enantioselective hydrogenation with racemic and enantiopure binap in the presence of a chiral ionic liquid. *Angew. Chem. Int. Ed.*, *120*, 7449–7451.

[61] Bao, W., Wang, Z. (2006). An effective synthesis of bromoesters from aromatic aldehydes using tribromide ionic liquid based on *L*-prolinol as reagent and reaction medium under mild conditions. *Green Chem.*, *8*, 1028–1033.

[62] Schulz, P. S., Müller, N., Bösmann, A., Wasserscheid, P. (2007). Effective chirality transfer in ionic liquids through ion-pairing effects. *Angew. Chem. Int. Ed.*, *46*, 1293–1295.

[63] Schneiders, K., Bösmann, A., Schulz, P. S., Wasserscheid, P. (2009). Chirality transfer in imidazolium camphorsulfonate ionic liquids through ion pairing effects. *Adv. Synth. Catal.*, *351*, 432–440.

[64] Kiss, L., Kurtan, T., Antus, S., Brunner, H. (2003). Further insight into the mechanism of Heck oxyarylation in the presence of chiral ligands. *Arkivoc*, *5*, 69–76.

[65] Wang, Z., Wang, Q., Zhang, Y., Bao, W. (2005). Synthesis of new chiral ionic liquids from natural acids and their applications in enantioselective Michael addition. *Tetrahedron Lett.*, *46*, 4657–4660.

[66] Mukherjee, D. K., Ghosh, N. (2008). Enantioselective phase transfer alkylation using orthopalladated complex in chiral ionic liquid. *Catal. Commun.*, *9*, 40–44.

[67] Van Buu, O. N., Aupoix, A., Vo-Thanh, G. (2009). Synthesis of novel chiral imidazolium-based ionic liquids derived from isosorbide and their applications in asymmetric aza-Diels-Alder reaction. *Tetrahedron*, *65*, 2260–2265.

[68] Biedron, T., Kubisa, P. (2003). Ionic liquids as reaction media for polymerization processes: atom transfer radical polymerization (ATRP) of acrylates in ionic liquids. *Polym. Int.*, *52*, 1584–1588.

[69] Biedron, T., Kubisa, P. (2005). Radical polymerization in a chiral ionic liquid: atom transfer radical polymerization of acrylates. *J. Polym. Sci. A Polym. Chem.*, *43*, 3454–3459.

[70] Ishida, Y., Sasaki, D., Miyauchi, H., Saigo, K. (2004). Design and synthesis of novel imidazolium-based ionic liquids with a pseudo crown-ether moiety: diastereomeric interaction of a racemic ionic liquid with enantiopure europium complexes. *Tetrahedron Lett.*, *45*, 9455–9459.

[71] Ding, J., Welton, T., Armstrong, D. W. (2004). Chiral ionic liquids as stationary phases in gas chromatography. *Anal. Chem.*, *76*, 6819–6822.

[72] Schuur B., Verkuijl B. J. V., Minnaard A. J., de Vries J. G., Heeres H. J., Feringa B. L. (2011). Chiral separation by enantioselective liquid–liquid extraction. *Org. Biomol. Chem.*, *9*, 36–51.

[73] Tang F., Zhang Q., Ren D., Nie Z., Liu Q., Yao S. (2010). Functional amino acid ionic liquids as solvent and selector in chiral extraction. *J. Chromatogr. A*, *1217*, 4669–4674.

[74] Zgonnik V., Gonella S., Mazières M.-R., Guillen F., Coquerel G., Plaquevent J.-C. Design and scalable synthesis of new chiral selectors. Part 2: chiral ionic liquids derived from DACH and Histidine, manuscript in preparation.

[75] Seebach, D., Oei, H.-A. (1975). Mechanism of electrochemical pinacolization. The first asymmetric synthesis in a chiral medium. *Angew. Chem. Int. Ed. Engl.*, *14*, 634–636.

[76] Di Furia, F., Modena, G., Curci, R. (1976). Chiral solvent-induced asymmetric synthesis of sulphoxides in the metal-catalysed oxidation of sulphides by *t*-butyl hydroperoxide. *Tetrahedron Lett.*, *50*, 4637–4638.

[77] Laarhoven, W. H., Cuppen, T. J. H. (1977). Chiral solvent-induced asymmetric synthesis; photosynthesis of optically enriched hexahelicene. *J. Chem. Soc. Chem. Commun.*, 47a.

[78] Laarhoven, W. H., Cuppen, T. J. H. (1978). Chiral solvent-induced asymmetric synthesis. Part 2. Photosynthesis of optically enriched hexahelicenes. *J. Chem. Soc. Perkin Trans. 2*, 315–318.

CHAPTER 8

ASYMMETRIC REACTIONS IN FLOW REACTORS

MUNAWWER RASHEED, SIMON C. ELMORE, AND THOMAS WIRTH

8.1. Introduction 346
8.2. Homogeneous asymmetric catalytic reactions 346
- 8.2.1. Enantioselective addition of trimethylsilyl cyanide and acyl cyanide to aldehydes and kinetic resolution 347
- 8.2.2. Enzyme-catalyzed cyanations 348
- 8.2.3. Asymmetric catalytic hydrogenation 349
- 8.2.4. Asymmetric catalytic aldol reaction 351
- 8.2.5. Asymmetric catalytic epoxidations 351
- 8.2.6. Asymmetric diethylzinc additions 352
- 8.2.7. Selective glycosylation 352
- 8.2.8. Photochemical asymmetric reactions 353

8.3. Heterogeneous asymmetric catalytic reactions 354
- 8.3.1. Kinetic resolutions 354
- 8.3.2. Role of diethylzinc in alkyl addition to aldehydes 356
- 8.3.3. Asymmetric cyclopropanations 358
- 8.3.4. Asymmetric hydrogenations 358
- 8.3.5. Asymmetric Michael additions 360
- 8.3.6. Asymmetric chlorination and β-lactam synthesis from acid chlorides 360
- 8.3.7. Stereoselective Diels–Alder and Ene reactions 362
- 8.3.8. Synthesis of chiral building blocks 363
- 8.3.9. Enantioselective addition of trimethylsilyl cyanide (TMSCN) and acetyl cyanide to benzaldehyde 364
- 8.3.10. Applications in multistep synthesis 364

References 367

Catalytic Methods in Asymmetric Synthesis: Advanced Materials, Techniques, and Applications, First Edition. Edited by Michelangelo Gruttadauria and Francesco Giacalone.

8.1. INTRODUCTION

Microreactor technology offers advantages to classical approaches by allowing miniaturization of structural features up to the micrometer regime. In recent years, chemists have recognized this as a very powerful tool and many reactions have been performed in such devices. These reactions benefit from the physical properties of microreactors, such as enhanced mass and heat transfer due to a very large surface/volume ratio as well as regular flow profiles leading to improved yields with increased selectivities. Catalytic reactions are becoming more and more important in chemistry, as increasing size and sustainability of processes require a careful use of limited resources. Strict control over thermal or concentration gradients within the microreactor allows new methods and technologies for efficient chemical transformations with high space–time yields. The mixing of substrates and reagents can be performed under highly controlled conditions leading to improved protocols. In last few years, a number of communications have reviewed this emerging field [1]. This chapter presents a selection of stereoselective reactions, which have been performed in microreactors. Microreactor technology seems to provide an additional platform for efficient homogeneous and heterogeneous catalytic reactions.

8.2. HOMOGENEOUS ASYMMETRIC CATALYTIC REACTIONS

Flow reactors and microreactors are potentially of very exciting interest to the field of homogeneous catalysis. Tending to give better thermal control and mixing over reactions, microreactors can give rise to enhanced yields, purities, and selectivities [2]. Homogeneous catalysis can be particularly appealing for use in flow, where issues such as catalyst insolubility and an even catalyst distribution are less of a concern than in heterogeneous examples. Until now, there is only a small number of homogeneous asymmetric catalyzed reactions that have been performed in microstructured devices.

While some organocatalysts are soluble in the reaction solvents, there are many cases where the catalyst must be adapted or otherwise changed chemically to be capable of performing selective homogeneous catalysis. Solubility in flow is a major issue, especially in microfluidic systems where even a small amount of insoluble catalyst or substrate can potentially jeopardize the entire reactor by causing unwanted and potentially dangerous blockages and leaks. As a result, significant work has been put into finding methods of dissolving catalysts for the flow environment, the most prominent of these being molecular weight enhancement [3]. In addition to the problem of actually dissolving the catalyst, it is also preferable if the catalyst can be recovered from the microreactor before workup or subsequent further reactions and, if possible, reused. Investigations into catalyst separation methods have also been performed and this chapter will be covering several types of these.

Separation of catalyst and substrate can often be quite challenging in homogeneous catalysis, especially if they have similar molecular weights. Molecular weight enhancement aims to help distinguish between the catalyst and the substrate by greatly increasing catalyst size and weight without altering its catalytic properties. This would then allow the filtration of the catalyst (potentially in flow), without losing the product to the filtration process. In a homogeneous system this would be accomplished by soluble polymer supports. This is highly advantageous as it means that the catalyst will be very thoroughly distributed throughout the system and the reactor blockage is not an issue due to both the catalyst and support being soluble in the reaction solvent [4].

The first polymer-bound active transition metal catalysts were synthesized in the early 1970s by Manassen [5], and since then a whole range of different polymers have been used to enhance the molecular weight of transition metal catalysts [6]. It has been found that linear polymer-supported catalysts are particularly good at keeping a similar reactivity to the low molecular weight version.

An alternative method to linear supports would be dendrimeric support, where the catalyst is attached to a dendrimer. The catalyst shows kinetic behavior and the transition metals on the outer sphere are directly accessible to the substrate, which allows for reaction rates that are comparable with other homogeneous systems. These types of catalyst contain many different reaction sites and ligands, resulting in a very high local catalyst and ligand concentration, which in turn can lead to catalyst destabilization or deactivation in some bimetallic systems.

8.2.1. Enantioselective Addition of Trimethylsilyl Cyanide and Acyl Cyanide to Aldehydes and Kinetic Resolution

Lewis acid and Lewis base-catalyzed reactions are quite common for the production of cyanohydrins from aldehydes. A T-shaped, electroosmotic-driven microreactor was used for the optimization of the enantioselective cyanation of benzaldehyde. Due to their robust nature [7], chiral bis(oxazolinyl)pyridine (Pybox) ligands were used to form the required lanthanide complexes used for the reaction [8]. Compared to a conventional batch procedure, higher conversion was observed within shorter reaction time. The microreactor process involving Lu(III) afforded essentially the same enantioselectivity as the batch process, whereas the enantioselectivity was lower in the microreactor for catalysts containing Yb(III). Ce(III)-based catalysts showed negligible selectivity in both processes (Scheme 8.1). The microreactor used for these experiments was a T-shaped borosilicate microreactor operating under electroosmotic flow where a positive voltage was applied across the two inlets and a negative voltage was applied across the outlet. Unfortunately, the initial results for the reaction catalyzed using $YbCl_3$ showed both a yield and selectivity lower than those obtained from the equivalent flask reaction. Further investigation showed that the catalyst loading tolerance of this reaction in flow

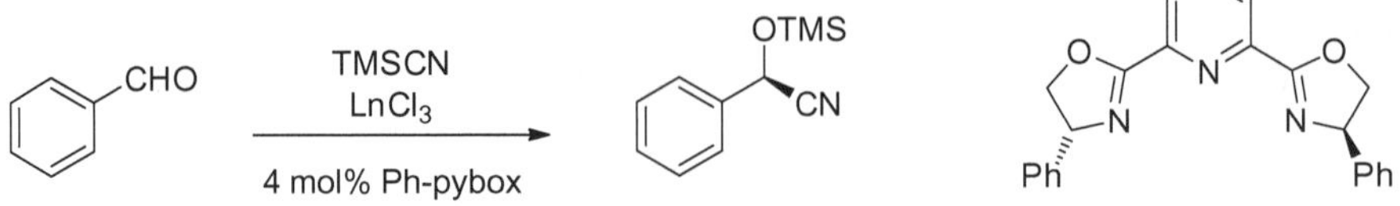

1 Ph-pybox

Ln:		
Lu	73% ee	(batch: 76% ee)
Yb	84% ee	(batch: 89% ee)
Ce	1% ee	(batch: 11% ee)

SCHEME 8.1. Asymmetric catalytic synthesis of cyanohydrin **1**.

was quite high, with little or no variation in selectivities observed when the catalyst loading was reduced from 10 to 2 mol%, potentially allowing for a more efficient process. Various attempts to improve the selectivities in the flow reactions were made. Tritylamine as an additive had a positive affect on the enantioselectivity of the reaction, affording the product with 66% and 81% enantiomeric excess (ee) in microreactor and batch processes, respectively, compared to 53% and 72% ee for reactions without any additive.

Moberg and co-workers demonstrated the possibilities of asymmetric silylcyanation in flow [8], initially using a literature protocol developed for Yb-catalyzed reactions [9]. While obtaining poorer selectivities and yields than in the batch process, the application of different metal catalysts (Lu-catalyzed reactions) resulted in both higher yields and selectivities in the corresponding flow processes for both the literature and Lu-catalyzed reactions.

Acylated cyanohydrins have been obtained from benzaldehyde in a microflow reaction with two different residence times, using various Lewis bases and a dimeric salen-Ti complex $(Salen\text{-}Ti)_2$ as chiral inductor. Enzymatic methods have been applied to determine the yield and purity of the enantiomeric acetylated cyanohydrins. A combination of enzymes was used for the analysis of reactants, target products, and side products: horse liver alcohol dehydrogenase (HLADH) for the reduction of the unreacted aldehyde, (*S*)-selective *Candida antarctica* lipase B for the selective hydrolysis of the (*S*)-ester, and nonselective pig liver esterase for the hydrolysis of the remaining ester (Scheme 8.2). The released acid, in turn, protonated the 4-nitrophenoxide to 4-nitrophenol, which was analyzed on a spectrophotometer. The hydrolyzed cyanohydrins remained in equilibrium with the corresponding aldehydes and therefore were reduced by nicotinamide adenine dinucleotide (NADH)/HLADH to corresponding alcohols. The excess NADH was analyzed to calculate the yield and purity. All methods of analysis were correlated and ee's of >80% were reported [10].

8.2.2. Enzyme-Catalyzed Cyanations

Similar to the synthesis of cyanohydrins described in Section 8.2.1, the enzyme-catalyzed asymmetric addition of hydrogen cyanide to aldehydes has been

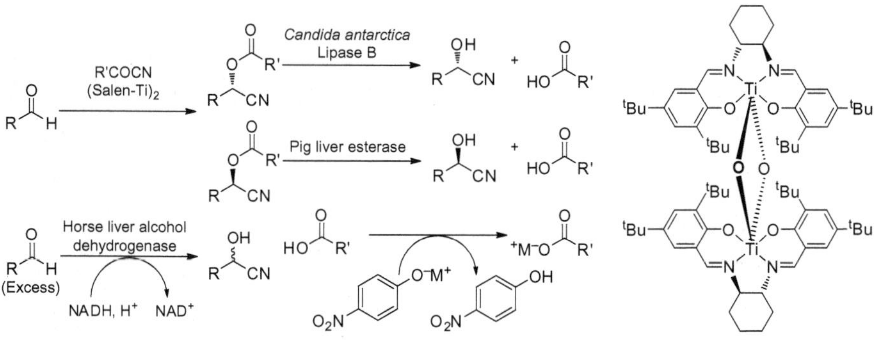

SCHEME 8.2. Asymmetric catalytic synthesis, resolution, and analysis of acetylated cyanohydrins.

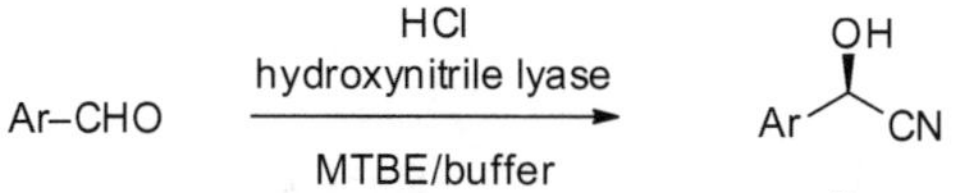

2

SCHEME 8.3. Enzyme-catalyzed synthesis of cyanohydrins **2**.

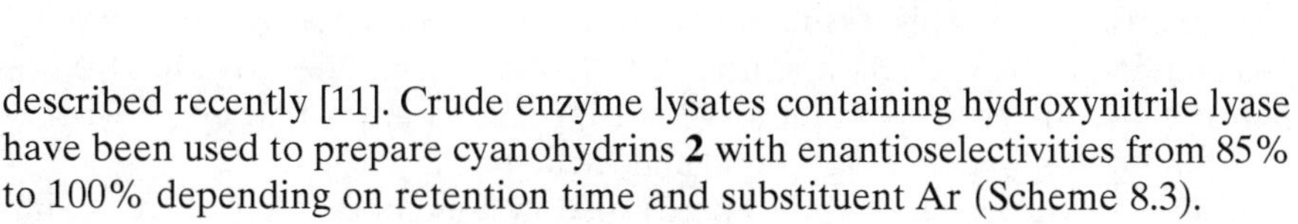

SCHEME 8.4. Catalytic stereoselective hydrogenation using rhodium catalysts.

described recently [11]. Crude enzyme lysates containing hydroxynitrile lyase have been used to prepare cyanohydrins **2** with enantioselectivities from 85% to 100% depending on retention time and substituent Ar (Scheme 8.3).

8.2.3. Asymmetric Catalytic Hydrogenation

High-throughput kinetic investigations of asymmetric hydrogenations using rhodium catalysts in microdevices have been investigated some time ago [12]. De Bellefon et al. have used a falling film microreactor in asymmetric hydrogenations (Scheme 8.4). Such falling film microreactors allow a continuous mixing of the aqueous and gaseous phase, which is passed through the reactor tubing in a well-defined segment. In this reaction, a residence time of 12 minutes and a reaction temperature of 70°C with no noticeable change to the

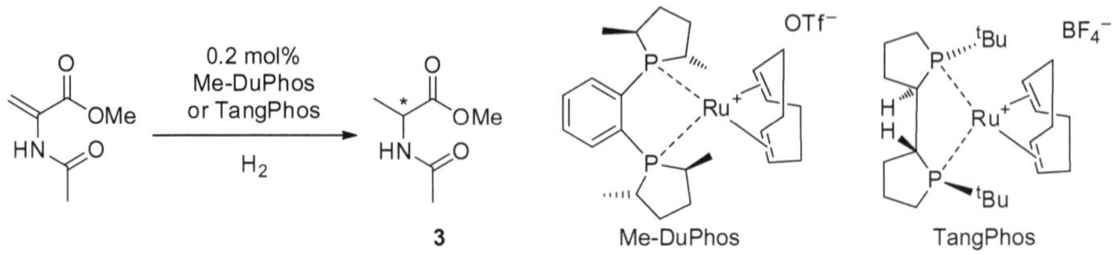

SCHEME 8.5. Hydrogenation in a falling film microreactor.

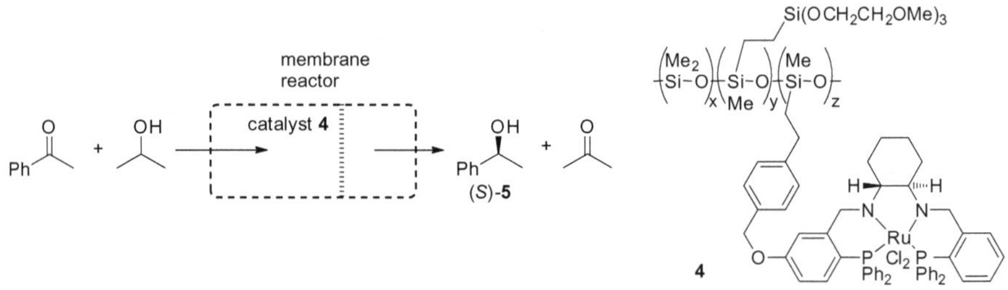

SCHEME 8.6. Transfer hydrogenation in a membrane reactor.

segment shape or composition give high yields (up to 94%) with ee's in line with those obtained from flask reactions (20% ee). Interestingly, the microreactor approach required a much lower catalyst loading than in other counterparts, needing only about 0.1 μg of Rh catalyst per run. In total, 17 different chiral phosphines were trialed, with the diphosphine ligand BDPPTS [(2*S*,4*S*)-2,4-bis(diphenylphosphino)pentane] proving the most effective.

This research was extended later in a study by Allwardt et al. using methyl-2-acetamido acrylate as substrate, 0.2 mol% of $Rh(cod)_2BF_4$ or $Rh(cod)_2SO_3CF_3$ as metal precursors, and a series of different chiral ligands in a falling film microreactor. This resulted in an array of observations. (*R*)-methyl 2-acetamidopropanoate **3** was obtained in 97.7% ee with the (*S,S*)-Me-DuPhos ligand while the corresponding (*S*)-**3** was produced in high ee by (*R,R*)-Me-DuPhos and (*S,S,R,R*)-TangPhos ligand (Scheme 8.5) [13].

Transfer hydrogenations were performed in a membrane reactor by using the soluble polysiloxane-based catalyst **4** (Scheme 8.6) [14]. A membrane reactor utilizes the size exclusion principle and is an alternative approach to phase transfer catalysis. The catalyst is homogeneous and therefore enables faster reaction kinetics, but is retained in the flow reactor by a membrane while the product can be removed easily. Compound (*S*)-**5** was obtained in 79% conversion with 91% ee. When compared with a similar enzymatic system, catalyst **4** was five times more productive and did not require an additional cofactor as a reducing agent; however, the enzyme could achieve an impressive 99% ee. Furthermore, careful exclusion of water and oxygen must be ensured, otherwise the catalyst **4** slowly decomposes.

8.2.4. Asymmetric Catalytic Aldol Reaction

Odedra and Seeberger successfully used 5-(pyrrolidin-2-yl)tetrazole **6** as an organocatalyst to synthesize a range of aldol products and it was claimed that both shorter reaction times and lower catalyst loadings can be achieved with this setup compared to the reaction conditions employed in traditional flask chemistry (Scheme 8.7) [15]. In addition to these advantages, similar or higher yields and better selectivities were also observed under microflow conditions. This point of view was recently challenged by a contribution from Blackmond and co-authors who provided a detailed thermodynamic and kinetic analysis of this reaction [16]. This led to their main conclusion that the profiles of a batch reaction and a reaction performed in a microreactor leading to product **7** are very similar with respect to yields and selectivities, with Ar = Ph compound **7** obtained in about 80% yield with 75% ee. The scope of this procedure was also widened to include some examples of the Mannich reaction, where high yields and selectivities were once again observed. A silicon-made microreactor consisting of a standard T-mixer with subsequent reaction channel has been used for these experiments.

8.2.5. Asymmetric Catalytic Epoxidations

A soluble polyethylene glycol poly-L-leucine catalyst has been used for the synthesis of chalcone epoxide **8** using hydrogen peroxide as oxidant (Scheme 8.8). Efficient premixing of the catalyst and hydrogen peroxide was necessary to obtain about 90% ee in the product **8** [17]. In this example, the catalyst has to be constantly provided together with the substrate for the flow reaction.

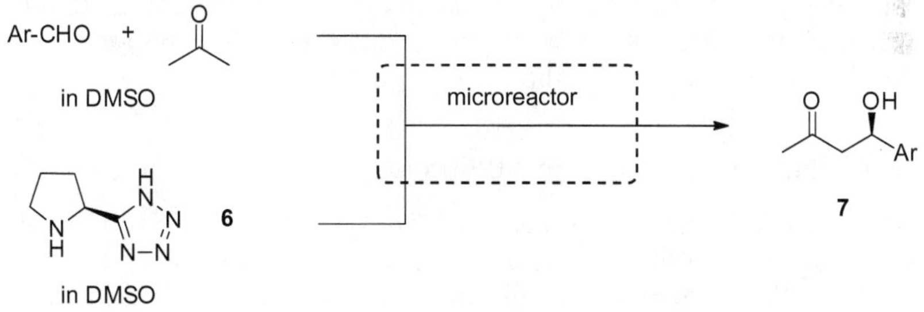

SCHEME 8.7. Asymmetric aldol reactions.

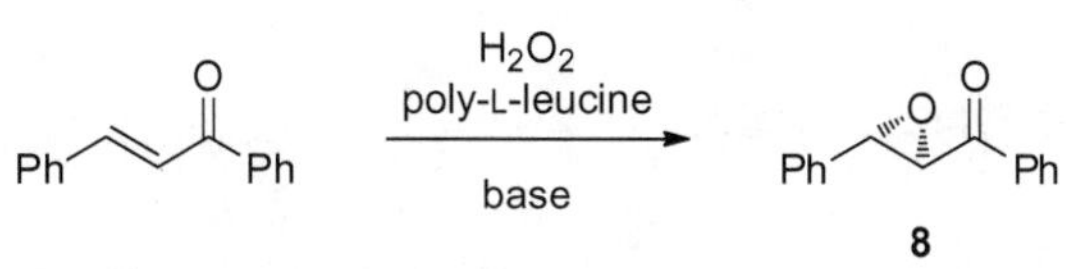

SCHEME 8.8. Stereoselective chalcone epoxidation.

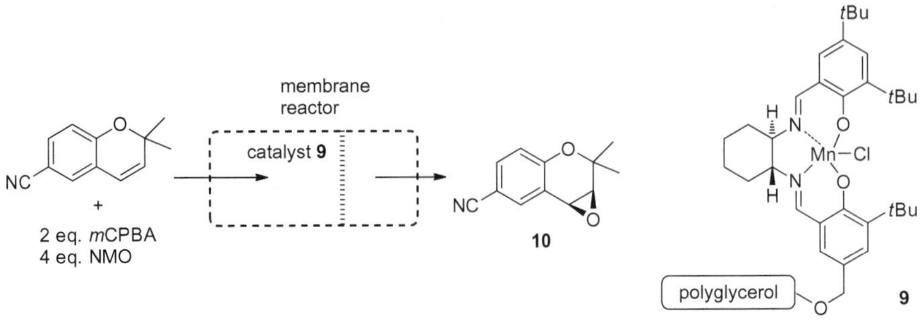

SCHEME 8.9. Asymmetric epoxidation in a membrane flow reactor.

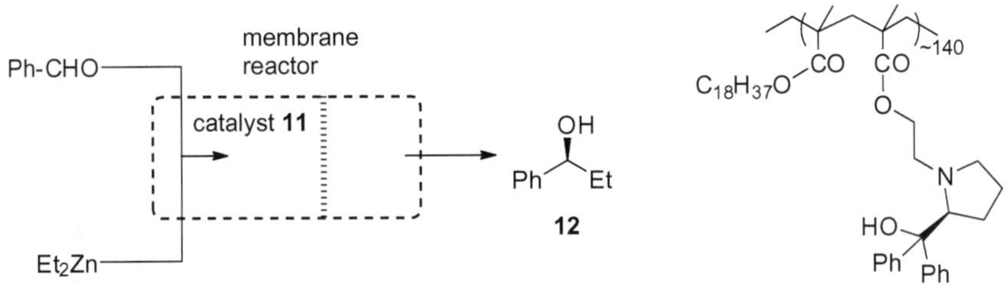

SCHEME 8.10. Diethylzinc addition to benzaldehyde in a membrane reactor.

A soluble polyglycerol-supported salen catalyst **9** was used in a continuous membrane flow reactor (Scheme 8.9) [18]. Continuous removal of the high molecular weight catalyst by a membrane allows an efficient flow epoxidation using *meta*-chloroperbenzoic acid (*m*CPBA) as the stoichiometric oxidant. Under these conditions, 70% conversion to the epoxide **10**, obtained in up to 92% ee, are achieved. The presence of 4 eq. *N*-methylmorpholine *N*-oxide (NMO) was found to increase the conversion by 10%.

8.2.6. Asymmetric Diethylzinc Additions

A soluble prolinol catalyst has been used for asymmetric diethylzinc additions to benzaldehydes [19]. The catalytically active prolinol moiety of the catalyst **11** is attached to a copolymer of different methacrylates (Scheme 8.10). The catalyst is retained in the membrane reactor by an ultrafiltration membrane because of its large molecular weight. The ee of alcohol **12** using catalyst **11** is 80%, lower than with the nonimmobilized ligand (97% ee) [20].

8.2.7. Selective Glycosylation

Glycosylations have been reported in microflow reactors, starting from a mixture of α/β-anomers of sialic acid derivatives **13** and **14**, resulting in an almost quantitative yield of exclusively α-sialylated product **15** (Scheme 8.11).

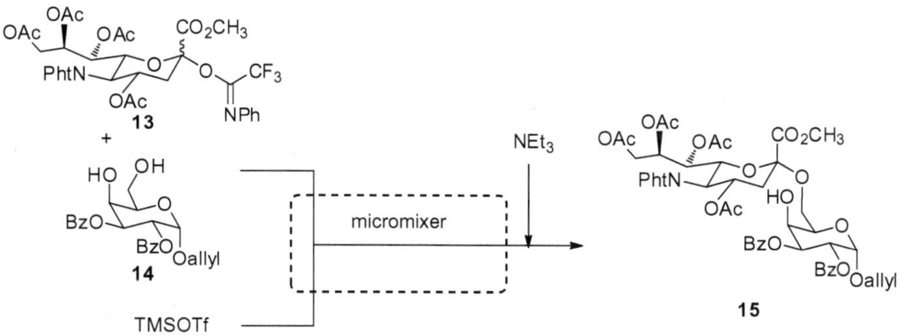

SCHEME 8.11. Highly selective α-(2,6)-sialylation.

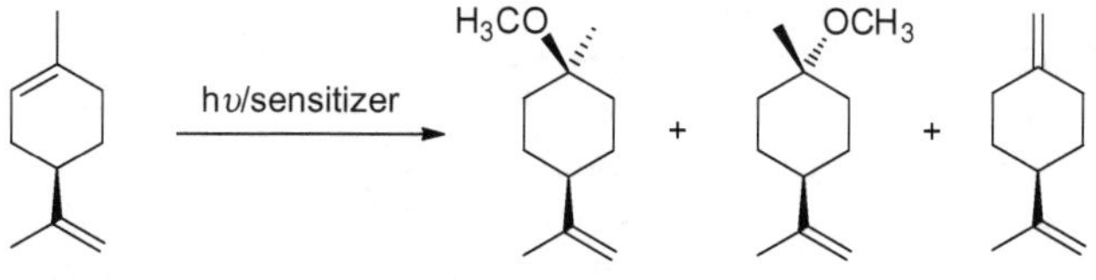

SCHEME 8.12. Photochemical stereoselective methanol addition.

The authors have used a micromixer for a fast and intense mixing of the donor and acceptor molecules with a Lewis acid [21]. A scale-up of the corresponding batch reaction was difficult, because a decomposition of the donor has been regarded as a severe problem. Efficient micromixing accelerated the reaction and the rapid heat transfer avoided the undesired hydrolysis and glycal formation. Similarly, selective β-mannosylated glucose has been obtained from a mixture of α/β-anomer of glucose [22]. α-Mannosylations have also been described [23]. Recently, microreactors have been used to establish kinetic parameters in the synthesis of unsymmetrical trehalose analogues [24].

8.2.8. Photochemical Asymmetric Reactions

Sakeda et al. have reported the addition of methanol to (*R*)-(+)-limonene **16** as shown in Scheme 8.12. In batch reactions, photochemical processes normally result in a larger number of uncontrolled side products as compared with the microflow reactions, where the formation of side products is suppressed. The formation of photoproducts increases with longer irradiation and the *trans*-product is formed in excess of the *cis*-product. A quartz microreactor with ultraviolet (UV) irradiation was used for this reaction [25].

[2+2] Photocycloaddition of cyclohexenone carboxylic acid (–)-8-(phenyl) menthyl ester **17** with cyclopentene **18** was performed in a microflow reactor and compared with a batch reaction (Scheme 8.13). The products formed in

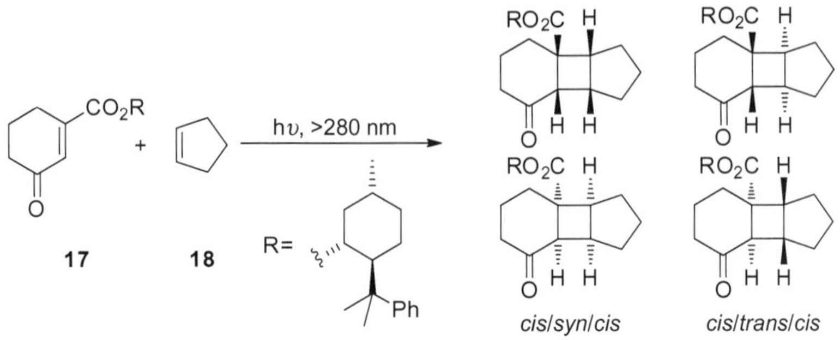

SCHEME 8.13. Asymmetric [2+2] photocycloaddition.

microflow had higher diastereoselectivities than the products from the batch reaction. In general the *cis–syn–cis* selectivity over *cis–trans–cis* decreases at higher temperature while the highest diastereoselectivities observed for the *cis–syn–cis* product was 82% in toluene at –40°C with a reaction time of 30 minutes. Under similar conditions the batch reaction needed an hour for completion, forming the *cis–syn–cis* product in 70% diastereomeric excess (de). The (–)-8-(phenyl)menthyl group acts as a chiral auxiliary in these reactions [26].

8.3. HETEROGENEOUS ASYMMETRIC CATALYTIC REACTIONS

Enantiomerically pure catalysts are often very valuable and many strategies have been developed to recover, recycle, or immobilize such compounds [27]. Complete separation from the reaction products can be challenging as contamination by side products or leached catalysts can be troublesome in large-scale syntheses. Polymer-supported catalysts are therefore of particular interest and several research groups have investigated their incorporation in flow chemistry. Such strategies allow an easy reuse and flow processes become more economically and environmentally friendly. However, the polymer support can have a strong influence on the catalytic activity. Access to the active site of a catalyst can be different, leading to microenvironments with catalytic behaviors different from homogeneous solution. Depending on the individual reaction, the polymer-supported catalyst can be floating in a solution and physically held in a defined compartment of the flow reactor. Other reactions use catalysts bound to monoliths or packed catalysts on polymers in a cartridge. In such cases, the stability of polymer-supported catalysts with sometimes higher pressures in flow chemistry can also effect their efficiency.

8.3.1. Kinetic Resolutions

A large range of immobilized catalysts has been employed for kinetic resolutions of various substrates using flow chemistry. Kirschning et al. have devel-

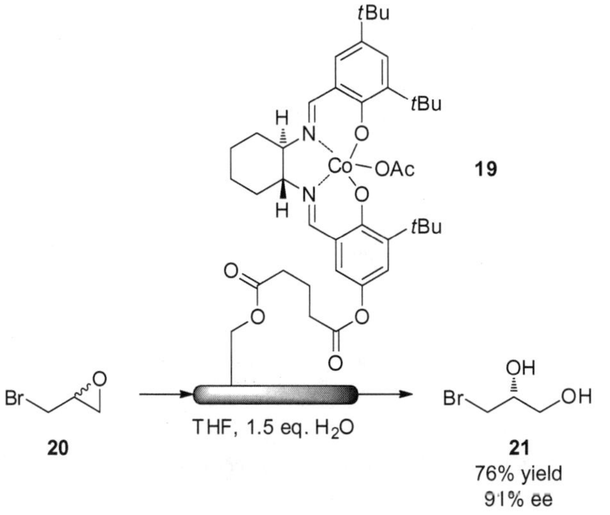

SCHEME 8.14. Enantioselective kinetic resolutions.

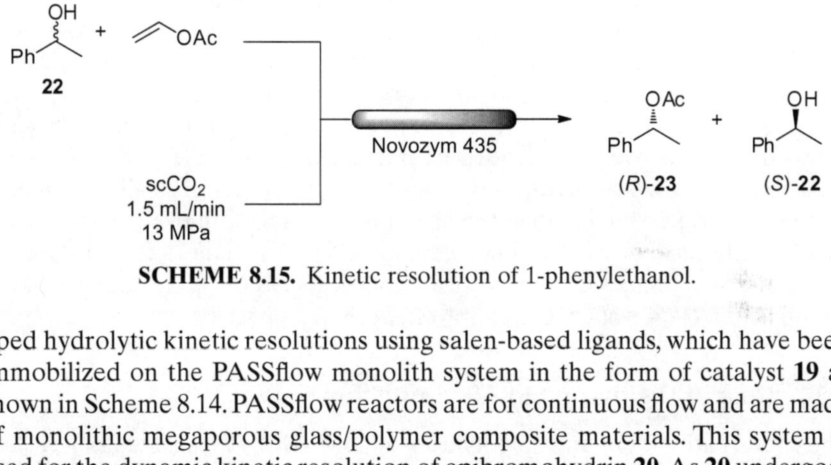

SCHEME 8.15. Kinetic resolution of 1-phenylethanol.

oped hydrolytic kinetic resolutions using salen-based ligands, which have been immobilized on the PASSflow monolith system in the form of catalyst **19** as shown in Scheme 8.14. PASSflow reactors are for continuous flow and are made of monolithic megaporous glass/polymer composite materials. This system is used for the dynamic kinetic resolution of epibromohydrin **20**. As **20** undergoes fast racemization under the reaction conditions, all starting material can be theoretically converted to the diol reaction product **21** [28]. Continuous large-scale runs have been performed without any loss of activity of the immobilized catalyst providing the diol **21** in similar yields and selectivities. Other terminal epoxides have been resolved using a salen catalyst bound to silica gel [29].

Different applications using immobilized enzymes as catalysts for kinetic resolutions using flow chemistry have been published recently. Some of these processes make beneficial use of supercritical solvents such as supercritical carbon dioxide ($scCO_2$). The kinetic resolution of 1-phenylethanol **22** using Novozym 435 as an immobilized lipase and vinyl acetate provided (*R*)-acetate **23** and the (*S*)-alcohol **22** in 47% yield each in almost enantiomerically pure forms (Scheme 8.15). Compared to a 10 mL batch reactor converting 0.83 mmol

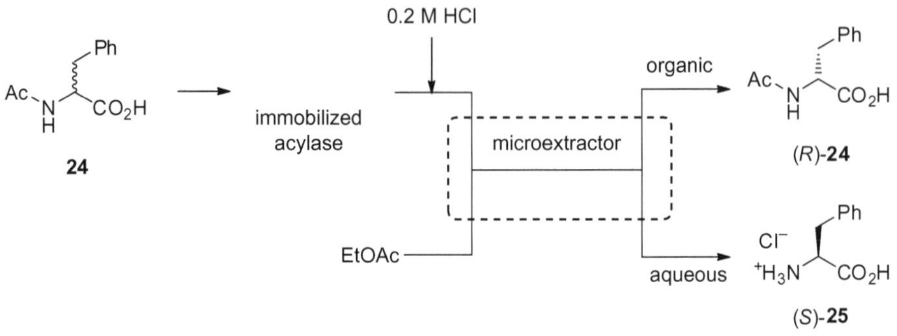

SCHEME 8.16. Kinetic resolution and separation of reaction products in continuous flow.

in 7 hours the flow process allowed a conversion rate of 25 mmol/h [30]. A similar investigation utilizes a Lipase immobilized on microporous silica was used without $scCO_2$ and in all cases the ee observed was >99% [31].

The products shown in the reaction in Scheme 8.15 are, however, still obtained as a (separable) mixture. A promising example of a kinetic resolution coupled with a microextractor for product separation was reported by Maeda and co-workers [32]. An immobilized acylase was used for the resolution of the racemic acetylated phenylalanine **24** dissolved in an aqueous buffer. The reaction mixture was acidified after the resolution and subsequently extracted with ethyl acetate using a microextractor as shown in Scheme 8.16. Chemically modified channels with hydrophobic and hydrophilic surfaces allowed a continuous separation of laminar biphasic flow. Under optimized reaction and flow conditions, the product phenylalanine (*S*)-**25** remained in the aqueous phase and was almost enantiomerically pure (99% ee), while the acetylated amino acid (*R*)-**24** was extracted into the organic phase.

8.3.2. Role of Diethylzinc in Alkyl Addition to Aldehydes

The coordination of metals to polymer-supported ligands is a general motif in heterogeneous catalysis, and many of such compounds have been employed in flow chemistry. The enantioselective addition of diethylzinc to benzaldehyde has been extensively investigated in flow chemistry [33–36]. Different catalysts, usually aminoalcohol derivatives, have been bound to polymer supports. The structures of the catalysts **27–31** are shown in Scheme 8.17. Mixtures of diethylzinc and aromatic aldehydes are used for the reactions, which are typically performed at temperatures below room temperature. Yields and selectivities of the different reactions are shown in Scheme 8.17. Catalyst **27** has also been used in an enantioselective reduction of valerophenone with borane in an early flow experiment [37]. For the first run with catalyst **29**, yields and selectivities were excellent, but subsequent runs using the same catalyst showed a drop in activity. This was related also to a breakdown of the polymeric mate-

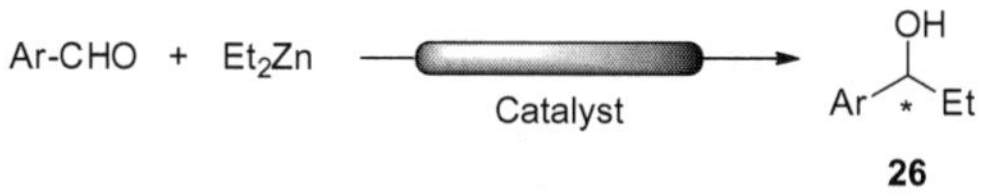

Catalyst	**26** (Ar=Ph)
27	99% *ee* (*S*)
28	81% yield, 81% ee (*R*)
29	94% yield, 97% ee (*R*)
30	85% yield, 99% ee (*R*)
31	99% yield, 93% ee (*S*)

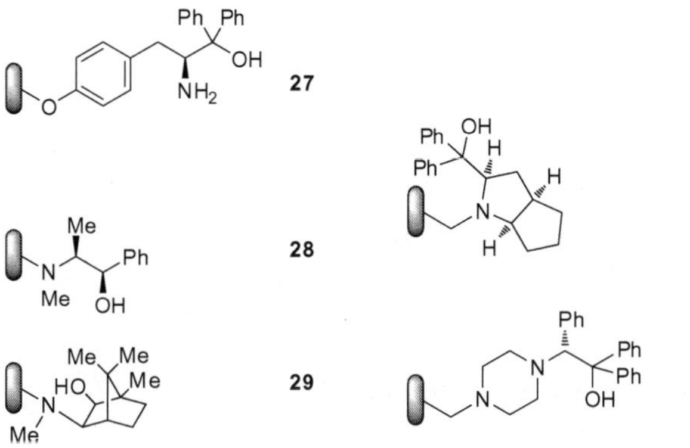

SCHEME 8.17. Diethylzinc addition to aromatic aldehydes.

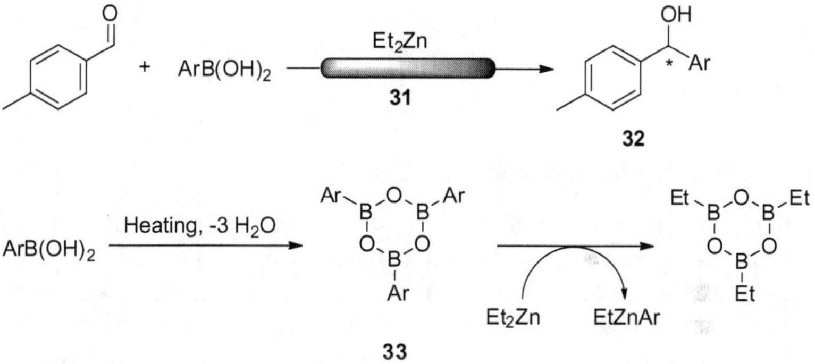

SCHEME 8.18. Role of diethylzinc in the addition to aromatic aldehydes to produce diarylmethanols.

rial, indicating that the choice of polymer support is important, which must be chemically resistant to the reaction conditions [34]. Catalyst **30** was attached to a monolith as supporting material using a chiral waste material from the synthesis of Ramipril [35]. This is especially interesting as the reaction with the homogeneous analog catalyst and also a heterogeneous analogue with the catalyst **31** bound to Merrifield resin provided the product in lower (87%–89% ee) selectivities. Several of these catalysts have also been used for the synthesis of gram amounts of the secondary alcohols **26**.

Catalyst **31**, immobilized on a polystyrene–divinylbenzene copolymer (PS-DVB) resin, has also been used in the stereoselective synthesis of diarylmethanols **32** from arylboronic acids and *p*-tolualdehyde (Scheme 8.18). Boroxin **33**

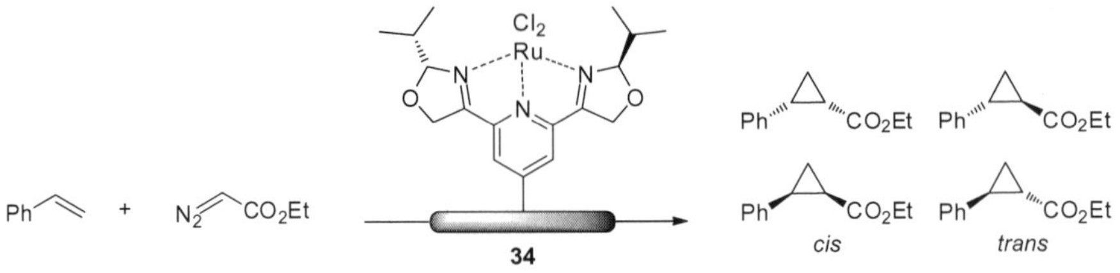

	yield	*cis*/*trans*	ee (*cis*)	ee (*trans*)
CH_2Cl_2 (rt, 35 minutes) or	53%	80:20	48%	79%
neat (rt, 35 minutes) or	72%	83:17	43%	79%
$scCO_2$ (8 MPa, 40°C, 1.3 minutes)	22%	85:15	56%	89%
homogeneous batch reaction	44%	87:13	18%	77%

SCHEME 8.19. Cyclopropanation in flow.

formed by the dehydration of arylboronic acids act as an arylating agent and ethylaryl zinc formed in flow via transmetallation by diethylzinc are the actual reactants. Adequate combination of arylating boroxin and aromatic aldehyde owing to their electron donating or withdrawing effect can produce the required diarylmethanols in almost quantitative yields and in enantioselectivities up to 83% [38]

8.3.3. Asymmetric Cyclopropanations

Asymmetric cyclopropanations have been performed in continuous flow, using ruthenium-based compounds bound to monoliths as potentially long-term stable catalysts [39]. The polymer-bound ruthenium complex **34** was initially generated by flowing a solution of dichlororuthenium(II) (*p*-cymene) through the monolith, followed by washing to remove uncomplexed ruthenium (Scheme 8.19). The cyclopropanation of styrene with ethyldiazoacetate has been subsequently catalyzed in flow using different solvents. Yields as well as *cis*/*trans* selectivities were found to be comparable to the reaction carried out under homogeneous reaction conditions. Under neat conditions, or when using $scCO_2$ as solvent, the productivity of the catalyst was superior to the use of dichloromethane. All reactions, however, produced a significant amount of ethyl fumarate and ethyl maleate as dimerized by-products.

8.3.4. Asymmetric Hydrogenations

The use of supercritical fluids such as $scCO_2$ is attractive for environmental reasons, but also because of the ease of product isolation by simple depressurization. Their miscibility with gases was used in an early example of continuous asymmetric hydrogenation using a fixed-bed reactor containing a platinum–aluminum oxide catalyst modified with cinchonidine alkaloids [40].

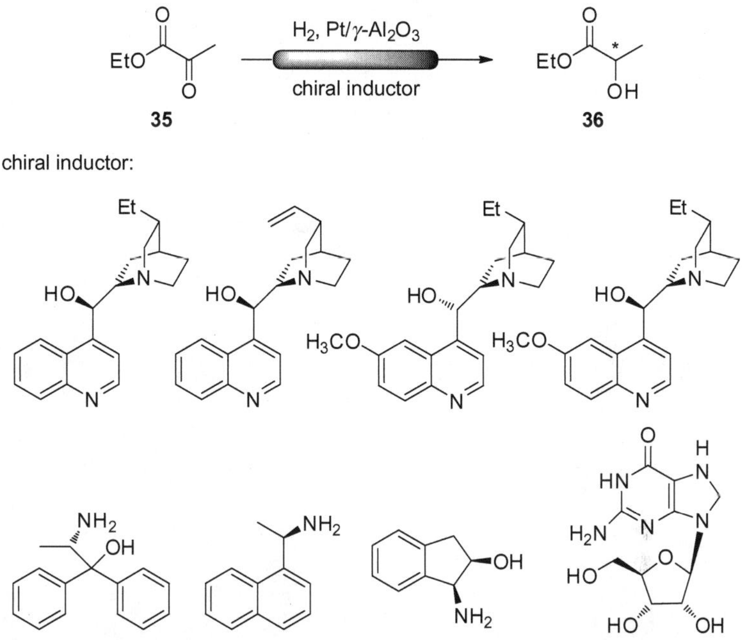

SCHEME 8.20. Asymmetric hydrogenation of ethyl pyruvate.

Improvement of this asymmetric approach has resulted in the development of a gas–liquid–solid microstructured hydrogenation reactor, which contain a nickel mesh contactor to separate gas–liquid phases. The solid phase catalyst, Pt/γ-Al_2O_3, was immobilized by depositing on a glass insert, followed by impregnation, calcination, and hydrogenation. The enantioselective reduction of ethyl pyruvate **35** with cinchonidine as the chiral inductor was best among the eight chiral inductors used in the study and led to product **36** in up to 63% ee; the conversion was just 21% (Scheme 8.20). The glass insert can be regenerated and adsorption and desorption of chiral inductors can be done on the catalytic surface without dismantling the reactor. The adsorption–desorption profile of chiral inductors was also discussed [41]. A similar catalyst modified with *O*-methyl-cinchonidine was used for the asymmetric hydrogenation of isopropyl-4,4,4-trifluoroacetoacetate [42].

Poliakoff and co-workers have used $scCO_2$ for the asymmetric hydrogenation of dimethyl itaconate using an immobilized rhodium-based catalyst, attached to alumina via a phosphotungstic linker, together with chiral bisphosphine ligands [43]. The ferrocenyl ligand shown in the complex **37** gave highest selectivities in the product **38** (up to 83% ee) in the asymmetric hydrogenation in flow (Scheme 8.21). Also α,β-unsaturated carboxylic acids have been used as substrates for asymmetric hydrogenations in flow [44].

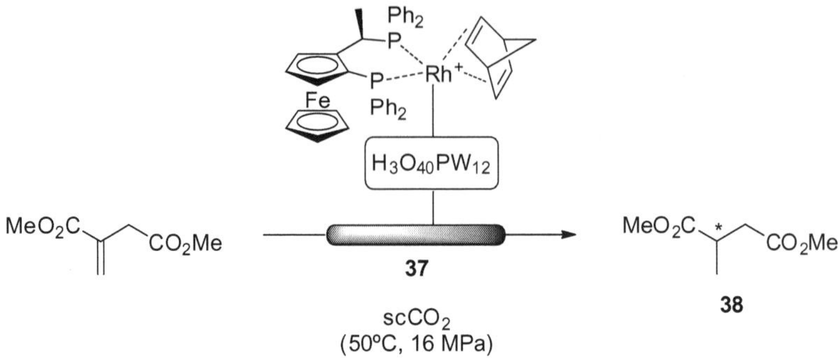

SCHEME 8.21. Continuous asymmetric hydrogenation in $scCO_2$.

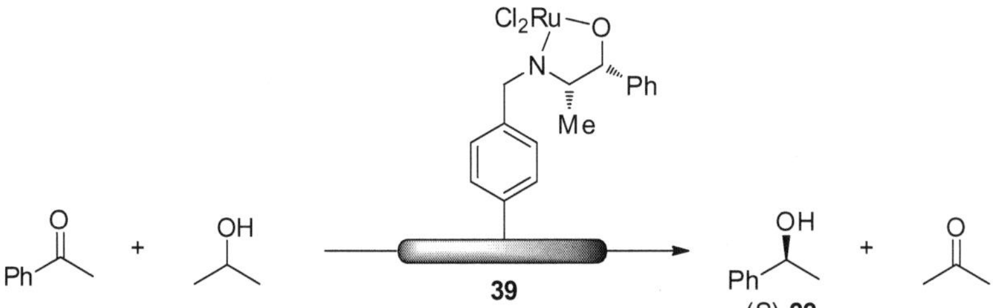

SCHEME 8.22. Transfer hydrogenation in flow using a ruthenium-based immobilized catalyst.

Transfer hydrogenations have also been adopted to flow chemistry. A ruthenium catalyst was complexed to a norephedrine ligand attached to modified silica **39** as shown in Scheme 8.22 [45]. Under optimized conditions, the transfer hydrogenation of acetophenone using 2-propanol provided a steady-state conversion of 95% and the alcohol (*S*)-**22** in about 90% ee. It was found that the supported catalyst is stable and active for at least 3 weeks, whereas the analogous homogeneous catalyst was active for only up to 20 hours. The extended stability could be a consequence of the isolation of the active sites in the immobilized catalyst.

8.3.5. Asymmetric Michael Additions

The Michael addition of methyl vinyl ketone and indanone derivative **40** was investigated in a flow reactor packed with polymer beads loaded with the catalyst **41** (Scheme 8.23) [46]. The use of a fluid bed is attractive as it allows the use of polymer-supported catalysts in a flow reactor. The use of the polymer-supported cinchonidine **41** led to the formation of the Michael product **42** in

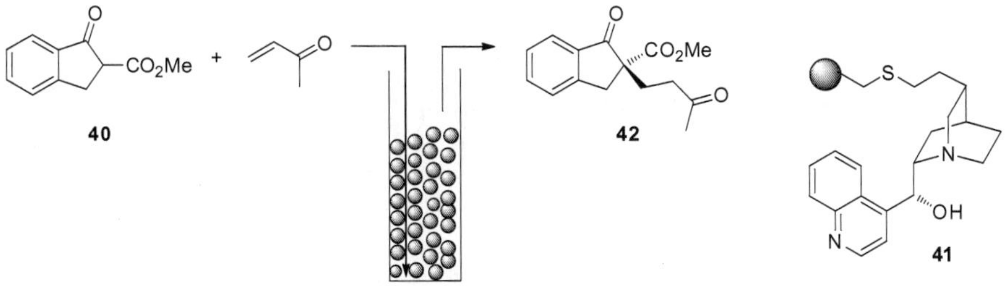

SCHEME 8.23. Asymmetric Michael reaction.

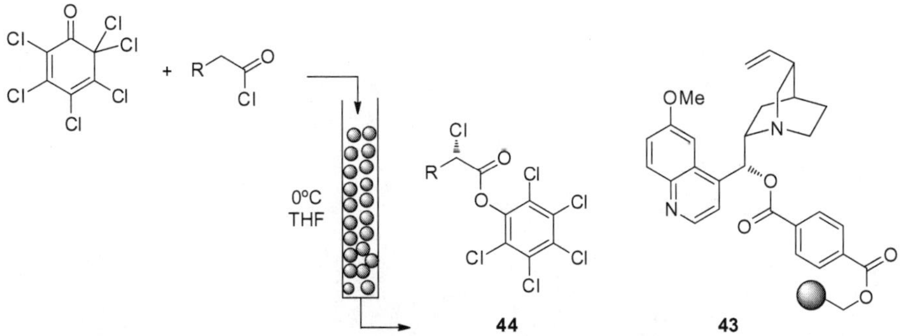

SCHEME 8.24. Asymmetric α-chlorination and active ester formation in flow.

high yields with selectivities (up to 51% ee) comparable to the homogeneously catalyzed reaction.

8.3.6. Asymmetric Chlorination and β-Lactam Synthesis From Acid Chlorides

Another natural product, a quinine derivative, was used for an asymmetric α-chlorination of acid chlorides in flow as shown in Scheme 8.24 [47]. The immobilized cinchona alkaloid **43** was used for the formation of an active ester and also served as a chiral inducer for the subsequent chlorination, which the authors proposed to proceed via the corresponding ammonium enolate species. The formation of a ketene intermediate and a subsequent pericyclic reaction with chlorine transfer as a mechanistic pathway cannot be excluded. The α-chlorinated esters **44** are formed in moderate yields (40%–60%), but with good enantioselectivities (up to 94% ee). One of these reaction products has been used as a key intermediate for the stereoselective synthesis of the peptidic metalloprotease inhibitor BMS-275291 as shown in Section 8.3.10 of this chapter [47b].

The same quinine-based polymer-supported catalyst **43** was used for the flow synthesis of β-lactams [48]. Ketenes were generated from acid chlorides

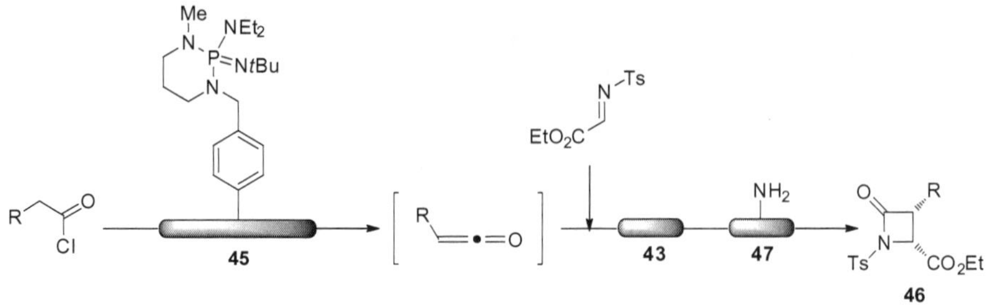

SCHEME 8.25. Three step β-lactam synthesis in flow.

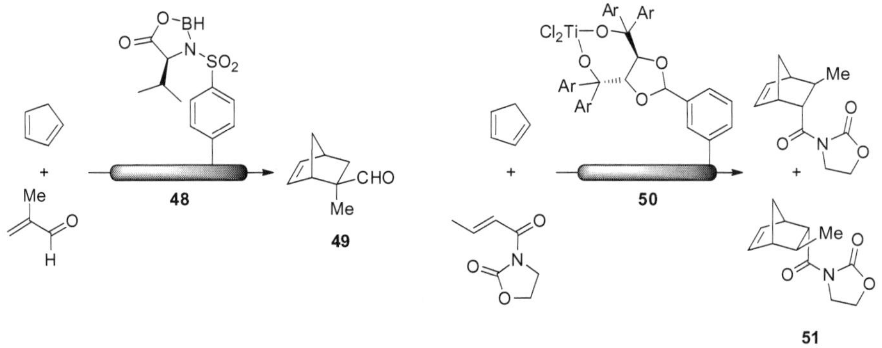

SCHEME 8.26. Enantioselective Diels–Alder reactions.

using polymer-supported BEMP (2-tert.-butylimino-2-diethylamino-1,3-dimethylperhydro-1,3,2-diazaphosphorine) **45**; the subsequent reaction with an imino ester was catalyzed by **43**, providing β-lactams **46** (Scheme 8.25). A subsequent nucleophilic scavenger resin with free amino groups was used in line with the polymer-supported reagent **47** to remove excess reagents and by-products. The three-step sequence was carried out in packed glass columns and proved to be robust enough to prepare gram quantities of the β-lactams **46**.

8.3.7. Stereoselective Diels–Alder and Ene Reactions

Chiral oxazaboroline compounds are well established for catalyzing enantioselective Diels–Alder reactions [49]. Copolymerization of a modified styrene monomer led to the immobilized PS-based reagent **48** (Scheme 8.26). The Diels–Alder reaction of cyclopentadiene and methacroleine was performed in a cooled (–30°C) flow reactor leading to the cyclization product **49** in 95% yield and 71% ee [50]. This result is comparable with the batch reaction. A different catalyst for Diels–Alder reactions has been prepared by attachment of a TADDOL (α,α,α,α-tetraaryl-1,3-dioxolane-4,5-dimethanol)

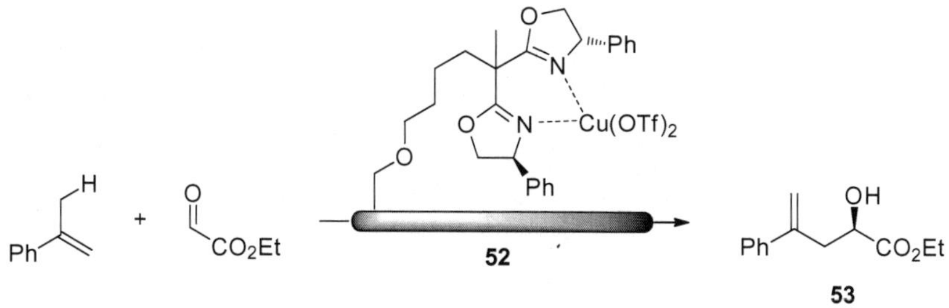

SCHEME 8.27. Glyoxylate-ene reaction in flow.

ligand to a monolith (**50**) [51]. The results obtained are different from a similar catalyst prepared by copolymerization as the ratio of the two isomers **51** changes. Enantioselectivities of up to 61% have been obtained. The monolith-based catalysts **50** need to be regenerated after use and their activity and stability is much lower. Titanium complexes no longer bound to the ligand are suspected to cause a background reaction to the racemate.

A polymer-bound bis(oxazoline) ligand **52** has been developed for the copper-catalyzed glyoxylate-ene reaction (Scheme 8.27) [52]. The copolymerized catalyst was filled into a high performance liquid chromatography (HPLC) column and a solution of the reagents passed through at 0°C. Although chromatographic purification of the product **53** was necessary, it was obtained in 78% yield and in 88% ee. Similar results have been obtained in the corresponding batch reactions.

8.3.8. Synthesis of Chiral Building Blocks

Flow chemistry has also been exploited in the synthesis of chiral building blocks. Using a protection–dehydrogenation–hydrogenation approach, chiral building blocks were produced. Solid supported reagents and scavengers were used and tartrate butane-2,3-diacetal **54** was obtained in shorter time (Scheme 8.28). Subsequent oxidation to **55** and reduction using flow chemistry generated the diastereomer **56**. In a similar fashion, butane-2,3-diacetal (BDA) of mannitol was produced and the C3–C4 bond of the protected mannitol was cleaved oxidatively using different sets of solid supported reagents and scavengers in flow. Depending on the oxidants, either BDA-protected methyl ether of glyceraldehyde or the corresponding acid, 2,3-dihydroxy propanoic acid, were obtained. Extending the protocol, the authors have also reported the production of an enantioenriched diketo derivative **58** using a chiral diol as starting material via *exo*-alkene **57** in flow. The initial oxidative cleavage of the *exo*-alkene **57** was followed by in situ infrared (IR) and the side product **58** was observed. It was then found that the formation of **58** is due to the reaction with sodium periodate, leading to enantioenriched **58** in good yields [53].

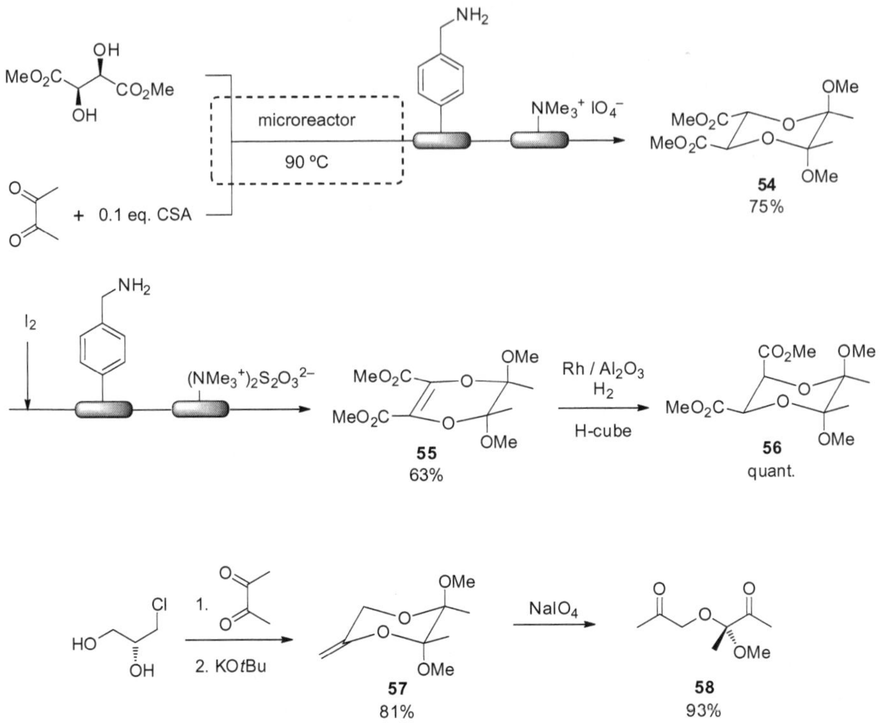

SCHEME 8.28. Synthesis of diacetals in flow.

8.3.9. Enantioselective Addition of Trimethylsilyl Cyanide (TMSCN) and Acetyl Cyanide to Benzaldehyde

The enantioselective synthesis of trimethylsilyl-protected (*S*)-mandelonitrile from benzaldehyde, as discussed in Section 8.2.1, has also been performed via heterogeneous asymmetric catalysis (Scheme 8.29). The hydrophobic surface of DVB beads were modified to hydrophilic surface by allowing the free vinyl groups to react with 4-(hydroxybutyl)vinyl ether in a free radical-mediated reaction. Pentandioic acid is then used as a linker to bridge the free hydroxy moieties and the phenolic residues, introduced to the salen ligand specifically for this purpose. The polymeric Ti(IV)-salen complex was formed by reaction with titanium tetrachloride. Acetylcyanide was also used, but the yield of acetylated (*S*)-mandelonitrile was lower (48% yield, 70% ee) as compared to the yields using TMSCN (92% yield, 72% ee) [54].

8.3.10. Applications in Multistep Synthesis

The drug discovery from nature most often results in molecules having a complex arrangement of a variety of functional groups. These natural products

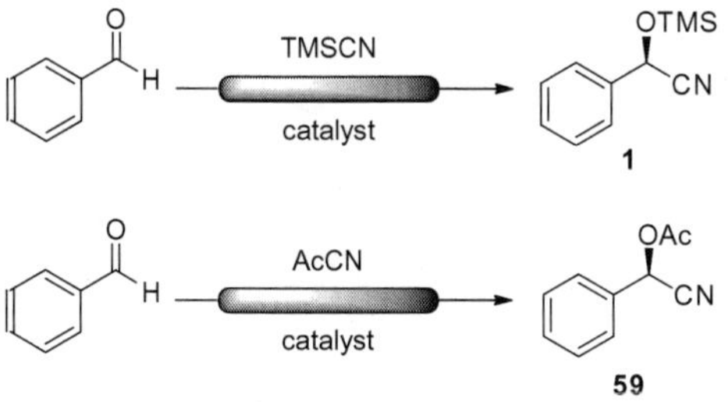

Flow (μL/min)	Product **1**
2.0	32% yield, 70% ee (*S*)
1.0	94% yield, 72% ee (*S*)
0.8	92% yield, 72% ee (*S*)

Flow (μL/min)	Product **59**
1.0[a]	35% yield, 32% ee (*S*)
0.6[a]	25% yield, 69% ee (*S*)
1.0[b]	66% yield, 31% ee (*S*)
0.6[b]	48% yield, 70% ee (*S*)

[a]20 and [b]40 mg of $(Salen\text{-}Ti)_2$ complex used

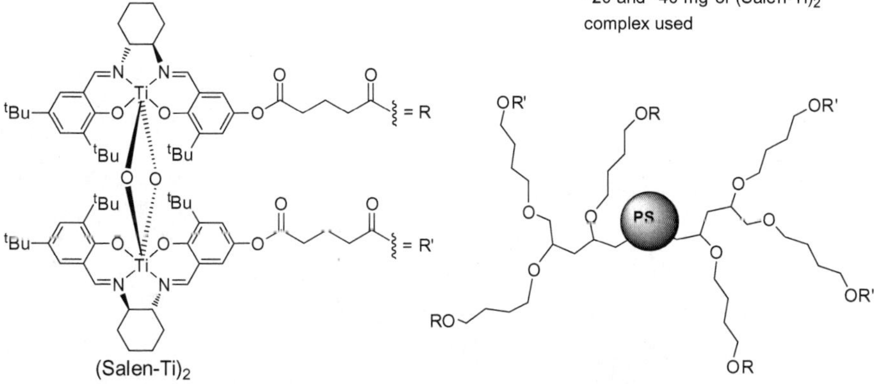

SCHEME 8.29. Heterogeneously catalyzed asymmetric synthesis of protected cyanohydrins.

with the potential as a future drug are present in minute quantity in nature, and often the ultimate solution is total synthesis.

Flow chemistry has been used in a simple way to synthesize oligomers with the required number of monomers using repetitive cycles. One example has appeared describing the synthesis of β-(1 $\rightarrow$ 6)-linked D-glucopyranoside homotetramers. This communication has also reported a perfluorinated linker at one of the anomeric positions to utilize the fluorous solid-phase extraction for automated purification and discussed chances for further functionalization [55]. On the other hand, microreactors have been used in the mulitstep synthesis of complex molecules. A potential anticancer metalloproteinase inhibitor BMS-275291 **61** (Scheme 8.30) has been synthesized using flow chemistry. The methodology included six different solid-supported packed columns including catalytic, reagent, diluent, and scavenger columns, arranged such that the target molecule is obtained at the final outlet. The report describes an α-clorination in 88% ee to obtain **60** as a key step in the synthetic sequence, detailed in Section 8.3.6, using a solid-supported quinine alkaloid **43** in the total synthesis of BMS-275291 **61** with 34% overall yield in 83% de [47b]. Furthermore, oxomaritidine **62** and grossamide **63** were some of the larger natural products synthesized exploiting microreactor chemistry in 2006 [56].

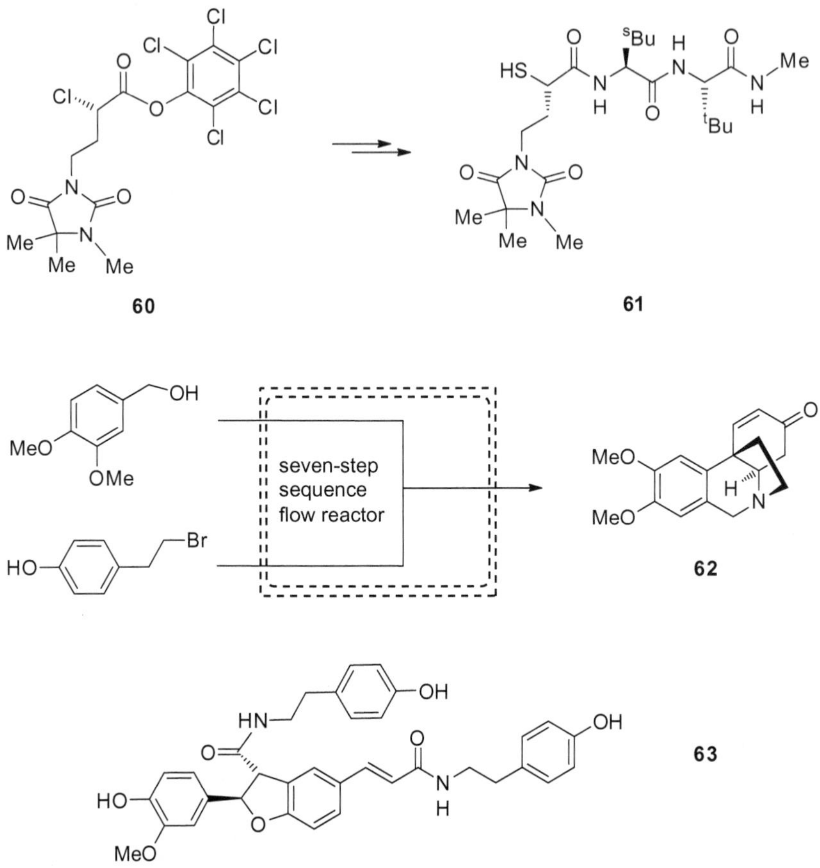

SCHEME 8.30. Multistep total syntheses.

A number of patents have also been published on asymmetric synthesis in microreactors. These include systems in which a reactant is brought in contact with a tethered chiral auxiliary [57] and the stereoselective production of (2*E*)-1-chloro-6,6-dimethyl-2-hepten-4-yne, an intermediate in the synthesis of terbinafine [58].

Stereoselective reactions have been performed under homogeneous as well as under heterogeneous reaction conditions in flow. A strict control of reaction parameters can be easily achieved. This is often crucial for asymmetric transformations where small changes of reaction conditions can have a large effect on the stereochemical outcome. Several promising examples of asymmetric synthesis in flow chemistry have already been reported and it is expected that many more procedures will be developed in the future.

REFERENCES

[1] (a) Zhang, S., Chen, J., Lykakys, I. N., Perchyonok, V. T. (2010). Streamlining organic free radical synthesis through modern molecular technology: from polymer supported synthesis to microreactors and beyond. *Curr. Org. Synth.*, *7*, 177–188; (b) Illg, T., Löb, P., Hessel, V. (2010). Flow chemistry using milli- and microstructured reactors. From conventional to novel process windows. *Bioorg. Med. Chem.*, *18*, 3707–3719; (c) Razzaq, T., Kappe, C. O. (2010). Continuous flow organic synthesis under high-temperature/pressure conditions. *Chem. Asian J.*, *5*, 1274–1289; (d) McMullen, P. J., Jensen, K. F. (2010). Integrated microreactors for reaction automation: new approaches to reaction development. *Annu. Rev. Anal. Chem.*, *3*, 19–42; (e) Moberg, C., Wingstrand, E. (2010). Enantioenriched acylated cyanohydrins: synthesis and analysis. *Synlett*, 355–367; (f) Verboom, W. (2009). Selected examples of high-pressure reactions in glass microreactors. *Chem. Eng. Technol.*, *32*, 1695–1701; (g) Hessel, V. (2009). Novel process windows—gate to maximizing process intensification via flow chemistry. *Chem. Eng. Technol.*, *32*, 1655–1681; (h) Hartman, R. L., Jensen, K. F. (2009). Microchemical systems for continuous-flow synthesis. *Lab Chip*, *9*, 2495–2507; (i) Geyer, K., Gustafsson, T., Seeberger, P. H. (2009). Developing continuous-flow microreactors as tools for synthetic chemists. *Synlett*, 2382–2391; (j) Weiler, A., Junkers, M. (2009). Using microreactors in chemical synthesis. Batch process versus continuous flow. *Pharm. Tech.*, *S6*, S10–S11; (k) Miyazaki, M., Maeda, H. (2009). Organic synthetic reaction using microspace. *Kagaku Kyoiku*, *57*, 364–367; (l) Liu, Q., Liu, Z. (2009). Application of photo-induced electron transfer reactions to organic synthesis. *Youji Huaxue*, *29*, 380–391; (m) Mak, X. Y., Laurino, P., Seeberger, P. H. (2009). Asymmetric reactions in continuous flow. *Beilstein J. Org. Chem.*, *5*, No. 19; (n) López, F. B., Egberink, R. J. M., Reinhoudt, D. N., Verboom, W. (2008). High pressure in organic chemistry on the way to miniaturization. *Tetrahedron*, *64*, 10023–10040.

[2] Wirth, T., Ed. (2008). *Microreactors in Organic Synthesis*. Wiley-VCH, Weinheim, Germany.

[3] Müller, C., Nijkamp, M. G., Vogt, D. (2005). Continuous homogeneous catalysis. *Eur. J. Inorg. Chem.*, 4011–4021.

[4] Cole-Hamilton, D. J. (2003). Homogeneous catalysis—new approaches to catalyst separation, recovery, and recycling. *Science*, *299*, 1702–1706.

[5] Manassen, J. (1971). Homogeneous catalysis with macromolecular ligands. *Platinum Met. Rev.*, *15*, 142–143.

[6] (a) Yao, Q. (2000). A soluble polymer-bound ruthenium carbene complex: a robust and reusable catalyst for ring-closing olefin metathesis. *Angew. Chem., Int. Ed.*, *39*, 3896–3898; (b) Bergbreiter, D. E. (2002). Using soluble polymers to recover catalysts and ligands. *Chem. Rev.*, *102*, 3345–3384; (c) Bolm, C., Derrien, N., Seger, A. (1996). Hyperbranched macromolecules in asymmetric catalysis. *Synlett*, 387–388; (d) Schwab, E., Mecking, S. (2001). Immobilization of a catalytically active rhodium complex by electrostatici of multiply charged phosphine ligands with a soluble polyelectrolyte and recovery by ultrafiltration. *Organometallics*, *20*, 5504–5506.

[7] Nishiyama, H. (1997). Chiral bis(oxazolinyl)pyridine (Pybox) ligands for asymmetric transition metal catalysis. *Adv. Catal. Proc.*, *2*, 153–188.

[8] Jönsson, C., Lundgren, S., Haswell, S. J., Moberg, C. (2004). Asymmetric catalysis in a micro reactor-Ce, Yb and Lu catalysed enantioselective addition of trimethylsilyl cyanide to benzaldehyde. *Tetrahedron*, *60*, 10515–10520.

[9] Gregory, R. (1999). Cyanohydrins in nature and the laboratory: biology, preparations, and synthetic applications. *Chem. Rev.*, *99*, 3649–3682.

[10] Hamberg, A., Lundgren, S., Wingstrand, E., Moberg, C., Hult, K. (2007). High-throughput synthesis and analysis of acylated cyanohydrins. *Chem. Eur. J.*, *13*, 4334–4341.

[11] (a) Koch, K., van den Berg, R. J. F., Nieuwland, P. J., Wijtmans, R., Schoemaker, H. E., van Hest, J. C. M., Rutjes, F. P. J. T. (2008). Enzymatic enantioselective C-C bond formation in microreactors. *Biotechnol. Bioeng.*, *99*, 1028–1033; (b) Koch, K., van den Berg, R. J. F., Nieuwland, P. J., Wijtmans, R., Wubbolts, M. G., Schoemaker, H. E., Rutjes, F. P. J. T., van Hest, J. C. M. (2008). Enzymatic synthesis of optically pure cyanohydrins in microchannels using a crude cell lysate. *Chem. Eng. J.*, *135S*, S89–S92.

[12] de Bellefon, C., Pestre, N., Lamouille, T., Grenouillet, P., Hessel, V. (2003). High throughput kinetic investigations of asymmetric hydrogenations with microdevices. *Adv. Synth. Catal.*, *345*, 190–193.

[13] Allwardt, A., Laue, S. H., Wendler, C., Stoll, N. (2008). A high parallel reaction system for efficient catalyst research. *Catal. Today*, *137*, 11–16.

[14] Laue, S., Greiner, L., Wöltinger, J., Liese, A. (2001). Continuous application of chemzymes in a membrane reactor: asymmetric transfer hydrogenation of acetophenone. *Adv. Synth. Catal.*, *343*, 711–720.

[15] Odedra, A., Seeberger, P. H. (2009). 5-(Pyrrolidin-2-yl)tetrazole-catalyzed aldol and Mannich reactions: acceleration and lower catalyst loading in a continuous-flow reactor. *Angew. Chem., Int. Ed.*, *48*, 2699–2702.

[16] Valera, F. E., Quaranta, M., Moran, A., Blacker, J., Armstrong, A., Cabral, J. T., Blackmond, D. G. (2010). The flow's the thing . . . or is it? Assessing the merits of homogeneous reactions in flask and flow. *Angew. Chem., Int. Ed.*, *49*, 2478–2485.

[17] Kee, S.-P., Gavriiliidis, A. (2007). Batch versus continuous mg-scale synthesis of chalcone epoxide with soluble polyethylene glycol poly-l-leucine catalyst. *J. Mol. Catal. A Chem.*, *263*, 156–162.

[18] Beigi, M., Haag, R., Liese, A. (2008). Continuous application of polyglycerol-supported Salen in a membrane reactor: asymmetric epoxidation of 6-cyano-2,2-dimethylchromene. *Adv. Synth. Catal.*, *350*, 919–925.

[19] Kragl, U., Dreisbach, C. (1996). Continuous asymmetric synthesis in a membrane reactor. *Angew. Chem., Int. Ed. Engl.*, *35*, 642–644.

[20] Soai, K., Niwa, S. (1992). Enantioselective addition of organozinc reagents to aldehydes. *Chem. Rev.*, *92*, 833–856.

[21] (a) Tanaka, K., Goi, T., Fukase, K. (2005). Highly efficient sialylation towards α(2-3)- and α(2-6)-Neu5Ac-Gal synthesis: significant "fixed dipole effect" of N-phthalyl group on α-selectivity. *Synlett*, 2958–2962; (b) Tanaka, S.-I., Goi, T., Tanaka, K., Fukase, K. (2007). Highly efficient α-sialylation by virtue of fixed dipole effects of N-phthalyl group: application to continuous flow synthesis of α(2-3)-and α(2-6)-Neu5Ac-Gal motifs by microreactor. *J. Carbohydr. Chem.*, *26*, 369–394.

[22] Tanaka, K., Fukase, K. (2009). Acid- mediated reactions under microfluidic conditions: a new strategy for practical synthesis of biofunctional natural products. *Beilstein J. Org. Chem.*, *5*, No. 40.

[23] Rathner, D. M., Murphy, E. R., Jhunjhunwala, M., Snyder, D. A., Jensen, K. F., Seeberger, P. H. (2005). Microreactor-based reaction optimization in organic chemistry-glycosylation as a challenge. *Chem. Commun.*, 578–580.

[24] Patel, M. K., Davis, B. G. (2010). Flow chemistry kinetic studies reveal reaction conditions for ready access to unsymmetrical trehalose analogues. *Org. Biomol. Chem.*, *8*, 4232–4235.

[25] Sakeda, K., Wakabayashi, K., Matsushita, Y., Ichimura, T., Suzuki, T., Wada, T., Inoue, Y. (2007). Asymmetric photosensitized addition of methanol to (R)-(+)-(Z)-limonene in a microreactor. *J. Photochem. Photobiol. A*, *192*, 166–171.

[26] Tsutsumi, K., Terao, K., Yamaguchi, H., Yoshimura, S., Morimoto, T., Kakiuchi, K., Fukuyama, T., Ryu, I. (2010). Diastereoselective [2+2] photocycloaddition of chiral cyclic enone and cyclopentene using a microflow reactor system. *Chem. Lett.*, *39*, 828–829.

[27] Trindade, A. F., Gois, P. M. P., Afonso, C. A. M. (2009). Recyclable stereoselective catalysts. *Chem. Rev.*, *109*, 418–514.

[28] (a) Solodenko, W., Kunz, U., Schönfeld, H., Kirschning, A. (2003). Polymer/carrier composites as materials and reactors for organic synthesis. *J. Chromatogr. A*, *1006*, 241–249; (b) Solodenko, W., Jas, G., Kunz, U., Kirschning, A. (2007). Continuous enantioselective kinetic resolution of terminal epoxides using immobilized chiral cobalt-Salen complexes. *Synthesis*, 583–589.

[29] Annis, D. A., Jacobsen, E. N. (1999). Polymer-supported chiral Co(Salen) complexes: synthetic applications and mechanistic investigations in the hydrolytic kinetic resolution of terminal epoxides. *J. Am. Chem. Soc.*, *121*, 4147–4150.

[30] Matsuda, T., Watanabe, K., Harada, T., Nakamura, K., Arita, Y., Misumi, Y., Ichikawa, S., Ikariya, T. (2004). High-efficiency and minimum-waste continuous kinetic resolution of racemic alcohols by using lipase in supercritical carbon dioxide. *Chem. Commun.*, 2286–2287.

[31] Kataoka, S., Takeuschi, Y., Harada, A., Yamada, M., Endo, A. (2010). Microreactor with mesoporous silica support layer for lipase catalyzed enantioselective transesterification. *Green Chem.*, *12*, 331–337.

[32] Honda, T., Miyazaki, M., Yamaguchi, Y., Nakamura, H., Maeda, H. (2007). Integrated microreaction system for optical resolution of racemic amino acids. *Lab Chip*, *7*, 366–372.

[33] Itsuno, S., Sakurai, Y., Ita, K., Maruyama, T., Nakahama, S., Fréchet, J. M. J. (1990). New solid-phase catalysts for asymmetric synthesis: cross-linked polymers containing a chiral Schiff base-zinc complex. *J. Org. Chem.*, *55*, 304–310.

[34] Hodge, P., Sung, D. W. L., Stratford, P. W. (1999). Asymmetric synthesis of 1-phenylpropanol using polymer-supported chiral catalysts in simple bench-top flow systems. *J. Chem. Soc. Perkin Trans. 1*, 2335–2342.

[35] Burguete, M. I., García-Verdugo, E., Vicent, M. J., Luis, S. V., Pennemann, H., von Keyserling, N. G., Martens, J. (2002). New supported β-amino alcohols as efficient catalysts for the enantioselective addition of diethylzinc to benzaldehyde under flow conditions. *Org. Lett.*, *4*, 3947–3950.

[36] Pericàs, M. A., Herrerías, C. I., Solà, L. (2008). Fast and enantioselective production of 1-aryl-1-propanols through a single pass, continuous flow process. *Adv. Synth. Catal.*, *350*, 927–932.

[37] Itsuno, S., Ito, K., Maruyama, T., Kanda, N., Hirao, A., Nakahama, S. (1986). Asymmetric reduction of aromatic ketone in a continuous flow system. *Bull. Chem. Soc. Jpn.*, *59*, 3329–3331.

[38] Rolland, J., Cambeiro, X. C., Rodríguez-Escrich, C., Pericàs, M. A. (2009). Continuous flow enantioselective arylation of aldehydes with ArZnEt using triarylboroxins as the ultimate source of aryl groups. *Beilstein J. Org. Chem.*, *5*, No. 56.

[39] (a) Burguete, M. I., Cornejo, A., García-Verdugo, E., García, J., Gil, M. J., Luis, S. V., Martínez-Merino, V., Mayoral, J. A., Sokolova, M. (2007). Bisoxazoline-functionalised enantioselective monolithic mini-flow-reactors: development of efficient processes from batch to flow conditions. *Green Chem.*, *9*, 1091–1096; (b) Burguete, M. I., Cornejo, A., García-Verdugo, E., Gil, M. J., Luis, S. V., Mayoral, J. A., Martínez-Merino, V., Sokolova, M. (2007). Pybox monolithic miniflow reactors for continuous asymmetric cyclopropanation reaction under conventional and supercritical conditions. *J. Org. Chem.*, *72*, 4344–4350.

[40] (a) Wandeler, R., Künzle, N., Schneider, M. S., Mallat, T., Baiker, A. (2001). Continuous platinum-catalyzed enantioselective hydrogenation in "supercritical" solvents. *Chem. Commun.*, 673–674; (b) Künzle, N., Solèr, J.-W., Baiker, A. (2003). Continuous enantioselective hydrogenation in fixed-bed reactor: towards process intensification. *Catal. Today*, *79–80*, 503–509; (c) Szöllösi, G., Cserényi, S., Balázsik, K., Fülöp, F., Bartók, M. (2009). New data in the enantioselective hydrogenation of ethyl pyruvate on Pt-cinchona chiral catalyst using continuous-flow fixed-bed reactor system: the origin of rate enhancement. *J. Mol. Catal. A Chem.*, *305*, 155–160.

[41] Abdallah, R., Fumey, B., Meille, V., de Bellefon, C. (2007). Micro-structured reactors as a tool for chiral modifier screening in gas–liquid–solid asymmetric hydrogenations. *Catal. Today*, *125*, 34–39.

[42] Künzle, N., Mallat, T., Baiker, A. (2003). Enantio selective hydrogenation of isopropyl-4,4,4-trifluoroacetoacetate in a continuous flow reactor. *Appl. Catal. A Gen.*, *238*, 251–257.

[43] (a) Stephenson, P., Licence, P., Ross, S. K., Poliakoff, M. (2004). Continuous catalytic asymmetric hydrogenation in supercritical CO_2. *Green Chem.*, *6*, 521–523; (b) Stephenson, P., Kondor, B., Licence, P., Scovell, K., Ross, S. K., Poliakoff, M. (2006). Continuous asymmetric hydrogenation in supercritical carbon dioxide using an immobilised homogeneous catalyst. *Adv. Synth. Catal.*, *348*, 1605–1610.

[44] Hermán, B., Szöllösi, G., Fülöp, F., Bartók, M. (2007). Enantioselective hydrogenation of α,β-unsaturated carboxylic acids in fixed-bed reactor. *App. Catal. A Gen.*, *331*, 39–43.

[45] Sandee, A. J., Petra, D. G. I., Reek, J. N. H., Kamer, P. C. J., van Leeuwen, P. W. N. M. (2001). Solid-phase synthesis of homogeneous ruthenium catalysts on silica for the continuous asymmetric transfer hydrogenation reaction. *Chem. Eur. J.*, *7*, 1202–1208.

[46] Bonfils, F., Cazaux, I., Hodge, P., Caze, C. (2006). Michael reactions carried out using a bench-top flow system. *Org. Biomol. Chem.*, *4*, 493–497.

[47] (a) Bernstein, D., France, S., Wolfer, J., Lectka, T. (2005). A column-based “flush and flow” system for the asymmetric α-chlorination of acid halides. *Tetrahedron Asymmetry*, *16*, 3481–3483; (b) France, S., Bernstein, D., Weatherwax, A., Lectka, T. (2005). Performing the synthesis of a complex molecule on sequentially linked columns: toward the development of a “synthesis machine”. *Org. Lett.*, *7*, 3009–3012.

[48] Hafez, A. M., Taggi, A. E., Dudding, T., Lectka, T. (2001). Asymmetric catalysis on sequentially-linked columns. *J. Am. Chem. Soc.*, *123*, 10853–10859.

[49] Kamahori, K., Ito, K., Itsuno, S. (1996). Asymmetric Diels-Alder reaction of methacrolein with cyclopentadiene using polymer-supported catalysts: design of highly enantioselective polymeric catalysts. *J. Org. Chem.*, *61*, 8321–8324.

[50] Altava, B., Burguete, M. I., García-Verdugo, E., Luis, S. V., Vicent, M. J. (2006). Functional monolithic resins for the development of enantioselective versatile catalytic minireactors with long-term stability: TADDOL supported systems. *Green Chem.*, *8*, 717–726.

[51] Altava, B., Burguete, M. I., Fraile, J. M., García, J. I., Luis, S. V., Mayoral, J. A., Vicent, M. J. (2000). How important is the inert matrix of supported enantiomeric catalysts? Reversal of topicity with two polystyrene backbones. *Angew. Chem., Int. Ed.*, *39*, 1503–1506.

[52] Mandoli, A., Orlandi, S., Pini, D., Salvadori, P. (2004). Insoluble polystyrene-bound bis(oxazoline): batch and continuous-flow heterogeneous enantioselective glyoxylate-ene reaction. *Tetrahedron Asymmetry*, *15*, 3233–3244.

[53] Carter, C. F., Baxendale, I. R., Pavey, J. B. J., Ley, S. V. (2010). The continuous flow synthesis of butane-2,3-diacetal protected building blocks using microreactors. *Org. Biomol. Chem.*, *8*, 1588–1595.

[54] Lundgren, S., Ihre, H., Moberg, C. (2008). Enantioselective cyanation of benzaldehyde. An asymmetric polymeric catalyst in a microreactor. *ARKIVOC*, *vi*, 73–80.

[55] (a) Carrel, F. R., Geyer, K., Codée, J. D. C., Seeberger, P. H. (2007). Oligosaccharide synthesis in microreactors. *Org. Lett.*, *9*, 2285–2288; (b) Kawakami, H., Goto, K., Mizuno, M. (2009). Multi-step synthesis of a protected monosaccharide unit by iterative reactions in microreactors and fluorous liquid-phase extractions. *Chem. Lett.*, *38*, 906–908.

[56] (a) Baxendale, I. R., Griffiths-Jones, C. M., Ley, S. V., Tranmer, G. K. (2006). Preparation of the neolignan natural product grossamide by a continuous-flow process. *Synlett*, 427–430; (b) Baxendale, I. R., Deeley, J., Griffiths-Jones, C. M., Ley, S. V., Saaby, S., Tranmer, G. K. (2006). A flow process for the multi-step synthesis of the alkaloid natural product oxomaritidine: a new paradigm for molecular assembly. *Chem. Commun.*, 2566–2568; (c) Kündig, P. (2006). The future of organic synthesis. *Science*, *314*, 430–431.

[57] Brophy, J. H., Jarosch, K. (2004). PCT Int. Appl., WO 2004016348.

[58] Koh, J. S., Hong, J. B., Kim, M. J., Park, J. K., Choi, J. T., Kim, C. Y., Kim, M. J. (2005). Repub. Korean Kongkae Taeho Kongbo., KR 2005081974.

CHAPTER 9

ASYMMETRIC CATALYTIC SYNTHESIS IN SUPERCRITICAL FLUIDS

TOMOKO MATSUDA

9.1. Introduction 374
9.2. Basic properties of $scCO_2$ for the application in organic synthesis 374
    9.2.1. Advantages of organic synthesis in $scCO_2$ 374
    9.2.2. Reactors 376
9.3. Enzyme-mediated asymmetric synthesis 377
    9.3.1. Lipase-catalyzed kinetic resolution of alcohols 378
    9.3.2. Lipase and chemical catalysts-catalyzed kinetic resolution and dynamic kinetic resolution of alcohols 379
    9.3.3. Control of enantioselectivity of lipase-catalyzed kinetic resolution 380
    9.3.4. Alcohol dehydrogenase-catalyzed asymmetric reductions 383
9.4. Metal complexes-mediated asymmetric synthesis 383
    9.4.1. Polymer-supported (*R,S*)-BINAPHOS-Rh(I) complex-catalyzed olefin hydroformylation 384
    9.4.2. Hydrogenation of olefin using aqueous/$scCO_2$ biphasic systems 385
    9.4.3. Hydrogenation using IL and $scCO_2$ systems 385
    9.4.4. Identification of catalyst surface species for platinum-catalyzed hydrogenation in supercritical ethane 386
    9.4.5. Control of enantioselectivity of photoaddition by temperature and pressure 387
9.5. Conclusions 388
References 388

Catalytic Methods in Asymmetric Synthesis: Advanced Materials, Techniques, and Applications, First Edition. Edited by Michelangelo Gruttadauria and Francesco Giacalone.

9.1. INTRODUCTION

Green methods for asymmetric synthesis are increasing in importance as the demand for the use of chiral compounds grows. Such products are useful for a number of purposes, such as drug synthesis, where the majority of drug candidate molecules possess more than one chiral center. To synthesize them in environmentally friendly methods, green solvents and green catalysts are necessary. For these purposes, supercritical fluids (fluids above their critical points) have been used for organic synthesis [1]. Among supercritical fluids, supercritical water (scH_2O) and supercritical carbon dioxide ($scCO_2$) are considered to be green because water and carbon dioxide exist naturally and abundantly. For asymmetric synthesis, carbon dioxide has been used due to the ambient critical temperature as shown in Table 9.1 and Figure 9.1.

As catalysts for green asymmetric synthesis, both chemical and biological catalysts need to be developed because they are complementary to each other. They are used separately for most of the cases, but the methods of using both in one reaction, such as dynamic kinetic resolutions of racemates and using enzyme for resolution and metal catalysts for racemization, have become increasingly important. These reactions have also been conducted in $scCO_2$.

Here, some representative reactions for asymmetric synthesis using $scCO_2$ by chemical catalysts and/or enzymes are shown. First, basic properties of $scCO_2$ are explained in Section 9.2. Then, enzyme-mediated asymmetric synthesis is described in detail in Section 9.3, and metal complexes-mediated asymmetric synthesis is introduced briefly in Section 9.4.

9.2. BASIC PROPERTIES OF $SCCO_2$ FOR THE APPLICATION IN ORGANIC SYNTHESIS

This section introduces the advantage of the use of $scCO_2$ for organic synthesis. The experimental apparatus are also explained briefly here.

9.2.1. Advantages of Organic Synthesis in $scCO_2$

The properties of $scCO_2$ and the advantages of organic synthesis using $scCO_2$ are listed in Table 9.2. $scCO_2$ is a high-density CO_2 solvent, so the use of $scCO_2$

TABLE 9.1. Critical Points of Some Materials Used for Organic Synthesis

	Critical Temperature (°C)	Critical Pressure (MPa)	Critical Density (g/mL)
CO_2	31	7.4 (73 atm)	0.47
Water	374	22 (218 atm)	0.32
Methanol	239	8.1 (80 atm)	0.27

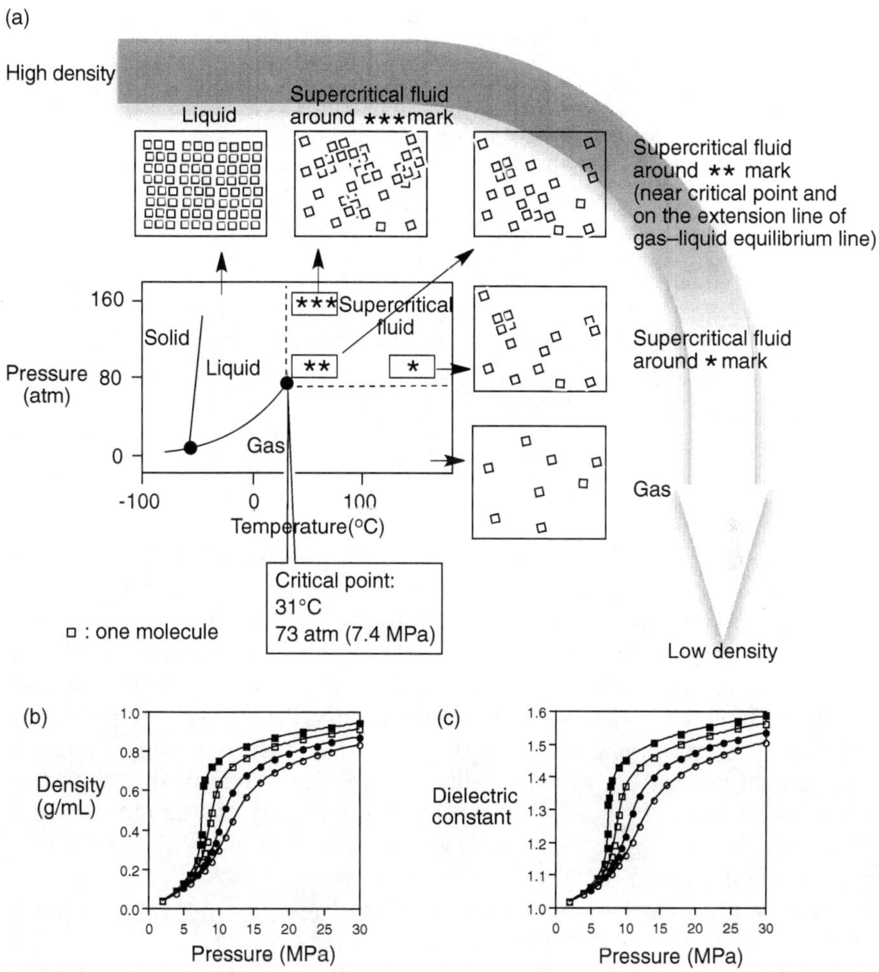

FIGURE 9.1. (a) Phase diagram of carbon dioxide and image of each state; (b) tunability of the density of CO_2 by the temperature and pressure; (c) tunability of dielectric constant (closed square: 32°C; open square: 40°C; closed circle: 50°C; open circle: 60°C) [2].

makes feasible the development of a reaction using carbon dioxide as a reactant. The diffusivity of $scCO_2$ is high like a gas, which makes the reaction rate faster. The high solubilizing power of $scCO_2$ (like that of liquids) to dissolve hydrophobic substrate also leads to a higher reaction rate. Since $scCO_2$ is a hydrophobic solvent like perfluorinated solvents (Fig. 9.1c), a cosolvent may be necessary to dissolve hydrophilic substances. $scCO_2$ becomes a gas after the reaction pressure reaches 1 atm, so the separation of the product from the

TABLE 9.2. Advantages of Organic Synthesis in $scCO_2$

Properties of $scCO_2$	Advantages
High-density CO_2	Development of reaction using CO_2 as a reactant.
High diffusivity	High reaction rate.
High solubilizing power	High substrate concentration to increase reaction rate.
Gas at ambient conditions	Easy separation of the product from the solvent.
CO_2 exists naturally	No waste problem. No problem with existing traces of solvent in the product.
Tunability	Control of selectivity by tuning the solvent properties. Examination of effect of solvent without changing the kind of molecule.

solvent is easy, and an extraction process is not necessary. For the enzymatic process in aqueous media, the extraction of product and treatment of the wastewater contaminated with extraction solvent may be difficult, but the use of $scCO_2$ solves these problems. CO_2 exists naturally and abundantly, so there is no waste problem and no problem with existing traces of solvent in the product. Finally, the tunability of $scCO_2$ clearly distinguishes supercritical fluids from conventional solvents. Its properties such as density, dielectric constant, diffusivity, viscosity, solubility, and so on, can be tuned by adjusting the pressure and temperature. For example, the density of $scCO_2$ can change from high (like a liquid) to low (like a gas) as shown in Figure 9.1a,b by adjusting the pressure and temperature [2]. Therefore, solvent property can be adjusted to be suitable to the desired reactions by adjusting the pressure and temperature. In some cases, the selectivity of the reaction can be controlled by tuning the solvent property by adjusting the pressure and temperature. Moreover, with supercritical fluids, solvent effects on the reaction can be examined without changing the kind of solvent. Therefore, continuous changes in the result—for example, enantioselectivity—can be expected, since the solvent properties can be changed continuously by manipulating the pressure and temperature.

9.2.2. Reactors

Both batch- and flow-type reactors have been used for asymmetric synthesis using $scCO_2$. A representative flow reactor [3] is shown in Figure 9.2. With flow reactors, the addition of a substrate to the column with a catalyst (chemical catalysts or enzymes) yields the product and CO_2, which is a gas at ambient pressure. With the batch reactor, however, extraction of the product from the

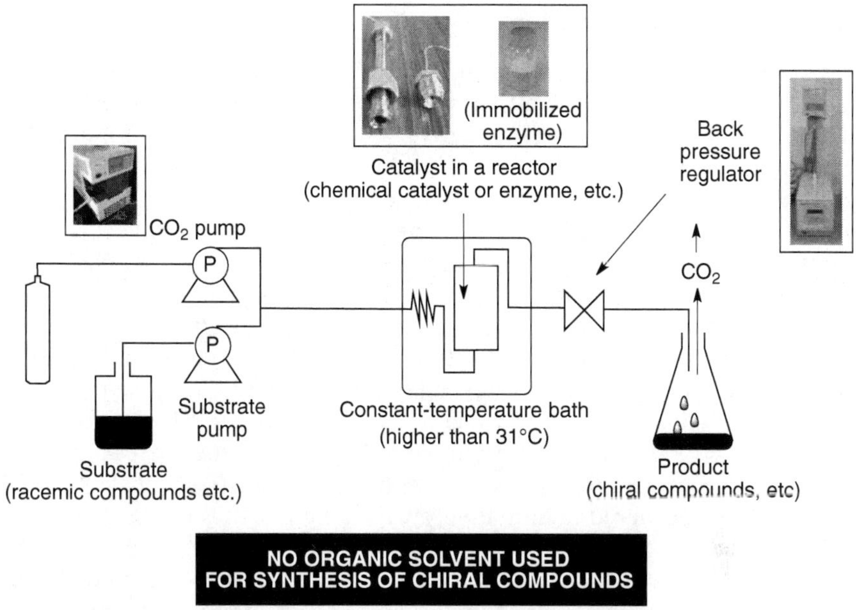

FIGURE 9.2. Simplified illustration of $scCO_2$ experimental apparatus.

biocatalyst is necessary after depressurization, and an organic solvent may be used. Therefore, the flow type is superior to the batch type for achieving virtually no solvent reaction. Moreover, the size of the reactors that use the flow process to generate an amount of product comparable with the corresponding batch reactors is smaller [4], which is particularly attractive for a supercritical fluid system.

9.3. ENZYME-MEDIATED ASYMMETRIC SYNTHESIS

The attraction of combining natural catalysts with a natural solvent has been the driving force behind a growing body of literature on the stability, activity, and specificity of enzymes in $scCO_2$ [5]. Since the first report on biotransformations in supercritical fluids in 1985 [6], the benefits of using supercritical fluids for biotransformations have been demonstrated, for example, in improved reaction rates, control of selectivities by pressure, and so on. The kind of biocatalytic reaction in the majority of reports has been hydrolysis due to the high stability and ease in handling. Oxidation and reduction have accounted for the second largest number of studies of biocatalysis. Here are some examples presented for kinetic resolution of racemic alcohol through esterification by lipase and asymmetric reduction.

9.3.1. Lipase-Catalyzed Kinetic Resolution of Alcohols

Heterogeneous, immobilized enzymes have been used for the synthesis of optically active compounds which have been virtually obtained without solvent using a flow reactor as shown in Figure 9.2 [3]. The use of the continuous flow reaction for the kinetic resolution of 1-phenylethanol by lipase Novozyme 435 resulted in a completely organic solvent-free process and in a significant improvement in the productivity for long periods of reaction times. The productivity of the optically active compounds, namely space–time yield, was improved by over 400 times compared to the corresponding batch reaction using $scCO_2$. The reaction of 1-phenylethanol and vinyl acetate, with the molar ratio of 1:0.5 at a flow rate of 0.70 mL/min, over the catalyst under 13 MPa of $scCO_2$ (1.5 mL/min), gave the corresponding acetate with 99.7% enantiomeric excess (ee) in 47% yield. The *E* value exceeds 1800 (Scheme 9.1a). The use of a slight excess of vinyl acetate resulted in an increase in the chemical yield of optically active acetate from 47% to 50%, in which the unreacted alcohol with 98.8% ee was recovered quantitatively. Changing the CO_2 pressure from 8.9 to 20 MPa did not significantly change the outcome of the reaction.

When the substrate specificity of this system was investigated, it was found that aliphatic alcohols, 2-undecanol, and 1-tetralol were kinetically resolved with the Novozym catalyst to give a mixture of the corresponding optically active (*R*)-ester and unreacted (*S*)-alcohol (Scheme 9.1b,c). This synthetic process is particularly useful for the large-scale production of optically active alcohols. The biocatalyst maintained its performance in terms of the reactivity and selectivity during 3 days of operation under supercriti-

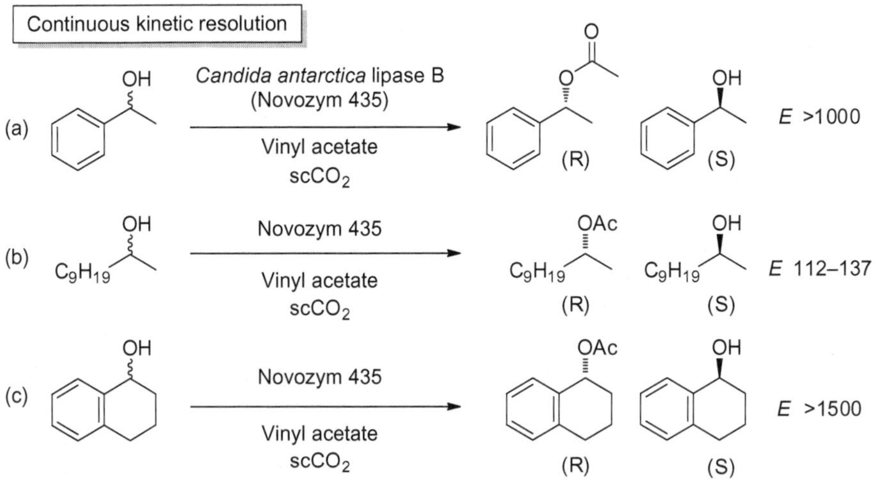

SCHEME 9.1. Asymmetric enzymatic reactions using $scCO_2$ flow system.

cal conditions (12.9–13 MPa at 42°C), and resulted in a quantitative transformation of (*R*/*S*)-1-phenylethanol (221 g) to (*S*)-phenylethanol with 99% ee and the corresponding (*R*)-acetate with 99% ee using 1.73 g of the immobilized enzyme.

9.3.2. Lipase and Chemical Catalysts-Catalyzed Kinetic Resolution and Dynamic Kinetic Resolution of Alcohols

Enzymes and chemical catalysts are complementary to each other, so they are used together in an $scCO_2$ flow system. For example, the two-step reaction shown in Scheme 9.2a was conducted successfully by linking two reactors in series [7]. The first reactor was used for the metal-catalyzed hydrogenation of acetophenone and the second reactor for lipase catalyzed the kinetic resolution of the resulting 1-phenylethanol.

Using chemical catalysts and lipase together, dynamic kinetic resolution (DKR) was also successfully conducted in $scCO_2$ [8]. For the above-described kinetic resolution, the maximum conversion in the reaction is only 50%, and the product has to be separated from the reactants. However, in DKR, it is possible to convert the racemic reactant to product with 100% yield and 100% ee theoretically, because the reactant is racemized to replenish the faster reacting enantiomer at the expense of the slower reacting enantiomer. Continuous DKR processes in ionic liquid (IL)/$scCO_2$ biphasic systems

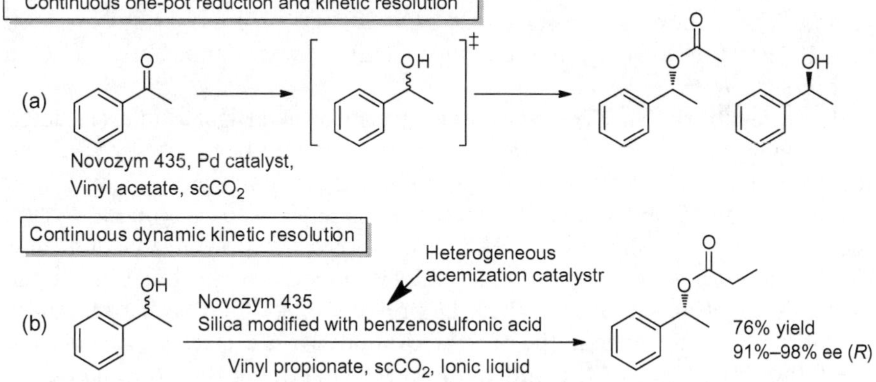

SCHEME 9.2. Asymmetric synthesis with lipase together with chemical catalysts using $scCO_2$ flow system: (a) hydrogenation of acetophenone over an achiral Pd catalyst to produce rac-1-phenylethanol (first step) and the kinetic resolution of (*R*/*S*)-1-phenylethanol to produce (*R*)-acetate and (*S*)-alcohol (second step); (b) DKR of rac-1-phenylethanol by transesterfication with vinyl propionate catalyzed by the combined action of immobilized lipase (Novozym) and silica modified with benzenosulphonic acid groups.

were carried out by simultaneously using both immobilized lipase (Novozym 435) and silica modified with benzenesulfonic acid (SCX) catalysts at 40°C and 10 MPa as shown in Scheme 9.2b. SCX racemized (*S*)-1-phenylethanol efficiently in different IL media ([emim][NTf_2], [btma][NTf_2], and [bmim] [PF_6]), and lipase catalyzed the kinetic resolution of (*R*/*S*)-(1)-phenylethanol. Coating both chemical and enzymatic catalysts with ILs greatly improved the efficiency of the process, providing a good yield (76%) of (*R*)-1-phenylethyl propionate product with excellent ee (ee = 91%–98%) in continuous operation.

9.3.3. Control of Enantioselectivity of Lipase-Catalyzed Kinetic Resolution

One of the most prominent characteristics of $scCO_2$ is the tunability of the solvent properties. Its properties such as density can be tuned by adjusting the pressure and temperature [2], so the property can be continuously changed. Here is an example of controlling the enantioselectivity of the lipase-catalyzed reaction by changing the pressure and temperature [9].

The enantioselective acetylation of racemic 1-(*p*-chlorophenyl)-2,2,2-trifluoroethanol with lipases and vinyl acetate in $scCO_2$ was examined in detail (Fig. 9.3a), and it was found that the enantioselectivity of the reaction catalyzed by lipase Novozym can be controlled by adjusting the pressure and the temperature of the $scCO_2$. First, various lipases were screened for the reaction (Table 9.3). In all but one case, the (*S*)-enantiomer reacted faster than the (*R*)-enantiomer, affording (*S*)-acetate and the remaining (*R*)-alcohol. The highest enantioselectivity ($E = 38$) was obtained using Novozym at 9.1 MPa. Interestingly, the enantioselectivity was significantly affected by pressure.

The effect of pressure on enantioselectivity was investigated in more detail by carrying out the esterification at pressures ranging from 8 to 19 MPa and for different reaction times, while maintaining the temperature at 55°C. As shown in Figure 9.3b, the E value decreased from 50 to 10 continuously when the pressure was changed from 8 to 19 MPa, regardless of the reaction time. The effect of pressure on enantioselectivity is indeed noteworthy, although the reason is not clear at present. When the pressure of $scCO_2$ was changed, there was no significant change in the polarity evaluated as a dielectric constant and log P (at 50°C; 1.4 at 8 MPa and 1.9 at 11 MPa) [10]. On the other hand, the density of $scCO_2$ does change from 0.20 to 0.42 g/mL when the pressure is changed from 8 to 11 MPa at 55°C [2]. It was proposed that the large change in density could significantly change the interaction between CO_2 and the enzyme by the formation of carbamates from CO_2 and the free amine groups on the surface of the enzyme. This can also occur by CO_2 adsorption on the enzyme and/or by CO_2 incorporation in the substrate-binding pocket of the enzyme, in analogy to the incorporation of organic molecules in enzymes. These interactions may gradually change the conformation of the enzyme in

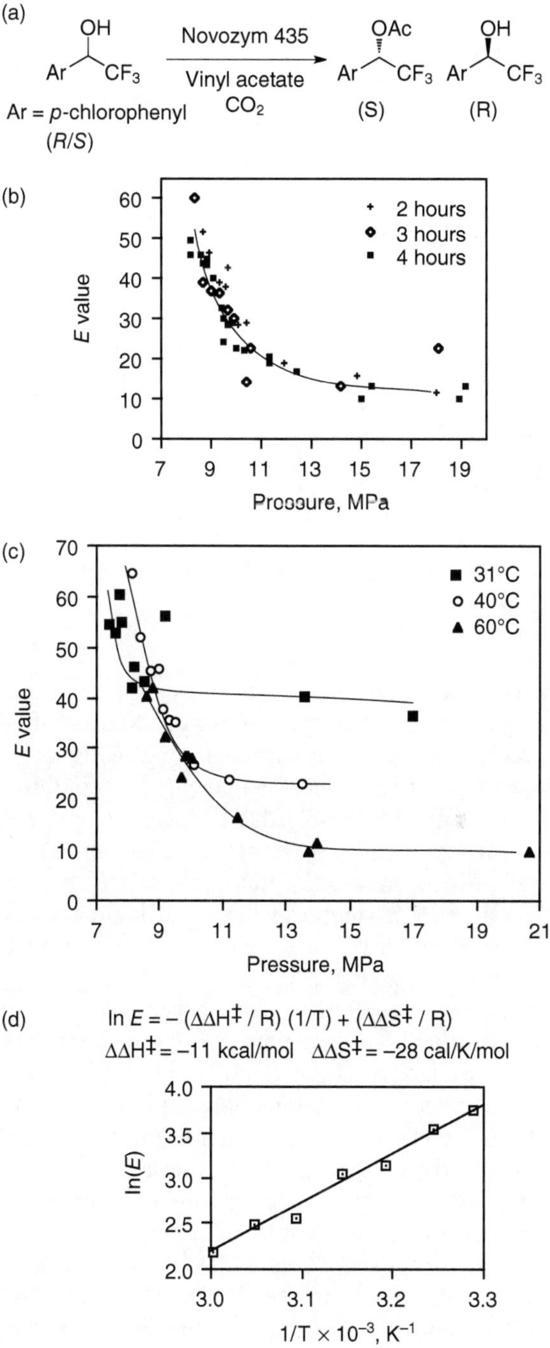

FIGURE 9.3. Control of enantioselectivity of lipase-catalyzed kinetic resolution by pressure and temperature.

TABLE 9.3. Screening of Lipases for Enantioselective Acetylation of 1-(*p*-Chlorophenyl)-2,2,2-trifluoroethanol in $scCO_2$

	Low Pressure Conditions (9.1 MPa)		High Pressure Conditions (14.5 MPa)	
Lipase	Yield (%)	E[a]	Yield (%)	E[a]
LPL (*Pseudomonas aeruginosa*)	52	12	38	16
AY (*Candida rugosa*)	8	1[b]	2	2
AH (*Pseudomonas cepacia*)	3	29	0	–
PS-D (*P. cepacia*)	0	–	0	–
PS-C (*P. cepacia*)	43	8	22	17
Lipozyme (*Rhizomucor miehei*)	0	–	0	–
Novozym 435 (*Candida antarctica*)	25	38	24	23

[a] Enantiomeric ratio, *E* value, was used to evaluate enantioselectivity. *E* = (VA/KA)/(VB/KB), where VA and KA, and VB and KB denote maximal velocities and Michaelis constants of the fast- and slow-reacting enantiomers, respectively. The (*S*)-enantiomer reacted faster than (*R*)-enantiomer.

[b] In this case, the (*R*)-enantiomer reacted slightly faster than the (*S*)-enantiomer.

response to pressure, resulting in a continuous change in enantioselectivity. The effect of pressure on the enantioselective acetylation of the alcohol with vinyl acetate in $scCO_2$ by Novozym was also investigated at 31, 40, and 60°C (Fig. 9.3c). As in the case at 55°C, the *E* value changed continuously according to the pressure. This is most probably due to the change of $scCO_2$ density, as described above. This explanation is in agreement with the observation that at lower temperatures (31 and 40°C), the decrease of the *E* value measured at pressures below 10 MPa is steeper, whereas at higher temperatures, the *E* values decrease more gradually. These changes correlate well with the change in density as shown in Figure 9.1b [2].

However, when *E* values obtained at the same density but at different temperatures were compared, a significant temperature effect became evident. Therefore, the enantioselectivity is determined not only by density but also by temperature. In a reaction under ambient conditions, the enantioselectivity of a kinetic resolution is temperature dependent and obeys a modified Eyring equation [11]. Sakai et al. provided the first experimental evidence supporting the theory of the effect of temperature on stereochemistry [11b,c]. Here, we examine whether the theory is applicable to the reaction in $scCO_2$. At a density of 0.75 g/mL (31°C at 9.5 MPa, 35°C at 11.2 MPa, 40°C at 13.2 MPa, 45°C at 15.3 MPa, 50°C at 17.5 MPa, 55°C at 19.6 MPa, and 60°C at 21.8 MPa), ln *E* was plotted against 1/T. As shown in Figure 9.3d, the Eyring plot was found to be linear throughout this range and thus indicates the conformational stability of the transition state. The differences in enthalpy and entropy values calculated from the above graph are given in Figure 9.3d.

9.3.4. Alcohol Dehydrogenase-Catalyzed Asymmetric Reductions

Most of the biocatalysts used in supercritical fluids are hydrolytic enzymes such as lipases and proteases, and very few reports on the use of alcohol dehydrogenases in supercritical fluids have been published, despite the fact that they are an important class of enzymes for the asymmetric reduction of ketones to produce chiral alcohols. So far, alcohol dehydrogenases from *Geotrichum candidum* and horse liver dehydrogenase have been used in $scCO_2$. The former was used for asymmetric synthesis, and the latter was used for reduction of butyraldehyde to butanol using fluorinated coenzyme, nicotinamide adenine dinucleotide (NADH).

Asymmetric reduction was realized by using an immobilized cell of *G. candidum* NBRC 5767 in 10 MPa $scCO_2$; high activities and excellent enantioselectivities were observed for the asymmetric reduction of aromatic and cyclic ketones [12]. For example, 2-fluoroacetophenone was reduced to the (*S*)-alcohol in 81% yield with >99% ee as shown in Figure 9.4. The immobilized cell was also used in the flow system [12b,c].

9.4. METAL COMPLEXES-MEDIATED ASYMMETRIC SYNTHESIS

Homogeneous and heterogeneous chemical catalysts have been used for asymmetric synthesis using supercritical fluid as a solvent [1]. Biphasic systems,

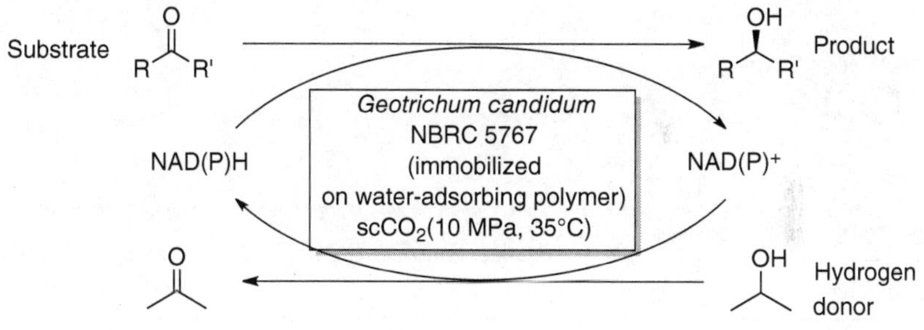

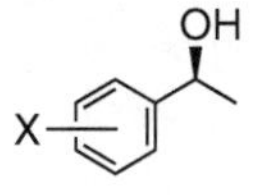

Product	X	Yield (%)	ee (%)
	H	51	>99%(*S*)
	o-F	81	>99%(*S*)
	m-F	53	>99%(*S*)
	p-F	11	97%(*S*)
	o-Me	22	>99%(*S*)

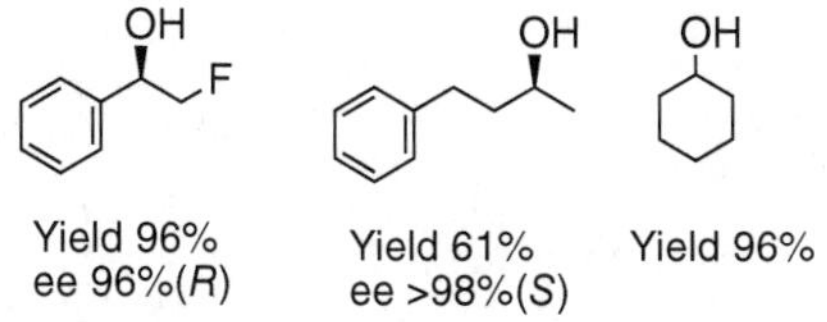

FIGURE 9.4. $scCO_2$ as a solvent for reduction of ketone by fungus, *G. candidum*.

$scCO_2$/aqueous, or ILs are also being developed for the promotion of green chemistry. Some of the interesting examples are shown in this section.

9.4.1. Polymer-Supported (*R,S*)-BINAPHOS-Rh(I) Complex-Catalyzed Olefin Hydroformylation

Asymmetric hydroformylation of olefins was performed using an (*R,S*)-BINAPHOS-Rh(I) catalyst that is covalently anchored to a highly cross-linked polystyrene support (Fig. 9.5) [13]. The polymer-supported catalyst was appli-

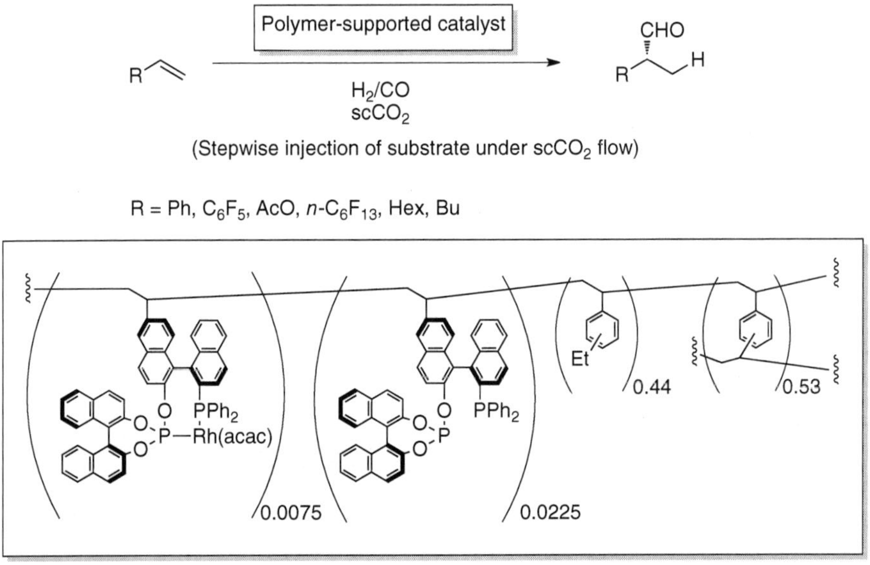

Reaction type	Olefin	Product	Iso/normal ratio	ee (%)
Batch	*cis*-2-Butene	(*S*)-2-Methylbutanal	100:0	82
Injection to vapor-flow column reactor	3,3,3-Trifluoropropene	(*S*)-2-Trifluoromethylpropanal	95:5	90

Sequential conversion

Cycle	Olefin	Conv (%)	Iso/normal ratio	ee (%)
1st	Styrene	49	82:18	77
2nd	Vinyl acetate	5	70:30	74
3rd	1-Octene	47	21:79	73
4th	1-Hexene	40	21:79	60
5th	Styrene	36	81:19	82
6th	2,3,4,5,6-Pentafluorostyrene	27	89:11	88
7th	$CF_3(CF_2)_5CH{=}CH_2$	21	91:9	78
8th	Styrene	54	80:20	80

FIGURE 9.5. Olefin hydroformylation by polymer-supported chiral catalyst.

cable to a continuous column reactor as well as to a batch reactor. Various olefins were injected and were successfully converted into the corresponding isoaldehydes with high enantioselectivity.

9.4.2. Hydrogenation of Olefin Using Aqueous/scCO_2 Biphasic Systems

Rh complex is also used in aqueous/scCO_2 biphasic systems, which allows the highly enantioselective hydrogenation of polar water-soluble substrates and efficient recycling of the CO_2-philic catalysts (Fig. 9.6) [14]. Chiral CO_2-philic catalysts were efficiently immobilized in scCO_2 as the stationary phase, while the polar substrates and products were contained in water as the mobile phase. Therefore, product separation and catalyst recycling were conducted without depressurization.

9.4.3. Hydrogenation Using IL and scCO_2 Systems

The IL/scCO_2 biphasic system has been used for the asymmetric hydrogenation of imine and olfine. The catalysts were immobilized in IL solution so

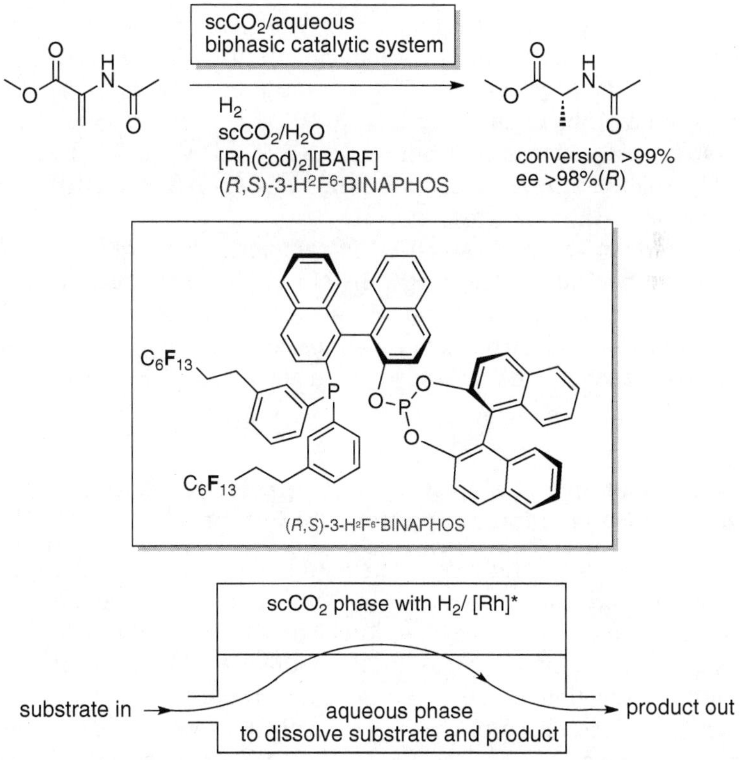

FIGURE 9.6. Hydrogenation of methyl 2-acetamidoacrylate using aqueous/scCO_2 biphasic system.

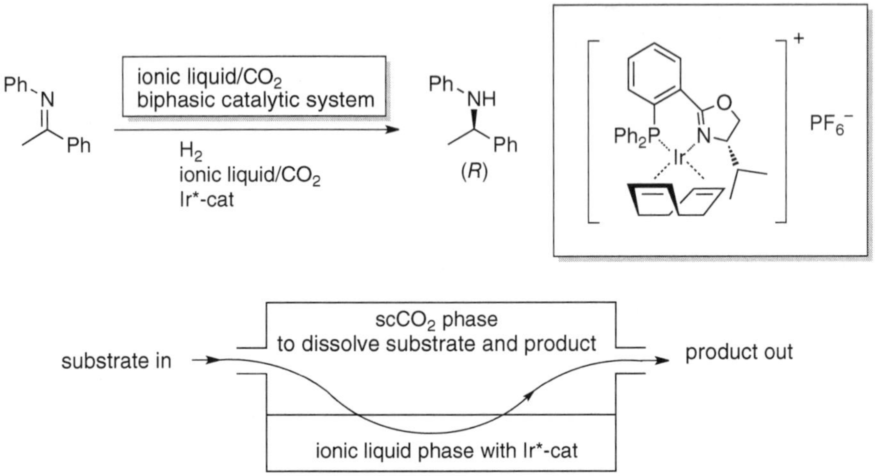

FIGURE 9.7. Iridium-catalyzed hydrogenation of imines using IL/$scCO_2$ biphasic system.

they were reused repeatedly without significant loss of enantioselectivity or conversion. For example, chiral iridium-catalyzed hydrogenation of imines was conducted first in $scCO_2$ in 1999 [15a] and then later using the IL/$scCO_2$ biphasic system [15b] as shown in Figure 9.7. The biphasic system leads to activation, tuning, and immobilization of the catalyst. Moreover, the products are readily isolated from the catalyst solution by CO_2 extraction without cross-contamination of IL or catalyst.

The IL and $scCO_2$ were also used in symmetric hydrogenation of tiglic acid catalyzed by $Ru(O_2CMe)_2$((*R*)-tolBINAP) [16]. The reaction in wet IL ([bmim]PF_6 with added water, bmim: 1-*n*-butyl-3-methylimidazolium) gave 2-methylbutanoic acid with high enantioselectivity and conversion. The product was extracted with $scCO_2$, giving a clean separation of product and catalyst.

9.4.4. Identification of Catalyst Surface Species for Platinum-Catalyzed Hydrogenation in Supercritical Ethane

There have been successfully conducted studies for the analysis of catalysts species in supercritical ethane. In situ attenuated total reflection infrared spectroscopy studies during the enantioselective hydrogenation of ethyl pyruvate in supercritical ethane over a chirally modified Pt/Al_2O_3 catalyst show the preferential adsorption of ethyl pyruvate as *cis*-conformer and indicate a hydrogen bond interaction of this species with the coadsorbed modifier cinchonidine as shown in Scheme 9.3 [17]. The coadsorbed species interact via hydrogen bonding, forming a diastereomeric complex.

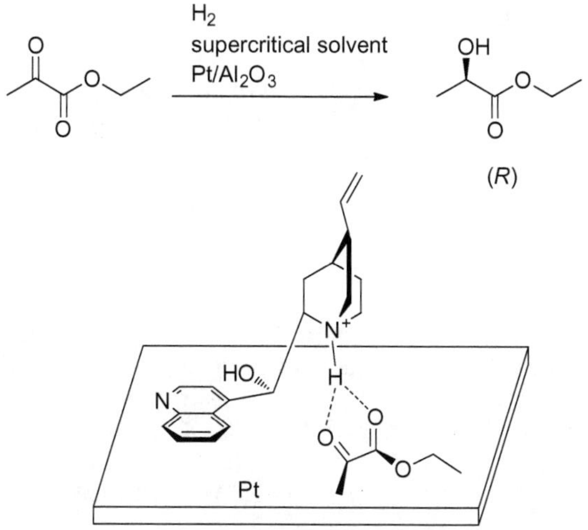

SCHEME 9.3. Proposed model for ethyl pyruvate adsorption during enantioselective hydrogenation in scC_2H_6 over Pt/Al_2O_3 chirally modified by CD (cinchonidine).

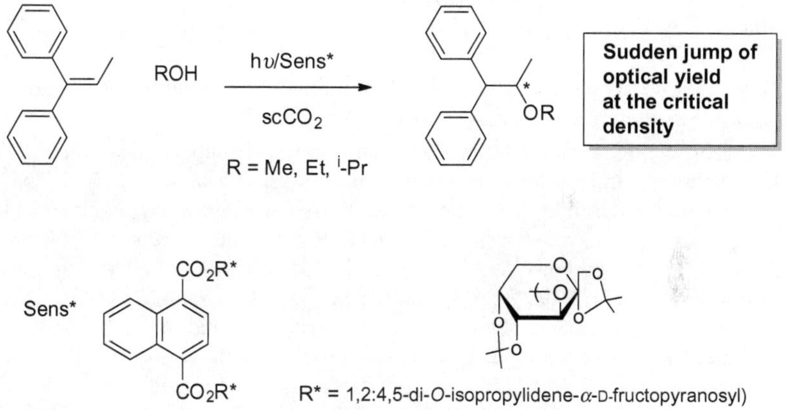

SCHEME 9.4. Control of enantioselectivity of the photoaddition by pressure and temperature.

9.4.5. Control of Enantioselectivity of Photoaddition by Temperature and Pressure

The enantioselectivity of anti-Markovnikov photo addition of methanol to diphenylpropene was controlled by temperature and pressure of CO_2 (Scheme 9.4) [18]. The enantioselectivity is enhanced by increasing alcohol size and pressure. Interestingly, the pressure dependence of ee is discontinuous at the critical density, accompanying a big jump caused most probably by enhanced clustering of the alcohol.

9.5. CONCLUSIONS

Some reactions for asymmetric synthesis using $scCO_2$ by chemical catalysts and/or enzymes are explained. These reactions may promote developing of green chemistry in the future.

REFERENCES

[1] (a) Jessop, P. G., Ikariya, T., Noyori, R. (1994). Homogeneous catalytic hydrogenation of supercritical carbon dioxide. *Nature*, *368*, 231–232; (b) Jessop, P. G., Ikariya, T., Noyori, R. (1999). Homogeneous catalysis in supercritical fluids. *Chem. Rev.*, *99*, 475–493; (c) Ikariya, T., Noyori, R. (2003). Carbon dioxide as a reagent and solvent in catalysis. In: J. M. DeSimone, W. Tumas (Eds.), *Green Chemistry Using Liquid and Supercritical Carbon Dioxide*. Oxford University Press, Oxford, pp. 48–63.

[2] Moriyoshi, T., Kita, T., Uosaki, Y. (1993). Static relative permittivity of carbon dioxide and nitrous oxide up to 30 MPa. *Ber. Bunsenges. Phys. Chem.*, *97*, 589–596.

[3] Matsuda, T., Watanabe, K., Harada, T., Nakamura, K., Arita, Y., Misumi, Y., Ichikawa, S., Ikariya, T. (2004). High-efficiency and minimum-waste continuous kinetic resolution of racemic alcohols by using lipase in supercritical carbon dioxide. *Chem. Commun.*, 2286–2287.

[4] Hitzler, M. G., Poliakoff, M. (1997). Continuous hydrogenation of organic compounds in supercritical fluids. *Chem. Commun.*, 1667–1668.

[5] (a) Hobbs, H. R., Thomas, N. R. (2007). Biocatalysis in supercritical fluids, in fluorous solvents, and under solvent-free conditions. *Chem. Rev.*, *107*, 2786–2820; (b) Matsuda, T., Harada, T., Nakamura, K., Ikariya, T. (2005). Asymmetric synthesis using hydrolytic enzymes in supercritical carbon dioxide. *Tetrahedron Asymmetry*, *16*, 909–915; (c) Mesiano, A. J., Beckman, E. J., Russell, A. J. (1999). Supercritical biocatalysis. *Chem. Rev.*, *99*, 623–634; (d) Mori, T., Okahata, Y. (1998). Effective biocatalytic transgalactosylation in a supercritical fluid using a lipid-coated enzyme. *Chem. Commun.*, 2215–2216; (e) Hobbs, H. R., Kirke, H. M., Poliakoff, M., Thomas, N. R. (2007). Homogeneous biocatalysis in both fluorous biphasic and supercritical carbon dioxide systems. *Angew. Chem. Int. Ed.*, *46*, 7860–7863.

[6] (a) Randolph, T. W., Blanch, H. W., Prausnitz, J. M., Wilke, C. R. (1985). Enzymatic catalysis in a supercritical fluid. *Biotechnol. Lett.*, *7*, 325–328; (b) Hammond, D. A., Karel, M., Klibanov, A. M., Krukonis, V. J. (1985). Enzymic reactions in supercritical gases. *Appl. Biochem. Biotechnol.*, *11*, 393–400; (c) Nakamura, K., Chi, Y. M., Yamada, Y., Yano, T. (1986). Lipase activity and stability in supercritical carbon dioxide. *Chem. Eng. Commun.*, *45*, 207–212.

[7] Hobbs, H. R., Kondor, B., Stephenson, P., Sheldon, R. A., Thomas, N. R., Poliakoff, M. (2006). Continuous kinetic resolution catalysed by cross-linked enzyme aggregates, "CLEAs," in supercritical CO_2. *Green Chem.*, *9*, 816–821.

[8] Lozano, P., Diego, T. D., Larnicol, M., Vaultier, M., Iborra, J. L. (2006). Chemoenzymatic dynamic kinetic resolution of rac-1-phenylethanol in ionic liquids and ionic liquids/supercritical carbon dioxide systems. *Biotechnol. Lett.*, *28*, 1559–1565.

[9] (a) Matsuda, T., Kanamaru, R., Watanabe, K., Harada, T., Nakamura, K. (2001). Control on enantioselectivity with pressure for lipase catalyzed esterification in supercritical carbon dioxide. *Tetrahedron Lett.*, *42*, 8319–8321; (b) Matsuda, T., Kanamaru, R., Watanabe, K., Kamitanaka, T., Harada, T., Nakamura, K. (2003). Control of enantioselectivity of lipase catalyzed esterification in supercritical carbon dioxide by tuning the pressure and temperature. *Tetrahedron Asymmetry*, *14*, 2087–2091.

[10] Nakaya, H., Miyawaki, O., Nakamura, K. (2001). Determination of log P for pressurized carbon dioxide and its characterization as a medium for enzyme reaction. *Enzyme Microb. Technol.*, *28*, 176–182.

[11] (a) Phillips, R. S. (1996). Temperature modulation of the stereochemistry of enzymatic catalysis: prospects for exploitation. *Trends Biotechnol.*, *14*, 13–16; (b) Sakai, T., Kawabata, I., Kishimoto, T., Ema, T., Utaka, M. (1997). Enhancement of the enantioselectivity in lipase-catalyzed kinetic resolutions of 3-phenyl-2H-azirine-2-methanol by lowering the temperature to –40°C. *J. Org. Chem.*, *62*, 4906–4907; (c) Sakai, T., Kishimoto, T., Tanaka, Y., Ema, T., Utaka, M. (1998). Low-temperature method for a dramatic improvement in enantioselectivity in lipase-catalyzed reactions. *Tetrahedron Lett.*, *39*, 7881–7884.

[12] (a) Matsuda, T., Harada, T., Nakamura, K. (2000). Alcohol dehydrogenase is active in supercritical carbon dioxide. *J. Chem. Soc. Chem. Commun.*, 1367–1368; (b) Matsuda, T., Watanabe, K., Kamitanaka, T., Harada, T., Nakamura, K. (2003). Biocatalytic reduction of ketones by a semi-continuous flow process using supercritical carbon dioxide. *J. Chem. Soc. Chem. Commun.*, *10*, 1198–1199; (c) Matsuda, T., Marukado, R., Mukouyama, M., Harada, T., Nakamura, K. (2008). Asymmetric reduction of ketones by *Geotrichum candidum*: immobilization and application to reactions using supercritical carbon dioxide. *Tetrahedron Asymmetry*, *19*, 2272–2275.

[13] Shibahara, F., Nozaki, K., Hiyama, T. (2003). Solvent-free asymmetric olefin hydroformylation catalyzed by highly cross-linked polystyrene-supported (*R*,*S*)-BINAPHOS-Rh(I) Complex. *J. Am. Chem. Soc.*, *125*, 8555–8560.

[14] (a) Burgemeister, K., Franciò, G., Hugl, H., Leitner, W. (2005). Enantioselective hydrogenation of polar substrates in inverted supercritical CO_2/aqueous biphasic media. *Chem. Commun.*, 6026–6028; (b) Burgemeister, K., Franciò, G., Gego, V. H., Greiner, L., Hugl, H., Leitner, W. (2007). Inverted supercritical carbon dioxide/aqueous biphasic media for rhodium-catalyzed hydrogenation reactions. *Chemistry*, *13*, 2798–2804.

[15] (a) Kainz, S., Brinkmann, A., Leitner, W., Pfaltz, A. (1999). Iridium-catalyzed enantioselective hydrogenation of imines in supercritical carbon dioxide. *J. Am. Chem. Soc.*, *121*, 6421–6429; (b) Solinas, M., Pfaltz, A., Cozzi, P. G., Leitner, W. (2004). Enantioselective hydrogenation of imines in ionic liquid/carbon dioxide media. *J. Am. Chem. Soc.*, *126*, 16142–16147.

[16] Brown, R. A., Pollet, P., McKoon, E., Eckert, C. A., Liotta, C. L., Jessop, P. G. (2001). Asymmetric hydrogenation and catalyst recycling using ionic liquid and supercritical carbon dioxide. *J. Am. Chem. Soc.*, *123*, 1254–1255.

[17] Schneider, M. S., Urakawa, A., Grunwaldt, J.-D., Burgi, T., Baiker, A. (2004). Identification of catalyst surface species during asymmetric platinum-catalysed hydrogenation in a "supercritical" solvent. *Chem. Commun.*, 744–745.

[18] (a) Nishiyama, Y., Kaneda, M., Saito, R., Mori, T., Wada, T., Inoue, Y. (2004). Enantiodifferentiating photoaddition of alcohols to 1,1-diphenylpropene in supercritical carbon dioxide: sudden jump of optical yield at the critical density. *J. Am. Chem. Soc.*, *126*, 6568–6569; (b) Nishiyama, Y., Wada, T., Mori, T., Inoue, Y. (2007). Critical control by temperature and pressure of enantiodifferentiating anti-Markovnikov photoaddition of methanol to diphenylpropene in near critical and supercritical carbon dioxide. *Chem. Lett.*, *36*, 1488–1489.

CHAPTER 10

MICROWAVE-ASSISTED TRANSITION METAL-CATALYZED ASYMMETRIC SYNTHESIS

LUKE R. ODELL AND MATS LARHED

10.1. Introduction 391
  10.1.1. Microwave heating 391
  10.1.2. Microwave-assisted transition metal-catalyzed asymmetric reactions 394
10.2. Allylic substitution reactions 395
  10.2.1. Palladium catalysis 395
  10.2.2. Molybdenum catalysis 398
10.3. The Heck reaction 399
10.4. Asymmetric reductions (or saturations) 401
10.5. Rhodium-catalyzed conjugate addition reactions 404
10.6. The Suzuki–Miyaura reaction 405
10.7. Epoxide opening 406
10.8. Miscellaneous transformations 407
10.9. Conclusion 407
Acknowledgment 408
References 408

10.1. INTRODUCTION

10.1.1. Microwave Heating

In principle, all metal-catalyzed organic transformations require an input of energy to proceed at a useful rate. In the correct amount and in the right form,

Catalytic Methods in Asymmetric Synthesis: Advanced Materials, Techniques, and Applications, First Edition. Edited by Michelangelo Gruttadauria and Francesco Giacalone.

energy drives the catalytic cycle forward, providing the desired product with high selectivity and in high yield. In the wrong amount or in the wrong form, energy can instead fully inhibit the catalytic process [1, 2]. Thus, the regulation of the energy input and the choice of the energy source are of vital importance in homogeneous catalysis. After the invention of the Bunsen burner in the middle of the nineteenth century, this heating device was used as a standard tool for rapid energy transfer in the daily work of chemists during the twentieth century. The heating techniques for organic reactions have not changed much in the last century, during which oil baths and heating mantles have dominated the organic chemistry landscape. Among the available alternatives, microwave irradiation has a huge potential to provide controlled thermal energy directly to the molecules of interest [2].

Microwaves are able to heat a reaction mixture by two general mechanisms, dipolar polarization and ionic conductance [3–5]. All noncrystalline solids and liquids that contain dipoles and/or charged species can absorb microwave energy and convert it into heat. This is because dipoles and ions are constantly trying to align themselves to the electric component of the oscillating electromagnetic vector of the microwave field, resulting in the rotation of molecules and the oscillation of ions. Hence, the microwave energy absorbed in this process is first converted to kinetic energy, which is then transferred into thermal energy through molecular friction (microwave in situ heating) [3–5]. To accomplish efficient heating, it is important that the frequency of the applied microwave radiation is within certain limits. If the frequency is too low, the dipoles will have time to align with the electric field and smoothly follow the field fluctuations, resulting in poor heating. If instead it is too high, the dipoles will not have time to realign themselves to the quickly alternating field, which means no molecular motion is created and therefore no heat is generated [2]. The frequency used by both microwave synthesizers and domestic microwave ovens (2450 MHz, $\lambda = 12.2$ cm) is located between these extremes where dipoles have time to partly realign with the oscillating electric field but are not quite able to follow the field fluctuations. The end result is effective heating [5].

Since the mid-1990s, an increasing number of research reports have indeed demonstrated that controlled microwave heating can be used to carry out almost all types of organic transformations that require thermal energy [6], including reactions catalyzed by homogeneous metal complexes [7]. Furthermore, with microwave in-situ heating, the thermal energy does not need to be transferred via the vessel wall, affording high temperatures quickly. Thus, this remote and volumetric heating method opens new ways to achieve fast and selective synthesis, which is especially important in the fine-tuning and optimization of sensitive transition metal-catalyzed reactions [7].

In the beginning of the twenty-first century, dedicated microwave synthesizers for controlled small-scale synthesis in sealed vessels became commercially available [5]. These "single-mode"[8] instruments enabled easy handling, programming, and documentation of time, temperature, and irradiation power during the reactions, while at the same time providing high microwave density, rapid heating, direct reaction control, good reproducibility, and safe operation. In addition, by using 0.2–20 mL septum-sealed purpose-made closed microwave-transparent vessels, which in general could accommodate at least 20 bar of pressure, it was possible to maintain reaction solutions at temperatures higher than their conventional reflux temperature. Together with high heating rates and efficient air jet cooling systems, this led to a dramatic reduction in the overall process times. Sealed vessels also permit the use of low-boiling solvents, allowing easy workup and reduction of solvent waste. Simultaneously, the multimode microwave reactor was improved, affording a possibility to safely conduct parallel synthesis and to increase the reaction scale up to liter volumes. Most likely, improved equipment for microwave-induced flow synthesis will be highly useful for further scaleup applications.

An additional environmentally friendly factor is the reduced energy consumption compared to the classical heating of lab-scale reactions [9, 10]. It is, however, not clear if pilot-scale or production-scale microwave processing share this advantage [11].

It is today fully established that early claims of the existence of "nonthermal microwave effects" should be reinterpreted as the effects of superheating [12]. For example, Raner and Strauss [13] and Kappe et al. [14] have conducted rigorous kinetic studies and demonstrated that reaction rates were the same regardless of heating method. An accurate comparison of microwave heating with traditional heating is possible only if the correct reaction temperature is known, and if the temperature is not uniform, then the exact temperature distribution has to be known. Importantly, Kaiser [15] and Kappe et al. [14] have shown that high microwave power and poor stirring furnish uneven heating, since the average rate of thermal diffusion of energy becomes less than the rate of absorption in a well-defined area of the reaction mixture.

It is somewhat difficult to measure temperatures during microwave irradiation, as ordinary thermometers relying on liquid expansion (mercury, ethanol) or standard thermocouples interact with the electromagnetic field. Correct temperature measurements can be achieved either by means of an immersed microwave-compatible temperature detector (e.g., a fiber optic sensor, a gas balloon, or a *p*-xylene thermometer) or on the outer surface of the reaction vessel by means of an infrared (IR) pyrometer [6]. As early as 1994, Whittaker and Mingos used a *p*-xylene thermometer, a fiber optic sensor, and an IR pyrometer for measuring the temperature of superheated boiling solvents during microwave irradiation [16]. The difference between the values found was at worst 24°C tetrahydrofuran (THF), which according to the Arrhenius equation, corresponds to a six-fold ambiguity in reaction rate (Ea = 84 kJ/mol) [17]. Thus, the fact that only local temperatures can be measured by these methods, unless a number of immersed probes are used simultaneously, might result in large rate estimation errors if the temperature distribution is not homogeneous [14].

10.1.2. Microwave-Assisted Transition Metal-Catalyzed Asymmetric Reactions

It is convenient to divide metal-catalyzed stereoselective transformations into two broad categories: reactions involving asymmetric catalysts producing enantioselective reactions, and those performed with symmetric catalysts and stereopure starting materials. The latter classification is outside the scope of this account and will not be discussed further.

Screening methods are finding increased use in the identification of efficient chiral catalysts for metal-catalyzed asymmetric synthesis [18]. To allow rapid screening of new ligands and optimization of reaction conditions, short process times are desirable. Modern single-mode microwave reactors have excellent potential for the integration of a high level of automation and rapid analysis. However, in most investigations on asymmetric catalysis, researchers strive to keep the temperature as low as possible while maintaining a sufficient reaction rate, since an increase in enantiomeric excess (ee) is often observed upon temperature decrease. High enantioselectivity is obtained for reactions having large differences in activation energy ($\Delta G^{\ddagger}$). Provided that the selectivity determining step remains the same, the higher the reaction temperature, the larger the difference in $\Delta G^{\ddagger}$ ($\Delta\Delta G^{\ddagger}$) that is required to achieve high enantioselectivity [1, 19]. Assuming the selectivity does not change during the reaction, plots of ee as a function of $\Delta\Delta G^{\ddagger}$ at 0, 100, and 200°C are depicted in Figure 10.1. As illustrated, for reactions with a difference in $\Delta\Delta G^{\ddagger}$ of 15 kJ/mol, the ee value is not affected to any large degree upon heating (given that the catalyst is not degraded).

Microwave heating has been used extensively in homogeneous catalysis since the first reports of microwave-assisted Heck, Suzuki, and Stille couplings by Larhed and Hallberg in 1996 [20]. The development of new methods for

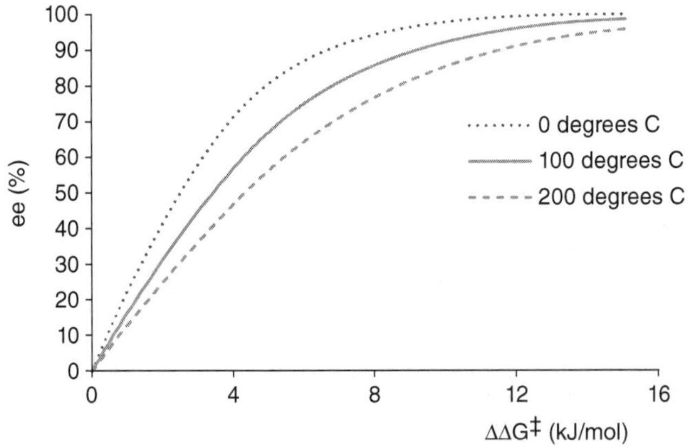

FIGURE 10.1. The ee as a function of the difference in activation energy for the formation of two enantiomers ($\Delta\Delta G^{\ddagger}$) [1].

the preparation of enantiopure building blocks, fine chemicals, and pharmaceuticals is of great importance. Nevertheless, the first report on the impact of microwave irradiation on an asymmetric transition metal-catalyzed reaction did not appear until 1999 [21]. Since then, the potential value of microwave technology for metal-catalyzed asymmetric reactions has been demonstrated in a number of small-scale studies. This short review will highlight some of the asymmetric transition metal-catalyzed reactions described to date with focus on small-scale applications carried out with dedicated single-mode reactors.

10.2. ALLYLIC SUBSTITUTION REACTIONS

10.2.1. Palladium Catalysis

The palladium(0)-catalyzed allylic substitution reactions, the Tsuji–Trost reaction, and in particular the asymmetric versions, have been extensively developed during the last 40 years [22, 23]. Interestingly, microwave heating has been found to promote these somewhat slow substitutions without any significant loss of the ee's. Thus, a highly enantioselective allylic substitution of *rac*-1,3-diphenyl-2-propenyl acetate was performed by Moberg, Hallberg, et al. with the thermostable palladium–phosphineoxazoline catalytic system, as shown in Scheme 10.1. This catalytic system produced higher ee's than, for example, the corresponding 2,2′-bis(diphenyl-phosphino)-1,1′binaphthyl (BINAP) system [19, 21]. Very high yields, ee's of 99% and up to 7000 turnovers/h could be obtained after 30 seconds of irradiation in an early single-mode applicator [24]. These substitutions were carried out in acetonitrile (b.p. 81–82°C) in sealed vessels with single-mode microwave heating up to 145°C, as determined with a fiber optic probe.

The standard reaction in Scheme 10.1 was also used by Braga et al. to test the potential of a number of cysteine-derived oxazolidine and thiazolidine ligands. Good yields and enantioselectivities of up to 72% ee were achieved after 2 minutes of microwave heating at 70°C [25]. By switching the ligand

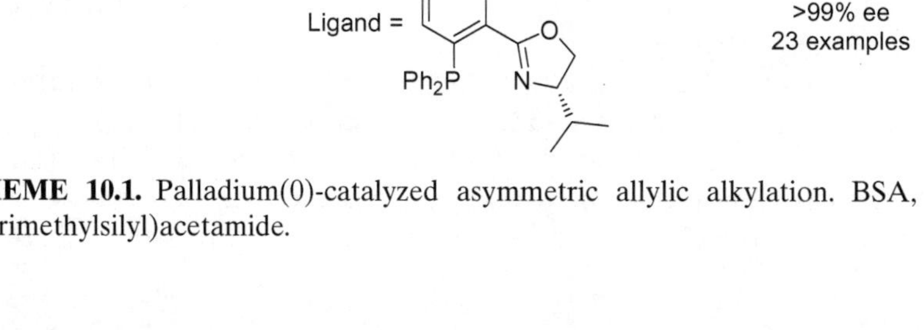

SCHEME 10.1. Palladium(0)-catalyzed asymmetric allylic alkylation. BSA, *N*,*O*-bis(trimethylsilyl)acetamide.

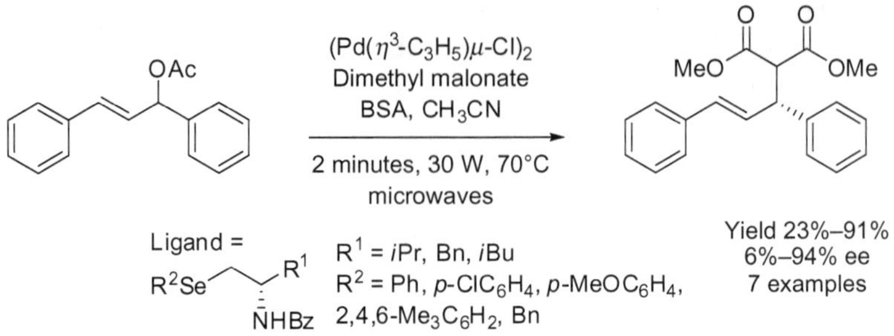

SCHEME 10.2. Palladium(0)-catalyzed asymmetric allylic alkylation using *β*-seleno amide ligands.

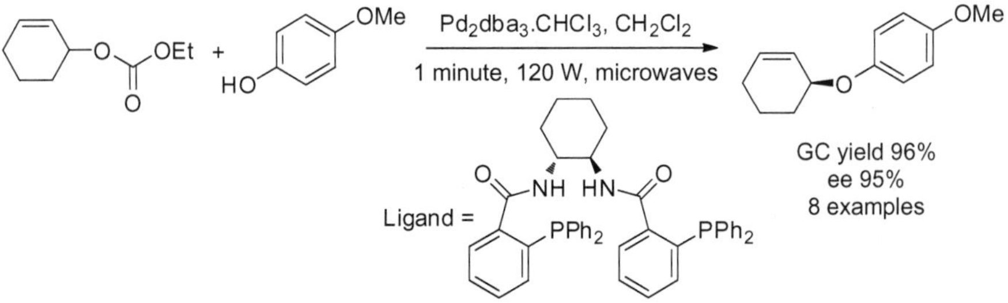

SCHEME 10.3. Asymmetric substitution of *rac*-ethyl-3-cyclohexenyl carbonate with *p*-methoxy phenol.

class to a set of different chiral *β*-seleno amides, improved enantioselectivities were obtained (Scheme 10.2) [26]. The conclusion by the authors was that the major advantage of high-density microwave radiation is the reduction in reaction time, from hours to minutes, enabling rapid feedback and efficient reaction optimization.

The palladium(0)-catalyzed substitution of the less reactive *rac*-ethyl 3-cyclohexenyl carbonate could, in a analogous fashion, be completed in 45 seconds to 1.5 minutes using a *P,P* ligand [19]. Dimethyl malonate, *p*-methoxy phenol, or phthalimide could be employed as the nucleophile furnishing satisfactory enantiomeric purities (94%–95% ee) and good yields (75%–96%). As an illustrative example, the outcome with the *O*-centered nucleophile is presented in Scheme 10.3. These microwave-assisted substitutions delivered similar yields and ee values to those conducted in sealed vials using classical oil-bath heating.

The use of microwave heating to quickly reach high temperatures and promote sluggish reactions is an attractive concept. Vidal-Ferran et al. investigated this approach to accelerate substitutions of hindered allylic substrates [27]. Microwave-assisted allylic alkylation turned out to be highly effective for this class of triphenyl-substituted substrates using a phosphinooxazoline

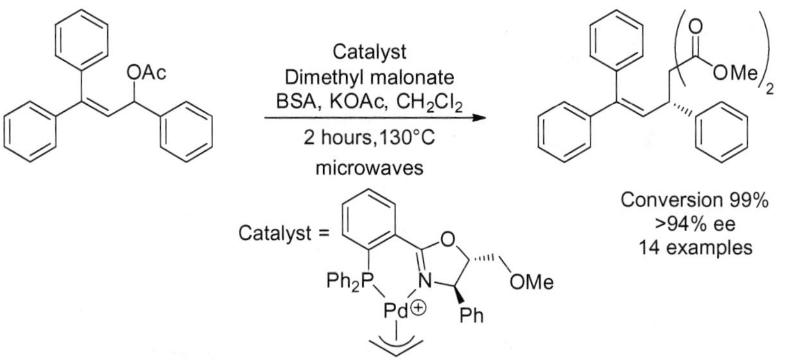

SCHEME 10.4. Palladium(0)-catalyzed asymmetric allylic alkylation of sterically hindered *rac*-1,3,3-triphenyl-2-propenyl acetate.

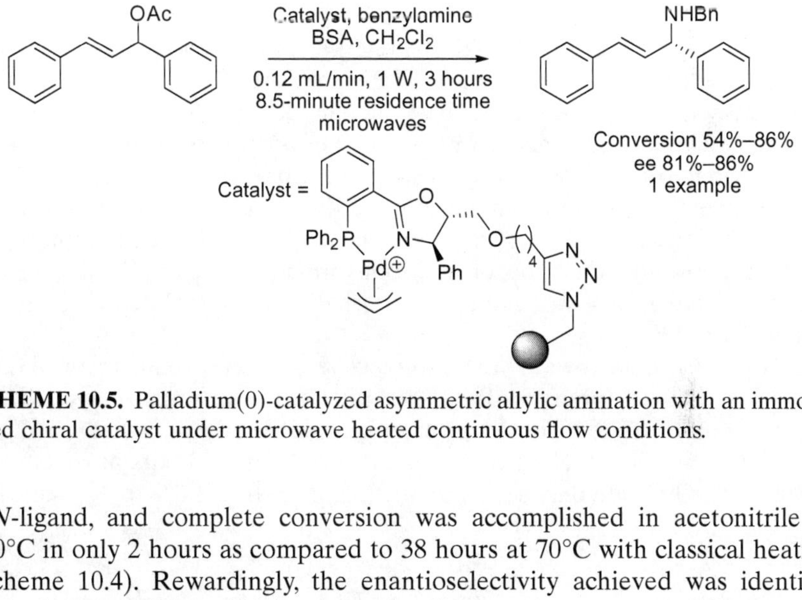

SCHEME 10.5. Palladium(0)-catalyzed asymmetric allylic amination with an immobilized chiral catalyst under microwave heated continuous flow conditions.

P,N-ligand, and complete conversion was accomplished in acetonitrile at 130°C in only 2 hours as compared to 38 hours at 70°C with classical heating (Scheme 10.4). Rewardingly, the enantioselectivity achieved was identical under both conditions (94% ee) despite the large temperature difference.

In order to investigate a continuous-flow system for asymmetric allylic aminations, a phosphinooxazoline ligand was anchored onto a cross-linked azidomethyl polystyrene resin via "click" chemistry. Palladium complexation was performed and the polymer-supported catalyst was transferred to a vertical, 1/4-inch Teflon tube and positioned in the microwave cavity. Despite using a very low and somewhat uncertain microwave power (1W), and with no temperature measurements of the reaction mixture in the microwave-heated tube reactor (temperature was only monitored outside the heated zone), up to 86% conversion and 86% ee was realized using a 0.12 mL/min flow rate for 3 hours (Scheme 10.5) [28]. This work represents, to the best of our knowledge, the first example of an enantioselective metal-catalyzed microwave-assisted continuous-flow process.

10.2.2. Molybdenum Catalysis

Molybdenum(0) catalysis in allylic alkylations give rise to regioselectivity favoring the most branched product (internal substitution), in contrast to allylic palladium(0) catalysis, which yield the terminal product [29]. The molybdenum(0)-catalyzed protocols also suffer from significantly lower reactivity. It was discovered by Trost and Hachiya that by using an unstable and sensitive catalyst such as $(EtCN)_3Mo(CO)_3$ together with a chiral bipyridine ligand, the low reactivity could be circumvented [30]. Instead of directly employing this procedure under microwave conditions, a noninert and operationally simple one-pot microwave protocol was developed, using inexpensive and stable $Mo(CO)_6$ as the metal source [31]. Despite the high reaction temperatures—up to 250°C with THF as solvent (b.p. 65–67°C)—high enantiopurities were achieved although the regioselectivity was reduced from 32:1 to 19:1 (internal/terminal alkylation) (Scheme 10.6) [31].

Recently, a series of new chiral bispyridylamido ligands were prepared and investigated by Moberg and co-workers. The 4-methoxypyridine derivative afforded an impressive 88% isolated yield of the branched product with >99% ee and an internal/terminal regioisomeric distribution of 41:1 after only 4 minutes of microwave irradiation (Scheme 10.7) [32]. Further optimization resulted in the identification of a carbohydrate-based pyridylamide ligand, which afforded a somewhat improved internal/terminal regioselectivity (49:1) and 99% ee [33].

In work directed toward a stereoselective synthetic route to the HIV-1 protease inhibitor Tipranavir, Trost and Andersen used a microwave-assisted molybdenum-catalyzed route to prepare one of the key intermediates (Scheme 10.8). The reaction time was reduced from 24 hours at 67°C to 20 minutes at 180°C with only a small reduction in ee (from 96% to 94% ee) [34].

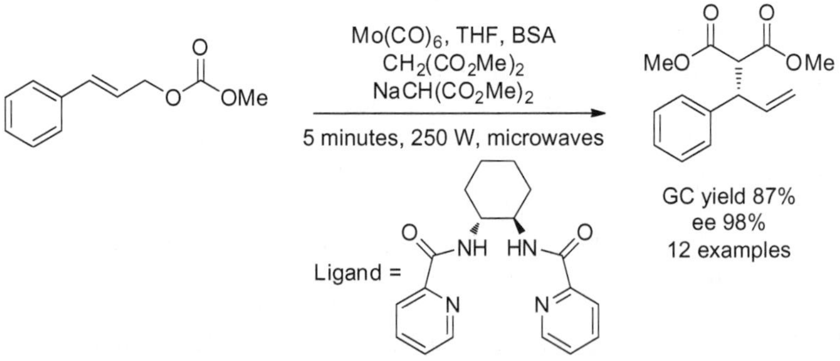

SCHEME 10.6. Internal asymmetric molybdenum(0)-catalyzed allylic alkylation.

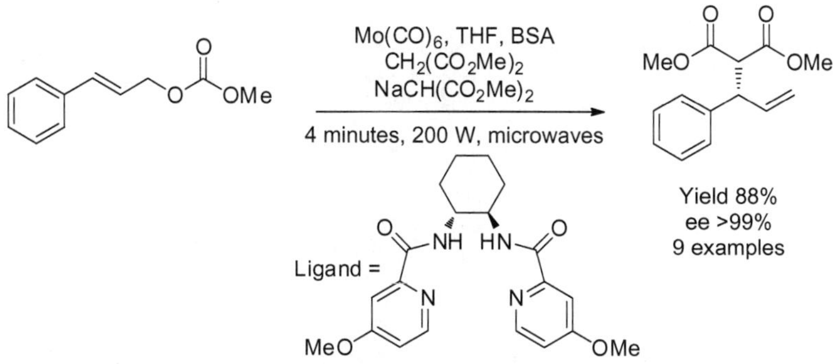

SCHEME 10.7. Internal asymmetric molybdenum-catalyzed allylic alkylation with an optimized ligand.

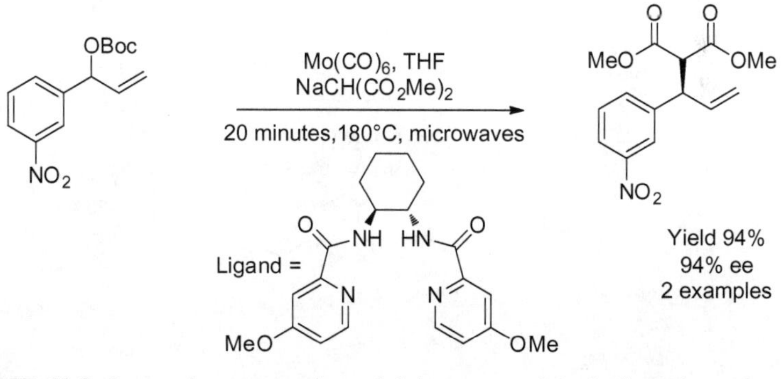

SCHEME 10.8. Internal asymmetric molybdenum-catalyzed alkylation of an allylic carbamate.

In addition, more convenient $Mo(CO)_6$ could be used as the molybdenum source instead of the sensitive catalyst complex $Mo(CO)_3(C_7H_8)$.

Finally, a resin-supported pyridylamide ligand was synthesized and evaluated (Scheme 10.9) by Moberg and co-workers. The polymer-attached ligand furnished the substitution product in 97% ee but in only 82% isolated yield, probably a result of the heterogeneous nature of the catalyst. The resin-supported ligand could be reused at least seven times with no significant change in the reaction outcome [35].

10.3. THE HECK REACTION

In the last 35 years, the palladium(0)-catalyzed vinylic substitution reaction, generally known as the Heck (or Heck–Mizoroki) reaction, has been extensively explored and used for the enantioselective preparation of natural

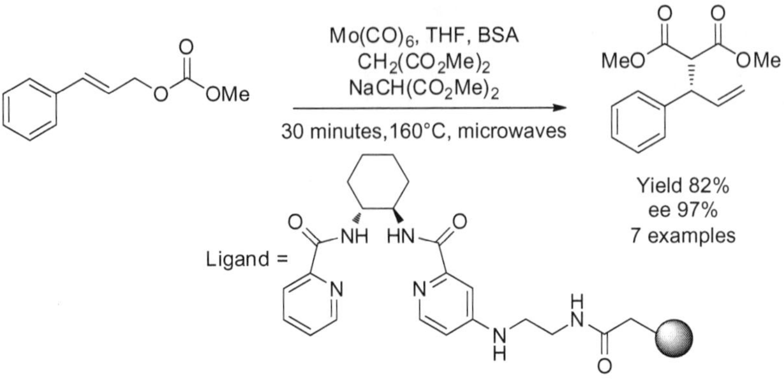

SCHEME 10.9. Asymmetric molybdenum-catalyzed allylic alkylation with a resin-supported ligand.

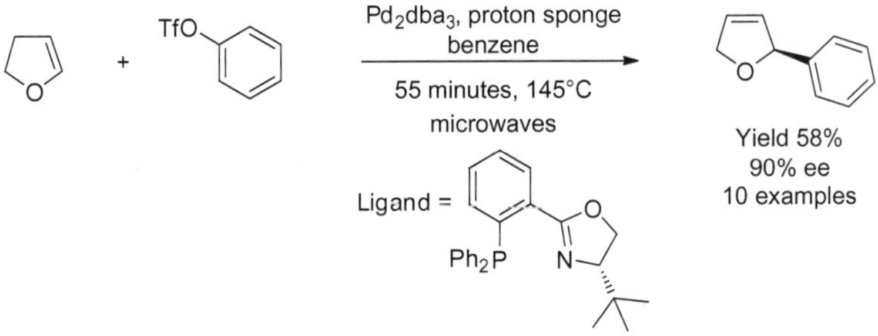

SCHEME 10.10. Asymmetric Heck arylation of prochiral 2,3-dihydrofuran with phenyl triflate and a phosphineoxazoline ligand.

products and pharmaceuticals, because of the mildness and high selectivity of the procedure. The Heck arylation presented in Scheme 10.10 was reported in 2002 and constitutes the first example of an asymmetric, microwave-promoted, intermolecular Heck reaction [36]. The power of the microwave methodology was manifested by the reduction of reaction time from 96 hours at 70°C to 55 minutes at 145°C. However, the increase in reaction rate was also coupled with a slight reduction in ee (from 97% to 90% ee).

Since then, a number of new ligands for asymmetric intermolecular Heck reactions have been developed, affording impressive results with high ee's and short reaction times [37]. In 2007, Andersson et al. found that a library of phosphine-thiazole ligands furnished excellent results in microwave-assisted arylation of 2,3-dihydrofuran with various organic triflates. The outcome using one of the best ligands is presented in Scheme 10.11 [38].

Pàmies et al. used a class of highly modular phosphiteoxazoline ligands to obtain full conversions in short reaction times with ee above 99%, employing not only the standard olefin 2,3-dihydrofuran but also the more sluggish

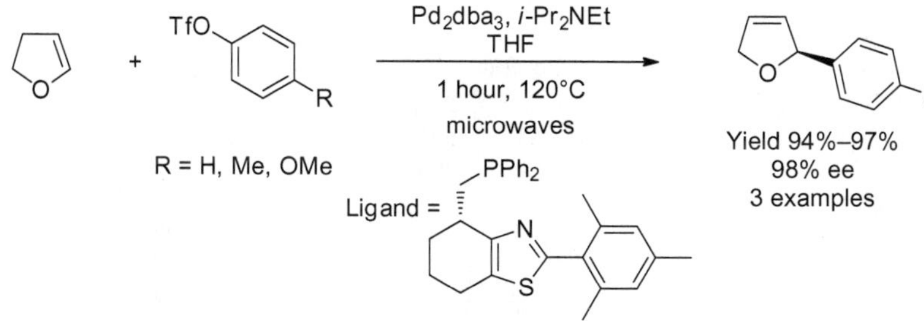

SCHEME 10.11. Asymmetric Heck arylation of prochiral 2,3-dihydrofuran with aryl triflates and a phosphine-thiazole ligand.

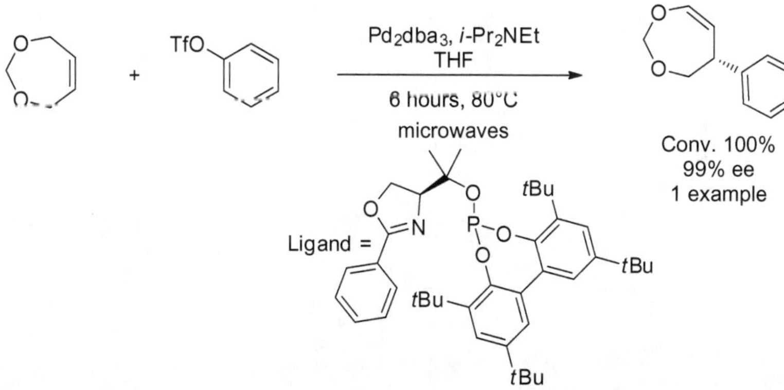

SCHEME 10.12. Asymmetric Heck arylation of prochiral 4,7-dihydro-1,3-dioxepin with phenyl triflate and a phosphiteoxazoline ligand.

olefins *N*-carbomethoxy-2,3-dihydropyrrole, cyclopentane, and 4,7-dihydro-1,3-dioxepin. In Scheme 10.12, we have depicted an impressive example with phenyl triflate and the seven-membered olefin, 4,7-dihydro-1,3-dioxepin [39].

Two examples of intramolecular microwave-assisted asymmetric intramolecular Heck cyclizations have been carried out by Overman et al. as part of the total synthesis of the *Strychnos* alkaloid (+)-Minfiensine. In order to increase the rate of the Heck cyclizations of the highly funtionalized triflate from 70 hours to 30–45 minutes, the reaction temperature was increased from 100 to 170°C using a single-mode microwave applicator (Scheme 10.13) [40]. Impressively, no erosion in yield or enantiopurity of the dihydrocarbazole product was detected (75%–87% yield, 99% ee).

10.4. ASYMMETRIC REDUCTIONS (OR SATURATIONS)

Synthetic protocols for the asymmetric saturation of sp^2-hybridized centers are highly valued by organic chemists and can be used to obtain optically

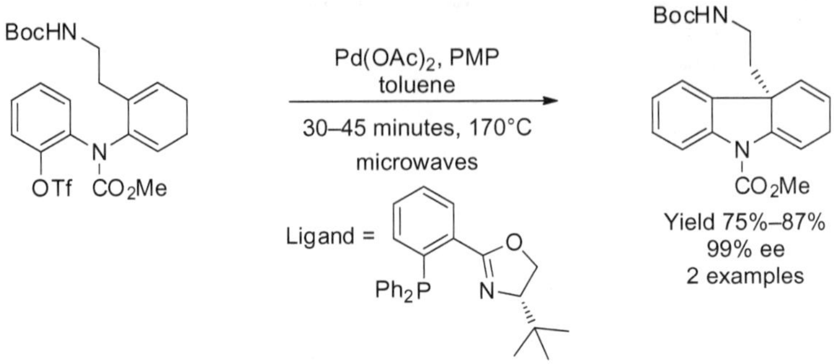

SCHEME 10.13. Intramolecular asymmetric Heck arylation of an aryl triflate using a phosphineoxazoline ligand.

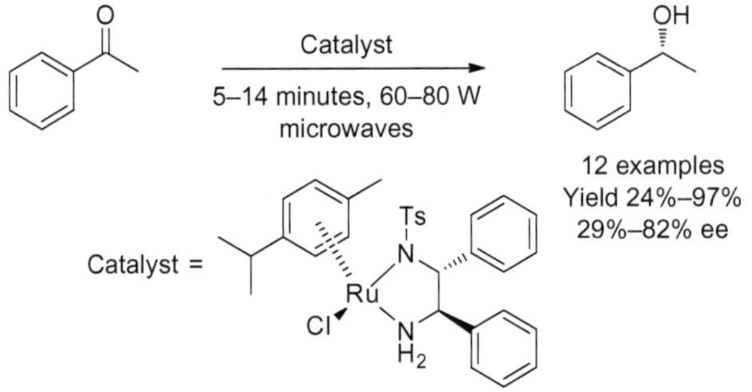

SCHEME 10.14. Ruthenium-catalyzed asymmetric reduction of acetophenone.

active alkanes, alcohols, and amines from simple prochiral precursors. In 2001, Lutsenko and Moberg reported the first microwave-assisted enantioselective reduction, via a ruthenium-catalyzed hydrogen transfer reaction (Scheme 10.14) [41]. In this study, Noyori's monotosylated (*R*,*R*)-diphenylethylenedaimine [42] was employed as the chiral ligand and propan-2-ol as the hydrogen source. Under standard conditions, the reaction takes 15 hours to reach completion, affording (*R*)-1-phenylethanol in 95% yield and 97% ee. Utilizing microwave heating, the authors could produce the same product in 90% yield and 82% ee after only 9 minutes of irradiation at 60 W. In addition, the sterically hindered substrate *t*-butylphenylketone was also efficiently reduced under these conditions (95% yield); however, the ee was poor (21%).

In a similar study, Jin and co-workers reported the ruthenium-catalyzed asymmetric reduction of various ketones using an SBA-15-supported ligand derived from (1*R*,2*R*)-diaminocyclohexane [43]. As depicted in Scheme 10.15, excellent conversions (>90%) and good enantioselectivities (61%–92%) were obtained using this catalytic system after only 20–40 minutes of microwave irradiation.

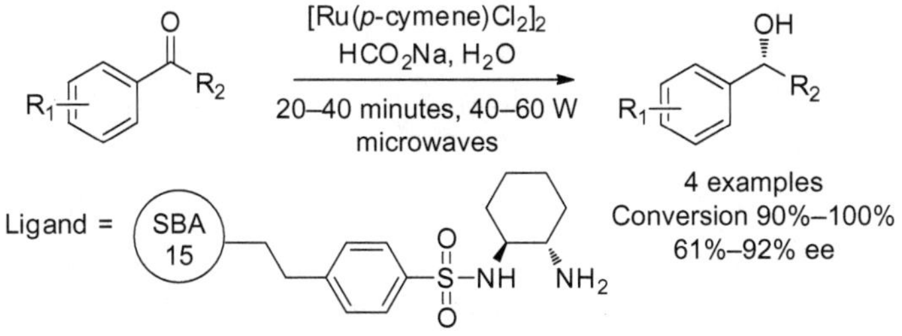

SCHEME 10.15. Ruthenium-catalyzed, solid-supported asymmetric ketone reduction.

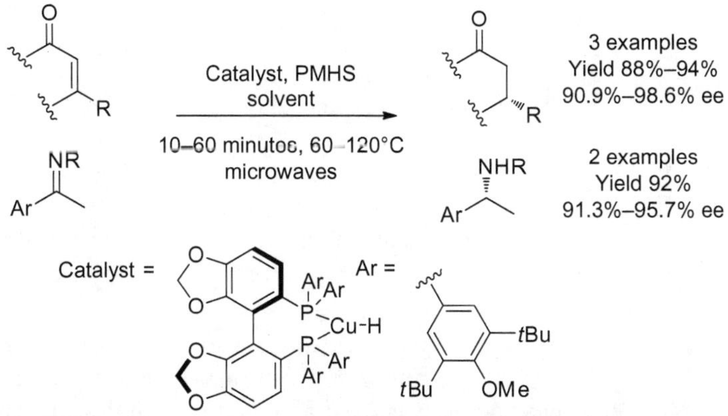

SCHEME 10.16. Asymmetric hydrosilylations with "copper hydride in a bottle." PMHS, polymethylhydrosiloxane.

Lipshutz and co-workers have described the asymmetric hydrolsilylation of various prochiral substrates using ligated copper hydride (CuH) [44, 45]. In particular, "CuH in a bottle" (see Scheme 10.16) derived from $Cu(OAc)_2{\cdot}H_2O$ was found to be highly effective and, more importantly, could be conveniently stored for weeks without compromising enantioselectivity (ee decreased from 99% to 94% over 2 months). Excellent enantioselectivities and yields were obtained for a variety of hydrolsilylations and, by using microwave heating, reaction times could be reduced from hours to minutes without a significant erosion in ee values.

Gustafsson et al. have employed a microwave-assisted asymmetric hydrogenation in the synthesis of a *C*-glycoside analogue of *β*-D-galactosyl hydroxylysine (Scheme 10.17) [46]. Exceptional selectivity (>99% de) and yields (95%) were obtained using Burk's catalyst [47], 2-propanol as solvent, and microwave heating (70°C for 40 minutes). Interestingly, increasing the reaction temperature by 10°C resulted in a threefold loss in selectivity, highlighting the difficulty in obtaining both fast reaction times and high stereoselectivities at elevated temperatures.

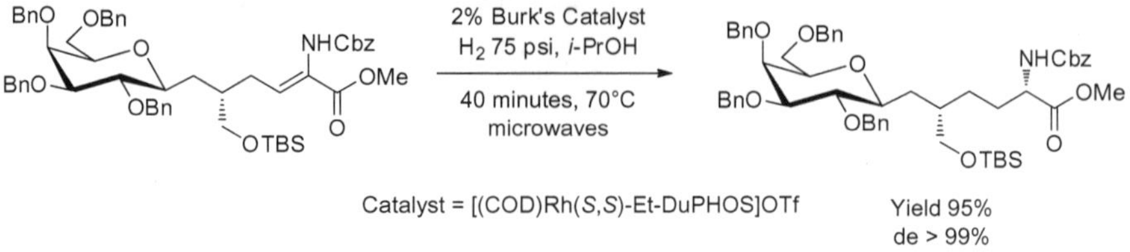

SCHEME 10.17. Asymmetric hydrogenation using Burk's catalyst. Cbz, carbobenzyloxy; TBS, *t*-butyldimethylsilyl.

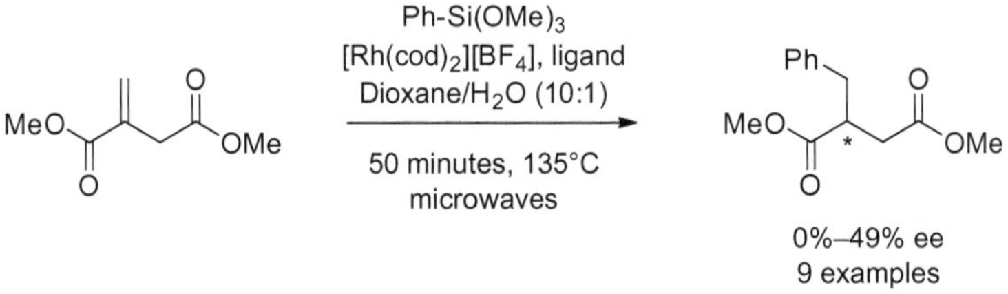

SCHEME 10.18. Rhodium-catalyzed enantioselective addition of phenyltimethoxysilane.

Recently, Toukoniitty and co-workers have studied the enantioselective hydrogenation of ethyl pyruvate under conventional and microwave heating conditions [48, 49]. Cinchondine was used as the chiral catalyst modifier and the reaction kinetics and selectivity were determined using ethanol and toluene as solvents. In the case of microwave transparent toluene, no significant differences (>75% ee) were observed between the two heating sources. However, when ethanol was used, the observed ee decreased from 60% to 40%. This was probably due to local superheating of ethanol in the microwave cavity or the higher bulk temperature of the ethanol reactions (36 vs. 23°C), leading to desorption of the catalyst modifier.

10.5. RHODIUM-CATALYZED CONJUGATE ADDITION REACTIONS

The rhodium-catalyzed conjugate addition of organometallic reagents to α,β-unsaturated carbonyl derivatives is a powerful method for the stereoselective construction of C–C bonds [50, 51]. In 2006, Frost and co-workers reported their preliminary investigations into the enantioselective addition of organotrialkoxysilanes to α-substituted acrylic esters (Scheme 10.18) [52]. The reaction proceeded smoothly in the absence of a ligand; however, upon the addition of (*R*)-BINAP, a sharp decrease in reactivity (70% conversion) was observed along with a disappointingly low ee (24%). Attempts to optimize the reaction (ligand screen, alterative proton, and fluoride sources) resulted in only a slight increase in enantioselectivity (49% ee) and thus, the authors chose not to pursue the transformation any further.

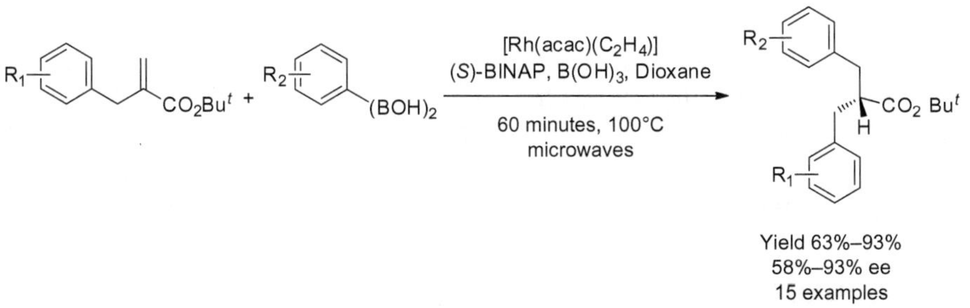

SCHEME 10.19. Rhodium-catalyzed enantioselective synthesis of α,α'-dibenzyl esters.

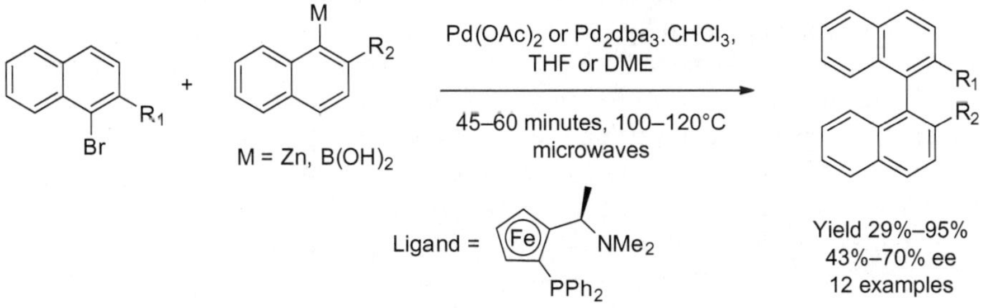

SCHEME 10.20. Asymmetric Suzuki–Miyaura and Negishi cross-couplings.

One year later, the same group reported a related protocol for the preparation of α,α'-dibenzyl esters from α-benzyl acrylates and boronic acids (Scheme 10.19) [53]. (*S*)-BINAP and boric acid were found to be the best chiral ligand/proton source combination for this tandem conjugate addition–enantioselective protonation process. A diverse range of α,α'-dibenzyl esters were obtained in good yields and with good levels of stereoselectivity.

10.6. THE SUZUKI–MIYAURA REACTION

The transition metal-catalyzed cross-coupling of an organoboron compound and an alkyl, vinyl or aryl halide or halide surrogate, known as the Suzuki–Miyaura reaction, is one of the most versatile and frequently utilized methods for C–C bond formation [54]. In 2007, Genov and co-workers disclosed the first microwave-assisted asymmetric Suzuki–Miyaura and Negishi cross-coupling reactions using palladium catalysis [$Pd(OAc)_2$ or $Pd_2dba_3 \cdot CHCl_3$] and the ferrocenyl monophosphane derivative (R,S_p)-[1-(2-diphenylphosphinoferrocenyl)ethyl]dimethylamine as the chiral source (Scheme 10.20) [55]. Microwave irradiation was found to afford equal or better yields and reduce the reaction times from several days to 1 hour when compared to the corresponding classical heating reactions. However, the enantioselectivity was somewhat lower under microwave irradiation, presumably due to the higher reaction temperatures (65 vs. 120°C). This study is an excellent example of the trade-off between higher temperatures/shorter reaction times/lower ee's and lower temperatures/longer reaction times/higher ee's, which is a common theme in the majority of investigations involving asymmetric catalysis under microwave irradiation.

10.7. EPOXIDE OPENING

The asymmetric ring opening of epoxides, with nitrogen nucleophiles, is a powerful method for the preparation of chiral β-amino alcohols. In 2005, Dioos and Jacobs reported the first microwave-assisted asymmetric epoxide ring opening catalyzed by a commercially available chiral (salen)Cr(III) complex (Scheme 10.21) [56]. Microwave irradiation reduced the reaction time from 18 hours to 30 seconds when compared to the corresponding room temperature reaction, without significantly comprising the enantioselectivity (84% vs. 78%) [57].

One year later, Kureshy et al. reported the asymmetric ring opening of *meso*-epoxides with various anilines, catalyzed by a Ti-(*S*)-BINOL (Ti-(*S*)-1,1′-

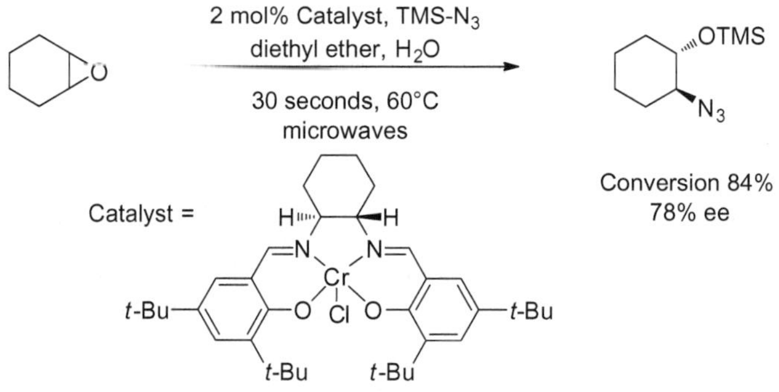

SCHEME 10.21. (Salen)Chromium(III)-catalyzed asymmetric ring opening of cyclohexene oxide with TMS-N_3.

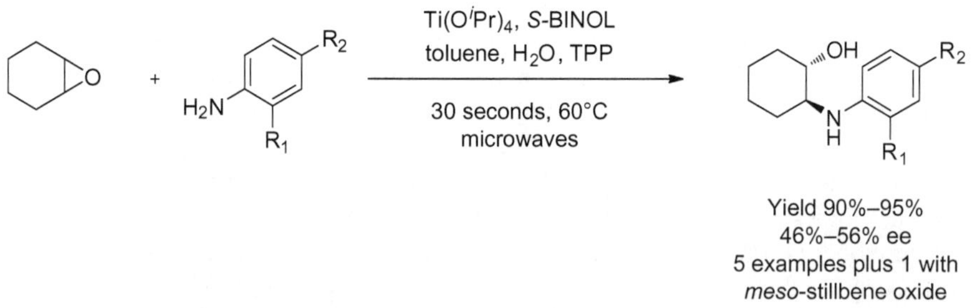

SCHEME 10.22. Titanium-(*S*)-BINOL catalyzed asymmetric ring opening of *meso*-epoxides with anilines.

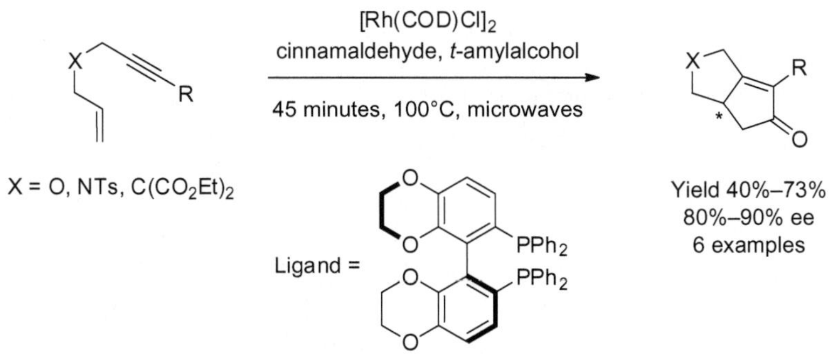

SCHEME 10.23. Rhodium-catalyzed asymmetric Pauson–Khand-type cyclizations.

bi-2-naphthol) complex (Scheme 10.22) [58]. Excellent yields (90%–95%) and reasonable enantioselectivities (46%–56% ee) were obtained after only 30 seconds of microwave irradiation at 60°C. The microwave-assisted reaction was 10 times and 235 times faster than classical oil-bath heating and room temperature reactions, respectively, with no significant decrease in ee values. Interestingly, a sharp decrease in enantioselectivity was observed when the reaction temperature was increased to 70°C, suggesting decomposition of the Ti-(*S*)-BINOL complex at this temperature.

10.8. MISCELLANEOUS TRANSFORMATIONS

In 2008, Kwong and co-workers disclosed a novel rhodium-catalyzed asymmetric Pauson–Khand-type cyclization (Scheme 10.23) [59]. Using a chiral Rh-(*S*)-bisbenzodioxanPhos complex, a range of O-, N-, and C-tethered enynes were transformed into the corresponding cyclopentenones, in good yields (40%–73%) and enantioselectivities (80%–90% ee). Lower enantioselectivities were observed when the reactions were performed under conventional heating, presumably due to complex decomposition over the prolonged reaction time (36 hours).

10.9. CONCLUSION

In this chapter we have presented and discussed some of the applications of microwave heating in asymmetric transition metal-catalyzed reactions, highlighting the enantioselective outcome, the increased reaction kinetics, the high reaction control, and the convenient handling. More and more chemists are becoming aware of the versatility of this energy source, and it is already clear that modern microwave synthesizers have much to offer the synthetic chemist. It is also likely that the widespread acceptance of this technique will be important for the screening of new asymmetric ligands, catalysts, and reaction systems, enabling fast optimization of new catalytic methods. In the opinion of the authors, controlled microwave heating will prove its main value at the reaction development stage, while a lower reaction temperature will probably be advantageous in the synthetic target application or production process since, in most cases, the stereoselectivity will decrease with higher temperature. Nevertheless, the development of thermostable chiral catalysts providing high enantioselectivities at high reaction temperatures remains an important research goal. The possibilities to simplify the preparative procedure by in situ generation of the active catalyst under microwave radiation should not be overlooked (i.e., molybdenum-catalyzed allylic substitutions).

A complementary and much desired future area of microwave chemistry will lie in the development of purpose-built commercially available flow equipment, allowing direct upscaling of pressurized "single-mode" small-scale reactions into kilogram-scale processes. Finally, we hope that despite not presenting a fully comprehensive review, we have provided some insight into the current state of the art of asymmetric transition metal-catalyzed microwave chemistry. We are convinced that many more applications will be developed in the near future.

ACKNOWLEDGMENTS

We thank Askan Fardost, Patrik Nordeman, and Dr. Malte Behrends for fruitful discussions and critical review of this chapter. We also thank Dr. Ulf Bremberg, Professor Christina Moberg, and Professor Anders Hallberg.

REFERENCES

[1] Larhed, M., Moberg, C., Hallberg, A. (2002). Microwave-accelerated homogeneous catalysis in organic chemistry. *Acc. Chem. Res.*, *35*, 717–727.

[2] Michael, D., Mingos, P. (2005). *Microwave Assisted Organic Synthesis*. Blackwell, Oxford, pp. 1–22.

[3] Strauss, C., Trainor, R. (1995). Developments in microwave-assisted organic chemistry. *Aust. J. Chem.*, *48*, 1665–1692.

[4] Lidstrom, P., Tierney, J., Wathey, B., Westman, J. (2001). Microwave assisted organic synthesis: a review. *Tetrahedron*, *57*, 9225–9283.

[5] Kappe, C. O. (2004). Controlled microwave heating in modern organic synthesis. *Angew. Chem. Int. Ed.*, *43*, 6250–6284.

[6] Kappe, C. O., Stadler, A. (Eds.) (2005). *Microwaves in Organic and Medicinal Chemistry*. Wiley-VCH, Weinheim, Germany.

[7] Nilsson, P., Olofsson, K., Larhed, M. (2006). Microwave-assisted and metal-catalyzed coupling reactions. *Top. Curr. Chem.*, *266*, 103–144.

[8] Robinson, J., Kingman, S., Irvine, D., Licence, P., Smith, A., Dimitrakis, G., Obermayer, D., Kappe, C. O. (2010). Understanding microwave heating effects in

single mode type cavities-theory and experiment. *Phys. Chem. Chem. Phys.*, *12*, 4750–4758.

[9] Gronnow, M. J., White, R. J., Clark, J. H., Macquarrie, D. J. (2005). Energy efficiency in chemical reactions: a comparative study of different reaction techniques. *Org. Process Res. Dev.*, *9*, 516–518.

[10] Razzaq, T., Kappe, C. O. (2008). On the energy efficiency of microwave-assisted organic reactions. *ChemSusChem.*, *1*, 123–132.

[11] Kremsner, J. M., Stadler, A., Kappe, C. O. (2006). The scale-up of microwave-assisted organic synthesis. *Top. Curr. Chem.*, *266*, 233–278.

[12] Obermayer, D., Gutmann, B., Kappe, C. O. (2009). Microwave chemistry in silicon carbide reaction vials: separating thermal from nonthermal effects. *Angew. Chem. Int. Ed.*, *48*, 8321–8324.

[13] Raner, K. D., Strauss, C. R. (1992). Influence of microwaves on the rate of esterification of 2,4,6-trimethylbenzoic acid with 2-propanol. *J. Org. Chem.*, *57*, 6231–6234.

[14] Herrero, M. A., Kremsner, J. M., Kappe, C. O. (2008). Nonthermal microwave effects revisited: on the importance of internal temperature monitoring and agitation in microwave chemistry. *J. Org. Chem.*, *73*, 36–47.

[15] Kaiser, N. F. (2001) Fast chemistry by microwave heating. Ph.D. Thesis, Uppsala University, Uppsala.

[16] Whittaker, A. G., Mingos, D. M. P. (1994). The application of microwave-heating to chemical syntheses. *JMPEE*, *29*, 195–219.

[17] Bremberg, U. (2000) Asymmetric catalysis: ligand design and microwave acceleration. Ph.D. Thesis, Royal Institute of Technology, Stockholm.

[18] Goudriaan, P. E., van Leeuwen, P., Birkholz, M. N., Reek, J. N. H. (2008). Libraries of bidentate phosphorus ligands; Synthesis strategies and application in catalysis. *Eur. J. Inorg. Chem.*, 2939–2958.

[19] Bremberg, U., Lutsenko, S., Kaiser, N. F., Larhed, M., Hallberg, A., Moberg, C. (2000). Rapid and stereoselective C-C, C-O, C-N and C-S couplings via microwave accelerated palladium-catalyzed allylic substitutions. *Synthesis*, 1004–1008.

[20] Larhed, M., Hallberg, A. (1996). Microwave-promoted palladium-catalyzed coupling reactions. *J. Org. Chem.*, *61*, 9582–9584.

[21] Bremberg, U., Larhed, M., Moberg, C., Hallberg, A. (1999). Rapid microwave-induced palladium-catalyzed asymmetric allylic alkylation. *J. Org. Chem.*, *64*, 1082–1083.

[22] Trost, B. M., VanVranken, D. L. (1996). Asymmetric transition metal-catalyzed allylic alkylations. *Chem. Rev.*, *96*, 395–422.

[23] Lu, Z., Ma, S. M. (2008). Metal-catalyzed enantioselective allylation in asymmetric synthesis. *Angew. Chem. Int. Ed.*, *47*, 258–297.

[24] Kaiser, N. F. K., Bremberg, U., Larhed, M., Moberg, C., Hallberg, A. (2000). Microwave-mediated palladium-catalyzed asymmetric allylic alkylation; An example of highly selective fast chemistry. *J. Organometallics Chem.*, *603*, 2–5.

[25] Braga, A. L., Silveira, C. C., de Bolster, M. W. G., Schrekker, H. S., Wessjohann, L. A., Schneider, P. H. (2005). Microwave-accelerated asymmetric allylations using

cysteine derived oxazolidine and thiazolidine palladium complexes. *J. Mol. Catal. A Chem.*, *239*, 235–238.

[26] Braga, A. L., Vargas, F., Sehnem, J. A., Wessjohann, L. A. (2006). Microwave-mediated palladium-catalyzed asymmetric allylic alkylation using chiral beta-seleno amides. *Eur. J. Org. Chem.*, 4993–4997.

[27] Popa, D., Puigjaner, C., Gomez, M., Benet-Buchholz, J., Vidal-Ferran, A., Pericas, M. A. (2007). Phosphinooxazolines derived from 3-amino-1,2-diols: highly efficient modular P-N ligands. *Adv. Synt. Catal.*, *349*, 2265–2278.

[28] Popa, D., Marcos, R., Sayalero, S., Vidal-Ferran, A., Pericas, M. A. (2009). Towards continuous flow, highly enantioselective allylic amination: ligand design, optimization and supporting. *Adv. Synt. Catal.*, *351*, 1539–1556.

[29] Belda, O., Moberg, C. (2004). Molybdenum-catalyzed asymmetric allylic alkylations. *Acc. Chem. Res.*, *37*, 159–167.

[30] Trost, B. M., Hachiya, I. (1998). Asymmetric molybdenum-catalyzed alkylations. *J. Am. Chem. Soc.*, *120*, 1104–1105.

[31] Kaiser, N. F. K., Bremberg, U., Larhed, M., Moberg, C., Hallberg, A. (2000). Fast, convenient, and efficient molybdenum-catalyzed asymmetric allylic alkylation under noninert conditions: an example of microwave-promoted fast chemistry. *Angew. Chem. Int. Ed.*, *39*, 3596–3598.

[32] Belda, O., Kaiser, N. F., Bremberg, U., Larhed, M., Hallberg, A., Moberg, C. (2000). Highly stereo- and regioselective allylations catalyzed by Mo-pyridylamide complexes: electronic and steric effects of the ligand. *J. Org. Chem.*, *65*, 5868–5870.

[33] Del Litto, R., Benessere, V., Ruffo, F., Moberg, C. (2009). Carbohydrate-based pyridine-2-carboxamides for Mo-catalyzed asymmetric allylic alkylations. *Eur. J. Org. Chem.*, 1352–1356.

[34] Trost, B. M., Andersen, N. G. (2002). Utilization of molybdenum- and palladium-catalyzed dynamic kinetic asymmetric transformations for the preparation of tertiary and quaternary stereogenic centers: a concise synthesis of tipranavir. *J. Am. Chem. Soc.*, *124*, 14320–14321.

[35] Belda, O., Lundgren, S., Moberg, C. (2003). Recoverable resin-supported pyridylamide ligand for microwave-accelerated molybdenum-catalyzed asymmetric allylic alkylations: synthesis of baclofen. *Org. Lett.*, *5*, 2275–2278.

[36] Nilsson, P., Gold, H., Larhed, M., Hallberg, A. (2002). Microwave-assisted enantioselective Heck reactions: expediting high reaction speed and preparative convenience. *Synthesis*, 1611–1614.

[37] Mata, Y., Pamies, O., Dieguez, M. (2007). Screening of a modular sugar-based phosphite-oxazoline ligand library in asymmetric Pd-catalyzed heck reactions. *Chem. Eur. J.*, *13*, 3296–3304.

[38] Kaukoranta, P., Kallstrom, K., Andersson, P. G. (2007). Microwave-assisted asymmetric intermolecular heck reaction Using phosphine-thiazole ligands. *Adv. Synt. Catal.*, *349*, 2595–2602.

[39] Mazuela, J., Pàmies, O., Diéguez, M. (2010). Biaryl phosphite-oxazoline ligands from the chiral pool: highly efficient modular ligands for the asymmetric Pd-catalyzed Heck reaction. *Chem. Eur. J.*, *16*, 3434–3440.

[40] Dounay, A. B., Humphreys, P. G., Overman, L. E., Wrobleski, A. D. (2008). Total synthesis of the Strychnos alkaloid (+)-minfiensine: tandem enantioselective

intramolecular Heck-iminium ion cyclization. *J. Am. Chem. Soc.*, *130*, 5368–5377.

[41] Lutsenko, S., Moberg, C. (2001). Microwave-mediated ruthenium-catalyzed asymmetric hydrogen transfer. *Tetrahedron Asymmetry*, *12*, 2529–2532.

[42] Hashiguchi, S., Fujii, A., Takehara, J., Ikariya, T., Noyori, R. (1995). Asymmetric transfer hydrogenation of aromatic ketones catalyzed by chiral ruthenium(II) complexes. *J. Am. Chem. Soc.*, *117*, 7562–7563.

[43] Jin, M.-J., Sarkar, M. S., Jung, J.-Y. (2007). Highly efficient microwave-assisted asymmetric transfer hydrogenation with SBA-15-supported TsCHDA chiral ligands. *Stud. Surf. Sci. Catal.*, *165*, 455–458.

[44] Lipshutz, B. H., Frieman, B. A., Unger, J. B., Nihan, D. M. (2005). Thermally accelerated asymmetric hydrosilylations using ligated copper hydride. *Can. J. Chem.*, *83*, 606–614.

[45] Lipshutz, B. H., Frieman, B. A. (2005). CuH in a bottle: a convenient reagent for asymmetric hydrosilylations. *Angew. Chem. Int. Ed.*, *44*, 6345–6348.

[46] Gustafsson, T., Hedenström, M., Kihlberg, J. (2006). Synthesis of a *C*-glycoside analogue of β-D-galactosyl hydroxylysine and incorporation in a glycopeptide from type II collagen. *J. Org. Chem.*, *71*, 1911–1919.

[47] Burk, M. J., Feaster, J. E., Nugent, W. A., Harlow, R. L. (1993). Preparation and use of C2-symmetric bis(phospholanes): production of α-amino acid derivatives via highly enantioselective hydrogenation reactions. *J. Am. Chem. Soc.*, *115*, 10125–10138.

[48] Toukoniitty, B., Roche, O., Mikkola, J.-P., Toukoniitty, E., Klingstedt, F., Eränen, K., Salmi, T., Murzin, D. Y. (2007). Ethyl pyruvate hydrogenation under microwave irradiation. *Chem. Eng. J.*, *126*, 103–109.

[49] Toukoniitty, B., Wärnå, J., Mikkola, J. P., Helle, M., Saxén, H., Murzin, D. Y., Salmi, T. (2009). Modelling of enantioselective and racemic hydrogenation of ethyl pyruvate on a Pt/Al_2O_3 catalyst in the presence of microwave irradiation. *Chem. Eng. Process Process Intensification*, *48*, 837–845.

[50] Hayashi, T., Yamasaki, K. (2003). Rhodium-catalyzed asymmetric 1,4-addition and its related asymmetric reactions. *Chem. Rev.*, *103*, 2829–2844.

[51] Fagnou, K., Lautens, M. (2002). Rhodium-catalyzed carbon–carbon bond forming reactions of organometallic compounds. *Chem. Rev.*, *103*, 169–196.

[52] Hargrave, J. D., Herbert, J., Bish, G., Frost, C. G. (2006). Rhodium-catalysed addition of organotrialkoxysilanes to [α]-substituted acrylic esters. *Org. Biomol. Chem.*, *4*, 3235–3241.

[53] Frost, C. G., Penrose, S. D., Lambshead, K., Raithby, P. R., Warren, J. E., Gleave, R. (2007). Rhodium-catalyzed conjugate addition–enantioselective protonation: the synthesis of α,α′-dibenzyl esters. *Org. Lett.*, *9*, 2119–2122.

[54] Miyaura, N., Suzuki, A. (1995). Palladium-catalyzed cross-coupling reactions of organoboron compounds. *Chem. Rev.*, *95*, 2457–2483.

[55] Genov, M., Almorín, A., Espinet, P. (2007). Microwave assisted asymmetric Suzuki-Miyaura and Negishi cross-coupling reactions: synthesis of chiral binaphthalenes. *Tetrahedron Asymmetry*, *18*, 625–627.

[56] Dioos, B. M. L., Jacobs, P. A. (2005). Microwave-assisted Cr(salen)-catalysed asymmetric ring opening of epoxides. *J. Catal.*, *235*, 428–430.

[57] Martinez, L. E., Leighton, J. L., Carsten, D. H., Jacobsen, E. N. (1995). Highly enantioselective ring opening of epoxides catalyzed by (salen)Cr(III) complexes. *J. Am. Chem. Soc.*, *117*, 5897–5898.

[58] Kureshy, R. I., Singh, S., Khan, N.-U. H., Abdi, S. H. R., Agrawal, S., Mayani, V. J., Jasra, R. V. (2006). Microwave-assisted asymmetric ring opening of meso-epoxides with aromatic amines catalyzed by a Ti-*S*-(−)-BINOL complex. *Tetrahedron Lett.*, *47*, 5277–5279.

[59] Lee, H. W., Lee, L. N., Chan, A. S. C., Kwong, F. Y., (2008). Microwave-assisted Rhodium-complex-catalyzed cascade decarbonylation and asymmetric Pauson-Khand-type cyclizations. *Eur. J. Org. Chem.*, 3403–3406.

PART II

RECENT ADVANCES IN ORGANOCATALYTIC, ENZYMATIC, AND METAL-BASED MEDIATED ASYMMETRIC SYNTHESIS

CHAPTER 11

RECENT ADVANCES ON STEREOSELECTIVE ORGANOCATALYTIC REACTIONS. ORGANOCATALYTIC SYNTHESIS OF NATURAL PRODUCTS AND DRUGS

MONIKA RAJ AND VINOD K. SINGH

11.1.	Introduction	416
11.2.	Enamine catalysis	417
	11.2.1. Aldol reaction	418
	11.2.2. α-Heterofunctionalization	424
	11.2.3. Conjugate addition	428
11.3.	Iminium catalysis	431
	11.3.1. Iminium–enamine reaction	433
	11.3.2. Conjugate additions	434
	11.3.3. Cycloaddition reactions	443
11.4.	Dienamine catalysis	445
11.5.	SOMO activation	449
11.6.	Brønsted acid and hydrogen bond catalysis	451
	11.6.1. Chiral phosphoric acids	452
	11.6.2. Monofunctional ureas and thioureas	454
11.7.	Brønsted base catalysis	458
11.8.	Bifunctional catalysis	458
	11.8.1. Bifunctional *Cinchona* alkaloids	459
	11.8.2. Bifunctional chiral urea and thiourea	459
11.9.	Phase transfer catalysis	465
11.10.	Conclusions and future perspective	467
	References	467

Catalytic Methods in Asymmetric Synthesis: Advanced Materials, Techniques, and Applications, First Edition. Edited by Michelangelo Gruttadauria and Francesco Giacalone.

11.1. INTRODUCTION

The need for chiral compounds, essentially as one enantiomer, has increased extensively in recent years. This is particularly important in pharmaceuticals because different enantiomers invariably have different biological activities [1]. A large number of drugs are increasingly becoming single enantiomer, and most of the precursors available in nature for bioactive compounds are chiral, such as amino acids (the building block of proteins) and sugars [2]. Chiral compounds are equally prevalent in the fields of agriculture, flavors, fragrances, and materials [3]. The properties of a material at times depend upon the chirality of polymers and liquid crystals [4]. The latter are used in thermometers to detect flaws in circuit board connections and to determine the condition of batteries. The chiral liquid crystals are used in the synthesis of "mood rings" (to determine the emotional state of the wearer) [5].

Historically, enantioenriched compounds were obtained from nature's chiral pool or by the resolution of a racemic mixture (an equimolar mixture of two enantiomers) [6]. Both of these methods have some limitations. Whereas the former requires stoichiometric use of a chiral precursor, the latter provides the desired enantiomer in only 50% theoretical yield. The efficient method to procure enantiomeric compounds is via asymmetric catalysis that allows us to construct stereocenters by using a substochiometric amount of a chiral substance, thus possessing significant potential advantages over the older procedures [6].

For a long time, enzymes [7] and metal complexes [8] have played important roles in asymmetric catalysis. Although metal-catalyzed reactions exhibit broad substrate scope, they are associated with the limitations of having a high cost and a toxicity related to metals, which are difficult to remove from the products. However, for the past few years, organocatalyzed reactions [9], which utilize substoichiometric amounts of organic compounds that do not contain even a small amount of metal or enzyme, have emerged as a new technique for asymmetric catalysis. Organic compounds, as compared to metals, are more stable, less expensive, nontoxic, readily available, and environmentally friendly [9]. Besides, organocatalytic reactions are less sensitive to the presence of water or air in comparison to metal-catalyzed reactions. Thus, the reproducibility and operational simplicity of these reactions are enhanced.

Organocatalysts provide better alternatives not only to metal catalysts but also to enzymes [9]. They provide a broad substrate scope in contrast to enzymes, which are highly substrate specific and cannot tolerate even the minor change in the structure of substrate molecules. In addition, they display another advantage over both metal catalysts and enzymes in that they are easily amenable to being attached to a solid support, leading to easy recovery of the catalyst and simplification of the reaction workup.

A pioneering organocatalyzed asymmetric reaction was reported independently by two groups over three decades ago: Eder et al. at Schering AG9 [10] and Hajos and Parrish at Hoffmann–La Roche Inc. [11]. They carried out the intramolecular aldol reaction of triketones by using L-proline. Although the reaction was known for a long time, its potential was realized less then a decade ago after List et al. discovered that L-proline can also catalyze the intermolecular aldol reaction [12]. During the same time, MacMillan et al. reported on the organocatalytic enantioselective Diels–Alder reaction [13]. After these discoveries, asymmetric organocatalysis has become a focal point of research in current years [9].

Organocatalysts are being used extensively for synthesizing drug molecules and natural products [14]. They display the characteristics of enzymes, which are known for high catalytic activity and stereospecificity [6]. Thus, the use of organic molecules enables chemists to understand the origin of chirality in the prebiotic environment.

The design of an organocatalyst and an understanding of its mechanism are very important for obtaining high stereoselectivity in a reaction. The catalyst, associated with different functional groups, catalyze the reaction by different mechanisms such as the enamine mechanism [9n, 15], iminium ion mechanism [16], dienamine mechanism [17], radical mechanism (singly occupied molecular orbital, SOMO) [18], Brønsted acid and H-bonding catalysis [19a], Brønsted base catalysis [19b], and phase transfer catalysis [20]. Organocatalysts activate a substrate molecule either by the formation of a covalent bond (enamine, iminium catalysis, and SOMO) or by the noncovalent interactions (Brønsted acid and H-bonding catalysis, Brønsted base catalysis, and phase transfer catalysis). The most recent contribution is by the activation of more than one substrate (bifunctional catalysis) [21].

With this in mind, this goal of this chapter is to critically describe the above eight distinct activation modes to detail the scope of asymmetric organocatalysis and its application in the synthesis of natural products and pharmaceutically relevant molecules [14]. The examples reported herein are a selection of the recent significant contributions that in our opinion have had a major impact in this area. Furthermore, this chapter demonstrates the ideas and challenges that are essential for progress in this field.

11.2. ENAMINE CATALYSIS

The α-functionalization of ketones involving the formation of enamine [15] as an intermediate has been termed "enamine catalysis" (Scheme 11.1) [22]. This proposal finds its genesis in the pioneering work of Stork and others [23]. The catalytic cycle commences with the condensation of a chiral aminocatalyst with a carbonyl compound, forming an iminium ion **1** that increases the acidity of α-protons and a deprotonation that leads to a reactive nucleophilic enamine **2**. The enamine **2** can either undergo an addition reaction (double bond containing an electrophile) or a substitution reaction (single bond containing an electrophile) depending on the nature of an electrophile. In either case, α-functionalized iminium ion **3** is formed, which on

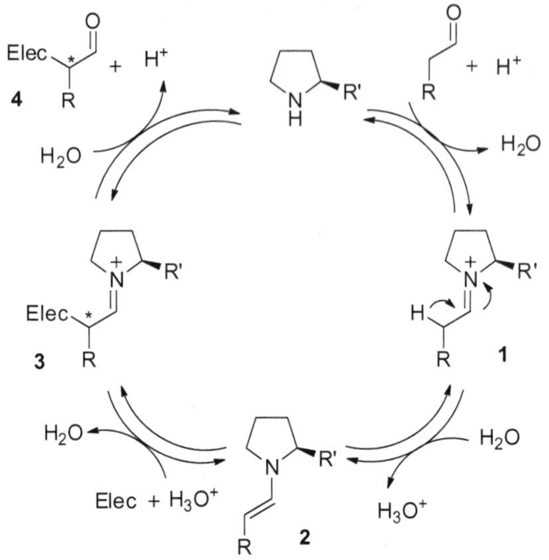

SCHEME 11.1. Enamine catalysis.

hydrolysis leads to product **4** and the amine catalyst that reenters the catalytic cycle.

It has been proposed that an incoming electrophile approaches the enamine by either of two different ways depending upon the nature of the catalyst. The catalysts having hydrogen bond donors direct the approach of the electrophile (H-bond acceptor) by the formation of a hydrogen bond as in the case of proline or prolinamide-catalyzed aldol, Mannich, hydroxylation, or amination reactions [24]. The hydrogen-bonding interaction accounts for the observed stereochemical outcome. Alternatively, the catalysts with bulky substituents orient electrophiles on the basis of steric factors as in the case of diarylprolinol catalyzed aldol, Mannich, amination, and halogenation reactions. The steric interactions account for the observed stereoinduction (Fig. 11.1). It is to be noted that these two directing effects deliver different enantiomers of the product even if both types of catalysts have the same absolute configuration (Fig. 11.1) [22].

Many asymmetric reactions proceed by enamine mechanisms, and they have been categorized in the following sections with application in the synthesis of natural products and biologically active compounds.

11.2.1. Aldol Reaction

Asymmetric organocatalytic aldol reaction is one of the most important methods for the formation of the C–C bond [25]. A wide variety of small organic molecules, including L-proline, are known to catalyze the asymmetric

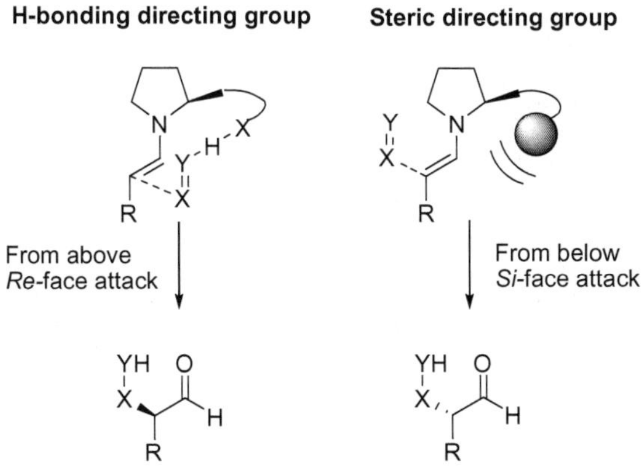

FIGURE 11.1. H-bond versus steric shielding of enamine intermediate.

aldol reaction. This methodology offers an efficient starting point for the synthesis of various complex scaffolds in natural products and biologically active drug molecules [14].

MacMillan et al. reported an impressive example where the enamine mechanism is applied twice in the total synthesis of callipeltoside C **15**, a cytotoxic marine macrolide (Scheme 11.2) [26]. The first step started with the double-differentiating aldol reaction of Roche ester derived from propanaldehyde **6** with aldehyde **7**, using L-proline **5** as a catalyst, yielding product **11** [27]. The product **11** further led to tetrahydropyran fragment **14**. The key stereocenter of the other fragment **12** was synthesized by an enantioselective α-oxyamination reaction of iodo-aldehyde **8** with nitroso benzene **9**, catalyzed by L-proline **5** [28]. The third step involves the aldol dimerization of **10** catalyzed by D-proline to give product **13** [29]. Finally, the coupling of all these fragments (**12**, **13**, and **14**) led to enanatiomeric sugar L-callipeltoside C **15**.

Convolutamydines A–E, isolated from the Floridian marine bryozoan *Amanthia convoluta* by Kamano and co-workers [30a], are members of a class of oxindole alkaloids containing 4,6-dibromo-3-hydroxy-oxindole as a common structural unit and different side-chain moieties at quaternary stereocenter on C-3. (*R*)-convolutamydine A **24** (Scheme 11.3) and B **27** (Scheme 11.4) exhibit potent inhibitor activity toward the differentiation of HL-60 human promyelocytic leukemia cells at 0.1 (convolutamydine A **24**) and 25 μg/mL (convolutamydine B 27) [30b,c]. Due to scarcity of convolutamydine E **26**, its biological activity has not yet been determined.

Tomasini et al. [31a,b] presented the first organocatalytic enantioselective synthesis of (*R*)-convolutamydine A **24** with 68% enantiomeric excess (ee). After this report, synthesis of convolutamydine A **24** was developed independently by four different groups. The key step involved in these procedures is direct enantioselective aldol reaction of isatin **16** with acetone **17** catalyzed by chiral amines **18–20** (Scheme 11.3) [32].

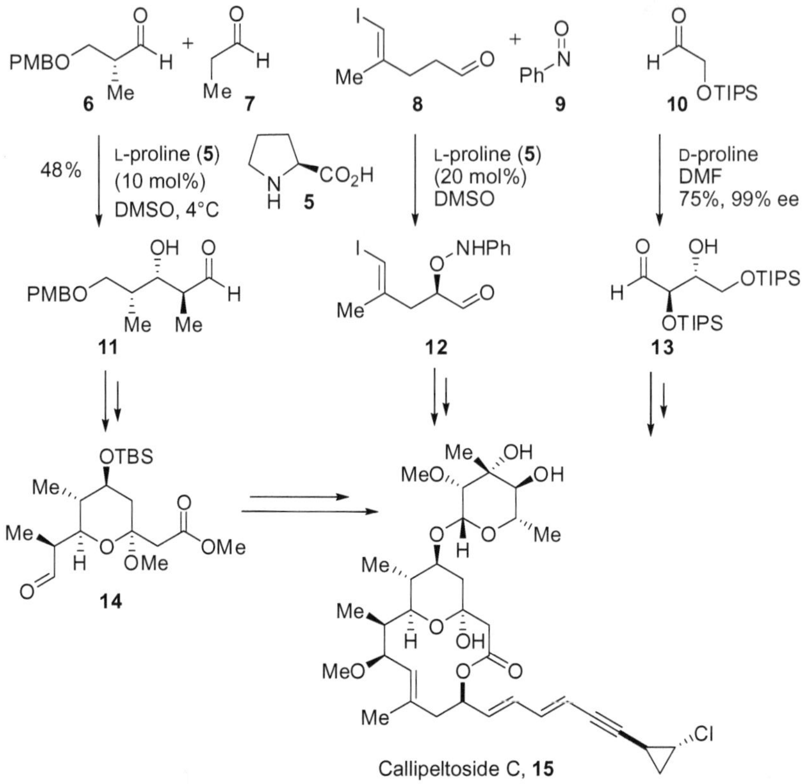

SCHEME 11.2. Total synthesis of callipeltoside C **15**.

Xiao and co-workers [32a] synthesized (*S*)-convolutamydine A **24** in 60% ee (87% ee after single recrystallization) by using a catalyst **18**. A plausible transition state **21** explains the reason for the observed stereochemical outcome.

Malkov et al. [32b] utilized D-leucinol **19** (20 mol%) for the synthesis of (*R*)-convolutamydine A **24** in 80% yield and with high enantioselectivity (94% ee). According to transition state model **22**, the reaction takes place through the *syn* form of an enamine intermediate. In 2008, Nakamura and co-workers [32c] carried out the synthesis of (*R*)-convolutamydine A **24** with 95% ee (enantiomerically pure after recrystallization) by using a novel *N*-heteroarylsulfonylprolinamide **20** organocatalyst (0.5–5 mol%). As studied by the authors, the hydrogen bond between the amide proton, 2-thienyl sulfur atom in chiral catalyst **20**, plays an important role in obtaining the high enantioselectivity as depicted in the transition state **23**.

Nakamura and co-workers utilized the same catalyst, *N*-heteroarylsulfonylprolinamide **20**, for carrying out the synthesis of (*R*)-

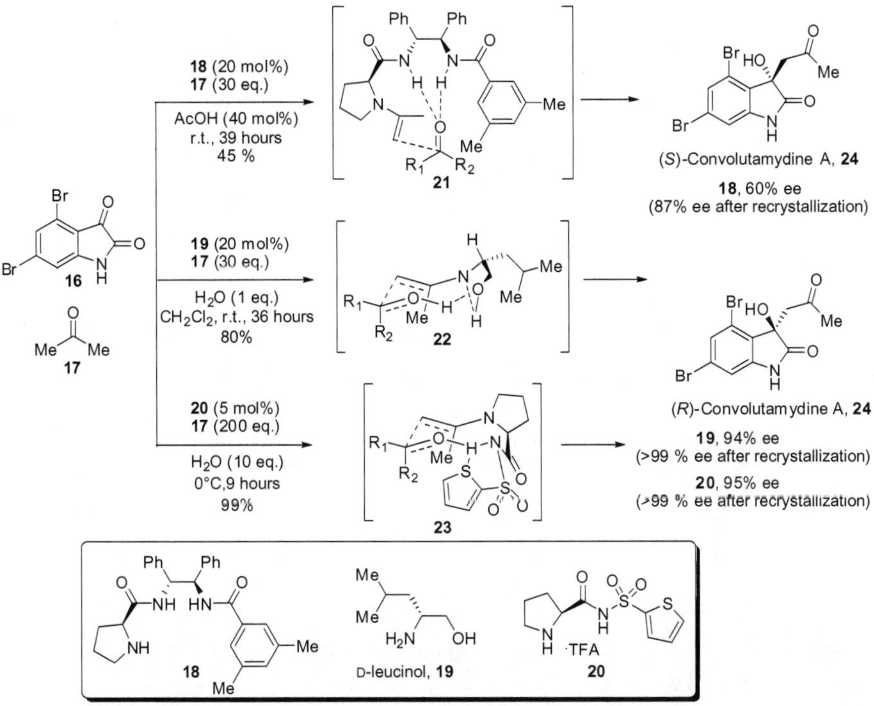

SCHEME 11.3. Syntheses of (*S*)- and (*R*)-convolutamydine A **24**.

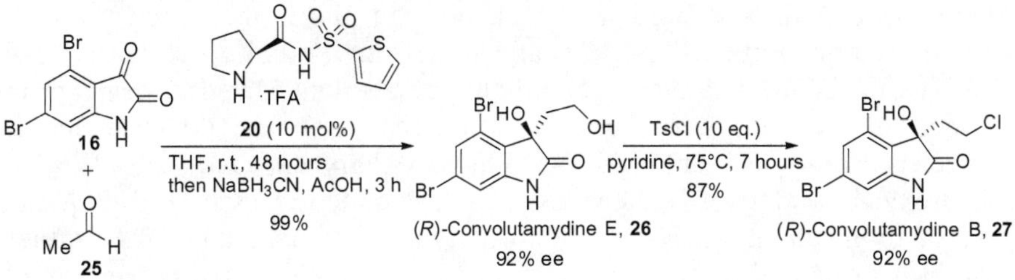

SCHEME 11.4. Syntheses of (*R*)-convolutamydine E **26** and B **27**.

convolutamydine E **26** and B **27** (Scheme 11.4) [33, 34]. The key step involves the direct asymmetric aldol reaction between acetaldehyde **25** and isatin **16** followed by reduction with sodium cyanoborohydride to give (*R*)-convolutamydine E **26** with high enantioselectivity (92% ee). Finally, the synthesis of (*R*)-convolutamydine B **27** in enantiomerically pure form was achieved by a single-step reaction from (*R*)-convolutamydine E **26**, followed by recrystallization.

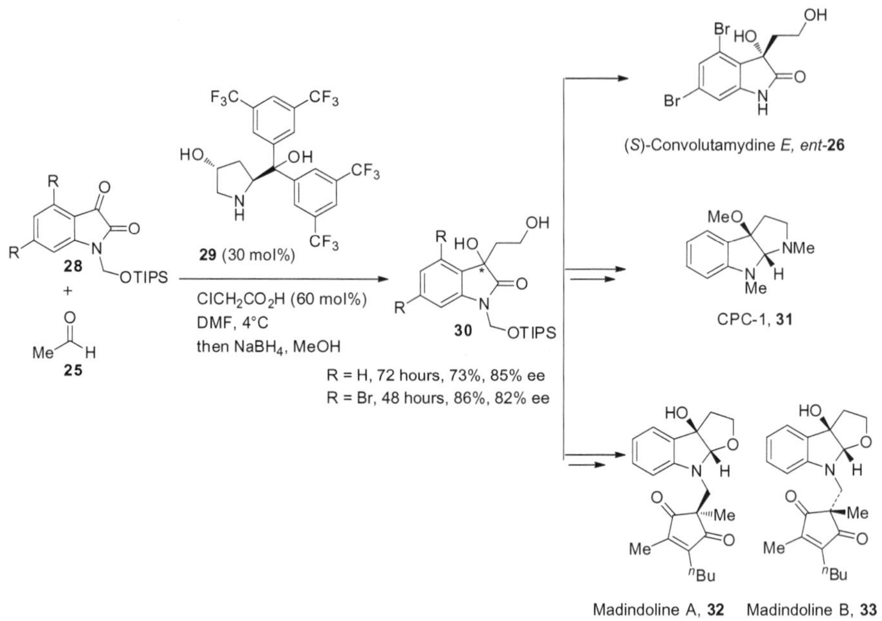

SCHEME 11.5. Syntheses of (*S*)-convolutamydine E *ent*-**26**, CPC-1 **31**, and madindoline A **32** and B **33**.

Recently, Hayashi and co-workers reported the synthesis of unnatural *ent*-convolutamydine E **26**, CPC-1 **31**, and madindoline A **32** and B **33** [35]. Madindoline A **32** and B **33** were isolated from the fermentation broth of streptomycesnitrosporeus K93-0771 and are selective inhibitors of interlukin-6 [36]. The CPC-1 **31** is a new pyrrolidine-type alkaloid [37]. The key step in synthesis involves the direct asymmetric aldol reaction of acetaldehyde **25** with protected isatin **28**, catalyzed by 4-hydroxydiaryl prolinol **29** afforded 3-hydroxyindole derivative **30** as a key intermediate (Scheme 11.5) [35]. The reversal of the enantiofacial selection when the substituent at C-5 is Br rather than H is explained on the basis of transition state models shown in Figure 11.2 [35]. The authors proposed that the proton of the hydroxy group of the catalyst **29** coordinates the carbonyl group of isatin **28** by two protonation modes, A and B, as shown in the figure. When the substituent at C-5 is small (H atom), then mode A is preferred, and the reaction proceeds through transition state (TS)-1 to give *R*-isomer as a major product. If the substituent at C-5 is large (Br atom), then mode B is favored because of steric hindrance in mode A. The steric interactions between the protecting group on nitrogen and aryl groups of the catalyst make the TS-2 unsuitable, thus reaction takes place through TS-3, affording the *S*-isomer predominantly.

W.-C. Yuan and co-workers recently reported on the synthesis of alkaloids; convolutamydine E **26** and B **27**. They also reported the synthesis of (–)-

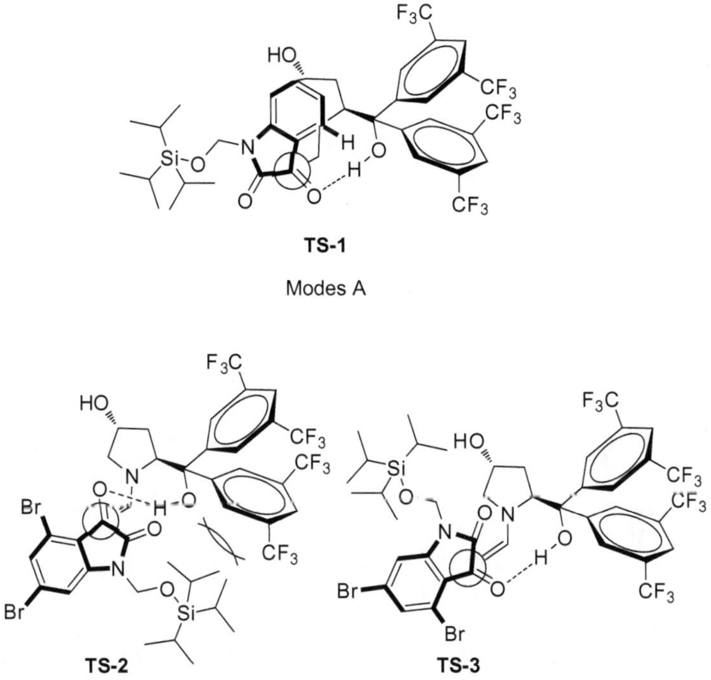

FIGURE 11.2. Transition state models.

donaxaridine **37** [38a], an alkaloid isolated from *Arundo donax*; and (*R*)-chimonamidine **38**, an alkaloid isolated from the seeds of *Chimonanthus praecox Link* [38b]. The key step involves the direct asymmetric aldol reaction of acetaldehyde **25** with isatin **34** and substituted isatin **16/35** catalyzed by 4-hydroxydiaryl prolinol **29** followed by $NaBH_4$ reduction, affording convolutamydine E **26** and 3-hydroxyindole derivative **36a/36b**. The aldol adduct convolutamydine E **26** was transformed into convolutamydine B **27** by following the sequence of tosylation and chlorination reactions. Further, 3-hydroxyindole derivatives **36a/36b** underwent a tosylation reaction followed by transamidation of methylamino intermediates that successfully generated (−)-donaxaridine **37** in 99% ee and (*R*)-chimonamidine **38** in 96% ee (Scheme 11.6) [39].

Singh et al. synthesized a diamine catalyst **41** for carrying out enantioselective aldol reactions with different substrates in both organic and aqueous media. This methodology was applied for the formal synthesis of (*S*)-oxybutynin **44** [40], a muscarinic receptor antagonist for the treatment of urinary infection, urgency, and urge incontinence. The synthesis commenced with the aldol reaction between cyclohexananone **39** and ketoester (ethyl phenyl glyoxylate **40a** or methyl phenyl glyoxylate **40b**) catalyzed by a diamine/Brønsted acid

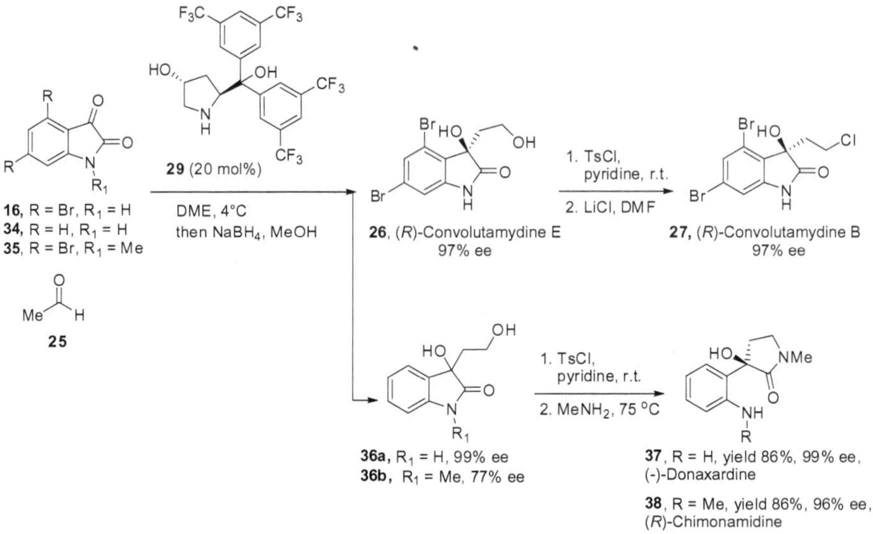

SCHEME 11.6. Syntheses of (*R*)-convolutamydine E **26**, (*R*)-convolutamydine B **27**, (–)-donaxaridine **37**, and (*R*)-chimonamidine **38**.

catalyst, and **41**/trifluoro acetic acid (TFA) gave the product **42a/42b** with quaternary chiral center in high ee's (Scheme 11.7). Additionally, by following series of reactions, the aldol adduct **42a/42b** was converted into (*S*)-2-cyclohexyl-2-phenylglycolic acid **43**, a key precursor for the synthesis of (*S*)-oxybutynin **44** [41].

Singh and co-workers utilized the same catalyst for the synthesis of 3-cyclohexanone-3-hydroxy-2-oxindoles **46a/46b** [42], which act as potential anticonvulsants in mice (Scheme 11.8) [43]. The key step of the synthesis was the asymmetric aldol reaction of cyclohexanone **39** with isatin **34/45** catalyzed by diamine/Brønsted acid **41**, which afforded the products **46a/46b** in ~99% ee.

11.2.2. α-Heterofunctionalization

The α-functionalization of carbonyl compounds is a versatile process. This process is conceptually similar to the previous reaction. The reactions of carbonyl compounds that have received much attention are the α-aminoxylation reaction with nitrosobenzene and the α-amination reaction with alkyl diazodicarboxylates. The simple methodology works well for a broad range of carbonyls, thus providing a convenient approach to natural products and drug molecules [24a, 44]. Some recent examples utilizing this methodology have been compiled in the succeeding paragraphs.

Varseev and Maier reported the enantioselective total synthesis of (+)-neosymbioimine **51** [45], a minor amphoteric metabolite of the symbiotic marine dinoflagellate *Symbiodinium* sp. [46]. This alkaloid is synthetically very chal-

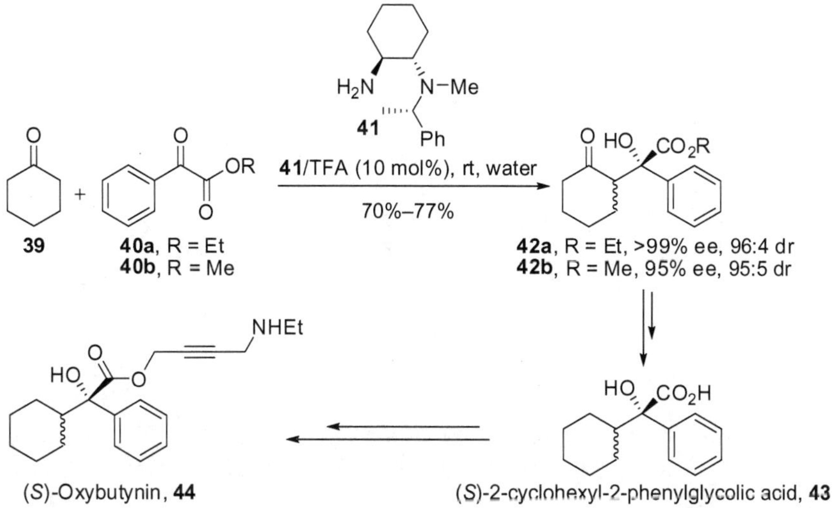

SCHEME 11.7. Formal synthesis of (*S*)-oxybutynin **44**.

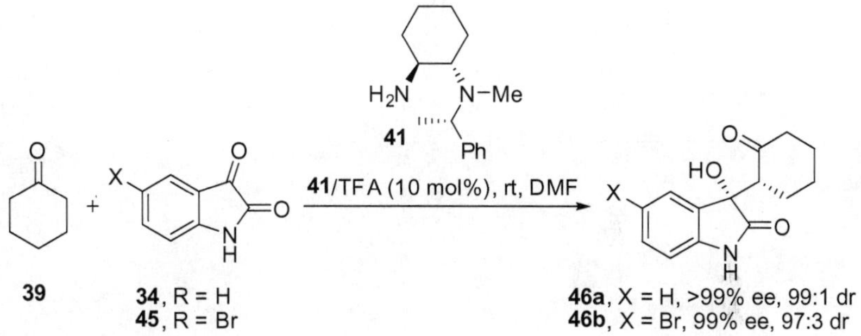

SCHEME 11.8. Synthesis of 3-cyclohexanone-3-hydroxy-2-oxindoles, **46a/46b**.

lenging, as it comprises a tricyclic iminium core with an attached resorcinol monosulfate unit [47]. This could be used in the development of efficient antiosteoporosis drugs. The key feature of the synthesis involves a one-pot three-step reaction including α-hydroxylation followed by the Horner–Wadsworth–Emmons (HWE) reaction, and a cleavage of the N–O bond originally devised by Mangion and MacMillan [48] (Scheme 11.9). The (*S*)-citronellol 47-derived aldehyde **48** underwent the one-pot MacMillan procedure with nitrosobenzene **8** in the presence of D-proline **5** followed by a Wittig reaction with **49** and gave 4-hydroxy-8-silyloxy-enolate **50** in 55% yield. Finally, synthesis of (+)-neosymbioimine **51** was accomplished from 4-hydroxy-8-silyloxy-enolate **50** by following the sequence of 18 steps in 10% overall yield and with >99% enantioselectivity.

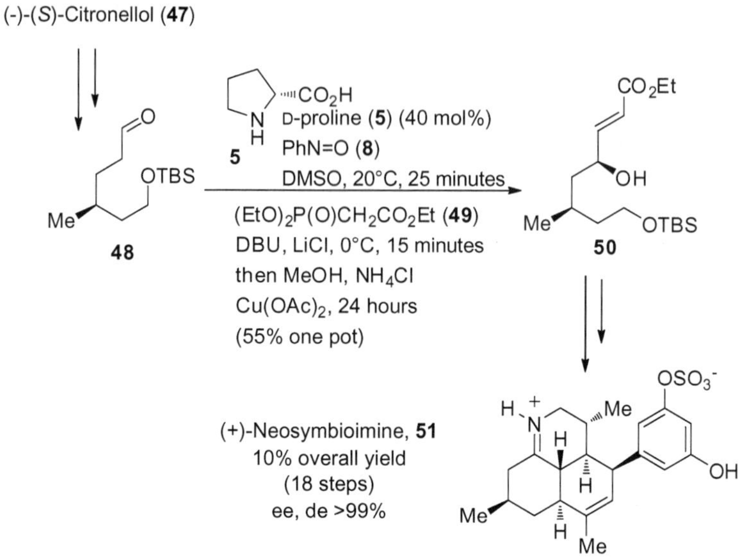

SCHEME 11.9. Synthesis of (+)-neosymbioimine **51**.

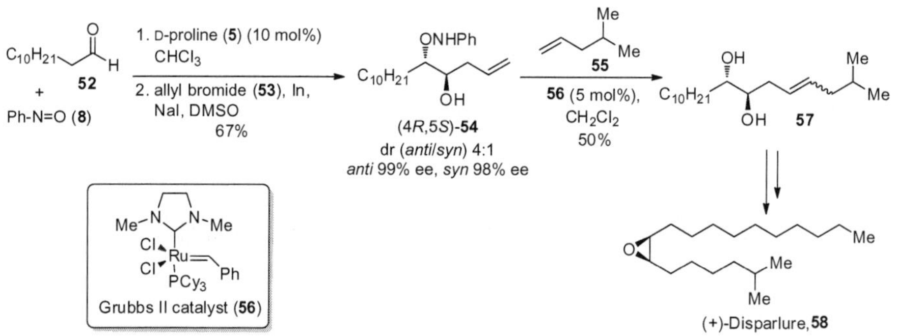

SCHEME 11.10. Total syntheses of (+)- and (−)-disparlure **58**, and its (7*S*,8*S*)- and (7*R*,8*R*)-*trans*-isomers.

Using similar methodology, the enantioselective organocatalytic synthesis of (+)-disparlure **58** [49] and its *trans*-isomer has been described by Kim and co-workers (Scheme 11.10) [49a]. (+)-Disparlure **58** is a sex pheromone of the female gypsy moth, *Porthetria dispar* (L) [50], while its *trans*-isomer (−) **58** antagonizes the effect of (+)-disparlure **58** and is itself slightly repellent [51]. The synthesis of (+)-disparlure **58** commenced with the organocatalytic enantioselctive α-aminoxylation of dodecanal **52** with nitrosobenzene **8** catalyzed by D-proline **5** and in situ indium-mediated allylation with allylbromide **53** affording the desired homoallylic alcohol **54** in 67% yield and high enantiose-

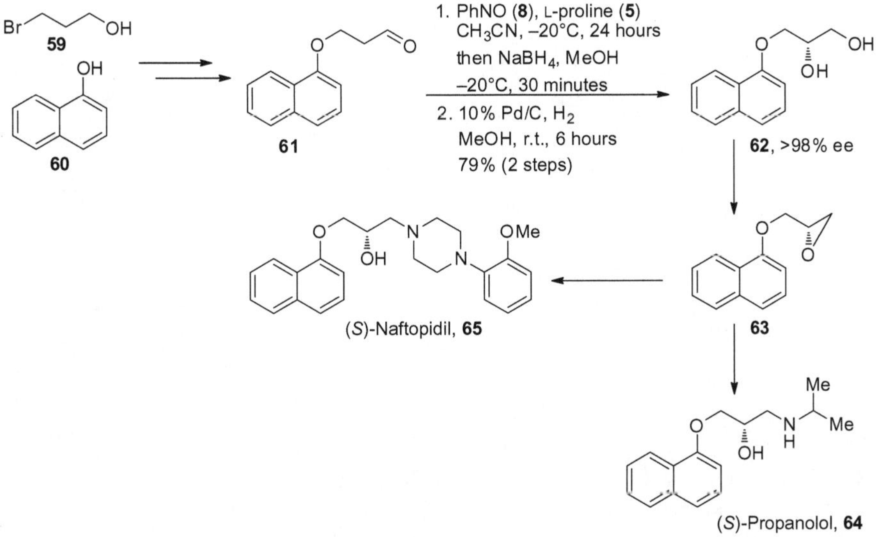

SCHEME 11.11. Organocatalytic enantioselective syntheses of *β*-blockers: (*S*)-propranolol **64** and (*S*)-naftopidil **65**.

lectivity (*anti*-99% ee). Furthermore, compound **54** underwent olefin cross-metathesis with alkene **55** in the presence of Grubb's catalyst **56** and yielded compound **57**, which on hydrogenation, followed by few standard reaction sequences, concluded the synthesis of (+)-disparlure **58**. The synthesis of *trans*-isomer (–) **58** was achieved in a similar manner by using L-proline in one of the key steps.

Applying similar methodology, Kalkote and co-workers reported the enantioselective synthesis of (*S*)-propranolol **64** and (*S*)-naftopidil **65**, which act as a *β*-adrenergic blocking agent (Scheme 11.11) [52]. The key step of synthesis involves the organocatalytic enantioselective *α*-aminoxylation [53] of aldehyde **61** by using L-proline and in situ reduction, followed by hydrogenation that gave chiral diol **62** in high yield (79%) and high enantioselectivity (>98% ee). The conversion of the chiral diol **62** to an epoxide **63** was achieved under Mitsonobu conditions [54]. Finally, the epoxide **63** was converted into (*S*)-propranolol **64** and (*S*)-naftopidil **65** by following the standard reaction sequence.

The *Lycopodium* genus of plants produce a number of alkaloids such as cernuine **75**, which comprises the fused tetracyclic ring system containing aminal moiety [55]. It was isolated by Marion and Manske in 1948, and the structure was determined by Ayer et al. in 1967 [56]. Kobayashi et al. in 2004 isolated cermizine D **74**, cermizine C **71**, and senepodine G **72**, in which **72** and **74** exhibited cytotoxicity against murine lymphoma L1210 cells with an half maximal inhibitory concentration (IC_{50}) of 7.8 and 7.5 g/mL, respectively

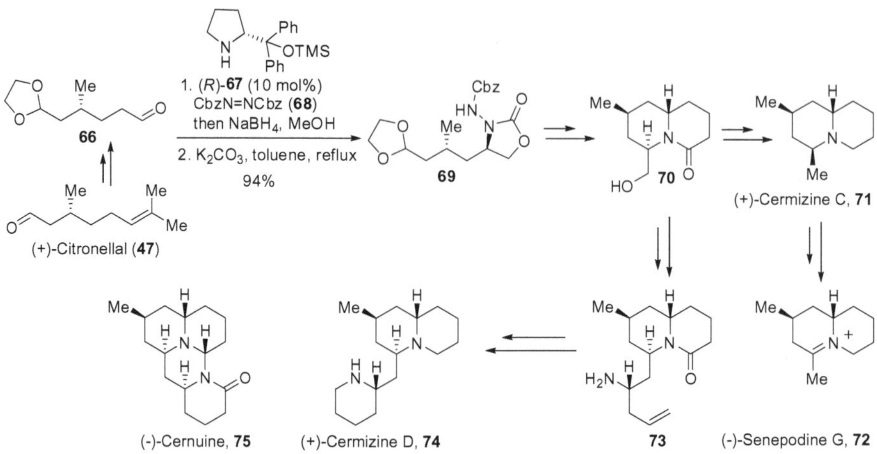

SCHEME 11.12. Asymmetric total syntheses of cernuane-type and quinolizidine-type *Lycopodium* alkaloids: (+)-cermizine C **71**, (−)-senepodine G **72**, (+)-cermizine D **74**, and (−)-cernuine **75**.

[57]. Takayama and co-workers reported the enantioselective total synthesis of cernuine **75**, cermizine D **74**, cermizine C **71**, and senepodine G **72** from a common intermediate **70** (Scheme 11.12) [58]. The key step of the synthesis is the α-amination reaction of aldehyde **66**, obtained from (+)-citronellal **47**, with dibenzyl azodicarboxylate 68 by using diaryl prolinol **67** as a catalyst. The α-amination product was converted into oxazolidinone **69** in 94% yield and 84% diastereoisomeric excess (de). After following a few other reaction sequences, including the ring-closing metathesis (RCM) with first-generation Grubb's catalyst **56** and a Wittig reaction, oxazolidinone **69** was converted into key intermediate **70**. Global reduction–deoxygenation afforded cermizine C (**71**), which was subsequently converted into senepodine G (**72**) via a regioselective Polonovsky–Potier reaction [59]. Finally, cernuine **75** and cermizine D **74** were obtained via intermediate **73** (comprising both an alkyl chain and amine functionality onto C-5) by use of an asymmetric transfer aminoallylation reaction developed by Kobayashi et al. [60].

11.2.3. Conjugate Addition

The conjugate addition-type reactions have held constant interest in the area of asymmetric organocatalysis [61]. Particularly, the addition of carbonyl compounds to nitroolefins, catalyzed by diarylprolinol, provides a convenient approach for the synthesis of complex scaffolds [62]. It is important to note that stereochemical induction in this reaction seems to be governed by steric properties of the catalyst rather than by H-bonding [63]. In the following section, some recent examples utilizing this approach are described.

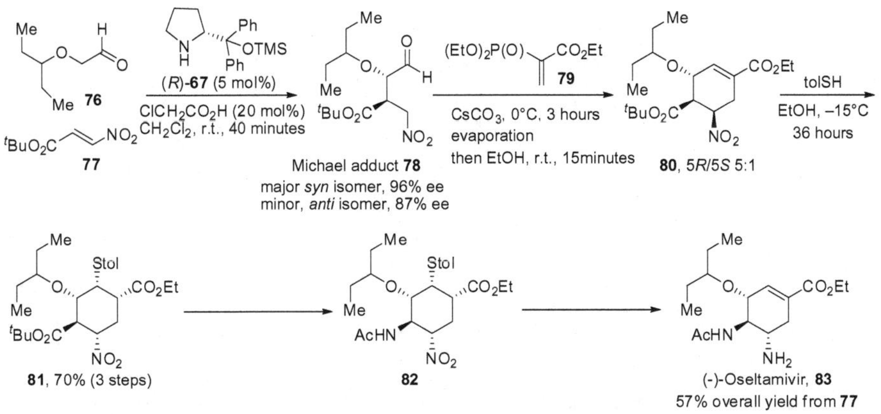

SCHEME 11.13. Synthesis of (–)-oseltamivir **83**.

The enamine mechanism is also involved in the organocatalytic synthesis of (–)-oseltamivir **83** (Tamiflu), a neuraminidase inhibitor used in the treatment of both type A and type B human influenza and is one of the most promising therapeutics [64]. Recently, Hayashi and co-workers reported the organocatalytic asymmetric total synthesis of (–)-oseltamivir **83** (Scheme 11.13) [65]. The key step of the synthesis involves the Michael addition [66, 67] of aldehyde **76** to nitroalkene **77** by using diaryl prolinol **67** as a catalyst to gave Michael adduct **78** with high enantioselectivity (major *syn* isomer, 96% ee). Further, Michael adduct **78** underwent Michael and HWE reaction in a domino fashion [62] to give intermediate **80**. The thiol-Michael reaction and base-catalyzed isomerization of intermediate (5*S*)-**80**/(5*R*)-**80** afforded the desired product (5*S*)-**81**/(5*R*)-**81** in high yield (70%). The next step involves the domino Curtius rearrangement and amide formation, affording **82**. Finally, the nitro group of **82** was reduced to an amine followed by the retro-Michael reaction of the thiol, and yielded the desired product, (–)-oseltamivir **83**. Summing up, nine reactions and three separate one-pot operations and one column chromatography are required for the synthesis of (–)-oseltamivir **83** in 57% overall yield.

Biyouyanagin A, **92** is isolated from the leaves of *Hypericum chinense* L. var. *salicifolium*, a *Hypericum* sp. used as a folk medicine in Japan for the treatment of female disorders. This compound exhibited significant and selective inhibitory activity against HIV replication in H9 lymphocytes and lipopolysaccharide (LPS)-induced cytokin production [68]. Thus, it acts as a potential biological tool and a drug discovery lead.

Nicolaou and co-workers reported the enantioselective synthesis of both **92a** (24*R*) and **92b** (24*S*) epimers of Biyouyanagin A (Scheme 11.14) [69]. The syntheses commenced with the asymmetric Michael addition of (*S*)- or (*R*)-citronellals **47** to methyl vinyl ketone (MVK) **84** in the presence of diaryl

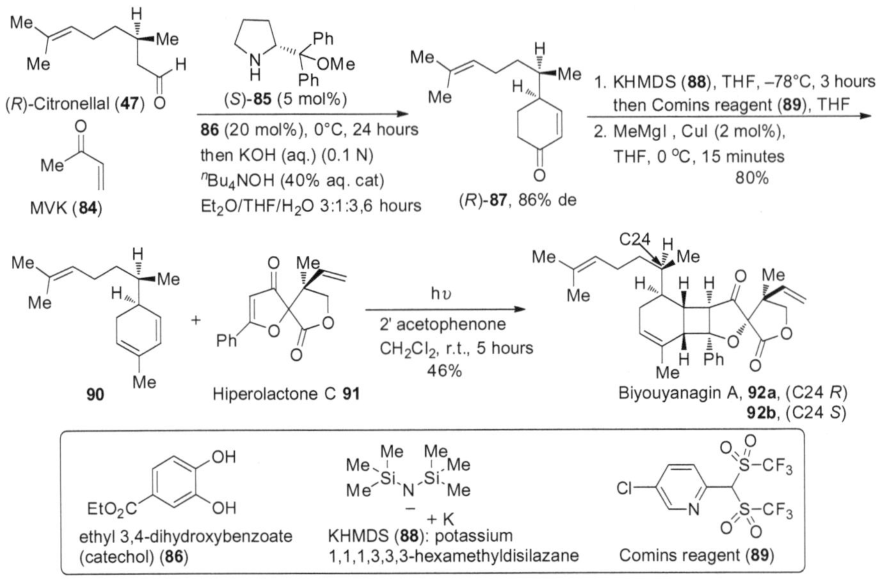

SCHEME 11.14. Total synthesis of biyouyanagin A **92**.

prolinol **85** as a catalyst and 3,4-dihydroxybenzonate **86** as a cocatalyst [70]. The resulting ketoaldehyde underwent intramolecular aldol condensation, giving enone (*S*)-**87a**/(*R*)-**87b** (72% yield and 93% de), which was converted to the desired terpenes (*S*)-**90a**/(*R*)-**90b** by Cu-catalyzed methylenation. Finally, the irradiation of the mixture of terpene (*S*)-**90a**/(*R*)-**90b** and **91**, synthesized from commercially available L-malic acid, led to regio-, chemo-, and stereoselective formation of cyclobutane through heterocoupling, and yielded the desired **92a** (24*S*) and **92b** (24 *R*) epimers of Biyouyanagin A.

Using a similar methodology, Chen and Baran carried out the synthesis of eudesmane terpenes **97–100** by the Michael addition of isovaleraldehyde **93** to MVK **84** followed by intramolecular aldol reaction of the intermediate **94** (92%–95% ee), which generated cryptone **95** in modest yield (63%) and good enantioselectivity (89% ee) (Scheme 11.15) [71]. Additionally, dihydrojunenol **96** was obtained from cryptone **95** by following a series of nine steps in overall 21% yield. Finally, the site-selective oxidation of dihydrojunenol **96** under different conditions concluded the syntheses of four eudesmane terpenes: 4-*epi*-ajanol **97**, pygmol **98**, dihydroxy eudesmane **99**, and eudesmantetraol **100**.

Campbell and Johnson described the total synthesis of (+)-polyanthellin A **105**, which acts as an antimalerial agent (Scheme 11.16) [72]. The synthesis commenced with the Michael addition reaction of aldehyde **101** with MVK **84** under the similar Michael addition conditions, and generated aldehyde **102** (90% yield and 94% ee). After following a sequence of three steps, the aldehyde **102** was converted into bicycle[4.1.0] heptanone **103**. Finally, [3+2] cyclo-

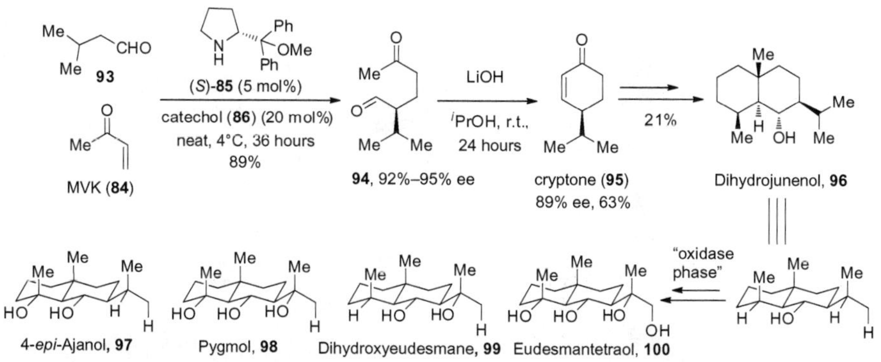

SCHEME 11.15. Total syntheses of eudesmane terpenes **97–100**.

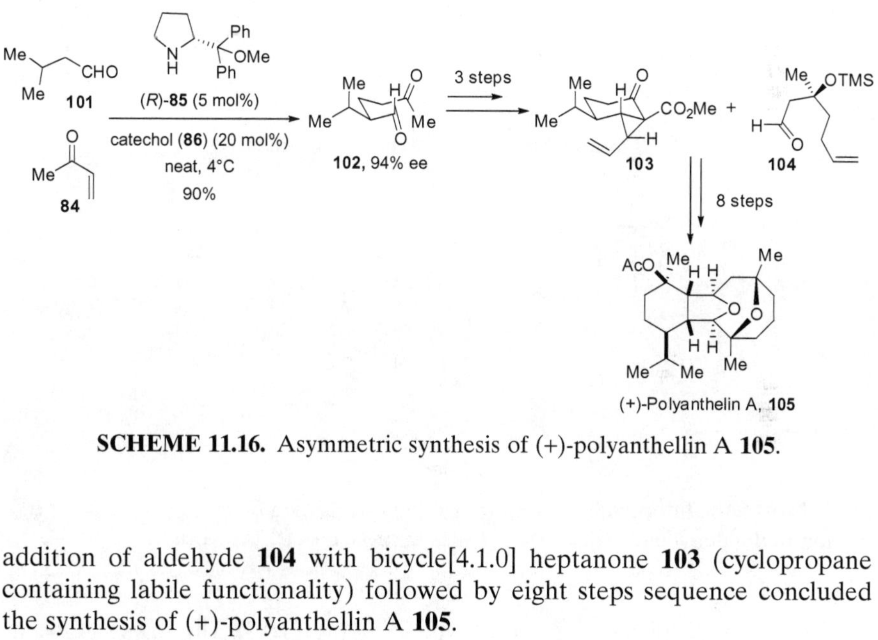

SCHEME 11.16. Asymmetric synthesis of (+)-polyanthellin A **105**.

addition of aldehyde **104** with bicycle[4.1.0] heptanone **103** (cyclopropane containing labile functionality) followed by eight steps sequence concluded the synthesis of (+)-polyanthellin A **105**.

11.3. IMINIUM CATALYSIS

The addition of a large number of different nonmetallic nucleophiles to the β-position of α,β-unsaturated carbonyl compounds via a key iminium ion intermediate (therefore termed "iminum catalysis") [16] has been very successfully developed [73]. It provides an operationally simple alternative to Lewis acid-catalyzed reactions of α,β-unsaturated carbonyl compounds. The initial condensation of α,β-unsaturated carbonyl compounds or enals with

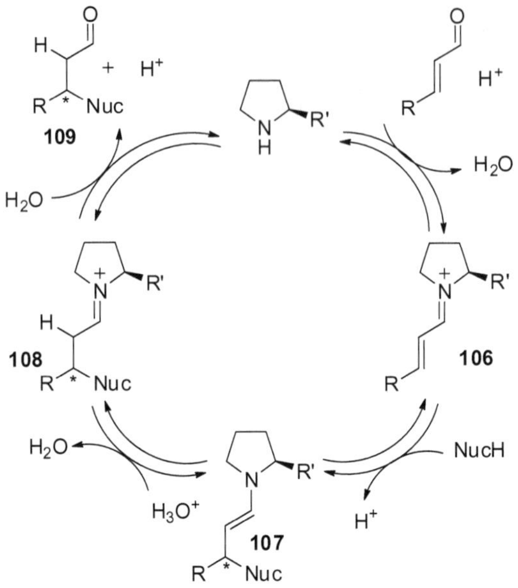

SCHEME 11.17. Iminium catalysis.

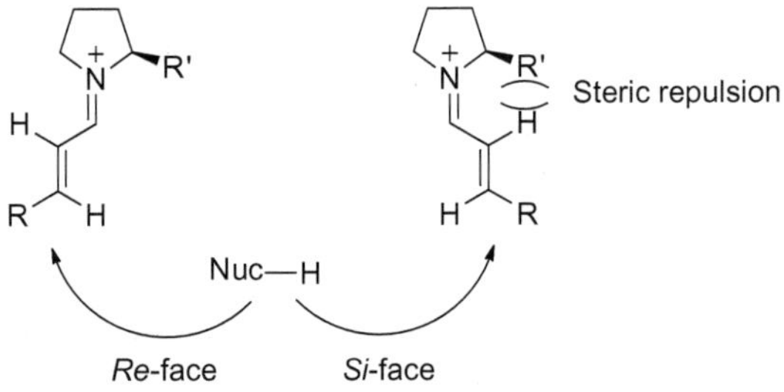

SCHEME 11.18. Intermediates in the pyrrolidine-catalyzed β-functionalization of α,β-unsaturated aldehydes in which the *Re*-face approach (left) is favored.

amine catalyst leads to iminium ion **106** (Scheme 11.17). The lowest unoccupied molecular orbital (LUMO) of the iminium ion **106** is lowered in energy, thereby acting as a reactive electrophilic partner for nucleophilic reactions such as conjugate addition reactions and pericyclic reactions [63]. The addition of nucleophile to the β-carbon atom forms a β-substituted enamine **107**, which tautomerizes to a corresponding iminium ion **108**. Finally, the hydrolysis of the iminium ion **108** releases the product **109** with concomitant regeneration of the catalyst.

The computational and theoretical studies [74] of the reaction have predicted that the *trans-trans* iminium ion is energetically more favorable both as a single molecule and in the transition state as compared to *cis-trans* iminium

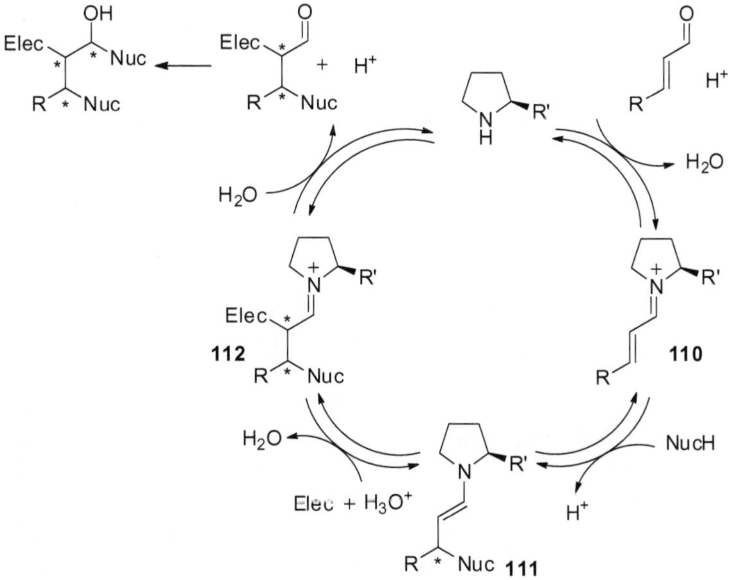

SCHEME 11.19. Pyrrolidine-catalyzed α,β-functionalization and nucleophilic addition to the carbonyl functionality of α,β-unsaturated aldehydes.

ion (Scheme 11.18). This conformation governs the *Re*-face approach of the nucleophile as the *Si*-face approach is unfavorable because of steric repulsions [22]. The reactions on the alkenyl portion of the iminium ion can be divided into two groups: conjugate additions and cycloaddition reactions. These reactions are operationally simple and provide versatile products with high ee, and thus are employed in the synthesis of complex natural products.

11.3.1. Iminium–Enamine Reaction

Besides the above fundamental transformations of carbonyl compounds, organocatalysts can also be used for one-pot, domino, and tandem reactions [75]. A closer look at the mechanism reveals that conjugate addition via iminium ion **110** forms an enamine **111** (Scheme 11.19) [22]. The further reaction of the resulting enamine **111** with suitable electrophile can lead to a second functionalization in a cascade manner. The nucleophile and electrophile involved in the iminium–enamine **112** double functionalization of an α,β-unsaturated carbonyl compound can be two different reagents or can be a part of the same molecule. As shown in Scheme 11.19, up to three stereocenters can be obtained by using this methodology. Depending upon the nature of the other reactive partners, multiple stereocenters can be generated in one reaction.

These organocatalytic tandem and domino reactions are characterized by high efficiencies. In addition, they avoid time-consuming, costly protection and deprotection steps, and purification of intermediates. Different types of asymmetric reactions involving the iminium ion intermediate have been categorized in the following section and their application in the synthesis of natural products and biologically active compounds are described in detail.

11.3.2. Conjugate Additions

Conjugate additions via the iminium ion intermediate largely depend upon the nature of the nucleophile. The matter of concern is the leaving ability of the group attached to the nucleophile as conjugate addition reactions are likely to be reversible. The other factor is competent 1,2 addition rather than the desired 1,4 addition [23]. This can be minimized (avoided) by using soft nucleophiles, which prefer to react in a 1,4-fashion, due to the softer electrophillic carbon at the 4-position as compared to the 2-position. A range of aromatic and heteroaromatic nucleophiles can be added to the iminium ion to form diverse functionalities that are common in drug molecules and natural products.

MacMillan and co-workers utilized organocascade catalysis [76] for organocatalytic enantioselective total synthesis of the sesquiterpene, (−)-aromadendranediol **120** [77] (Scheme 11.21), isolated from the marine coral *Sinularia mayi* [78] and from leaves of the Amazonian tree *Xylopia brasiliensis* [79]. (−)-Aromadendranediol **120** is also synthetically very challenging as it comprises a tricyclic framework containing six stereocenters (five of which are contiguous), two tertiary hydroxy groups, and a 1,1-dimethylated cyclopropane ring.

The key feature involves the triple-cascade-catalysis sequence that utilizes Grubb's catalyst **56** (catalyzing cross-metathesis reaction), imidazolidinone catalyst **117** (catalyzing Mukaiyama–Michael reaction), and proline **5** (catalyzing intramolecular aldol reaction) (Schemes 11.20 and 11.21). The synthesis commenced by the treatment of crotonaldehyde **113** with 5-hexene-2-one **114** in the presence of Grubb's catalyst **56** generated the keto-enal **115**. The introduction of silyloxyfuran **116** at this stage leads to a Mukaiyama–Michael iminium reaction with keto-enal **113** in the presence of imidazolidinone catalyst **117** that gave keto-aldehyde **118** with two contiguous chiral centers. The resulting keto-aldehdye **118** further underwent intramolecular aldol reaction via enamine-catalytic cycle in the presence of proline **5** to give butenolide **119** [64% yield, 95% ee with 5:1 diastereoisomeric ratio (dr)] containing a *trans*-substituted pentyl ring, tertiary alcohol, and third and fourth stereocenters. Finally, butenolide **119** underwent a sequence of eight-step reactions that yielded synthetic (−)-aromadendranediol **120** (Scheme 11.21).

List et al. have reported that tandem functionalization can be carried out intramolecularly via conjugate iminium reduction followed by interception of the resulting enamine by intramolecualr cyclization to give cyclic products

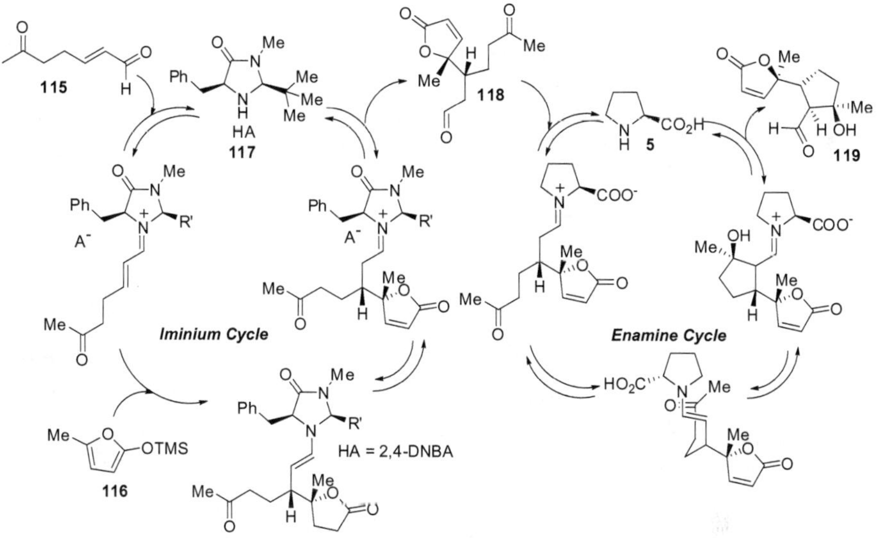

SCHEME 11.20. Proposed mechanism of cascade catalysis to generate the stereochemical core of (−)-aromadendranediol **120**.

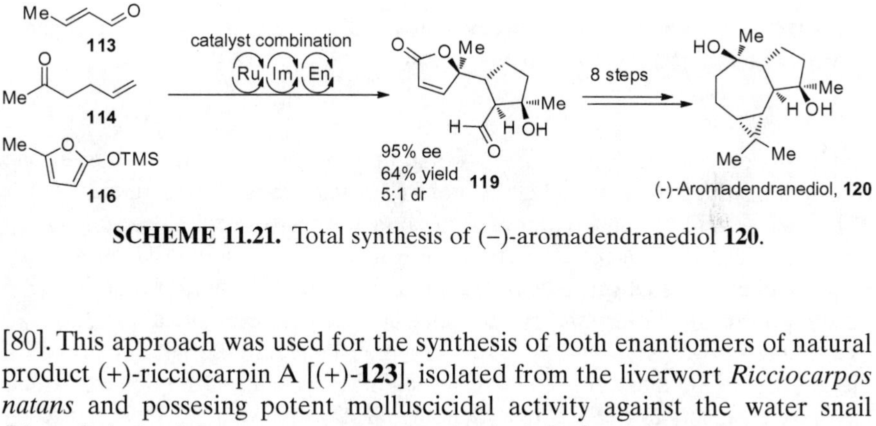

SCHEME 11.21. Total synthesis of (−)-aromadendranediol **120**.

[80]. This approach was used for the synthesis of both enantiomers of natural product (+)-ricciocarpin A [(+)-**123**], isolated from the liverwort *Ricciocarpos natans* and possesing potent molluscicidal activity against the water snail *Biomphalaria glabrata* vector of schistosomiasis [81]. Among human parasitic diseases, schistosomiasis ranks second in terms of its effects on public health in tropical and subtropical areas after malaria. Therefore, the synthesis of highly potent (+)-ricciocarpin A [(+)-**123**] has attracted considerable attention of various synthetic chemists.

The key features in synthesis include one-pot conjugate reduction and Michael cyclization cascade followed by $Sm(O^iPr)_3$-catalyzed isomerization and highly diastereoselective Tishchenko reaction [82]. In this synthesis, aldehyde **121** forms an iminium ion with MacMillan's catalyst **117**, which undergoes conjugate reduction by the hydride transfer from the Hantzsch ester **122**,

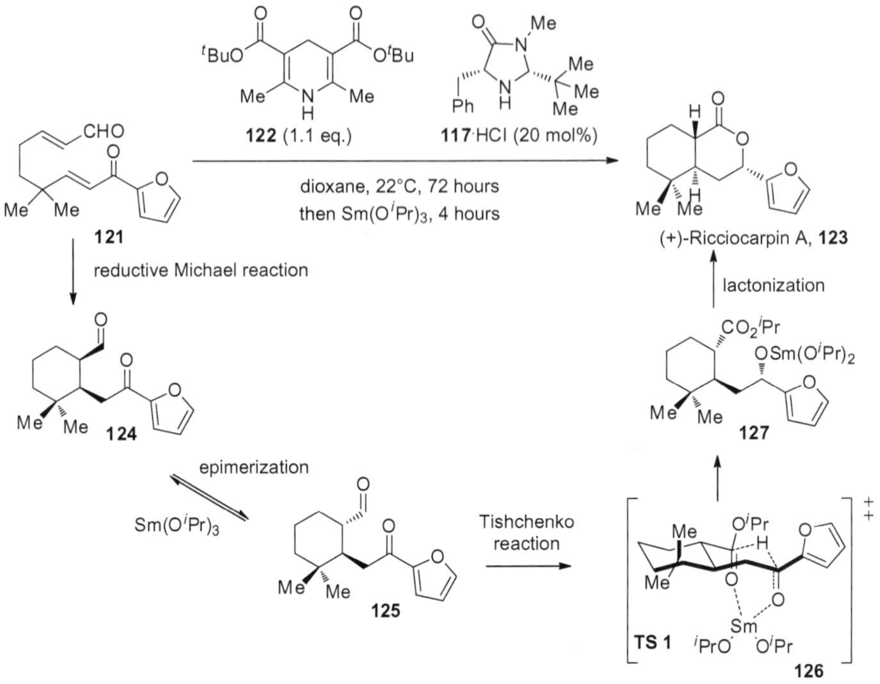

SCHEME 11.22. Synthesis of (+)-ricciocarpin A **123** by a one-pot, three-step, organocatalytic reductive Michael–Tishchenko cascade.

giving an enamine intermediate. This enamine intermediate adds on to enone to give ketoaldehydes *cis*-**124** (99% ee) and *trans*-**125** (97% ee) in a 2:1 ratio. The reaction mixture on treatment with samarium triisopropoxide [Sm(O^{i}Pr)$_3$] resulted in rapid *cis–trans* isomerization followed by a highly diastereoselective Tishchenko reaction. It was assumed that the Tishchenko reaction [83] occurs preferably from *trans*-**125** to generate the desired natural product **123** as a single *trans*-diastereomer in excellent enantioselectivity (99% ee and 48% yield). The highly diastereoselective Tishchenko cyclization may be explained by the transition state **TS-1 126** (Scheme 11.22), in which the two carbonyl groups are effectively chelated by the samarium complex in the critical hydride-transfer step.

Palomo et al. utilized the organocatalyzed Michael reaction for the synthesis of (*S*)-rolipram (**133**), a type IV phosphodiesterase inhibitor (Scheme 11.23) [84]. The key feature includes Michael addition followed by hydrogenation and spontaneous lactamization (Scheme 11.23). The reaction sequence commenced with the Michael addition of nitromethane **129** to the iminium ion generated by the reaction of substituted cinnamaldehyde **128** with the water-compatible pyrrolidine catalyst **130** [85]. The reaction afforded the corresponding Michael adduct in good yield and high enantioselectivity. The

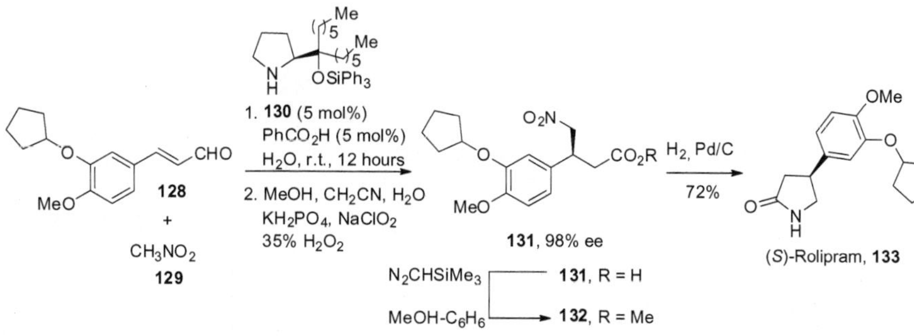

SCHEME 11.23. Synthesis of (*S*)-rolipram **133**.

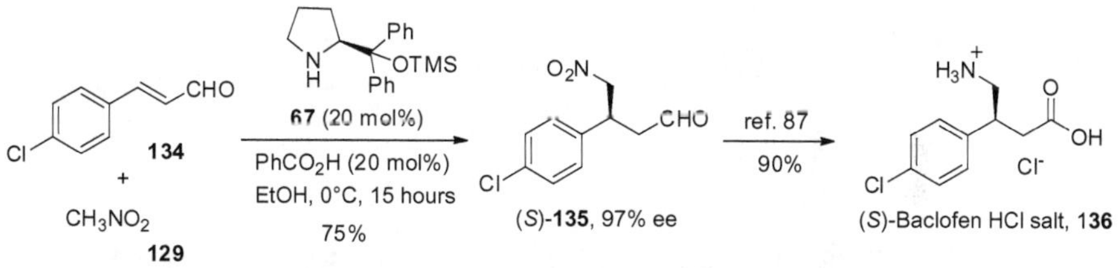

SCHEME 11.24. Synthesis of baclofen **136**.

Michael adduct on oxidation gave corresponding acid **131** (98% ee and 62% yield), which further underwent esterification and yielded **132**. Finally, the reaction proceeded by the hydrogenation of the nitro group to corresponding amine followed by lactamization afforded (*S*)-rolipram **133**, in 72% yield.

Another application of organocatalyzed Michael addition was described by Wang and co-workers for the synthesis of both enantiomers of baclofen (*S*)- and (*R*)-136, a potent gamma-aminobutyric acid (GABA) receptor agonist (Scheme 11.24) [86]. The reaction proceeded with Michael addition of nitromethane **129** to cinnamaldehyde derivative **134**, catalyzed by trimethyl silyl (TMS)-protected diphenyl prolinol (*S*)-**67**. The reaction gave the desired product **135** in high enantioselectivity (~97% ee). The Michael adduct **135** was converted into corresponding acid followed by hydrogenation of nitro group, and afforded baclofen (*S*)-**136**. The other enantiomer baclofen (*R*)-**136** was synthesized by using TMS-protected diphenyl prolinol (*R*)-**67** as a catalyst in a similar manner.

Using similar methodology, Kadas et al. reported the synthesis of (−)-7-deoxy-*trans*-dihydronarciclasine **140** (Scheme 11.25) [88]. The synthesis proceeded with the highly enantioselective Michael addition of nitromethane **129** to benzylideneacetone **137** in the presence of an amine **138** as a catalyst and yielded **139** (>99% ee). Furthermore, **139** underwent Claisen condensation with ethyl formate, followed by an intramolecular nitroaldol reaction and protection of keto group as dioxolane, and gave **140** [89]. Finally, the

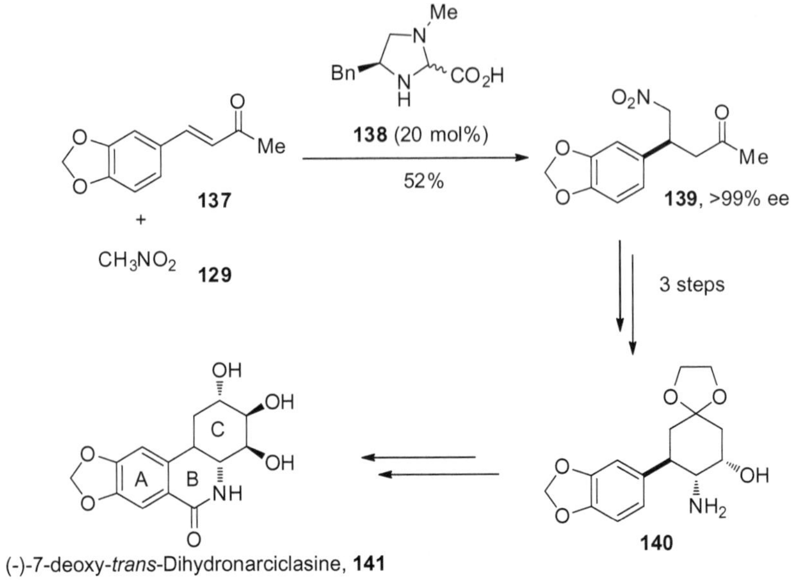

SCHEME 11.25. Synthesis of (–)-7-deoxy-*trans*-dihydronarciclasine **141**.

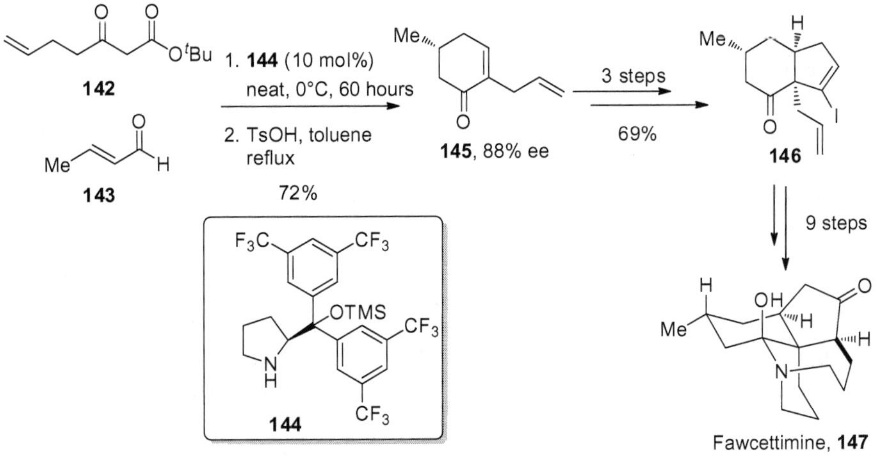

SCHEME 11.26. Total synthesis of (+)-fawcettimine **147**.

generation of the B-ring was achieved by the Banwell modification of the Bischler–Napieralski reaction, and afforded (–)-7-deoxy-*trans*-dihydronarciclasine **141** [90].

Another application of iminium catalysis in combination with an aldol elimination reaction has been used by Toste and co-workers [92a] for the total synthesis of (+)-fawcettimine **147**, a class of *Lycopodium* alkaloids having over 60 natural products [91] (Scheme 11.26) [55e, 92]. This class of compounds is synthetically very challenging as it comprises tetracyclic structure with four stereocenters, including a quaternary one.

The authors used secondary amine catalyst **144** for carrying out the enantioselective conjugate addition of ketoester **142** to crotonaldehyde **143** [93], followed by an intramolecular aldol reaction and decarboxylation of the ester, giving the desired 2-allylcyclohexenone **145** (72% yield and 88% ee). This dienone **145** was converted into allyl-substituted vinyl iodide **146** (as a single diastereomer in 69% yield) by using a sequence of conjugate propargylation, acetylene iodination, and Au(I)-catalyzed cyclization. Finally, the formation of (+)-fawcettimine **147** was completed in nine steps from allyl-substituted vinyl iodide **146**.

Applying the similar methodology, Rios and co-workers described the formal synthesis of (−)paroxetine **152**, an antidepressive drug [94]. One of the key steps involves the organocatalytic enantioselective conjugate addition of amidomalonate **148** to α,β-unsaturated aldehyde **149** followed by intramolecular cyclization of the Michael adduct, giving piperidinone **150** (90% ee). In the presence of borane in THF, the *ent*-**150** was converted into **151**, a key precursor for the synthesis of (−)paroxetine **152** (Scheme 11.27) [95]. The stereochemical outcome in the above reaction can be explained on the basis of a transition state model shown in Scheme 11.28 [94]. According to this scheme, the attack of the amidomalonate **148** to the β-carbon of the iminium ion **153** preferentially takes place from the *Si*-face as the *Re*-face approach is sterically hindered by the bulky aryl substituents of the catalyst to give enamine intermediate **154**. The enamine intermediate **154** isomerizes to iminium ion **155**, followed by hydrolysis to give hemiacetal **156**. It should be noted that epimerization of

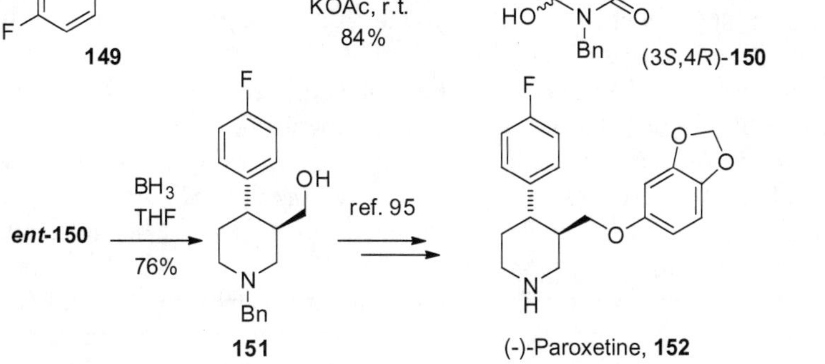

SCHEME 11.27. Formal synthesis of (−)-paroxetine **152**.

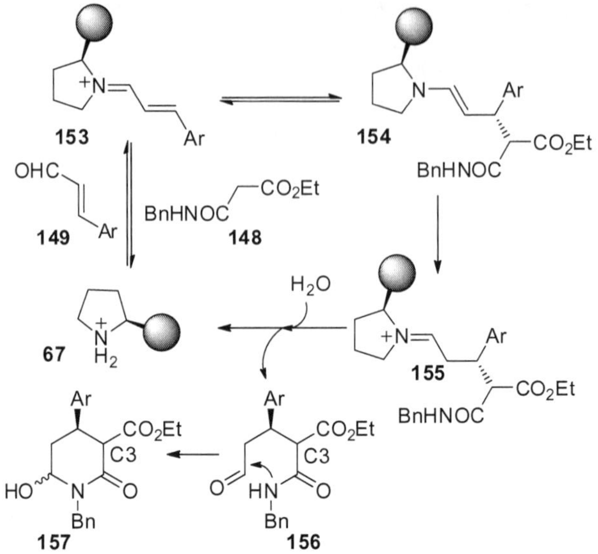

SCHEME 11.28. Proposed mechanism of the organocatalytic reaction.

stereochemically labile stereocenter at C-3 establishes the more stable (3*S*,4*R*)-*trans* configuration in the final hemiacetal **157**.

Fustero and co-workers reported the total synthesis of three different piperidine alkaloids: (+)-sedamine **162** [96], (+)-allosedamine **163**, and (+)-coniine·HCl **165** [97]. The key feature of the synthesis includes organocatalytic enantioselective intramolecular aza-Michael reaction (Scheme 11.29) [98]. The synthesis proceeded with the cross-metathesis reaction of Boc-**158** with **159** and Cbz-**158** with **159**, catalyzed by Grubb's catalyst **56**, providing Boc-**160** and Cbz-**160**. Additionally, Boc-**160** and Cbz-**160** underwent an enantioselective intramolecular aza-Michael reaction, catalyzed by amine **144**, and gave Boc-**161** (94% ee) and Cbz-**161** (94% ee). Finally, Boc-**161** underwent a Grignard reaction followed by reduction, giving (+)-sedamine **162** and (+)-allosedamine **163** in a 3:2 ratio with 60% yield. On the other hand, Cbz-**164** gave (+)-coniine·HCl **165** in a two-step sequence of reactions.

Similar methodology is used for the total synthesis of the biologically active tetrahydroquinoline alkaloid, (+)-angustureine **169** (Scheme 11.30) [99]. The synthesis commenced by the cross-metathesis reaction of **166** and aldehyde **159**, catalyzed by Grubb's catalyst **56**, to give product **167** in 73% yield. The key step involves an enantioselective intramolecular aza-Michael cyclization of **167** by using **144** as a catalyst to give **168** (92% ee) followed by a three-step sequence of reactions, affording (+)-angustureine **169** in 52% yield. The stereochemical outcome of the reaction can be explained on the basis that attack of the amine takes place from the *Re*-face of the iminium ion (**170**, **171**; Scheme 11.31).

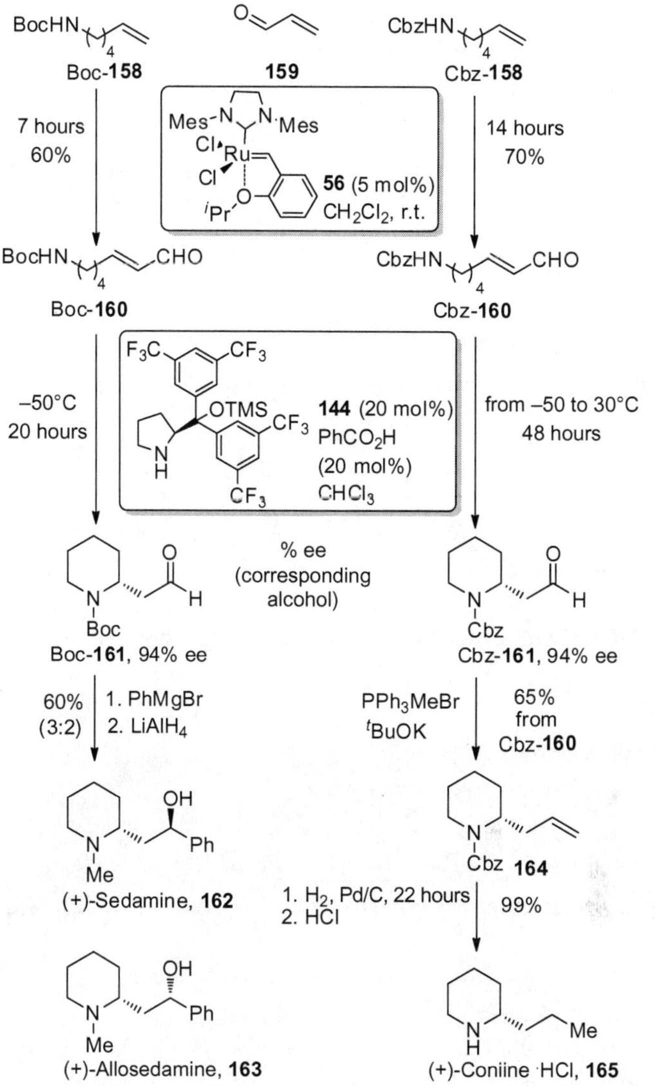

SCHEME 11.29. Concise syntheses of (+)-sedamine **162**, (+)-allosedamine **163**, and (+)-coniine·HCl **165**.

Marigo et al. exploited the iminium–enamine tandem reaction for the asymmetric oxidation of enal by urea peroxide [100]. Nicolaou and co-workers applied this methodology for the total synthesis of the fungal metabolite hirsutellone B **176** [101] (Scheme 11.32) [102]. The synthesis of the tricyclic core of hirsutellone B **176** proceeds with the conversion of (*R*)-(+)-citronellal **47** to iodo enal **172** by the standard reaction sequence. Further, iodo enal **172**

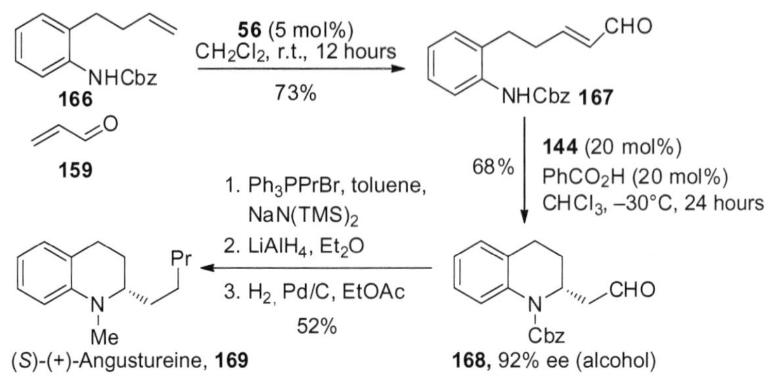

SCHEME 11.30. Enantioselective total synthesis of (+)-angustureine **169**.

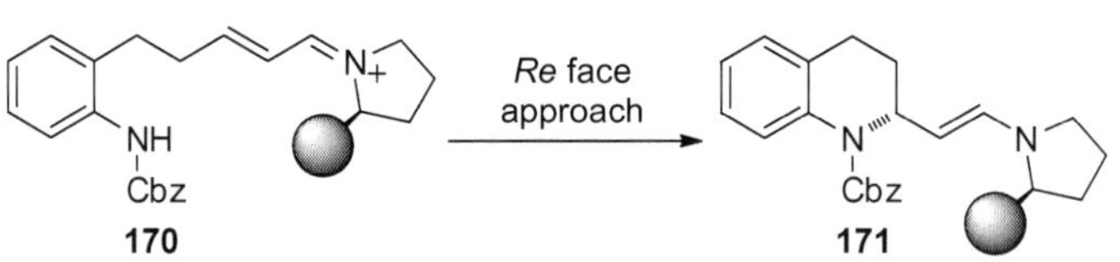

SCHEME 11.31. Stereochemical outcome for the cyclization of **167**.

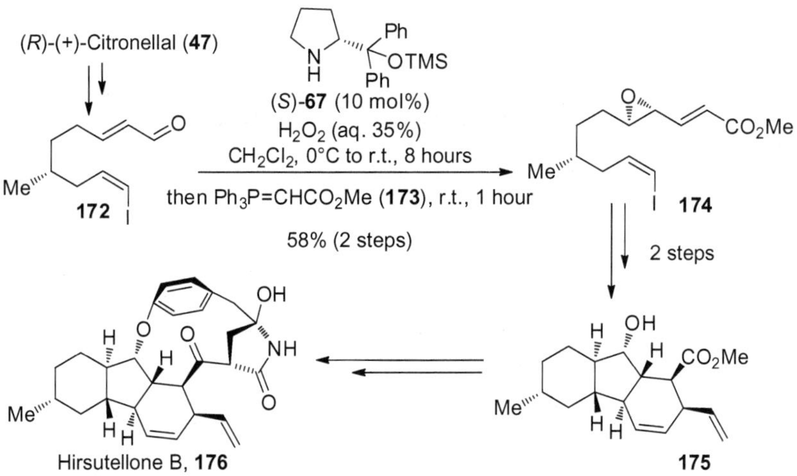

SCHEME 11.32. Synthesis of hirsutellone B **176**.

undergoes asymmetric Jørgensen epoxidation [100] by using H_2O_2, catalyzed by diaryl prolinol catalyst (*S*)-**67** followed by a Wittig reaction with ylide **173**, giving epoxy ester iodide **174** (58% yield). Then, epoxy ester iodide **174** underwent a sequence of novel cascade reactions, yielding hirsutellone B **176** via **175**.

11.3.3. Cycloaddition Reactions

The large variety of cycloadditions that can take place via iminium catalysis include [4+2] [103], [3+2] [104], and [4+3] [105] reactions as well as intramolecular versions of these reactions. The reactions can tolerate a range of dienes and dinophiles, thus exhibiting broad substrate scope and affording products in high yields and high stereoselectivities. Cycloaddition via intramolecular Diels–Alder reaction forms a complex scaffold with high ee. This process is used for the synthesis of natural products and biologically active molecules [106].

Recently, MacMillan and co-workers described a total synthesis of (+)-minfiensine **185** [107] (Scheme 11.33) [108]. The key steps of the synthesis include organocatalytic enantioselective Diels–Alder reaction followed by amine cyclization. The synthesis commenced with the formation of **178** from indole derivative **177** by following a three-step sequence. Additionally, the product **178** with propynal **179** underwent a cascade cyclization reaction, catalyzed by the imidazolodinone catalyst **180** that gave pyrroloindoline tetracycle **181** in good yield (87%) and excellent enantioselectivity (96% ee). Then, after following two successive steps, intermediate **183** was formed. The synthesis of the piperidine ring in **184** was achieved by alkyne radical cylization of intermediate **183** by using *t*Bu_3SnH. Finally, a partial (though regio- and diastereoselective) allene hydrogenation of compound **184** gave (+)-minfiensine **185**. The observed stereochemical outcome is explained from the transition state model shown in Figure 11.3. The cycloaddition step is initiated by the iminium ion intermediate formed by the reaction of secondary amine catalyst **180** with propynal **179**. The *endo*-selectivity is explained by the transition state in which alkyne group is pointing away from the bulky *t*-butyl group of the catalyst. In

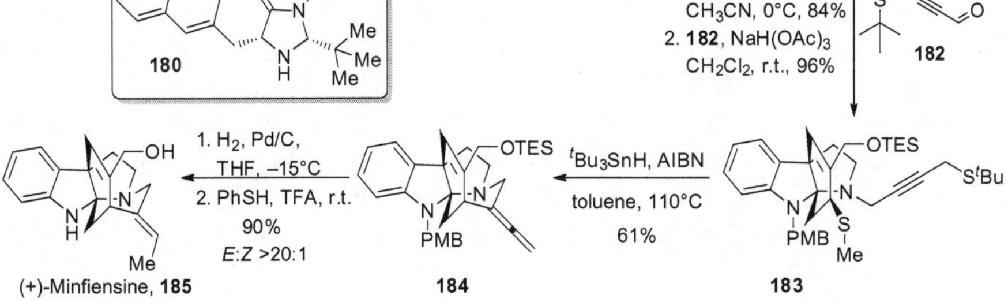

SCHEME 11.33. Nine-step total synthesis of (+)-minfiensine **185**.

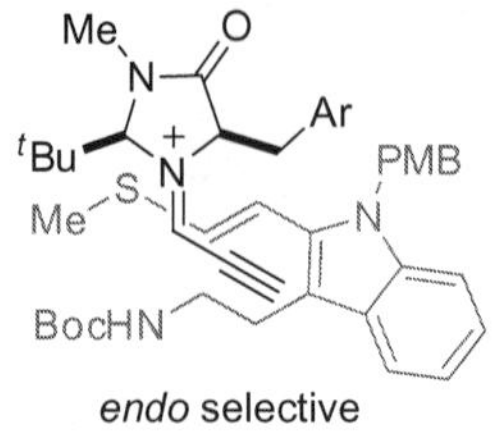

FIGURE 11.3. Proposed transition state for the Diels–Alder reaction.

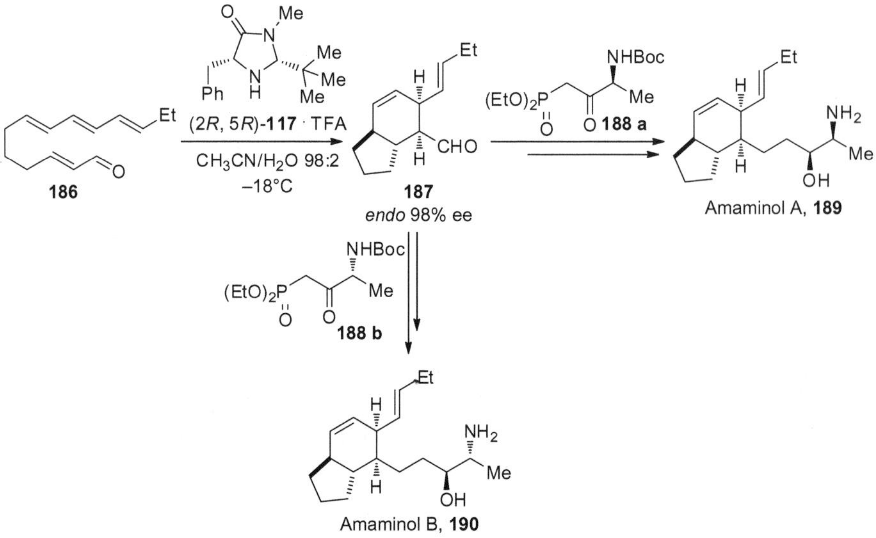

SCHEME 11.34. Organocatalytic syntheses of amaminol A **189** and B **190**.

this conformation, the aryl group shields the top face of the alkyne and thus facilitates the *endo*-selective Diels–Alder cycloaddition with 2-vinyl indole **178** in a regioselective manner.

Using a similar methodology, Kumpulainen and co-workers reported the enantioselective total synthesis of amaminol A **189** and B **190** (Scheme 11.34) [109], isolated by Sata and Fusetani from an unidentified tunicate from the Amami Islands [110]. These bicyclic amino alcohols show remarkable resemblance to crucigasterin and possess moderate cytotoxicity against P388 murine leukemia cells. The synthesis proceeded with intramolecular Diels–Alder reaction of **186**, catalyzed by imidazolidinone **117**, giving *endo* **187** with very high enantioselectivity (98% ee). Furthermore, the coupling of aldehyde **187** with *β*-ketophosphonates **188a** and **188b** in an HWE reaction followed by reduction gave desired products amaminol A **189** and B **190**.

11.4. DIENAMINE CATALYSIS

A closer inspection of the mechanism of β-functionalization in α,β-unsaturated carbonyl compounds reveal the presence of an intermediate, dienamine. The deprotonation at the γ-position of iminium ion gives dienamine intermediate **192** (Scheme 11.35) [17] (which has led to this type of catalysis being termed "dienamine catalysis") that can react with electrophiles, giving γ-functionalization. This discovery represents an example of the umpolung of iminium catalysis. This transformation allows the electrophilic attack to what was considered electrophilic enal in β-functionalization [22]. The catalytic cycle follows through the iminium ion **191**, arising by the condensation of amine catalyst with an α,β-unsaturated carbonyl compound (Scheme 11.35). The deprotonation at the γ-position of iminium ion **191** gives *trans*-dienamine intermediate **192**, which isomerizies to *cis*-dienamine **193**, followed by the [4+2] cycloaddition, a lowest energy reaction path, giving cyclic hemiaminal **194**. The hydrolysis of this hemiaminal **194** releases γ-functionalized *cis*-α,β-unsaturated product **195**, which isomerizes to more stable *trans* product.

The γ-functionalization of α,β-unsaturated carbonyl compounds can provide various chiral carbocycles [111] with complex networks in natural products and pharmaceutical molecules. Some of the recent examples utilizing this methodology are reported in the following section.

Woggon and co-workers reported the asymmetric synthesis of biologically the most significant member of the vitamin E family, α-tocopherol **199** (Scheme 11.36) [112]. It is well known as a very efficient radical chain-breaking

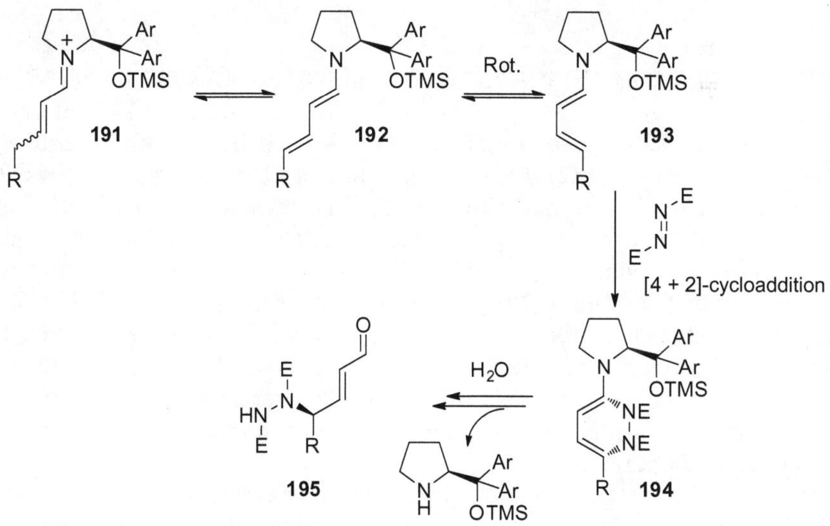

SCHEME 11.35. [4+2]-cycloaddition leading to α-amination of α,β-unsaturated aldehydes (E = CO_2Et).

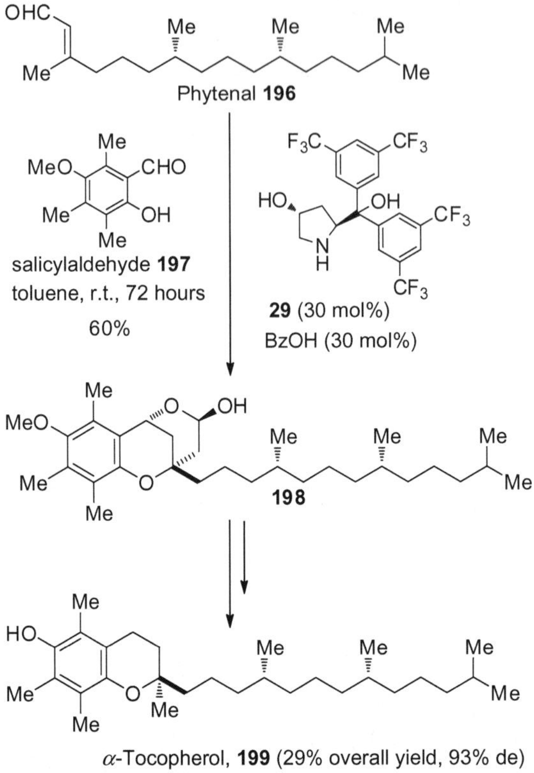

SCHEME 11.36. Synthesis of α-tocopherol **199**.

antioxidant in tissues. The key feature of synthesis involves the domino aldol/oxa-Michael reaction [113] of phytenal **196** with salicylaldehyde **197**, catalyzed by diaryl prolinol **29** to give hemiacetal 12*S*-**198**. Then, by following a sequence of reactions, hemiacetal 12*S*-**198** was converted into 15*S* α-tocopherol **199** with an overall yield of 29% and with 93% de. The formation of lactol 12*S*-**198** follows the mechanism shown in Scheme 11.37. The iminium ion **200** formed between diaryl prolinol catalyst **29** and the enone functionality of phytenal **196** isomerizes to dienamine **201**, which reacts with salicylaldehyde **197** to yield aldol adduct **202**. The aldol adduct **202** then cyclizes diastereoselectivity by intramolecular oxa-Michael reaction followed by hydrolysis, which closes the ring to give 12*S*-**198** and releases the catalyst **29**. Here, the aldol reaction is a key step, which controls the stereoselectivity in the formation of *syn* arrangement of six-membered lactol.

Hong and co-workers developed the highly enantioselective organocatalytic Robinson annulation of α,β-unsaturated aldehydes in the presence of L-proline or diarylpyrrolinol silyl ethers and trialkylamine as a catalyst, affording [4+2] cycloaddition adducts [114]. This methodology was applied for the

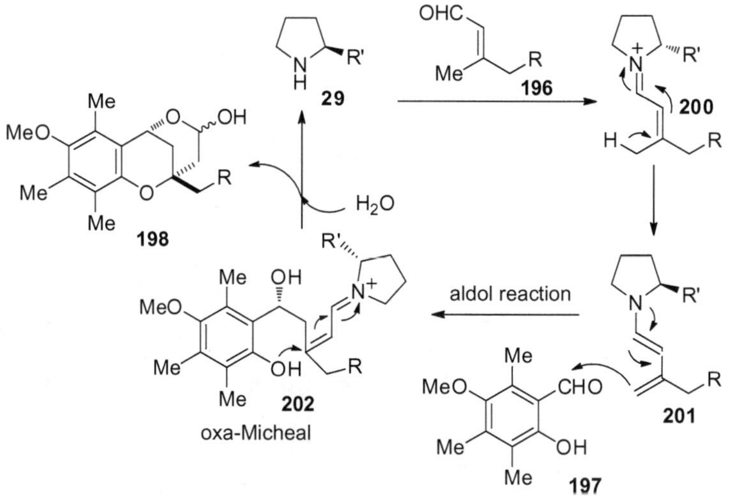

SCHEME 11.37. Proposed mechanism for the domino aldol–oxa-Michael reaction.

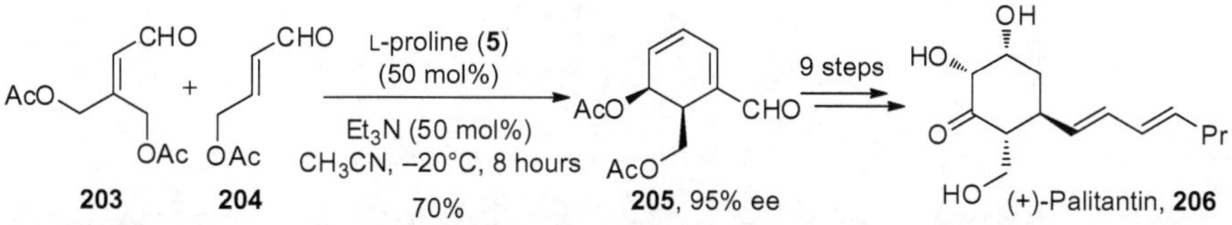

SCHEME 11.38. Application of the organocatalytic asymmetric Robinson annulation of *α*,*β*-unsaturated aldehydes to the total synthesis of (+)-palitantin **206**.

total synthesis of (+)-palitantin **206** (Scheme 11.38) [115], a polyketide metabolite isolated from *Penicillium palitans* [116] and *Pencillium brefeldianum* [117] that exhibits antifungal [118] and antibiotic activity [119], and acts as an HIV-I integrase inhibitor [120]. The key step involves the [4+2] cycloaddition of dialkyl substituted enone **203** and monoalkyl substituted enone **204**, catalyzed by L-proline and triethylamine as a cocatalyst, and giving cyclic adduct **205** in high yield (70%) and high enantioselectivity (95% ee). The cyclic adduct **205** further underwent other nine series of steps to yield (+)-palitantin **206**.

The authors proposed the mechanism for the key [4+2] cycloaddition step (Scheme 11.39). The reaction takes place by the formation of dienamine between proline **5** and *β*-dialkyl-*α*,*β*-unsaturated aldehyde **203**, as they are more easily enolized as compared to *β*-monoalkyl-*α*,*β*-unsaturated aldehyde **204**. The Michael reaction takes place through E-transition state **207** followed by intramolecular Mannich reaction activated by the formation of iminium ion **208** to afford the desired adduct **205**.

Another example of the application of dienamine catalysis in the total synthesis was reported by Christmann et al. [121]. They carried out synthesis

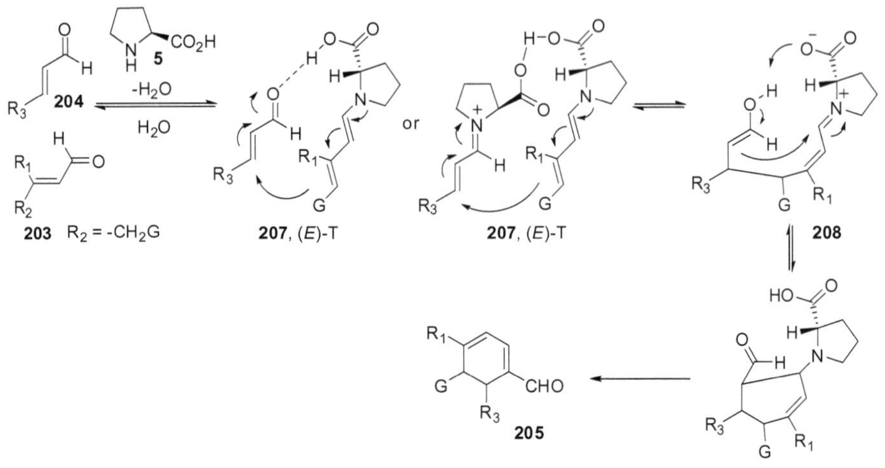

SCHEME 11.39. Mechanism of asymmetric Robinson annulation.

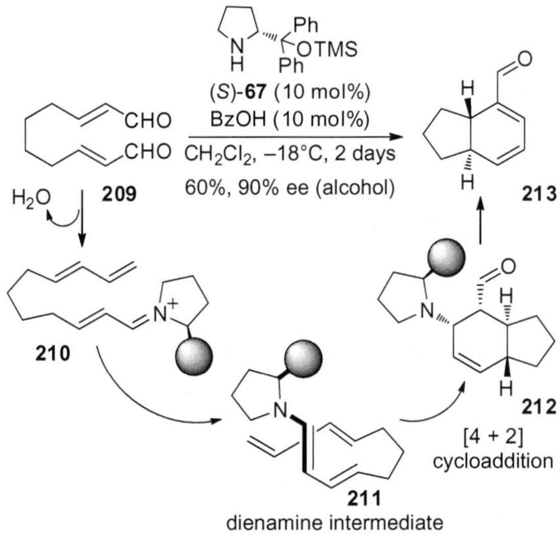

SCHEME 11.40. Synthesis of **213**, a pungent constituent of black cardamom.

of compound (+)-**213**, which was identified as a pungent constituent of black cardamom by using diaryl prolinol **67** as a catalyst (Scheme 11.40) [122]. The synthesis commenced with the formation of iminium ion **210** between diaryl prolinol **67** and α,β-unsaturated aldehyde **209**, which isomerizes to dienamine intermediate **211**. This intermediate presumably adopts the conformation that minimizes the repulsions with the bulky aryl substituents of the catalyst **67** and exposes the unshielded face of the π-system for an *endo* approach of the enal, providing **212**. Finally, β-hydride elimination regenerated the catalyst **67** and

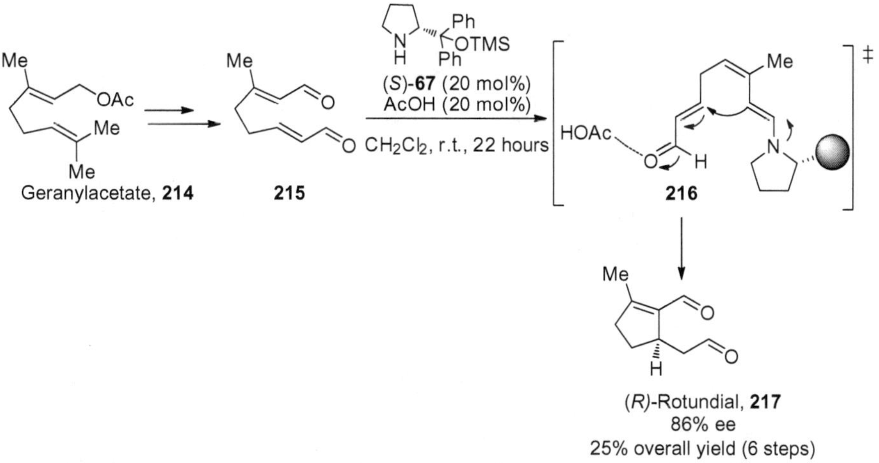

SCHEME 11.41. Synthesis of (*R*)-rotundial **217**, via an organocatalytic Rauhut–Currier-type reaction.

gave product **213**. It is proposed that a second molecule of the catalyst may assist this elimination by the formation of iminium ion with the remaining aldehyde moiety.

The same group reported the other application of dienamine catalysis for carrying out an intramolecular Rauhut–Currier reaction. This methodology was applied for the total synthesis of mosquito repellent (*R*)-rotundial **217** [123] (Scheme 11.41) [124], isolated from *Vitex rotundifolia* [125]. The synthesis commenced with the formation of an iminium ion between diaryl prolinol catalyst **67** and dialdehyde **215**, obtained from commercially available geranyl acetate **214**. The iminium ion isomerizes to dienamine **216**, which intramolecularly cyclizes to give Rauhut–Currier adduct **217**. Acetic acid assists this cyclization step by activating the acceptor.

11.5. SOMO ACTIVATION

The stereoselective α-alkylation of the aldehydes can be now be achieved through organocatalytic enamines by a new activation method involving a radical cation species, with a SOMO [126]. This is an interesting example of the umpolung of enamine catalysis devised recently by MacMillan and coworkers [127]. This reaction allows the nucleophilic attack to what was considered nucleophilic enamine in α-functionalization. The catalytic cycle follows by the condensation of amine catalyst with carbonyl compound to give an enamine intermediate (Scheme 11.42) [22]. The addition of an oxidant removes one electron from the enamine's π-system, giving a radical cation **218**, which reacts with nucleophiles, rich in π-electrons to form a new radical intermediate

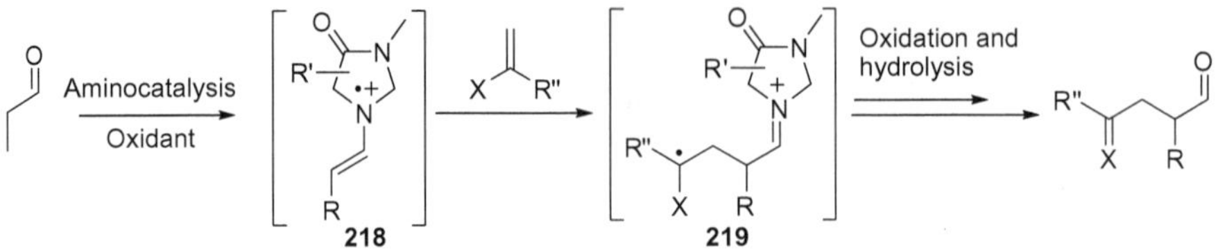

SCHEME 11.42. SOMO catalysis: the reaction of carbocation radical intermediate with two oxidations.

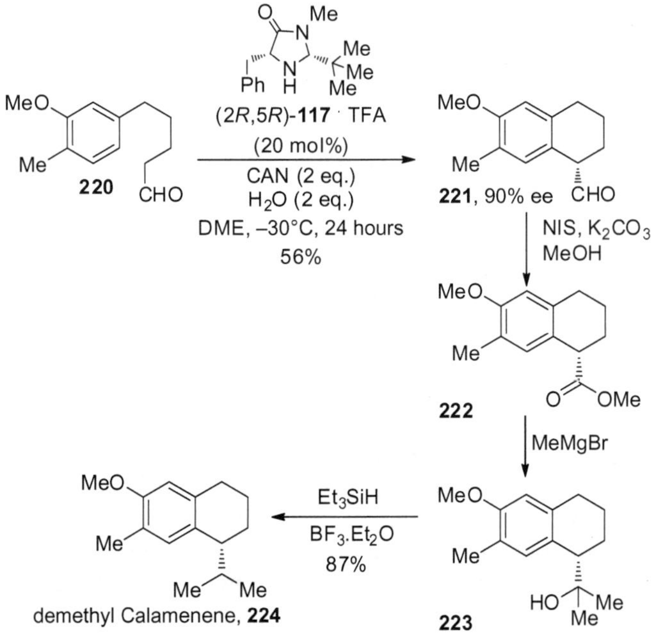

SCHEME 11.43. Total synthesis of demethyl calamenene **224**.

219. A second oxidation followed by hydrolysis gives the desired α-functionalized product with concomitant regeneration of the catalyst. Nowadays, this methodology is exploited in the synthesis of various natural products and biologically active molecules.

Nicolaou and co-workers utilized the SOMO catalysis to synthesize potent cytotoxic demethyl calamenene **224** [128] (Scheme 11.43) [129]. The synthesis proceeded with the reaction of an aryl iodide with 4-penten-1-ol and afforded aldehyde **220** [19b, 130]. Further, the aldehyde **220** underwent intramolecular Friedel–Crafts-type reaction [131] by using imidazolidinone **117** as a catalyst under the SOMO catalysis conditions to give a bicyclic aldehyde **221** (56% yield, 90% ee). The one-pot oxidation–esterification of the aldehyde **221** furnished a bicyclic ester **222**, followed by reaction with a Grignard reagent and

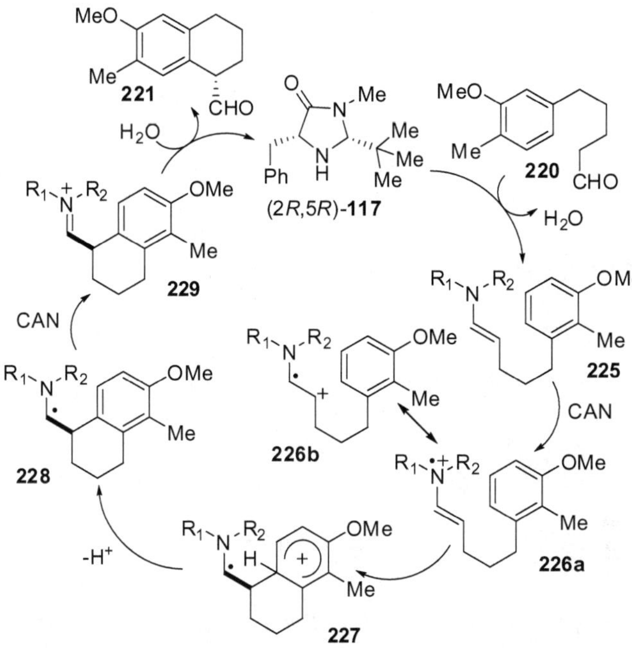

SCHEME 11.44. Proposed catalytic cycle of the intramolecular Friedel–Crafts α-arylation .

yielded an alcohol **223** in 65% yield. Finally, reductive deoxygenation using Et_3SiH in the presence of $BF_3{\cdot}Et_2O$ gave the desired demethyl calamenene **224** in 87% yield. The authors have proposed the catalytic cycle for an intramolecular Friedel–Crafts reaction (Scheme 11.44). The enamine **225** formed from the aldehyde **220** and imidazolidinone catalyst **117** on single-electron transfer (SET) reaction with cerium(IV) ammonium nitrate (CAN) generated a reactive radical cation intermediate **226a/226b**. This reactive intermediate underwent an intramolecular reaction to generate γ-complex **227**, which loses protons to form intermediate **228** followed by another SET reaction using CAN to give iminium ion **229**. After the hydrolysis of iminium ion **229**, the catalyst reenters the cycle and the desired bicyclic product **221** is obtained.

11.6. BRØNSTED ACID AND HYDROGEN BOND CATALYSIS

The Brønsted acids and chiral hydrogen bond donors play a very crucial role in asymmetric catalysis [19a, 132]. Brønsted acid catalyzes a reaction by forming a hydrogen bond to an electrophile, thus decreasing its electron density and activating it toward the nucleophilic attack (Fig. 11.4). This H-bond activation of electrophile is in direct analogy to Lewis acid activation.

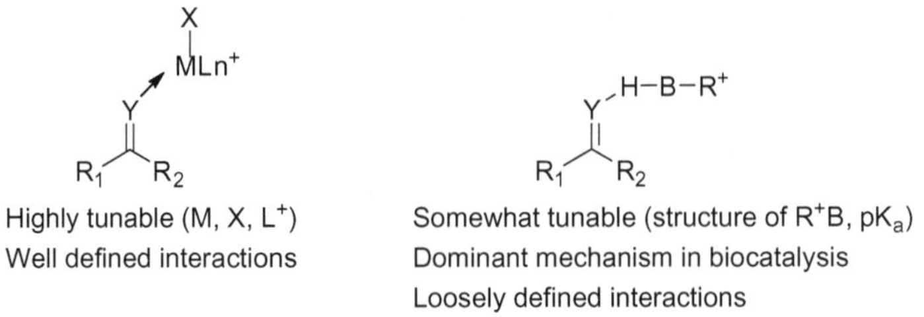

FIGURE 11.4. Brønsted acid and hydrogen bond donor catalysis.

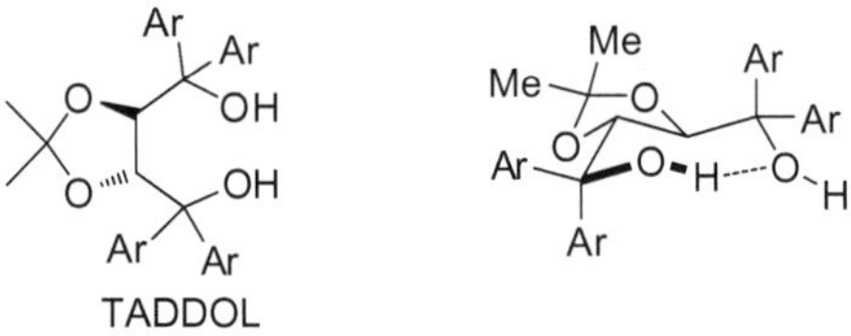

FIGURE 11.5. TADDOL and its transition state model.

Depending on the mode of activation, Brønsted acids have been divided into two different categories:

- *Single Hydrogen Bond Donors.* Examples are TADDOL [19b, 132a,b, 133], Binols, and Binol-derived chiral phosphoric acid [134]. These species catalyze the reaction by the formation of a single hydrogen bond with an electrophile. For highly enantioselective catalysis, attaining the suitable rigid catalyst–substrate complex by isolating a hydrogen bond is very challenging. The mechanistic aspects show that rigid transition state is achieved via the formation of an intramolecular hydrogen bond, which defines the orientation and increases the acidity of the OH bond not engaged in hydrogen bonding (Fig. 11.5). In addition, aromatic moieties surrounding the hydroxy groups further define the position of bound substrate through steric and electronic interactions.
- *Double Hydrogen Bond Donors.* Examples are thiourea, urea [132c, 135], and guanidinium and amidinium ions [136]. These species activate the electrophile by forming two hydrogen bonds simultaneously. In contrast to single hydrogen bond donors, two hydrogen bond interactions benefit from increased strength and directionality. These Brønsted acids and H-bond donors have been used for catalyzing asymmetric reactions and applied for the synthesis of natural products and pharmaceutical active compounds as described in the following sections.

11.6.1. Chiral Phosphoric Acids

Axially chiral phosphoric acid derived from binapthol is an efficient organocatalyst in various asymmetric reactions [134]. Notably, these chiral phosphoric

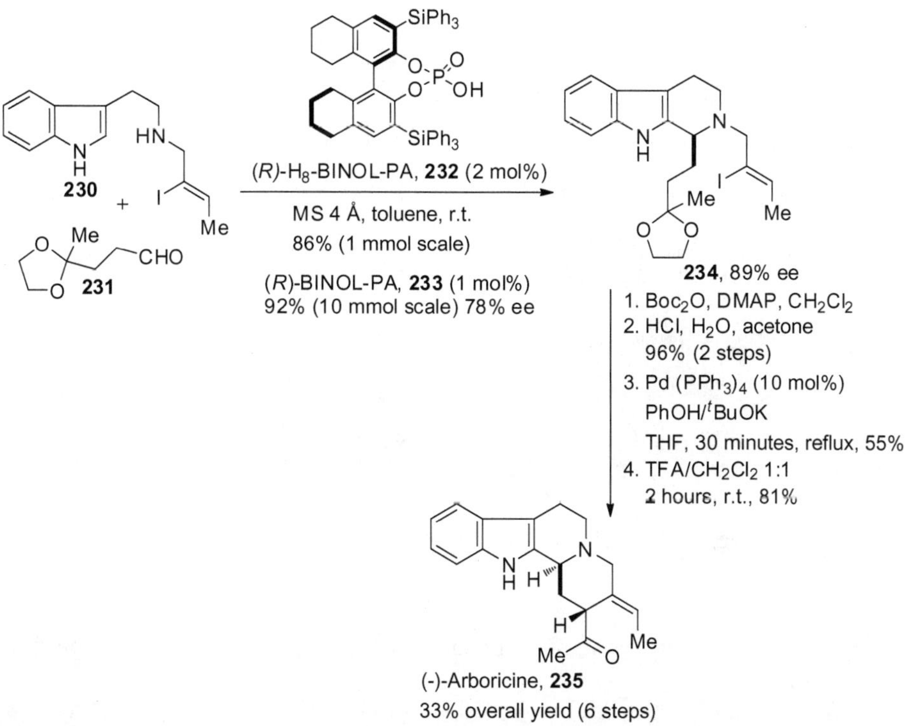

SCHEME 11.45. Organocatalytic enantioselective total synthesis of (–)-arboricine (**235**).

acids can be considered to be bifunctional. In addition to the hydroxy group (which activates the electrophile), they also contain a carbonyl group, which serves as a Lewis base (which activates the nucleophile).

Hiemstra and co-workers have reported the synthesis of deplancheine-type tetracyclic indole alkaloid (–)-arboricine **235** (Scheme 11.45) [137], isolated by Kam and co-workers from the leaves of *Kopsia arborea* [138]. The alkaloid (–)-arboricine **235** exhibits the moderate ability to reverse multidrug resistance in vincristine-resistant KB (VJ300) cells. One of the key steps involves the asymmetric Pictet–Spengler reaction [139] of protected 4-oxo-pentanal **231** and known tryptamine derivative **230** [140] by using phosphoric acid **232** as a catalyst, which gave **234** (86% yield and 89% ee). Finally, the diastereoselective Pd(0)-catalyzed iodoalkane–enolate cyclization of **234** yielded the desired (–)-arboricine **235** after deprotection. The authors utilized the methodology devised by Sole et al. [141], in which reaction occurs in the presence of potassium phenoxide, thus avoiding migration of the *exo*-cyclic double bond in conjugation with the keto-carbonyl group. It is to be noted that only 1 mol% of catalyst (*R*)-BINOL-PA **233** was required to carry out the reaction on a 10-mmol scale to generate the compound **234** in 92% yield and 78% ee. The absolute configuration of the product (3*S*, 15*R*) was determined by X-ray

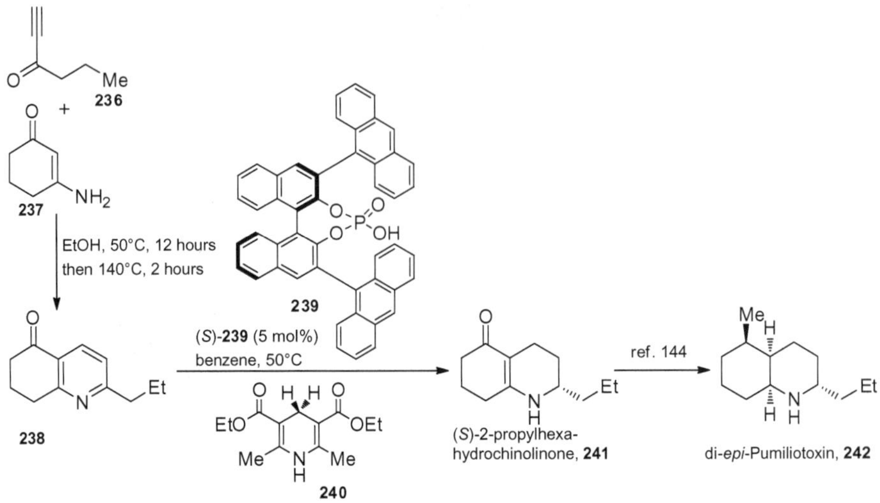

SCHEME 11.46. Synthesis of di-*epi*-pumiliotoxin C **242**.

crystallographic analysis and appeared to be opposite to the published structure.

Rueping and Antonchick reported a formal synthesis of di-*epi*-pumiliotoxin C **242** (Scheme 11.46) [142]. They described the synthesis of (*S*)-2-propylhexahydrochinolinone **241**, a key precursor of di-*epi*-pumiliotoxin C **242**. The synthesis of pyridine **238** was achieved by following the procedure of Bohlmann and Rahtz [143]. The key step involves the organocatalytic enantioselective reduction of pyridine **238** by using Hantzsch ester **240** as a hydride source and Brønsted acid **239** as the catalyst, giving (*S*)-2-propylhexahydrochinolinone **241**. Additionally, compound **241** underwent a series of standard reaction sequences, affording di-*epi*-pumiliotoxin C **242** [144]. The authors devised the catalytic cycle for this transformation as shown in Scheme 11.47. They proposed that pyridine ring **238** is activated by proton transfer from the Brønsted acid **239** to form the chiral on pair **243**. Then, the chiral ion pair **243** takes the hydride from Hantzsch ester **240**, which resulted in adduct **244**. The acid-catalyzed isomerization of the adduct **244** gives iminium ion **245**, which further takes the hydride ion from Hantzsch ester **240** to generate desired pyridine **241** with concomitant regeneration of the Brønsted acid catalyst **239**.

11.6.2. Monofunctional Ureas and Thioureas

A number of unique features of the urea structure enable it to act as a catalyst for diverse asymmetric reactions. The high catalytic activity is due to the ability of thiourea to activate electrophiles by double H-bonding with tenability [135].

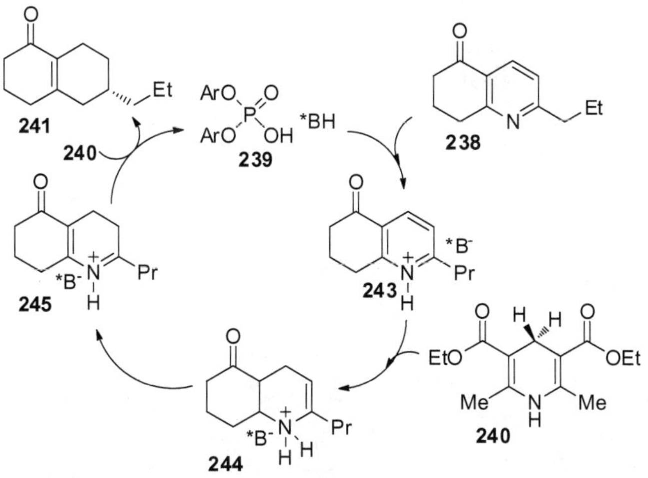

SCHEME 11.47. The postulated mechanism for the enantioselective organocatalytic reduction of pyridine **238**.

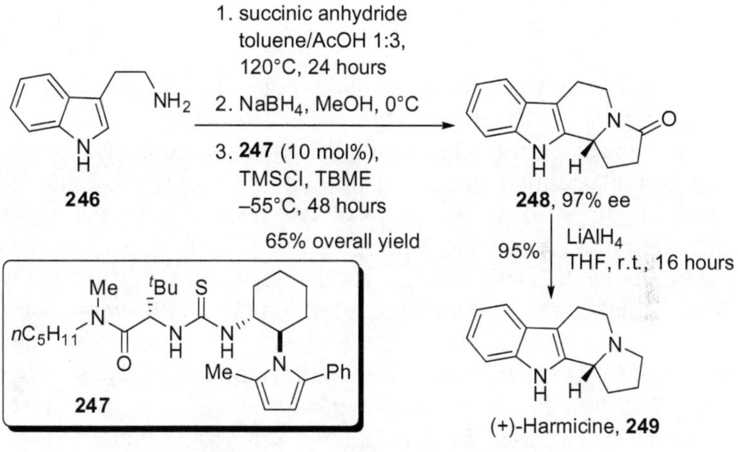

SCHEME 11.48. Synthesis of (+)-harmicine **249**.

The variation of substituents at the nitrogen atom permits a high degree of tuning of steric and electronic properties of the catalyst.

Jacobsen and co-workers reported a highly enantioselective organocatalytic synthesis of indolizidinones and quinolizidinones by using thiourea catalyst **247**. The key step of the reaction is the Pictet–Spengler cyclization of β-indolyl ethyl hydroxylactams. To demonstrate the application of this methodology, total synthesis of (+)-harmicine **249** [145] was carried out (Scheme 11.48) [146]. The synthesis commenced with highly enantioselective Pictet–Spengler cyclization of tryptamine **246** by using thiourea **247** as a catalyst, affording lactam

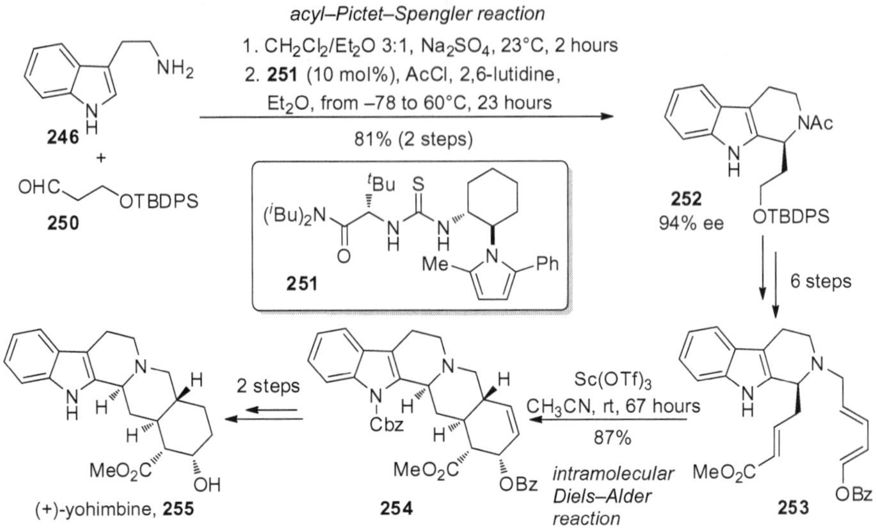

SCHEME 11.49. Synthesis of (+)-yohimbine **255**.

248 with 97% ee. Finally, the simple reduction of lactam **248** with $LiAlH_4$ gave the desired product (+)-harmicine **249**.

Similar methodology was utilized for the total synthesis of (+)-yohimbine **255**, an important member of the monoterpenoid indole alkaloids (Scheme 11.49) [147, 148]. One of the key steps of synthesis involves the acyl-Pictet–Spengler reaction of tryptamine **246** with aldehyde **250** by using thiourea **251** as a catalyst, giving **252** with 94% ee. The authors proposed that catalyst **251** controls the attack at the acyl iminium intermediate. Furthermore, incorporation of diene in the side chain was achieved in six steps by reductive amination, which yielded **253**. Then, **253** underwent an intramolecular Diels–Alder reaction, giving **254** with the simultaneous generation of four new stereogenic centers [149]. The synthesis of (+)-yohimbine **255** was concluded in two steps by following the reduction and deprotection of the hydroxy group of **254**.

Barbas and co-workers developed the novel, highly enantioselective organocatalytic synthesis of pyrrolidinoindolines comprising crucial chiral quaternary C-3 stereocenters by using thiourea-catalyzed 1,4-addition of oxindoles to nitroalkenes [150]. This methodology was applied for the synthesis of (+)-esermethole **261**, a key precursor to (+)-physostigmine **262** (Scheme 11.50) [151]. The 1,4-conjugate addition of oxindole **256** to nitroethylene **257**, catalyzed by thiourea **258**, gave the Michael adduct **259** with 65% yield and 83% ee (96% ee after crystallization). The reduction of Michael adduct **259** followed by methyl carbamate formation afforded a key intermediate **260**. Finally, intramolecular reductive amination/cyclization of **260** concluded the synthesis of (+)-esermethole **261** (72% yield).

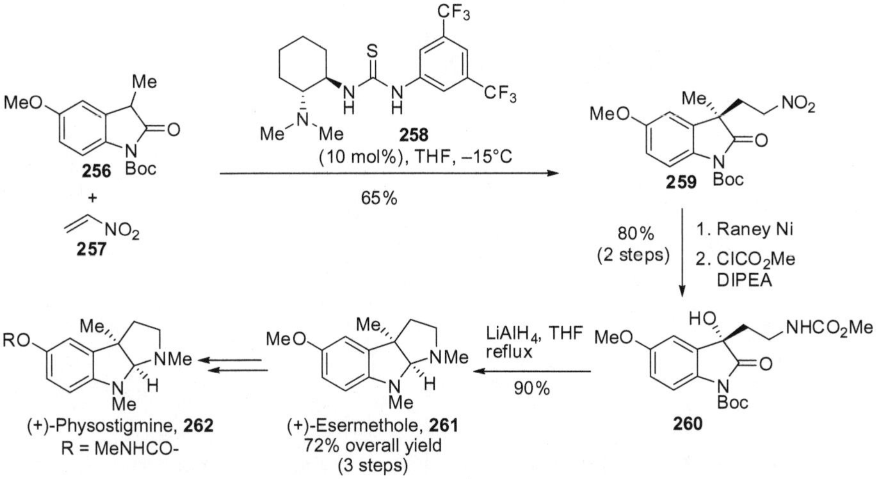

SCHEME 11.50. Synthesis of (+)-esermethole **261**.

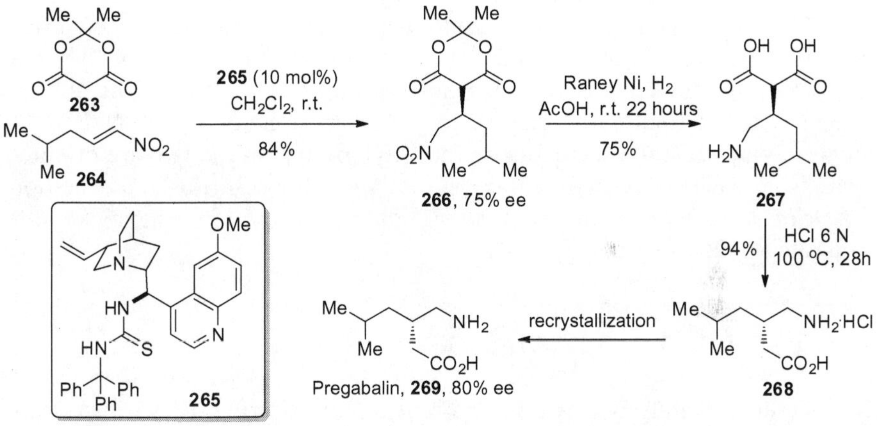

SCHEME 11.51. Organocatalytic enantioselective synthesis of pregabalin **269**.

Koskinen and co-workers reported the total synthesis of pregabalin **269** [152a,b], an important anticonvulsant drug (Scheme 11.51) [152c]. The synthesis commenced with the enantioselective Michael addition of Meldrum's acid **263** to nitroalkene **264**, catalyzed by quinidine-derived trityl-substituted thiourea **265**, and afforded nitro compound *S*-**266** in high yield and good enantioselectivity (75% ee). Finally, the hydrogenation of the compound **266** followed by decarboxylation of di-acid **267** afforded pregabalin hydrochloride **268** in 59% overall yield. The recrystallization of pregabalin hydrochloride **268** concluded synthesis of pregabalin **269** with enantioselectivity of 80%. The authors

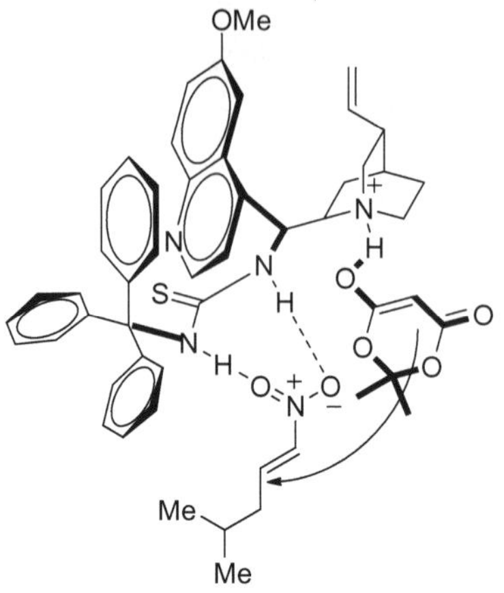

FIGURE 11.6. Proposed transition state leading to the major enantiomer (*S*)-**266**.

proposed that the stereochemical outcome of the reaction is mediated from the complex formed between Meldrum's acid **263**, nitro alkene **264**, and catalyst **265** by hydrogen bonding interactions (Fig. 11.6). The two oxygen atoms of the nitro group form hydrogen bonds with both NH groups of thiourea, thus activating the electron-deficient nitro-group-bearing olefin. The authors proposed that quinidine is oriented in the same direction as the N–H bond and also placed on the same face as the trityl group, but the two are oriented away from each other to avoid any steric interactions. In this conformation, attack of enolate preferably occurs from the top face of *S-cis* nitroalkene to generate (*S*)-**266**.

11.7. BRØNSTED BASE CATALYSIS

Brønsted base catalysts play an important role in carrying out various asymmetric reactions [153]. In particular, *Cinchona* alkaloids, isolated from the bark of the *Cinchona* and *Remijia* species, catalyze many useful processes with high enantioselectivities [154]. *Cinchona* alkaloids act as a catalyst by deprotonating substrates with acidic protons, forming the contact ion pair between the resulting anion and protonated amine of the catalyst (Fig. 11.7). This interaction provides a chiral environment around the anion and permits enantioselective reaction with electrophiles.

11.8. BIFUNCTIONAL CATALYSIS

In bifunctional catalysis, organocatalysts operate by simultaneous activation of both the electrophile and the nucleophile [155]. In recent years, species

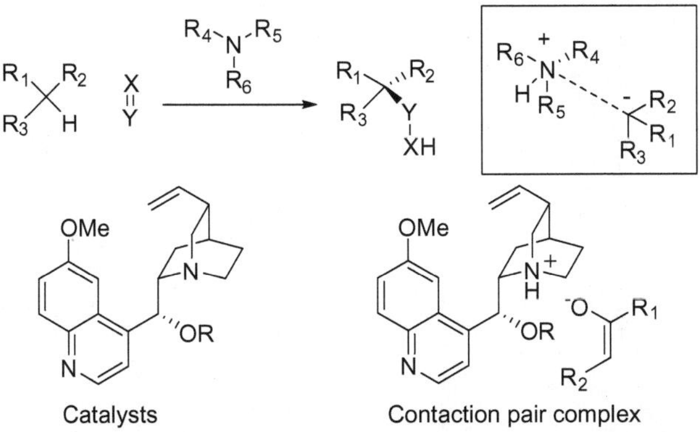

FIGURE 11.7. *Chinchona* alkaloids as catalysts.

incorporating H-bond donors along with other Brønsted bases, nucleophilic, or acidic functional groups in a chiral scaffold constitute a highly successful class of catalysts. An H-bond donor group activates the electrophile, whereas the Brønsted base activates the nucleophile. These include thiourea or urea derivatives and *Cinchona* alkaloid derivatives.

11.8.1. Bifunctional *Cinchona* Alkaloids

Several of these *Cinchona* alkaloids posses a free acidic hydroxy group in addition to basic tertiary amine moiety [154]. These alkaliods catalyze the reaction by activation of nucleophile by general base mechanism and electrophile by H-bonding interaction.

11.8.2. Bifunctional Chiral Urea and Thiourea

The urea or thiourea can easily be tuned to bifunctional catalysts by incorporating additional basic and acidic functional groups at the amine side chain [135]. The incorporation of thiourea moiety into a naturally, privileged chiral skeleton of axially binaphthyl and *Cinchona* alkaloids generate a novel class of efficient bifunctional organocatalysts. The use of derivatized bifunctional chiral *Cinchona* alkaloids, chiral urea, and thiourea for generating chiral stereogenic centers in natural products and biologically active compounds are described below.

Falck and co-workers reported the enantioselective organocatalytic oxa-Michael reaction of achiral γ/δ-hydroxy-α-β-enones by using boronates **272/273** as chiral hydroxide donors, catalyzed by bifunctional catalyst **274**. This methodology was applied for the synthesis of acetate **275** [156], a potent antifungal/hepatic protective agent isolated from avocados, and (+)-(S)-streptenol A **276** [157] (Scheme 11.52) [158]. The synthesis involves oxa-Michael addition of phenyl boronoic acid **272** and **273** to enone **270** and **271**, respectively, catalyzed

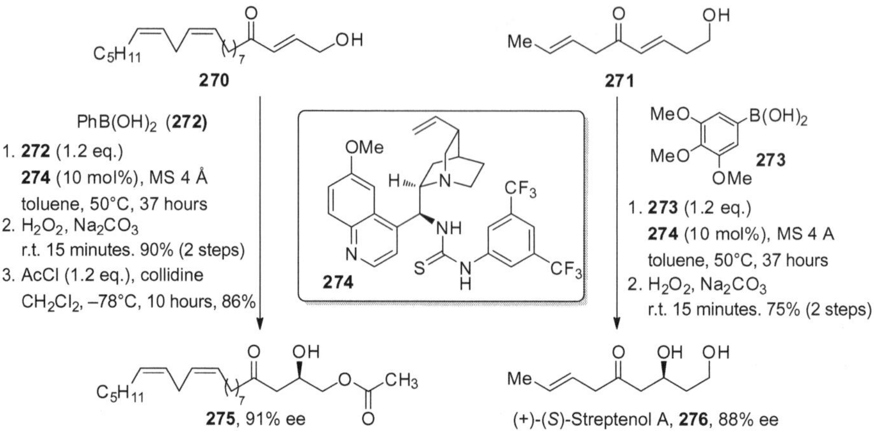

SCHEME 11.52. Organocatalytic synthesis of acetate **275** and (+)-(*S*)-streptenol A **276**.

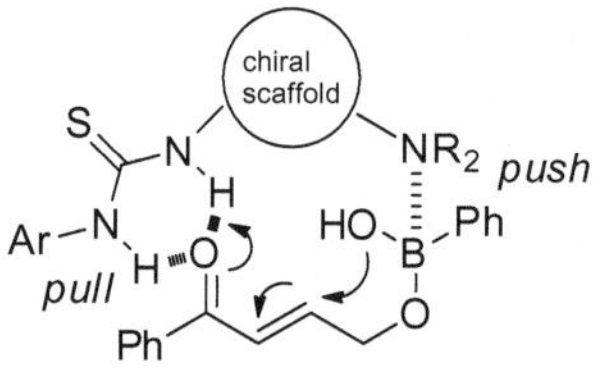

FIGURE 11.8. Proposed asymmetric catalysis based on push/pull-type bifunctional organocatalysis.

by bifunctional organocatalyst **274** followed by a standard reaction sequence. The reaction is believed to take place through a complex formed between a boronic acid hemiester, generated in situ from γ/δ-hydroxy-α,β-enones and a chiral amine catalyst by hydrogen bonding interactions. The authors devised a model, shown in Figure 11.8, to explain the induction of chirality, which is based on a push/pull-type mechanism of the bifunctional organocatalyst [150a, 159]. According to the model, the carbonyl group of the boronic acid hemiester forms a hydrogen bond with thiourea (the pull) and the tertiary nitrogen forms a complex with boron (the push). This complexation increases the nucleophilicity of the boronate oxygen as well as increasing the electrophilicity of the enone in a chiral environment [160].

Schaus and co-workers have reported two different strategies (Scheme 11.53) for the organocatalytic enantioselective total synthesis of SNAP-7941 285, a chiral dihydroprimidone (DHPM) [161]. SNAP-7941 **285** acts as a small molecule inhibitor of Melanin concentrating hormone (MCH1-R) in a G protein-coupled receptors biased library screening. As identified by social interaction studies, inhibition of MCH1-R promotes weight loss in obese rats

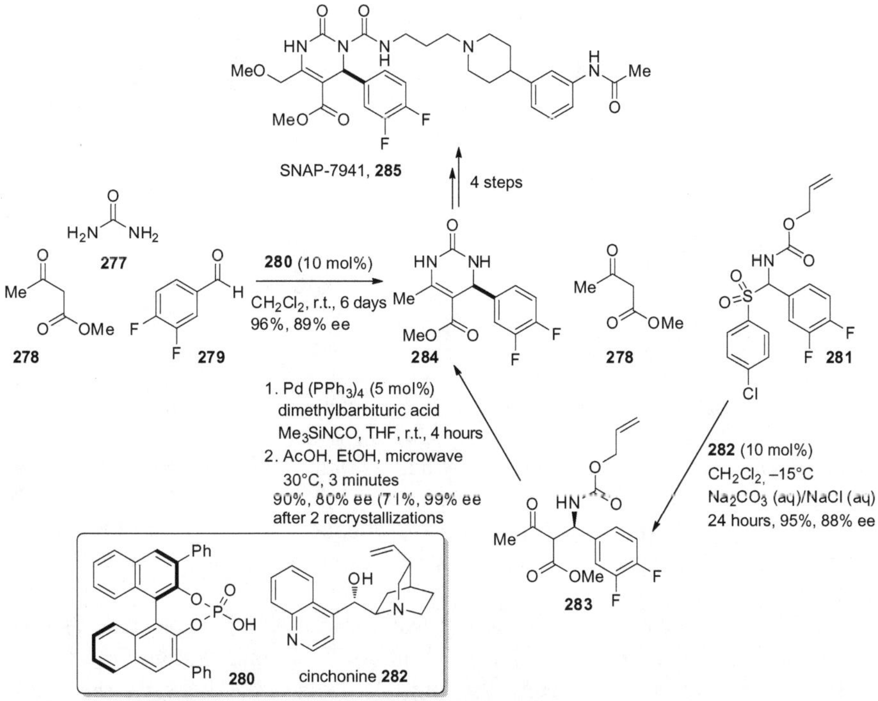

SCHEME 11.53. Synthesis of SNAP-7941 **285**, by an asymmetric Biginelli reaction and an asymmetric Mannich reaction.

and decreases depression and anxiety in both rats and guinea pigs [162]. The key step in one of the synthesis is the enantioselective organocatalytic Mannich reaction [163] of β-ketoester **278** and α,α-allyl-carbamate-3,4-difluorophenyl sulfone **281**, catalyzed by *Cinchona* alkaloid **282**. Further, the chiral amine **283**, obtained in 88% ee, underwent a two-step sequence, generating SNAP-7941 core **284**. The key step in the second strategy involves the Biginelli reaction [164] of urea **277**, methyl acetoacetate **278**, and aldehyde **279** in multicomponent reactions, catalyzed by chiral phosphoric acid **280**, giving the desired SNAP-7941 core **284**. In the first strategy, an additional step was required for the synthesis of Mannich substrate, thus favoring the Biginelli reaction method.

Deng et al. reported highly enantioselective organocatalytic tandem conjugate addition–protonation reactions, catalyzed by bifunctional catalyst **288**, generating compounds having a quaternary and tertiary stereocenter in a 1,3-relationship with high diastereoselectivities [165]. The diastereoselectivities obtained from bifunctional catalyst **288** is complementary to that obtained from nonbifunctional organocatalysts. To demonstrate the application of this methodology, the asymmetric formal synthesis of manzacidins C **290** was carried out (Scheme 11.54) [166]. The synthesis commenced with the conjugate

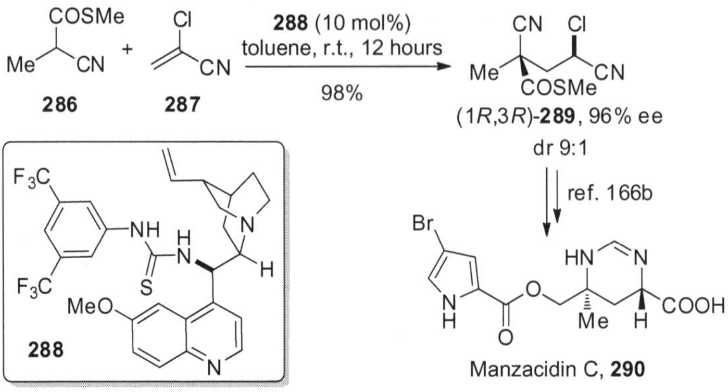

SCHEME 11.54. Formal synthesis of manzacidin C **290**.

addition of reactive **286** to electron-deficient olefin **287** catalyzed by bifunctional catalyst **288**, followed by protonation, which afforded (1*R*, 3*R*) **289** (comprising quaternary and tertiary stereocenters in a 1,3-relationship). Afterwards, following the standard 12-step sequence, synthesis of manzacidins C **290** was achieved.

Dixon and co-workers reported the enantioselective total synthesis of (–)-nakadomarin A **299** [84b, 167], an alkaloid of the manzamine family with a unique hexacyclic framework (Scheme 11.55) [168]. The synthesis commenced with the formation of azabicyclo[6.3.0]undecanone **292** by an intramolecular Julia–Kocienski olefination reaction from pyroglutamate-derived lactam **291** in six steps. The key feature of the synthesis is the conjugate addition of azabicyclo[6.3.0]undecanone **292** to nitroalkene **293** catalyzed by chiral urea catalyst **294**. It afforded γ-nitroester **295** in 57% yield with a 91:9 dr. Further, a critical three-component coupling generated the A-ring of δ-lactam **296** in a single operation from γ-nitroester **295** by in situ formaldimine formation, nitro-Mannich addition, and terminal lactamization [169]. The reduction of the nitro group of **296** under radical conditions followed by $LiAlH_4$ reduction of δ-lactam generated **297**. The one-pot operation involving the partial reduction of the pyrrolidinone followed by furan–iminium ion cyclization generated ABCDE pentacycle **298** of nakadomarin A. Finally, the RCM reaction of **298** in the presence of Grubbs' catalyst **56**, yielded the desired product (–)-nakadomarin A **299** with a Z/E isomer ratio of 63:37. The separation of the desired Z isomer from the E isomer was achieved by high performance liquid chromatography (HPLC), thus affording pure (–)-nakadomarin A **299** in 30% yield.

Hatakeyama and co-workers discovered that β-isocupreidine **302** [170], bearing the acidic phenolic hydroxy group, efficiently catalyzes asymmetric Baylis–Hillman reactions [171] of hexafluoroisopropyl acrylate with aldehydes. This methodology was applied for the formal asymmetric synthesis of (+)-fostriecin **307** and (+)-phoslactomycin B **308**, potentially as lead compounds for anticancer drugs [172]. The asymmetric Baylis–Hillman reaction of

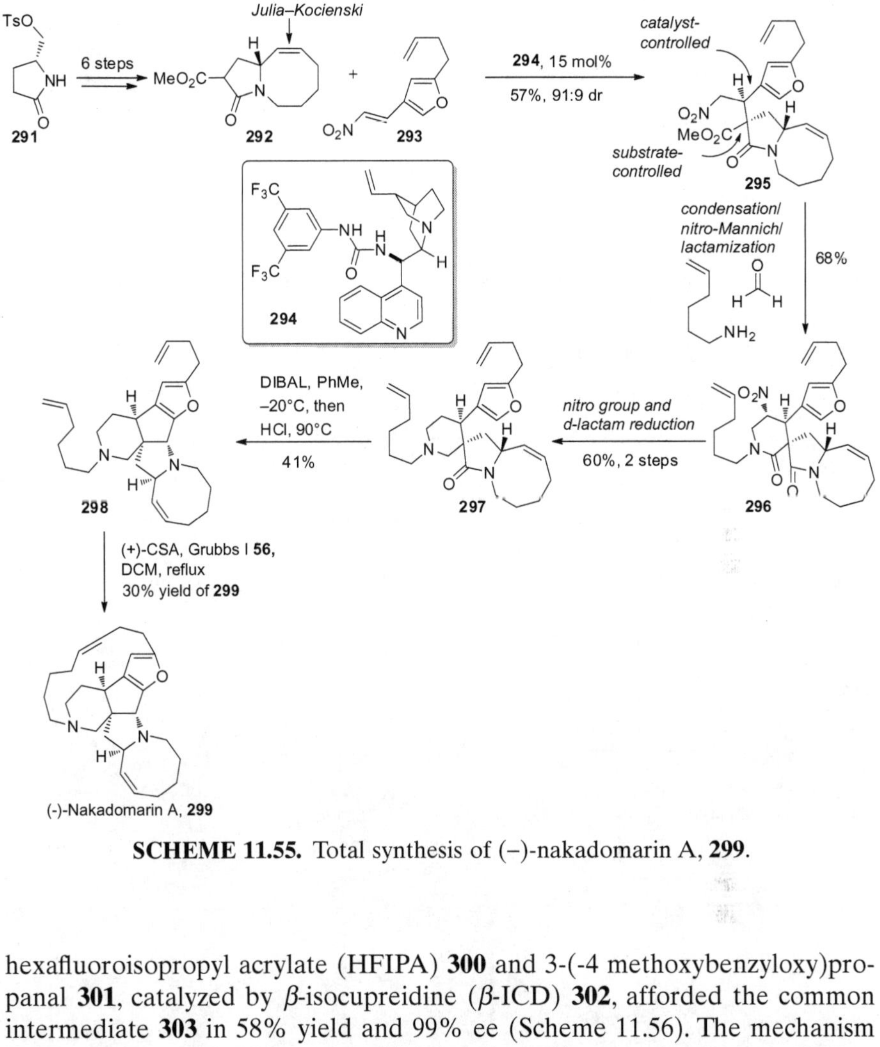

SCHEME 11.55. Total synthesis of (−)-nakadomarin A, **299**.

hexafluoroisopropyl acrylate (HFIPA) **300** and 3-(-4 methoxybenzyloxy)propanal **301**, catalyzed by *β*-isocupreidine (*β*-ICD) **302**, afforded the common intermediate **303** in 58% yield and 99% ee (Scheme 11.56). The mechanism involves a reversible aldol-type addition followed by enantioselective elimination, in which a key H-bond between catalyst and intermediate was proposed. Further, synthesis of (+)-fostriecin **307** and (+)-phoslactomycin B **308** was achieved via intermediate **303** by following a series of reactions.

Another efficient bifunctional organocatalyst, Yang's catalyst **310**, was reported by Seki for the asymmetric organocatalytic synthesis of methyl (2*R*,3*S*)-3-(-4-methoxyphenyl) glycidate **312**, a key precursor for the synthesis of the cardiovascular drug diltiazem hydrochloride **313** (Scheme 11.57) [173]. The key feature of the synthesis includes the dioxirane-mediated asymmetric epoxidation of methyl (E)-4-methoxycinnamate **309** [174], catalyzed by Yang's catalyst **310** [20a, 175], giving the desired product (−)-(2*R*,3*S*)-**311** in high yield (87%) and high enantioselectivity (78% ee). Further, the ring opening of

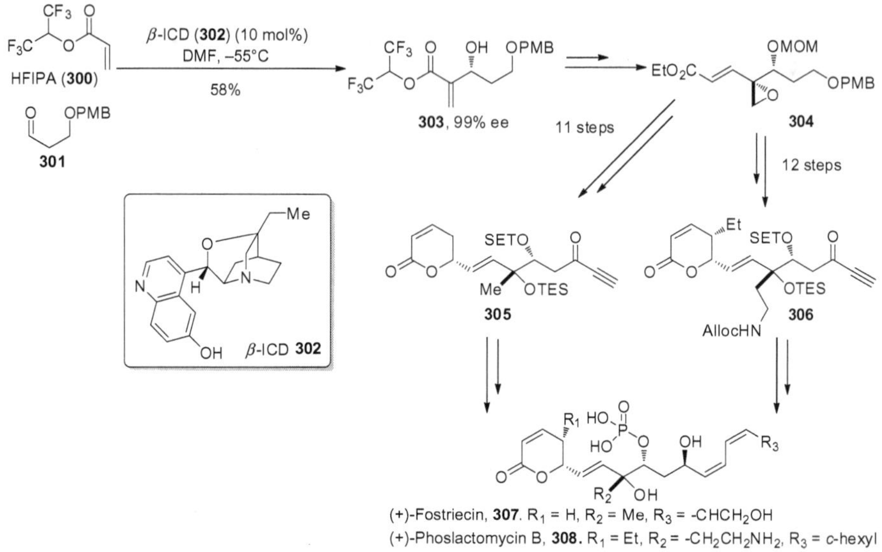

SCHEME 11.56. Syntheses of (+)-fostriecin **307** and (+)-phoslactomycin B **308**.

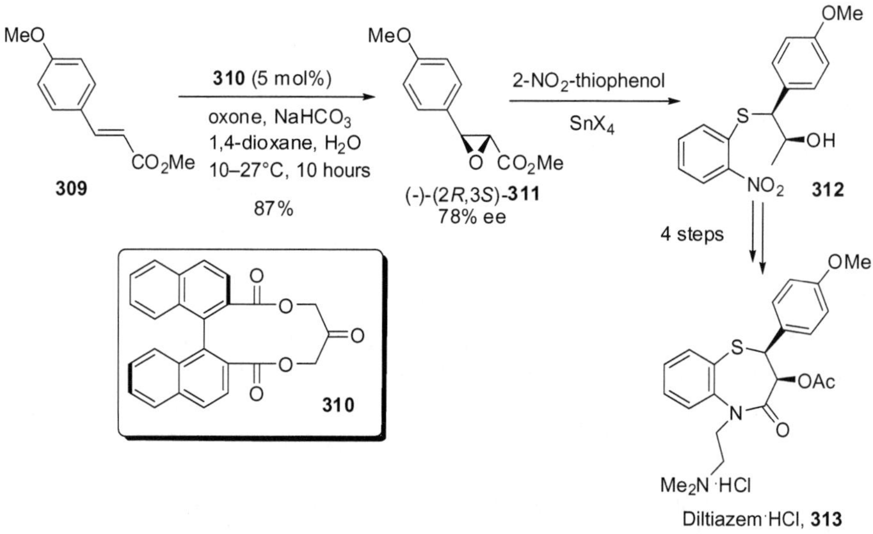

SCHEME 11.57. Synthesis of compound **312**, a key intermediate of diltiazem **313**.

epoxide (−)-(2*R*,3*S*)-**311** by 2-NO_2-thiophenol yielded **312**, a key precursor for the synthesis of diltiazem hydrochloride **313**. The stereochemical outcome of the reaction was explained by the transition state model shown in Fig. 11.9 [174]. The authors proposed that electrostatic interactions between ester groups and steric interactions between 3,3′ hydrogens decrease the stability of

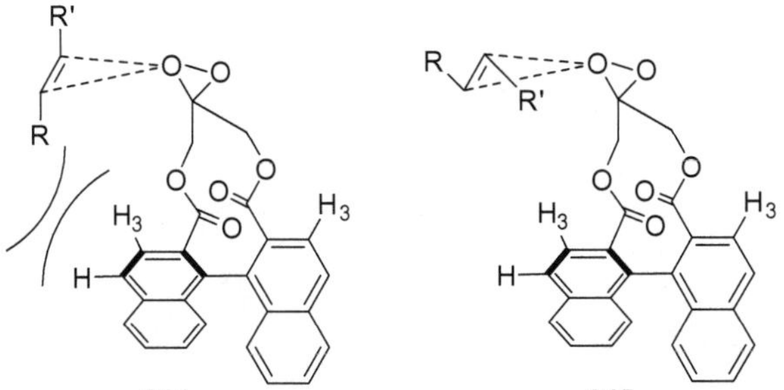

FIGURE 11.9. Transition state models.

transition state **314**. In order to avoid these repulsions, reaction takes place through the more stable transition state **315**. This explains the reason for the observed configuration of the product **311**.

11.9. PHASE TRANSFER CATALYSIS

The use of phase transfer catalysts (PTCs) for asymmetric synthesis offers further advantages because it involves mild conditions, simple reaction procedures, safe and inexpensive reagents and solvents, and greatly simplifies product isolation [176]. The most common PTCs are quaternary ammonium salts. The stereochemical communication between a chiral cation and anionic substrate is achieved by an ion-pairing interaction [177]. The most common PTCs are *N*-alkylated *Cinchona* alkaloids, pioneered by scientists at Merck [178]. There are many excellent reviews in this area but herein we describe some of recent advances in PTCs and their application in the synthesis of natural products and drug molecules.

Shibasaki and co-workers developed catalytic asymmetric phase-transfer Mannich-type reactions of glycine Shiff's base by utilizing TaDiAS **318** as a catalyst [179]. This methodology was applied for the total synthesis of (+)-nemonapride **323** (Scheme 11.58) [180]. The key step involves the asymmetric Mannich-type reaction of Schiff's base **316** with alkyl amine **317**, catalyzed by TaDiAS **318**, affording the desired product **319** in high yield (72%) and with moderate enantioselectivity (65% ee). Then, compound **319** underwent a sequence of six steps and yielded product **320** in 64% yield. Finally, the condensation of compound **321** with compound **322** concluded the synthesis of (+)-nemonapride **323**.

The authors proposed that enantioselectivity is mediated through the complex **324** formed between *Z*-enolate of Schiff's base **316** and two ammonium cations of the catalyst TaDiAS **318** (Fig. 11.10) by the hydrogen bonding networks. This model is based on preliminary computational calculations. In this transition state, benzyl moieties surrounding the catalyst **318** shielded the

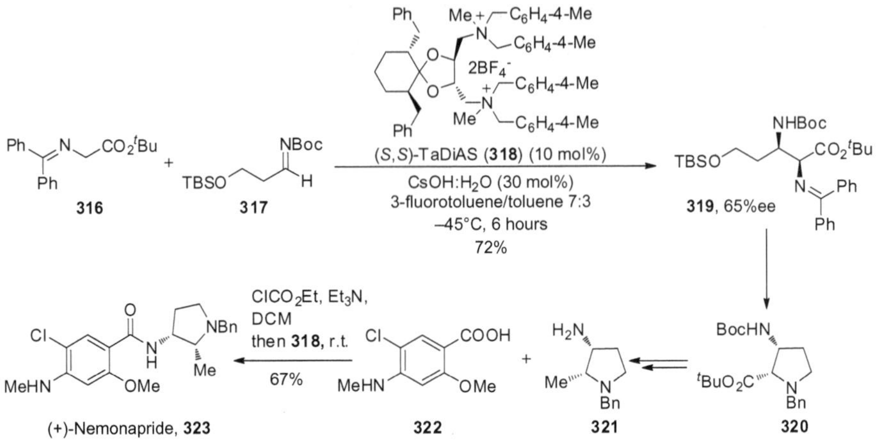

SCHEME 11.58. Synthesis of (+)-nemonapride **323**.

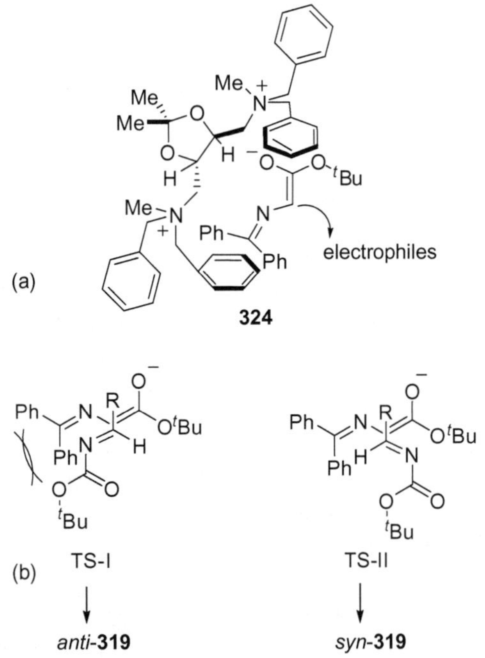

FIGURE 11.10. (a) Proposed transition state model **319**; (b) transition state models of the Mannich-type reaction.

Si-face of the *Z*-enolate of the Schiff's base **316**; thus, the approach of the electrophile can take place only from the *Re*-face, yielding an (*S*)-configured product. High diastereoselectivities observed in the Mannich-type reactions is independent of the catalyst effects (Fig. 11.10b). The results indicate that the reaction proceeds through the nonchelate, acyclic transition state. The authors

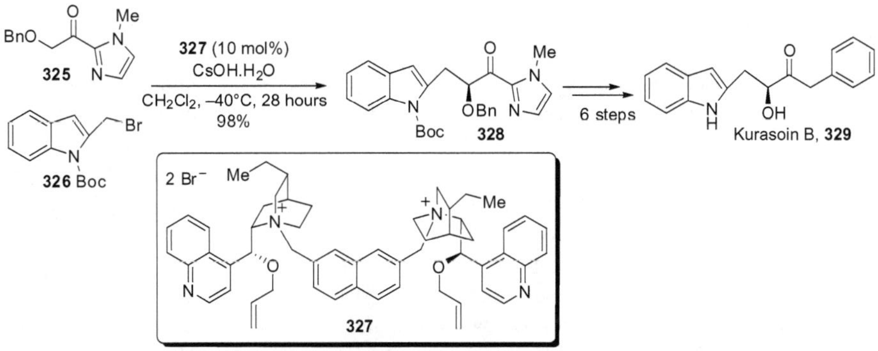

SCHEME 11.59. Synthesis of kurasoin B **329** by PTC.

proposed that steric interaction between the *N*-Boc and *N*-diphenyl methylene groups of enolate decreases the stability of transition state TS-I. In order to avoid these repulsions, reaction takes place through the more stable transition state TS-II to afford the *syn*-product.

Andrus and co-workers reported the organocatalytic asymmetric total synthesis of hydroxy ketone, kurasoin B **329**, a protein farnesyltransferase inhibitor (Scheme 11.59) [181]. The kurasoin B **329** was isolated from the fungus *Paecilomyces* sp. [182] and shows potential as a cancer drug [183]. The synthesis commenced with the alkylation reaction of benzyloxyacetyl imidazole **325** with 3-methylbromoindole **326**, catalyzed by biscinchonidinium dimethylnaphthalene catalyst **327**, affording the desired (*S*)-product **328** in high yield and with high enantioselectivity (99% ee). Finally, kurasoin B **329** was achieved by following the sequence of six steps with 34% overall yield.

11.10. CONCLUSIONS AND FUTURE PERSPECTIVE

Enantioselective organocatalysis has played a major role in the synthesis of complex natural products and drug molecules. The organocatalytic reactions are operationally simple and amenable to sequential transformation and domino processes. But most of these protocols still have some limitations such as high catalyst loading (10–20 mol%). Improvement is required in catalysts for broad applicability and large-scale synthesis of molecules. We hope that future organocatalysis will bring new endeavors in the field of enantioselective synthesis.

REFERENCES

[1] (a) Nguyen, L. A., He, H., Pham-Huy, C. (2006). Chiral drugs. An overview. *Int. J. Biomed. Sci.*, *2*, 85–100; (b) Eliel, E. L., Wilen, S. H. (1994). *The Stereochemistry of Organic Compounds*. Wiley-VCH, Weinheim, Germany.

[2] Halpern, J., Trost, B. M. (2004). Asymmetric catalysis. *Proc. Natl. Acad. Sci. U.S.A.*, *101*, 5347.

[3] Armstrong, D. W., Chang, C. D., Li, W. Y. (1990). Relevance of enantiomeric separations in food and beverage analyses. *J. Agric. Food Chem.*, *38*, 1674–1677.

[4] (a) de Gennes, P. G., Prost, J. (1993). *The Physics of Liquid Crystals*. Clarendon Press, Oxford; (b) Dierking, F. (2003). *Textures of Liquid Crystals*. Wiley-VCH, Weinheim, Germany; (c) Collings, P. J., Hird, M. (1997). *Introduction to Liquid Crystals*. Taylor & Francis, Bristol, PA.

[5] (a) Collings, P. J. (1990). *Liquid Crystals: Nature's Delicate Phase of Matter*. Princeton University Press, Princeton, NJ; (b) Tannas, L. E., Jr. (1994). Evolution of flat-panel displays. Proceedings of the IEEE 82, 4, April 1994, 499–509; (c) Sluckin, T. J., Dunmur, D. A., Stegemeyer, H. (2004). *Crystals, That Flow—Classic Papers from The History of Liquid Crystals*. Taylor & Francis, London.

[6] (a) Andersson, S., Allenmark, S. G. (2002). Preparative chiral chromatographic resolution of enantiomers in drug discovery. *J. Biochem. Biophys. Methods.*, *54*, 11–23; (b) Nugent, W. A., RajanBabu, T. V., Burk, M. J. (1993). Beyond nature's chiral pool: enantioselective catalysis in industry. *Science*, *259*, 479–483.

[7] (a) Breslow, R. (1982). Artificial enzymes. *Science*, *218*, 532–537; (b) Kirby, A. J. (1996). Enzyme mechanisms, models, and mimics. *Angew. Chem. Int. Ed. Engl.*, *35*, 706–724; (c) Heine, A., De Santis, G., Luz, J. G., Mitchell, M., Wong, C.-H., Wilson, I. A. (2001). Observation of covalent intermediates in an enzyme mechanism at atomic resolution. *Science*, *294*, 369–374, and references therein.

[8] (a) Fubini, B., Areán, C. O. (1999). Chemical aspects of the toxicity of inhaled mineral dusts. *Chem. Soc. Rev.*, *28*, 373–381; (b) Noyori, R. (1989). Chemical multiplication of chirality: science and applications. *Chem. Soc. Rev.*, *18*, 187–208; (c) Davies, S. G. (1982). *Organotransition Metal Chemistry Applications to Organic Synthesis*. Pergamon, Oxford.

[9] (a) Dalko, P. I. (2007). *Enantioselective Organocatalysis*. Wiley-VCH, Weinheim, Germany; (b) Berkessel, A., Groger, H. (2005). *Asymmetric Organocatalysis–From Biomimetic Concepts to Applications in Asymmetric Synthesis*. Wiley-VCH, Weinheim, Germany; (c) Berkessel, A., Groger, H. (2004). *Metal-Free Organic Catalysis in Asymmetric Synthesis*. Wiley-VCH, Weinheim, Germany; Reviews: (d) Barbas, C. F., III (2008). Organocatalysis lost: modern chemistry, ancient chemistry, and an unseen biosynthetic apparatus. *Angew. Chem. Int. Ed.*, *47*, 42–47; (e) Seayad, J., List, B. (2005). Asymmetric organocatalysis. *Org. Biomol. Chem.*, *3*, 719–724; (f) Dalko, P. I., Moisan, L. (2004). In the golden age of organocatalysis. *Angew. Chem. Int. Ed.*, *43*, 5138–5175; (g) List, B. (2002). Proline-catalyzed asymmetric reactions. *Tetrahedron*, *58*, 5573–5590; (h) Dalko, P. I., Moisan, L. (2001). Enantioselective organocatalysis. *Angew. Chem. Int. Ed.*, *40*, 3726–3748; (i) Notz, W., Tanaka, F., Barbas, C. F., III (2004). Enamine-based organocatalysis with proline and diamines: the development of direct catalytic asymmetric aldol, Mannich, Michael, and Diels–Alder reactions. *Acc. Chem. Res.*, *37*, 580–591; (j) List, B. (2001). Asymmetric aminocatalysis. *Synlett*, 1675–1686; (k) Melchiorre, P., Marigo, M., Carlone, A., Bartoli, G. (2008). Asymmetric aminocatalysis—gold rush in organic chemistry. *Angew. Chem. Int. Ed.*, *47*, 6138–6171; (l) Emsley, J. (1980). Very strong hydrogen bonding. *Chem. Soc. Rev.*, *9*, 91–124; (m) Kleiner, C. M., Schreiner, P. R. (2006). Hydrophobic amplification of noncovalent organocatalysis. *Chem. Commun.*, 4315–4317; (n) Mukherjee, S.,

Yang, J. W., Hoffmann, S., List, B. (2007). Asymmetric enamine catalysis. *Chem. Rev.*, *107*, 5471–5569; (o) Gruttadauria, M., Giacalone, F., Noto, R. (2009). Water in stereoselective organocatalytic reactions. *Adv. Synth. Catal.*, *351*, 33–57; (p) List, B. (2007). Introduction: organocatalysis. *Chem. Rev.*, *107*, 5413–5415; (q) Doyle, A. G., Jacobsen, E. N. (2007). Small-molecule H-bond donors in asymmetric catalysis. *Chem. Rev.*, *107*, 5713–5743; (r) Dondoni, A., Massi, A. (2008). Asymmetric organocatalysis: from infancy to adolescence. *Angew. Chem. Int. Ed.*, *47*, 4638–4660.

[10] Eder, U., Sauer, G., Wiechert, R. (1971). New type of asymmetric cyclization to optically active steroid cd partial structures. *Angew. Chem. Int. Ed. Engl.*, *10*, 496–497.

[11] Hajos, Z. G., Parrish, D. R. (1974). Asymmetric synthesis of bicyclic intermediates of natural product chemistry. *J. Org. Chem.*, *39*, 1615–1621.

[12] List, B., Lerner, R. A., Barbas, C. F., III (2000). Proline-catalyzed direct asymmetric aldol reactions. *J. Am. Chem. Soc.*, *122*, 2395–2396.

[13] Ahrendt, K. A., Borths, C. J., MacMillan, D. W. C. (2000). New strategies for organic catalysis: the first highly enantioselective organocatalytic Diels–Alder reaction. *J. Am. Chem. Soc.*, *122*, 4243–4244.

[14] Marques-Lopez, E., Herrera, R. P., Christmann, M. (2010). Asymmetric organocatalysis in total synthesis – a trial by fire. *Nat. Prod. Rep.*, *27*, 1–41.

[15] Houk, K. N., Cheong, P. H.-Y. (2008). Computational prediction of small-molecule catalysts. *Nature*, *455*, 309–313.

[16] Erkkila, A., Majander, I., Pihko, P. M. (2007). Iminium catalysis. *Chem. Rev.*, *107*, 5416–5470.

[17] Bertelsen, S., Marigo, M., Brandes, S., Dinér, P., Jørgensen, K. A. (2006). Dienamine catalysis: organocatalytic asymmetric γ-amination of α,β-unsaturated aldehydes. *J. Am. Chem. Soc.*, *128*, 12973–12980.

[18] Nicewicz, D., MacMillan, D. W. C. (2008). Merging photoredox catalysis with organocatalysis: the direct asymmetric alkylation of aldehydes. *Science*, *332*, 77–80.

[19] (a) Akiyama, T., Itoh, J., Fuchibea, K. (2006). Recent progress in chiral Brønsted acid catalysis. *Adv. Synth. Catal.*, *348*, 999–1010; (b) Taylor, M. S., Jacobsen, E. N. (2006). Asymmetric catalysis by chiral hydrogen-bond donors. *Angew. Chem. Int. Ed.*, *45*, 1520–1543; (c) Yu, X., Wang, W. (2008). Hydrogen-bond-mediated asymmetric catalysis. *Chem. Asian J.*, *3*, 516–532; (d) France, S., Guerin, D. J., Miller, S. J., Lectka, T. (2003). Nucleophilic chiral amines as catalysts in asymmetric synthesis. *Chem. Rev.*, *103*, 2985–3012.

[20] (a) Maruoka, K., Ooi, T. (2003). Enantioselective amino acid synthesis by chiral phase-transfer catalysis. *Chem. Rev.*, *103*, 3013–3028; (b) Halpern, M. E. (1997). *Phase Transfer Catalysis, Mechanisms and Synthesis*. American Chemical Society, Washington, DC; (c) Sasson, Y., Neumann, R. (Eds.) (1997). *Handbook of Phase Transfer Catalysis*. Blackie, London; (d) Starks, C. M., Liotta, C. L., Halpern, M. C. (1994). *Phase-Transfer Catalysis: Fundamentals, Applications, and Industrial Perspectives*. Chapman & Hall, New York; (e) Dehmlow, E. V., Dehmlow, S. S. (1993). *Phase Transfer Catalysis*. VCH, Weinheim, Germany; (f) Lygo, B. (2001). *Phase-Transfer Reactions. Rodd's Chemistry of Carbon Compounds*, Vol. V. Asymmetric Catalysis; Elsevier Science Ltd., Oxford, pp. 101–149; (g) Jones, R. A.

(2001). *Quaternary Ammonium Salts. Their Use in Phase-Transfer Catalysis.* Academic Press, London; (h) Hashimoto, T., Maruoka, K. (2007). Recent development and application of chiral phase-transfer catalysts. *Chem. Rev.*, *107*, 5656–5682.

[21] Bhadury, P. S., Song, B.-A., Yang, S., Hu, D.-Y., Xue, W. (2009). Bifunctional chiral organocatalysts in organic transformations. *Curr. Org. Synth.*, *6*, 380–399, and references therein.

[22] Bertelsen, S., Jørgensen, K. A. (2009). Organocatalysis—after the gold rush. *Chem. Soc. Rev.*, *38*, 2178–2189.

[23] Stork, G., Terrell, R., Szmuszkovicz, J. (1954). A new synthesis of 2-alkyl and 2-acyl ketones. *J. Am. Chem. Soc.*, *76*, 2029–2030.

[24] (a) Franzén, J., Marigo, M., Fielenbach, D., Tobias, W. C., Kjærsgaard, A., Jørgensen, K. A. (2005). A general organocatalyst for direct α-functionalization of aldehydes: stereoselective C–C, C–N, C–F, C–Br, and C–S bond-forming reactions. Scope and mechanistic insights. *J. Am. Chem. Soc.*, *127*, 18296–18304; (b) Schmid, M. B., Zeitler, K., Gschwind, R. M. (2010). The elusive enamine intermediate in proline-catalyzed aldol reactions: NMR detection, formation pathway, and stabilization trends. *Angew. Chem. Int. Ed.*, *49*, 4997–5003; (c) Maya, V., Singh, V. K. (2007). Highly enantioselective water-compatible organocatalyst for Michael reaction of ketones to nitroolefins. *Org. Lett.*, *9*, 1117–1119; (d) Raj, M., Maya, V., Ginotra, S. K., Singh, V. K. (2008). Highly enantioselective direct aldol reaction catalyzed by organic molecules. *Org. Lett.*, *8*, 4097–4099; (e) Maya, V., Raj, M., Singh, V. K. (2007). Highly enantioselective organocatalytic direct aldol reaction in an aqueous medium. *Org. Lett.*, *9*, 2593–2595; (f) Raj, M., Maya, V., Singh, V. K. (2009). Highly efficient small organic molecules for enantioselective direct aldol reaction in organic and aqueous media. *J. Org. Chem.*, *74*, 4289–4297; (g) Gandhi, S., Singh, V. K. (2008). Synthesis of chiral organocatalysts derived from aziridines: application in asymmetric aldol reaction. *J. Org. Chem.*, *73*, 9411–9416.

[25] (a) Borodin, A. (1873). Ueber einen neuen Abkömmling des *Valerals. Ber. Dtsch. Chem. Ges.*, *6*, 982–985; (b) Kane, R. (1838). Ueber die Zusammensetzung einiger ätherischen *Oele. J. Prakt. Chem.*, *15*, 129–155; (c) Mahrwald, R. (2004). *Modern Aldol Reactions*, Vol. 1–2. Wiley-VCH, Weinheim, Germany; (d) Geary, L. M., Hultin, P. G. (2009). The state of the art in asymmetric induction: the aldol reaction as a case study. *Tetrahedron Asymmetry*, *20*, 131–173; (e) Brodmann, T., Lorenz, M., Schaeckel, R., Simsek, S., Kalesse, M. (2009). Highly stereoselective aldol reactions in the total syntheses of complex natural products. *Synlett*, 174–192; (f) Mlynarski, J., Paradowska, J. (2008). Catalytic asymmetric aldol reactions in aqueous media. *Chem. Soc. Rev.*, *37*, 1502–1511.

[26] Carpenter, J., Northrup, A. B., Chung, D. M., Wiener, J. J. M., Kim, S.-G., MacMillan, D. W. C. (2008). Total synthesis and structural revision of callipeltoside C. *Angew. Chem. Int. Ed.*, *47*, 3568–3572.

[27] Northrup, A. B., MacMillan, D. W. C. (2002). The first direct and enantioselective cross-aldol reaction of aldehydes. *J. Am. Chem. Soc.*, *124*, 6798–6799.

[28] (a) Brown, S. P., Brochu, M. P., Sinz, C. J., MacMillan, D. W. C. (2003). The direct and enantioselective organocatalytic α-oxidation of aldehydes. *J. Am. Chem. Soc.*, *125*, 10808–10809; (b) Zhong, G. (2003). A facile and rapid route to highly enan-

tiopure 1,2-diols by novel catalytic asymmetric α-aminoxylation of aldehydes. *Angew. Chem. Int. Ed.*, *42*, 4247–4250.

[29] (a) Northrup, A. B., Mangion, I. K., Hettche, F., MacMillan, D. W. C. (2004). Enantioselective organocatalytic direct aldol reactions of α-oxyaldehydes: step one in a two-step synthesis of carbohydrates. *Angew. Chem. Int. Ed.*, *43*, 2152–2154; (b) Northrup, A. B., MacMillan, D. W. C. (2004). Two-step synthesis of carbohydrates by selective aldol reactions. *Science*, *305*, 1752–1755.

[30] (a) Zhang, H.-P., Shigemori, H., Ishibashi, M., Kosaka, T., Pettit, G. R., Kamano, Y., Kobayashi, J. (1994). Convolutamides A-F, novel γ-lactam alkaloids from the marine bryozoan *Amathia convoluta*. *Tetrahedron*, *50*, 10201–10206; (b) Kamano, Y., Zhang, H.-P., Ichihara, Y., Kizu, H., Komiyama, K., Pettit, G. R. (1995). Convolutamydine A, a novel bioactive hydroxyoxindole alkaloid from marine bryozoan *Amathia convoluta*. *Tetrahedron Lett.*, *36*, 2783–2784; (c) Kamano, Y., Kotake, A., Hashima, H., Hayakawa, I., Hiraide, H., Zhang, H.-P., Kizu, H., Komiyama, K., Hayashi, M., Pettit, G. R. (1999). Three new alkaloids, convolutamines F and G, and convolutamydine E, from the Floridian marine bryozoan *Amathia convoluta*. *Collect. Czech. Chem. Commun.*, *64*, 1147–1153.

[31] (a) Luppi, G., Cozzi, P. G., Monari, M., Kaptein, B., Broxterman, Q. B., Tomasini, C. (2005). Dipeptide-catalyzed asymmetric aldol condensation of acetone with (N-alkylated) isatins. *J. Org. Chem.*, *70*, 7418–7421; (b) Luppi, G., Monari, M., Corrêa, R. J., Violante, F. A., Pinto, A. C., Kaptein, B., Broxterman, A. B., Garden, S. J., Tomasini, C. (2006). The first total synthesis of (*R*)-convolutamydine A. *Tetrahedron*, *62*, 12017–12024.

[32] (a) Chen, J.-R., Liu, X.-P., Zhu, X.-Y., Li, L., Qiao, Y.-F., Zhang, J.-M., Xiao, W.-J. (2007). Organocatalytic asymmetric aldol reaction of ketones with isatins: straightforward stereoselective synthesis of 3-alkyl-3-hydroxyindolin-2-ones. *Tetrahedron*, *63*, 10437–10444; (b) Malkov, A. V., Kabeshov, M. A., Bella, M., Kysilka, O., Malyshev, D. A., Pluhá ková, K., Koovský, P. (2007). Vicinal amino alcohols as organocatalysts in asymmetric cross-aldol reaction of ketones: application in the synthesis of convolutamydine A. *Org. Lett.*, *9*, 5473–5476; (c) Nakamura, S., Hara, N. . H., Kubo, K., Shibata, N., Toru, T. (2008). Enantioselective synthesis of (*R*)-convolutamydine A with new *N*-heteroarylsulfonylprolinamides. *Chem. Eur. J.*, *14*, 8079–8081.

[33] Hara, N., Nakamura, S., Shibata, N., Toru, T. (2009). First enantioselective synthesis of (*R*)-convolutamydine B and E with *N*-(heteroarenesulfonyl) prolinamides. *Chem. Eur. J.*, *15*, 6790–6793.

[34] For previous syntheses of convolutamydines B and E using a diastereoselective vinylogous Mukaiyama aldol reaction as a key step, see: Nakamura, T., Shirokawa, S.-I., Hosokawa, S., Nakazaki, A., Kobayashi, S. (2006). Enantioselective total synthesis of convolutamydines B and E. *Org. Lett.*, *8*, 677–679.

[35] Itoh, T., Ishikawa, H., Hayashi, Y. (2009). Asymmetric aldol reaction of acetaldehyde and isatin derivatives for the total syntheses of ent-convolutamydine E and CPC-1 and a half fragment of madindoline A and B. *Org. Lett.*, *11*, 3854–3857.

[36] (a) Hayashi, M., Rho, M.-C., Enomoto, A., Fukami, A., Kim, Y.-P., Kikuchi, Y., Sunazuka, T., Hirose, T., Komiyama, K., Omura, S. (2002). Suppression of bone resorption by madindoline A, a novel nonpeptide antagonist to gp130. *Proc. Natl. Acad. Sci. U.S.A.*, *99*, 14728–14733. For several asymmetric syntheses of madindoline A and B, see: (b) Sunazuka, T., Hirose, T., Shirahata, T., Harigaya, Y.,

Hayashi, M., Komiyama, K., Mura, S., Smith, A. B., III (2000). Total synthesis of (+)-madindoline A and (−)-madindoline B, potent, selective inhibitors of interleukin 6. Determination of the relative and absolute configurations. *J. Am. Chem. Soc.*, *122*, 2122–2123; (c) Hirose, T., Sunazuka, T., Shirahata, T., Yamamoto, D., Harigaya, Y., Kuwajima, I., Mura, S. (2002). Determination of the absolute stereochemistry and asymmetric total synthesis of madindolines A and B: a practical improvement to a second-generation approach from the first-generation. *Org. Lett.*, *4*, 501–503; (d) Hirose, T., Sunazuka, T., Yamamoto, D., Kojima, N., Shirahata, T., Harigaya, Y., Kuwajima, I., Mura, S. (2005). Novel stereoselective construction of a quaternary carbon: application to synthesis of the cyclopentenedione moiety of madindolines. *Tetrahedron*, *61*, 6015–6039; (e) Hosokawa, S., Sekiguchi, K., Enomoto, M., Kobayashi, S. (2000). Total synthesis of madindoline A. *Tetrahedron Lett.*, *41*, 6429–6433; (f) Hosokawa, S., Sekiguchi, K., Hayase, K., Hirukawa, Y., Kobayashi, S. (2000). Synthesis of (±)-madindolines and chemical models. Studies of chemical reactivity. *Tetrahedron Lett.*, *41*, 6435–6439.

[37] (a) Kitajima, M., Mori, I., Arai, K., Kogure, N., Takayama, H. (2006). Two new tryptamine-derived alkaloids from *Chimonanthus praecox f. concolor*. *Tetrahedron Lett.*, *47*, 3199–3202. For the synthesis of a precursor of CPC-1 by a phase-transfer-catalytic method, see: (b) Sano, D., Nagata, K., Itoh, T. (2008). Catalytic asymmetric hydroxylation of oxindoles by molecular oxygen using a phase-transfer catalyst. *Org. Lett.*, *10*, 1593–1595.

[38] (a) Khuzhaev, V. U. (2004). Alkaloids of *Arundo donax*. XVIII. Nitrogenous bases in flowers of cultivars. *Chem. Nat. Compd.*, *40*, 516–517; (b) Takayama, H., Matsuda, Y., Masubuchi, K., Ishida, A., Kitajima, M., Aimi, N. (2004). Isolation, structure elucidation, and total synthesis of two new chimonanthus alkaloids, chimonamidine and chimonanthidine. *Tetrahedron*, *60*, 893–900.

[39] Chen, W.-B., Du, X.-L., Cun, L.-F., Zhang, X.-M., Yuan, W.-C. (2010). Highly enantioselective aldol reaction of acetaldehyde and isatins only with 4-hydroxydiarylprolinol as catalyst: concise stereoselective synthesis of (*R*)-convolutamydines B and E, (−)-donaxaridine and (*R*)-chimonamidine. *Tetrahedron*, *66*, 1441–1446.

[40] Raj, M., Parashari, G. S., Singh, V. K. (2009). Highly enantioselective organocatalytic *syn*- and *anti*-aldol reactions in aqueous medium. *Adv. Synth. Catal.*, *351*, 1284–1288.

[41] Tokuda, O., Kano, T., Gao, W.-G., Ikemoto, T., Maruoka, K. (2005). A practical synthesis of (*S*)-2-cyclohexyl-2-phenylglycolic acid via organocatalytic asymmetric construction of a tetrasubstituted carbon center. *Org. Lett.*, *7*, 5103–5105.

[42] Raj, M., Veerasamy, N., Singh, V. K. (2010). Highly enantioselective synthesis of 3-cycloalkanone-3-hydroxy-2-oxindoles, potential anticonvulsants. *Tetrahedron Lett.*, *51*, 2157–2159.

[43] (a) Popp, F. D., Donigan, B. E. (1979). Synthesis of 3-hydroxy-3-phenacyloxindole analogs. *J. Pharm. Sci.*, *68*, 519–520; (b) Pajouhesh, H., Parson, R., Popp, F. D. (1983). Potential anticonvulsants VI: condensation of isatins with cyclohexanone and other cyclic ketones. *J. Pharm. Sci.*, *72*, 318–321; (c) Joshi, K. C., Dandia, A., Sanan, S. (1989). Studies in spiroheterocycles. Part XVI. Synthesis of new fluorine-containing spiro-3-indoline derivatives. *J. Fluorine Chem.*, *44*, 59–72.

[44] (a) Marigo, M., Jørgensen, K. A. (2006). Organocatalytic direct asymmetric alpha-heteroatom functionalization of aldehydes and ketones. *Chem. Commun.*,

2001–2011; (b) Guillena, G., Ramón, D. J. (2006). Enantioselective α-heterofunctionalisation of carbonyl compounds: organocatalysis is the simplest approach. *Tetrahedron: Asymmetry*, *17*, 1465–1492; (c) de Figueiredo, R. M., Christman, M. (2007). Organocatalytic synthesis of drugs and bioactive natural products. *Eur. J. Org. Chem.*, 2575–2600.

[45] Varseev, G. N., Maier, M. E. (2007). Enantioselective total synthesis of (+)-neosymbioimine. *Org. Lett.*, *9*, 1461–1464.

[46] Kita, M., Ohishi, N., Washida, K., Kondo, M., Koyama, T., Yamada, K., Uemura, D. (2005). Symbioimine and neosymbioimine, amphoteric iminium metabolites from the symbiotic marine dinoflagellate *Symbiodinium* sp. *Bioorg. Med. Chem.*, *13*, 5253–5258.

[47] (a) Kita, M., Kondo, M., Koyama, T., Yamada, K., Matsumoto, T., Lee, K.-H., Woo, J.-T., Uemura, D. (2004). Symbioimine exhibiting inhibitory effect of osteoclast differentiation, from the symbiotic marine dinoflagellate *Symbiodinium* sp. *J. Am. Chem. Soc.*, *126*, 4794–4795; (b) Kita, M., Uemura, D. (2005). Iminium alkaloids from marine invertebrates: structure, biological activity, and biogenesis. *Chem. Lett.*, *34*, 454–459; (c) Herlt, A., Mander, L., Rombang, W., Rumampuk, R., Soemitro, S., Steglich, W., Tarigan, P., von Nussbaum, F. (2004). Alkaloids from marine organisms. Part 8: isolation of bisdemethylaaptamine and bisdemethylaaptamine-9-O-sulfate from an Indonesian *Aaptos* sp. marine sponge. *Tetrahedron*, *60*, 6101–6104; (d) Gray, C. A., de Lira, S. P., Silva, M., Pimenta, E. F., Thiemann, O. H., Oliva, G., Hajdu, E., Andersen, R. J., Berlinck, R. G. S. (2006). Sulfated meroterpenoids from the Brazilian sponge *Callyspongia* sp. are inhibitors of the antileishmaniasis target adenosine phosphoribosyl transferase. *J. Org. Chem.*, *71*, 8685–8690.

[48] Mangion, I. K., MacMillan, D. W. C. (2005). Total synthesis of brasoside and littoralisone. *J. Am. Chem. Soc.*, *127*, 3696–3697.

[49] (a) Kim, S. G. (2009). Concise total synthesis of (+)-disparlure and its trans-isomer using asymmetric organocatalysis. *Synthesis*, 2418–2422. For previous synthesis reviews, see: (b) Mori, K., Tashiro, T. (2004). Useful reactions in modern pheromone synthesis. *Curr. Org. Synth.*, *1*, 11–29; (c) Mori, K. (2000). Organic synthesis and chemical ecology. *Acc. Chem. Res.*, *33*, 102–110.

[50] Bierl, B. A., Beroza, M., Collier, C. W. (1970). Potent sex attractant of the gypsy moth: its isolation, identification, and synthesis. *Science*, *170*, 87–90.

[51] Miller, J. R., Mori, K., Roelofs, W. L. (1977). Gypsy moth field trapping and electroantennogram studies with pheromone enantiomers. *J. Insect Physiol.*, *23*, 1447–1453.

[52] Panchgalle, S. P., Gore, R. G., Chavan, S. P., Kalkote, U. R. (2009). Organocatalytic enantioselective synthesis of β-blockers: (*S*)-propranolol and (*S*)-naftopidil. *Tetrahedron Asymmetry*, *20*, 1767–1770.

[53] Hayashi, Y., Yamaguchi, J., Hibino, K., Shoji, M. (2003). Direct proline catalyzed asymmetric α-aminooxylation of aldehydes. *Tetrahedron Lett.*, *44*, 8293–8296.

[54] (a) Mitsunobu, O., Yamada, M. (1967). Preparation of esters of carboxylic and phosphoric acid via quaternary phosphonium salts. *Bull. Chem. Soc. Jpn.*, *40*, 2380–2382. For a recent review, see: (b) Kumara Swamy, K. C., Bhuvan Kumar, N. N., Balaraman, E., Pavan Kumar, K. V. P. (2009). Mitsunobu and related reactions: advances and applications. *Chem. Rev.*, *109*, 2551–2651.

[55] For recent reports on the total synthesis of *Lycopodium* alkaloids, see: (a) Yang, H., Carter, R. G., Zakharov, L. N. (2008). Enantioselective total synthesis of lycopodine. *J. Am. Chem. Soc.*, *130*, 9238–9239; (b) Kozak, J. A., Dake, G. R. (2008). Total synthesis of (+)-fawcettidine. *Angew. Chem. Int. Ed.*, *47*, 4221–4223; (c) Bisai, A., West, S. P., Sarpong, R. (2008). Unified strategy for the synthesis of the "miscellaneous" *Lycopodium* alkaloids: total synthesis of (±)-lyconadin A. *J. Am. Chem. Soc.*, *130*, 7222–7223; (d) Kozaka, T., Miyakoshi, N., Mukai, C. (2007). Stereoselective total syntheses of three *Lycopodium* alkaloids, (−)-magellanine, (+)-magellaninone, and (+)-paniculatine, based on two Pauson–Khand reactions. *J. Org. Chem.*, *72*, 10147–10154; (e) Linghu, X., Kennedy-Smith, J. J., Toste, F. D. (2007). Total synthesis of (+)-fawcettimine. *Angew. Chem. Int. Ed.*, *46*, 7671–7673; (f) Beshore, D. C., Smith, A. B., III (2007). Total syntheses of (+)-lyconadin A and (−)-lyconadin B. *J. Am. Chem. Soc.*, *129*, 4148–4149; (g) Snider, B. B., Grabowski, J. F. (2007). Total synthesis of (−)-senepodine G and (−)-scermizine. *C. J. Org. Chem.*, *72*, 1039–1042.

[56] (a) Ayer, W. A., Jenkins, J. K., Valverde-López, S. (1964). The alkaloids of *Lycopodium* cernuum. *Tetrahedron Lett.*, *5*, 2201–2222; (b) Ayer, W. A., Jenkins, J. K., Valverde-López, S., Burnell, R. H. (1967). The alkaloids of *Lycopodium* cernuum L. I. The structures of cernuine and lycocernuine. *Can. J. Chem.*, *45*, 433–443; (c) Ayer, W. A., Jenkins, J. K., Piers, K., Valverde-López, S. (1967). The alkaloids of *Lycopodium* cernuum L. II. The stereochemistry of cernuine and lycocernuine. *Can. J. Chem.*, *45*, 445–450; (d) Ayer, W. A., Piers, K. (1967). The alkaloids of *Lycopodium* cernuum L. III. The synthesis of dihydrodeoxyepiallocernuine. *Can. J. Chem.*, *45*, 451–459.

[57] Morita, H., Hirasawa, Y., Shinzato, T., Kobayashi, J. (2004). New phlegmarane-type, cernuane-type, and quinolizidine alkaloids from two species of *Lycopodium*. *Tetrahedron*, *60*, 7015–7023.

[58] (a) Nishikawa, Y., Kitajima, M., Takayama, H. (2008). First asymmetric total syntheses of cernuane-type *Lycopodium* alkaloids, cernuine, and cermizine D. *Org. Lett.*, *10*, 1987–1990; (b) Nishikawa, Y., Kitajima, M., Kogure, N., Takayama, H. (2009). A divergent approach for the total syntheses of cernuane-type and quinolizidine-type *Lycopodium* alkaloids. *Tetrahedron*, *65*, 1608–1617; (c) Shigeyama, T., Katakawa, K., Kogure, N., Kitajima, M., Takayama, H. (2007). Asymmetric total syntheses of two phlegmarine-type alkaloids, lycoposerramines-v and -w, newly isolated from *Lycopodium serrratum*. *Org. Lett.*, *9*, 4069–4072; (d) Katakawa, K., Nozoe, A., Kogure, N., Kitajima, M., Hosokawa, M., Takayama, H. (2007). Fawcettimine-related alkaloids from *Lycopodium serratum*. *J. Nat. Prod.*, *70*, 1024–1028; (e) Katakawa, K., Kitajima, M., Aimi, N., Seki, H., Yamaguchi, K., Furihata, K., Harayama, T., Takayama, H. (2005). Structure elucidation and synthesis of lycoposerramine-B, a novel oxime-containing *Lycopodium* alkaloid from *Lycopodium serratum* thunb. *J. Org. Chem.*, *70*, 658–663; (f) Takayama, H., Katakawa, K., Kitajima, M., Yamaguchi, K., Aimi, N. (2002). Seven new *Lycopodium* alkaloids, lycoposerramines-C, -D, -E, -P, -Q, -S, and -U, from *Lycopodium serratum* thunb. *Tetrahedron Lett.*, *43*, 8307–8311; (g) Takayama, H., Katakawa, K., Kitajima, M., Seki, H., Yamaguchi, K., Aimi, N. (2002). A new type of *Lycopodium* alkaloid, lycoposerramine-A, from *Lycopodium serratum* thunb. *Org. Lett.*, *4*, 1244–1244.

[59] Cave, A., Kan-Fan, C., Potier, P., Le Men, J. (1967). Modification de la reaction de Polonovsky: action de l'anhydride trifluoroacetique sur un aminoxyde. *Tetrahedron*, *23*, 4681–4689.

[60] Sugiura, M., Mori, C., Kobayashi, S. (2006). Enantioselective transfer aminoallylation: synthesis of optically active homoallylic primary amines. *J. Am. Chem. Soc.*, *128*, 11038–11039.

[61] (a) List, B., Pojarliev, P., Martin, H. J. (2001). Efficient proline-catalyzed Michael additions of unmodified ketones to nitro olefins. *Org. Lett.*, *3*, 2423–2425; (b) Betancort, J. M., Sakthivel, K., Thayumanavan, R., Tanaka, F., Barbas, III, C. F. (2004). Catalytic direct asymmetric Michael reacions: addition of unmodified ketone and aldehyde donors to alkylidene malonates and nitro olefins. *Synthesis.*, 1509–1521; (c) Cobb, A. J. A., Shaw, D. M., Longbottom, D. A., Gold, J. B., Ley, S. V. (2005). Organocatalysis with proline derivates: improved catalysts for asymmetric Mannich, nitro-Michael and aldol reactions. *Org. Biomol. Chem.*, *3*, 84–96.

[62] Hayashi, Y., Gotoh, H., Hayashi, T., Shoji, M. (2005). Diphenylprolinol silyl ethers as efficient organocatalysts for the asymmetric Michael reaction of aldehydes and nitroalkenes. *Angew. Chem. Int. Ed.*, *44*, 4212–4215.

[63] Gaunt, M. J., Johansson, C. C. C., McNally, A., Vo, N. T. (2007). Enantioselective organocatalysis. *Drug Discovery Today*, *12*, 8–27.

[64] Kim, C. U., Lew, W., Williams, M. A., Liu, H., Zhang, L., Swaminathan, S., Bischofberger, N., Chen, M. S., Mendel, D. B., Tai, C. Y., Laver, W. G., Stevens, R. C. (1997). Influenza neuraminidase inhibitors possessing a novel hydrophobic interaction in the enzyme active site: design, synthesis, and structural analysis of carbocyclic sialic acid analogues with potent anti-influenza activity. *J. Am. Chem. Soc.*, *119*, 681–690.

[65] (a) Ishikawa, H., Suzuki, T., Hayashi, Y. (2009). High-yielding synthesis of the anti-influenza neuramidase inhibitor (−)-oseltamivir by three "one-pot" operations. *Angew. Chem. Int. Ed.*, *48*, 1304–1307. For selected efficient preparations of (−)-oseltamivir **76** and of its derivatives, see: (b) Yeung, Y.-Y., Hong, S., Corey, E. J. (2006). A short enantioselective pathway for the synthesis of the anti-influenza neuramidase inhibitor oseltamivir from 1,3-butadiene and acrylic acid. *J. Am. Chem. Soc.*, *128*, 6310–6311; (c) Fukuta, Y., Mita, T., Fukuda, N., Kanai, M., Shibasaki, M. (2006). De novo synthesis of Tamiflu via a catalytic asymmetric ring-opening of meso-aziridines with TMSN3. *J. Am. Chem. Soc.*, *128*, 6312–6313; (d) Cong, X., Yao, Z.-J. (2006). Ring-closing metathesis-based synthesis of (3*R*,4*R*,5*S*)-4-acetylamino-5-amino-3-hydroxy- cyclohex-1-ene-carboxylic acid ethyl ester: a functionalized cycloalkene skeleton of GS4104. *J. Org. Chem.*, *71*, 5365–5368; (e) Farina, V., Brown, J. D. (2006). Tamiflu: the supply problem. *Angew. Chem. Int. Ed.*, *45*, 7330–7334; (f) Satoh, N., Akiba, T., Yokoshima, S., Fukuyama, T. (2007). A practical synthesis of (−)-oseltamivir. *Angew. Chem. Int. Ed.*, *46*, 5734–5736; (g) Shie, J.-J., Fang, J.-M., Wang, S.-Y., Tsai, K.-C., Cheng, Y.-S. E., Yang, A.-S., Hsiao, S.-C., Su, C.-Y., Wong, C.-H. (2007). Synthesis of Tamiflu and its phosphonate congeners possessing potent anti-influenza activity. *J. Am. Chem. Soc.*, *129*, 11892–11893; (h) Zutter, U., Iding, H., Spurr, P., Wirz, B. (2008). New, efficient synthesis of oseltamivir phosphate (Tamiflu) via enzymatic desymmetrization of a meso-1,3-cyclohexanedicarboxylic acid diester. *J. Org. Chem.*, *73*, 4895–4902; (i) Shie, J.-J., Fang, J.-M., Wong, C.-H. (2008). A concise

and flexible synthesis of the potent anti-influenza agents Tamiflu and Tamiphosphor. *Angew. Chem. Int. Ed.*, *47*, 5788–5791; (j) Trost, B. M., Zhang, T. (2008). A concise synthesis of (−)-oseltamivir. *Angew. Chem. Int. Ed.*, *47*, 3759–3761.

[66] (a) Michael, A. (1887). Ueber die Addition von Natriumacetessig- und natriummalonsäureäthern zu den aethern ungesättigter säuren. *J. Prakt. Chem.*, *35*, 349–356; (b) Michael, A. (1894). Ueber die addition von natriumacetessig- und natriummalonsäureäther zu den aethern ungesättigter säuren. *J. Prakt. Chem.*, *49*, 20–29.

[67] For a review of organocatalytic enantioselective Michael and hetero-Michael reactions, see: Vicario, J. L., Badía, D., Carrillo, L. (2007). Organocatalytic enantioselective Michael and hetero-Michael reactions. *Synthesis.*, 2065–2092.

[68] Tanaka, N., Okasaka, M., Ishimaru, Y., Takaishi, Y., Sato, M., Okamoto, M., Oshikawa, T., Ahmed, S. U., Consentino, L. M., Lee, K.-H. (2005). Biyouyanagin A, an anti-HIV agent from *Hypericum chinense* L. var. *salicifolium*. *Org. Lett.*, *7*, 2997–2999.

[69] Nicolaou, K. C., Sarlah, D., Shaw, D. M. (2007). Total synthesis and revised structure of biyouyanagin A. *Angew. Chem. Int. Ed.*, *46*, 4708–4711.

[70] Chi, Y., Gellman, S. H. (2005). Diphenylprolinol methyl ether: a highly enantioselective catalyst for Michael addition of aldehydes to simple enones. *Org. Lett.*, *7*, 4253–4256.

[71] Chen, K., Baran, P. S. (2009). Total synthesis of eudesmane terpenes by site-selective C–H oxidations. *Nature*, *459*, 824–828.

[72] Campbell, M. J., Johnson, J. S. (2009). Asymmetric synthesis of (+)-polyanthellin A. *J. Am. Chem. Soc.*, *131*, 10370–10371.

[73] Bertelsen, S., Dinér, P., Johansen, R. L., Jørgensen, K. A. (2007). Asymmetric organocatalytic β-hydroxylation of α,β-unsaturated aldehydes. *J. Am. Chem. Soc.*, *129*, 1536–1537, and references therein.

[74] (a) Dinér, P., Nielsen, M., Marigo, M., Jørgensen, K. A. (2007). Enantioselective organocatalytic conjugate addition of N-heterocycles to α,β-unsaturated aldehydes. *Angew. Chem. Int. Ed.*, *46*, 1983–1987; (b) Seebach, D., Grosělj, U., Badine, D. M., Schweizer, W. B., Beck, A. K. (2008). Isolation and X-ray structures of reactive intermediates of organocatalysis with diphenylprolinol ethers and with imidazolidinones: a survey and comparison with computed structures and with 1-acyl-imidazolidinones: the 1,5-repulsion and the geminal-diaryl effect at work. *Helv. Chim. Acta.*, *91*, 1999–2034.

[75] (a) Enders, D., Hüttl, M. R. M., Grondal, C., Raabe, G. (2006). Control of four stereocentres in a triple cascade organocatalytic reaction. *Nature*, *441*, 861–863; (b) Grondal, C., Jeanty, M., Enders, D. (2010). Organocatalytic cascade reactions as a new tool in total synthesis. *Nat. Chem.*, *2*, 167–178; (c) Zhang, X., Zhang, S., Wang, W. (2010). Iminium-allenamine cascade catalysis: one-pot access to chiral 4*H*-chromenes by a highly enantioselective Michael-Michael sequence. *Angew. Chem. Int. Ed.*, *49*, 1481 –11484.

[76] (a) MacMillan, D. W. C., Walji, A. M. (2007). Strategies to bypass the taxol problem. Enantioselective cascade catalysis, a new approach for the efficient construction of molecular complexity. *Synlett*, 1477–1489; (b) Huang, Y., Walji,

A. M., Larsen, C. H., MacMillan, D. W. C. (2005). Enantioselective organo-cascade catalysis. *J. Am. Chem. Soc.*, *127*, 15051–15053; (c) Zhou, J., List, B. (2005). Catalytic asymmetric reductive Michael cyclization. *J. Am. Chem. Soc.*, *127*, 15035–15037; (d) Marigo, M., Schulte, T., Franzén, J., Jørgensen, K. A. (2005). Asymmetric multicomponent domino reactions and highly enantioselective conjugated addition of thiols to α,β-unsaturated aldehydes. *J. Am. Chem. Soc.*, *127*, 15710–15711; (e) Enders, D., Grondal, C., Hüttl, M. R. M. (2007). Asymmetric organocatalytic domino reactions. *Angew. Chem. Int. Ed.*, *46*, 1570–1581; (f) Guillena, G., Ramon, D. J., Yus, M. (2007). Organocatalytic enantioselective multicomponent reactions (OEMCRs). *Tetrahedron Asymmetry*, *18*, 693–700; (g) Guo, H., Ma, J. (2006). Catalytic asymmetric tandem transformations triggered by conjugate additions. *Angew. Chem. Int. Ed.*, *45*, 354 –3366.

[77] Simmons, B., Walji, A. M., MacMillan, D. W. C. (2009). Cycle-specific organocascade catalysis: application to olefin hydroamination, hydro-oxidation, and amino-oxidation, and to natural product synthesis. *Angew. Chem. Int. Ed.*, *48*, 4349–4353.

[78] Beechan, C. M., Djerassi, C., Eggert, H. (1978). Terpenoids-LXXIV: the sesquiterpenes from the soft coral *Sinularia mayi*. *Tetrahedron*, *34*, 2503–2508.

[79] Moriera, I. C., Lago, J. H. G., Young, M. C. M., Roque, N. F. (2003). Antifungal aromadendrane sesquiterpenoids from the leaves of *Xylopia brasiliensis*. *J. Braz. Chem. Soc.*, *14*, 828–831.

[80] Yang, J. W., Hechavarria Fonseca, M. T., List, B. (2005). Catalytic asymmetric reductive Michael cyclization. *J. Am. Chem. Soc.*, *127*, 15036–15037.

[81] (a) Wurzel, G., Becker, H. (1990). Sesquiterpenoids from the liverwort *Ricciocarpos natans*. *Photochemistry*, *29*, 2565–2568; (b) Wurzel, G., Becker, H., Eicher, H. T., Tiefensee, K. (1990). Molluscicidal properties of constituents from the liverwort *Ricciocarpos natans* and/or synthetic lunalaric acid derivates. *Planta Med.*, *56*, 444–445; (c) Zinsmeister, H. D., Becker, H., Eicher, T. (1991). Bryophytes, a source of biologically active, naturally occurring material? *Angew. Chem. Int. Ed.*, *30*, 130–147.

[82] Michrowska, A., List, B. (2009). Concise synthesis of ricciocarpin A and discovery of a more potent analogue. *Nat. Chem.*, *1*, 225–228.

[83] Seki, T., Nakajo, T., Onaka, M. (2006). The Tishchenko reaction: a classic and practical tool for ester synthesis. *Chem. Lett.*, *35*, 824–829.

[84] (a) Palomo, C., Landa, A., Mielgo, A., Oiarbide, M., Puente, A., Vera, S. (2007). Water-compatible iminium activation: organocatalytic Michael reactions of carbon-centered nucleophiles with enals. *Angew. Chem. Int. Ed.*, *46*, 8431–8435. For another organocatalytic asymmetric total synthesis of (*R*)-rolipram, see: (b) Hynes, P. S., Stupple, P. A., Dixon, D. J. (2008). Organocatalytic asymmetric total synthesis of (*R*)-rolipram and formal synthesis of (3*S*,4*R*)-paroxetine. *Org. Lett.*, *10*, 1389–1391.

[85] For pertinent information on this subject, see: García, A. L. L., Carpes, M. J. S., de Oca, A. C. B. M., dos Santos, M. A. G., Santana, C. C., Correia, C. R. D. (2005). Synthesis of 4-aryl-2-pyrrolidones and β-aryl-γ-amino-butyric acid (GABA) analogues by Heck arylation of 3-pyrrolines with arenediazonium tetrafluoroborates. Synthesis of (±)-rolipram on a multigram scale and chromatographic

resolution by semipreparative chiral simulated moving bed chromatography. *J. Org. Chem.*, *70*, 1050–1053.

[86] Zu, L., Xie, H., Li, H., Wang, J., Wang, W. (2007). Highly enantioselective organocatalytic conjugate addition of nitromethane to α,β-unsaturated aldehydes: three-step synthesis of optically active baclofen. *Adv. Synth. Catal.*, *349*, 2660–2664.

[87] For other organocatalytic preparation of chiral baclofen using *Cinchona* alkaloid derivatives, see: Ji, L., Ma, Y., Li, J., Zhang, L., Zhang, L. (2009). An efficient synthesis of (*R*)- and (*S*)-baclofen via desymmetrization. *Tetrahedron Lett.*, *50*, 6166–6168.

[88] (a) Szántó, G., Hegedus, L., Mattyasovszky, L., Simon, A., Simon, A., Bitter, I., Tóth, G., Tõke, L., Kádas, I. (2009). An expedient total synthesis of *ent*-(−) -7-deoxy-*trans*-dihydronarciclasine. *Tetrahedron*, *65*, 8412–8417. For previous syntheses, see: (b) Isobe, K., Taga, J., Tsuda, Y. (1978). Syntheses of (±)-dihydrolycoricidines from the common intermediate to ethanophenanthridine alkaloids. *Heterocycles*, *9*, 625–630; (c) Chida, N., Jitsuoka, M., Yamamoto, Y., Ohtsuka, M., Ogawa, S. (1996). Total synthesis of (+)-7-deoxypancratistatin and (+)-7-deoxy-*trans*-dihydronarciclasine. *Heterocycles*, *43*, 1385–1390; (d) Fujimura, T., Shibuya, M., Ogasawara, K., Iwabuchi, Y. (2005). An enantio- and diastereocontrolled synthesis of (+)-7-deoxy-*trans*-dihydronarciclasine. *Heterocycles*, *66*, 167–173.

[89] For reviews, see: (a) Boruwa, J., Gogoi, N., Saikia, P. P., Barua, N. C. (2006). Catalytic asymmetric Henry reaction. *Tetrahedron Asymmetry*, *17*, 3315–3326; (b) Palomo, C., Oiarbide, M., Mielgo, A. (2004). Unveiling reliable catalysts for the asymmetric nitroaldol (Henry) reaction. *Angew. Chem. Int. Ed.*, *43*, 5442–5444.

[90] Bischler, A., Napieralski, B. (1893). Zur kenntniss einer neuen *Isochinolinsynthese*. *Ber. Dtsch. Chem. Ges.*, *26*, 1903–1908.

[91] Ma, X., Gang, D. R. (2004). The *Lycopodium* alkaloids. *Nat. Prod. Rep.*, *21*, 752–772.

[92] (a) Heathcock, C. H., Blumenkopf, T. A., Smith, K. M. (1989). Total synthesis of (±)-fawcettimine. *J. Org. Chem.*, *54*, 1548–1562; (b) Harayama, T., Takatani, M., Inubushi, Y. (1979). Stereoselective syntheses of *Lycopodium* alkaloids, (±)-fawcettimine and (±)-8-deoxyserratinine. *Tetrahedron Lett.*, *20*, 4307–4310; (c) Harayama, T., Takatani, M., Inubushi, Y. (1980). Total syntheses of the *Lycopodium* alkaloids (±)-fawcettimine and (±)-8-deoxyserratinine. *Chem. Pharm. Bull.*, *28*, 2394–2402; (d) Heathcock, C. H., Smith, K. M., Blumenkopf, T. A. (1986). Total synthesis of (±)-fawcettimine (Burnell's base A). *J. Am. Chem. Soc.*, *108*, 5022–5024.

[93] For an organocatalytic addition of 1, 3-dicarbonyl compounds as key step for preparation of warfarin, see: (a) Xie, J.-W., Yue, L., Chen, W., Du, W., Zhu, J., Deng, J.-G., Chen, Y.-C. (2007). Highly enantioselective Michael addition of cyclic 1,3-dicarbonyl compounds to α,β-unsaturated ketones. *Org. Lett.*, *9*, 413–415; (b) Kristensen, T. E., Vestli, K., Hansen, F. K., Hansen, T. (2009). New phenylglycine-derived primary amine organocatalysts for the preparation of optically active warfarin. *Eur. J. Org. Chem.*, 5185–5191; (c) Dong, Z., Wang, L., Chen, X., Liu, X., Lin, L., Feng, X. (2009). Organocatalytic enantioselective Michael addition of 4-hydroxycoumarin to α,β-unsaturated ketones: a simple synthesis of warfarin. *Eur. J. Org. Chem.*, 5192–5197.

[94] (a) Valero, G., Schimer, J., Cisarova, I., Vesely, J., Moyano, A., Rios, R. (2009). Highly enantioselective organocatalytic synthesis of piperidines. Formal synthesis of (−)-paroxetine. *Tetrahedron Lett.*, *50*, 1943–1946. For an excellent review of syntheses of paroxetine, see: (b) de Risi, C., Fanton, G., Pollini, G. P., Trapella, C., Zanirato, V. (2008). Recent advances in the stereoselective synthesis of trans-3,4-disubstituted-piperidines: applications to (−)-paroxetine. *Tetrahedron Asymmetry*, *19*, 131–155.

[95] For the last step in the synthesis, see: (a) Brandau, S., Landa, A., Franzén, J., Marigo, M., Jorgensen, K. A. (2006). Organocatalytic conjugate addition of malonates to α,β-unsaturated aldehydes: asymmetric formal synthesis of (−)-paroxetine, chiral lactams, and lactones. *Angew. Chem. Int. Ed.*, *45*, 4305–4309; (b) De Ferra, L., Massardo, P., Piccolo, O., Cigarillo, G. Rubidium and potassium salts of proline have been used in the synthesis of piperidines. U.S. Pat. 6, 444, 822, B1.

[96] (a) Josephsohn, N. S., Snapper, M. L., Hoveyda, A. H. (2004). Ag-Catalyzed asymmetric Mannich reactions of enol ethers with aryl, alkyl, alkenyl, and alkynyl imines. *J. Am. Chem. Soc.*, *126*, 3734–3735; (b) Zheng, G., Dwoskin, L. P., Crooks, P. A. (2004). Indirect trapping of the retroconjugate addition reaction intermediate involved in the epimerization of lobeline: application to the synthesis of (−)-sedamine. *J. Org. Chem.*, *69*, 8514–8517; (c) Angoli, M., Barilli, A., Lesma, G., Passarella, D., Riva, S., Silvani, A., Danielli, B. (2003). Remote stereocenter discrimination in the enzymatic resolution of piperidine-2-ethanol. Short enantioselective synthesis of sedamine and allosedamine. *J. Org. Chem.*, *68*, 9525–9527.

[97] Fustero, S., Jiménez, D., Moscardó, J., Catalán, S., del Pozo, C. (2007). Organocatalytic conjugate addition of malonates to α,β-unsaturated aldehydes: asymmetric formal synthesis of (−)-paroxetine, chiral lactams, and lactones. *Org. Lett.*, *9*, 5283–5286.

[98] For other applications of this organocatalytic strategy, see: (a) Carlson, E. C., Rathbone, L. K., Yang, H., Collett, N. D., Carter, R. G. (2008). Improved protocol for asymmetric, intramolecular heteroatom Michael addition using organocatalysis: enantioselective syntheses of homoproline, pelletierine, and homopipecolic acid. *J. Org. Chem.*, *73*, 5155–5158. For reviews of organocatalytic asymmetric aza-Michael additions, see: (b) Enders, D., Wang, C., Liebich, J. X. (2009). Organocatalytic asymmetric aza-Michael additions. *Chem. Eur. J.*, *15*, 11058–11076; (c) Krishna, P. R., Sreeshailam, A., Srinivas, R. (2009). Recent advances and applications in asymmetric aza-Michael addition chemistry. *Tetrahedron*, *65*, 9657–9672.

[99] (a) Fustero, S., Moscardó, J., Jiménez, D., Pérez-Carrión, M. D., Sánchez-Roselló, M., del Pozo, C. (2008). Organocatalytic approach to benzofused nitrogen-containing heterocycles: enantioselective total synthesis of (+)-angustureine. *Chem. Eur. J.*, *14*, 9868–9872. For the previous enantioselective organocatalytic synthesis by means of chiral Brønsted acids, see: (b) Rueping, M., Antonchick, A. P., Theissmann, T. (2006). A highly enantioselective Brønsted acid catalyzed cascade reaction: organocatalytic transfer hydrogenation of quinolines and their application in the synthesis of alkaloids. *Angew. Chem. Int. Ed.*, *45*, 3683–3686.

[100] Marigo, M., Franzén, J., Poulsen, T. B., Zhuang, W., Jørgensen, K. A. (2005). Asymmetric organocatalytic epoxidation of *α,β*-unsaturated aldehydes with hydrogen peroxide. *J. Am. Chem. Soc.*, *127*, 6964–6965.

[101] (a) Isaka, M., Rugseree, N., Maithip, P., Kongsaeree, P., Prabpai, S., Thebtaranonth, Y. (2005). Hirsutellones A–E, antimycobacterial alkaloids from the insect

pathogenic fungus *Hirsutella nivea* BCC 2594. *Tetrahedron*, *61*, 5577–5583; (b) Isaka, M., Prathumpai, W., Wongsa, P., Tanticharoen, M. (2006). Hirsutellone F, a dimer of antitubercular alkaloids from the seed fungus *Trichoderma* species BCC. *Org. Lett.*, *8*, 2815–2817.

[102] Nicolaou, K. C., Sarlah, D., Wu, T. R., Zhan, W. (2009). Total synthesis of hirsutellone B. *Angew. Chem. Int. Ed.*, *48*, 6870–6874.

[103] Northrup, A. B., MacMillan, D. W. C. (2002). The first general enantioselective catalytic Diels–Alder reaction with simple α,β-unsaturated ketones. *J. Am. Chem. Soc.*, *124*, 2458–2460.

[104] Jen, W. S., Wiener, J. J. M., MacMillan, D. W. C. (2000). New strategies for organic catalysis: the first enantioselective organocatalytic 1,3-dipolar cycloaddition. *J. Am. Chem. Soc.*, *122*, 9874–9875.

[105] Harmata, M., Ghosh, S. K., Hong, X., Wacharasindhu, S., Kirchhoefer, P. (2003). Asymmetric organocatalysis of 4 + 3 cycloaddition reactions. *J. Am. Chem. Soc.*, *125*, 2058–2059.

[106] Wilson, R. M., Jen, W. S., MacMillan, D. W. C. (2005). Enantioselective organocatalytic intramolecular Diels–Alder reactions. The asymmetric synthesis of solanapyrone D. *J. Am. Chem. Soc.*, *127*, 11616–11617.

[107] Massiot, G., Thépenier, P., Jacquier, M., Le Men-Olivier, L., Delaude, C. (1989). Normavacurine and minfiensine, two new alkaloids with $C_{19}H_{22}N_2O$ formula from strychnos species. *Heterocycles*, *29*, 1435–1438.

[108] (a) Jones, S. B., Simmons, B., MacMillan, D. W. C. (2009). Nine-step enantioselective total synthesis of (+)-minfiensine. *J. Am. Chem. Soc.*, *131*, 13606–13607. For previous total syntheses, see: (b) Dounay, A. B., Overman, L. E., Wrobleski, A. D. (2005). Sequential catalytic asymmetric Heck–iminium ion cyclization: enantioselective total synthesis of the *Strychnos* alkaloid minfiensine. *J. Am. Chem. Soc.*, *127*, 10186–10187; (c) Dounay, A. B., Humphreys, P. G., Overman, L. E., Wrobleski, A. D. (2008). Sequential catalytic asymmetric Heck–iminium ion cyclization: enantioselective total synthesis of the *Strychnos* alkaloid minfiensine. Total synthesis of the strychnos alkaloid (+)-minfiensine: tandem enantioselective intramolecular Heck–iminium ion cyclization. *J. Am. Chem. Soc.*, *130*, 5368–5377; (d) Shen, L., Zhang, M., Wu, Y., Qin, Y. (2008). Efficient assembly of an indole alkaloid skeleton by cyclopropanation: concise total synthesis of (±)-minfiensine. *Angew. Chem. Int. Ed.*, *47*, 3618–3621.

[109] Amaminol A: (a) Kumpulainen, E. T. T., Koskinen, A. M. P., Rissanen, K. (2007). Total synthesis of amaminol A: establishment of the absolute stereochemistry. *Org. Lett.*, *9*, 5043–5045. For initial synthetic studies toward a total synthesis of amaminol A and B, see: (b) Selkälä, S. A., Koskinen, A. M. P. (2005). Preparation of bicyclo[4.3.0]nonanes by an organocatalytic intramolecular Diels-Alder reaction. *Eur. J. Org. Chem.*, 1620–1624; and (c) Jacobs, W. C., Christmann, M. (2008). Synthesis of amaminol B. *Synlett*, 247–251, respectively.

[110] Sata, N. U., Fusetani, N. (2000). Amaminols A and B, new bicyclic amino alcohols from an unidentified tunicate of the family *Polyclinidae*. *Tetrahedron Lett.*, *41*, 489–492.

[111] (a) Nozaki, K., Sato, N., Ikeda, K., Takaya, H. (1996). Synthesis of highly functionalized gamma-butyrolactones from activated carbonyl compounds and

dimethyl acetylenedicarboxylate. *J. Org. Chem.*, *61*, 4516–4519; (b) Mielgo, A., Palomo, C. (2008). α,α-Diarylprolinol ethers: new tools for functionalization of carbonyl compounds. *Chem. Asian J.*, *3*, 922–948.

[112] Liu, K., Chougnet, A., Woggon, W.-D. (2008). A short route to α-tocopherol. *Angew. Chem. Int. Ed.*, *47*, 5827–5829.

[113] For selected examples, see: (a) Li, H., Wang, J., Nunu, T. E., Zu, L. S., Jiang, W., Wei, S. H., Wang, W. (2007). One-pot approach to chiral chromenes via enantioselective organocatalytic domino oxa-Michael–aldol reaction. *Chem. Commun.*, 507–509; (b) Sundén, H., Ibrahem, I., Zhao, G.-L., Eriksson, L., Córdova, A. (2007). Catalytic enantioselective domino oxa-Michael/aldol condensations: asymmetric synthesis of benzopyran derivatives. *Chem. Eur. J.*, *13*, 574–581; (c) Rueping, M., Sugiono, E., Merino, E. (2008). Asymmetric iminium ion catalysis: an efficient enantioselective synthesis of pyranonaphthoquinones and β-lapachones. *Angew. Chem. Int. Ed.*, *47*, 3046–3049. For excellent reviews of domino reactions, see Reference [76d] and (d) Yu, X., Wang, W. (2008). Organocatalysis: asymmetric cascade reactions catalysed by chiral secondary amines. *Org. Biomol. Chem.*, *6*, 2037–2046. For reviews of oxa-Michael additions, see: (e) Berkessel, A., Gröger, H. (2005). *Asymmetric Organocatalysis*. Wiley-VCH, Weinheim, Germany, Chapter 4; (f) Nising, C. F., Bräse, S. (2008). The oxa-Michael reaction: from recent developments to applications in natural product synthesis. *Chem. Soc. Rev.*, *37*, 1218–1228.

[114] Hong, B.-C., Wu, M.-F., Tseng, H.-C., Liao, J.-H. (2006). Enantioselective organocatalytic formal [3 + 3]-cycloaddition of *α,β*-unsaturated aldehydes and application to the asymmetric synthesis of (–)-isopulegol hydrate and (–)-cubebaol. *Org. Lett.*, *8*, 2217–2220.

[115] (a) Hong, B.-C., Wu, M.-F., Tseng, H.-C., Huang, G.-F., Su, C.-F., Liao, J.-H. (2007). Organocatalytic asymmetric Robinson annulation of α,β-unsaturated aldehydes: applications to the total synthesis of (+)-palitantin. *J. Org. Chem.*, *72*, 8459–8471. For previous syntheses, see: (b) Ichihara, A., Ubukata, M., Sakamura, S. (1980). Stereoselective synthesis of (±)-palitantin. *Tetrahedron*, *36*, 1547–1550; (c) Ichihara, A., Ubukata, M., Sakamura, S. (1977). Stereoselective synthesis of (±)-palitantin. *Tetrahedron Lett.*, *18*, 3473–3476; (d) Hanessian, S., Sakito, Y., Dhanoa, D., Baptistella, L. (1989). Synthesis of (+)-palitantin. *Tetrahedron*, *45*, 6623–6630; (e) Deruyttere, X., Dumortier, L., Van der Eycken, J., Vandewalle, M. (1992). The total synthesis of (+)-palitantin. *Synlett*, 51–52; (f) Hareau, G., Koiwa, M., Hanazawa, T., Fumie, S. F. (1999). Asymmetric synthesis of palitantin from the (5*R*)-*tert*-butyldimethylsiloxy-2-cyclohexenone. *Tetrahedron Lett.*, *40*, 7493–7496; (g) Angelaud, R., Babot, O., Charvat, T., Landais, Y. (1999). Desymmetrization of cyclohexadienylsilanes. Regio-, diastereo-, and enantioselective access to sugar mimics. *J. Org. Chem.*, *64*, 9613–9624.

[116] (a) Birkinshaw, J. H., Raistrick, H. (1936). Studies in the biochemistry of micro-organisms. XLIX. Palitantin, C14H22O4, a hitherto undescribed metabolic product of *Penicillium* palitans Westling. *Biochem. J.*, *30*, 801–808; (b) Birkinshaw, J. H. (1952). Studies in the biochemistry of micro-organisms. 88. Palitantin. Part 2. Further derivatives and degradation products. *Biochem. J.*, *51*, 271–276.

[117] Demetriadou, A., Laue, E. D., Staunton, J. (1988). Biosynthesis of palitantin, a polyketide metabolite of *Penicillium brefeldianum*: ^{13}C n.m.r. assignment and

incorporation of ^{13}C-, ^{2}H-, and $^{18}O_2$-labelled acetates. *J. Chem. Soc. Perkin Trans. 1.*, 773–778.

[118] Yamaji, K., Fukushi, Y., Hashidoko, Y., Yoshida, T., Tahara, S. (1999). Characterization of antifungal metabolites produced by *Penicillium* species isolated from seeds of *Picea glehnii*. *J. Chem. Ecol.*, *25*, 1643–1653.

[119] Curtis, P. J., Hemming, H. G., Smith, W. K. (1951). Frequentin: an antibiotic produced by some strains of penicillium frequentans westling. *Nature*, *167*, 557–558.

[120] Hong, H., Neamati, N., Wang, S., Nicklaus, M. C., Mazumder, A., Zhao, H., Burke, T. R., Jr., Pommier, Y., Milne, G. W. A. (1997). Discovery of HIV-1 integrase inhibitors by pharmacophore searching. *J. Med. Chem.*, *40*, 930–936.

[121] de Figueiredo, R. M., Fröhlich, R., Christmann, M. (2008). Amine-catalyzed cyclizations of tethered α,β-unsaturated carbonyl compounds. *Angew. Chem. Int. Ed.*, *47*, 1450–1453.

[122] Starkenmann, C., Mayenzet, F., Brauchli, R., Wunsche, L., Vial, C. (2007). Structure elucidation of a pungent compound in black cardamom: *Amomum tsao-ko* Crevost et Lemarié (Zingiberaceae). *J. Agric. Food Chem.*, *55*, 10902–10907.

[123] For a previous synthesis of rotundial, see: Takikawa, H., Yamazaki, Y., Mori, K. (1998). Synthesis of mono- and sesquiterpenoids, part XXVI. Synthesis and absolute configuration of rotundial, a mosquito repellent from the leaves of *Vitex rotundiforia*. *Eur. J. Org. Chem.*, 229–232.

[124] (a) Marqués-López, E., Herrera, R. P., Marks, T., Jacobs, W. C., Könning, D., de Figueiredo, R. M., Christmann, M. (2009). Crossed intramolecular Rauhut–Currier-type reactions via dienamine activation. *Org. Lett.*, *11*, 4116–4119. For seminal work on the Rauhut–Currier reaction, see: (b) Rauhut, M. M., Currier, H. (1963). US Pat. 3,074,999. For an excellent review, see: (c) Aroyan, C. E., Dermenci, A., Miller, S. J. (2009). The Rauhut–Currier reaction: a history and its synthetic application. *Tetrahedron*, *65*, 4069–4084.

[125] Watanabe, K., Takada, Y., Matsuo, N., Nishimura, H. (1995). Rotundial, a new natural mosquito repellent from the leaves of *Vitex rotundifolia*. *Biosci., Biotechnol., Biochem.*, *59*, 1979–1980.

[126] Sibi, M. P., Hasegawa, M. (2007). Organocatalysis in radical chemistry. Enantioselective α-oxyamination of aldehydes. *J. Am. Chem. Soc.*, *129*, 4124–4125.

[127] (a) Jang, H., Hong, J., MacMillan, D. W. C. (2007). Enantioselective organocatalytic singly occupied molecular orbital activation: the enantioselective α-enolation of aldehydes. *J. Am. Chem. Soc.*, *129*, 7004–7005; (b) Kim, H., MacMillan, D. W. C. (2008). Enantioselective organo-SOMO catalysis: the α-vinylation of aldehydes. *J. Am. Chem. Soc.*, *130*, 398–399; (c) Conrad, J. C., Kong, J., Laforteza, B. N., MacMillan, D. W. C. (2009). Enantioselective α-arylation of aldehydes via organo-SOMO catalysis. An ortho-selective arylation reaction based on an open-shell pathway. *J. Am. Chem. Soc.*, *131*, 11640–11641; (d) Um, J. M., Gutierrez, O., Schoenebeck, F., Houk, K. N., MacMillan, D. W. C. (2010). Nature of intermediates in organo-SOMO catalysis of α-arylation of aldehydes. *J. Am. Chem. Soc.*, *132*, 6001–6005; (e) Rendler, S., MacMillan, D. W. C. (2010). Enantioselective polyene cyclization via organo-SOMO catalysis. *J. Am. Chem. Soc.*, *132*, 5027–5029.

[128] Bohlmann, F., Zdero, C., Robinson, H., King, R. M. (1979). Neue cadinen- und norcadinen-derivate aus *Heterotheca grandiflora*. *Phytochem*, *18*, 1675–1680.

[129] For a related transformation, see: (a) Nicolaou, K. C., Reingruber, R., Sarlah, D., Bräse, S. (2009). Enantioselective intramolecular Friedel–Crafts-type α-arylation of aldehydes. *J. Am. Chem. Soc.*, *131*, 2086–2087; (b) Alemán, J., Cabrera, S., Maerten, E., Overgaard, J., Jørgensen, K. A. (2007). Asymmetric organocatalytic α-arylation of aldehydes. *Angew. Chem. Int. Ed.*, *46*, 5520–5523. For previous enantioselective total syntheses, see: (c) Tietze, L. F., Raschke, T. (1995). Enantioselective total synthesis of a natural norsesquiterpene of the calamenene group by a silane-terminated intramolecular Heck reaction. *Synlett*, 597–598; (d) Schmalz, H.-G., Kiehl, O., Korell, U., Lex, J. (2003). An enantioselective approach to cytotoxic norcalamenenes via electron-transfer-driven benzylic umpolung of an arene tricarbonyl chromium complex. *Synthesis.*, 1851–1855.

[130] (a) Mizoroki, T., Mori, K., Ozaki, A. (1971). Arylation of olefin with aryl iodide catalyzed by palladium. *Bull. Chem. Soc. Jpn.*, *44*, 581–581; (b) Heck, R. F., Nolley, J. P., Jr. (1972). Palladium-catalyzed vinylic hydrogen substitution reactions with aryl, benzyl, and styryl halides. *J. Org. Chem.*, *37*, 2320–2322.

[131] (a) Friedel, C., Crafts, J. M., Hebd, C. R. (1877). *Seances Acad. Sci.*, *84*, 1392–1395. For reviews, see: (b) Rueping, M., Nachtsheim, B. J. (2010). A review of new developments in the Friedel–Crafts alkylation—from green chemistry to asymmetric catalysis. *Beilstein J. Org. Chem.*, *6*, 1–24; (c) Marqués-López, E., Diez-Martinez, A., Merino, P., Herrera, R. P. (2009). The role of the indole in important organocatalytic enantioselective Friedel-Crafts alkylation reactions. *Curr. Org. Chem.*, *13*, 1585–1609, and references cited therein.

[132] For reviews: (a) Schreiner, P. R. (2003). Metal-free organocatalysis through explicit hydrogen bonding interactions. *Chem. Soc. Rev.*, *32*, 289–296; (b) Pihko, P. M. (2004). Activation of carbonyl compounds by double hydrogen bonding: an emerging tool in asymmetric catalysis. *Angew. Chem. Int. Ed.*, *43*, 2062–2064; (c) Bolm, C., Rantanen, T., Schiffers, I., Zani, L. (2005). Protonated chiral catalysts: versatile tools for asymmetric synthesis. *Angew. Chem. Int. Ed.*, *44*, 1758–1763; (d) Yamamoto, H., Futatsugi, K. (2005). "Designer acids": combined acid catalysis for asymmetric synthesis. *Angew. Chem. Int. Ed.*, *44*, 1924–1942.

[133] Toda, F. (1995). Solid state organic chemistry: efficient reactions, remarkable yields, and stereoselectivity. *Acc. Chem. Res.*, *28*, 480–486.

[134] For a review about chiral phosphoric acid organocatalysts, see: Connon, S. J. (2006). Chiral phosphoric acids: powerful organocatalysts for asymmetric addition reactions to imines. *Angew. Chem. Int. Ed.*, *45*, 3909–3912.

[135] For reviews about chiral thiourea organocatalysts, see: (a) Takemoto, Y. (2005). Recognition and activation by ureas and thioureas: stereoselective reactions using ureas and thioureas as hydrogen-bonding donors. *Org. Biomol. Chem.*, *3*, 4299–4306; (b) Connon, S. J. (2006). Organocatalysis mediated by (thio)urea derivatives. *Chem. Eur. J.*, *12*, 5418–5427; (c) Miyabe, H., Takemoto, Y. (2008). Discovery and application of asymmetric reaction by multi-functional thioureas. *Bull. Chem. Soc. Jpn.*, *81*, 785–795.

[136] (a) Terada, M., Ube, H., Yaguchi, Y. (2006). Axially chiral guanidine as enantioselective base catalyst for 1,4-addition reaction of 1,3-dicarbonyl compounds with conjugated nitroalkenes. *J. Am. Chem. Soc.*, *128*, 1454–1455; (b) Schuster, T.,

Kurz, M., Göbel, M. W. (2000). Catalysis of a Diels–Alder reaction by amidinium ions. *J. Org. Chem.*, *65*, 1697–1701; (c) Lao, J.-H., Zhang, X.-J., Wang, J.-J., Li, X.-M., Yan, M., Luo, H.-B. (2009). The effect of hydrogen bond donors in asymmetric organocatalytic conjugate additions. *TetrahedronAsymmetry*, *22*, 2818–2822.

[137] Wanner, M. J., Boots, R. N. A., Eradus, B., de Gelder, R., van Maarseveen, J. H., Hiemstra, H. (2009). Organocatalytic enantioselective total synthesis of (–)-arboricine. *Org. Lett.*, *11*, 2579–2581.

[138] Lim, K.-H., Komiyama, K., Kam, T.-S. (2007). Arboricine and arboricinine, unusual tetracyclic indole regioisomers from Kopsia. *Tetrahedron Lett.*, *48*, 1143–1145.

[139] (a) Pictet, A., Spengler, T. (1911). Über die Bildung von Isochinolin-derivaten durch Einwirkung von Methylal auf Phenyl-äthylamin, Phenyl-alanin und *Tyrosin*. *Ber. Dtsch. Chem. Ges.*, *44*, 2030–2036. For reviews, see: (b) Cox, E. D., Cook, J. M. (1995). The Pictet-Spengler condensation: a new direction for an old reaction. *Chem. Rev.*, *95*, 1797–1842; (c) You, S.-L., Cai, Q., Zeng, M. (2009). Chiral Brønsted acid catalyzed Friedel–Crafts alkylation reactions. *Chem. Soc. Rev.*, *38*, 2190–2201.

[140] Takayama, H., Watanabe, F., Kitajima, M., Aimi, N. (1997). A radical cyclization strategy for the concise total synthesis of (±)-geissoschizine. *Tetrahedron Lett.*, *38*, 5307–5310.

[141] (a) Sole, D., Urbaneja, X., Bonjoch, J. (2004). Palladium-catalyzed intramolecular coupling of amino-tethered vinyl halides with ketones, esters, and nitriles using potassium phenoxide as the base. *Adv. Synth. Catal.*, *346*, 1646–1650. For a similar reaction, see also: (b) Yu, J. M., Wang, T., Liu, X. X., Deschamps, J., Flippen-Anderson, J., Liao, X. B., Cook, J. M. (2003). General approach for the synthesis of sarpagine indole alkaloids. Enantiospecific total synthesis of (+)-vellosimine, (+)-normacusine B, (–)-alkaloid Q3, (–)-panarine, (+)-Na-methylvellosimine, and (+)-Na-methyl-16-epipericyclivine. *J. Org. Chem.*, *68*, 7565–7581.

[142] Rueping, M., Antonchick, A. P. (2007). Organocatalytic enantioselective reduction of pyridines. *Angew. Chem. Int. Ed.*, *46*, 4562–4565.

[143] (a) Bohlmann, F., Rahtz, D. (1957). Über eine neue Pyridinsynthese. *Chem. Ber.*, *90*, 2265–2272; (b) Bagley, M. C., Brace, C., Dale, J. W., Ohnesorge, M., Phillips, N. G., Xiong, X., Bower, J. (2002). Synthesis of tetrasubstituted pyridines by the acid-catalysed Bohlmann–Rahtz reaction. *J. Chem. Soc., Perkin Trans. 1.*, 1663–1671.

[144] Sklenicka, H. M., Hsung, R. P., McLaughlin, M. J., Wie, L.-L., Gerasyuto, A. I., Brennessel, W. B. (2002). Stereoselective formal [3 + 3] cycloaddition approach to cis-1-azadecalins and synthesis of (–)-4*a*,8*a*-*diepi*-pumiliotoxin c. Evidence for the first highly stereoselective 6π-electron electrocyclic ring closures of 1-azatrienes. *J. Am. Chem. Soc.*, *124*, 10435–10442.

[145] Kam, T.-S., Sim, K.-M. (1998). Alkaloids from *Kopsia griffithii*. *Phytochem*, *47*, 145–147.

[146] (a) Raheem, I. T., Thiara, P. S., Peterson, E. A., Jacobsen, E. N. (2007). Enantioselective Pictet–Spengler-type cyclizations of hydroxylactams: H-bond donor catalysis by anion binding. *J. Am. Chem. Soc.*, *129*, 13404–13405. For alter-

native synthetic approaches to harmicine, see: (b) Szawkao, J., Czarnocki, S. J., Zawadzka, A., Wojtasiewicz, K., Leniewski, A., Maurin, J. K., Czarnockia, Z., Drabowicz, J. (2007). Enantioselective synthesis of some tetrahydroisoquinoline and tetrahydro-β-carboline alkaloids. *Tetrahedron Asymmetry*, *18*, 406–413, and references cited therein.

[147] For reviews, see: (a) Brown, R. T. (1983). The Yohimbine group. In: J. E. Saxton (Ed.), *The Chemistry of Heterocyclic Compounds*. John Wiley and Sons, New York, pp. 147–199; (b) Baxter, E. W., Mariano, P. S. (1992). Recent advances in the synthesis of Yohimbine alkaloids. In: S. W. Pelletier (Ed.), *Alkaloids: Chemical and Biological Perspectives*. Springer-Verlag, New York. pp. 197–319; (c) Szántay, C., Honty, K. (1994). Monoterpenoid indole alkaloids. In: J. E. Saxton (Ed.), *The Chemistry of Heterocyclic Compounds*. John Wiley and Sons, New York. pp. 161–216; (d) Goldberg, M. R., Robertson, D. (1983). A pharmacological probe for study of α_2 adrenoreceptor. *Pharmacol. Rev.*, *35*, 143–180.

[148] For the first total syntheses of yohimbine, see: van Tamelen, E. E., Shamma, M., Burgstahler, A. W., Wolinsky, J., Tamm, R., Aldrich, P. E. (1969). Total synthesis of yohimbine. *J. Am. Chem. Soc.*, *91*, 7315–7333.

[149] Mergott, D. J., Zuend, S. J., Jacobsen, E. N. (2008). Catalytic asymmetric total synthesis of (+)-yohimbine. *Org. Lett.*, *10*, 745–748.

[150] (a) Okino, T., Hoashi, Y., Takemoto, Y. (2003). Enantioselective Michael reaction of malonates to nitroolefins catalyzed by bifunctional organocatalysts. *J. Am. Chem. Soc.*, *125*, 12672–12673; (b) Okino, T., Hoashi, Y., Furukawa, T., Xu, X., Takemoto, Y. (2005). Enantio- and diastereoselective Michael reaction of 1,3-dicarbonyl compounds to nitroolefins catalyzed by a bifunctional thiourea. *J. Am. Chem. Soc.*, *127*, 119–125.

[151] (a) Bui, T., Syed, S., Barbas, C. F., III (2009). Thiourea-catalyzed highly enantio- and diastereoselective additions of oxindoles to nitroolefins: application to the formal synthesis of (+)-physostigmine. *J. Am. Chem. Soc.*, *131*, 8758–8759. For selected previous syntheses of (–)-physostigmines, see: (b) Matsuura, T., Overman, L. E., Poon, D. J. (1998). Catalytic asymmetric synthesis of either enantiomer of the calabar alkaloids physostigmine and physovenine. *J. Am. Chem. Soc.*, *120*, 6500–6503; (c) Asakawa, K., Noguchi, N., Takashima, S., Nakada, M. (2008). Preparation of a new chiral building block containing a benzylic quaternary stereogenic center and a formal total synthesis of (–)-physostigmine. *Tetrahedron Asymmetry*, *19*, 2304–2309; (d) Trost, B. M., Zhang, Y. (2006). Molybdenum-catalyzed asymmetric allylation of 3-alkyloxindoles: application to the formal total synthesis of (–)-physostigmine. *J. Am. Chem. Soc.*, *128*, 4590–4591; (e) Lee, T. B. K., Wong, G. S. K. (1991). Asymmetric alkylation of oxindoles: an approach to the total synthesis of (–)-physostigmine. *J. Org. Chem.*, *56*, 872–875.

[152] (a) Silverman, R. B., Andruszkiewicz, R., Nanavati, S. M., Taylor, C. P., Vartanian, M. G. (1991). 3-Alkyl-4-aminobutyric acids: the first class of anticonvulsant agents that activates L-glutamic acid decarboxylase. *J. Med. Chem.*, *34*, 2295–2298; (b) Silverman, R. B. (2008). From basic science to blockbuster drug: the discovery of lyrica. *Angew. Chem. Int. Ed.*, *47*, 3500–3504; (c) Bassas, O., Huuskonen, J., Rissanen, K., Koskinen, A. M. P. (2009). A simple organocatalytic enantioselective synthesis of pregabalin. *Eur. J. Org. Chem.*, 1340–1351.

[153] (a) Tian, S.-K., Chen, Y., Hang, J., Tang, L., McDaid, P., Deng, L. (2004). Asymmetric organic catalysis with modified *Cinchona* alkaloids. *Acc. Chem. Res.*, *37*, 621–631;

(b) Marcelli, T., van Maarseveen, J. H., Hiemstra, H. (2006). Cupreines and cupreidines: an emerging class of bifunctional *Cinchona* organocatalysts. *Angew. Chem. Int. Ed.*, *45*, 7496–7504.

[154] (a) Hartikka, A., Modin, S. A., Andersson, P. G., Arvidsson, P. I. (2003). *Cinchona* alkaloid derived ligands in catalytic asymmetric transfer hydrogenation. *Org. Biomol. Chem.*, *1*, 2522–2526; (b) Gaunt, M. J., Johansson, C. C. C. (2007). Recent developments in the use of catalytic asymmetric enolates in chemical synthesis. *Chem. Rev.*, *107*, 5596–5605; (c) Bartoli, G., Melchiorre, P. (2008). A novel organocatalytic tool for the iminium activation of α,β-unsaturated ketones. *Synlett*, 1759–1772.

[155] (a) Tanaka, S., Nagasawa, K. (2009). Guanidine-urea bifunctional organocatalyst for asymmetric epoxidation of 1,3-diarylenones with hydrogen peroxide. *Synlett*, 667–670; (b) McCooey, S. H., Connon, S. J. (2005). Urea- and thiourea-substituted *Cinchona* alkaloid derivatives as highly efficient bifunctional organocatalysts for the asymmetric addition of malonate to nitroalkenes: inversion of configuration at C9 dramatically improves catalyst performance. *Angew. Chem. Int. Ed.*, *44*, 6367–6370.

[156] MacLeod, J. K., Schäffeler, L. (1995). Interestingly, the (*S*)-enantiomer is inactive. *J. Nat. Prod.*, *58*, 1270–1273.

[157] Dollt, H., Hammann, P., Blechert, S. (1999). Synthesis of (+)-(*S*)-streptenol A and biomimetic synthesis of (2*R*,4*S*)- and (2*S*,4*S*)-2-(pent-3-enyl)piperidin-4-ol. *Helv. Chim. Acta.*, *82*, 1111–1122 , and references cited therein.

[158] Li, D. R., Murugan, A., Falck, J. R. (2008). Enantioselective, organocatalytic oxy-michael addition to γ/δ-hydroxy-α,β-enones: boronate-amine complexes as chiral hydroxide synthons. *J. Am. Chem. Soc.*, *130*, 46–48.

[159] Vakulya, B., Varga, S., Csámpai, A., Soós, T. (2005). Highly enantioselective conjugate addition of nitromethane to chalcones using bifunctional cinchona organocatalysts. *Org. Lett.*, *7*, 1967–1969.

[160] Shi, M., Lei, Z.-Y., Zhao, M.-X., Shi, J.-W. (2007). A highly efficient asymmetric Michael addition of anthrone to nitroalkenes with cinchona organocatalysts. *Tetrahedron Lett.*, *48*, 5743–5746.

[161] Goss, J. M., Schaus, S. E. (2008). Enantioselective synthesis of SNAP-7941: chiral dihydropyrimidone inhibitor of MCH1-R. *J. Org. Chem.*, *73*, 7651–7656.

[162] (a) Borowsky, B., Durkin, M. M., Ogozalek, K., Marzabadi, M. R., DeLeon, J., Heurich, R., Lichtblau, H., Shaposhnik, Z., Daniewska, I., Blackburn, T. P., Branchek, T. A., Gerald, C., Vaysse, P. J., Forray, C. (2002). Antidepressant, anxiolytic and anorectic effects of a melaninconcentrating hormone-1 receptor antagonist. *Nat. Med.*, *8*, 825–830; (b) Basso, A. M., Bratcher, N. A., Gallagher, K. B., Cowart, M. D., Zhao, C., Sun, M., Esbenshade, T. A., Brune, M. E., Fox, G. B., Schmidt, M., Collins, C. A., Souers, A. J., Iyengar, R., Vasudevan, A., Kym, P. R., Hancock, A. A., Rueter, L. E. (2006). Lack of efficacy of melanin-concentrating hormone-1 receptor antagonists in models of depression and anxiety. *Eur. J. Pharmacol.*, *540*, 115–120.

[163] (a) Mannich, C., Krösche, W. (1912). Uber ein kondensationsprodukt aus formaldehyde, ammoniak und antipyrin. *Arch. Pharm.*, *250*, 647–667. For recent reviews, see: (b) Ting, A., Schaus, S. E. (2007). Organocatalytic asymmetric

Mannich reactions: new methodology, catalyst design, and synthetic applications. *Eur. J. Org. Chem.*, 5797–5815; (c) Verkade, J. M. M., van Hemert, L. J. C., Quaedfliegb, P. J. L. M., Rutjes, F. P. J. T. (2008). Organocatalysed asymmetric Mannich reactions. *Chem. Soc. Rev.*, *37*, 29–41; (d) Arrayas, R. G., Carretero, J. C. (2009). Catalytic asymmetric direct Mannich reaction: a powerful tool for the synthesis of α β-diamino acids. *Chem. Soc. Rev.*, *38*, 1940–1948.

[164] (a) Biginelli, P. (1893). The condensation reaction described by Biginelli. *Gazz. Chim. Ital.*, *23*, 360–416. For reviews, see: (b) Kappe, C. O. (1993). 100 years of the Biginelli dihydropyrimidine synthesis. *Tetrahedron*, *49*, 6937–6963; (c) Kampe, C. O. (2005). The Biginelli reaction. In: J. Zhu, H. Bienaymé (Eds.), *Multicomponent Reactions*. Wiley-VCH, Weinheim, Germany, pp. 95–120; (d) Gong, L.-Z., Chen, X.-H., Xu, X.-Y. (2007). Asymmetric organocatalytic Biginelli reactions: a new approach to quickly access optically active 3,4-dihydropyrimidin-2-(1*H*)-ones. *Chem. Eur. J.*, *13*, 8920–8926.

[165] Wang, B., Wu, F., Wang, Y., Liu, X., Deng, L. (2007). Control of diastereoselectivity in tandem asymmetric reactions generating nonadjacent stereocenters with bifunctional catalysis by *Cinchona* alkaloids. *J. Am. Chem. Soc.*, *129*, 768–769.

[166] Wang, Y., Liu, X., Deng, L. (2006). Dual-function cinchona alkaloid catalysis: catalytic asymmetric tandem conjugate addition–protonation for the direct creation of nonadjacent stereocenters. *J. Am. Chem. Soc.*, *128*, 3928–3930.

[167] (a) Jakubec, P., Cockfield, D. M., Dixon, D. J. (2009). Total synthesis of (–)-nakadomarin A. *J. Am. Chem. Soc.*, *131*, 16632–16633; (b) Martin, D. B. C., Vanderwal, C. D. (2010). Concise synthesis of (–)-nakadomarin A. *Angew. Chem. Int. Ed.*, *49*, 2830–2832.

[168] Kobayashi, J., Watanabe, D., Kawasaki, N., Tsuda, M. (1997). Nakadomarin A, a novel hexacyclic manzamine-related alkaloid from *Amphimedon* sponge. *J. Org. Chem.*, *62*, 9236–9239.

[169] Jakubec, P., Helliwell, M., Dixon, D. J. (2008). Cyclic imine nitro-Mannich/lactamization cascades: a direct stereoselective synthesis of multicyclic piperidinone derivatives. *Org. Lett.*, *10*, 4267–4270.

[170] (a) Shibahara, S., Fujino, M., Tashiro, Y., Takahashi, K., Ishihara, J., Hatakeyama, S. (2008). Asymmetric total synthesis of (+)-phoslactomycin B. *Org. Lett.*, *10*, 2139–2142; (b) Morita, K.-I., Suzuki, Z., Hirose, H. (1968). A tertiary phosphine-catalyzed reaction of acrylic compounds with aldehydes. *Bull. Chem. Soc. Jpn.*, *41*, 2815.

[171] For seminal work, see: (a) Morita, K.-I., Suzuki, Z., Hirose, H. (1968). A tertiary phosphine-catalyzed reaction of acrylic compounds with aldehydes. *Bull. Chem. Soc. Jpn.*, *41*, 2815–2815; (b) Baylis, A. B., Hillman, M. E. D. (1972) Offenlegungsschrift 21155113, US Pat. 3,743,669 (*Chem. Abstr.*, 1972, *77*, 34174q). For selected reviews, see: (c) Basavaiah, D., Rao, K. V., Reddy, R. J. (2007). The Baylis-Hillman reaction: a novel source of attraction, opportunities, and challenges in synthetic chemistry. *Chem. Soc. Rev.*, *36*, 1581–1588; (d) Masson, G., Housseman, C., Zhu, J. (2007). The enantioselective Morita-Baylis-Hillman reaction and its aza counterpart. *Angew. Chem. Int. Ed.*, *46*, 4614–4602; (e) Ma, G.-N., Jiang, J.-J., Shi, M., Wei, Y. (2009). Recent extensions of the Morita-Baylis-Hillman reaction. *Chem. Commun.*, 5496–5514.

[172] Sarkar, S. M., Wanzala, E. N., Shibahara, S., Takahashi, K., Ishihara, J., Hatakeyama, S. (2009). Enantio- and stereoselective route to the phoslactomcin family of antibiotics: formal synthesis of (+)-fostriecin and (+)-phoslactomycin B. *Chem. Commun.*, 5907–5909.

[173] Seki, M. (2008). A practical synthesis of a key chiral drug intermediate via asymmetric organocatalysis. *Synlett*, 164–176.

[174] Frohn, M., Shi, Y. (2000). Chiral ketone-catalyzed asymmetric epoxidation of olefins. *Synthesis.*, 1979–2000.

[175] (a) Yang, D. (2004). Ketone-catalyzed asymmetric epoxidation reactions. *Acc. Chem. Res.*, *37*, 497–505; (b) Electrostatic repulsion between the oxygen-containing functional groups plays an important role in the stereoselectivity of the dioxirane-mediated asymmetric epoxidation; see: Matsumoto, K., Tomioka, K. (2002). Chiral ketone-catalyzed asymmetric epoxidation of olefins with Oxone®. *Tetrahedron Lett.*, *43*, 631–633.

[176] (a) Dehmlow, E. V., Dehmlow, S. S. (1993). *Phase Transfer Catalysis*, 3rd ed. VCH, Weinheim, Germany; (b) Starks, C. M., Liotta, C. L., Halpern, M. (1994). *Phase-Transfer Catalysis*. Chapman & Hall, New York; (c) Sasson, Y., Neumann, R. (Eds.) (1997). *Handbook of Phase-Transfer Catalysis*. Blackie Academic & Professional, London; (d) Halpern, M. E. (Ed.) (1997). *Phase-Transfer Catalysis, ACS Symposium Series 659*. American Chemical Society, Washington, DC.

[177] (a) Shioiri, T. (1997). Chiral phase transfer catalysis. In: Y. Sasson, R. Neumann (Eds.), *Handbook of Phase-Transfer Catalysis*. Blackie Academic & Professional, London. Chapter 14; (b) Nelson, A. (1999). Asymmetric phase-transfer catalysis. *Angew. Chem. Int. Ed.*, *38*, 1583–1585; (c) Shioiri, T., Arai, S. (2000). Asymmetric phase transfer catalysis. In: F. Vögtle, J. F. Stoddart, M. Shibasaki (Eds.), *Stimulating Concepts in Chemistry*. Wiley-VCH, Weinheim, Germany, p. 123; (d) O'Donnell, M. J. (2000). Asymmetric phase transfer reactions. In: I. Ojima (Ed.), *Catalytic Asymmetric Syntheses*, 2nd ed. Wiley-VCH, New York. Ch. 10; (e) O'Donnell, M. J. (2001). The preparation of optically active α-amino acids from the benzophenone imines of glycine derivatives. *Aldrichim. Acta.*, *34*, 3–15; (f) O'Donnell, M. J. (2004). The enantioselective synthesis of α-amino acids by phase-transfer catalysis with achiral Schiff base esters. *Acc. Chem. Res.*, *37*, 506–517; (g) Lygo, B., Andrews, B. I. (2004). Asymmetric phase-transfer catalysis utilizing chiral quaternary ammonium salts: asymmetric alkylation of glycine imines. *Acc. Chem. Res.*, *37*, 518–525; (h) Vachon, J., Lacour, J. (2006). Recent developments in enantioselective phase transfer catalysis using chiral ammonium salts. *Chimia*, *60*, 266–275.

[178] (a) Dolling, U.-H., Davis, P., Grabowski, E. J. J. (1984). Efficient catalytic asymmetric alkylations. 1. Enantioselective synthesis of (+)-indacrinone via chiral phase-transfer catalysis. *J. Am. Chem. Soc.*, *106*, 446–447; (b) Hughes, D. L., Dolling, U.-H., Ryan, K. M., Schoenewaldt, E. F., Grabowski, E. J. J. (1987). Efficient catalytic asymmetric alkylations. 3. A kinetic and mechanistic study of the enantioselective phase-transfer methylation of 6,7-dichloro-5-methoxy-2-phenyl-1-indanone. *J. Org. Chem.*, *52*, 4745–4752.

[179] (a) Shibuguchi, T., Fukuta, Y., Akachi, Y., Sekine, A., Ohshima, T., Shibasaki, M. (2002). Development of new asymmetric two-center catalysts in phase-transfer reactions. *Tetrahedron Lett.*, *43*, 9539–9543; (b) Ohshima, T., Shibuguchi, T.,

Fukuta, Y., Shibasaki, M. (2004). Catalytic asymmetric phase-transfer reactions using tartrate-derived asymmetric two-center organocatalysts. *Tetrahedron*, *60*, 7743–7754; (c) Okada, A., Shibuguchi, T., Ohshima, T., Masu, H., Yamaguchi, K., Shibasaki, M. (2005). Enantio- and diastereoselective catalytic Mannich-type reaction of a glycine Schiff base using a chiral two-center phase-transfer catalyst. *Angew. Chem. Int. Ed.*, *44*, 4564–4567; (d) Fukuta, Y., Ohshima, T., Ganadesikan, V., Shibuguchi, T., Nemoto, T., Kisugi, T., Okino, T., Shibasaki, M. (2004). Asymmetric catalysis special feature part I: enantioselective syntheses and biological studies of aeruginosin 298-A and its analogs: application of catalytic asymmetric phase-transfer reaction. *Proc. Natl. Acad. Sci. U. S. A.*, *101*, 5433–5438.

[180] (a) Shibuguchi, T., Mihara, H., Kuramochi, A., Ohshima, T., Shibasaki, M. (2007). Catalytic asymmetric phase-transfer Michael reaction and Mannich-type reaction of glycine Schiff bases with tartrate-derived diammonium salts. *Chem. Asian J.*, *2*, 794–808. For previous racemic syntheses, see: (b) Murakami, K., Takahashi, M., Hirata, Y., Takashima, M., Iwanami, S., Hasegawa, O., Nozaki, Y., Tashikawa, S., Takeda, M., Usuda, S. (1978) U.S. Pat. 4,097,487; (c) Takashima, M., Iwanami, S., Usuda, S. (1980) U.S. Pat. 4,210,660; (d) Iwanami, S., Takashima, M., Hirata, Y., Hasegawa, O., Usuda, S. (1981). *Synthesis* and neuroleptic activity of benzamides. Cis-N-(1-benzyl-2-methylpyrrolidin-3-yl)-5-chloro-2-methoxy-4-(methylamino) benzamide and related compounds. *J. Med. Chem.*, *24*, 1224–1230. For an asymmetric formal synthesis, see: (e) Huang, P. Q., Wang, S. L., Zheng, H., Fei, X. S. (1997). First asymmetric synthesis of (2*R*, 3*R*)-3-amino-1-benzyl-2-methyl-pyrrolidine via a highly diastereoselective reductive alkylation. *Tetrahedron Lett.*, *38*, 271–274; (f) Huang, P. Q., Wang, S. L., Ye, J. L., Ruan, Y. P., Huang, Y. Q., Zheng, H., Gao, J. X. (1998). An easy access to protected (4*S*, 5*R*)-5-alkyl-4-hydroxy-2-pyrrolidinones and their use as versatile synthetic intermediates. *Tetrahedron.*, *54*, 12547–12560.

[181] (a) Christiansen, M. A., Butler, A. W., Hill, A. R., Andrus, M. B. (2009). Synthesis of kurasoin B using phase transfer-catalyzed acylimidazole alkylation. *Synlett*, 653–657. For an efficient seven-step (24% overall yield) synthesis of kurasoin A using the same reaction as the key step to install the (*S*)-hydroxy functionality, see: (b) Andrus, M. B., Hicken, E. J., Stephens, J. C., Bedke, D. K. (2006). Total Synthesis of the hydroxyketone kurasoin A using asymmetric phase-transfer alkylation. *J. Org. Chem.*, *71*, 8651–8654. For another example [(–)-ragaglitazar, a known diabetes drug], see: (c) Andrus, M. B., Hicken, E. J., Stephens, C. J., Bedke, D. K. (2005). Asymmetric phase-transfer catalyzed glycolate alkylation, investigation of the scope, and application to the synthesis of (–)-ragaglitazar. *J. Org. Chem.*, *70*, 9470–9479. For previous syntheses, see: (d) Sunazuka, T., Hirose, T., Zhi-Ming, T., Uchida, R., Shiomi, K., Harigaya, Y., Omura, S. (1997). Syntheses and absolute structures of novel protein farnesyltransferase inhibitors, kurasoins A and B. *J. Antibiot.*, *50*, 453–455; (e) Uchida, R., Shiomi, K., Sunazuka, T., Inokoshi, J., Nishizawa, A., Hirose, T., Tanaka, H., Iwai, Y., Omura, S. (1996). Kurasoins A and B, new protein farnesyltransferase inhibitors produced by Paecilomyces sp. FO-3684. II. Structure elucidation and total synthesis. *J. Antibiot.*, *49*, 886–889.

[182] Uchida, R., Shiomi, K., Inokoshi, J., Masuma, R., Kawakubo, T., Tanaka, H., Iwai, Y., Omura, S. (1996). Kurasoins A and B, new protein farnesyltransferase inhibi-

tors produced by *Paecilomyces* sp FO-3684. I. Producing strain, fermentation, isolation, and biological activities. *J. Antibiot.*, *49*, 932–934.

[183] (a) Kohl, N. E., Wilson, F. R., Mosser, S. D., Giuliani, E., DeSolms, S. J. (1994). Protein farnesyltransferase inhibitors block the growth of ras-dependent tumors in nude mice. *Proc. Natl. Acad. Sci. U.S.A.*, *91*, 9141–9145; (b) Kohl, N. E., Omer, C. A., Conner, M. W., Anthony, N. J., Davide, J. P. (1995). Inhibition of farnesyltransferase induces regression of mammary and salivary carcinomas in ras transgenic mice. *Nat. Med.*, *1*, 792–797.

CHAPTER 12

RECENT ADVANCES IN BIOCATALYSIS APPLIED TO ORGANIC SYNTHESIS

GONZALO DE GONZALO, IVÁN LAVANDERA, AND VICENTE GOTOR

12.1. Biocatalysis: Introduction 491
12.2. Oxidoreductases 492
12.3. Transferases 505
12.4. Use of hydrolases in biocatalytic processes 507
12.5. Recent biocatalyzed methodologies employing lyases 516
12.6. Isomerases 520
References 521

12.1. BIOCATALYSIS: INTRODUCTION

For decades, synthetic organic chemists have devoted their efforts to developing new and versatile strategies in order to selectively modify similar reacting moieties in the same molecule or one enantiomer from a racemate. In the past several years, the need of (chiral) precursors for fine chemicals in, for example, the flavor and fragrance, agrochemical, and pharmaceutical industries has increased [1]. But along with this need is the necessity to develop sustainable

Catalytic Methods in Asymmetric Synthesis: Advanced Materials, Techniques, and Applications, First Edition. Edited by Michelangelo Gruttadauria and Francesco Giacalone.

processes, due to the creation of literally hundreds of programs and governmental initiatives related to the concept of "green chemistry" [2]. Thus, nowadays, the "process efficiency" concept is not only related to a high chemical yield but also to the minimized use of large amounts of harmful organic solvents and production of chemical waste. At this point, the use of biological methods applied to the organic synthesis has become a powerful tool for highly stereo- and regioselective transformations due to the intrinsic chirality of enzymes, and nowadays is a commonly employed strategy in industrial and academic laboratories [3–6].

As Hudlicky and Reed recently pointed out [7], while *biotransformation* can be defined as "the reaction of chemical compounds in a living system, and it need not be a process defined by the organism's metabolism," *biocatalysis* is more broadly defined as "the mediation of chemical reactions by means of biological systems, including isolated enzymes, whole cells, or cell-free extracts." Thus, the use of biological catalysts has several advantages over other catalysts because of the great enhancement of rate reactions, the usual exquisite regio-, chemo-, and stereoselectivities obtained using mild reaction conditions, and that they are kinder to the environment. Furthermore, the development of novel biochemical and genetic tools in recent years has opened the possibilities of obtaining, for example, overexpressed enzymes in host cells that increase the efficacy of these processes, thereby lowering total costs. On the other hand, some drawbacks still have to be overcome, such as the usual instability of isolated enzymes and the high costs of special equipment and cofactors employed with redox biocatalysts.

In this chapter we want to show some relevant examples from 2006 concerning biocatalysis applied to the synthesis of interesting organic compounds like pharmaceutical derivatives, while also emphasizing in some cases the importance of the development of new systems to obtain these compounds in a more efficient and clean manner. We have subdivided this chapter depending on the enzyme type following the standard classification of biocatalysts [6].

12.2. OXIDOREDUCTASES

Oxidoreductases are an important class of enzymes that catalyze redox processes by transferring electrons from a reductant to an oxidant [6, 8]. These biocatalysts are widely applied due to their usually exquisite chemo-, regio-, and stereoselectivities through mild and environmentally friendly protocols.

The oxidoreductases most often employed are the alcohol dehydrogenases [ADHs, EC 1.1.1.X., also called ketoreductases (KRED) or carbonyl reductases], which are able to perform stereoselective carbonyl reductions or enantioselective alcohol oxidations [9]. These biocatalysts have been historically employed as multienzymatic systems as whole cells (bacteria, yeasts, fungi, or plants) [10], but usually low selectivities have been obtained in these processes due to the action of several enzymes with different and/or opposite selectivities. Due to this, since the mid-1990s, the tendency has been to employ purified or overexpressed ADHs [9–11]. These biocatalysts work in aqueous media, although in the past few years, novel enzymes that can work in nonconventional media have been described [12]. In any regard, besides all the advantages that biocatalyzed oxidations present over chemical methods, the requirement of the expensive nicotinamide adenine dinucleotide phosphate (NADPH) cofactor has precluded, to some extent, their use on a large scale. Therefore, several chemical, electrochemical, photochemical, or enzymatic methods have been designed to successfully regenerate the cofactor [13, 14]. Due to these novel techniques, ADHs have been recently employed as adequate catalysts to obtain important pharmaceutical and high-added-value compounds.

Enantioenriched diols are important building blocks of many natural compounds, such as pheromones or antitumor agents like discodermolide. As a result, they are used as chiral precursors for fine chemicals in the flavor and fragrance, agrochemical, and pharmaceutical industries. The α-hydroxy ketones (also called acyloins) constitute well-known derivatives for the synthesis of 1,2-amino alcohols through diastereoselective reductive amination. Furthermore, short-chain diols can be employed as starting materials for chiral polymers or as a backbone for chiral ligands applied to asymmetric transition metal catalysis. In a work described by Kroutil and co-workers [15], the asymmetric reduction of symmetrical and nonsymmetrical diketones (**1**) as well as the stereoselective oxidation of various diols (**2**) by biocatalytic hydrogen transfer was investigated by employing lyophilized cells of *Rhodococcus ruber* DSM 44541 containing ADH ADH-"A," to obtain final products hydroxyketones (**3**) with high stereopreference at the (ω-1)- and (ω-2)-positions, while sterically more demanding ketone moieties—for example, those at the (ω-3)-position—remain unchanged (Scheme 12.1). For the oxidation mode, differentiation between primary and secondary alcohols was achieved, with the (*S*)-configured secondary alcohols preferentially oxidized while the Prelog alcohols were obtained with very high enantiomeric excesses (ee's) and diastereomeric excesses (de's) for the reduction mode. All experiments were performed using lyophilized cells of *R. ruber* DSM 44541 as the catalyst while 2-propanol or acetone was used as the hydrogen donor or acceptor source, respectively, to recycle the cofactor and as cosolvent to improve the solubility of the substrates.

In a more recent example, we presented a stereo- and regioselective ADH-catalyzed reduction of several 1,2-, 1,3- and 1,4-diketones of interest with

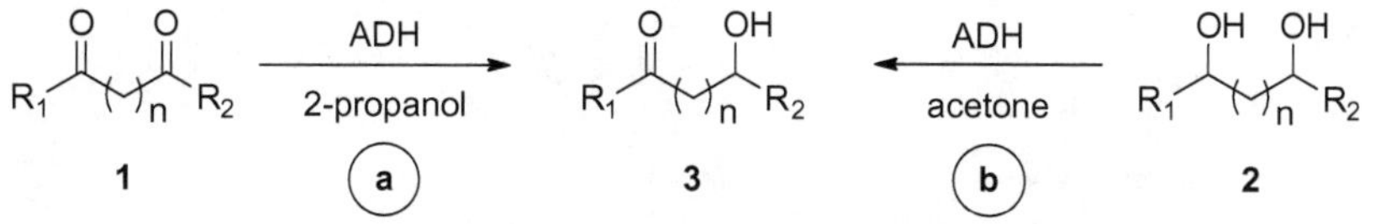

SCHEME 12.1. (a) Reduction of diketones **1** and (b) oxidation of diols **2** by biocatalytic hydrogen transfer to obtain enantioenriched hydroxyketones **3**.

excellent ee's and de's, using again an internal recycling of cofactor using purified enzymes and 2-propanol. *Lactobacillus brevis* ADH (LBADH), *Thermoanaerobium* sp. ADH (ADH-T), and *R. ruber* ADH (ADH-"A") were successfully employed in these bioreductions [16]. Katzberg et al. presented a new synthetic route starting from 2,5-hexanedione to synthesize (5*S*)-hydroxy-2-hexanone with high enantioselectivity (>99% ee) [17]. (5*S*)-Hydroxy-2-hexanone is a valuable bifunctional chiral building block for, for example, pharmaceuticals or aroma compounds. A major issue has appeared because this compound can be further reduced to furnish (2*S*,5*S*)-hexanediol. Among the tested biocatalysts, the whole-cell system *Saccharomyces cerevisiae* L13 surpassed the purified bacterial dehydrogenase ADH-T in terms of chemoselectivity. Thus, the use of whole cells of *S. cerevisiae* L13 afforded (*S*)-hydroxyketone from prochiral 2,5-hexanedione with 85% yield, which was 21% more than the value obtained with ADH-T.

Optically active α-alkyl-β-hydroxyketones and their analogs represent an important class of synthons that have been used in the synthesis of many natural products and pharmaceuticals and have been identified in a broad spectrum of biologically active compounds including polyketides. Smonou and co-workers reported a chemoenzymatic route to a great number of these compounds by starting from prochiral 2-alkyl-1,3-diketones **4** [18]. Thus, the regio- and stereoselective synthesis of different diastereomers of a large number of α-alkyl-β-hydroxyketones (**5**) was achieved in good to excellent optical purities, from readily available diketones **4** using a kit of 34 commercially available KREDs enzymes (Scheme 12.2a). At least two, and in many cases three out of the four possible diastereomers, were selectively produced. With only two exceptions, all enzymes gave the same absolute configuration for the reduction of the first keto group regardless of the α-alkyl substitution, although the stereoselectivity of this second chiral center was not always high and was dependent on the enzyme and the substrate. In these cases, glucose dehydrogenase (GDH) with glucose was used to recycle the nicotinamide cofactor. Applying this methodology, these authors were able to synthesize (+)-sitophilure **6** (Scheme 12.2b), the aggregation pheromone of the pests rice weevil (*Sitophilus oryzae* L.) and maize weevil (*Sitophilus zeamais* M.), in two

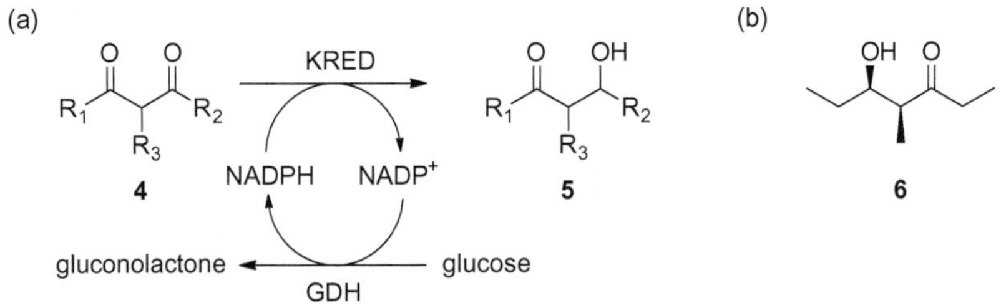

SCHEME 12.2. (a) Enzymatic reduction of α-alkyl-1,3-diketones with NADPH-dependent KREDs and GDH for cofactor recycling; (b) structure of (+)-sitophilure **6**.

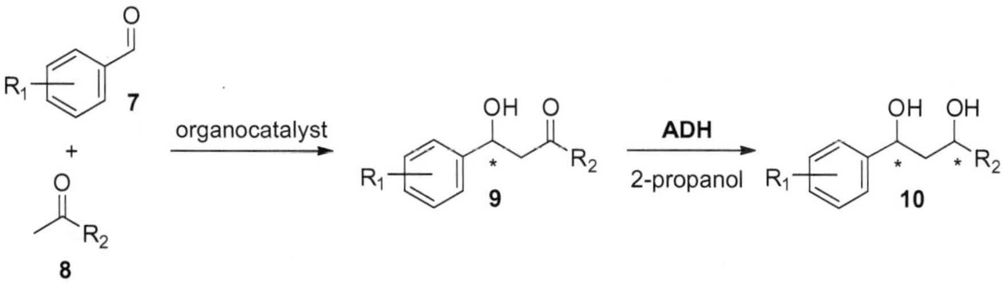

SCHEME 12.3. Synthesis of enantioenriched 1,3-diols combining organo- and biocatalysis.

steps with an overall yield of 81%, starting from commercially available 3,5-heptanedione [19].

Widely used as building blocks in the synthesis of pharmaceutically active compounds are 1,3-diols with two stereogenic centers. In addition, numerous natural products possess a chiral 1,3-diol subunit. In a recent report, a modular chemoenzymatic process for the selective synthesis of all four stereoisomers of 1,3-diols through a combination of asymmetric organocatalysis and biocatalysis was shown (Scheme 12.3) [20]. Thus, each center could be constructed selectively by means of a catalyst that was specifically suitable for this purpose in a two-step, one-pot synthesis. Aldol reaction between aldehydes **7** and ketones **8** followed by the enzymatic reduction of 1,3-hydroxyketones **9** afforded final 1,3-diols **10** with excellent enantioselectivities and very high de's. Some of these processes were subsequently carried out on a preparative scale according to the concept of the substrate-coupled regeneration of the cofactor employing an excess of 2-propanol.

In 2006, Gröger and co-workers developed an efficient "tailor-made" whole-cell catalyst for the desired asymmetric reduction of several ketones of interest coexpressing the genes of an ADH [from *Lactobacillus kefir* (LKADH) or from *Rhodococcus erythropolis* (ReADH)] with a GDH used to internally recycle the coenzyme in a host organism (*Escherichia coli*) [21]. Thus, very efficient catalysts (>1000 U/mg) for a range of substrates were achieved. Furthermore, this practical biocatalytic reduction concept was suitable for upscaling since it proceeded at concentrations of substrate >100 g/L.

Optically active *β*-hydroxy nitriles are a class of important compounds because they can be converted into chiral 1,3-amino alcohols and *β*-hydroxy carboxylic acids. Optically active 1,3-amino alcohols have been widely used as precursors in the synthesis of *β*-blocker drugs and as chiral ligands in asymmetric reactions. The *β*-hydroxy carboxylic acids are key intermediates for the synthesis of a variety of pharmaceutically important compounds such as *β*-amino acids, *β*-lactams, *β*-lactones, and pheromones, and are the key components of polyketide natural products. In a recent report, an ADH from *Candida magnoliae* (CMCR) with *anti*-Prelog selectivity was applied to the reduction of a series of aromatic *β*-keto nitriles **11** to obtain the enantiopure

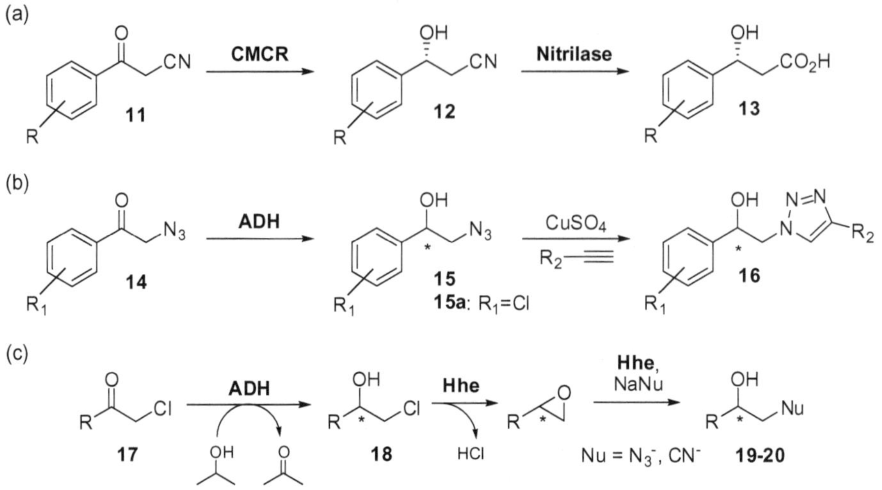

SCHEME 12.4. (a) ADH- plus nitrilase-catalyzed synthesis of enantiopure β-hydroxy carboxylic acids; (b) ADH- plus click-catalyzed reaction to obtain β-adrenergic receptor blocker analogs; (c) ADH- plus Hhe-catalyzed cascade process to obtain β-azidoalcohols and β-hydroxy nitriles.

(*R*)-β-hydroxy nitriles (*R*)-**12** with very high isolated yields using GDH to recycle the cofactor [22]. In addition, (*R*)-**12** was treated with a nitrilase from the cyanobacterium *Synechocystis* sp. strain PCC 6803 (NIT6803) in order to achieve their enzymatic hydrolysis to obtain the (*R*)-β-hydroxy acids (*R*)-**13** with excellent yields (Scheme 12.4a). In another report by the same authors [23], starting from the corresponding α-azidoacetophenone derivatives **14**, both antipodes of 2-azido-1-arylethanols **15** were synthesized with excellent optical purity via enzymatic reduction employing CMCR or an ADH from *S. cerevisiae* (Ymr226c). These derivatives are important intermediates of several compounds such as aziridines, important building blocks in organic synthesis, and key components in bioactive molecules (e.g., aziridine-based cysteine protease inhibitors) and 1,2-amino alcohols, widely used as chiral ligands for asymmetric catalysis and in the construction of biologically active compounds. Employing click chemistry, **15** can be converted into β-adrenergic receptor blocker analogs with 1,2,3-triazole moiety. Here, (*S*)-2-azido-1-(*p*-chlorophenyl) ethanol **15a** reacted with alkynes to afford high yields of optically pure triazole-containing β-adrenergic receptor blocker analogs **16** with potential biological activity (Scheme 12.4b).

In a very recent contribution, a three-step, two-enzyme, one-pot cascade sequence starting from prochiral α-chloro ketones **17** leading to enantiopure β-azidoalcohols **19** and β-hydroxy nitriles **20** was described [24]. In this case, the asymmetric bioreduction of **17** by hydrogen transfer catalyzed by an ADH established the stereogenic center in the first step to furnish enantiopure chlorohydrin intermediates **18**. Subsequent biocatalyzed ring closure to the epoxide and nucleophilic ring opening with azide, N_3^-, or cyanide, CN^-, both catalyzed by a nonselective halohydrin dehalogenase (Hhe), proceeded with full retention of configuration to give enantiopure **19** and **20** derivatives, respectively (Scheme 12.4c). By selecting the adequate ADH, both enantiomers of various optically pure *β*-azidoalcohols and *β*-hydroxy nitriles were synthesized in moderate to excellent yields. For the reduction of **17** the two stereocomplementary ADHs—ADH-"A" from *R. ruber* and LBADH from *L. brevis*—were used with 2-propanol, while for the epoxide ring-closure and ring-opening reactions, two different Hhes were investigated: HheB from *Mycobacterium* sp. GP1 as well as HheC from *Agrobacterium radiobacter* AD1.

Kaneka alcohol **22** is an essential building block for 3-hydroxy-3-methylglutaryl-CoA (HMG-CoA) reductase inhibitors that present two stereogenic centers (Scheme 12.5). In 2006, Patel and co-workers reported an investigation concerning the microbial reduction of several diketoesters by *Acinetobacter* sp. SC13874 and other microbial systems. Thus, a cell suspension of *Acinetobacter* sp. afforded the desired *syn*-(3*R*,5*S*)-dihydroxy ester **21**, precursor of alcohol **22**, with complete conversion and high de [25]. Enantioenriched 3-chloro-2-hydroxyalkanoic esters are very useful in the synthesis of chiral glycidic esters and 2,3-epoxialcohols, which are key intermediates for the synthesis of important bioactive molecules. Milagre et al. described the asymmetric bioreduction of ethyl 3-halo-2-oxo-4-phenylbutanoate compounds with several immobilized microorganisms, especially with *S. cerevisiae*, optimizing the matrix of immobilization and the concentration of substrate [26]. Thus, high enantio- (especially in the case of the *anti*-derivative) and diastereoselectivities (up to 70% favoring *anti* over *syn*) were achieved.

Tamsulosin hydrochloride **24** is employed in the symptomatic treatment of benign prostatic hyperplasia. A simple approach to (*R*)-**24** was reported by Brenna and co-workers (Scheme 12.5) [27]. Thus, incubation of *p*-anisaldehyde with fermenting baker's yeast cells in the presence of glucose yielded a 7:3 mixture of anisic alcohol and of enantiopure diol **23**, a precursor of **24**. 5-Hydroxyhept-6-enoates have been used as key intermediates for a variety of physiologically active compounds, for example, prostaglandins, leukotrienes, or isoprostanes. In 2007, both enantiomers of ethyl 5-hydroxyhept-6-enoate were achieved by ADH-catalyzed reduction of the corresponding ketone. To obtain the *R* enantiomer, LBADH ($c = 95\%$, ee $= 98\%$) was selected as the best enzyme, while for the *S* enantiomer, ADH-T ($c = 94\%$, ee $= 95\%$) was the best choice. In both cases, 2-propanol was employed for cofactor recycling [28].

Naturally occurring 10-membered ring lactones from fungal metabolites present a wide variety of bioactive substances. Among them, modiolide A (**27**) is an antibacterial and antifungal substance from marine-origin microorganisms. Sugai and co-workers [29] achieved the asymmetric reduction of a silylated propargyl ketone intermediate (**25**) mediated by whole cells of *Pichia minuta* IAM 12215 in 88% yield and 96% ee (Scheme 12.5). Another synthon

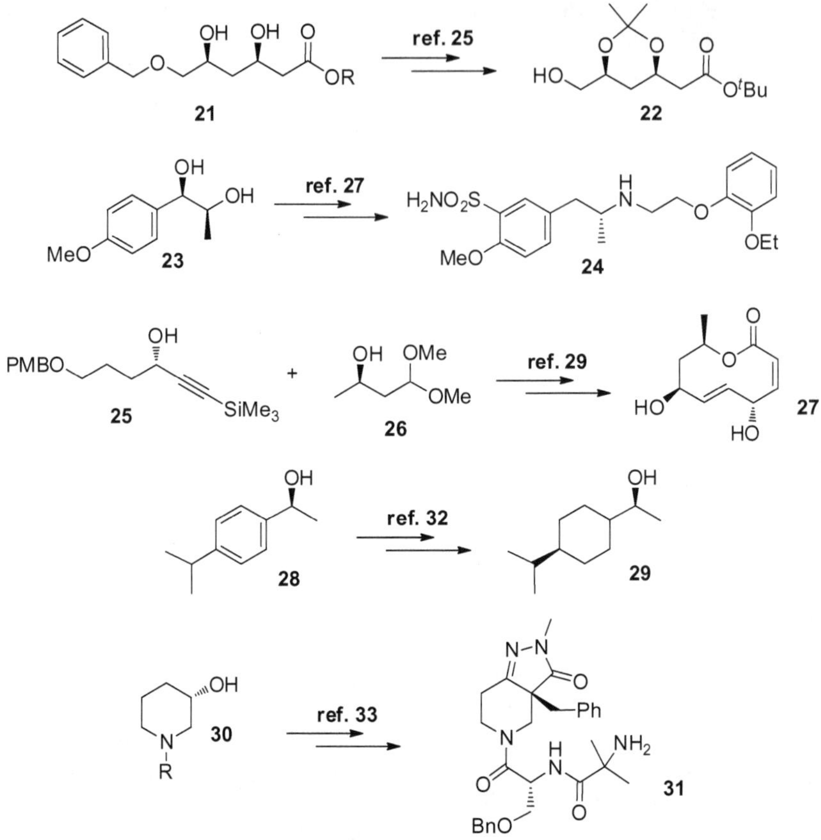

SCHEME 12.5. Examples of pharmaceutical intermediates obtained through ADH-catalyzed processes applied to the synthesis of drugs, fragrances, and biologically active derivatives.

(**26**) needed to perform the total synthesis of modiolide A was prepared ($c = 76\%$, ee $= 92\%$) via *Yamadazyma farinosa* NBRC 10896-mediated asymmetric reduction on a large scale. Clopidogrel is a platelet aggregation inhibitor widely administered to atherosclerotic patients with the risk of a heart attack or stroke that are caused by the formation of a clot in the blood. Methyl (*R*)-*o*-chloromandelate, which is an intermediate of clopidogrel, was obtained in >99% ee by the asymmetric reduction of methyl *o*-chlorobenzoylformate on a 2-g scale (1.0M) with a recombinant *E. coli* coproducing a carbonyl reductase from *S. cerevisiae* and a GDH [30].

Berkowitz and co-workers showed an interesting work applied to the obtainment of several pharmaceutical building blocks employing purified ADHs and EtOH as a four-electron reductant. For this, apart from the ADH that oxidized ethanol into acetaldehyde, an aldehyde dehydrogenase (AIDH)

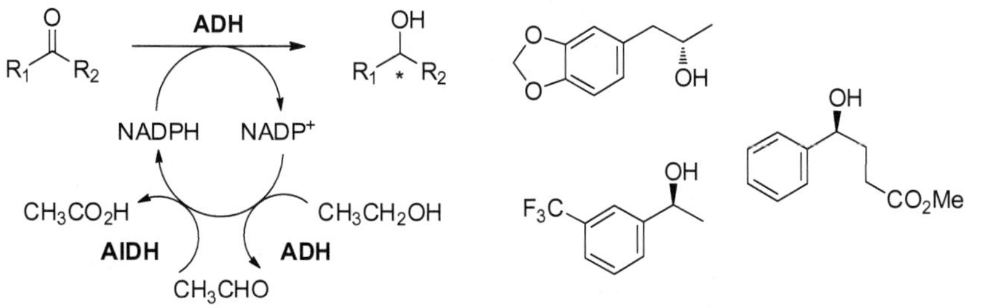

SCHEME 12.6. Entries to several pharmaceutical building-blocks employing EtOH as cosubstrate.

was used to convert acetaldehyde into acetic acid (Scheme 12.6) [31]. Thus, viable KRED-based entries into secondary alcohol building blocks for Dolastatin, Prozac, and Strattera were presented. In some cases the reactions could be easily scaled-up at 1 mmol scale.

Mugetanol [1-(4-isopropylcyclohexyl)ethanol] (**29**) (Scheme 12.5) is a fragrance that possesses four diasteroisomeric forms, with the (−)-1-(*S*)-*cis* isomer giving the highest odor intensity. In a paper by Gotor and co-workers, the chemoenzymatic stereoselective synthesis of Mugetanol from commercially available 4-isopropylacetophenone using different biocatalysts was reported [32]. Thus, three approaches were considered depending on the enzyme used in the biocatalytic key step. In the case of purified ADHs, ADH-T, *Candida parapsilosis* ADH (CP-ADH), and ADH-"A" showed a complete Prelog selectivity affording the enantiopure precursor of (*S*)-**28**, while LBADH and PR2-ADH provided the (*R*)-enantiomer. Numerous natural and nonnatural products with interesting biological activities contain in their structure a piperidine core possessing a quaternary stereogenic center at C3, such as capromorelin (**31**) or isonitramine and sibirine. In a very recent contribution, Cossy and co-workers used *Daucus carota* in aqueous media as a mild, economically viable, and eco-compatible reductive agent for the preparation of optically active cyclic 3-hydroxyamines **30**. Usually, the yields obtained were high and the ee's were higher than 90%. Furthermore, this methodology was applied for the enantioselective synthesis of an advanced precursor of **31** [33].

The goal of obtaining a highly valuable enantiomerically pure product in 100% yield from a cheap racemic substrate in a one-pot process is currently an important topic in catalysis. In one-pot sequential catalysis, the reaction conditions can be adjusted for each step, and although concurrent catalysis is more demanding, such processes circumvent the often time-intensive and yield-reducing isolation and purification of intermediates in multistep syntheses. Kroutil and co-workers described an elegant protocol to deracemize secondary alcohols employing whole cells of *Alcaligenes faecalis* DSM 13975 as a very active and enantioselective catalyst for the oxidation step that simply

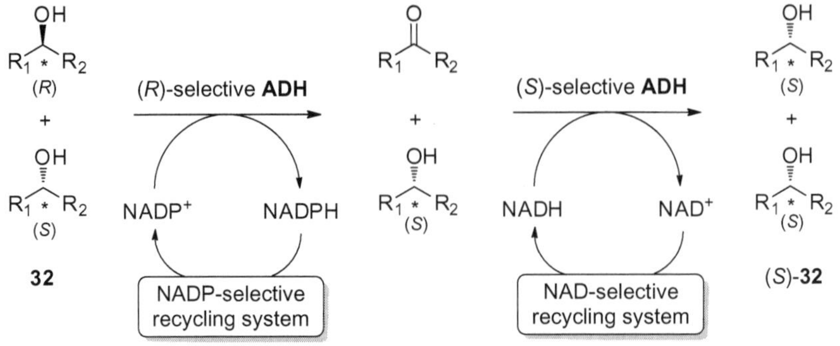

SCHEME 12.7. One-pot deracemization of *sec*-alcohols combining simultaneous concurrent tandem oxidation and reduction cycles with opposite stereo- and cofactor preference.

required molecular oxygen as a mild oxidant for the oxidation step, and a commercially available stereoselective ADH for the reduction step (ADH-"A" or ReADH) [34]. Thus, enantiopure *S* alcohols could be obtained with excellent conversions in many cases. Furthermore, it was shown that the opposite enantiomers were accessible with a similar system employing whole cells of *R. erythropolis* DSM 43066 and LKADH. As an extension of this work, these authors showed that independent oxidative cofactor recycling of $NADP^+$ and reductive regeneration of NADH could be performed simultaneously in the same compartment by two different ADHs with opposite stereo- and cofactor preference in order to deracemize several *sec*-alcohols (Scheme 12.7) [35]. By careful selection of the appropriate enzymes, stereoinversion processes can also be achieved representing a "green" equivalent to the chemical Mitsunobu inversion. The generality of the system was exemplified for several *rac-sec*-alcohols (**32**) yielding the optically pure (*R*)- as well as (*S*)-enantiomers by choosing the matching pairs of ADHs.

The asymmetric reduction of C=C bonds can create up to two novel chiral centers and is one of the most widely employed strategies for the synthesis of enantioenriched compounds. Although metal-based methodologies to achieve this transformation have been impressively developed, the biocatalytic counterpart is catalyzed by enoate reductases (EC 1.3.1.X), members of the "old yellow enzyme" family [36]. These ubiquitous enzymes are widely distributed in microorganisms, particularly in bacteria, lower fungi, and plants, and in the past few years more synthetic applications have appeared that use them. These common enzymes act through the transfer of a hydride ion, derived from a flavin cofactor ($FMNH_2$), onto the β-carbon atom of an α,β-unsaturated carbonyl compound, while a proton, derived from the solvent, is added from the opposite side onto the α-carbon atom. As a consequence of this mechanism, the hydrogenation occurs in a *trans* fashion (Scheme 12.8a). The reduced flavin is recovered by the action of a nicotinamide cofactor. Thus, Hall et al. showed

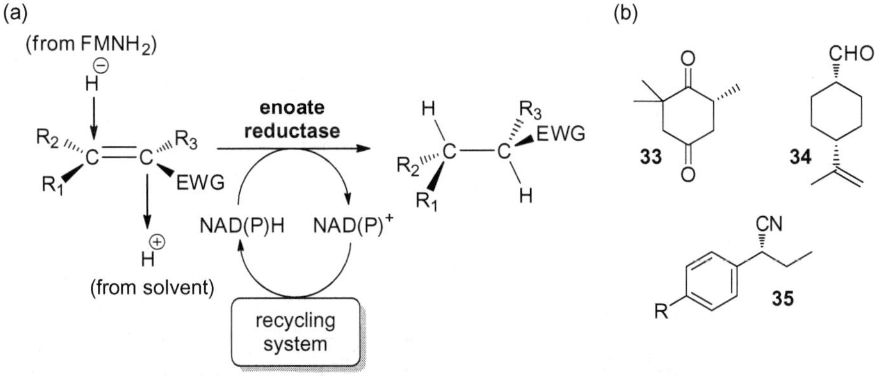

SCHEME 12.8. (a) Asymmetric bioreduction of activated alkenes bearing activating electron-withdrawing groups (EWG) using enoate reductases; (b) compounds obtained using this methodology.

that the 12-oxophytodienoate reductase isoenzymes OPR1 and OPR3 from tomato displayed a remarkably broad substrate range for the stereoselective asymmetric bioreduction of α,β-unsaturated enals, enones, dicarboxylic acids, *N*-substituted maleimides, and nitroalkenes with excellent conversions and ee's in many cases [37]. In a subsequent paper, these authors also demonstrated that these enzymes and YqjM from *Bacillus subtilis* (BS2) displayed a remarkably broad substrate spectrum by reducing electron-deficient alkenes. The reduction of ketoisophorone to levodione **33** (Scheme 12.8b) was achieved, an important building block for the synthesis of zeaxanthin on an industrial scale [38].

Recently, it was described that leukotriene B_4 12-hydroxydehydrogenase (Ltb4dh) reduced both enantiomers of perillaldehyde to afford the same *cis*-product (**34**) (Scheme 12.8b) with high selectivity [39]. Kosjek and co-workers applied commercial isolated enoate reductases to convert a series of α,β-unsaturated nitriles to the optically active nitrile products such as **35** (Scheme 12.8b) in high yields and excellent enantioselectivities (up to 99% ee) [40]. In particular, enantiopure nitriles are important synthetic precursors for the preparation of pharmacologically important building blocks.

Other oxidoreductases that have been employed with synthetic purposes are the oxygenases, able to perform the oxidation of substrates by inserting one (monooxygenases) or two (dioxygenases) oxygen atoms from molecular oxygen [41]. Cytochrome P450 monooxygenases (P450s or CYP, EC 1.14.X.X) catalyze the regio- and often stereoselective introduction of oxygen into a huge range of substrates. In the presence of the cofactor NAD(P)H and a corresponding electron transfer system or alternatively with hydrogen peroxide using the peroxy-shunt pathway, C–H aliphatic and aromatic bonds can be hydroxylated and C=C can be epoxidized [42]. Recently, Glieder and co-workers obtained a set of bacterial and fungal P450 enzymes by protein sequence alignments, expressed in *E. coli* and characterized. Notably, a fungal self-sufficient cytochrome P450 (CYP) from *Aspergillus fumigatus* turned out to be especially stable also in the presence of organic cosolvents. Selected biocatalysts were used as lyophilized lysates for the synthesis of 4′-hydroxydiclofenac and 6-hydroxychlorzoxazone, the two metabolites of active pharmaceutical compounds diclofenac and chlorzoxazone as generated by human P450s [43].

Baeyer–Villiger monooxygenases (BVMOs) (EC 1.14.X.X) are a class of flavin-containing monooxygenases that catalyze with high regio- and/or enantioselectivities the Baeyer–Villiger oxidation as well as the oxygenation of different heteroatoms [44, 45]. The low stability of most of the Baeyer-Villiger monooxygenases as well as the requirement of an expensive cofactor (generally NADPH) has limited their application. To circumvent these bottlenecks, recombinant whole-cell overexpression systems have been introduced recently. Mihovilovic et al. have developed the Baeyer–Villiger oxidation of different prochiral ketones presenting a cyclobutanone structural motif to obtain the corresponding chiral butyrolactones (Scheme 12.9a), interesting synthons for the preparation of natural and bioactive molecules, like analgesics, GABA receptor inhibitors, β-amino acids, or structurally related lignans. Eight BVMOs from different bacterial origin (*Acinetobacter*, *Arthrobacter*, *Brachymonas*, *Brevibacterium*, *Comamonas*, and *Rhodococcus*) were overexpressed in *E. coli* and employed as whole-cells biocatalysts in the oxidative processes. Prochiral cyclobutanones **36** were prepared by Zn-coupled mediated [2+2] cyclization from the corresponding alkene. Depending on the BVMO used, two possible

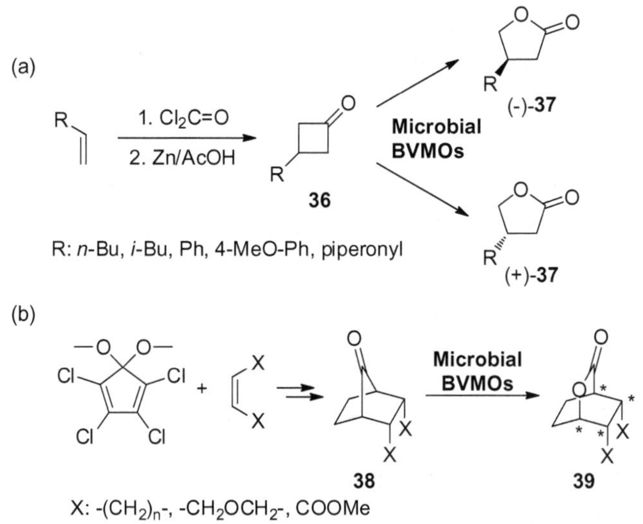

SCHEME 12.9. (a) BVMO-catalyzed oxidation of prochiral cyclobutanones; (b) BVMO-catalyzed oxidation of prochiral bridged cyclic ketones.

stereochemical outcomes were observed, with it being possible to achieve both enantiomers of the corresponding butyrolactones **37** with different optical purities by simple tuning of the biocatalyst. Yields and selectivities obtained were also strongly dependent on the substrate structure [46]. The same library of BVMOs has been employed by these authors for the biooxidation of polycyclic substrates with the ketone moiety located at the bridge of the bicyclic compounds **38** (Scheme 12.9b). BVMO-catalyzed oxidation of prochiral ketones **38** can afford the generation of four to six stereogenic centers, and the final lactones **39** can be easily transformed in other functionalities, giving access, for example, to optically pure bicycle[4.2.0]octan-2-ol and bicycle[4.2.0]octan-2-one. The starting ketones have been synthesized with good overall yields in a multistep procedure starting from 5,5-dimethoxy-1,2,3,4-tetrachlorocyclopentadiene and the appropriate cyclic olefin. The biocatalyzed oxidations were performed in a 100-mg substrate scale, achieving in all the cases yields between 50% and 80%. For most of the substrates, both enantiomers of the final lactone can be achieved with high optical purities depending on the BVMO employed [47].

BVMOs have also been employed for kinetic resolution processes. Thus, a set of aromatic benzylketones have been oxidized by the recently cloned and overexpressed in *E. coli* enzymes phenylacetone monooxygenase (PAMO) from *Thermobifida fusca*, its M446G mutant, and 4-hydroxyacetophenone monooxygenase (HAPMO) from *Pseudomonas fluorescens* ACB. Good to excellent enantioselectivities ($E \geq 200$) and conversions can be retrieved when oxidizing racemic α-benzylketones presenting a methyl or ethyl group, observing a drop in these parameters when employing substrates containing longer alkyl chains. The presence in the aromatic ring of different substituents in the oxidation of (±)-3-phenylbutan-2-one derivatives led to a decrease in the BVMOs activity, while good selectivities can be obtained for almost all of the substrates by choosing properly the BVMO. Reactions have been scaled up to 50–100 mg in order to isolate the corresponding chiral α-benzylketones **40** and α-benzylesters **41** with good yields. The main drawback when applying this methodology is that purified BVMOs are used, so the expensive cofactor NADPH is required for performing the biocatalytic activity, and a second enzymatic system formed by glucose-6-phosphate with glucose-6-phosphate dehydrogenase was employed for regenerating this cofactor [48, 49]. In order to minimize the quantity of reagents employed and maximize the redox economy of the oxidative processes, some BVMO-biocatalyzed reactions have been coupled to the biocatalyzed oxidation of *sec*-alcohols **32** employing ADHs. Thus, in a parallel and concurrent manner, the BVMO is able to oxidize selectively racemic ketones or sulfides employing NADPH that is oxidized to $NADP^+$, and this cofactor will be reduced again to NADPH in the selective oxidation of the racemic alcohol catalyzed by different ADHs. This approach, called parallel interconnected kinetic asymmetric transformations (PIKATs) (Scheme 12.10), has been applied when using isolated PAMO and HAPMO in the preparation of chiral sulfoxides and in the selective Baeyer–Villiger

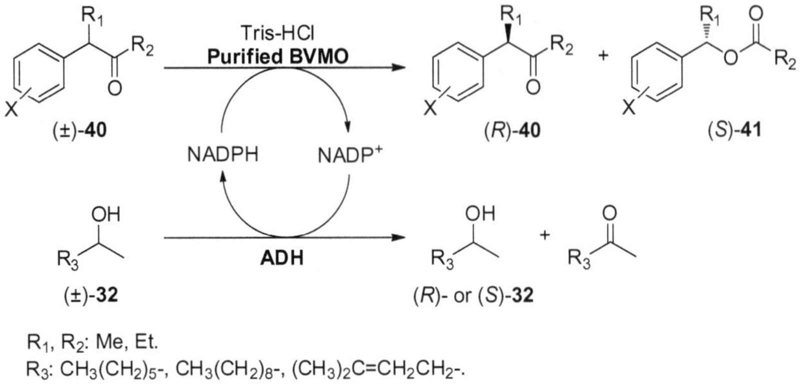

SCHEME 12.10. PIKATs by combining BVMOs and ADHs.

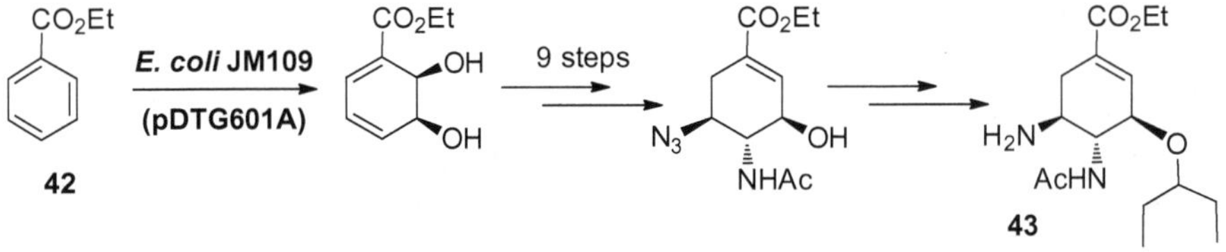

SCHEME 12.11. Formal synthesis of Tamiflu starting from ethyl benzoate **42**.

oxidation of racemic benzylketones, while LBADH and ADH-T have been used for the resolution of alkyl secondary alcohols (±)-**32** [50].

One of the main advantages that biocatalysis can offer with regard to more conventional synthetic procedures is that it is possible to achieve transformations that cannot be performed using nonbiological catalysts. In this sense, one of the best examples is the enzymatic *cis*-dihydroxylation of substituted benzenes catalyzed by dioxygenases (EC 1.13.11.X) [51]. Hudlicky and co-workers showed a short formal 10-step synthesis (10% overall yield) of oseltamivir (Tamiflu, **43**) from ethyl benzoate (**42**). The first stage of this sequential synthesis was the *cis*-dihydroxylation of **42** in a 1.0 g scale using whole cells of *E. coli* JM109 overexpressing a dioxygenase (Scheme 12.11) [52]. Tamiflu is one of the compounds that have been found to be effective as an inhibitor of H5N1 influenza virus neuroaminidase.

Oxidases catalyze the oxidation of substrates through the electron transfer to molecular oxygen, which is reduced to water or hydrogen peroxides. Asymmetric methods to prepare optically active α-chiral primary amines are in high demand in asymmetric synthesis due to the biological/pharmacological activity of these compounds. Among all the possible biocatalytic methods used to obtain them, one of the most promising is the use of amino oxidases (EC 1.4.3.X) [53]. These enzymes are responsible for the transformation of amines

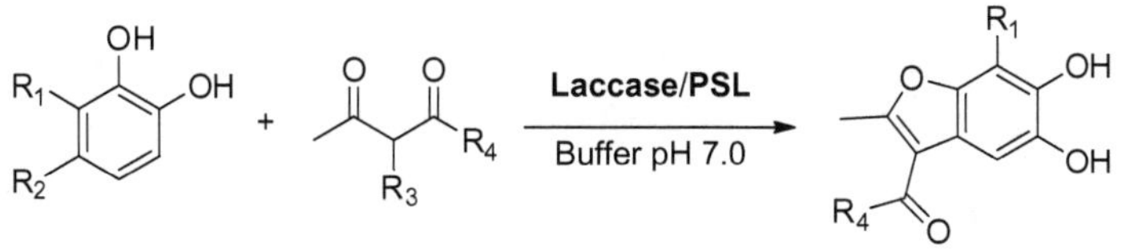

44

45

R_1, R_2: H, Cl, OMe, Me.
R_3: H, Cl.
R_4: Me, OMe.

SCHEME 12.12. Biocatalyzed domino reaction preparation of benzofuran derivatives **45** by combining laccase catalyzed oxidation with a Michael addition performed by the lipase from *P. cepacea.*

into imines and have been applied to deracemization processes of racemic amines by means of an enantioselective oxidation in combination with a non selective reductive agent (e.g., ammonia borane). In 2007, Turner et al. used a monoamine oxidase from *Aspergillus niger* to successfully obtain the alkaloid (*R*)-crispine A in 97% ee [54].

Laccase (EC 1.10.3.2) is a multinuclear copper-containing oxidase that catalyzes the oxidation of various aromatic compounds, including phenols, *o*- and *p*-diphenols, and lignin derivatives [55]. This enzyme often displays high selectivity in aqueous media, providing a green chemistry solution for organic synthesis. Laccase has been employed to catalyze the oxidation of catechols **44** to the corresponding *o*-quinones, which can react in situ with nucleophilic reagents as 1,3-dicarbonyl compounds and aromatic amines via a Michael addition to generate the corresponding benzofurans **45**, as shown in Scheme 12.12. Although lipases catalyze the hydrolysis of esters and the esterification of acids, they have demonstrated a catalytic promiscuity by catalyzing other reactions, such as the Michael addition [56]. Thus, a novel one-step procedure using two biocatalysts [laccase and the lipase from *Pseudomonas cepacea* (PSL)] in aqueous medium was tested for the oxidation of catechols and further Michael addition with 1,3-dicarbonyl compounds [57]. The yields obtained strongly depended on the reactivity of the reaction substrates. Addition of PSL was also shown to improve to a 40% extent the yield obtained in the Michael addition between *o*-quinones obtained from laccase-catalyzed oxidation and aromatic amines.

12.3. TRANSFERASES

This type of enzyme is responsible of the transference of one group into another one. Among them, probably the most often employed are the aminotransferases or transaminases (TAs, EC 2.6.1.X), which transfer an amino group (from an amino donor) onto a carbonyl moiety (amino acceptor). These

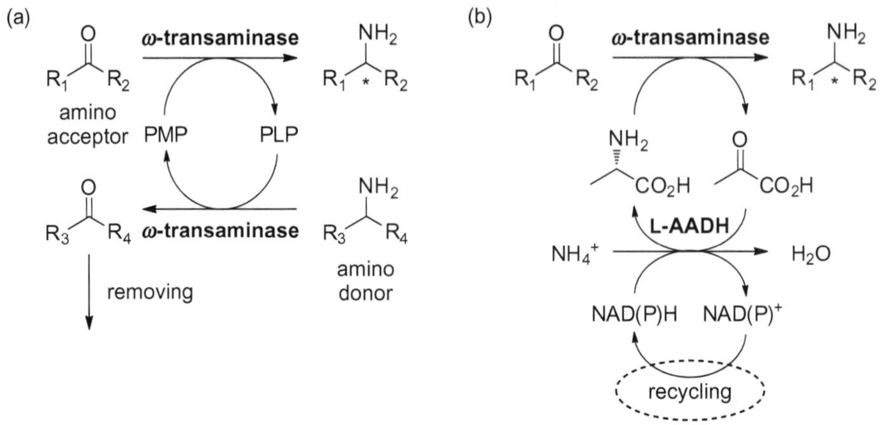

SCHEME 12.13. (a) Employment of ω-TAs to synthesize enantiopure primary amines using an amino donor or acceptor and PLP; (b) formal reductive amination concept for the preparation of optically pure amines.

biocatalysts are divided in α-TAs, responsible for obtaining α-amino acids from α-keto acids, or in ω-TAs, which transfer the amino group into a carbonyl moiety that is not an α-keto acid (Scheme 12.13a) [58]. ω-TAs offer a unique opportunity for the asymmetric synthesis of bioactive compounds that possess a chiral amine moiety starting from simple prochiral ketones through a stereoselective amination process or from low-cost racemic amines through a kinetic oxidative resolution. In the past few years, more and more examples have appeared in the literature that make use of these novel biocatalysts. They require as coenzyme pyridoxal-5′-phosphate (PLP), which serves as a molecular shuttle for ammonia and electrons between the amino donor and the amino acceptor. During the reaction, PLP is reversibly interconverted to pyridoxamine (PMP). Usually, in processes of kinetic resolution of *rac*-amines, pyruvate is employed as an amino acceptor in stoichiometric amounts while for asymmetric aminations, an excess of alanine is used as an amino donor, although these last transformations are usually impeded due to a disfavored thermodynamic equilibrium and inhibitory problems. Recently, a cascade methodology employing three purified enzymes and a catalytic amount of alanine to perform the efficient synthesis of several enantiopure primary amines by means of an ω-TA, an amino acid dehydrogenase (AADH), and a formate or a GDH (Scheme 12.13b) has been described [59].

Koszelewski et al. obtained (*S*)- and (*R*)-mexiletine [1-(2,6-dimethylphenoxy)-2-propanamine], an antiarrhythmic, antimyotonic, and analgesic oral drug in its racemic form, by deracemization starting from the commercially available racemic amine using purified ω-TAs in up to >99% ee and conversion with 97% isolated yield by a one-pot, two-step procedure at a 100-mg scale. The cosubstrate pyruvate needed in the first oxidative step was recycled by using an amino acid oxidase [60]. These enzymes have also been employed in cascade processes combined with transketolases (TKs). Thus, Lye and co-workers constructed a biocatalyst overexpressing a TK and a TA to obtain 2-aminobutane-1,3,4-triol, a building block for statins used in the synthesis of protease inhibitors such as Nelfinavir or a detoxifying agent used for reducing the toxicity of the antibiotic treatment for rice blast disease like detoxinine. While TK achieved the aldol reaction between hydroxypyruvate and glycolaldehyde to form L-erythrulose, the TA catalyzed the synthesis of the final derivative [61].

12.4. USE OF HYDROLASES IN BIOCATALYTIC PROCESSES

Hydrolases (EC 3.X.X.X) have been the most widely employed class of enzymes in the field of biocatalysis. These enzymes can catalyze the hydrolysis of esters, peptides, glycerides, and other structures, and have been shown to be very useful for the resolution of racemic mixtures and the desymmetrization of *meso* and prochiral compounds with the generation of enantiomerically pure compounds [62]. Among all the types of hydrolases, lipases (EC 3.1.1.3) are the most deeply investigated and employed group of biocatalysts [63]. The natural function of these enzymes consists in catalyzing the hydrolysis of triglycerides into fatty acids and glycerol. A large range of lipases are produced by fungi or bacteria and are excreted as extracellular enzymes, which makes their large-scale production easy and inexpensive. Other advantages of these biocatalysts are that they do not require cofactors and are highly stable in organic solvents, and they are able to catalyze a wide set of reactions apart from the natural one. In the past few years, several examples of chemoenzymatic processes using lipases have been described.

Thus, these enzymes have been employed for the preparation of *N*-substituted-*β*-prolines, compounds of high interest due to their biological properties. Two different enzymatic reactions were analyzed. In the biocatalyzed hydrolysis of the pyrrolidine derivative 1-*tert*-butyl-3-methylpyrrolidine-1,3-dicarboxylate, the best results were achieved using lipase from *A. niger* [64]. As lipases can catalyze different reactions in addition to the natural one, racemic *N*-Boc-proline was also tested in biocatalyzed esterifications in the presence of methanol. Lipase AS was chosen for the scaling up due to its availability and affordability after optimization of the reaction conditions. Esterification was carried out in phosphate buffer pH 7.5, starting from 1.0 kg of substrate using 20% w/w of enzyme at 30°C in order to obtain after 24 hours enantiopure (*S*)-ester with 43% yield and the (*R*)-*N*-Boc-(*S*)-*β*-proline with 55% yield and 85% ee. Hydrolysis of the ester in the presence of LiOH 2.0 N in tetrohydrofuran (THF) led to enantiopure *N*-Boc-(*S*)-*β*-proline with complete conversion.

Cephalosporins (**47**) are one of the most important classes of antibiotics. Several derivatives have been prepared through modification at the 3′-hydroxymethyl group. However, the deprotection of the 3′-acetoxy precursor **46** in the presence of cephalosporanic acid still remains a difficult process

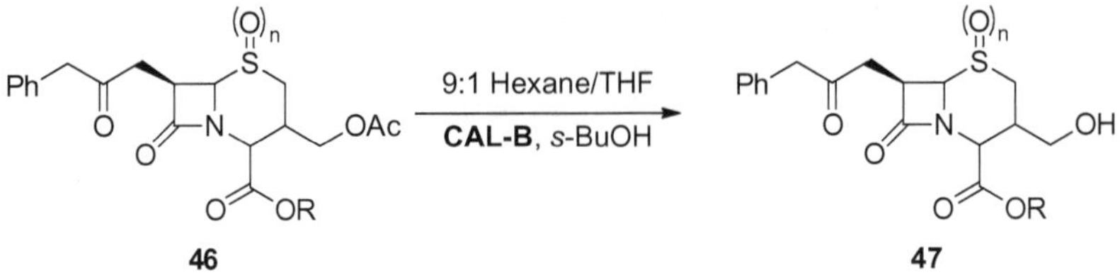

n: 0, 1, 2
R: H, benzhydryl, p-methoxybenzyl, *t*-butyl

SCHEME 12.14. Enzymatic synthesis of cephalosporins employing the lipase from CAL-B.

(Scheme 12.14) [65]. *Candida antarctica* B (CAL-B)-catalyzed deacylation of different cephalosporins in the presence of *sec*-butanol occurred with moderate to good yields (52%–70%) depending on the substrate structure, revealing a useful methodology for preparing cephalosporins analogs that require further modification at the allylic alcohol.

Bornscheuer et al. have employed one esterase present in BS2 for the preparation of chiral tertiary alcohols. These compounds are important building blocks in organic synthesis. Furthermore, tertiary alcohols containing fluorine present an increasing interest as ferroelectric liquid crystals and drugs. BS2 has been identified as a GGG(A)X-hydrolase type, a subclass of hydrolases that show activity through esters of sterically more demanding tertiary alcohols due to their bigger active site when compared with the more common GX-hydrolases. Thus, wild-type BS2 biocatalyzed the kinetic resolution of 1,1,1-trifluoro-2-phenylbut-3-yn-1-yl acetate leading to (*R*)-1,1,1,-trifluoro-2-phenylbut-3-yn-1-ol with moderate enantioselectivity. Mutagenesis studies have demonstrated that two BS2 mutants (G105A and E188D) were able to catalyze this kinetic resolution with excellent enantioselectivity ($E > 100$) [66]. Recently, the BS2 double mutant E188W/M193C has demonstrated to be highly enantioselective in the enzymatic hydrolysis of this substrate presenting a reverse enantioselectivity, obtaining both (*S*)-alcohol and (*R*)-ester with high optical purities [67].

The α,α-dialkyl-α-hydroxycarboxylic acids (**49**) are important building blocks in organic synthesis. These compounds can be prepared in enantiomerically pure form by employing a two-step procedure from commercially available molecules (Scheme 12.15). The first step consisted of the Passerini reaction starting from isocyanide, a ketone, and a carboxylic acid, leading to the corresponding α,α-dialkyl-α-hydroxycarboxylic esters **48**, which were employed as substrates in hydrolytic reactions catalyzed by different hydrolases. Two esterases bearing the GGG(A)X-motif in their actives sites and a protease were active for this reaction, having achieved a moderate enantioselectivity ($E = 42$) in the presence of esterase 8, an enzyme obtained from metagenome [68].

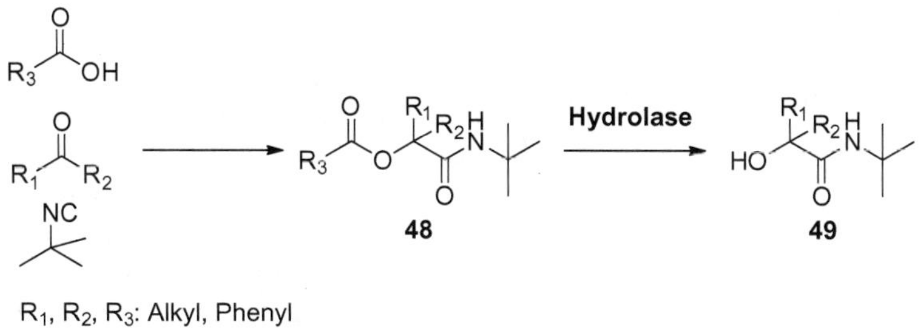

SCHEME 12.15. Two-step synthesis of α,α-dialkyl-α-hydroxycarboxylic acids **49**.

Enantiomerically pure β-amino acids and their derivatives have attracted attention due to their pharmaceutical applications and to their enhanced resistance toward proteolitic enzymes, so they can be used as building blocks for preparing peptides with increased activity and stability. β-Amino acids are also components in many bioactive natural and synthetic products, such as the antitumor agents Taxol and cryptophycin, antibiotics, and different insecticidal and antifungal compounds. Synthetic β-amino acids are also precursors of β-lactams. In past years, several procedures for the preparation of these compounds that employ hydrolases have been developed. Li et al. have described the enzymatic resolution of racemic β-phenylalanine (±)-**50** employing penicillin G acylase (PGA) from *E. coli* [69]. This enzyme was able to catalyze the highly enantioselective acylation of this β-amino acid with phenylacetamide (**51**) in aqueous solution at pH 10.0 in order to obtain (*S*)-β-phenylalanine and (*R*)-*N*-phenylacetyl-β-phenylalanine, (*R*)-**52**. After acidification and separation of the two compounds, enantiopure (*R*)-β-phenylalanine can also be achieved by the amide hydrolysis catalyzed by PGA in water at pH 7.5 (Scheme 12.16a). Both enantiomers of β-phenylalanine can be obtained with optical purities higher than 98%, indicating the potential of this reaction for industrial applications.

β-Aminopeptidases BapA from *Sphingosinicella xenopeptidilytica* and from *S. microcystinivorans* as well as DmpA from *Ochrobractum anthropi* are able to catalyze the hydrolysis of *N*-terminal β-amino acids residues from amides and peptides, so they have been employed in the selective hydrolysis of racemic aliphatic β-amino acids amides to produce the corresponding chiral β-amino acids. These three biocatalysts have shown an excellent enantioselectivity ($E > 200$, ee $\geq 98\%$) when hydrolyzing substrates that possess a methyl or propyl group in the β-carbon, while the use of bulkier substituents as cyclohexyl or *tert*-butyl led to slight decrease in the selectivity ($E > 100$, ee $> 97\%$). For all the enzymes, the L-β-amino acids and the corresponding D-β-amino acid amides were obtained [70].

Alicyclic β-amino acids **54** can be prepared by the biocatalyzed hydrolysis of racemic *cis*- and *trans*-β-amino acid esters (±)-**53**. The use of CAL-B with

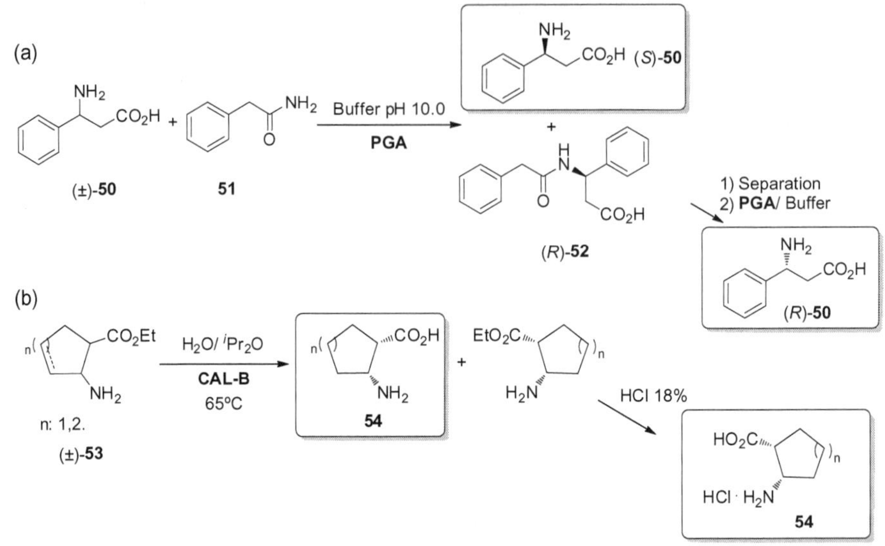

SCHEME 12.16. Synthesis of *β*-amino acids employing hydrolases.

0.5 eq. of water in ^{i}Pr$_2$O at 65°C led to enantiopure *cis*-(1*S*,2*R*)- and *trans*-(1*R*,2*R*)-*β*-amino acids with high enantioselectivities. Transformation of the nonreactive enantiomers with aqueous HCl resulted in the formation of the *cis*-(1*R*,2*S*) and *trans*-(1*S*,2*S*)-*β*-amino acids hydrochlorides (Scheme 12.16b). The recycling of the biocatalyst in the hydrolysis of racemic ethyl *cis*-aminocyclohexane-1-carboxylate was analyzed, and it was revealed that there was a slight decrease in the activity after three cycles, while no changes in the selectivity were observed. Enzymatic hydrolyses were scaled up to gram scale with good yields, achieving conversions close to 50% between 1 and 3 days [71]. The same methodology has been applied for the enzymatic hydrolysis of *β*-aryl-*β*-amino acid esters to obtain chiral *β*-amino acids. For these substrates, the highest performance was observed by employing the lipase from *P. cepacea* and 0.5 eq. of water in ^{i}Pr$_2$O at 45°C. (*S*)-*β*-aryl-*β*-amino acids were recovered in highly selective processes, while the corresponding (*R*)-*β*-aryl-*β*-amino acids hydrochlorides were accessible by treatment with HCl [72].

Hydrolases have also been employed in the desymmetrization of symmetric compounds. These processes consist of an asymmetric synthetic transformation and can therefore give a quantitative yield. Recently, Zutter et al. have described the chemoenzymatic preparation of oseltamivir phosphate (Tamiflu, **43**), starting from the cheap and commercially available 2,6-dimethoxyphenol **55**, as indicated in Scheme 12.17 [73]. This 10-step synthesis provided enantiomerically pure Tamiflu in ~30% overall yield without any chromatographic purification. The two key steps of this synthesis are the *cis*-hydrogenation of the 5-(1-ethylpropoxy)-4,6-dimethoxyisopthalic acid diethyl ester **56** and the

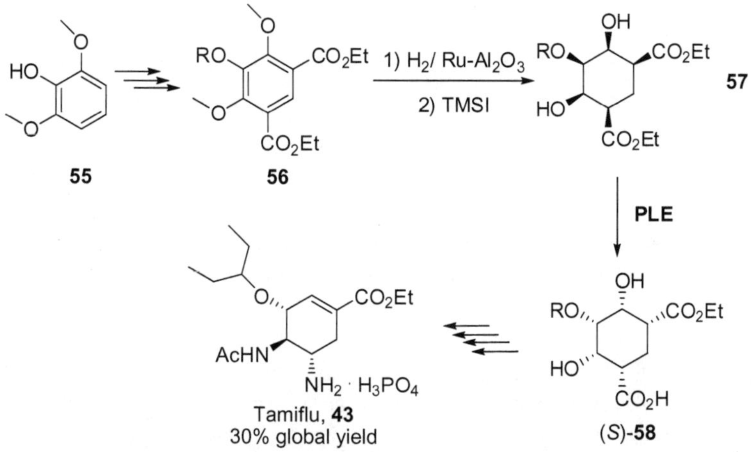

SCHEME 12.17. Chemoenzymatic synthesis of Tamiflu.

desymmetrization of the resultant all-*cis meso*-diester **57**. After an extensive enzymatic screening, the most selective hydrolysis was performed by pig liver esterase (PLE), affording the desired (*S*)-monoacid (*S*)-**58** with high ee (96%–98% ee) and quantitative yield. The esterase was able to tolerate 10% substrate concentration even at 35°C.

P. cepacea lipase has been employed as desymmetrization reagent in the biocatalyzed alkoxycarbonylation of prochiral 2-substituted propane 1,3-diamines with diallyl carbonate for the synthesis of chiral (*R*)-amino carbamates with high yields [74]. Enzyme activity and selectivity were highly dependent on the 2-substituent bonded to the diamine core. Final products were obtained with moderate yields and good to excellent optical purities. Chiral amino carbamates can be easily deprotected leading the corresponding (*R*)-diamines, compounds with high interest in pharmaceutical, surface, bioconjugate chemistry, and also employed as bidentate ligands and linkers in solid-phase synthesis.

Nucleosides and derivatives are relevant compounds of great interest as chemotherapeutic agents. In the past few years, several methodologies have been developed in order to synthesize derivatives presenting pharmacological properties [75]. Two cytotoxic nucleosides—1-β-arabinofuranosyl uracil and 9-β-arabinofuranosyl adenosine—were selectively acylated with hexanoic anhydride and vinyl esters by different lipases [76]. Processes took place regioselectively at the hydroxyl group in the C-5′ position. The best results in all cases were achieved with CAL-B, while the use of *P. fluorescens* lipase led to lower yields and higher amounts of the diacyl derivative. Reactions were performed on THF and in 2-methyltetrahydrofuran (MeTHF), an aprotic solvent that is being used in industrial processes due to its properties. It is partially soluble in water, gives clean water phase separations, has a

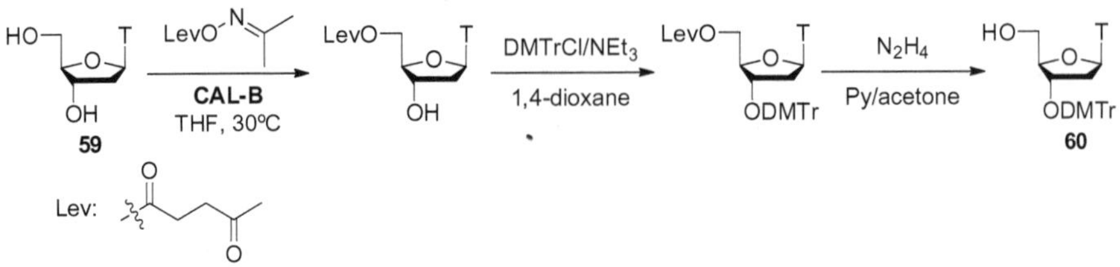

SCHEME 12.18. Chemoenzymatic preparation of 3′-*O*-dimethoxytrityl-protected nucleosides employing lipase B from *Candida antarctica*.

boiling point of 89°C, and presents a low environmental impact as it is obtained from furfural, isolated from corn, sugar cane, and oat. Both solvents gave similar results.

3′-*O*-Dimethoxytrityl (DMTr)-protected nucleosides **60** are valuable building blocks for the assembly of oligonucleotides. Until now, no efficient protocols have been developed for their preparation. Gotor and co-workers have described their synthesis by a chemoenzymatic methodology starting from free nucleosides, which were regioselectively protected with the levulinyl group (Scheme 12.18) [77]. Effective 5′-*O*-levulinylation was performed by CAL-B-catalyzed acylation in an organic solvent using acetonoxime levulinate as an acylating agent, obtaining yields of 71%–80%. The dimethoxytritilation of the resultant nucleosides with DMTrCl in 1,4-dioxane occurred with high yields (higher than 80%). Finally, the 5′-*O*-levulinyl group was chemoselectively cleaved by treatment with hydrazine in pyridine and acetic acid with yield varying from 68% to 77%. The synthesis of the 3′-*O*-DMTr-thymidine (**60**) was developed on large scale, starting from 2.5 g of thymidine **59** and using CAL-B (1:1 w/w). The overall yield for this three-step synthesis was 73%.

Enzymatic kinetic resolutions are straightforward and satisfactory in terms of optical purities, but suffer from the intrinsic drawback of being limited to a maximum theoretical yield for a desirable enantiomer of 50%. Many efforts have been developed to overcome this limitation in order to obtain enantiopure products with higher yields. By introducing an efficient catalyst for the racemization of the substrate coupled to the biocatalyzed resolution, dynamic kinetic resolution (DKR) processes can be performed. Theoretically, DKR can provide one single enantiomeric product in 100% yield in the case of combining an efficient racemization catalyst with a highly enantioselective enzyme (it is recognized that good DKR can be performed when $E > 20$) [78, 79]. Racemization of the starting material can be performed by employing a metal-based catalyst, another biocatalyst, or by using other compounds, such as zeolites, acids or bases, S_N2 displacements, or aldehydes. Among all these methodologies, ruthenium-based organometallic complexes have been most intensively tested as the racemization catalyst.

A cholinesterase inhibitor described for the treatment of Alzheimer's disease, (*S*)-rivastigmine, (*S*)-**64**, has been synthesized with 29% global

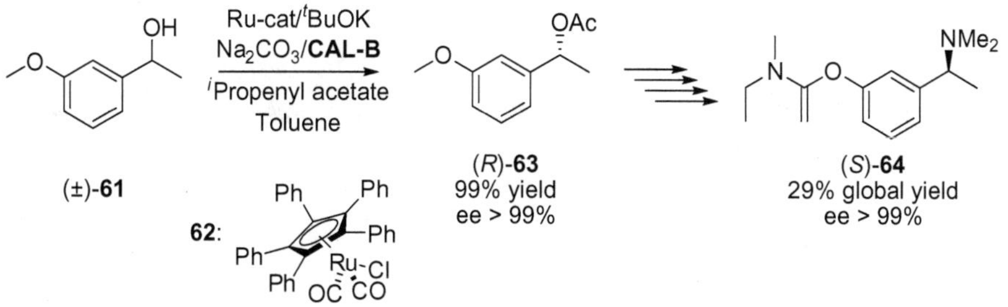

SCHEME 12.19. Synthesis of (−)-rivastigmine **64** using a biocatalyzed DKR process as the key step.

yield by a six-step chemoenzymatic methodology starting from 3-methoxyacetophenone (Scheme 12.19) [80]. Chirality can be introduced in the synthesis at different stages. First, the lipase-catalyzed acylation of (±)-1-(3-methoxyphenyl)ethanol (**61**) was studied at different reaction conditions, and excellent enantioselectivities with both PSL-C and CAL-B in the presence of vinyl acetate and *tert*-butyl methyl ether as solvent were obtained. CAL-B showed its robustness for this process as it was able to catalyze this reaction during five cycles without loss in its biocatalytic properties, and with only 75 mg of crude biocatalyst, 1.5 g of starting material could be converted after 24 hours. In order to improve the yield of this resolution step, a DKR was tested by combining the lipase with a ruthenium complex (**62**) in the presence of sodium carbonate and potassium *tert*-butoxide. In these conditions, enantiopure (*R*)-1-(3-methoxyphenyl)ethyl acetate (*R*)-**63** was recovered with complete yield. As the acetate obtained presents an opposite configuration to the desired final product, two additional steps were required: (1) hydrolysis of the (*R*)-acetate and (2) Mitsunobu inversion leading to the (*S*)-amine to complete the synthesis with a global yield of 29%. Racemic amine can also be subjected to kinetic resolution catalyzed by CAL-B with excellent enantioselectivity in order to obtain the amine of the right configuration to complete the synthesis of (*S*)-rivastigmine.

Chiral amines are of great interest in the pharmaceutical and agrochemical industries. DKRs have also been proposed to prepare these compounds; however, few practical procedures have been developed. Over the past few years, Kim et al. described the DKR of aromatic and aliphatic amines and over the amino acid phenylalanine. The CAL-B-catalyzed acylation of the starting amines has been combined with a metal-catalyzed racemization that employs Pd nanocatalysts [Pd/AlO(OH)] prepared by trapping palladium nanoparticles in aluminum hydroxide [81]. Since both catalysts are thermostable, the DKR could be performed in toluene at 100°C during multiple cycles without loss of activity and selectivity. All the amides were obtained with high conversions and optical purities. Bäckvall and co-workers have also developed the DKR of several amines in biocatalyzed acetylations with isopropyl acetate and

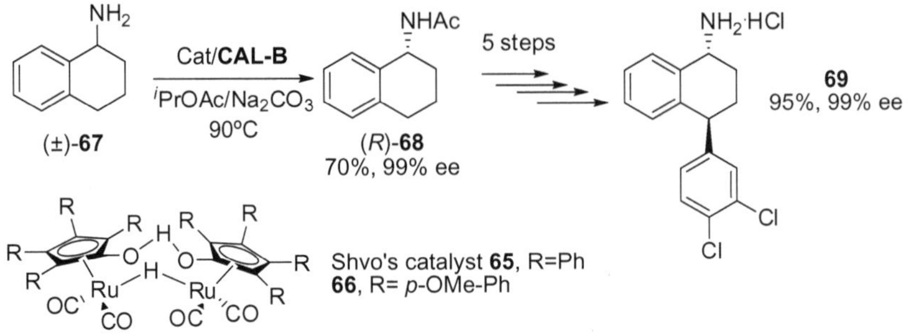

SCHEME 12.20. Chemoenzymatic preparation of norsertraline employing a CAL-B-catalyzed DKR as a key step.

CAL-B employing ruthenium complexes as racemization catalysts [82]. The best results concerning the racemization step were obtained with the Shvo's catalyst (**65**) and its *p*-methoxy substituted derivative (**66**). Benzylic amines can be converted into the corresponding amides with yields higher than 70% and optical purities between 95% and 99%. The system was able to tolerate different functional groups as nitrile, nitrate, or ether. Aliphatic primary amides were obtained in high yields and ee's. The enzymatic alkoxycarbonylation of the racemic amines employing as resolution agent dibenzyl carbonate has also been studied [83]. Resultant chiral carbamates led to the free amines after treatment with Pd on charcoal under hydrogen atmosphere. The amines were obtained with quantitative yield and with no loss in optical purity. This Ru-based DKR methodology has been employed as a key step in the chemoenzymatic preparation of norsertraline (**69**), a selective serotonin reuptake inhibitor, starting from (±)-1,2,3,4-tetrahydro-1-naphthylamine, **67** (Scheme 12.20). The DKR led to a 70% yield of enantiopure (*R*)-amide **68**. After a five-step procedure, this amide led to norsertraline.

A biocatalyzed transformation of a mixture of diols would give, in the best case, a theoretical yield of 25% of one enantiomer. The development of the DKR methodologies for secondary alcohols can be extended to the simultaneous resolution of two chiral *sec*-alcohols centers by combining the asymmetric transformation with racemization/epimerization processes in order to lead to a dynamic kinetic asymmetric transformation (DYKAT) [84]. Thus, after optimizing the reaction conditions and the biocatalyst, the epimerization reagent (catalysts **65** or **66**), and the acyl donor, it was possible to selectively diacetylate C_2-symmetric diols, the source of chiral auxiliaries, and ligands, as well as the precursors of nitrogen heterocycles. The CAL-B- and PSL-C-catalyzed acetylation of 2,5-hexanediol with isopropenyl acetate using a ruthenium catalyst activated by a catalytic amount of potassium *tert*-butoxide led to the corresponding (2*R*,5*R*)-hexanediol diacetate in quantitative yields with excellent ee (≥99%) and a high (*R*,*R*)/*meso* ratio (94:6). DYKAT of 2,4-pentanediol by using CAL-B and the same ruthenium catalyst was also performed, achieving

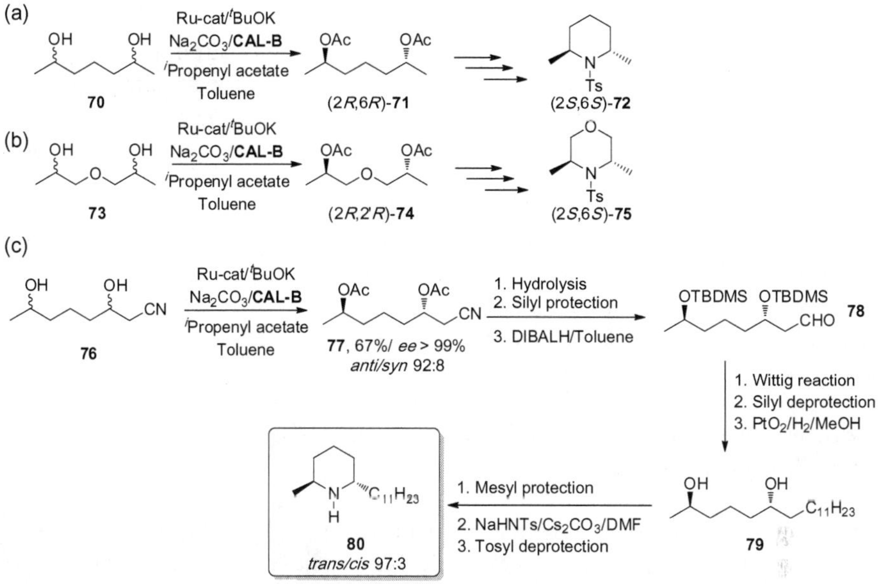

SCHEME 12.21. Some examples of DYKAT processes on 1,5-diols in order to obtain the corresponding heterocycles.

the enantiopure (2*R*,4*R*)-diacetate in a 96% extent with an excellent (*R,R*)/*meso* ratio (97:3). This reaction was performed at 50°C, because at this temperature epimerization outruns possible acyl migrations [85].

The DYKAT methodology has been also applied to the resolution of different 1,5-diols. These compounds are important synthons for the preparation of optically active 2,6-disubstituted and 3,5-disubstituted six-membered heterocycles. The CAL-B or PSL-C type II biocatalyzed acetylation of 1,5-diols employing isopropyl acetate as acyl donor and a ruthenium catalyst for racemization/epimerization leads to the corresponding diacetates with high yields, optical purities, and diastereoselectivities (*anti*/*syn* ratios higher than 8:2) in almost all cases [86]. Obtained from heptanediol (**70**), (2*R*,6*R*)-heptanediol acetate (**71**) was hydrolyzed to the (2*R*,6*R*)-**70**, which was subsequently mesylated (Scheme 12.21a). Reaction of the protected diol with sodium tosylamide in *N*,*N*-dimethylformamide (DMF) at 50°C led to the corresponding enantiopure (2*S*,6*S*)-*N*-tosyl-2,6-dimethylpiperidine **72** (*trans*/*cis* 95:5). In the same way, optically pure (2*R*,2′*R*)-1,1′-oxybis-2-propanol-2,2′-diacetate **74** obtained from diol **73** was employed to obtain enantiopure (2*S*,6*S*)-*N*-tosyl-2,6-dimethylmorpholine **75** (*trans*/*cis* 95:5), as shown in Scheme 12.21b. The selective CAL-B-catalyzed acylation of racemic 3,7-dihydroxyoctanenitrile **76** using isopropenyl acetate in the presence of the same ruthenium catalyst afforded the corresponding enantiopure (3*S*,7*R*)-diacetate **77** in 67% yield with high diastereoselectivity (*anti*/*syn* 92:8). The diacetate was hydrolyzed in 89% yield and subsequently protected with silyl groups. Reduction of the nitrile moiety with diisobutylaluminium hydride (DIBALH) led to the aldehyde **78**, followed by a Wittig reaction with (1-nonyl)triphenylphosphonium bromide. Removal of the silyl protective groups and reduction of the double bond using Adam's catalyst afforded **79**, which was protected as dimesylate. This compound was treated with sodium tosylamide to afford the *N*-tosyl piperidine, which was detosylated with sodium naphthalide to give (+)-solenopsin A95 (**80**) in a total yield of 8% (Scheme 12.21c) [87]. This alkaloid, the unnatural enantiomer of one of the constituents of the venom of the fire ant *Solenopsis invicta*, presents cytotoxic, antibacterial, antifungal, and anti-HIV properties.

12.5. RECENT BIOCATALYZED METHODOLOGIES EMPLOYING LYASES

Lyases (EC 4.X.X.X) have been employed for several years as biocatalysts in the addition of small molecules to double bonds. These enzymes present a high importance in organic synthesis, as they are able to catalyze the carbon–carbon bond formation, a reaction of great importance in organic synthesis. Hydroxynitrile lyases or oxynitrilases (EC 4.1.2.X) are the most widespread lyases, catalyzing the reversible addition of hydrogen cyanide (HCN) to carbonylic compounds such as aldehydes and methyl or ethyl ketones in order to obtain cyanohydrins [88, 89]. Depending on the oxynitrilase stereochemistry, the cyanide attack can be performed by one of the carbonyl compound faces, making it possible to afford (*R*)- or (*S*)-cyanohydrins. These molecules, which possess two functional groups, can be converted into valuable building blocks of a huge number of useful compounds.

The recently discovered (*R*)-oxynitrilase from white apricot [*Prunus armeniaca* (ParsHNL)] has been employed for the synthesis of chiral cyanohydrins starting from bulky aldehydes. This enzyme was able to convert biphenyl, napthtyl, anthracyl, and phenoxyphenyl aldehydes as well as substrates in which the aromatic aldehyde was substituted by a pyridil, pyrimidiyl, or a piperazinyl group with good to high yields. In most cases, excellent enantioselectivities were obtained [90].

The (*S*)-oxynitrilase from *Manihot esculenta* (MeHNL) has been recently combined with a nitrile hydratase (NHase) from *Nitriliruptor alkaliphilus* to obtain the corresponding α-hydroxycarboxylic amides from aldehydes in a one-pot process. Both enzymes were immobilized as cross-linked enzyme aggregates (CLEAs). The NHase was sensitive to water-immiscible solvents as well as to aldehydes and hydrogen cyanide, but showed good activity in acidic media with a pH of 4.0–5.5. By using a portionwise feed of HCN (a maximum of 5 mM) and moderate aldehyde concentrations (35–45 mM), after optimization of the pH and enzymes activity ratio, it was possible to transform

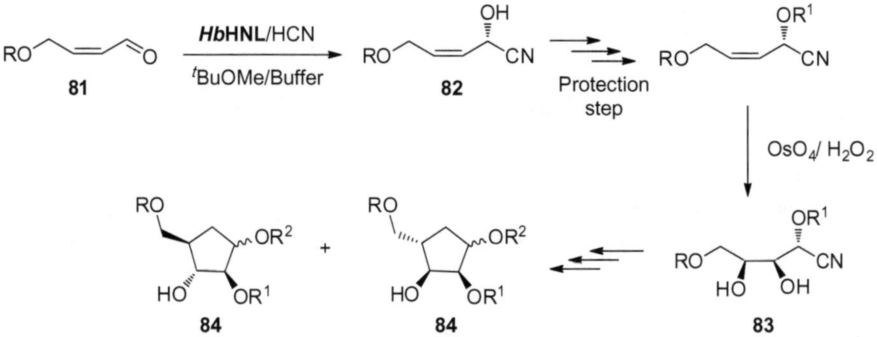

R: Bn
R^1: H, TBDPS
R^2: H, Me.

SCHEME 12.22. Chemoenzymatic procedure for the biocatalyzed formation of pentoses employing the (*S*)-oxynitrilase from *H. brasiliensis*.

five aliphatic aldehydes into the corresponding (*S*)-amides with high yields (higher than 80%) and optical purities (ee > 85%) [91].

Hevea brasiliensis hydroxynitrile lyase (*Hb*HNL) has been used in the chemoenzymatic preparation of different pentoses (**84**) starting from (*Z*)-2-buten-1,4-diol (Scheme 12.22). The key steps of this chemoenzymatic synthesis were the transformation of different *O*-protected 4-hydroxybut-2-enals (**81**) into the corresponding (*S*)-cyanohydrins **82** and the subsequent asymmetric dihydroxylation of the *O*-protected cyanohydrin to obtain the corresponding diol **83**. After testing a series of adequate *O*-protected aldehydes for the purpose of synthesizing the final pentoses, only the *O*-allyl-protected derivative was converted with sufficient high selectivity. The dihydroxylation step was also influenced for the different protecting groups, with the best results achieved when employing a catalytic amount of OsO_4 in the presence of H_2O_2 [92].

Hhes (also called haloalcohol dehalogenases or hydrogen-halide lyases, EC 4.5.1.X) catalyze the conversion of vicinal halohydrins into their corresponding epoxides, while releasing halide ions [93]. They can be found in several bacteria that use halogenated alcohols or compounds that are degraded via halohydrins as a carbon source for growth. In the epoxide ring opening, different nucleophiles can be accepted, including azide, nitrite, and cyanide. This remarkable catalytic promiscuity allows the enzymatic production of a broad range of *β*-substituted alcohols from epoxides. Apart from the example already described in combination with an ADH (Scheme 12.4c) [24], Janssen and coworkers have demonstrated the great potential of these biocatalysts in several examples. Thus, by employing a mutant from Hhe from *A. radiobacter* AD1 (HheC), the synthesis of optically pure methyl 4-cyano-3-hydroxybutanoate became feasible [94]. This compound is a versatile building block containing three functional groups. The *R* enantiomer can be applied as an intermediate in the production of cholesterol-lowering drugs of the statin type. This procedure provided (*S*)-methyl 4-cyano-3-hydroxybutanoate in good yield (81% from the kinetic resolution) and with very high ee (>95%). In a later contribution [95], HheC catalyzed the kinetic nucleophilic ring-opening resolution of epoxides with cyanide and azide. In the case of 2,2-disubstituted epoxides, this reaction proceeded with excellent enantioselectivity ($E > 200$), affording both enantiopure epoxides and β-cyano and β-azido tertiary alcohols, important building blocks for various optically active products that are difficult to prepare. In 2008, an elegant chemoenzymatic DKR procedure leading to aromatic terminal epoxides was published [96]. Thus, coupling an Ir catalyst which promoted the racemization of the halohydrins with a double mutant of HheC, several (*R*)-oxiranes could be achieved with moderate to high conversions (60%–90%) and high enantioselectivities (usually higher than 90%).

Carbon–carbon bond formation is a reaction of great importance in organic synthesis. Catalytic C–C coupling is one of the most useful synthetic methods in asymmetric synthesis. In nature, these reactions are catalyzed by aldolases (EC 4.1.2.X), a subclass of lyases that are able to catalyze the addition of carbonucleophiles to carbonyl groups [97, 98]. More than 40 aldolases have been characterized to date, offering biocatalytic access to different hydroxycarbonyl compounds. However, the application of these enzymatic processes is still low due to thermodynamic limitations (the reaction involves an equilibrium) as well as to kinetic limitations (the rates of formation of the different enantiomers of the product are often not big enough to achieve high selectivities). Nevertheless, in the past few years, some nice examples of the use of these enzymes in organic synthesis have been developed. Threonine aldolases catalyze the reversible aldol addition of glycine to a variety of acceptor aldehydes with complete control of the stereochemistry. Recently, a serine hydroxymethyltransferase (SHMT) with threonine aldolase activity from *Streptococcus thermophilus* and an L-threonine aldolase (L-TA) from *E. coli* have been overexpressed in *E. coli* M15 and employed as biocatalysts for the aldol addition of glycine to different *N*-benzyloxycarbonyl (Cbz)-protected aminoaldehydes [99], as (*R*)- and (*S*)-*N*-Cbz-alaninal, *N*-Cbz-3-aminopropanal, *N*-Cbz-2-aminoethanal, and benzyloxyacetaldehyde. Low temperatures (4°C) and higher excesses of glycine lead to the higher conversions, showing higher SHMT activity in DMF/water mixtures while emulsions were the best media for L-TA. The aldol reactions allowed preparing *N*-Cbz-γ-amino-β-hydroxy-α-amino acid as final products in useful yields on a preparative scale, but the isolated yields are still low for an industrial application. Treatment of the *N*-Cbz-γ-amino-β-hydroxy-α-amino acid with potassium hydroxide led to the spontaneous formation of the 2-oxazolidinone derivatives in quantitative yields.

L-*allo*-Threonine aldolase from *Aeromonas jandaei* and D-threonine aldolase from *Pseudomonas* sp. have been employed as biocatalysts in the aldol

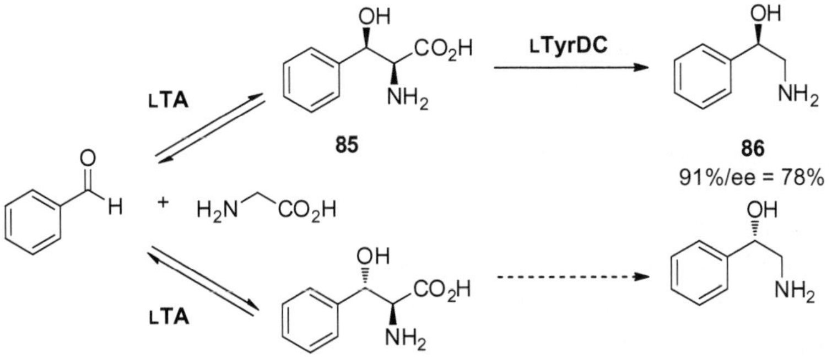

SCHEME 12.23. Enzymatic preparation of (*R*)-2-aminophenylethanol by employing two biocatalysts in a DYKAT procedure.

reaction between different aldehydes as acceptors and several donors as D-alanine, D-serine, and D-cysteine in order to obtain the corresponding L- or D-β-hydroxy-α,α-dialkyl-α-amino acids with moderate to good yields and diastereoselectivities depending on the substrate structure [100]. These non-natural occurring compounds are employed as building blocks, as enzyme inhibitors, and as conformational modifiers of physiologically active peptides.

In order to improve some of the thermodynamic limitations that the use of aldolases presents, the aldol reaction catalyzed by L-threonine aldolase (L-TA) from *Pseudomonas putida* between benzaldehyde and glycine leading to L-phenylserine (**85**) has been coupled to the subsequent decarboxylation catalyzed by the L-tyrosine decarboxylase (L-TyrDC) from *Enterococcus faecalis* V583 to obtain (*R*)-2-amino-1-phenylethanol (**86**) with moderate selectivity (ee = 78%) in a one-pot reaction, as these two enzymes have been overexpressed in *E. coli* (Scheme 12.23). The aldol equilibrium is shifted to the products, thus resulting in high conversions. The formal global reaction that is produced is the aminomethylation of an aldehyde, being the absolute configuration retained in the hydroxyl substituent stereogenic center while disappearing in the second one. Thus, the moderate C_β selectivity that the aldolase exhibits in the formation of L-phenylserine was improved by the high *syn*/*anti* selectivity of the decarboxylase. This reaction is the first reported bienzymatic resolution of diastereomers, the so-called DYKAT previously described for lipases coupled to organometallic compounds in the resolution of diols [101]. The conversion of benzaldehyde into the intermediate phenylserine (8%) and the final aminoalcohol (91%) was completed after 58 hours.

D-Fructose-6-phosphate aldolase (FSA) has been used as a biocatalyst in the direct *syn* cross-aldol addition of glycoaldehyde to different aldehydes in aqueous media. As glycoaldehyde presents a higher affinity as a donor for FSA, good to poor aldehyde acceptors can be employed in the synthesis by maintaining a low concentration of glycoaldehyde. These biocatalyzed processes have been employed to obtain 1-deoxyiminosugars and precursors of L-xylose and D-arabinose [102].

12.6. ISOMERASES

Isomerases (EC 5.X.X.X) are enzymes that catalyze the structural rearrangement of isomers. Due to their high substrate specificity, few examples of isomerases with application in organic synthesis have been described. In most cases, these enzymes are employed as racemization agents in order to perform DKR reactions catalyzed by other enzymes [103]. Phenylalanine aminomutase (PAM) from *Taxus chinensis* is an enzyme that catalyzes the conversion of α-phenylalanine into β-phenylalanine. Recently, it has been observed that the PAM-catalyzed reaction on α-phenylalanine led to (*E*)-cinnamic acid and β-phenylalanine, indicating that this biocatalyst presented ammonia lyase activity [104]. This enzymatic promiscuity was exploited in order to synthesize a set of chiral α- and β-amino acids by treating different substituted (*E*)-cinnamic acids **87** with ammonia (Scheme 12.24) [105]. Addition of ammonia to cinnamic acids yielded a mixture 1:1 of α- and β-phenylalanines (**88** and **89**, respectively) with excellent enantioselectivities (ee > 99% for both isomers). PAM was able to accept substrates with halogen, alkoxy, and alkyl substituents in the aromatic ring. For all the substrates tested, enantiopure β-amino acids were produced while the enantioselectivity for the α-isomers was excellent in almost all cases. The presence of substituents in the *ortho* position resulted in almost exclusive formation of the α-phenylalanine analogs. The PAM-catalyzed addition of ammonia to *meta*-substituted cinnamic acids led to a mixture of compounds, while the transformation of acids with different groups in the *para* position occurred with lower catalytic rates but with higher affinities than the *o*- and *m*-substituted substrates. In general, affinity of PAM for *p*-substituted cinnamic acids was higher with lipophilic groups; however, this trend was disturbed by steric effects. It was also observed that there was a clear correlation between the initial amount of the β-phenylalanine derivative and the Hammet constant of the substituent [106], with higher percentages of β-product with more electron-donating groups being achieved.

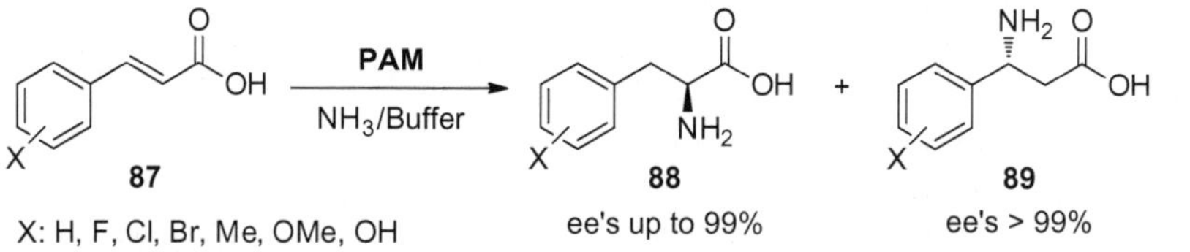

SCHEME 12.24. PAM-catalyzed conversion of cinnamic acids into α- and β-amino acids.

REFERENCES

[1] Breuer, M., Ditrich, K., Habicher, T., Hauer, B., Kesseler, M., Stürmer, R., Zelinski, T. (2004). Industrial methods for the production of optically active intermediates. *Angew. Chem. Int. Ed.*, *43*, 788–824.

[2] Anastas, P., Eghbali, N. (2010). Green chemistry: principles and practice. *Chem. Soc. Rev.*, *39*, 301–312.

[3] Fessner, W. D., Anthonsen, T. (Eds.) (2009). *Modern Biocatalysis: Stereoselective and Environmentally Friendly Reactions*. Wiley-VCH, Weinheim, Germany.

[4] Gotor, V., Alfonso, I., García-Urdiales, E. (Eds.) (2008). *Asymmetric Organic Synthesis with Enzymes*. Wiley-VCH, Weinheim, Germany.

[5] Patel, R. N. (Ed.) (2007). *Biocatalysis in the Pharmaceutical and Biotechnology Industry*. CRC Press, Boca Raton, FL.

[6] Faber, K. (2004). *Biotransformations in Organic Chemistry*, 5th ed., Springer, Berlin, Germany.

[7] Hudlicky, T., Reed, J. W. (2009). Applications of biotransformations and biocatalysis to complexity generation in organic synthesis. *Chem. Soc. Rev.*, *38*, 3117–3132.

[8] Schmid, R. D., Urlacher, V. B. (Eds.) (2007). *Modern Biooxidation. Enzymes, Reactions and Applications*, Wiley-VCH, Weinheim, Germany.

[9] Huisman, G. W., Liang, J., Krebber, A. (2010). Practical chiral alcohol manufacture using ketoreductases. *Curr. Opin. Chem. Biol.*, *14*, 122–129.

[10] Matsuda, T., Yamanaka, R., Nakamura, K. (2009). Recent progress in biocatalysis for asymmetric oxidation and reduction. *Tetrahedron: Asymmetry*, *20*, 513–557.

[11] de Wildeman, S. M. A., Sonke, T., Schoemaker, H. E., May, O. (2007). Biocatalytic reductions: from lab curiosity to "first choice." *Acc. Chem. Res.*, *40*, 1260–1266.

[12] de Gonzalo, G., Lavandera, I., Faber, K., Kroutil, W. (2007). Enzymatic reduction of ketones in "micro-aqueous" media catalyzed by ADH-A from *Rhodococcus ruber*. *Org. Lett.*, *9*, 2163–2166.

[13] Hildebrand, F., Lütz, S. (2009). Stable electroenzymatic processes by catalyst separation. *Chem. Eur. J.*, *15*, 4998–5001.

[14] Bisogno, F. R., Lavandera, I., Kroutil, W., Gotor, V. (2009). Tandem concurrent processes: one-pot single-catalyst biohydrogen transfer for the simultaneous preparation of enantiopure secondary alcohols. *J. Org. Chem.*, *74*, 1730–1732.

[15] Edegger, K., Stampfer, W., Seisser, B., Faber, K., Mayer, S. F., Oerhrlein, R., Hafner, A., Kroutil, W. (2006). Regio- and stereoselective reduction of diketones and oxidation of diols by biocatalytic hydrogen transfer. *Eur. J. Org. Chem.*, 1904–1909.

[16] Kurina-Sanz, M., Bisogno, F. R., Lavandera, I., Orden, A. A., Gotor, V. (2009). Promiscuous substrate binding explains the enzymatic stereo- and regiocontrolled synthesis of enantiopure hydroxy ketones and diols. *Adv. Synth. Catal.*, *351*, 1842–1848.

[17] Katzberg, M., Wechler, K., Müller, M., Dünkelmann, P., Stohrer, J., Hummel, W., Bertau, M. (2009). Biocatalytic production of (5*S*)-hydroxy-2-hexanone. *Org. Biomol. Chem.*, *7*, 304–314.

[18] Kalaitzakis, D., Rozzell, J. D., Kambourakis, S., Smonou, I. (2006). Synthesis of valuable chiral intermediates by isolated ketoreductases: synthesis of α-alkyl-β-hydroxyketones and 1,3-diols. *Adv. Synth. Catal.*, *348*, 1958–1969.

[19] Kalaitzakis, D., Rozzell, J. D., Kambourakis, S., Smonou, I. (2006). A two-step chemoenzymatic synthesis of the natural pheromone (+)-sitophilure utilizing isolated, NADPH-dependent ketoreductases. *Eur. J. Org. Chem.*, 2309–2313.

[20] Baer, K., Kraußer, M., Burda, E., Hummel, W., Berkessel, A., Gröger, H. (2009). Sequential and modular synthesis of chiral 1,3-diols with two sterogenic centers: access to all four stereoisomers by combination of organo- and biocatalysis. *Angew. Chem. Int. Ed.*, *48*, 9355–9358.

[21] Gröger, H., Chamouleau, F., Orologas, N., Rollmann, C., Drauz, K., Hummel, W., Weckbecker, A., May, O. (2006). Enantioselective reduction of ketones with "designer cells" at high substrate concentrations: highly efficient access to functionalized optically active alcohols. *Angew. Chem. Int. Ed.*, *45*, 5677–5681.

[22] Zhu, D., Ankati, H., Mukherjee, C., Yang, Y., Biehl, E. R., Hua, L. (2007). Asymmetric reduction of *β*-ketonitriles with a recombinant carbonyl reductase and enzymatic transformation to optically pure *β*-hydroxy carboxylic acids. *Org. Lett.*, *9*, 2561–2563.

[23] Ankati, H., Yang, Y., Zhu, D., Biehl, E. R., Hua, L. (2008). Synthesis of optically pure 2-azido-1-arylethanols with isolated enzymes and conversion of triazole-containing β-blocker analogues employing click chemistry. *J. Org. Chem.*, *73*, 6433–6436.

[24] Schrittwieser, J. H., Lavandera, I., Seisser, B., Mautner, B., Kroutil, W. (2009). Biocatalytic cascade for the synthesis of enantiopure β-azidoalcohols and β-hydroxynitriles. *Eur. J. Org. Chem.*, 2293–2298.

[25] Guo, Z., Chen, Y., Goswami, A., Hanson, R. L., Patel, R. N. (2006). Synthesis of ethyl and *t*-butyl (3*R*,5*S*)-dihydroxy-6-benzyloxy hexanoates via diastereo- and enantioselective microbial reduction. *Tetrahedron: Asymmetry*, *17*, 1589–1602.

[26] Milagre, H. M. S., Milagre, C. D. F., Moran, P. J. S., Santana, M. H. A., Rodrigues, J. A. R. (2006). Asymmetric bioreduction of ethyl 3-halo-2-oxo-4-phenylbutanoate by *Saccharomyces cerevisiae* immobilized in Ca-alginate beads with double gel layer. *Org. Process Res. Dev.*, *10*, 611–617.

[27] Acetti, D., Brenna, E., Fuganti, C. (2007). A new enzymatic approach to (*R*)-tamsulosin hydrochloride. *Tetrahedron: Asymmetry*, *18*, 488–492.

[28] Fischer, T., Pietruszka, J. (2007). Efficient synthesis of either enantiomer of ethyl 5-hydroxyhept-6-enoate. *Adv. Synth. Catal.*, *349*, 1533–1536.

[29] Matsuda, M., Yamazaki, T., Fuhshuku, K., Sugai, T. (2007). First total synthesis of modiolide A, based on the whole-cell yeast-catalyzed asymmetric reduction of a propargyl ketone. *Tetrahedron*, *63*, 8752–8760.

[30] Ema, T., Ide, S., Okita, N., Sakai, T. (2008). Highly efficient chemoenzymatic synthesis of methyl (*R*)-*o*-chloromandelate, a key intermediate for clopidogrel, *via* asymmetric reduction with recombinant *Escherichia coli*. *Adv. Synth. Catal.*, *350*, 2039–2044.

[31] Broussy, S., Cheloha, R. W., Berkowitz, D. B. (2009). Enantioselective ketoreductases-based entry into pharmaceutical building blocks: ethanol as tunable nicotinamide reductant. *Org. Lett.*, *11*, 305–308.

[32] Vieira, G. A. B., Lemos, T. L. G., de Mattos, M. C., de Oliveira, M. C. F., Melo, V. M. M., de Gonzalo, G., Gotor-Fernández, V., Gotor, V. (2009). Chemoenzymatic synthesis of optically active mugetanol isomers: use of lipases and oxidoreductases in fragrance chemistry. *Tetrahedron: Asymmetry*, *20*, 214–219.

[33] Lacheretz, R., Gomez Pardo, D., Cossy, J. (2009). *Daucus carota* mediated-reduction of cyclic 3-oxo-amines. *Org. Lett.*, *11*, 1245–1248.

[34] Voss, C. V., Gruber, C. C., Kroutil, W. (2008). Deracemization of secondary alcohols through tandem biocatalytic oxidation and reduction. *Angew. Chem. Int. Ed.*, *47*, 741–745.

[35] Voss, C. V., Gruber, C. C., Faber, K., Knaus, T., Macheroux, P., Kroutil, W. (2008). Orchestration of concurrent oxidation and reduction cycles for stereoinversion and deracemisation of *sec*-alcohols. *J. Am. Chem. Soc.*, *130*, 13969–13972.

[36] Stuermer, R., Hauer, B., Hall, M., Faber, K. (2007). Asymmetric bioreduction of activated C=C bonds using enoate reductases from the old yellow enzyme family. *Curr. Opin. Chem. Biol.*, *11*, 203–213.

[37] Hall, M., Stueckler, C., Kroutil, W., Macheroux, P., Faber, K. (2007). Asymmetric bioreduction of activated alkenes using cloned 12-oxophytodienoate reductase isoenzymes OPR-1 and OPR-3 from *Lycopersicum esculentum* (tomato): a striking change of stereoselectivity. *Angew. Chem. Int. Ed.*, *46*, 3934–3937.

[38] Hall, M., Stueckler, C., Ehammer, H., Pointner, E., Oberdorfer, G., Gruber, K., Hauer, B., Stuermer, R., Kroutil, W., Macheroux, P., Faber, K. (2008). Asymmetric bioreduction of C=C bonds using enoate reductases OPR1, OPR3 and YqjM: enzyme-based stereocontrol. *Adv. Synth. Catal.*, *350*, 411–418.

[39] Bougioukou, D. J., Stewart, J. D. (2008). Opposite stereochemical courses for enzyme-mediated alkene reductions of an enantiomeric pair substrate. *J. Am. Chem. Soc.*, *130*, 7655–7658.

[40] Kosjek, B., Fleitz, F. J., Dormer, P. G., Kuethe, J. T., Devine, P. N. (2008). Asymmetric bioreduction of α,β-unsaturated nitriles and ketones. *Tetrahedron: Asymmetry*, *19*, 1403–1406.

[41] Urlacher, V. B., Eiben, S. (2006). Cytochrome P450 monooxygenases: perspectives for synthetic application. *Trends Biotechnol.*, *24*, 324–330.

[42] Bernhardt, R. (2006). Cytochromes P450 as versatile biocatalysts. *J. Biotechnol.*, *124*, 128–145.

[43] Weis, R., Winkler, M., Schittmayer, M., Kambourakis, S., Vink, M., Rozzell, J. D., Glieder, A. (2009). A diversified library of bacterial and fungal bifunctional cytochrome P450 enzymes for drug metabolite synthesis. *Adv. Synth. Catal.*, *351*, 2140–2146.

[44] Alphand, V., Fraaije, M. W., Mihovilovic, M. D., Ottolina, G. (2009). Second generation Baeyer-Villiger biocatalysts. In: W. D. Fessner, T. Anthonsen (Eds.) *Modern Biocatalysis: Stereoselective and Environmentally Friendly Reactions*, Wiley-VCH, Weinheim, Germany, pp. 339–368.

[45] Kayser, M. M. (2009). "Designer reagents" recombinant microorganisms: new and powerful tools for organic synthesis. *Tetrahedron*, *65*, 947–974.

[46] Rudroff, F., Rydz, J., Ogink, F. H., Fink, M., Mihovilovic, M. D. (2007). Comparing the stereoselective biooxidation of cyclobutanones by recombinant strains expressing bacterial Baeyer-Villiger monooxygenases. *Adv. Synth. Catal.*, *349*, 1436–1444.

[47] Snajdrova, R., Braun, I., Bach, T., Mereiter, K., Mihovilovic, M. D. (2007). Biooxidation of bridged cycloketones using Baeyer-Villiger monooxygenases of various bacterial origin. *J. Org. Chem.*, *72*, 9597–9603.

[48] Rodríguez, C., de Gonzalo, G., Fraaije, M. W., Gotor, V. (2007). Enzymatic kinetic resolution of racemic ketones catalyzed by Baeyer-Villiger monooxygenases. *Tetrahedron: Asymmetry*, *18*, 1338–1344.

[49] Rodríguez, C., de Gonzalo, G., Torres Pazmiño, D. E., Fraaije, M. W., Gotor, V. (2009). Baeyer-Villiger monooxygenase-catalyzed kinetic resolution of racemic α-alkyl benzyl ketones: enzymatic synthesis of α-alkyl benzylketones and α-alkyl benzylesters. *Tetrahedron: Asymmetry*, *20*, 1168–1173.

[50] Rioz-Martínez, A., Bisogno, F. R., Rodríguez, C., de Gonzalo, G., Lavandera, I., Torres Pazmiño, D. E., Fraaije, M. W., Gotor, V. (2010). Biocatalysed concurrent production of enantioenriched compounds through parallel interconnected kinetic asymmetric transformations. *Org. Biomol. Chem.*, *8*, 1431–1437.

[51] Hudlicky, T., Reed, J. W. (2009). On the merits of biocatalysis and the impact of arene *cis*-dihydrodiols on enantioselective synthesis. *Synlett*, 685–703.

[52] Sullivan, B., Carrera, I., Drouin, M., Hudlicky, T. (2009). Symmetry-based design for the chemoenzymatic synthesis of oseltamivir (Tamiflu) from ethyl benzoate. *Angew. Chem. Int. Ed.*, *48*, 4229–4231.

[53] Turner, N. J. (2010). Deracemisation methods. *Curr. Opin. Chem. Biol.*, *14*, 115–121.

[54] Bailey, K. R., Ellis, A. J., Reiss, R., Snape, T. J., Turner, N. J. (2007). A template based mnemonic for monoamine oxidase (MAO-N) catalyzed reactions and its application to the chemo-enzymatic deracemisation of the alkaloid (±)-crispine A. *Chem. Commun.*, 3640–3642.

[55] Riva, S. (2006). Laccases: blue enzymes for green chemistry. *Trends Biotechnol.*, *24*, 219–226.

[56] Torre, O., Alfonso, I., Gotor, V. (2004). Lipase catalyzed Michael addition of secondary amines to acrylonitrile. *Chem. Commun.*, 1724–1726.

[57] Witayakran, S., Ragauskas, A. J. (2009). Cocatalytic enzyme system for the Michael reaction in-situ generated *ortho*-quinones. *Eur. J. Org. Chem.*, 358–363.

[58] Koszelewski, D., Tauber, K., Faber, K., Kroutil, W. (2010). ω-Transaminases for the synthesis of non-racemic α-chiral primary amines. *Trends Biotechnol.*, *28*, 324–332.

[59] Koszelewski, D., Lavandera, I., Clay, D., Guebitz, G. M., Rozzell, D., Kroutil, W. (2008). Formal asymmetric biocatalytic reductive amination. *Angew. Chem. Int. Ed.*, *47*, 9337–9340.

[60] Koszelewski, D., Pressnitz, D., Clay, D., Kroutil, W. (2009). Deracemization of mexiletine biocatalyzed by ω-transaminases. *Org. Lett.*, *11*, 4810–4812.

[61] Ingram, C. U., Bommer, M., Smith, M. E. B., Dalby, P. A., Ward, J. M., Hailes, H. C., Lye, G. J. (2007). One-Pot synthesis of amino-alcohols using a de-novo transketolase and β-alanine:pyruvate transaminase pathway in *Escherichia coli. Biotechnol. Bioeng.*, *96*, 559–569.

[62] Bornscheuer, U. T., Kaszlauskas, R. S. (2005). *Hydrolases in Organic Synthesis: Regio- and Stereoselective Biotransformations*, Wiley-VCH, Weinheim, Germany.

[63] Ghanem, A. (2007). Trends in lipase-catalyzed asymmetric access to enantiomerically pure/enriched compounds. *Tetrahedron*, *63*, 1721–1754.

[64] Mendiola, J., García-Cerrado, S., de Frutos, O., Puente, M. L., Gu, R. L., Khau, V. V. (2009). Enzymatic resolution of *N*-substituted-β-prolines. *Org. Process Res. Dev.*, *13*, 292–296.

[65] Patterson, L. D., Miller, M. J. (2010). Enzymatic deprotection of the cephalosporin 3′-acetoxy group using *Candida antarctica* lipase B. *J. Org. Chem.*, *75*, 1289–1292.

[66] Henke, E., Bornscheuer, U. T., Schmid, R. D., Pleiss, J. (2003). A molecular mechanism of enantiorecognition of tertiary alcohols by carboxylesterases. *ChemBioChem*, *4*, 485–493.

[67] Bartsch, S., Kourist, R., Bornscheuer, U. T. (2008). Complete inversion of enantioselectivity towards acetylated tertiary alcohols by a double mutation of a *Bacillus subtilis* esterase. *Angew. Chem. Int. Ed.*, *47*, 1508–1511.

[68] Kourist, R., Nguyen, G.-S., Strübing, D., Böttcher, D., Liebeton, K., Naumer, C., Eck, J., Bornscheuer, U. T. (2008). Hydrolase-catalyzed stereoselective preparation of protected α,α-dialkyl-α-hydroxycarboxylic acids. *Tetrahedron: Asymmetry*, *19*, 1839–1843.

[69] Li, D., Cheng, S., Wei, D., Ren, Y., Zhang, D. (2007). Production of enantiomerically pure (*S*)-β-phenylalanine and (*R*)-β-phenylalanine by penicillin G acylase from *Escherichia coli* in aqueous medium. *Biotechnol. Lett.*, *29*, 1825–1830.

[70] Heck, T., Seebach, D., Osswald, S., ter Wiel, M. K. J., Kholer, H.-P. E., Geueke, B. (2009). Kinetic resolution of aliphatic β-amino acid amides by β-aminopeptidases. *ChemBioChem.*, *10*, 1558–1561.

[71] Forró, E., Fülöp, F. (2007). The first directed enzymatic hydrolysis of alicyclic β-amino acid esters: a route to enantiopure *cis*- and *trans*-β-amino acids. *Chem. Eur. J.*, *13*, 6397–6401.

[72] Tasnádi, G., Forró, E., Fülöp, F. (2008). An efficient new enzymatic method for the preparation of β-aryl-β-amino acid enantiomers. *Tetrahedron: Asymmetry*, *19*, 2072–2077.

[73] Zutter, U., Iding, H., Spurr, P., Wirz, B. (2008). New, efficient synthesis of oseltamivir phosphate (Tamiflu) via enzymatic desymmetrization of a *meso*-1,3-cyclohexanedicarboxylic acid diester. *J. Org. Chem.*, *73*, 4895–4902.

[74] Ríos-Lombardía, N., Busto, E., García-Urdiales, E., Gotor-Fernández, V., Gotor, V. (2009). Enzymatic desymmetrization of prochiral 2-substituted-1,3-diamines: preparation of valuable nitrogenated compounds. *J. Org. Chem.*, *74*, 2571–2574.

[75] Gotor, V. (2009). Biocatalysis applied to the synthesis of nucleoside analogs. In: W. D. Fessner, T. Anthonsen (Eds.) *Modern Biocatalysis: Stereoselective and Environmentally Friendly Reactions*, Wiley-VCH, Weinheim, Germany, pp. 135–152.

[76] Simeó, Y., Sinisterra, J. V., Alcántara, A. R. (2009). Regioselective enzymatic acylation of pharmacologically interesting nucleosides in 2-methyltetrahydrofuran, a greener substitute for THF. *Green Chem.*, *11*, 855–862.

[77] Díaz-Rodríguez, A., Fernández, S., Sanghvi, Y. S., Ferrero, M., Gotor, V. (2006). Novel chemoenzymatic protocol for the synthesis of 3′-*O*-dimethoxytrityl-2′-deoxynucleoside derivatives as building blocks for oligonucleotide synthesis. *Org. Process Res. Dev.*, *10*, 581–587.

[78] Lee, J. H., Han, K., Kim, M.-J., Park, J. (2010). Chemoenzymatic dynamic kinetic resolution of alcohols and amines. *Eur. J. Org. Chem.*, 999–1015.

[79] Kamal, A., Azhar, M. A., Krishnaji, T., Malik, M. S., Azeeza, S. (2008). Approaches based on enzyme mediated kinetic to dynamic kinetic resolutions: a versatile route for chiral intermediates. *Coord. Chem. Rev.*, *252*, 569–592.

[80] Suárez-Mangas, J., Rodríguez-Mata, M., Busto, E., Gotor-Fernández, V., Gotor, V. (2009). Chemoenzymatic synthesis of rivastigmine based on lipase-catalyzed processes. *J. Org. Chem.*, *74*, 5304–5310.

[81] Kim, M.-J., Kim, W.-H., Han, K., Choi, Y. K., Park, J. (2007). Dynamic kinetic resolution of primary amines with a recyclable Pd nanocatalyst for racemization. *Org. Lett.*, *9*, 1157–1159.

[82] Thalen, L. K., Zhao, D., Sortais, J.-B., Paetzold, J., Hoben, C., Bäckvall, J.-E. (2009). A chemoenzymatic approach to enantiomerically pure amines using dynamic kinetic resolution: application to the synthesis of norsertraline. *Chem. Eur. J.*, *15*, 3403–3410.

[83] Hoben, C. E., Kanupp, L., Bäckvall, J.-E. (2008). Practical chemoenzymatic dynamic kinetic resolution of primary amines via transfer of readily removable benzyloxycarbonyl group. *Tetrahedron Lett.*, *49*, 977–979.

[84] Stenreiber, J., Faber, K., Griengl, H. (2008). De-racemization of enantiomers versus de-epimerization of diastereomers—classification of dynamic kinetic asymmetric transformations (DYKAT). *Chem. Eur. J.*, *14*, 8060–8072.

[85] Martín-Matute, B., Edin, M., Bäckvall, J.-E. (2006). Highly efficient synthesis of enantiopure diacetylated C_2-symmetric diols by ruthenium- and enzyme-catalyzed dynamic kinetic asymmetric transformation (DYKAT). *Chem. Eur. J.*, *12*, 6053–6061.

[86] Leijondahl, K., Borén, L., Braun, R., Bäckvall, J.-E. (2008). Enantiopure 1,5-diols from dynamic kinetic asymmetric transformation. Useful synthetic intermediates for the preparation of chiral heterocycles. *Org. Lett.*, *10*, 2027–2030.

[87] Leijondahl, K., Borén, L., Braun, R., Bäckvall, J.-E. (2009). Enzyme- and ruthenium-catalyzed dynamic kinetic asymmetric transformation of 1,5-diols. Application to the synthesis of (+)-Solenopsin A. *J. Org. Chem.*, *74*, 1988–1993.

[88] Holt, J., Hanefeld, U. (2009). Enantioselective enzyme-catalysed synthesis of cyanohydrins. *Curr. Org. Synth.*, *6*, 15–37.

[89] Andexer, J. N., Langermann, J. V., Kragl, U., Pohl, M. (2009). How to overcome limitations in biotechnology processes—examples from hydroxynitrile lyase applications. *Trends Biotechnol.*, *27*, 599–607.

[90] Bhunya, R., Mahapatra, T., Nanda, S. (2009). *Prunus armeniaca* hydroxynitrile lyase (ParsHNL)-catalyzed asymmetric transformation of cyanohydrins from sterically demanding aromatic aldehydes. *Tetrahedron: Asymmetry*, *20*, 1526–1530.

[91] van Pelt, S., van Rantwijk, F., Sheldon, R. (2009). Synthesis of aliphatic (*S*)-α-hydroxycarboxylic amides using a one-pot bienzymatic cascade immobilised oxynitrilase and nitrile hydratase. *Adv. Synth. Catal.*, *351*, 397–404.

[92] Avi, M., Gaisberger, R., Feichtenhofer, S., Griengl, H. (2009). *De novo* synthesis of pentoses via cyanohydrins as key intermediates. *Tetrahedron*, *65*, 5418–5426.

[93] Janssen, D. B., Majerić Elenkov, M., Hasnaoui, G., Hauer, B., Lutje Spelberg, J. H. (2006). Enantioselective formation and ring-opening of epoxides catalysed by halohydrin dehalogenases. *Biochem. Soc. Trans.*, *34*, 291–295.

[94] Majerić Elenkov, M., Tang, L., Hauer, B., Janssen, D. B. (2006). Sequential kinetic resolution catalyzed by halohydrin dehalogenase. *Org. Lett.*, *8*, 4227–4229.

[95] Majerić Elenkov, M., Hoeffken, H. W., Tang, L., Hauer, B., Janssen, D. B. (2007). Enzyme catalyzed nucleophilic ring opening of epoxides for the preparation of enantiopure tertiary alcohols. *Adv. Synth. Catal.*, *349*, 2279–2285.

[96] Haak, R. M., Berthiol, F., Jerphagnon, T., Gayet, A. J. A., Tarabiono, C., Postema, C. P., Ritleng, V., Pfeffer, M., Janssen, D. B., Minnaard, A. J., Feringa, B. L., de Vries, J. G. (2008). Dynamic kinetic resolution of racemic *β*-haloalcohols: direct access to enantioenriched epoxides. *J. Am. Chem. Soc.*, *130*, 13508–13509.

[97] Clapes, P., Sprenger, G. A., Joglar, J. (2009). Novel strategies in aldolase-catalyzed synthesis of iminosugars. In: W. D. Fessner, T. Anthonsen (Eds.) *Modern Biocatalysis: Stereoselective and Environmentally Friendly Reactions*, Wiley-VCH, Weinheim, Germany, pp. 299–312.

[98] Clapes, P., Fessner, W. D., Sprenger, G. A., Samland, A. K. (2010). Recent progress in stereoselective synthesis with aldolases. *Curr. Opin. Chem. Biol.*, *14*, 154–167.

[99] Gutiérrez, M. L., Garrabou, X., Agosta, E., Servi, S., Parella, T., Joglar, J., Clapés, P. (2008). Serine hydromethyl transferase from *Streptococcus thermophilus* and L-threonine aldolase from *Escherichia coli* as stereocomplementary biocatalysts for the synthesis of β-hydroxy-α,ω-amino acids derivatives. *Chem. Eur. J.*, *14*, 4647–4656.

[100] Fesko, K., Uhl, M., Stenreiber, J., Gruber, K., Griengl, H. (2010). Biocatalytic access to α,α-dialkyl-α-amino acids by a mechanism-based approach. *Angew. Chem. Int. Ed.*, *49*, 121–124.

[101] Steinreiber, J., Schürmann, M., Wolberg, M., van Assema, F., Reisinger, C., Fesko, K., Mink, D., Griengl, H. (2007). Overcoming thermodynamic and kinetic limitations of aldolase catalyzed reactions by applying multienzymatic dynamic kinetic asymmetric transformations. *Angew. Chem. Int. Ed.*, *46*, 1624–1626.

[102] Garrabou, X., Castillo, J. A., Guérard-Hélaine, C., Parella, T., Joglar, J., Clapés, P. (2009). Asymmetric self- and cross-aldol reactions of glycoaldehyde catalyzed by D-fructose-6-phosphate aldolase. *Angew. Chem. Int. Ed.*, *48*, 5521–5525.

[103] Glueck, S. M., Pirker, M., Nestl, B. M., Ueberbacher, B. T., Larissegger-Schnell, B., Csar, K., Hauer, B., Stuermer, R., Kroutil, W., Faber, K. (2005). Biocatalytic racemization of aliphatic, arylaliphatic, and aromatic α-hydroxycarboxylic acids. *J. Org. Chem.*, *70*, 4028–4032.

[104] Wu, B., Szymanski, W., Wietzes, P., de Wildeman, S., Poelarends, G. J., Feringa, B. L., Janssen, D. B. (2009). Enzymatic synthesis of enantiopure α- and β-amino acids by phenylalanine aminomutase-catalyzed amination of cinnamic acid derivatives. *ChemBioChem*, *10*, 338–344.

[105] Szymanski, W., Wu, B., Weiner, B., de Wildeman, S., Feringa, B. L., Janssen, D. B. (2009). Phenylalanine aminomutase-catalyzed addition of ammonia to substituted cinnamic acids: a route to enantiopure α- and β-amino acids. *J. Org. Chem.*, *74*, 9152–9157.

[106] Hansch, H., Leo, A., Taft, R. W. (1971). A survey of Hammett substituent constants and resonance and field parameters. *Chem. Rev.*, *91*, 165–195.

CHAPTER 13

PEPTIDES FOR ASYMMETRIC CATALYSIS

MATTHIAS FREUND AND SVETLANA B. TSOGOEVA

13.1. Introduction 530
13.2. Cyanation of aldehydes and Strecker reaction 530
13.3. Peptide-catalyzed asymmetric 1,4-conjugate addition reactions 532
13.3.1. *N*-alkyl imidazole-based peptides 532
13.3.2. *N*-terminal prolyl peptides 534
13.3.3. *N*-terminal primary amino peptides 541
13.4. Peptide-catalyzed asymmetric aldol reactions 543
13.4.1. *N*-terminal prolyl peptides 543
13.4.2. *N*-terminal primary amino peptides 549
13.5. Peptide-catalyzed asymmetric Morita–Baylis–Hillman reactions 553
13.6. Peptide-catalyzed Stetter reaction 555
13.7. Peptide-catalyzed regioselective acylation reactions 555
13.8. Peptide-catalyzed asymmetric α-functionalizations 557
13.9. Peptide-catalyzed desymmetrization reaction 559
13.10. Peptide-catalyzed kinetic resolutions 560
13.11. Peptide-catalyzed asymmetric protonation reactions 570
13.12. Peptide-catalyzed asymmetric transfer hydrogenation reactions 571
13.13. Summary 573
References 573

Catalytic Methods in Asymmetric Synthesis: Advanced Materials, Techniques, and Applications, First Edition. Edited by Michelangelo Gruttadauria and Francesco Giacalone.

13.1. INTRODUCTION

The employment of short peptides and peptide-like molecules as catalysts with an enzyme-like character in asymmetric syntheses is a subject of considerable current interest and continues to receive increasing attention from chemists [1]. To understand the driving forces of this research one has to look at the main advantages of small peptide catalysts with respect to enzymes as outlined in Figure 13.1. Features such as straightforward accessibility from nature's toolbox and modularity render peptides excellent asymmetric catalysts for different important transformations and attractive alternatives to other organocatalysts [2]. The structural diversity available with short peptide sequences makes this class of molecules particularly promising for the development of a broad spectrum of small peptide catalysts that mimic various qualities of enzymes. In addition, the structural simplicity of the short peptides contrasts with the complexity of the enzymes, making the mechanistic investigations easier. Furthermore, it is easy to prepare the peptide sequence that can produce the opposite enantiomer.

In this chapter, representative examples of peptide catalysts are described. The material is ordered according to the type of reaction catalyzed.

13.2. CYANATION OF ALDEHYDES AND STRECKER REACTION

The first examples of asymmetric peptide catalysis emerged in the 1980s. An important milestone for peptide-catalyzed asymmetric reactions in general was reported by Inoue and co-workers in 1981 and 1982 in the context of an asymmetric cyanation of benzaldehyde to a cyanohydrine [3, 4]. In the pres-

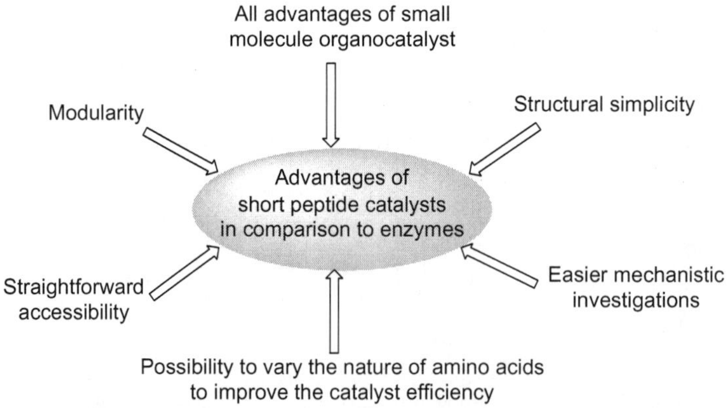

FIGURE 13.1. Main advantages of small peptide catalysts with respect to enzymes.

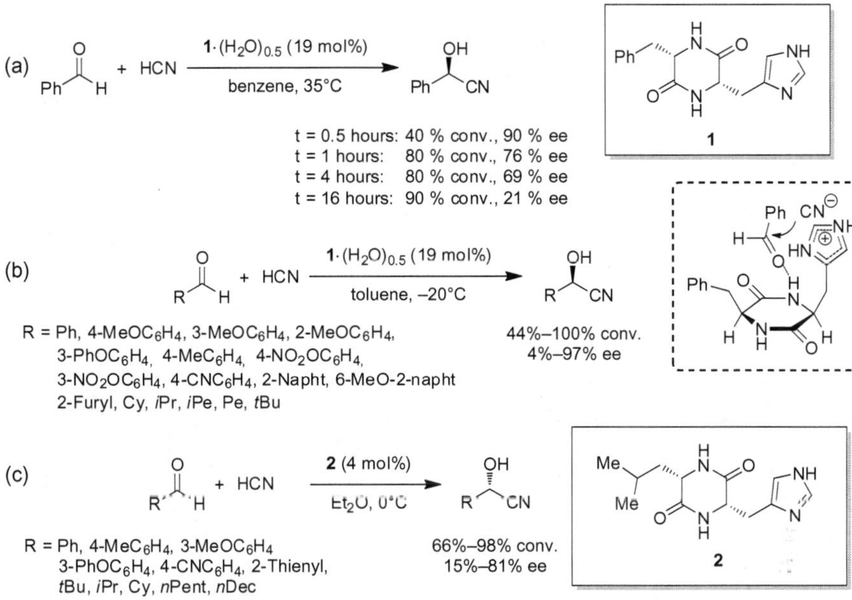

SCHEME 13.1. Asymmetric cyanation of aldehydes.

ence of the cyclic dipeptide *cyclo*[(*S*)-His-(*S*)-Phe] (**1**, Scheme 13.1a) as its hydrate form, a remarkably high enantiomeric excess (ee) of 90% was measured in the early stage of this reaction but decreased with ongoing reaction because of racemization. This, in fact, was the first reported case of a highly enantioselective catalysis by small peptides. Regarding the mechanism, the imidazole in the catalyst was protonated by hydrogen cyanide (HCN), and the cyanide ion is transferred to the aldehyde, which is also located at the dipeptides by hydrogen bonding (Scheme 13.1b).

Encouraged by these results, Inoue and co-workers investigated the aldehyde scope of this catalyst at lower temperatures to prevent racemization of the cyanohydrines [5]. A variety of aromatic, heteroaromatic, and aliphatic aldehydes was converted to the cyanated products with 44%–100% conversion and 4%–97% ee.

During this time, the same group reported another cyclic dipeptide, *cyclo*[(*S*)-His-(*S*)-Leu] (**2**, Scheme 13.1c), as the catalyst for this type of reaction [6]. Various aromatic and aliphatic aldehydes were cyanated to cyanhydrines with 66%–98% conversion and 15%–81% ee. It is worth noting that **2** furnished the opposite enantiomer in comparison to **1**.

In the 1990s, Lipton and co-workers also became aware of such cyclic structures and applied the closely related dipeptide **3** for asymmetric Strecker reactions of aromatic and aliphatic *N*-benzhydryl imines (Scheme 13.2), furnishing the products with 71%–97% conversions and <10%–>99% ee [7].

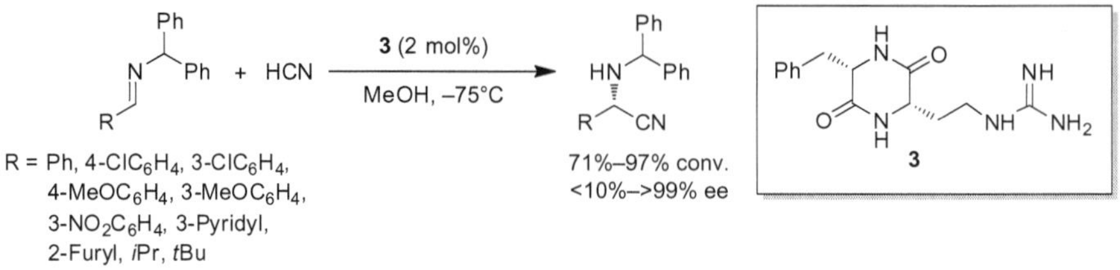

SCHEME 13.2. Asymmetric Strecker reactions of aromatic and aliphatic *N*-benzhydryl imines.

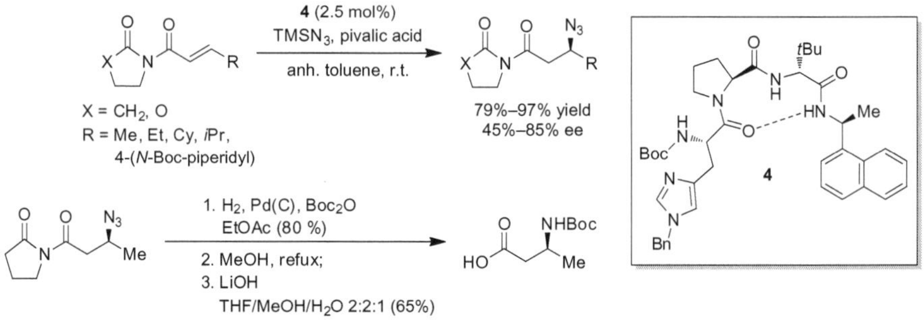

SCHEME 13.3. Asymmetric conjugate addition of azide to enoates catalyzed by tripeptide **4**.

13.3. PEPTIDE-CATALYZED ASYMMETRIC 1,4-CONJUGATE ADDITION REACTIONS

13.3.1. *N*-Alkyl Imidazole-Based Peptides

The asymmetric conjugate addition of azide to enoates is an important key step in a synthetic pathway to chiral enantiopure β-amino acids. Such unnatural building blocks are often incorporated in peptidic drugs in order to circumvent their degradation in biological systems.

Miller and co-workers, who employed small peptides in kinetic resolution studies of racemic alcohols, expanded the application of such catalysts to this area of research [8]. A survey of various tripeptides revealed the catalyst structure **4** (Scheme 13.3), which promoted the azidation of pyrrolidinone-derived imides with $TMSN_3$ in 79%–97% yields and with 45%–85% ee. In the enoate, γ-branching provided the conversion to the β-azido products with better enantioselectivity than the unbranched substrates. An oxazolidinone ring instead of pyrrolidinone was found to decrease the reactivity and the enantioselectivity. To demonstrate the synthetic value of such β-azido alkanoates, one of the catalytic products was converted to the *N*-Boc-protected β-amino acid.

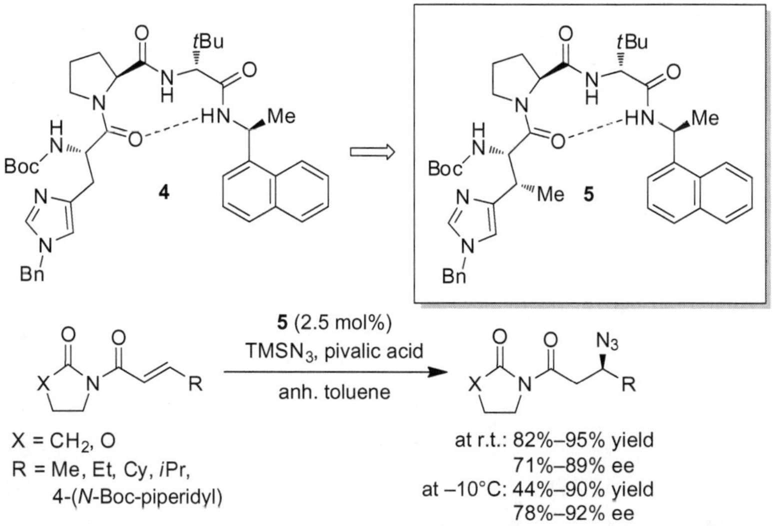

SCHEME 13.4. Asymmetric conjugate addition of azide to enoates catalyzed by tripeptide **5**.

Two years later, Miller et al. was further improving their original peptide-catalyzed azidation reaction [8] by introducing a β-branch in the histidine part of the catalyst (Scheme 13.4) [9]. In particular, a methyl group in the β-position was found to increase the enantioselectivity of the azidation reaction, especially for pyrrolidinone substrates, which were unbranched in the γ-position. The respective products were isolated in 82%–95% yield and with 71%–89% ee.

Further improvement of enantioselectivity was achieved by decreasing the reaction temperature from ambient to −10°C, albeit at cost of some reactivity, and the β-azido alkanoates were obtained in 44%–90% yield and with 78%–92% ee.

Besides the previously mentioned synthetic application of such β-azido alkanoates in the preparation of β-amino acids, another potential transformation of azides was studied: the 1,3-dipolar cycloaddition with dienophilic alkines and alkenes, a pathway to enantioenriched triazoles (Scheme 13.5).

Linton et al. [10] investigated peptides in enantioselective Michael additions of α-nitro ketones and esters to α,β-unsaturated ketones. A screening survey revealed the structural motif **6** (Scheme 13.6) with the Brønsted basic *N*-benzyl imidazole as an ultimate requirement not only for catalytic activity, but also for stereoselectivity, as the basic side chain operates in the chiral environment of the peptide. To probe the qualities and the limitations of this catalyst, a series of α-nitro-substituted ketones and esters were converted with α,β-unsaturated ketones to the Michael products in rather heterogeneous results, as 29%–99% yields and 0%–74% ee were obtained.

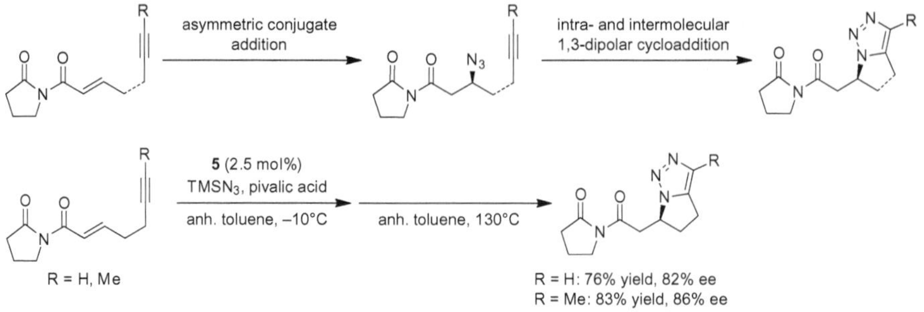

SCHEME 13.5. Enantioenriched triazoles via 1,3-dipolar cycloaddition of *β*-azido alkanoates.

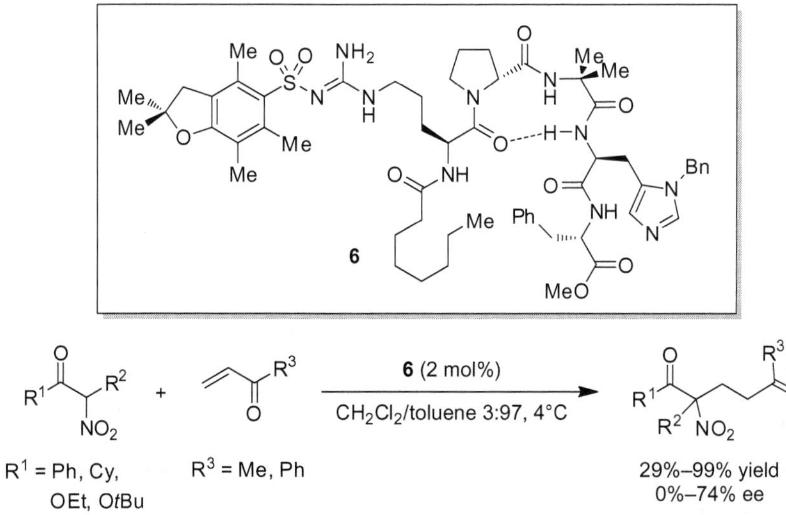

SCHEME 13.6. Enantioselective Michael addition reactions of *α*-nitro ketones and esters catalyzed by **6**.

By comparing **6** with the achiral and structurally much simpler *N*-methyl imidazole in kinetic studies, a participation of the peptide side chains (especially the sulfonated guanidin function) seems to be involved, as the peptide catalyst was promoting the model reaction significantly stronger. Based on this observation, a transition state arrangement was proposed (Fig. 13.2).

13.3.2. *N*-Terminal Prolyl Peptides

The products of 1,4-addition of nitroalkanes to *α,β*-unsaturated enones are useful precursors for a variety of structures such as aminocarbonyl compounds, aminoalkanes, and pyrrolidines. Efforts toward achieving asymmetric conju-

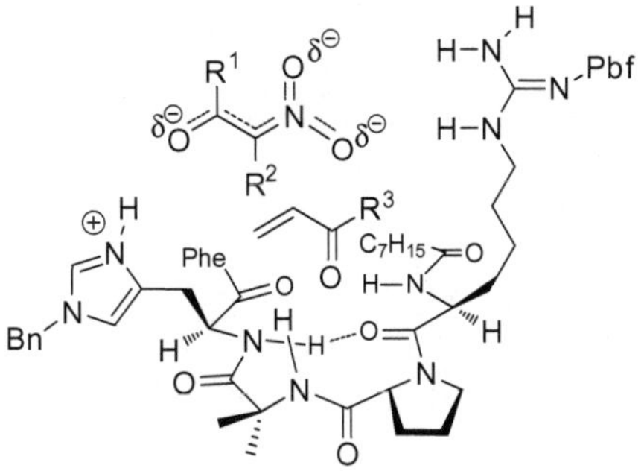

FIGURE 13.2. Transition state model proposed for the Michael addition reaction catalyzed by **6**.

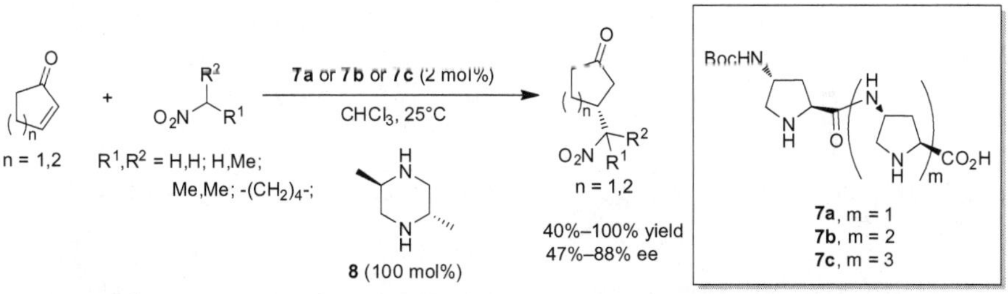

SCHEME 13.7. Asymmetric 1,4-conjugate addition reactions catalyzed by di-, tri-, and tetrapeptides **7a–c**.

gate addition of nitroalkanes to *α,β*-unsaturated ketones have been the subject of several recent reports.

Hanessian and Pham [11] first described the catalytic asymmetric conjugate addition of various nitroalkanes to cyclic enones in the presence of L-proline as a catalyst and *trans*-2,5-dimethylpiperazine (**8**) as an additive.

In a study by Tsogoeva et al., short peptides based on *trans*-4-amino-L-proline were the subject of studies concerning this asymmetric 1,4-conjugate addition reaction [12, 13]. The di-, tri- and tetrapeptides **7a–c** (Scheme 13.7) were investigated in the addition of different nitroalkanes to cyclic *α,β*-unsaturated ketones in the presence of achiral amine base **8** as an additive. Two 4-*trans*-amino-proline residues were shown to be sufficient enough to catalyze the conjugate addition reactions with up to 88% ee and up to 100% yield [13].

In a study by Palomo et al., two *trans*-4-hydroxy-L-proline amides (**9**, **10**) were introduced in the asymmetric 1,4-conjugate addition of aldehydes to aromatic and aliphatic nitroalkenes (Scheme 13.8) [14]. The authors also showed the synthetic efficiency of their methodology as they converted two of their catalytic products to *γ*-butyrolactones. This organocatalytic motif influences the arrangement of the reactants in two ways. On the one hand, the

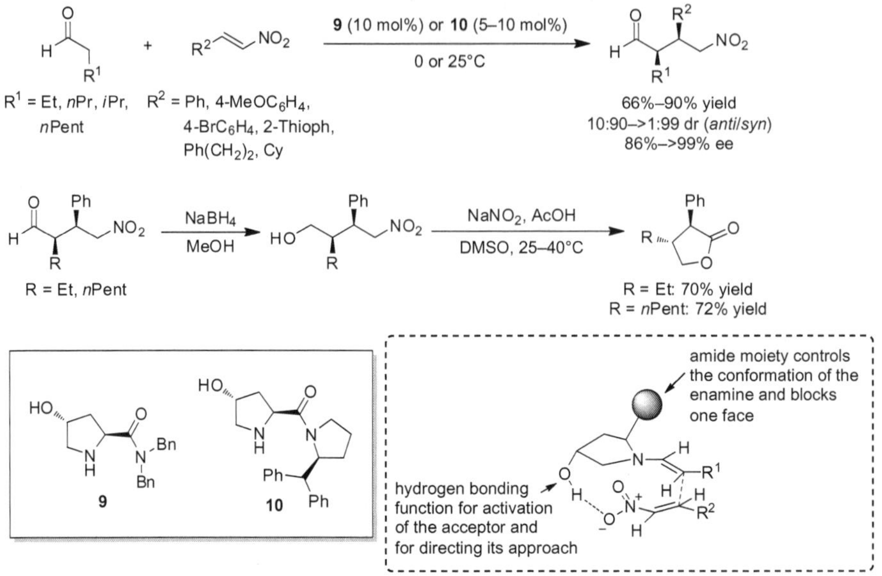

SCHEME 13.8. Asymmetric Michael addition reactions catalyzed by L-proline derivatives **9** and **10**.

bulky amide function controls the conformation of the enamine, while on the other, the hydroxy function directs the approach by the acceptor from the less hindered side.

The tripeptide **11a** (Scheme 13.9) was designed and applied by Wennemers and co-workers in asymmetric Michael additions of aldehydes to nitroolefins [15]. A range of aliphatic aldehydes was converted with both aliphatic and aromatic nitroalkenes to Michael products in up to quantitative yields and in 4:1–>99:1 diastereomeric ratio (dr) (*syn*/*anti*) and in 81–>99% ee. The absolute configuration of the amino-terminal proline was found to dictate the stereochemical outcome, as the diastereomer of **11a** (tripeptide **11b**) produced the opposite configuration in the Michael adduct. This observation can be rationalized by the generally well-accepted mechanism for this type of reaction (Scheme 13.9), as the *N*-terminal proline moiety is responsible for the enamine activation. The acceptor is approaching the donor in a chiral environment, which is influenced largely by the absolute configuration at the *N*-terminus.

To further understand this behavior, both possible transition state arrangements for the *syn* stereoisomers were subjected to molecular modeling (Fig. 13.3). Wennemers and co-workers focused particularly on nitroethylene as a Michael acceptor employed in an interesting, alternative pathway to γ^2-amino acids (Scheme 13.10a) [16].

Initially, an exchange of the L-asparagine amide in **11a** by L-glutamic acid amide resulted in a more soluble tripetide (**12**) (Scheme 13.10c,d), which showed slightly better results in this context. A set of aldehydes was trans-

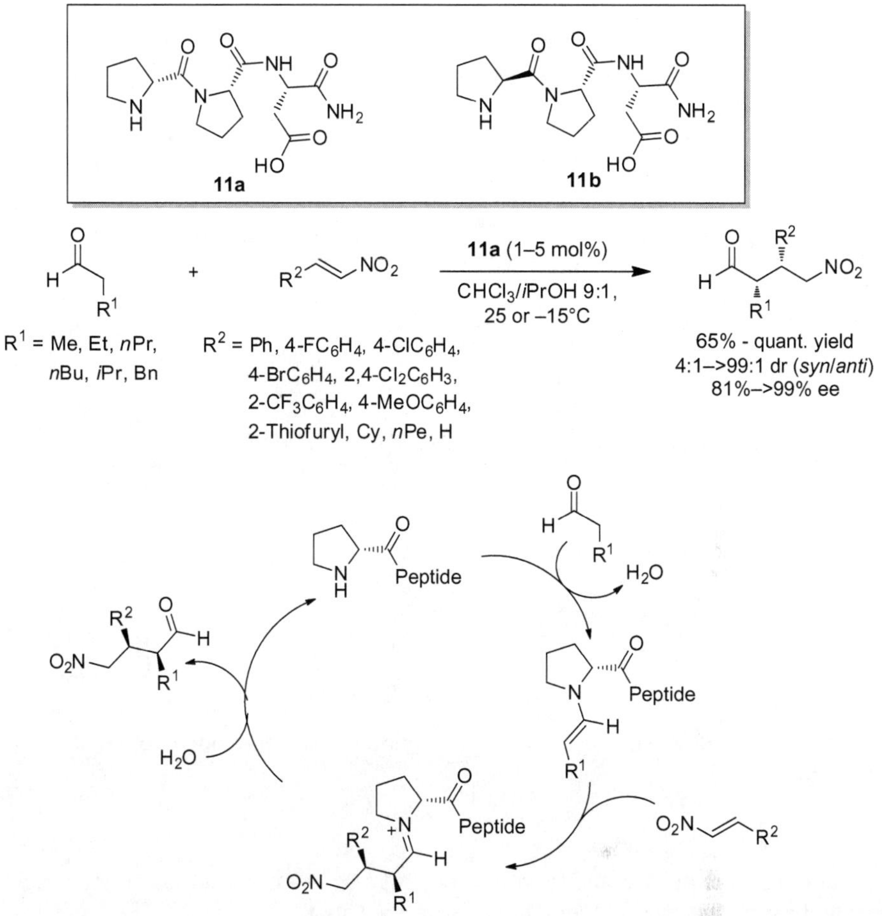

SCHEME 13.9. Asymmetric Michael addition reactions catalyzed by tripeptide **11a**.

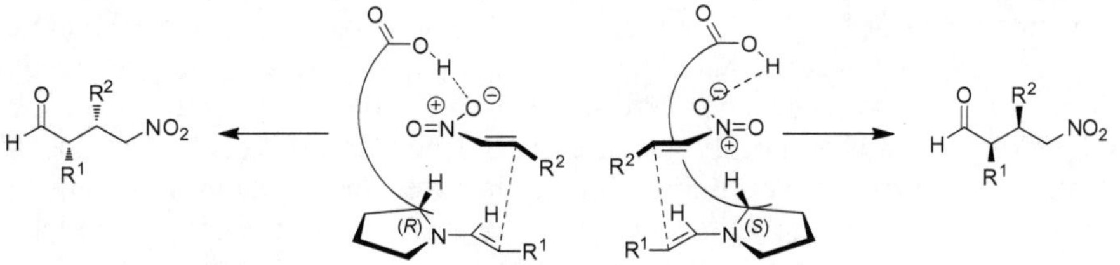

FIGURE 13.3. Possible transition state arrangements for the *syn* stereoisomers.

formed to γ-nitro aldehydes, which were reduced in situ to alcohols to circumvent racemization.

In order to achieve their original goal—the enantioselective preparation of γ^2-amino acids—an exemplary procedure was also reported (Scheme 13.10a).

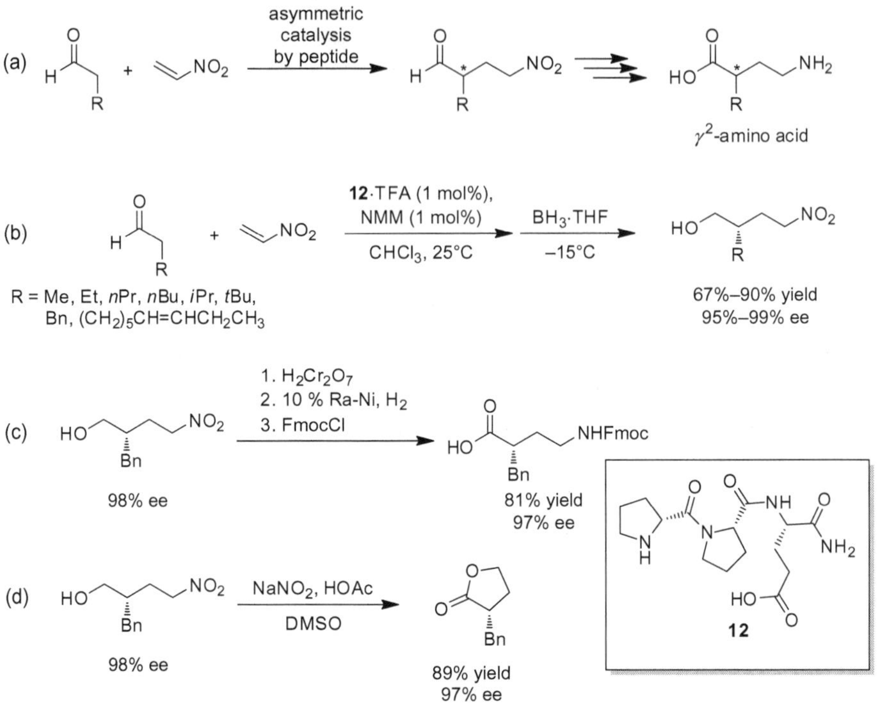

SCHEME 13.10. Asymmetric Michael addition reactions on nitroethylene as acceptor.

Not only amino acids could be prepared, but also γ-butyrolactones were accessible. In both cases, no racemization was observed.

Wennemers and co-workers characterized the structural requirements of their tripeptidic catalysts based on *N*-terminal H-D-Pro-L-Pro-OH, as such catalyst motifs were presented recently [15–18] as effective promoters of asymmetric aldol and Michael reactions. For H-D-Pro-L-Pro-L-Asp-NH_2 (**11a**, Fig. 13.4), a single-crystal X-ray structure analysis was presented, in which an intramolecular hydrogen bond between the primary amide and one of the peptide bonds was found. By this interaction, the catalyst is conformationally restricted to a β-turn. This finding was further solidified by catalysis results of analogous structures to **11a**, which lack the ability to form such a hydrogen bond. Besides this result, the importance of the carboxylic function was shown by screening derivatives of **11a**, in which the acid function was either amidated or esterified. Finally, by homologation of the asparaginic acid side chain in **11a**, a remarkable degree of flexibility could be achieved without decreasing the catalyst's qualities.

In 2009, Tsogoeva et al. described a first study of unmodified proline-based di- and tripeptides as enantioselective catalysts for Michael additions of

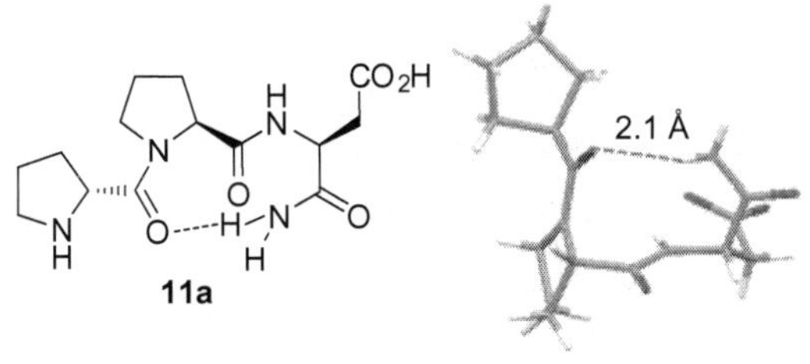

FIGURE 13.4. Single-crystal X-ray structure analysis of H-D-Pro-L-Pro-L-Asp-NH_2 **11a**.

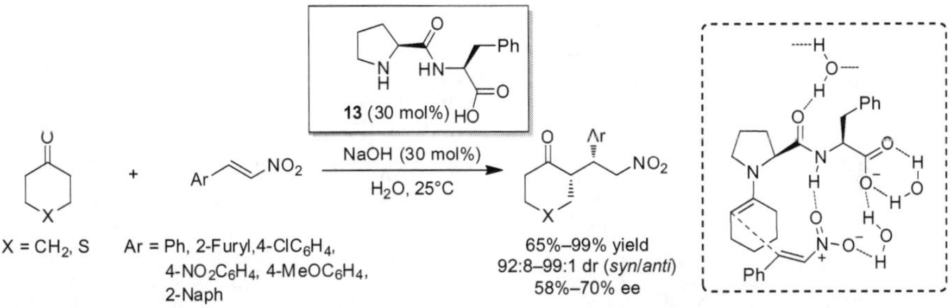

SCHEME 13.11. Asymmetric Michael addition reactions catalyzed by L-proline derivative **13**.

ketones to nitrostyrenes on water and without any organic cosolvents, providing access to valuable building blocks such as γ-nitroketones (Scheme 13.11) [19]. It was found that a base (e.g., NaOH) is necessary for catalytic activity of unmodified dipeptides on water; in its absence, no conversion was observed. Additionally, acidic additives like acetic acid were unsuitable. A series of aromatic nitroalkenes was converted with six-membered cyclic ketones to Michael products in 65%–99% yield, with 92:8-99:1 dr (*syn*/*anti*) and in good enantioselectivity (up to 70% ee) in the presence of H-Pro-Phe-OH (**13**).

Kudo and co-workers used the methodology of asymmetric iminium activation in the context of stereoselective conjugate 1,4-additions of indoles to α,β-unsaturated aldehydes in an aqueous system [20]. As a catalytic motif, the authors were using a solid supported peptide (**14**, Scheme 13.12). In the presence of water, which had a beneficial effect on the reaction rate, a series of unsaturated aldehydes was converted with *N*-methyl-indole and indole to the respective adducts and was reduced in situ with 44%–88% yield and 52%–94% ee. The resin, loaded with the catalyst, was recycled up to five times, without losing its catalyzing properties.

Here the mechanism involves at first the acid-catalyzed formation of an iminium intermediate (Fig. 13.5a), which is approached by the indole from the less-hindered *Si*-face of the C=C double bond. In this step, the Brønsted acidic

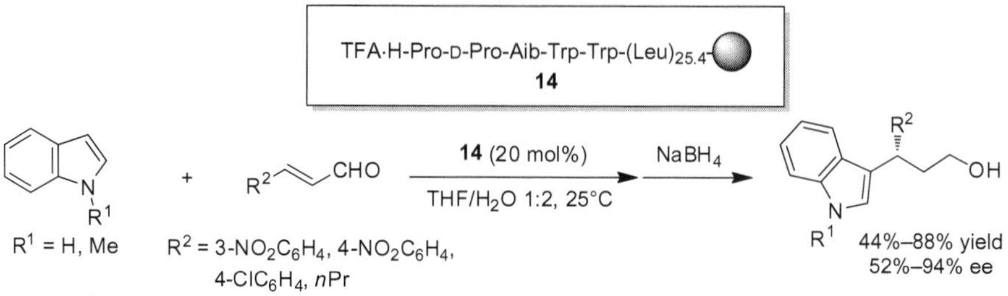

SCHEME 13.12. Asymmetric conjugate 1,4-additions of indoles to α,β-unsaturated aldehydes.

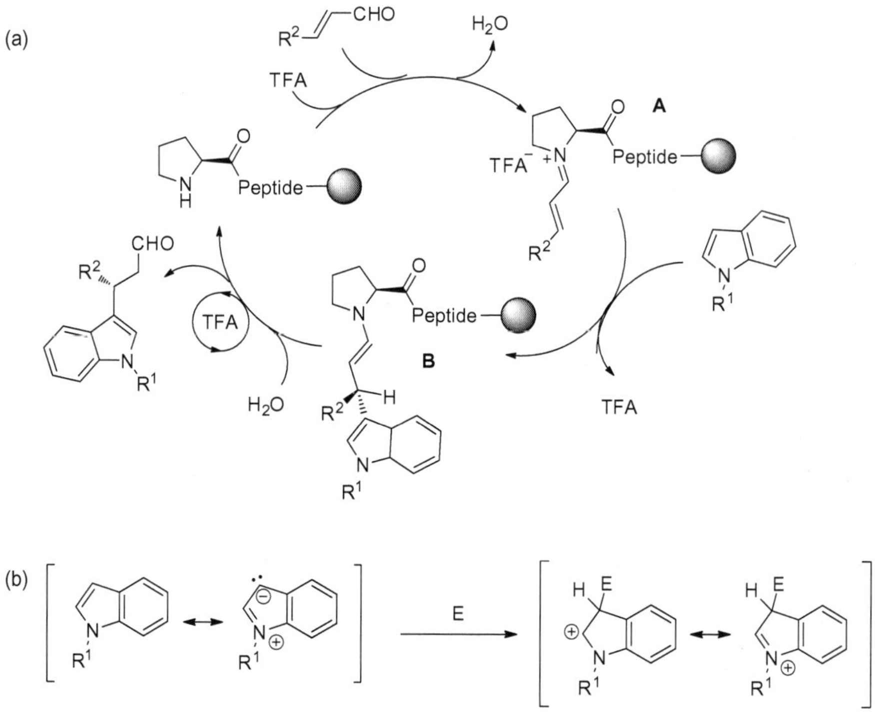

FIGURE 13.5. Proposed mechanism for the asymmetric conjugate 1,4-additions of indoles to α,β-unsaturated aldehydes.

cocatalyst is regenerated and further used in the following hydrolysis of the enamine **B** (Fig. 13.5a) to the product and the restored catalyst.

Indoles, which show aromatic properties over both cycles, can be also considered as quite active enamines, with their 3-position being of pronounced nucleophilicity. This behavior can be described by resonance structures: During electrophilic substitutions (E = electrophile, Fig. 13.5b), the aromatic system of the benzene ring stays intact, while the positive charge is still delocalized and therefore stabilized.

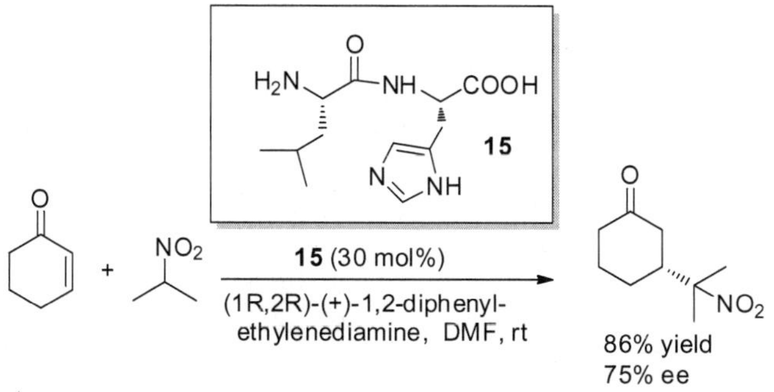

SCHEME 13.13. Asymmetric addition of 2-nitropropane to cyclohexanone catalyzed by *N*-terminal primary amino dipeptide **15**.

13.3.3. *N*-Terminal Primary Amino Peptides

The use of chiral primary amines as organocatalysts has particular appeal because of their known occurrence in the catalytic sites of several enzymes, such as type I aldolases, decarboxylases, and dehydratases [21]. In the area of organocatalysis, chiral primary amines have recently emerged as new and powerful catalysts for many important organic transformations [22].

The addition of nitroalkanes to α,β-unsaturated carbonyl compounds has become an area of intense interest for investigations involving asymmetric catalysis. In 2004, Tsogoeva and co-workers became interested in the possibility of *N*-terminal primary amino peptides as potential catalysts for the addition of 2-nitropropane to cyclohexenone (Scheme 13.13) [23]. After examination of various *N*-terminal primary amino dipeptides along with chiral amine additives, H-Leu-His-OH (**15**), in combination with (1*R*,2*R*)-1,2-diphenylethylenediamine (0.3 eq.) as a cocatalyst, was found to be a new catalytic system that provided the product in 86% yield and 75% ee. This constituted the first example of *N*-terminal primary amine-based unmodified dipeptide being used in catalytic asymmetric 1,4-conjugate addition reactions; previously, only proline and proline-based catalysts were thought capable of this chemical feat.

This finding stimulated work of other groups employing primary amine-based amino acids and dipeptides as catalysts in other 1,4-conjugate addition reactions.

Córdova and co-workers reported that the dipeptide **16a** and its diastereomer **16b** (Scheme 13.14) are effective catalysts for the asymmetric 1,4-conjugate addition of ketones and aldehydes to nitroalkenes [24]. A set of aromatic nitroalkenes and cyclic, as well as some acyclic ketones and aldehydes were converted to the respective products in 30%–95% yield, 1:2–36:1 dr (*syn*/*anti*), and 29%–98% ee of the *syn* product. The addition of a rather large amount of water was found to have positive effect on the catalysis.

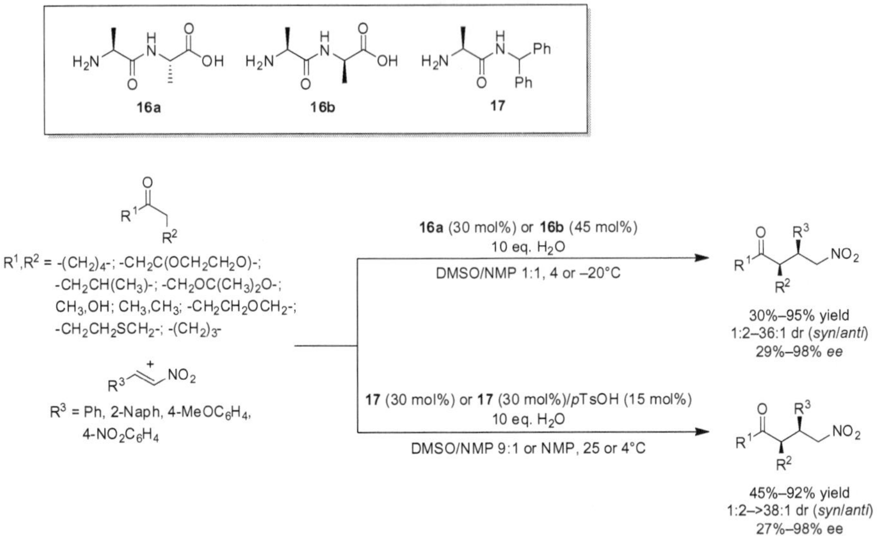

SCHEME 13.14. Asymmetric 1,4-conjugate addition of ketones and aldehydes to nitroalkenes catalyzed by dipeptides **16** and **17**.

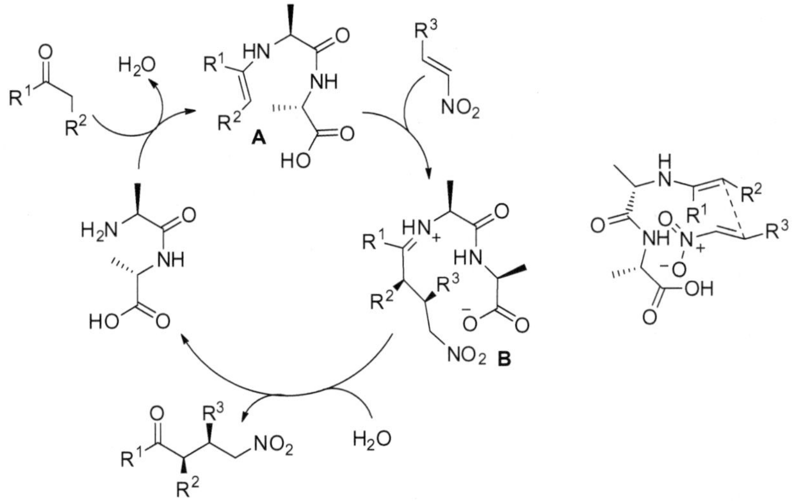

FIGURE 13.6. Proposed mechanism for the Michael reactions catalyzed by dipeptides **16** and **17**.

In the course of their explanations, the authors also gave insights into a proposed mechanism (Fig. 13.6), with an enamine **A** and an iminium ion **B** playing central roles.

In a similar fashion and with comparable catalytic qualities [45%–92% yield, 1:2–>38:1 dr (*syn*/*anti*) and 27–9% ee (*syn*)], the L-alanine amide **17** was employed as an organocatalyst in this reaction, again in the presence of water [25].

13.4. PEPTIDE-CATALYZED ASYMMETRIC ALDOL REACTIONS

13.4.1. *N*-Terminal Prolyl Peptides

A different vista has been opened up by some recent contributions from several researchers who discovered that L-proline amides and L-proline-based short peptides acted as efficient catalysts for asymmetric direct aldol reactions [26–28]. The first such successful example of using L-proline amino alcohol amides as catalysts for highly enantioselective reactions of aldehydes with acetone has been reported by Gong et al. [26, 29]. With L-prolinamide **18**, prepared from L-proline and (1*S*,2*S*)-diphenyl-2-aminoethanol, high yields (up to 93%) and enantioselectivities of up to 93% ee for aromatic aldehydes and up to >99% ee for aliphatic aldehydes under −25°C (Scheme 13.15) were obtained.

The enantioselectivities observed for both aromatic and aliphatic aldehydes were higher [26] than those attained by using L-proline as the catalyst [30]. Additionally, the theoretical studies performed on the transition state structures disclosed that the amide N-H and the terminal hydroxyl group could form hydrogen bonds with the benzaldehyde substrate, thereby reducing the activation energy and causing high enantioselectivity. These results suggested a new approach in the design of new organic catalysts for direct asymmetric aldol reactions and motivated researchers for further investigations. Based on these observations and results obtained with L-proline-based dipeptides [27], Gong et al. next anticipated that larger L-proline-based peptides might also be useful as organic catalysts for the direct aldol reaction, not only because of their structural similarity to L-proline amides but also because they contain more amide units, which are the same building blocks that constitute enzymes [31]. An examination of methyl ester-protected di-, tri-, tetra-, and pentapeptides has shown that the increase in the peptide size led to an increase in ee's.

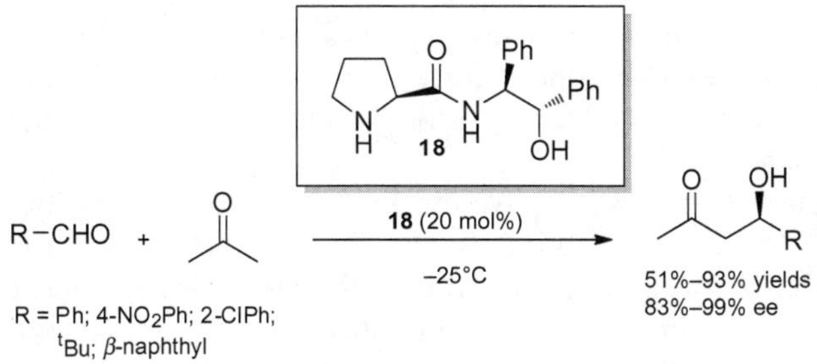

SCHEME 13.15. Asymmetric aldol reactions catalyzed by L-prolinamide **18**.

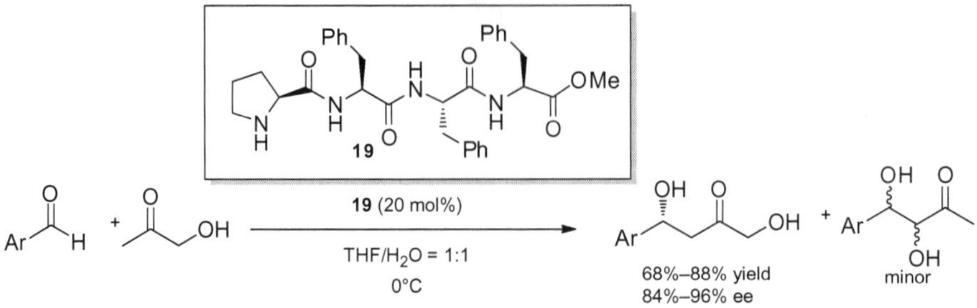

SCHEME 13.16. Asymmetric aldol reactions catalyzed by tetrapeptide **19**.

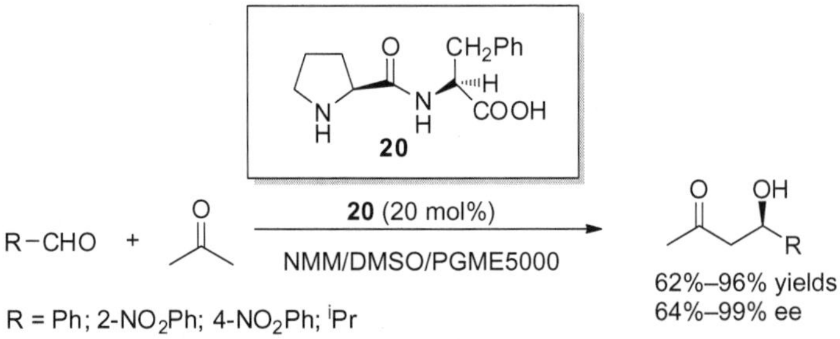

SCHEME 13.17. Asymmetric aldol reactions catalyzed by dipeptide **20**.

These results demonstrated that the peptide size also played an important role in the stereo- and regiocontrol. L-Proline-based tetra- and pentapeptides (H-Pro-Phe-Phe-Phe-OMe and H-Pro-Phe-Phe-Phe-Phe-OMe) have thus been developed as efficient catalysts for the asymmetric direct aldol reactions of hydroxyacetone with aldehydes. Chiral 1,4-diols, which are disfavored products in similar aldol reactions catalyzed by L-proline, were obtained in high yields and enantioselectivities with tetrapeptide **19** in aqueous media (Scheme 13.16).

As an extension of the work published by Martin and List [27], unprotected dipeptide H-Pro-Phe-OH (**20**) was recently shown by Li and co-workers [32] to be an efficient catalyst for direct asymmetric aldol reactions between acetone and various aldehydes in a dimethyl sulfoxide (DMSO)-N-methylmorpholine (NMM)-polyethylene glycol monomethyl ether (PGME) 5000 system at 0°C and gave the aldol product in high yields and with up to 99% ee (Scheme 13.17).

In previous results found by Wennemers et al. [33], the tripeptide H-L-Pro-L-Pro-L-Asp-NH_2 was identified as an effective catalyst for organocatalytic aldol reactions between acetone and aldehydes. Based on this, the catalytic system was further improved by immobilization of the tripeptide on a solid support [34]. A series of different resins was investigated and covalent binding of the tripeptide to TentaGel (**21**, Scheme 13.18) was chosen as the most suit-

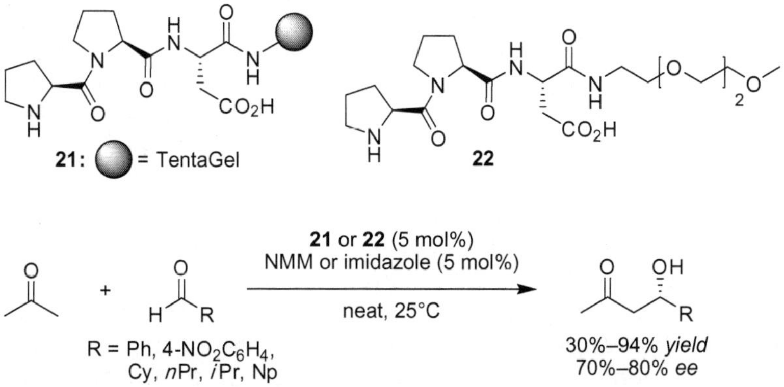

SCHEME 13.18. Asymmetric aldol reactions catalyzed by supported tripeptides **21** and **22**.

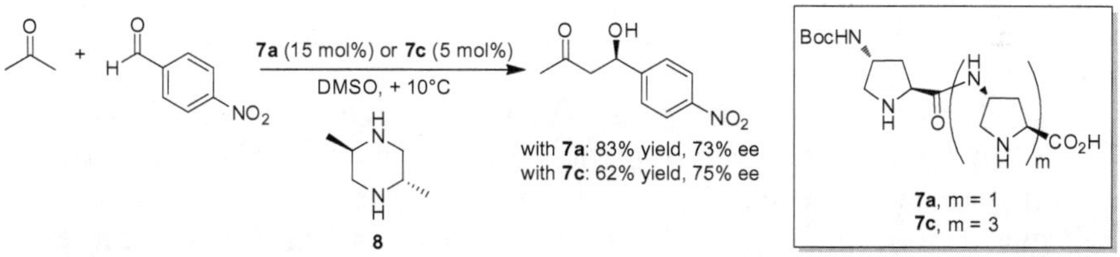

SCHEME 13.19. Asymmetric aldol reactions catalyzed by peptides **7a** and **7c**.

able support. Using this catalyst, a series of aromatic and aliphatic aldehydes was converted to the respective aldol adducts with acetone in yields ranging up to 94%, and in 70%–80% ee. The solid support of the catalyst enabled recycling up to three times, without affecting catalytic activity and stereoselectivity.

Another improvement of their original catalytic system involved poly(ethylene) glycolation of the C-terminus in H-L-Pro-L-Pro-L-Asp-NH_2 (**22**) in order to improve the solubility in organic solvents. Also in this case, a series of aldol products was isolated in comparable yields and in likewise stereoselectivities.

Two *trans*-4-amino-L-proline-based peptides **7a** and **7c** were further evaluated by Tsogoeva and co-workers [13] in the aldol reaction of acetone and an aromatic aldehyde (Scheme 13.19). While both structures were comparable in their enantioselectivity, the dipeptide **7a** was more reactive, resulting in a better yield of the aldol adduct. Interestingly, peptides **7a** and **7c** in 15 and 5 mol% loading, respectively, showed higher or similar yields (83% and 62%) and approximately the same enantioselectivities (73% and 75% ee) as L-proline (68% and 76% ee) at 30 mol% [30].

Proline-based dipeptide-catalyzed aldol reaction of acetone with several *N*-alkylated isatins was described by Tomasini et al. [35]. The desired compound was obtained in quantitative yield and with good enantioselectivities up to 77%. The best results were obtained with 10 mol% H-D-Pro-L-β^3-hPhg-OBn as a catalyst, resulting in the preferential formation of the (*R*)-enantiomer.

Peptides with prolyl N-termini, attached to a PEG–polystyrene (PEG-PS) (TentaGel, TG) synthesis resin, have been tested by Davis et al. as heterogeneous catalysts for the aldol reaction between acetone and *p*-nitrobenzaldehyde [36]. Proline directly attached to TG showed good activity but poor enantioselectivity. However, in combination with serine or threonine, the selectivity improved considerably. At −25°C, the dipeptide H-Pro-Ser-NH-TG yielded 82% ee.

The PEG-PS resin-supported proline-based tripeptide/zinc chloride catalyst system has been developed by Kudo et al. for use in the direct asymmetric aldol reaction of acetone with aldehydes in aqueous media [37]. The peptide catalyst could be separated from the reaction mixture by filtration and was reusable at least five times without significant change in its activity and selectivity.

Contributing a methodology for the enantioselective synthesis of tertiary alcohols by an aldol reaction of methyl ketones and α,β-unsaturated trifluoromethyl ketones, Liu and co-workers introduced the L-proline-derived sulfonylamide **23** as a catalyst [38]. After a preliminary screening phase of a set of L-proline derivatives, this structural motif was applied in the aldol reaction of various acceptors with small ketones, and the respective products were isolated in 76%–99% yields, 81%–95% ee (Scheme 13.20). A very good regioselectivity toward the 1,2-addition was observed. Quite common for enamine

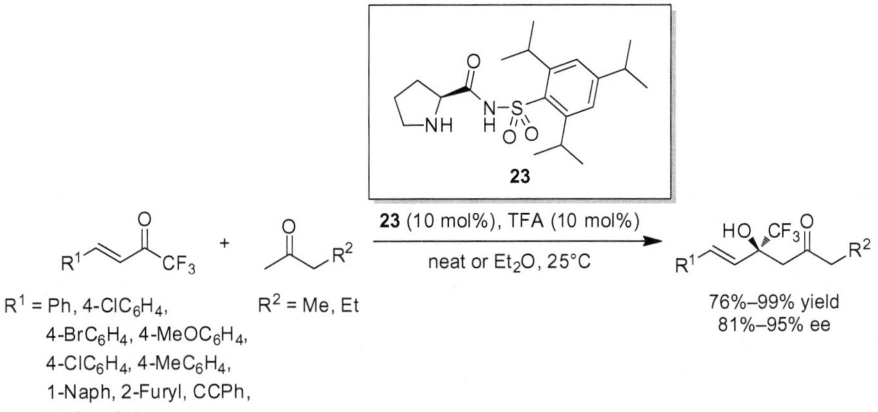

SCHEME 13.20. Asymmetric aldol reaction of methyl ketones and α,β-unsaturated trifluoromethyl ketones catalyzed by L-proline-derived sulfonylamide **23**.

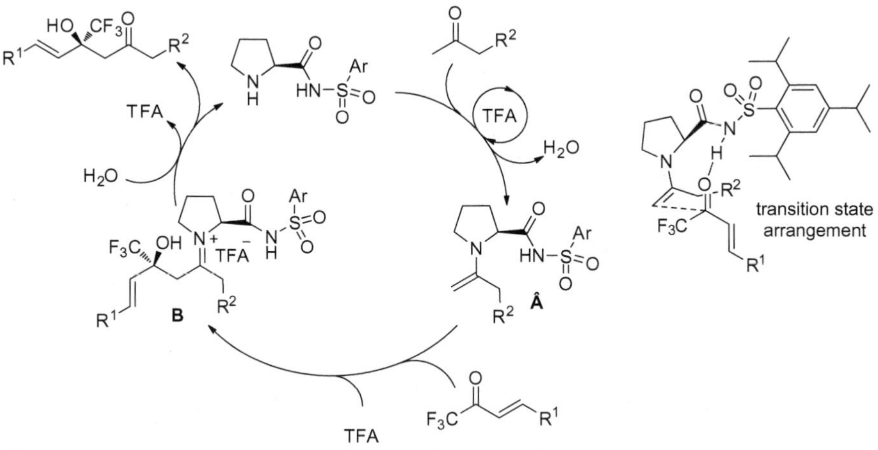

FIGURE 13.7. Proposed mechanism for the aldol reaction catalyzed by sulfonylamide **23**.

catalysis, the addition of a Brønsted acid such as trifluoroacetic acid had a positive effect on the stereoselectivity.

A rationalization for the observed stereochemistry was also presented by a proposed transition state and mechanism (Fig. 13.7). The enamine **A** intermediate is approached from the *Si*-side under direction by the sulfone amide. Here, a stabilization of the transition state by hydrogen bonding was proposed. The resulting iminium compound **B** is finally hydrolyzed to the product and the regenerated catalyst, again under acid catalysis.

In a study by Yan and Wang, a series of silica-supported peptides based on *cis*-4-amino L-proline and L-proline, respectively, was synthesized and evaluated in the aldol reaction of *p*-nitrobenzaldehyde and acetone [39]. As a privileged structure, **24** was identified, containing a dipeptide motif as its central active site. From various aromatic aldehydes, the aldol products with acetone were produced by this immobile promoter in 55%–96% yield and in 64%–96% ee (Scheme 13.21). The ease of recyclability was also demonstrated, as recovering the catalyst five times did not result in a decay of activity.

The unnatural building block β-aminocyclopropane α,γ-dicarboxylic acid (**25**) was incorporated in small peptides by Reiser and co-workers to evaluate the potential of such structures in asymmetric aldol reactions (Scheme 13.22) [40].

Basically, two promising catalyst motifs were identified from a series of screened small peptides (**26**, **27**) and applied in the aldol reaction of acetone with a set of aromatic and one aliphatic aldehyde. Interestingly, both enantiomers of the aldol products were accessible with these catalysts, with **27** being much more selective in the case of the aliphatic aldehyde. Furthermore, cyclic ketones were also inspected in this context, as was an intramolecular version of the aldol reaction (Scheme 13.23).

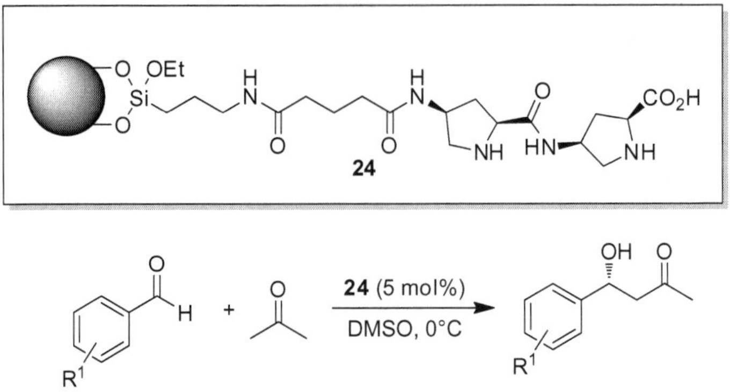

R^1 = 4-NO_2, 3-NO_2, 2-NO_2, 4-CF_3, 2-Cl-5-NO_2, 3-NO_2-4-Cl, 2-Cl, 4-Cl, 2-Br, 4-Br, 2,4-di-Cl, 2,6-di-Cl, 4-Me

55%–97% yield
64%–96% ee

SCHEME 13.21. Asymmetric aldol reactions catalyzed by silica-supported catalyst **24**.

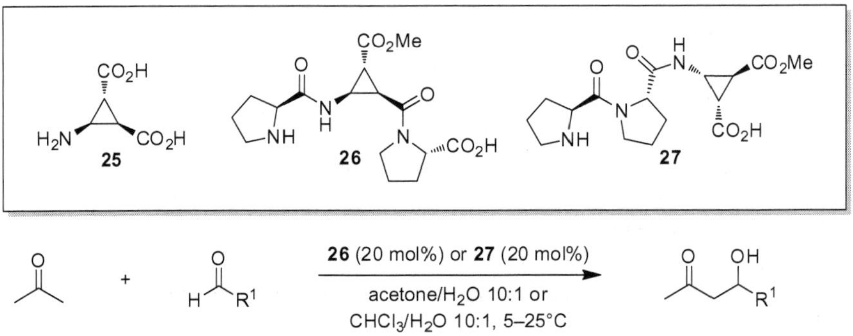

R^1 = Ph, 2-ClC_6H_4, 4-ClC_6H_4, 2-BrC_6H_4, 2-$NO_2C_6H_4$, 4-$NO_2C_6H_4$, Cy

with **26**: 43%–89% yield, 41%–91% ee (*R*)

with **27**: 42%–91% yield, 79%–88% ee (*S*)

SCHEME 13.22. Asymmetric aldol reactions catalyzed by L-proline derivatives **26** and **27**.

Miravet and co-workers developed a hydrogel-forming prolinyl-valine-based peptide (**28**, Scheme 13.24) for the asymmetric direct aldol reaction of cyclohexanone and *p*-nitrobenzaldehydes [41]. The hydrogel was prepared by sudden cooling of a hot aqueous solution of **28**, followed by sonification. Structure analyses (scanning electron microscopy, X-ray powder diffraction) revealed a staggered laminar supramolecular arrangement, in which hydrophobic and hydrogen-bonding interactions are expressed. With respect to catalysis, after optimization of the chosen aldol reaction, the product was iso-

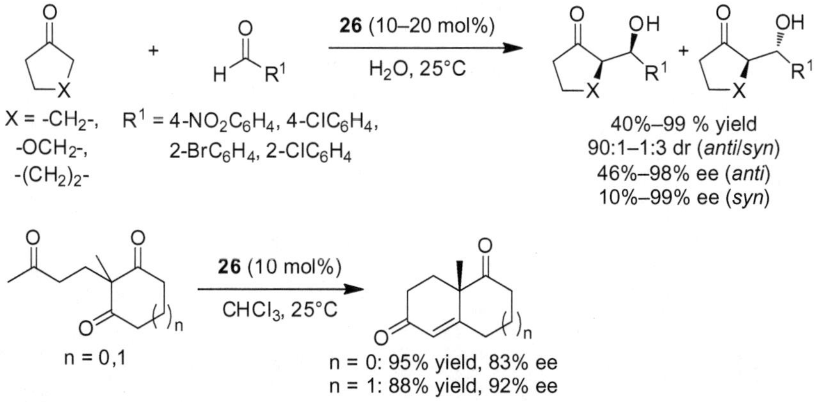

SCHEME 13.23. Asymmetric aldol reactions catalyzed by L-proline derivative **26**.

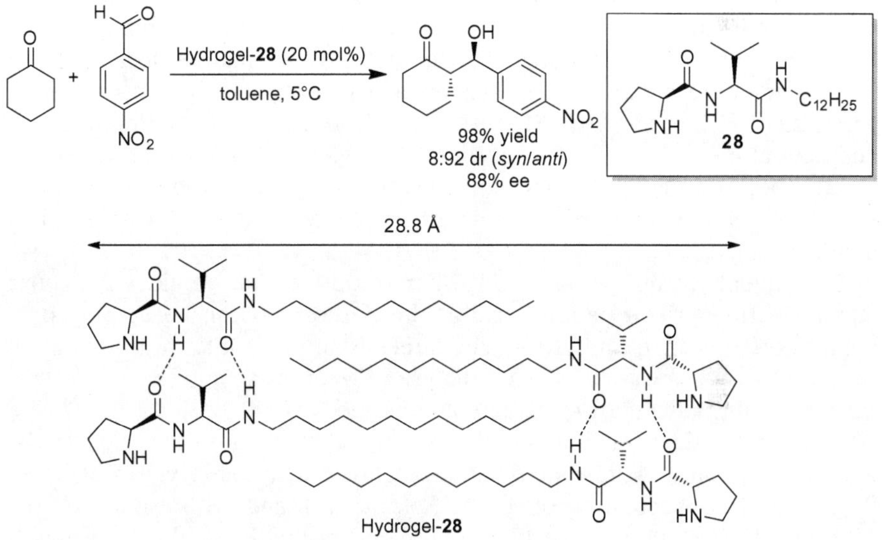

SCHEME 13.24. Asymmetric aldol reaction catalyzed by L-proline hydrogel derivative **28**.

lated in 98% yield, 8:92 dr (*syn*/*anti*) and 88% ee. The catalyst was recycled up to three times without any decreases in catalytic efficiency.

13.4.2. *N*-Terminal Primary Amino Peptides

The capability of different unmodified *N*-terminal primary amino dipeptides, containing histidine (e.g., H-His-Phe-OH, H-Phe-His-OH, H-Lys-His-OH, H-Leu-Phe-OH, H-Leu-His-OH, and H-His-Leu-OH) to act as effective

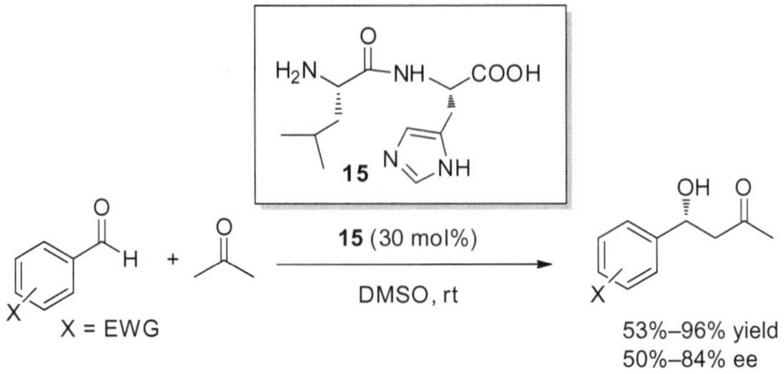

SCHEME 13.25. Asymmetric aldol reaction catalyzed by dipeptide **15**.

catalysts for aldol reactions of acetone with aromatic aldehydes, has been shown by Tsogoeva and co-workers [42]. The reactivities and stereoselectivities are shown to be dependent upon the intramolecular cooperation of side-chain functional groups and the presence of a suitable combination and sequence of amino acids. Good yields (up to 96%) and enantioselectivities (up to 76% ee) were obtained with electron-deficient aromatic aldehydes in the presence of H-Leu-His-OH (Scheme 13.25). The influence of different chiral and achiral cocatalysts on the reaction rates, yields, and enantioselectivities has also been evaluated. A significant increase in the rate of the reaction, most remarkably for achiral *trans*-2,5-dimethylpiperazine (**8**) as cocatalyst, was demonstrated.

Subsequently, Córdova and co-workers reported that peptides and their analogs with a primary amino acid at the *N*-terminus can be employed as highly stereoselective catalysts for the direct asymmetric intermolecular aldol reaction [43, 44]. It was shown that the use of water as an additive was necessary to obtain the highest levels of enantioselectivity.

A set of dipetides was found to catalyze the aldol reaction of various cyclic and acylic ketones with aromatic and aliphatic aldehydes in yields of 20%–93%, diastereoselectivities of 1:1–1:13 (*syn*/*anti*) and with 51%–99% ee (Scheme 13.26a) [44]. Regarding the mechanism, the authors proposed enamine catalysis (Scheme 13.26b) and a chair-like transition state (**A**). Also in this case, water plays a central role in the mechanism as its addition to the reaction mixture has positive influences. Finally, this group also observed the absence of nonlinear effects, as a correlation between the ee in the catalyst and the product was studied. This fact suggests the participation of one catalyst molecule in the stereoselectivity determining step.

Regarding a prebiotic enantioselective formation of sugars, Weber and Pizzarello used a water-based model in which glycolaldehyde was self-condensating to D-configured tetroses (threose, erythrose) under catalysis of homochiral L-configured dipetides [45]. A survey of various dipeptides showed, generally, that the catalyst only influences the enantioselectivity of erythrose; threose was more or less unaffected. Among the dipeptides tested, β-branched

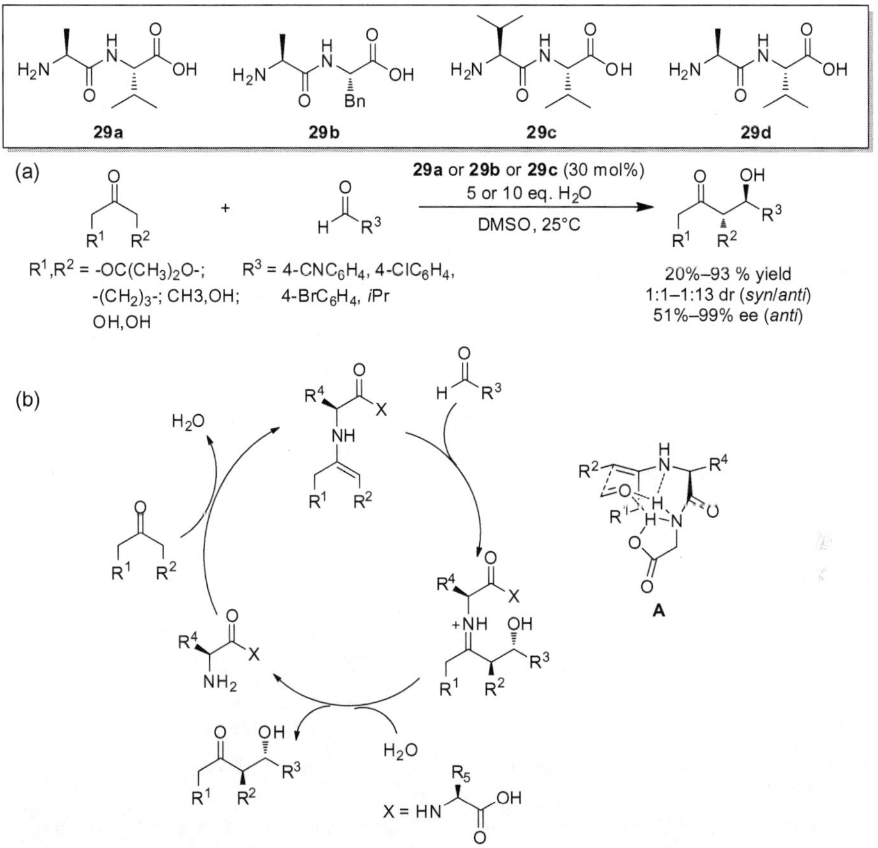

SCHEME 13.26. (a) Asymmetric aldol reactions catalyzed by dipeptides **29a–c**; (b) proposed mechanism for the above reaction.

structures like L-Val-L-Val (**30**) were the most useful catalysts, as 82% ee in the formation of erythrose was observed (Scheme 13.27).

Most published organocatalytic diastereoselective aldol reactions result in *anti* arrangement of both stereocenters; *syn* selectivity is observed more rarely. Finding more *syn* selective catalysts is still the subject of challenging research, and Gong et al. were pleased to contribute a new type of aldol catalyst **31** (Scheme 13.28) [46]. Besides hydroxy acetone, other small ketones were converted—with both aromatic and aliphatic aldehydes—to aldol adducts in 45% and 97% yields. The observed diastereoselectivities were between 10:1 and more than 20:1 (*syn*/*anti*); the enantioselectivities of the *syn* products ranged between 80% and 99% ee.

According to the authors, the *syn* selectivity is a result of a decreased steric repulsion, compared to another established pyrrolidine-containing catalyst, which results in *anti* products (Fig. 13.8).

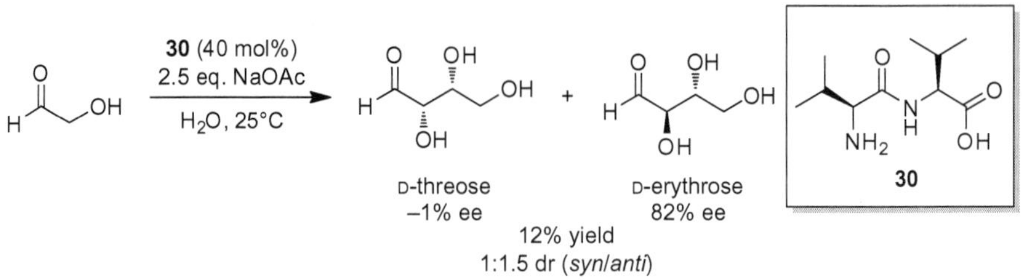

SCHEME 13.27. Asymmetric erythrose synthesis catalyzed by dipeptide **30**.

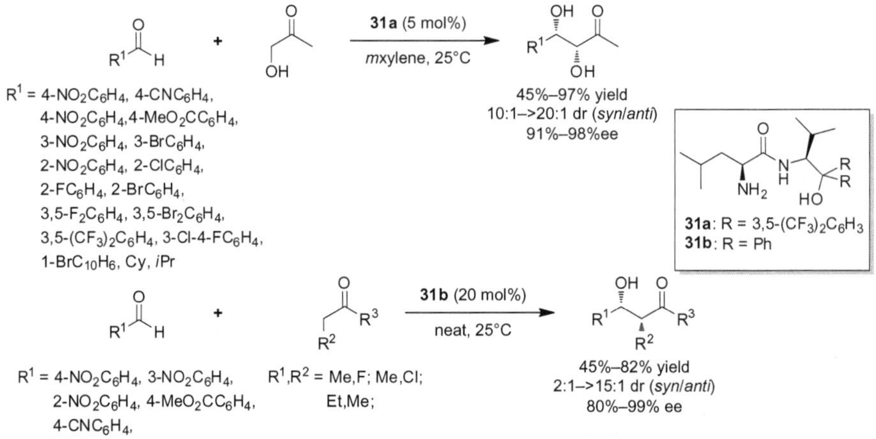

SCHEME 13.28. Asymmetric *syn* aldol reactions catalyzed by amides **31a,b**.

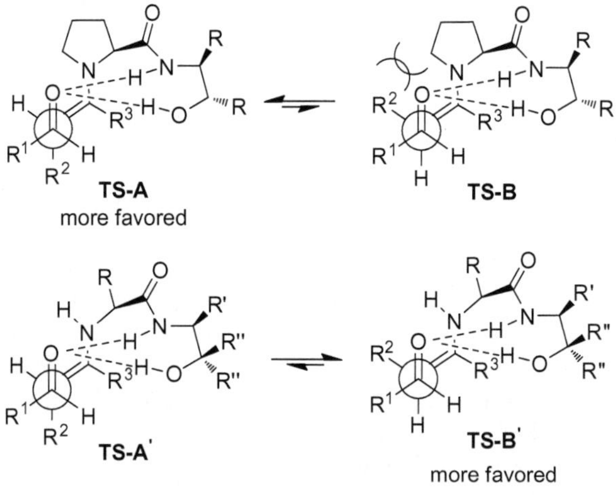

FIGURE 13.8. Proposed transition state structures for *anti* selectivity (**TS-A**) and *syn* selectivity (**TS-A′**).

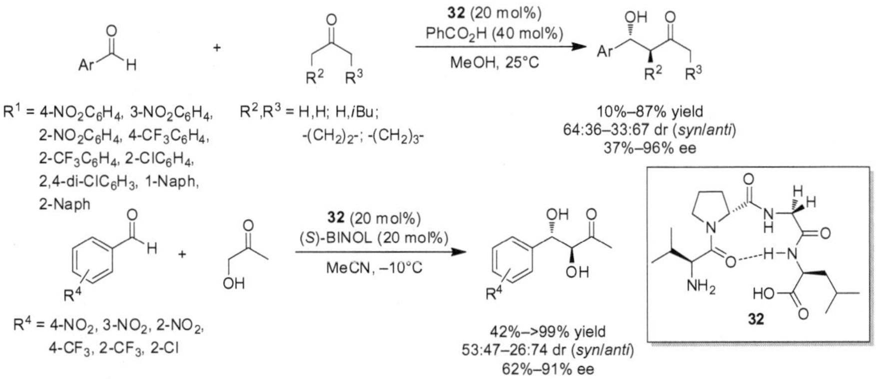

SCHEME 13.29. Asymmetric aldol reactions catalyzed by tetrapeptide **32**.

A series of conformationally restricted peptides was prepared by Da et al. and applied as catalysts in the context of asymmetric aldol reactions [47]. A structural motif such as **32** (Scheme 13.29) turned out to be an especially fruitful promoter of this type of enamine-catalyzed reaction. Studies concerning cooperative effects of catalyst and additives were investigated, with benzoic acid being the most beneficial agent, as the aldol reactions of varying aromatic aldehydes with acyclic and cyclic ketones furnished the corresponding products in 10%–87% yields, 64:36–33:67 dr (*syn*/*anti*), and in 37%–96% ee of the *anti*-configured adduct. By circular dichroism and nuclear Overhauser effect spectroscopy (NOESY) analyses in methanol, a β-turn-like conformation was supported. Furthermore, hydroxy acetone was also employed with this catalyst; in this case, (*S*)-1,1′-bi-2-naphthol (BINOL) was used as an additive.

13.5. PEPTIDE-CATALYZED ASYMMETRIC MORITA–BAYLIS–HILLMAN REACTIONS

Given that short peptides provide extensive opportunities for catalyst tuning, Miller and co-workers extended the scope of these catalysts to enantioselective Morita–Baylis–Hillman reactions and explored the possibility of synergistic effects between two distinct cocatalytic entities [48].

The octapeptide **33**, in combination with L-proline as cocatalyst, was identified by extensive screening studies in the group of Miller as an effective promoter of asymmetric Morita–Baylis–Hillman reactions between α,β-unsaturated ketones and aldehydes. This catalytic methodology was quite suitable for the conversion of a set of aromatic aldehydes and methyl vinyl ketone to the respective adducts in 52%–95% yields and in 41%–81% ee (Scheme 13.30a) [49].

With the goal of clarifying the catalyst–cocatalyst interaction, further studies showed that proline is a vital component in the catalysis, with its configuration

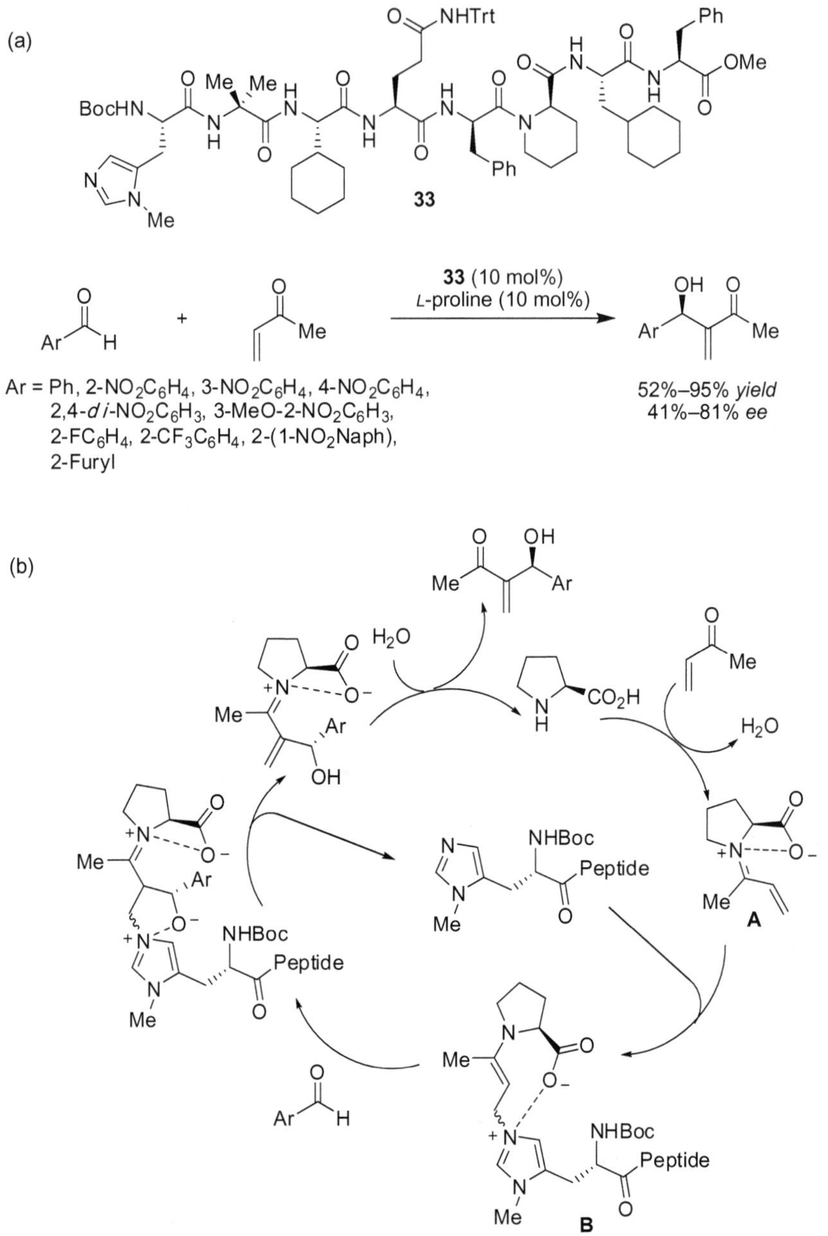

SCHEME 13.30. (a) Asymmetric Morita–Baylis–Hillman reaction catalyzed by peptide **33** and L-proline; (b) proposed mechanism for the above reaction.

having a direct influence on the products' absolute configuration. Besides, the amine, as well as the carboxyl function and geometry of proline, seems to be important.

Based on those facts, a dual mechanism, such as what follows, is quite plausible (Scheme 13.30b). At the beginning, the enone is activated by the formation of an iminium intermediate **A**. Next, a nucleophilic attack by the *N*-methyl-imidazole-containing peptide generates the enamine **B**, which approaches the aldehyde from its *Re*-side. Finally, the imidazole peptide is eliminated, and the L-proline regenerated by hydrolysis.

In the context of asymmetric additions of allenoates to *N*-acyl imines, a conversion reminiscent of the aza-Baylis–Hillman reaction, small peptides have also found application. Miller et al. introduced a peptide based on pyridylalanine as a catalyst in this case [50]. After initially screening a set of such closely related peptides, the structure **34** was identified, which promoted the conversion of *N*-benzoyl imines with two different allenoates in up to 88% yield and in up to 89% enantioselectivities (Scheme 13.31a).

The mechanistic principle lies in a nucleophilic attack of the catalyst's pyridyl moiety on the allenoate (Scheme 13.31b). The resulting adduct **A** shows nucleophilic properties at the α-position to the ester function, which is then approached by the *N*-acyl imine as the acceptor in the chiral environment of the peptide backbone to a second adduct **B**. Finally, **B** is decomposing to the product and the regenerated catalyst by an intramolecular deprotonation.

13.6. PEPTIDE-CATALYZED STETTER REACTION

The range of reactions that may be catalyzed by short peptides and peptide-like molecules is expanding continuously. In 2005, Miller and co-workers found that thiazolylalanine-based peptide **35** functioned as an enantioselective catalyst for an intramolecular Stetter reaction [51]. A new family of catalysts that promote the cyclization of the test substrates with up to 81% ee was developed (Scheme 13.32). Further studies might pave the way to more effective peptide-based catalysts for this enantioselective Stetter reaction.

13.7. PEPTIDE-CATALYZED REGIOSELECTIVE ACYLATION REACTIONS

In a study by Miller et al., peptide-based catalysts were used for regioselective modifications of the natural product erythromycine A (Scheme 13.33), which exhibits antibiotic properties [52]. In detail, a controlled regioselective esterification of this complex substrate was desired; as an acyl transfer agent, histidine-based peptidic catalysts were chosen.

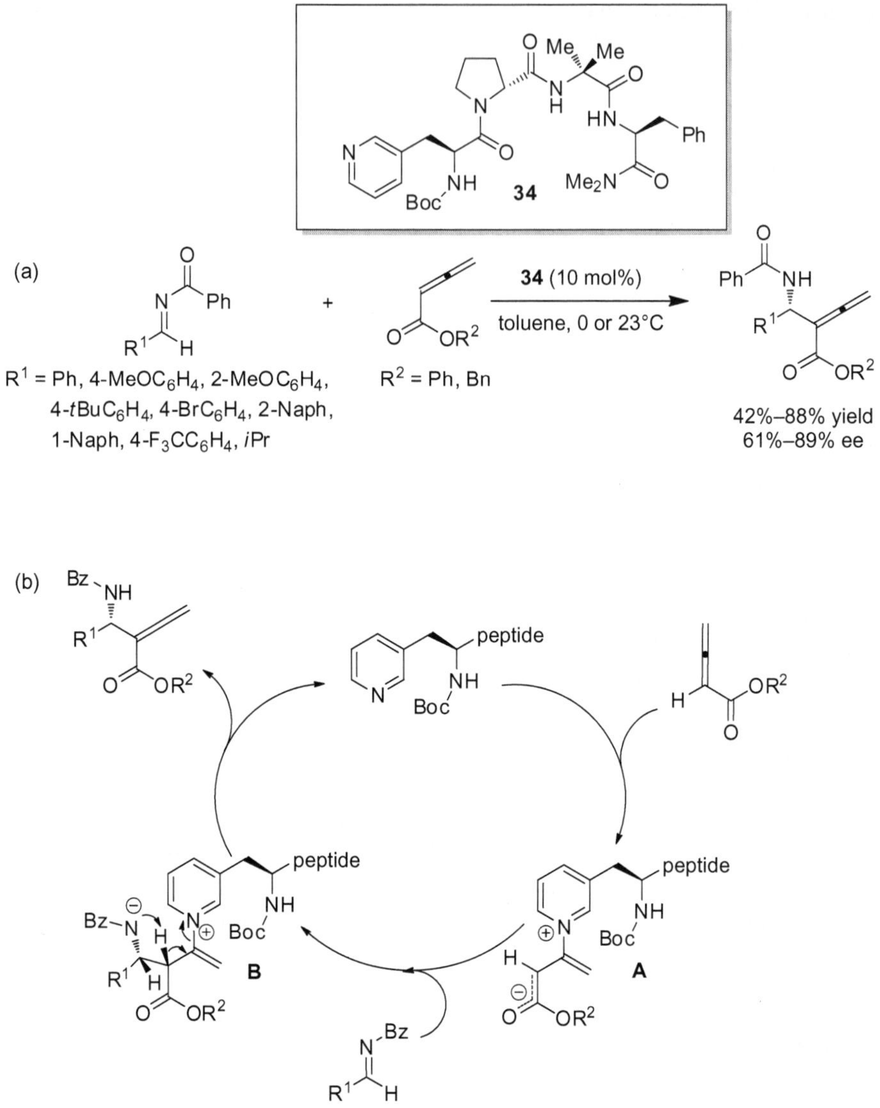

SCHEME 13.31. (a) Asymmetric additions of allenoates to *N*-acyl imines catalyzed by peptide **34**; (b) proposed mechanism for the above reaction.

In a preceding achiral study with *N*-methylimidazole (NMI) acylation catalyst, the intrinsic reactivities of the single hydroxyl groups were clarified. The tertiary hydroxyl groups were generally less reactive than the secondary ones; among them, the 2′-position was the most reactive, followed by the 4″- and the 11-positions.

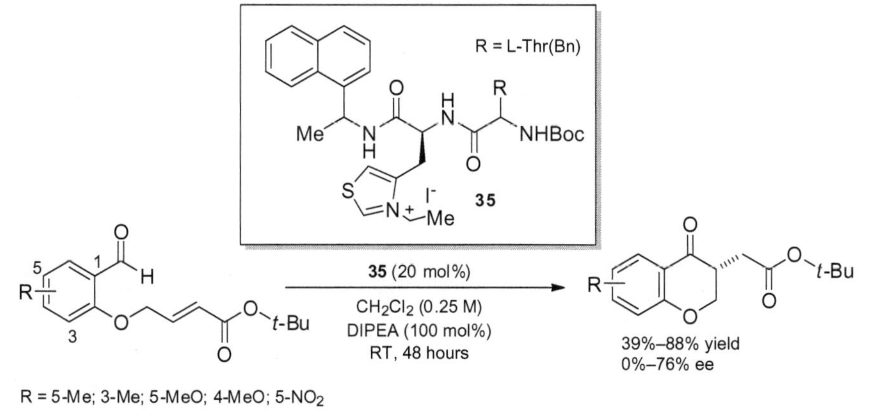

SCHEME 13.32. Asymmetric Stetter reaction catalyzed by thiazolylalanine-based peptide **35**.

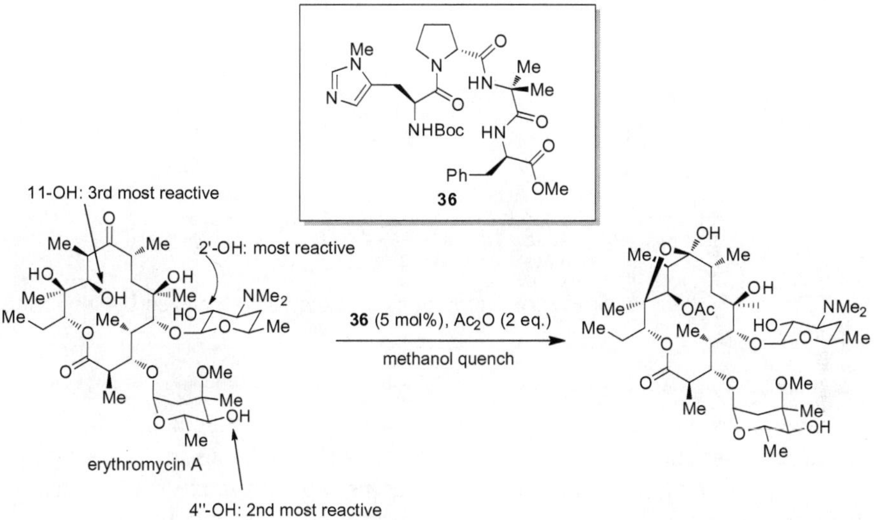

SCHEME 13.33. Regioselective acylation catalyzed by peptide **36**.

Through an extensive screening of peptides, a preferential acylation of the position 11 was observed with the *N*-methyl L-histidine-based peptide **36** (Scheme 13.33), which was also not only limited to acetic anhydride.

13.8. PEPTIDE-CATALYZED ASYMMETRIC α-FUNCTIONALIZATIONS

Kudo and co-workers extended their solid-phase-supported peptide catalyst **37**, which was also used as a promoter for Friedel–Crafts alkylations [20], to

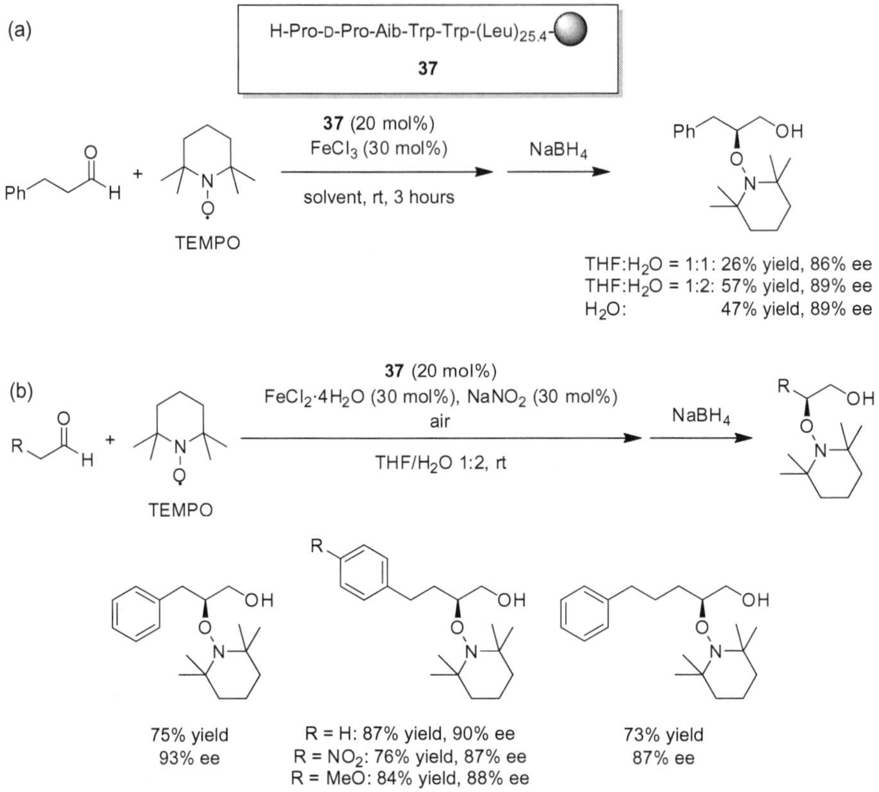

SCHEME 13.34. Asymmetric α-oxyaminations of aldehydes catalyzed by supported peptide **37**.

asymmetric α-oxyaminations of aldehydes, which were reduced in situ to alcohols to circumvent the difficulties associated with aldehydes in analysis [53].

During their optimization studies with this reaction, mixtures of tetrahydrofuran (THF) and water as solvents were found to positively influence the catalysis, as the yield of oxyamination products increased with more water present (Scheme 13.34a). This can be primarily ascribed to increased hydrophobic interactions between the substrates and the catalyst. As the authors investigated a series of structurally different peptides in analogy to previous studies, the poly-L-leucine part of the catalyst's backbone was shown to be a vital prerequisite for the yield and the enantioselectivity.

Considering the scope of this catalysis, various derivatives and homologs of 3-phenylpropionic aldehyde were successfully oxyaminated (Scheme 13.34b), in all cases producing yields of 73%–87% and enantioselectivities of 87%–93% ee. As the aromatic system of these substrates is in a quite remote region with respect to the reacting α-position, their electronic properties had a rather minor and not really systematic influence on the catalysis.

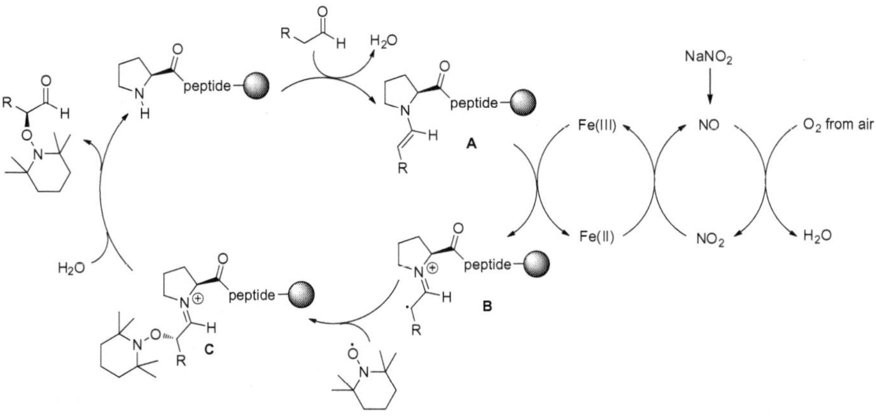

FIGURE 13.9. Proposed mechanism for the asymmetric α-oxyaminations of aldehydes catalyzed by peptide **37**.

The mechanism of such an α-oxyamination is quite interesting (Fig. 13.9). While the formation of the aldehyde's enamine **A** is an often-encountered common feature of *N*-terminal prolinyl peptides, the oxidation of such enamines by a single-electron transfer (SET) process to an iminium radical **B** is a rather rare mechanistic step. As an electron transferring reagent, an inorganic oxidizing system, $FeCl_2 \cdot 4H_2O/NaNO_2$/air, was used for this purpose in catalytic amounts. Basically, this redox system ensures a steady concentration of iron(III) in the catalysis system, which is the actual SET reagent (Fig. 13.9). The radical intermediate **B** is then associating with 2,2,6,6-tetramethylpiperidine-1-oxyl (TEMPO) to a second iminium species **C** under enantioface discrimination because of the chirality in the catalyst, which is finally hydrolyzed to the product and the regenerated catalyst.

13.9. PEPTIDE-CATALYZED DESYMMETRIZATION REACTION

In 2001, Sculimbrene and Miller published their first results on peptide-catalyzed regio- and enantioselective monophosphorylation of *myo*-inositol to D-*myo*-inositol-1-phosphate as a kinase mimic [54]. Their strategy involved first benzylating three of the possible six hydroxyl groups, followed by a peptide-catalyzed desymmetrizing phosphorylation (Scheme 13.35). During the screening of a peptide library including 39 members, the structural motif **38** was found to be an effective promoter for this kind of regio- and enantioselective transformation, as the monophosphorylated intermediate was isolated in 65% yield with >98% ee. Finally, the desired target compound, D-*myo*-inositol-1-phosphate (D-I-1P), was obtained by a dissolving metal reduction using lithium.

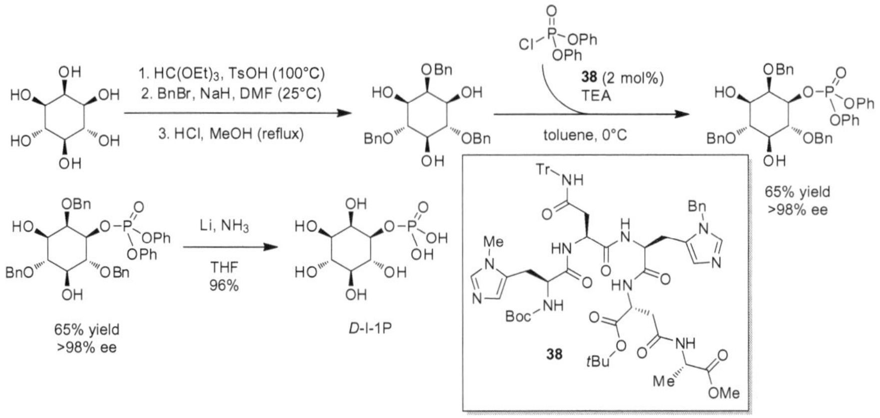

SCHEME 13.35. Desymmetrization reaction catalyzed by peptide **38**.

Miller and co-workers reported a desymmetrization of a prochiral bisphenole by monoacetylation with *N*-methylhistidine containing peptides [55]. An extensive screening of peptides with varying sizes revealed the species **39** as the lead structure of choice, which catalyzed the enantioselective monoesterification in 80% yield and with 95% ee (Scheme 13.36a). The acyl transfer function in this type of catalyst is the *N*-methyl imidazole moiety, which at first gets acylated and plays its role as a nucleophilic catalyst in the vicinity of the peptides chirality (Scheme 13.36b).

In order to gain synthetic access to a series of enantiopure polyphosphatidyl inositoles, Miller et al. developed a synthetic strategy, which is centered on a site-selective, desymmetrizing, peptide-catalyzed phosphorylation of a *meso*-inositole precursor (**A**, Scheme 13.37) [56,57]. Two suitable *N*-methyl histidine-based peptide catalysts, **40** and **41**, were identified for this kind of transformation, each providing one of both enantiomers (**B**, *ent*-**B**) in 97% and 98% ee and in 53% yields in both cases. Afterward, those desymmetrized precursors were transformed to different *myo*-inositol polyphosphates by functionalization and deprotection steps.

13.10. PEPTIDE-CATALYZED KINETIC RESOLUTIONS

Miller and co-workers reported a series of peptides, containing alkylated histidine residues, that are capable of effecting kinetic resolutions of functionalized secondary alcohols [58]. Octapeptide **42** (Scheme 13.38) was shown to function as an effective asymmetric acylation catalyst for the kinetic resolution of a variety of racemic secondary alcohol substrates.

In 2004, Miller et al. described an experimental study of this effective peptide catalyst that shed light on the mechanistic basis for its stereoselectivity

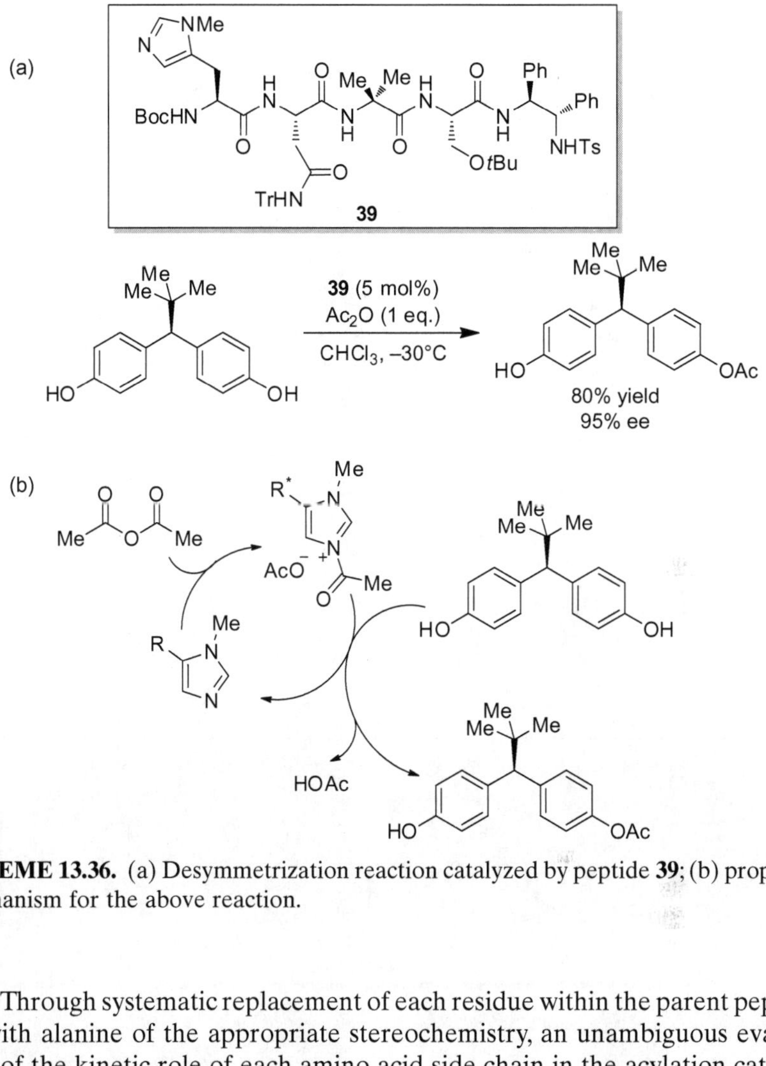

SCHEME 13.36. (a) Desymmetrization reaction catalyzed by peptide **39**; (b) proposed mechanism for the above reaction.

[59]. Through systematic replacement of each residue within the parent peptide **42** with alanine of the appropriate stereochemistry, an unambiguous evaluation of the kinetic role of each amino acid side chain in the acylation catalyst was carried out and the bifunctional mechanism of action was confirmed. While a hydrogen bond between the imidazole π-nitrogen and a backbone NH group might contribute to secondary structural stabilization, it may also serve to transmit heightened basicity to the corresponding backbone carbonyl oxygen, which could then serve as a general base (secondary nucleophile) within the bifunctional catalyst [11]. In addition, the results of the alanine scan underlined the importance of a combination of both of the two His residues to create a highly active and selective peptide catalyst.

Miller and co-workers designed catalyst **43** (Scheme 13.39), with the 1-imidazolyl group as acyl transfer moiety being crucial for general activity and with the chiral peptide backbone including a hydrogen-bond stabilized

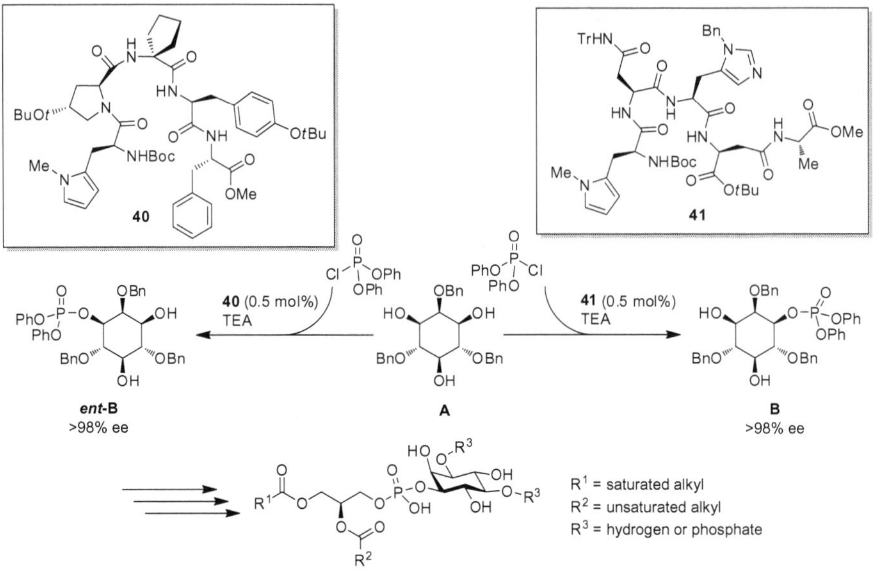

SCHEME 13.37. Desymmetrization reaction catalyzed by peptides **40** and **41**.

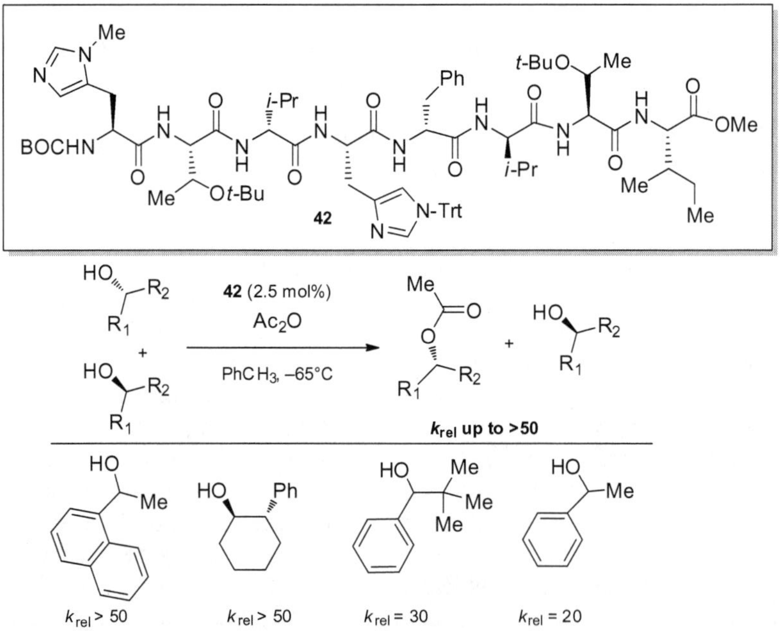

SCHEME 13.38. Kinetic resolutions of functionalized secondary alcohols catalyzed by octapeptide **42**.

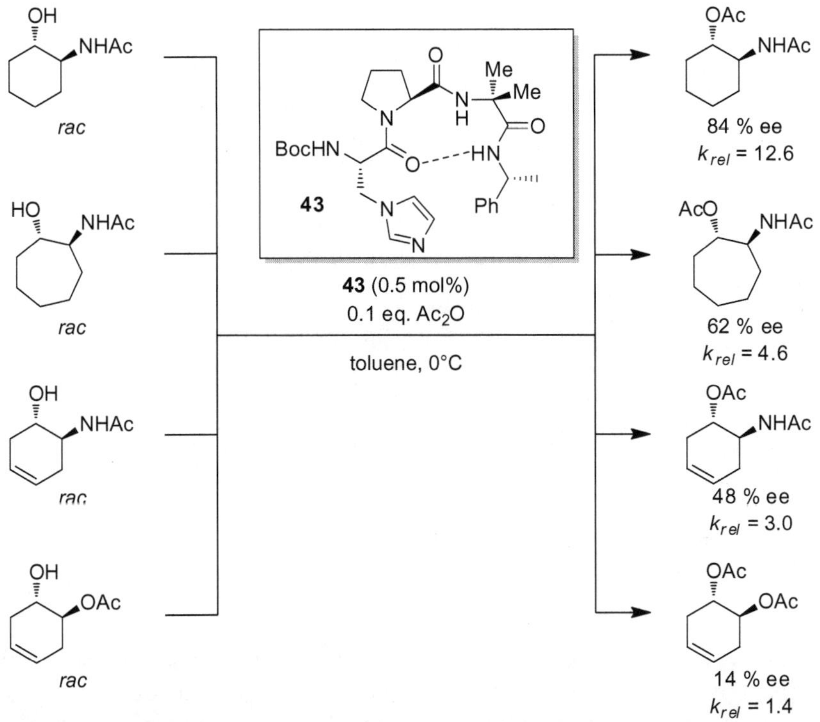

SCHEME 13.39. Kinetic resolutions of functionalized secondary alcohols catalyzed by peptide **43**.

ß-turn being necessary for satisfying the enantiodiscrimination of the substrates [60]. Using this peptidic scaffold, a set of racemic cyclic α-acetamido and α-acetato alcohols was acylated enantiodiscriminatively to the respective *S,S*-configured diacetyl products in 14%–84% ee and with 1.4–12.6 relative rate constants (k_{rel}).

After this initial success, Miller and co-workers continued their kinetic resolution studies by expanding their original catalytic motif to tetrapeptidic structures of type **44** (Scheme 13.40) [61]. A peptide library with 10 members was evaluated with a cyclic α-acetamidoalcohol, furnishing important structural prerequisites for an efficient resolution. It was found that the absolute configuration in the proline part (region *i*+1) of the catalyst dictates the absolute configuration of the preferentially acylated stereoisomer (Scheme 13.40): incorporation of L-proline leads to acylation of the (*S*,*S*)-enantiomer, D-proline to acylation of the (*R*,*R*)-enantiomer. Moreover, the absolute configuration in the C-terminal amino acid (region *i*+3) in relation to one of the proline parts had an influence on the selectivity: matching pairs were L-proline in (*i*+1) and D-configuration in (*i*+3) and vice versa; of lower selectivity were such structures with homochirality in the (*i*+1) and (*i*+3) regions. On basis of nuclear

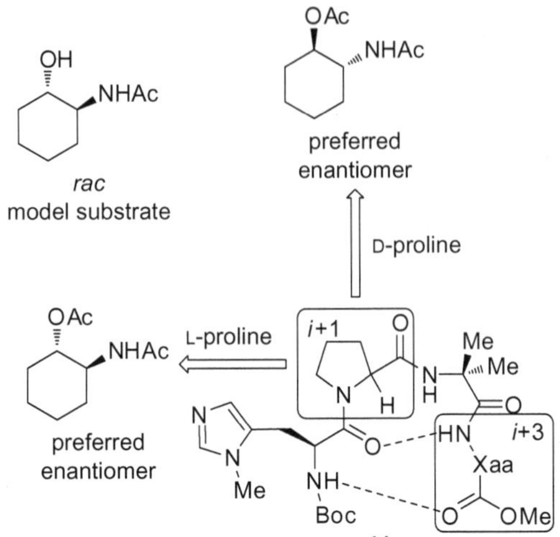

SCHEME 13.40. Structural prerequisites for an efficient resolution of a cyclic α-acetamidoalcohol.

magnetic resonance (NMR) measurements, the authors came to the conclusion that epimerization of the proline part in the tetrapeptide leads to approximate mirror images, additional evidence for the observed empiric data.

Such tetrapeptidic catalyst structures exhibit a β-hairpin motif, which is stabilized by two intramolecular hydrogen bonds.

Miller and co-workers were further expanding the peptidic framework to the octapeptides **45** and **46** (Fig. 13.10) [62]. The introduction of an additional four amino acids in the previous tetrapeptide motif resulted in a longer β-hairpin structure. In the case of the D-proline containing catalyst **45**, four intramolecular hydrogen bonds are formed and make the conformation quite rigid. In contrast, the L-proline containing molecule **46** is more flexible, as only one hydrogen bond is formed. Together with the tetrapeptide **47** (Fig. 13.10), the asymmetric acylation properties have been evaluated and compared to each other in the kinetic resolution of a variety of racemic acetamido alcohols (Fig. 13.10). In most of the chosen racemates, the rigid octapeptide **45** was more enantiodiscriminative than its more flexible diastereomere **46**, and was also superior to its smaller relative **47**.

A structural model, to explain the sense and the degree of the observed enantiodiscrimination for different peptide structures, was also presented by Miller et al. [63]. NMR spectroscopic techniques [correlation spectroscopy (COSY), rotating-frame Overhauser effect spectroscopy [ROESY]), in combination with solvent titration in order to study the hydrogen-bonded network in the peptide, confirmed both hydrogen bonds in **47**, which were already mentioned and studied previously (Fig. 13.11) [61]. A significant Nuclear Overhauser Effect (NOE) between the methine proton of D-proline part and

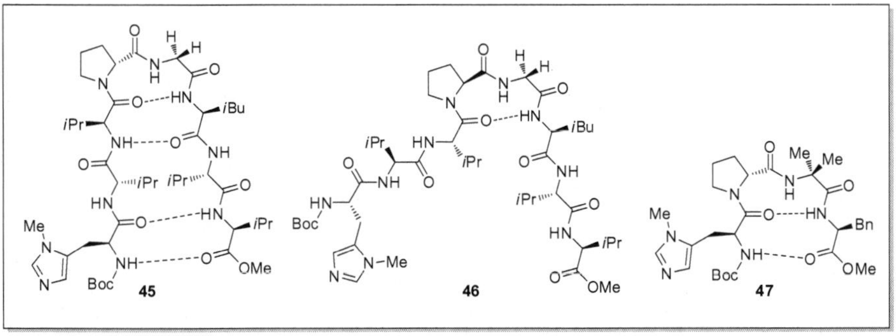

with **45**: 50% conv., k_{rel} = 51	with **45**: 45% conv., k_{rel} = 15	with **45**: 49% conv., k_{rel} = 27	with **45**: 35% conv., k_{rel} = 1
with **46**: 46% conv., k_{rel} = 7	with **46**: 56% conv., k_{rel} = 2	with **46**: 45% conv., k_{rel} = 3	with **46**: 45% conv., k_{rel} = 1
with **47**: 49% conv., k_{rel} = 28	with **47**: 51% conv., k_{rel} = 17	with **47**: 56% conv., k_{rel} = 6	with **47**: 49% conv., k_{rel} = 4

FIGURE 13.10. Kinetic resolution of racemic acetamido alcohols catalyzed by peptides **45–47**.

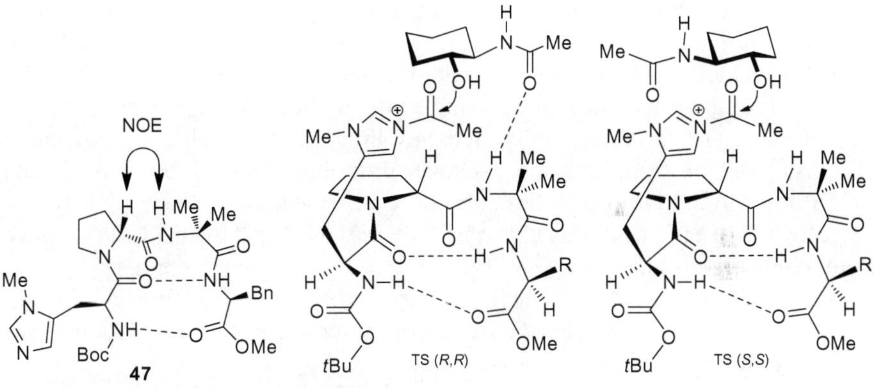

FIGURE 13.11. Transition states of acylation reaction catalyzed by peptide **47**.

the NH proton of the Aib part was also determined and served as additional evidence for the proposed rigid *β*-hairpin structure.

In order to describe the arrangement of the D-proline-containing catalyst and of the fast-reacting enantiomer [the (*R*,*R*)-stereoisomer], a stabilizing hydrogen bonding interaction between the acetamido group and the D-Pro-Aib peptide bond was proposed [see TS(*R*,*R*) in Fig. 13.11]. In the case of the slower-reacting enantiomer [the (*S*,*S*)-stereoisomer], such a beneficial interaction is not possible without major conformational changes in the catalyst and the substrate [see TS(*S*,*S*) in Fig. 13.11].

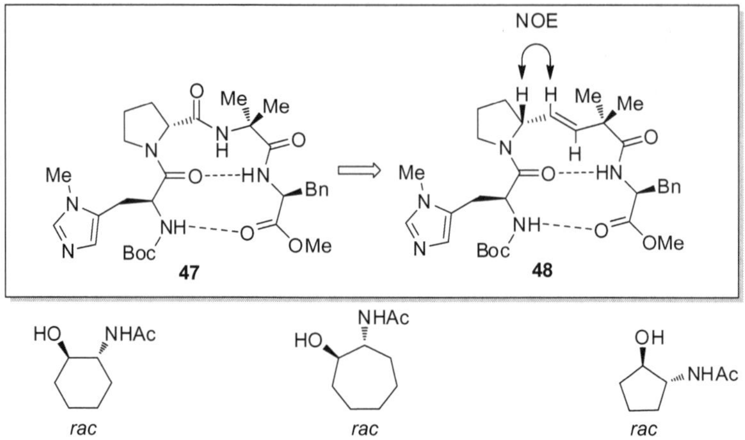

with **47**: 49% conv., k_{rel} = 28 | with **47**: 51% conv., k_{rel} = 17 | with **47**: 56% conv., k_{rel} = 6
with **48**: 50% conv., k_{rel} = <1.5 | with **48**: 52% conv., k_{rel} = <1.5 | with **48**: 50% conv., k_{rel} = <1.5

FIGURE 13.12. Kinetic resolution of racemic acetamido alcohols catalyzed by peptides **47** and **48**.

In order to prove this postulate, a derivative of **47** has been synthesized, with an olefin moiety (peptide **48**) instead of the D-Pro-Aib peptide bond (Fig. 13.12). This methodology is known from the area of peptidomimetics, in order to prepare analogs with nearly the same conformation lacking the polar hydrogen bonding amide functionality. Because **48** is no longer able to coordinate the faster-reacting substrate by hydrogen bonding, a strong decrease in the enantiodiscriminative properties should be observed. Again using NMR, the same two hydrogen bonds as in its parent compound and the same NOE have been confirmed [61].

The performance of this olefin isoster in the kinetic resolution of three racemic α-acetamido alcohols indeed confirmed Miller's expectations (Fig. 13.12). In all cases, a $k_{rel} < 1.5$ was observed.

The same comparison of the octapetide **45** (Fig. 13.13) with its olefine isoster **49**, however, stands in sharp contrast to the above-mentioned explanations, as both catalysts resolved the same racemates with comparable enantiodiscrimination. Obviously, in the case of **45**, no hydrogen-bonding interaction between the acetamido group and the D-Pro-Gly peptide bond is involved in the transition state arrangement.

In the context of a kinetic resolution of tertiary alcohols by peptide-catalyzed acylation, Angione and Miller investigated two peptidic catalysts (**50**, **51**), differing only in a β-methyl group of the histidine residue [64]. It was found that introduction of a β-branch in **50**, **51** results in improvements of both the enantioselectivity and activity (Scheme 13.41). The authors proposed that the β-branched catalyst is conformationally more restricted in comparison to the unbranched structure.

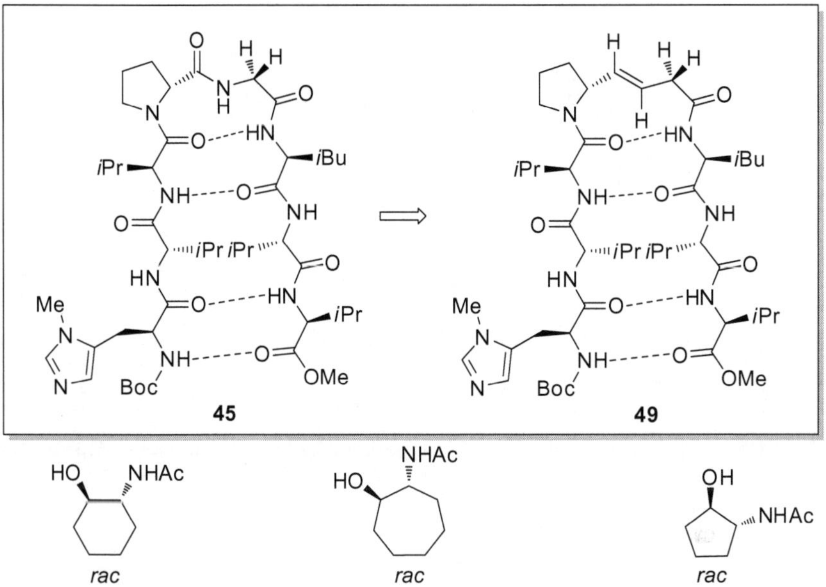

with **45**: 50% conv., k_{rel} = 51 with **45**: 45% conv., k_{rel} = 15 with **45**: 49% conv., k_{rel} = 27
with **49**: 53% conv., k_{rel} = 50 with **49**: 47% conv., k_{rel} = 31 with **49**: 53% conv., k_{rel} = 26

FIGURE 13.13. Kinetic resolution of racemic acetamido alcohols catalyzed by peptides **45** and **49**.

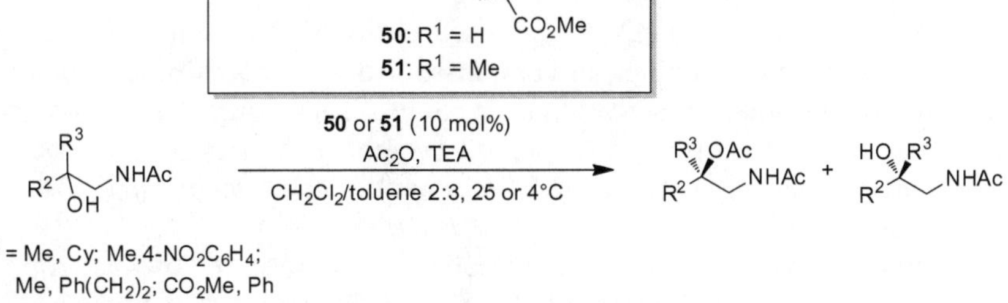

SCHEME 13.41. Kinetic resolution of racemic acetamido alcohols catalyzed by peptides **50–51**.

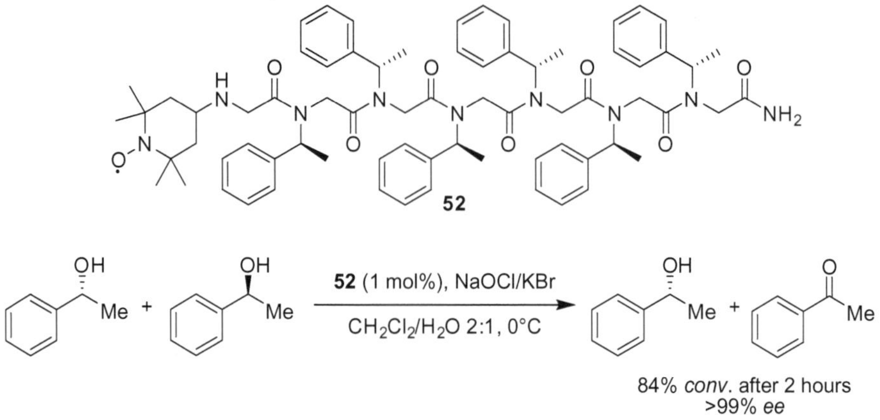

SCHEME 13.42. Asymmetric oxidative kinetic process of 1-phenylethanol catalyzed by **52**.

Ward et al. were synthesizing and evaluating peptoid oligomers in an asymmetric oxidative kinetic process of 1-phenylethanol (Scheme 13.42) [65]. The catalyst structures consisted of a polyglycine backbone, in which the amide functions were substituted by three different moieties: (*S*)- and (*R*)-1-phenylethylamine, benzyl, and TEMPO. By varying the amounts of chiral and achiral substituents, as well as the absolute configurations in the chiral parts, the secondary structure was varied systematically. Besides these structure-defining elements, the radical moiety was needed as the active site. Collectively, structure **52** was identified as the most selective promoter for such selective oxidation. Mimicking the enantiodiscriminating nature of enzymes, only one stereoisomer of the substrate was oxidized to acetophenone, which furnished the remaining stereoisomer in excellent enantioselectivity of more than 99% ee. To reoxidize the active site in the catalyst, an inorganic oxidizing agent (NaOCl/KBr) was employed.

The kinetic resolution of racemic amines, as a potential pathway to enantiopure amines, was the subject of research by Miller et al. [66]. Generally, kinetic resolution of amines, in comparison to alcohols, is more challenging, as amines are more reactive than alcohols. Therefore, the mechanism of kinetic resolution is often accompanied by an achiral background reaction of the derivatizing reagent with both enantiomers, which lead to lower enantioselectivities.

In analogy to the kinetic resolution of alcohols presented by this group [64], small *N*-methyl L-histidine-based peptides **53** were used (Scheme 13.43). The authors aimed their research at the following conceptual design: simple racemic formamides could be *O*-acylated enantioselectively with a peptide catalyst using Boc_2O under kinetic control to a carbonate **A**, which then would rear-

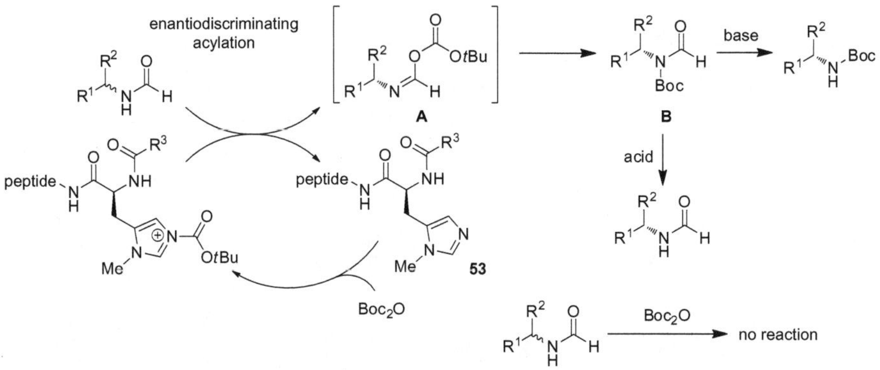

SCHEME 13.43. General scheme for the kinetic resolution of racemic formamides catalyzed by small *N*-methyl L-histidine-based peptides **53**.

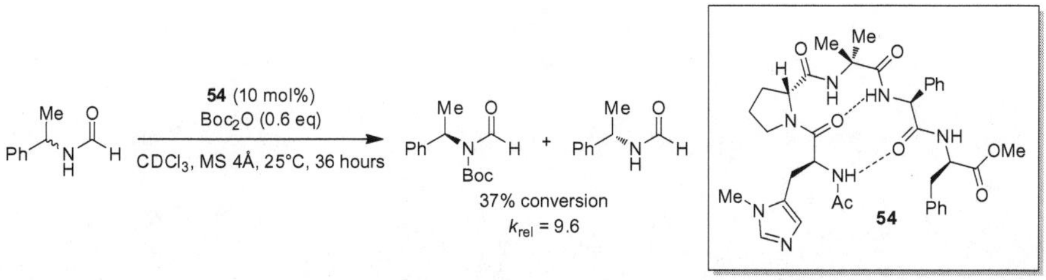

SCHEME 13.44. Kinetic resolution of racemic formamides catalyzed by *N*-methyl L-histidine-based peptide **54**.

range to an imide **B**. In the absence of the catalyst, the acylation would not proceed, which is good for an effective resolution. Further transformation of such products using acids or bases would lead to enantiopure formamides and carbamates, respectively.

Studies dealing with structure optimization of the catalyst revealed the lead motif **54** (Scheme 13.44), in which the structure is fixed by intramolecular hydrogen bonding to a *β*-hairpin conformation. However, in all cases, a rather low conversion of 37% during 36 hours was observed by NMR, owing to the rather low nucleophilicity of the formyl oxygen.

In comparison to the oxygen of formamides, the sulfur of thioformamides is more nucleophilic. A competition experiment (Scheme 13.45) between both types of formamides clearly confirmed this: A thioformamide was strongly favored over a formamide in such a catalysis with respect to reactivity. Working from this background, a kinetic resolution experiment of a thioformamide was also studied. The product with the same absolute configuration was formed in a considerably shorter time period of 12 hours. Relatively good enantioselectivity of 73% ee was observed in this case.

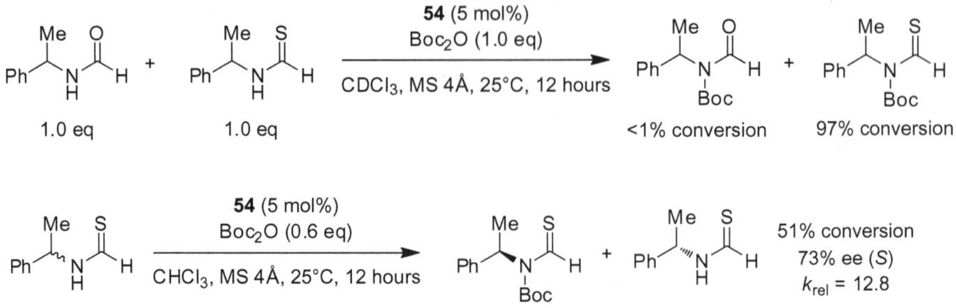

SCHEME 13.45. Comparison of kinetic resolution between thioformamides and formamides catalyzed by *N*-methyl L-histidine-based peptide **54**.

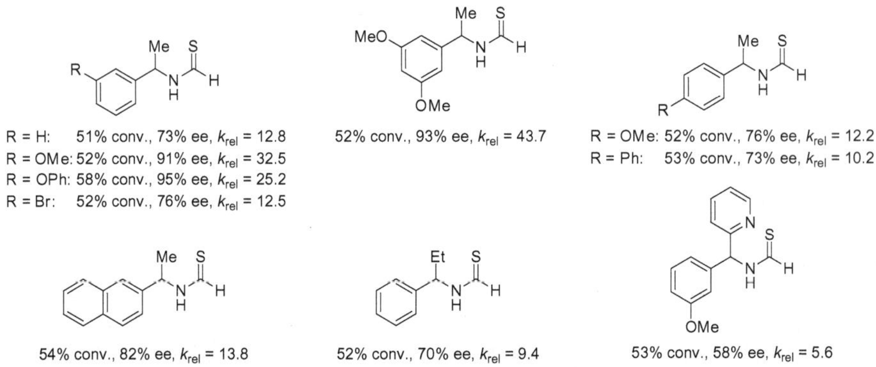

SCHEME 13.46. Kinetic resolution of racemic thioformamides catalyzed by *N*-methyl L-histidine-based peptide **54**.

Using this catalysis method, a series of racemic thioformamides was successfully resolved (Scheme 13.46). Various derivatives of 1-phenyl-1-ethyl-thioformamide were good substrates, because *meta*- and *para*-substitution as well as disubstitutions in these positions were tolerated. Enantioselectivities were especially good in the case of electron-donating substituents in the *meta*-position [MeO: 91% ee; PhO: 95% ee, 3,5-*di*$(MeO)_2$] and for the 2-naphthyl substrate (82% ee).

13.11. PEPTIDE-CATALYZED ASYMMETRIC PROTONATION REACTIONS

Asymmetric synthesis of ketones bearing a tertiary chiral center in the α-position was the subject of studies by Yanagisawa et al. [67]. Lithium enolates, prepared in situ from silylated enol ethers, were protonated in an enantioselective manner by the dipeptide **55** (Scheme 13.47). An additional achiral proton

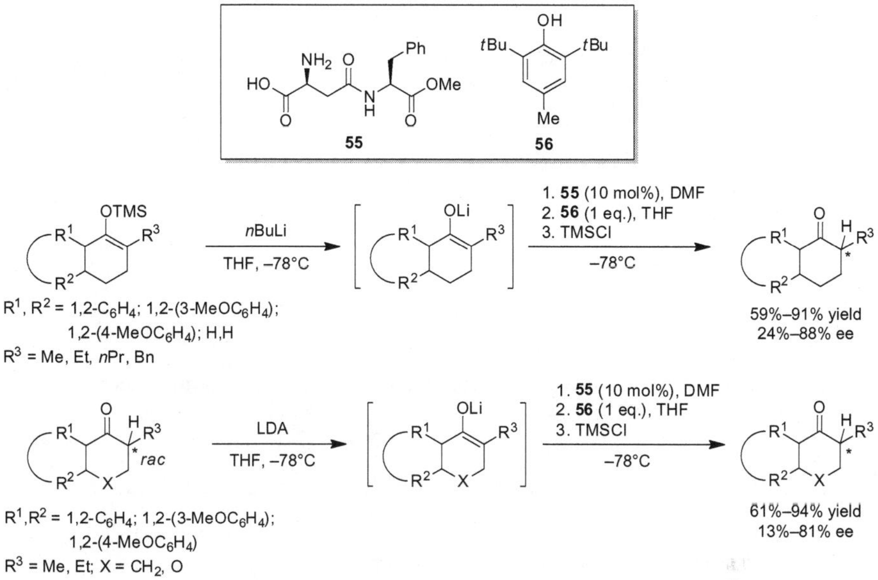

SCHEME 13.47. Asymmetric protonation reactions catalyzed by the dipeptide **55**.

source (**56**) was found to be crucial for any useful levels of enantioselectivity. During their screening experiments of various amino acids and peptides, **55** was found to be the most effective catalyst motif, as a series of lithium enolates was converted to the corresponding ketones in 59%–91% yield and in 24%–88% ee. The amino function in **55** was found to be crucial for the asymmetric induction, as its constitutional isomer aspartame showed no enantioselectivity. To increase the efficiency and atomic economy, this group also investigated another methodology, in which the lithium enolate was generated by α-deprotonation of ketones prior to the asymmetric protonation step. In this case, comparable enantioselectivities (13–81% ee) and yields (61–94%) were obtained, except for cases where sterically more demanding ketones were employed.

According to theYanagisawa's proposals, the catalyst as the chiral proton source (**55**) (Fig. 13.14) stereoselectively protonates the lithium enolate. After this step, this chiral proton source is reprotonated by the achiral proton source (**56**). Generally, it is a vital necessity for this catalytic system that the achiral proton source does not participate in the protonation of the lithium enolate.

13.12. PEPTIDE-CATALYZED ASYMMETRIC TRANSFER HYDROGENATION REACTIONS

A peptidic catalyst **14**, anchored to PEG and attached to polystyrene, was reported as an organocatalyst from Kudo et al. for asymmetric transfer

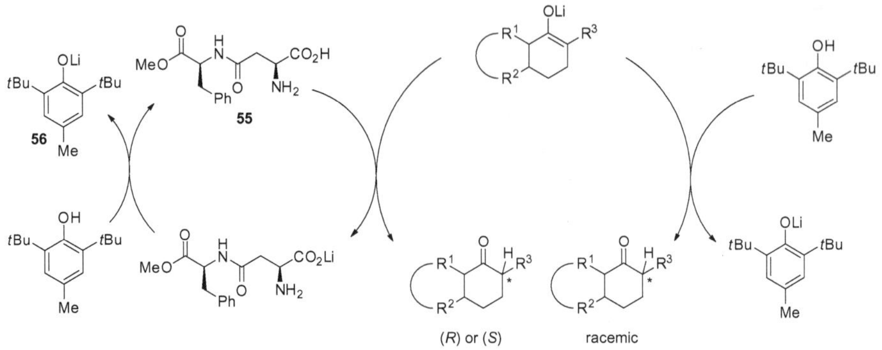

FIGURE 13.14. Proposed mechanism for the asymmetric protonation reactions catalyzed by the dipeptide **55**.

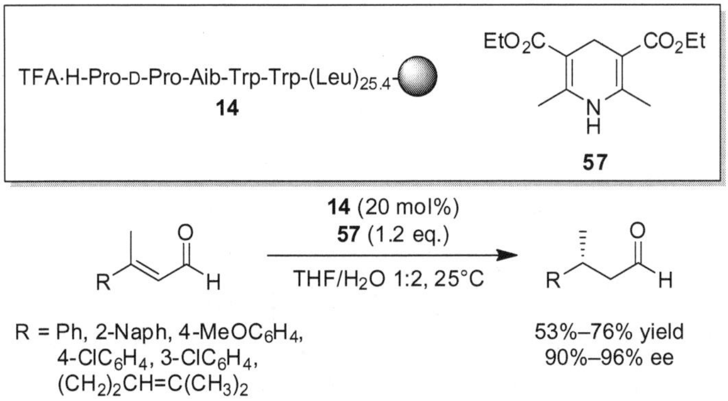

SCHEME 13.48. Asymmetric transfer hydrogenation reactions catalyzed by supported peptide **14**.

hydrogenations in an aqueous medium, affording the asymmetric reduction of a series of *β*-methyl *α,β*-unsaturated aldehydes in 53%–76% yield and in 90%–96% ee (Scheme 13.48) [68]. However, if *R* was an *ortho*-substituted phenyl moiety, no conversion was observed.

The amino acid sequence of **14** was the consequence of extensive studies concerning structure–activity relationships. It was found that a certain degree of hydrophobicity around the active site (the *N*-terminal L-proline) in the catalyst is a prerequisite for both the activity and also for the stereoselectivity. However, not only the structure had an influence on the transfer hydrogenation, but also the amount of water in the solvent was significant for the catalysis, as water also increases the interactions between the hydrophobic reactants.

The mechanism of such reactions is based on iminium catalysis (Fig. 13.15). The unsaturated aldehyde is converted to an iminium species **A**, attached to

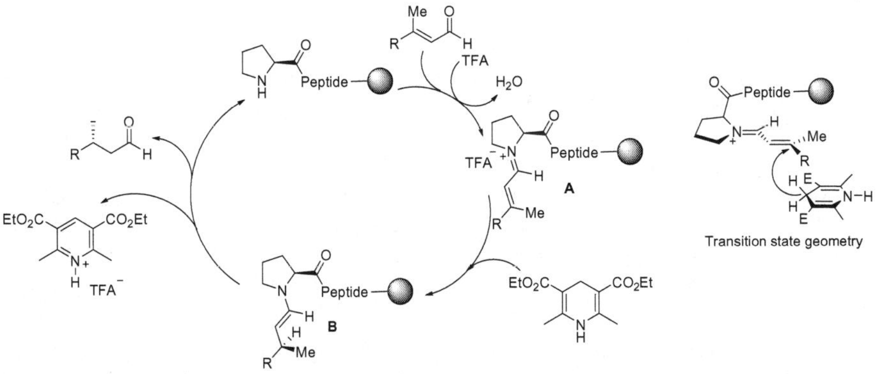

FIGURE 13.15. Proposed mechanism for the asymmetric transfer hydrogenation reactions catalyzed by supported peptide **14**.

the solid phase. This intermediate is then hydrogenated from the sterically less demanding side, the *Si*-face of the C=C double bond, to an enamine **B**. The geometry of this step could be described as proposed by List et al. [69] in their pioneering work. Finally, the catalytic cycle is completed by hydrolysis.

13.13. SUMMARY

In conclusion, the ever-expanding contributions in the field of asymmetric synthesis with short peptides as modular organocatalysts and enzyme mimics undoubtedly confirm that in many cases, peptide catalysts provide highly selective and active alternatives to enzymes and metal catalysts.

Over the past years, a remarkable number of new enantioselective reactions subject to peptide catalysis have been identified for a wide range of substrates; further exciting discoveries of new, unpredicted, and unprecedented industrially attractive peptide catalysts are to be expected in the near future.

Without question, a more detailed mechanistic understanding of these versatile catalysts is needed to realize the full potential of peptide catalysis and to pave the way to new organic transformations.

REFERENCES

[1] For reviews on peptides as catalysts, see: (a) Jarvo, E. R., Miller, S. J. (2002). Amino acids and peptides as asymmetric organocatalysts. *Tetrahedron*, *58*, 2481–2495; (b) Berkessel, A. (2003). The discovery of catalytically active peptides through combinatorial chemistry. *Curr. Opin. Chem. Biol.*, *7*, 409–419; (c) Gröger, H., Wilken, J., Berkessel, A. (2003). Simple amino acids and short-chain peptides as efficient

metal-free catalysts in asymmetric syntheses. In: H.-G. Schmalz, T. Wirth (Eds.), *Organic Synthesis Highlights V*, Wiley-VCH, Weinheim, Germany, pp. 178–186; (d) Miller, S. J. (2004). In search of peptide-based catalysts for asymmetric organic synthesis. *Acc. Chem. Res.*, *37*, 601–610; (e) Tsogoeva, S. B. (2005). Short peptides and peptide-like enzyme mimics—efficient organic catalysts in asymmetric synthesis. *Lett. Org. Chem.*, *2*, 208–213; (f) Darbre, T., Reymond, J.-L. (2006). Peptide dendrimers as artificial anzymes, receptors, and drug-delivery agents. *Acc. Chem. Res.*, *39*, 925–934; (g) Revell, J. D., Wennemers, H. (2007). Peptidic catalysts developed by combinatorial screening methods. *Curr. Opin. Chem. Biol.*, *11*, 269–278; (h) Davie, E. A. C., Mennen, S. M., Xu, Y., Miller, S. J. (2007). Asymmetric catalysis mediated by synthetic peptides. *Chem. Rev.*, *107*, 5759–5812.

[2] For general reviews on asymmetric organocatalyis, see: (a) Dalko, P. I., Moisan, L. (2001). Enantioselective organocatalysis. *Angew. Chem. Int. Ed.*, *40*, 3726–3748; (b) Dalko, P. I., Moisan, L. (2004). In the golden age of organocatalysis. *Angew. Chem. Int. Ed.*, *43*, 5138–5175; (c) Houk, K. N., List, B. (2004). Asymmetric organocatalysis. *Acc. Chem. Res.*, *37*, 487–631; (d) Berkessel, A., Gröger, H. (2005). *Asymmetric Organocatalysis*, Wiley-VCH, Weinheim, Germany; (e) List, B., Yang, J. W. (2006). The organic approach to asymmetric catalysis. *Science*, *313*, 1584- 1586; (f) Dalko, P. I. (2007). *Enantioselective Organocatalysis*, Wiley-VCH, Weinheim, Germany; (g) Pellissier, H. (2007). Asymmetric organocatalysis. *Tetrahedron*, *63*, 9267–9331; (h) Dondoni, A., Massi, A. (2008). Asymmetric organocatalysis: from infancy to adolescence. *Angew. Chem. Int. Ed.*, *47*, 4638–4660.

[3] Oku, J.-I., Inoue, S. (1981). Asymmetric cyanohydrin synthesis catalysed by a synthetic cyclic dipeptide. *J. Chem. Soc. Chem. Commun.*, *5*, 229–230.

[4] Oku, J.-I., Ito, N., Inoue, S. (1982). Asymmetric cyanohydrin synthesis catalyzed by synthetic dipeptides, 2. *Makromol. Chem.*, *183*, 579–586.

[5] Tanaka, K., Mori, A., Inoue, S. (1990). The cyclic dipeptide *cyclo*[(*S*)-Phenylalanyl-(*S*)-histidyl] a catalyst for asymmetric addition of hydrogen cyanide to aldehydes. *J. Org. Chem.*, *55*, 181–185.

[6] Mori, A., Ikeda, Y., Kinoshita, K., Inoue, S. (1989). Cyclo-((*S*)-leucyl-(*S*)-histidyl). A catalyst for asymmetric addition of hydrogen cyanide to aldehydes. *Chem. Lett.*, *18*, 2119–2122.

[7] Iyer, M. S., Gigstad, K. M., Namdev, N. D., Lipton, M. (1996). Asymmetric catalysis of the Strecker amino acid synthesis by a cyclic dipeptide. *J. Am. Chem. Soc.*, *118*, 4910–4911.

[8] Horstmann, T. E., Guerin, D. J., Miller, S. J. (2000). Asymmetric conjugate addition of azide to α,β-unsaturated carbonyl compounds catalyzed by simple peptides. *Angew. Chem. Int. Ed.*, *39*, 3635–3638.

[9] Guerin, D. J., Miller, S. J. (2002). Asymmetric azidation-cycloaddition with open-chain peptide-based catalysts. A sequential enantioselective route to triazoles. *J. Am. Chem. Soc.*, *124*, 2134–2136.

[10] Linton, B. R., Reutershan, M. H., Aderman, C. M., Richardson, E. A., Brownell, K. R., Ashley, C. W., Evans, C. A., Miller, S. J. (2007). Asymmetric Michael addition of α-nitro-ketones using catalytic peptides. *Tetrahedron Lett.*, *48*, 1993–1997.

[11] Hanessian, S., Pham, V. (2000). Catalytic asymmetric conjugate addition of nitroalkanes to cycloalkenones. *Org. Lett.*, *2*, 2975–2978.

[12] Tsogoeva, S. B., Jagtap, S. B., Ardemasova, Z. A., Kalikhevich, V. N. (2004). Trends in asymmetric Michael reactions catalysed by tripeptides in combination with an achiral additive in different solvents. *Eur. J. Org. Chem.*, 4014–4019.

[13] Tsogoeva, S. B., Jagtap, S. B., Ardemasova, Z. A. (2006). 4-*trans*-Amino-proline based di- and tetrapeptides as organic catalysts for asymmetric C-C bond formation reactions. *Tetrahedron Asymmetry*, *17*, 989–992.

[14] Palomo, C., Vera, S., Mielgo, A., Gómez-Bengoa, E. (2006). Highly efficient asymmetric Michael addition of aldehydes to nitroalkenes catalyzed by a simple *trans*-4-hydroxyprolylamide. *Angew. Chem. Int. Ed.*, *45*, 5984–5987.

[15] Wiesner, M., Revell, J. D., Wennemers, H. (2008). Tripeptides as efficient asymmetric catalysts for 1,4-addition reactions of aldehydes to nitroolefins—a rational approach. *Angew. Chem. Int. Ed.*, *47*, 1871–1874.

[16] Wiesner, M., Revell, J. D., Tonazzi, S., Wennemers, H. (2008). Peptide catalyzed asymmetric conjugate addition reactions of aldehydes to nitroethylene—a convenient entry into γ^2–amino acids. *J. Am. Chem. Soc.*, *130*, 5610–5611.

[17] Revell, J. D., Wennemers, H. (2008). Investigating sequence space: how important is the spatial arrangement of functional groups in the asymmetric aldol reaction catalyst H-Pro-Pro-Asp-NH2? *Adv. Synth. Catal.*, *350*, 1046–1052.

[18] Wiesner, M., Neuburger, M., Wennemers, H. (2009). Tripeptides of the type H-D-Pro-Pro-Xaa-NH2 as catalysts for asymmetric 1,4-addition reactions: structural requirements for high catalytic efficiency. *Chem. Eur. J.*, *15*, 10103–10109.

[19] Freund, M., Schenker, S., Tsogoeva, S. B. (2009). Enantioselective nitro-Michael reactions catalyzed by short peptides on water. *Org. Biomol. Chem.*, *7*, 4279–4284.

[20] Akagawa, K., Yamashita, T., Sakamoto, S., Kudo, K. (2009). Friedel-Crafts-type alkylation in aqueous media using resin-supported peptide catalyst having polyleucine. *Tetrahedron Lett.*, *50*, 5602–5604.

[21] Hupe, D. J. (1984). Enzyme reactions involving imine formations. In: M. I. Page (Ed.), *New Comprehensive Biochemistry*, Vol. 6. Elsevier, Amsterdam, The Netherlands, pp. 271–301.

[22] For reviews, see: (a) Peng, F., Shao, Z. (2008). Advances in asymmetric organocatalytic reactions catalyzed by chiral primary amines. *J. Mol. Catal. A.*, *285*, 1–13; (b) Xu, L.-W., Luo, J., Lu, Y. (2009). Asymmetric catalysis with chiral primary amine-based organocatalysts. *Chem. Commun.*, 1807–1821.

[23] Tsogoeva, S. B., Jagtap, S. B. (2004). Dual catalyst control in the chiral diamine-dipeptide-catalyzed asymmetric Michael addition. *Synlett*, 2624–2626.

[24] Xu, Y., Zou, W., Sundén, H., Ibrahem, I., Córdova, A. (2006). Small peptide-catalyzed enantioselective addition of ketones to nitroolefins. *Adv. Synth. Catal.*, *348*, 418–424.

[25] Xu, Y., Córdova, A. (2006). Simple highly modular acyclic amine-catalyzed direct enantioselective addition of ketones to nitro-olefins. *Chem. Commun.*, 460–462.

[26] Tang, Z., Jiang, F., Yu, L.-T., Cui, X., Gong, L.-Z., Mi, A.-Q., Jiang, Y.-Z., Wu, Y.-D. (2003). Novel small organic molecules for a highly enantioselective direct aldol reaction. *J. Am. Chem. Soc.*, *125*, 5262–5263.

[27] Martin, H. J., List, B. (2003). Mining sequence space for asymmetric aminocatalysis: N-terminal prolyl-peptides efficiently catalyze enantioselective aldol and Michael reactions. *Synlett*, 1901–1902.

[28] Kofoed, J., Nielsen, J., Reymond, J.-L. (2003). Discovery of new peptide-based catalysts for the direct asymmetric aldol reaction. *Bioorg. Med. Chem. Lett.*, *13*, 2445–2447.

[29] Tang, Z., Jiang, F., Cui, X., Gong, L.-Z., Mi, A.-Q., Jiang, Y.-Z., Wu, Y.-D. (2004). Asymmetric catalysis special feature, Part II: enantioselective direct aldol reactions catalyzed by L-prolinamide derivatives. *Proc. Natl. Acad. Sci. U.S.A.*, *101*, 5755–5760.

[30] List, B., Lerner, R. A., Barbas, C. F., III (2000). Proline-catalyzed direct asymmetric aldol reactions. *J. Am. Chem. Soc.*, *122*, 2395–2396.

[31] Tang, Z., Yang, Z.-H., Cun, L.-F., Gong, L.-Z., Mi, A.-Q., Jiang, Y.-Z. (2004). Small peptides catalyze highly enantioselective direct aldol reactions of aldehydes with hydroxyacetone: unprecedented regiocontrol in aqueous media. *Org. Lett.*, *6*, 2285–2287.

[32] Shi, L.-X., Sun, Q., Ge, Z.-M., Zhu, Y.-Q., Cheng, T.-M., Li, R.-T. (2004). Dipeptide-catalyzed direct asymmetric aldol reaction. *Synlett*, 2215–2217.

[33] Krattiger, P., Kovasky, R., Revell, J. D., Ivan, S., Wennemers, H. (2005). Increased structural complexity leads to higher activity: peptides as efficient and versatile catalysts for asymmetric aldol reactions. *Org. Lett.*, *7*, 1101–1103.

[34] Revell, J. D., Gantenbein, D., Krattiger, P., Wennemers, H. (2006). Solid-supported and pegylated H-Pro-Pro-Asp-NHR as catalysts for aymmetric aldol reactions. *Biopolymers (Peptide Science)*, *84*, 105–113.

[35] Luppi, G., Cozzi, G. P., Monari, M., Kaptein, B., Broxterman, Q. B., Tomasini, C. (2005). Dipeptide-catalyzed asymmetric aldol condensation of acetone with (N-Alkylated) isatins. *J. Org. Chem.*, *70*, 7418–7421.

[36] Andreae, M. R. M., Davis, A. P. (2005). Heterogeneous catalysis of the asymmetric aldol reaction by solid-supported proline-terminated peptides. *Tetrahedron Asymmetry*, *16*, 2487–2492.

[37] Akagawa, K., Sakamoto, S., Kudo, K. (2005). Direct asymmetric aldol reaction in aqueous media using polymer-supported peptide. *Tetrahedron Lett.*, *46*, 8185–8187.

[38] Wang, X.-J., Yan, Z., Liu, J.-T. (2007). Regiospecific organocatalytic asymmetric aldol reaction of methyl ketones and α,β-unsaturated trifluoromethyl ketones. *Org. Lett.*, *9*, 1343–1345.

[39] Yan, J., Wang, L. (2009). Asymmetric aldol reactions catalyzed by efficient and recyclable silica-supported proline-based peptides. *Chirality*, *21*, 413–420.

[40] D'Elia, V., Zwicknagl, H., Reiser, O. (2008). Short α,β-peptides as catalysts for intra- and intermolecular aldol reactions. *J. Org. Chem.*, *73*, 3262–3265.

[41] Rodríguez-Llansola, F., Miravet, J. F., Escuder, B. (2009). A supramolecular hydrogel as a reusable heterogeneous catalyst for the direct aldol reaction. *Chem. Commun.*, 7303–7305.

[42] Tsogoeva, S. B., Wei, S.-W. (2005). (*S*)-Histidine-based dipeptides as organic catalysts for direct asymmetric aldol reactions. *Tetrahedron Asymmetry*, *16*, 1947–1951.

[43] Zou, W., Ibrahem, I., Dziedzic, P., Sundén, H., Córdova, A. (2005). Small peptides as modular catalysts for the direct asymmetric aldol reaction: ancient peptides with aldolase enzyme activity. *Chem. Commun.*, 4946–4948.

[44] Córdova, A., Zou, W., Dziedzic, P., Ibrahem, I., Reyes, E., Xu, Y. (2006). Direct asymmetric intermolecular aldol reactions catalyzed by amino acids and small peptides. *Chem. Eur. J.*, *12*, 5383–5397.

[45] Weber, A. L., Pizzarello, S. (2006). The peptide-catalyzed stereospecific synthesis of tetroses: a possible model for prebiotic molecular evolution. *Proc. Natl. Acad. Sci. U.S.A.*, *103*, 12713–12717.

[46] Xu, X.-Y., Wang, Y.-Z., Gong, L.-Z. (2007). Design of organocatalysts for asymmetric direct *syn*-aldol reactions. *Org. Lett.*, *9*, 4247–4249.

[47] Wu, F.-C., Da, C.-S., Du, Z.-X., Guo, Q.-P., Li, W.-P., Yi, L., Jia, Y.-N., Ma, X. (2009). N-Primary-amine-terminal β-turn tetrapeptides as organocatalysts for highly enantioselective aldol reaction. *J. Org. Chem.*, *74*, 4812–4818.

[48] Imbriglio, J. E., Vasbinder, M. M., Miller, S. J. (2003). Dual catalyst control in the amino acid-peptide-catalyzed enantioselective Baylis-Hillman reaction. *Org. Lett.*, *5*, 3741–3743.

[49] Vasbinder, M. M., Imbriglio, J. E., Miller, S. J. (2006). Amino acid-peptide-catalyzed enantioselective Morita–Baylis–Hillman reactions. *Tetrahedron*, *62*, 11450–11459.

[50] Cowen, B. J., Saunders, L. B., Miller, S. J. (2009). Pyridylalanine (Pal)-peptide catalyzed enantioselective allenoate additions to N-acyl imines. *J. Am. Chem. Soc.*, *131*, 6105–6107.

[51] Mennen, S. M., Blank, J. T., Tran-Dubé, M. B., Imbriglio, J. E., Miller, S. J. (2005). A peptide-catalyzed asymmetric Stetter reaction. *Chem. Commun.*, 195–197.

[52] Lewis, C. A., Miller, S. J. (2006). Site-selective derivatization and remodeling of Erythromycin A by using simple peptide-based chiral catalysts. *Angew. Chem. Int. Ed.*, *45*, 5616–5619.

[53] Akagawa, K., Fujiwara, T., Sakamoto, S., Kudo, K. (2010). Efficient asymmetric α-oxyamination of aldehydes by resin-supported peptide catalyst in aqueous media. *Org. Lett.*, *12*, 1804–1807.

[54] Sculimbrene, B. R., Miller, S. J. (2001). Discovery of a catalytic asymmetric phosphorylation through selection of a minimal kinase mimic: a concise total synthesis of D-*myo*-inositol-1-phosphate. *J. Am. Chem. Soc.*, *123*, 10125–10126.

[55] Lewis, C. A., Chiu, A., Kubryk, M., Balsells, J., Pollard, D., Esser, C. K., Murry, J., Reamer, R. A., Hansen, K. B., Miller, S. J. (2006). Remote desymmetrization at near-nanometer group separation catalyzed by a miniaturized enzyme mimic. *J. Am. Chem. Soc.*, *130*, 16358–16365.

[56] Morgan, A. J., Komiya, S., Xu, Y., Miller, S. J. (2006). Unified total syntheses of the inositol polyphosphates: D-I-3,5,6P3, D-I-3,4,5P3, D-I-3,4,6P3, and D-I-3,4,5,6P4 via catalytic enantioselective and site-selective phosphorylation. *J. Org. Chem.*, *71*, 6923–6931.

[57] Xu, Y., Sculimbrene, B. R., Miller, S. J. (2006). Streamlined synthesis of phosphatidylinositol (PI), PI3P, PI3,5P2, and deoxygenated analogues as potential biological probes. *J. Org. Chem.*, *71*, 4919–4928.

[58] Sculimbrene, B. R., Morgan, A. J., Miller, S. J. (2003). Nonenzymatic peptide-based catalytic asymmetric phosphorylation of inositol derivatives. *Chem. Commun.*, 1781–1785.

[59] Fierman, M. B., O'Leary, D. J., Steinmetz, W. E., Miller, S. J. (2004). Structure-selectivity relationships and structure for a peptide-based enantioselective acylation catalyst. *J. Am. Chem. Soc.*, *126*, 6967–6971.

[60] Miller, S. J., Copeland, G. T., Papaioannou, N., Horstmann, T. E., Ruel, E. M. (1998). Kinetic resolution of alcohols catalyzed by tripeptides containing the N-alkylimidazole substructure. *J. Am. Chem. Soc.*, *120*, 1629–1630.

[61] Copeland, G. T., Jarvo, E. R., Miller, S. J. (1998). Minimal acylase-like peptides. Conformational control of absolute stereospecificity. *J. Org. Chem.*, *63*, 6784–6785.

[62] Jarvo, E. R., Copeland, G. T., Papaioannou, N., Bonitatebus, P. J., Jr., Miller, S. J. (1999). A biomimetic approach to asymmetric acyl transfer catalysis. *J. Am. Chem. Soc.*, *121*, 11638–11643.

[63] Vasbinder, M. M., Jarvo, E. R., Miller, S. J. (2001). Incorporation of peptide isosteres into enantioselective peptide-based catalysts as mechanistic probes. *Angew. Chem. Int. Ed.*, *40*, 2824–2827.

[64] Angione, M. C., Miller, S. J. (2006). Dihedral angle restriction within a peptide-based tertiary alcohol kinetic resolution catalyst. *Tetrahedron*, *62*, 5254–5261.

[65] Maayan, G., Ward, M. D., Kirshenbaum, K. (2009). Folded biomimetic oligomers for enantioselective catalysis. *Proc. Natl. Acad. Sci. U.S.A.*, *106*, 13679–13684.

[66] Fowler, B. S., Mikochik, P. J., Miller, S. J. (2010). Peptide-catalyzed kinetic resolution of formamides and thioformamides as an entry to nonracemic amines. *J. Am. Chem. Soc.*, *132*, 2870–2871.

[67] Mitsuhashi, K., Ito, R., Arai, T., Yanagisawa, A. (2006). Catalytic asymmetric protonation of lithium enolates using amino acid derivatives as chiral proton sources. *Org. Lett.*, *8*, 1721–1724.

[68] Akagawa, K., Akabane, H., Sakamoto, S., Kudo, K. (2008). Organocatalytic asymmetric transfer hydrogenation in aqueous media using resin-supported peptide having a polyleucine tether. *Org. Lett.*, *10*, 2035–2037.

[69] Woon Yang, J., Hechavarria Fonseca, M. T., List, B. (2004). A metal-free transfer hydrogenation: organocatalytic conjugate reduction of α,β-unsaturated aldehydes. *Angew. Chem. Int. Ed.*, *43*, 6660–6662.

CHAPTER 14

SILICATE-MEDIATED STEREOSELECTIVE REACTIONS CATALYZED BY CHIRAL LEWIS BASES

MAURIZIO BENAGLIA, STEFANIA GUIZZETTI, AND SERGIO ROSSI

14.1. Introduction 580
14.2. Stereoselective C–H bond formation 584
14.2.1. Reactions catalyzed by *N*-formyl derivatives 584
14.2.2. Reactions catalyzed by chiral picolinamides 592
14.2.3. Reductions catalyzed by other chiral Lewis bases 597
14.3. Stereoselective C–C bond formation 600
14.3.1. Allyltrichlorosilane addition to C=O and C=N bonds 601
14.3.1.1. Allylation of C=N group 601
14.3.1.2. Allylation of C=O group 602
14.3.2. Aldol condensation reaction 608
14.3.3. Lewis base-catalyzed Lewis acid-mediated reaction 610
14.4. Ring-opening reaction of epoxides 614
14.5. Miscellaneous 615
14.6. Outlook and perspectives 616
References 617

Catalytic Methods in Asymmetric Synthesis: Advanced Materials, Techniques, and Applications, First Edition. Edited by Michelangelo Gruttadauria and Francesco Giacalone.

14.1. INTRODUCTION

The chemistry of penta and/or hexavalent silicon compounds has recently attracted much attention because of the possibility to develop organocatalyzed enantioselective reactions in the presence of cheap, low-toxic, and environmentally friendly species such as hypervalent silicates [1]. Even if the discovery of silicon compounds with a coordination number greater than four dates back to 1809, when the adduct $SiF_4 \cdot 2\ NH_3$ was reported by Gay-Lussac [2], only in the last 40 years has the distinctive reactivity displayed by penta- and hexavalent silicon compounds been increasingly studied [3]. More recently, the possibility to develop organocatalytic silicon-based methodologies has given new impulse to the studies in the field. The tremendous growth of interest in what is currently referred to as the "organocatalytic" approach toward enantioselective synthesis is strongly indicative of the general direction toward which modern stereoselective synthesis is moving. After scattered reports on this methodology in the 1970s, organic catalysis now represents an established method of using an organic molecule of relatively low molecular weight, simple structure, and low cost to promote a given transformation in substoichiometric quantity, in the absence of any metal, and under nonstringent reaction conditions that are typical of organometallic catalysis [4].

The organocatalytic approach fulfills many of the requirements listed in the well-known 12 principles of green chemistry [5]. A process based on a catalytic methodology is already "green" by definition, since it is clearly stated that "catalysts are preferable to stoichiometric reagents" because they minimize waste and increase energy efficiency; a catalyst often allows a reaction to be run in milder experimental conditions, once again improving the efficiency of a process from the economic and energetic point of view. Specifically, organocatalysts may lead to the design of "safer chemicals and products," as expected by modern synthetic chemists, also with the goal of using less hazardous solvents or reaction conditions. The replacement of metal-based catalysts with equally efficient metal-free counterparts is very appealing in view of possible applications in the future of nontoxic, low-cost, and more environmentally friendly promoters on an industrial scale with obvious advantages from the environmental and economic points of view [6].

In this framework, chiral organocatalysts present other fundamental characteristics of potential enormous interest and, if possible, are even more appealing for both academic research and, in particular, for industrial-scale applications. Recent market analysis showed that global revenues from chiral technology soared from \$6.63 billion in 2000 to \$16.03 billion in 2007, growing at a compound annual rate of 13.4% during that period. Approximately 80% of all products currently in development for the pharmaceutical industry are based on chiral building blocks; this clearly demonstrates why chiral technology has become of fundamental importance for pharmaceutical companies and others [7]. In this context, the extraordinary efforts made by several groups in studying and developing novel and alternative synthetic organocata-

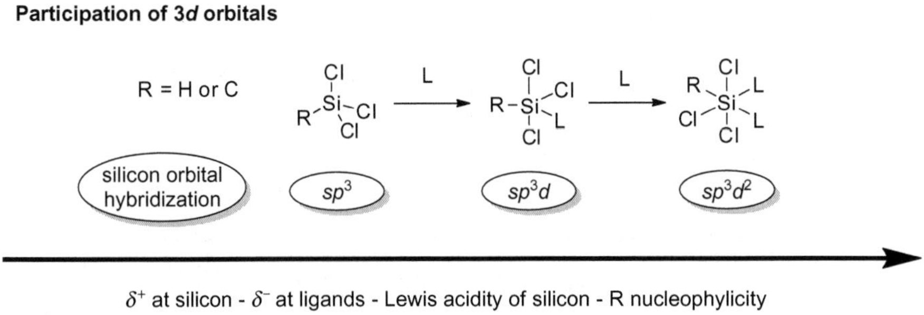

FIGURE 14.1. Hypercoordinated silicon species. (Charges on silicon atom are omitted for clarity.)

lytic stereoselective methodologies is understandable. And it is not surprising that in the last few years, stereoselective versions of several reactions promoted by silicon-based catalysts have been developed [8], particularly those promoted by hypervalent silicate intermediates used as chiral Lewis bases [9].

Before entering in the discussion of different reactions, a general picture of the mechanism of this chemistry should be given [10].

Contrary to carbon (its first row group 14-analog), silicon displays the ability to form more bonds than the four necessary for fulfilling the octet rule: in the presence of donor molecules or ions it is possible for the formation of five-, six-, and even seven-coordinated silicon species, some of which have been isolated and/or characterized. To explain this behavior, two main different theories have been formulated: the first invokes the participation of the silicon 3*d* orbitals in the expansion of the coordination sphere (Fig. 14.1); in the five-coordinated species, the silicon orbitals would have an sp^3d hybridization (with trigonal bipyramidal geometry), while in the six-coordinated species the hybridization would be sp^3d^2 (with octahedral geometry). The reduced *s*-character of the silicon orbitals in the hypercoordinated species would explain their increased Lewis acidity and the transfer of electron density to the ligands.

The second theoretical approach (Fig. 14.2), in contrast, rules out the participation of the 3*d* orbitals in the bonding process and hypothesizes instead a so-called hypervalent bonding [11]. The ability of main-group elements to form compounds that appear to break the Langmuir–Lewis octet rule was originally explained by invoking an availability of *d* orbitals (such as 3*d* for silicon) by using an analogy to transition-metal complexation. However, silicon is not a transition metal, and it is now generally accepted that the 3*d* orbitals on silicon are too diffuse to engage in meaningful bonding [10].

The ability of silicon to expand its coordination sphere (to engage in hypervalent bonding) is due to the ability of the silicon 3*p* orbitals to engage in electron-rich, three-center, four-electron bonding. Therefore, the formation of a penta- or hexacoordinated silicon species would involve, respectively, one or two three-center, four-electron molecular bonds, each formed by a silicon

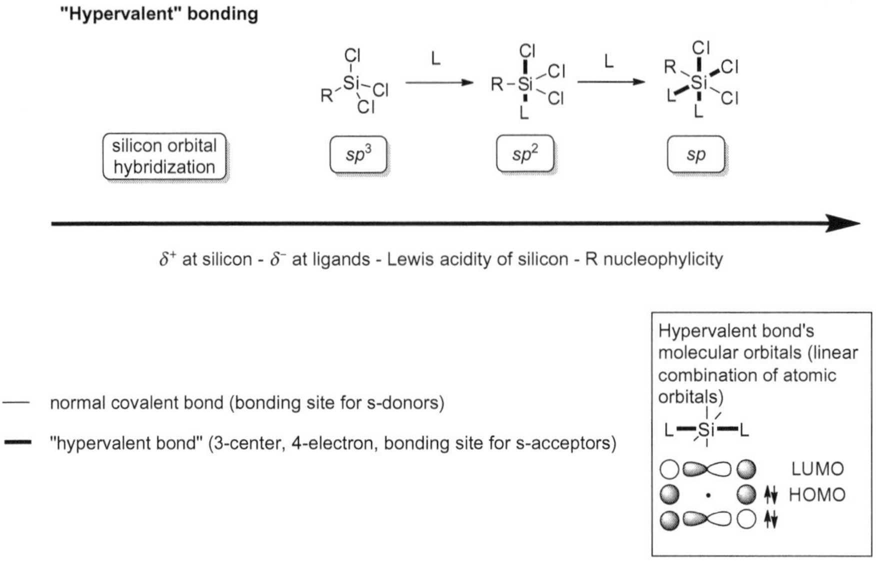

FIGURE 14.2. Hypervalent bonding in silicon species. LUMO, lowest unoccupied molecular orbital; HOMO, highest occupied molecular orbital.

p-orbital and two *p*-orbitals of electronegative ligands featuring a relative *trans*-disposition. An important consequence is the nonequivalence of the ligand positions in five- and six-coordinated silicon species, the σ-acceptor ligands preferring "hypervalent" bonds and the σ-donors forming preferentially normal covalent bonds with the sp^2 (for pentacoordianted compounds) or *sp* (for hexacoordinated compounds) silicon orbitals. The presence of hypervalent bonds imposes some stereochemical constraints (like the *trans*-disposition of the most electronegative ligands) and it allows the formulation of predictions about the positions of the other ligands on the base of their electronic properties. Accordingly, the number of possible configurations of the silicon ligands to be considered in the elaboration of a stereoselection model is actually restricted, as shown in a recent paper by Denmark and coworkers [12].

Both theories are helpful in the interpretation of the fundamental properties of hypervalent silicon species that clearly distinguish their reactivity from that of four-coordinated compounds, such as the increased Lewis acidity of the silicon atom and the transfer of electronic density to the ligands, which confer to silicon-bound R groups (carbanion or hydride equivalent) marked nucleophilic properties. The hypervalent silicon species involved in synthetically useful processes are generally formed in situ by reaction between a four-coordinated species and a Lewis base in what is often called the "activation step" [8, 11]. The so-formed five- or six-coordinated silicon species is able to promote the desired reaction in a catalytic process if the base can dissociate from the silicon after the product is formed.

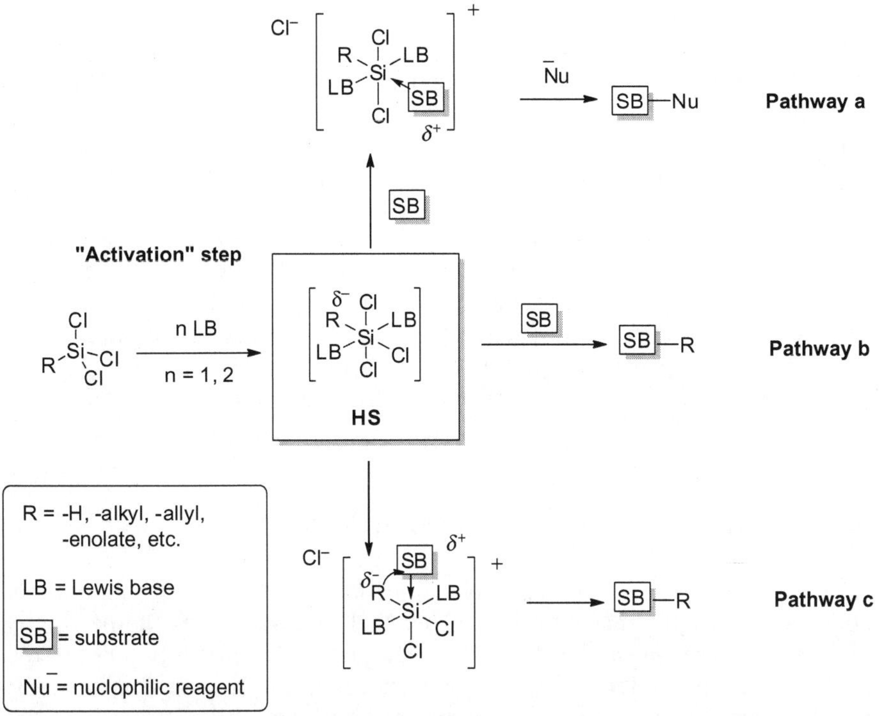

FIGURE 14.3. Reactions involving hypervalent silicon species.

Three general kinds of reaction mechanism can be envisaged depending on the role played by the hypervalent species (HS) (Fig. 14.3): (1) the HS may act as a Lewis acid coordinating the substrate and activating it toward the attack of an external nucleophile (Fig. 14.3, pathway a); (2) as in pathway b of Figure 14.3, a nucleophilic silicon ligand is transferred to the substrate that is not coordinated by silicon; and (3) the HS coordinates the substrate transferring at the same time one of its ligands to it (Fig. 14.3, pathway c). In the last case, both of the peculiar properties of hypervalent silicon species are so exploited at the same time. When a mechanism of type C is operating, the cyclic transition state (TS) allows an efficient control of the relative stereochemistry of the product.

This classification should be helpful for an immediate comprehension of the mechanistic details that are discussed in the following sections. However, it is not used as the criterion of organization of this chapter, since the actual mechanism of several reactions is still a matter of discussion. For example, in the trichlorosilane-mediated reduction of carbon–nitrogen double bond, the activation of a chiral Lewis base is needed in order to activate the reducing agent and to control the stereochemistry of the process. However, a cationic silicon species has been never proposed so far, even if an octahedrally coordinated silicon species is usually invoked to explain the sense of steoreselection. A cation silicon species is believed to be the catalytically active species in most of the C–C bond formation reactions. In the present chapter, we prefer to report on the more important transformations promoted by hypervalent silicon species, organized by type of reactions: first reductions will be discussed, followed by carbon–carbon bonds synthesis; opening of epoxides and other miscellaneous reactions will conclude the survey. Trimethylsilyl cyanide addition to carbon–nitrogen double bonds will not be discussed, since it is still under debate if all of these reactions actually involve a hyperconjugated silicate species. However, enantioselective versions of silyl cyanide reactions with carbonyl and imine derivatives have been described and recently reviewed, with a special attention to the addition of cyanide anione to imines to afford α-amino nitriles, key intermediates for the synthesis of α-amino acids [10, 11].

14.2. STEREOSELECTIVE C–H BOND FORMATION

Since chiral amines are finding applications in an always increasing number of fields, such as pharmaceutical, agrochemical, and fragrances, the possibility to develop an organocatalytic approach has gained much attention because it might represent a solution to the problems related to the presence of toxic metal, whose leaching could contaminate the product [13].

Among the metal-free methodologies recently developed [14], the use of trichlorosilane as a reducing agent is particularly attractive. This cheap reagent is a colorless liquid, easily prepared by the silicon industry, which has already been employed in large scale for transforming phosphine oxide to phosphine and *N*-acyliminium ion to *N*-acylamine. Trichlorosilane needs to be activated by coordination with Lewis bases, such as *N,N*-dimethylformamide (DMF), acetonitrile, and trialkylamines, to generate hexacoordinated hydridosilicate, the real active reducing agent that operates under mild conditions. The use of chiral Lewis bases offer the possibility to control the absolute stereochemistry of the process and it has been widely explored in the last few years, leading to the development of some really efficient catalysts. The catalytic systems may be classified as (1) *N*-formyl derivatives, which may be historically considered the first class of compounds developed as chiral activators of trichlorosilane, (2) chiral picolinamides, deeply investigated in the very last few years, and (3) other Lewis basic compounds.

14.2.1. Reactions Catalyzed by *N*-Formyl Derivatives

The first example of stereoselective catalytic reduction with $HSiCl_3$ was reported in 1999: Matsumura and co-workers reported *N*-formyl cyclic amine compounds [basically (*S*)-proline derivatives] to be effective activators to reduce ketones in the presence of trichlorosilane [15a]. A catalytic amount of

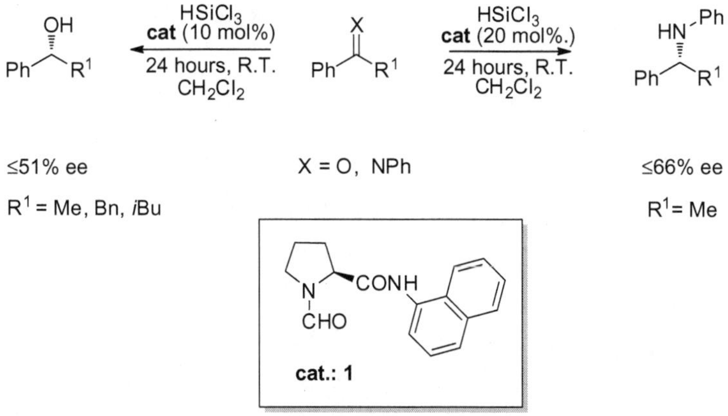

SCHEME 14.1. *N*-formyl proline as chiral promoter of $HSiCl_3$-mediated reactions.

these Lewis bases was used for obtaining enantiomerically enriched secondary alcohols [up to 51% enantiomeric excess (ee), Scheme 14.1]. Two years later the same group found that triclorosilane, activated with the same chiral Lewis basic *N*-formylproline, is an effective reagent for chemo- and stereoselective reduction of imines (Scheme 14.1). The corresponding amines were isolated in moderate yields with up to 66% ee [15b].

Matsumura et al.'s contribution in designing *N*-formyl pyrrolidine derivatives as $HSiCl_3$ activators can be considered as a milestone for the asymmetric reduction of ketones and imines using $HSiCl_3$ as a reducing agent, and paved the road to the synthesis of other related systems. Since then, considerable efforts have been devoted to the development of efficient catalysts for the reduction of carbon–nitrogen double bonds, and remarkable progress has been made.

One important breakthrough in the field was achieved by Malkov and et al. in 2004, when they developed the first highly stereochemical efficient catalyst; the compounds of choice are *N*-methyl-(*S*)-valine-derived Lewis basic organocatalysts type 2, commercially available since 2009 (Scheme 14.2) [16].

Two years later [17], the same authors reported a detailed investigation of the reduction of imines with $HSiCl_3$ catalyzed by *N*-methyl-(*S*)-aminoacid derivatives. A library of chiral *N*-formylated aminoacids was designed and synthesized, with structural variations at the carboxyamide group, with either aromatic or aliphatic substituents. The reaction was carried out in nonpolar solvents; toluene was chosen for its relatively low environmental impact. Different substituted *N*-aryl ketimines have been tested as substrates.

After the screening of a variety of *N*-methyl-(*S*)-amino acid in the reduction of ketomines, valine was selected as the chiral element of choice to perform stereocontrol and a few conclusions were proposed: (1) the *N*-methyl formamide moiety of the catalyst is fundamental for the enantioselectivity;

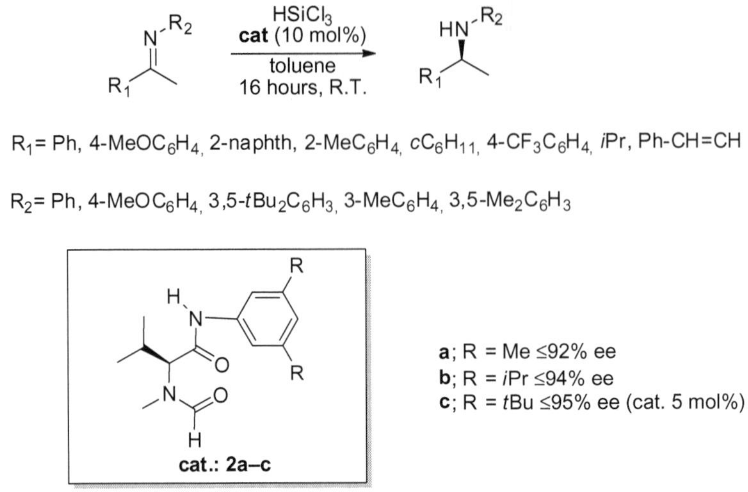

SCHEME 14.2. *N*-formyl derivative of *N*-methyl-(*S*)-valine.

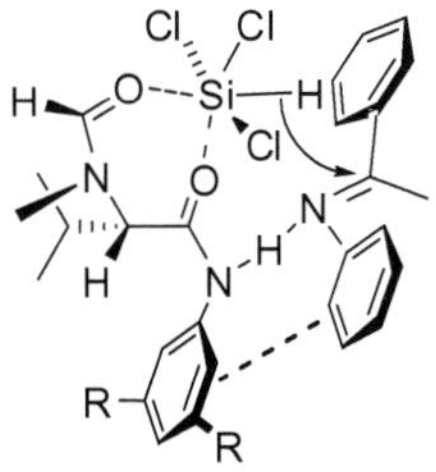

FIGURE 14.4. Model of stereoselection for (*S*)-valine-derived organocatalyst.

(2) arene–arene interactions may play an important role in determining the stereoselectivity of the catalyst; (3) the anilide moiety of the catalyst has to be a secondary amide (i.e., with an NH group); (4) the silicon atom is activated by coordination with formamide moiety; (5) the configuration of the resulting product depends on the nature of the aminoacid side chain; and (6) bulkier groups in the 3,5-positions of the aromatic ring (diisopropyl and di-*tert*-butyl) determine an increase of enantioselectivity in the reduction of aromatic and nonaromatic ketoimines.

Catalyst–substrate hydrogen bonding and coordination of the silicon atom by the two carboxyamide groups were suggested to play a fundamental role in determining the stereoselectivity of the catalyst. In the proposed TS, an additional element of stereocontrol is the formation of the hydrogen bond between the amide group of the catalyst and the substrate (Fig. 14.4).

The general applicability of catalyst **2c** (called Sigamide, Scheme 14.3) was then investigated in the reduction of multifunctionalized ketoimines bearing

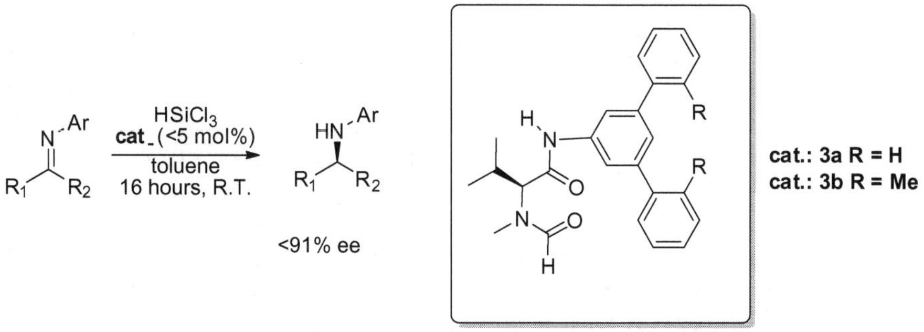

SCHEME 14.3. New (*S*)-valine-derived organocatalysts.

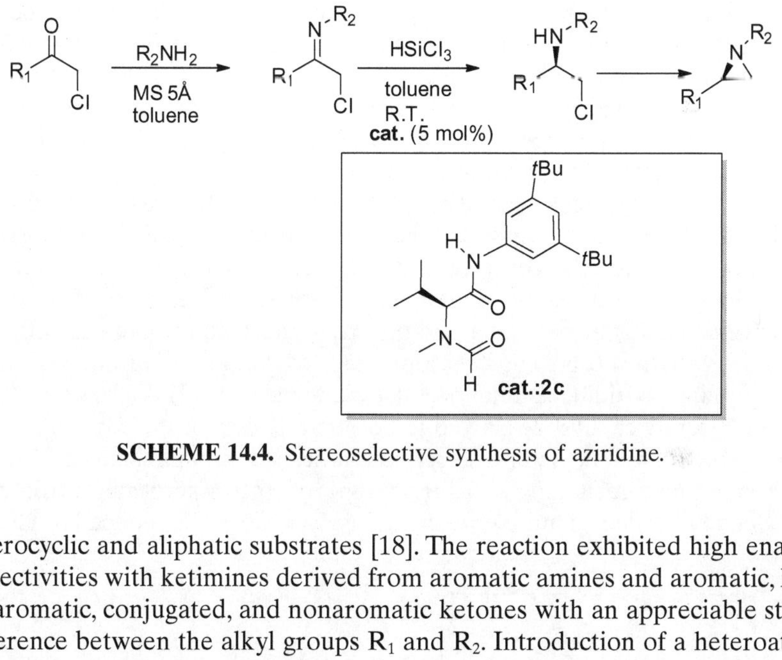

SCHEME 14.4. Stereoselective synthesis of aziridine.

heterocyclic and aliphatic substrates [18]. The reaction exhibited high enantioselectivities with ketimines derived from aromatic amines and aromatic, heteroaromatic, conjugated, and nonaromatic ketones with an appreciable steric difference between the alkyl groups R_1 and R_2. Introduction of a heteroatom into the aromatic system (pyridyl derivatives) afforded the products with almost no enantioselection, probably due to the competition of the substrate pyridine nitrogen with the catalyst in the coordinating of the silicon atom of $HSiCl_3$.

Recently, two new (*S*)-valine-derived organocatalysts (**3a**,**b** Scheme 14.3) bearing a bulky aromatic substituent at the amidic nitrogen were synthesized [19]. The efficiency of the new compounds was tested in the model reduction of ketoimines derived from aryl methyl ketones and it was found to be slightly inferior compared with that of Sigamide **2c** (Scheme 14.4).

The recoverability of this family of catalysts has also been studied (see also Chapter 2). A fluorous-tagged catalyst was shown to operate in solution, with very little difference in terms of catalytic efficiency with respect to the untagged version [20a]. The products were separated from the catalyst by filtration though a pad of fluorous silica; the catalyst was easily recovered and recycled.

Recently, another study on the development of a polymer-anchored version of the same catalyst was reported; different supports were investigated, such as Merrifield, and extended Merrifield, Wang, TentaGel, and Marshall resins were all employed to immobilize the organocatalysts through an ethereal bond [20b]. The enantioselective reduction of *N*-aryl ketoimines in the presence of trichlorosilane was performed by employing typically 15–25 mol% amount of the supported catalyst, a higher loading than that one used with the nonsupported system (typically 5%–10% mol cat.). The immobilized catalysts showed a remarkable dependence on the reaction solvent; while the nonsupported organocatalysts work well in toluene, the polymer-anchored species behave much better in chloroform. By operating in the best experimental condition, with the Merrifield-anchored catalyst, the product was isolated in good yield and in 82% ee, which is, however 10% lower than the enantioselectivity obtained with the nonsupported catalyst. After filtration of the immobilized organocatalyst, it was possible to reuse it five times while always maintaining the same level of stereoselectivity, but a reactivation step was required.

The use of (*S*)-valine-derived formamide was extended also to the reduction of α-chloro-imines. These compounds were generated in situ from corresponding α-chloro-ketones and aniline derivatives. The reduction in the presence of $HSiCl_3$ at room temperature gave the α-chloro-amines with high enantioselectivities (up to 96% ee) and good yields, and after cyclization the corresponding aziridine as final product (Scheme 14.5) [21].

Very recently catalyst **2c** proved to be suitable also for the development of a new protocol for the enantioselective synthesis of β-aminoacids derivatives from enamine precursors [22]. Treatment of the β-ketoester/nitrile with p-anisidine afforded enamines, which themselves cannot be reduced by Cl_3SiH.

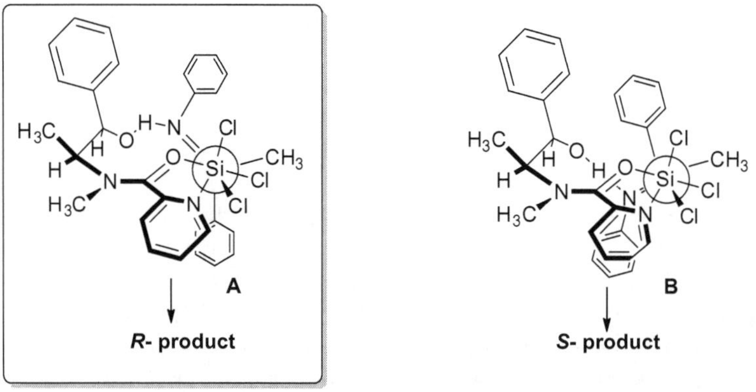

FIGURE 14.5. Proposed model of stereoselection for reduction of *N*-aryl ketoimines.

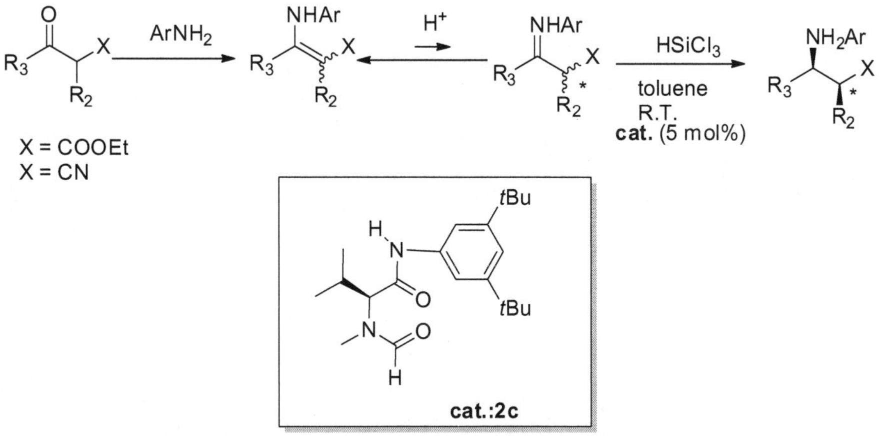

SCHEME 14.5. Stereoselective reduction of enamines.

Because the enamine–imine equilibration is facilitated by Brønsted acids, a number of acid additives were examined, among which AcOH (1 eq. of which was used) emerged as a good compromise between reactivity and selectivity. Enamine was reduced to give the amino ester in high yield and 89% ee. The enantiomerically pure product could be obtained by a single crystallization. Nitriles exhibited the same behavior as the esters in terms of reactivity. The authors then focused on the synthesis of β-2,3-amino acids. In this case, since the starting achiral enamines are in fast equilibrium with the corresponding chiral racemic imines, the reaction can be considered as a dynamic kinetic resolution (Scheme 14.5). The corresponding amino esters and amino nitriles were prepared in good yields and high enantioselectivities (≤90% ee) and diastereoselectivities [≤99% diastereoisomeric excess (de)].

Following their previous studies with prolinamides, in 2006 Matsumura et al. reported the activity of *N*-formyl proline derivatives in the reduction of ketones in the presence of trichlorosilane [23]. Secondary alcohols could be synthesized with high enantioselectivity (up to 97%, Scheme 14.6), employing a catalytic amount of *N*-formyl-α'-(2,4,6-triethylphenyl)-(*S*)-proline (catalyst **4**). The selection of the best performing compound was the result of the screening of a series of α'-arylproline derivatives. Both the carbonyl group at the α-position and a 2,4,6-triethylphenyl group at the 5-position in the proline ring play an important role for determining the high enantioselectivity.

More recently the use of *N*-formyl L-pipecolinic acid derivatives as organocatalysts was also explored in the reduction of aromatic and aliphatic ketones [24]. The best catalyst for the reduction of carbonyl compounds was found to be **5**, characterized by the presence of the methoxy group on carbon C2′ of the chiral aminoalcohol moiety (Scheme 14.6). The methoxy functional group on that position turned out to be crucial for obtaining alcohols with high enantioselectivity. Indeed, the replacement of this mojety with either a bigger

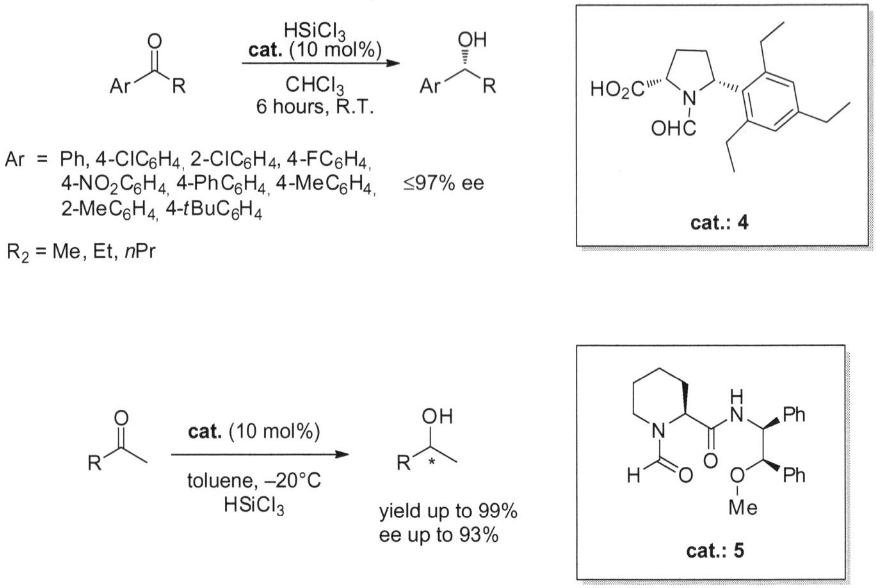

SCHEME 14.6. Stereoselective reduction of ketones.

alkoxy group or a group with a less electron-rich 2-oxygen led to a decreased reactivity and/or enantioselectivity of the catalyst. According to the authors, catalyst **5** works as a tridentate activator and promotes the hydrosilylation of ketones through a heptacoordinate silicon transition structure.

In 2007, Tsogoeva et al. [25] reported the use of new chiral formamides in the reduction of ketoimines in the presence of trichlorosilane. A second element of stereocontrol, such as a chiral amine, was added to the proline mojety. Catalyst **6**, the *N*-formyl prolinamide of (*R*)-α-methyl amine, activated trichlorosilane in the ketoimine reduction, affording the product in 75% yield and 81% ee in the presence of an additive. Hexamethylphosphoric triamide (HMPA) and p-nitrobenzoic acid were tested as additives, and, quite surprisingly, the former turned out to be the more effective one.

In the same year, Sun et al. [26] reported the (*S*)-proline-derived C_2-symmetric chiral tetraamide **7** (Scheme 14.7) as a novel catalyst in the enantioselective hydrosilylation of ketomines. The choice of catalyst **7** is the result of the screening of a series of C_2-symmetric chiral tetraamide derivatives where the linkage of the two proline diamide units proved to have a significant impact on the enantioselectivity. Either a shorter or longer linkage and aromatic linkages provided products with much lower enantioselectivities.

The reaction products were isolated in high yields (up to 95%) and moderate to high enantioselectivities (up to 86% ee) for a broad range of substrates, including aromatic and aliphatic imines. The two diamide units in the chiral Lewis base work cooperatively, showing a synergistic effect.

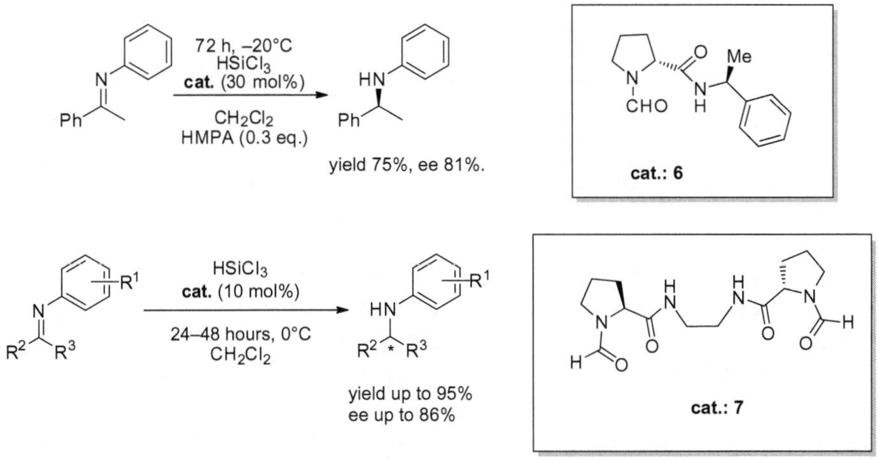

SCHEME 14.7. Novel *N*-formyl chiral Lewis bases.

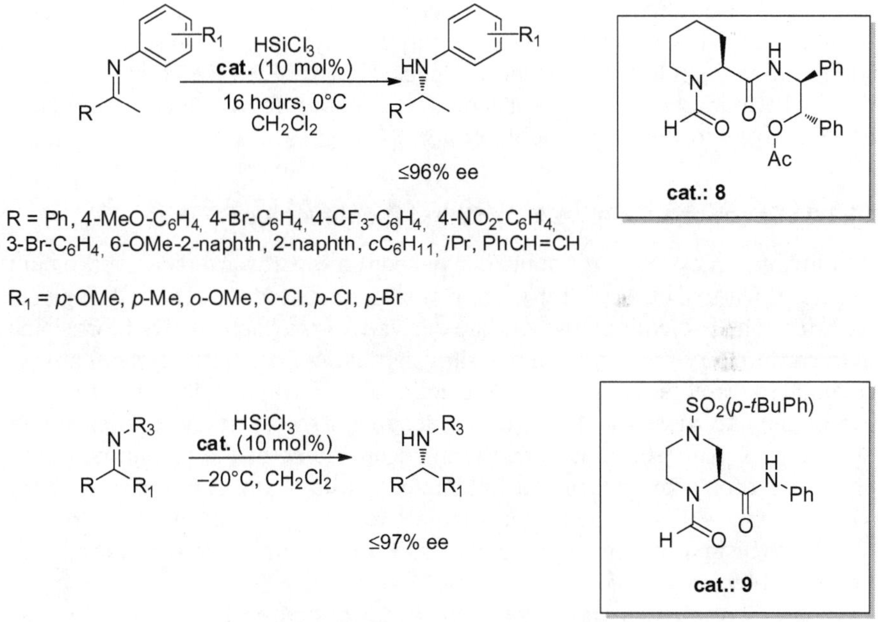

SCHEME 14.8. Chiral pipecolinic acid derivatives.

In 2006, Sun et al. reported that switching from the five-membered ring of proline to a six-membered ring had a beneficial effect on the enantioselectivity. The first catalyst [27] derived from L-pipecolinic acid, compound **8**, promoted the reduction of *N*-aryl ketoimines with trichlorosilane with high yields and good enantioselectivities (Scheme 14.8).

Later, a piperazinyl backbone was employed as a building block for the construction of a new catalyst [28]. The arenesulfonyl group on the 4-position group has been shown to be a key element for obtaining a high level of enantiocontrol (Scheme 14.8). Catalyst **9** promoted the reduction of a broad range of imines with good yields and enantioselectivities.

Very recently, Schreiner et al. have published a detailed investigation of the influence that nonaromatic groups in N-formylprolinamide may have on the ee's of ketimine reductions, by also employing computational methods in the attempt to get some mechanistic insights of the process [29]. By working with a series of novel chiral organocatalysts derived from proline, valine, and pipecolinic acid, the dominant role of the amino acids scaffold for the mode of action in the enantiodifferentiating step was demonstrated. Mechanistic studies by means of density functional theory (DFT) computations seem to confirm that the catalyst not only coordinates to trichlorosilane, but also reacts as a proton donor in the crucial transition structure; indeed, the importance of the presence of acidic NH proton of a secondary amide group, able to bind with the basic nitrogen of the reacting imine, has been demonstrated. Although the authors suggest that enantiodifferentiating steps for proline, pipecolinic acid, and valine-derived catalysts may be different, based on the computational studies they propose a general picture for the catalytic reduction of ketimines with trichlorosilane, which could be described as a formal H^+/H^-transfer to the C=N double bond.

14.2.2. Reactions Catalyzed by Chiral Picolinamides

A contribution by Matsumura in 2006 paved the way toward the development of a novel class of catalysts for trichlorosilane-mediated reductions. His group reported that *N*-picolinoylpyrrolidine derivatives may activate trichlorosilane in the reduction of aromatic imines, showing that *N*-formyl group is not always essential for catalytic activity [30]. *N*-Picolinoyl-(2*S*)-(diphenylhydroxymethyl)-pyrrolidine **10** gave the best results, leading to enantioselectivities up to 80%. It was proposed that both the nitrogen atoms of the picolinoyl group and the carbonyl oxygen play a fundamental role in the coordination of the silicon atom, while the hydrogen of the hydroxy group is supposed by the authors to be involved in a hydrogen bond with the nitrogen atom of the imine (Scheme 14.9).

Based on these seminal works, our group has recently focused onto the design and synthesis of a wide class of catalysts prepared by simple condensation of a chiral aminoalcohol with picolinic acid or its derivatives. While our investigation led to a patent deposit [31], at the same time Zhang et al. independently reported in a preliminary communication the use of ephedrine and pseudoephedrine-derived picolinamides in the reduction of *N*-aryl and *N*-benzyl ketimines promoted by trichlorosilane [32]. With catalyst **11** easily prepared from 2-picolinic acid and (1*R*,2*S*)-ephedrine, a variety of *N*-aryl ketimines and *N*-benzyl ketimines were reduced with trichlorosilane in high

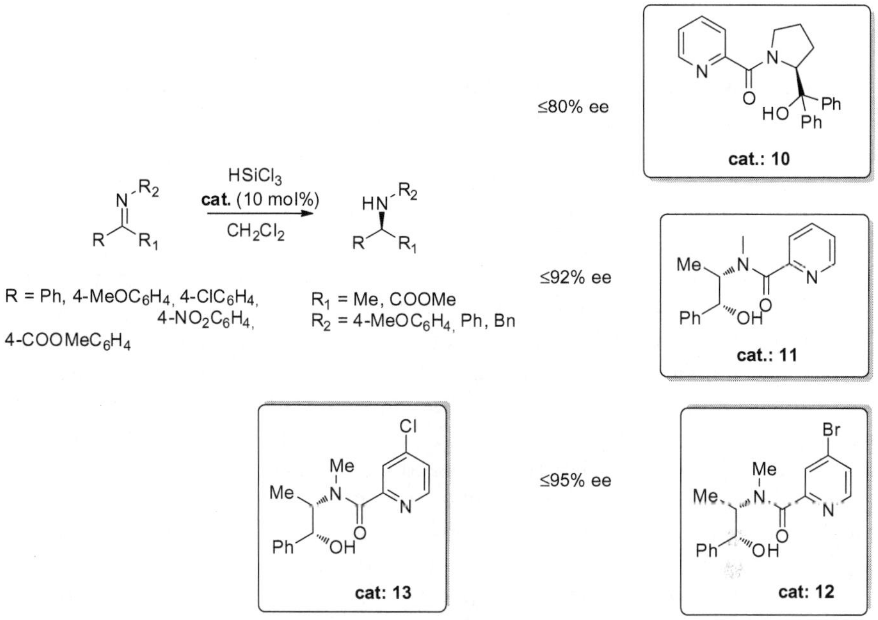

SCHEME 14.9. Chiral *N*-picolinoyl derivatives.

yields (<93%) and moderate to excellent ee values (<92%) under mild conditions (Scheme 14.9).

Our group systematically investigated this class of organocatalysts [33]. In a single-step procedure, several derivatives were synthesized simply by reaction of picolinic acid and different enantiomerically pure amino alcohols mediated by condensing agents or by reaction of picolinoyl chloride and the amino alcohol. The pyridine ring, the free hydroxyl group, and *N*-alkyl substitution in the aminoalcohol portion were identified as key structural elements, necessary to secure good stereocontrol; the effect of different substituents at the nitrogen atom and at the two stereocenters were also studied. By studying several, differently substituted derivatives it was shown that the introduction of a proper substituent in the 4-position of the pyridine moiety could improve catalyst efficiency. Indeed, 4-bromo and 4-chloro picolinic derivatives **12** and **13** showed especially remarkable catalytic properties. Working at 0°C in dichloromethane (DCM) with catalyst **13** the chiral amine was obtained in quantitative yield and 83% ee; enantioselectivity was increased up to 88% by working in chloroform. A further improvement was observed by performing the reaction at –20°C when enantioselectivity reached 95% with no erosion of the chemical yield, being the reduction product isolated, basically, in quantitative yield. Even by working with 1 mol% amount, catalyst **13** promoted the reduction in 90% yield after only 2 hours.

The screening of systematically modified organocatalysts of this family led to identifying the key structural factors that influence their catalytic properties and to proposing a tentative model of stereoselection observed in the reaction promoted by picolinamide derivatives. In this model, pyridine nitrogen and the CO amidic group of picolinamide activate trichlorosilane by coordination; the hydrogen atom of hydroxyl group plays a fundamental role in coordinating the imine through hydrogen bonding. The presence of two stereogenic centers on the aminoalcohol moiety with the correct relative configuration such as in (1*R*, 2*S*)-(−) ephedrine is necessary to stereodirect the imine attack by trichlorosilane. The methyl groups on the amide nitrogen and on the stereocenter in position 2 of the aminoalcohol chain apparently have the optimum size for maximizing the enantiodifferentiation of the process. In the proposed stereoselection model **A**, leading to the major enantiomer, the steric interaction between the pyridine ring and *N*-aryl group is much less significant than that observed in adduct **B**, which is thus disfavored.

Good results were also obtained in the enantioselective reduction of *N*-alkyl imines [34], a transformation that only recently has been accomplished organocatalytically [35]. Ephedrine-based picolinamides promoted the reaction of *N*-butyl imine of acetophenone in excellent yields and high stereoselectivities: under the best conditions (chloroform, 0°C, 24 hours), 4-chloropicolinic derivative **13** promoted the reduction in 98% yield and 91% ee.

The organocatalysts have several positive features: they are easily prepared, by a single condensation step, from commercially available compounds; they are low-cost catalysts, the source of stereocontrol being a very cheap and largely available aminoalcohol, such as ephedrine; and the reduction of carbon–nitrogen double bond is performed under very mild reaction conditions and with an extremely simple experimental procedure that allows to obtain, after an aqueous workup, a highly pure product. A very convenient enantioselective organocatalytic three-component methodology was also developed; the reductive amination process, starting simply by a mixture of a ketone and an aryl amine, opens an easy access to chiral amines with a straightforward experimental methodology. All these positive features make the present catalytic method, in principle, also suitable for large-scale applications; its synthetic potentiality was indeed demonstrated by successfully employing the present metal-free catalytic procedure in the preparation of (*S*)-metolachlor, a potent and widely used herbicide [33].

In order to further improve the selectivity of the process, the trichlorosilane-mediated reduction was accomplished on ketoimines derived from (*R*)-1-phenyl-ethyl amine (Scheme 14.10). It was found that a catalytic amount of *N,N*-DMF was able to promote $HSiCl_3$ addition with good stereoselectivity, although in low yield [36]. By optimizing the reaction conditions it was shown that best results were obtained at −50°C in chlorinated solvents by performing the reduction with 6 eq. of DMF. In these conditions, *N*-α-methyl benzyl imines of methyl aryl ketones of different electronic properties were effectively reduced to the corresponding secondary amines in quantitative yields, with 90%–99% diastereoselectivity.

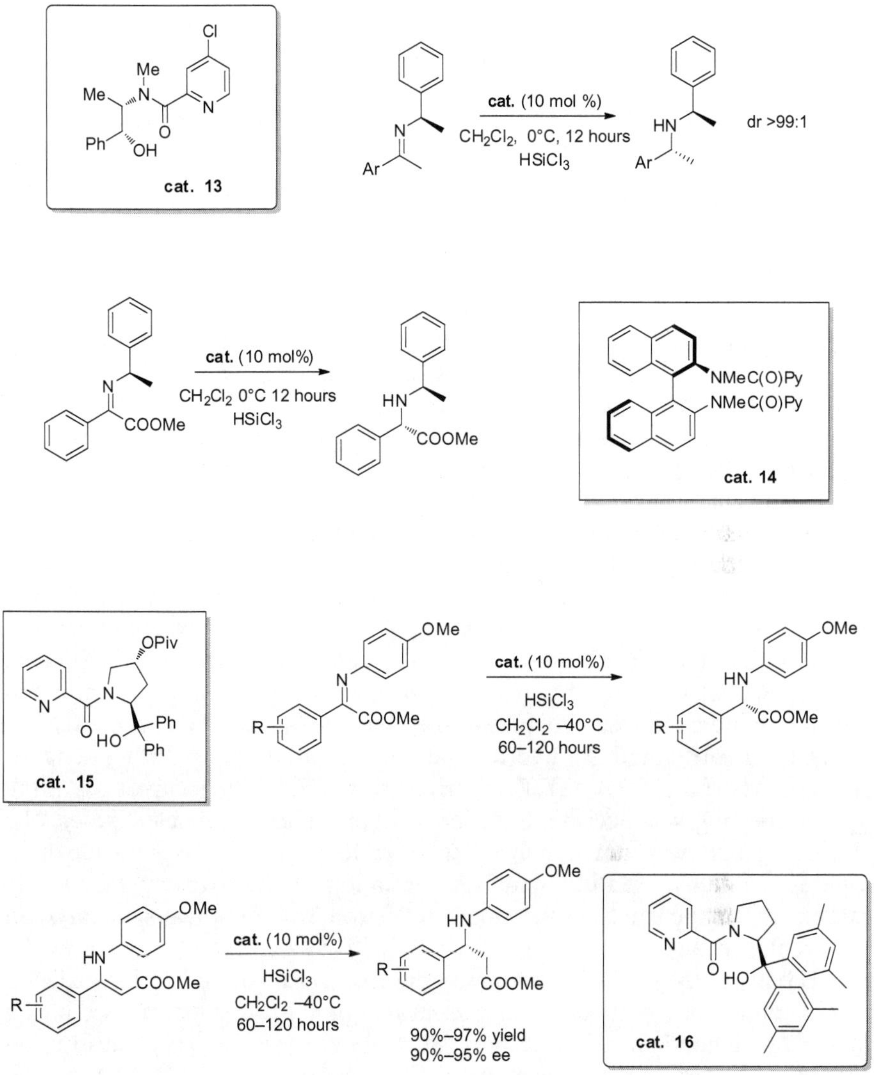

SCHEME 14.10. Stereoselective catalytic reduction of chiral imines and imino esters.

However, when chiral picolinamide **13** was employed as the catalyst, the control of the stereoselectivity was total, as a demonstration of the presence of a cooperative effect of Lewis basic catalyst with the (*R*)-methyl benzyl residue at the imine nitrogen [37].

The methodology was extended to the synthesis of an enantiomerically pure secondary amine of both C_1 and C_2 symmetry. Also, imine derived from methyl isobutyl ketone was readily reduced in >98% yield in the presence of catalyst **13** to afford an enantiomerically pure direct precursor of (*R*)-isopropyl methyl amine. The methodology is attractive since the combination of low-cost, easy to make metal-free catalyst and an inexpensive chiral auxiliary allowed to reduce ketimines with different structural features, often with total control of the stereoselectivity.

The wide applicability was also demonstrated in the preparation of α-amino esters (Scheme 14.10). Catalyst **13** promoted the reduction of *N*-benzyl iminoester in quantitative yield and up to 71% ee. However, with the same catalyst the reduction of *N*-α-methyl benzyl imines of methyl phenylglyoxilate at 0°C in DCM afforded the corresponding chiral aminoester in 73% yield and 91% de.

Our group also developed a second class of chiral picolinamides as efficient chiral organocatalysts for trichlorosilane-mediated reactions. Picolinic acid was condensed with (*R*)-*N*,*N'*-dimethyl amino binaphthyl diamine to afford catalyst **14** in 73% yield after chromatographic purification. Noteworthy binaphthyldiamine-derived bis-picolinamides showed a remarkable activity in performing the reduction of *N*-aryl (up to 83% ee), *N*-benzyl (up to 87% ee), and *N*-alkyl ketoimines (up to 87% ee), by working typically at 0°C [38]. Furthermore, catalyst **14** was employed also in the imine reduction of *N*-benzyl imine of ketoesters, although with less success (71% ee) [39].

Very recently, Zhang et al. also reported an efficient protocol for the organocatalytic synthesis of α-amino esters [40]. A novel chiral Lewis base organocatalysts derived from *trans*-4-hydroxy-L-proline was developed (Scheme 14.10); it is worth noting that the catalyst of choice, compound **15**, exhibited only moderate enantioselectivities in the hydrosilylation of *N*-aryl β-enamino esters, but it promoted the hydrosilylation of imino esters with high enantioselectivities (up to 93% ee). The introduction of a bulky group at C4 of the pyrrolidine ring was decisive in order to obtain high stereoselectivities. The hydroxy group was functionalized with various bulky groups; while benzylation, trimethylsilylation, and isovalerylation of the hydroxy group only caused marginal changes in the enantioselection, when *O*-pivaloyl catalyst **15** was employed, an increase in enantioselectivity was observed. In exploring the applicability of the catalyst to imines of differently substituted aryl glyoxilates, it was found that both *para*- and *meta*-functionalized substrates could be reduced with good enantioselectivity (80%–93% ee), while *ortho* substitution caused a decrease of stereoselection (50%–60% ee).

The same group has employed picolinamide derivatives of prolinol to also realize an efficient reduction of enamine [41]. Chiral *N*-picolinoylpyrrolidine derivatives and *N*-picolinoylephedrine were evaluated in hydrosilylation of (*Z*)-methyl 3-phenyl-3-(phenylamino)acrylate, leading to the corresponding reduction product in good enantioselectivities in chloroform at 0°C. The enantioselectivity increased slightly along with the increase of the size of the aryl groups in the catalysts. The best yield and enantioselectivity were obtained with catalyst **16** at –30°C for 48 hours. Under the optimized conditions, the generality of the Lewis base organocatalyzed hydrosilylation of various β-enamino esters were examined. In the presence of 10 mol% of Lewis base, β-enamino esters were reduced in high yields and enantioselectivities typically ranging from 90% to 95% ee. It is worth mentioning that *N*-acyl *β*-enamino esters were totally inactive in the present organocatalytic system. The reaction is supposed to proceed through the imine tautomer rather than the enamine tautomer. In the proposed mechanism, the nitrogen atom of the pyridine ring and the carbonyl oxygen atom of catalyst are coordinated to Cl_3SiH, while the imine is activated by the hydroxy group of the Lewis base through hydrogen bonding. It has also been hypothesized that a stabilization due to arene–arene interactions between the aromatic systems of the catalyst and the substrate may occur.

14.2.3. Reductions Catalyzed by Other Chiral Lewis Bases

In 2006, a novel chiral Lewis basic system was reported by Malkov et al.: chiral oxazolines bearing an isoquinoline fragment. Catalyst **17** has been employed in the reduction of aromatic ketones and imines with trichlorosilane providing the products with a good level of enantioselectivity (Scheme 14.11) [42].

The maximum level of enantioselectivity reached in the reduction of ketones was 87%, while even better results were achieved in the ketoimines reduction, where 92% enantioselectivity was reached. The authors hypothesized that coordination of the trichlorosilane by the catalyst would generate a chiral hexacoordinated silicon species that would be the actual reducing

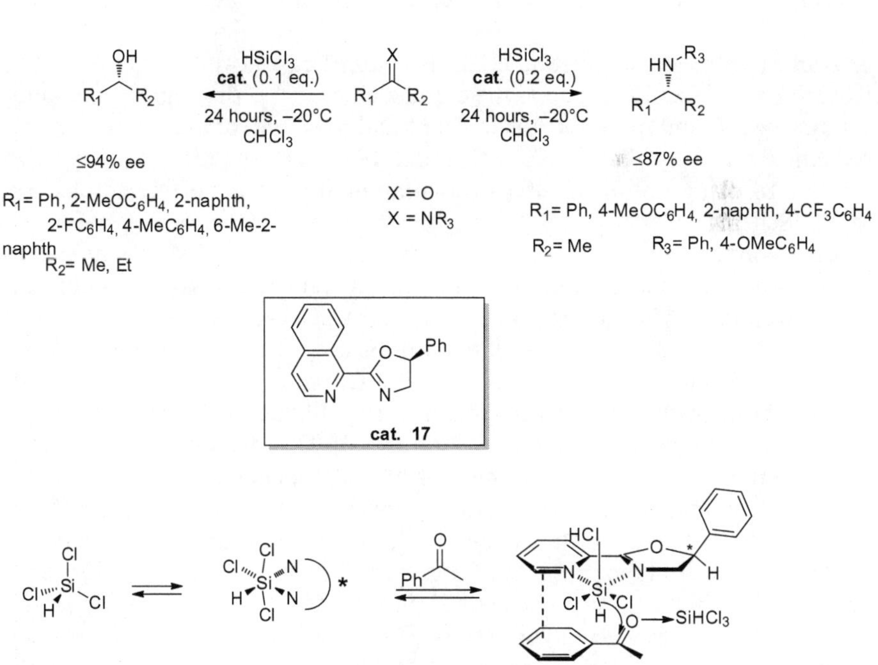

SCHEME 14.11. Stereoselective catalytic reduction promoted by oxazoline-based chiral catalyst.

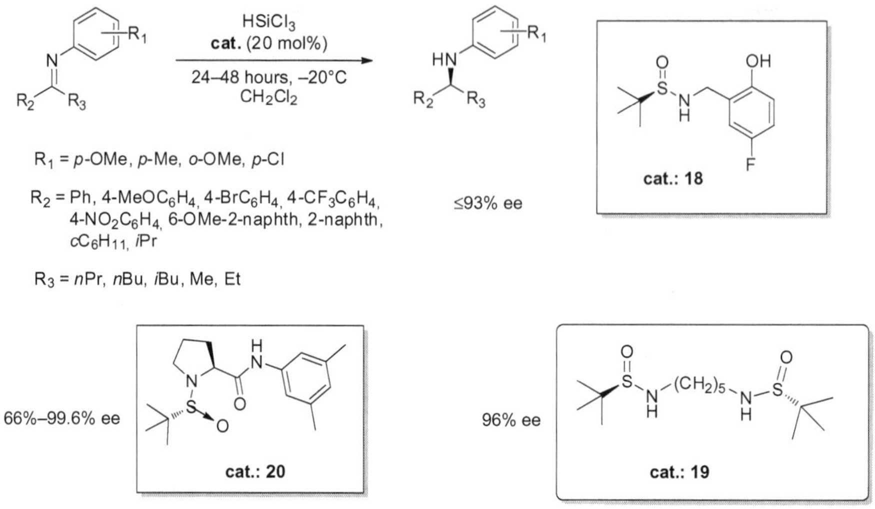

SCHEME 14.12. Chiral sulfinamide as promoter of imine reduction.

species. When a ketone is the reactive substrate, further activation would be provided by coordination of a molecule of trichlorosilane by the carbonyl oxygen.

Almost at the same time Sun et al. also published a novel designed catalyst featuring a sulfinamide group as the chiral element [43]. This family of organocatalysts was found to be able to activate trichlorosilane for the stereoselective reduction of *N*-aryl ketimines with good yield and enantioselectivity, with catalyst **18** being the most successful compound in terms of stereoselection levels (Scheme 14.12).

Based on the assumption that the mechanism would involve two molecules of Lewis base for the activation of $HSiCl_3$, a novel chiral bis-sulfinamide was then developed. After a screening of different derivatives, the compound of choice was found to be a bis-sulfinamide bearing a five-methylene linkage. Catalyst **19** promoted the reduction of the model substrate, *N*-phenyl imine of acetophenone with 96% ee (Scheme 14.12). Sulfinamide **20** represents the most recent catalyst prepared by Sun et al. [44]. This derivative incorporates two different elements responsible for the stereochemical control of the process: a sulfinamide group with a stereogenic sulfur atom and an *N*-aryl prolinamides. This system has been employed in the reduction of aromatic *N*-alkyl ketimines in the presence of trichlorosilane, providing the corresponding amines with good enantioselectivities (up to 99.6% ee) and high yields.

A clearly innovative catalytic system was reported by Nakajima et al., who introduced chiral phosphine oxides as suitable Lewis bases for activating trichlorosilane in stereoselective transformations. Indeed, trichlorosilane has been used in the conjugate reduction of α,β-unsaturated ketones in the

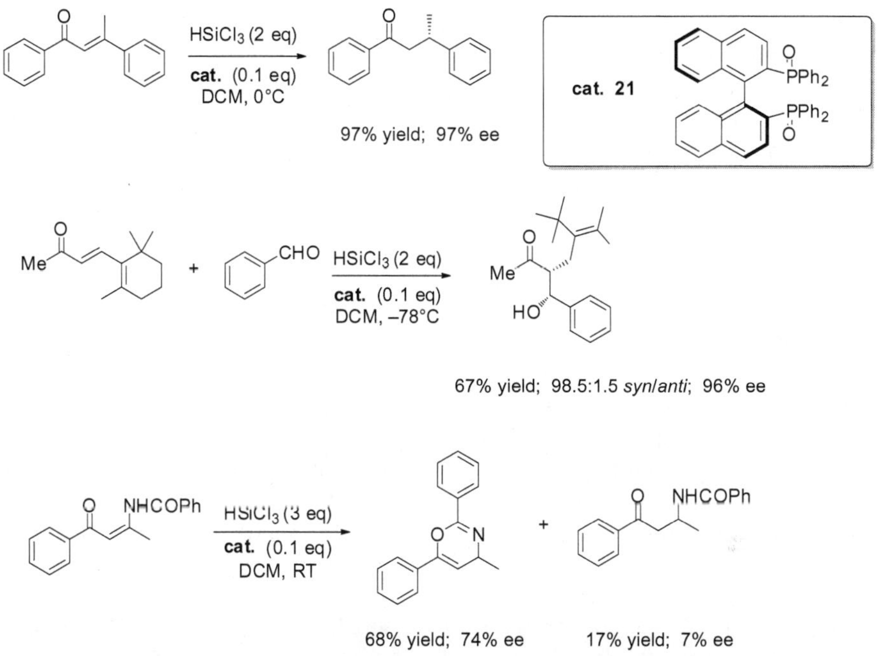

SCHEME 14.13. Chiral phosphine oxide-catalyzed reductions.

presence of a catalytic amount of a chiral Lewis base. The reduction of 1,3-diphenylbutenone promoted by catalytic amounts of 2,2′-Bis(diphenylphosphanyl)-1,1′-binaphthyl dioxide **21** (BINAPO) at 0°C was successfully accomplished, leading to the corresponding saturated compound in 97% yield and a somehow surprising, but very good, 97% ee [45].

An alternative methodology for organocatalytic conjugate reduction of enones and subsequent reaction with aldehydes to realize a reductive aldol reaction was developed. The idea was to activate the silane with a suitable Lewis base to perform the 1,4-reduction via a six-membered TS; then, with the assistance of the same Lewis base, the generated trichlorosilyl enolate should react with the electrophilic aldehyde (Scheme 14.13).

Triphenylphosphine oxide was shown to be able to catalyze the three-component reaction of chalcone, benzaldehyde, and trichlorosilane (reductive aldol reactions) to afford the corresponding aldol product in 78% yield. Preliminary experiments with (*S*)-BINAPO as chiral Lewis base were very promising in terms of stereocontrol. 2,2′-Bis(diphenylphosphanyl)-1,1′-binaphthyl dioxide **21** gave even more appealing results in the reductive aldol reduction of *β*-ionone with benzaldehyde, where a very high *syn* stereoselectivity was observed along with 96% enantioselectivity for the *syn* isomer.

A great variety of Lewis base-catalyzed stereoselective transformations are currently under investigation, and novel synthetic methodologies have been developing as well. An example comes from a very recent report where, by studying the stereoselective synthesis of *N*-acylated *β*-amino ketones, it was unexpectedly found that optically active 4H-1,3-oxazines could be directly obtained via reductive cyclization of *N*-acylated *β*-amino enones using trichlorosilane and chiral Lewis base catalysts [46]. The reaction of trichlorosilane in the presence of catalytic amounts of BINAPO with (Z)-*N*-benzoyl enone derived from 3-amino-1-phenylbutane-1,3-dione surprisingly afforded the 4H-1,3-oxazine as a major product in 56% yield and 71% enantioselectivity (Scheme 14.13). Similar yields and stereoselectivity (up to 81% ee) were obtained by extending the reaction to other five substrates; among different chiral phosphine oxides investigated, BINAPO was found to secure the best performances. From some preliminary experiments, it was observed that trichlorosilane does not act only as a reductant, but also as a dehydrating agent. In the reaction, different ratios of oxazine and the expected *β*-keto amide were formed, depending on the variation of experimental conditions. Interestingly, it was observed that the two products were obtained with different levels of stereoselection, and sometimes even with a different absolute configuration.

The result was tentatively explained by assuming that the oxazine was not derived from the ketoamide by simple dehydration. It was proposed that 4H-1,3-oxazine was generated via the conjugate reduction of *N*-acylated *β*-amino enone, followed by cyclization of the resulting enolate and elimination of $HOSiCl_3$, whereas the ketoamide originates from the 1,2-reduction of the *N*-acyl imine were generated via equilibration of the enamide. Further studies will be necessary to fully understand the reaction mechanism, in order to design more efficient catalysts.

14.3. STEREOSELECTIVE C–C BOND FORMATION

The coordination of a Lewis base to a tetracoordinated silicon atom leads to hypervalent silicate species of increased Lewis acidity at the silicon center. As a consequence, such extracoordinated organosilicon compounds become very reactive carbon nucleophiles or hydride donors with a strong electrophilic character at silicon and an enhanced capability to transfer a formally negative charged group to an acceptor. (See Figs. 14.1 and 14.2.) Somehow, when a hypervalent silicon atom is involved as the reactive site in a transformation, carbon–carbon as well as carbon–heteroatom bond formation can occur. On the contrary, when a tetracoordinated silicon atom is exclusively involved in the reaction mechanism, a carbon–silicon as well as heteroatom–silicon bond formation may occur but not a carbon–carbon formation. Along these lines, several asymmetric catalytic systems have been explored in order to develop new stereoselective substoichiometric methodologies for carbon–carbon bond construction.

14.3.1. Allyltrichlorosilane Addition to C=O and C=N Bonds

14.3.1.1. Allylation of C=N Group. The synthesis of enantiomerically enriched homoallylic amines is a topic of paramount importance since they represent useful synthetic intermediates that may be converted in different functional groups. However, while the catalytic enantioselective allyl addition to carbonyl compounds is well developed, only a few examples of the analogous reaction with imines and imino esters are known, despite their utility in organic synthesis [47].

In 2004, Kobayashi et al. reported the first example of allyltrichlorosilane addition to *N*-benzoyl hydrazones, a reaction promoted by a chiral phosphine oxide as chiral Lewis base [48]. In that case, bis-(diphenylphosphanyl)-binaphthyl dioxide (BINAPO) was employed in the reaction of α-hydrazono esters, obtained from ethyl glyoxylate and benzhydrazide, with allyltrichlorosilane, affording the product in high yields and enantioselectivities at –78°C in DCM (Scheme 14.14). The reaction is stereospecific, in that (*E*)-crotyltrichlorosilanes

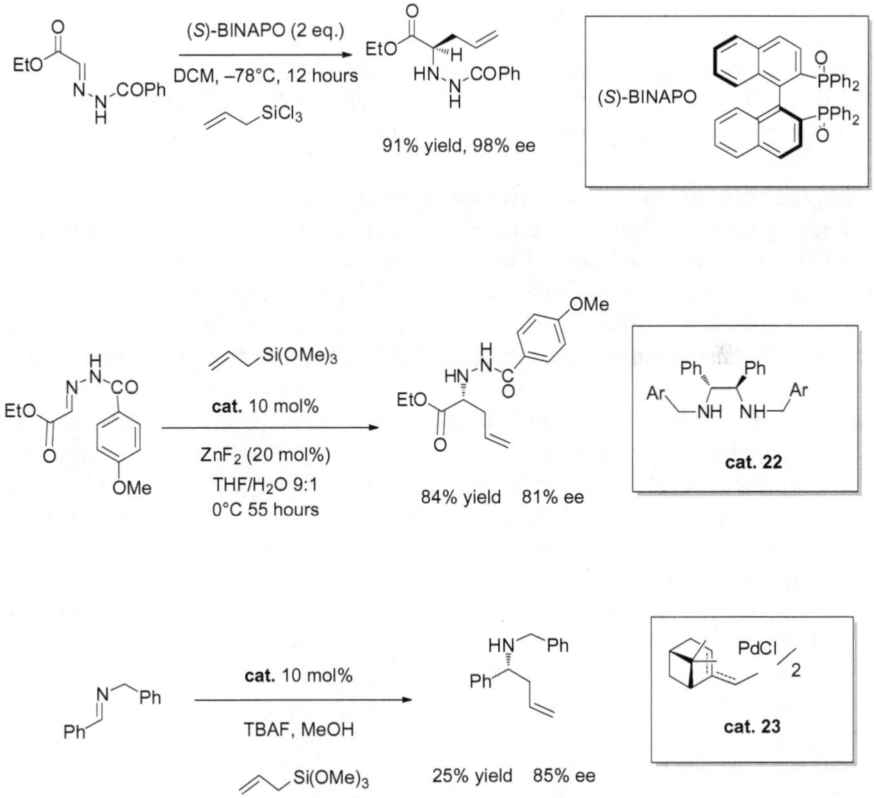

SCHEME 14.14. Stereoselective addition of carbon-nitrogen double bond.

afford the *syn* isomers, and (*Z*)-crotyltrichlorosilanes afforded their *anti* counterparts.

It must be noted that a more than stoichiometric amount of what was called *NCO* (neutral coordinate-organocatalyst) was necessary in order to achieve high stereoselectivities, while 0.2 eq. of BINAPO catalyzed the reaction in only 11% yield and 56% ee. Another drawback of the methodology is represented by the reductive cleavage of the N–N bond, accomplished by using SmI_2, required in order to obtain synthetically useful compounds. Even if the chiral source could be recovered without loss of stereochemical integrity, it is obvious that the reaction cannot be considered "organocatalyzed"; however, it has some merit since it represents the first example, and still one of the few cases, of enantioselective allylation of a carbon–nitrogen double bond involving a metal-free "promoter" [49].

Recently, Kobayashi et al. has also developed a zinc fluoride-catalyzed addition of allyltrimethoxy silane [50] to acylhydrazono esters, in the presence of a chiral diamine ligand **22** (Scheme 14.14) [51]. In the reaction, water plays a determinant role in affording the product of reaction, which suffers anyway of substrate limitations. Recently, Fernandes and Yamamoto have reported the addition of allyltrimethoxysilane to simple imines mediated by a dual activation/promotion process that involves the use of tetrabutyl ammonium fluoride (TBAF) and a chiral complex of palladium **23**; the product is isolated in 84% ee but with very low yields.(Scheme 14.14) [52].

14.3.1.2. Allylation of C=O Group. Previously promoted by chiral Lewis acids, this reaction that may lead to the formation of two new stereocenters can currently be carried out in the presence of a variety of organic Lewis bases as catalysts. Since a few reviews have recently covered the topic [10, 11], in the present section only the more important contributions in the field will be discussed as representative examples of different classes of developed catalysts; in addition, the more recent achievements in the allylation reaction of carbonyl compounds will be included.

In 1994, Denmark et al. reported the first enantioselective, noncatalytic addition of allyltrichlorosilane to aldehydes promoted by chiral phosphorotriamides [53]. A series of detailed studies demonstrated that two pathways were possible: one involving an octahedral cationic silicon atom, coordinated by two Lewis bases molecules leading to a good selectivity [54a], and another less selective one where only one phosphoroamide was bound to a pentacoordinated silicon center [54b]. In view of these mechanistic considerations, several chiral bidentate phosphoroamides were prepared and studied in the test allylation of benzaldehyde; a catalyst loading as low as 5 mol% of compound **24** was found to promote the reaction affording the product in high yield and enantioselectivity up to 72% (Scheme 14.7, eq. a) [55].

Based on these results, which clearly indicated the beneficial effect of combining two phosphoramide units through a diamminoalkyl chain, new bidentate catalysts derived from 2,2′-bispyrrolidine and 2,2′-bis piperidine units

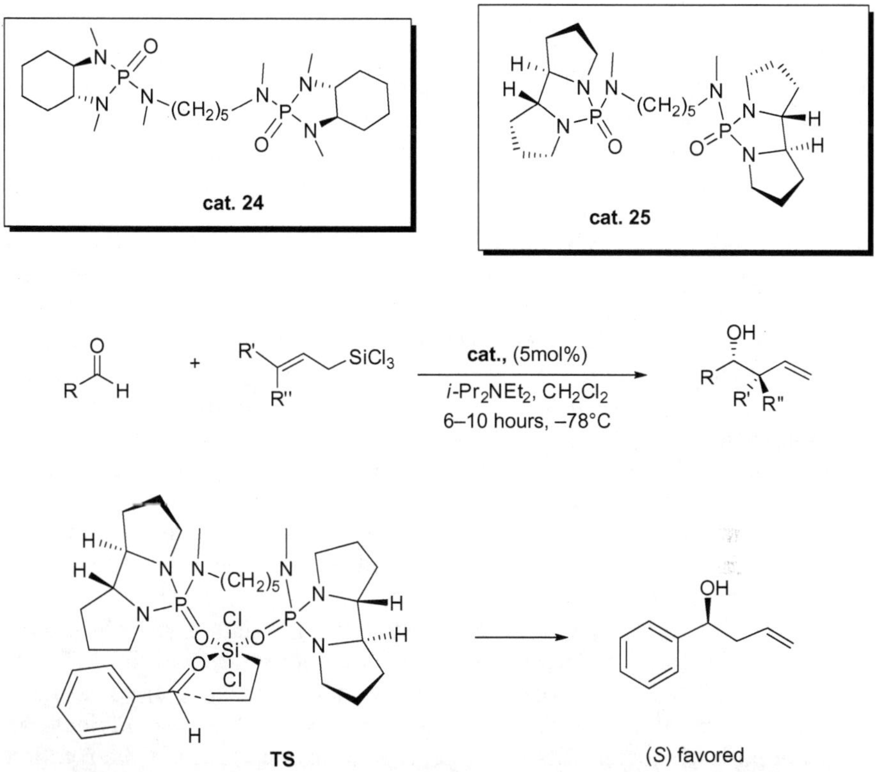

SCHEME 14.15. Allylation reactions promoted by chiral phosphoroamides.

were investigated. Compound **25** was found to be a really efficient promoter for the allylation reaction, and by addition of allyltrichlorosilane to benzaldehyde, afforded the omoallylic alcohol in 85% yield and 87% ee [54]. Various γ-substituted allyltrichlorosilanes were employed, leading to the products in high yields and up to 96% ee, showing a good correlation between the configuration of the C=C double bond in the reagent and the *syn*/*anti* diastereoisomeric ratio of the products. A rationalization of the behavior of catalyst **25** was also proposed (Scheme 14.15). In the chairlike, cyclic TS **A** the aldehyde ring is located in an unfavorable position occupied by a forward-pointing pyrrolidine ring, creating destabilizing steric interactions. In the diastereoisomeric chairlike arrangement of TS **B**, the aldehyde ring does not have any unfavorable interaction with the reward-pointing pyrrolidine unit, leading to the experimentally observed product of *S*-configuration.

Among Lewis basic catalysts, another class of compounds that deserve special attention are amine *N*-oxides [56]. The high nucleophilicity of the oxygen in *N*-oxides, coupled with a high affinity of silicon for oxygen, represents ideal properties for the development of synthetic methodology based on nucleophilic activation of organosilicon reagents. The first asymmetric addition of trichlorosilane to aldehyde catalyzed by biquinoline *N,N′*-dioxides **26** was reported in 1998 by Nakajima et al. [57]. The reaction was accelerated by the addition of diisopropylethylamine and afforded the products in high yields and enantioselectivities (up to 92%) with aromatic and heteroaromatic aldehydes, but lower yields and stereocontrol were observed with nonconjugated aldehydes (Scheme 14.16).

Later, Hayashi et al. developed another chiral catalyst, **27**, with a stereogenic axis as key element of stereocontrol, leading to comparable enantioselection (56%–98% ee) [58]. Remarkably, the Hayashi catalyst was found to be effective at the 0.1 mol% level (–40°C, acetonitrile) and retains moderate activity even at 0.01 mol% loading, which makes this organocatalyst the most reactive one reported to date. More recently, a simple synthesis of unsymmetric atropoisomeric bipyridine *N,N′*-dioxides in three steps from commercially available material was reported [59]. The key step of this reaction sequence is cobalt-catalyzed heterocyclotrimerization of 1-pyridyl-1,7-octadiynes with nitriles to provide unsymmetrical bipyridines, followed by oxidation and resolution into enantiomers. Catalyst **28** promoted the addition of allyltrichlorosilane to aromatic aldehydes in up to 80% ee.

Another class of catalysts was actively studied by Malkov et al., which have shown that the terpene-derived bipyridine *N*-monoxides, Me_2PINDOX, **29** (cat. 10 mol%, –78°C, CH_2Cl_2), was extremely enantioselective (up to 98% ee), although the reaction was somehow slow [60]. Such catalyst combines the effects of both stereoelements, stereogenic centers and axis, since the rotation about the bond connecting the two pyridine moieties is restricted by the two methyl groups and the N–O residue. In analogy to the previously proposed model, chelation of the silicon in allyltrochlorosilane by O and N was also proposed for **29** (Scheme 14.16).

In another important contribution Malkov et al. showed that two *N*-oxide groups are not necessary, but one *N*-oxide and a second coordination element are enough to guarantee high levels of stereocontrol, such as in the new developed catalyst **30** [61]. The proposed transition structure for the mono-*N*-oxide derivatives **A** is very similar to that proposed for bis-*N*-oxide compounds, **B** (Scheme 14.17). In catalyst **30** arene–arene interactions between the catalyst and the substrate have been suggested to account for the high reactivity and selectivity. Furthermore, it was shown that the axial stereogenicity, whether predetermined or induced during the reaction, is not an absolute prerequisite for attaining high enantioselectivity in the allylation reaction [62]. As further demonstration of these considerations, Hoveyda et al. developed the *N*-oxide **31**—the only representative of aliphatic tertiary amine *N*-oxides so far reported in this series—that presents a stereogenic center at the nitrogen [63]. It is pertinent to note that catalysts **30** and **31** secure high enantioselectivity even at room temperature. The only other example of chiral nonpyiridinic *N*-oxide used as a promoter of the allylation reaction has been recently developed by our group [64]. A new class of amine *N*-oxides derived from

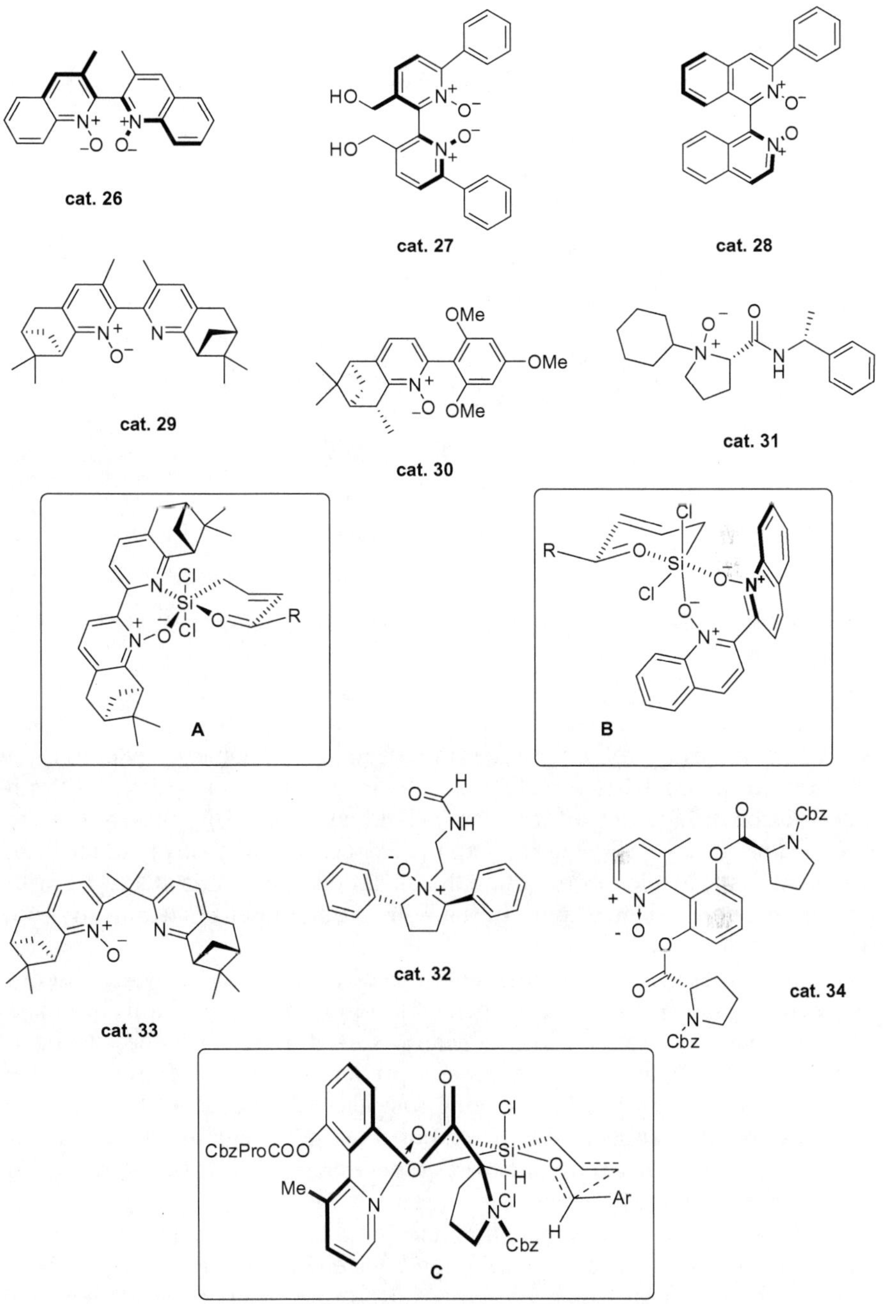

SCHEME 14.16. Chiral *N*-oxides as catalysts for allylation of carbonyl compounds.

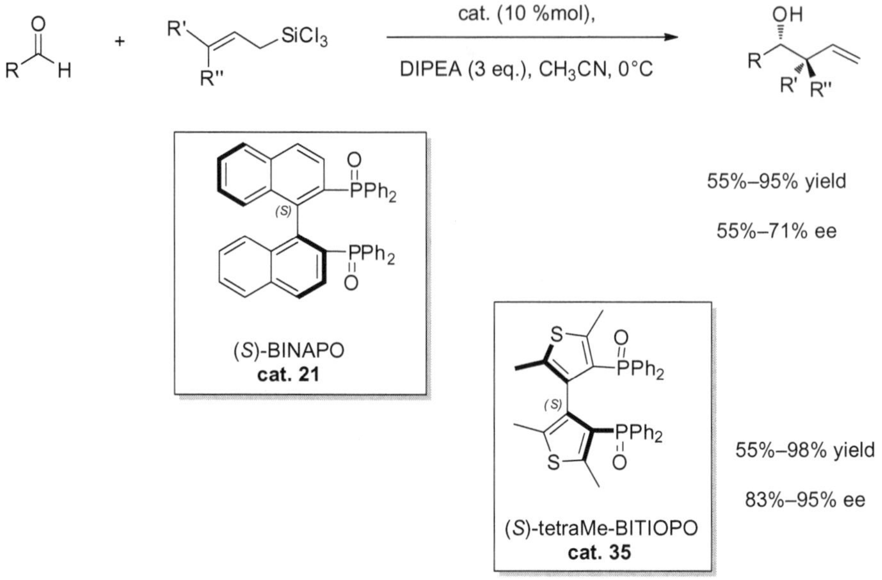

SCHEME 14.17. Chiral phosphine-oxide-catalyzed addition of allyltrichlorosilane to aldehydes.

trans-2,5-diphenylpyrrolidine were synthesized in enantiomerically pure form and tested as metal-free catalysts in the reaction of aldehydes with allyl(trichloro)silane to afford homoallylic alcohols. The products were obtained in fair to good yields and up to 85% ee. It is noteworthy to mention that a catalyst capable of promoting the allylation of aliphatic aldehydes with an almost unprecedented and unusually high enantioselectivity, up to 85%, was identified in **32**.

Based on these studies, other systems recently characterized by the absence of the stereogenic axis were developed [65]. For example, new chiral dipyridine *N*-monoxides and *N,N′*-dioxides, which possess an isopropylidene backbone between two pyridine rings, have been prepared from naturally occurring monoterpenes, the more efficient being compound **33** (Scheme 14.16) [66]. Its utility as organocatalysts has been demonstrated in the enantioselective addition of allyltrichlorosilane to aldehydes, where enantioselectivities up to 85% ee have been obtained.

A series of structurally simple pyridine *N*-oxides have readily been assembled from inexpensive amino acids and tested as organocatalysts in the allylation of aldehydes with allyltrichlorosilane to afford homoallylic alcohols [67]. (*S*)-Proline based catalyst **34** afforded the products derived from aromatic aldehydes in fair to good yields and up to 84% ee. By implementing the results of conformational analysis with those of a few control experiments, transition structure **C** shown in Scheme 14.16 can be proposed to tentatively explain the stereochemical result of the allylation reaction. In this model, the hypervalent silicon atom is coordinated by the pyridine *N*-oxide oxygen and the phenolic oxygen of one side arm. The bulky proline residue effectively blocks one side of the adduct and accommodates the aldehyde better than the sterically more requiring allyl residue as its *cis* substituent.

In 2005, for the first time, it was demonstrated that chiral phosphine oxides, such as BINAPO, can also act as organocatalysts in the enantioselective addition of allyltrichlorosilane to aldehydes [68]. In the presence of 10% mol amount of (*S*)-BINAPO **21** the homoallylic alcohol was obtained in DCM in only 32% yield and 36% ee (Scheme 14.17). By employing a proper additive such as $Bu_4N^+I^-$and 5 eq. of DIPEA (*N*,*N*-diisopropyl ethyl amine), the yield was improved to 92% after only 4 hours at room temperature, even if with still modest enantioselectivity (43% ee), definitely lower than that obtained with the best phosphoroamide-derived catalysts. The reduced chemical activity might be ascribed to the different electronic properties of the ligands. However, for this kind of reactions, it had been already suggested that the effectiveness of a catalyst is determined not only by the donor properties of the Lewis base but also by the steric hindrance at the oxygen atom. For example, dimethylphosphinic *N*,*N*-dimethylamide is a better promoter for the allylation of benzaldehyde than methylphosphonic di-(*N*,*N*-dimethylamide), which, in its turn, is better than HMPA. However, if in dimethylphosphinic *N*,*N*-dimethylamide a methyl group is replaced by a isopropyl group, the chemical efficiency of the catalyst dramatically decreases, clearly pointing at the importance of the steric accessibility of the oxygen atom.

A marked improvement of the catalytic efficiency of chiral phosphine oxides was obtained by our group exploring the characteristics of heteroaromatic systems [69]. The advantages offered by the biheteroaryldiphosphine oxides, with respect to carbocyclic aromatic derivatives, reside in their greater synthetic accessibility and in the possibility of testing a series of catalysts displaying different electronic properties, where the influence of both the electronic availability of the heterocyclic system and of the position of the phosphorus atoms on the latter may be investigated.

Electron-deficient diphosphine oxide did not promote the reaction in appreciable yields, while more electron-rich compounds showed a significant catalytic activity, promoting the addition of allyl(trichloro)silane to benzaldehyde at 0°C in higher yield than medium electron-rich diphosphine oxide BINAPO. Biheteroaromatic diphosphine oxides also showed an extraordinary ability in determining the stereochemical outcome of the reaction. The catalyst of choice was found to be (*S*)-tetramethyl-bithiophene phosphine oxide, (*S*)-TetraMe-BITIOPO, **35**, which promoted the allylation of benzaldehyde in 93% ee, a very high level of enantioselectivity, comparable to those obtained with the best known organocatalysts. The same catalyst efficiently promoted the addition of allyltrichlorosilane to aromatic aldehydes bearing electron-withdrawing groups as well as electron-donating groups with enantioselectivities constantly higher than 90%.

14.3.2. Aldol Condensation Reaction

Since the structure and the reaction mode of allylsilane may recall that of silyl enol ether (C–Si bond cleavage vs. O–Si bond cleavage), the addition of trichlorosilyl enol ethers to carbonyl derivatives catalyzed by Lewis bases was studied. However, since silyl enol ethers have a higher nucleophilicity compared to the corresponding allylsilanes, the aldol addition of trichlorosilyl enol ethers to aldehydes proceeds readily at room temperature without a catalyst and it exhibits simple first-order kinetics in each component. Nevertheless, the reaction is substantially accelerated by Lewis bases, which set the scene for the development of an asymmetric variant. Denmark et al. introduced a range of efficient chiral phosphoramides as nucleophilic activators for enantioselective C–C bond aldol formation and also carried out a detailed mechanistic investigation [70]. In 1996, the first example of aldol condensation of trichlorosilyl enol ethers was reported [71]. The chiral phosphoroamide **36** derived from 1,2-diphenyl-ethylendiamine successfully promoted in 10 mol% the addition of the trichlorosilyl enol ether of cyclohexanone to benzaldehyde in 95% yield, 65:1 *syn/anti* ratio, and 93% ee (Scheme 14.18).

However it was demonstrated that the diastereoselectivity was largely dependent on the structure of the chiral catalyst. After carrying a detailed

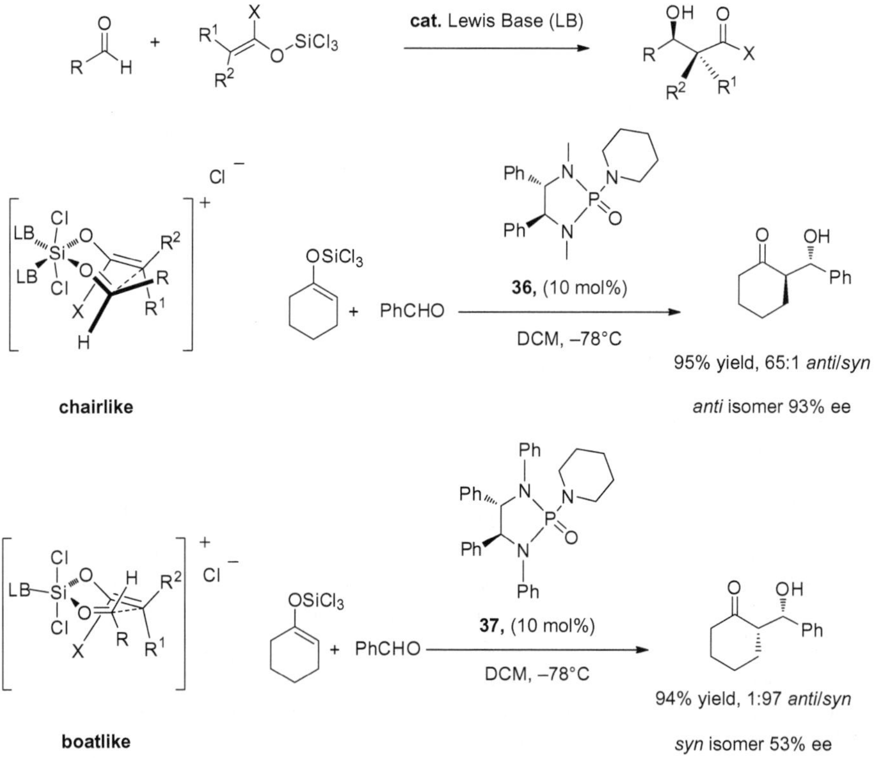

SCHEME 14.18. Aldol condensation promoted by chiral phosphoroamides.

mechanistic study, bidentate and smaller monodentate catalysts were shown to react through a cationic chairlike TS, similar to that usually proposed for the allylation reaction, with octahedral extracoordinate silicon. According to this scheme, (*Z*)-enol ethers produced *syn* adducts, whereas (*E*) derivatives furnished *anti* diastereoisomers. In the case of a bulky monodentate activator, where coordination of the second catalyst molecule is precluded by steric factors, the diastereoselectivity of the reaction was reversed. Here, the reaction presumably proceeds via the cationic boatlike TS, in which the silicon is pentacoordinate. According to this scheme, the cyclohexanone-derived enol ether with a fixed (*E*) configuration of the double bond gave rise to the *syn* product with sterically demanding catalyst **36** and to the *anti* isomer with catalyst **37** [72].

Pyridine *N*-oxides were also demonstrated to work as catalysts in the aldol reaction. In the absence of an activator, addition of trichlorosilyl ketene acetal to acetophenone slowly takes place at 0°C, but it can be accelerated by a Lewis base (Scheme 14.19). Bis-*N*-oxide **38** emerged as the most promising in terms

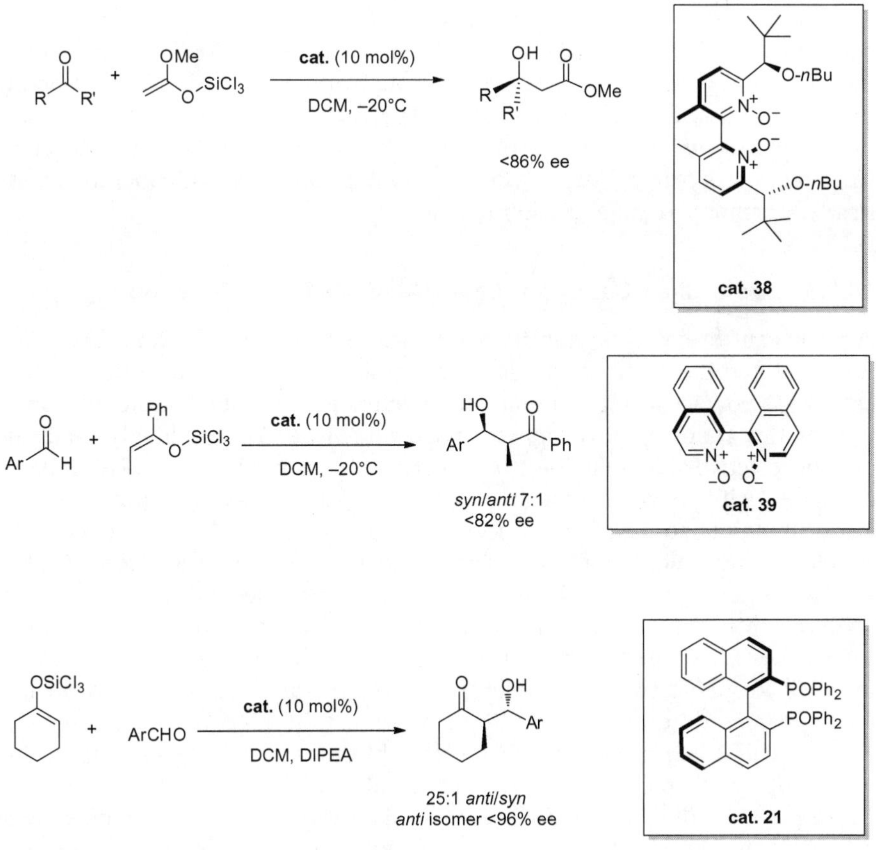

SCHEME 14.19. Aldol reactions catalyzed by chiral *N*-oxides and phosphine-oxides.

of reactivity and enantioselectivity (cat. 10 mol%, –20°C, CH_2Cl_2), affording the β-hydroxy ester, with a tertiary stereocenter in 94% yield and 84% ee [73]. A new procedure for the synthesis of atropoisomeric bis-*N*-oxide has also been developed. An X-ray crystal structure of the complex between a catalyst and silicon tetrachloride has been obtained. Extensive computational analyses were conducted to propose a stereochemical rationale for the observed trends in enantioselectivities.

Other chiral *N*-oxides were employed, but with less success; for example, Nakajima et al. reported that catalyst **39** promoted the addition of trichlorosilyl enol ethers to aromatic aldehydes in decent diastereoselectivity and enantioselection up to 82% (Scheme 14.19) [74]. Also, phosphine oxide **21** was able to catalyze the addition of cyclohexanone-derived silyl enol ether with activated aromatic aldehydes in high stereoselectivity [75]. The aldol adduct was obtained in moderate yield and diastereoselectivity, with 82% of enantioselectivity for the *anti* isomer (Scheme 14.19). With the addition of diisoproyl ethyl amine (DIPEA), both chemical and stereochemical efficiency were increased, the product being isolated at –78°C in DCM in 94% yield, 86% of *anti* diastereoselectivity, and 87% ee for the *anti* isomer. It was proposed that the tertiary amines additive not only may work as an acid scavenger to neutralize the hydrogen chloride produced by adventitious hydrolysis of trichlorosilyl enolethers, but it also accelerates the reaction rate by promoting the dissociation of phosphine oxide from the silicon atom. The Lewis base-promoted condensation afforded *anti* adducts starting from (*E*)-silanes and *syn* adducts from (*Z*)-silanes, confirming the hypothesis that a cyclic, six-membered TS is involved, similar to the allylation reaction.

14.3.3. Lewis Base-Catalyzed Lewis Acid-Mediated Reaction

A significant breakthrough in the field was accomplished by Denmark et al. who explored the possibility to develop chiral hypervalent silicates to be used as Lewis acids, according to the mode of activation described in the introduction of this chapter and proposed in Scheme 14.20 [76]. A highly efficient enantioselective aldol reaction of silyl keteneacetals catalyzed by a Lewis base activated with tetrachlorosilane was reported. Catalytic amounts of a binaphthyldiamino-based phosphoramide **40** (1 mol%) in the presence of a stoichiometric amount of tetrachlorosilane promoted the addition of silyl keteneacetals to aromatic aldehydes in high enantioselectivities.

In the proposed mechanism a basic ligand is employed to enhance the activity of a Lewis acid. The coordination of a Lewis base to a Lewis acid makes it more electrophilic; since a cationic species is generated, the result is a significantly increased Lewis acidity of the new adduct. In this aspect the combination of a chiral Lewis base and silicon tetrachloride to generate a strong Lewis acid is different from most of the other chiral Lewis acid-promoted reactions. Lewis base coordination to $SiCl_4$ activates the Lewis acid, while complexation of a basic chiral ligand to a Lewis acid precursor normally decreases the reac-

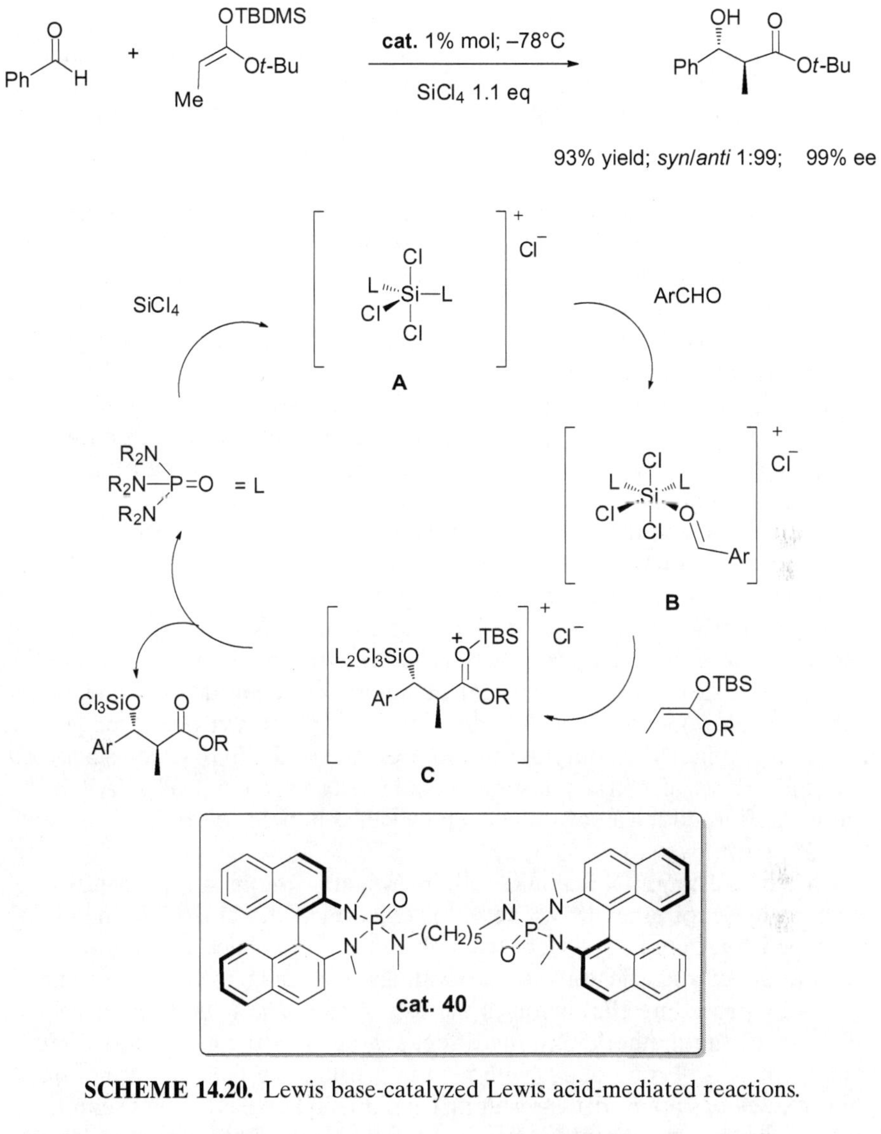

SCHEME 14.20. Lewis base-catalyzed Lewis acid-mediated reactions.

tivity of the chiral complex. It is correct to say that these are not Lewis acid-catalyzed reactions; the fact that the aldol products are trichlorosilyl ethers, as demonstrated by nuclear magnetic resonance (NMR), is clear evidence that each molecule of tetrachlorosilane participating in the catalytic cycle is incorporated into the product.

By exploiting the same concept, and always by using catalyst **40**, other reactions were performed, such as the addition of silyl enol ethers (in the presence of tetrabutyl ammonium salt), allylation with allyltributylstannane, and vinylogous aldol reactions [77]. Recently, the vinylogous aldol addition of conjugated

N,O-silyl ketene acetals to aldehydes was also described [78]. In the course of the last years it was shown that an in situ-generated chiral phosphoroamide-bound trichlorosilyl cation is an active catalyst for different transformations [79], according to a general mechanism for these reactions that can be properly defined as phosphoroamide-catalyzed and $SiCl_4$-mediated transformations, as proposed by Denmark et al. [10].

In the case of reaction of silyl keteneacetal, the hypothesized catalytic cycle involves the chiral trichlorosilyl cation **A** that binds the aldehyde to give adduct **B**, attacked by the silyl ketene acetal to afford the intermediate **C**, which, after dissociation from the catalyst, leads to the product as trichlorosilyl ether (Scheme 14.20). Not only the reaction is *anti* selective, but is also diastereoconvergent, affording the same stereoisomer independently from the geometry of the starting enolate. The behavior was tentatively rationalized by proposing that the decisive factor responsible for the observed trend in diastereoselectivity is the interaction between the α-substituent and the bound silyl cation complex in an open, acyclic transition structure. The analysis to explain the sense of the enantioselectivity of the process was less conclusive, but it was shown that the catalyst pocket is quite congested and the stereo and enantiocontrol was possibly dominated by steric factors.

Very recently it was demonstrated that chiral phosphine oxides are also able to promote the silyl ketene acetals addition to aromatic aldehydes in the presence of a stoichiometric amount of silicon tetrachloride (Scheme 14.21) [80]. The reaction of benzaldehyde with the trimethylsilyl ketene acetal, derived from methyl isobutyrate in the presence of a 1.5 eq. of tetrachlorosilane and 0.1 eq. of chiral phosphine oxide BINAP dioxide (BINAPO **21**), smoothly afforded the aldol adduct in high yield but moderate enantioselectivity (52% ee).

More interestingly, Nakajima et al. investigated the in situ preparation of trichlorosilyl enol ether [81]. The aldol reaction of trichlorosilyl enolethers developed by Denmark suffers from the major drawback of requiring the preparation of the trichlorosilyl derivatives according to an environment-unfriendly procedure that involves the use of mercuric salts. To improve the efficiency of the methodology, the direct synthesis of the enolethers directly from the carbonyl compounds with tetrachlorosilane in the presence of phosphine oxides was realized; the resulting trichlorosilyl enol ether was simultaneously activated by phosphine oxide to react with aldehydes, to afford a β-hydroxy ketone in a direct aldol-type reaction of two carbonyl compounds (Scheme 14.21). After the usual screening of several experimental conditions, propionitrile was found to be the solvent of choice; under the best conditions, the adduct was isolated in high yield (0°C, 2 hours, 81% yield) with a good diasteroselectivity (*syn*/*anti* = 17:83) but moderate enantioselectivity (54% ee for *anti* isomer).

Also, our group has been deeply involved in the development of new reactions promoted by catalytic amounts of chiral phosphine oxides derived from biheteroaromatic systems, already employed with success in the allyltrichlo-

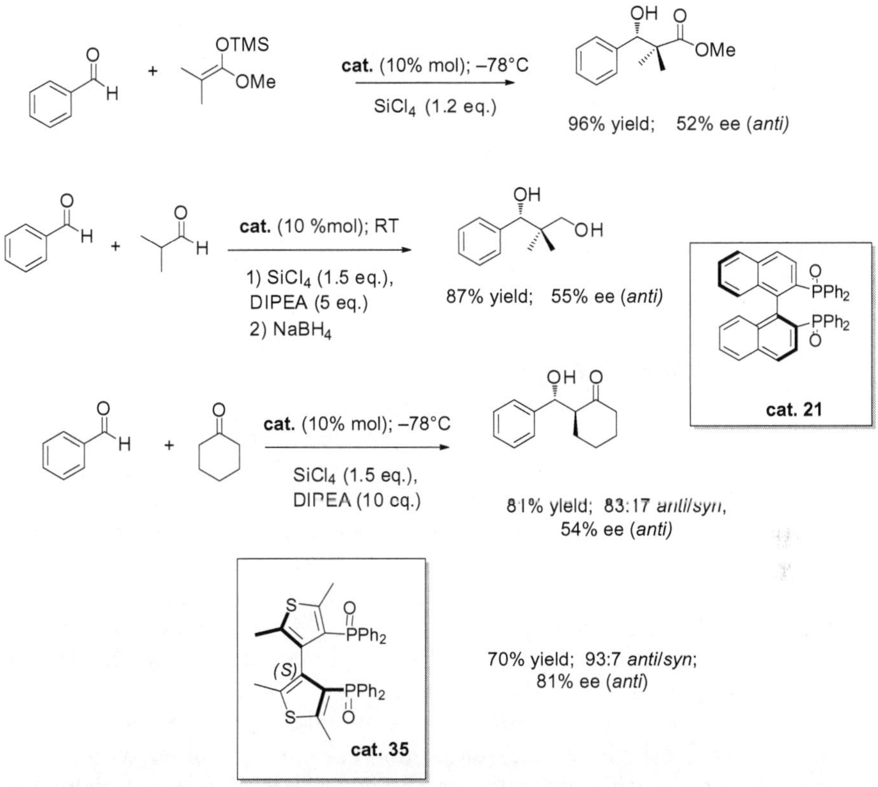

SCHEME 14.21. Phosphine oxide-catalyzed $SiCl_4$-mediated reactions.

rosilane addition. Since electron-rich (*S*)-TetraMe-BITIOPO had already shown higher performances than (*S*)-BINAPO in the allylation, we thought that such a Lewis base could successfully produce, by reaction with $SiCl_4$, the chiral cationic hypervalent silicon species, an active catalyst for the addition of several nucleophiles to carbonyl compounds. Once again, (*S*)-TetraMe-BITIOPO **35** was demonstrated to perform better than (*S*)-BINAPO, and higher levels of stereocontrol were reached (Scheme 14.21) [82]. For example, the direct condensation of cyclohexanone with benzaldehyde at 0°C in DCM, in the presence of 3 eq. of $SiCl_4$, 10 eq. of DIPEA and 0.1 eq. of (*S*)-TetraMe-BITIOPO, afforded the aldol product in 12:88 *syn/anti* stereoselectivity and 75% ee for the *anti* isomer. At a lower temperature (–25°C), both the diasteoreoselectivity and the enantioselectivity were improved up to 81% ee for the *anti* isomer.

Taking advantage of the poor reactivity of aliphatic aldehydes as electrophiles, a direct aldol-type reaction between two different aldehydes was also positively accomplished [81]. The reaction between benzaldehyde and isobutyraldehyde in the presence of (*S*)-BINAPO **21** leads to the expected

β-hydroxy aldehyde that was reduced to the corresponding diol in order to facilitate the isolation of the product. Also, in this case, the enantioselectivity was not really satisfactory (55% ee).

The very promising results of these studies pave the way toward the application of this methodology to other nucleophilic attacks to activated C=O, C=N, and even C=C promoted by chiral hypervalent silicon Lewis acids. It is even too easy to predict that further studies in the field and more detailed investigations for the development of innovative synthetic methodologies will appear soon.

14.4. RING-OPENING REACTION OF EPOXIDES

The first catalytic enantioselective opening of *meso*-epoxides with tetrachlorosilane in the presence of the chiral phosphoramide **41** as a Lewis base (Scheme 14.23) was reported by Denmark et al. [83]. The mechanistic hypothesis is that the Lewis base reacts with tetrachlorosilane, forming a pentacoordinate silicate complex that coordinates the oxygen atom of the substrate, activating it toward nucleophilic substitution. The attack of the chloride ion proceeds in an S_N2 fashion. The enantioselectivity depends on the substrate's structure. A higher level of ee was obtained for acyclic substrates, whereas for cyclic substrates, it also depends on ring size.

Denmark et al. investigated different aspects of the reaction [84] with the use of different chlorosilane sources and with different stoichiometry, catalyst loading, internal quench, and kinetic and nonlinear effect. This survey highlights that only silicon tetrachloride affords the product with a high level of stereocontrol: only one chlorine is released, and in the course of the reaction, the nature of the silicon reagent does not change; the selectivity doesn't change with catalyst loadings ranging from 100 to 4 mol%, but it decreases with a 2 mol% loading. This behavior suggests that a single pathway mechanism is active in the 100–4 mol% range. The authors conclude that probably more than one molecule of the catalyst is involved in the stereochemistry determining step.

Fu and co-workers reported [85] a new family of chiral catalysts capable of promoting the opening of *meso*-epoxides with high enantioselectivity (up to 98% ee) in the presence of tetrachlorosilane via a hexacoordinate silicate (Scheme 14.22). The choice of catalyst **42** is the result of the screening of a series of enantiopure pyridine *N*-oxides where the authors show that increased steric hindrance increases the level of stereocontrol. An electron-poor aromatic group of the substrate leads to the chlorohydrin with the highest selectivity: the ee depends on electronic effects. A positive nonlinear correlation between the ee of the catalyst and the ee of the product was also observed.

Later, Nakajima's group showed [86] that also the chiral phosphine oxide (*S*)-BINAPO provided the product of ring opening of *meso*-epoxides with high enantioselectivity (up to 90% ee) in the presence of tetrachlorosilane and diisopropylethylamine, whose presence seems to be necessary in order to obtain a good level of stereocontrol. More recently, a novel catalytic system

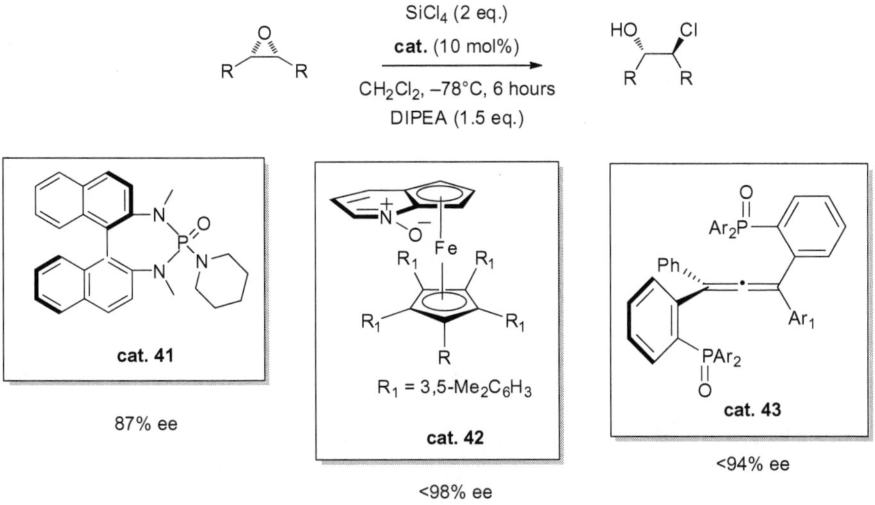

SCHEME 14.22. Catalytic stereoselective epoxide-opening reactions.

has been reported; optically active mono- and bisphosphine oxides containing an allene backbone were prepared in enantiomerically pure form and used as organocatalysts (Scheme 14.22) [87]. The novel chiral allenes were tested in the opening of *meso* epoxides by the addition of silicon tetrachloride. It was observed that bisphosphine oxides performed much better than monophosphine oxides. It must be said that in the solid state, bisphosphine oxide adopts a conformation characterized by π-stacking interactions involving two phenyl rings of the phosphine oxides and one of the backbone phenyl rings. Because of this arrangement, the two oxygen atoms project in roughly the same direction. The selected catalyst **43** was found to be a really effective promoter that could be employed at 0.1% mol cat. loading, clearly indicating that a completely unexplored novel class of chiral phosphine oxide, based on the allene backbone, are now available as suitable catalysts.

14.5. MISCELLANEOUS

Propargyl trichlorosilane, prepared by CuCl-catalyzed reaction between propargyl chloride and $HSiCl_3$; and allenyl trichlorosilane, synthesized analogously in the presence of Ni(acac)$_2$, were shown to react with aromatic aldehydes under activation of a Lewis base, similar to the addition of allyltrichlorosilane to carbonyl compounds [88]. The addition of propargyl trichlorosilane to aldehydes leads to allenyl alcohols (Scheme 14.23), while the reaction of allenyl trichlorosilane affords the corresponding homopropargyl alcohol. An asymmetric version has been reported by Nakajima et al. [89], who employed the chiral biquinoline bis-*N*-oxide **17** as catalyst (10 mol%), but the enantioselectivities observed were rather modest (40%–62% ee).

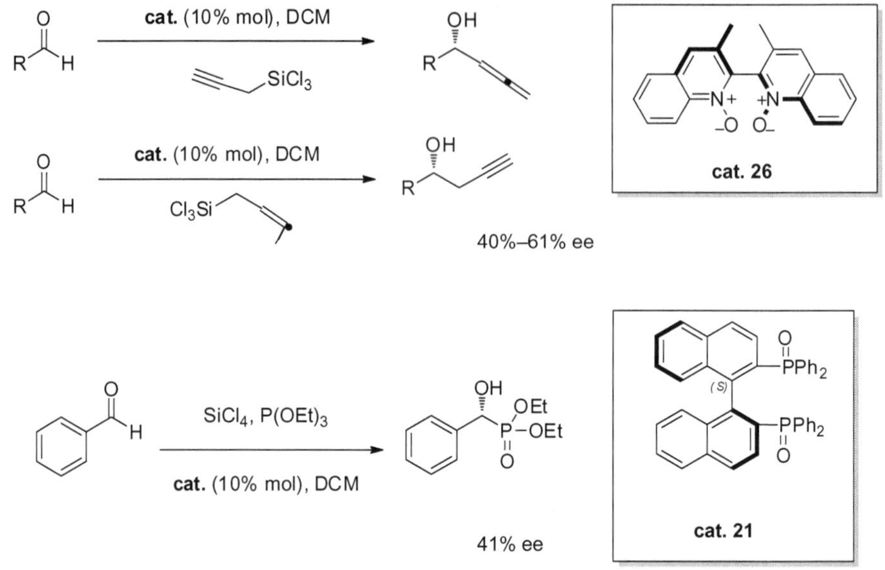

SCHEME 14.23. Organocatalytic miscellaneous reactions.

In the attempt to further explore new synthetic methodologies based on the use of trichlorosilyl derivatives activated by chiral Lewis bases, the enantioselective addition of trialkylphosphites to achiral aldehydes was investigated [80]. However, the catalytic Abramov-type phosphonylation of carbonyl compounds promoted by tetrachlorosilane in the presence of catalytic amounts of (*S*)-BINAPO was met with limited success, affording the corresponding α-hydroxyphosphonate in 86% yield but only 41% enantioselectivity (Scheme 14.5). Once again, the role of diisopropylethylamine as additive was decisive to achieve high chemical yields, but did not improve the stereocontrol. Even if the level of stereoselectivity is unsatisfactory, the work is worth mentioning because it represents the first organocatalytic enantioselective Abramov-type phosphonylation of aldehydes and the first report where different classes of chiral phosphine oxides other than the bis-(diarylphosphanyl)-binaphthyl dioxides were taken in consideration, although without success in this kind of reaction.

14.6. OUTLOOK AND PERSPECTIVES

Since hypervalent silicon species may work through different activation mechanisms, they have recently attracted much attention for their versatility and for the possibility to develop several catalytic processes. Indeed, tuning the chemistry of penta- and/or hexavalent silicon compounds by the design and the synthesis of chiral organocatalytic species is not only feasible, but highly desirable, with the goal to always develop new enantioselective reactions in the presence of cheap, low-toxic, and environmental friendly species such as silicon-based reagents.

The development of Lewis-bases catalyzed reactions for regio- and stereochemical bond formation is a topic of primary importance in modern organic chemistry. The design of innovative methodologies for increasing molecular diversity and complexity is another field where enantiomerically chiral Lewis bases may find extensive application.

The study of novel chiral Lewis bases able to generate hypervalent silicon species and catalyze stereoselective transformations is another topic of great perspective. Quite unexpectedly, among the different classes of Lewis bases currently employed in organocatalysis, phosphine oxides have received little attention so far. Their use as chiral Lewis basic metal-free catalysts has been limited to relatively few reactions. Chiral phosphine oxides have been studied in the polyhalosilanes chemistry, as clearly shown above, but a variety of different reactions with different mechanism are suitable to be explored. Another point worthy of consideration is that relatively few classes of phosphine oxides have been investigated; basically, only BINAPO have been used. However, it should be noted that not only carbocyclic aromatic but also heteroaromatic-based phosphine oxides may be considered. These compounds offer new possibilities of development, since the electronic and steric properties of the ligands could be modulated by a proper choice of substituents. It is also remarkable that alkyl phosphine oxides have been basically unexplored and, to best of our knowledge, no studies on their use in organocatalysis have been reported so far.

Finally, the design and development of multifunctional organocatalytic systems where the Lewis basic group is present, together with another metal-free catalytically active residue, typically an acidic group, is almost totally unexplored [90]. Only examples of bifunctional catalysts where a metal Lewis acid is combined with a Lewis basic site have been investigated. However, it is clear how many more options and possibilities could be offered by the availability of properly designed multifunctional catalysts.

On the basis of these considerations it is easy to predict that we will see a continuously increasing interest in the field of stereoselective reactions promoted by chiral Lewis bases; hopefully, this survey will stimulate further research in a very exciting area, where hypervalent silicate species will play a decisive role in the invention of new, highly chemical and stereochemical efficient catalytic systems of low environmental impact.

REFERENCES

[1] Reviews about the formation and structure of hypervalent silicon compounds: (a) Tandura, S. N., Voronkov, M. G., Alekseev, N. V. (1986). Molecular and electronic structure of penta- and hexacoordinate silicon compounds. *Top. Curr.*

Chem., *131*, 99–189; (b) Shklover, V. E., Struchov, Y. T., Voronkov, M. G. (1989). Organic compounds of silicon with non-standard types of coordination. *Russ. Chem. Rev.*, *58*, 211–229; (c) Lukevics, E., Pudova, O., Sturkovich, R. (Eds.) (1989). *Molecular Structure of Organosilicon Compounds*, Ellis Horwood, Chichester, Germany; (d) Holmes, R. R. (1996). Comparison of phosphorus and silicon: hypervalency, stereochemistry, and reactivity. *Chem. Rev.*, *96*, 927–950.

[2] Gay-Lussac, J. L., Thenard, L. J. (1809). Sur le rapport qui existe entre l'oxidation des métaux et leur capacité de saturation pour les acides. *Mémoires de Physique et de Chimie de la Société d'Arcueil*, *2*, 159–175.

[3] (a) Chuit, C., Corriu, R. J. P., Reye, C., Young, J. C. (1993). Reactivity of penta- and hexacoordinate silicon compounds and their role as reaction intermediates. *Chem. Rev.*, *93*, 1371–1448; (b) Hosomi, A. (1988). Characteristics in the reactions of allylsilanes and their applications to versatile synthetic equivalents. *Acc. Chem. Res.*, *21*, 200–206; (c) Furin, G. G., Vyazankina, O. A., Gostevsky, B. A., Vyazankin, N. S. (1988). Synthetic aspects of the use of organosilicon compounds under nucleophilic catalysis conditions. *Tetrahedron*, *44*, 2675–2749. For a discussion about the possible involvement of heptacoordinated silicon species in nucleophilic addition to hexacoordinated silicon species, see Reference [3a], p. 1240.

[4] (a) Pellissier, H. (2007). Asymmetric organocatalysis. *Tetrahedron*, *63*, 9267–9331; (b) Dalko, P. I. (Ed.), (2007). *Enantioselective Organocatalysis. Reactions and Experimental Procedures*. Wiley VCH, Weinheim, Germany; (c) Recent reviews:Dondoni, A. Massi, A. (2008). Asymmetric organocatalysis: from infancy to adolescence. *Angew. Chem. Int. Ed.*, *47*, 4638–4660; (d) Melchiorre, P., Marigo, M., Carlone, A., Bartoli, G. (2008). Asymmetric aminocatalysis. Gold rush in organic chemistry. *Angew. Chem. Int. Ed.*, *47*, 6138–6171.

[5] Anastas, P. T., Warner, J. C. (1998). *Green Chemistry: Theory and Practice*, Oxford University Press, New York.

[6] Blaser, H. U. (2003). Enantioselective catalysis in fine chemicals production. *Chem. Commun.*, 293–296.

[7] (a) Federsel, H.-J. (2008). Handing over the baton: connecting medicinal chemistry with process R&D. *Drug News Perspect.*, *21*, 193.[7] (b) See also Federsel, H.-J. (2009). Chemical process research and development in the 21st century: challenges, strategies, and solutions from a pharmaceutical industry perspective. *Acc. Chem. Res.*, *42*, 671–680, and references cited.

[8] Rendler, S., Oestreich, M. (2005). Hypervalent silicon as a reactive site in selective bond-forming processes. *Synthesis*, 1727–1747.

[9] Orito, Y., Nakajima, M. (2006). Lewis base catalyzed asymmetric reactions involving hypervalent silicate intermediates. *Synthesis*, 1391–1401.

[10] For a profound and very detailed analysis of different kinds of Lewis base-catalysis see the fundamental review, see: Denmark, S. E., Beutner, G. L. (2008). Lewis base catalysis in organic synthesis. *Angew. Chem. Int. Ed.*, *47*, 1560–1638.

[11] For a recent review on hypervalent silicates-mediated reactions, see: Benaglia, M., Guizzetti, S., Pignataro, L. (2008). Stereoselective reactions involving hypervalent silicate complexes. *Coord. Chem. Rev.*, *252*, 492–512.

[12] (a) Denmark, S. E., Wynn, T., Beutner, G. L. (2002). Lewis base activation of Lewis acids. Addition of silyl ketene acetals to aldehydes. *J. Am. Chem. Soc.*, *124*, 13405–13407; (b) Denmark, S. E., Fan, Y., Eastgate, M. D. (2005). Lewis base catalyzed,

enantioselective aldol addition of methyl trichlorosilyl ketene acetal to ketones. *J. Org. Chem.*, *70*, 5235–5248.

[13] For a recent review on chiral amine synthesis, see: Nugent, T. C., El-Shazly, M. (2010). Chiral amine synthesis—recent developments and trends for enamide reduction, reductive amination, and imine reduction. *Adv. Synth. Catal.*, *352*, 753–819.

[14] For chiral phosphoric acids-mediated reductions, see: (a) Connon, S. J. (2007). Asymmetric organocatalytic reductions mediated by dihydropyridines. *Org. Biomol. Chem.*, *5*, 3407–3417; (b) Akiyama, T. (2007). Stronger Brønsted acids. *Chem. Rev.*, *107*, 5744–5758; (c) Terada, M. (2008). Binaphthol-derived phosphoric acid as a versatile catalyst for enantioselective carbon-carbon bond forming reactions. *Chem. Commun.*, 4097–4112.

[15] (a) Iwasaki, F., Onomura, O., Mishima, K., Maki, T., Matsumura, Y. (1999). Catalytic activation of trichlorosilane for efficient and stereoselective reduction of ketones. *Tetrahedron Lett.*, *40*, 7507–7511; (b) Iwasaki, F., Onomura, O., Mishima, K., Kanematsu, T., Maki, T., Matsumura, Y. (2001). First chemo- and stereoselective reduction of imines using trichlorosilane activated with *N*-formylpyrrolidine derivatives. *Tetrahedron Lett.*, *42*, 2525–2527.

[16] Malkov, A. V., Mariani, A., MacDougal, K. N., Kočovský, P. (2004). Role of noncovalent interactions in the enantioselective reduction of aromatic ketimines with trichlorosilane. *Org. Lett.*, *6*, 2253–2256.

[17] Malkov, A. V., Stončius, S., MacDougal, K. N., Mariani, A., McGeoch, G. D., Kočovský, P. (2006). Formamides derived from *N*-methyl amino acids serve as new chiral organocatalysts in the enantioselective reduction of aromatic ketimines with trichlorosilane. *Tetrahedron*, *62*, 264–284.

[18] Malkov, A. V., Figlus, M., Stončius, S., Kočovský, P. (2009). Asymmetric reduction of imines with trichlorosilane, catalyzed by sigamide, an amino acid-derived formamide: scope and limitations. *J. Org. Chem.*, *74*, 5839–5849.

[19] Malkov, A. V., Vranková, K., Sigerson, R. C., Stončius, S., Kočovský, P. (2009). New organocatalysts for the asymmetric reduction of imines with trichlorosilane. *Tetrahedron*, *65*, 9481–9486.

[20] (a) Malkov, A. V., Figlus, M., Stoncius, S., Kočovský, P. (2007). Organocatalysis with a fluorous tag: asymmetric reduction of imines with trichlorosilane catalyzed by amino acid-derived formamides. *J. Org. Chem.*, *72*, 1315–1325; (b) Malkov, A. V., Figlus, M., Kočovský, P. (2008). Polymer-supported organocatalysts: asymmetric reduction of imines with trichlorosilane catalyzed by an amino acid-derived formamide anchored to a polymer. *J. Org. Chem.*, *73*, 3985–3995; (c) for another recoverable system, see: Malkov, A. V., Figlus, M., Prestly, M. R., Rabani, G., Cooke, G., Kočovský, P. (2009). Soluble polymer-supported organocatalysts: asymmetric reduction of imines with trichlorosilane catalyzed by an amino acid derived formamide anchored to a soluble polymer. *Chem. Eur. J.*, *15*, 9651–9654.

[21] Malkov, A. V., Stončius, S., Kočovský, P. (2007). Enantioselective synthesis of 1,2-diarylaziridines by the organocatalytic reductive amination of α-chloroketones. *Angew. Chem. Int. Ed.*, *46*, 3722–3724.

[22] Malkov, A. V., Stončius, S., Vranková, K., Arndt, M., Kočovský, P. (2008). Dynamic kinetic resolution in the asymmetric synthesis of β-amino acids by organocatalytic reduction of enamines with trichlorosilane. *Chem. Eur. J.*, *14*, 8082–8085.

[23] Matsumura, Y., Ogura, K., Kouchi, Y., Iwasaki, F., Onomura, O. (2006). New efficient organic activators for highly enantioselective reduction of aromatic ketones by trichlorosilane. *Org. Lett.*, *17*, 3789–3792.

[24] Zhou, L., Wang, Z., Wei, S., Sun, J. (2007). Evolution of chiral Lewis basic *N*-formamide as highly effective organocatalyst for asymmetric reduction of both ketones and ketimines with an unprecedented substrate scope. *Chem. Commun.*, 2977–2979.

[25] Baudequin, C., Chaturvedi, D., Tsogoeva, S. B. (2007). Organocatalysis with chiral formamides: asymmetric allylation and reduction of imines. *Eur. J. Org. Chem.*, 2623–2629.

[26] Wang, Z. Y., Wei, S., Wang, C., Sun, J. (2007). Enantioselective hydrosilylation of ketimines catalyzed by Lewis basic C_2-symmetric chiral tetraamide. *Tetrahedron Asymmetry*, *18*, 705–709.

[27] Wang, Z., Ye, X., Wei, S., Wu, P., Zhang, A., Sun, J. (2006). A highly enantioselective Lewis basic organocatalyst for reduction of *N*-aryl imines with unprecedented substrate spectrum. *Org. Lett.*, *5*, 999–1001.

[28] Wang, Z., Cheng, M., Wu, P., Wei, S., Sun, J. (2006). L-Piperazine-2-carboxylic acid derived *N*-formamide as a highly enantioselective Lewis basic catalyst for hydrosilylation of *N*-aryl imines with an unprecedented substrate profile. *Org. Lett.*, *14*, 3045–3048.

[29] Zhang, Z., Rooshenas, P., Hausmann, H., Schreiner, P. R. (2009). Asymmetric transfer hydrogenation of ketimines with trichlorosilane: structural studies. *Synthesis*, *9*, 1531–1544.

[30] Onomura, O., Kouchi, Y., Iwasaki, F., Matsumura, Y. (2006). New organic activators for the enantioselective reduction of aromatic imines with trichlorosilane. *Tetrahedron Lett.*, *47*, 3751–3754.

[31] Guizzetti, S., Benaglia, M. (2009). European Patent Application November 30 2007; PCT/EP/2008/010079, Nov. 27, 2008. WO2009068284 (A2), 2009-06-04.

[32] Zheng, H., Deng, J., Lin, W., Zhang, X. (2007). Enantioselective hydrosilylation of ketimines with trichlorosilane promoted by chiral *N*-picolinoylaminoalcohols. *Tetrahedron Lett.*, *48*, 7934–7937.

[33] Guizzetti, S., Benaglia, M., Annunziata, R., Cozzi, F. (2009). Chiral Lewis base promoted trichlorosilane reduction of ketimines. An enantioselective organocatalytic synthesis of chiral amines. *Tetrahedron*, *65*, 6354–6363.

[34] Guizzetti, S., Benaglia, M. (2009). European Patent Appl. n. EP07023240.0, September 22, 2008.

[35] Wang, C., Wu, X., Zhou, L., Sun, J. (2008). A highly enantioselective organocatalytic method for reduction of aromatic *N*-alkyl ketimines. *Chem. Eur. J.*, *14*, 8789–8792.

[36] Guizzetti, S., Biaggi, C., Benaglia, M., Celentano, G. (2010). A convenient, highly stereoselective, metal-free synthesis of chiral amines. *Synlett*, 134–136.

[37] Guizzetti, S., Benaglia, M., Rossi, S. (2009). Highly stereoselective metal-free catalytic reduction of imines: an easy entry to enantiomerically pure amines and natural and unnatural α-amino esters. *Org. Lett.*, *11*, 2928–2931.

[38] Guizzetti, S., Benaglia, M., Cozzi, F., Rossi, S., Celentano, G. (2009). Enantioselective catalytic reduction of ketoimines with trichlorosilane promoted by readily available chiral Lewis bases. *Chirality*, *21*, 233–238.

[39] Guizzetti, S., Benaglia, M., Celentano, G. (2009). 1,1′-Binaphthyldiamine-based Lewis bases as readily available and efficient organocatalysts for the reduction of *N*-aryl and *N*-alkyl ketimines. *Eur. J. Org. Chem.*, 3683–3687.

[40] Xue, Z.-Y., Jiang, Y., Yuan, W.-C., Zhang, X.-M. (2010). The first highly enantioselective Lewis base organocatalyzed hydrosilylation of α-imino esters. *Eur. J. Org. Chem.*, 616–619.

[41] Zheng, H.-J., Chen, W.-B., Wu, Z.-J., Deng, J.-G., Lin, W.-Q., Yuan, W.-C., Zhang, X.-M. (2008). Highly enantioselective synthesis of β-amino acid derivatives by the Lewis base catalyzed hydrosilylation of β-enamino esters. *Chem. Eur. J.*, *14*, 9864–9867.

[42] Malkov, A. V., Liddon, A. J. P. S., Ramírez-López, P., Bendová, L., Haigh, D., Kočovský, P. (2006). Remote chiral induction in the organocatalytic hydrosilylation of aromatic ketones and ketimines. *Angew. Chem. Int. Ed.*, *45*, 1432–1435.

[43] Pei, D., Wang, Z., Wei, S., Zhang, Y., Sun, J. (2006). *S*-Chiral sulfinamides as highly enantioselective organocatalysts. *Org. Lett.*, *25*, 5913–5915.

[44] Pei, D., Wang, Z., Wei, S., Wang, M., Sun, J. (2008). Rationally-designed *S*-chiral bissulfinamides as highly enantioselective organocatalysts for reduction of ketimines. *Adv. Synth. Catal.*, *350*, 619–623.

[45] Sugiura, M., Sato, N., Kotani, S., Nakajima, M. (2008). Lewis base-catalyzed conjugate reduction and reductive aldol reaction of α,β-unsaturated ketones using trichlorosilane. *Chem. Commun.*, 4309–4311.

[46] Sugiura, M., Kumahara, M., Nakajima, M. (2009). Asymmetric synthesis of 4*H*-1,3-oxazines: enantioselective reductive cyclization of *N*-acylated β-amino enones with trichlorosilane catalyzed by chiral Lewis bases. *Chem. Commun.*, 3585–3587.

[47] Colombo, F., Benaglia, M., Annunziata, R. (2007). Catalytic, enantioselective allylation of α-iminoesters promoted by silver(I) complexes of chiral imines. *Tetrahedron Lett.*, *46*, 2687–2690, and references cited there.

[48] Ogawa, C., Sugiura, M., Kobayashi, S. (2004). Stereospecific, enantioselective allylation of α-hydrazono esters by using allyltrichlorosilanes with BINAP dioxides as neutral-coordinate organocatalysts. *Angew. Chem. Int. Ed.*, *43*, 6491–6493.

[49] (a) The adddition of allyltrichlorosilane to *N*-acylhydrazones promoted by stoichiometric amount of a chiral sullfoxide was also reported; see: (a) Kobayashi, S., Ogawa, C., Konishi, H., Sugiura, M. (2003). Chiral sulfoxides as neutral coordinate-organocatalysts in asymmetric allylation of N-acylhydrazones using allyltrichlorosilanes. *J. Am. Chem. Soc.*, *125*, 6610–6611; (b) Garcia-Flores, F., Flores-Michel, L. S., Juaristi, E. (2006). Asymmetric allylation of N-benzoylhydrazones promoted by novel C2-symmetric bis-sulfoxide organocatalysts. *Tetrahedron Lett.*, *47*, 8235–8238; for the use of a chiral allyl silane reagents see: (c) Rabbat, P. M. A., Corey Valdez, S., Leighton, J. L. (2006). Phenol-directed enantioselective allylation of aldimines and ketimines. *Org. Lett.*, *8*, 6119–6121, and references cited there.

[50] For a review on silver catalyzed asymmetric allylation see: Yamamoto, H., Wadamoto, M. (2007). Silver-catalyzed asymmetric allylation: allyltrimethoxysilane as a remarkable reagent. *Chem. Asian J.*, *2*, 692–698.

[51] Kiyohara, H., Nakamura, Y., Matsubara, R., Kobayashi, S. (2006). Enantiomerically enriched allylglycine derivatives through the catalytic asymmetric allylation of iminoesters and iminophosphonates with allylsilanes. *Angew. Chem. Int. Ed.*, *45*, 1615–1617.

[52] Fernandes, R. A., Yamamoto, H. (2004). The first catalytic asymmetric allylation of imines with the tetraallylsilane–TBAF–MeOH system, using the chiral bis-π-allylpalladium complex. *J. Org. Chem.*, *64*, 735–738.

[53] Denmark, S. E., Coe, D. M., Pratt, N. E., Griedel, B. D. (1994). Asymmetric allylation of aldehydes with chiral Lewis bases. *J. Org. Chem.*, *59*, 6161–6163.

[54] (a) Denmark, S. E., Fu, J., Coe, D. M., Su, X., Pratt, N. E., Griedel, B. D. (2006). Chiral phosphoramide-catalyzed enantioselective addition of allylic trichlorosilanes to aldehydes. Preparative and mechanistic studies with monodentate phosphorus-based amides. *J. Org. Chem.*, *71*, 1513–1522, and references cited there; (b) Denmark, S. E., Fu, J. (2000). On the mechanism of catalytic, enantioselective allylation of aldehydes with chlorosilanes and chiral Lewis bases. *J. Am. Chem. Soc.*, *122*, 12021–12022.

[55] Denmark, S. E., Fu, J., Lawler, M. J. (2006). Chiral phosphoramide-catalyzed enantioselective addition of allylic trichlorosilanes to aldehydes. Preparative studies with bidentate phosphorus-based amides. *J. Org. Chem.*, *71*, 1523–1526 and references cited there.

[56] For a microreview about chiral *N*-oxides in asymmetric catalysis, see: Malkov, A. V., Kočovský, P. (2007). Chiral N-oxides in asymmetric catalysis. *Eur. J. Org. Chem.*, 29–36.

[57] Nakajima, M., Saito, M., Shiro, M., Hashimoto, S. (1998). (*S*)-3,3′-Dimethyl-2,2′-biquinoline *N,N*′-dioxide as an efficient catalyst for enantioselective addition of allyltrichlorosilanes to aldehydes. *J. Am. Chem. Soc.*, *120*, 6419–6420.

[58] Shimada, T., Kina, A., Ikeda, S., Hayashi, T. (2002). A novel axially chiral 2,2′-bipyridine *N,N*′-dioxide. Its preparation and use for asymmetric allylation of aldehydes with allyl(trichloro)silane as a highly efficient catalyst. *Org. Lett.*, *4*, 2799–2801.

[59] Hrdina, R., Valterová, I., Hodačová, J., Cisařová, I., Kotora, M. (2007). A simple approach to unsymmetric atropoisomeric bipyridine *N,N*′-dioxides and their application in enantioselective allylation of aldehydes. *Adv. Synth. Catal.*, *349*, 822–826.

[60] (a) Malkov, A. V., Orsini, M., Pernazza, D., Muir, K. W., Langer, V., Meghani, P., Kočovský, P. (2002). Chiral 2,2′-bipyridine-type *N*-monoxides as organocatalysts in the enantioselective allylation of aldehydes with allyltrichlorosilane. *Org. Lett.*, *4*, 1047–1049; (b) Malkov, A. V., Bell, M., Orsini, M., Pernazza, D., Massa, A., Herrmann, P., Meghani, P., Kočovský, P. (2003). New Lewis-basic *N*-oxides as chiral organocatalysts in asymmetric allylation of aldehydes. *J. Org. Chem.*, *68*, 9659–9668.

[61] Malkov, A. V., Bell, M., Castelluzzo, F., Kočovský, P. (2005). METHOX: a new pyridine *N*-oxide organocatalyst for the asymmetric allylation of aldehydes with allyltrichlorosilanes. *Org. Lett.*, *7*, 3219–3222.

[62] Malkov, A. V., Dufkovà, A., Farrugia, L., Kočovský, P. (2003). Quinox, a quinoline-type *N*-oxide, as organocatalyst in the asymmetric allylation of aromatic aldehydes with allyltrichlorosilanes: the role of arene-arene interactions. *Angew. Chem. Int. Ed.*, *42*, 3674–3677.

[63] Traverse, J. F., Zhao, Y., Hoveyda, A. H., Snapper, M. L. (2005). Proline-based *N*-oxides as readily available and modular chiral catalysts. Enantioselective reactions of allyltrichlorosilane with aldehydes. *Org. Lett.*, *7*, 3151–3154.

[64] Simonini, V., Benaglia, M., Guizzetti, S., Pignataro, L., Celentano, G. (2008). A new class of chiral Lewis basic metal-free catalysts for stereoselective allylations of aldehydes. *Synlett*, 1061–1065.

[65] (a) Pignataro, L., Benaglia, M., Cinquini, M., Cozzi, F., Celentano, G. (2005). Readily available pyridine- and quinoline-*N*-oxides as new organocatalysts for the enantioselective allylation of aromatic aldehydes with allyl(trichloro)silane. *Chirality*, *17*, 396–403; (b) Wong, W.-L., Lee, C.-S., Leung, H.-K., Kwong, H.-L. (2004). The first series of chiral 2,2′:6′,2″-terpyridine tri-*N*-oxide ligands for Lewis base-catalyzed asymmetric allylation of aldehydes. *Org. Biomol. Chem.*, *2*, 1967–1969.

[66] Chelucci, G., Belmonte, N., Benaglia, M., Pignataro, L. (2007). Enantioselective allylation of aldehydes with allyltrichlorosilane promoted by new chiral dipyridylmethane *N*-oxides. *Tetrahedron Lett.*, *48*, 4037–4041.

[67] Pignataro, L., Benaglia, M., Annunziata, R., Cinquini, M., Cozzi, F. (2006). Structurally simple pyridine *N*-oxides as efficient organocatalysts for the enantioselective allylation of aromatic aldehydes. *J. Org. Chem.*, *71*, 1458–1463.

[68] Nakajima, M., Kotani, S., Ishizuka, T., Hashimoto, S. (2005). Chiral phosphine oxide BINAPO as a catalyst for enantioselective allylation of aldehydes with allyltrichlorosilanes. *Tetrahedron Lett.*, *46*, 157–159.

[69] Simonini, V., Benaglia, M., Benincori, T. (2008). Novel chiral biheteroaromatic diphosphine oxides for Lewis base activation of Lewis acids in enantioselective allylation and epoxide opening. *Adv. Synth. Catal.*, *350*, 561–564.

[70] For an overview, see Denmark, S. E., Stavenger, R. A. (2000). Asymmetric catalysis of aldol reactions with chiral Lewis bases. *Acc. Chem. Res.*, *33*, 432–440.

[71] Denmark, S. E., Winter, S. B., Su, X., Wong, K.-T. (1996). Conformational study of solid polypeptides by ^{1}H combined rotation and multiple pulse spectroscopy NMR. *J. Am. Chem. Soc.*, *118*, 7604–7607.

[72] Denmark, S. E., Pham, S. M. (2003). Stereoselective aldol additions of achiral ethyl ketone-derived trichlorosilyl enolates. *J. Org. Chem.*, *68*, 5045–5055, and references cited there.

[73] Denmark, S. E., Fan, Y. (2002). Catalytic, enantioselective aldol additions to ketones. *J. Am. Chem. Soc.*, *124*, 4233–4235.

[74] Nakajima, M., Yokota, T., Saito, M., Hashimoto, S. (2004). Enantioselective aldol reactions of trichlorosilyl enol ethers catalyzed by chiral *N*,*N*′-dioxides and monodentate *N*-oxides. *Tetrahedron Lett.*, *45*, 61–64.

[75] Kotani, S., Hashimoto, S., Nakajima, M. (2006). Enantioselective aldol reactions of trichlorosilyl enol ethers catalyzed by the chiral phosphine oxide BINAPO. *Synlett*, 1116–1118.

[76] Denmark, S. E., Wynn, T., Beutner, G. L. (2002). Lewis base activation of Lewis acids. Addition of silyl ketene acetals to aldehydes. *J. Am. Chem. Soc.*, *124*, 13405–13407.

[77] Denmark, S. E., Beutner, G. L., Wynn, T., Eastgate, M. D. (2005). Lewis base activation of Lewis acids: catalytic, enantioselective addition of silyl ketene acetals to aldehydes. *J. Am. Chem. Soc.*, *127*, 3774–3789.

[78] Denmark, S. E., Heemstra, J. R. Jr. (2006). Lewis base activation of Lewis acids. Vinylogous aldol addition reactions of conjugated *N,O*-silyl ketene acetals to aldehydes. *J. Am. Chem. Soc.*, *128*, 1038–1039.

[79] (a) For selected recent works in the field see: Denmark, S. E., Eklov, B. M. (2008). Neutral and cationic phosphoramide adducts of silicon tetrachloride: synthesis and characterization of their solution and solid-state structures. *Chem. Eur. J.*, *14*, 234–239; (b) Denmark, S. E., Chung, W.-J. (2008). Lewis base activation of Lewis acids: catalytic enantioselective glycolate aldol reactions. *Angew. Chem. Int. Ed.*, *47*, 1890–1892.

[80] Shimoda, Y., Tando, T., Kotani, S., Sugiura, M., Nakajima, M. (2009). Enantioselective aldol reaction of silyl ketene acetals promoted by a Lewis base-activated Lewis acid catalyst. *Tetrahedron Asymmetry*, *20*, 1369–1370.

[81] Kotani, S., Shimoda, Y., Sugiura, M., Nakajima, M. (2009). Novel enantioselective direct aldol-type reaction promoted by a chiral phosphine oxide as an organocatalyst. *Tetrahedron Lett.*, *50*, 4602–4605.

[82] Rossi, S., Benaglia, M., Benincori, T., Celentano, G. (2011). *Tetrahedron*, *67*, 158–166.

[83] Denmark, S. E., Barsanti, P. A., Wong, K.-T., Stavenger, R. A. (1998). Enantioselective ring opening of epoxides with silicon tetrachloride in the presence of a chiral Lewis base. *J. Org. Chem.*, *63*, 2428–2429.

[84] Denmark, S. E., Barsanti, P. A., Beutner, G. L., Wilson, T. W. (2007). Enantioselective ring opening of epoxides with silicon tetrachloride in the presence of a chiral Lewis base: mechanism studies. *Adv. Synth. Catal.*, *349*, 567–582.

[85] Tao, B., Lo, M. M.-C., Fu, G. C. (2001). Planar-chiral pyridine *N*-oxides, a new family of asymmetric catalysts: exploiting an η^5-C_5Ar_5 ligand to achieve high enantioselectivity. *J. Am. Chem. Soc.*, *123*, 353–354.

[86] Tokuoka, E., Totani, S., Matsunaga, H., Ishizuka, T., Hashimoto, S., Nakajima, M. (2005). Asymmetric ring opening of *meso*-epoxides catalyzed by the chiral phosphine oxide BINAPO. *Tetrahedron Asymmetry*, *16*, 2391–2392.

[87] Pu, X., Qi, X., Ready, J. M. (2009). Allenes in asymmetric catalysis: asymmetric ring opening of *meso*-epoxides catalyzed by allene-containing phosphine oxides. *J. Am. Chem. Soc.*, *131*, 10364–10365.

[88] Kobayashi, S., Nishio, K. (1995). Selective formation of propargylsilanes and allenylsilanes and their reactions with aldehydes for the preparation of homopropargylic and allenic alcohols. *J. Am. Chem. Soc.*, *117*, 6392–6393.

[89] Nakajima, M., Saito, M., Hashimoto, S. (2002). Selective synthesis of optically active allenic and homopropargylic alcohols from propargyl chloride. *Tetrahedron Asymmetry*, *13*, 2449–2452.

[90] See Cowen, B. J., Miller, S. J. (2009). Enantioselective catalysis and complexity generation from allenoates. *Chem. Soc. Rev.*, *38*, 3102–3116, and references cited there.

CHAPTER 15

RECENT ADVANCES IN THE METAL-CATALYZED STEREOSELECTIVE SYNTHESIS OF BIOLOGICALLY ACTIVE MOLECULES

CATALINA FERRER, XAVIER VERDAGUER, AND ANTONI RIERA

15.1.	Introduction	625
15.2.	Enantioselective reductions	626
15.3.	Enantioselective oxidations	631
15.4.	Asymmetric additions	634
15.5.	Asymmetric cycloadditions	646
15.6.	Cyclizations and rearrangements	654
	References	658

15.1. INTRODUCTION

In this chapter, we highlight recent reports on asymmetric organometallic reactions for synthesizing biologically active compounds. Asymmetric catalysis has matured as a field, to the point that synthetic chemists can now choose from myriad methods that provide optically active products in high yields and high enantiomeric excesses (ee) using only substoichiometric amounts of a chiral compound. Over the past decade, complexes of almost every metal on the periodic table plus hundreds of ligands have been explored, illustrating

Catalytic Methods in Asymmetric Synthesis: Advanced Materials, Techniques, and Applications, First Edition. Edited by Michelangelo Gruttadauria and Francesco Giacalone.

the extraordinary diversity of organometallic chemistry. Although the field is rapidly expanding, and new methodologies are being uncovered daily, this chapter deals exclusively with those reactions that have been used in the total synthesis of biologically active complex molecules: Instead of reviewing the latest methodologies reported, we focus on reliable and robust methods that are frequently selected by chemists other than those that developed the technique. Application of these methodologies in the total synthesis of complex natural products or in the pilot production of pharmaceuticals is the definitive proof of their reliability and efficiency.

Given the immense number of syntheses described to date, an exhaustive report would have been impossible. Therefore, we have chosen the most representative examples from an overview of the literature from 2005 to 2009.

15.2. ENANTIOSELECTIVE REDUCTIONS

Catalytic hydrogenation was the second organometallic reaction for which an asymmetric version was developed (the first was the cyclopropanation of alkenes, developed by Nozaki et al. [1]). The pioneering work of Knowles et al. [2] and Kagan et al. [3] on rhodium-catalyzed hydrogenation of enamides provided the foundation for a reliable methodology that remains widely used. In addition to the phosphorous-stereogenic diphosphine DIPAMP, myriad chiral diphosphines featuring different types of chiral backbones have been developed [4]. Figure 15.1 shows examples of the structural diversity of chiral diphosphines used in asymmetric hydrogenation.

The rhodium complexes of many chiral phosphines can be used in asymmetric hydrogenation of enamides with excellent ee's [5]. The C2-symmetric bis(phospholane) Et-DuPHOS, developed by Burk [6], is among the most popular of these phosphines, and has become a synthetic standard tool. Several recent syntheses of natural compounds such as Aphanorphine, Mucronine E, and Houamide A, have employed *β*-aryl alanine esters as starting chiral materials and have prepared them by asymmetric hydrogenation using rhodium complex [Rh(COD)(Et-DuPHOS)]OTf (**1**) as a catalyst (COD, cyclooctadi-

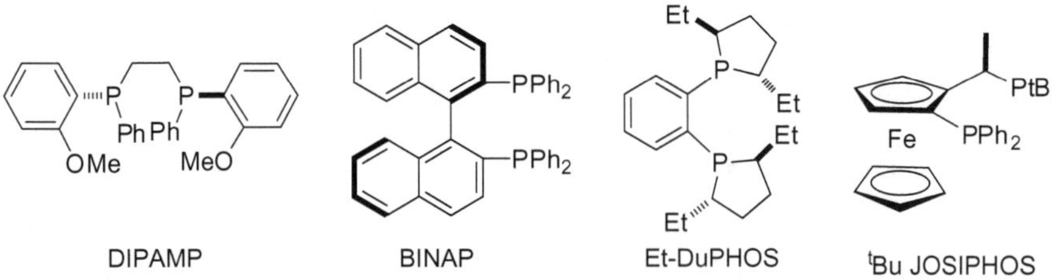

FIGURE 15.1. Different structural types of chiral diphosphines used in asymmetric hydrogenation.

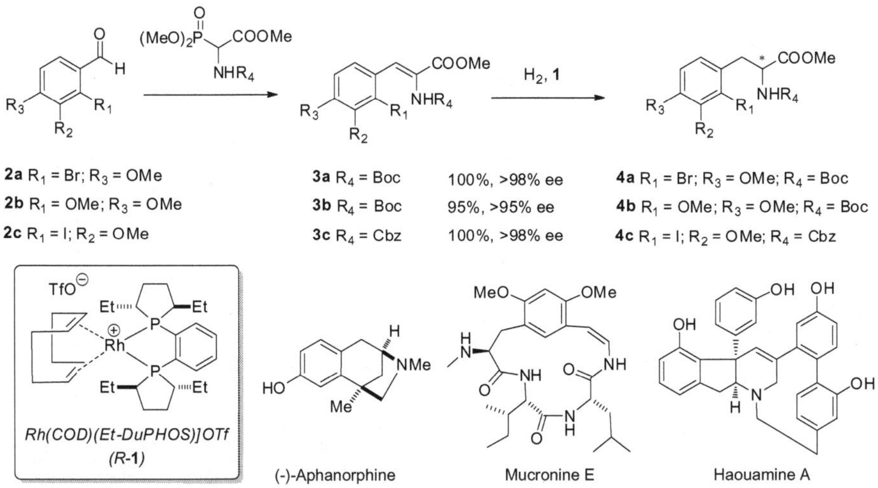

SCHEME 15.1. Enantioselective preparation of the *β*-aryl alanine esters **4a–c**, used as starting materials to synthesize the alkaloids Aphanorphine, Mucronine E, and Haouamide A, respectively.

ene) (Scheme 15.1). The dehydroaryl alanine precursors **3a–c** were easily prepared in good yields by Horner–Emmons olefination of the corresponding aldehydes **2a–c**. Starting from 2-bromo-4-methoxybenzaldehyde (**2a**), Gallagher et al. [7] prepared the amino ester **4a** with excellent enantioselectivity, and then used it in the synthesis of the marine alkaloid Aphanorphine. Using an analogous sequence, Evano et al. [8] prepared *N*-Boc-dimethoxyphenyl alanine (**4b**) as the chiral starting material, for the synthesis of the cyclopeptide alkaloid Mucronine E. Finally, Fürstner and Ackerstaff [9] also used this catalytic hydrogenation to prepare amino ester **4c**, a key starting material for the total synthesis of the cytotoxic alkaloid Haouamine A.

The reliability of the asymmetric hydrogenation, plus its high turnover rates and easy workup, have made it ideal for scaling up; indeed, from its earliest stages of development, it has garnered enormous industrial interest. Although most asymmetric hydrogenations described to date involve chiral diphosphines as ligands, some recently reported methodologies based on chiral monophosphines have also given excellent results [10] after careful optimization. A nice example is the preparation of the key intermediate **6**, used for the industrial synthesis of the renin inhibitor Aliskiren [11]. Chemists from the chemical company DSM (Koninklijke DSM N.V.), in collaboration with Feringa's group, found that a mixture of phosphoramidite **7** and an achiral phosphine gives excellent enantioselectivity as well as a high turnover number (TON) in the hydrogenation of the unsaturated acid **5** (Scheme 15.2).

Ruthenium catalysts (Fig. 15.2) have expanded the scope of the asymmetric hydrogenation to encompass carbonyl reduction [12]. The hydrogenation of

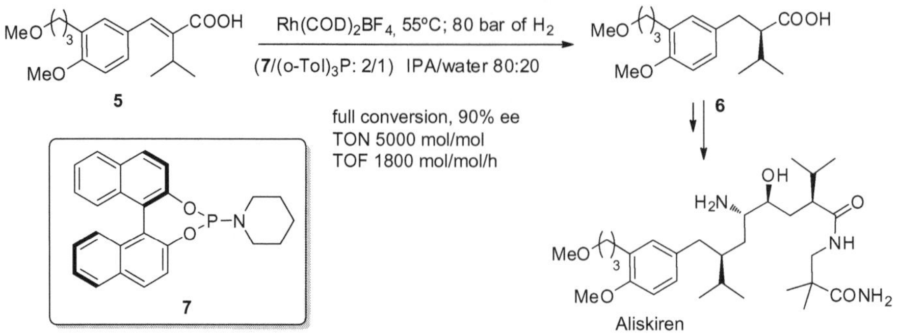

SCHEME 15.2. Industrial synthesis of Aliskiren.

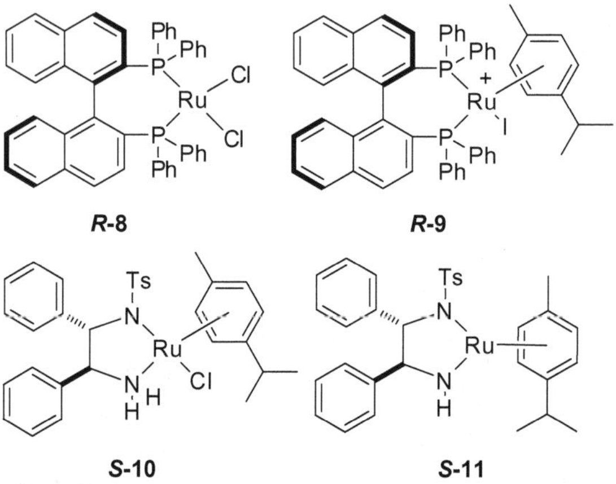

FIGURE 15.2. Ruthenium catalysts for carbonyl asymmetric hydrogenation.

β-keto ester catalyzed by the BINAP ruthenium complexes **8** and **9** (developed by Noyori) is a well-established asymmetric methodology. Transfer hydrogenation using the ruthenium catalysts **10** and **11**, derived from 1,2-diphenylethane-1,2-diamine (DPEN), is extremely reliable for the reduction of ketones and imines [13].

As a representative example of applications of such methodologies, we have selected the preparation of certain chiral building blocks used in the total syntheses of Myxovirescin A1 and (–)-Pseudolaric acid (Scheme 15.3). Large quantities of hydroxy ester **13** were readily available in high optical purity (96% ee) by asymmetric hydrogenation of keto ester **12**. Substitution of the chloride by azide, protection of the alcohol as methoxymethyl (MOM) ether, and subsequent reduction of the ester afforded the aldehyde **14**, which is one of the starting materials selected by Fürstner et al. in their synthesis of the antibiotic macrolide Myxovirescin A_1 [14]. Hydroxylactone **16** is another useful chiral starting material that can be prepared following Noyori's procedure. Hydrogenation of 2-acetyl butyrolactone (**15**) using the BINAP complex

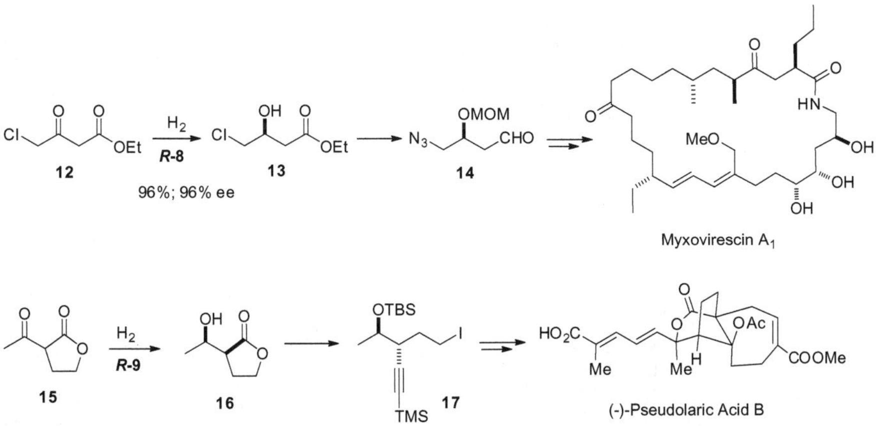

SCHEME 15.3. Reduction of *β*-keto esters by ruthenium catalyzed hydrogenation to prepare starting materials for the synthesis of Myxovirescin A1 and Pseudolaric acid.

***R*-9** enabled the generation of the two adjacent stereocenters of **16** in >95:5 diastereomeric ratio (dr) and >90% ee. Trost et al. [15] transformed compound **16** into the iodide **17** for use as an intermediate in the total synthesis of (−)-Pseudolaric acid, a diterpene with strong potential as a new anticancer agent.

A recent example of industrial asymmetric hydrogenation can be found in the synthesis of Sitagliptin, a dipeptidyl peptidase IV (DPP-4) inhibitor approved by the U.S. Food and Drug Administration (FDA) for the treatment of type 2 diabetes. In 2005, Merck chemists developed the first-generation route to provide large quantities (>100 kg) of Sitagliptin. This process is based on the asymmetric hydrogenation of the *β*-keto ester **18** using the BINAP ruthenium complex ***S*-8** as catalyst (Scheme 15.4) [16]. The enantiomerically enriched hydroxy acid **19** was then transformed into Sitagliptin in six steps. In collaboration with Solvias, Merck chemists recently improved this chemistry by developing an environmentally friendly, cost-effective version based on rhodium-catalyzed enamine hydrogenation [17]. The key intermediate dehydrositagliptin (**21**) was prepared in an efficient three-step, one-pot process in 82% yield. Dehydrositagliptin was enantioselectively hydrogenated using loadings of Rh(I)/^{t}Bu-JOSIPHOS (see Fig. 15.1) as low as 0.15% to afford Sitagliptin in 95% yield and 95% ee (Scheme 15.4).

Reduction of simple ketones is an important transformation for which several methodologies have been developed. Asymmetric transfer hydrogenation is very convenient for preparing starting materials. For example, the enantiomerically enriched alcohols **23** and **26** were used to construct key intermediates for the synthesis of the anticancer agent Iejimalide A, by Fürstner et al. [18], and the potent histone deacetylase (HDAC) inhibitor FK 228, by Williams et al., respectively (Scheme 15.5) [19]. In both cases, the targets were conveniently prepared by asymmetric transfer hydrogenation of

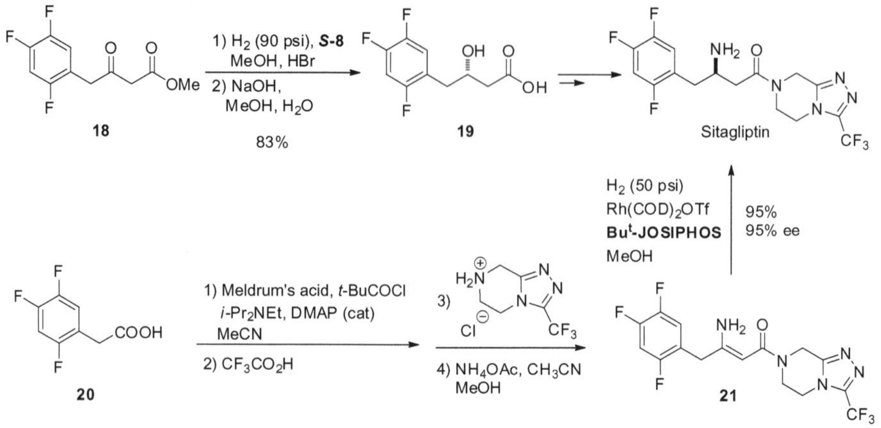

SCHEME 15.4. Industrial syntheses of Sitagliptin based on asymmetric hydrogenations.

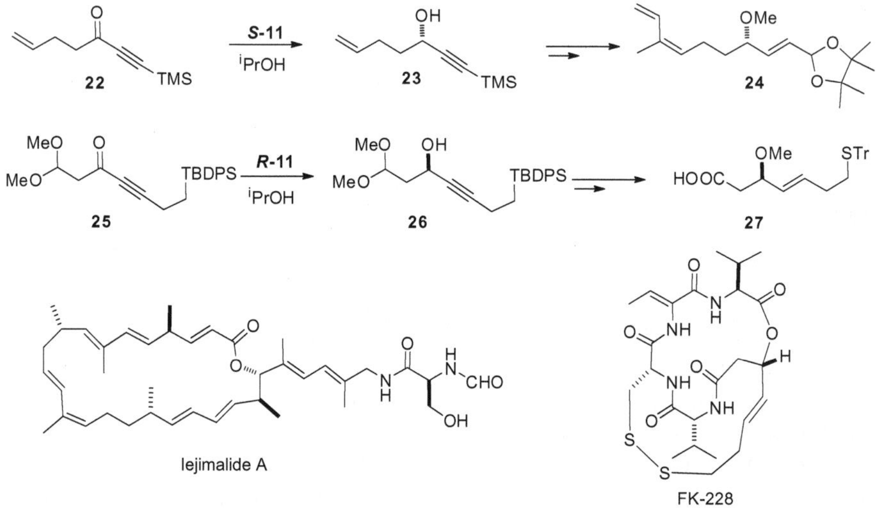

SCHEME 15.5. Preparation of chiral starting materials in the syntheses of Iejimalide A and FK-228 via asymmetric reduction of acetylenic ketones.

acetylenic ketones (**22** and **25**, respectively), using the two enantiomers of the ruthenium catalyst **11**.

Simple methyl ketones can also be reduced using Noyori's catalyst. The endiyne Uncialamycin is a promising lead for cancer and infectious diseases drug discovery. In the first asymmetric synthesis of Uncialamycin, Nicolaou et al. [20] reduced the prochiral methyl ketone **28** to the secondary alcohol **29** in 95% yield and 98% ee (Scheme 15.6).

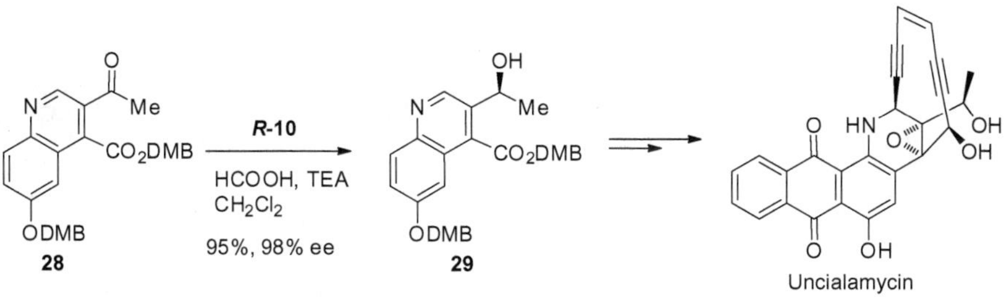

SCHEME 15.6. Noyori's asymmetric transfer hydrogenation, used to prepare a key precursor of Uncialamycin. DMB, 3,4-dimethoxybenzyl.

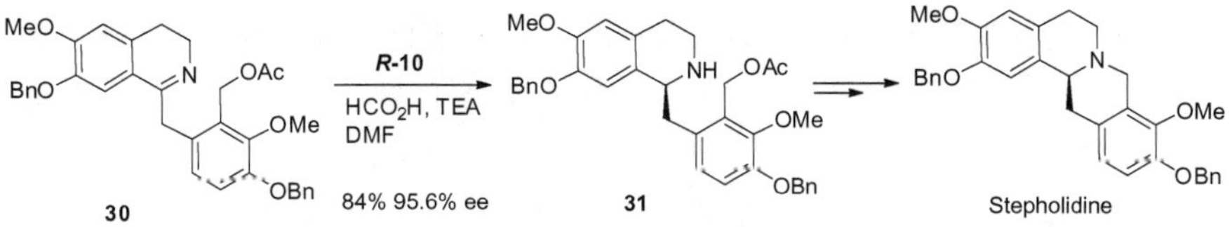

SCHEME 15.7. Asymmetric imine reduction via hydrogen transfer, used to prepare a key precursor of Stepholidine.

Cheng and Yang employed the same ruthenium catalyst (***R*-10**) in the synthesis of (*S*)-Stepholidine, a promising antipsychotic drug candidate [21]. They reduced the cyclic imine **30** to the enantiomerically enriched tetrahydroisoquinoline **31**, a key intermediate for the synthesis of Stepholidine, with concomitant formation of the chiral center found in the product (Scheme 15.7).

15.3. ENANTIOSELECTIVE OXIDATIONS

Alkene oxidations, together with hydrogenations and cyclopropanations, are among the first reactions for which enantioselective versions were developed. Sharpless asymmetric epoxidations [22] and dihydroxylations [23] have become standard methodologies, and are sometimes taught in general organic chemistry textbooks. For instance, the Sharpless epoxidation of allylic alcohols is widely used to prepare optically active starting materials; it offers reliability, a readily available catalyst, facile prediction of the product configuration, and the versatility of the epoxy alcohol functionality. Furthermore, among the different types of epoxy alcohols accessible by Sharpless epoxidation, those bearing unsaturated side chains have the additional benefit of the extraordinary development of alkene metathesis [24]. A representative recent example is the synthesis of *ent*-Clavilactone B by Barrett et al. (Scheme 15.8) [25].

Sharpless asymmetric dihydroxylation [26] has been used to prepare enantiomerically enriched fragments, for example, to prepare the intermediate **39** for the total synthesis of (+)-Amphidinolide W by Ghosh and Gong (Scheme

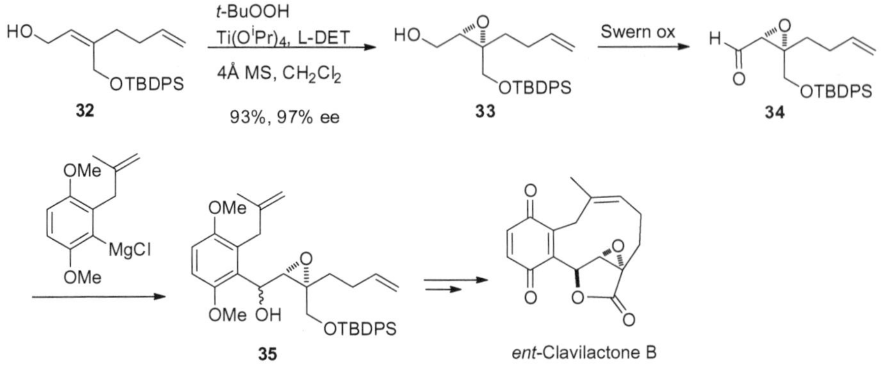

SCHEME 15.8. Sharpless epoxidation an epoxy alcohol, used to prepare a key precursor of *ent*-Clavilactone.

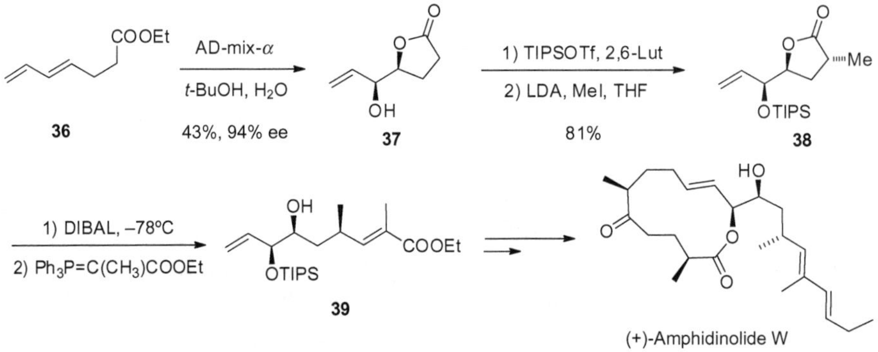

SCHEME 15.9. Preparation of a synthetic precursor of (+)-Amphidinolide W via Sharpless asymmetric dihydroxylation.

15.9) [27]. (E)-Hepta-4,6-dienoate (**36**), readily available from Claisen rearrangement of penta-1,4-dienol, was submitted to Sharpless asymmetric dihydroxylation to give the lactone **37** in high enantiomeric purity but only moderate regioselectivity. Protection and subsequent diastereoselective alkylation afforded the lactone **38**, which, after reduction and Wittig olefination, gave the intermediate **39**.

Asymmetric oxidations of functional groups other than alkenes are far less common. Kinetic resolution of secondary alcohols by aerobic oxidation is a methodology recently developed independently by Stoltz et al. [28] and by Sigman et al. [29]. This palladium-catalyzed oxidation provides excellent ee's when sparteine is used as chiral ligand. Stoltz et al. [30] used this chemistry in the total synthesis of several alkaloids, including (−)-Aurantioclavine (Scheme 15.10). The process can be made more efficient by recycling the ketone product **42** via reduction with lithium aluminum hydride.

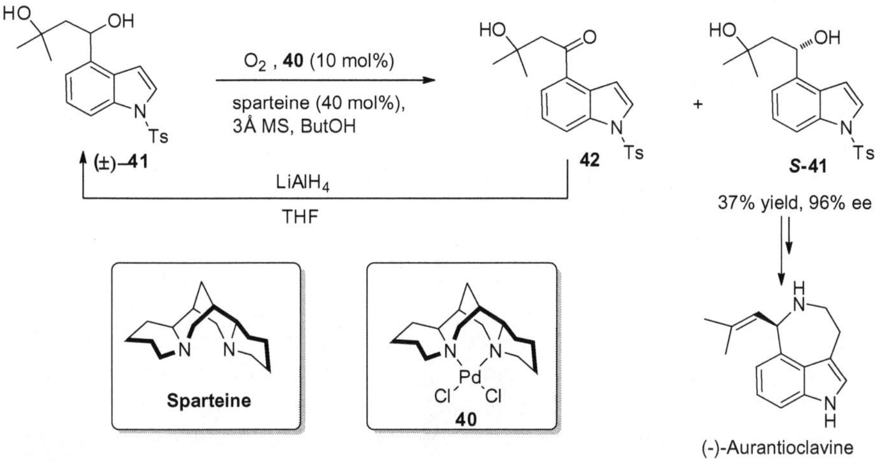

SCHEME 15.10. Secondary alcohol resolution via asymmetric oxidation, used in the synthesis of Aurantioclavine.

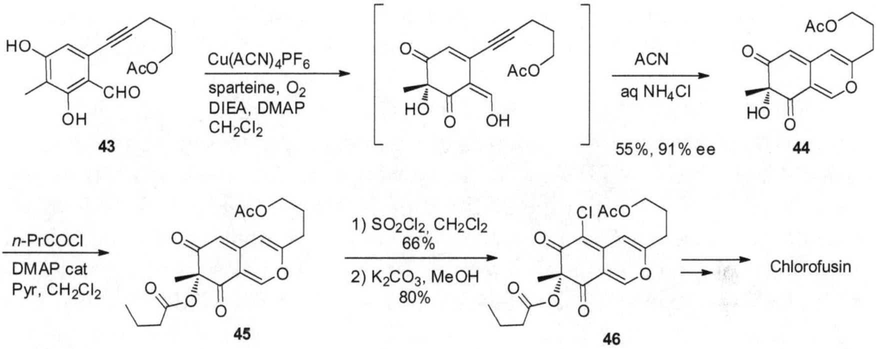

SCHEME 15.11. Synthesis of the nonpeptidic fragment of Chlorofusin by asymmetric deoxigenation.

Oxidative dearomatization of phenols is another type of asymmetric oxidation catalyzed by sparteine metal complexes. Copper-mediated oxidative hydroxylation of phenols affords highly enantiomerically enriched intermediates that can be used as advanced starting materials [31]. The synthesis of Chlorofusin [32], a novel fungal peptide with promising anticancer activity, reflects the value of this asymmetric methodology. The copper-mediated oxidation of bis-phenol **43** enabled preparation of the enantiomerically enriched chiral quinone **44**, which was ultimately converted into the key intermediate Azaphilone **46** (Scheme 15.11). Coupling of **46** to a cyclopeptide completed the synthesis.

15.4. ASYMMETRIC ADDITIONS

Enantioenriched chiral secondary alcohols can also be prepared by enantioselective alkylation of aldehydes. This methodology is advantageous over the enantioselective reduction of ketones in that the chiral center and the molecule skeleton are assembled simultaneously in the same reaction. Dialkylzinc reagents are typically used, as they do not usually react with aldehydes in the absence of catalyst. The pioneering work of Soai et al. and Noyori et al. [33] showed that chiral amino alcohols can be used as Lewis bases to enantioselectively add simple alkyl groups to aldehydes. Since few dialkylzinc compounds are commercially available, the main difficulty is often the preparation of the organozinc reagents. Ohno et al. [34] introduced the use of chiral Lewis acids derived from the C2 bis-sulfonamide ligand **47**, as catalysts for the nucleophilic addition of organozinc reagents. Some representative ligands and a chiral Lewis acid used in this reaction are shown in Figure 15.3.

In the total synthesis of (−)-Gloeosporone [35], Trenkle and Jamison added a pentyl side chain to aldehyde **51** by enantioselective addition. In this case, generation of the dialkylzinc (using Knochel's method) was critical for obtaining high enantioselectivity. Esterification of the resulting secondary alcohol **52**, followed by nickel-catalyzed, epoxide-alkyne reductive macrocyclization, afforded an advanced precursor (**54**) of the macrolide natural product (Scheme 15.12).

Enantioselective alkynyl addition to carbonyl or imino groups is a very important transformation. Chemists from Merck pioneered this chemistry during the syntheses of HIV-1 reverse transcriptase inhibitors, by using amino alcohols as chiral ligands in the addition of lithium acetylides to ketones [36]. Recently, Trost developed a dinuclear Zn-catalyzed asymmetric alkynylation using the proline-derived amino alcohol **48**. The alkynylation of several aromatic aldehydes using this bimetallic catalyst system proceeded efficiently with high enantioselectivities. This methodology was employed in the total synthesis of Adociacetylene (Scheme 15.13).

Allyl addition is usually an easier transformation. Several methodologies for metal-catalyzed enantioselective allylation of carbonyl compounds to give optically active homoallylic alcohols have been described to proceed in high

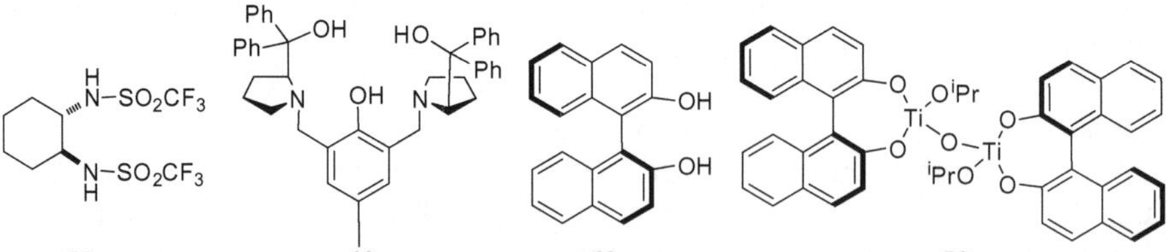

47 **48** **49** **50**

FIGURE 15.3. Chiral ligands used to prepare Lewis acids catalysts and Maruoka's catalyst (**50**) for enantioselective additions to aldehydes.

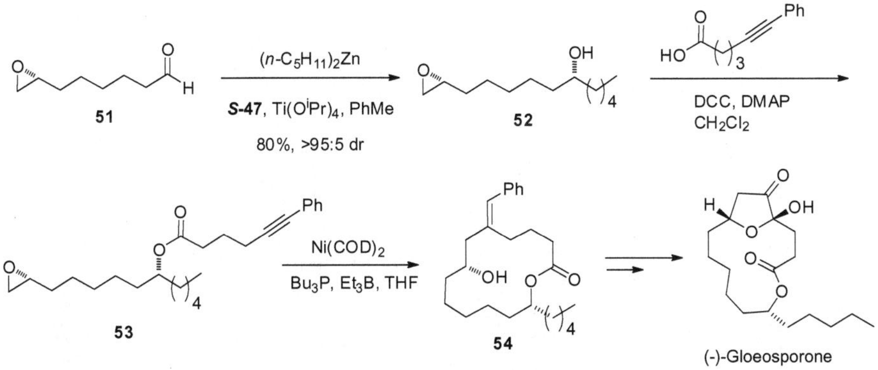

SCHEME 15.12. Enantioselective addition of a pentyl group to an aldehyde and nickel-catalyzed macrocyclization in the synthesis of Gloeosporone.

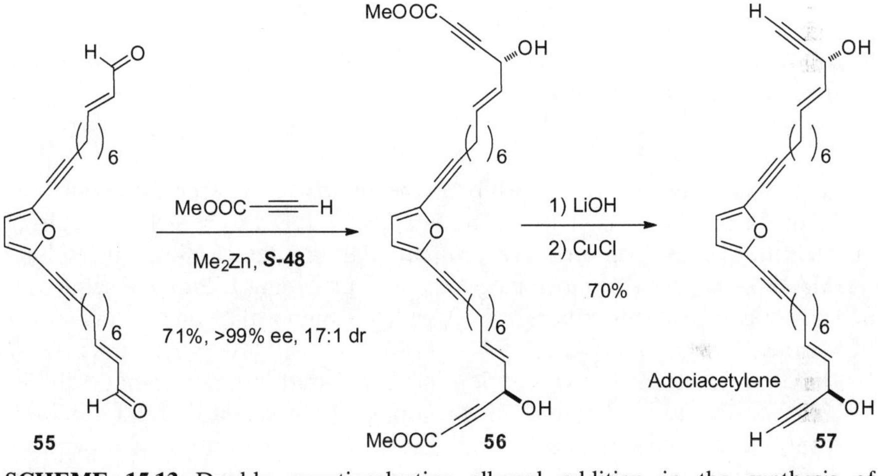

SCHEME 15.13. Double enantioselective alkynyl addition in the synthesis of Adociacetylene.

selectivity [37]. The titanium-catalyzed [38] version of this reaction using allyltributyltin as a nucleophile has been applied to the synthesis of (−)-Incarvillateine [39]. In this case, (*S*)-BINOL (**49**) was the chiral ligand of choice to carry out the reaction. Enantioselective addition to aldehyde **58** gave compound **59** in excellent yield and ee. Protection of the secondary alcohol, and cross-metathesis with methacrolein afforded aldehyde **60**. The intramolecular alkylation of α,β-unsaturated esters via Rh-catalyzed C–H activation developed by Ellman et al. [40] was one of the key steps in this synthesis. Reaction of the imine **61** in the presence of $[RhCl(coe)_2]_2$ (coe, cyclooctene) results in an olefinic C–H activation followed by *syn*-alkene insertion and reductive elimination to exclusively provide the desired exocyclic double bond

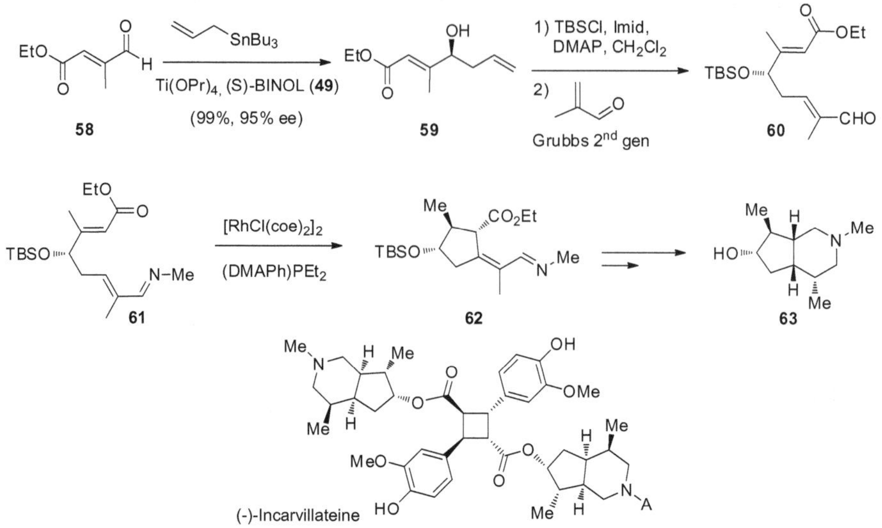

SCHEME 15.14. Enantioselective allyl addition in the early stages of the synthesis of Incarvillateine and diastereoselective Rh-catalyzed cycloisomerization.

geometry and the *anti* relationship of the methyl and ester functionalities (Scheme 15.14). Though many of the ligands explored were active, resulting in quantitative cyclization of **61**, dimethylaminophenyl diethylphosphine [(DMAPh)-PEt_2] was the most selective ligand, providing a 5:1 dr. Subsequently, **62** was transformed into piperidine **63**, which is one of the constituting motifs of (−)-Incarvillateine (Scheme 15.14).

The enantioselective allyl addition has been used in many total syntheses to generate chirality early on. For example, in the synthesis of the natural anticancer compound Blepharocalyxin D [41], the catalyst used for promoting allyl addition to the aldehyde **64** was the BINOL derivative **50** (described by Maruoka et al. [42]). As in the preceding example, the powerful combination of allyl addition and cross-metathesis was employed to construct compound **66**. The secondary alcohol was then submitted to a Prins cyclization to give the pyrane **67**, an advanced precursor of Blepharocalyxin D (Scheme 15.15).

Similar transformations can be achieved using the silver/BINAP catalyst developed by Yamamoto et al. [43]. Reaction of aldehyde **68** with methallyl-tributylstannane in the presence of catalytic amounts of AgOTf and BINAP gave the homoallylic alcohol **69** in good yield and high enantioselectivity (Scheme 15.16). Compound **70** was further functionalized into the lateral alkylic chain of (*R*,*R*,*R*)-α-Tocopherol [44].

The past few years have witnessed considerable effort toward the development of efficient catalytic systems for asymmetric conjugated addition using organometallic reagents [45]. The broadest scope in this chemistry has been reached with complexes of copper salts and chiral ligands. Recent advances

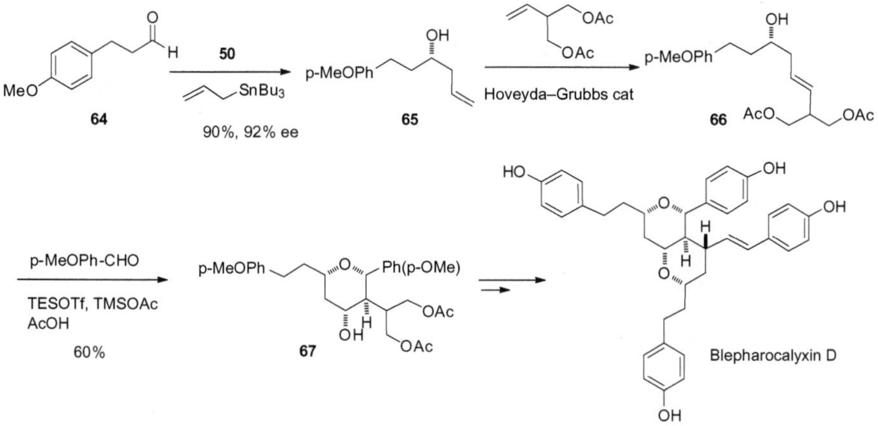

SCHEME 15.15. Enantioselective allyl addition in the early stages of the synthesis of Blepharocalyxin D.

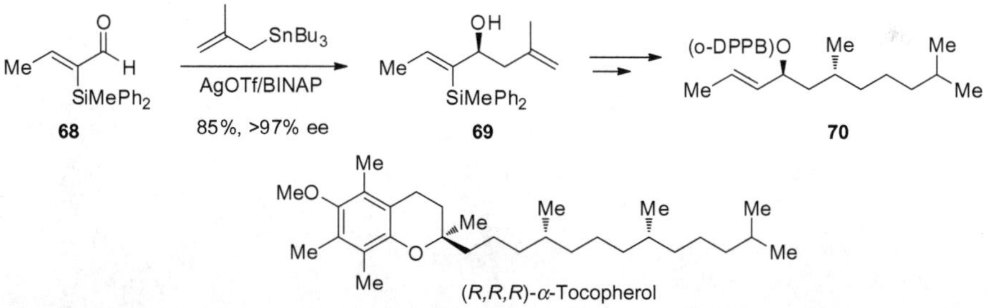

SCHEME 15.16. Enantioselective allyl addition in the synthesis of Tocopherol. *o*-DPPB, *ortho*-diphenylphosphanylbenzoate.

have enabled the preparation of several natural compounds [46]. For example, the highly enantioselective conjugated addition of Grignard reagents to α,β-unsaturated thioesters catalyzed by a copper bromide complex of Cy-JOSIPHOS (**71**) (developed by Feringa et al. [47]), has been employed in the synthesis of a *Phytophthora* mating hormone (Scheme 15.17) [48]. The sequence included three similar conjugated additions. The enantioselective copper-catalyzed reaction of the thioesther **72** with methyl magnesium bromide gave **73**, which was converted into dithiane **74** in a few steps. The same protocol was used to synthesize **76**, which was obtained in very good yield and ee. Thioester **76** was transformed into **77**, which was subjected to an asymmetric conjugated addition to give **78** in very good diastereomeric excess. Finally, reduction of the thioesther in **78**, followed by reaction with **74**, and protecting group manipulation, gave the desired hormone.

Hoveyda et al. have done substantial work to develop catalytic methods for asymmetric conjugate additions. Some of the successful ligands are shown in Figure 15.4. The peptide phosphane **79** gave excellent yields and ee's in the

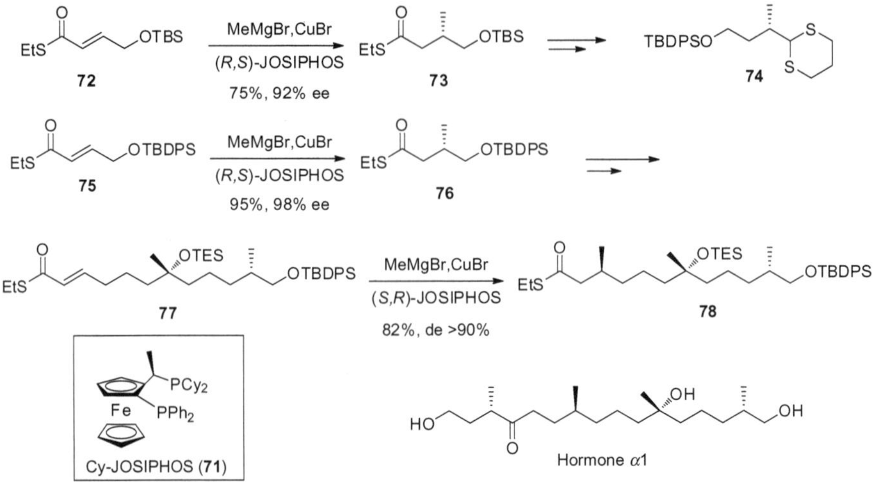

SCHEME 15.17. Enantioselective conjugate methyl additions of cuprate reagents, used in the synthesis of the hormone α1.

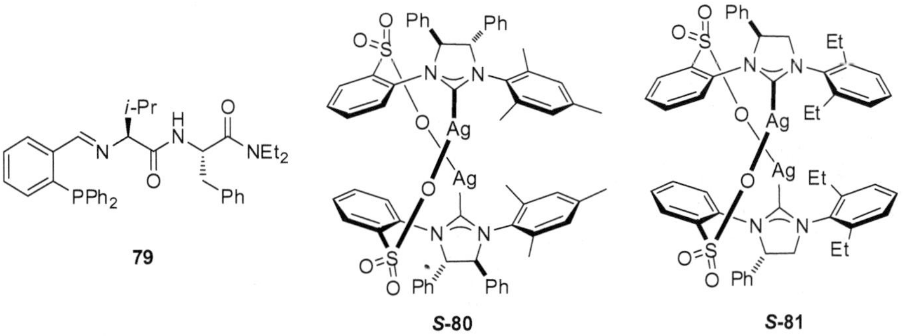

FIGURE 15.4. Chiral ligands used in enantioselective copper-catalyzed conjugate additions.

copper-catalyzed conjugate addition of dialkylzincs to unsaturated furanones and pyranones [49]. The same group used this methodology in the synthesis of Clavirolide C [50]. Thus, reaction of the unsaturated lactone **82** and dimethylzinc with $(CuOTf)_2$·toluene in the presence of the chiral ligand **79** generated **83** in 86% yield and 98% ee. Compound **83** was then converted into aldehyde **84** in a few steps (Scheme 15.18). In contrast, when β-substituted cyclopentenones were used as substrate, a chiral copper complex generated in situ from the N-heterocyclic carbene sulfonates **80** or **81** had to be used [51]. This conjugate addition was used in the preparation of the quaternary centre of Clavirolide C. Conjugate addition of trimethylaluminium to cyclopentenone

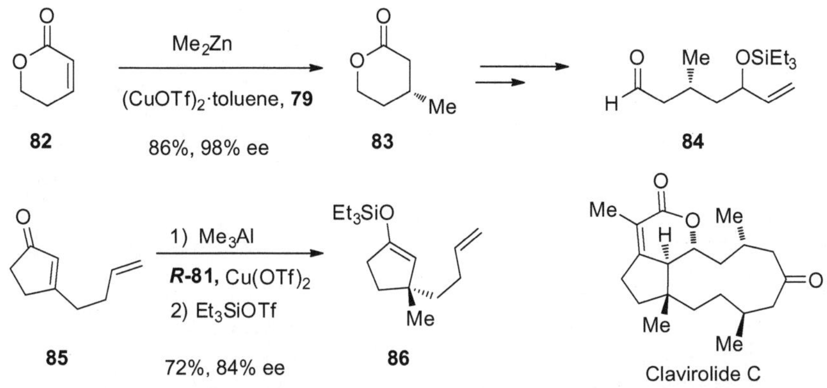

SCHEME 15.18. Enantioselective conjugate additions in the synthesis of Clavirolide C.

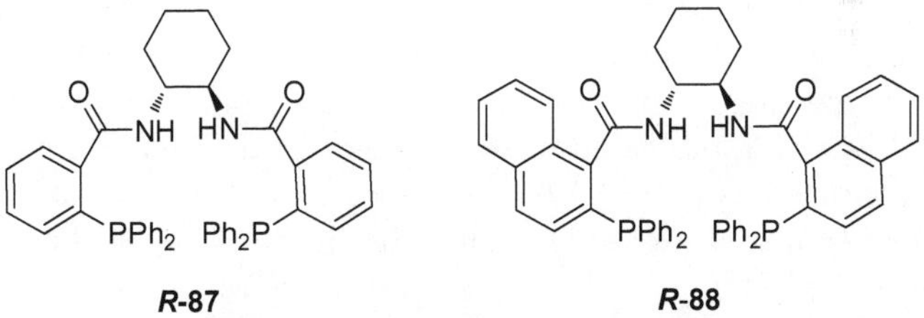

FIGURE 15.5. Chiral ligands for AAA.

85 promoted by a copper complex derived from the optimized sulfonate ***R*-81**, followed by quenching with Et_3SiOTf, gave compound **86** in good yield and ee. Generation of the lithium enolate derived from **86**, and subsequent addition of aldehyde **84** in the presence on BEt_3, gave highly selective formation of an aldol product, which was then converted into the final dolabellane diterpene (Scheme 15.18).

Asymmetric allylic alkylation (AAA) is a very powerful synthetic tool for complex natural and biologically important targets, due to its broad substrate scope and its ability to form different types of bonds (e.g., C–C, C–O, C–N, and C-S) [52]. The reaction involves metal-promoted ionization, which leads to the formation of π-allyl system, followed by nucleophilic addition. The newly formed double bond is a point for further functionalization, making this reaction very attractive for total synthesis [53]. Trost et al. have extensively studied this reaction, for which they developed the chiral ligands **87** and **88** (Fig. 15.5). They have also prepared diverse chiral building blocks with high diastereo- and enantioselectivities [54].

An elegant example of the advantages of AAA is the synthesis of (−)-Terpestacin [55]. The synthesis started with the Pd-catalyzed AAA between

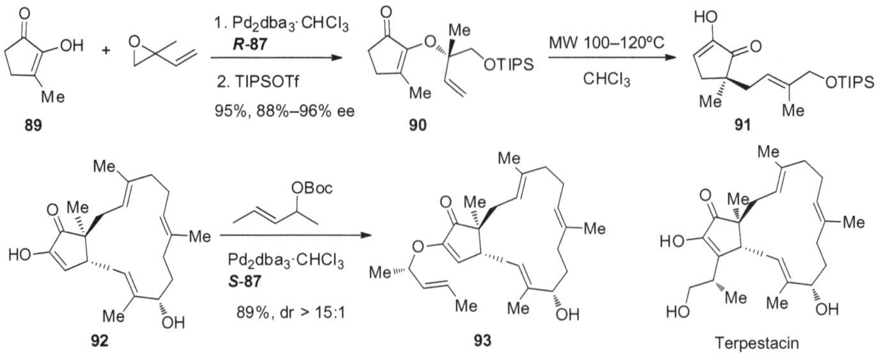

SCHEME 15.19. Catalytic asymmetric allyl alkylations used in the synthesis of Terpestacin.

cyclopentenone **89** and isoprene monoepoxide, which proceeded in excellent yield and ee to give adduct **90** (Scheme 15.19). Compound **90** was subjected to a Claisen rearrangement to form **91**; this also proceeded with excellent selectivity, illustrating the value of the double-bond formation in AAA. The same strategy of Pd-catalyzed AAA followed by Claisen rearrangement was later repeated to introduce the requisite chirality on the side chain. The same ligand, but with different configuration, was used to obtain the desired stereochemistry. The reaction was also done using a more elaborate substrate to give compound **93**, an advanced precursor of Terpestacin, with excellent selectivity (Scheme 15.19).

The same group also used ligand ***R*-87** in the synthesis of Agelastatin A [56]. They employed a double Pd-catalyzed AAA, using a pyrrole and an *N*-alkoxyamide as nucleophile. By varying the functional groups at the 2-position of the pyrrole, they performed the AAA stepwise or in a cascade sequence. In a four-step sequence, they obtained compound **94** in 92% ee. Alternatively, in a cascade process through a double AAA pathway, they obtained compound **95** (a pseudoenantiomer of **94**) in 97.5% ee. Further transformations of **94** and **95** ultimately afforded (+)-Agelastatin A and (−)-Agelastatin A, respectively (Scheme 15.20).

Xie et al. employed AAAs to generate two vicinal quaternary carbon centers in short and efficient syntheses of Hyperolactone C and Biyouyanagin [57]. AAA of the *β*-ketoester **96** had to be promoted by using the more bulky ligand (*R*,*R*)-**88** to give the hydroxy ester **97** in good enantio- and diastereoselectivities. Compound **97** was converted into its *tert*-butyldimethylsilyl (TBS) derivative **98** to avoid an undesired intramolecular carbonate migration. Alternatively, **97** was lactonized by treatment with *p*-toluenesulfonic acid to give Hyperolactone C. Biomimetic [2 + 2] cycloaddition of Hyperolactone C with *ent*-Zingiberene afforded the natural product (−)-Biyouyanagin. (Scheme 15.21).

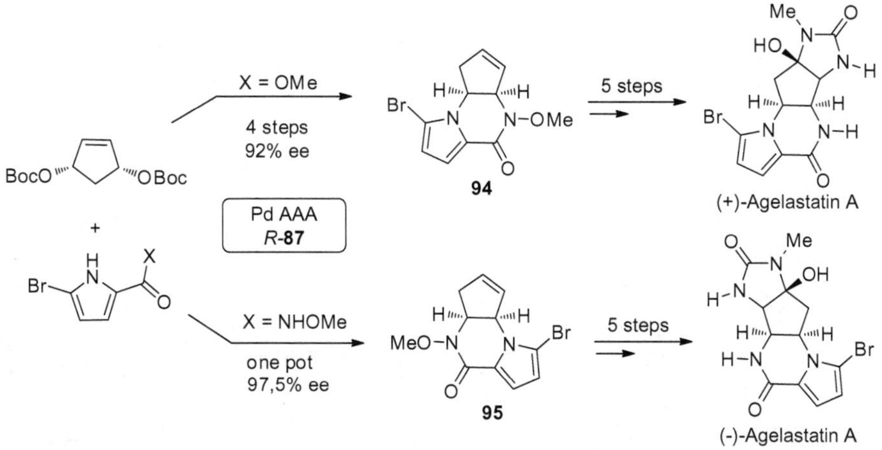

SCHEME 15.20. Synthesis of both enantiomers of Agelastatin A via AAAs: the (+) enantiomer was formed by a sequential reaction (bottom), and the (−) enantiomer was formed by a cascade reaction (top).

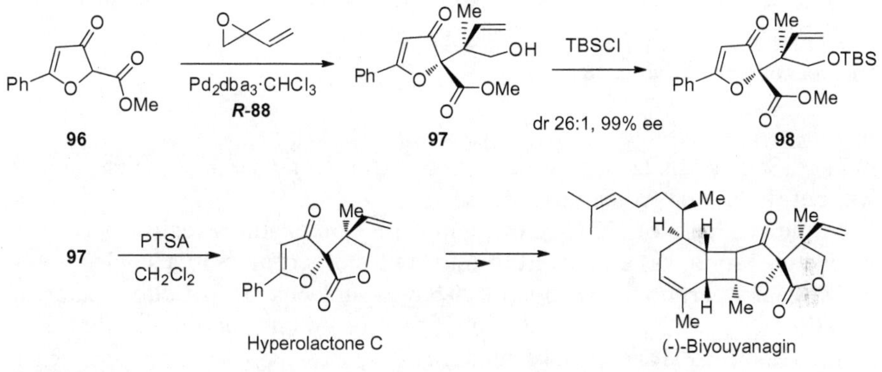

SCHEME 15.21. Synthesis of Hyperolactone and Biyouyanagin by asymmetric allyl alkylation.

The enantioselective version of the decarboxylative alkylation of allyl enol carbonates to the corresponding α-allylcyclohexanone derivatives, described originally by Tsuji and Minami [58], has been expanded by Behenna and Stoltz over the past few years [59]. This mild, operationally straightforward and stereoselective reaction produces chiral cycloalkanones with quaternary stereocenters at the α-position in high enantiopurity and excellent yield.

This methodology was applied for the first time in total synthesis in the preparation of (+)-Dichroanone [60]. In the key step of the asymmetric Tsuji allylation, the carbonate **99** was treated with catalytic $Pd_2(dba)_3$ and (*S*)-*tert*-Bu-PHOX (**100**) to afford the quaternary allyl ketone **101** in 83% yield and

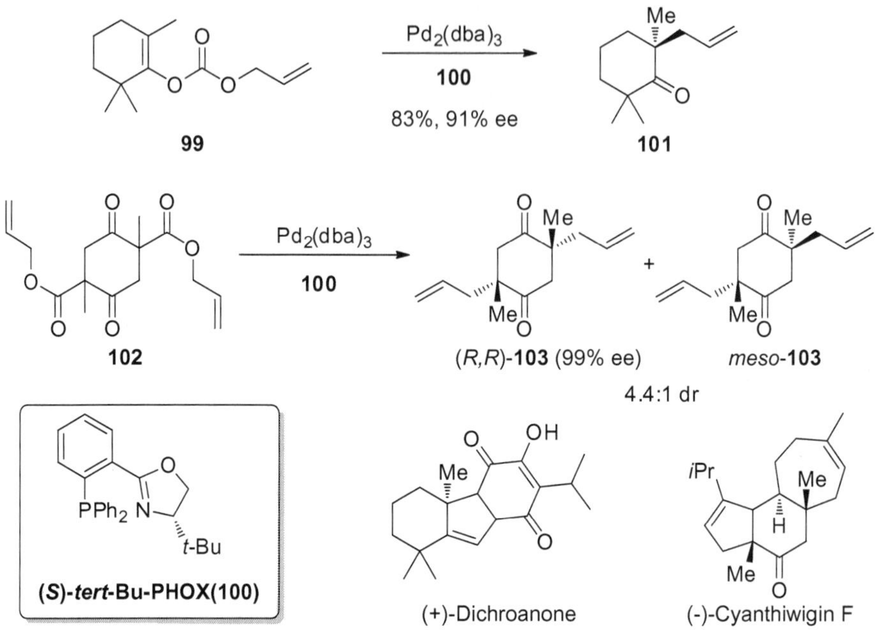

SCHEME 15.22. Asymmetric decarboxylative allylations used in the syntheses of Dichroanone and Cyanthiwigin.

91% ee (Scheme 15.22). Compound **101** was transformed into (+)-Dichroanone without the use of protecting groups.

In the synthesis of (−)-Cyanthiwigin F [61], an enantioselective decarboxylative allylation was employed to form two stereocenters in a single stereoconvergent operation. Compound **102** was subjected to a double catalytic enantioselective alkylation by treatment with catalytic amounts of $Pd_2(dba)_3$ and (*S*)-*tert*-Bu-PHOX. A 4.4:1 mixture of the bis-alkylated products (*R*,*R*)-**103** (in 99% ee) and *meso*-**103** were obtained. Although this reaction could progress through 16 different pathways to afford any of the three different stereoisomers, the high degree of selectivity imparted by the catalytic system predominantly favors the formation of the desired diketone (*R*,*R*)-**103** (Scheme 15.22).

Hydroboration is one of the most valuable synthetic transformations, as the initially formed C–B bond can be efficiently transformed into C–C, C–O, or C–N bonds. The asymmetric version of this transformation is especially attractive, as it occurs with retention of the configuration in the carbon bond. Although hydroboration of alkenes with chiral hydroboration agents remains very useful and reliable, metal-catalyzed hydroboration reactions have become an attractive alternative for the functionalization of double bonds [62]. The most widely used metal in this chemistry is rhodium, although copper has also been employed. These metals are generally used in combination with bidentate P,P or P,N chiral ligands. For example, a rhodium-catalyzed enantioselective

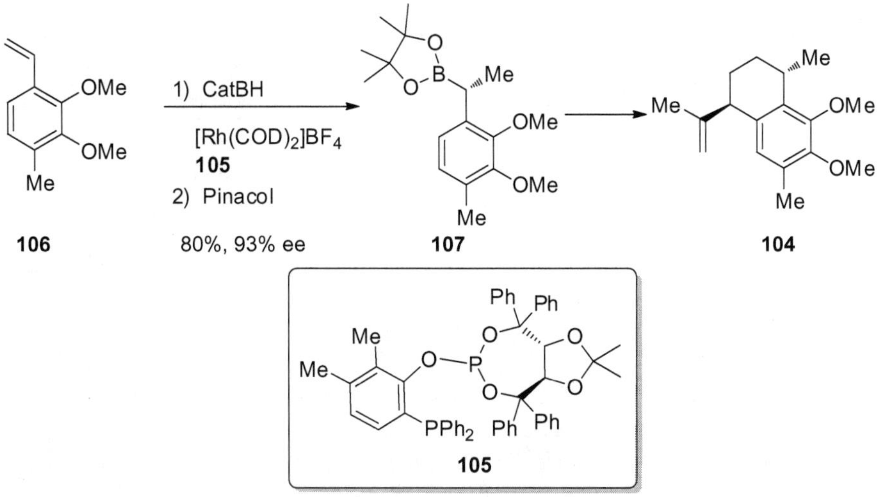

SCHEME 15.23. Rhodium-catalyzed enantioselective hydroboration used to prepare compound **104**, a synthetic intermediate of marine Serrulatane and Amphilectane diterpenes.

hydroboration has been used in the asymmetric synthesis of *trans*-7,8-dimethoxycalamenene (**104**), a projected intermediate for the total synthesis of marine Serrulatane and Amphilectane diterpenes (Scheme 15.23) [63]. Using $[Rh(COD)_2]BF_4$ as catalyst, the $\alpha,\alpha,\alpha',\alpha'$-tetraphenyl-1,3-dioxolan-4,5-dimethanol (TADDOL)-derivative chiral ligand **105**, and catecholborane as the boron source, and following addition of pinacol, the styrene derivative **106** was converted into the borane derivative **107** in 80% yield and with 93% ee.

The zirconium-catalyzed asymmetric carboalumination of alkenes, known as the *ZACA reaction*, was discovered and has been mainly developed by Negishi et al. [64]. In this reaction, after the carboalumination of an alkene catalyzed by Zr, a new carbon–carbon bond is formed, and the newly generated organoaluminium species can either be oxidized (typically, with oxygen) to give 2-alkyl-1-alkanols or treated with iodine to give 2-alkyl-1-iodo compounds. Chiral zirconium catalysts [generally, $(NMI)_2ZrCl_2$ dichlorobis(1-neomenthylindenyl)zirconium (**108**)] provide good ee's.

The utility of this reaction has been shown in the synthesis of (−)-Spongidepsin [65]. The two key chiral fragments **109** and **110** were efficiently synthesized using ZACA reactions (Scheme 15.24). The acid fragment **109** was prepared from allyl alcohol in various steps, including two ZACA reactions. The first one afforded compound **111** (82% ee), which was then subjected to a palladium-catalyzed vinylation to give **112**. The second ZACA reaction on **112** generated the alcohol **113** in 76% yield and 5.5:1 dr. After chromatographic purification, **113** was obtained in 45% yield and > 40:1 dr, and was then converted by standard functional group manipulation into **109** (Scheme 15.24). Fragment **110** was also prepared via a ZACA reaction, from compound **115**, to give **116**

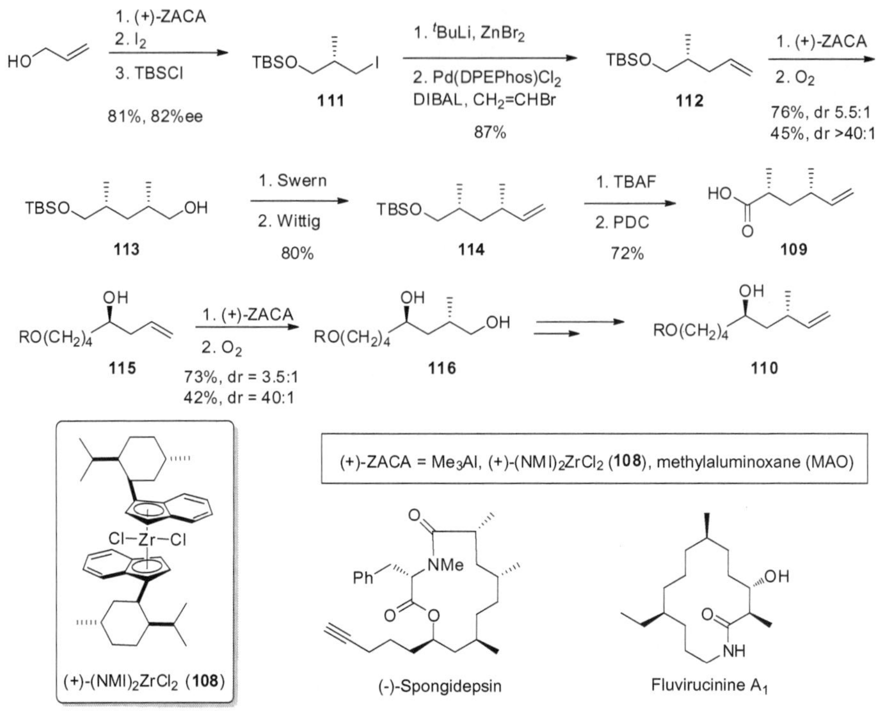

SCHEME 15.24. ZACA reaction used in the synthesis of Spongidepsin. TBSCl, *tert*-butyldimethylsilyl chloride; DPEPhos, bis(*o*-diphenylphosphinophenyl) ether; DIBAL, diisobutylaluminium hydride; TBAF, tetra-butylammonium fluoride; PDC, pyridinium dichromate.

in 73% yield and 3.5:1 dr (after chromatographic purification: 42% yield and 40:1 dr). Interestingly, unprotected ω-vinyl secondary alcohols were used in the ZACA reaction. Subsequent reactions and functional group interconversions completed the total synthesis of (−)-Spongidepsin (Scheme 15.24). More recently, Negishi et al. reported the combination of the ZACA reaction followed by the lipase-catalyzed acetylation with vinyl acetate to synthesize stereomerically pure (>99% ee) 2-methyl-1-alkanols [66]. This methodology has been applied to the synthesis of Fluvirucinine A_1 (Scheme 15.24) [67].

The Heck reaction, defined as the Pd(0)-catalyzed coupling of an aryl or vinyl halide or triflate with an alkene, is among the most important carbon–carbon bond-forming processes [68]. It is advantageous in that its substrates are not limited to activated alkenes, but also include simple olefins. Moreover, the Heck reaction is compatible with other functional groups in the reactive substrates, such as ketones, esters, amides, ethers, or heterocycles. Since the first reported examples of the asymmetric version of this reaction [69] in the late 1980s, major effort has been devoted to its development and it has been applied to various complex natural product syntheses.

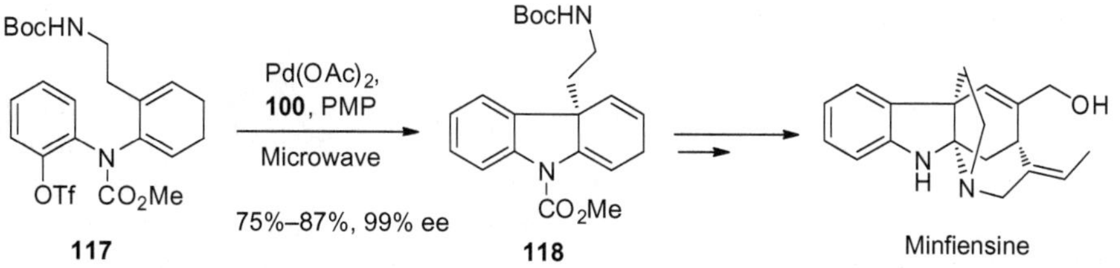

SCHEME 15.25. Enantioselective Heck reaction in the synthesis of Minfiensine.

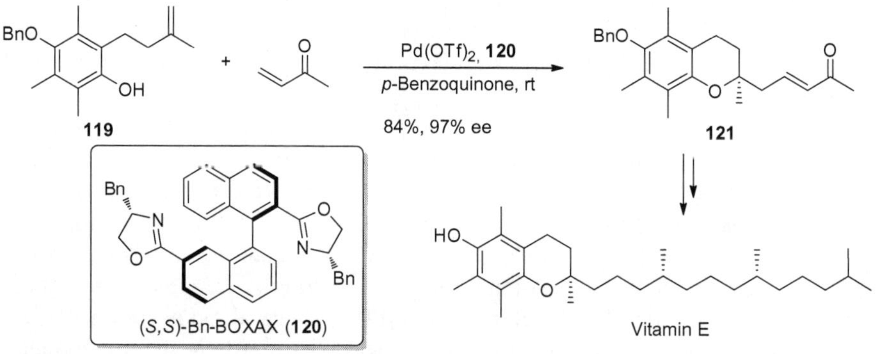

SCHEME 15.26. Enantioselective Wacker–Heck tandem reaction in the synthesis of vitamin E.

Overman et al. have used an asymmetric Heck reaction for the total synthesis of Minfiensine, a member of the *Strychnos* alkaloids family [70]. The intramolecular asymmetric Heck reaction of the triflate **117** using $Pd(OAc)_2$ as catalyst and the ^{t}Bu-PHOX ligand (**100**), in the presence of 1,2,2,6,6-pentamethylpiperidine (PMP) gave the dihydrocarbazole **118** in good yield and with excellent enantioselectivity (Scheme 15.25). Using microwave conditions enabled the shortening of the reaction time at 100°C from 70 hours to 30–45 minutes. Compound **118** was then transformed into Minfiensine in several steps.

Tietze et al. harnessed a domino process comprising a Wacker reaction followed by a Heck reaction in the synthesis of vitamin E [71]. Reaction of phenol **119** with methyl vinyl ketone in the presence of catalytic amounts of $Pd(OTf)_2$, the chiral ligand (*S*,*S*)-Bn-BOXAX (**120**), and p-benzoquinone as a reoxidant for the Pd, afforded the key intermediate **121** in 84% yield and with 97% ee (Scheme 15.26). In this cascade process, the initial intramolecular Wacker reaction defines the stereochemistry of the new chiral center, and the resulting palladium intermediate then undergoes Heck reaction with methyl vinyl ketone. Compound **121** was ultimately converted into vitamin E.

15.5. ASYMMETRIC CYCLOADDITIONS

Since its discovery in 1928 [72], the Diels–Alder reaction (DA) has fascinated organic chemists to the point that it may be considered our favorite reaction. The love story between synthetic chemists and the *diene synthesis*, as it was initially known, can be explained by many factors. First, and central to this chapter, is its capacity to construct molecular complexity in a single reaction step. The [4 + 2] cycloaddition enables simple and straightforward construction of six-membered rings with up to four stereocenters in an experimentally simple and straightforward way. Moreover, the regio- and stereoselectivity rules that govern this process enable chemists to make predictions and retrosynthetic plans. Given the profusion of six-membered rings in nature, the DA was clearly stated to find broad application in the total synthesis of natural products. [73]

A historic turning point in the development of asymmetric DA reactions was the discovery that Lewis acids accelerate the cycloaddition, which enabled running the reaction at lower temperatures than ever before, with concomitant enhancements in stereoselectivity. Initially, as in many other transformations, the chiral auxiliary approach showed that excellent stereoselectivities could be obtained provided that the dieneophile was correctly activated and aligned. This set the stage for the development of efficient catalytic asymmetric DA reactions in which the chiral information was furnished by the Lewis acid, rather than by the substrates. Thus, from the 1990s to present, this has been a very active field that has progressed markedly [74, 75].

An overview of applications of metal-catalyzed asymmetric DA chemistry in natural product synthesis over the past 5 years reveals a handful of popular methodologies. These include the use of Jacobsen's Schiff base chromium complex (**122**) or Rawal–Jacobsen's salen complexes (**123** and **124**), which have become excellent reagents for enantioselective DA or hetero Diels–Alder (HDA) reactions [76–78] (Fig. 15.6).

The HDA process between Danishefsky-type dienes and aldehydes has become a highly efficient route to chiral six-membered pyrans, a ubiquitous structural motif in nature. Gademann et al. employed Jacobsen's catalyst (***R*-122**) to prepare the dihydropyranone fragment of Anguinomycin C, a potent

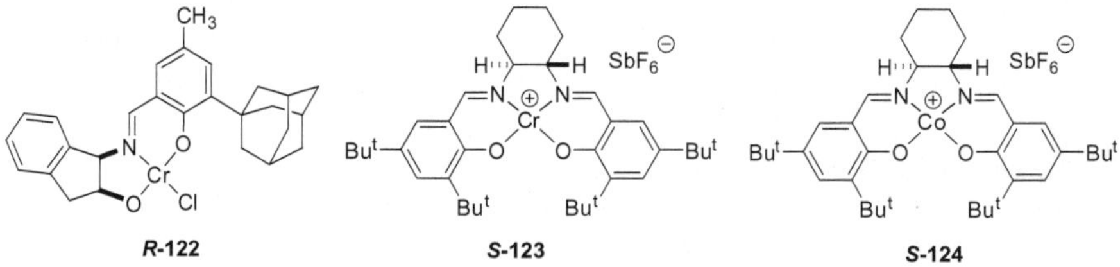

FIGURE 15.6. Jacobsen's Schiff base-chromium catalyst (**122**) and Rawal–Jacobsen's salen complexes (**123** and **124**) for asymmetric DA reactions.

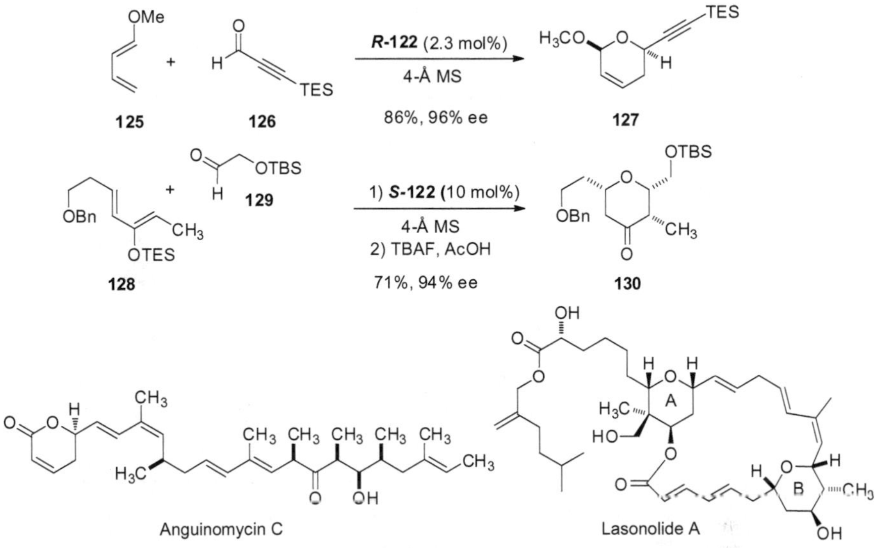

SCHEME 15.27. Synthesis of the pyranyl fragments in Anguinomycin C and in Lasonolide A by asymmetric HDA reaction.

and selective antitumor agent [79]. Reaction of methoxybutadiene (**125**) and the protected propargylic aldehyde **126** catalyzed by ***R*-122** at room temperature provided 2-methoxy dihydropyrane **127** in 86% yield and with 96% ee (Scheme 15.27). As often occurs, this reaction was run neat (i.e., without solvent) and with molecular sieves, which are necessary for high conversions and selectivities. Acid-catalyzed isomerization of the acetal functionality, and oxidation in a later stage, provided the dihydropyranone moiety of Anguinomycin C. Ghosh et al., in their synthesis of the marine antitumor agent Lasonolide A, also used Jacobsen's catalyst (***S*-122**) to construct the tetrahydropyran ring B [80, 81]. Reaction of diene **128** with the protected hydroxyaldehyde **129** afforded, after acidic workup, the corresponding tetrahydropyrone **130** (Scheme 15.27). Further reduction of the ketone group and subsequent functionalization of the side arms gave the desired target compound. The same group exploited this chemistry in a similar HDA reaction as part of a concise approach to the central core of Brevisamide, a biosynthetic precursor of Brevitoxins [82].

Paterson and Miller harnessed asymmetric HDA transformation to build the central core of (+)-Neopeltolide [83]. The tetrahydropyran ring was assembled (with up to three chiral centers) using a chiral aldehyde as dienophile (Scheme 15.28). Thus, the reaction of the chiral aldehyde **131** with the diene **132** catalyzed by Jacobsen's catalyst (***R*-122**), followed by acidic workup, afforded the tetrahydropyrone **133** with complete control over the C3 and C7 chiral centers. Reduction of the ketone with $NaBH_4$ at a later stage provided the chiral center at C5.

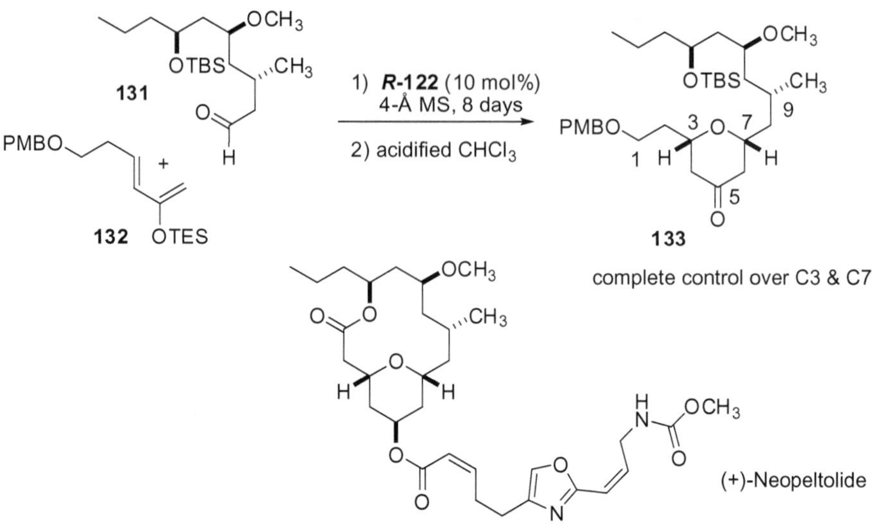

SCHEME 15.28. Synthesis of the pyrane ring of Neopeltolide by asymmetric HDA.

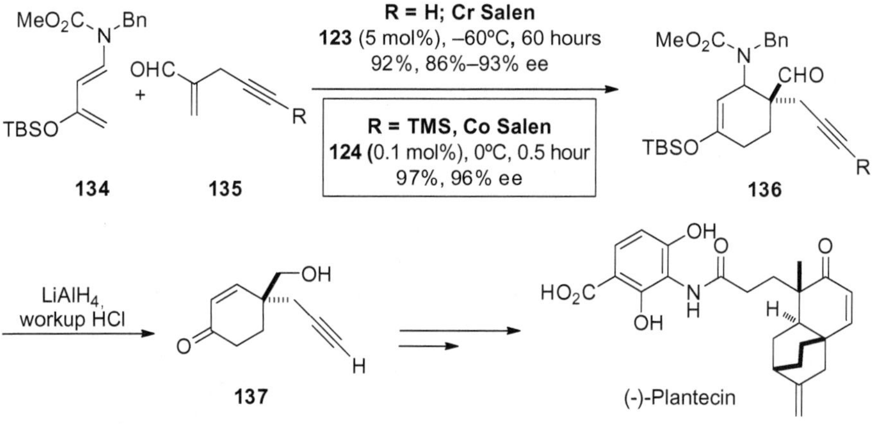

SCHEME 15.29. Asymmetric DA reactions in the synthesis of Plantecin.

Chromium salen complex **123** was first reported by Jacobsen et al. as an efficient catalyst for the enantioselective epoxidation of olefins or the hydrolytic kinetic resolution of racemic epoxides [84]. Rawal et al. showed its utility in the DA reactions of 1-amino substituted dienes and α-substituted acroleins [85], and later developed the even more efficient Co(III) salen **124** [86, 87]. Nicolaou et al. employed these catalysts to synthesize (–)-Plantecin, an antibiotic isolated from a soil sample collected in Spain [88, 89]. Their strategy relied on an asymmetric DA reaction between the diene **134** and compound **135** to prepare the enone **137** early on (Scheme 15.29). Optimization of the reaction conditions showed that at 0°C, the cobalt catalyst **124** gave better

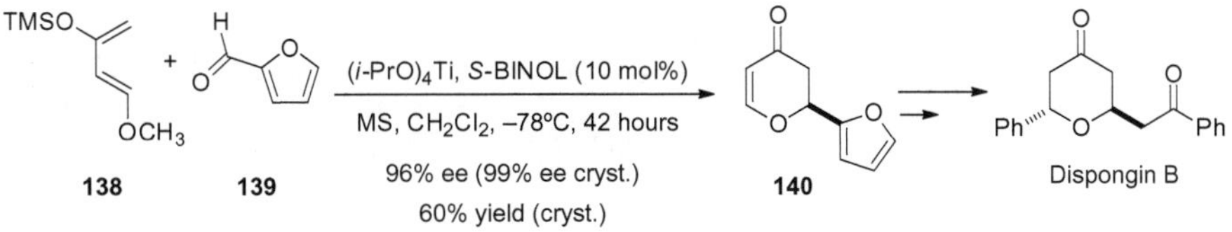

SCHEME 15.30. Asymmetric HDA in the synthesis of Dispongines.

rates and selectivity than did its chromium analog **123**. Use of the trimethylsilyl (TMS)-protected dienophile facilitated purification of **135**, thereby accelerating the subsequent chemistry.

Other catalytic systems that were developed during the 1990s and that remain widely used in total synthesis include the Ti/BINOL Lewis acids, employed in various experimental procedures developed in different laboratories. For example, Keck et al. developed the $Ti(O^iPr)_4$/BINOL reagent pair for use in a formal HDA reaction between Danishefsky's diene (**138**) and aldehydes [90]. A two-step Mukaiyama aldol–Michael sequence provided the corresponding dihydropyrones in excellent ee. Kumaraswamy et al. applied this methodology to the synthesis of Dispongines A and B (Scheme 15.30) [91].

Mikami et al. devised another variation of the Ti/BINOL system, which is based on using $Cl_2Ti(O^iPr)_4$ as metal source [92]. Combination of $Cl_2Ti(O^iPr)_4$ and BINOL (with or without molecular sieves), known as *Mikami's catalyst*, has proven invaluable in DA reactions of quinones. Ward and Shen utilized this catalyst to prepare diterpenoid Cyathin A_3, a nerve growth factor [93]. Screening of various catalysts showed that Mikami's catalyst afforded the best results for the unusual diene **142** (Scheme 15.31). Addition of silica and magnesium to the acidic reaction medium increased the stability of **142** and were crucial for consistently obtaining high yields. The Mikami catalyst was also used by Yu and Danishefsky, in their synthesis of the antibiotic Fluostatin C [94]. Reaction of vinylindene diene **144** and quinoneketal **145** provided an unexpected regiochemistry, demonstrating that the directing group within the diene was the benzylic methylene moiety (see asterisk) rather than the aromatic group (Scheme 15.31). The chiral centers generated upon the formation of **146** ultimately disappeared in Fluostatin C, but were exploited to direct the formation of the α-hydroxy epoxide in the final product.

Copper complexes can also catalyze asymmetric DA reactions. Evans used his own chiral copper-bisoxazolidine catalyst **147** to prepare the central tetrahydropyran (ring E) of (+)-Azaspiracid, a toxin isolated from mussels [95]. The hydrated complex **147** is a shelf-stable compound that must be dehydrated with molecular sieves prior to use (Scheme 15.32). Inverse-electron-demand HDA between a *cis*-propenyl ether and the diketoenone **148** provided a mixture of *cis* and *trans* isomers of the dihydropyran **149**. The desired product was isolated in 84% yield as a single isomer on a 10–20g scale. In the same synthesis, the copper complex **150**, derived from the same bis-oxazolidine

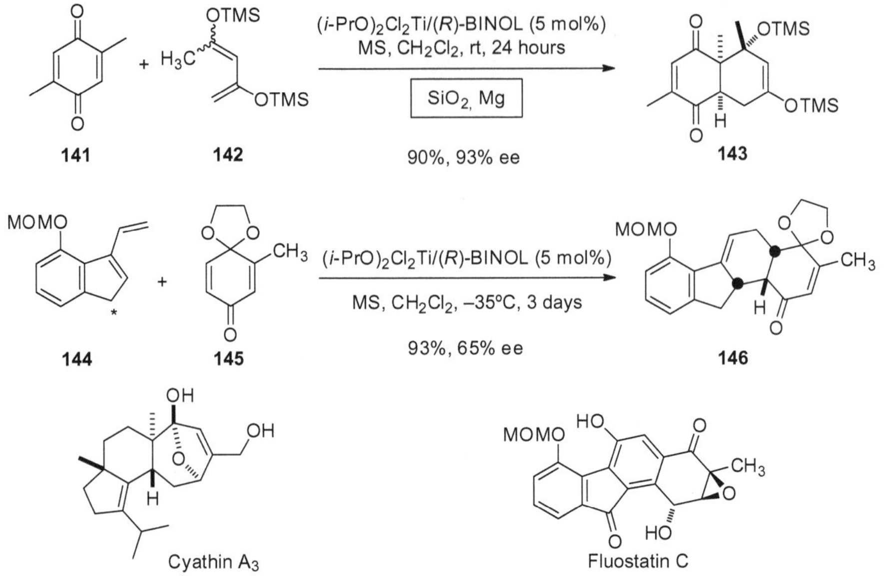

SCHEME 15.31. Asymmetric DA reactions of quinones, used in syntheses of Cyathin and Fluostatin C.

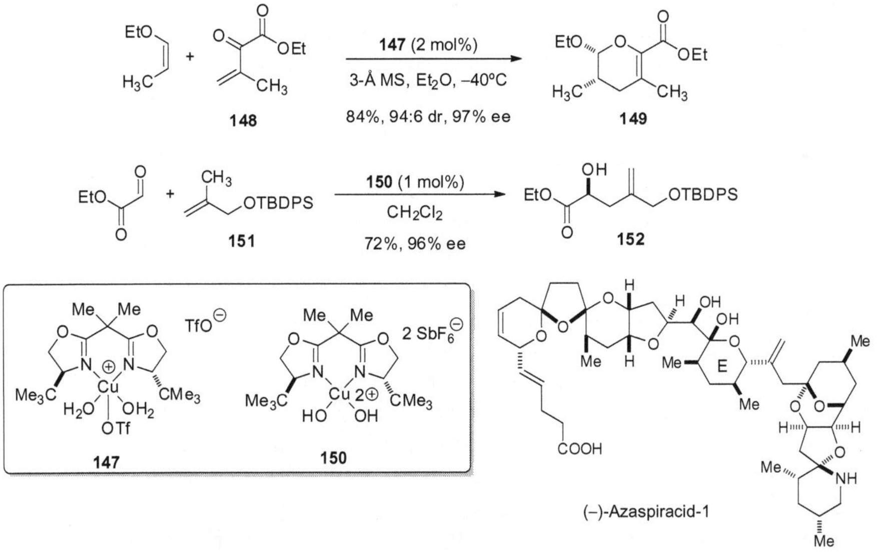

SCHEME 15.32. Asymmetric DA and DA-ene reactions used to synthesize Azaspiracid.

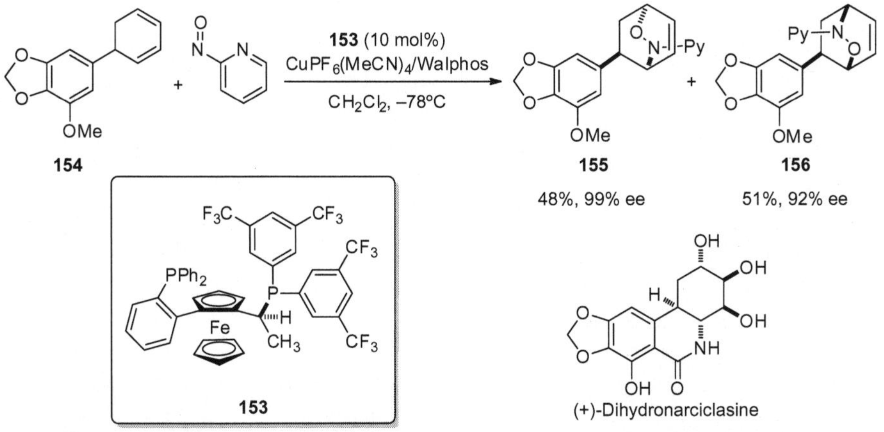

SCHEME 15.33. Nitroso-DA reaction used to prepare dihydronarciclasine.

ligand, was used to catalyze an asymmetric glyoxylate-ene reaction to afford hydroxy ester **152**.

Other recently reported copper complexes have also been employed in total syntheses. Typically, the research group that develops the catalyst also develops the synthetic methodology, as proof of principle for the methodology. For example, in 2007, Jana and Studer developed an asymmetric nitroso-DA reaction [96]. Their catalyst is based on the ferrocene diphosphine ligand Walphos (**153**) and copper. The reaction is marked by the fact that it provides almost equal quantities of two regioisomers, with high enantioselectivity for both. Thus, reaction of 2-nitroso pyridine with the diene **154** provides the regioisomers **155** and **156** with 99% and 92% ee, respectively (Scheme 15.33). Isomer **155** was utilized for the synthesis of the antitumor agent Dihydronarciclasine [97].

Corey et al. have been very active in the field of catalytic asymmetric DA chemistry, for which they developed very selective oxazaborolidine catalysts [98]. One of their latest advances is the acid-activated oxazaborolidinium cation **157**, which they found to be a general catalyst for poorly reactive dienes and dienophiles [99]. Using **157** as catalyst, they reported a remarkable macrobicyclization as the key step in the total synthesis of the marine natural product Palominol [100]. Diene **158** was cyclized in the presence of **157** (20 mol%) at low temperature (Scheme 15.34). The *endo* stereochemistry for intermediate **159** was confirmed by X-ray crystallography. Most interestingly, only **157** promoted the intramolecular cycloaddition; exposure of diene **158** to standard achiral Lewis acids ($MeAlCl_2$, Me_2AlCl) or to heat led to deprotection or polymerization of the DA precursor.

Enantioselective cyclopropanation employing α-diazo esters has been widely studied using Cu, Rh, Ru, and Co as catalyst metals [101]. The

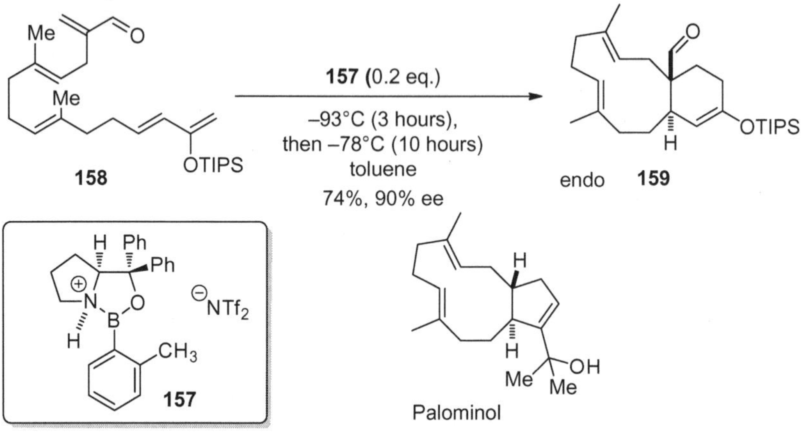

SCHEME 15.34. Asymmetric DA reaction catalyzed by an oxazaborolidine salt.

copper-based catalysts are probably the most effective for a wide scope of substrates. The seminal work of Pfaltz et al. in this area, on semicorrin-type ligands [102], was quickly followed by numerous reports of ligands with improved scope, including the copper(I) bis(oxazoline) ligands reported by Evans et al. in the early 1990s, which are now routinely used [103].

Reiser et al. used the isopropyl-bis(oxazoline) copper complex **160** to catalyze a cyclopropanation between ethyl diazoacetate and **161** to afford **162**, an early intermediate in the synthesis of the antitumor guaianolide Arglabin (Scheme 15.35). As the initial step in the synthesis, the reaction was carried out on a 50-g scale. Crystallization with pentane afforded optically pure **162** in 38% yield [104]. Honma and Nakada used a similar catalyst to prepare (+)-Digitoxigenin (Scheme 15.35) [105]. They used an intramolecular cyclopropanation of α-diazo-β-keto sulfone **163** to tricyclic sulfone **164** [106] developed in-house to construct rings A and B of the steroid system.

The enantioselective Simmons–Smith cyclopropanation of allyl alcohols, developed by Charette et al. [107], is among the most reliable methods for preparing cyclopropane compounds. Trost et al. used this reaction to prepare the starting material for the key diastereoselective [5 + 2] cycloaddition in a synthesis of (+)-Frondosin A [108]. The allyl alcohol **165** was submitted to cyclopropanation using Charette's ligand (**166**) to give the enantiomerically enriched cyclopropane **167**. Compound **167** was transformed into enantiomerically pure **168**, which, in the presence of catalytic amounts of $CpRu(CH_3CN)_3PF_6$, gave **169** as a single diastereomer in very good yield (Scheme 15.36). This Ru-catalyzed [5 + 2] cycloaddition of alkynes with vinylcyclopropanes, independently developed by Wender et al. [109] and Trost et al. [110], has been exploited in other syntheses of natural products [111]. Compound **169**, which already contains the carbocyclic core of Frondosin A, was ultimately converted into the desired product.

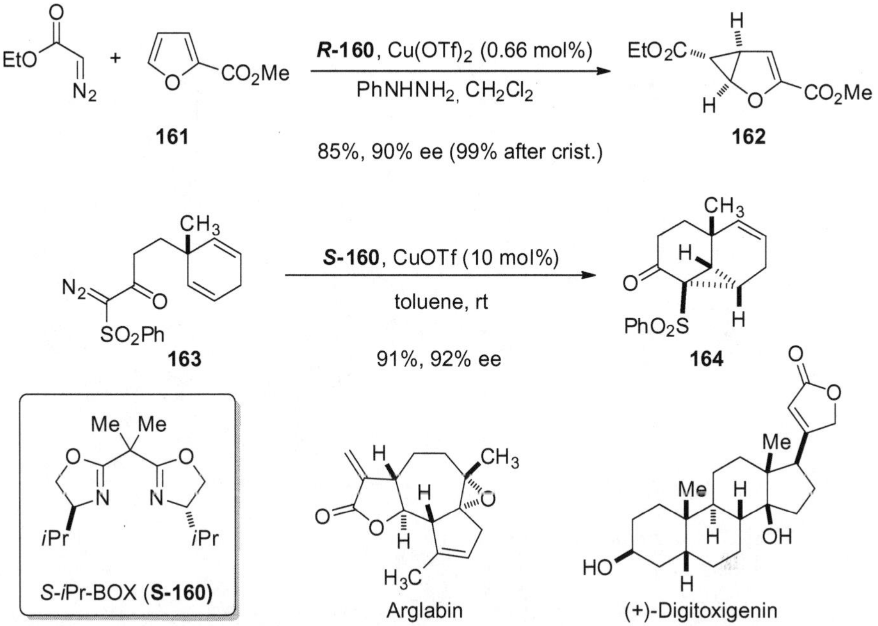

SCHEME 15.35. Copper-catalyzed enantioselective cyclopropanations used in syntheses of Arglabin and Digitoxigenin.

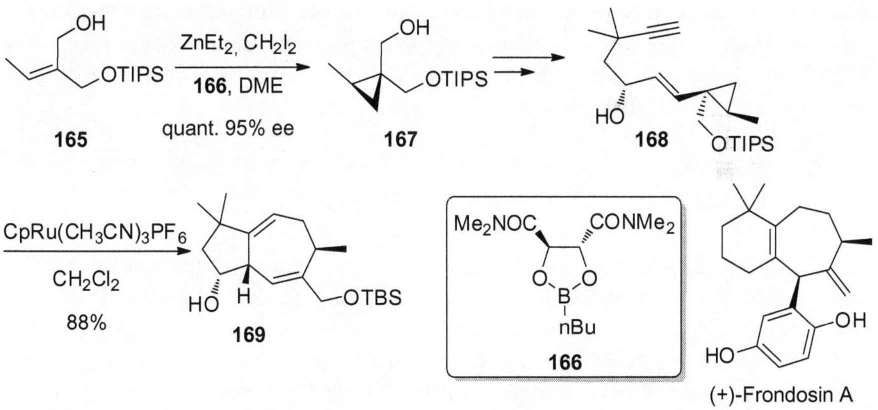

SCHEME 15.36. Enantioselective Simmons–Smith reaction used in the preparation of a vinyl cyclopropane compound needed for a Ru-catalyzed [5 + 2] cycloaddition.

Nucleoside analogs are of interest for their antiviral and antitumor activities. Most of the reported syntheses of carbonucleosides rely on enzymatic and kinetic resolution or on the use of sugars as starting materials. Riera et al. synthesized Carbovir and Abacavir using a highly enantioselective intermolecular Pauson–Khand reaction (Scheme 15.37) [112]. They used the chiral

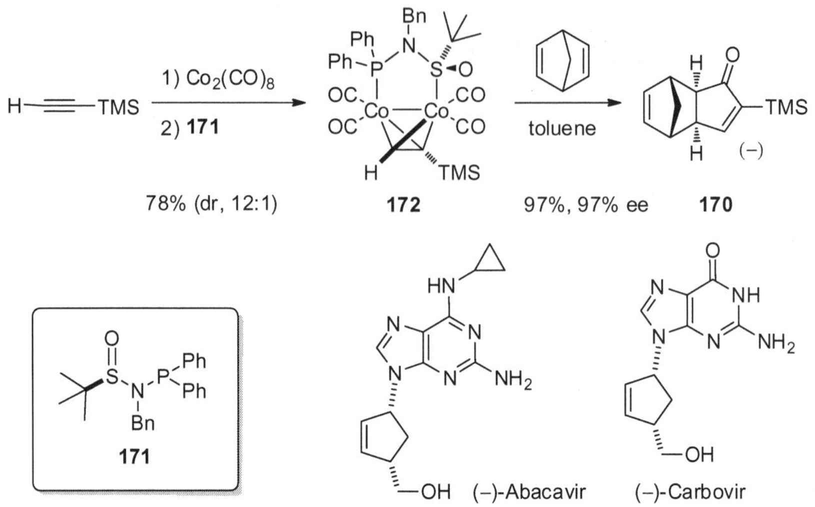

SCHEME 15.37. Asymmetric intermolecular Pauson–Khand reaction used in the syntheses of Abacavir and Carbovir.

P,S ligand **171** to generate the diastereomerically pure bridged dicobalt complex **172**, which, in turn, provided the key intermediate, the tricyclic compound **170**, in high yield and with 97% ee [113]. Further crystallization of **170** afforded optically pure product, which, upon subsequent stereoselective transformations, gave the desired carbonucleosides. Notably, the norbornene moiety in **170** serves both as chiral inductor and as protecting group of the double bond present in the final product.

15.6. CYCLIZATIONS AND REARRANGEMENTS

Metal-catalyzed cycloisomerizations, especially those of enynes, are among the most important transformations for synthesizing functionalized carbocycles and heterocycles [114]. They are also a good example of atom economy [115] in synthesis: all of the atoms from the starting materials end up in the final product. Despite major advances in this chemistry, for total synthesis, it has been mainly restricted to diastereoselective processes. One of the few examples of an enantioselective enyne cycloisomerization used in a total synthesis is found in the synthesis of Platensimycin by Nicolaou et al. [116]. To develop an efficient and straightforward synthetic route, they devised a new method for enantioselective cyclization of 1,6-enynes that have a terminal alkyne. They first used the method described by Zhang et al. [117], for the rhodium-catalyzed enantioselective cyclization of 1,6-enynes containing an internal alkyne, but this led to a more elaborate synthesis [118], as achieving a highly enantioselective cyclization required the additional step of removing the carboxylic group

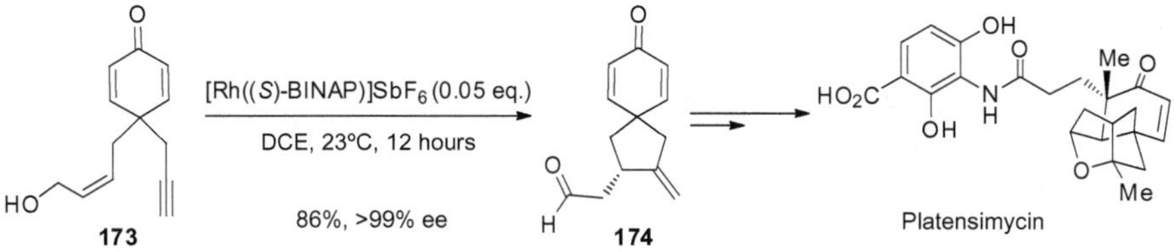

SCHEME 15.38. Enantioselective enyne cycloisomerization developed in the synthesis of Platensimycin.

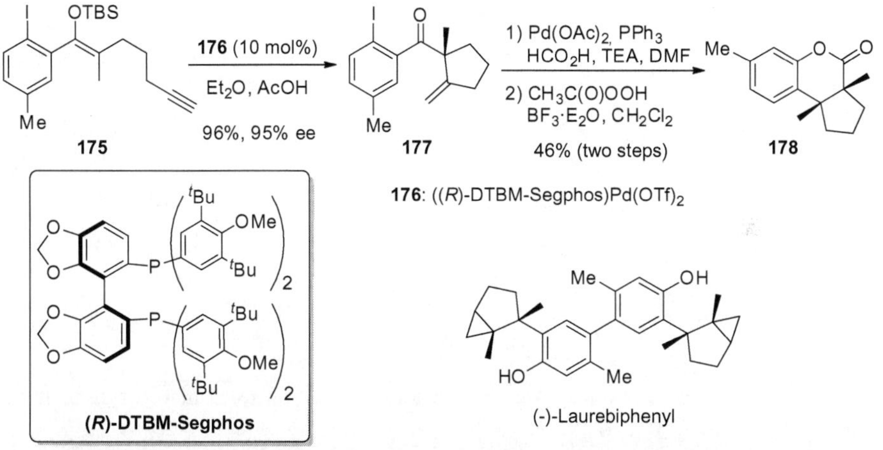

SCHEME 15.39. Palladium-catalyzed enantioselective cyclization of silyloxy 1,6-enynes as a key reaction in the synthesis of Laurebiphenyl.

from the alkyne. The new and improved conditions comprised the use of a preformed rhodium catalyst [Rh((*S*)-BINAP)]SbF_6 instead of the combination of $[Rh(COD)Cl]_2$, $AgSbF_6$, and (*S*)-BINAP employed by Zhang with internal alkynes. Under these new conditions, cyclization of 1,6-enyne **173**, which contains a terminal alkyne, furnished the spiro-dienone aldehyde **174** in 86% yield and with >99% ee (Scheme 15.38). Compound **174** was later transformed into the complex molecule Platensimycin.

Corkey and Toste used the enantioselective cyclization of 1,6-enynes in an asymmetric synthesis of the dimeric compound (−)-Laurebiphenyl [119]. Palladium-catalyzed cyclization of silyloxy-substituted olefins with alkynes provided rapid access to highly functionalized, enantioenriched methylene cyclopentane adducts. Thus, treatment of enol ether **175** with 10% ((*R*)-DTBM-Segphos)Pd$(OTf)_2$ (**176**) gave clean conversion to the cyclopentyl ketone **177** in 96% yield and with 95% ee. The ketone **177** was then subjected to Heck cyclization and Baeyer–Villiger oxidation to afford lactone **178** (Scheme 15.39). Subsequent transformations completed the synthesis of (−)-Laurebiphenyl.

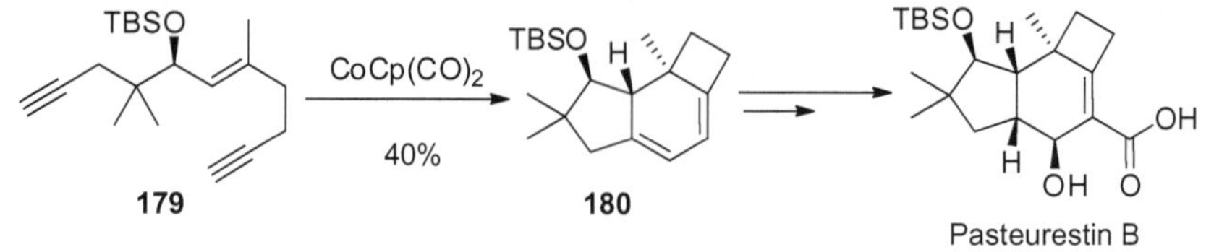

SCHEME 15.40. Cobalt-catalyzed diastereoselective [2 + 2 + 2] used in the synthesis of Pasteurestin B.

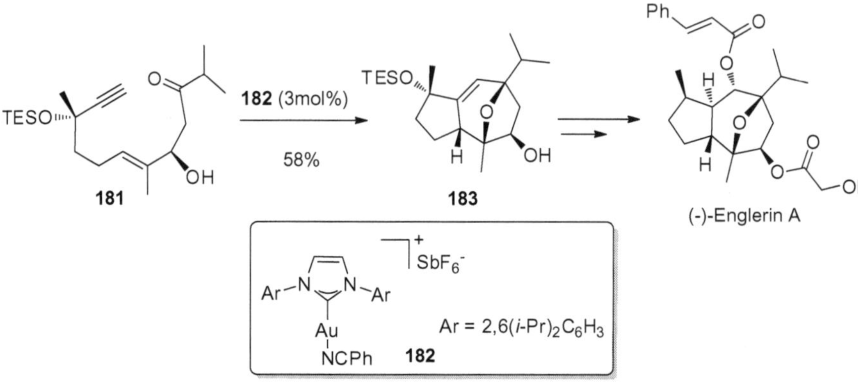

SCHEME 15.41. Gold-catalyzed enyne cycloisomerization used in the synthesis of Englerin A.

As previously mentioned, most of the metal-catalyzed cyclizations that have been used in total synthesis are diastereoselective processes. The utility of these reactions is demonstrated by the speed with which they are applied in total synthesis after their development. For example, Mulzer et al. employed a cobalt-catalyzed [2 + 2 + 2] cyclization of enediynes (developed by Johnson and Vollhardt [120]) to the synthesis of Pasteurestin B [121]. Treatment of compound **179** with catalytic amounts of a cobalt catalyst gave clean formation of **180** as a single diastereoisomer. Interestingly, **179** contains virtually the full carbon skeleton of Pasteurestin B (Scheme 15.40).

Gold has recently emerged as a very powerful metal for catalytic cycloisomerizations [122]; indeed, several new gold-mediated transformations have been reported. Its catalytic utility is intimately associated with the ability to selectively activate, under mild conditions, π-systems to the addition of a wide range of nucleophiles. These new transformations are generally highly stereoselective and have been used in several total syntheses.

Echavarren et al. [123] and Ma et al. [124] simultaneously reported total syntheses of (–)-Englerin A. Both routes are based on stereoselective gold-catalyzed enyne cyclization. In Echavarren et al.'s approach, cyclization of enyne **181** catalyzed by the cationic gold complex **182** afforded the oxatricyclic derivative **183** as a single diastereomer in 58% yield (Scheme 15.41). In con-

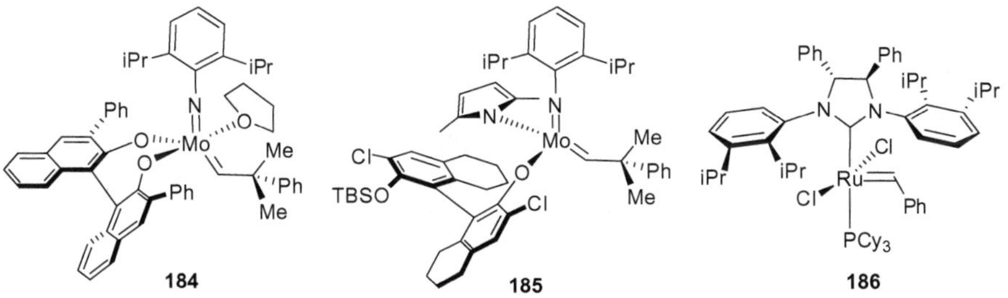

FIGURE 15.7. Chiral catalysts for asymmetric olefin metathesis.

trast, Ma et al.'s approach uses an enyne that is similar to **181**, but lacks the hydroxyl group on the propargylic position and uses AuCl as catalyst.

Olefin metathesis has been extensively studied and broadly used in the total synthesis of complex molecules [125]. However, asymmetric olefin metathesis [126] is still in development; thus, only a few examples of its application in total synthesis have been published. A pioneering example is Schrock and Hoveyda's synthesis of (+)-Africanol using the chiral molybdenum catalyst **184**, which they developed in-house (Fig. 15.7) [127].

Major advances in this field have been achieved through the search for more effective and straightforward syntheses of enantiomerically pure complex molecules. A nice example of this development is Hoveyda et al.'s synthesis of (+)-Quebrachamine [128], in which the need for an efficient catalyst for the enantioselective ring-closing metathesis (RCM) of substrate **187** led to the introduction of a new class of enantiomerically pure molybdenum catalysts represented by **185** (Fig. 15.7). Complex **185** was generated in situ by reaction of the corresponding achiral molybdenum bis pyrrolide and an equivalent of the chiral aryl alcohol. Using this catalyst, the authors obtained the desired tetracyclic diene **188**, containing a quaternary stereogenic center, in good yield and with excellent enantioselectivity by RCM of **187**. Finally, hydrogenation of **188** gave (+)-Quebrachamine.

Despite the fact that most of the research on asymmetric olefin metathesis has been focused on high oxidation state molybdenum complexes, chiral ruthenium alkylidene complexes have also been shown to catalyze these reactions efficiently and to provide good ee's [129]. For example, in the preparation of 5-epi-Citreoviral [130], the use of less than 1 mol% of ruthenium catalyst **186** (based on a chiral *N*-heterocyclic carbene) in the RCM of the silyl ether **189** gave compound **190** with 92% ee; this compound (**190)** was then oxidized to diol **191**. It is worth noting that all remaining stereocenters in the final product were established from the single chiral center generated in this process (Scheme 15.42).

In summary, asymmetric transformations based on organometallic catalysis are being widely used in complex total synthesis. The variety of asymmetric reactions in the modern organic chemist's toolbox is enormous, encompassing

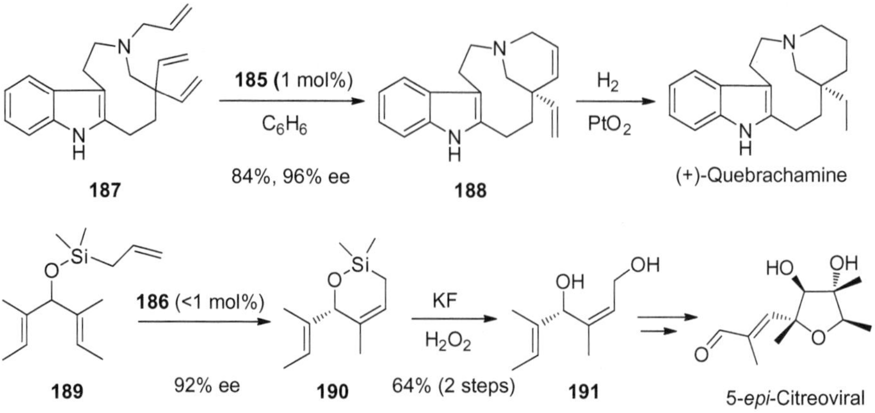

SCHEME 15.42. Asymmetric olefin metathesis used in syntheses of Quebrachamine and 5-epi-Citreoviral.

dozens of metals and hundreds of chiral ligands that enable fine tuning of the desired transformation. In an attempt to give a general overview of the state of the art, we have shown here only a selection of the transformations used in total synthesis in the last 5 years.

REFERENCES

[1] Nozaki, H., Moriuti, S., Takaya, H., Noyori, R. (1966). Asymmetric induction in carbenoid reactions by means of a dissymmetric copper chelate. *Tetrahedron Lett.*, *43*, 5239–5244.

[2] (a) Knowles, W. S., Sabacky, M. J., Vineyard, B. D. (1972). Catalytic asymmetric hydrogenation. *J. Chem. Soc. Chem. Commun.*, 10–11; (b) Knowles, W. S. (1983). Asymmetric hydrogenation. *Acc. Chem. Res.*, *16*, 106–112; (c) Vineyard, B. D., Knowles, W. S., Sabacky, M. J., Bachman, G. L., Weinkauff, D. J. (1977). Asymmetric hydrogenation. Rhodium chiral bisphosphine catalyst. *J. Am. Chem. Soc.*, *99*, 5946–5952.

[3] Kagan, H. B., Dang-Tuan-Phat (1972). Asymmetric catalytic reduction with transition metal complexes. I. catalytic system of rhodium(I) with (−)-2,3-0-isopropylidene-2,3-dihydroxy-1,4-bis(diphenylphosphino)butane, a new chiral diphosphine. *J. Am. Chem. Soc.*, *94*, 6429–6433.

[4] Börner, A. (Ed.) (2008). *Phosphorus ligands in Asymmetric Catalysis*, Wiley-VCH, Weinheim, Germany.

[5] Recent reviews: (a) Gridnev, I. D., Imamoto, T. (2009). Mechanism of enantioselection in Rh-catalyzed asymmetric hydrogenation. The origin of utmost catalytic performance. *Chem. Commun.*, *48*, 7447–7464; (b) Shimizu, H., Nagasaki, I., Matsumura, K., Sayo, N., Saito, T. (2007). Developments in asymmetric hydrogenation from an industrial perspective. *Acc. Chem. Res.*, *40*, 1385–1393.

[6] (a) Burk, M. J. (2000). Modular phospholane ligands in asymmetric catalysis. *Acc. Chem. Res.*, *33*, 363–372; (b) Burk, M. J., Feaster, J. E., Nugent, W. A., Harlow, R. L. (1993). Preparation and use of C2-symmetric bis(phospholanes): production of α -amino acid derivatives via highly enantioselective hydrogenation reactions. *J. Am. Chem. Soc.*, *115*, 10125–10138; (c) Burk, M. J. (1991). C2-symmetric bis(phospholanes) and their use in highly enantioselective hydrogenation reactions. *J. Am. Chem. Soc.*, *113*, 8518–8519.

[7] (a) Bower, J. F., Szeto, P., Gallagher, T. (2007). Cyclic sulfamidates as versatile lactam precursors. an evaluation of synthetic strategies towards (−)-Aphanorphine. *Org. Biomol. Chem.*, *5*, 143–150; (b) Bower, J. F., Szeto, P., Gallagher, T. (2005). Cyclic sulfamidates as lactam precursors. An efficient asymmetric synthesis of (−)-Aphanorphine. *Chem. Commun.*, 5793–5794.

[8] Toumi, M., Couty, F., Evano, G. (2008). Eight-step total synthesis of the cyclopeptide alkaloid Mucronine E. *Synlett.*, 29–32.

[9] Fürstner, A., Ackerstaff, J. (2008). Formal total synthesis of (−)-Haouamine A. *Chem. Commun.*, 2870–2872.

[10] (a) Teichert, J. F., Feringa, B. L. (2010). Phosphoramidites: privileged ligands in asymmetric catalysis. *Angew. Chem., Int. Ed.*, *49*, 2486–2528; (b) Minnaard, A. J., Feringa, B. L., Lefort, L., de Vries, J. G. (2007). Asymmetric hydrogenation using monodentate phosphoramidite ligands. *Acc. Chem. Res.*, *40*, 1267–1277.

[11] Boogers, J. A. F., Felfer, U., Kotthaus, M., Lefort, L., Steinbauer, G., De Vries, A. H. M., De Vries, J. G. A. (2007). A mixed-ligand approach enables the asymmetric hydrogenation of an α -isopropylcinnamic acid en route to the renin inhibitor aliskiren. *Org. Process Res. Dev.*, *11*, 585–591.

[12] (a) Noyori, R., Hashiguchi, S. (1997). Asymmetric transfer hydrogenation catalyzed by chiral ruthenium complexes. *Acc. Chem. Res.*, *30*, 97–102; (b) Mashima, K., Kusano, K., Ohta, T., Noyori, R., Takaya, H. (1989). Synthesis of new cationic BINAP-ruthenium(II) complexes and their use in asymmetric hydrogenation [BINAP = 2,2′-bis(diphenylphosphino)-1,1′-binaphthyl]. *J. Chem. Soc. Chem. Commun.*, 1208–1210.

[13] (a) Matsumura, K., Hashiguchi, S., Ikariya, T., Noyori, R. (1997). Asymmetric transfer hydrogenation of α,β-acetylenic ketones. *J. Am. Chem. Soc.*, *119*, 8738–8739; (b) Fujii, A., Hashiguchi, S., Uematsu, N., Ikariya, T., Noyori, R. (1996). Ruthenium(II)-catalyzed asymmetric transfer hydrogenation of ketones using a formic acid-triethylamine mixture. *J. Am. Chem. Soc.*, *118*, 2521–2522; (c) Uematsu, N., Fujii, A., Hashiguchi, S., Ikariya, T., Noyori, R. (1996). Asymmetric transfer hydrogenation of imines. *J. Am. Chem. Soc.*, *118*, 4916–4917.

[14] Fürstner, A., Bonnekessel, M., Blank, J. T., Radkowski, K., Seidel, G., Lacombe, F., Gabor, B., Mynott, R. (2007). Total synthesis of Myxovirescin A1. *Chem.- Eur. J.*, *13*, 8762–8783.

[15] (a) Trost, B. M., Waser, J., Meyer, A. (2008). Total synthesis of (−)-Pseudolaric acid B. *J. Am. Chem. Soc.*, *130*, 16424–16434; (b) Trost, B. M., Waser, J., Meyer, A. (2007). Total synthesis of (−)-Pseudolaric acid B. *J. Am. Chem. Soc.*, *129*, 14556–14557.

[16] Hansen, K. B., Balsells, J., Dreher, S., Hsiao, Y., Kubryk, M., Palucki, M., Rivera, N. R., Steinhuebel, D., Armstrong, J. D., III, Askin, D., Grabowski, E. J. J. (2005).

First generation process for the preparation of the DPP-IV inhibitor Sitagliptin. *Org. Process Res. Dev.*, *9*, 634–639.

[17] Hansen, K. B., Hsiao, Y., Xu, F., Rivera, N., Clausen, A., Kubryk, M., Krska, S., Rosner, T., Simmons, B., Balsells, J., Ikemoto, N., Sun, Y., Spindler, F., Malan, C., Grabowski, E. J., Armstrong, J. D., III (2009). Highly efficient asymmetric synthesis *of Sitagliptin. J. Am. Chem. Soc.*, *131*, 8798–8804.

[18] Fürstner, A., Nevado, C., Waser, M., Tremblay, M., Chevrier, C., Teply, F., Aïssa, C., Moulin, E., Müller, O. (2007). Total synthesis of Iejimalide A-D and assessment of the remarkable actin-depolymerizing capacity of these polyene macrolides. *J. Am. Chem. Soc.*, *129*, 9150–9161.

[19] Greshock, T. J., Johns, D. M., Noguchi, Y., Williams, R. M. (2008). Improved total synthesis of the potent HDAC inhibitor FK228 (FR-901228). *Org. Lett.*, *10*, 613–616.

[20] Nicolaou, K. C., Chen, J. S., Zhang, H., Montero, A. (2008). Asymmetric synthesis and biological properties of Uncialamycin and 26-epi-uncialamycin. *Angew. Chem. Int. Ed.*, *47*, 185–189.

[21] Cheng, J., Yang, Y. (2009). Enantioselective total synthesis of (−)-(S)-Stepholidine. *J. Org. Chem.*, *74*, 9225–9228.

[22] (a) Gao, Y., Klunder, J. M., Hanson, R. M., Masamune, H., Ko, S. Y., Sharpless, K. B. (1987). Catalytic asymmetric epoxidation and kinetic resolution: modified procedures including in situ derivatization. *J. Am. Chem. Soc.*, *109*, 5765–5780; (b) Katsuki, T., Martin, V. S. (1996). Asymmetric epoxidation of allylic alcohols: the Katsuki-Sharpless epoxidation reaction. *Org. React.*, *48*, 1–299.

[23] (a) Jacobsen, E. N., Marko, I., Mungall, W. S., Schroeder, G., Sharpless, K. B. (1988). Asymmetric dihydroxylation via ligand-accelerated catalysis. *J. Am. Chem. Soc.*, *110*, 1968–1970; (b) Xu, D., Crispino, G. A., Sharpless, K. B. (1992). Selective asymmetric dihydroxylation (AD) of dienes. *J. Am. Chem. Soc.*, *114*, 7570–7571; (c) Wang, L., Sharpless, K. B. (1992). Catalytic asymmetric dihydroxylation of *cis*-disubstituted olefins. *J. Am. Chem. Soc.*, *114*, 7568–7570.

[24] Riera, A., Moreno, M. (2010). Synthetic applications of chiral unsaturated epoxy alcohols prepared by Sharpless asymmetric epoxidation. *Molecules*, *15*, 1041–1073.

[25] Larrosa, I., Da Silva, M. I., Gomez, P. M., Hannen, P., Ko, E., Lenger, S. R., Linke, S. R., White, A. J., Wilton, D., Barrett, A. G. (2006). Highly convergent three component benzyne coupling: the total synthesis of ent-Clavilactone B. *J. Am. Chem. Soc.*, *128*, 14042–14043.

[26] (a) Noe, M. C., Letavic, M. A., Snow, S. L. (2005). Asymmetric dihydroxylation of alkenes. *Org. React.*, *66*, 109–625; (b) Kolb, H. C., VanNieuwenhze, M. S., Sharpless, K. B. (1994). Catalytic asymmetric dihydroxylation. *Chem. Rev.*, *94*, 2483–2547.

[27] Ghosh, A. K., Gong, G. (2006). Total synthesis and revision of C6 stereochemistry of (+)-Amphidinolide W. *J. Org. Chem.*, *71*, 1085–1093.

[28] (a) Trend, R. M., Stoltz, B. M. (2004). An experimentally derived model for stereoselectivity in the aerobic oxidative kinetic resolution of secondary alcohols by (sparteine)$PdCl_2$. *J. Am. Chem. Soc.*, *126*, 4482–4483; (b) Ferreira, E. M., Stoltz, B. M. (2001). The palladium-catalyzed oxidative kinetic resolution of secondary alcohols with molecular oxygen. *J. Am. Chem. Soc.*, *123*, 7725–7726.

[29] Jensen, D. R., Pugsley, J. S., Sigman, M. S. (2001). Palladium-catalyzed enantioselective oxidations of alcohols using molecular oxygen. *J. Am. Chem. Soc.*, *123*, 7475–7476.

[30] Krishnan, S., Bagdanoff, J. T., Ebner, D. C., Ramtohul, Y. K., Tambar, U. K., Stoltz, B. M. (2008). Pd-catalyzed enantioselective aerobic oxidation of secondary alcohols: applications to the total synthesis of alkaloids. *J. Am. Chem. Soc.*, *130*, 13745–13754.

[31] (a) Dong, S., Zhu, J., Porco, J. A., Jr. (2008). Enantioselective synthesis of bicyclo[2.2.2]octenones using a copper-mediated oxidative dearomatization/[4 + 2] dimerization cascade. *J. Am. Chem. Soc.*, *130*, 2738–2739; (b) Zhu, J., Grigoriadis, N. P., Lee, J. P., Porco, J. A.,Jr (2005). Synthesis of the Azaphilones using copper-mediated enantioselective oxidative dearomatization. *J. Am. Chem. Soc.*, *127*, 9342–9343.

[32] Qian, W., Wei, W., Zhang, Y., Yao, Z. (2007). Total synthesis, assignment of absolute stereochemistry, and structural revision of Chlorofusin. *J. Am. Chem. Soc.*, *129*, 6400–6401.

[33] (a) Soai, K., Ookawa, A., Kaba, T., Ogawa, K. (1987). Catalytic asymmetric induction. Highly enantioselective addition of dialkylzincs to aldehydes using chiral pyrrolidinylmethanols and their metal salts. *J. Am. Chem. Soc.*, *109*, 7111–7115; (b) Kitamura, M., Suga, S., Kawai, K., Noyori, R. (1986). Catalytic asymmetric induction. highly enantioselective addition of dialkylzincs to aldehydes. *J. Am. Chem. Soc.*, *108*, 6071–6072.

[34] (a) Takahashi, H., Kawakita, T., Yoshioka, M., Kobayashi, S., Ohno, M. (1989). Enantioselective alkylation of aldehyde catalyzed by a disulfonamide-titanium tetraisopropoxide-dialkylzinc system. *Tetrahedron Lett.*, *30*, 7095–7098; (b) Yoshioka, M., Kawakita, T., Ohno, M. (1989). Asymmetric induction catalyzed by conjugate bases of chiral proton acids as ligands. enantioselective addition of dialkylzinc-orthotitanate complex to benzaldehyde with catalytic ability of a remarkable high order. *Tetrahedron Lett.*, *30*, 1657–1660; (c) Takahashi, H., Kawakita, T., Ohno, M., Yoshioka, M., Kobayashi, S. (1992). A catalytic enantioselective reaction using a C2-symmetric disulfonamide as a chiral ligand: alkylation of aldehydes catalyzed by disulfonamide-Ti(O-iso-pr)$_4$-dialkyl zinc system. *Tetrahedron*, *48*, 5691–5700.

[35] Trenkle, J. D., Jamison, T. F. (2009). Macrocyclization by nickel-catalyzed, ester-promoted, epoxide-alkyne reductive coupling: total synthesis of (−)-Gloeosporone. *Angew. Chem. Int. Ed.*, *48*, 5366–5368.

[36] Thompson, A., Corley, E. G., Huntington, M. F., Grabowski, E. J. J., Remenar, J. F., Collum, D. B. (1998). Lithium ephedrate-mediated addition of a lithium acetylide to a ketone: solution structures and relative reactivities of mixed aggregates underlying the high enantioselectivities. *J. Am. Chem. Soc.*, *120*, 2028–2038.

[37] Denmark, S. E., Fu, J. (2003). Catalytic enantioselective addition of allylic organometallic reagents to aldehydes and ketones. *Chem. Rev.*, *103*, 2763–2793.

[38] Keck, G. E., Tarbet, K. H., Geraci, L. S. (1993). Catalytic asymmetric allylation of aldehydes. *J. Am. Chem. Soc.*, *115*, 8467–8468.

[39] Tsai, A. S., Bergman, R. G., Ellman, J. A. (2008). Asymmetric synthesis of (−)-Incarvillateine employing an intramolecular alkylation via Rh-catalyzed olefinic C-H bond activation. *J. Am. Chem. Soc.*, *130*, 6316–6317.

[40] Colby, D. A., Bergman, R. G., Ellman, J. A. (2006). Stereoselective alkylation of α,β-unsaturated imines via C-H bond activation. *J. Am. Chem. Soc.*, *128*, 5604–5605.

[41] Ko, H. M., Lee, D. G., Kim, M. A., Kim, H. J., Park, J., Lah, M. S., Lee, E. (2007). Total synthesis of (–)-Blepharocalyxin D. *Org. Lett.*, *9*, 141–144.

[42] Hanawa, H., Uraguchi, D., Konishi, S., Hashimoto, T., Maruoka, K. (2003). Catalytic asymmetric allylation of aldehydes and related reactions with bis[((S)-binaphthoxy) (isopropoxy)titanium] oxide as a micro -oxo-type chiral Lewis acid. *Chem.-Eur. J.*, *9*, 4405–4413.

[43] Yanagisawa, A., Nakashima, H., Ishiba, A., Yamamoto, H. (1996). Catalytic asymmetric allylation of aldehydes using a chiral silver(I) complex. *J. Am. Chem. Soc.*, *118*, 4723–4724.

[44] Rein, C., Demel, P., Outten, R. A., Netscher, T., Breit, B. (2007). Reagent directing group controlled organic synthesis: total synthesis of (*R,R,R*)-(α)-Tocopherol. *Angew. Chem. Int. Ed.*, *46*, 8670–8673.

[45] (a) Harutyunyan, S. R., den Hartog, T., Geurts, K., Minnaard, A. J., Feringa, B. L. (2008). Catalytic asymmetric conjugate addition and allylic alkylation with Grignard reagents. *Chem. Rev.*, *108*, 2824–2852; (b) Alexakis, A., Benhaim, C. (2002). Enantioselective copper-catalyzed conjugate addition. *Eur. J. Org. Chem.*, 3221–3236; (c) Krause, N., Hoffmann-Roder, A. (2001). Recent advances in catalytic enantioselective Michael additions. *Synthesis*, 171–196.

[46] Some recent examples: (a) Chin, Y., Wang, S., Loh, T. (2009). Synthesis of Iriomoteolide-1a C13-C23 fragment via asymmetric conjugate addition and Julia-Kocienski coupling reaction. *Org. Lett.*, *11*, 3674–3676; (b) Casas-Arce, E., Ter Horst, B., Feringa, B. L., Minnaard, A. J. (2008). Asymmetric total synthesis of PDIM A: a virulence factor of mycobacterium tuberculosis. *Chem. Eur. J.*, *14*, 4157–4159; (c) Vintonyak, V. V., Maier, M. E. (2008). Formal total synthesis of Neopeltolide. *Org. Lett.*, *10*, 1239–1242; (d) Ter Horst, B., Feringa, B. L., Minnaard, A. J. (2007). Catalytic asymmetric synthesis of Phthioceranic acid, a heptamethyl-branched acid from mycobacterium tuberculosis. *Org. Lett.*, *9*, 3013–3015.

[47] Des Mazery, R., Pullez, M., Lopez, F., Harutyunyan, S. R., Minnaard, A. J., Feringa, B. L. (2005). An iterative catalytic route to enantiopure deoxypropionate subunits: asymmetric conjugate addition of Grignard reagents to α,β-unsaturated thioesters. *J. Am. Chem. Soc.*, *127*, 9966–9967.

[48] Harutyunyan, S. R., Zhao, Z., den Hartog, T., Bouwmeester, K., Minnaard, A. J., Feringa, B. L., Govers, F. (2008). Biologically active Phytophthora mating hormone prepared by catalytic asymmetric total synthesis. *Proc. Natl. Acad. Sci. U.S.A.*, *105*, 8507–8512.

[49] Brown, M. K., Degrado, S. J., Hoveyda, A. H. (2005). Highly enantioselective Cu-catalyzed conjugate additions of dialkylzinc reagents to unsaturated furanones and pyranones: preparation of air-stable and catalytically active Cu-peptide complexes. *Angew. Chem. Int. Ed.*, *44*, 5306–5310.

[50] Brown, M. K., Hoveyda, A. H. (2008). Enantioselective total synthesis of Clavirolide C. applications of Cu-catalyzed asymmetric conjugate additions and Ru-catalyzed ring-closing metathesis. *J. Am. Chem. Soc.*, *130*, 12904–12906.

[51] Brown, M. K., May, T. L., Baxter, C. A., Hoveyda, A. H. (2007). All-carbon quaternary stereogenic centers by enantioselective Cu-catalyzed conjugate additions

promoted by a chiral N-heterocyclic carbene. *Angew. Chem. Int. Ed.*, *46*, 1097–1100.

[52] (a) Pfaltz, A., Lautens, M. (1999). Allylic substitution reactions. *Compr. Asymmetric Catal. I-III*, *2*, 833–884; (b) Trost, B. M., Lee, C. (2000). Asymmetric allylic alkylation reactions. In: Ojima I. (Ed.), *Catalytic Asymmetric Synthesis*, 2nd ed. Wiley-VCH, New York, pp. 593–649; (c) Trost, B. M. (2002). Pd asymmetric allylic alkylation (AAA). A powerful synthetic tool. *Chem. Pharm. Bull.*, *50*, 1–14; (d) Trost, B. M., Crawley, M. L. (2003). Asymmetric transition-metal-catalyzed allylic alkylations: applications in total synthesis. *Chem. Rev.*, *103*, 2921–2943; (e) Trost, B. M., Machacek, M. R., Aponick, A. (2006). Predicting the stereochemistry of diphenylphosphino benzoic acid (DPPBA)-based palladium-catalyzed asymmetric allylic alkylation reactions: a working model. *Acc. Chem. Res.*, *39*, 747–760.

[53] Some recent examples of use of AAA in total synthesis: (a) Cook, C., Guinchard, X., Liron, F., Roulland, E. (2010). Total synthesis of (–)-Exiguolide. *Org. Lett.*, *12*, 744–747; (b) Farwick, A., Helmchen, G. (2010). Enantioselective total synthesis of (–)-(α)-Kainic acid. *Org. Lett.*, *12*, 1108–1111; (c) Nemoto, T., Yamamoto, E., Franzen, R., Fukuyama, T., Wu, R., Fukamachi, T., Kobayashi, H., Hamada, Y. (2010). Catalytic asymmetric total synthesis of Tangutorine. *Org. Lett.*, *12*, 872–875; (d) González-Gálvez, D., García-García, E., Alibés, R., Bayón, P., de March, P., Figueredo, M., Font, J. (2009). Enantioselective approach to Securinega alkaloids. total synthesis of Securinine and (–)-Norsecurinine. *J. Org. Chem.*, *74*, 6199–6211; (e) Gnamm, C., Krauter, C. M., Broedner, K., Helmchen, G. (2009). Stereoselective synthesis of 2,6-disubstituted piperidines using the iridium-catalyzed allylic cyclization as configurational switch: asymmetric total synthesis of (+)-241 D and related piperidine alkaloids. *Chem. Eur. J.*, *15*, 2050–2054; (f) Gnamm, C., Franck, G., Miller, N., Stork, T., Broedner, K., Helmchen, G. (2008). Enantioselective iridium-catalyzed allylic aminations of allylic carbonates with functionalized side chains. asymmetric total synthesis of (S)-Vigabatrin. *Synthesis*, 3331–3350; (g) Bian, J., Van Wingerden, M., Ready, J. M. (2006). Enantioselective total synthesis of (+)- and (–)-Nigellamine A2. *J. Am. Chem. Soc.*, *128*, 7428–7429.

[54] (a) Trost, B. M., Osipov, M., Dong, G. (2010). A concise enantioselective synthesis of (–)-Ranirestat. *Org. Lett.*, *12*, 1276–1279; (b) Trost, B. M., Aponick, A., Stanzl, B. N. (2007). A convergent Pd-catalyzed asymmetric allylic alkylation of dl- and meso-divinylethylene carbonate: enantioselective synthesis of (+)-Australine hydrochloride and formal synthesis of Isoaltholactone. *Chem. Eur. J.*, *13*, 9547–9560; (c) Trost, B. M., Dong, G. (2007). Asymmetric annulation toward pyrrolopiperazinones. concise enantioselective syntheses of pyrrole alkaloid natural products. *Org. Lett.*, *9*, 2357–2359; (d) Trost, B. M., Stiles, D. T. (2007). Total synthesis of Spirotryprostatin B via diastereoselective prenylation. *Org. Lett.*, *9*, 2763–2766; (e) Trost, B. M., Horne, D. B., Woltering, M. J. (2006). Palladium-catalyzed DYKAT of butadiene mono-epoxide: enantioselective total synthesis of (+)-DMDP, (–)-Bulgecinine, and (+)-Broussonetine G. *Chem- Eur J.*, *12*, 6607–6620; (f) Trost, B. M., Machacek, M. R., Faulk, B. D. (2006). Sequential Ru-Pd catalysis: a two-catalyst one-pot protocol for the synthesis of N- and O-heterocycles. *J. Am. Chem. Soc.*, *128*, 6745–6754.

[55] Trost, B. M., Dong, G., Vance, J. A. (2007). A diosphenol-based strategy for the total synthesis of (–)-Terpestacin. *J. Am. Chem. Soc.*, *129*, 4540–4541.

[56] Trost, B. M., Dong, G. (2009). A stereodivergent strategy to both product enantiomers from the same enantiomer of a stereoinducing catalyst: Agelastatin A. *Chem.-Eur. J.*, *15*, 6910–6919.

[57] Du, C., Li, L., Li, Y., Xie, Z. (2009). Construction of two vicinal quaternary carbons by asymmetric allylic alkylation: total synthesis of Hyperolactone C and (−)-Biyouyanagin A. *Angew. Chem. Int. Ed.*, *48*, 7853–7856.

[58] Tsuji, J., Minami, I. (1987). New synthetic reactions of allyl alkyl carbonates, allyl Î²-keto carboxylates, and allyl vinylic carbonates catalyzed by palladium complexes. *Acc. Chem. Res.*, *20*, 140–145.

[59] Behenna, D. C., Stoltz, B. M. (2004). The enantioselective Tsuji allylation. *J. Am. Chem. Soc.*, *126*, 15044–15045.

[60] McFadden, R. M., Stoltz, B. M. (2006). The catalytic enantioselective, protecting group-free total synthesis of (+)-Dichroanone. *J. Am. Chem. Soc.*, *128*, 7738–7739.

[61] Enquist, J. A., Jr, Stoltz, B. M. (2008). The total synthesis of (−)-Cyanthiwigin F by means of double catalytic enantioselective alkylation. *Nature*, *453*, 1228–1231.

[62] Carroll, A., O'Sullivan, T. P., Guiry, P. J. (2005). The development of enantioselective rhodium-catalyzed hydroboration of olefins. *Adv. Synth. Catal.*, *347*, 609–631.

[63] Werle, S., Fey, T., Neudoerfl, J. M., Schmalz, H. (2007). Enantioselective synthesis of a *trans*-7,8-Dimethoxycalamenene. *Org. Lett.*, *9*, 3555–3558.

[64] (a) Negishi, E. (2000). Asymmetric carbometallations. In Ojima I. (Ed.), *Catalytic Asymmetric Synthesis* II. Wiley-VCH, New York, pp. 165–189; (b) Negishi, E., Huo, S. (2002). Synthesis and reactivity of zirconocene derivatives. In Marek I. (Ed.), *Titanium and Zirconium in Organic Synthesis*. Wiley-VCH, Weinheim, pp. 1–49; (c) Kondakov, D. Y., Negishi, E. (1996). Zirconium-catalyzed enantioselective alkylalumination of monosubstituted alkenes proceeding via noncyclic mechanism. *J. Am. Chem. Soc.*, *118*, 1577–1578; (d) Kondakov, D. Y., Negishi, E. (1995). Zirconium-catalyzed enantioselective methylalumination of monosubstituted alkenes. *J. Am. Chem. Soc.*, *117*, 10771–10772.

[65] Zhu, G., Negishi, E. (2007). Fully reagent-controlled asymmetric synthesis of (−)-Spongidepsin via the Zr-catalyzed asymmetric carboalumination of alkenes (ZACA reaction). *Org. Lett.*, *9*, 2771–2774.

[66] Huang, Z., Tan, Z., Novak, T., Zhu, G., Negishi, E. (2007). Zirconium-catalyzed asymmetric carboalumination of alkenes: ZACA-lipase-catalyzed acetylation synergy. *Adv. Synth. Catal.*, *349*, 539–545.

[67] Liang, B., Negishi, E. (2008). Highly efficient asymmetric synthesis of Fluvirucinine A1 via Zr-catalyzed asymmetric carboalumination of alkenes (ZACA)-lipase-catalyzed acetylation tandem process. *Org. Lett.*, *10*, 193–195.

[68] Oestreich, M. (Ed.) (2009). *The Mirozoki-Heck Reaction*. Wiley-VCH, Weinheim.

[69] (a) Shibasaki, M., Vogl, E. M., Ohshima, T. (2004). Asymmetric Heck reaction. *Adv. Synth. Catal.*, *346*, 1533–1552; (b) Dounay, A. B., Overman, L. E. (2003). The asymmetric intramolecular Heck reaction in natural product total synthesis. *Chem. Rev.*, *103*, 2945–2963; (c) Shibasaki, M., Boden, C. D. J., Kojima, A. (1997). The asymmetric Heck reaction. *Tetrahedron*, *53*, 7371–7395.

[70] (a) Dounay, A. B., Humphreys, P. G., Overman, L. E., Wrobleski, A. D. (2008). Total synthesis of the strychnos alkaloid (+)-Minfiensine: tandem enantioselective intramolecular Heck-iminium ion cyclization. *J. Am. Chem. Soc.*, *130*, 5368–5377; (b) Dounay, A. B., Overman, L. E., Wrobleski, A. D. (2005). Sequential catalytic asymmetric Heck-iminium ion cyclization: enantioselective total synthesis of the strychnos alkaloid Minfiensine. *J. Am. Chem. Soc.*, *127*, 10186–10187.

[71] Tietze, L. F., Stecker, F., Zinngrebe, J., Sommer, K. M. (2006). Enantioselective palladium-catalyzed total synthesis of vitamin E by employing a domino Wacker-Heck reaction. *Chem. Eur. J.*, *12*, 8770–8776.

[72] Diels, O., Alder, K. (1928). Syntheses in the hydroaromatic series. I. Addition of "diene" hydrocarbons. *Liebigs Ann.*, *460*, 98–122.

[73] Nicolaou, K. C., Snyder, S. A., Montagnon, T., Vassilikogiannakis, G. (2002). The Diels-Alder reaction in total synthesis. *Angew. Chem. Int. Ed.*, *41*, 1668–1698.

[74] Jorgensen, K. A. (2000). Catalytic asymmetric hetero-Diels-Alder reactions of carbonyl compounds and imines. *Angew. Chem. Int. Ed.*, *39*, 3558–3588.

[75] Pellissier, H. (2009). Asymmetric hetero-Diels-Alder reactions of carbonyl compounds. *Tetrahedron*, *65*, 2839–2877.

[76] Dossetter, A. G., Jamison, T. F., Jacobsen, E. N. (1999). Highly enantio- and diastereoselective hetero-Diels-Alder reactions catalyzed by new chiral tridentate chromium(III) catalysts. *Angew. Chem. Int. Ed.*, *38*, 2398–2400.

[77] Gademann, K., Chavez, D. E., Jacobsen, E. N. (2002). Highly enantioselective inverse-electron-demand hetero-Diels-Alder reactions of α,β-unsaturated aldehydes. *Angew. Chem. Int. Ed.*, *41*, 3059–3061.

[78] Jarvo, E. R., Lawrence, B. M., Jacobsen, E. N. (2005). Highly enantio- and regioselective quinone Diels-Alder reactions catalyzed by a tridentate [(Schiff base) CrIII] complex. *Angew. Chem. Int. Ed.*, *44*, 6043–6046.

[79] Bonazzi, S., Guttinger, S., Zemp, I., Kutay, U., Gademann, K. (2007). Total synthesis, configuration, and biological evaluation of Anguinomycin C. *Angew. Chem. Int. Ed.*, *46*, 8707–8710.

[80] Ghosh, A. K., Gong, G. (2008). Total synthesis of potent antitumor agent (−)-Lasonolide A: a cycloaddition-based strategy. *Chem.-As. J.*, *3*, 1811–1823.

[81] Ghosh, A. K., Gong, G. (2007). Enantioselective total synthesis of macrolide antitumor agent (−)-Lasonolide A. *Org. Lett.*, *9*, 1437–1440.

[82] Ghosh, A. K., Li, J. (2009). An asymmetric total synthesis of Brevisamide. *Org. Lett.*, *11*, 4164–4167.

[83] Paterson, I., Miller, N. A. (2008). Total synthesis of the marine macrolide (+)-Neopeltolide. *Chem. Commun.*, 4708–4710.

[84] (a) Jacobsen, E. N. (2000). Asymmetric catalysis of epoxide ring-opening reactions. *Acc. Chem. Res.*, *33*, 421–431; (b) Tokunaga, M., Larrow, J. F., Kakiuchi, F., Jacobsen, E. N. (1997). Asymmetric catalysis with water: efficient kinetic resolution of terminal epoxides by means of catalytic hydrolysis. *Science*, *277*, 936–938; (c) Schaus, S. E., Brnalt, J., Jacobsen, E. N. (1998). Asymmetric hetero-Diels-Alder reactions catalyzed by chiral (salen)chromium(III) complexes. *J. Org. Chem.*, *63*, 403–405.

[85] Huang, Y., Iwama, T., Rawal, V. H. (2002). Broadly effective enantioselective Diels-Alder reactions of 1-amino-substituted-1,3-butadienes. *Org. Lett.*, *4*, 1163–1166.

[86] Huang, Y., Iwama, T., Rawal, V. H. (2002). Design and development of highly effective Lewis acid catalysts for enantioselective Diels-Alder reactions. *J. Am. Chem. Soc.*, *124*, 5950–5951.

[87] Kozmin, S. A., Iwama, T., Huang, Y., Rawal, V. H. (2002). An efficient approach to aspidosperma alkaloids via [4 + 2] cycloadditions of aminosiloxydienes: stereocontrolled total synthesis of (±)-Tabersonine. gram-scale catalytic asymmetric syntheses of (+)-Tabersonine and (+)-16-methoxytabersonine. asymmetric syntheses of (+)-Aspidospermidine and (−)-Quebrachamine. *J. Am. Chem. Soc.*, *124*, 4628–4641.

[88] Nicolaou, K. C., Tria, G. S., Edmonds, D. J. (2008). Total synthesis of Platencin. *Angew. Chem. Int. Ed.*, *47*, 1780–1783.

[89] Nicolaou, K. C., Tria, G. S., Edmonds, D. J., Kar, M. (2009). Total syntheses of (±)-Platencin and (−)-Platencin. *J. Am. Chem. Soc.*, *131*, 15909–15917.

[90] Keck, G. E., Li, X., Krishnamurthy, D. (1995). Catalytic enantioselective synthesis of dihydropyrones via formal hetero Diels-Alder reactions of "Danishefsky's diene" with aldehydes. *J. Org. Chem.*, *60*, 5998–5999.

[91] Kumaraswamy, G., Ramakrishna, G., Naresh, P., Jagadeesh, B., Sridhar, B. (2009). A flexible enantioselective total synthesis of Diospongins A and B and their enantiomers using catalytic hetero-Diels-Alder/Rh-catalyzed 1,4-addition and asymmetric transfer hydrogenation reactions as key steps. *J. Org. Chem.*, *74*, 8468–8471.

[92] Mikami, K., Motoyama, Y., Terada, M. (1994). Asymmetric catalysis of Diels-Alder cycloadditions by an MS-free binaphthol-titanium complex: dramatic effect of MS, linear versus positive nonlinear relationship, and synthetic applications. *J. Am. Chem. Soc.*, *116*, 2812–2820.

[93] Ward, D. E., Shen, J. (2007). Enantioselective total synthesis of Cyathin A3. *Org. Lett.*, *9*, 2843–2846.

[94] Yu, M., Danishefsky, S. J. (2008). A direct route to Fluostatin C by a fascinating Diels-Alder reaction. *J. Am. Chem. Soc.*, *130*, 2783–2785.

[95] Evans, D. A., Kvaerno, L., Dunn, T. B., Beauchemin, A., Raymer, B., Mulder, J. A., Olhava, E. J., Juhl, M., Kagechika, K., Favor, D. A. (2008). Total synthesis of (+)-Azaspiracid-1. an exhibition of the intricacies of complex molecule synthesis. *J. Am. Chem. Soc.*, *130*, 16295–16309.

[96] Jana, C. K., Studer, A. (2007). Divergent reactions for racemates: catalytic, enantioselective, and regio-divergent nitroso Diels-Alder reactions. *Angew. Chem. Int. Ed.*, *46*, 6542–6544.

[97] Jana, C. K., Studer, A. (2008). Total synthesis of (+)-*trans*-Dihydronarciclasine by a catalytic enantioselective regiodivergent nitroso Diels-Alder reaction. *Chem.-Eur. J.*, *14*, 6326–6328.

[98] Corey, E. J. (2002). Catalytic enantioselective Diels-Alder reactions: methods, mechanistic fundamentals, pathways, and applications. *Angew. Chem. Int. Ed.*, *41*, 1650–1667.

[99] Ryu, D. H., Corey, E. J. (2003). Triflimide activation of a chiral oxazaborolidine leads to a more general catalytic system for enantioselective Diels-Alder addition. *J. Am. Chem. Soc.*, *125*, 6388–6390.

[100] Snyder, S. A., Corey, E. J. (2006). Concise total syntheses of Palominol, Dolabellatrienone, β-Araneosene, and Isoedunol via an enantioselective Diels-Alder macrobicyclization. *J. Am. Chem. Soc.*, *128*, 740–742.

[101] Lebel, H., Marcoux, J., Molinaro, C., Charette, A. B. (2003). Stereoselective cyclopropanation reactions. *Chem. Rev.*, *103*, 977–1050.

[102] Fritschi, H., Leutenegger, U., Pfaltz, A. (1986). Chiral copper semicorrin complexes as enantioselective catalysts for the cyclopropanation of olefins with diazo compounds. *Angew. Chem. Int. Ed.*, *25*, 1005.

[103] Evans, D. A., Woerpel, K. A., Hinman, M. M., Faul, M. M. (1991). Bis(oxazolines) as chiral ligands in metal-catalyzed asymmetric reactions. catalytic, asymmetric cyclopropanation of olefins. *J. Am. Chem. Soc.*, *113*, 726–728.

[104] Kalidindi, S., Jeong, W. B., Schall, A., Bandichhor, R., Nosse, B., Reiser, O. (2007). Enantioselective synthesis of Arglabin. *Angew. Chem. Int. Ed.*, *46*, 6361–6363.

[105] Honma, M., Nakada, M. (2007). Enantioselective total synthesis of (+)-Digitoxigenin. *Tetrahedron Lett.*, *48*, 1541–1544.

[106] Honma, M., Sawada, T., Fujisawa, Y., Utsugi, M., Watanabe, H., Umino, A., Matsumura, T., Hagihara, T., Takano, M., Nakada, M. (2003). Asymmetric catalysis on the intramolecular cyclopropanation of α-diazo-β-keto sulfones. *J. Am. Chem. Soc.*, *125*, 2860–2861.

[107] (a) Charette, A. B., Juteau, H., Lebel, H., Molinaro, C. (1998). Enantioselective cyclopropanation of allylic alcohols with dioxaborolane ligands: scope and synthetic applications. *J. Am. Chem. Soc.*, *120*, 11943–11952; (b) Charette, A. B., Juteau, H. (1994). Design of amphoteric bifunctional ligands: application to the enantioselective Simmons-Smith cyclopropanation of allylic alcohols. *J. Am. Chem. Soc.*, *116*, 2651–2652.

[108] Trost, B. M., Hu, Y., Horne, D. B. (2007). Total synthesis of (+)-Frondosin A. Application of the Ru-catalyzed [5 + 2] cycloaddition. *J. Am. Chem. Soc.*, *129*, 11781–11790.

[109] (a) Wender, P. A., Williams, T. J. (2002). $[(Arene)Rh(COD)]^{+}$ complexes as catalysts for [5 + 2] cycloaddition reactions. *Angew. Chem. Int. Ed.*, *41*, 4550–4553; (b) Wender, P. A., Takahashi, H., Witulski, B. (1995). Transition metal catalyzed [5 + 2] cycloadditions of vinylcyclopropanes and alkynes: a homolog of the Diels-Alder reaction for the synthesis of seven-membered rings. *J. Am. Chem. Soc.*, *117*, 4720–4721.

[110] (a) Trost, B. M., Shen, H. C., Horne, D. B., Toste, F. D., Steinmetz, B. G., Koradin, C. (2005). Syntheses of seven-membered rings: ruthenium-catalyzed intramolecular [5 + 2] cycloadditions. *Chem.-Eur. J.*, *11*, 2577–2590; (b) Trost, B. M., Toste, F. D., Shen, H. (2000). Ruthenium-catalyzed intramolecular [5 + 2] cycloadditions. *J. Am. Chem. Soc.*, *122*, 2379–2380.

[111] (a) Trost, B. M., Waser, J., Meyer, A. (2008). Total synthesis of (−)-Pseudolaric acid B. *J. Am. Chem. Soc.*, *130*, 16424–16434; (b) Trost, B. M., Waser, J., Meyer, A. (2007). Total synthesis of (−)-Pseudolaric acid B. *J. Am. Chem. Soc.*, *129*, 14556–14557.

[112] Vazquez-Romero, A., Rodriguez, J., Lledo, A., Verdaguer, X., Riera, A. (2008). Enantioselective syntheses of carbanucleosides from the Pauson-Khand adduct of trimethylsilylacetylene and norbornadiene. *Org. Lett.*, *10*, 4509–4512.

[113] Solà, J., Revés, M., Riera, A., Verdaguer, X. (2007). *N*-phosphino sulfinamide ligands: an efficient manner to combine sulfur chirality and phosphorus coordination behavior. *Angew. Chem. Int. Ed.*, *46*, 5020–5023.

[114] (a) Michelet, V., Toullec, P. Y., Genet, J. (2008). Cycloisomerization of 1,n-enynes: challenging metal-catalyzed rearrangements and mechanistic insights. *Angew. Chem., Inter. Ed.*, *47*, 4268–4315; (b) Zhang, L., Sun, J., Kozmin, S. A. (2006). Gold and platinum catalysis of enyne cycloisomerization. *Adv. Synth. Catal.*, *348*, 2271–2296; (c) Echavarren, A. M., Nevado, C. (2004). Non-stabilized transition metal carbenes as intermediates in intramolecular reactions of alkynes with alkenes. *Chem. Soc. Rev.*, *33*, 431–436; (d) Lloyd-Jones, G. C. (2003). Mechanistic aspects of transition metal catalysed 1,6-diene and 1,6-enyne cycloisomerisation reactions. *Org. Biomol. Chem.*, *1*, 215–236; (e) Trost, B. M., Toste, F. D., Pinkerton, A. B. (2001). Non-metathesis ruthenium-catalyzed C-C bond formation. *Chem. Rev.*, *101*, 2067–2096.

[115] Trost, B. M. (1991). The atom economy: a search for synthetic efficiency. *Science*, *254*, 1471–1477.

[116] (a) Nicolaou, K. C., Li, A., Edmonds, D. J., Tria, G. S., Ellery, S. P. (2009). Total synthesis of Platensimycin and related natural products. *J. Am. Chem. Soc.*, *131*, 16905–16918; (b) Nicolaou, K. C., Li, A., Ellery, S. P., Edmonds, D. J. (2009). Rhodium-catalyzed asymmetric enyne cycloisomerization of terminal alkynes and formal total synthesis of (–)-Platensimycin. *Angew. Chem. Int. Ed.*, *48*, 6293–6295.

[117] (a) Cao, P., Wang, B., Zhang, X. (2000). Rh-catalyzed enyne cycloisomerization. *J. Am. Chem. Soc.*, *122*, 6490–6491; (b) Cao, P., Zhang, X. (2000). The first highly enantioselective Rh-catalyzed enyne cycloisomerization. *Angew. Chem. Int. Ed.*, *39*, 4104–4106.

[118] Nicolaou, K. C., Edmonds, D. J., Li, A., Tria, G. S. (2007). Asymmetric total syntheses of Platensimycin. *Angew. Chem. Int. Ed.*, *46*, 3942–3945.

[119] Corkey, B. K., Toste, F. D. (2007). Palladium-catalyzed enantioselective cyclization of silyloxy-1,6-enynes. *J. Am. Chem. Soc.*, *129*, 2764–2765.

[120] Johnson, E. P., Vollhardt, K. P. C. (1991). Stereoselective, one-step assembly of the strained protoilludane framework by cobalt-mediated cyclization of an acyclic enediyne precursor. A total synthesis of Illudol. *J. Am. Chem. Soc.*, *113*, 381–382.

[121] (a) Koegl, M., Brecker, L., Warrass, R., Mulzer, J. (2008). Novel protoilludane lead structure for veterinary antibiotics: total synthesis of Pasteurestins A and B and assignment of their configurations. *Eur. J. Org. Chem.*, 2714–2730; (b) Koegl, M., Brecker, L., Warrass, R., Mulzer, J. (2007). Total synthesis and configurational assignment of Pasteurestin A and B. *Angew. Chem. Int. Ed.*, *46*, 9320–9322.

[122] (a) Belmont, P., Parker, E. (2009). Silver and gold catalysis for cycloisomerization reactions. *Eur. J. Org. Chem.*, 6075–6089; (b) Soriano, E., Marco-Contelles, J. (2009). Mechanistic insights on the cycloisomerization of polyunsaturated precursors catalyzed by platinum and gold complexes. *Acc. Chem. Res.*, *42*, 1026–1036; (c) Gagosz, F. (2009). Recent developments in gold catalysis. *Tetrahedron.*, *65*, 1757; (d) Hashmi, A. S. K., Rudolph, M. (2008). Gold catalysis in total synthesis. *Chem. Soc. Rev.*, *37*, 1766–1775; (e) Krause, N., Belting, V., Deutsch, C., Erdsack, J., Fan, H., Gockel, B., Hoffmann-Röder, A., Morita, N., Volz, F. (2008).

Golden opportunities in catalysis. *Pure Appl. Chem.*, *80*, 1063–1069; (f) Li, Z., Brouwer, C., He, C. (2008). Gold-catalyzed organic transformations. *Chem. Rev.*, *108*, 3239–3265; (g) Jimenez-Nunez, E., Echavarren, A. M. (2007). Molecular diversity through gold catalysis with alkynes. *Chem. Commun.*, 333–346; (h) Stephen, A., Hashmi, K., Hutchings, G. J. (2006). Gold catalysis. *Angew. Chem. Int. Ed.*, *45*, 7896–7936; (i) Ma, S., Yu, S., Gu, Z. (2006). Gold-catalyzed cyclization of enynes. *Angew. Chem. Int. Ed.*, *45*, 200–203; (j) Zhang, L., Sun, J., Kozmin, S. A. (2006). Gold and platinum catalysis of enyne cycloisomerization. *Adv. Synth. Catal.*, *348*, 2271–2296.

[123] Molawi, K., Delpont, N., Echavarren, A. M. (2010). Enantioselective synthesis of (−)-Englerins A and B. *Angew. Chem. Int. Ed.*, *49*, 3517–3519.

[124] Zhou, Q., Chen, X., Ma, D. (2010). Asymmetric, protecting-group-free total synthesis of (−)-Englerin A. *Angew. Chem. Int. Ed.*, *49*, 3513–3516.

[125] Hoveyda, A. H. (2003). *Handbook of Metathesis*, Wiley-VCH, Weinheim, Germany.

[126] Some reviews on asymmetric olefin metathesis: (a) Hoveyda, A. H., Malcolmson, S. J., Meek, S. J., Zhugralin, A. R. (2010). Catalytic enantioselective olefin metathesis in natural product synthesis. chiral metal-based complexes that deliver high enantioselectivity and more. *Angew. Chem. Int. Ed.*, *49*, 34–44; (b) Schrock, R. R., Hoveyda, A. H. (2003). Molybdenum and tungsten imido alkylidene complexes as efficient olefin-metathesis catalysts. *Angew. Chem. Int. Ed.*, *42*, 4592–4633; (c) Hoveyda, A. H., Schrock, R. R. (2001). Catalytic asymmetric olefin metathesis. *Chem.- Eur. J.*, *7*, 945–950.

[127] Some examples: (a) Gillingham, D. G., Hoveyda, A. H. (2007). Chiral N-heterocyclic carbenes in natural product synthesis: application of Ru-catalyzed asymmetric ring-opening/cross-metathesis and Cu-catalyzed allylic alkylation to total synthesis of Baconipyrone C. *Angew. Chem. Int. Ed.*, *46*, 3860–3864; (b) Sattely, E. S., Cortez, G. A., Moebius, D. C., Schrock, R. R., Hoveyda, A. H. (2005). Enantioselective synthesis of cyclic amides and amines through Mo-catalyzed asymmetric ring-closing metathesis. *J. Am. Chem. Soc.*, *127*, 8526–8533; (c) Weatherhead, G. S., Cortez, G. A., Schrock, R. R., Hoveyda, A. H. (2004). Mo-catalyzed asymmetric olefin metathesis in target-oriented synthesis: enantioselective synthesis of (+)-Africanol. *Proc. Natl. Acad. Sci. U.S.A.*, *101*, 5805–5809; (d) Cefalo, D. R., Kiely, A. F., Wuchrer, M., Jamieson, J. Y., Schrock, R. R., Hoveyda, A. H. (2001). Enantioselective synthesis of unsaturated cyclic tertiary ethers by Mo-catalyzed olefin metathesis. *J. Am. Chem. Soc.*, *123*, 3139–3140.

[128] (a) Sattely, E. S., Meek, S. J., Malcolmson, S. J., Schrock, R. R., Hoveyda, A. H. (2009). Design and stereoselective preparation of a new class of chiral olefin metathesis catalysts and application to enantioselective synthesis of Quebrachamine: catalyst development inspired by natural product synthesis. *J. Am. Chem. Soc.*, *131*, 943–953; (b) Meek, S. J., Malcolmson, S. J., Li, B., Schrock, R. R., Hoveyda, A. H. (2009). The significance of degenerate processes to enantioselective olefin metathesis reactions promoted by stereogenic-at-Mo complexes. *J. Am. Chem. Soc.*, *131*, 16407–16409; (c) Malcolmson, S. J., Meek, S. J., Sattely, E. S., Schrock, R. R., Hoveyda, A. H. (2008). Highly efficient molybdenum-based catalysts for enantioselective alkene metathesis. *Nature*, *456*, 933–937.

[129] (a) Seiders, T. J., Ward, D. W., Grubbs, R. H. (2001). Enantioselective ruthenium-catalyzed ring-closing metathesis. *Org. Lett.*, *3*, 3225–3228; (b) Van Veldhuizen, J. J., Gillingham, D. G., Garber, S. B., Kataoka, O., Hoveyda, A. H. (2003). Chiral Ru-based complexes for asymmetric olefin metathesis: enhancement of catalyst activity through steric and electronic modification. *J. Am. Chem. Soc.*, *125*, 12502–12508; (c) Van Veldhuizen, J. J., Garber, S. B., Kingsbury, J. S., Hoveyda, A. H. (2002). A recyclable chiral Ru catalyst for enantioselective olefin metathesis. Efficient catalytic asymmetric ring-Opening/Cross metathesis in air. *J. Am. Chem. Soc.*, *124*, 4954–4955.

[130] Funk, T. W. (2009). Enantioselective synthesis of 5-epi-Citreoviral using ruthenium-catalyzed asymmetric ring-closing metathesis. *Org. Lett.*, *11*, 4998–5001.

CHAPTER 16

STEREOSELECTIVE NITROGEN HETEROCYCLE SYNTHESIS MEDIATED BY CHIRAL METAL CATALYSTS

SHERRY R. CHEMLER

16.1. Introduction 672
16.2. Metal-catalyzed enantioselective additions of amines and amine derivatives to unsaturated carbon–carbon bonds 672
16.2.1. Aziridination of alkenes 672
16.2.2. Intramolecular hydroamination of alkenes, alkynes, and allenes 672
16.2.2.1. Hydroamination of alkenes 672
16.2.2.2. Hydroamination of alkynes 674
16.2.2.3. Hydroamination of allenes 675
16.2.3. Oxidative amination of alkenes 676
16.2.3.1. Pd-catalyzed asymmetric aza-Wacker cyclizations 676
16.2.4. Carboamination and aminocarbonylation of alkenes 677
16.2.4.1. Alkene carboamination 677
16.2.4.2. Aminocarbonylations 679
16.2.5. Aminooxygenation of alkenes 679
16.2.5.1. Enantioselective intramolecular alkene aminooxygenation 679
16.2.5.2. Enantioselective intermolecular alkene aminooxygenation 680
16.2.6. Diamination of alkenes and dienes 682
16.2.6.1. Intramolecular alkene diamination 682
16.2.6.2. Diamination of dienes 682
16.3. Conclusions and future directions 684
References 684

Catalytic Methods in Asymmetric Synthesis: Advanced Materials, Techniques, and Applications, First Edition. Edited by Michelangelo Gruttadauria and Francesco Giacalone.

16.1. INTRODUCTION

The prevalence of chiral nitrogen heterocycles in small molecule and natural product-derived therapeutics as well as their robust use as ligands and catalysts in chemical processes has stimulated a high demand for synthetic methods that can produce them efficiently and enantioselectively. Reactions mediated by chiral transition metals play an important role in enantioselective nitrogen heterocycle synthesis. Two common strategies for the synthesis of chiral nitrogen heterocycles are: (1) enantioselective additions to carbon–nitrogen π-bonds (e.g., imines) [1–5], including cycloadditions [6–9], and (2) enantioselective addition of amines to carbon–carbon π-bonds. These addition reactions are catalyzed by a range of chiral transition metal complexes. This chapter will highlight important examples in the second category.

16.2. METAL-CATALYZED ENANTIOSELECTIVE ADDITIONS OF AMINES AND AMINE DERIVATIVES TO UNSATURATED CARBON–CARBON BONDS

16.2.1. Aziridination of Alkenes

The smallest nitrogen heterocycle is the aziridine. The catalytic asymmetric aziridination of alkenes has been recently reviewed [10, 11]. The most enantioselective reactions are aziridination of styrenes (predominantly α,β-unsaturated styrenes and chalcones) with chiral copper complexes [12–15], but chiral rhodium [16] and ruthenium [17] complexes have also been used to catalyze the aziridination of alkenes. The ruthenium complex-catalyzed aziridination has demonstrated the widest substrate scope, including styrene, 1-hexene, and α,β-unsaturated esters [17]. Table 16.1 summarizes the aziridination of alkenes with various copper complexes [12–14].

Table 16.2 summarizes the aziridination of alkenes with a chiral ruthenium complex [17]. Despite these advances, a protocol for the aziridination of a broader range of aliphatic alkenes is lacking.

16.2.2. Intramolecular Hydroamination of Alkenes, Alkynes, and Allenes

The catalytic asymmetric hydroamination/cyclization of alkenes (AHA), alkynes, and allenes is an atom economical process that generally provides five- and six-membered nitrogen heterocycles in good to excellent enantioselectivity. This topic has been reviewed extensively [18–23], so only highlights and updates of each reaction will be provided in this chapter.

16.2.2.1. Hydroamination of Alkenes. A broad range of main group, group 4, lanthanide and transition metals catalyze the hydroamination of alkenes. Of

TABLE 16.1. (Cu)-Catalyzed Enantioselective Alkene Aziridination

$$\text{Alkene} + \text{PhI} = \text{NTs} \xrightarrow[\text{solvent, temp}]{\text{catalyst (mol\%)}} \text{Aziridine}$$

Entry	Alkene	Catalyst	Solvent/Temp	Product	Yield (%)	ee (%)
1	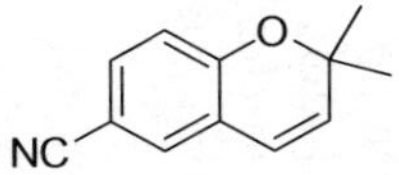	**A**/CuOTf (10 mol%)	CH_2Cl_2/–78°C	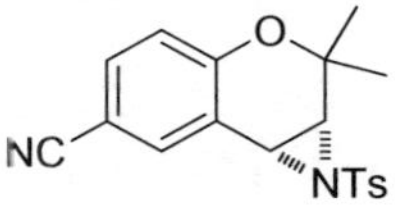	75	>98
2	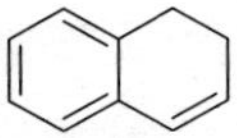	**A**/CuOTf (10 mol%)	CH_2Cl_2/–78°C	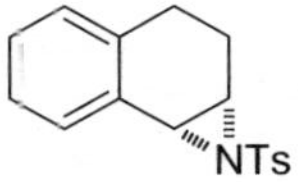	70	87
3	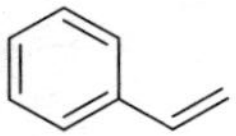	**A**/CuOTf (10 mol%)	CH_2Cl_2/–78°C	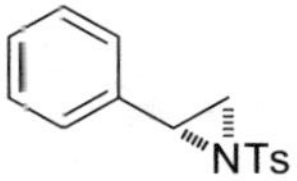	76	66
4	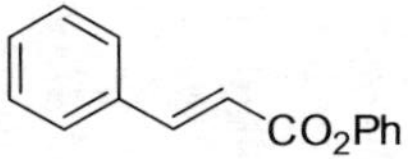	**B**/CuOTf (5 mol%)	styrene/0°C	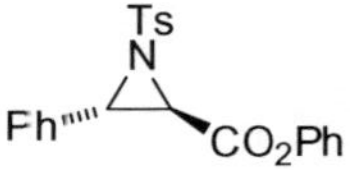	89	63
5	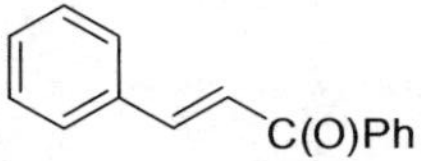	**C**/CuOTf (5 mol%)	C_8H_6/21°C	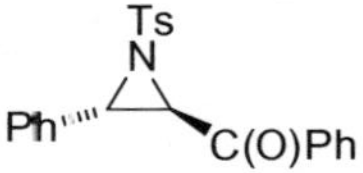	64	97
6		**D**/CuOTf (5 mol%)	CH_2Cl_2/24°C		56	91

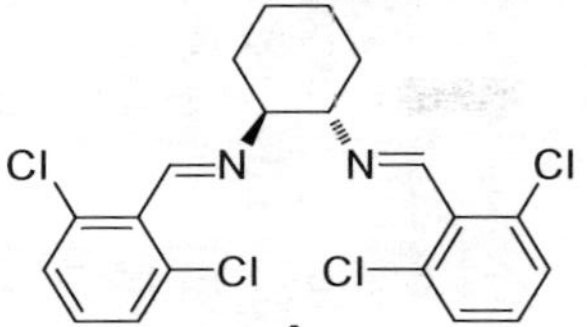

A

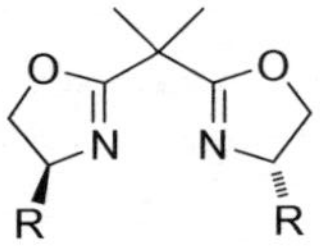

B, R = *t*-Bu
C, R = Ph

D 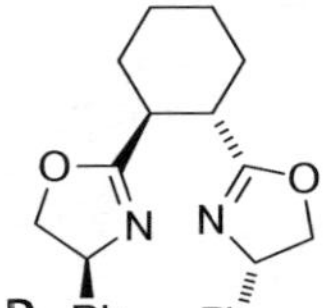

TABLE 16.2. (Ru)-Catalyzed Enantioselective Alkene Aziridination

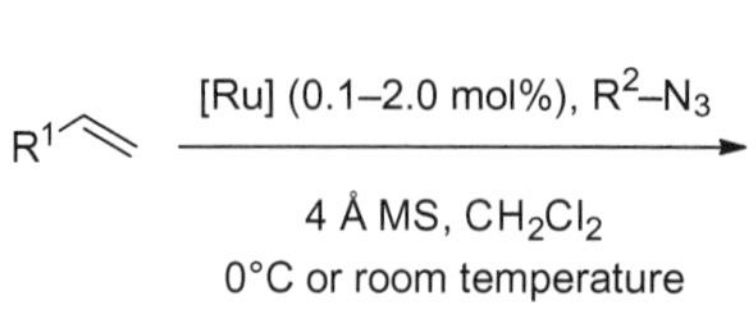

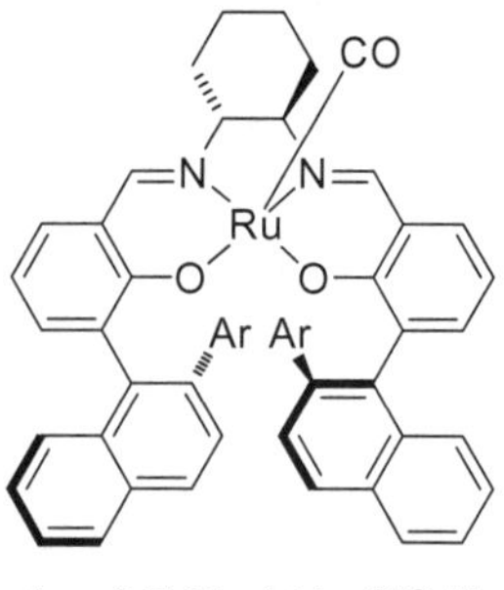

Entry	R^1	R^2	Yield (%)	%ee
1	Ph	*p*-Ns	70	81
2	Ph	SES	99	92
3	CO_2Bn	SES	81	99
4	nC_6H_{13}	Ts	64	84

Ar = 3,5-Cl_2-4-$Me_3SiC_6H_2$

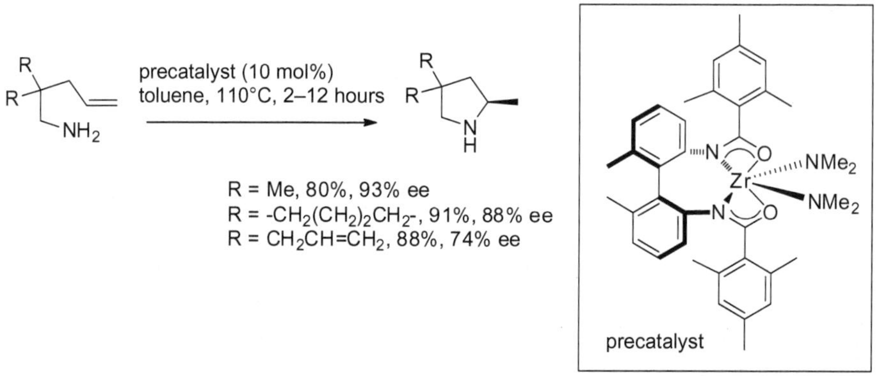

SCHEME 16.1. (Zr)-catalyzed enantioselective alkene hydroamination.

the group 4 metals, chiral zirconium amidate catalysts have provided relatively high yields and enantioselectivities in the synthesis of pyrrolidines from γ-alkenyl amines (Scheme 16.1) [24, 25]. The mechanism is thought to involve a cyclic transition state with a stereochemistry-determining [2+2] *cis* aminozirconation step.

A rhodium-catalyzed enantioselective hydroamination provides competitive levels of enantioselectivity for secondary γ-alkenyl amines and aniline (Scheme 16.2) [26]. This reaction provides high enantioselectivity for substrates without alkyl backbone substitution (R = H), which is less common.

16.2.2.2. Hydroamination of Alkynes. The palladium-catalyzed enantioselective hydroamination/cyclization of alkynes provides 2-alkenyl pyrrolidines and piperidines in good yield and enantioselectivity (Table 16.3) [27–29]. The asymmetry is provided by the chiral (*R*,*R*)-methyl Norphos ligand and a Pd-π-allyl intermediate is thought to be involved in the stereochemistry-determining step.

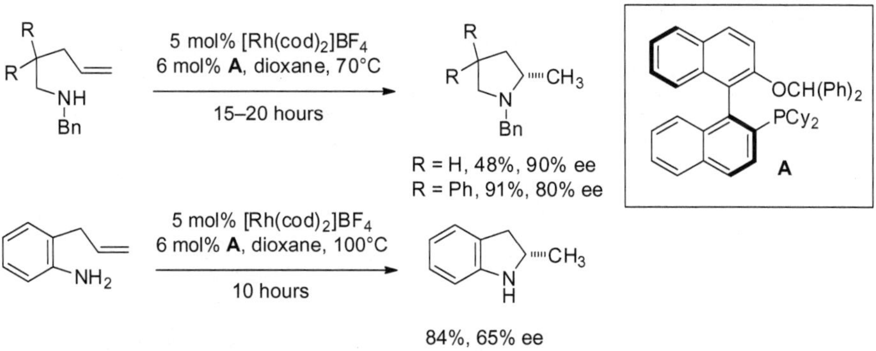

SCHEME 16.2. (Rh)-catalyzed enantioselective alkene hydroamination.

TABLE 16.3. (Pd)-Catalyzed Enantioselective Alkyne Hydroamination

Entry[a]	Substrate / Product	Yield (%)	ee (%)
1	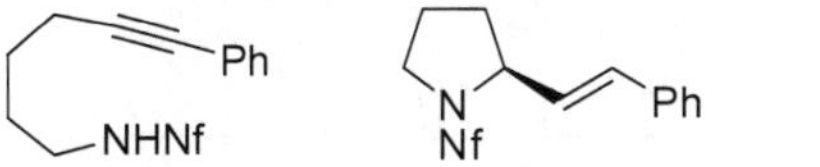	93	92
2	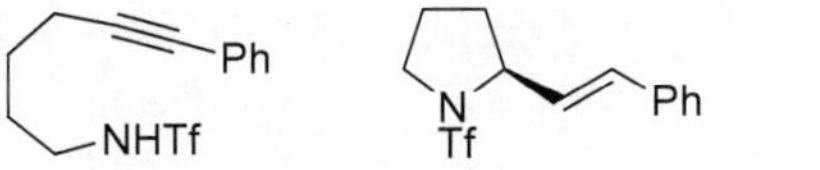	90	88
3	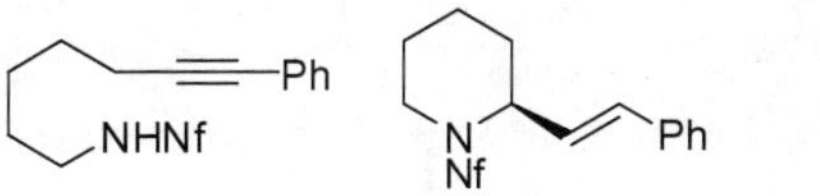	92	79
4	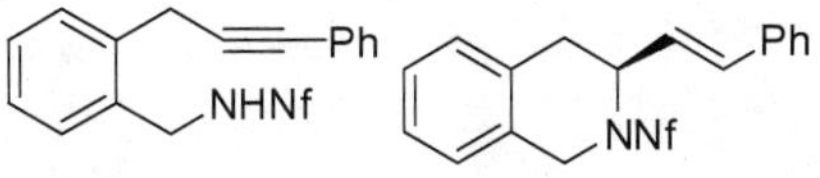	90	95

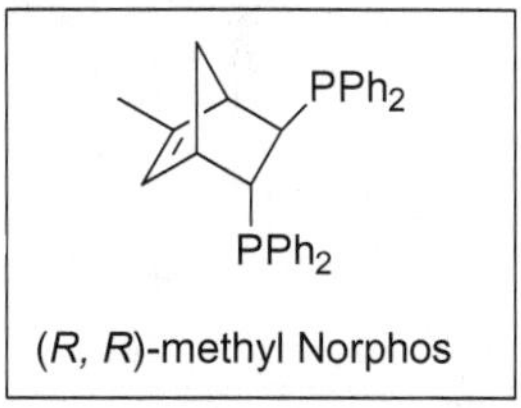

[a]Conditions: 5 mol% $Pd_2(dba)_3 \cdot CHCl_3$, 10 mol% $PhCO_2H$, 20 mol% (*R*,*R*)-methyl Norphos in PhH at 100°C, 48 hours.

16.2.2.3. Hydroamination of Allenes. The gold-catalyzed enantioselective hydroamination/cyclization of γ- and δ-aminoallenes provides α-alkenyl pyrrolidines, piperidines, pyrazolidines, and isoxazolidines with high yield and enantioselectivity [30–33]. Higher substitution on the termal carbon of the allene generally provides higher levels of enantioselectivity. Table 16.4 illustrates the gold-catalyzed hydroamination/cyclization of various allenes for the formation of five- and six-membered ring nitrogen heterocyles [30, 32].

TABLE 16.4. (Au)-Catalyzed Enantioselective Allene Hydroamination

Entry[a]	Substrate	Conditions	Product	Yield (%)	ee (%)
1	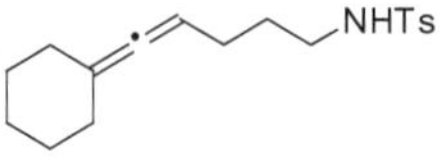	A	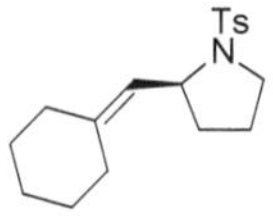	88	98
2	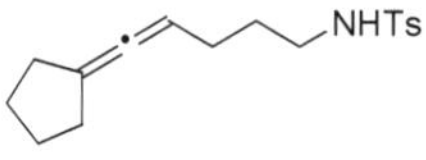	A	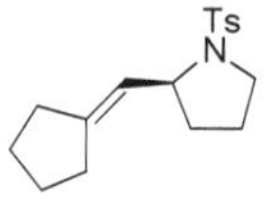	75	83
3	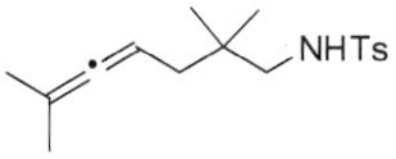	B	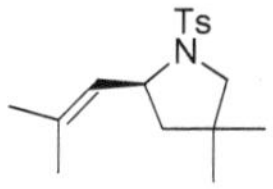	94	93
4		C	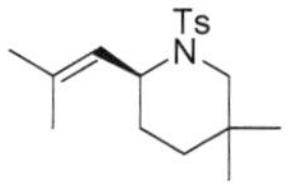	70	98
5	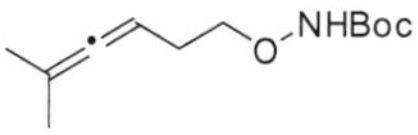	A	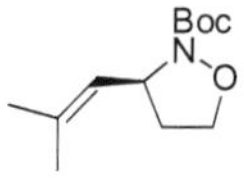	91	98
6	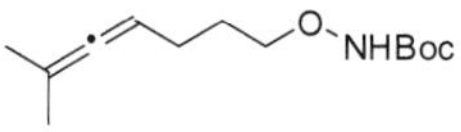	C[b]	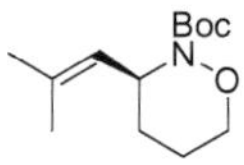	63	89

[a] Conditions: A = 3 mol% of 5, 0.3 M in DCE, 23°C; B = 5 mol% of **6**, 0.3 M in $MeNO_2$, 50°C; C = 5 mol% of **7**, 0.3 M in $MeNO_2$, 50°C.
[b] 36 hours, 65°C.

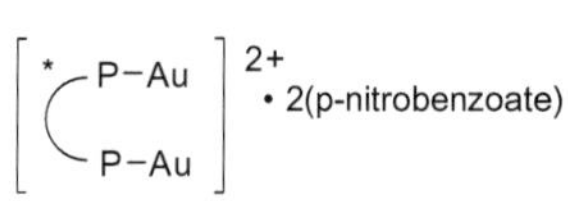

5 = (R)-xylyl-BINAP(AuOPNB)$_2$
6 = (R)-SEGPHOS(AuOPNB)$_2$
7 = (R)-ClMeOBiPHEP(AuOPNB)$_2$

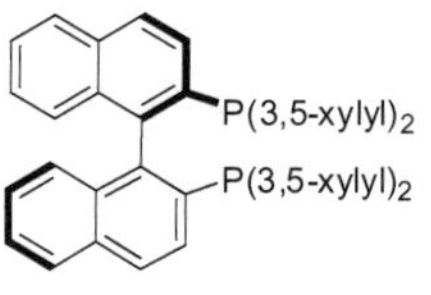

(*R*)-xylyl-BINAP

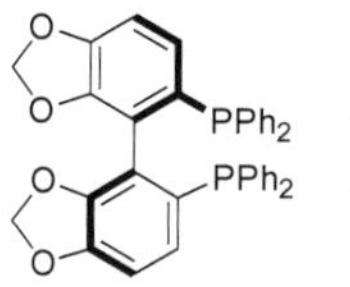

(*R*)-SEGPHOS

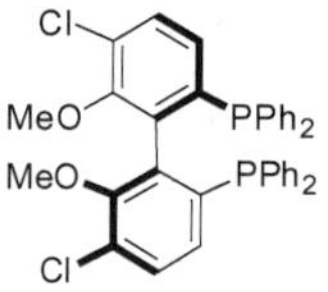

(*R*)-ClMeOBIPHEP

16.2.3. Oxidative Amination of Alkenes

16.2.3.1. Pd-Catalyzed Asymmetric Aza-Wacker Cyclizations. The palladium-catalyzed oxidative amination/cyclization of alkenes has been optimized extensively in racemic form [34], but few enantioselective variants have been reported [35, 36]. Scheme 16.3 summarizes the most promising report where the 4-methoxy-2-crotyl-N-tosylaniline provided the highest selectivity [74% enantiomeric excess (ee)] with (*S*)-*i*-Pr-quinolineoxazoline ligand [36].

The oxidatively neutral Pd-catalyzed enantioselective intramolecular S_N2' of terminal allylic acetates also provides vinyl-substituted nitrogen heterocycles with excellent yield and enantioselectivity (not shown) [37, 38].

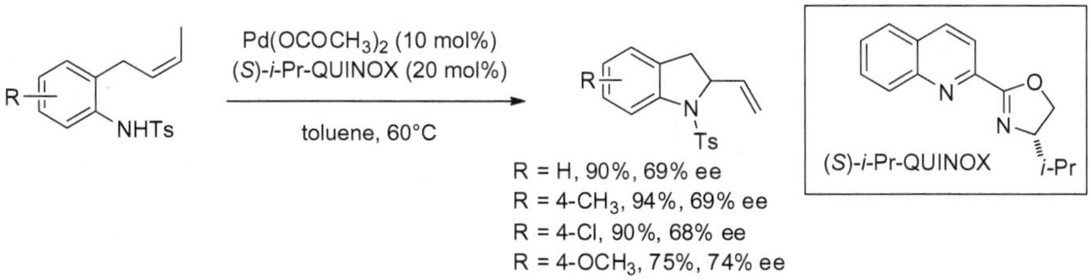

SCHEME 16.3. (Pd)-catalyzed enantioselective alkene oxidative amination.

16.2.4. Carboamination and Aminocarbonylation of Alkenes

The catalytic enantioselective carboamination of alkenes for the synthesis of nitrogen heterocycles involves intramoleular metal-catalyzed alkene amination followed by C–C bond formation. Copper and palladium catalysts have demonstrated the most significant advances in this area [23, 39–41]. In addition to the alkene carboamination and aminocarbonylation reactions summarized below, the enantioselective synthesis of nitrogen heterocycles via enantioselective domino Heck-allylic amination of dienes [42] and enantioselective hetero- and carboannulation of allenes have also been reported [43].

16.2.4.1. Alkene Carboamination. Enantioselective alkene carboamination/cyclizations are relatively rare reactions that have been reported only recently. The reactions form N-bearing stereocenters in the alkene amination step. To date, enantioselective alkene carboamination reactions have been reported only with chiral copper and palladium complexes. Interestingly, all of the reported reactions are thought to proceed via *cis*-aminometallation, concerted intramolecular addition of N-[Met] across the alkene [44–46]. Such a mechanism allows for a tight, cyclic transition state that can facilitate induction of stereochemistry from the chiral metal ligands.

The chiral $Cu(OTf)_2$·(*R*)-Ph-Box complex catalyzes the doubly intramolecular carboamination of γ-alkenylsulfonamides, providing chiral pyrrolidine sultams [44, 47] and hexahydrobenz[*f*]indoles [48] in good to excellent yield and good to excellent enantioselectivity. MnO_2 is used as the stoichiometric oxidant. The mechanism is thought to involve enantioselective *cis*-aminocupration followed by C–Cu homolysis. Addition of the resulting carbon radical intermediate to a nearby aryl ring, the C–C bond formation, is a net C–H functionalization. If the aryl ring is tethered through the nitrogen, a sultam product results (Table 16.5) [44]. The sulfonamide moiety is optimal for both reactivity and enantioselectivity. The pyrrolidine can be obtained by removal of SO_2 under dissolving metal conditions (Li, NH_3) [44].

If the aryl ring undergoing radical addition is in the allylic position, a chiral hexahydrobenz[*f*]indole can result (Table 16.6) [48]. This reaction is highly enantioselective, diastereoselective, and chemoselective, forming the *cis*-fused hexahydrobenz[*f*]indole exclusively.

TABLE 16.5. (Cu)-Catalyzed Enantioselective Alkene Carboamination

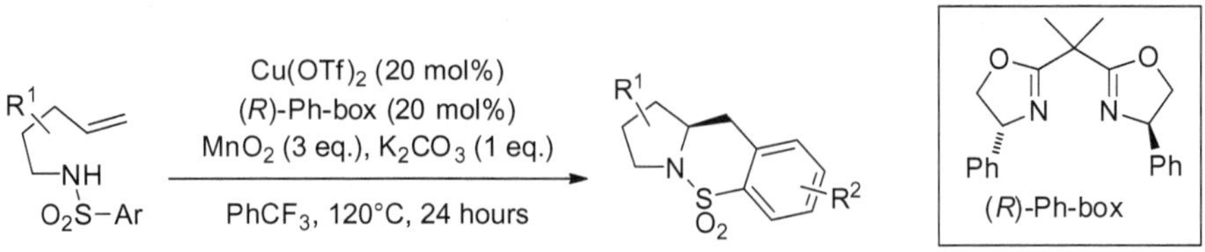

Entry	Substrate	Product	Yield (%)	ee (%)[a]
	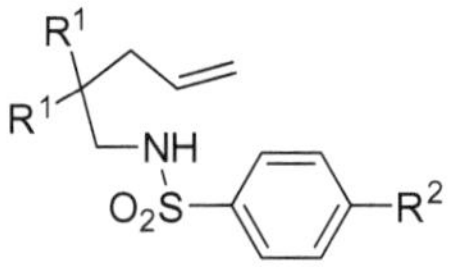	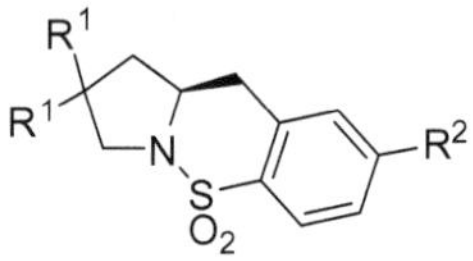		
1	R^1 = Me, R^2 = Me		85	92
2	R^1 = Me, R^2 = H		73	92
3	R^1 = Me, R^2 = CI		45	92
4	R^1 = Me, R^2 = OMe		75	94
5	R^1 = Ph, R^2 = Me		78	94
6	R^1 = H, R^2 = H		77	82
7	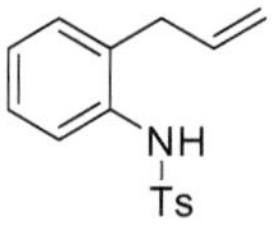	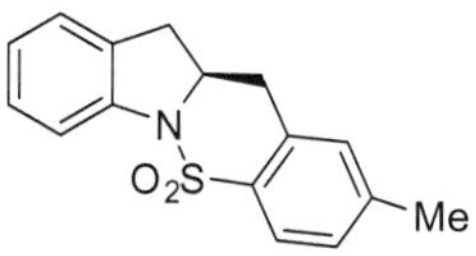	78	46
8d	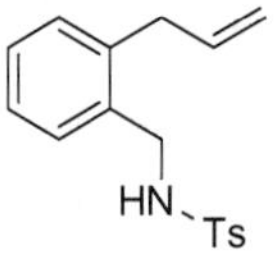	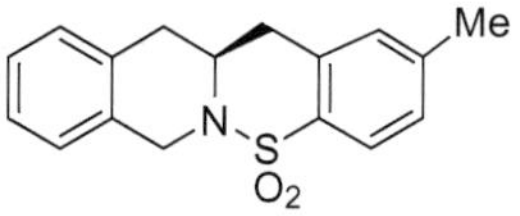	63 (96 hours)	72

[a]Enantiomeric excess was determined by chiral high performance liquid chromatography (HPLC) analysis.

Doubly intramolecular enantioselective carboaminations have also been catalyzed by chiral Pd-complexes. 2-Allylbenzamides undergo tandem cyclization to provide chiral indolines with good yields and excellent enantioselectivities (Scheme 16.4) [45, 49].

A Pd-catalyzed intra/intermolecular alkene carboamination that produces 2-benzyl and 2-alkenylpyrrolidines enantioselectively using the chiral phosphoramidite ligand (*R*)-Siphos-PE [50] has been reported recently [46]. A summary of substrates and products is presented in Table 16.7 [46]. The reaction initiates with oxidative addition into an arylhalide or vinyl halide followed by enantioselective aminopalladation/cyclization of the γ-alkenyl-*tert*-butylcarbamate and reductive elimination to give C–C bond formation.

TABLE 16.6. (Cu)-Catalyzed Hexahydrobenz[*f*]indole Synthesis

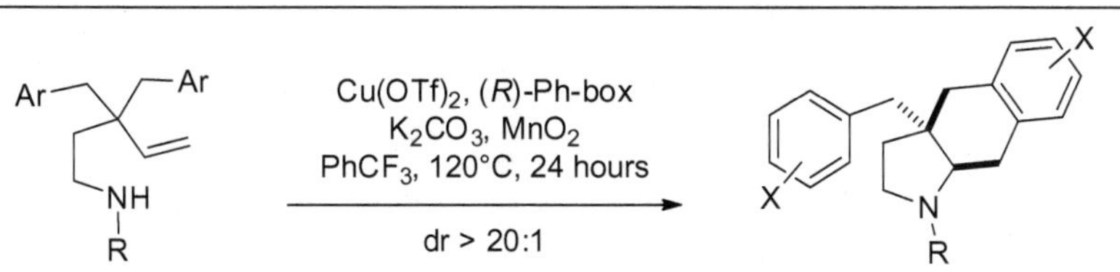

Entry	Ar	R	Yield (%)	ee (%)[a]
1	Ph	Ms	99	82
2	Ph	SES	93	83
3	Ph	Bs	99	94
4	Ph	Ts	96	96
5	Ph	Ns	99	97
6	4-F-C_6H_4	Ts	89	96
7	4-MeS-C_6H_4	Ts	91	97
8	4-MeO-C_6H_4	Ts	89	96

[a] Enantiomeric excess was determined by chiral HPLC analysis.

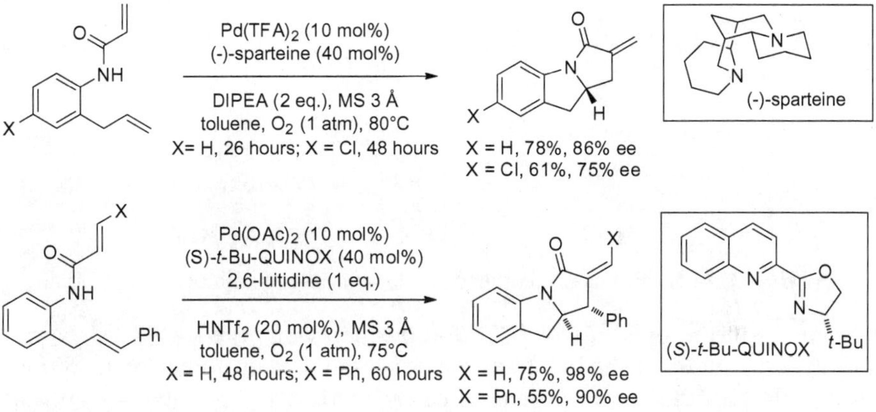

SCHEME 16.4. (Pd)-catalyzed enantioselective tandem cyclization.

16.2.4.2. Aminocarbonylations. Palladium-catalyzed enantioselective intramolecular aminocarbonylations of γ-pentenylureas provide tetrahydropyrrolo (1,2-*c*)pyrimidine-1,3-diones in moderate to good yields and enantioselectivities (Scheme 16.5) [51, 52]. These reactions are performed under CO (1 atm) using the tetraisopropyl spiro bis(isooxazoline) ligand (*i*-Pr-SPRIX) as chirality source.

16.2.5. Aminooxygenation of Alkenes

16.2.5.1. Enantioselective Intramolecular Alkene Aminooxygenation. Chiral 2-methyleneoxy indolines and pyrrolidines have been synthesized via $Cu(OTf)_2$·(4*R*,5*S*)-di-Ph-box catalyzed intramolecular cyclization/amination

TABLE 16.7. (Pd)-Catalyzed Enantioselective Alkene Carboamination

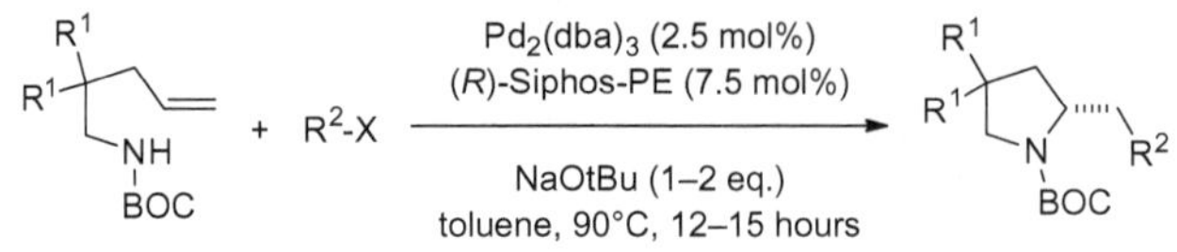

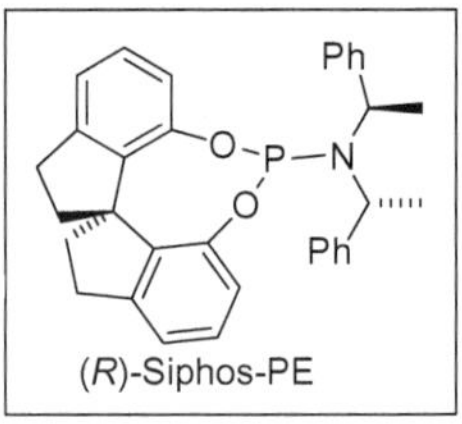

Entry	R^1	R^2-X	Yield (%)	ee (%)[a]
1	H	4-$MeOC_6H_4Br$	72	86
2	H	4-FC_6H_4Br	66	80
3	Me	4-$Me_2NC_6H_4Br$	70	92
4	Me	m-$CF_3C_6H_4I$	71	91
5	OEt	4-PhC_6H_4Br	64	75
6	H	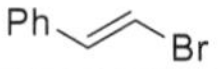	76	93

[a] Enantiomeric excess was determined by chiral HPLC analysis.

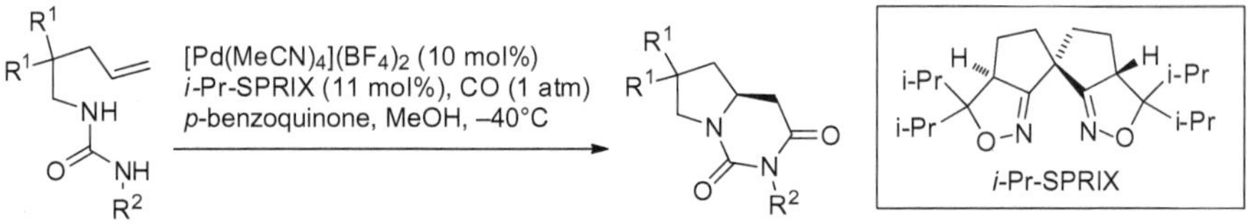

R^1 = Me, R^2 = Ts, 89%, 88% ee
R^1 = Me, R^2 = Bs, 89%, 87% ee
R^1 = H, R^2 = Ts, 28%, 61% ee
R^1 = Ph, R^2 = Ts, 80%, 30% ee
R^1 = CO_2Et, R^2 = Ts, 50%, 51% ee

SCHEME 16.5. (Pd)-catalyzed enantioselective alkene aminocarbonylation.

of alkenes in the presence of (2,2,6,6-tetra-methylpiperidin-1-yl)oxyl radical (TEMPO) [53]. While the indolines can be synthesized on a small scale (50 mg) using only TEMPO as the oxygen source and oxidant, the pyrrolidine synthesis requires an O_2 (1 atm) atmosphere to go to completion (Table 16.8). O_2 (1 atm) is also required for larger scale indoline synthesis reactions (unpublished results).

The resulting hydroxylamine products can be converted to the corresponding chiral alcohol or aldehyde by either reduction (Zn, MeOH, NH_4Cl) or oxidation (meta-chloro-perbenzoic acid, MCPBA), respectively (Scheme 16.6).

16.2.5.2. Enantioselective Intermolecular Alkene Aminooxygenation. Chiral copper complexes have also been used to catalyze the enantioselective intermolecular aminooxygenation of styrenes [54]. Benzenesulfonylphenyloxaziridine serves as the heteroatom source, providing diastereomeric mixtures of 1,3-oxazolines with moderate levels of enantioselectivity and yield (Table 16.9). The aminal moiety can be hydrolyzed under acidic conditions to reveal the aminoalcohol, and recrystallization can enrich the aminoalcohol to >99% ee.

TABLE 16.8. (Cu)-Catalyzed Enantioselective Intramolecular Aminooxygenation

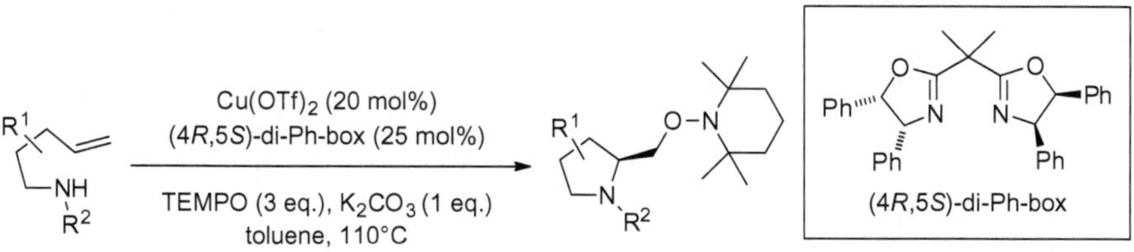

Entry	Substrate	Product	Yield (%)	ee (%)[a]
	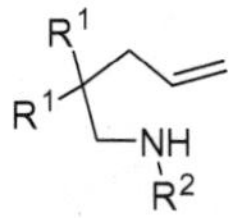	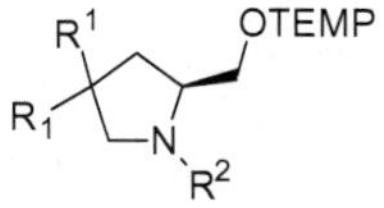		
1[b]	R^1 = Me, R^2 = Ts		97	88
2[b]	R^1 = Ph, R^2 = Ts		97	92
3[b,c]	R^1 = H, R^2 = Ts		74	75
4[b,c]	R^1 = Me, R^2 = Ns		86	89
	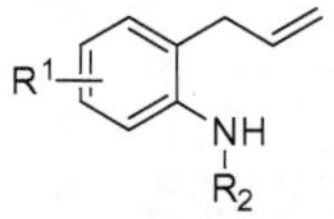	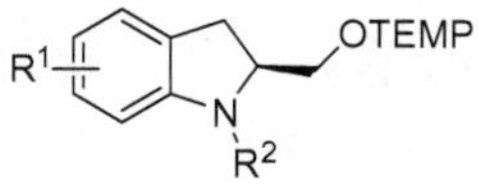		
5	R^1 = H, R^2 = Ts		97	90
6	R^1 = p-F, R^2 = Ts		83	89
7	R^1 = p-Cl, R^2 = Ts		73	91
8	R^1 = p-OMe, R^2 = Ts		82	86
9[b,c]	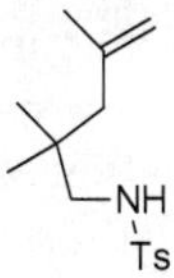	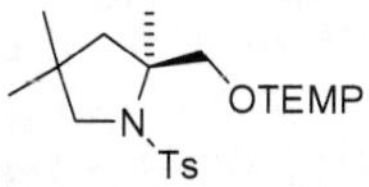	61	82

[a]Enantiomeric excess was measured by chiral HPLC analysis.
[b]Reaction was run at 120°C under O_2 (1 atm, balloon).
[c]40 mol% $Cu(OTf)_2$ and 50 mol% ligand was used.

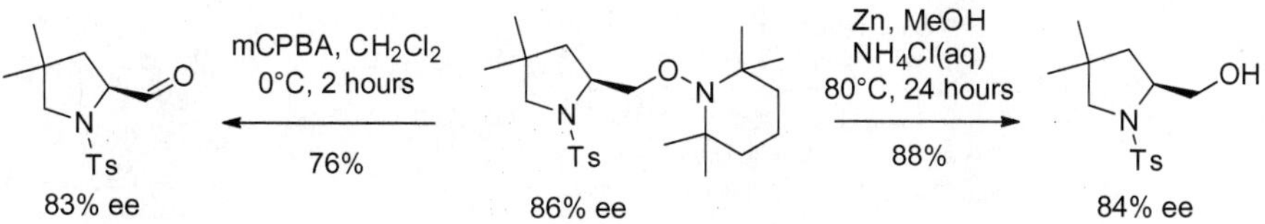

SCHEME 16.6. Oxidation or reduction of the hydroxylamine products.

TABLE 16.9. (Cu)-Catalyzed Enatioselective Intermolecular Aminooxygenation

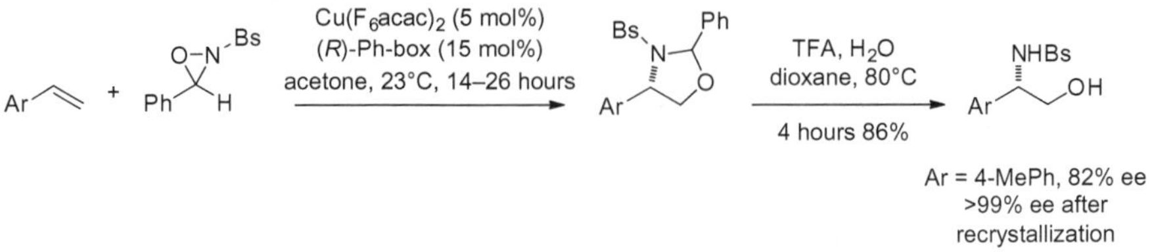

Entry	Ar	Yield (%)	dr (*cis/trans*)	%ee[a]
1	Ph	81	1.5:1	81
2	4-MePh	72	2.2:1	82
3	3-MePh	78	1.6:1	79
4	4-BrPh	61	1:1	65
5	4-ClPh	70	0.8:1	69
6	4-FPh	57	1.4:1	78

[a] ee values reported for weighted average of the *cis/trans* isomers.

16.2.6. Diamination of Alkenes and Dienes

The diaminination of alkenes is a particularly active area of research that has seen substantial progress in methods development in recent years [55–57]. This transformation results in the formation of vicinal diamines, particularly common moieties found in biologically active compounds, as well as ligands and catalysts for organic reactions. This chapter highlights only the chiral metal-catalyzed alkene diamination reactions.

16.2.6.1. Intramolecular Alkene Diamination. A singular example of a promising but unoptimized $Cu(OTf)_2$·(*R*)-Ph-box-catalyzed intramolecular alkene diamination for the formation of chiral indoline diamine has been reported (Scheme 16.7) [58]. This reaction is mechanistically similar to the copper-catalyzed alkene carboamination and aminooxygenation reactions discussed in Sections 16.2.4.1 and 16.2.5.1.

16.2.6.2. Diamination of Dienes. The regioselective, catalytic enantioselective diamination of dienes for the formation of cyclic ureas has been achieved using both palladium and copper chiral complexes [59–61]. The palladium-catalyzed reaction [59] is selective for diamination of the internal alkene of

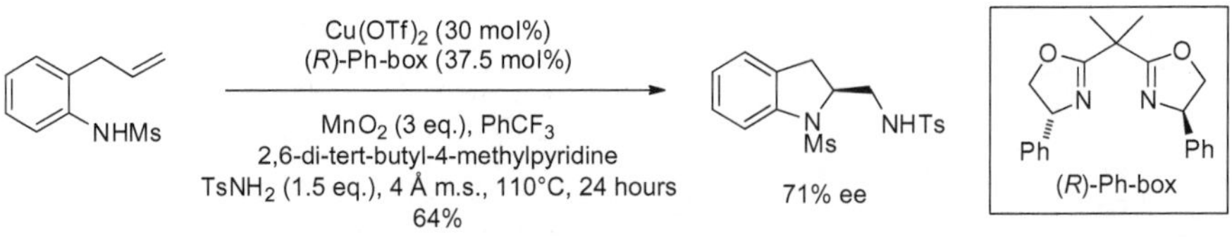

SCHEME 16.7. (Cu)-catalyzed enantioselective intramolecular alkene diamination.

the diene while the copper-catalyzed reaction [61] is selective for diamination of the terminal alkene of the diene. In these reactions, di-*tert*-butyldiaziridinone is the nitrogen source [62].

A range of dienes underwent enantioselective diamination of the internal alkene to provide chiral cyclic ureas in excellent yields and enantioselectivity using $Pd_2(dba)_3$ and chiral phosphoramidite ligand **A** (Table 16.10) [59].

TABLE 16.10. (Pd)-Catalyzed Enantioselective Intermolecular Alkene Diamination

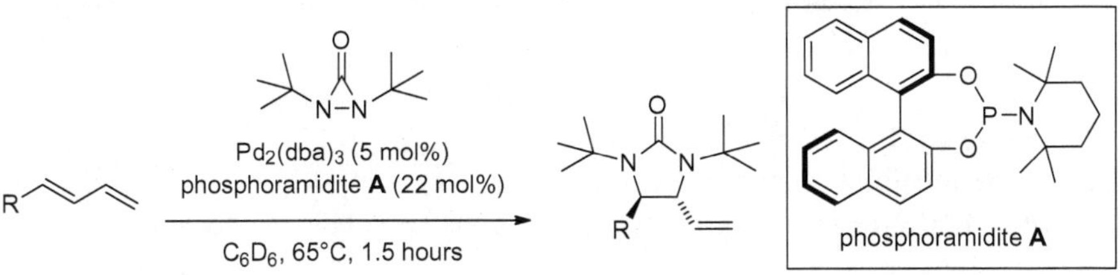

Entry	Substrate	Product	Yield (%)	ee (%)[a]
	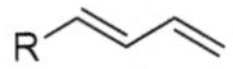	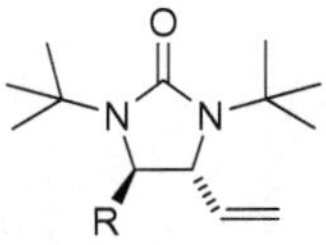		
1	R = Me		91	91
2	R = C_5H_{11}		90	92
3	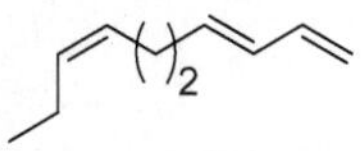	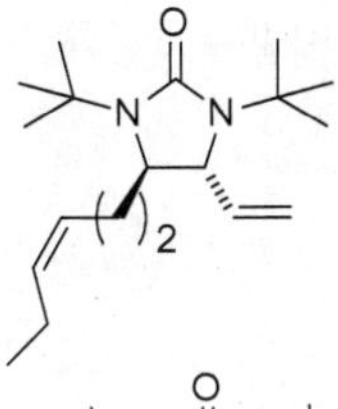	90	93
4	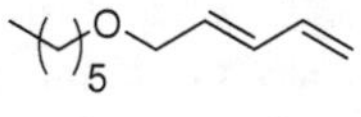	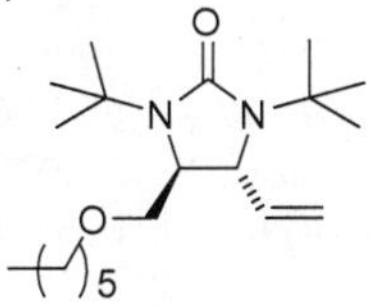	86	92
5	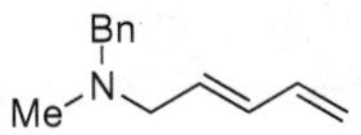	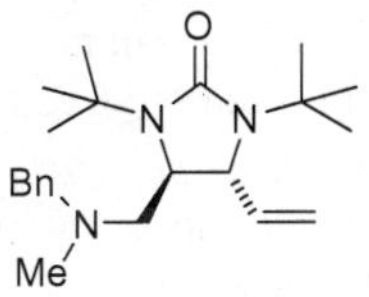	70	92
6	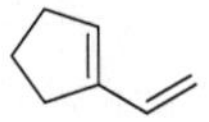	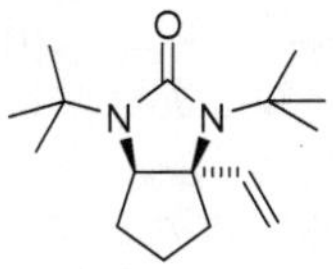	83	87

[a] Enantioselectivity was determined by chiral gas chromatography (GC) analysis.

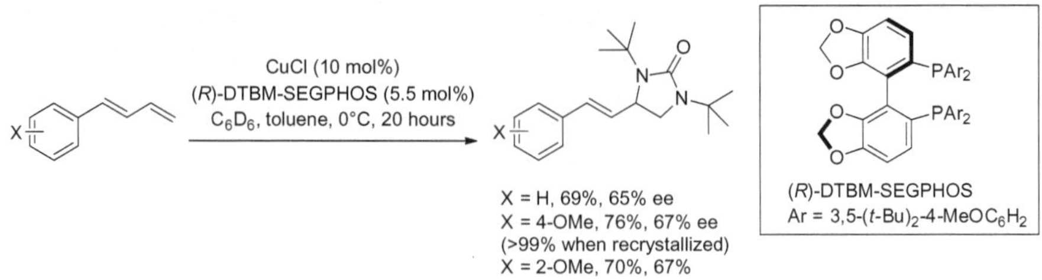

SCHEME 16.8. (Cu)-catalyzed enatioselective intermolecular alkene diamination.

The copper(I)-catalyzed enantioselective diamination is selective for reaction at the terminal alkene of the diene and occurs in moderate to good yield and with moderate enantioselectivity (Scheme 16.8) [61].

16.3. CONCLUSIONS AND FUTURE DIRECTIONS

The advent of asymmetric catalysis has modernized nitrogen heterocycle synthesis. Nitrogen heterocycles are rich sources of bioactive compounds and their enantioselective synthesis is no longer limited by chiral pool starting materials. Further application of the methods described above in drug discovery will undoubtedly be forthcoming, and these applications will, in turn, drive the further development and optimization of these methods as well as related methods.

REFERENCES

[1] Noyori, R., Hashiguchi, S. (1997). Asymmetric transfer hydrogenation catalyzed by chiral ruthenium complexes. *Acc. Chem. Res.*, *30*, 97–102.

[2] Tietze, L. F., Zhou, Y., Topken, E. (2000). Synthesis of simple enantiopure tetrahydro-β-carbolines and tetrahydroisoquinolines. *Eur. J. Org. Chem.*, 2247–2252.

[3] Taylor, A. M., Schreiber, S. L. (2006). Enantioselective addition of terminal alkynes to isolated isoquinoline iminiums. *Org. Lett.*, *8*, 143–146.

[4] Itoh, T., Miyazaki, M., Fukuoka, H., Nagata, K., Ohsawa, A. (2006). Formal total synthesis of (−)-emetine using catalytic asymmetric allylation of cyclic imines as a key step. *Org. Lett.*, *8*, 1295–1297.

[5] Wang, S., Onaran, B., Seto, C. T. (2010). Enantioselective synthesis of 1-aryltetrahydroisoquinolines. *Org. Lett.*, *12*, 2690–2693.

[6] Stanley, L. M., Sibi, M. P. (2008). Enantioselective copper-catalyzed 1,3-dipolar cycloadditions. *Chem. Rev.*, *108*, 2887–2902.

[7] Kobayashi, S., Ishitani, H. (1999). Catalytic enantioselective addition to imines. *Chem. Rev.*, *99*, 1069–1094.

[8] Jorgensen, K. A. (2000). Catalytic asymmetric hetero-Diels-Alder reactions of carbonyl comopunds and imines. *Angew. Chem. Int. Ed.*, *39*, 3558–3588.

[9] Dalton, D. M., Oberg, K. M., Yu, R. T., Lee, E. E., Perreault, S., Oinen, M. E., Pease, M. L., Malik, G., Rovis, T. (2009). Enantioselective rhodium-catalyzed [2+2+2] cycloadditions of terminal alkynes and alkenyl isocyanates: mechanistic insights lead to a unified model that rationalizes product selectivity. *J. Am. Chem. Soc.*, *131*, 15717–15728.

[10] Pellissier, H. (2010). Recent developments in asymmetric aziridination. *Tetrahedron*, *66*, 1509–1555.

[11] Muller, P., Fruit, C. (2003). Enantioselective catalytic aziridinations and asymmetric nitrene insertions into CH bonds. *Chem. Rev.*, *103*, 2905–2920.

[12] Evans, D. A., Faul, M. M., Bilodeau, M. T., Anderson, B. A., Barnes, D. M. (1993). Bis(oxazoline)-copper complexes as chiral catalysts for the enantioselective aziridinaiton of olefins. *J. Am. Chem. Soc.*, *115*, 5328–5329.

[13] Li, Z., Conser, K. R., Jacobsen, E. N. (1993). Asymmetric alkene aziridination with readily available chiral diimine-based catalysts. *J. Am. Chem. Soc.*, *115*, 5326–5327.

[14] Ma, L., Du, D.-M., Xu, J. (2005). Rational tuning chelate size of bis-oxazoline ligands to improve enantioselectivity in the asymmetric aziridination of chalcones. *J. Org. Chem.*, *70*, 10155–10158.

[15] Wang, X., Ding, K. (2006). One-pot enantioselective aziridination of olefins catalyzed by a copper(I) complex of a novel diimine ligand by using $PhI(OAc)_2$ and sulfonamide as nitrene precursors. *Chem. Eur. J.*, *12*, 4568–4575.

[16] Liang, J.-L., Yuan, S.-X., Chan, P. W. H., Che, C.-M. (2003). Chiral rhodium(II,II) dimers catalyzed enantioselective intramolecular aziridination of sulfonamides and carbamates. *Tetrahedron Lett.*, *44*, 5917–5920.

[17] Kawabata, H., Omura, K., Uchida, T., Katsuki, T. (2007). Construction of robust ruthenium(salen)(CO) complexes and asymmetric aziridination with nitrene precursors in the form of azide compounds that bear easily removable N-sulfonyl groups. *Chem. Asian. J.*, *2*, 248–256.

[18] Roesky, H. W., Muller, T. E. (2003). Enantioselective catalytic hydroamination of alkenes. *Angew. Chem. Int. Ed.*, *42*, 2708–2710.

[19] Hultzsch, K. C. (2005). Transition metal-catalyzed asymmetric hydroamination of alkenes (AHA). *Adv. Synth. Catal.*, *347*, 367–391.

[20] Hultzsch, K. C. (2005). Catalytic asymmetric hydroamination of non-activated olefins. *Org. Biomol. Chem.*, *3*, 1819–1824.

[21] Hong, S., Marks, T. J. (2004). Organolanthanide-catalyzed hydroamination. *Acc. Chem. Res.*, *37*, 673–686.

[22] Muller, T. E., Hultzsch, K. C., Yus, M., Foubelo, F., Tada, M. (2008). Hydroamination: direct addition of amines to alkenes and alkynes. *Chem. Rev.*, *108*, 3795–3892.

[23] Chemler, S. R. (2009). The enantioselective intramolecular aminative functionalization of unactivated alkenes, dienes, allenes and alkynes for the synthesis of chiral nitrogen heterocycles. *Org. Biomol. Chem.*, *7*, 3009–3019.

[24] Wood, M. C., Leitch, D. C., Yeung, C. S., Kozak, J. A., Schafer, L. (2007). Chiral neutral zirconium amidate complexes for the asymmetric hydroamination of alkenes. *Angew. Chem. Int. Ed.*, *46*, 354–358.

[25] Wood, M. C., Leitch, D. C., Yeung, C. S., Kozak, J. A., Schafer, L. (2009). Chiral neutral zirconium amidate complexes for the asymmetric hydroamination of alkenes. *Angew. Chem. Int. Ed.*, *48*, 6937.

[26] Shen, X. Q., Buchwald, S. L. (2010). Rhodium-catalyzed asymmetric intramolecular hydroamination of unactivated alkenes. *Angew. Chem. Int. Ed.*, *49*, 564–567.

[27] Patil, N. T., Lutete, L. M., Wu, H. Y., Pahadi, N. K., Gridnev, I. D., Yamamoto, Y. (2006). Palladium-catalyzed intramolecular asymmetric hydroamination, hydroalkoxylation, and hydrocarbonation of alkynes. *J. Org. Chem.*, *71*, 4270–4279.

[28] Lutete, L. M., Kadota, I., Yamamoto, Y. (2004). Palladium-catalyzed intramolecular asymmetric hydroamination of alkynes. *J. Am. Chem. Soc.*, *126*, 1622–1623.

[29] Narsireddy, M., Yamamoto, Y. (2008). Catalytic asymmetric intramolecular hydroamination of alkynes in the presence of a catalyst system consisting of Pd(0)-methyl Norphos (or tolyl Renorphos)-benzoic acid. *J. Org. Chem.*, *73*, 9698–9709.

[30] LaLonde, R. L., Sherry, B. D., Kang, E. J., Toste, F. D. (2007). Gold(I)-catalyzed enantioselective intramolecular hydroamination of allenes. *J. Am. Chem. Soc.*, *129*, 2452–2453.

[31] Hamilton, G. L., Kang, E. J., Mba, M., Toste, F. D. (2007). A powerful chiral counterion strategy for asymmetric transition metal catalysis. *Science*, *317*, 496–499.

[32] LaLonde, R. L., Wang, Z. J., Mba, M., Lackner, A. D., Toste, F. D. (2010). Gold(I)-catalyzed enantioselective synthesis of pyrazolidines, isoxazolidines, and tetrahydrooxazines. *Angew. Chem. Int. Ed.*, *49*, 598–601.

[33] Zhang, Z. B., Bender, C. F., Widenhoefer, R. A. (2007). Gold(I)-catalyzed enantioselective hydroamination of N-allenyl carbamates. *Org. Lett.*, *9*, 2887–2889.

[34] Kotov, V., Scarborough, C. C., Stahl, S. S. (2007). Palladium-catalyzed aerobic oxidative amination of alkenes: development of intra- and intermolecular aza-Wacker reactions. *Inorg. Chem.*, *46*, 1910–1923.

[35] Scarborough, C. C., Bergant, A., Sazama, G. T., Guzei, I. A., Spencer, L. C., Stahl, S. S. (2009). Synthesis of Pd-II complexes bearing an enantiomerically resolved seven-membered N-heterocyclic carbene ligand and initial studies of their use in asymmetric Wacker-type oxidative cyclization reactions. *Tetrahedron*, *65*, 5084–5092.

[36] Jiang, F., Wu, Z., Zhang, W. (2010). Pd-catalyzed asymmetric aza-Wacker-type cyclization reaction of olefinic tosylamides. *Tetrahedron Lett.*, *51*, 5124–5126.

[37] Overman, L. E., Remarchuk, T. P. (2002). Catalytic asymmetric intramolecular aminopalladation: enantioselective synthesis of vinyl-substituted 2-oxazolidinones, 2-imidazolidinones, and 2-pyrrolidinones. *J. Am. Chem. Soc.*, *124*, 12–13.

[38] Kirsch, S. F., Overman, L. E. (2005). Catalytic asymmetric intramolecular aminopalladation: improved palladium(II) catalysts. *J. Org. Chem.*, *70*, 2859–2861.

[39] Tietze, L. F., Ila, H., Bell, H. P. (2004). Enantioselective palladium-catalyzed transformations. *Chem. Rev.*, *104*, 3453–3516.

[40] Chemler, S. R., Fuller, P. H. (2007). Heterocycle synthesis by copper facilitated addition of heteroatoms to alkenes, alkynes and arenes. *Chem. Soc. Rev.*, *36*, 1153–1160.

[41] Chemler, S. R. (2011). Evolution of copper(II) as a new alkene amination promoter and catalyst. *J. Organomet. Chem.*, *696*, 150–158.

[42] Flubacher, D., Helmchen, G. (1999). Enantioselective domino Heck-allylic amination reactions. *Tetrahedron Lett.*, *40*, 3867–3868.

[43] Zenner, J. M., Larock, R. C. (1999). Palladium-catalyzed, asymmetric hetero- and carboannulation of allenes using functionally-substituted aryl and vinylic iodides. *J. Org. Chem.*, *64*, 7312–7322.

[44] Zeng, W., Chemler, S. R. (2007). Copper(II)-catalyzed enantioselective intramolecular carboamination of alkenes. *J. Am. Chem. Soc.*, *129*, 12948–12949.

[45] He, W., Yip, K.-T., Zhu, N.-Y., Yang, D. (2009). Pd(II)/tBu-quinolineoxazoline: an air-stable and modular chiral catalyst system for enantioselective oxidative cascade cyclization. *Org. Lett.*, *11*, 5626–5628.

[46] Mai, D. N., Wolfe, J. P. (2010). Asymmetric palladium-catalyzed carboamination reactions for the synthesis of enantiomerically enriched 2-(arylmethyl)- and 2-(alkenylmethyl)pyrrolidines. *J. Am. Chem. Soc.*, *132*, 12157–12159.

[47] Zeng, W., Chemler, S. R. (2008). Total synthesis of (S)-(+)-Tylophorine via enantioselective intramolecular alkene carboamination. *J. Org. Chem.*, *73*, 6045–6047.

[48] Miao, L., Haque, I., Manzoni, M. R., Tham, W. S., Chemler, S. R. (2010). Diastereo- and enantioselective copper-Catalyzed intramolecular carboamination of alkenes for the synthesis of hexahydro-1H-benz[f]indoles. *Org. Lett.*, *12*, 4739–4741.

[49] Yip, K.-T., Yang, M., Law, K.-L., Zhu, N.-Y., Yang, D. (2006). Pd(II)-catalyzed enantioselective oxidative tandem cyclization reactions. Synthesis of indolines through C-N and C-C bond formation. *J. Am. Chem. Soc.*, *128*, 3130–3131.

[50] Guo, X.-X., Xie, J.-H., Hou, G.-H., Shi, W.-J., Wang, L.-X., Zhou, Q.-L. (2004). Asymmetric palladium-catalyzed hydrosilylation of styrenes using efficient chiral spiro phosphoramidite ligands. *Tetrahedron Asymmetry*, *15*, 2231–2234.

[51] Tsujihara, T., Shinohara, T., Takenaka, K., Takizawa, S., Onitsuka, K., Hatanaka, M., Sasai, H. (2009). Enantioselective intramolecular oxidative aminocarbonylation of alkenylureas catalyzed by palladium-spiro bis(isoxazoline) complexes. *J. Org. Chem.*, *74*, 9274–9279.

[52] Shinohara, T., Arai, M. A., Wakita, K., Arai, T., Sasai, H. (2003). The first enantioselective intramolecular aminocarbonylation of alkenes promoted by Pd(II)-spiro bis(isoxazoline) catalyst. *Tetrahedron Lett.*, *44*, 711–714.

[53] Fuller, P. H., Kim, J. W., Chemler, S. R. (2008). Copper catalyzed enantioselective intramolecular aminooxygenation of alkenes. *J. Am. Chem. Soc.*, *130*, 17638–17639.

[54] Michaelis, D. J., Williamson, K. S., Yoon, T. P. (2009). Oxaziridine-mediated enantioselective aminohydroxylation of styrenes catalyzed by copper(II) bis(oxazoline) complexes. *Tetrahedron*, *65*, 5118–5124.

[55] Lucet, D., Le Gall, T., Mioskowski, C. (1998). The chemistry of vicinal diamines. *Angew. Chem. Int. Ed.*, *37*, 2580–2627.

[56] Cardona, F., Goti, A. (2009). Metal-catalysed 1,2-diamination reactions. *Nat. Chem.*, *1*, 269–275.

[57] de Figueiredo, R. M. (2009). Transition metal-catalyzed diamination of olefins. *Angew. Chem. Int. Ed.*, *48*, 1190–1193.

[58] Sequeira, F. C., Turnpenny, B. W., Chemler, S. R. (2010). Copper-promoted and copper-catalyzed intermolecular alkene diamination. *Angew. Chem. Int. Ed.*, *49*, 6365–6368.

[59] Du, H., Yuan, W., Zhao, B., Shi, Y. (2007). Catalytic asymmetric diamination of conjugated dienes and trienes. *J. Am. Chem. Soc.*, *129*, 11688–11689.

[60] Du, H., Zhao, B., Shi, Y. (2008). Catalytic asymmetric allylic and homoallylic diamination of terminal olefins via formal C-H activation. *J. Am. Chem. Soc.*, *130*, 8590–8591.

[61] Du, H., Zhao, B., Yuan, W., Shi, Y. (2008). Cu(I)-Catalyzed asymmetric diamination of conjugated dienes. *Org. Lett.*, *10*, 4231–4234.

[62] Greene, F. D., Stowell, J. C., Bergmark, W. R. (1969). Diaziridinones (2,3-diazacyclopropanes). II. Synthesis, properties, and reactions. *J. Org. Chem.*, *34*, 2254–2262.

INDEX

Abacavir 653, 654
α-acetamido cinnnamic acid 7
α-acetamidoalcohol 563, 654
acetoacetates
  hydrogenation of 9–12
acetone
  in aldol reaction 108–112, 114–119, 122–123, 127, 129–131, 187, 191, 419, 543–547, 550
  in Michael reaction 137, 141, 145–147
acetophenone
  in aldol reaction 609
  oxidation to 568
  reduction of 12, 52–56, 216, 232, 262, 263, 292, 297, 298, 360, 379, 402
acetyl cyanide
  addition of 364
acidic aluminosilicate 7
Acinetobacter 497, 502
acrolein 61, 138, 159, 648
3-acryloyl-2-oxazolidinone 32
acylation 101, 215, 237, 239, 555
  lipase-catalysed 509, 512, 513, 515
acylhydrazono esters 602
Adam's catalyst 516
Adociacetylene 634
(+)-Africanol 657
Agelastatin A 640
4-*epi*-ajanol 430
L-alaninol 301
Alcaligenes faecalis 499
alcohol dehydrogenases (ADH) 383, 492–500, 503, 517
aldol reaction 418
  biocatalyzed 518, 519
  peptide-catalyzed 543
  in flow reactors 351
  Mukaiyama-aldol reactions 30, 236, 609, 610
  supported catalysts 102, 191, 222, 239
aldolases 518, 541
Aliskiren 627

Catalytic Methods in Asymmetric Synthesis: Advanced Materials, Techniques, and Applications, First Edition. Edited by Michelangelo Gruttadauria and Francesco Giacalone.

alkene hydrogenation 5, 264
alkene metathesis 631
γ-alkenyl amines 674
γ-alkenylsulfonamides 677
alkoxycarbonylation 511, 514
alkylation of glycine and alanine Schiff bases 92–99
alkylimidazoles 327
alkyne 443, 496, 634, 652–655, 674
 in supported catalysts 219–225
alkynylation 634
allenoates 555
allenyl trichlorosilane 615
(+)-Allosedamine 440
allylation 24, 26, 63, 159, 427, 601, 611, 613, 634, 641
allylic acetate 16, 676
allylic alkylation 16, 395, 396, 398, 639
allylic amination 16, 397, 677
allylic substitution 15, 395
allylsilanes 608
allyltributylstannane 611, 635
allyltrichlorosilane 158, 601
allyltrimethoxy silane 602
alumina 7, 117, 180, 298, 359
alumina-supported Pt 294
Amaminol A and B 444
amidoacrylic acids 7
α-amination reaction 424, 428
amino acid dehydrogenase 506
α-amino acids 92, 95, 100, 124, 192, 292, 297, 326, 328, 330, 334, 337, 339, 506, 519, 520, 571, 585, 607
β-amino acids 495, 502, 509, 510, 520, 533, 589
γ^2-amino acids 536, 537
1,2-amino alcohols 51, 58, 239, 283, 294, 406, 444, 493, 496, 593, 634
amino carbamates 511
amino esters 589, 596, 597
amino ketones 600
aminoallenes 675
aminoallylation 428
2-aminobutane-1,3,4-triol 507
aminocarbonylation reaction 677
trans-4-amino-*L*-proline-based peptides 546
aminooxygenation 679
aminopalladation 678
β-aminopeptidases 509
3-aminopropyltriethoxysilane 37, 89
aminotransferases 505
α-aminoxylation 104, 107, 424, 426, 427
(+)-Amphidinolide W 631
Anguinomycin C 646, 647
(+)-Angustureine 440
anion metathesis 327–329
Aphanorphine 626, 627
9-β-arabinofuranosyl adenosine 511
1-β-arabinofuranosyl uracil 511
(−)-Arboricine 453
Arglabin 653
Argopore resin 104
(−)-Aromadendranediol 434
arylacrylic acids 9
Aspergillus fumigatus 502
Aspergillus niger 505
atom transfer radical polymerization (ATRP) 228, 336
atomic absorption spectroscopy 204
ATR-IR spectroscopy 292, 304
(−)-Aurantioclavine 632
aza-Baylis-Hillman reaction 159, 332, 339, 555
aza-Diels-Alder reaction 336
aza-Michael reaction 440
aza-Morita-Baylis-Hillman reaction *see* aza-Baylis-Hillman reaction
(+)-Azaspiracid 649
azidation 532, 533
β-azido alkanoates 532
β-azido tertiary alcohols 518
α-azidoacetophenone 496
β-azidoalcohols 496
4-azido-L-proline 104, 121
azidomethyl polystyrene 397
aziridination of alkenes 672
azomethine ylides 15

Bacillus subtilis 501
Baclofen 437
Baeyer-Villiger monooxygenases 502
Baeyer-Villiger oxidation 502, 504, 655
baker's yeast 497
Baylis-Hillman reaction 105, 332, 462, 553

bentonite 119, 182, 192
benzalacetophenone
  epoxidation of 89
benzaldehydes
  aldol reaction 104, 112, 115, 117, 122, 126, 129, 191, 543, 608
  allylation 34, 602
  bromoester synthesis 335
  diethylzinc addition *see* diethylzinc
  in Strecker reaction *see* Strecker reaction
  silylcyanation of *see* silylcyanation
benzofurans 505
benzylation of glycine and alanine Schiff bases *see* alkylation of glycine and alanine Schiff bases
α-benzyl acrylates 405
α-benzylketones 503
Biginelli reaction 461
BINAP 214
  $AgClO_4$ complex 15
  PEG-supported 11
  Polystyrene-supported 15
BINAPO 599–602, 606–607, 612–617
BINOL 5, 59, 157, 191, 215, 218, 231, 268, 270, 271, 274, 277–281, 406, 452, 553, 635, 649
bipyridines 156, 336, 398
bipyridine *N,N′*-dioxides 604
bipyridine *N*-monoxides, Me_2PINDOX 604
2,2′-bis(diphenylphosphanyl)-1,1′-binaphthyl dioxide *see* BINAPO
Bischler-Napieralski reaction 438
bisoxaolines ligands
  Bn-BOXAX 645
  Aza-BOX 24, 33, 34
  BOX 26
  Phe-BOX 33
  Py-BOX 24, 219, 236, 347
Biyouyanagin 429, 640
Blepharocalyxin D 636
borane, reduction of ketones 55, 150, 356
boronates 459
boronic acid 149, 357, 405, 460
borosilicate microreactor 347
Brevisamide 647
Brevitoxins 647
Brønsted acid 326
Burkholderia cepacia see lipase
1-butylimidazole 331
trans-4-(4-*tert*-butylphenoxy)-proline 129
t-butylphenylketone 402
butylpyridinium cation 12
2-(tert-butylsulfonyl)iodosylbenzene 35, 273
butyraldehyde, reduction of 383
butyrolactones 502, 535, 538

Callipeltoside C 419
camphorsulfonato anion 331
Candida antarctica B (CAL-B) *see* Novozym 435
Candida magnoliae (CMCR) 495
Candida parapsilosis 499
Candida rugosa 49, 382
Capromorelin 499
1-carbadodecaborate anion 12
Carbamates 380, 511, 514, 569
carboalumination of alkenes *see* ZACA reaction
carboamination of alkenes 677
carboannulation of allenes 677
carbon nanotubes 7, 43, 184
carbonyl-ene reaction 28, 61, 278
Carbovir 653
N-carboxyanhydride of L-Leucine 89–91
catecholborane 643
catechols, oxidation of 505
cephalosporins 507
cerium(IV) ammonium nitrate 451
Cermizine C 427
Cermizine D 427
C-H insertions 218, 235, 332
chalcone 89–91, 271, 351, 599, 672
(R)-Chimonamidine 423
α-chlorination of acid chlorides 361
α-chloro ketones 496, 588
chlorofusin 633
(–)-chloromethylmenthylether 327
chloroperoxidase 50
chlorzoxazone 502
chromanones 109

cinchona 18–23, 86, 92, 94, 100, 135, 149, 220, 224–226, 237–239, 292- 307, 361, 459, 465
(DHQ)$_2$PHAL ligand 19
(DHQD)$_2$PHAL ligand 21
(DHQD) ligand 20
(DHQD)$_2$PYR ligand 21
(QN)$_2$PHAL ligand 22
(DHD)$_2$PHAL ligand 20
cinnamaldehyde 133, 150, 153, 155, 407, 436
5-epi-Citreoviral 657
(+)-citronellal 428, 441
(R)-citronellol 327
(S)-citronellol 425
Claisen condensation 437
Claisen rearrangement 632, 640
Claisen–Schmidt condensation 63
ent-Clavilactone B 631
Clavirolide C 638
clays 4, 38, 182, 191, 199, 202, 205
click chemistry 26, 34, 103, 104, 146, 225, 397, 496
Clopidogrel 498
(+)-Coniine•HCl 440
Convolutamydines A-E 419
copper
Cu(I)- and Cu(II)-PhPyBox triflate complex 25
Cu-Inda-BOX 31
Cu-BOX 26
copper bromide complex of Cy-JOSIPHOS 637
copper catalyzed azide-alkyne cycloaddition 219–221
copper triflate Cu(OTf)$_2$•(*R*)-Ph-BOX complex 29, 31, 236, 334, 653, 677, 679, 681, 682
copper-catalyzed benzoylation 34
copper-catalyzed methylenation 430
Corey-Bakshi-Shibata (CBS) asymmetric reduction 149, 222
(R)-Crispine A 505
crotonaldeyde 152
crotyltrichlorosilanes 601
Cryptophycin 509
cumene hydroperoxide 270
Curtius rearrangement 429
β-cyano tertiary alcohols 518
cyanohydrins 44, 45, 189, 347, 348, 349, 516, 517, 530, 531
cyanosilylation 26, 47, 162, 185, 189, 192, 197, 285
cyanation 530
enzyme-catalyzed 348
(−)-Cyanthiwigin F 642
Cyathin A_3 649
cyclic voltammetry 198
[2+2+2] cyclization of enediynes 656
cycloaddition 14, 32, 61, 186, 433, 443, 446, 646, 672
azide-alkyne cycloaddition 219, 221, 223, 225
[2+2] cycloaddition 502, 640
[4+2] cycloaddition 445
[5+2] cycloaddition 652
1,3-dipolar cycloaddition 15, 103, 533
[2+2] photocycloaddition 353
cyclobutanones 502
cycloctene 10
β-cyclodextrin 126, 129
cyclohexane-1,2-dione 294
cyclohexanecarboxaldehyde 123
cyclohexanone 136, 138, 277, 334, 353, 439, 541
cyclohexenone 136, 138, 277, 334, 353, 541
2-allyl- 439
cyclopentadiene 10, 31, 61, 150–153, 362, 503
cyclopentanone 104, 116, 123, 137, 141, 145–147, 334
cyclopentene 353
cyclopropanation 24, 26–28, 33–34, 218, 235–236, 358, 626, 631, 652

DABCO 232, 322
DABCO-H_2O_2 89
Danishefsky's diene 14, 15, 45, 649
Daucus carota 499
deacylation 508
decarboxylases 541
dehydratases 541
α-dehydroamino acid methyl esters 6
dehydrobromation 328
demethyl calamenene 450

dendrimer 53, 97, 115, 141, 347
(−)-7-deoxy-*trans*-dihydronarciclasine 437
α-deprotonation of ketones 571
deracemisation 523, 524
desymmetrization 100, 144, 507, 510, 511, 525, 559, 577
diamination 682
diamine ligand 51, 602
diamines 51, 120, 123, *468*, 511, *525*, 682, *687*
diaminocyclohexane 189, 285, 402
1,3-diaryl-1,3-propanedione 149
diarylprolinols 131, 133, 135, 139, 149, 222, 239–242, 418, 428, 437, 442, 446, 448
α-diazoesters 651
dibenzyl azodicarboxylate 428
α,α′-dibenzyl esters 405
dicarboxylic acids 275, 282, 501, 547
(+)-Dichroanone 641
diclofenac 502
dicyclopentadiene 10
Diels-Alder reaction 15, 31, 45, 84, 150, 153, 243, 325, 329, 336, 340, 341, 343, 362, 417, 443, 444, 456, dienamine mechanism 417, 445
di-*epi*-pumiliotoxin C 454
diethylsylilchloride polystyrene 16
diethylzinc 57, 58, 60, 61, 271, 278, 334, 352, 356
(+)-Digitoxigenin 652
2,3-dihydrofuran 400
Dihydrojunenol 430
Dihydronarciclasine 437, 651
dihydroxy eudesmane 430
3,4-dihydroxybenzonate 430
dihydroxylation reaction 17, 22, 186, 239, 332, 504, 517, 631
di-isocyanates 10
diketoesters, microbial reduction of 497
1,2-, 1,3- and 1,4-diketones, reduction of 493
Diltiazem 20, 463
dimethyl itaconate 5, 7, 8, 359
2,2-dimethyl-2*H*-chromene 35, 272
2,4-dimethylimidazole 328
dimethylmalonate 16
trans-2,5-dimethylpiperazine 535, 550
1,1-diphenyl ethylene 33
(*E*)-1,3-diphenyl-3-acetoxyprop-1-ene 16
(1*S*,2*S*)-diphenylethylenediamine *see* DPEN
diphenylpropene 387
dirhodium catalysts *see* rhodium catalysts
Discodermolide 493
(+)-Disparlure 426
dispersion polymerization 105, 229
Dispongines A and B 649
divinylbenzene 92, 135, 157, 230, 233, 236, 241, 243, 357, 364
dodecanal, α-aminoxylation of 426
domino reaction 160, 434, 505
(−)-Donaxaridine 422
double layered hydroxides (LDH) 20, 38, 182, 191, 192, 199, 202, 205
DPEN 9, 10, 216, 232, 233, 245, 262, 263, 298, 628
dynamic kinetic asymmetric transformation (DYKAT) 514, 515, 519
dynamic kinetic resolution (DKR) 47, 48, 50, 379, 512–514, 518, 520

electron paramagnetic resonance (EPR) or (ESR) 198
electrospinning 133, 240, 242
emulsion polymerization 229
enamides, hydrogenation of 9, 268, 626
enamine mechanism 161, 417, 550
enamines hydrogenation 5, 6
enantioselective liquid-liquid extraction (ELLE) 337
Enders cascade-reaction 135
(−)-Englerin A 656
enoates 266, 269, 497
  azide conjugate addition 532
  enoate reductase 500
enynes 407, 654, 655
Ephedra alkaloids 224, 332
ephedrine 56, 92, 327, 360, 592, 594, 596
ephedrinium 337
epibromohydrin 355
epichlorohydrin 41, 42
2,3-epoxialcohols 497

epoxidation
of α,β-unsaturated aldehydes 442
of α,β-unsaturated esters 463
of α,β-unsaturated ketones 88, 90, 91, 270, 271, 351
of alkenes 35–40, 190, 193, 199, 233, 271, 273, 648
of allylic alcohol 631
epoxide ring opening 232, 282, 406, 463, 517, 614
epoxides kinetic resolution 40–42, 217, 355, 518, 648
Erythromycine A 555
L-erythrulose 507
(+)-Esermethole 456
esterification 218, 219, 259, 437, 450, 505
biocatalyzed 377, 507
peptide-catalyzed 556
ethyl 3-halo-2-oxo-4-phenylbutanoate 497
ethyl 5-hydroxyhept-6-enoate 497
ethyl diazoacetate 25, 28, 33, 652
ethyl formate 12, 437
ethyl lactate 292, 327
ethyl pyruvate *see* hydrogenation of
ethyl vinyl ketone 159
ethyleneglycol dimethacrylate (EGDMA) 133, 230–234, 239, 241
ethylglyoxalate 30
eudesmantetraol 430
extended X-ray absorption fine structure (EXAFS) 196, 204

(+)-Fawcettimine 438
(S)-Fenoprofen 90
ferrocenyl derivatives 8, 359, 405, 651
flavin cofactor 500
fluorous solid-phase (F-SPE) extraction 61, 126, 148, 153
fluorous-tagged 34, 155, 156, 587
Fluostatin C 649
(S)-Fluoxetine 216
Fluvirucinine A_1 644
N-formyl derivatives 584
(+)-Fostriecin 462
Friedel-Crafts reaction 152, 450, 451, 557
(+)-Frondosin A 652
furanones 638

glass transition temperature (T_g) 329
(−)-Gloeosporone 634
glucose dehydrogenase (GDH) 494–496, 498, 506
glycidic esters 497
glycosylations 352
gold nanoparticles 29, 186, 195
Grignard reaction 241, 440, 450, 637
grossamide 365
Grubbs' catalyst 427, 428, 434, 440, 462, 463, 636
guanidinium cation 22, 332, 452

Halohydrin dehalogenases 497
halohydrins 517, 518
haloysite 192
Hantzsch ester 153, 157, 244, 435, 454
(+)-Harmicine 455
H-bonding catalysis 417
Heck reactions 336, 399, 644, 645, 677
hectorite 182, 191, 192, 202
Henry reaction 45
Hirsutellone B 441
histidine 101, 328, 533, 549, 555, 557, 560, 566, 568
homoallylic alcohols 426, 606, 607, 636
homochiral metal-organic polymer 258–260, 274, 275, 285–286
Horner-Wadsworth-Emmons reaction 425, 627
horse liver alcohol dehydrogenase 348
Houamide A 626
Huisgen cyclization 329
hydroamination 672
hydroboration 225, 642, 643
hydroformylation 5, 13, 14, 384
hydrogen peroxide 40, 50, 51, 88, 351, 501, 504
hydrogenation 5, 250, 293, 298, 301, 349, 358, 386, 572, 626
of acetylenic ketones 630
of acrylates 334
of aromatic ketones 10, 12, 52–56, 261–263, 297, 597
of dehydro-α-amino acid derivatives 264, 266
of dimethyl itaconate 7, 355, 359
of enamides 9, 626
of ethyl formate 12

of ethyl pyruvate 292, 294, 359, 386, 404
of fluorinated aromatic ketones 297
of imine 385
of isophorone 300, 501
of isopropyl-4,4,4-trifluoroacetoacetate 359
of itaconic acid 194, 267
of ketopantolactone 294
of methyl 2-acetamidoacrylate 7, 265, 385
of methyl acetoacetate 261
of methyl α-acetamidobut-2-enoate 266, 269
of pyrones 300
of quinolines 12
ruthenium-mediated 10–12, 52, 53
of tiglic acid 386
of (*Z*)-methyl 2-acetamidobut-2-enoate 266
of (Z)-methyl-α-acetamidocinnamate 193
of α,β-unsaturated carboxylic acids 298–299, 359
of α,β-unsaturated ketones 300, 598
of α-diketones 293–297, 493
of α-fluoro ketones 292
of α-hydroxyketones 303
of α-keto esters 292
of α-ketoacetals 293, 296
of α-ketoacids 294, 296
of α-ketoamides 294
of β-diketones 297, 493, 494
of β-keto esters 260, 297, 628
of β-ketoalcohols 297
of β-ketoethers 297
of β-ketosulfones 297
hydrolases 507
hydrolytic kinetic resolution 40–42, 217, 355, 648
hydrosilylation
of imino esters 596
of ketomines 596
of ketones 11, 590
of enamino esters 596
hydrotalcite 38, 191, 202
β-hydroxy aldehydes 614
β-hydroxy carboxylic acids 495
α-hydroxy ketones 467, 493
β-hydroxy ketone 102, 612
β-hydroxy nitriles 48, 495–497
hydroxyacetone 112, 114, 116, 187, 544
hydroxyapatite 298
α-hydroxycarboxylic amides 516
hydroxyethyl methacrylate (HEMA) 230
hydroxynitrile lyases 349, 516, 517
α-hydroxyphosphonate 616
4-hydroxyproline 86, 102, 110, 114, 122, 221, 241, 329
hydroxypyruvate 507
Hyperolactone C 640

iminium catalysis 161, 431, 572
(−)-Incarvillateine 635
indium
In(III) Py BOX 26
indium mediated-allylation 24, 426
indoles 154, 161, 422, 443, 453, 456, 467, 539, 540, 677
indolines 678–682
inductively coupled plasma atomic emission spectroscopy 203
infrared (IR) spectroscopy 198
inositoles 560
ion exchange 4, 7, 18, 20, 99, 121, 130, 146, 152, 191- 193, 202, 223, 329
ionic liquid
[bmim]Cl 329
[bdmim]BF_4 49
[beim]PF_6 43
[bmim]BF_4 116–118, 122, 127, 131, 143, 147
[bmim]Cl 329
[bmim]OH 330
[bmim]PF_6 21, 38, 40, 42, 49, 61, 115, 116, 153, 380, 386
[bmim]NTf_2 48, 123, 127
[bmim]TfO 123
[bmim]SbF_6 32
[bmpy]NTf_2 123
[bmpy]TfO 123
[btma]NTf_2 380
[emim]NTf_2 31, 48, 380
[emim]OTf 33
[emim]PF_6 43
[hmim]PF_6 26
[PPMIM]PF_6 48

ionic liquid-coated enzyme 48
ionic liquid-tagged organocatalyst 120
iridium catalyst 12
iridium-catalyzed hydrogenation of imines 386
isatin, aldol reaction of 130, 419–424, 546
β-isocupreidine 462
isomerases 520
isonitramine 499
isosorbide 327

Jacobsen's catalyst 40–42, 46, 647
Jacobsen's Schiff base-chromium complex 646
JandaJel 92, 100, 151, 218, 223
Jørgensen epoxidation 442
Juliá-Colonna epoxidation 163–164, 224
Julia-Kocienski olefination reaction 462

α-keto acids 293–296, 506
ketoisophorone 501
ketopantolactone 294, 302, 305
ketoreductases 492
kinase mimic 559
kinetic resolution
  of thioformamide 569, 570
  of 1,2-diols 34
  of 1-phenyl-2-propanol 260
  of amines 505, 506, 514, 568
  of secondary alcohols 100, 499, 504, 514, 560, 632
  of terminal epoxides 40–42, 217, 355, 518
  of β-hydroxynitriles 48
  Kröhnke condensation 328
Kurasoin B 467

laccase 505
β-lactams 137, 361, 495, 509
lactic acid 22, 275
Lactobacillus brevis 494
Lactobacillus kefir 495
lactones 495, 497; *see also* butyrolactones
laponite 28, 33
Lasonolide A 647
(–)-Laurebiphenyl 655
layered double hydroxides 182
L-leucine 158, 330
  N-carboxyanhydride 89–91
D-leucinol 420
leukotriene 497, 501
Levodione 501
Lewis acid 14, 24, 35, 59, 155, 158, 162, 233, 347, 353, 431, 451, 581–583, 600, 610, 634, 646, 649, 651
Lewis base 133, 159, 332, 347, 453, 580, 634
(*R*)-(+)-limonene 353
lipase 48, 49, 336, 348, 355, 356, 379, 380, 499, 505, 507, 510–513, 519, 644
LUMO 432, 582
lyases 349, 516

MacMillan catalyst 86, 150, 152, 153, 161, 223, 239, 242, 243, 417, 419, 435
macrocyclization 634
Madindoline A and B 422
magnetic nanoparticles 19, 41, 51
magnetite 34, 100, 186
maleimides 15, 501
L-malic acid 430
Mannich reaction 109, 114, 162, 351, 447, 461
mannitol 363
Manzacidins C 461
MCM-41 7, 35–38, 52, 60, 112, 138, 179–181, 184–189, 197, 202–205, 297
*m*CPBA 217, 352, 680
Meldrum's acid 457
membrane reactor 39, 56, 350, 352
mercaptomethyl polystyrene 92, 104
Merrifield resin 26, 34, 58, 94, 103–104, 109, 155, 212- 218, 221–225, 357, 588
mesocellular silica foam 19, 28, 151, 223
meso-carboxylic anhydrides 285
meso-epoxides 285, 406, 614
metal organic frameworks (MOF) 180, 188, 190, 258, 271–284
metalloproteinase inhibitor 365
methacroleine 362, 635
methallyltributylstannane 636
1-methoxy-1-trimethylsilyloxypropene 31
methyl (*R*)-*o*-chloromandelate 498
(*R*)-methyl 2-acetamidopropanoate 350
methyl methacrylate 133, 239, 241

methyl *o*-chlorobenzoylformate 498
methyl pyruvate 31
methyl vinyl ketone 159, 161, 230, 360, 429, 553, 645
methyl α-acetamido cinnamate 5
methyl α-acetamidoacrylate 5
methyl-2-acetamido acrylate 350
α-methylbenzylamine 92, 328
2-methylbutanoic acid 386
α-methylstyrene 23, 30, 36–38, 278
(*S*)-Metolachlor 594
Mexiletine 506
Michael addition 46, 61, 131, 277, 336, 339, 360, 429, 436, 457, 459, 505, 533–539
microencapsulation 4, 18, 26
microwave 50, 112, 138, 391, 645
Mikami's catalyst 649
Minfiensine 401, 443, 645
Mitsunobu reaction 238
molybdenum
  $Mo(CO)_3(C_7H_8)$ 399
  $Mo(CO)_6$ 398–400
  $(EtCN)_3Mo(CO)_3$ 398
Modiolide A 497
montmorillonite 126, 152, 182, 191, 199, 202
Morita-Baylis-Hillman reaction *see* Baylis-Hillman reaction
Mucronine E 626
Mugetanol 499
Mukaiyama aldol reaction 30, 236
Mukaiyama-Michael reaction 434
Myxovirescin A1 628

N-(2-thiophenesulfonyl)prolinamide 130
N_2 adsorption-desorption isotherms 203
N-acryloyloxazolidinone 31
NADH 348, 383, 500
NADPH 493, 499–504
nafion-silica 33
(*S*)-Naftopidil 427
(–)-Nakadomarin A 462
1-naphthyl-1,2-ethanediol 294
naphthyl-ethylamine 292
N-arylations of *N*-heterocycles 186
N-benzhydryl imines 531
Negishi cross-coupling reactions 405
Nelfinavir 507
(+)-Nemonapride 465
(+)-Neopeltolide 647
(+)-Neosymbioimine 424
N-heterocyclic carbene 638, 658
(S)-Nicotine 327
nitrilase 496
nitrile hydratase 516
nitroalkenes 148, 456, 458, 501, 535, 536, 539, 541
α-nitro ketones and esters 533
4-nitrobenzaldehyde 105, 108, 109, 112–115, 119, 123–126, 130, 131
γ-nitroketones 539
nitrosobenzene 107, 424–426
nitrostyrene 132–140, 144–147, 153, 339, 539
nitroxide-mediated polymerization (NMP) 228, 231, 240, 242
N-methylmorpholine *N*-oxide (NMO) 18, 352
Nontronite 192
Norsertraline 514
novasyn TG amino resin 108
Novozym 435 48–50, 348, 355, 378–382, 508–515
Noyori's catalyst 630
N-substituted maleimides 501
N-toluenesulfonyl-proline 116
nuclear magnetic resonance (NMR) 104, 196, 337, 564, 611
nucleosides 511, 512, 653

O-acetylcyanohydrins 44
olefin metathesis 657
oligonucleotides 211, 512
(–)-Oseltamivir *see* Tamiflu
osmium tetroxide 18–24, 332, 517
oxa-Michael reaction 447
oxazaborolidines 56, 651
oxidoreductases 492
Oxomaritidine 365
oxophytodienoate reductase isoenzymes 501
α-oxyamination reaction 419, 558
(*S*)-oxybutynin 423
(*R*)-oxynitrilase 516
(*S*)-oxynitrilase 516

(+)-Palitantin 447
palladium catalyst 15–17, 20, 50, 220, 292, 336, 395, 399, 405, 513, 602, 632, 643, 645, 655, 674, 676, 679, 682
Palominol 651
pantoyl-naphthylethylamine (PNEA) 297
parallel interconnected kinetic asymmetric transformations (PIKAT) 503
(−)-Paroxetine 439
Passerini reaction 508
Pasteurestin B 656
Pauson-Khand reaction 653
Penicillin G acylase 509
pentandioic acid 364
perfluorinated linker 365
perfluorinated solvents 4, 375
perfluorophase 22
perillaldehyde 501
periodic mesoporous organosilicas 188, 189, 197, 210–203
phase transfer catalysis (PTC) phase-transfer catalyst 92, 97, 465
(4*S*)-phenoxyproline 126
1-phenyl-1,2-propanedione 294
phenylacetamide 509
phenylalanine 7, 243, 356, 509, 513
phenylalanine aminomutase 520
phenylalaninol 301
1-phenylethanol
  kinetic resolution 48, 356, 378, 379
  oxidative kinetic process 568
  reduction of acetophenone 402
(*R*)-1-phenylethyl propionate 49, 380
phosphorylation 559
(+)-Phoslactomycin B 462
phosphine ligand
  BDPP 7, 349
  BINAP 7, 9, 11, 12, 14, 15, 194, 214–218, 231, 232, 260, 262, 335, 386, 395, 404, 628, 636, 655
  BINAPHOS 13
  diAm-BINAP 10
  DIOP 194, 202, 205, 214, 230
  DIPAMP 215, 626
  Et-DuPhos 9, 404, 626
  Josiphos 9, 626, 629, 637
  Mandyphos 9
  Me-DuPhos 7, 350
  Monophos 5, 264–267
  P-Phos 12
  PPM 216, 230
  Pyrphos 7
  TangPhos 350
  Taniaphos 9
  Walphos 9, 651
phosphine oxides
  BINAPO 599, 601, 607, 612–616
  tetraMe-BITIOPO 606, 607, 613
phosphinooxathiane ligand 16
phosphinooxazoline ligand 397
phosphiteoxazoline ligand 400
phosphonioxazolidine ligand 16
phosphonylation 616
phosphoramide 159, 602, 608, 610, 614
phosphoramidite 264, 627, 678, 683
phosphoramidite ligand (*R*)-Siphos 678
phosphorates 14
phosphoroamides 602, 608
phosphorotriamides 602
phosphotungstic acid 7
[2+2] photocycloaddition 353
(+)-Physostigmine 456
Pichia minuta 497
picolinamides 592
Pictet-Spengler reaction 453–456
pinane-fused ionic liquid 329
(+)-Pinocarvone 328
piperidines 443, 499, 515, 602, 636, 674, 675
planar chirality 328
(−)-Plantecin 648
Platensimycin 654
platinum 213, 291, 292, 297, 298, 301, 358, 386
Polonovsky-Potier reaction 428
poly(ethylene glycol) dimethyl ether 12
poly-(L)-leucine 87–91, 153, 158, 351, 588
poly(norbonene) polymer 46
(+)-Polyanthellin A 430
poly(ethelyene glycol) 4, 9, 11, 16, 17, 20, 21, 23, 30, 33, 34, 45, 51–55, 90, 95, 114, 187, 213, 218–223, 241, 243, 351, 545, 571

poly(ethelyene glycol) dimethyl ether 12
poly(oxyethyleneglycol)alkyl sulfate 49
polyethyleneglycol-polystyrene copolymer 92
polyoxomethalate 120
polystyrene 9, 16, 25–27, 33, 37, 40, 52, 56–59, 63, 88, 92, 100, 102, 193, 213, 215–222, 225, 277, 357, 384, 397, 546, 571
powder X-ray diffraction (XRD) 196, 202, 205
Pregabalin 457
Prins cyclisation 636
prolinamides 86, 110–116, 125, 130, 420, 543, 589
prolinate anion 219, 337
D-proline 239, 419, 426
  containing peptides 563–565
N-Boc-proline 507
proline-*N*-sulfonamide 115
prolinethioamide 131
prolinol derivative 56, 58, 147, 149, 327, 352, 596; *see also* diaryl prolinols
propargyl trichlorosilane 615
propargylation 439
(*S*)-propranolol 427
(*S*)-2-propylhexahydrochinolinone 454
protonation reactions 461, 570
(–)-Pseudolaric acid 628
Pseudomonas cepacea 505, 510, 511; *see also* lipase
Pseudomonas fluorescens 503, 511; *see also* lipase
Pt/Al_2O_3 294, 298, 386
Pygmol 430
pyranones 638, 647
pyrazolidines 675
pyridine
  as ligands 10, 24, 58, 259, 273; *see also* bisoxaolines ligands, Py-BOX
  derivatives 398, 454
pyridine-4-carbaldehyde 328
pyridine-2-nitroso 651
pyridine N-oxides 606, 614
pyridinium cations 12, 24, 328
pyridoxal-5′-phosphate 506
pyridoxamine 506
5-(pyrrolidin-2-yl)tetrazole 351
pyrrolidines 100, 116, 120, 131, 133, 135, 137, 138, 140, 144–147, 153, 161, 218, 297, 327, 422, 432, 436, 507, 534, 551, 585, 592, 596, 602, 674–680

(+)-Quebrachamine 657
quinic acid 22
quinolines 12, 52, 226, 244, 300

Ramipril 357
Raney Ni 50, 297, 457
Rauhut-Currier reaction 449
Rawal-Jacobsen's salen 646
reduction
  of α,β-unsatured aldehydes 572
  of enones 599
  of imines 155, 585, 588, 592–598
  of ketones 10, 51, 55, 215, 383, 585, 589, 598, 628, 634
  of *N*-benzyl iminoester 596
reductive elimination 635, 678
reversible addition-fragmentation chain transfer 228
rhodacarborane 12
rhodium catalyst
  (*R,S*)-BINAPHOS-Rh(I) 13, 384
  [(BPPM)Rh(cod)Cl] 194
  $[(PNNP)\text{-}Rh(cod)]^+$ 192
  [(DIOP)Rh(cod)Cl)] 194
  [(R,R)-Me-DuPHOS)Rh(COD)] 7
  $[Rh((S)\text{-}BINAP)]SbF_6$ 655
  [Rh(COD)(Et-DuPHOS)]OTf
  $[Rh(COD)(Ph\text{-}\beta\text{-}glup)]^+BF_4^-$
  $[Rh(cod)_2]BF_4$ 5, 7, 264, 267, 269, 334, 350, 628, 643, 675
  $[Rh(COD)L]^+$ 193
  $[RhCl(coe)_2]_2$ 635
  $Rh(cod)_2SO_3CF_3$ 350
  $Rh(I)/^tBu$-JOSIPHOS 629
  Rh-Phe-BOX 34
  Rh-(*S*)-bisbenzodioxanPhos 407
  dirhodium catalysts
    $Rh_2(S\text{-}DOSP)_4$ 219
    $Rh_2(5S\text{-}MEPY)_4$ 218
    $Rh_2(S\text{-}PTTL)_4$ 235
rhodium-catalyzed conjugate addition 404
rhodium-catalyzed hydrogenation 9, 213, 349

rhodium-catalyzed C-H activation 635
Rhodococcus erythropolis 495
Rhodococcus ruber 493, 497
(+)-Ricciocarpin A 435
ring closing metathesis 428, 462, 657
ring-opening olefin methathesis polymerization (ROMP) 10, 235
(*S*)-Rivastigmine 512
Robinson annulations 109, 160, 446
(*S*)-Rolipram 436
(*R*)-Rotundial 449
ruthenium catalyst
  (COD)Ru(bis-methallyl) 215
  (*R,S*)- BINAPHOS-Rh(I) 13, 384
  [(BINAP)Ru(p-cumene)Cl]Cl 194
  $[RuCl_2(p\text{-cymene})]_2$ 236, 358
  $CpRu(CH_3CN)_3PF_6$ 652
  $Ru(4,4'\text{-}(PO_3H_2)_2\text{-BINAP})(DPEN)Cl_2$ 10
  Ru-(BINAP)(dpea) 7
  $Ru(O_2CMe)_2((R)\text{-tolBINAP})$ 386
  Ru-BINAP-DPEN 262
  Ru-Py-BOX 25
  trans-$RuCl_2(Py)_2((R,R)\text{-Norphos})$ 10
ruthenium complexes as racemization catalysts 514, 515
ruthenium-mediated hydrogenation 9, 628
ruthenium for asymmetric reduction of ketones 215, 402
ruthenium mediated hydrogenation of acetoacetates 10–12
ruthenium mediated reduction of acetophenone 52, 53
ruthenium-catalyzed cyclopropanations 24, 236, 358

Salen complexes
  AlCl(salen) 46
  Co(salen) 41, 217
  Cr(III)(salen) 203, 406
  Ti(IV)(salen) 44, 348, 365
  Mn(III)(salen) 35–39, 47, 190, 195, 203, 205, 217, 271- 274
  VO(salen) 43–45, 189, 299
samarium triisopropoxide 436
secondary alcohols 11, 45–50, 100, 280, 357, 493, 500–504, 514, 560, 585, 589, 630–635, 644
(+)-Sedamine 440
L-Selectride 226
α-selenenylation 105
β-seleno amides 396
Senepodine G 427
single-electron transfer 451
Shiff's base 465
Shvo's catalyst 514
sialic acid derivative 352
Sibirine 499
Sigamide 586, 587
Silica
  MCM-41 7, 35–38, 52, 60, 112, 138, 179, 181, 184–189, 197, 202–205, 297
  MCM-48 37, 38, 181
  SBA-15 30, 36–40, 52, 60, 63, 154, 181, 184, 402
  SBA-16 40, 52, 181
silsesquioxanes 188
silyl enol ethers 608–612
silyl keteneacetals 610, 612
silylcyanation 24, 26, 43, 45, 185, 348
silylketene 30
silyloxy-L-serine 116
Simmons-Smith cyclopropanation 652
single electron transfer 451
Sitagliptin 629
(+)-Sitophilure 494
(+)-Solenopsin A95 516
Solid-state NMR spectroscopy 196
SOMO 417, 449
Sonogashira coupling 219, 266
Sparteine 632, 679
(−)-Spongidepsin 643
(S)-Stepholidine 631
Stetter reaction 555
trans-stilbene dihydroxylation 19, 23
Strecker reaction 157, 530
(+)-(*S*)-Streptenol A 459
styrene
  aminooxidation 680
  aziridation 672
  cyclopropanation 25, 27, 28, 33, 34, 358
  dihydroxylation 18, 23